# 3rd International Conference on Energy Materials and Environment Engineering (ICEMEE 2017)



IOP Conference Series: Earth and Environmental Science
Volume 61

Bangkok, Thailand
10 – 12 March 2017

ISBN: 978-1-5108-3957-1
ISSN: 1755-1307

**Printed from e-media with permission by:**

Curran Associates, Inc.
57 Morehouse Lane
Red Hook, NY 12571

**Some format issues inherent in the e-media version may also appear in this print version.**



Printed by Curran Associates, Inc. (2017)

For permission requests, please contact the Institute of Physics
at the address below.

Institute of Physics
Dirac House, Temple Back
Bristol BS1 6BE UK

Phone: 44 1 17 929 7481
Fax: 44 1 17 920 0979

techtracking@iop.org

**Additional copies of this publication are available from:**

Curran Associates, Inc.
57 Morehouse Lane
Red Hook, NY 12571 USA
Phone: 845-758-0400
Fax: 845-758-2633
Email: curran@proceedings.com
Web: www.proceedings.com

# Table of contents

## Volume 61

**3rd International Conference on Energy Materials and Environment Engineering**

**10–12 March 2017, Bangkok, Thailand**

**Accepted papers received: 31 March 2017**
**Published online: 21 April 2017**

### Preface



### Papers

IOP Conf. Series: Earth and Environmental Science **61** (2017) 011001 doi:10.1088/1755-1315/61/1/011001

# Preface

The 2017 3rd International Conference on Energy Materials and Environment Engineering (ICEMEE 2017) was successfully held from March 10-12, 2017 in Bangkok, Thailand. The conference addressed recent development in Energy Materials & Environment Engineering and the relevant applications. The technical program consisted of 4 international keynote speakers, 19 invited speakers, 30 oral presentations, and a poster session with more that 70 posters.

This volume of IOP Conference Series: Earth and Environmental Science (EES) is a compilation of the accepted papers in ICEMEE2017 and represents important results that were presented in the conference. The electronic submission and handling of the manuscripts via the conference website, including the selection of reviewers and evaluation of manuscripts, were identical to the procedures applied to manuscripts submitted as regular contributions for publication in peer reviewed journals.

The organization of this conference and the preparation of proceedings volume would have been impossible without the tremendous efforts and dedication of many individuals. Prof. Sivakumar Manickam oversaw the organization of the Conference. We would like to acknowledge a large team of reviewers for their timely submission of quality reports. Without this support, the conference could not have been successful. We sincerely hope that ICEMEE2017 will serve as an ideal forum for academia, industrial practitioners and government agents.

Prof. Sivakumar Manickam
University of Nottingham, Malaysia
General Chairman of ICEMEE2017

IOP Conf. Series: Earth and Environmental Science **61** (2017) 011001 doi:10.1088/1755-1315/61/1/011001

# Organizing Committees

## Co-Chair

Prof.Avi Friedman, McGill University, Canada
Prof.SIVAKUMAR MANICKAM, UNIVERSITY OF NOTTINGHAM, MALAYSIA

## Organizing Committee

Prof.Dr.Jinlong Zhang, East China University of Science and Technology, China
Dr.Paolo Beccarelli, The University of Nottingham, UK
Dr.Azharul karim, Queensland University of Technology, Australia
A.Prof.Fredy Kurniawan, Institut Teknologi Sepuluh Nopember, INDONESIA
A.Prof.Hakan CALISKAN, Usak University, TURKEY
Prof.Chris Bowen, University of Bath, UK
Dr.Masa Noguchi, Faculty of Architecture, Building and Planning, Australia
Lecturer.Alper KEREM, Osmaniye Korkut Ata University, Turkey
Dr.Xingxing Zhang, University of Nottingham Ningbo, China
Er. Uttam Roy, National Institute of Technology Agartala, India
Dr.Tze Ching Ong, Politeknik Kuching Sarawak, Malaysia
Prof.Dr.Jarot S. Suroso, Bina Nusantara University, Indonesia
Dr.George Filis, Bournemouth University, UK
Prof.Dr.Bingqing Wei, University of Delaware, USA
Dr.Nathan Vinod Kumar, National Institute of Oceanography, India
Dr.Narottam Das, University of Southern Queensland, Australia
Prof.Marcello De Falco, University "Campus Bio-Medico" of Rome, Italy
Dr.Lluc Canals Casals, Universitat Politecnica de Catalunya, Spain
Prof.Alan J Simson, Leeds Beckett University, UK
Prof.Laurent Ruhlmann, University of Strasbourg, France
Mr.Tochukwu A.S. Ugwoke, Michael Okpara University of Agriculture, Nigeria
Dr.Ali Mansour LAGOUN, Scientific and Technical Research Center On Physical And Chemical Analyses, CRAPC, Bousmail, Algeria
Assistant Prof.Abdul Lateef Olanrewaju, Universiti Tunku Abdul Rahman, Malaysia
Dr.Diana Iruretagoyena Ferrer, Department of Chemical Engineering, Imperial College London, United Kingdom
Dr.Yanhui Huang, Rensselaer Polytechnic Institute, USA
Prof.Luigi Fortuna, University of Catania Department of Electrical, Italy
Dr.Laura Castro-Santos, Department of Naval and Ocean enxeñaría, Spain
Dr.Qingbin Song, Macau University of Science and Technology, China

IOP Conf. Series: Earth and Environmental Science **61** (2017) 011002 doi:10.1088/1755-1315/61/1/011002

# Peer review statement

All papers published in this volume of *IOP Conference Series: Earth and Environmental Science* have been peer reviewed through processes administered by the proceedings Editors. Reviews were conducted by expert referees to the professional and scientific standards expected of a proceedings journal published by IOP Publishing.

# Research about impacts of specific pollutants like herbicides upon microbial activity of activated sludge systems in wastewater treatment plants

**Wei Hong[1], Shi-qing Liu[2], Cong Han[2], Gong-fa Chang[1*], Bo Liu[1] and Xiao-feng Zou[1]**

[1]Shandong Academy of Environmental Science, Jinan 250013, China
[2]Binzhou Municipal Bureau of Total Pollutant Discharge Control,Binzhou 256600,China
* Corresponding author's e-mail: changgongfa@126.com, First author's e-mail: honw@163.com

**Abstract**. The impacts of five types of typical specific pollutants like herbicides (including prometon, propazine, acetochlor, metolachlor and bentazone) upon microbial activity of activated sludge systems in wastewater treatment plants were studied by quantitatively feeding specific pollutants into these systems and measuring changes to specific oxygen uptake rate (SOUR) of the systems. The research results suggest that all these five types of specific pollutants significantly inhibit microbial and aerobic metabolisms of the activated sludge systems. The impacts of these herbicides are arranged as follows according to their degrees: bentazone<prometon≈ propazine≈metolachlor<acetochlor. When the concentration of specific pollutants was 20mg/L in the wastewater, bentazone, prometon, propazine and metolachlor decreased the SOUR of the activated sludge by 17.0%, 28.4%, 25.8% and 31.1% respectively. The SOUR declined by 37.1% when the concentration of acetochlor was 10mg/L. Useful lessons may be drawn from this study for controlling concentration of specific pollutants like herbicides in influents of the wastewater treatment plants in pesticide parks.

## 1. Introduction

With striking features such as low operating costs, wide scope of application and remarkable effects of treatment, aerobic activated sludge treatment process has become the most widespread and leading one for treating industrial wastewater at present [1]. However, some highly toxic specific pollutants of influents often have significant impacts upon microbial activity of the activated sludge systems when the aerobic activated sludge treatment process is used. As a result, the degradation efficiency of the activated sludge systems is decreased and the effects of wastewater treatment are affected.

In this study, a wastewater treatment plant of an industrial pesticide park in Binzhou of Shandong Province was exemplified. It mainly receives pretreated wastewater from manufacturers of pesticides like herbicides. Owing to frequent fluctuations in wastewater discharge of pesticide manufacturers, biochemical system of the wastewater treatment plant was impacted in the industrial park. To specify the impacts of specific pollutants in discharged wastewater of pesticide enterprises upon microorganisms of the activated sludge system in the wastewater treatment plant of the industrial park, five types of typical specific pollutants like herbicides were determined based on preliminary research, including prometon, propazine, acetochlor, metolachlor and bentazone, in order to explore their effects on these microorganisms and thereby provide references for controlling specific pollutants of the influents for the wastewater treatment plant of this park.

## 2. Experimental Materials

The activated sledge was taken from the aeration basin of the wastewater treatment plant in an industrial pesticide park of Binzhou.

The raw water of the wastewater was taken from the equalization basin of the wastewater treatment plant of an industrial pesticide park in Binzhou.

Prometon, propazine, acetochlor, metolachlor and bentazone are purchased from a pesticide shop of Jinan.

## 3. Experimental Methods

SOUR was one of the important parameters which characterize microbial activity of activated sludge systems for wastewater treatment,it reflects physiological state, mechanisms and metabolic state of microorganisms in activated sludge from the perspective of the respiration rate [2-3]. As a parameter for detecting microbial activity of activated sludge, SOUR has been generally recognized and widely used by scholars home and abroad for analyzing, evaluating and predicting operating conditions and treatment capacity of the systems [4-7].

The experimental methods were introduced as follows: 5L activated sludge was settled for 15min and 2L supernatant was taken out by siphonage. Then, 2L raw water of wastewater was added into it. After stirring and mixing, specific pollutants, including prometon, propazine, acetochlor, metolachlor and bentazone, were added at the prescribed doses. Sgitator was started to keep mixing the muddy water, and the air pump was turned on for aeration. The dissolved oxygen of the sludge treatment system was measured online, and the aerationwas suspended when the concentration of the dissolved oxygen was approximately 6.0 mg/L. Subsequently, initial and final numerical values of dissolved oxygen (DO) and time were recorded with a dissolved oxygen meter and a timer. Next, the concentration of mixed liquor volatile suspended solids (MLVSS) was measured for the activated sludge, and the SOUR of the sludge was calculated according to the following formula.

$$SOUR=(DO_{initial}-DO_{final})\ (mg/L)\ /\ (t_{initial}-t_{final})\ (h)\times MLVSS\ (g/L)$$

## 4. Results and Discussions

### *4.1. Impacts of Prometon upon SOUR of the Activated Sludge System*

The tested system of SOUR was shown in Fig 1 as follows when no prometon was fed and it was fed at different dosages, including 5mg/L, 10mg/L and 20mg/L.

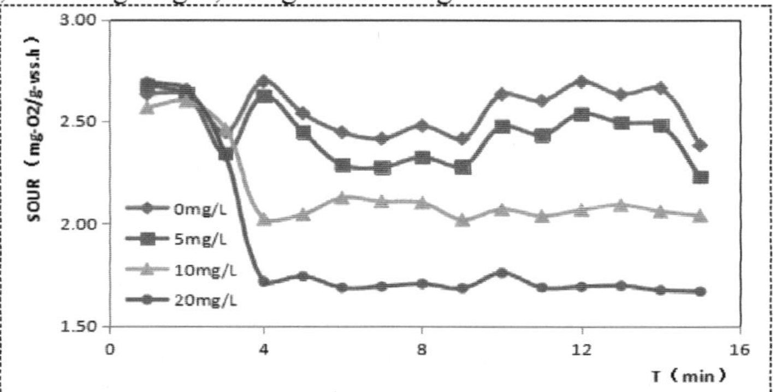


**Figure 1**. Impacts of Prometon upon SOUR of the Activated Sludge System

From Fig 1, it has been discovered that without prometon, the SOUR of the activated sludge system was about 2.5mg- $O_2$/g-vss.h. When prometon was fed at 5mg/L, 10mg/L and 20mg/L, the SOUR got to decline drastically at the 4th minute and generally kept stable thereafter. On average, it was declined by 7.62%, 11.3% and 25.8% respectively at these doses. This suggests that the metabolic activity of microorganisms would be inhibited in the activated sludge system if prometon exists in the wastewater.

### *4.2. Impacts of Propazine upon SOUR of Activated Sludge System*

The tested system of SOUR was respectively measured when propazine was fed at 5mg/L, 10mg/L and 20mg/L, as shown in Fig 2 as follows.

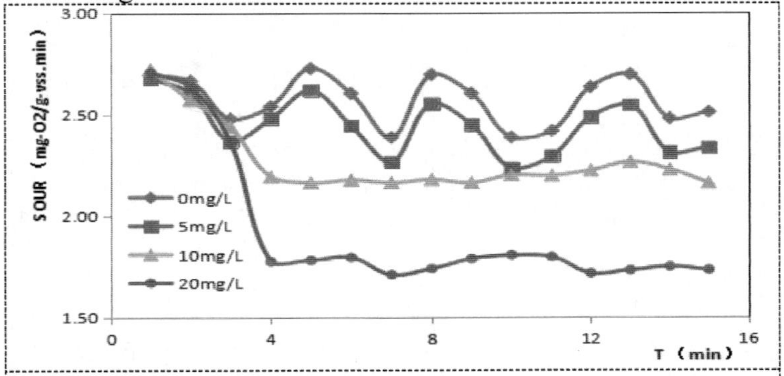


**Figure 2**. Impacts of Propazine upon SOUR of the Activated Sludge System

It has been observed from Fig 2 that when propazine was fed at 5mg/L, the SOUR of the activated sludge system has been slightly impacted and declined by about 3.7%. As it was fed at 10mg/L and 20mg/L respectively, it could decreased by 12.3% and 31.1% respectively. This indicates that the metabolic activity of microorganisms would be inhibited in the activated sludge system when there was propazine in the wastewater.

### *4.3. Impacts of Metolachlor upon SOUR of Activated Sludge System*

The tested system of SOUR was respectively measured when the metolachlor was fed at 5mg/L, 10mg/L and 20mg/L, as shown in Fig 3 as follows.

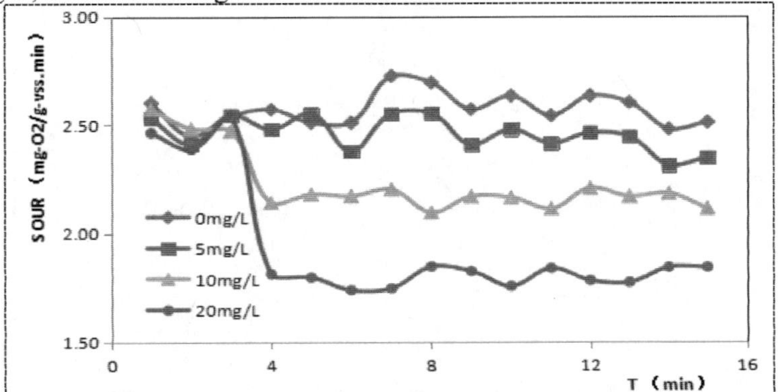


**Figure 3**. Impacts of Metolachlor upon SOUR of Activated Sludge System

It has been discovered from Fig 3 that SOUR of the activated sludge system was affected a little and declines by about 4.2% when metolachlor was fed at 5mg/L. It dropped by 10mg/L and 20mg/L respectively when the metolachlor was fed at 10mg/L and 20mg/L respectively. This reveal that the metabolic activity of microorganisms would be inhibited in the activated sludge system when metolachlor exist in the wastewater.

### *4.4. Impacts of Acetochlor upon SOUR of Activated Sludge System*

The SOUR was respectively measured when the acetochlor was fed at 3mg/L, 5mg/L and 10mg/L, as shown in Fig 4 as follows.

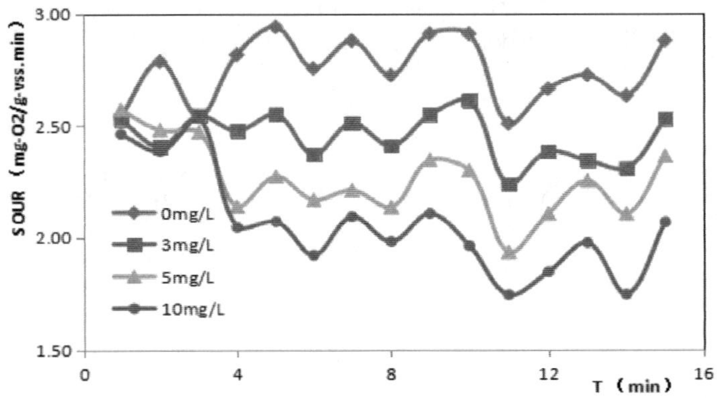


**Figure 4**. Impacts of Acetochlor upon SOUR of Activated Sludge System

It has been observed from Fig 4 that the SOUR of the activated sludge system declined significantly by 7.4%, 14.8% and 37.1% respectively when acetochlor was fed at 3mg/L, 5mg/L and 10mg/L. The metabolic activity of microorganisms was even inhibited to a great extent in the activated sludge system when acetochlor was fed at a low dose. Thus, it was significantly much more effective for inhibiting toxicity of microorganisms in the activated sludge system than prometon, propazine and metolachlor.

### *4.5. Impacts of Bentazone upon SOUR of Activated Sludge System*

The tested system of SOUR was respectively measured when the bentazone was fed at 5mg/L, 10mg/L and 20mg/L, as shown in Fig 5 as follows.

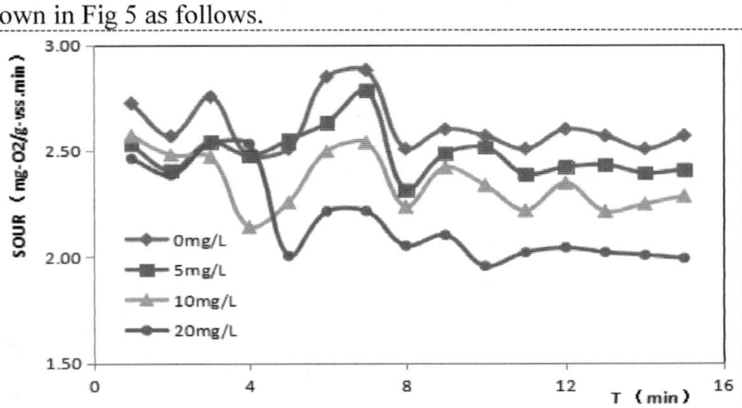


**Figure 5**. Impacts of Bentazone upon SOUR of Activated Sludge System

It has been observed from Fig 5 that SOUR of the activated sludge system declined by 4.8%, 10.1% and 17.0% on average respectively when the bentazone was fed at 5mg/L, 10mg/L and 20mg/L. Bentazone was effective for inhibiting metabolic activity of microorganisms in the activated sludge system. However, it was less effective for inhibiting such activity than prometon, propazine, acetochlor and metolachlor in case of the same dose.

## 5. Conclusions

- Specific pollutants like herbicides, including prometon, propazine, acetochlor, metolachlor and bentazone, may significantly inhibit microbial activity of microorganisms in activated sludge systems of wastewater treatment plants, and become increasingly more effective for inhibiting the activity with the increase of their concentration.

- All five types of specific pollutants inhibit anaerobic metabolism of microorganisms in the activated sludge system and their impacts are arranged as follows in view of their degree: bentazone<prometon≈propazine≈metolachlor<acetochlor. In case that the concentration of specific pollutants was 20mg/L in the wastewater, bentazone, prometon, propazine and acetochlor contribute to a decline of SOUR by 17.0%, 28.4%, 25.8% and 31.1% respectively in the activated sludge system. When the concentration of acetochlor was up to 10mg/L, the SOUR of the activated sludge system would decrease by 37.1%. Useful lessons may be drawn from this study for controlling the concentration of specific pollutants like herbicides in the influents of wastewater treatment plants of pesticide parks.

**References**

[1] WANG Zhi-hai, WEI Hong-bin, JIA Zhi-yu, et al. Activated sludge process for treatment of form aldehyde wastewater[J]. China water & wastewater. 2009,25(1): 86-88.

[2] GAO Chun-di, PENG Yong-zhen, GAO Kai, et al. On-line detection of oxygen uptake rate in SBR process and error analysis[J]. Industrial Water and Wastewater, 2001, 32( 2) : 4-7.

[3] SUN Yan, LI Ruo-gu, ZHANG Yan-qiu. Determination of the our in wastewater treatment plants with a manometric respiromete[J].Energy Environmental Protection. 2011(01): 19-22.

[4] PAMBRUN V, PAUL E, SPERANDIO M. Control and modelling of partial nitrification of effluents with high ammonia concentrations in sequencing batch reactor[J].Chemical Engineering and Processing: Process Intensification, 2008, 47( 3) :323-329.

[5] WANG Shu-tao, LI Su-ping, WANG Wei-qing, et al. Impace of ZnO nanoparticles on the activity of the activated sludge in SBR[J].China Environmental Science. 2014,34(10): 2575-2580.

[6] SONG Ying-hao, WANG Kai-jun, NI Wen, et al. Study on measurement of inhibition kinetic constant based on respiration rate in activated sludge process[J].Chinese Journal of Environmental Engineering. 2008,2(04): 493-497.

[7] SONG Ying-hao, WANG Kai-jun, NI Wen, et al. Study on judgment of inhibition type based on respiration rate in activated sludge process[J].Environmental protection Science. 2008,34(05): 7-9.

IOP Conf. Series: Earth and Environmental Science **61** (2017) 012002 doi:10.1088/1755-1315/61/1/012002

# Optimization of self-study room open problem based on green and low-carbon campus construction

**Baoyou Liu**

Key Laboratory of Pollution Prevention and Control in Hebei Province, College of Environmental Science and Engineering, Hebei University of Science and Technology, 050018 Shijiazhuang, China

* Corresponding author: liuzhang544@163.com

**Abstract.** The optimization of self-study room open arrangement problem in colleges and universities is conducive to accelerate the fine management of the campus and promote green and low-carbon campus construction. Firstly, combined with the actual survey data, the self-study area and living area were divided into different blocks, and the electricity consumption in each self-study room and distance between different living and studying areas were normalized. Secondly, the minimum of total satisfaction index and the minimum of the total electricity consumption were selected as the optimization targets respectively. The mathematical models of linear programming were established and resolved by LINGO software. The results showed that the minimum of total satisfaction index was 4055.533 and the total minimum electricity consumption was 137216 W. Finally, some advice had been put forward on how to realize the high efficient administration of the study room.

## 1 Introduction

In recent years, electricity waste has become a very serious problem in colleges and universities, which is incarnated in the evening self study. One situation is there are a handful of students in the study room, but the lights are all turned on; the other situation is there are a few students who want to study at night, but lots of classrooms are opened. Sometimes, there is even no student in a classroom while the lights are all still lit. Therefore, it is necessary to establish an economical and reasonable management method, which is not only conducive to promote the fine management of the campus, but also conducive to promote the green and low-carbon campus construction [1-2]. Our present research interests are focused on the optimization of typical environmental problems and their effective solution methods [3-5]. In this paper, a typical self-study room open arrangement problem was analyzed and resolved by lingo software. This paper may provide a reference for the relevant departments to intensify the management solutions in colleges and universities.

## 2 PROBLEM DESCRIPTIONS

A classroom administrator in a university needs to open some classrooms every night from 17:00 to 22:00. If a classroom is opened, then all the lights in this classroom are assumed to be opened please complete the following questions:

(1) There are about 19,500 students in the school. Sample survey indicates that each student is independent in self study and the probability of self-study is 0.9. Assuming that students who need to self-study can be satisfied by the degree of not less than 95%, so the number of self-study students is $19500 \times 0.9 \times 95\% = 16672$. In order to ensure the occupancy rate of classroom is not less than 4/5,

while not more than 90%, please try to answer which classrooms should be opened, so as to achieve the purpose of saving electricity?

(2) According to the specific location of the dormitories, the living area was divided into seven blocks (S1-S7). There are 135 classrooms available for evening study. Due to the geographical location, we divided them into four areas: Z1, Z2, Z3 and Z4. Z1 represents south public teaching building, Z2 represents Lecture group and library building, Z3 represents main public teaching building B and Z4 represents main public teaching building C and D. The main public teaching building is divided into Z3 and Z4 areas, because there are a large number of self-study rooms, and the classrooms in main public teaching building B are irregular, so it is helpful for the following targeted analysis. It is assumed that the spent time from the dormitory area to a classroom is the same as that of any classrooms in the same self-study area. Give a reasonable degree of satisfaction measurement, and re-consider how to arrange the classroom, can not only achieve the purpose of saving electricity, but also improve the student satisfaction index. In addition, try to arrange to open the same district of the classrooms.

## 3 PROBLEM ASSUMPTIONS

The power supply is normal; there is no power outage; the probability of self-study is not affected by the weather; the light bulbs, tables and chairs are intact, and can be used normally; all the lights and seats are distributed evenly; the probability of self-study has nothing to do with the distance between the living area and studying area.

## 4 MODEL AND SOLUTIONS

### *4.1 Mathematical model for question 1 and its solution*

First of all, electricity consumption in each self-study room and distances between different areas were normalized, the results are showed in Table 1 - Table 2. Then, the satisfactory function was constructed, which uses unit electricity consumption *Pn*(n=1-4) and distance *Lmn* as the dependent variable of the function. For unit electricity consumption and distance, we assume that the unit electricity consumption is slightly more important than the distance, so when given the weight value, we take *W*= (0.6, 0.4). Satisfactory function was showed as: *F(Pn, Lmn)*=0.6*pn*0.4*Lmn* (m=1, 2, 3, 4, n=1, 2...7). The data in Table 1 and Table 2 were substituted into the above formula, satisfactory function values form the dormitory area to the various study areas were calculated. For each opened classroom, we require its occupancy rate is not less than 4/5, while not more than 90%. Four study areas should meet the above requirements, so the number of students who need to study in four study areas was obtained; the results were shown in Table 3.

Assuming the number of the students is same in each dormitory building, the students who needed to self-study were evenly assigned to each building, and then the number of students can be obtained according to the number of buildings in each dormitory area, the results was shown in Table 4.

**Table 1.** Normalized electricity consumption of self study room in different areas.

| Area code | Z1 | Z2 | Z3 | Z4 |
|---|---|---|---|---|
| Unit electricity consumption pn /W | 0.399 | 0.155 | 0.243 | 0.202 |

**Table 2.** Normalized distance between different dormitory area and the study room area.

| Distance Lmn/km | Z1 | Z2 | Z3 | Z4 |
|---|---|---|---|---|
| S1 | 0.2391 | 0.2804 | 0.2402 | 0.2402 |
| S2 | 0.2487 | 0.2841 | 0.2336 | 0.2336 |
| S3 | 0.2082 | 0.2780 | 0.2569 | 0.2569 |
| S4 | 0.2220 | 0.2824 | 0.2478 | 0.2478 |

| | | | | |
|---|---|---|---|---|
| S5 | 0.1777 | 0.2713 | 0.2812 | 0.2812 |
| S6 | 0.1680 | 0.2700 | 0.2810 | 0.2810 |
| S7 | 0.1606 | 0.2688 | 0.2853 | 0.2853 |

**Table 3.** Number of students calculated in the 4 self-study areas according to requirements.

| | Z1 | Z2 | Z3 | Z4 | Sum |
|---|---|---|---|---|---|
| Total Seats | 2964 | 3716 | 4194 | 10440 | 21314 |
| *Mn*4/5 | 2371 | 2973 | 3355 | 8352 | 17050 |
| *Mn*90% | 2668 | 3345 | 3774 | 9396 | 19183 |

**Table 4.** The number of students in seven dormitory areas

| Dormitory areas | S1 | S2 | S3 | S4 | S5 | S6 | S7 |
|---|---|---|---|---|---|---|---|
| Number | 3449 | 2874 | 3874 | 3449 | 1149 | 1725 | 1149 |

Then let $M$ represent the sum of the satisfactory function:

$$M=\sum_{m=1}^{4}\sum_{n=1}^{7}XnmF(Pn,Lmn)\quad(m=1-4,n=1-7) \tag{1}$$

The linear functions on the objective function M and the constraints can be established. The following standard form was obtained:

Objective function

$$minM=\sum_{m=1}^{4}\sum_{n=1}^{7}XnmF(Pn,Lmn)\quad(m=1-4,n=1-7)=0.2391X11+0.2804X12+\ldots+0.2853X74 \tag{2}$$

Constraints:

*X11+X12+X13+X14*<=3449;*X21+X22+X23+X24*<=2874;*X31+X32+X33+X34*<=3874;*X41+X42+X43+X44*<=3449;*X51+X52+X53+X54*<=1149;*X61+X62+X63+X64*<=1725;*X71+X72+X73+X74*<=1149;2371<=*X11+X21+X31+X41+X51+X61+X71*<=2668;2973<=*X12+X22+X32+X42+X52+X62+X72*<=3354; 3355<=*X13+X23+X33+X43+X53+X63+X73*<=3774;8352<=*X14+X24+X34+X44+X54+X64+X74*<=9396 (3)

The above linear programming problem can be resolved by lingo 11.0. The minimum value of the satisfactory functions index is 4055.533.

*4.2 Mathematical Model 2 for question 2 and its solution*

The total number of seats required for each study room area can be calculated from the results of Model 1, which is *Z1*=2371, *Z2*=2973, *Z3*=3854, *Z4*=8854. In order to obtain the minimum value of the total electricity consumption, the mathematical model of the four study areas can be established respectively. For mathematical model for *Z1* area, the linear programming model was established as follows,

Objective function

$$minZ1=\sum_{i=n202}^{n413}PiXi=1104Xn202+\ldots+Xn413 \tag{4}$$

Constraints:

178*Xn202*+178*Xn205*+126*Xn206*+126*Xn207*+126*Xn208*+172*Xn212*+126*Xn302*+178*Xn305*+126*Xn306*+126*Xn307*+126*Xn308*+126*Xn309*+126*Xn310*+172*Xn313*+178*Xn403*+178*Xn406*+126*Xn407*+126*Xn408*+172*Xn412*+172*Xn413*>=2371 (5)

*Xi*=0 or 1（*i*=n202, n205...n413） (6)

The results were obtained by LINGO 11.0 software: the classrooms in the south public teaching building area should all be opened except 208, 212, 302, 306 and those classrooms on fifth floor, when the minimum electricity consumption of Z1 area is 32240 W. Similarly established were the

Mathematical Models for Z2, Z3 and Z4 area, the minimum electricity consumptions were 12672, 31104 and 61200 W respectively. The total minimum electricity consumption can be calculated through the above optimized results, which was 32240 +12672 +31104 +61200 = 137216W.

## 5 CONCLUSIONS AND PROSPECT

Based on the above optimized procedure, the open response system software of self-study room open arrangement can be designed. Once college students input their preferred time and location of self-study into the computer, the system can automatically output the optimal results to meet the requirement based on the software calculation. The operations of the software will be helpful to improve the utilization efficiency of the classrooms, thereby promote low-carbon campus construction. Additionally, energy-saving lamps should be vigorously popularized, which may solve the campus waste of electric power greatly. As far as we know, energy-saving lights cost less money in the process of using than normal fluorescent lamp. What's more, reasonable campus planning, humanized distribution and distance of teaching building and dormitories can not only make the learning more comfortable, but also promote campus energy conservation.

## ACKNOWLEDGMENT

The author gratefully acknowledges environment engineering key subject construction of Hebei Province for the financial supports.

## REFERENCES

[1] F. Jiang, Y.H. Fan, H.Y. Ding, L. She, Management Observation **20**, 107 (2009)
[2] T. Liu, X.L. Liu, Journal of Ankang University, 25, 9(2013)
[3] H.W. Zhang, Z.G. Niu, LINGO8.0 and its application in environmental systems optimization (Tianjin University Press, Tianjin, 2005)
[4] B.Y. Liu, N.X. Jin, Applied Mechanics and Materials, 744, 2126(2015)
[5] B.Y. Liu, Int. J. Modelling, Identification and Control, 16, 285(2012)

# Unsteady heat transfer performance of heat pipe with axially swallow-tailed microgrooves

**R P Zhang**

School of Materials Science and Engineering, Jingdezhen Ceramic Institute, Jingdezhen Jiangxi 333001, China
E-mail: rpzhang@126.com

**Abstract.** A mathematical model is developed for predicting the transient heat transfer and fluid flow of heat pipe with axially swallow-tailed microgrooves. The effects of liquid convective heat transfer in the microgrooves, liquid-vapor interfacial phase-change heat transfer and liquid-vapor interfacial shear stress are accounted for in the present model. The coupled non-linear control equations are solved numerically. Mass flow rate at the interface is obtained from the application of kinetic theory. Time variation of wall temperature is studied from the initial startup to steady state. The numerical results are verified by experiments. Time constants for startup and shutdown operation are defined to determine how fast a heat pipe responds to an applied input heat flux, which slightly decreases with increasing heat load.

## 1. Introduction

Heat pipes with axially swallow-tailed microgrooves as shown in Fig. 1 are highly efficient thermal systems due to phase change heat transfer of the internal working fluid. It has demonstrated excellent thermal performance, high degree of temperature uniformity and reliable operation under microgravity conditions, which are being used in the area of efficient and reliable heat removal, such as spacecraft thermal control systems and microelectronic cooling systems.

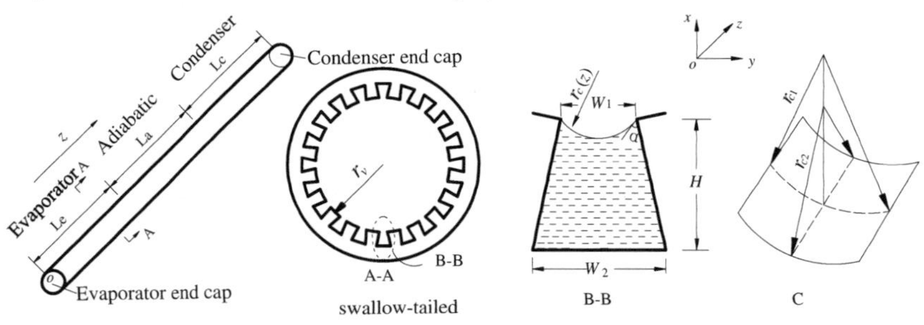


Fig. 1 Schematic diagram of heat pipe with axial swallow-tailed microgrooves

In many thermal control applications, heat pipes typically do not operate at steady state conditions. Grarimella [1] pointed out that many heat pipes operate in transient conditions to manage heat dissipation problems and heat fluctuations of electronic components. Colwell and Chang [2] made a series of tests on an 80-cm long Freon(R11) stainless-steel heat pipe under normal and supercritical startup. Wang [3,4]performed experimental investigation on a flat plate copper heat pipe during startup and shutdown cycles, experimental results were compared with the analytical model of Zhu [5] and were found to be in very good agreement. Most of the available researches on transient characteristics of heat pipes with axially microgrooves are focused on the triangular microgrooves [6-9], while there have been very few reports on heat pipes with axially swallow-tailed microgrooves. Unsteady investigation made on the heat pipe with the axially swallow-tailed microgrooves not only understand the startup / shutdown and load variation characteristics of heat pipe, also contribute to preliminary transient analyses of loop heat pipes and capillary pumped loops. Therefore, in the current study, a one-dimensional fluid flow and heat/mass transfer unsteady model is developed for heat pipe with axially swallow-tailed microgrooves and coupled nonlinear governing equations are solved numerically.

## 2. Theoretical model

In the evaporator section, the heat imposed on the heat pipe wall together with the difference of the heat entering and leaving the control element by conduction is partly used to raise the heat pipe wall temperature while the remaining heat is absorbed by the working liquid in the grooves through convection. Therefore, the unsteady energy balance equation of the heat pipe wall in the evaporator section can be given by

$$\rho_w C_w A_w dz \frac{\partial T_w}{\partial t} = A_w dz K_w \frac{\partial^2 T_w}{\partial z^2} + \frac{Q_{in} dz}{L_e} - h_{wl}(T_w - T_l) w_{wl} dz \quad (1)$$

In the adiabatic section, variation of wall temperature as a function of time not only comes from the condensation of vapour in the fin and swallow-tailed microgrooves, but also comes from the working liquid heat convection in the swallow-tailed microgrooves and the difference of the heat entering and leaving the control element by conduction, the unsteady state energy balance of the heat pipe wall can be expressed as:

$$\rho_w C_w A_w dz \frac{\partial T_w}{\partial t} = A_w dz K_w \frac{\partial^2 T_w}{\partial z^2} - m_i h_{fg} R_{wv} dz - h_{wl}(T_w - T_l) w_{wl} dz \quad (2)$$

In the condenser section, the heat taken up by the cooling liquid and the difference of the heat entering and leaving the control element by conduction not only comes from the condensation of vapour in the fin and swallow-tailed microgrooves, but also comes from the working liquid heat convection in the swallow-tailed microgrooves , so the unsteady state energy balance of the heat pipe wall is written as

$$\rho_w C_w A_w dz \frac{\partial T_w}{\partial t} = A_w dz K_w \frac{\partial^2 T_w}{\partial z^2} - h_{wl}(T_w - T_l) w_{wl} dz - m_i h_{fg} R_{wv} dz - h_\infty (T_w - T_\infty) \pi d_{out} dz \quad (3)$$

The energy entering the control volume includes two parts, one is the sensible heat of working liquid flowing into the element, the other is the heat taken up by the working liquid from the heat pipe wall. The energy leaving the control volume also include two portion, one is the sensible heat of working liquid flowing out of the element, the other is the heat leaving the element by evaporation. Therefore, the unsteady energy conservation equation of the liquid in the microgrooves in differential form is expressed as

$$\rho_l C_l A_l dz \frac{\partial T_l}{\partial t} = \rho_l C_l V_l A_l dz \frac{\partial T_l}{\partial z} + h_{wl}(T_w - T_l) w_{wl} dz - m_i h_{fg} R_{lv} dz \quad (4)$$

The rate of accumulation of the convective momentum is equal to the momentum change due to liquid flow in the microgrooves, pressure force acting on the element, the wall shear force and liquid-vapour interfacial force. So the differential form of unsteady state momentum balance of liquid can be expressed as

$$\frac{\partial(\rho_l A_l V_l)}{\partial t} = \rho_l A_l V_l \frac{\partial V_l}{\partial z} + A_l \frac{\partial P_l}{\partial z} - w_{wl} \tau_{wl} - R_{lv} \tau_{il} \quad (5)$$

The variation of mass as a function of time is equal to the difference of change of liquid mass due to convection and the mass evaporated from the control volume. So the differential form of mass balance at unsteady state can be expressed as

$$\frac{\partial(\rho_l A_l)}{\partial t} = \frac{\partial(\rho_l A_l V_l)}{\partial z} - \frac{m_i h_{fg} R_{lv}}{h_{fg}} \quad (6)$$

## 3. Experimental validation

The present mathematical model is verified by an experiment on the heat transfer performance of a swallow-tailed heat pipe with the geometry shown in Table 1. The schematic of the experimental setup is shown in Fig.2. Heat removal at the condenser was provided using a clamped cold water jacket, a Thin-film heater was mounted on the evaporator region to simulate the heat source. Adequate contact between the water jacket and the condenser region of the heat pipe, as well as the thin-film heater and evaporator region, was ensured using pads of thermal interface material. The thin-film heater was insulated on top using a thick piece of acrylic. Nine thermocouples (type-T), which had uncertainty of ±0.1°C， are attached on the outside wall of the heat pipe. All the thermocouples are calibrated against any variation in the room temperature. Before each test, heat pipe was allowed to reach a steady-state temperature condition by turning on the flow through the cooling jacket. This provided the initial condition for the experiments as well as for the model. The heater was then turned on, and temperature measurements continued until a new steady state was reached. Temperature measurements were recorded during the transient heating period as well as at steady state. The axial temperatures and the water inlet and outlet temperatures are recorded at an interval of 1 second using the data acquisition Instrument. Steady state temperatures reported at each power level were averaged over a period of one minute. Working temperature is regarded as the mean value of these measured data of the adiabatic segment. Given a heat load, $Q_{in}$, the working temperature of the heat pipe can be controlled by regulating the temperature and velocity of the cooling water supplied by the constant temperature water bath.

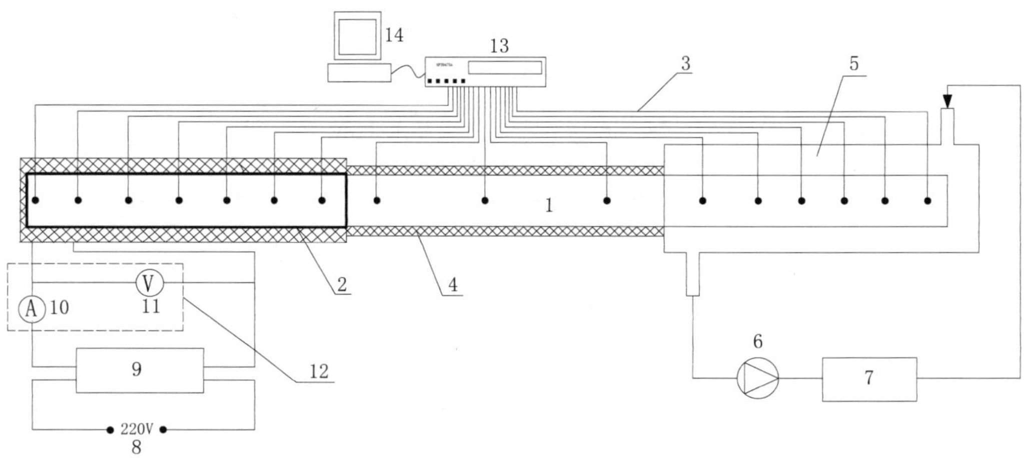


Fig.2 Schematic diagram of experimental setup

1- Heat pipe, 2-Thin-film electrical heater, 3-Thermocouple, 4-Thermal insulation, 5-Tank,6-Pump, 7-Constant temperature water bath, 8-power meter, 9-Voltage regulator, 10-Amperemeter, 11-Voltmeter, 12-Power supply, 13-Data acquisition instrument, 14-Computer

## 4. Results and discussion

To make the comparison meaningful, the transient behaviour of an aluminium-ammonia axially grooved heat pipe was studied using the experimental working condition. Heat pipe started at ambient temperature, T=293.15, constant heat load of 200W was applied and condenser section was cooled by convection. Both the experimental and numerical computational temporal temperature response for the evaporator section of heat pipe are presented in Fig.3. In view of the uncertainty in the internal dimensions and wick properties of the heat pipe tested, the agreement between the numerical prediction and experiment is reasonable.

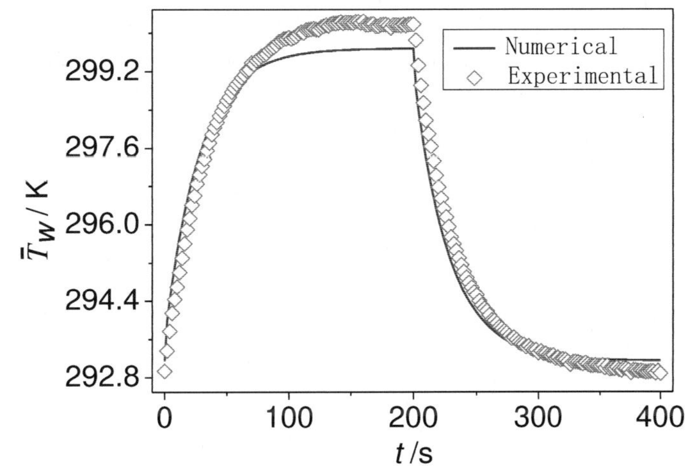


Fig.3 Comparison of the experimental and numerical wall temperature of the evaporator section

Model predictions for transient variation of axial wall temperature distribution are shown in Fig.4. It can be seen from the figure, the temperature distribution at each section (i.e. the evaporator, adiabatic and condenser sections ) during the startup process is almost uniform, and the major temperature difference occurs at the junctions among neighbouring sections. The uniformity of the temperature distribution at the evaporator section is vital for the heat pipe to be used to thermal control of electronic product. It can be seen from the figure that temperature increase gets more and more slowly with time. This phenomenon also indicated that, during the initial startup, heat transported by phase change play less important role in the whole heat transfer, and become more and more important as the process continues. Once the heat pipe reaches the steady state, the input heat is completely transported through evaporation and condensation. In this context, heat transport device consisting of grooved heat pipe can reduce weight of electronic cooling system, which is of significance to aero- and space flight. In addition, the wall temperature of heat pipe maintain at a relatively low level for the high heat flux imposed.

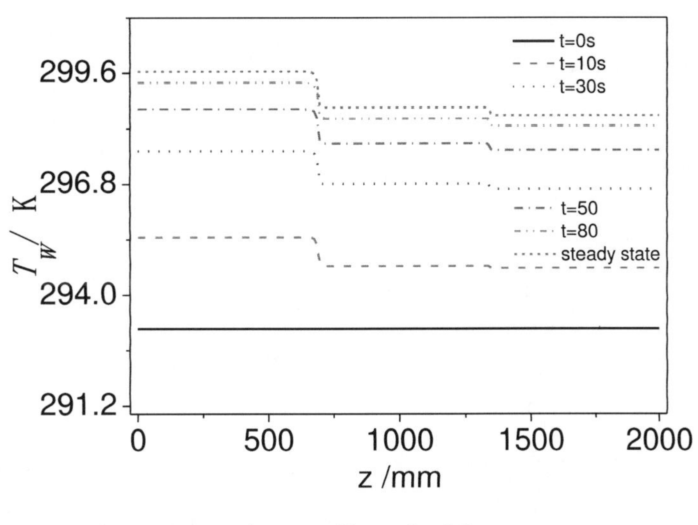


Figure 4 Transient profiles of axial temperature

The time required to reach steady state is an important parameter for the startup/shutdown of a heat pipe. The concept of the time constant, τ, as proposed by El-Genk and Huang [10] is utilized in the analysis to comprehensively simulate startup/shutdown operation.

Figure 5 presents time constants for both the startup and shutdown operation for different heat inputs. As can be seen from the figure, heat flux minimally affects the time constants of heat pipe. Increasing the heat flux slightly lowers the time constants. The results also show that time constant for the shutdown operation is a little more than that for the startup operation.

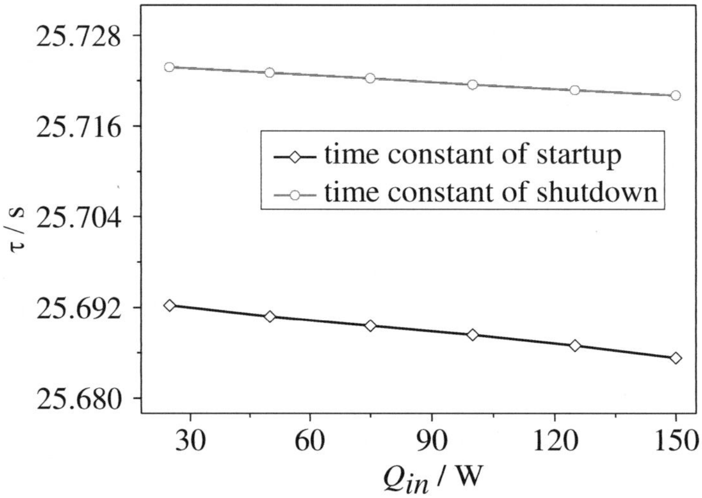


Fig. 5 Startup/shutdown time constant profiles as a function of heat inputs

Fig.6 shows the temporal variation for the effective thermal conductivity. As can be seen from the figure, the effective thermal conductivity increases gradually until the heat pipe achieve the steady state during the startup operation, and retains a constant value during the shutdown operation. In addition, there is a sharp increase in the value of the effective thermal conductivity when the heat pipe operate from the steady state to shutdown operation.

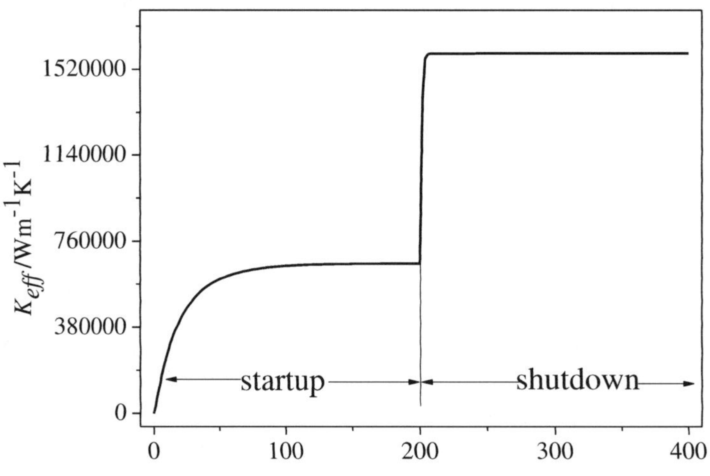


Fig. 6 Transient variation of the effective thermal conductivity

**5. Conclusions**

A one-dimensional analysis model for a heat pipe with axial swallow-tailed microgrooves is developed and analyzed numerically to predict the transient heat transfer and fluid flow characteristics. The transient thermal performance of heat pipe is investigated by the experiment. The results obtained from the proposed model are in close agreement with the experimental data in terms of startup and shutdown temperature response. The conclusions can be summarized as: Temperature distribution at each section (i.e. the evaporator, adiabatic and condenser sections) during the startup process is almost uniform, and the major temperature difference occurs at the junctions among neighbouring sections. Heat flux minimally affects the time constants of the heat pipe. Increasing the heat flux slightly lowers the time constants for wall temperature response of heat pipe. The results also show that the time constant for the shutdown operation is a little more than that for the startup operation. Effective thermal conductivity increases gradually until heat pipe achieve the steady state during the startup operation, and retains a constant value during the shutdown operation.

**References**

[1] Sobhan C B Garimella S V and Unnikrishnan V V 2000 A Computational Model for the Transient Analysis of Flat Heat Pipes *Inter Society Conference on Thermal Phenomena* 106-113
[2] Colwell G T and Chang W S 1984 Measurements of the Transient Behaviour of a Capillary Structure Under Heavy Thermal Loading International Journal of Heat and Mass Transfer 27 541-551.
[3] Wang Y and Vafai K 2000 An Experimental Investigation of the Transient Characteristics of a Flat Plate Heat Pipe During Startup and Shutdown Operations ASME Journal of Heat Transfer 122 525-535
[4] Wang Y and Vafai K 2000 An Experimental Investigation of the thermal Performance of an Asymmetrical Flat Plate Heat Pipe International Journal of Heat and Mass Transfer 43 2657-2668
[5]Sobhan C B Huang X Y and Liu C Y 2000 Investigations on Transient and Steady-State Performance of a Micro Heat Pipe Journal of Thermophysics and Heat Transfer 14(2) 161-169
[6] Suman B Sirshendu D and Sunando D Transient modeling of micro-grooved heat pipe International Journal of Heat and Mass Transfer 48 1633-1646
[7] Suman B 2007 On the Transient Analysis of a V-Shaped Microgrooved Heat Pipe Journal of Heat Transfer 129 1584-1591
[8] Anand s De s and Dasgupta s 2002 Experimental and Theoretical Study of Axial Dryout Point for Evaporation from V-shaped Micro Heat Pipe International Journal of Heat and Mass Transfer 45 1535-1543
[9] Suman B 2006 On the fill charge and the sensitivity analysis of a V-shaped micro heat pipe AICE Journal 52 (9) 3041-3054
[10] El-Genk M S and Huang L 1993 An experimental investigation of the transient response of a water heat pipe International Journal of Heat and Mass Transfer 36 3823-3830

IOP Conf. Series: Earth and Environmental Science **61** (2017) 012004 doi:10.1088/1755-1315/61/1/012004

# Effects of trade openness and market scale on different regions

**Renqu Tian**[1]**, Zisheng Yang**[2,*]

[1] Institute of economics, Yunnan University of Finance and Economics, Kunming, Yunnan, China
[2] Institute of Land & Resources and Sustainable Development, Yunnan University of Finance and Economics, Kunming, Yunnan, China
goldfishgogo@163.com*

**Abstract.** This paper revisits the relationship between growth, trade openness and market scale. Empirical studies have provided that area develops lopsided problem in China is increasingly serious, while large trade openness and market scale bring about more economic growth. We use a number of data set from province-level's gross domestic product and socio-economic, as well as statistical methods panel ordinary least squares and instrumental variables estimation techniques to explore the effects of trade openness and regional market scale on the three major economic regions. The results indicate: Firstly, the impact of market scale and trade openness on economic growth is found to be positive. Secondly, the overall regional disparity is owing to the trade openness, market scale and macroeconomic policies. Thirdly, midland and western region should take advantage of regional geographical location and resource to expand exports and narrow the regional difference.

## 1. Introduction

Over the past several decades, foreign trade has become the most active part in China's economy. In 1978, import and export goods of China amounted to $20 billion 500 million, which in the world trade only ranked thirty-second, accounting for less than 1%. In 2013, Chinese total import and export of goods amounted to $4 trillion and 160 billion, of which exports amounted to $2 trillion and 210 billion, imports amounted to $1 trillion and 950 billion. According to the World Trade Organization Secretariat preliminary statistics, China has surpassed the United States to become the world's largest trade country.

At the meantime, unbalanced growth of regional economy has been one of the key development characteristics considered across China. The economic disparities is even more discernible between eastern coastal areas and western regions. For example, per capita GDP in Yunnan province in 1999 was 4,478 yuan, while per capita GDP in Shanghai was 27,561 yuan, there is a difference of 23,083 yuan. However, in 2013 per capita GDP of Yunnan province increased to 27,264 yuan, per capita GDP of Shanghai at this time already increased to 97,343 yuan, the gap between them is 72,731 yuan, which is 3.04 times in 1999. It means per capita GDP in China has increased year by year, inequality among individuals is growing among most regions.

Are there exist some connections between growth and openness? Theoretically, the influence of trade openness on economic fluctuation is uncertain. Openness will increase the external impact of economic shocks. Meanwhile, the degree of specialization of the industry makes it more vulnerable in the face of shocks from specific trade sectors.

To reflect the social and economic development of different regions and to provide the basic regional development policy for the national development, Chinese Seventh Five-Year Plan divided country into three economic regions, respectively they are east region, central and western region. We contributes to the literature by using province-level information to study and try to solve (1) the main factors behind unbalanced growth in China; (2) the effects of trade openness and regional market scale on the three major economic regions.

The remainder of this paper proceeds as follows: Section 2 briefly reviews the literature and introduces the instrumental variables; Section 3 describes the dataset and method; Section 4 presents the empirical results of three economic regions; Section 5 investigates the determinants of regional disparity; followed by conclusions and policy implication in Section 6.

## 2. Literature review

There have been rich discussions on effects of trade openness and market scale. According to the thoughts of early scholars, openness and trade scale these potential determinants of growth might be interlinked (see Magdalene, 2016). However, due to the different situation in regions and different research methods that adopted by scholars, there are several arguments of interaction effect between trade openness and market scale in explaining economic growth. On the one hand, most studies support the positive effect of trade openness on economic growth. Geoffrey (2014) suggested that open economies grow faster than closed economies. Kim (2016) showed that on average, greater international trade, promotes economic growth and amplifies growth volatility in the long-run, in the short-run, causes a negative short-run correlation between growth and growth volatility. Ma (2012) claimed that trade openness has a positive effect on economic growth and in the meanwhile it baffles the development of labour-intensive industries. On the other hand, as emphasized in Menyah (2014), trade liberalization and financial development may not have made a significant impact on growth.

Recent empirical work on Chinese research also focuses on relationship of openness, growth and other influencing factors. Xu and Meng (2015) estimated the relationship between human capital and growth though panel threshold model. The results indicated that there was negative association between openness and growth with a lower level of human capital, while there was significant positive association between these variables if with a higher level of human capital. By analysing panel data of 31 provinces in China from 1998 to 2012, Xu and Guo (2014) demonstrated that the energy consumption plays an important role in promoting regional economic growth, but the effect of trade liberalization on economic growth is not important in China's central and western area. Wang (2015) and Li (2011) drew a conclusion that carbon dioxide emissions is increasing due to trade openness in China, which means international trade has a negative impact on China's environment.

Previous cross-country studies on the openness-growth connection are subject to fundamental methodological and econometric problems, as pointed out by Mohammad et al.(2016). However the endogenous problem of trade openness may exist in various aspects of country: the more affluent provinces and autonomous regions, the greater the regional advantage to expand exports. Therefore researchers use the geographical distance from the provincial capital to the nearest coastline to measure the index of trade openness (Huang and Li, 2006), although in a certain extent this method can reflect the transportation cost, there still has a certain deviation between geographical distance and the actual transport distance.

So far there remains a lack of satisfactory measure of openness, and very little empirical evidence on the impact of the integrated economy on regional economic growth. Motivated by this idea, we investigate the research methods of Huang and Li (2006), using road distance from the provincial capital city to the nearest port which can more appropriately represent the transportation cost. Moreover, the geographical distance of the city at the prefecture level or above is taken as the instrumental variable, which can effectively solve the endogenous problem of the empirical study.

## 3. Model and data

Based on the model of Huang and Li (2006), we use the following empirical framework to investigate long-run growth. Accordingly, the author puts forward the corresponding with China's actual situation of countermeasures, to narrow the regional economic disparities exist among regions. With a number of data set from province-level GDP and socio-economic, this paper introduces fixed effects into regressions to control the endogenous problem of trade openness and market scale. In general form, this model can be characterized as

$$g_{it} = \beta_0 + \beta_1 FMA_{it} + \beta_2 RMS_{it} + \beta_3 FMA_{it} \times RMS_{it} + \beta_4 logy_{i,t-\tau} + \beta'_5 Z_{it} + \varepsilon_{it} \quad (1)$$

The data consists of i=1,... , 28 provinces, and t=1, 2, ... ,4 period. Where $\beta_0$ and $\varepsilon_{it}$ respectively for constant and error term. Explained variables $g_{it}$ is the logarithm of the average growth rate of per provinces capita income in period t. $logy_{it}$ is the logarithm of the initial capita income per province in period t. $FMA_{it}$ is the trade openness of province in period t. $RMS_{it}$ is a calculated measure of regional market scale:

$$RMS_{it} = \Sigma_{j \neq i}(Y_{ij} / D_{ij} + Y_{it} / D_{ii}) \quad (2)$$

Where $D_{ij}$ is the shortest road distance between provinces i and j, $D_{ii}$ is the provincial internal distance:

$$D_{ii} = \frac{2}{3}\sqrt{S_i \pi} \quad (3)$$

Where $S_i$ is the land area of province i; $Y_{it}$ is the per capita GDP, $Y_{ij}$ is the gross regional domestic product. $FMA_{it} \times RMS_{it}$ is the multiplicative interaction between trade openness and regional market scale, parameter $\beta_3$ is expected to be negative if the foreign and regional markets exhibit substitutes for each other. The variable $Z_{it}$ represents a vector of control and environmental variables that are primarily determined by investment rate ($INV_{it}$), literacy rate ($LR_{it}$), as well as the proportion of the total area of the government consumption ($GOV_{it}$).

Data on literacy rate, used to measure the level of literacy in a country, currently we use the proportion of people who over school age (15 years old) could read and write in the corresponding population, also come from the National Bureau of Statistics of China. Data for geographical factors, distance inside and between provinces, are taken from electronic map. Considering the topography (such as the mountain plateau and desert, etc.), we use road distance from the provincial capital to the nearest port instead of straight-line distance.

The annual data for the period 1970-2010 is calculated using in this paper comes from the National Bureau of Statistics of China. We divide this period into four sections, and in the regression we take the arithmetic mean of variables in each time period. The cross-sectional data totals 28 provinces (not include Xizang, Chongqing and Hainan), hence the total data consist of 112 provincial-level observations of province GDP for four periods, every ten years as a time period. Notice that there are only 84 observations for literacy rate because of incomplete information in 1970s.

## 4. Empirical analysis

Following the definition of the variables and the methodology, we now turn to the empirical analysis. Table 1 reports the summary statistics for both complexity and the actual volatility. According to the regression Equation (1), the author uses the geographical distance of China provincial capital city from the nearest port (variable FMA) as the instrumental variable instead actual trade openness, obtained by two stage least squares results as shown in Table 2. The return of the three major economic zones of the east region, midland, and west region, the estimated results of different regions are obtained as shown in Table 3.

**Table 1.** Summary statistics

| Variable | Obs | Mean | Std. | Min | Max |
|---|---|---|---|---|---|
| $g_{it}$ | 112 | 0.088 | 0.038 | 0.019 | 0.198 |
| **rms** | 112 | 0.297 | 0.394 | 0.008 | 1.847 |
| **fma** | 112 | 1.993 | 3.310 | 0.050 | 23.877 |
| **fma*rms** | 112 | 0.934 | 2.540 | 0.000 | 16.401 |

IOP Conf. Series: Earth and Environmental Science **61** (2017) 012004 doi:10.1088/1755-1315/61/1/012004

| | | | | | |
|---|---|---|---|---|---|
| **lny** | 112 | 5.712 | 1.417 | 1.977 | 9.134 |
| **gov** | 112 | 0.114 | 0.044 | 0.035 | 0.248 |
| **inv** | 112 | 0.442 | 0.160 | 0.172 | 1.008 |
| **lr** | 84 | 0.804 | 0.151 | 0.443 | 0.984 |

In Table 2 as a benchmark in column (1), we use the augmented model that explains differences in average growth rate of per provinces capita income across China by the geographical level in the previous period to mitigate the endogeneity problem, including the regional market scale ($RMS_{it}$), trade openness ($FMA_{it}$) and interaction $FMA_{it} \times RMS_{it}$, we observe that the effect of $FMA_{it}$ and $RMS_{it}$ are both positive and highly significant, and the interaction between trade openness and regional market scale $FMA_{it} \times RMS_{it}$ is negative across all formulas, and in column (3) and (5) significant at the five percent level. In column (2) (3) (4) we introduce a range of other possible control variables. Finally in column (5) we consider all of the variables, we can see that the results are still stable.

**Table 2.** Panel two-stage least squares estimation results

| | **(1)** | **(2)** | **(3)** | **(4)** | **(5)** |
|---|---|---|---|---|---|
| **fma*rms** | -0.007** | -0.005* | 0.005*** | -0.004** | 0.005*** |
| | (-2.387) | (-1.757) | (-3.095) | (-2.196) | -0.956 |
| **rms** | 0.070*** | 0.050*** | 0.039*** | 0.024* | -0.009 |
| | -6.109 | -3.528 | -3.419 | -1.894 | (-0.559) |
| **fma** | 0.007*** | 0.006** | 0.004** | 0.003* | 0.001 |
| | -3.224 | -2.621 | -2.496 | -1.894 | -0.981 |
| **lny** | 0.010*** | -0.012*** | -0.005* | -0.007** | 0.013*** |
| | (-3.994) | (-4.692) | (-2.019) | (-2.673) | (-5.491) |
| **gov** | | 0.286*** | | 0.239*** | 0.042 |
| | | -3.579 | | -3.455 | -0.508 |
| **inv** | | | 0.135*** | 0.126*** | 0.088*** |
| | | | -5.582 | -5.531 | -3.563 |
| **lr** | | | | | 0.039 |
| | | | | | -1.282 |
| **_cons** | 0.118*** | 0.106*** | 0.043** | 0.037* | 0.098*** |
| | -7.926 | -6.169 | -2.061 | -1.884 | -4.241 |
| **N** | 112 | 112 | 112 | 112 | 84 |
| **R-sq** | 0.653 | 0.695 | 0.741 | 0.769 | 0.847 |

t statistics in parentheses,* p<0.1, ** p<0.05, *** p<0.01

And we consider to what extent do these variables matter for understanding differences in growth across different regions? Table 3 provides an analysis of economic growth focusing on the three economic regions of China for selected years.

**Table 3.** District data

| | **eastern area** | **midland** | **western area** | **all** |
|---|---|---|---|---|
| **fma*rms** | -0.005*** | -0.034 | 0.037* | -0.004** |
| | (-4.329) | (-1.066) | -1.88 | (-2.196) |
| **rms** | 0.042* | 0.074 | 0.092*** | 0.024* |
| | -2.177 | -1.798 | -5.182 | -1.894 |
| **fma** | 0.002 | 0.025 | 0.007* | 0.003* |
| | -1.763 | -1.891 | -1.872 | -1.894 |
| **lny** | 0.010** | -0.002 | -0.014*** | -0.007** |
| | -2.692 | (-0.514) | (-4.385) | (-2.673) |
| **gov** | -0.145 | 0.08 | 0.229** | 0.239*** |
| | (-0.875) | -0.655 | -2.309 | -3.455 |
| **inv** | 0.107** | 0.123** | -0.018 | 0.126*** |

| | | | | |
|---|---|---|---|---|
| | -2.257 | -3.489 | (-0.465) | -5.531 |
| **_cons** | -0.019 | 0.009 | 0.115*** | 0.037* |
| | (-0.967) | -0.392 | -3.6 | -1.884 |
| **N** | 44 | 28 | 40 | 112 |
| **R-sq** | 0.775 | 0.87 | 0.894 | 0.769 |

t statistics in parentheses,* p<0.1, ** p<0.05, *** p<0.01

## 5. Results

The empirical result confirms the existence of research that market scale and trade openness have a positive impact on economic growth in the long run. For the regions with higher export openness, its growth will be faster in the provinces closer to overseas market. For the regions with higher dependence on domestic demand, its growth will be faster in the provinces with greater market scale.

We further evaluate $FMA_{it} \times RMS_{it}$ in different period and region, a remarkable substitutive effect is found to exist between markets at home and abroad.

Table 2 shows that investment rate and government consumption are significant at the 1% level across column (1)-(4). Since these variables are government controlled this result suggests that investment rate and government consumption is also important in understanding differences across regions. It can be seen however that literacy rate is not significant, one possibility is that there is migration across regions, while most internal trade in China is in agricultural products and processed products, this result is not unexpected.

In Table 3 the association of multiplicative interaction between trade openness and market scale is particularly significant in eastern area. One way to interpret this is that eastern area enjoys lower transport costs and experiences a high influx of foreign capital, hence it has a large substitution effect of domestic and foreign market.

In summary, unbalanced economic development is very common no matter with or between countries. To narrow the regional difference and promote the regional sustainable development is an important challenge at the present stage. Further research may be conducted to ascertain the following factors: (1) the relationship between spatial evolution mode and regional industrial structure; (2) the quality of infrastructure and how does it influence per capita income in China.

## 6. Policy suggestions

Our empirical research confirms that the growth is determined jointly by the trade openness and market scale. According to the above conclusions, this paper puts forward following policy suggestions:

(1) The difference of openness degree is one important reason for the difference of economic growth among regions. In a more open economy, the market mechanism may play a more obvious role, resulting in higher efficiency of resource allocation. Speeding up the reform and open-up in middle and western regions may also be an important lever for facilitating their economic growth. However, the openness is largely dependent upon exterior system, policy convergence is needed for the convergence of economic growth among regions.

(2) High-quality market environments should be established, backed by an improved legal system, with surveillance by social networks, the media, and individuals. For instance, our results suggest that border areas like Yunnan, a relatively closed economy by the measure of trade to GDP, would probably improve its overall market competitiveness if it moved from its current fixed regime to a more flexible one, such as be endowed with greater administrative and financial autonomy, etc. Only by achieving the transformation of primary processing products exports to high value-added exports, can these backward area expend its market scale.

(3) Provinces and regions in mainland China shall further improve transport infrastructure and weaken the geographical disadvantage in terms of international market association. Meanwhile, western provinces and regions shall give play to their own comparative advantages and expand their trade with neighbouring countries. Also, government could increase the intensity of economic support

and improve the local infrastructure construction, to be tilted to a certain extent in the tax system, which will accelerate the development of central and western provinces.

**Acknowledgments**
Authors would like to thank Miao Hongwei who kindly provided his ideas and suggestions. Authors also wish to thank anonymous reviewers for their valuable comments to improve the paper.

**References**

[1] Ferreira, F., Gignoux, J. The measurement of inequality of opportunity: Theory and an application to Latin America. *Review of Income and Wealth*, 2011. 57(4), pp.622–657.
[2] Geoffrey J.D. Hewings, Spatially blind trade and fiscal impact policies and their impact on regional economies. *The Quarterly Review of Economics and Finance*, 2014.54, pp.590–602.
[3] Hering L., S. Poncet, "Market access and individual wages: evidence from China", *Review of Economics and Statistics*, forthcoming. 2016.
[4] Huang Jl., Li Kw., Export openness, regional market size and economic growth, *Economic research*, 2006. 6, pp.27-38.
[5] Henry, M., Kneller, R., Milner, C., Girma, S.. Do Natural Barriers Affect the Relationship Between Trade Openness and Growth. *Oxford Bulletin of Economics and Statistics*. 2012.74(1), pp.1-19.
[6] Kim, Dh., Lin, Sc., Suen, Yb.. Trade, growth and growth volatility: New panel evidence. *International Review of Economics and Finance*. 2016.45, pp.384–399.
[7] Li, K., Qi, Sz,. Trade Openness, Economic Growth and Carbon Dioxide Emission in China. *Economic Research Journal*. 2011.11 (4), pp.60–72.
[8] Mehmet N. Eris, Ulasan, B.. Trade openness and economic growth: Bayesian model averaging estimate of cross-country growth regressions, *Economic Modelling*, 2013.33, pp.867–883.
[9] Menyah, K., Nazlioglu, S.. Wolde-Rufael, Y., Financial development, trade openness and economic growth in African countries: New insights from a panel causality approach. *Economic Modelling*. 2014.37, pp.386–394.
[10] Silberbergera, M., Könige, J., Regulation, Trade and economic growth, *Economic Systems*. 2016. 40, pp.308–322.
[11] Tian, X., Zhang, X., Zhou, Y., Yu, X. Regional income inequality in China revisited: A perspective from club convergence. *Economic Modelling*, 2016. 56, pp.50–58.
[12] Wen, Dw., FDI、Export openness and Chinese provincial industry growth. *Journal of Financial Research*, 2013. (6), pp.104-117.
[13] Wang, Mc., Xu, Kn.,Trade Openness, Economic Growth and CO2 Emission in China: A GVAR Analysis. *China population, resources and environment*. 2015. 25(11), pp.52-58.
[14] Xu J., Meng, J,. Trade openness, economic growth and human capital. *World Economy Studies*, 2015.6(25), pp.84–91.
[15] Xu, Yz., Guo, J., Energy consumption, trade liberalization and economic growth. *Finance and Trade Economics*, 2014. 32(12), pp.99–109.

IOP Conf. Series: Earth and Environmental Science **61** (2017) 012005 doi:10.1088/1755-1315/61/1/012005

# Gas migration model based on overburden strata fracture evolution law in three dimensional mine-induced stress conditions

**Xing LI** [1,2], **Guangzhi YIN**[1,2]

[1] State Key Laboratory of Coal Mine Disaster Dynamics and Control, Chongqing University, Chongqing 400030, China;

[2] College of Resources and Environmental Sciences, Chongqing University, Chongqing 400030, China

Email: lixingmail@yeah.net

**Abstract**: Based on the fracture features obtained from similarity simulation excavation experiment in 3D mine-induced stress condition, the gas migration model in gob area was set up, and numerical simulation using COMSOL Multiphysics software on gas migration law was studied. The research results show that: Gas pressure variety gradient and flow velocity are obviously influenced by fracture distribution shape. Gas concentration distribution presents the rounded rectangle shape in the lower strata, while in higher strata it tends to be a "O shape" morphology which is similar to the fracture form in strata above the gob. Besides, the pressure relief gas mainly accumulates in the working face fracture field, especially in the higher overlying strata. In the compaction field, gas concentration distribution form presents a "saddle" shape morphology in the vertical section, and gas concentration in boundary fracture zone is higher than in the compaction zone. It is remarkable that gas concentration in start-up fractured zone is also relatively high, and with the advance of working face, its gas enrichment degree decreases. The results of the research have an important guiding significance for coal and gas simultaneous extraction.

## 1. Introduction

During the extraction of coal seam, the cracks, deformation, shear failure and collapse of the overburden strata caused by the pressure relief action in the gob area would lead to the occurrence of the bed-separated fissures and vertical cross-measure cracks [1-3]. The fractures network greatly increases the permeability of overlying strata, which is the main area of coal mine gas drainage. Many researchers have made plenty of research on the law of gas migration in mined-out area. Hu and Jin [4-6] regarded the gob as porous medium and used CFD software for numerical simulation to study the gob gas flow and distribution. Li[7] established a double-layered seepage flow model by studying the cave-off medium delamination and wind flow stratified flow law. Based on the simplification of the mined-out area to porous medium, Jiang[8] proposed a three-dimensional study of fully mechanized top-coal caving face with porous media hydrodynamics theory. However, due to randomness of the fracture structure and accumulation mode of the mining fractured zone, the fracture features, the scales and the continuity from the micro-point of view are no longer valid, thus the fracture structures should not be regarded as an isotropic continuous medium. Plenty of site practices have proved that the law of the methane migration in the fractured zone is quite complex, including the mechanical dispersion, floating, diffusion and the coupling of the fracture and gas flow, the overlying strata evolution law is actually a problem of the space structure evolution. The intrinsic relation between the fracture structures and the gas migration under 3D stress conditions can be more accurate to reveal the gas migration law. Hence, in this paper, firstly, the characteristics of spatial fractured fields of overlying strata are described under 3D mining stress, and then the gas migration mathematical model is established. Finally, gas migration law by using the numerical simulation method is elaborated. The gas migration law in fractured field could provide theoretical supports for the parameter optimization of pressure-relief gas drainage system.

## 2. Evolution law of fracture field under 3D stress condition

Under the 3D stress conditions, the rock strata will bend, shear and collapse when the working face is advanced for a certain period of time. The deformation and incongruity caused by different flexural rigidity of the strata lead to the occurrence of the bed-separated and cross-measure fractures. While the fractures in the overburden strata present a "trapezoid body" form, and the fractures in the start-up and the working face area are more developed. Furthermore, the cross-measure fractures and bed-separated fissures always interpenetrate each other. In the extension plane of coal seam, the middle area of the gob is gradually compacted away from the start-up. In the lower area of the overlying strata, the fractured zone presents a "rounded rectangle" form in the parallel plane, and with the increase of the height, the fracture distribution of the middle and high-level strata tends to be an "O-shape" form, as shown in Fig.1.

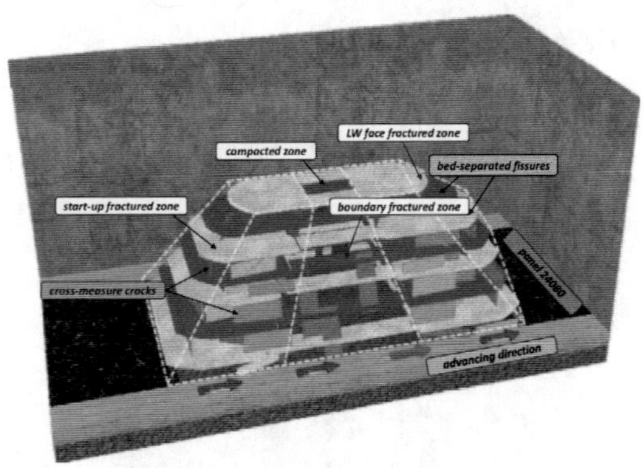


Fig.1 Three-dimensional space model of overburden strata fractured zone

## 3. Numerical calculation of gas migration law and analysis

Due to the random distribution of the fractures, the fracture forms and parameters are different. Because of the different properties of the rock mass and fractures, the gas flow in the gob is non-linear. In this paper, the RNG k-ε turbulence equation is adopted in the fractures. and the linear Darcy law is adopted in the rock mass. Besides, the following formula are used to deal with the interface problem between the unbroken rock and fractures, : $\mathbf{u}_{k-\varepsilon} = \mathbf{u}_{darcy}$ , $P_{k-\varepsilon} = P_{darcy}$ , as shown in Fig 2.

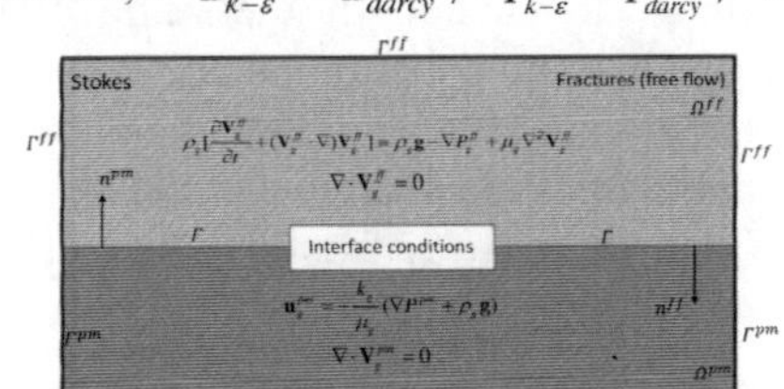


Fig.2 Schematic depiction of gob gas migration

*3.1 Engineering conditions and computational model*

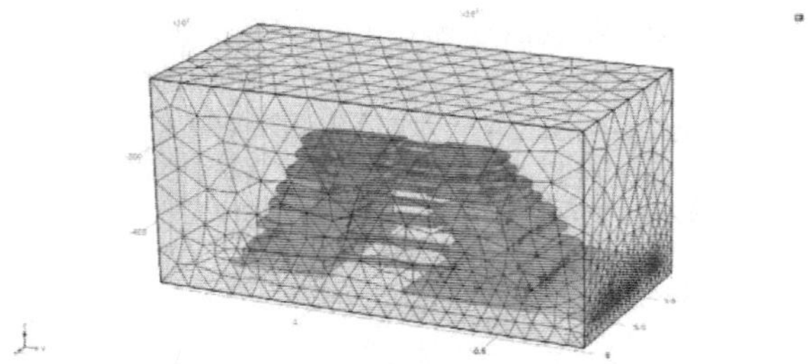

Fig.3 Mesh generation of gob fracture field

The depth of coal seam is about 900 meters, and the length and height of the working face are respectively 190m and 3.5m. Besides, the average gas pressure is 2.6MPa. Based on the results of three-dimensional excavation experiment, the key parameters such as the positions, opening sizes and lengths of fractures at different heights and longitudinal sections are statistically counted. The parameters of the physical model are as follows: The whole size of the computational model is 200m height, 100m in width and 100m length, while the heights of the caving and the fractured zones are 8m and 36m. Furthermore the mining fracture angles are separately 70° in start-up side and 64° in working face side. The gas migration law is simulated by using COMSOL Multiphysics software. Figure 3 shows the gob area model after meshing.

*3.2 Methane migration and enrichment law in overlying strata*

In order to study the law of gas migration and enrichment law in the overburden strata, ZXQM01-03 and QXQM01-03 planes are intercepted respectively along the strike and inclination direction of coal seam, see Figure 4. Features such as gas pressure gradient, concentration and velocity will be displayed in these planes.

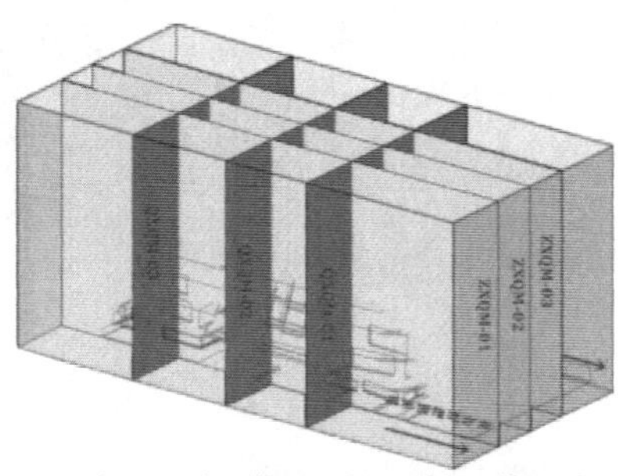


Figure 4 Cross-sections schematic diagram of inclination and strike directions

Figure 5 shows the distribution law of pressure with time in the fractured field. Figure 6 shows the gas pressure and velocity field features in the ZXJM-02 plane at t = 10000 s. We can see that both the pressure gradients and velocity are greatly affected by fracture morphology. The fractures in the working face are developed and the velocity of gas through this area is relatively large. Since the fractures are fully compacted under the action of the mine pressure, the gas is hard to flow in compacted zone. While there are still some fractures in the boundary areas near the upper and lower roadways, the gas could flow easily and the pressure gradient decreases significantly. Though plenty of fractures still exist in the start-up area, the gas pressure is smaller compared with the gas pressure nearby working face due to the far distance away from the gas gushing source.

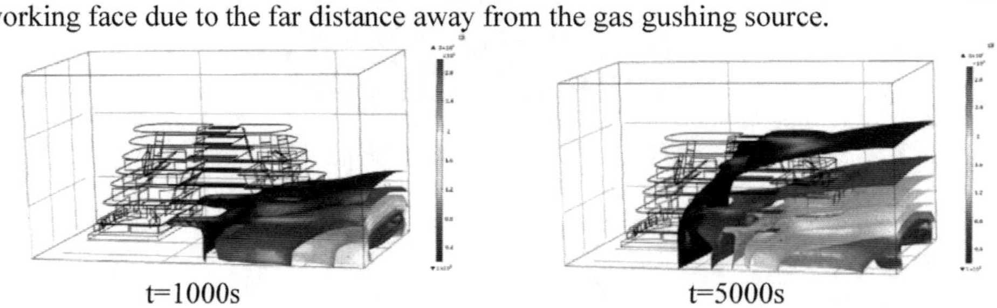

t=1000s t=5000s

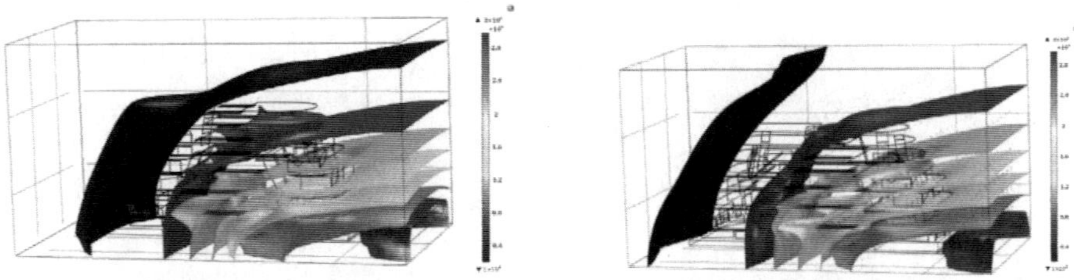

t=9000s t=13000s

Figure.5 Features of gas pressure changing with time in fractured zones

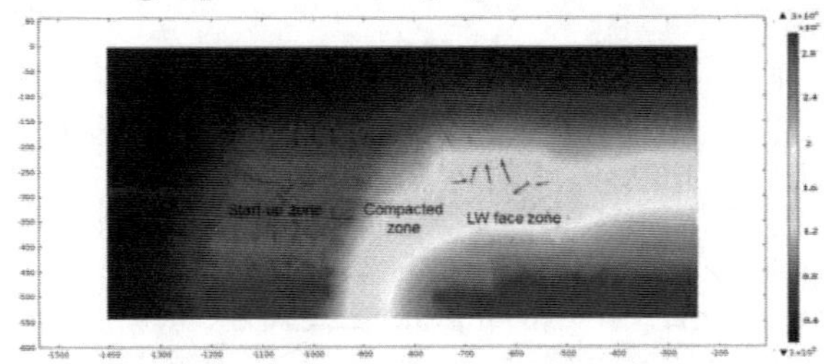


Fig.6 Gas pressure and velocity distribution at t = 10000s in ZXQM-02 plane

From the top of working face, cloud maps of the gas concentration distribution law in different height levels are drawn at t = 100000s. As shown in Fig. 7, the distribution patterns of gas concentration are in the form of a "rounded rectangle" shape in the lower strata and an "O" shape in the higher strata. This is consistent with the distribution law of the fractures in the overburden strata. While, when the gas flows from the working face to the compaction zone, most of gas flows across the cracks in the boundary area. The gas concentration distribution form in this area always presents a "□" type, and gas concentration in the boundary zone is higher than that in the compacted zone.

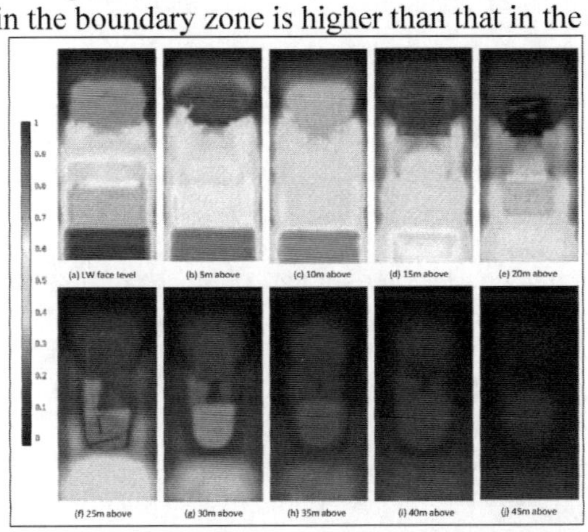


Fig. 7 Gas concentration distributions at different level planes above the LW face

Fig. 8 is the cloud maps of the velocity field features in different areas of the overburden strata. In working face fractured zone, plenty of high-pressure gas rush into the fractures, and then gradually migrate to the higher level, and the rest of gas flow through the unbroken rock as the Darcy's pattern. In the compacted zone, the fractures are almost fully compressed by the stress, thus the gas is difficult to run through the compacted space. While cracks in the boundary fractured zone provide convenient channels for gas migration. Therefore, gas concentration distribution feature in this vertical section present a "saddle" type. In the start-up fractured zone, the gas accumulates in the bed-separated and cross-measure fractures, and the gas concentration is relatively higher.

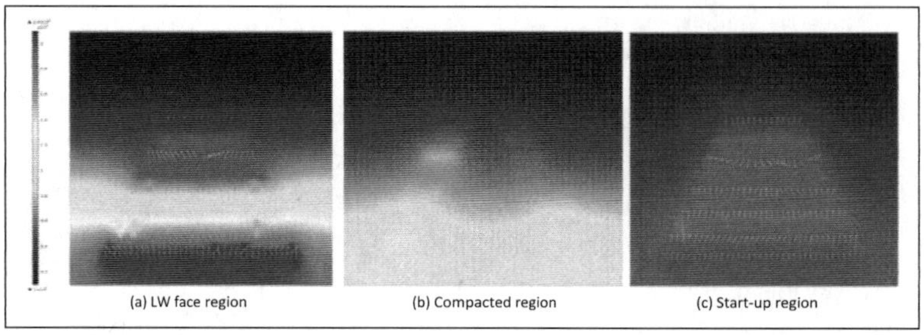


Figure 8 Methane pressure and flow distribution in different fractured zones above gob

## 4 Conclusions

(1) The gas pressure gradient and velocity field are greatly affected by the fractures. Both the gas pressure and velocity are higher in the working face fractured zone. Then, in the compacted zone, the fissures are fully compacted with the action of the mine pressure, thus the gas is more difficult to flow across this area. Though plenty of fractures still exist in the start-up area, the gas pressure is smaller compared with the gas pressure nearby working face due to the far distance away from the gas gushing source.

(2) The distribution law of gas concentration presents respectively a "rounded rectangle" and an "O-shape" in the lower and higher strata, which is consistent with the fractures distribution law in the overburden strata.

(3) The pressure relief gas is mainly gathered in the working face fractured zone, boundary fractured zone and start-up fractured zone, and the gas concentration gradually decreases with the increase of height. Besides, with the advance of the working face, the degree of gas enrichment in start-up fractured zone is gradually reduced.

## Acknowledgments

This study is financially supported by National Natural Science Foundation of China (51434003,51374256), and Fundamental Research Funds for the Central Universities (CDJXS12241103). Authors are also grateful to Pingdingshan 10th Mine for their data sharing.

## References

[1] Qian Minggao, Li Hongchang. The movement of overlying strata in longwall mining and its effect on ground pressure [J]. Journal of China Coal Society, 1982,2:1-12.

[2] Lin Haifei, Li Shugang, Cheng Lianhua, et,al. Model experiment of evolution pattern of mining-induced fissure in overlying strata[J].Journal of Xi'an University of Science and Technology. 2010,30(5):507-512.

[3] S.K. Das. Observations and classification of roof strata behavior over longwall coal mining panels in India [J]. International Journal of Rock Mechanics and Mining Sciences, 2000, 37:585–597.

[4] Hu Qianting, Liang Yunpei, Liu Jianzhong. CFD simulation of goaf gas flow patterns[J]. Journal of China Coal Society, 2007,32(7):719-723.

[5] Jin Longzhe, Yao Wei, Zhang Jun. CFD simulation of gas seepage regularity in goaf[J]. Journal of China Coal Society, 2010,35(9):1476-1480.

[6] Lan Zequan, Zhang Guoshu. Numerical simulation of gas concentration field in multi-source and multi-congruence goaf[J]. Journal of China Coal Society, 2007,32(4):396-401.

[7] Li Zongxiang, Wu Qiang, Pan Liming. Double layered seepage model and oxygen consumption-temperature rising characteristics in goaf[J]. Journal of China University of Mining & Technology, 2009,38(2):182-186.

[8] Jiang Shuguang, Zhang Renwei. Mathematical model and numerical calculation of air flow field in sub-level caving workings[J]. Journal of China Coal Society, 1998,23(3):258-261.

IOP Conf. Series: Earth and Environmental Science **61** (2017) 012006 doi:10.1088/1755-1315/61/1/012006

# Research for full-term forecasting model of production rate with multiple life cycle

**Yan Wende[1], Yuan Yingzhong[*1], Liu Jun[2], Qi Zhilin[1], Li Jiqiang[1], Xiao Qianhua[1]**

1. Chongqing Key Laboratory of Complex Oil & Gas Fields Exploration and Development, Chongqing University of Science & Technology, Shapingba, Chongqing, China
2. Petro China Southwest Oil and Gas Field Company, Chengdu, China
*Corresponding Author, e-mail: yuanyingzh0001@126.com

**Abstract**. Based on generalized production forecasting model, in the case of oil production with multiple production peak value and multiple life cycle, a multi-peak forecasting model is proposed, which can describe full course of rise, growth, maturity and decline for every life cycle. Linear trial-and-error method, binary regression method and ternary regression method are used to resolve the parameters of production forecasting model. The results indicate that parameters in linear trial-and-error method are not resolved simultaneously and the solutions are usually not the best matching. Parameters in multi-variant regression method are resolved simultaneously with the best correlation factor, and the solution is more accurate than linear trial-and-error method. Growth factor including cumulative production rate term and exponential function term of time are considered simultaneously in ternary regression method, and the solution is more accurate than binary regression method. For actual example with three-peak production, ternary regression method are used to resolve the parameters of multi-peak production forecasting model, and matching accuracy is greatly improved.

## 1. Introduction

Conventional Arps decline curve is a monotonically decreasing function. Seeing from the shape of function, there is no peak. Since Weng's cycle prediction model[1] is used to predict oil and gas field production rate, nearly 20 kinds of other full-term production rate prediction models including Logistic, Log Normal Distribution, Rayleigh are proposed in succession[2-6]. Seeing from the shape of the function, Weng's cycle model is a single-peak function. However, actual oil production rate curve may display the characteristics of increase-decrease-increase-decrease, which is a form of multiple-peak or multiple life cycle. Analyzing from mathematical, probability density function with shape of multiple-peak[7-8] is used to describe production rate curve of multiple life cycle. However, the application of multiple-peak function is complex and the parameters are difficult to get. In this paper, based on generalized production rate prediction model, the production rate curve of multiple life cycle is processed with piecewise function, which realizes accurate fitting of oil production rate with multiple life cycle.

## 2. Generalized formula of production rate prediction model

In the generalized production prediction model, the decline function term essentially reflects the inherent decline form of oil production rate under different driving types and percolation characteristics, and the increasing function term is an expression form of oil production rate which reflects the external influence people exerting on oil field including increasing production well number, stimulation of oil layer and application of new technology[9]. At present, many scholars attempt to combine empirical formula of production prediction and parameters of reservoir rock or fluid[10-11], which face many difficulties and uncertainties. It is main research method to analyze characteristics of production rate change from the perspective of mathematics and propose high accuracy calculation model[12].

In the generalized production rate prediction model, production rate is multiplication of an increasing function and a decline function, whose formula is

$$Q = a \cdot g(t) \cdot f(t) \tag{1}$$

Seeing from the actual process of crude oil production, cumulative oil production rate $N_p$ increases along the time. Seeing from pure mathematical characteristics, exponential function $t^b$ (the time $t$ is base number) also increases along the time. Following formula can be used to describe increasing function term $g(t)$ using cumulative oil production rate $N_p$ and the time $t$.

$$g(t) = N_p^k \cdot t^b \tag{2}$$

The decline function term can be expressed with following general formula.

$$f(t) = \exp[-c_1 t^{m_1} - n \ln^P (c_2 t^{m_2} + c_3)] \tag{3}$$

Substituting Eq. (2) and Eq.(3) into Eq.(1), following formula can be got.

$$Q = a \cdot N_p^k \cdot t^b \cdot \exp[-c_1 t^{m_1} - n \ln^P (c_2 t^{m_2} + c_3)] \tag{4}$$

When $n$=0, Eq.(3) can be simplified to be $f(t) = \exp(-c_1 t^{m_1})$, which is Weibull production rate decline equation.

When $c_1 = 0$, $m_2 = p = c_3 = 1$, Eq.(3) can be simplified to be $f(t) = \exp[-n \ln(c_2 t + 1)]$, which is hyperbolic decline equation.

When $c_1 = 0$, $m_2 = p = n = c_3 = 1$, Eq.(3) can be simplified to be $f(t) = \exp[-\ln(c_2 t + 1)]$, which is harmonic decline equation.

When $c_1 = c_3 = 0$, $m_2 = c_2 = 1$, $p = 2$, Eq.(3) can be simplified to be $f(t) = \exp(-n \ln^2 t)$, which is decline function with lognormal distribution.

In the elastic displacement oilfield or depletion development oilfield, usually $m_1 = 0$, $n$=0, then decline function $f(t)$ can be expressed as $f(t) = \exp(-c_1 t)$, therefore

$$Q = a \cdot N_p^k \cdot t^b \cdot \exp(-ct) \tag{5}$$

## 3. Determination of parameters in generalized production rate prediction model

### *3.1 Improvement of conventional linear trial-and-error method*

When $k$=0, Eq.(5) is Weng's cycle model

$$Q = a t^b e^{-ct} \tag{6}$$

In Eq.(6), there are three parameters $a$, $b$, $c$ need to be resolved. A simple linear trial-and-error method is proposed by Chen Yuanqian[13], and Eq.(6) can be rewritten as

$$\ln \frac{Q}{t^b} = \ln a - ct \tag{7}$$

According to data of actual production rate $Q$ and corresponding production time $t$, given different value of $b$, Eq.(7) can be resolved with linear trial-and-error method of semi-logarithm, then the value of $b$ with the highest correlation coefficient and the best linear relationship can be got, which is the correct value to seek. The intercept and slope can be got by linear regression, then value of $a$ and $c$ can be obtained.

It is indicated that linear trial-and-error method is some defective in actual application, although the method is effective somewhere. This is because in the process of resolving the value of $b$ in Eq.(7)

with linear trial-and-error method of semi-logarithm, although the linear relationship is the best, values of $a$, $c$ and $b$ are not got simultaneously. Therefore, when the value of $b$ is the best, parameter combination of $a$, $b$ and $c$ are not sure to be optimal.

For relatively simple expression, if it can be converted to multivariate linear form, multiple linear regression method could be used directly instead of linear trial-and-error method, Eq.(6) can be written as

$$\ln Q = \ln a + b \ln t - ct \quad (8)$$

Using binary linear regression analysis method, parameters $a$, $b$ and $c$ with the highest correlation coefficient can be obtained simultaneously through Eq.(8), which is more accurate than linear trial-and-error method (Figure 1). In Fig.1, fitting equation is $Q = 0.2048t^{2.65}e^{-0.1402t}$ with linear trial-and-error method, whose average error is 22.4%. Fitting equation with binary linear regression method is $Q = 0.40708t^{2.1126}e^{-0.0963t}$, $R^2 = 0.7642$, whose average error is 19.55%.

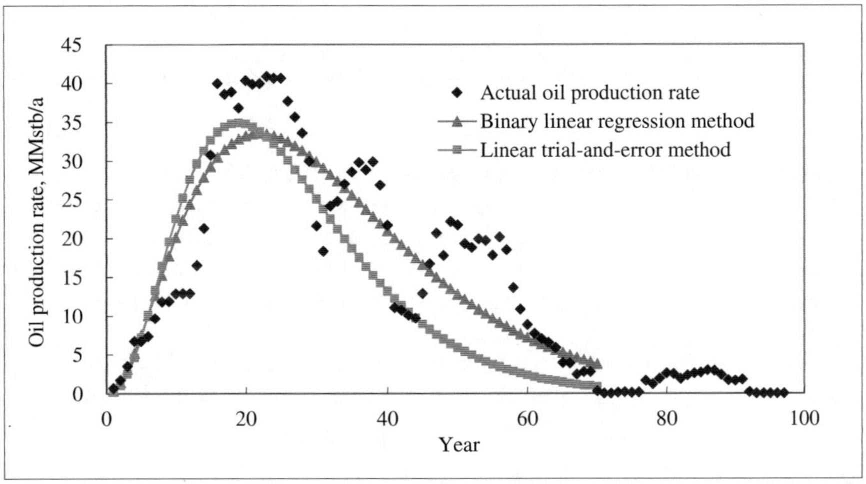


**Figure 1.** Comparison of linear trial-and-error method and binary linear regression method for an oil field

*3.2 Improvement of fitting accuracy with ternary linear regression method*

Data of production rate is fitted through Eq.(8) with binary linear regression method, whose average error is 19.55%. The reason of relatively big error is that increasing function term of production rate $g(t) = t^b$ only considers issue of time, yet issue of cumulative oil production rate which increases along the time is not considered.

Considering $g(t) = N_p^k \cdot t^b$, then $Q = a \cdot N_p^k \cdot t^b \cdot \exp(-ct)$, therefore

$$\ln Q = \ln a + k \ln N_p + b \ln t - ct \quad (9)$$

Parameters $a$, $b$ and $c$ can be got through ternary linear regression method, whose expression is $Q = 1.1353 N_p^{1.4124} t^{-1.2634} e^{-0.05226t}$, $R^2 = 0.7853$, with average error of 17.59%.

It is indicated that fitting accuracy of ternary linear regression method is higher than binary linear regression method (Figure 2).

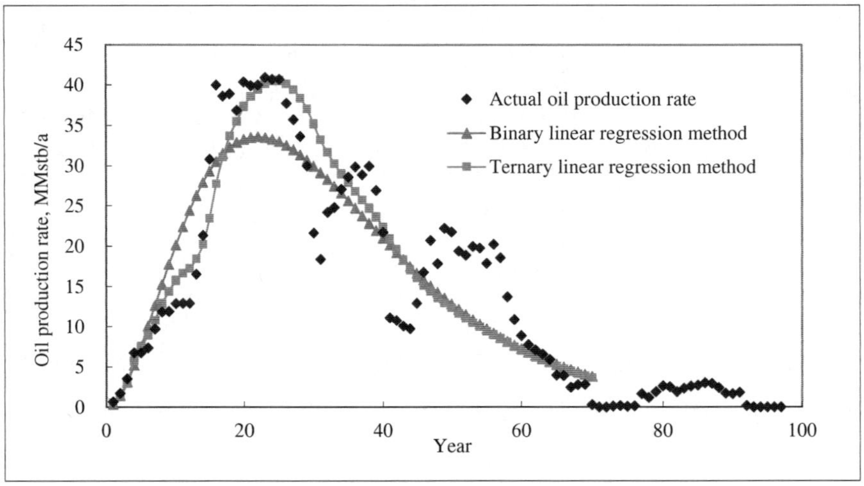


**Figure 2**. Comparison of binary linear regression method and ternary linear regression method for an oil field

## 4. Full-term forecasting model of production rate with multiple life cycle

For actual oilfield, stage drilling and commissioning may exist, therefore there are multiple production peak value and life cycle for actual production rate. Different life courses including rise, growth, maturity and decline exist in different life cycle. For double-peak, tri-peak or multi-peak oil field development model, it is difficult to get reliable prediction result using conventional single peak prediction model[14-15].

Based on generalized single peak production rate prediction model with four parameters, a multi-peak production rate prediction model is proposed using ternary linear regression method.

$$Q_n(t) = a_n \cdot [N_p(t) - Np(t_{0(n-1)})]^{k_n} \cdot [t - t_{0(n-1)}]^{b_n} \cdot \exp[-c_n(t - t_{0(n-1)})] \quad (10)$$

Where, $Q_n(t)$——production rate in different peak period at the time $t$, $10^4$t/a;

$t_{0(n-1)}$——initial production time for different peaks（when n=1, $t_{0(n-1)}= t_{00}=0$）, a;

$N_p(t)$——cumulative oil production rate at the time $t$, $10^4$t;

$N_p(t_{0(n-1)})$——cumulative oil production rate at the time $t_{0(n-1)}$, $10^4$t.

For the condition of three-peak production, the formula of production rate in different peak period is that

$$Q_1(t) = a_1 \cdot [N_p(t) - Np(t_{00})]^{k_1} \cdot [t - t_{00}]^{b_1} \cdot \exp[-c_1(t - t_{00})] \quad (11)$$

$$Q_2(t) = a_2 \cdot [N_p(t) - Np(t_{01})]^{k_2} \cdot [t - t_{01}]^{b_2} \cdot \exp[-c_2(t - t_{01})] \quad (12)$$

$$Q_3(t) = a_3 \cdot [N_p(t) - Np(t_{02})]^{k_3} \cdot [t - t_{02}]^{b_3} \cdot \exp[-c_3(t - t_{02})] \quad (13)$$

In Figure 1 and Figure 2, because a general life cycle is used to calculate the production rate, the error between calculated and actual production rate is big. For the actual oilfield, there are three life cycle and production peak value, where the first life cycle is from the first year to thirty year, the second life cycle is from thirty-one year to forty-three year, and the third life cycle is from forty-four year to seventy year.

### *4.1 Life cycle of One*

For the first life cycle from the first year to thirty year, if the data after 30 years is the real production rate, fitting equation is that

$$Q_1 = 1.1353 N_p^{1.4124} t^{-1.2634} e^{-0.05226t}, \; R^2 = 0.7853$$

If the data after 30 years is extrapolated production rate according to natural decline (Figure 3), fitting equation is that

$$Q_1 = 2.0207 N_p^{2.5211} t^{-3.5107} e^{-0.071t}, \; R^2 = 0.9952$$

*4.2 Life cycle of Two*

For the second life cycle from the thirty-first year to forty-three year, if the data after 43 years is the real production rate, fitting equation is that

$$Q_2 = 0.1239[N_p - N_p(t=30)]^{1.5983}(t-30)^{-1.1338} e^{-0.07546(t-30)}, \; R^2 = 0.6566$$

If the data after 43 years is extrapolated production rate according to natural decline, fitting equation is that

$$Q_2 = 0.1721[N_p - N_p(t=30)]^{1.6277}(t-30)^{-1.1115} e^{-0.1770(t-30)}, \; R^2 = 0.9985$$

*4.3 Life cycle of Three*

For the third life cycle from the forty-four year to seventy year, all the data is real production rate, fitting equation is that

$$Q_3 = 0.2377[N_p - N_p(t=43)]^{1.6614}(t-43)^{-1.1893} e^{-0.1445(t-43)}, \; R^2 = 0.8736$$

The results indicate that the fitting accuracy (average error of 1.19%) for extrapolated production rate is higher than fitting accuracy (average error of 10.04%) for real production rate in every life cycle (Figure 4). This is because if production rate in the decline period is extrapolated according to production decline rule, full course of rise, growth, maturity and decline for every life cycle can be constructed, which can reduce the influences of other life cycle.

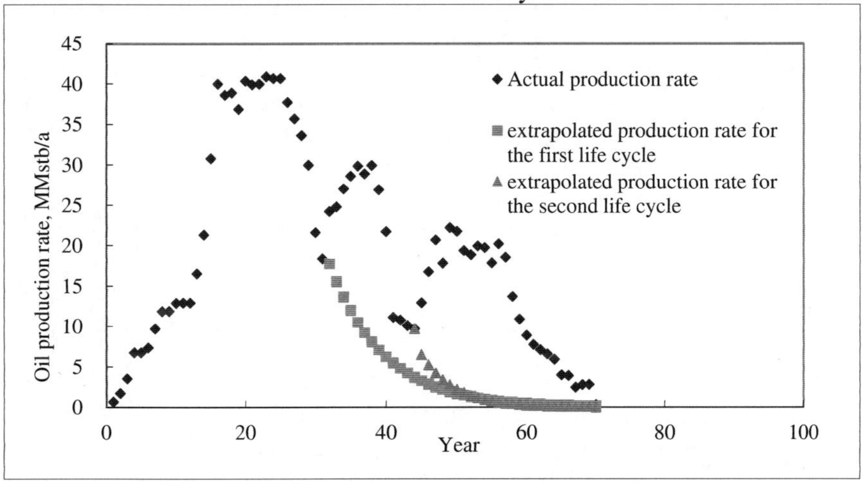


**Figure 3**. Actual and extrapolated production rate

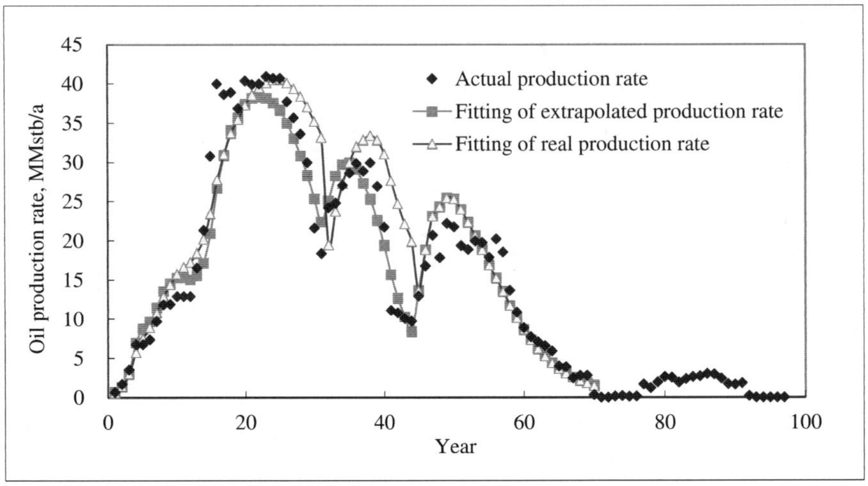


**Figure 4**. Full-term prediction of multiple life cycle for actual and extrapolated production rate

## 5. Conclusions

(1) In the production rate prediction model with several parameters, parameters are not resolved simultaneously in the linear trial-and-error method, therefore the solutions are usually not the best matching parameters. Parameters with the highest correlation coefficient are resolved simultaneously in the multiple linear regression method, whose accuracy is higher than linear trial-and-error method. If multivariate linear correlation can be easily constructed, it is best to use multiple linear regression method other than linear trial-and-error method.

(2) Fitting accuracy of ternary linear regression method is higher than binary linear regression method because the former simultaneously considers two factor that increase over time including cumulative oil production rate term and time exponential function term.

(3) For the oil production with multiple production peak value and multiple life cycle, if production rate in the decline period is extrapolated according to production decline rule, full course of rise, growth, maturity and decline for every life cycle can be constructed, then ternary linear regression method can be used to resolve four parameters in the generalized production rate prediction model, which can greatly improve fitting accuracy for production rate with multiple life cycle.

## Acknowledgments

Authors wishing to acknowledge assistance of National Natural Science Foundation of China (51574052,51374269), Chongqing Basic Science and Advanced Technology Research Project (cstc2016jcyjA0607), Science and Technology Research Project of Chongqing Municipal Education Committee (KJ1601319), University Innovation Team Project of Chongqing Municipal (CXTDX201601033).

## References

[1] Weng Wenbo 1984 *Fundamentals of forecasting theory* (Beijing: Petroleum Industry Press).
[2] Gao Wenjun, Zhang Yuanping, Yang Xinghua 2001 Percolation theory of logistic production decline equation. Xinjiang Petroleum Geology 22(6) 506-508.
[3] Clark A J, Petrobras L, Larry W, 2011 Production forecasting with logistic growth models. SPE 144790.
[4] Liu Pengcheng 2000 Generalized logistic cyclic model for predicting oilfield recovery percent. Oil & Gas Recovery Technology 7(3) 31-34.
[5] Hu Jianguo, Pan Zengyao 2000 Improvement and simplification of logarithmic normal distribution prediction model. Xinjiang Petroleum Geology 21(5) 412-414.
[6] Yu Qitai 1994 A study of features for seven decline curves. Xinjiang Petroleum Geology 15(1) 49-55.

[7] Jiang Renyan 1997 One new probability density function with bimodal shape. Journal of Changsha Communications University 13(1) 1-8.
[8] Wang Chen, Gao Chengtai 1999 The determination of double peak pattern of heterogeneity reservoir beds. Petroleum Exploration and Development 26(4) 57-59.
[9] Sun Yukai, Gao Wenjun 2007 Improvement and application of common reservoir engineering methods (Beijing: Petroleum Industry Press).
[10] Kegang L, Xingru W, He Z 2013 Tactics and pitfalls in production decline curve analysis. SPE 164503.
[11] Mosobalaje O, Tiab D, Lu J 2014 Theories of arps' decline curve exponent and loss ratio, for saturated reservoirs. SPE 172128.
[12] Ali T, Sheng J 2015 Production decline models: a comparison study. SPE 177300.
[13] Chen Yuanqian 1999 Practical methods of petroleum reservoir engineering (Beijing: Petroleum Industry Press).
[14] Chen Yuanqian, Hao Mingqiang 2013 Establishment and application of the multi-peak forecasting model. Xinjiang Petroleum Geology 34(3) 296-299.
[15] Yan Yongtao, Lei Dengsheng, Yuan Yingzhong 2016 Research and application of Weibull model for production prediction with bimodal distribution characteristics. Reservoir Evaluation and Development 6(2) 14-18.

IOP Conf. Series: Earth and Environmental Science **61** (2017) 012007 doi:10.1088/1755-1315/61/1/012007

# The application of geological advanced prediction in tunnel construction

**LU Youzhong**

Physical Science and Technology College of YICHUN UNIVERSITY, Xuefu Road, Yuanzhou District, Yichun, JiangXi 336000, China
E-mail: yz_lu@126.com

**Abstract**: This article studied the theoretical composition of tunnel geological prediction (TGP), put forth the concepts of broad-sense TGP, narrow-sense TGP and comprehensive TGP and established a technical solution of broad-sense TGP on this basis, including three steps: macroscopic prediction of unfavorable geologies in the tunnel area, prediction of unfavorable geological bodies in the tunnel (i.e.: narrow-sense TGP) and alarm on approaching of geological hazards in the tunnel. The article also researched the methods and technical means of short-term geological prediction and long-term geological prediction of narrow-sense TGP and proposed a fault parameter prediction method, a surface geological body projection prediction method and a precursor prediction method of unfavorable geological bodies in the tunnel. In addition, the article also introduced the concept of alarm on approaching of geological hazards in tunnel construction and established a method for alarm on approaching of geological hazards in the tunnel. In the end, the article concluded that the implementation of comprehensive prediction is the most effective means to achieve expected prediction effect and increase prediction accuracy.

## 1 Introduction

The 21st century is a century of exploitation of underground space as a resource, also a century of Western Development of China. China is a mountainous country. Two thirds of its national territorial area is mountains and plateaus. The large-scale development of railway and highway transport projects, hydropower projects, South-to-North Water Diversion Project, mine projects and military and protective projects all involves construction of enormous mountain tunnels, long tunnels in particular. We can see this century will witness great progress in Chinese tunnel construction.

Tunnels usually refer to the engineering works used as underground passages. By characteristics of wall rock, tunnels are classified into two categories in general. One category is rock tunnels, built in rock strata. As rock tunnels are mostly built in mountains, they are also called mountain tunnels; the other category is soft soil tunnels, built in soil strata or quasi-soil strata. As soft soil tunnels are often built under water or urban overpasses, they are also called underwater tunnels or urban road tunnels. Due to the complexity and variability of geological conditions in the longitudinal direction, railway and highway tunnels bear the characteristics of both rock tunnels and soft soil tunnels. As tunnels are also located in underground rock masses with complex hydrogeology and engineering geology, the precondition for safe and high-quality tunnel construction without future troubles is to get to know and foresee surrounding hydrogeological and engineering geological conditions. Following enormous construction of mountain tunnels, geological work of tunnel construction as a necessary step of tunnel construction certainly will be widely popularized.

IOP Conf. Series: Earth and Environmental Science **61** (2017) 012007 doi:10.1088/1755-1315/61/1/012007

Geological work of tunnel construction is conducted after design submission, in tunnel construction stage and throughout construction process.

With regard to geological work of tunnel construction, in foreign countries, general work has been done universally and academic research also has reached a certain depth, but all-round and systematic geological work of construction and academic research is yet to be done. In China, general work is conducted only in a minority of tunnels and basically is still in a starting stage, and academic research is seldom reported and not deep. In order to avoid the blindness of tunnel excavation and construction, we must do well in geological work of tunnel construction.

Tunnel geological prediction (TGP) is a key link of geological work of tunnel construction. Its importance goes without saying. Following the increase of quantity and the enlargement of scale of tunnels to be built, how to implement TGP with "high quality and high accuracy" becomes an urgent and important issue. A major measure for fast construction is elimination of geological hazards in construction. The most effective means to eliminate geological hazards in construction is to do well in TGP [2].

## 2 Concept and Technical Solution of TGP

TGP includes broad-sense TGP and narrow-sense TGP.

Broad-sense TGP is a generic term of prediction of geological hazards inside and outside the tunnel during tunnel construction, and consists of three parts of content or three steps: macroscopic prediction of unfavorable geologies in the tunnel area, prediction of unfavorable geological bodies in the tunnel (i.e.: narrow-sense TGP) and alarm on approaching of geological hazards in the tunnel.

Macroscopic prediction of unfavorable geologies in the tunnel area is a precondition and basis of prediction of geological hazards during tunnel construction, and the first step. From macroscopic prediction of unfavorable geologies of a tunnel on the basis of analyzing geological condition in the tunnel area, we may know the type, scale and possible location and direction of unfavorable geologies that might be encountered during tunnel construction, and may also know the possibility that these unfavorable geologies induce geological hazards during construction, the type and intensity of the geological hazards and their impact on tunnel construction. Therefore, only under the guidance of the principle of macroscopic prediction of unfavorable geologies of the tunnel, can we more accurately and more effectively predict unfavorable geological bodies in the tunnel and geological hazards during construction. The former is the work basis of the latter two.

The basis of macroscopic prediction of unfavorable geologies in the tunnel area is analysis of geological conditions of the tunnel. This is because most of the unfavorable geological bodies of a tunnel constitute a part of stratum, geological structure or karstic geological body in the tunnel area. For example, cores of folds, fault fracture zones, steeply inclined rock strata and interlayer sliding faults are parts of a geological structure; karst caves, underground rivers and karst silt zones are parts of a karst geological body. Whether they will meet with the tunnel not is closely related to the erosion basis to the tunnel area; high pressure high concentration gassy seams, old coal kilns and goafs are parts of a coal seam stratum.

Precondition of analysis of geological conditions of a tunnel: on the basis of the tunnel design specification and design drawings submitted by a design institute, the geological technicians for tunnel construction conduct in-depth and detailed surface geological review or survey on the strata (coal seam strata in particular), geological structure, karst and hydrologic regime of the tunnel area again. As the accuracy required in the design stage is limited, the tunnel design specification and design drawings submitted by the design institute only roughly describe major unfavorable geologies and construction-related geological hazards of the tunnel. As far as the analysis of geological conditions of the tunnel and the analysis of geological conditions of the tunnel needed by macroscopic prediction are concerned, the submitted data is far from enough in research depth, and also has a gap in accuracy.

## 3 Prediction of Unfavorable Geological Bodies in the Tunnel

Prediction of unfavorable geological bodies in the tunnel (narrow-sense TGP, "TGP in the tunnel" for short, the same below) is the second step of broad-sense TGP.

At present, enormous work has been done on TGP in the tunnel in construction practice at home and abroad, but scientific research is still in a state of "piling" of scattered and sporadic raw data or results. It not only does not ascend to a certain height of theoretic research but also lacks in-depth and systematic research findings.

*3.1 Long-Term TGP*

A rock strength reduction zone is shown in Fig. 1. Practice has proved that in every fault influenced zone, there is a change zone of physical and mechanical properties of rocks, which faces the direction of the fault and shows reduction of rock strength. From the perspective of geomechanics, the mechanical mechanism for existence of this change zone is as follows: in the process of generation of the fault, the instant violent displacement of the rocks on the two walls inevitably generates a local stress field in the walls. The strong internal stress not only may form folds, joints (minor faults) and other macroscopic structures inside the fault influenced zone but also will form numerous microscopic structures in a certain width of rock mass near the fault, particularly a large number of microscopic fissures. The latter is the fundamental reason for strength reduction of the rocks in a certain range on the two sides of the fault.

Rock strength reduction zone consists of a gradual reduction zone and a dramatic reduction zone.

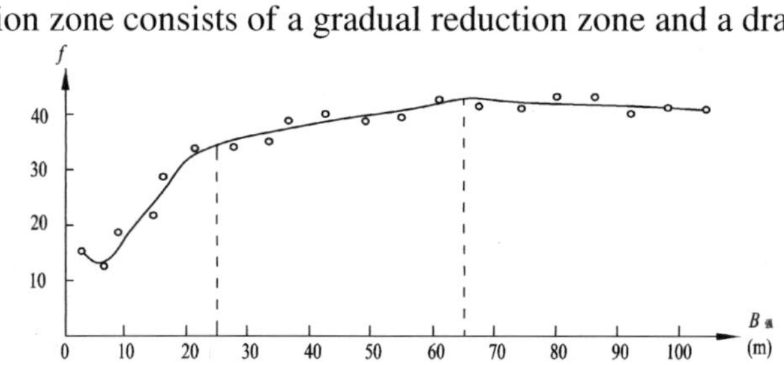


Fig. 1 Two inflection points of the rock strength reduction zone

The dramatic reduction zone of rock strength is almost in the range of the fault fracture zone and is meaningless to both long-term and short-term TGP.

According to the rough distribution characteristics of discrete points, the type of a functional equation of an approximation curve that can represent scatter distribution law is preliminarily determined. Data fitting method is used to repeatedly compare the preliminarily determined curve equation with other types of curve equations to get an optimal curve equation.

The least square method is used to determine the coefficients and constants of the optimal curve equation.

Expression of the functional equation:

$$f(x) = a_0 + a_1 x + a_2 x^2 + \ldots + a_n x^n$$

At point $x_i$, the difference between function value and measured value $y_i$ is:

$$r_i = f(x_i) - y_i$$

where, $r_i$-residual.

Residual sum of squares at every measuring point:

$$s = \sum_{i=1}^{n} r_i^2 = \sum_{i=1}^{n} [f(x_i) - y_i]^2$$

Residual sum of squares should be reduced to minimum. If $s$ is regarded as a function of polynomial constant coefficient, then according to the principle of function value in mathematical analysis, in order to minimize $s$, the following condition must be met:

$$\frac{\partial s}{\partial a_i} = 0(i = 1,2,3,\ldots n)$$

i.e.: $$\frac{\partial \left\{ \sum_{i=1}^{n} [f(x_i) - y_i]^2 \right\}}{\partial a_i} = 0$$

From the above expression, the following system of linear equations may be obtained:

$$\begin{bmatrix} S_0 & S_1 & S_2 \ldots & S_n \\ S_1 & S_2 & S_3 \ldots & S_{n+1} \\ S_2 & S_3 & S_4 \ldots & S_{n+2} \\ M & M & M \ldots & M \\ S_n & S_{n+1} & S_{n+2} \ldots & S_{2n} \end{bmatrix} \times \begin{bmatrix} a_0 \\ a_1 \\ a_2 \\ M \\ a_n \end{bmatrix} = \begin{bmatrix} T_0 \\ T_1 \\ T_2 \\ M \\ T_n \end{bmatrix}$$

where, $S_k = \sum_{i=1}^{n} x_i^k (k = 1,2,3,\ldots,2n+1)$.

$$T_k = \sum_{i=1}^{n} y_i x_i^k$$

$a_0$、$a_1$、$a_2$、$a_3 \ldots a_n$ obtained from the above expression are constant coefficients of the optimal curve equation.

A lot of practice proves that when stratigraphic separation $N$ is $\leq$5m, the scatters in the scatter diagram are roughly in linear distribution, the optimal curve is a straight line and the optimal functional equation is a linear equation in one unknown.

If the optimal curve equation is a linear equation in one unknown, then the general expression will be $y = a_0 + a_1 x$, $\frac{\partial S}{\partial a_i} = 0$, get

$$\begin{bmatrix} S_0 & S_1 \\ S_1 & S_2 \end{bmatrix} \times \begin{bmatrix} a_0 \\ a_1 \end{bmatrix} = \begin{bmatrix} T_0 \\ T_1 \end{bmatrix}, \text{ i.e.: } \begin{cases} S_0 a_0 + S_1 a_1 = T_0 \\ S_1 a_0 + S_2 a_1 = T_1 \end{cases},$$

so

$$\begin{cases} a_0 = \dfrac{S_2 T_0 - S_1 T_1}{S_2 S_0 - S_1^2} \\ a_1 = \dfrac{S_1 T_0 - S_0 T_1}{S_1^2 - S_0 S_2} \end{cases}$$

where, $S_0 = \sum_{i=1}^{n} x_i^0 = n$; $S_1 = \sum_{i=1}^{n} x_i^1$; $S_2 = \sum_{i=1}^{n} x_i^2$; $T_0 = \sum_{i=1}^{n} y_i x_i^0$; $T_1 = \sum_{i=1}^{n} y_i x_i^1$.

In order to combine with TGP, the expression of linear equation is rewritten into $B = a_1 N + a_0$, then

IOP Conf. Series: Earth and Environmental Science **61** (2017) 012007 doi:10.1088/1755-1315/61/1/012007

$$\begin{cases} a_0 = \dfrac{\sum_{i=1}^{n} B_i \times \sum_{i=1}^{n} N_i^2 - \sum_{i=1}^{n} N_i B_i \times \sum_{i=1}^{n} N_i}{\sum_{i=1}^{n} N_i^2 \times n - \left(\sum_{i=1}^{n} N_i\right)^2} \\ a_1 = \dfrac{\sum_{i=1}^{n} N_i B_i \times n - \sum_{i=1}^{n} B_i \times \sum_{i=1}^{n} N_i}{\sum_{i=1}^{n} N_i^2 \times n - \left(\sum_{i=1}^{n} N_i\right)^2} \end{cases}$$

The above formula is applied in the practice of engineering prediction. Its accuracy may meet the requirements of engineering construction.

### *3.2 Short-Term TGP*

At present, only one detection means is adopted in most cases at home and abroad, such as: geological radar, infrared [14] and sound wave detector. In China, a face compiling record method and advance drilling are also used for short-term TGP. This article established a method for prediction of unfavorable geological bodies in the tunnel and studied the face compiling record method in depth. Similarly, in prediction practice, two or more than two methods and means are comprehensively used to achieve the purpose of verifying each other, making the best of the both worlds and improving prediction effect. Particularly, during short-term prediction applying geological radar and other geophysical prospecting means, it is more suitable to combine them with at least another short-term geological analysis method because the result of geophysical prospecting often has multiple solutions and limitation, and only the combination of multiple means can turn multiple solutions of the result of geophysical prospecting into a single solution.

Short-term TGP is prediction conducted on the basis of long-term TGP, so the main task of short-term TGP is to more accurately predict the following information in a range of 15~20m, maximum 30m in advance of face: possible rock stratum, nature of adjoining unfavorable geological bodies, possible nature of underground water bodies, extension of unfavorable geological bodies at or near face towards the front of face, and grade of wall rock. Different prediction methods and technical means may result in different prediction distance, accuracy and effect. If at least two of the abovementioned methods and technical means are adopted to implement comprehensive short-term TGP, the general technical indexes that should be achieved include:

Prediction of distance: It may reach 20m in advance of face; prediction of the nature of fault fracture zone and most unfavorable geological bodies: Full accuracy may be achieved; prediction of the positions of unfavorable geological bodies: The accuracy may exceed 95%; prediction of the scale of unfavorable geological bodies: The accuracy may exceed 90%; it may predict both the existence of water enriched zones and the nature of underground water bodies; both the relative quality and accurate grade of wall rock.

Short-term TGP focuses on the sections with complex geological conditions during tunnel excavation and advancement on the basis of long-term TGP.

Basic principle:

In most cases, the position, sequence, thickness, lithology and stratal combination of surface rock strata correspond to the position, sequence, thickness, lithology and stratal combination of the rock strata in the tunnel. By applying this principle, we may compare the position, sequence, thickness, lithology and stratal combination of surface rock strata obtained from surface geological review and survey with the rock strata at and near the face. When it is confirmed that the rock strata exposed at and near tunnel face are identical to the characteristics of a rock stratum on ground surface, we may deduce the position, thickness, lithology and strata combination of the rock stratum that will appear in a certain range in advance of face as well as its position in the tunnel, as shown in Fig. 2.

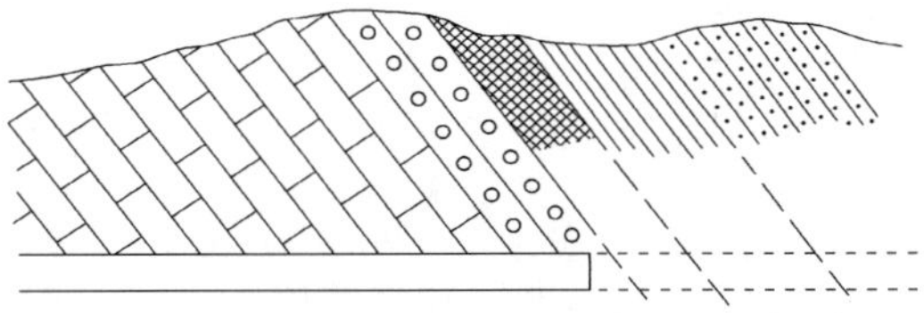

Fig. 2 Short-term prediction of rock strata in the tunnel

## 4 Alarm on approaching of Geological Hazards in the Tunnel

The technology for alarm on approaching of geological hazards in the tunnel, i.e.: the technology for monitoring and judgment on approaching geological hazards in the tunnel, is a core task of broad-sense TGP, also its third step. It is to judge whether geological hazards will occur in front of the working face during tunnel advancing when the unfavorable geological bodies stated in short/long-term TGP are drawing near, and to make further confirmation when the unfavorable geological bodies are drawing near so as to work out a construction improvement plan and reinforcement measures in time, and avoid occurrence of geological hazards. Therefore, we may say all other geological work of tunnel construction, such as: macroscopic prediction of unfavorable geologies in the tunnel area, long-term and short-term TGP, and wall rock stability evaluation, serves alarm on approaching of geological hazards in the tunnel and centers on it in the final analysis. It assumes a very important position in geological work of tunnel construction. Alarm on approaching of geological hazards in the tunnel is based on long-term and short-term TGP, mainly by method of monitoring and judging approaching unfavorable geological bodies.

## 5 Conclusion

TGP includes broad-sense TGP and narrow-sense TGP. Broad-sense TGP is a generic term of prediction of geological hazards inside and outside the tunnel during tunnel construction; narrow-sense TGP is geological prediction of unfavorable geological bodies inside the tunnel during tunnel construction.

The combination of long-term and short-term TGP and alarm on approaching of geological hazards during construction is called comprehensive TGP. A lot of construction practice proves that only combining geological methods with physical methods, inside the tunnel with outside the tunnel, and long-term prediction with near-term prediction and implementing comprehensive prediction can achieve the purpose of verifying each other, making the best of the both worlds and improving prediction effect.

## References

[1] Qihu QIAN, Xiaoli RONG. State, issues and relevant recommendations for security risk management of China's underground engineering [J]. Chinese Journal of Rock Mechanics and Engineering, 2008, 27 (4): 649–654. (in Chinese)

[2] Mengshu WANG. Hydrological and geological forecast of tunnel construction in the karst district [J].Railroad Survey, 2004, (1): 7-10. (in Chinese)

[3] Shucai LI, Yiguo XUE, Qingsong ZHANG et al. Key technology study on comprehensive prediction and early-warning of geological hazards during tunnel construction in high-risk karst areas [J]. Chinese Journal of Rock Mechanics and Engineering, 2008, 27 (7): 1 297–1 306. (in Chinese)

[4] Zhigang LIU and Lianzhong WANG. Applied geomechanics [M]. Beijing: China Coal Industry Publishing House, 1993: 1–45. (in Chinese)

[5] BUTTON E, BRETTEREBNER H, SCHWAB P. The application of TRT-true reflection tomography at the Unterwald tunnel in Felsbau [J]. Geophysics, 2002, 20 (2): 51–56.

[6] Zhigang LIU and Yong ZHAO. Geological technique for tunnel construction [M]. Beijing: China Railway Publishing House, 2001. (in Chinese)

[7] Chengzong CHEN and Faliang HE. Engineering geological and sonic detection technique of the

tunnel [M]. Chengdu: Southwest Jiaotong University Press, 2005. (in Chinese)
[8] Zhigang LIU and Xiufeng LIU. TSP application and development in tunnel forecast [J]. Chinese Journal of Rock Mechanics and Engineering, 2003, 22 (8): 1 399–1 402. (in Chinese)
[9] Faliang HE and Cangsong LI. Development of geological forecast in tunnel construction [J]. Modern Tunnelling Technology, 2001, 38 (3): 12–15. (in Chinese)
[10] KLOSE C D. Fuzzy rule-based expert system for short-range seismic prediction [J]. Computers and Geosciences, 2002, 28 (3): 377–386.
[11] Jianfeng CHEN. Comparison of techniques for geological prediction of tunnels [J]. Underground Space, 2003, 23 (1): 5–9. (in Chinese)
[12] Ying WANG, Qiang CHEN, Youyi WEI et al. Application of infrared acquisition technology in prediction of water gushing in Yuanliangshan tunnel [J]. Chinese Journal of Rock Mechanics and Engineering, 2003, 22 (5): 855–857. (in Chinese)
[13] Jun WU, Haihe MAO, Song YING et al. Application of ground probing radar to short-term geological forecast for tunnel construction [J]. Rock and Soil Mechanics, 2003, 24 (Supp.1): 154–157. (in Chinese)
[14] K.-J.Shou. A Three-Dimensional Hybrid Boundary Element Method for Non–Linear Analysis of a Weak Plane Near an Under-ground Excavation [J]. Tunneling and Underground Space Technology, 2005, 15 (2): 215-226
[15] D. Sari, A. G. Pasamehmetoglu. Proposed support design, Kaletepe tunnel [J]. Turkey.: Engineering Geology, 2004, (72): 201-216
[16] CHRISTIAN D. Klose fuzzy rule-based expert system for short-range seismic prediction [J]. Computers and Geosciences, 2002, 28 (3): 377-386
[17] BUTTON E, BRETTEREBNER H, SCHWAB P. The application of TRT-truereflection tomography at the Unterwald tunnel in Felsbau [J]. Geophysics, 2002, 20 (2): 51-56
[18] Sweeting, N.Warth. Predicting ahead of the Face [J].Tunnels＆Tunneling, 1998, (4): 14-20
[19] A. Ozsan, C. Karpuz. Preliminary support design for Ankara subway extension tunnel [J]. Engineering Geology, 2001, (59): 161-172

# International comparison on the management system of end-of-life vehicles

**Jiaxi WANG[1], Anqi ZHU[2], Yaodong ZHOU[1]**

1.Beijing Jiaotong University, Haidian District, Beijing 100044, China.
2.Worcester Polytechnic Institute, Worcester, MA, 01609, U.S.A
16125250@bjtu.edu.cn

**Abstract**. With the rapid development of economy, Chinese vehicle ownership has risen significantly. At the same time, the end-of-life vehicles put a huge pressure to energy and environment, which will finally have an extremely bad effect on sustainable development of economy. The paper takes the United States, the European Union and Japan as samples, compares and analyses these countries' experiences in law, administration, operation. The key of developed country governance to solve 'trust' problem in recycling is its diversity within a complete legal system. At last, the paper gives some suggestions on the issue of Chinese end-of-life vehicles recycling.

## 1. Introduction

'Trust' problem is the key problem in the reproduction and remanufacture of end-of-life car. These cars must be effectively degraded into renewable substances before they are reproduced and remanufactured. In this way, they can provide a reliable basis for quality of new products. Developed country establishes a strong system to solve such problems. America is most productive and consumption nation of automobiles all over the world. They scrap more than 10 million vehicles every year. As the third economic heavyweight all over world, Japan is also a productive and consumption nation of automobiles. They scrap about 5 million vehicles every year All the European Union has been in 'Vehicle Society'. Just in 2002, they scrapped almost 9 million vehicles. China end-of-life vehicle market started quietly and the quantity of cars was expected to reach from 12 to 16 million at 2020.

At Special Press Conference on Automobile Circulation Development Promoting Green & Circular Consumption, spokesman for Ministry of Commerce Yao Jian pointed out that by the end of 2012, the vehicle ownership in China had exceeded 0.12 billion. He estimated that by the end of 2015 and 2020, the number would increase to 0.15 billion and 0.2 billion. Such huge inventory and demand size indicates that automobile market is gradually forging into the peak time of renewal. According to China statistical yearbook, the growth tread of China privately owned automobile from 1995 to 2014 is shown as Fig. 1.1. From the diagram it can be concluded that over the 20 past year China private vehicle ownership increased sharply. The yearly growth rate exceeded 15 percent. With the sharp increase of vehicle ownership, the amount of end-of-life vehicle is continuously increasing.

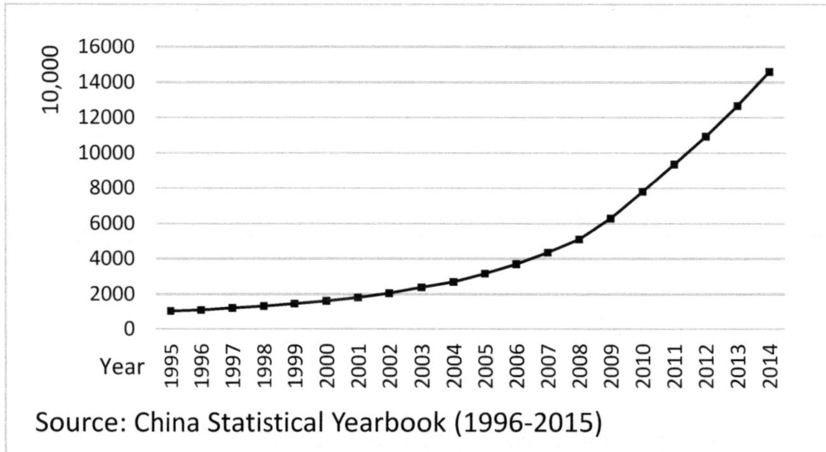


Fig. 1.1 Trend of China's motor vehicle`s number 1995~2014

The countries which have developed automobile industry met problems of end-of-life vehicles in advance. After many years' discovery and practice, they have a lot of experience of how to use macro-economic control and market allocating resource to manage end-of-life vehicles effectively. The core problem of recycling and managing end-of-life vehicles is how to select management measures. Since there are multiple ways to facilitate end-of-life vehicles recycle and management, we can classify them with different standards[1]. In this part, we discuss the problem through legal and policy measure.

## 2. Legislation restricition

### *2.1. United States – Indirect Legislation Restrictions*

With the increased harm caused by end-of-life vehicle, American government tried to take serious actions since the nineteen nineties. They legislated complete laws and regulations about environment protection and production joint liability. But they didn't have special national end-of-life recycling regulations. End-of-life recycling regulations should obey the following government laws, The Ministry of transportation vehicle recalling law, the Ministry of Commerce car insurance usage law, the Ministry of energy average fuel consumption law, EPA emissions law and so on.

In 1991, America firstly legislated the law to facilitate recycling wasted tyre. In 1994, government ruled that asphalt highway which was funded by government must contain 5% rubber particles. These rubber particles were made by old tyres. After implementing the regulation, asphalt highways contained rubber particles in America have already exceeded 11 thousand kilometers. Asphalt Pavement can be anti-aging by using rubber particles. Besides, it can make road surface be more elastic and extend lifespan. Moreover, according to the related law, the vehicle accessory can be use again only if it doesn't reach the years that to be completely abandoned and affect normal use. All of these laws facilitate the development of American end-of-life vehicle recycling.

With the strict environmental protection regulations, American consumers can't abandon cars anywhere. Usually, they will give end-of-life car to car dismantling enterprises voluntarily. Production joint liability means that in a manufacturing chain, some or all of the producers and sellers should share associative responsibility for the losses caused by the production that is dangerous in manufacture and sale or defective products[2]. Such product liability regulations promote the entire automotive production supply chain to participate in the research of end-of-life car recycling, making the end-of-life cars become a higher residual value of material resources in the United States. End-of-life can be recycled in market-oriented mode.

### *2.2. Japan – Completed recycling regulation and effective processed operating procedure*

Since 1970s, Japan began to build a recycling economy system to get rid of mass production, consumption and waste-based development model. They use the end-of-life products as resources to

IOP Conf. Series: Earth and Environmental Science **61** (2017) 012008 doi:10.1088/1755-1315/61/1/012008

recycle them effectively and establish a basic legal system to promote recycling-oriented society. They also clearly defined the obligation of automobile manufacturers, importers and vehicle owners.

In May 1997, the METI promulgated the "Rules for the Regeneration of Used Vehicles" on the basis of the Report of the Council on Economic and Industrial Structure. The main objective of the law is to reduce the use of hazardous substances, reduce debris and improve the reusability. Meanwhile, it also plays the following roles.

Improve the original processing channels and make it more efficient. For example, in order to prevent illegal discard, they established the upper and lower processes to manage note and strengthened the penalty of illegal discard.

They complete the information exchange between relevant departments and make the work more efficient. They also clarified the obligations of the automobile manufacturers and the relevant departments of the government, local governments, distributors, disintegration and crusher employees. In addition, it also emphasized the following two points: Establish end-of-life vehicle recycling support center. This center can not only strengthen information exchange with relevant department, but also can improve the recycling rate of technology development. In new car design respect, fully considering the easy decomposition, easy recycling of the structure and selection and improvement of plastic design, which is aim at the waste plastics that take up 50% of the current debris.

In 2004, Japan revised the Law of Registration of Vehicle Cancellation, and strengthened information management of updated vehicles from all aspects of vehicle registration and cancellation. It also promoted the recycling, dismantling and comprehensive utilization of end-of-life vehicles.

### *2.3. European Union – Legislation from the root*

On 18 September 2000, the European Parliament and the European Council passed legislation to implement the "Directive on End-of-Life Vehicles (2000/53 / EC)", which was used as a legislation of Member States in 2001.

Directive made a series of regulations on the disposal of end-of-life vehicles, recycling rate, vehicle free recycling, etc. to promote the development of end-of-life vehicles recycling. The EU was also released "Directive 2005/64/EC on the type-approval of motor vehicles with regard to their reusability, recyclability and recoverability" in 2005. The directive was based on the "Directive 2000/53/EC on end-of life vehicles" and brought category M1 or N1 into administration and technical regulation of EU type certification system. The Directive explicitly requires individual vehicle manufacturers to collect information from the entire supply chain as well.

The Directive has become the basis for the EU members to develop end-of-life vehicles recycling regulations. The relevant regulations legislated by the Member States are based on Directives (Table 2.1).

Table 2.1 Relevant Regulations of End-of-Life Vehicles Recycling in EU Member States

| Country | Release / implementation time | Relevant laws and regulations |
|---|---|---|
| Germany | 1972<br>1996<br>21 June 2002 | "Waste Disposal Law"<br>"Circular Economy and Waste Management Law"<br>"End-Of-Life Vehicle Regulation" |
| Britain | 2003 | "End-Of-Life Vehicle Regulation"<br>"End-Of-Life Vehicle Regulation 2003" |
| | 3 March 2005 | "End-of-Life Vehicle (Producer Responsibility) Regulations"<br>"End-Of-Life Vehicle Regulation 2005" |
| France | August 1, 2003 | "Decree on the construction of vehicles and the disposal of motor vehicles"<br>"Decree No.2003-727" |
| Netherlands | 24 May 2002 | "The Scrap Car Management Act"<br>"End-Of-Life Vehicle Management Decree" |
| Spain | 20 December 2002 | "ROYAL DECREE 1382/2002" |

According to the above directive or law, the main responsibility of the end-of-life vehicle recycling chain is the automobile manufacturer, that is, car manufacturers or importers. The automobile manufacturer connects upstream (parts suppliers) of end-of-life vehicles recycling chain to its downstream (recyclers, dismantlers and crushers), which is the key to end-of-life vehicles recycling. When they produce cars, they are required to achieve the following objectives at least: low energy consumption, easy dismantling, suitable for recycling, reduce toxic substances emissions, etc. To achieve these goals, manufacturers need to understand the appropriate technical and economic means to manufacture vehicle that is easy to be dismantled, recycled and is low-harmful. In addition, the cooperation of the recycler, the dismantler, and the shredder is basis for achieving the Directive.

## 3. Operation system management

### *3.1. United States – Fully develop private and business autonomy*

In 1994, the first Cooperative Research and Development Agreement (CRAPA) was signed by the Vehicle Recycling Consortium (GM, Ford, DaimlerChrysler), the US Department of Energy's Argonne National Laboratory at the University of Chicago and the American Plastics Institute (APC). Automotive fuel economy, vehicle safety and lightweight materials are listed as mainly studied objectives in the agreement and they reach consensus to share cost and research results. The dismantling of end-of-life vehicle and the recycling of seat foam have made phased research results at the first cooperative research and development agreement.

With the leadership of three American major automobile companies, the US Department of Energy, the Office of the car pre-technology and Argonne National Laboratory jointly developed a "Future End-of-Life Recycling Guide" in 2001, which clearly defined that the US end-of-life recycling rate should reach 95% before 2020. The vehicles that reach the end of life in 2020 are products produced between 2005 and 2010. Steel and aluminum are still the main material components of these vehicles, and thermoplastics are still included in their plastic parts. The Guide assumed that the materials for end-of-life vehicles in 2020 should include 75% metal, 15% plastic and 10% other materials such as glass, liquid and so on. The guide predicted that 10% of end-of-life vehicles would be disassembled, 5% of metal parts would be re-used and remanufactured, and 5% would be recycled. Crushing and material sorting recycling would account for 90% of end-of-life cars, while landfill would only account for5%.

### *3.2. Japan – strengthen recycling information system establishment*

Japan's management of end-of-life vehicles recycling is carried out through both government and non-governmental organizations. Mainly including the following three points: 1. Increase awareness of consumer protection by levying recycling fees. 2. Establish a fair and transparent electronic inventory system, and increase information management efforts[3]. 3. Establish a recycling system which is based on the principle of expanding the responsibility of producers.

For the definition of consumer and producer responsibility, the Ministry of Economy, Trade and Industry (METI) examined a wide range of questionnaires on who should take responsibility for the cost of recycling end-of-life vehicles. The result showed that consumers should take this responsibility. Japan adopt levying recycling fee approach to promote the recycling of end-of-life vehicles, which means the money levied from vehicle producers or consumers will pay to end-of-life vehicles recyclers. This will enhance consumers' awareness of resource conservation and environmental protection.

Besides, to the producers, establish recycling mechanism which is based on the principle of expanding producers' responsibility. The system expands the responsibility of car manufacturers. They are not only responsible for the recovery and disposal of ASR, freon and airbags, but also need to ensure that the operation of the entire recycling system should be successful.

### *3.3. The European Union – Strengthen the control of producers*

After the European Commission issued the first directive, the "Directive on End-of-Life Vehicles (2000/53 / EC)" (referred to as "ELV Directive"), in May 2000, they also issued the "Directive 2005/64/EC on the type-approval of motor vehicles with regard to their reusability, recyclability and

IOP Conf. Series: Earth and Environmental Science **61** (2017) 012008 doi:10.1088/1755-1315/61/1/012008

recoverability" (referred to as "RRR type certification"), "Motor vehicles with trailers: EC type-approval (70/156/EEC)" revision. The ELV Directive also become an independent directive within the framework of the 70/156 / EEC, the 2009/1 / EC Directive revision and the RRR Type Approval Regulations of 2005/64 / EC, which clearly requires the vehicle companies to collect information from the entire supply chain.

The ELV Directive defines the two-phase target of end-of-life vehicles' "re-use and re-utilize" and "re-use and recycling" in the EU, prohibits or limits the use of four heavy metals (lead, mercury, cadmium and hexavalent chromium) as well. The ELV Directive is composed of 13 parts, which is aim to prevent vehicle waste. It establishes mechanisms for the collection, treatment and reuse of end-of-life vehicles to reduce waste disposal. Besides, it improves the environmental protection effectiveness of all operators involved in the vehicle life cycle.

## 4. Suggestion to China

As of 2014, China end-of-life vehicle recycling dismantling industry develops steadily. There are 597 companies acquired dismantling qualification in China, which means that end-of-life vehicle recycling network is taking shape. The number and overall quality of recycling dismantling company are increasing gradually. But there are still some problems in our country, such as low recovery, low level enterprise, heavy tax burden and disorder. Besides, the problems are caused by the enterprises' management. It's not suited to social development. And relevant laws also still need to be improved. We have following suggestions:

### *4.1. Complete for the recycling of end-of-life vehicles standards and regulations*

End-of-life vehicle recycling standards and regulatory system should be combined with the characteristics of our country from the design and production of new cars to the maintenance and repair of motor vehicles scrap dismantling and recycling. The scope should be involved in domestic and imported cars. In addition, Chinese auto products and accessories' import and export also need to comply with international laws and regulations.

### *4.2. As soon as possible to improve the recovery of automotive product standards*

Develop automotive product recycling standards as soon as possible, emphasis on environmental design, quantify the use of hazardous substances in product design, and clarify the ban on hazardous substances and the standards that automobile and accessory enterprises that must implement. This can improve the recycling rate of end-of-life vehicles, protect the environment and promote the realization of stage objectives.

### *4.3. Regulate the end-of-life vehicle recycling management system*

Further improve the management system of scrap car recycling, standardize the operation process is focus we need to improve in the recovery of end-of-life vehicles. First of all, we need to clarify that the car manufacturer is the responsibility of the end-of-life vehicle recycling system. The car manufacturers do not only have responsibility for the design of automotive products, but also should be responsible for recycling end-of-life vehicles and disclose vehicle dismantling information. Secondly, the various enterprises in the scrapped car recycling system (Automobile and accessories enterprises recycling dismantling enterprises) should strengthen cooperation to achieve common goals. Besides, we should use foreign advanced technology in the recycling of end-of-life vehicles for reference, such as the EU end-of-life vehicle directive, to further refine the recycling process with a view of China's special circumstances.

### *4.4. Encourage automotive product recycling technology research*

Recycling, dismantling and reuse of end-of-life vehicles is actually an important part of the whole life cycle of product design, production, consumption, and recycling. To improve the product recycling rate, on the one hand, we need to use scientific methods to regulate the scientific recovery dismantling process; On the other hand, to start from the source. That is, at the beginning of product design, we need to take the performance of the product cost index into account to reduce environmental and energy consumption of components and other recycling problem, so as to design more pollution-free and reusable automobile. Save resources and protect the environment while meet people's needs.

### *4.5. Identify the responsibilities of all the parties involved in the end-of-life vehicle*

Not matter the EU's 'Who manufacture, who is responsible for it', or the Japanese consumer taking whole responsibility, they all clearly defined the responsibility of every party. This avoid shirking the responsibility at the most extent, thus promoting the recovery of end-of-life vehicles. China should strengthen the definition of end-of-life vehicle responsibility so that the end-of-life vehicles can be handled according to the standard procedure.

**Acknowledgement**

This paper is supported by the Beijing Education Commission project "An empirical study on governance mechanism of environment externality in Beijing urban traffic (2016LJJD-2)" and the Beijing Social Science Association project "Research on the coordination of urban vehicle restriction policy under the integration of BTH(B13H100010)".

## References

[1] GENG Yong-sheng. Tail Gas Pollution from the Vehicles and It`s Control Technoloty[J]. Environmental Science Survey,2010,06:62-69.
[2] Xuan Xiangyang Ge Jinxi Cheng Ying etc. Discussion on the operation mode of China's automobile resources recycling [J]. Shanghai Auto,2008,09:28-31.
[3] ZHOU Dan HAI Reti etc.Implementary Measure of Extended Producer Responsibility in Recovery and Management of Used Automobile[J]. Environmental Science & Technology,2007,09:61-64+118-119.
[4] Kanari N, Pineau J L, Shallari S. End-of-life vehicle recycling in the european union[J]. JOM, 2003, 55(8):15-19.
[5] CHEN Ming;WANG Jun-jun. Recycling of End-of-Life Vehicle: Refulation,Management and Prospects [J]. Journal of Shanghai Jiaotong University, 2014(01):125-131.
[6] Mazzanti M, Zoboli R. Economic instruments and induced innovation: The European policies on end-of-life vehicles[J]. Ecological Economics, 2006, 58(2):318-337.
[7] Sawyer-Beaulieu S S, Stagner J A, Tam E K L. Sustainability Issues Affecting the Successful Management and Recycling of End-of-Life Vehicles in Canada and the United States[M]// Environmental Issues in Automotive Industry. 2014:223-245.

# Adsorption and desorption of agricultural waste-derived DOMs in soil

**H L Li, G H Yang, G Y Zhu, Z Q Sun and X Y Yu**

School of Resources and Environment, University of Jinan, Jinan 250022, China
ujn_lihl@163.com

**Abstract.** The sorption and desorption of two forms of dissolved organic matter (DOM) extracted from agricultural wastes were studied by batch experiments. The adsorption of the two DOMs on the soil were well fitted to the Linear and Freundlich isotherms. DOM extracted from cow manure (MDOM) shows higher affinity to the soil than that extracted from wheat straw (SDOM). Significant desorption hysteresis was observed for both DOMs. Due to the desorption of some aromatic substances with larger molecular weight from the soil, the average molecular weight and aromaticity of the DOMs increased at sorption equilibrium compared with those before sorption.

## 1. Introduction

The incorporation of organic amendments such as straw and animal manure are common farming management practices in China and many other countries around the world [1], thus DOMs originated from these amendments will be released into soil environments. DOM can be bound to the soil via electrostatic adsorption, ligand exchange, complexation, hydrophobic interaction, hydrogen bonding and cation bridging mechanisms [2]. Since DOMs play an important role in the transport of pollutants in soils, their fate has been a subject of considerable interest in recent years.

There are several studies on the sorption of DOMs in soils. For example, Kaiser et al. [3] investigated the sorption of total, hydrophilic, and hydrophobic DOM on 125 forest soil samples and found that the majority of the soils preferentially sorbed hydrophobic dissolved organic carbon. Shen [4] evaluated some important physical-chemical factors determining the sorption of DOM onto natural soil and demonstrated that DOM sorption on soil is largely due to ligand exchange between DOM and hydroxyl groups on the soil mineral surface. He also found that soil sorption capacity of DOM is positively correlated to the soil clay content as well as the soil water solution ionic strength and pH. The sorption of DOMs on soils can decrease the mobility of the associated pollutants. Furthermore, the adsorbed DOMs can change the physicochemical properties of the soil surface, which will have an effect on the fate of some other contaminants. Thus, it is essential to better understand the processes that alter the concentration and properties of DOMs in soils.

The aims of this study was to explore the sorption and desorption behaviors of two DOMs derived from agricultural wastes and the changes in their physiochemical characteristics before and after sorption.

## 2. Materials and methods

### *2.1. Materials*

$CaCl_2$ and $NaN_3$ were analytical grade purchased from Sinopharm Chemical Reagent Co. ltd and Sigma-Aldrich, respectively.

An uncontaminated soil was collected from the suburb of Jinan city, China. The soil was air dried and sieved through a 2 mm mesh. The $K_2Cr_2O_7$ titration method was used to determine the soil total organic carbon (TOC). The soil pH was determined in 1:5 soil - water slurries using a glass electrode. The cation-exchange capacity (CEC) of the soil was determined by the $NH_4^+$–acetate method. Selected soil properties are shown in Table 1.

Table 1. Selected soil properties

| pH | Grain size distribution (%) | | | CEC (cmol/kg) | TOC (g/kg) |
|---|---|---|---|---|---|
| | <0.002 mm | 0.002-0.02 mm | 0.02-2 mm | | |
| 6.8 | 53.2 | 24.4 | 22.4 | 6.84 | 7.19 |

*2.2. Dissolved organic matter extraction and characterization*
Two forms of dissolved organic matter were used. One form of DOM was extracted from wheat straw (SDOM) and the other from cow manure (MDOM). The DOMs were extracted as described in our previous study [1]. Briefly, wheat straw and cow manure were extracted with deionized water using a solid/water ratio of 1:10 and 1:5 (w/v, dry weight basis), respectively, in a reciprocal shaker at 200 rpm and 25 °C for 24 h. The suspensions were centrifuged at 5000 rpm for 30 min, then, vacuum filtered through a filter paper (Whatman 20-25 μm pore size), and a 0.45 μm cellulous acetate filter (Whatman, Maidstone, UK) successively. The filtrates were freeze dried as powders and stored in a desiccator for later use. The concentrations of DOMs were measured at mg $L^{-1}$ dissolved organic carbon (DOC) using a total organic carbon (TOC) analyzer (Shimadzu TOC-L CPH/CPN, Japan). The electrical conductivity (EC) of the DOM solution was measured at 25 °C by a YSI 3100 conductivity instrument (YSI, US). The salt ions in the DOM solution were determined using an inductive coupled plasma atomic emission spectroscopy (ICP-AES) (PerkinElmer, Optima 7000 DV). The molecular weights were measured using the high-pressure size exclusion chromatography method as described by Yue et al. [5]. Specific UV absorbance at 254 nm was used to evaluate the aromaticity of DOM samples [6]. Absorbance of DOM solution at 465 and 665 nm was also measured to calculate the E4/E6 ratio, which is negatively correlated with molecular size [7]. The $^{13}C$ CP/MAS NMR spectra were obtained using a Varian Infinity Plus-300 spectrometer equipped with a 4 mm double-resonance magic angle spinning probe head, operating at $^{13}C$ frequency of 75.4 MHz. The aliphatic carbon and aromatic carbon were determined by quantifying the peak area in the 0-109 ppm [8] and 110-160 ppm [9] chemical shift bands, respectively. The polar carbon of the DOMs was determined by quantifying the peak area in the 50-109 ppm and 145-220 ppm chemical shift band [8].

*2.3. Adsorption and desorption experiments*
Batch adsorption experiments were conducted in 30-mL glass vials with teflon-lined caps. Soil sample (3.00 g) was mixed with 30 mL of DOM solution in 0.01 M $CaCl_2$ as the background electrolyte and 0.2 g $L^{-1}$ $NaN_3$ as a biocide. The initial concentrations of DOMs were in the range of 0–150 mg $L^{-1}$. The mixtures were shaken for 48 h at 25 °C on a rotary shaker (200 rpm) and afterwards centrifuged for 30 min at 5000 rpm. After the sorption period, desorption was conducted by withdrawing 28 mL supernatant and replenishing the samples with the same volume of background electrolyte solution. The vials were sealed, shaken and centrifuged as described in the sorption experiments. The supernatant of sorption and desorption experiments was filtered through a 0.45 μm cellulous acetate filter (Whatman, Maidstone, UK). A portion of the filtrate was used for the determination of TOC and the remaining was used for UV detection at 254 nm, 465 nm and 665 nm.

Sorbed DOM concentration in the sorption experiments was calculated by the following mass balance equation

$$Q_e = \frac{(C_0 - C_e + C_{ck})V}{W} \tag{1}$$

where $C_e$ is the liquid phase equilibrium concentration in mg $L^{-1}$ and $Q_e$ is the concentration on the solid phase in mg $kg^{-1}$, $C_{ck}$ is the DOM concentration that released from the soil to the solution when no SDOM and MDOM was added, $V$ (mL) is the volume of solution and $W$ (g) is the mass of soil used.

For the desorption experiments, the desorption amount of DOM can be calculated with the following equation

$$Q_d = \frac{(C_e' - C_{ck}')V - C_e V_1}{W} \quad (2)$$

where $Q_d$(mg $kg^{-1}$) is the desorption amount of DOM at equilibrium, $C_e'$ (mg $L^{-1}$) is the equilibrium concentration in the liquid phase, $C_{ck}'$ is the DOM concentration that released from the soil to the solution in the desorption treatment with no SDOM and MDOM was added initially, $V_1$ is the residual solution volume before the desorption experiments.

### 2.4. Adsorption isotherm models

Linear, Langmiur, Freundlich and Temkin isotherm which have the following forms were used to fit the adsorption data

Linear isotherm: $Q_e = aC_e + b$ (3)

Langmiur isotherm: $\frac{C_e}{Q_e} = \frac{C_e}{Q_m} + \frac{1}{kQ_m}$ (4)

Freundlich isotherm: $\ln Q_e = lnK_f + nlnC_e$ (5)

Temkin isotherm: $Q_e = A + B\ln C_e$ (6)

where $Q_m$ is the Langmuir parameter related to maximum adsorption capacity (mg $kg^{-1}$), $k$ is the Langmuir parameter related to energy of adsorption, $K_f$ is the Freundlich affinity coefficient [(mg $kg^{-1}$)/(mg $L^{-1}$)$^n$], n is the Freundlich nonlinearity factor, A and B are constants.

## 3. Results and discussion

### 3.1. Characterization of DOMs

The major properties of the two DOM samples were reported previously [1] and are shown in Table 2. Higher $A_{254}$/DOC values for MDOM indicate that MDOM has relatively higher aromaticity than SDOM. The E4/E6 ratios of SDOM and MDOM were 8.83 and 7.66, respectively, indicating that MDOM had a relatively higher molecular weight. The differences between SDOM and MDOM might affect their adsorption on the soil, thus their interaction with other pollutants. Table 3 presents the molecular weights and structural composition of the two DOMs. It can be seen that MDOM has a higher molecular weight than SDOM. The contents of aliphatic and polar carbon for SDOM are higher than those for MDOM, while the content of aromatic carbon for SDOM are lower than that for MDOM.

**Table 2.** Properties of dissolved organic matters extracted from wheat straw (SDOM) and cow manure (MDOM) [a]

| Sample | pH | EC (μS/cm) | $K^+$ (mg $L^{-1}$) | $Na^+$ (mg $L^{-1}$) | $Ca^{2+}$ (mg $L^{-1}$) | $Mg^{2+}$ (mg $L^{-1}$) | $A_{254}$/DOC (L·$mg^{-1}$·$cm^{-1}$) | E4/E6 |
|---|---|---|---|---|---|---|---|---|
| SDOM | 8.86 | 254 | 12.4 | 1.68 | 1.19 | 0.79 | 1.6 | 8.83 |
| MDOM | 9.06 | 192 | 9.49 | 2.94 | 0.36 | 0.57 | 1.8 | 7.66 |

[a] Data measured at DOM concentration of 50 mg/L DOC.

### 3.2. Sorption-desorption isotherms of DOMs

The DOM concentration at sorption/desorption equilibrium and the sorption/desorption amounts are shown in table 4. In the sorption experiments, when no SDOM or MDOM was added to the solution, a concentration of 18.4 mg $L^{-1}$ was measured at equilibrium, implying that DOM was released from the soil. The desorption amount is 184 mg $kg^{-1}$. Significant desorption hysteresis was observed for both

**Table 3.** Molecular weight, polydispersity, aliphatic carbon, aromatic carbon and polar carbon of DOM samples

| Sample | Mw [a] | Mn [b] | Polydispersity (ρ) | Aliphatic C (%) | Aromatic C (%) | Polar C (%) |
|---|---|---|---|---|---|---|
| SDOM | 2632 | 1255 | 2.1 | 75.6 | 3.2 | 77.0 |
| MDOM | 3542 | 1964 | 1.8 | 71.1 | 3.8 | 69.1 |

[a] Weight-averaged molecular weight; [b] Number-averaged molecular weight.

DOMs. For SDOM, the percentage that can be desorbed is 9.86%-50%. While for MDOM, the percentage is only 2.52%-25.2%. This means that MDOM are more difficult to be desorbed than SDOM. As can be seen from table 5, the linear equation and the Freundlich equation can well fit the adsorption of the two DOMs in the soil. The sorption isotherms were also shown in figure 1. In the concentration range studied, the adsorption amount of DOM was linearly correlated with the

**Table 4.** Sorption/desorption amount of DOMs

| Treatment | Initial concentration ($mg \cdot L^{-1}$) | Concentration at sorption equilibrium ($mg \cdot L^{-1}$) | Sorption amount ($mg \cdot kg^{-1}$) | Concentration at desorption equilibrium ($mg \cdot L^{-1}$) | Desorption amount ($mg \cdot kg^{-1}$) |
|---|---|---|---|---|---|
| DOM-0 | 0 | 18.4 | -184 [a] | 2.98 | 17.5 |
| SDOM-25 | 25 | 36.6 | 72.5 | 8.01 | 38.2 |
| SDOM-50 | 50 | 55.5 | 142 | 8.67 | 32.2 |
| SDOM-75 | 75 | 73.9 | 216 | 10.2 | 34.8 |
| SDOM-100 | 100 | 99.1 | 223 | 10.6 | 22.3 |
| SDOM-150 | 150 | 127 | 455 | 16.6 | 63.2 |
| MDOM-25 | 25 | 34.2 | 88.5 | 4.37 | 3.38 |
| MDOM-50 | 50 | 51.8 | 164 | 9.45 | 42.5 |
| MDOM-75 | 75 | 66.8 | 265 | 6.88 | 6.77 |
| MDOM-100 | 100 | 83.0 | 354 | 9.82 | 25.4 |
| MDOM-150 | 150 | 113 | 554 | 11.2 | 19.6 |

[a] Negative number means the desorption amount from soil.

**Table 5.** Fitting parameters of adsorption isotherms

| | SDOM | | | MDOM | | |
|---|---|---|---|---|---|---|
| Linear isotherm | $a$ | $b$ | $R^2$ | $a$ | $b$ | $R^2$ |
| | 4.25 | -8.02 | 0.998 | 5.97 | -17.8 | 0.995 |
| Langmiur isotherm | $Q_m$ | $k$ | $R^2$ | $Q_m$ | $k$ | $R^2$ |
| | 1111 | 0.006 | 0.862 | 769 | 0.006 | 0.972 |
| Freundlich isotherm | $K_f$ | $n$ | $R^2$ | $K_f$ | $n$ | $R^2$ |
| | 4.06 | 0.999 | 0.998 | 5.34 | 0.989 | 0.992 |
| Temkin isotherm | $A$ | $B$ | $R^2$ | $A$ | $B$ | $R^2$ |
| | -579 | 211 | 0.905 | -652 | 250 | 0.887 |

equilibrium concentration in the solution phase, indicating that the adsorption of DOM onto soil may not reach saturation state, and there are still vacant adsorption sites on the soil surface. The higher the value of parameter $a$ in the linear adsorption equation, the easier the adsorption of DOM by soil. The value of parameter $a$ for MDOM was higher than that for SDOM, indicating that MDOM was more easily adsorbed by soil. This might be originated from the difference in composition and structure of the two DOMs. The aromaticity of MDOM is higher than that of the SDOM, implying it is more hydrophobic and more easily to be adsorbed by soil. This is also verified by the higher $K_f$ value of MDOM.

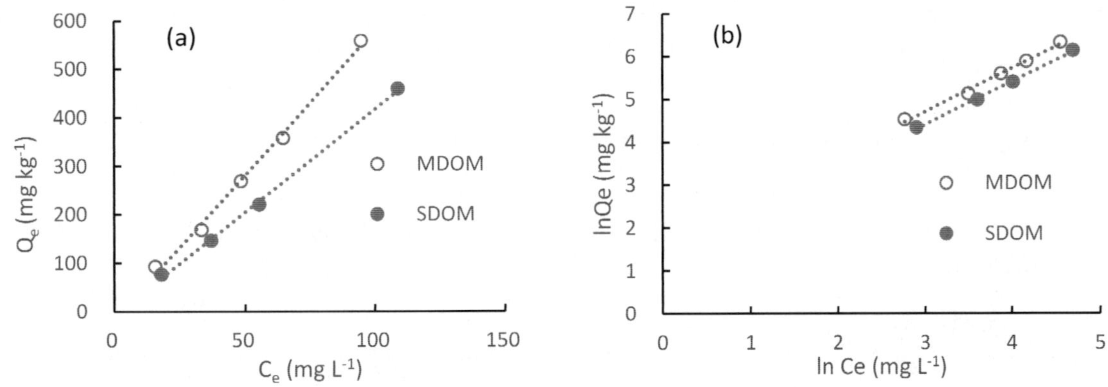


**Figure 1.** Adsorption isotherms of DOM: (a) Linear isotherm and (b) Freundlich isotherm

*3.3. Aromaticity and molecular weight characteristics before and after sorption*

Table 6 shows the aromaticity and molecular weight characteristics at sorption equilibrium. The E4/E6 values of SDOM and MDOM before adsorption were 8.83 and 7.66 (Table 2), respectively. At adsorption equilibrium, the E4/E6 values of SDOM and MDOM were 2.29-2.53 and 2.07-3.26, respectively, which both decreased significantly. It is indicated that the average molecular weight of DOMs greatly increased. Some researchers [3] have reported the substance with large molecules in DOMs were preferentially adsorbed on the surface of the soil, thus the molecular weight would decrease if only the sorption process being considered, which contradicted the result of this study. The increase in molecular weight might be attributed to desorption of some substances with large molecules from the soil used in this study, which proved by the lower E4/E6 (1.89) value of DOM-0 treatment (Table 6). Comparing the data in Table 1 and Table 6, it can be found that the $A_{254}$/DOC values at equilibrium was higher than those before sorption, indicating an increase in aromaticity. This might be explained by desorption of some aromatic substances from the soil, which was proved by the higher $A_{254}$/DOC value (34.8) of DOM-0 treatment.

**Table 6.** Aromaticity and molecular weight characteristics at sorption equilibrium

| Treatment | $A_{254}$/DOC (L·mg$^{-1}$·cm$^{-1}$) | E4/E6 |
|---|---|---|
| DOM-0 | 34.8 | 1.89 |
| SDOM-25 | 19.9 | 2.47 |
| SDOM-50 | 17.3 | 2.28 |
| SDOM-75 | 16.7 | 2.58 |
| SDOM-100 | 14.4 | 2.53 |
| SDOM-150 | 16.0 | 2.98 |
| MDOM-25 | 24.7 | 2.07 |
| MDOM-50 | 17.3 | 2.28 |
| MDOM-75 | 14.7 | 3.10 |
| MDOM-100 | 13.5 | 2.63 |
| MDOM-150 | 14.0 | 3.26 |

## 4. Conclusions

The adsorption of the two DOMs on the soil were well fitted to the Linear and Freundlich isotherms. Significant desorption hysteresis was observed for both DOMs. Compared with SDOM, MDOM are more easily to be adsorbed by soil and more difficult to be desorbed. Due to the desorption of some

IOP Conf. Series: Earth and Environmental Science **61** (2017) 012009 doi:10.1088/1755-1315/61/1/012009

aromatic substances with larger molecular weight from the soil, the average molecular weight and aromaticity of the DOMs increased at sorption equilibrium compared with those before sorption.

**Acknowledgments**
Financial supports are from the National Natural Science Foundation of China (21207049).

**References**

[1] Li H L, Qiu Y H, Wang X N, Liu W H, Chen G C, Ma Y B and Xing B S 2016 *Environ. Pollut.* **210** 323
[2] Mavi M S, Sanderman J, Chittleborough D J, Cox J W and Marschner P 2012 *Sci. Total Environ.* **435-436** 337
[3] Kaiser K, Guggenberger G and Zech W 1996 *Geoderma* **74** 281
[4] Shen Y H 1999 *Chemosphere* **38** 1505
[5] Yue L, Wu F, Liu C, Li W, Wang J and Mei Y 2004 *Chin. J. Geochem.* **23** 275
[6] Tusseau-Vuillemin M H, Gourlay C, Lorgeoux C, Mouchel J M, Buzier R, Gilbin R, Seidel J L and Elbaz-Poulichet F 2007 *Sci. Total Environ.* **375** 244
[7] Yu H, Huang G H, An C j and Wei J 2011 *J. Hazard Mater.* **190** 883
[8] Wang X, Shu L, Wang Y, Xu B, Bai Y, Tao S and Xing B 2011 *Environ. Sci. Technol.* **45** 9276
[9] Chin Y P, Aiken G R and Danielsen K M 1997 *Environ. Sci. Technol.* **31** 1630

# The development and application of bi-functional clay inhibitor of CYY-2

**Yongbin Li[1, a], Fengqun Li [2], Jiliang Yu[1], Changsheng Han[3], Hui Gong[2],Jing Wang[1],Yingde Wu[1],Rui Wang[1]**

[1]Petroleum Production Engineering Research Institute, Huabei Oilfield Company, Renqiu ,Hebei062552, china
[2]The Third Exploit Factory of Huabei Oilfield Company, Hejian ,Hebei062450, china
[3]The Erlian Filiale of Huabei Oilfield Company of petrochina, Xilin Hot026017,china
[a]email:cyy_liyb@petrochina.com.cn

**Abstract**. with the constant discovery and gradual development of low-permeability and water-sensitivity formation, reservoir protection is especially important for improving water-injection development efficiency and enhancing working level of low-permeability reservoir, which will be a critical sector in enhancing development quality of low-permeability and water-sensitivity reservoir. Combined with the understanding of clay mineral expansion theory, starting from hydratability, polarity and oxidability of adopted materials, one type of bi-functional clay inhibitor, CYY-2, is developed which has the function of swelling prevention and swelling shrinkage, and property evaluation is carried out by adopting different methods and measures. Test results indicated that: by using this bi-functional clay inhibitor, the swelling prevention rate of magcogel and natural core powder could reach to 91.7% and 89.8%, swelling shrinkage of which could reach to 61.3% and 81.5%; in the natural flowing experiment, core damage ratio was declined from 86.3% to 12.8%, and the core permeability could be raised by 3 times averagely; in the field experiment of 5 wells, effective rate of 80%, the well effect, is reached. With the extension of injection time of injecting wells, injection pressure rose constantly, injection allocation request was not finished, which influenced the overall reservoir development effect. So reservoir protection is important to deal with the above problems.

## 1. Introduction

Reservoir protection research is one of important technologies in the oilfield development as well as an important element to improve exploration and development quality. As the extension of develop time, some injection wells in the seriously water-sensitive reserves had many conditions such as rising injection pressure, decreasing injection volume and being difficult to accomplish injection allocation, which had bad influence on the reservoir development effect. The reservoir clay content is high and easy to be water-sensitive. In the water injection development, the salinity of injection water is lower than that of formation water, so clay mineral will be hydrate and swell, break and migrate which make oil layer damaged, injection pressure raised, injection allocation decreased. Thus, aiming at the conditions that clay content is high and clay is easy to hydrate and swell with fresh water and to migrate in the water-sensitive reserves of Huabei oilfield, on the base of taking example by domestic and overseas experiences of swelling prevention measures, bi-functional clay inhibitor of CYY-2 is developed. The CYY-2 has the functions of preventing clay from hydrating and swelling and dredging formation hole of low-permeable and water-sensitive reserves, realizing decompression and stimulation. The CYY-2 can achieve well effects through the field pilot tests.

## 2. The development of clay inhibitor formula

### *2.1. Main materials molecular formula*

The molecular formula for N-propyl-N,N,N-three methyl ammonium chloride is $C_9H_{19}ClN_2O$, The molecular formula for γ-propyl three methoxy silane is $C_9H_{21}NO_3Si$, molecular formula for PEG is $HO(CH_2CH_2O)_nH$.

### *2.2. Main materials filtering*

Through literature researches, four chemical raw materials are chose of which main materials are easy to get and price is reasonable. By using centrifugation evaluation methods, swelling prevention effect is confirmed.

Swelling prevention percentage and swelling shrinkage percentage is measured by centrifugation methods, experiment methods are as follows:

Swelling prevention percentage: The mass of bentonite power is 0.50g (montmorillonite content is higher than 80 percent), accurate to 0.01g. The power is put into centrifuge tube of 10ml, added antiswelling agent solution of 10ml to the tube and shaking enough. The mixture is placed for 2 hours in the indoor temperature and then put into centrifugal machine. After being centrifugalized for 15minutes in the rotate speed of 1500r/min, the expanding volume $V_1$ of bentonite power is read. The antiswelling agent solution is replaced by deionized water of 10ml and then the expanding volume $V_2$ in water of bentonite power is measure. The antiswelling agent solution is replaced by kerosene of 10ml and then the expanding volume $V_0$ in kerosene of bentonite power is measure.

The equation of swelling prevention percentage is $B=(V_2-V_1)/(V_2-V_0)\times100\%$

Swelling shrinkage ratio: According to the above method, deionized water and injecting water of 10ml are added to the centrifuge tube and then the volume $V_3$ is read. Clay inhibitor solution of different concentrations are added into the above centrifuge tubes, placed for 4 hours in the indoor temperature and then put into centrifugal machine. After being centrifugalized for 15minutes in the rotate speed of 1500r/min, bentonite power volume $V_4$ is read after swelling shrinkage.

The equation of swelling shrinkage ratio is $P=(V_3-V_4)/V_3\times100\%$

**Table 1 The filtering results of main materials for clay inhibitor**

| Material name | alkylolamides | N-propyl-N,N,N-three methyl ammonium chloride | Potassium polyacrylate | Aluminium polychlorid |
|---|---|---|---|---|
| Swelling prevention percentage, % | 83.9 | 88.2 | 80.4 | 79.7 |
| Swelling shrinkage ratio, % | 46.8 | 51.4 | 49.3 | 47.9 |

From the experiment results, all the filtered main materials have higher swelling shrinkage percentage and swelling shrinkage ratio. By contrast, the swelling prevention and swelling shrinkage effect of N-propyl-N,N,N-three methyl ammonium chloride is excellent, so N-propyl-N,N,N-three methyl ammonium chloride is chose as main material of clay inhibitor.

**Table 2 the influence of N-propyl-N,N,N-three methyl ammonium chloride dosage on the swelling prevention and swelling shrinkage effect**

| N-propyl-N,N,N-three methyl ammonium chloride dosage, % | 1 | 2 | 3 | 4 | 5 | 6 | 7 | 8 | 9 | 10 |
|---|---|---|---|---|---|---|---|---|---|---|
| Swelling prevention percentage, % | 66.5 | 80.9 | 84.1 | 90.7 | 92.1 | 92.3 | 92.2 | 92.0 | 92.4 | 92.8 |
| Swelling shrinkage ratio, % | 35.6 | 45.4 | 48.2 | 50.1 | 50.6 | 50.4 | 50.6 | 50.2 | 50.7 | 50.6 |

From Table 2, the influence of N-propyl-N,N,N-three methyl ammonium chloride dosage on the swelling prevention and swelling shrinkage effect is: when dosage is lower than 3 percent, swelling prevention percentage and swelling shrinkage ratio rise obviously, but when dosage is higher than 3 percent, the property is stable. So, the range of N-propyl-N,N,N-three methyl ammonium chloride dosage is from 3 to 5 percent.

### 2.3. *Functional agents filtering*

By using centrifugation evaluation methods, the swelling prevention effect of 3 functional agents is evaluated.

**Table 3 The filtering results of functional agents**

| Material name | 0 | DMC | γ-propyl three methoxy silane | cetyl trimethyl ammonium bromide |
|---|---|---|---|---|
| Swelling prevention percentage, % | 54.2 | 87.9 | 91.6 | 84.6 |
| Swelling shrinkage ratio, % | 9.6 | 43.8 | 52.3 | 45.9 |

From Table 3, the addition of functional agents could improve swelling prevention effect of clay inhibitor. Among them, the addition of functional agent γ-propyl three methoxy silane improves swelling prevention percentage obviously, so it is chose as functional agent.

**Table 4 the influence of γ-propyl three methoxy silane dosage change on the swelling prevention and swelling shrinkage effect**

| γ-propyl three methoxy silane dosage, % | 0 | 0.5 | 1.0 | 1.5 | 2.0 | 2.5 | 3.0 | 3.5 | 4.0 | 5.0 |
|---|---|---|---|---|---|---|---|---|---|---|
| Swelling prevention percentage, % | 44.8.2 | 73.3 | 86.2 | 90.5 | 92.3 | 92.4 | 92.6 | 92.1 | 92.4 | 92.3 |
| Swelling shrinkage ratio, % | 9.4 | 38.3 | 42.7 | 50.8 | 54.1 | 54.8 | 54.2 | 54.6 | 54.5 | 54.4 |

From the experiment results, along with the addition of γ-propyl three methoxy silane and the increase of dosage, the swelling prevention percentage of core sample improves obviously. When dosage is lower than 2 percent, swelling prevention percentage improves from 44.8 to 92.5 percent and swelling shrinkage ratio improves from 9.4 to 54.1 percent, but when dosage is higher than 2 percent, the property is stable. So, the range of γ-propyl three methoxy silane dosage is from 1.5 to 2 percent.

### 2.4. *Additives filtering*

In order to improve swelling prevention effect of clay inhibitor, additive is needed. Table 5 is filtering experiment of different additives.

**Table 5 The filtering results of additives**

| Material name | 0 | 2,3-propylene oxide three methyl ammonium chloride | PEG | secaline |
|---|---|---|---|---|
| Swelling prevention percentage, % | 52.2 | 88.4 | 91.5 | 90.7 |
| Swelling shrinkage ratio, % | 18.6 | 44.4 | 51.9 | 47.9 |

From Table 5, all the additives are polymeric substances in strong-polarity which owns bigger molecular clew, diffusion ability and proper molecular shape. Additives are absorbed to several clay particles to form the bridge through hydrogen bond force and electrostatic force, in order to prevent clay particles from dispersing. Thus, PEG is chose as additive.

**Table 6 the influence of** PEG **dosage change on the swelling prevention and swelling shrinkage effect**

| PEG dosage, % | 0 | 0.1 | 0.2 | 0.3 | 0.4 | 0.5 | 0.6 | 0.7 | 0.8 | 0.9 |
|---|---|---|---|---|---|---|---|---|---|---|
| Swelling prevention percentage, % | 56.4 | 69.2 | 72.6 | 80.1 | 90.3 | 92.8 | 92.2 | 92.4 | 92.7 | 92.5 |
| Swelling shrinkage ratio, % | 18.4 | 29.8 | 37.6 | 46.7 | 53.6 | 54.4 | 54.1 | 53.9 | 54.5 | 54.1 |

From the experiment results, long with the addition of PEG and the increase of dosage, the swelling prevention percentage of core sample improves obviously. When dosage is lower than 0.5 percent, swelling prevention percentage improves from 56.4 to 92.8 percent and swelling shrinkage ratio improves from 18.4 to 54.4 percent, but when dosage is higher than 0.5 percent, the property is stable. So, the range of PEG dosage is from 0.4 to 0.5 percent.

### 2.5. *Confirm the basic formula*

On the basis of optimizing every component, the basic formula of clay inhibitor CYY-2 is confirmed. The main components are:

N-propyl-N,N,N-three methyl ammonium chloride($C_9H_{19}ClN_2O$), the dosage is from 3 to 5 percent, the best dosage is 4 percent.

γ-propyl three methoxy silane($C_9H_{21}NO_3Si$), the dosage is from 1.5 to 2 percent, the best dosage is 2 percent.

PEG ($HO(CH_2CH_2O)_nH$), the dosage is from 0.4 to 0.5 percent, the best dosage is 0.5 percent.

## 3. Performance evaluation of clay inhibitor

### 3.1. *Swelling prevention effect evaluation*

*3.1.1. The examine on interplanar spacing change.* By using X-ray diffractometer evaluation method, the spacing of calcium montmorillonite sample is determined by diffraction analysis. The interplanar spacing, *d* value, of calcium montmorillonite in completely hydrated state is regarded as upper limit ratio to estimate swelling prevention effects of clay inhibitor.

Eight clay inhibitors and the developed clay inhibitor are prepared to solutions of different concentration respectively, and then calcium montmorillonite of a certain quality is put into different solutions. After 24 hours, they are centrifuged and sedimentated, then prepared to samples for X-ray diffraction analysis. The spacing of calcium montmorillonite of different clay inhibitors under different concentrations is measured. The results are shown in Table 7.

**Table 7 The spacing of calcium montmorillonite of different clay inhibitors under different concentrations**

| Clay inhibitor | Interplanar spacing *d* (in different concentration) | | | | | | |
|---|---|---|---|---|---|---|---|
| | 0.5% | 1.0% | 1.5% | 2.0% | 2.5% | 3.0% | 4.0% |
| KCL | 21.3635 | 18.6419 | 16.8321 | 16.2745 | 16.1852 | 15.8871 | 15.7982 |
| $NH_4CL$ | 22.6550 | 16.1852 | 16.0968 | 16.6420 | 16.0968 | 15.7874 | 15.6541 |
| BC-61 | 19.7623 | 18.5241 | 15.8371 | 14.5761 | 14.5834 | 14.8284 | 14.7124 |
| BC-81 | 19.8571 | 18.0704 | 16.8575 | 14.6541 | 14.4984 | 14.3651 | 14.7524 |
| XK-2 | 19.6542 | 18.5436 | 16.7425 | 15.6332 | 14.7965 | 14.6547 | 14.7125 |
| HT-1 | 19.7852 | 18.5541 | 16.8124 | 15.8126 | 14.8274 | 14.7691 | 14.6873 |
| HYZ-g | 21.3214 | 19.2456 | 18.2125 | 16.4531 | 15.6547 | 14.7723 | 14.8142 |
| JDB | 22.1251 | 19.8823 | 18.5641 | 16.7652 | 15.6642 | 14.2387 | 14.5124 |
| CYY-2 | 17.8155 | 16.2446 | 14.5596 | 14.2728 | 14.3732 | 14.4785 | 14.5562 |

Some conclusions can be drawn from the test results. For the clay inhibitors BC-81 and BC-61, when the concentration is more than 2 percent, interplanar spacing diminishes and the swelling prevention effect is best. For the clay inhibitors XK-2, HT-1, HYZ-g and JDB, when the concentration is more than 2.5 percent, the swelling prevention effect is best. For the developed clay inhibitors CYY-2, when the concentration is more than 1.5 percent, the swelling prevention effect is best. For the inorganic salt KCl, $NH_4Cl$, their swelling prevention effect is poor, because the ionic adsorption of inorganic salt is reversible.

*3.1.2. Stable effect evaluation.* Washing experiment: The dry samples treated with clay inhibitors are rinsed with deionized water for 3 times, then the spacing *d* changes before and after washing are analyzed and measured.

The water immersion test: The calcium montmorillonite samples were soaked in the clay inhibitor solution of best concentration. They are placed for 24 hours after fully stirring. Then clay inhibitor solution is emptied and deionized water is added. After stirring and 24 hours of fully immersing, the samples is centrifuged and sampled to analyze. Test results are shown in Table 8.

**Table 8 The spacing changes of calcium montmorillonite samples before and after washing and water immersion**

| Clay inhibitors | Dosage % | The spacing of dry sample $\times 10^{-1}$um | The spacing of sample after washing $\times 10^{-1}$um | The spacing of sample before water immersion $\times 10^{-1}$um | The spacing of sample after water immersion $\times 10^{-1}$um |
|---|---|---|---|---|---|
| KCL | 3 | 13.7639 | 19.7754 | 16.0854 | 16.6234 |
| $NH_4CL$ | 3 | 13.5586 | 21.0346 | 16.0966 | 16.5217 |
| BC-61 | 2 | 14.1412 | 14.9532 | 14.5561 | 14.6532 |
| BC-81 | 2 | 14.2426 | 14.5213 | 14.6231 | 14.7156 |
| XK-2 | 2.5 | 14.1953 | 14.5741 | 14.8421 | 14.9781 |
| HT-1 | 2.5 | 14.8652 | 14.6213 | 14.3215 | 14.4452 |
| HYZ-g | 2.5 | 14.8032 | 14.5214 | 14.2254 | 14.5416 |
| JDB | 2.5 | 14.8621 | 14.5126 | 14.6213 | 14.8791 |
| CYY-2 | 1.5 | 14.1034 | 14.2236 | 14.1567 | 14.2213 |

Except that the swelling prevention effect and stable effect of inorganic salt KC1 and $NH_4C1$ is poor, other clay inhibitors are better. Among them the propery of clay inhibitor CYY-2 is excellent.

### *3.2. Dynamic evaluation of swelling prevention effect*

*3.2.1. Dilatometer test method.* By using JHPZ-II intelligent dilatometer bearing high temperature and pressure, at the temperature of 60°C, pressure of 5MPa conditions, bentonite powders and natural core powders crushed to below 0.9mm in different reservoirs are added to different clay inhibitors solutions. The expansion rate of them is determined. Table 9 is the measured swelling prevention percentage and expansion rate.

The specific experimental method is: the core powder of 5.0g is put into core barrel of dilatometer. Suppress for 5 minutes under the pressure of 5MPa. The expanding height $H_0$ is measured by using deionized water. The expanding height after antiswelling agent solution is added is $H_1$. The calculation formula is: F= $(H_0-H_1)/H_0\times100\%$.

**Table 9 The results of expansion rate for bentonite powder and core power**

| | Antiswelling agent dosage | Original height | Stopping height | The growth | Swelling prevention percentage | Swelling shrinkage ratio |
|---|---|---|---|---|---|---|
| Bentonite | / | 8.197 | 16.749 | 8.552 | 104.33 | / |
| | 1.5% CYY-2 | 8.144 | 9.104 | 0.87 | 10.7 | 89.74 |
| The well Tong 47X | / | 14.256 | 14.422 | 0.166 | 1.3 | / |
| | 1.5% CYY-2 | 14.261 | 14.288 | 0.027 | 0.19 | 83.6 |
| The well Ba 38-2 | / | 14.278 | 14.448 | 0.17 | 1.19 | / |
| | 1.5% CYY-2 | 14.001 | 14.03 | 0.029 | 0.21 | 82.9 |

The results show that for bentonite powder and core powder from different reservoirs, the screened clay inhibitors CYY-2 have better swelling prevention effect under the condition of 60°C and 5MPa. The swelling prevention percentage on bentonite can reach 90 percent, for different reservoir core powder it can reach 80 percent.

*3.2.2. The ability tests on preventing core particles from spalling and dispersing.* Take reservoir core particles of 20 to 40 meshes from well Zhou 16-41 of 3.00g. They are put into sieve of 4cm diameter and 60 meshes. Deionized water of 100ml and different concentrations of clay inhibitor CYY-2 and antiswelling agent XK-2 solution is put for 4 hours of soak. They are weighed after sieving and analyzing. The weight loss ratio is calculated. Test results are shown in Table 10.

**Table 10 The weight loss ratio of core particles in different chemicals**

| Chemical name and dosage | Deionized water | CYY-2 | | | XK-2 | | |
|---|---|---|---|---|---|---|---|
| | | 1.0 | 1.5 | 2.0 | 2.0 | 2.5 | 3.0 |
| weight loss ratio, % | 81.6 | 14.7 | 10.2 | 8.5 | 42.1 | 38.7 | 35.9 |

Test results show that if the dosage of filtered clay inhibitor in solution is 1.0 to 2.0 percent, the weight loss ratio of core particles used in test is less than 15 percent. If the dosage of antiswelling agent XK-2 in solution is 2.0 to 3.0 percent, the weight loss ratio of core particles used in test is more than 35 percent. The weight loss ratio is 81 percent in deionized water. The above conclusions show that clay inhibitor CYY-2 has better ability to inhibit reservoir clay minerals from hydrating, expanding and dispersing in the static condition.

*3.2.3. Evaluate prevention effect of clay inhibitor by core tests.* Using washing oil core from block Tong 47 with well Tong 12X, block Yan 9 with well Yan 9-5X and block Ba 38 with well Ba 38-2, adopting core evaluation test method, prevention effect of screened clay inhibitor in 1.5 percent concentration is examined. Test results are shown in Table 11.

**Table 11 Test results of clay inhibitor by core evaluation**

| Well | Core NO. | Permeability of injecting saline $\times 10^{-3}\mu m^2$ | Permeability of injecting deionized water $\times 10^{-3}\mu m^2$ | Damage rate of permeability % | Injection order |
|---|---|---|---|---|---|
| Tong 12X | 12-1 | 6.8 | 0.89 | 86.9 | Inject saline and then deionized water |
| | 12-2 | 6.45 | 5.63 | 12.8 | Inject saline and then clay inhibitor, deionized water |
| Yan 9-5X | Y5-1 | 3.29 | 0.45 | 86.3 | Inject saline and then deionized water |
| | Y5-2 | 2.93 | 2.57 | 12.3 | Inject saline and then clay inhibitor, deionized water |
| Ba 38-2 | 38-1 | 3.29 | 0.45 | 87.3 | Inject saline and then deionized water |
| | 38-2 | 2.93 | 2.57 | 12.6 | Inject saline and then clay inhibitor, deionized water |

Core test results show that due to prevention effect of clay inhibitor, core damage rate of each block is greatly reduced after being treated with clay inhibitor.

## 4. Field test and effect

As the extension of developing time, some injection wells had many problems such as rising injection pressure, decreasing injection volume and being difficult to accomplish injection allocation, which had bad influence on the reservoir development effect. Since 2006, the field tests of 5 wells were carried out and 4 wells effective which measurement success rate is 80 percent, achieving good effect. For example, well Quan 241-9 was constructed on April 30 2006, and then water flooding pressure was rising from 20.5MPa to 18.5MPa, injection water volume was increased from 13.3$m^3$ to 19.7$m^3$, the cumulative injection water increased by 3146 $m^3$ in validity period, the validity period is 165 days. The cumulative increased oil of connected well, Quan 241-12, is 2064 tons.

## 5. Conclusions

Antiswelling agent matching water sensitive reservoir is used for water injection well treatment. It is not only necessary for low- permeability and water- sensitive reservoir development, but also can effectively control or slow down pressure rising rate of injection wells, thus improving exploration effect.

The developed bi-functional clay inhibitor not only could achieve good performance of 90 percent swelling shrinkage percentage and 90 percent swelling shrinkage ratio, but also obtained well effect in field application. The measurement success rate was 80 percent, the average decreasing water injection pressure was 3.2MPa, the average improving injection volume was 14.43$m^3$, and cumulative oil increment of oil wells corresponding to effective wells was 2680.9 tons.

IOP Conf. Series: Earth and Environmental Science **61** (2017) 012010 doi:10.1088/1755-1315/61/1/012010

**Acknowledgements**
Key S&T Special Projects of PetroChina Company Ltd. "Key technology research and application for increasing production and stable yields of 8 million tons in Huabei oilfield" (Project number 2014E-3507).

**References**
[1] SONG DouGui, ZHU GuiLin, LI JinFa, et al. 2014 *Elementary Research and Practice of Heavy Crude Oil Recovery from a Sensible Sandstone Reservoir by the Way of Clay Stabilization and Water Flood (Oilfield Chemistry vol 4)* chapter 21 pp 320–323.
[2] LU Zhenhua, HAN Xiaoqiang, LIU Jing. 2006 *Influencing Factors Research on Evaluation Method of Anti-swelling Agent Performance(Xinjiang Petroleum Science & Technology vol 3)* chapter 26 pp 39–41.
[3] Cui Fuyi, Ma Xingqin, Zhang Yunzhi, et al. 2005 *To Optimize Clay Stabilizing Agent for Water Injection Treatment(Oil & Gas Recovery Technology vol 1)* chapter 12 pp 75–80.
[4] JIN Baojun, MA Xingqin. 2003 *Optimization of Clay Inhibitor for Water Injection Pre-treatment in Low Permeability Oilfield (Special Oil & Gas Reservoirs vol 6)* chapter 10 pp 75–78.
[5] ZHU Weiqun, ZHANG Rui, LIU Guangqeng. 2003 *Property of a Clay Stabilizer-CETA(Chinese Journal of Applied Chemistr vol 10)* chapter 20 pp 955–958.
[6] Zhang Xuan, Huang Zhiyu, Li Wen. 2012 *Synthesis and Evaluation of a Novel Gemini Quaternary Ammonium as an Anti-swelling Chemical(Chemical Engineering of Oil & Gas vol 1)* chapter 41 pp 90–91.
[7] Huang Zhiyu, He Yan, Yu Nian, et al. 1995 *Study of FP-1 Anti-clayswelling Agent(Oil Drilling & Production Technology vol 3)* chapter 17 pp 55–57.
[8] JIN Baojun, LI Shanggui, MA Xingqin, et al. 2001 *Preparation and Use of Low Molecular Mass Clay Stabilizer J-1(Oilfield Chemistry vol 2)* chapter 18 pp 127–138.
[9] HE Jinlong. 1997 *Study and Use of Clay Stabilizer CT12-1(Drilling & Production Technology vol 3)* chapter 20 pp 56–59.
[10] ZHU Jian. 2008 *Performance Properties of Clay Stabilizer KS-1 for Low Permeability and Water Sensitive Oil Reservoirs(Oilfield Chemistry vol 4)* chapter 25 pp 336–339.

# Mineral migration of alkali metal during the combustion of coal and sewage sludge

**Peng Ling[1], Liang Liu[1], Jinjin Ai[1], Yanshan Yin[1], Hong Tian[1] and Fei Xie[1]**

[1] School of Power and Energy Engineering, Changsha University of Science and Technology, 960, 2nd Section, Wanjiali South RD, Changsha, Hunan, 410114, China
The corresponding author's e-mail address: liuliang_hn@126.com

**Abstract**. The mineral migration of alkali metal during the combustion of coal and sewage sludge was studied by X-ray fluorescence (XRF), X-ray diffraction (XRD) phase analysis and Factsage 7.0 simulation. It showed that the migration process of potassium(K)during the combustion of coal-sludge blends mainly manifested the process from muscovite ($KAl_3Si_3O_{10}(OH)_2$) to sanidine ($KAlSi_3O_8$), which included the transforming process of escaped atom K combined with $Na_2SO_4$ from glaserite or aphthitalite to sanidine ($KAlSi_3O_8$).The mutual transformation of ardite ($Na_2SO_4$) and albite ($NaAlSi_3O_8$) was mainly in the migration process of element Na with low-albite and high-albite appeared occurring in the process. The mutual transformation of anhydrite ($CaSO_4$) and anorthite ($CaAl_2Si_2O_8$) was mainly in the migration of element Ca.

## 1. Introduction

Municipal sludge is considered as a useful biological resource because of its rich organic matters and a large number of trace elements [1,2]. However, environmental pollution can be caused by improper disposal because of its microbes, eggs, bacteria and heavy metals [3]. The pollution and the waste which brings huge environmental and economic benefits can be reduced by the energy regeneration of sludge, effective use of sludge. As a new method of sludge incineration, the power generation boiler of the co-firing of coal and sludge can not only improve the harmless disposal of sludge, but also generate power with the calorific value of sludge [4] to reduce the coal consumption, which effectively alleviates the current problems of large amount of sludge hard to be handled.

Nevertheless, many experts and scholars have found that the alkali metal at high temperature is easy to react with silicon to produce alkaline silicate particles and a large amount of alkaline silicate particles bound in grate furnace wall and heating surface easily formed slag [5]. The alkali metal is easy to react with the bed material and generate the low melting point eutectic compounds causing particles agglomeration[6,7]. Chloride salt, sulfate, hydrogen chloride and other volatile phase materials which cause high temperature corrosion of boiler materials can be easily formed by the alkali metals, Cl, S, and other elements at high temperature. At the same time, low temperature corrosion to the heat exchanger surface can be caused by the condensation of these substances in the low temperature transferring surface [8-10]. In addition, it is a seriously threat to safe operation of the gas turbine. The alkali metal deposition can block the flow channels, suppressing the flow through and reducing the turbine efficiency[11,12]. The gas turbine blade corrosion and reducing the thermal efficiency of gas turbine can be caused by alkali metal components which are not completely cooled with the gasification gas flowing into the gas turbine [13,14]. Therefore, it is worth paying attention to the problems of the migration and transformation of alkali metals in the mixed combustion of sludge and coal. The mineral migration process of alkali metals during combustion of coal mixed with municipal sewage sludge by chemical thermodynamic software is simulated in this paper. A certain reference value in the study of micro compound in the combustion process of sludge mixed coal and to a certain guiding role in solving the problems of corrosion and slagging are provided.

## 2. Materials and methods

Lanba coal from China and sewage sludge obtained from an urban wastewater treatment plant situated in Changsha, China are chosen for the sample. The mixtures of sewage sludge and Lanba coal through dry processing are grinded in a ball milling into particles which are then passed through a 200-mesh miller sieve.

Different mass fractions of sludge on a dry basis (0, 30, 50, 70 and 100%) are mixed with coal (drying basis) to prepare the blended samples. In order to accomplish the homogenization, the mixture is then milled by a mechanical agitator with 400 rpm lasting 20 minutes. Consequently, series of coal-sludge blends (W1, W2, W3, W4 and W5) are made. The contents of sewage sludge are showed by the proportion between wet sewage sludge and coal.

A D/MAX-2200 type rotating anode XRD is used for identifying mineral morphologies for different samples. The rated power of XRD analyzed is 12kW, and its angular accuracy is $\Delta 2\theta \leq \pm 0.02°$. The data of XRD are analyzed by software Jade6. The ashes obtained at 815°C grounded at 74μm are heated at a rate of 15°C/min in air atmosphere from a room temperature to an end temperature of 1100°C.

## 3. Results and discussion

### *3.1. Characteristics of materials and ashes*

The proximate analyses of coals and sewage sludge are listed in Table 1. As was showed in Table 1, sewage sludge has some special characteristics compared with Lanba coal. The moisture of sludge is much higher than the coal while the fixed carbon is only 5.13(wt.%). Therefore, The heating values of sludge are lower than the coal because it has a lot of organic matters in the coal. Moreover, the ash yields of samples follow the sequence W1<W2<W3<W4<W5 because it has a lot of inorganic compounds in the sewage sludge.

**Table 1.** Proximate and ultimate analysis of samples

| Samples | Proximate analysis (wt.%) | | | | Ultimate analysis (wt.%) | | | | |
|---|---|---|---|---|---|---|---|---|---|
| | $M_{ad}$ | $A_{ad}$ | $V_{ad}$ | $FC_{ad}$ | $C_{ad}$ | $H_{ad}$ | $O_{ad}$ | $N_{ad}$ | $S_{t,ad}$ |
| W1 | 3.09 | 15.77 | 18.57 | 62.57 | 68.10 | 4.28 | 11.03 | 9.50 | 7.09 |
| W2 | 4.65 | 29.6 | 18.73 | 47.02 | 53.61 | 3.89 | 29.19 | 10.18 | 3.13 |
| W3 | 6.24 | 38.79 | 19.22 | 35.75 | 45.16 | 3.12 | 41.85 | 7.41 | 2.46 |
| W4 | 6.74 | 49.37 | 19.86 | 24.04 | 32.81 | 3.72 | 57.85 | 3.9 | 1.72 |
| W5 | 9.34 | 63.77 | 21.76 | 5.13 | 11.94 | 3.13 | 80.58 | 3.97 | 0.38 |

ad, air-dried basis; M, moisture; A, ash; V, volatile; FC, fixed carbon

The compositions of major and minor elements by XRF at 815°C are shown in Table 2. As can be seen in Table 2, most of the minerals in the sludge exist in the form of oxide. And the sludge ash has much higher $SiO_2$ contents than coal ash.

IOP Conf. Series: Earth and Environmental Science **61** (2017) 012011 doi:10.1088/1755-1315/61/1/012011

**Table 2.** Ash chemical composition (wt%) of samples

| Samples | $SiO_2$ | $Al_2O_3$ | CaO | MgO | $Fe_2O_3$ | $Na_2O$ | $K_2O$ | $SO_3$ |
|---|---|---|---|---|---|---|---|---|
| W1 | 48.40 | 23.59 | 1.39 | 0.43 | 22.85 | 0.20 | 0.92 | 2.20 |
| W2 | 52.34 | 23.65 | 1.41 | 0.64 | 18.72 | 0.20 | 1.25 | 1.62 |
| W3 | 54.13 | 23.51 | 1.48 | 0.72 | 16.33 | 0.22 | 1.99 | 1.41 |
| W4 | 57.60 | 23.80 | 1.43 | 0.91 | 13.26 | 0.21 | 0.20 | 0.85 |
| W5 | 61.58 | 23.91 | 1.51 | 1.17 | 9.16 | 0.23 | 2.04 | 0.30 |

### *3.2. Minerals identified in ashes*

The major minerals identified by XRD in the ashes obtained at 815°C are shown in Fig.1 and Fig.2. From the X-ray diffractometry, several minerals including kaolinite, quartz, szomolnokite, rozenite, heulandite and boehmite are shown in the coal ash and several minerals including quartz, muscovite, illite, kaolinite and orthoclase are shown in the sludge. The main minerals in coal , sludge and blends are shown in Table 3 and Table 4.

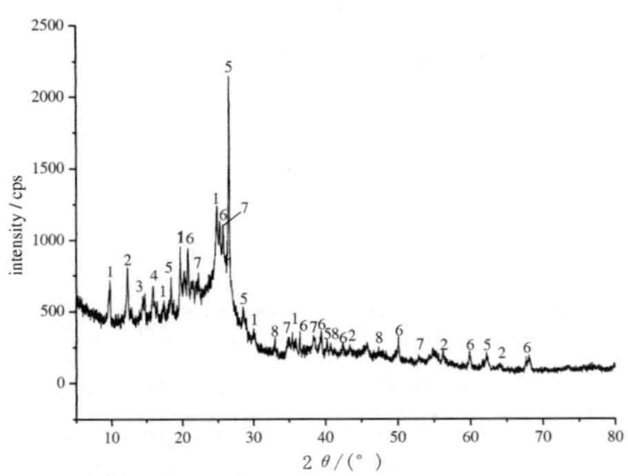


1—heulandite（Ca $(Al_2Si_7O_{18})\cdot 6H_2O$）；

2—kaolinite（$Al_2(Si_2O_5)(OH)_4$）；

3—boehmite（$AlO(OH)$）；

4—Dawsonite（$NaAl(CO_3)(OH)_2$）；

5—szomolnokite（$Fe(SO_4)\cdot H_2O)$）；

6—quartz（$SiO_2$）；

7—rozenite（$Fe(SO_4)\cdot 4H_2O$）；

8—pyrite（$FeS_2$）

**Figure 1.** XRD spectrum of W1

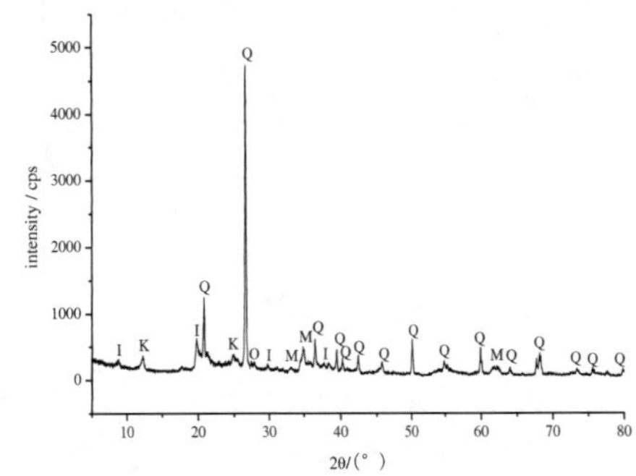


I—illite（$K(Al_4Si_2O_9(OH)_3)$）；

K—kaolinite（$Al_2(Si_2O_5)(OH)_4$）；

Q—quartz（$SiO_2$）；

M—muscovite（$KAl_3Si_3O_{10}(OH)_2$）；

O—orthoclase（$KAlSi_3O_8$）

**Figure 2.** XRD spectrum of W5

**Table 3.** Mineral content in the W1 and W5

| Sample | Name | Wt/% | R |
|---|---|---|---|
| W1 | Kaolinite 1A | 18.6 | 13.19% |
| | Szomolnokite | 8.9 | |
| | Rozenite | 2.9 | |
| | Heulandite | 9.1 | |
| | Quartz | 42.9 | |
| | Boehmite | 7.1 | |

| | | | |
|---|---|---|---|
| | Pyrite | 1.3 | |
| | Dawsonite | 8.1 | |
| W5 | Quartz | 47.0 | 13.61% |
| | Muscovite 2M1 | 17.3 | |
| | Illite | 21.0 | |
| | Kaolinite 1A | 10.9 | |
| | Orthoclase | 3.8 | |

**Table 4.** Main mineral content of samples（wt.%）

| Sample | quartz | muscovite | illite | kaolinite | orthoclase | szomolnokite | rozenite | heulandite | boehmite | pyrite | dawsonite |
|---|---|---|---|---|---|---|---|---|---|---|---|
| W1 | 42.9 | - | - | 18.6 | - | 8.9 | 2.9 | 9.1 | 7.1 | 1.3 | 8.1 |
| W2 | 44.13 | 5.19 | 6.30 | 16.29 | 1.14 | 6.23 | 2.03 | 6.37 | 4.97 | 0.91 | 5.67 |
| W3 | 44.95 | 8.65 | 10.50 | 14.75 | 1.90 | 4.45 | 1.45 | 4.55 | 3.55 | 0.65 | 4.05 |
| W4 | 45.77 | 5.19 | 14.7 | 13.21 | 2.66 | 2.67 | 0.87 | 2.73 | 2.13 | 0.39 | 2.43 |
| W5 | 47 | 17.3 | 21.0 | 10.9 | 3.8 | - | - | - | - | - | - |

Equilib of Factsage 7.0 software whose calculation principle is mainly based on the ChemSage Gibbs minimum free energy algorithm is used to calculate. The function is to find the system according to Gibbs energy global minimization of the different phase composition and content [15].

The 100kg various samples of the mineral mole are calculated through W1, W2, W3, W4 and W5 of the main mineral mass fraction data and mineral type amount of data. In order to simulate the real combustion environment, the temperature is set to 200-2100°C and the pressure is set to normal pressure (101325KPa) assuming that the air is composed of 79%$N_2$+21%$O_2$. The excess air factor is set to 1.2. The air volume is theoretical calculation from Eq.1-1 as shown blow:

$$V_0 = 0.0889(C_{ar} + 0.375S_{ar}) + 0.265H_{ar} - 0.0333O_{ar} \quad (1\text{-}1)$$

The mineral composition and air quantity of the sample are shown in the chemical thermodynamic balance of Equilib. As was shown in Table 5, the database ideal gas(g), solid state(s), liquid (l), and the use of normal algorithm are selected in the species Compound. As was shown in Fig 3 to Fig 7, the results of the calculation of alkali metal minerals are selected out to draw the alkali metal mineral phase change through the calculation of Equilib module in Factsage.

**Table 5.** Data input in Factsage software

| Sample | $SiO_2$ | $KAl_3Si_3O_{10}(OH)_2$ | $KAl_4Si_2O_9(OH)_3$ | $Al_2(Si_2O_5)(OH)_4$ | $KAlSi_3O_8$ | $FeSO_4(H_2O)$ |
|---|---|---|---|---|---|---|
| W1 | 715.00 | - | - | 70.99 | - | 52.35 |
| W2 | 735.50 | 12.98 | 15.71 | 62.18 | 4.10 | 36.65 |
| W3 | 749.17 | 21.63 | 26.18 | 56.30 | 6.83 | 26.18 |
| W4 | 762.83 | 30.28 | 36.66 | 50.42 | 9.57 | 15.71 |
| W5 | 783.33 | 43.25 | 52.37 | 41.6 | 13.67 | - |

| Sample | $Fe(SO_4)\cdot 4H_2O$ | $Ca(Al_2Si_7O_{18})\cdot 6H_2O$ | $AlO(OH)$ | $FeS_2$ | $NaAl(CO_3)(OH)_2$ |
|---|---|---|---|---|---|
| W1 | 12.95 | 13.09 | 118.3 | 10.83 | 56.25 |

| | | | | | |
|---|---|---|---|---|---|
| W2 | 9.06 | 9.16 | 82.83 | 7.58 | 39.38 |
| W3 | 6.47 | 6.54 | 59.17 | 5.42 | 28.13 |
| W4 | 3.88 | 3.93 | 35.5 | 3.25 | 16.88 |
| W5 | - | - | - | - | - |

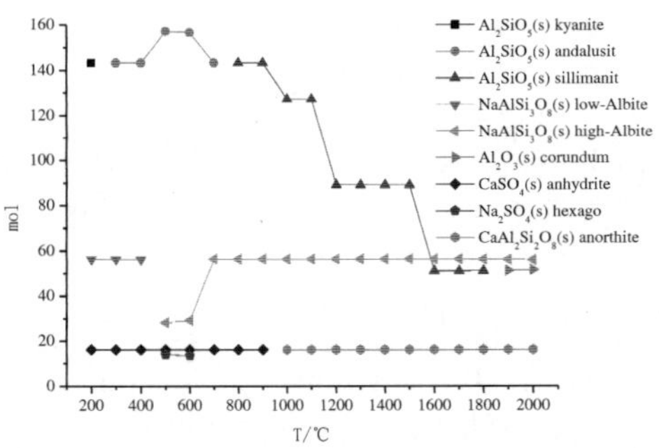


**Figure 3.** Mineral transformation of sample W1

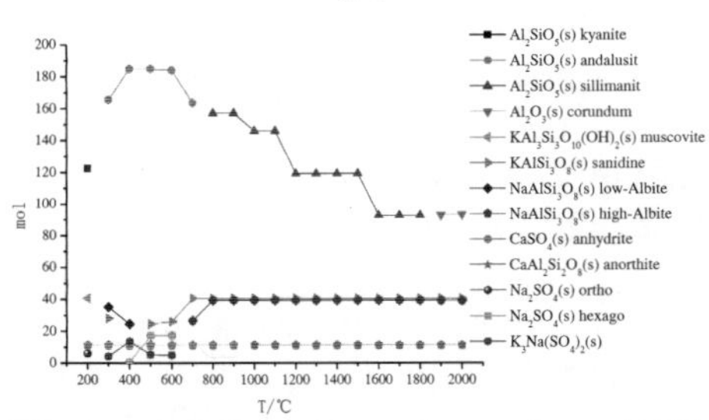


**Figure 4.** Mineral transformation of W2

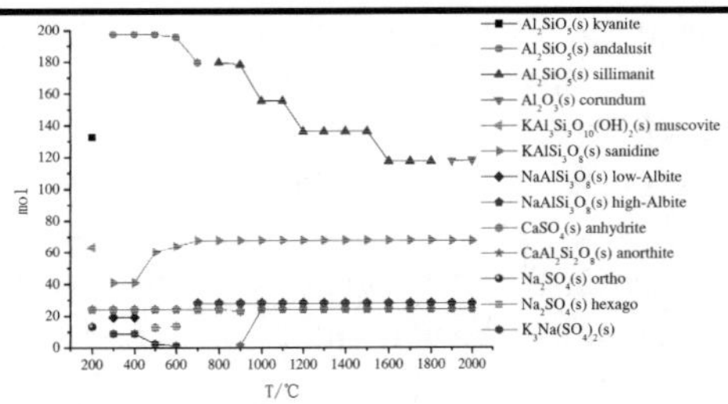


**Figure 5.** Mineral transformation of W3

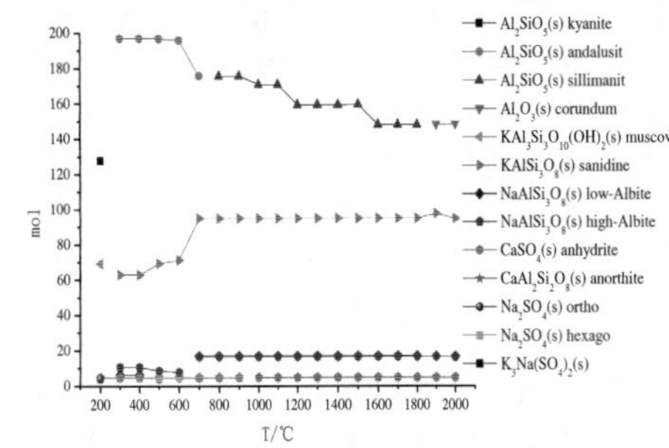


**Figure 6.** Mineral transformation of W4

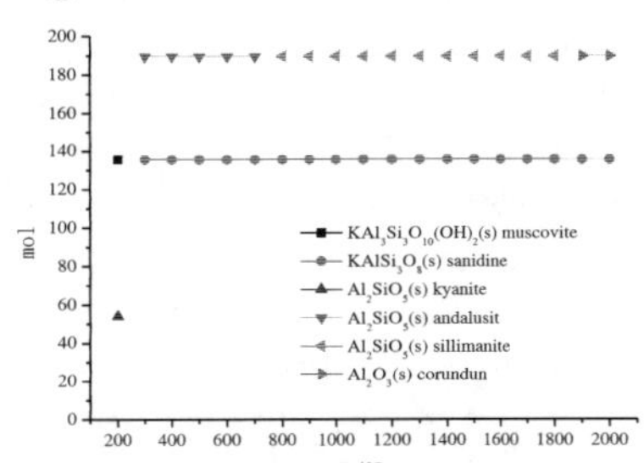


**Figure 7.** Mineral transformation of W5

The samples of mineral species and content are analyzed by the XRD experiment and Jade 6.0 software. As was shown in W2, W3 and W4, element K is contained mainly in muscovite, illite, kaolinite and orthoclase while element Ca is mainly in heulandite and element Na is mainly in dawsonite. The products calculated by the Factsage7.0 software at the specific temperature through the principle of the single metallic element in the reactants and the products are also analyzed in this thesis.

### *3.3. Phase transfer of minerals containing elements K*

From Fig 3 and Fig 7, the muscovite in the samples disappears in the temperature of 200-300°C. Therefore, the transformation of muscovite in this range can be found. Element K is contained only in muscovite in the samples of CN. The muscovite converted tosanidine in the temperature range can be found. In Fig 7, the molar amount of muscovite and sanidine are almost equal and the molar amount of andalusite, sillimanite and corundum are also almost equal so that muscovite in W1is only converted to sanidine.

There are various mineral species in W2, W3 and W4. Muscovite disappears and sanidine appears at 200-300°C in each sample. It proves that the muscovite transforms into sanidinein in this temperature range. At the same time, escaped atom K and $Na_2SO_4$ (ortho) are combined into

IOP Conf. Series: Earth and Environmental Science **61** (2017) 012011 doi:10.1088/1755-1315/61/1/012011

$K_3Na(SO_4)_2(s)$ so that the molar amount of sanidine is significantly smaller than that of muscovite. $K_3Na(SO_4)_2(s)$ is a kind of complex salt, which is formed by two kinds of simple salt crystals whose chemical composition are still the simplest elements [16]. $K_3Na(SO_4)_2(s)$ is identified as the glaserite or the aphthitalite all belonging to the same kind of salt with the same chemical formula [17]. In this process, muscovite itself not only transforms to sanidine, but also interactes with other substances in the reaction system that appeares glaserite (or aphthitalite), which fully proves the existence of synergistic effect between various mineral potential.

With the increasing of temperature, the content of sanidine gradually increases. When the reaction system temperature reaches 700°C, the molar amount of sanidine tends to be stable with the potassium sulfate (or potassium sulfate stone) disappearing. According to the migration of element K, atom K transforms to sanidine with the crystal structure of $K_3Na(SO_4)_2(s)$ changing. In term of the molar quantity, the quantity of the feldspar increases significantly in the temperature of 600-700°C.

### *3.4. Phase transfer of minerals containing elements Na*

Na is contained mainly in dawsonite which is a kind of orthorhombic carbonate minerals containing water, sodium and aluminium. With the increasing temperature of the reaction system, the product category of W5 combustion process is analyzed. The Dawsonite transforms to albite in the reaction system. The carbonate mineral decomposition, the combination process of Na and silicate Albite exists in this process because albite belongs to the aluminosilicate. However, the reaction mechanism remains to be further studied.

In the temperature range of 400-700°C, there is a small amount of $Na_2SO_4$ (hexago) in the reaction system. $Na_2SO_4$ (hexago) is one of the six hexagonal crystal. According to the molecular formula and crystal type, it can be presumed to be thenardite [18]. With the temperature rising, thenardite in the reaction system disappears and the molar quantity of albite increases dramatically, which shows that Na of thenardite is regressed to albite.

The mineral species of W2, W3 and W4 are complex. Atom Na all concentrates in the orthorhombic thenardite at 200°C. With the temperature rising, atomic vibration amplitude becomes larger in Fig4. Orthorhombic thenardite transformed into low-albite [19] at 200-300°C.With the temperature further increasing, the low-albite transforms into hexagonal thenardite at 400-500°C, which shows that the thenardite of hexagonal crystal structure is more stable than orthorhombic crystal structure. Then hexagonal thenardite transforms into High-Albit at the temperature of 600-700°C. At the temperature of 700-2000°C, the molar amount of sodium high-albite remains constant. In Fig 5 and Fig 6, the migration of element Na is in conformity with the process of that in Fig 4.

### *3.5. Phase transfer of minerals containing elements Ca*

The heulandite is formed in a low temperature environment [20]. it has converted to anhydrite when the reaction system temperature rises to 200°C. Therefore, element Ca is mainly contained in the anhydrite in Fig 4-7 at the temperature of 200°C. As the temperature of the reaction system rising, the anhydrite is converted to the anorthite at the temperature of 900-1000°C. Anhydrite is one of sulfate minerals while calcium feldspar is one of aluminosilicate minerals. Therefore, there must be products of anhydrite decomposition which reacts with aluminosilicate with anorthite formed. It also shows that there is a mutual synergy between the mineral material to achieve a state of equilibrium and stability. In the temperature of 1000-2000°C, there is no transformation of anorthite. In this process, the molar amounts of anhydrite and anorthite are equal substantially. The element Ca of the migration is completely consistent with it in Fig 4-7, which shows that the preceding analysis of the element Ca derived from the migration process is reliable.

## 4. Conclusions

The Through the simulating study on the migration of alkali metal minerals in combustion process of sludge and coal, we can get following conclusions:

(1) Quartz, muscovite, illiteilerite, kaolinite and orthoclase are contained in the municipal sewage sludge. Kaolinite, szomolnokite, rozenite, heulandite, quartz, boehmite, pyrite, dawsonite are contained in the Lanba coal. The content of quartz is very high in the coal and sludge.

(2) The migration of K in process of sludge mixed coal combustion is a transformation of muscovite, glaserite (or aphthitalite) and sanidine. At the temperature of 200-300°C, muscovite transforms into sanidine. At the same time, atom K escaped from muscovite is combined with $Na_2SO_4$ (ortho) with glaserite (or aphthitalite) formed. At the temperature of 600-700°C, glaserite (or aphthitalite) disappears and atom K returns to sanidine. At the temperature of 700-2000°C, mineral containing potassium is sanidine.

(3) The migration of element Na during the combustion of municipal sewage sludge mixed with coal has four main processes. At the temperature of 200-400°C, orthorhombic thenardite transforms to low-albite. At the temperature of 400-500°C, low-albite transforms to hexagonal thenardite. At the temperature of 600-700°C, hexagonal thenardite transforms to high albite minerals. At the temperature of 700-2000°C, mineral containing element Na is albite.

(4) The migration of Ca in process of city sludge mixed with coal combustion is a transformation of anhydrite and anorthite. At the temperature of 200-900°C, element Ca mainly exists in the anhydrite. At the temperature of 900-1000°C, anhydrite transformed into anorthite. At the temperature of 1000-2000°C, mineral containing element Ca is anorthite.

**Acknowledgments**

The authors gratefully acknowledge the support provided bythe National Natural Science Foundation of China (No.51206012).

**References**

[1] Tao J, Wu S, Sun L, Tan X, Yu S and Zhang Z 2012 Procedia Environ Sci 12 964-71
[2] Folgueras M.B, DíAz, R.M, Xiberta J and Prieto I 2003 Fuel 82 1939–48
[3] Azizi A.B, Lim M.P, Noor Z.M and Abdullah N 2013 Ecotoxicol. Environ. Saf 90 13–20
[4] Bianchini A 2015 Waste Manage 42 159–65
[5] Okuno T, Sonoyama N, Hayashi J, Li C.Z, Sathe C and Chiba T 2005 Energ Fuel 19 2164-71
[6] Zhang Jun, Lin X.F, Yin J.M 2005 J Eng Thermophys 3 537-9 (in Chinese)
[7] Yang W, Ji X, Dong K, Song X and Bie R 2015 Energiae Solaris Sinica 36 3072-5 (in Chinese)
[8] KAI Xingping,YANG Tianhua,SUN Yang, et al. Study on migration mechanism of alkali metals during co-firing of rice straw and coal[J]. Proceedings of the CSEE, 2012, 32(8): 133-8 (in Chinese)
[9] Kijo-Kleczkowska A, Środa K, Kosowska-Golachowska M, Musiał T and Wolski K 2016 Waste Manage 53 165-81.
[10] Kijo-Kleczkowska A, Środa K, Kosowska-Golachowska M, Musiał T and Wolski K 2016 Fuel 170 141-60
[11] Capablo J 2016 Fuel Process Techno 153 58-73
[12] Gilbe C, Öhman M, Lindstrom E, Dan B, Backman R., & Samuelsson R and Burvall J 2008 Energ Fuel 22 3536-43
[13] Fatehi H, Li Z.S, Bai X.S and Aldén M 2016 Proc Combust Inst
[14] Mao H.Z, Yin Z.L, Xu X.H 2006 Exp Technol Manage 23 30-2
[15] C.W. Bale, E. Bélisle, P. Chartrand, S.A. Decterov, G. Eriksson, and A.E. Gheribi 2016 Calphad 54 35-53
[16] MahadevaNatarajan and Etalo A. Secco 1983 J Solid State Chem 47 231-35
[17] A EICHNER, M KACZMARSKI, M WIESNER and B MROZ 2004 Ferroelectrics 1 31-4
[18] Weidemann H.G and Smykatz-Kloss W 1981 ThermochimicaActa 50 17-29
[19] Post J.E and Burnham C.W Amer Miner 72 507-14
[20] Zhao D, Cleare K, Oliver C, Ingram C, Cook D and Szostak R 1998 Microporous Mesoporous Mat 21 371-9

IOP Conf. Series: Earth and Environmental Science **61** (2017) 012012 doi:10.1088/1755-1315/61/1/012012

# Preparation of Sm doped cerium dioxide film by anodization

**Xiaozhen LIU[1, a*], Junhua YANG[1, b], Xiaozhou LIU[2, c], Letian XIA[1, d], Jie CHEN[3,e] and Ying ZHU[1, f]**

[1]School of Chemical and Environmental Engineering, Shanghai Institute of Technology Shanghai, China 201418
[2]GanNan Normal University Jiangxi, China 341000
[3]Regenia AB, Stockholm, Sweden 10691
[a]liuxiaozhen5291@163.com, [b]yjh@sit.edu.cn, [c]liuxiaozhou@yahoo.com, [d]xlt@sit.edu.cn, [e]cj@yahoo.com, [f]zy@sit.edu.cn

**Abstract**. The Sm doped cerium dioxide films were prepared with cerium foils as raw materials by anodization in $Sm(NO_3)_3$-$Na_2C_2O_4$-$NH_3{\cdot}H_2O$-$H_2O$-$(CH_2OH)_2$ electrolyte. The anodic Sm doped cerium oxide film was heat treated at 550°C. The Sm doped cerium dioxide films were characterized with X-ray diffraction (XRD), energy-dispersive analyses of X-ray (EDAX), Fourier transform infrared (FTIR) techniques and scanning electron microcopy (SEM), respectively. The anodic Sm doped cerium oxide film is semi crystalline film. The heat treated anodic Sm doped cerium oxide film at 550°C has a structure of cubic fluorite. The doping of Sm is replacement doping or caulking doping. The Sm doped cerium dioxide film is porous film. The water, ethylene glycol and $CO_2$ are adsorbed in the anodic Sm doped cerium oxide film. The adsorbing water, ethylene glycol and $CO_2$ in the anodic Sm doped cerium oxide film are removed at 550°C. The Sm doped cerium dioxide film has strong absorption in the range of 1200 ~ 4000cm$^{-1}$.

## 1. Introduction

Doped cerium dioxide has recently been attracting much attention in the oxidative catalysis and solid oxide fuel cell (SOFC) research communities, due to its high oxygen ion mobility and its multiple valence states [1-3]. The porous doped cerium dioxide films are potentially suitable for efficient catalytic treatment of reactants because such structures do not aggregate like particles which have diminished effective catalytic surface areas when used in bulk forms. They can also serve as media for separating reactants and products on different sides of the films such as in SOFC. The common preparation methods of the porous doped cerium dioxide films are ball-milling and sintering methods. There are still some problems, such as controllability of thickness and density of film, defects by calcinations and high cost. If the porous doped cerium dioxide film is prepared with cerium foils as raw materials, we may solve the above problems.

In this paper, the Sm doped cerium dioxide films were prepared with cerium foils as raw materials by anodization in $Sm(NO_3)_3$-$Na_2C_2O_4$-$NH_3{\cdot}H_2O$-$H_2O$-$(CH_2OH)_2$ electrolyte, and the anodic Sm doped cerium oxide film was heat treated at 550°C. The Sm doped cerium dioxide films were characterized with X-ray diffraction (XRD), energy-dispersive analyses of X-ray (EDAX), Fourier transform infrared (FTIR) techniques and scanning electron microcopy (SEM), respectively.

## 2. Experimental

### *2.1. Preparation of the Sm doped cerium dioxide films*

The Sm doped cerium dioxide films were prepared by anodization in $Sm(NO_3)_3$-$Na_2C_2O_4$-$NH_3 \cdot H_2O$-$H_2O$-$(CH_2OH)_2$ electrolyte according to [4]. High purity cerium foils (99.99% purity, 15mm×15 mm×0.25 mm) were used to grow anodic porous layers. Before anodization the cerium foils were degreased in absolute ethyl alcohol by using ultrasonic cleaner and annealed at 500°C for 3 h to remove mechanical stresses and recrystallize of the cerium foils under a nitrogen atmosphere, and then polished with diamond spray polishing agent, followed in absolute ethyl alcohol by using ultrasonic cleaner. An electrochemical reactor was designed and built to carry out anodizing experiments, and got the anodic Sm doped cerium oxide film. The cerium foil and lead plate were the anode and the cathode respectively. The electrolyte were 1.4 mmol/L $Sm(NO_3)_3$, 0.02 mol/L $Na_2C_2O_4$, 1 mol/L $NH_3 \cdot H_2O$ aqueous ethylene glycol solution, ethylene glycol : $H_2O$ = 10 : 1. The anodizing parameters: temperature 20°C, current density 0.5 mA/cm$^2$, time 10 h. After anodizing, the anodized sample was washed thoroughly with distilled water, dried with an air jet, the anodized sample was heat treated at 550°C for 2 h, and the Sm doped cerium dioxide film was got.

### *2.2. Testing instruments and methods*

X-ray powder diffraction patterns of the anodic and the heat treated anodic Sm doped cerium oxide films were acquired with a Rigaku D/max 2550 VB/PC X-ray diffractometer, using Cu Kα radiation, respectively. A continuous scan mode was used to collect 2θ data from 20 ~ 90° with a 0.02 sampling pitch and a 2° min$^{-1}$ scan rate. X-ray tube voltage and current were set at 40 kV and 30 mA, respectively.

The energy-dispersive analysis of X-ray of the heat treated anodic Sm doped cerium oxide film was acquired with S-3400N scanning electron microcopy.

Infrared absorption spectra of the anodic and the heat treated anodic Sm doped cerium oxide films were determined by Avatar 360 FTIR infrared spectrophotometer by potassium bromide disc method respectively.

The morphology of the Sm doped cerium dioxide films were examined by HITACHI S3400 scanning electron microcopy.

## 3. Results and discussion

### *3.1. XRD studies*

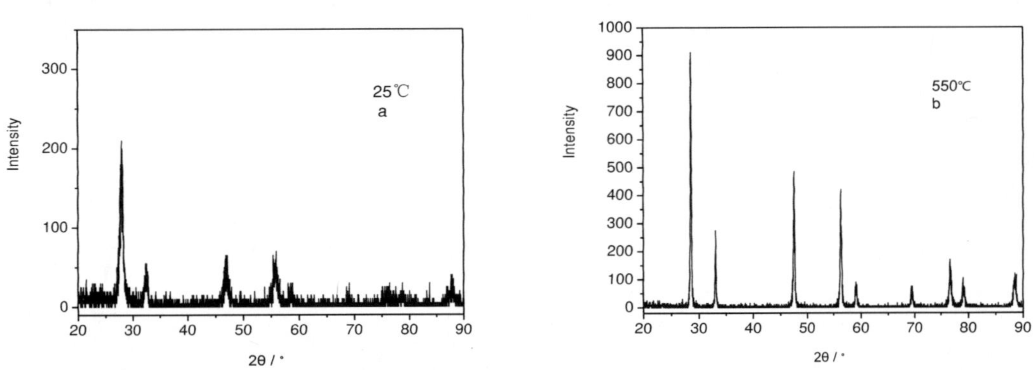


Fig.1 XRD spectra of the anodic and the heat treated anodic Sm doped cerium oxide films

The X-ray diffraction patterns of the anodic and the heat treated anodic Sm doped cerium oxide films are shown in Fig.1. The diffraction peaks 2θ of the films are shown in Table 1. As shown in Fig.1a and listed in Table 1, the anodic Sm doped cerium oxide film has the diffraction patterns at 28.52°, 32.94°, 47.56°, 56.34° respectively, the diffraction peaks correspond to cubic $CeO_2$(111), $CeO_2$(200),

$CeO_2(220)$, $CeO_2(311)$, respectively. The anodic Sm doped cerium oxide film is semi crystalline film. Comparing Fig.1b to JCPDS Standard card (PDF#43-1002), the diffraction peaks in the XRD spectrum of the heat treated anodic Sm doped cerium oxide film at 550°C can be recognized as the characteristic ones of $CeO_2$, which demonstrates that the samples have a structure of cubic fluorite. Moreover, they give the diffraction peaks a small migration, which demonstrates that the doping can be achieved better by anodization method and be recognized as replacement doping or caulking doping. Comparing Fig.1b to Fig.1a, after the heat treatment of the anodic Sm doped cerium oxide film at 550°C, new diffraction peaks are observed at 59.10°, 69.40°, 76.70°, 79.06°, 88.48°, respectively, the diffraction peaks correspond to cubic $CeO_2(222)$, $CeO_2(400)$, $CeO_2(331)$, $CeO_2(420)$, $CeO_2(422)$, respectively, and the crystal structure of the samples becomes complete.

Table 1 Diffraction peaks 2θ of films

| (hkl) | $CeO_2$(111) | $CeO_2$(200) | $CeO_2$(220) | $CeO_2$(311) | $CeO_2$(222) | $CeO_2$(400) | $CeO_2$(331) | $CeO_2$(420) | $CeO_2$(422) |
|---|---|---|---|---|---|---|---|---|---|
| 25°C | 28.52 | 32.94 | 47.56 | 56.34 | | | | | |
| 550 °C | 28.54 | 33.06 | 47.50 | 56.34 | 59.10 | 69.40 | 76.70 | 79.06 | 88.48 |

## 3.2. EDAX analysis

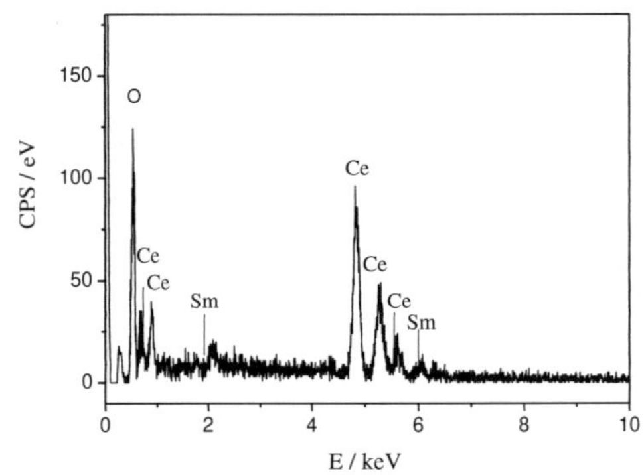


Fig.2 EDAX spectrum of the heat treated anodic Sm doped cerium oxide film at 550°C

Fig.2 shows the EDAX spectrum of the heat treated anodic Sm doped cerium oxide film at 550°C. The contents of Ce(at%), Sm(at%) and O(at%) are 31.67%, 1.09% and 67.24% respectively, and Ce(at%) + Sm (at%) : O(at%) =1:2. The heat treated anodic cerium oxide film at 550°C is the Sm doped cerium dioxide film. The result is consistent with the XRD analysis result.

## 3.3. Infrared spectra

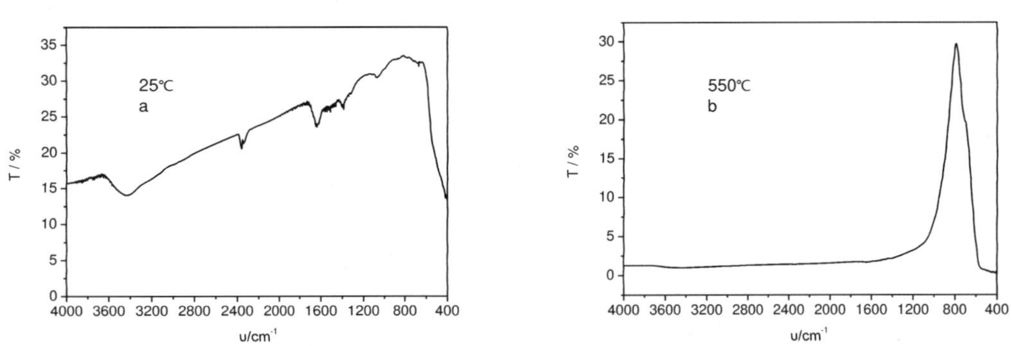


Fig.3 FTIR spectra of the anodic and the heat treated anodic Sm doped cerium oxide films

IOP Conf. Series: Earth and Environmental Science **61** (2017) 012012 doi:10.1088/1755-1315/61/1/012012

Fig.3 show FTIR spectra of the anodic and the heat treated anodic Sm doped cerium oxide films. According to Fig.3a, there are νO-H vibration peak at 3423.17cm$^{-1}$, ν$CO_2$ vibration peak at 2360.54 cm$^{-1}$, νO-H vibration peak at 1632.03 cm$^{-1}$, δ$CH_2$ vibration peak at 1544.28 cm$^{-1}$, δO-H vibration peak at 1383.73 cm$^{-1}$, νC-O vibration peak at 1059.34 cm$^{-1}$, $CeO_2$ vibration peak at 400.00 cm$^{-1}$[5], demonstrating the presence of adsorbing water, ethylene glycol and $CO_2$ in the anodic Sm doped cerium oxide film. According to Fig.3b, there are $CeO_2$ vibration peaks at 600.00 ~ 400.00 cm$^{-1}$, demonstrating the heat treated anodic Sm doped cerium oxide film at 550°C being Sm doped cerium dioxide film. The adsorbing water, ethylene glycol and $CO_2$ in the anodic Sm doped cerium oxide film are removed at 550°C. The result is consistent with the XRD analyses and EDAX analyses result respectively. The Sm doped cerium dioxide film has strong absorption in the range of 1200 ~ 4000cm$^{-1}$.

Table 2 The locations of infrred absorption spectra of the peaks (cm$^{-1}$)

| | | | | | | | |
|---|---|---|---|---|---|---|---|
| Anodic Sm doped cerium oxide film | 3423.17 | 2360.54 | 1632.03 | 1544.28 | 1383.73 | 1059.34 | 400.00 |
| Heat treated film(550°C) | | | | | | | 400.00 |

### *3.4. SEM image*

Fig.4 shows SEM surface morphology of the Sm doped cerium dioxide film. It is seen in Fig.4 that the Sm doped cerium dioxide film is the porous film.

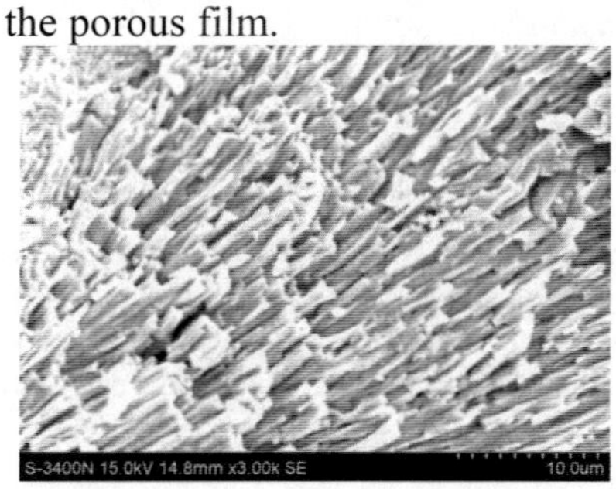


Fig.4 SEM of the Sm doped cerium dioxide film (×3000)

## 4. Conclusions

The Sm doped cerium dioxide film is prepared by anodization in $Sm(NO_3)_3$-$Na_2C_2O_4$-$NH_3 \cdot H_2O$-$H_2O$-$(CH_2OH)_2$ electrolyte. The anodic Sm doped cerium oxide film is semi crystalline film. The heat treated anodic Sm doped cerium oxide film at 550°C has a structure of cubic fluorite. The doping of Sm is replacement doping or caulking doping. The Sm doped cerium dioxide film is porous film. The water, ethylene glycol and $CO_2$ are adsorbed in the anodic Sm doped cerium oxide film. The adsorbing water, ethylene glycol and $CO_2$ in the anodic Sm doped cerium oxide film are removed at 550°C. The Sm doped cerium dioxide film has strong absorption in the range of 1200 ~ 4000cm$^{-1}$.

**Acknowledgements**

This work was financially supported by the project supported by the National Natural Science Foundation of China(51372153, 51572176), Shanghai Municipal Education Commission (Plateau Discipline Construction Program, 0817), Leading Academic Discipline Project of Shanghai Municipal Education Commission(J51503).

**References**

[1] H.P. Dasari, K. Ahn, S.Y. Park, et al., Journal of Alloys and Compounds. 672 (2016) 397.
[2] S.Y. Gomez, D. Hotza, Renewable and Sustainable Energy Reviews. 61 (2016) 155.
[3] L.L. Guan, S. Le, X.D. Zhu, et al., Electrochimica Acta. 161 (2016) 129.
[4] X.Z. Liu, L.T. Xia, J. Chen, C.N. 201510503095.3.(2015).
[5] C.H. Wang, C. H. Lu, Mater Res Bull. 37 (2002) 783.

IOP Conf. Series: Earth and Environmental Science **61** (2017) 012013 doi:10.1088/1755-1315/61/1/012013

# Effect of heat treatment on the crystal structure and FTIR spectra of Sm doped cerium dioxide film

**Xiaozhou LIU[1, a], Xiaozhou LIU[2, b*], Letian XIA[1, c], Jie CHEN[3, d] and Xiaoyu WANG[2, e]**

[1]GanNan Normal University Jiangxi, China 341000
[2]School of Chemical and Environmental Engineering, Shanghai Institute of Technology Shanghai, China 201418
[3]Regenia AB, Stockholm, Sweden 10691
[a]liuxiaozhou@yahoo.com, [b]liuxiaozhen5291@163.com, [c]xlt@sit.edu.cn, [d]cj@yahoo.com, [e]wxy@sit.edu.cn

**Abstract**. The Sm doped cerium dioxide films were prepared with cerium foils as raw materials by anodization in $Sm(NO_3)_3$-$Na_2C_2O_4$-$NH_3 \cdot H_2O$-$H_2O$-$(CH_2OH)_2$ electrolyte. The anodic Sm doped cerium oxide films were heat treated in 100°C ~ 400°C. The heat treated anodic Sm doped cerium oxide films were characterized with X-ray diffraction (XRD) and Fourier transform infrared (FTIR) techniques respectively. The heat treated anodic Sm doped cerium oxide film at 100°C is the semi crystalline film. As heat treatment temperatures being in 200°C ~ 400°C, the heat treated anodic Sm doped cerium oxide films have a structure of cubic fluorite respectively. The doping of Sm can be achieved well by anodization method and be recognized as replacement doping or caulking doping. The crystal structure of Sm doped cerium dioxide films become more complete with the increase of heat treatment temperature in 200 ~ 400 °C. The doping of Sm can improve the crystallinity of the cerium dioxide film. The presence of adsorbing water, ethylene glycol and $CO_2$ in the heat treated anodic Sm doped cerium oxide film at 100°C. The adsorbing ethylene glycols and water, $CO_2$ in the anodic Sm doped cerium oxide film are removed at 200°C and 300°C respectively.

## 1. Introduction

Because of doped cerium dioxide having high oxygen ion mobility and its multiple valence states, doped cerium dioxide has recently been attracting much attention in the oxidative catalysis and solid oxide fuel cell (SOFC) research communities [1-9]. The porous doped cerium dioxide films are potentially suitable for efficient catalytic treatment of reactants because such structures do not aggregate like particles which have diminished effective catalytic surface areas when used in bulk forms. They can also serve as media for separating reactants and products on different sides of the films such as in SOFC. The porous doped cerium dioxide film is prepared with cerium foils as raw materials, we may solve the existing problems, such as the controllability of thickness and density of film, defects by calcinations and high cost. The composition and crystal structure of the film have influence on the performance of the film.

In this paper, The Sm doped cerium dioxide films were prepared with cerium foils as raw materials by anodization in $Sm(NO_3)_3$-$Na_2C_2O_4$-$NH_3 \cdot H_2O$-$H_2O$-$(CH_2OH)_2$ electrolyte. The anodic Sm doped cerium oxide films were heat treated in 100 ~ 400°C. The heat treated anodic Sm doped cerium oxide films were characterized with X-ray diffraction (XRD) and Fourier transform infrared (FTIR) techniques respectively.

## 2. Experimental

### *2.1. Preparation of the Sm doped cerium dioxide films*

The Sm doped cerium dioxide films were prepared by anodization in $Sm(NO_3)_3$-$Na_2C_2O_4$-$NH_3 \cdot H_2O$-$H_2O$-$(CH_2OH)_2$ electrolyte according to [10]. The electrolyte were 1.4 mmol/L $Sm(NO_3)_3$, 0.02 mol/L $Na_2C_2O_4$, 1 mol/L $NH_3 \cdot H_2O$ aqueous ethylene glycol solution, ethylene glycol : $H_2O$ = 10 : 1. The anodizing parameters: temperature 20°C, current density 0.5 mA/cm$^2$, time 10 h. The anodic Sm doped cerium oxide films were heat treated at 100°C, 200°C, 300°C, 350°C, 400°C, respectively for 2 h, and the Sm doped cerium dioxide film was got. The cerium dioxide film was prepared as contrast.

### *2.2. Testing instruments and methods*

X-ray powder diffraction patterns of the heat treated anodic Sm doped cerium oxide films were acquired with a Rigaku D/max 2550 VB/PC X-ray diffractometer, using Cu Kα radiation, respectively. A continuous scan mode was used to collect 2θ data from 20 ~ 90° with a 0.02 sampling pitch and a 2° min$^{-1}$ scan rate. X-ray tube voltage and current were set at 40 kV and 30 mA, respectively.

Infrared absorption spectra of the heat treated anodic Sm doped cerium oxide films were determined by Avatar 360 FTIR infrared spectrophotometer by potassium bromide disc method respectively.

## 3. Results and discussion

### *3.1. XRD studies*

The X-ray diffraction patterns of the anodic Sm doped cerium oxide films at different heat treating temperature are shown in Fig.1. The diffraction peaks 2θ of the films are shown in Table 1. As shown in Fig.1a and listed in Table 1, the heat treated anodic Sm doped cerium oxide film at 100°C has the

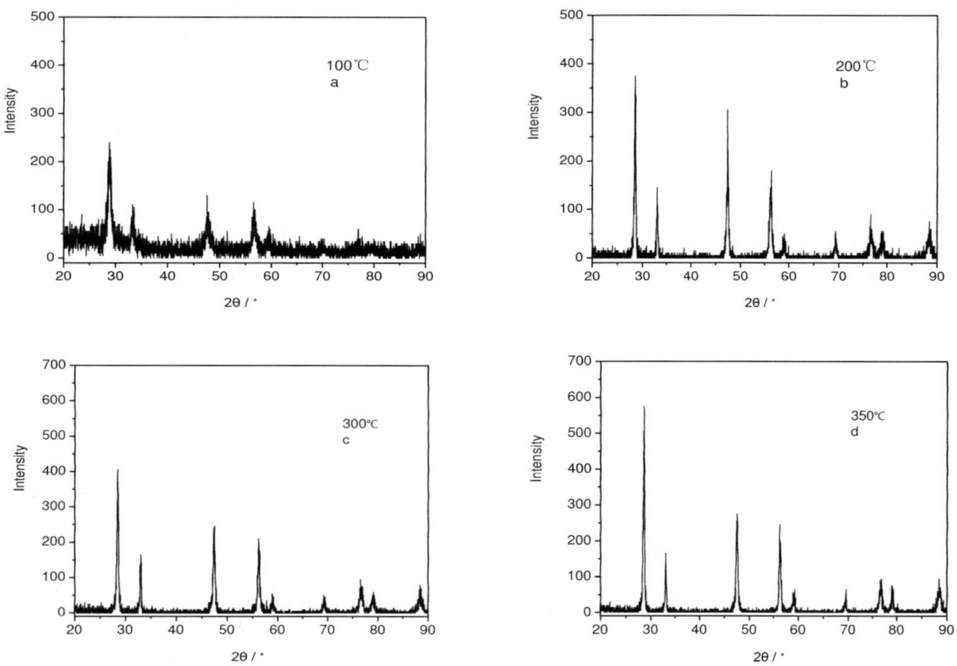

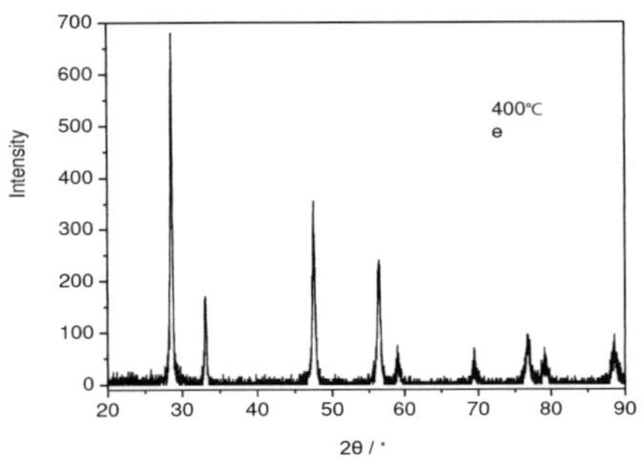


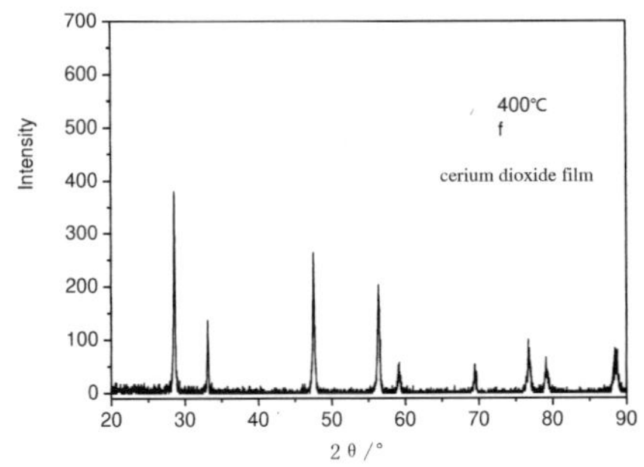


**Fig.1** XRD spectra of the anodic Sm doped cerium oxide films at different heat treating temperature

diffraction patterns at 28.54°, 33.08°, 47.54°, 56.32° respectively, the diffraction peaks correspond to cubic $CeO_2$(111), $CeO_2$(200), $CeO_2$(220), $CeO_2$(311), respectively, and the sample is the semi crystalline film. Comparing Fig.1b, Fig.1c, Fig.1d, Fig.1e, to Fig.1f, the diffraction peaks in the XRD spectra of the heat treated anodic Sm doped cerium oxide films at 200°C, 300°C, 350°C, 400°C, respectively, can be recognized as the characteristic ones of $CeO_2$, which demonstrates that the samples have a structure of cubic fluorite respectively. Moreover, they give the diffraction peaks a small migration, which demonstrates that the doping of Sm can be achieved well by anodization method and be recognized as replacement doping or caulking doping. The ionic radius of $Ce^{4+}$ (92pm) and $Sm^{3+}$ (96.4pm) is about the same, the site replacement of $Ce^{4+}$ by $Sm^{3+}$. As shown in Fig.1b, Fig.1c, Fig.1d, Fig.1e, the intensity of diffraction peaks increase with the increase of heat treatment temperature in 200 ~ 400 °C, therefore, the crystal structure of Sm doped cerium dioxide films become more complete with the increase of heat treatment temperature in 200 ~ 400 °C. Comparing Fig.1e to Fig.1f, as heat treatment temperature being 400 °C, the intensity of diffraction peaks of Sm doped cerium dioxide film is stronger than that of the cerium dioxide film, which demonstrates that the doping of Sm can improve the crystallinity of the cerium dioxide film.

**Table 1** Diffraction peaks 2θ of films

| (hkl) | $CeO_2$(111) | $CeO_2$(200) | $CeO_2$(220) | $CeO_2$(311) | $CeO_2$(222) | $CeO_2$(400) | $CeO_2$(331) | $CeO_2$(420) | $CeO_2$(422) |
|---|---|---|---|---|---|---|---|---|---|
| 100°C | 28.54 | 33.08 | 47.54 | 56.32 | | | | | |
| 200°C | 28.54 | 33.08 | 47.44 | 56.34 | 59.12 | 69.42 | 76.68 | 79.08 | 88.46 |
| 300°C | 28.56 | 33.10 | 47.44 | 56.30 | 59.12 | 69.42 | 76.70 | 79.08 | 88.42 |
| 350°C | 28.56 | 33.12 | 47.42 | 56.34 | 59.12 | 69.40 | 76.66 | 79.08 | 88.44 |
| 400°C | 28.52 | 33.10 | 47.44 | 56.36 | 59.12 | 69.42 | 76.66 | 79.06 | 88.42 |
| $CeO_2$ | 28.54 | 33.10 | 47.50 | 56.30 | 59.10 | 69.42 | 76.68 | 79.10 | 88.42 |

### 3.2. *Infrared spectra*

Fig.2 show FTIR spectra of the anodic Sm doped cerium oxide films at different heat treating temperature. According to Fig.2a, there are νO-H vibration peak at 3434.26 $cm^{-1}$, $\nu CO_2$ vibration peak at 2358.13 $cm^{-1}$, νO-H vibration peak at 1648.91 $cm^{-1}$, $\delta CH_2$ vibration peak at 1542.36 $cm^{-1}$, δO-H vibration peak at 1382.29 $cm^{-1}$, νC-O vibration peak at 1059.74 $cm^{-1}$, $CeO_2$ vibration peak at 400.00 $cm^{-1}$[11], demonstrating the presence of adsorbing water, ethylene glycol and $CO_2$ in the heat treated anodic Sm doped cerium oxide film at 100°C. According to Fig.2b, there are νO-H vibration peak at 3447.28 $cm^{-1}$, $\nu CO_2$ vibration peak at 2359.10 $cm^{-1}$, νO-H vibration peak at 1648.13 $cm^{-1}$, demonstrating the presence of adsorbing water and $CO_2$ in the heat treated anodic Sm doped cerium oxide film at 200°C. The adsorbing ethylene glycols in the anodic Sm doped cerium oxide film are removed at 200°C. According to Fig.2c and Fig.2d, there are $CeO_2$ vibration peaks at 610.42 ~ 400.00 $cm^{-1}$, demonstrating the heat treated anodic Sm doped cerium oxide film at 300°C and

400°Crespectively being Sm doped cerium dioxide film. The adsorbing water and $CO_2$ in the anodic Sm doped cerium oxide film are removed in 300°C ~ 400°C. The Sm doped cerium dioxide film has strong absorption in the range of 1200 ~ 4000cm$^{-1}$.

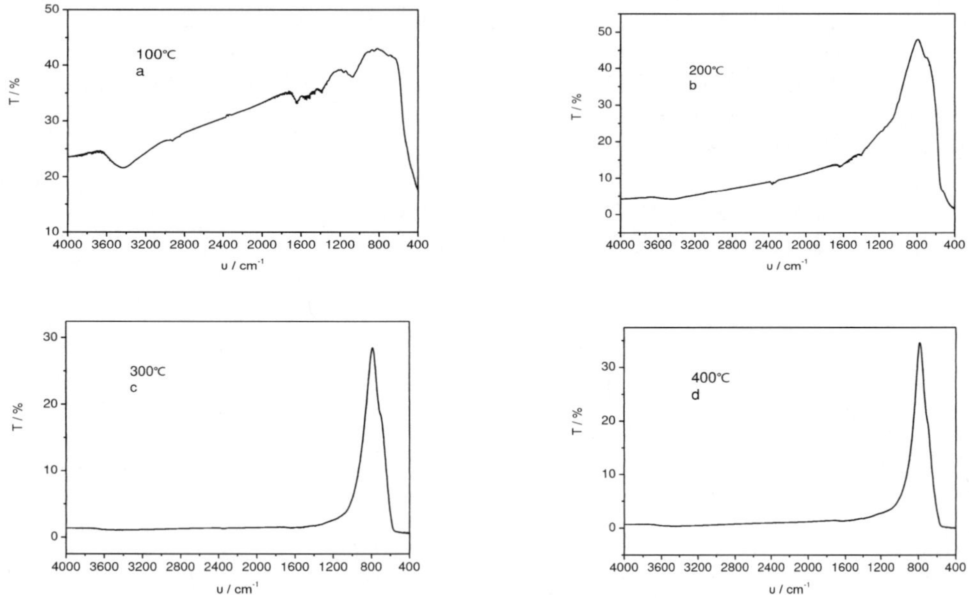


**Fig.2** FTIR spectra of the anodic Sm doped cerium oxide films at different heat treating temperature

**Table 2** The locations of infrred absorption spectra of the peaks (cm$^{-1}$)

| | | | | | | | |
|---|---|---|---|---|---|---|---|
| Heat treated film(100°C) | 3434.26 | 2358.13 | 1648.91 | 1542.36 | 1382.29 | 1059.74 | 400.00 |
| Heat treated film(200°C) | 3447.28 | 2359.10 | 1648.13 | | | | 400.00 |
| Heat treated film(300°C) | | | | | | 610.42 | 400.00 |
| Heat treated film(400°C) | | | | | | 610.23 | 400.00 |

## 4. Conclusions

The heat treated anodic Sm doped cerium oxide film at 100°C is the semi crystalline film. As heat treatment temperatures being in 200°C~ 400°C, the heat treated anodic Sm doped cerium oxide films have a structure of cubic fluorite respectively. The doping of Sm can be achieved well by anodization method and be recognized as replacement doping or caulking doping. The crystal structure of Sm doped cerium dioxide films become more complete with the increase of heat treatment temperature in 200 ~ 400 °C. The doping of Sm can improve the crystallinity of the cerium dioxide film. The presence of adsorbing water, ethylene glycol and $CO_2$ in the heat treated anodic Sm doped cerium oxide film at 100°C. The adsorbing ethylene glycols and water, $CO_2$ in the anodic Sm doped cerium oxide film are removed at 200°C and 300°C respectively.

**Acknowledgements**
This work was financially supported by the project supported by the National Natural Science Foundation of China (51372153, 51572176), Shanghai Municipal Education Commission (Plateau Discipline Construction Program, 0817), Leading Academic Discipline Project of Shanghai Municipal Education Commission (J51503).

## References

[1] H.P. Dasari, K. Ahn, S.Y. Park, et al., Journal of Alloys and Compounds. 672 (2016) 397.

[2] M.L. Faro, S. Trocino, S.C. Zignani, et al., International Journal of Hydrogen Energy. 41 (2016) 5741.
[3] B. Timurkutluk, C. Timurkutluk, M.Y. Kaplan, Renewable and Sustainable Energy Reviews. 56 (2016) 1101.
[4] S.Y. Gomez, D. Hotza, Renewable and Sustainable Energy Reviews. 61 (2016) 155.
[5] W.L. Li, C.Y. Xiong, Q. Zhang, et al., Electrochimica Acta. 190 (2016) 531.
[6] S.Y. Gomez, D. Hotza, Renewable and Sustainable Energy Reviews. 61 (2016) 155.
[7] L.L. Guan, S. Le, X.D. Zhu, et al., Electrochimica Acta. 161 (2016) 129.
[8] T. Somekawa, Y. Matsuzaki, Y.Tachikawa, et al., Solid State Ionics. 282 (2015) 1.
[9] B. Matovic, M. Stojmenovic, J. Pantic, et al., Journal of Asian Ceramic Societies. 2 (2014) 117.
[10] X.Z. Liu, L.T. Xia, J. Chen, C.N. 201510503095.3.(2015).
[11] C.H. Wang, C. H. Lu, Mater Res Bull. 37 (2002) 783.

IOP Conf. Series: Earth and Environmental Science **61** (2017) 012014 doi:10.1088/1755-1315/61/1/012014

# Enhanced thermopower of $YBaCo_4O_{7+\delta}$ by Na-for-Ba substitution

**S L Leng[1], X X Wang[2] and H Z Song[2, *]**

[1] College of Material and Chemical Engineering, Tongren University, Tongren 554300, China.
[2] Key Laboratory of Material Physics of Ministry of Education, and Physical Engineering College, Zhengzhou University, Zhengzhou 450052, China.
* star@zzu.edu.cn

**Abstract**. The electrical resistivity and Seebeck coefficients of $YBa_{1-x}Na_xCo_4O_{7+\delta}$ (x = 0.0, 0.1, 0.2, 0.3, 0.4, 0.5, 0.6) ceramic samples prepared by the solid-state reaction method were measured in the temperature range 390-973 K. The results show that with the increase of the Na doping amount, both the electrical resistivity and Seebeck coefficients increased, and that the maximum value of the Seebeck coefficient can reach 170 $\mu VK^{-1}$. According to the power factors, the samples with Na doping amount x = 0.4 and 0.5 have higher power factors at higher temperature, and the maximum value is 71 $\mu Wm^{-1}K^{-2}$ when x = 0.4 at 973 K, being 11% higher than that of the sample without Na doping.

## 1. Introduction

Thermoelectric (TE) materials can convert thermal energy directly into electricity mutually, making it becoming increasingly important in the field of cooling, heating, and recovering waste heat. However, these materials are incapable of wide commercialization because of their poor thermalelectric performance, so enhancing the thermal electrical performance has long been the researchers' study focus. The performance of a TE material is governed by the dimensionless figure-of-merit $ZT = S^2T/\rho\kappa$, where $S$ is the Seebeck coefficient, ρ the electric resistivity, κ the thermal conductivity and $T$ the absolute temperature. The parameter $S^2/\rho$ is called power factor, which represents the electric transport properties of TE materials. Therefore, good TE materials should possess large power factor as well as low thermal conductivity $\kappa$. Reducing electrical resistivity is a common way we take to optimize the electric transport properties, while improving Seebeck coefficient is also another way that should not be ignored. At present, some layered cobalt oxides, such as $Na_xCoO_2$, $Ca_3Co_4O_9$ and $Bi_2Sr_2Co_2O_y$,[1-3] exhibit better thermoelectric performance due to the layered structure, mixed valence of Co ion and the low spin of $Co^{3+}$,[4-6] attracting researchers' great attention.

$RBaCo_4O_{7+\delta}$ (R = Y, Dy–Lu, In, Ca; abbreviated as R114) synthesized by Valldor et al.[7] is another type of the layered cobalt oxides. Due to their unique structure, favorable physical and chemical properties, R114 compounds have been studied in many fields such as solid oxide fuel cells (SOFCs) and oxygen permeation membranes.[8-15] Especially, much work has been done to investigate their electric transport properties and thermal electric performance.[16-19] It was found that the $ZT$ value of R114 is much lower than that of conventional non-oxide thermoelectric materials due to their ceramic nature. In our previous study, the influence of R site or Co site doping on the electrical transport properties of R114 has been widely investigated,[19-24] but the investigations of Ba site doping has barely been reported. In this paper, the influence of Ba site doping on the electric transport properties of $YBaCo_4O_{7+\delta}$ was studied, and the $Na^+$ was selected as the doping element.

## 2. Experimental

$YBa_{1-x}Na_xCo_4O_{7+\delta}$ (x=0.0, 0.1, 0.2, 0.3, 0.4, 0.5, 0.6) ceramic samples were synthesized by the traditional solid-state reaction method. Stoichiometric amounts of $Y_2O_3$, $Na_2CO_3$, $BaCO_3$ and $Co_3O_4$ raw materials were mixed thoroughly in an agate mortar. The mixed powder was heated to 1000°C in a box furnace in air and kept at this temperature for 10 h in order to get rid of carbon. After regrinding, the power was consolidated into pellets and rectangular bars, and then sintered at 1140°C in air for 20 h again. The electrical resistivity and Seebeck coefficient were measured simultaneously under He atmosphere from 390 K to 973 K using a LSR-3/800 Seebeck coefficient/electrical resistance measuring system (Linseis, Germany). The rectangular bars with length ~15 mm, width ~4.5 mm, and height ~2.0 mm were prepared for electrical measurements.

## 3. Results and discussion

The relationship between Ln（$\rho/T$）and 1000/$T$ was plotted in Figure 1, showing the temperature dependence of the electrical resistivity of $YBa_{1-x}Na_xCo_4O_{7+\delta}$ samples from 390 K to 973 K. It was clearly seen that the electrical resistivity of all samples decreases with increase of the temperature, demonstrating semiconducting-like behavior in the measured temperature region. Unfortunately, the electrical resistivity increases with the increase of Na doping amount, which is contrary to our expectation. The phenomena may result from the lattice distortion even the presence of impurity phase leading by the reasonable high doping amount.

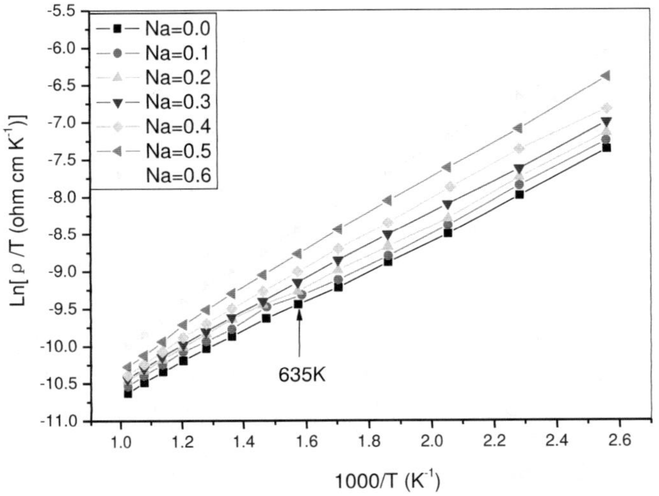


**Figure 1.**Temperature dependence of the electrical resistivity (ln($\rho$/T) vs 1000/$T$) of $YBa_{1-x}Na_xCo_4O_{7+\delta}$ samples.

It is also noticed that all the samples have good linear fitting results from 390 K to 635 K. Therefore the conduction mechanism should belong to small polaron hopping conduction mechanism, consistent with our previous study.[19,21,23] The behavior of electrical resistivity versus temperature is expressed as $\rho = \rho_0 T \exp(E_\sigma / k_B T)$ where $\rho$ is the electrical resistivity, $\rho_0$ is a constant, $T$ is the absolute temperature, $k_B$ is the Boltzmann constant, and $E_\sigma$ is the conductive activation energy. However, for temperatures beyond 635 K, the temperature dependence of the electrical resistivity deviates from a straight line, which can be attributed to oxygen absorption/desorption of the samples.

Oxygen absorption cannot exist because the electrical resistivity was measured under He atmosphere. According to previous report,[25] at temperature around 635 K, it begins to appear oxygen desorption from the lattice, resulting in a drop of the carrier concentration, and thus a rise of the resistivity. This can be understood from the defect chemistry where the resistivity of $RBaCo_4O_{7+\delta}$ is related to the oxygen vacancy concentration $[V_O^{\bullet\bullet}]$ in the ceramic. The reaction during oxygen desorbing can be represented as $O_O^x \Leftrightarrow \frac{1}{2}O_2 + [V_O^{\bullet\bullet}] + 2e'$. Thus, oxygen desorption increases the oxygen vacancy concentration, along with the reduction of hole carriers, resulting in the increment of the resistivity because R114 is *p*-type semiconductor, It seems that this behavior becomes insignificant even disappears as the Na doping amount increases. The reason may be that the increasing Na concentration results in a drop in oxygen adsorption/desorption ability, as shown by the influence of element doping on the oxygen adsorption/desorption properties.[26]

Figure 2 shows the temperature dependence of the Seebeck coefficients of the $YBa_{1-x}Na_xCo_4O_{7+\delta}$ samples. It can be seen that the Seebeck coefficients of all the samples are positive for the entire measured temperature range, indicating a *p*-type semiconductor and the conduction mechanism predominantly governed by holes. Moreover, at higher temperature above 635 K, the Seebeck coefficients increase markedly with the increase of the Na concentration corresponding to the evolution of electrical resistivity, which can be attributed to two competing factors. One is the carrier mobility: the carrier mobility should be decreased due to the enhanced scattering effect of carriers in the crystal lattice leading by the Na doping and according to conventional theory for the semiconductor, the smaller carrier mobility leads to smaller Seebeck coefficients. Another is the electron correlation and election-phonon interactions: the ionic bond of $Na^+$ is so strong that its presence makes more and more carriers localization even in a "self- trapped-state", which will lead to the enhancement of Seebeck coefficient. Moreover, in layered cobaltates it is realized that the electron correlation and election-phonon interactions will play a very important role in the determination of the Seebeck coefficient.[27] As a result, the Seebeck coefficients increase with the increase of the Na concentration resulting from the predominate contribution of the latter factor. When x = 0.6 at 830 K, the Seebeck coefficient arrives to the maximum value 170 $\mu VK^{-1}$ in the study range, which is benefit for the improvement of electric transport properties.

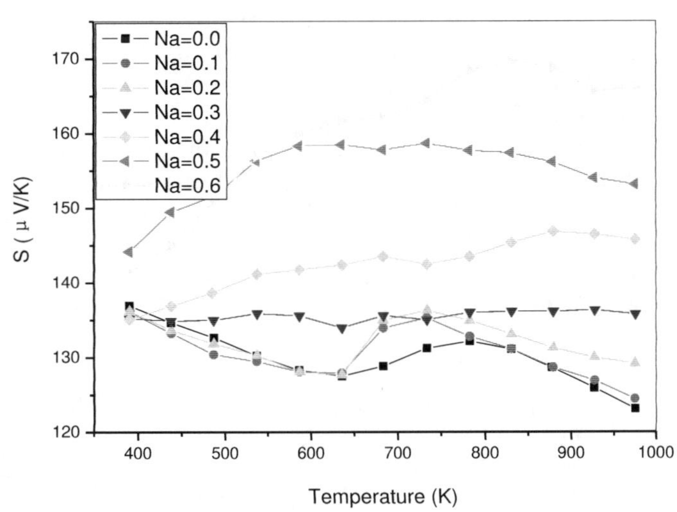


**Figure 2.** Temperature dependence of the Seebeck coefficients ($S$) of $YBa_{1-x}Na_xCo_4O_{7+\delta}$ samples.

With the electrical resistivity and Seebeck coefficient, power factors (PF= $S^2/\rho$) have been calculated and the variation with temperature as a function of Nb content has been represented in Figure 3. It is found that, for all the samples, the power factor increases with increase of the temperature, indicating that Y114 exhibit better electrical transport properties at higher temperature. Moreover, at higher temperature range, when x = 0.4 or 0.5, it appears that Na addition produces an important increase of power factor due to the relatively large enhancement of Seebeck coefficient. The sample with x = 0.4 has the maximum PF = 71μW m$^{-1}$ K$^{-2}$ at about 973 K, which is the highest value in the temperature range of our study and is 11% higher than the undoped sample.

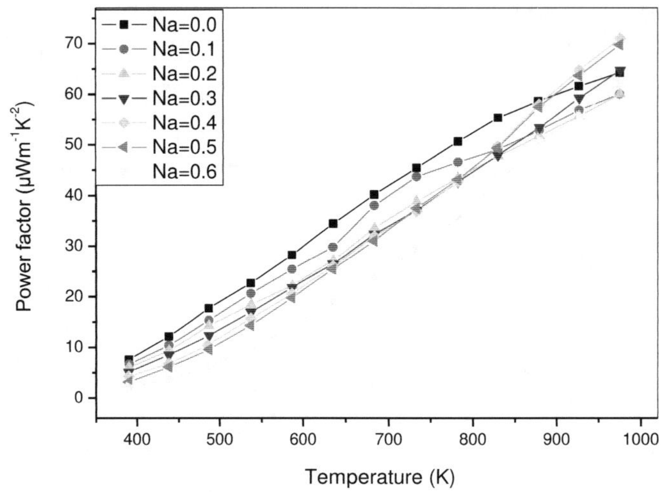


**Figure 3.** Temperature dependence of the power factors of $YBa_{1-x}Na_xCo_4O_{7+\delta}$ samples.

## 4. Conclusions

We have synthesized $YBa_{1-x}Na_xCo_4O_{7+\delta}$ (x=0.0, 0.1, 0.2, 0.3, 0.4, 0.5, 0.6) ceramic samples and investigated their high temperature electrical transport properties. The study demonstrates that the presence of Na produces significant effects on the electric resistivity and Seebeck coefficient. With the increase of the Na doping amount, the electrical resistivity and Seebeck coefficient remarkably increase simultaneously. Moreover, when x = 0.4 or 0.5, the increase magnitude of Seebeck coefficients is higher than that of electric resistivity at higher temperature, resulting an enhancement of the power factor. This phenomenon provides an idea of investigating the influence of double element doping on the electrical transport properties of R114 materials, i.e. Ag-for-Y or Pb-for-Y substitution improves the conductivity[21,24,28] and at the same time Na-for-Ba substitution enhance the thermopower. Thus, the power factor of co-doped $YBaCo_4O_{7+\delta}$ may be enhanced remarkably.

## Acknowledgments

This work was supported by the Foundation of Henan Educational Committee (Grant No. 16A140036)

## References

[1] Terasaki I, Sasago Y and Uchinokura K 1997 *Phys. Rev. B* vol 56 p R12685
[2] Funahashi R, Matsubara I, Ikuta H, Takeuchi T, Mizutani U and Sodeoka S 2000 *Jpn. J. Appl. Phys.* vol 39 p L1127
[3] Funahashi R, Matsubara I and Sodeoka S 2000 *Appl. Phys. Lett.* vol 76 p 2385
[4] He J, Liu Y and Funahashi R 2011 *J. Mater. Res.* vol 26 p 1762

[5] Zebarjadi M, Esfarjani K, Dresselhaus M S, Ren Z F and Chen G 2012 *Energy Environ. Sci.* vol 5 p 5147
[6] Doumerc J P, Blangero M, Pollet M, Carlier D, Darriet J, Berthelot R, Delmas C and Decourt R 2009 *J. Electron. Mater.* vol 38 p 1078
[7] Fergus J W 2012 *J. Euro. Ceram. Soc.* vol 32 p 525
[8] Valldor M 2004 *Solid State Sci.* vol 6 p 251
[9] Maignan A, Hébert S, V Caignaert, Pralong V and Pelloquin D 2008 *Solid State Commun.* vol 47 p 470
[10] Markina M, Vasiliev A N, Nakayama N, Mizota T and Yeda Y 2010 *J. Magn. Magn. Mater.* vol 322 p 1249
[11] Sarkar T, Caignaert V, Pralong V and Raveau B 2010 *Chem. Mater.* vol 22 p 6467
[12] Tsipis E, Kharton V and Frade J 2006 *Solid State Ionic* vol 177 p 1823
[13] Kim Y N, Kim J H, Huq A, Paranthaman M P and Manthiram A 2012 *J. Power Source* vol 214 p 7
[14] Zhu B, Hao H, Zhang Y, Jia J and Hu X 2010 *J. Rare Earths* vol 28 p 84
[15] Zhang K, Zhu Z, Ran R, Shao Z, Jin W and Liu S 2010 *J. Alloy. Compd.* vol 492 p 552
[16] Wong-Ng W, Xie W, Yan Y, Liu G, Kaduk J, Thomas E and Tritt T 2011 *J. Appl.Phys.* vol 110 p 113706
[17] Tsipis E V, Kharton V V, Frade J R and Núñez P 2005 *J. Solid State Electronchem.* vol 95 p 547
[18] Hao H S, Chen C Q, Pan L J, Gao J X and Hu X 2007 *Phys. B* vol 387 p 98
[19] Gao F, He Q L, Wu F, Yang D L, Hu X and Song H Z 2015 *Mod. Phys. Lett. B* vol 29 p 1550154
[20] Hao H S, Zhang X T, He Q L, Chen C Q and Hu X 2007 *Solid State Commun.* vol 141 p 591
[21] He Q L, Zhang J, Gao F, Wu F, Hu X and Song H Z 2013 *J. Electron. Mater.* vol 42 p 2570
[22] Hao H, He Q, Cheng Y and Zhao L 2014 *Mater. Res. Bull.* vol 53 p 84
[23] He Q L, Gao F, Hu X and Song H Z 2013 *Adv. Mater. Res.* vol 724-725 p 1029
[24] Chen Y B, Ma RX, Wang K L, Gao F, Hu X and Song HZ 2015 *Int. J. Mod. Phys. B* vol 29 p 1550082
[25] Song H, Jia J, Zhang S, Yang D, Sun H and Hu X 2012 *Mater. Res. Bull.* vol 47 p 518
[26] Wang S, Hao H, Zhu B, Jia J and Hu X 2008 *J. Mater. Sci.* vol 43 p 5385
[27] Schrade M, Fjeld H, Finstad T G and Norby T 2014 *J. Phys. Chem. C* vol 118 p 2915
[28] Chen Y B, Cao X L, Ma R X, Gao F, Hu X and Song H Z 2015 *J. Electron. Mater.* vol 44 p 3545

# Study on treatment of aquaculture wastewater using a hybrid constructed wetland

**Jinzhao Hu, Rui Hu, Dan Qi, Xujie Lu**

School of Ocean Science and Technology
Hainan Tropical Ocean University, Sanya, China
E-mail: xujie_lu@163.com

**Abstract.** This paper reported the pollutant removal performances of a hybrid wetland system for the treatment of aquaculture wastewater. The system consisted of two treatment stages: a subsurface vertical flow (VF) wetland, followed by a horizontal flow (HF). The aquaculture wastewater with the different concentrations such as eutrophy and mesotrophy was treated using hybrid constructed wetland. The experimental results showed that the removal efficiencies of eutrophy aquaculture wastewater achieved 56%, 71%, 73% for nitrite, phosphate and nitrate, respectively. At the same conditions, it can be found that the removal efficiencies of mesotrophy aquaculture wastewater achieved 39%, 74%, 73% for nitrite, phosphate and nitrate, respectively.

## 1. Introduction

Aquaculture has become the fastest growing food production sector and may be practiced in extensive, semi-intensive or intensive system. Aquaculture has gained prominence due to the stabilization of the fish captures and the increase of fish consumption in the last years.Aquaculture has been a fast-growing industry because of significant increases in demand for fish and seafood throughout the world. Total aquaculture production (including aquatic plants) in 2000 was 45.7 million tonnes by weight and US$ 56.5 billion by value [1]. Aquaculture is growing more rapidly than any other segment of the animal culture industry.

Aquaculture wastewater exerts adverse environmental impacts when the effluents from these systems are discharged to receiving waters. The organic matter loading reduces dissolved oxygen levels and contributes to the buildup of bottom sediments and high nutrient loading impairs water quality by simulating excessive phytoplankton production [2].Aquaculture systems produce large quantities of organic matter and nutrients( nitrogen, phosphorus and other elements) that require treatment and/or disposal. The production if 1 ton (1000kg) of live channel catfish releases 1190 kg of dry matter, 60kg of nitrogen and 12kg of phosphorus to the culture water as metabolic wastes [3]. The major contaminants of in aquaculture wastewater can be broadly classified into three categories: organic carbon, nitrogen and phosphorus. Nitrogen compounds such as ammonium and nitrite can be toxic to aquatic life if present at sufficiently high concentration, while nitrate is known to cause 'blue babe syndrome' and is therefore a potential public health threat. Furthermore, nutrients such as nitrogen are well known to stimulate growth of algae and other photosynthetic aquatic life. These nutrients were also found to lead to excessive eutrophication, excessive loss of oxygen resources, and undesirable changes in aquatic ecosystems, which becomes a serious environmental problem [4].Consequently, aquaculture industries look for appropriate and better methods in treating wastewater prior to recirculation or discharge into the receiving water.

Constructed wetlands are the constructed systems which utilize natural processes for treating wastewater that are using soil, vegetation, and microbial communities. They are quite similar to the natural wetlands using natural treatment processes but perform in a controlled environment [5]. Conducted wetland (CW) technology has grown in popularity for wastewater treatment since the early 1970s [6]. The use of constructed wetlands has several advantages over conventional wastewater treatment methods. Initial cost of constructed wetlands for use as primq or secondary treatments is considerably lower than conventional treatments. Annual cost of operation is also considerably less expensive. A paper issued by the Tennessee Valley Authority estimates the savings from constructed wetlands built for small publicly owned treatment works nationwide exceed $2 billion [7]. While constructed wetlands efficiently remove many pollutants from wastewater streams, constructed wetlands also greatly enhance the environment by providing a habitat for vegetation, fish, birds, and other wildlife. Many projects have added nature trails and picnic areas to encourage visitors. Constructed wetland treatment systems are engineered systems that have been designed and constructed to utilize the natural processes but do so within a more controlled environment. Constructed wetlands may be categorized according to the various design parameters, but the three most important criteria are hydrology (open water-surface flow and subsurface flow), type of macrophytic growth (emergent, submerged, free-floating) and flow path (horizontal and vertical). Different types of constructed wetlands may be combined with each other (i.e., hybrid or combined systems) to utilize the specific advantages of the different systems [8-10].One of the major problems with efficient performance of constructed wetlands is clogging of the filtration substrate. Therefore, it is necessary to select the filtration material carefully, distribute the wastewater evenly across the wetland surface, and also select the optimum hydraulic loading rate. Some studies[111] have demonstrated that constructed wetlands can effectively remove the major pollutants from catfish, shrimp and milkfish effluents, including organic matter ,SS, N, and P. CW is not only easy to operate, but also could be built at a relatively lower cost. The results of constructed wetland depend on its microbial activity, hydraulic retention, time, load temperature, and types of vegetation [12]. However, the disadvantages of wetland cannot be ignored[9], the wetland require more land area than other alternative systems, besides, it is not very stable because it requires an extra start-up time until vegetation is well grown, and seasonal uncertainties could damage the wetland for decrease of sunlight and temperature[13]. Although, all of the mentioned systems are efficient in removing contaminants and pathogens from wastewater, however being possessing an elevated evaporation rate as compared to lagoons and ponds, the potential of reusable water in these systems is very limited. To overcome this concern, a configuration named as "hybrid CWs (combination of vertical and horizontal flow)" could be an appropriate alternate by having minimum water loss [14].and improved effluent quality with less total-N concentrations [15].

The main objective of the work was to investigate treatment efficiencies of aquaculture wastewater using hybrid constructed wetland (combination of vertical and horizontal flow).

## 2. Materials and methods

### *2.1. The lab-scale wetland systems*

The lab-scale hybrid systems were built on campus (outdoors) at Hainan Tropical Ocean University. The hybrid systems consisted of two treatment stages: a HF wetland (A), followed by a VF wetland (B), as shown in Fig. 1.The available materials, organic sugarcane bagasse (size 3.4-8.5 mm) and sylhet sand (size 300.0-600.0 μm), were used as the main media in the VF and HF wetlands, respectively. The bagasse was a by-product of sugarcane processing, which contained 40% cellulose, 24% hemicellulose, and 25% lignin. The packed porosity of the media was measured as 65% and 30% for sugarcane bagasse and sylhet sand, respectively.

The volume of each VF and HF wetland is 57cm*37cm*30cm. The same type of locally available *mangrove* was planted into the wetlands. Experimental arrangement of hybrid wetland systems is shown in Fig.1.

**Fig. 1** Diagram of hybrid wetland systems.

### *2.2. Wastewater sampling and analysis*

The seawater sample were collected from Egret park(Sanya), making the water become eutrophic by adding the certain amount of monopotassium phosphate, ammonia chloride, potassium nitrate, sodium nitrite. Table.1 demonstrates two kinds of eutrophic water with different concentration.

**Table.1** Eutrophic body of water of different concentration（mg/mL）

| Nutrient Content | Mesotrophy | Eutrophy |
|---|---|---|
| Total Nitrogen | 0.5 | 1.0 |
| Total Phosphorus | 0.05 | 0.1 |

During experimental analyses, wastewater samples were collected on a weekly basis from the inlet and outlet of each wetland reactor. The water quality parameters measured included COD, total nitrogen (TN), nitrate nitrogen ($NO_3$-N), nitrite nitrogen ($NO_2$-N), ammonia nitrogen ($NH_4$-N), total phosphorus (TP), phosphate ($PO_4$-P), and turbidity. Sample analyses were conducted in accordance with the methods described in 'Water and Wastewater Monitoring and Analysis Methods' (State Environmental Protection Administration of the People's Republic of China, 2002).

## 3. Results and discussion

Nitrate ($NO_3$-N) and nitrite ($NO_2$-N) are the major sources of wastewater contamination and their excessive amount in conjunction with phosphorus can cause eutrophication. In the present work, treatment efficiencies of aquaculture wastewater using hybrid constructed wetland were investigated under the same conditions of mesotrophy and eutrophy.

IOP Conf. Series: Earth and Environmental Science **61** (2017) 012015 doi:10.1088/1755-1315/61/1/012015

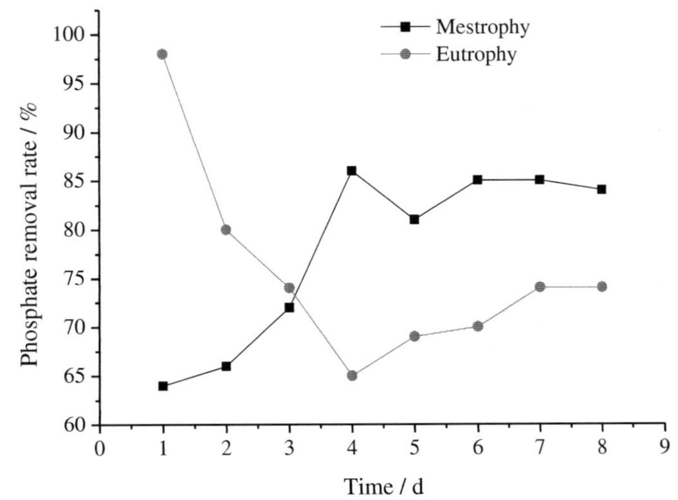


**Fig.2** The removal rate of phosphate

It is obvious that the phosphate removal efficiency has decreased dramatically at first, and then it went up slowly from 5[th] day. By contrast, the removal rate in mesotrophy jumped from 64% to 87%. As shown in Fig.2, it can be found that the phosphate removal efficiency in the condition of mesotrophication is relatively better than that in eutropication.

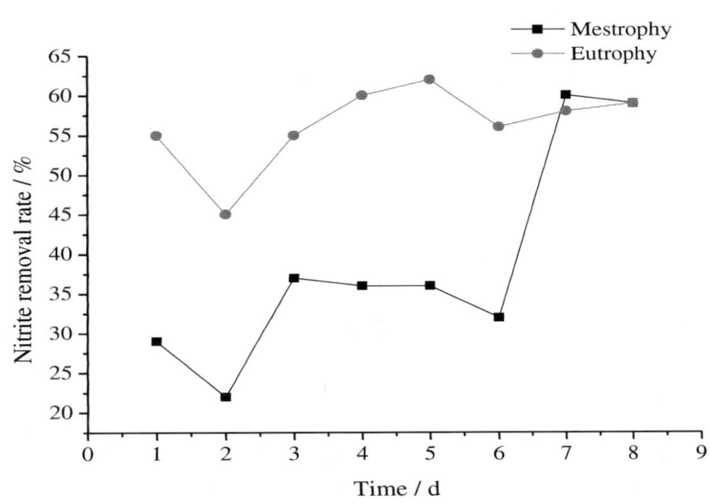


**Fig.3** The removal rate of nitrite

During the process of removing nitrite, the removal rate in eutrophication started with 55%, and then it fluctuated from 2[nd] day to 7[th] day and finally reached 57%, Compared with mesotophication, the removal rate jumped from 33% to 57%. Fig.3 indicated that the nitrite removal rates both in mesotrophication and in eutropication gained good results.

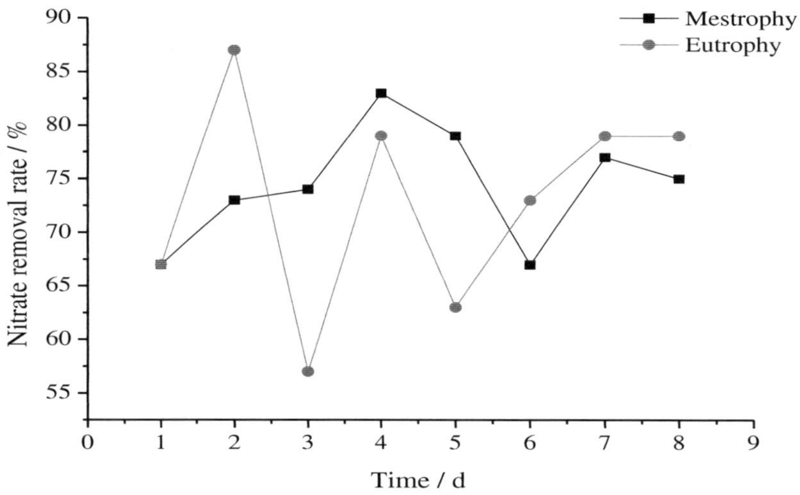


**Fig.4** The removal rate of nitrate

As shown in Fig.4, the removal rates of two kinds of aquaculture wastewater have experienced a fluctuating period from day 1 to day 7 and finally reached the same level of around 77%. It can found that the removal rate of eutrophy is slightly higher than that of mesotrophy.

## 4. Conclusions

Among current wastewater treatment technologies, the constructed wetland technology is considered as an eco-friendly, low cost technology with some distinct advantages such as: low operation and maintenance cost, as well as provide aesthetic value, generate usable plant biomass, and help support wildlife habitat. In the present work, a hybrid wetland system was built to treat two different concentrations of aquaculture wastewater.The experimental results showed that the removal efficiencies of eutrophy aquaculture wastewater achieved 56%, 71%, 73% for nitrite, phosphate and nitrate, respectively. At the same conditions, it can be found that the removal efficiencies of mesotrophy aquaculture wastewater achieved 39%, 74%, 73% for nitrite, phosphate and nitrate, respectively.

## Acknowledgments

The authors would like to greatly acknowledge financial support from Special Project of Hainan Social Development (Grant No.2015SF09).

## References

[1] Food and Agriculture Organizations of the United Nations (FAO), 2014. The State of World Fisheries and Aquaculture, 2002. FAO Fisheries Department, Rome
[2] Nora'Aini A, Mohammad A W, Jusoh A, et al 2005 Treatment of aquaculture wastewater using ultra-low pressure asymmetric polyethersulfone (PES) membrane Desalination 185:317-326
[3] Ghaly A E, Kamal M K and Mahmoud N S 2005 Phytoremediation of aquaculture wasterwater for water recycling and production of fish feed Environmental International 31:1-13
[4] Jang J D, Barford J P, Rennegerg L R 2004 Application of biochemical oxygen demand(BOD) biosensor for optimization of biological carbon and nitrogen removal from synthetic wastwater in a sequencing bacth reactor system Biosensor & Bioelectronics 19:805-812
[5] Vymazal J 2010 Constructed Wetlands for Wastewater Treatment, Water 2:530-549
[6] International Water Association.2002 Constructed Wetlands for Pollution Control. Processes, Performance, Design and Operation; IWA Publishing: London, UK, p. 156

[7] Jia W, Zhang J, Wu J, Xie H, Zhang B 2010 Effect of intermittent operation on contaminant removal and plant growth in vertical flow constructed wetlands: a microcosm experiment Desalination 262:202-208

[8] Vymazal J 2005 Horizontal sub-surfaceflowand hybrid constructed wetlands systems for wastewater treatment Ecol. Eng. 25:478-490

[9] Vymazal J 2008 Constructed wetlands, surface flow. In Encyclopedia of Ecology, Elsevier BV: Amsterdam.

[10] Li H Z, Ye J F, Wang S, Fu W and Xing S W 2008 Applicance and development ofvertical-flow constructed wetland technology for rural wastewater treatment in Shanghai. Environ. Pollut. Control. 30:84-89

[11] Lin YF, Jing S R, Lee DY and Wang T W 2002 Nutrient removal from aquaculture wastewater using a constructed wetlands system. Aquaculture 209: 169-184

[12] El-Khateeb M A, Al-Herrawy A Z, Kamel M M, et al.2009 Use of wetlands as post-treatment of anaerobically treated effluent Desalination. 245(1):50-59

[13] Shutes R B E 2001 Artificial wetlands and water quality improvement Environment International 26: 441-447

[14] Miller D and Semmens K 2002 Waste Management in Aquaculture Agricultural and Resource Economics Program Division of Resource Management College of Agriculture, Forestry, and Consumer Sciences West Virginia University Aquaculture Information Series Publication

[15] Masi F and Martinuzzi N 2007 Constructed wetlands for the Mediterranean countries: hybrid systems for water reuse and sustainable sanitation Desalination. 215: 44-55

[16] Sayadi M H, Kargar R, Doosti M R and Salehi H 2012 Hybrid constructed wetlands for wastewater treatment: a worldwide review. Proc. Int.l Acad. Ecol. Environ. Sci. 2:204-222

IOP Conf. Series: Earth and Environmental Science **61** (2017) 012016 doi:10.1088/1755-1315/61/1/012016

# Distribution characteristics of organochlorine pesticide in the water environment in Lanzhou section of Yellow River

**L Yang[1], X Zhao[*1], J M Shen[2], Z L Chen[2], X C Wang[2] and H R Qiu[1]**

[1]College of Petrochemical Technology, Lanzhou University of Technology, Lanzhou 730050, China;
[2]State Key Laboratory of Urban Water Resource and Environment, School of Municipal and Environmental Engineering, Harbin Institute of Technology, Harbin 150090, China
[*]zhaoxia@lut.cn

**Abstract.** Surface water, surface sediments and suspended particles in the Lanzhou section of Yellow River were collected. After the samples were lyophilised, extracted, concentrated, purified and separated, organochlorine pesticides in the samples were analysed by GC-MS. Results showed that organochlorine pesticide contents in surface water, surface sediments and suspended particles ranged from 28.63 ng/L to 123.2 ng/L, from 0.86 ng/g to 4.51 ng/g and from 23.29 ng/g to 126.14 ng/g, respectively. HCHs, DDTs and HCB were high; among these contents, HCH contents ranged from 1.49 ng/L to 18.1 ng/L, from 0.04ng/g to 1.53 ng/g and from 2.74ng/g to 25.64 ng/g, respectively. DDT contents ranged from 1.49 ng/Lto 18.1 ng/L, from 0.04 ng/g to 1.53 ng/g and from 2.74 ng/g to 25.64 ng/g, respectively. Component analysis results showed that organochlorine pesticide in the Lanzhou section of Yellow River was mainly from early residues or soil after pesticides were applied and long-term weathering occurred. Correlation analysis results showed that total organic carbon was an important factor affecting the distribution of organochlorine pesticide in sediments. Moderate organochlorine pesticide contents were detected in surface water in Lanzhou section of Yellow River compared with other rivers in our country and in other countries. Furthermore, the ecological risk of organochlorine pesticide in surface sediments was low.

## 1.Introduction

As the second longest river in China, Yellow River flows through nine provinces in North China with a length of 5,464km, a basin of 795,000 $km^2$ and average sediment concentration of 2.83 $kg/m^3$. Yellow River is one of the most important water sources of industrial and agricultural applications, as well as human consumption[1]. The Lanzhou section of Yellow River belongs to the upstream of Yellow River with a length of 219.7 km. Although the Lanzhou section is a source of water supply for local urban residents, as well as for industrial and agricultural production, this section is a pollutiondischarge-concentrated area. Approximately 40 km of the river is located in Lanzhou City. As a consequence of concentrated industrial and mining enterprises, as well as dense population, the Lanzhou section of Yellow River receives a large amount of production and household wastewater, which becomes one of the main pollution sources of Yellow River. Wastewater contains a large number of persistent organic pollutants, including organochlorine pesticide[2].

Organochlorine pesticide is a refractory toxic organic pollutant with stable physicochemical properties. With hydrophilic lipophilicity, organochlorine pesticide penetrating the environment easily accumulates in sediments, thereby damaging the ecosystem[3,4]. As such, China has gradually prohibited the use of organochlorine pesticide since 1983. However, organochlorine pesticide can still be detected in several environmental media because of low biodegradability[5].

Further research has been conducted regarding organochlorine pesticide in rivers, estuaries and coastal water, but no report has been presented regarding the distribution of organochlorine pesticide in the Lanzhou section of Yellow River. In this study, contents, components and distribution characteristics of organochlorine pesticide were investigated in surface water, suspended particles and surface sediments of the Lanzhou section of Yellow River; possible sources of these substances were also analysed. Furthermore, the pollution level of organochlorine pesticide in water environment was evaluated. This study aimed to provide a scientific basis for the treatment of organochlorine pesticide in Yellow River.

## 2.Materials and methods

### *2.1. Chemicals*

The following chemicals were used: standard sample of organochlorine pesticide (Supelco Bellefonte, USA), *n*-hexane, dichloromethane, acetone (HPLC grade; J. T. Baker, USA), anhydrous sodium sulphate (heated at 450 °C for 8 h; Chenguang, Tianjin) and silica gel (heated at 150 °C for 4 h; Merck, USA).

### *2.2. Sample collection and study area*

The samples were collected in October 2015. Sampling point distribution is shown in Fig.1 and latitudes and longitudes of the sampling points are listed in Table 1. Water sample (as surface water) was collected from 0.5 m below the surface and placed in 1 L sampling bottle. Surface sediment sample was collected using a grab sampler. Suspended particles were obtained by filtering a largevolume of water sample through 0.45 μm glass fibre filtration film[6]. After the sample was transported to the lab, surface water was stored at 4°C; surface sediments and suspended particles were stored at −20°C[7].

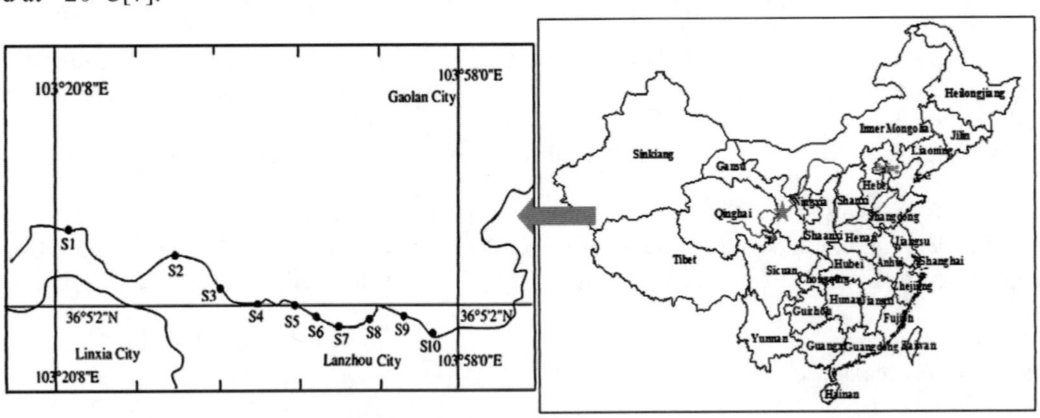


**Fig. 1** Spatial distributions of the 10 sampling sites in Lanzhou reach of the Yellow River in Northwest China

**Table 1** Description of sampling locations in detail

| Station | Location | | | | | |
|---|---|---|---|---|---|---|
| S1 | N36° | 10′ | 10.6″ | E103° | 29′ | 07.8″ |
| S2 | N36° | 08′ | 12.8″ | E103° | 36′ | 45.7″ |
| S3 | N36° | 05′ | 58.9″ | E103° | 40′ | 39.1″ |
| S4 | N36° | 05′ | 15.0″ | E103° | 43′ | 11.6″ |
| S5 | N36° | 04′ | 57.7″ | E103° | 45′ | 56.5″ |
| S6 | N36° | 04′ | 13.1″ | E103° | 47′ | 39.8″ |
| S7 | N36° | 03′ | 47.9″ | E103° | 49′ | 08.4″ |

IOP Conf. Series: Earth and Environmental Science **61** (2017) 012016 doi:10.1088/1755-1315/61/1/012016

| | | | | | | |
|---|---|---|---|---|---|---|
| S8 | N36° | 04′ | 25.3″ | E103° | 51′ | 33.8″ |
| S9 | N36° | 04′ | 07.3″ | E103° | 54′ | 10.4″ |
| S10 | N36° | 03′ | 10.1″ | E103° | 56′ | 52.8″ |

*2.3. Sample extraction*

For surface water sampling, 1 L of water sample was placed in a separatory funnel and 30 mL of methylene chloride was added. The funnel was shaken for 2 min and left to stand to allow the mixture to separate into layers. The organic phase in the sublayer was dried with anhydrous sodium sulphate and then collected in a round-bottomed flask. This procedure was repeated twice. The extracting solutions were combined and purified through silica gel column chromatography. The solution was concentrated to 2 mL to 3 mL.n-Hexane (10 mL) was added to replace the solvent. The concentration of the solution was adjusted until the final volume was 1 mL.For surface sediment and suspended particle sampling, the samples were placed in a Soxhlet extractor. The mixture (200 mL) containing acetone and n-hexane with a volume ratio of 1:1 was added to the extractor and then subjected to extraction for 48 h. After extraction was completed, the extracted solution was purified by silica gel column chromatography; the concentration was adjusted to 2 mL to 3 mL.n-Hexane (10 mL) was added to replace the solvent. The concentration of the solution was further adjusted until the final volume reached 1 mL[8].

*2.4. Analytical procedure*

Samples were detected with an Agilent 6890N-5975B mass spectrometer coupled to a gas chromatograph. An HP-5MS capillary chromatograph column (60 m × 0.25 mm × 0.25 μm) was also used. The injection temperature was 250 °C with asplitless injection mode. The flow rate of high-purity helium as carrier gas was 0.8 mL/min. The temperature of the transmission line was 280 °C. The temperature program of gas chromatography was set as follows: initial column temperature, 80 °C for 1 min; increase to 160 °C at a speed of 10 °C/min; increase to 250 °C at a speed of 2.0 °C/min; and final increase to 300 °C at a speed of 30.0 °C/min; temperature was then maintained at 300 °C for 5 min. No split injection of 2.0 μL was set for current flow. The temperature of ion source was 150 °C. The electron energy was 70 eV. The scanning quality scope was m/z 50–500[9,10]. The retention times of chromatogram peaks of standard samples were applied to qualitatively analyse organochlorine pesticide in actual samples[11].

The total organic carbon (TOC) of surface sediments was determined using a TOC-5000A analyser.

*2.5. Quality control and assurance*

The samples were monitored to determine recovery rate by using the following quality references and parameters: method blank, spike blank, duplicate and standard material; data were then corrected on the basis of recovery rate.

## 3. Results and discussion

*3.1. Organochlorine Pesticide in Surface Water*

Fig. 2 shows the total content distribution of organochlorine pesticide in surface water of 10 sampling points in the Lanzhou section of Yellow River. The organochlorine pesticide content ranged from 28.63 ng/Lto 123.2 ng/L with an average of 72.13 ng/L. Furthermore, the maximum content was detected at S8 and the minimum content was found at S1 (Fig. 2). The organochlorine pesticide content at S4 to S8 was high possibly because these sections were close to the urban centre, where urban wastewater and industrial pollution were the main sources of high organochlorine pesticide contents.

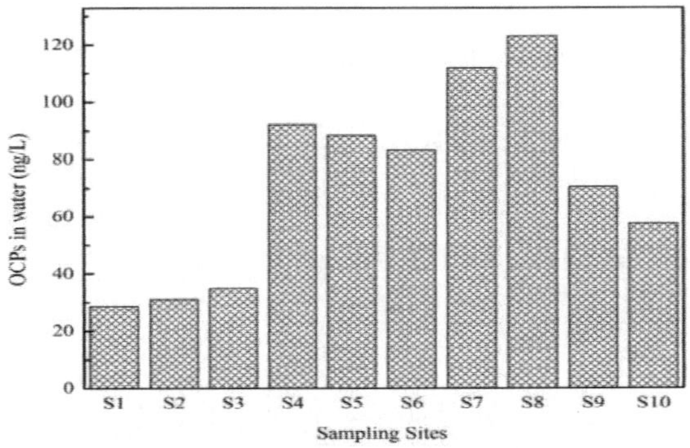


**Fig.2** Concentration of OCPs in water in Lanzhou Reach of Yellow River

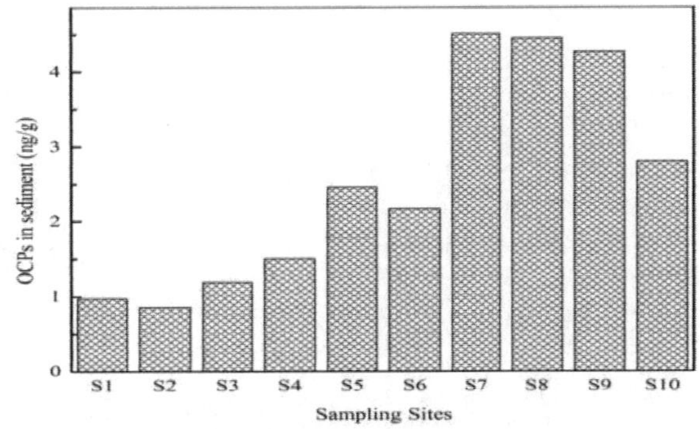


**Fig.3** Concentration of OCPs in sediment in Lanzhou Reach of Yellow River

*3.2. Organochlorine Pesticide in Surface Sediments*

Fig. 3 shows the total content distribution of organochlorine pesticide in surface sediments of 10 sampling points in the Lanzhou section of Yellow River. The organochlorine pesticide content ranged from 0.86 ng/Lto 4.51 ng/L with an average of 2.52 ng/g. The maximum content was found at S7 and the minimum value was detected at S2. This result showed a gradual increasing trend from upstream to downstream. This observation occurred because a large amount of organochlorine pollutants flowed into Yellow River via industrial and urban wastewater.

*3.3. Organochlorine Pesticide in Suspended Particles*

Fig. 4 shows the total content distribution of organochlorine pesticide in suspended particles of 10 sampling points in the Lanzhou section of Yellow River. The organochlorine pesticide content ranged from 23.29 ng/Lto 126.14 ng/L with an average of 65.76 ng/g. The maximum content was found at S6 and the minimum content was detected at S1. Overall, the content of organochlorine pesticide in suspended particles was similar to that in surface water but higher than that in surface sediments. This result was possible because suspended particles were characterised by a large surface area, high viscosity and high organic matter content; thus, organochlorine pesticide exhibits strong absorption ability[12].

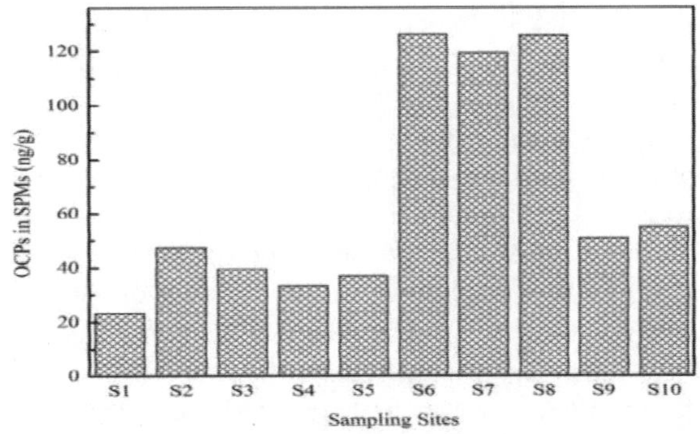


**Fig.4** Concentration of OCPs in SPMs in Lanzhou Reach of Yellow River

*3.4. Comparison of Organochlorine Pesticide in Yellow River with that in Other Rivers and Water Bodies*

Table 2 lists the total HCH and DDT residues in surface water and surface sediments in the Lanzhou section of Yellow River were compared with those of other rivers in our country and in other countries. The HCH content in surface water in Yellow River was higher than that in Liao River but lower than that in JiulongRiver; the HCH content in surface water in Yellow River was equal to that in Suzhou River. Likewise, the DDT content in surface water in Yellow River was higher than that in

Liao River but lower than that in Suzhou River and JiulongRiver. HCH and DDT contents in rivers in our country were higher than those in rivers in other countries. The HCH content in surface sediments was lower than that in ParramataRiver, KaveriRiver, HaiheRiver and Pearl River but almost equivalent to that in Huangpu River. Likewise, the DDT content was low. In general, organochlorine pesticide content in surface water in the Lanzhou section of Yellow River was moderate and that in surface sediments was low.

**Table 2** Comparisons of HCHs and DDTs in water and sediment from rivers in other areas

| Medium | Location | HCHs | DDTs | Reference |
|---|---|---|---|---|
| Water (ng/L) | Ebro River, Spain | 3.1 | 3.4 | Fernandez et al. (1999) |
| | St.Lawrence River, Canada | 0.06 | 0.9-22 | Pham et al. (1993) |
| | Suzhou River, China | 17-90 | 17-99 | Hu et al. (2005) |
| | Liao River, China | 7.6-31 | 0.1-4.2 | Zhang et al. (2002) |
| | Jiulong River, China | 32-130 | 19-97 | Zhang et al. (2001) |
| | Lanzhou Reach of Yellow River, China | 11-88 | 1.5-18 | This study |
| Sediment(ng/g) | Parramata River, Australia | 7.7 | 26 | Iwata et al. (1994) |
| | Kaveri River, India | 4.4-158 | 0.7-4.9 | Rajen et al. (1999) |
| | Haihe River, China | 7.8-11 | 9.5-12 | Wu et al. (1999) |
| | Pearl River, China | 1.4-3.8 | 2.2-5 | Luo et al. (2005) |
| | Huangpu River, China | 0.14-0.77 | 0.7-4.4 | Hu et al. (2005) |
| | Lanzhou Reach of Yellow River, China | 0.18-1.1 | 0.4-1.5 | This study |

### *3.5. Composition Characteristics of Organochlorine Pesticide*

*3.5.1. Composition Characteristics of HCHs.* The HCH contents in surface water, surface sediments and suspended particles in the Lanzhou section of Yellow River ranged from 1.12 ng/Lto 88.43 ng/L, from 0.18 ng/g to 1.1 ng/g and from 2.47 ng/g to 52.08 ng/g with averages of 42.55 ng/L, 0.54 and 20.26 ng/g, respectively. The average percentages of α-HCH, β-HCH and γ-HCH in surface water were 40%, 27% and 33%, respectively. The average percentages of α-HCH, β-HCH and γ-HCH in surface sediments were 14%, 46% and 40%, respectively. The average percentages of α-HCH, β-HCH and γ-HCH in suspended particles were 5%, 19% and 76%, respectively. The ratios of β-HCH to γ-HCH at most sampling points of surface water were ≤1; by contrast, the ratios of all sampling points of suspended particles were <1, indicating the presence of a new pollutant in the Lanzhou section of Yellow River or the application of lindane pesticide[13]. The ratios of β-HCH to γ-HCH in surface sediments were mostly between 1 and 2, indicating that HCHs in surface sediments was mainly from the residues of organochlorine pesticide after these substances undergo long-term degradation and lindane pesticide is applied[14]. The distribution characteristics of HCHs in surface water, surface sediments and suspended particles were compared. Our results showed that α-HCH content in surface water was higher than that in surface sediments and suspended particles; likewise, β-HCH content in surface sediments was higher than that in surface water and suspended particles. Furthermore, γ-HCH content was the highest in suspended particles. This condition was observed because α-HCH is highly volatile and easily present in water phase; β-HCH attaches to and accumulates in surface sediments with stable properties; γ-HCH easily attaches to microporous structures of suspended particles[15,16].

*3.5.2. Component Characteristics of DDTs.* The DDT contents in surface water, surface sediments and suspended particles in the Lanzhou section of Yellow River ranged from 1.49 ng/Lto 18.1 ng/L, from 0.04 ng/g to 1.53 ng/g and from 2.74 ng/g to 25.64 ng/g with averages of 6.38 ng/L, 0.7 ng/g and 14.33 ng/g, respectively. Under anaerobic condition, DDT is mainly degraded into DDD; under aerobic condition, DDT is mainly degraded into DDE. Therefore, DDT content is likely maintained at a high level if DDT is continuously applied; without additional DDT, DDT content is possibly reduced, but the corresponding degradation product is likely increased in a continuous pattern. Thus, the ratios of DDT to DDD + DDE and DDD to DDE can be used to determine whether new DDT input is added and to trace the degrading environment of DDTs. At a DDT-to-DDD + DDE ratio of <2, early residues or long-term weathering residues are retained after pesticides are applied[17-19]. Table 3 shows the ratios of DDT to DDD + DDE of all sampling points in surface water and surface sediments. All of the ratios were <2, indicating that DDTs were mainly obtained from early residues or long-term weathering residues after pesticides were applied and no new DDTs were added. At the same time, the ratios of DDD to DDE in surface sediments were mostly <1 and DDE was the main degradation product; this result indicated that DDTs in surface sediments underwent aerobic biodegradation, which was related to a large flow and great flow disturbance in this area.

**Table 3** The ratios of DDT / (DDD + DDE) in each sampling point of water and surface sediments

| Locations | S1 | S2 | S3 | S4 | S5 | S6 | S7 | S8 | S9 | S10 |
|---|---|---|---|---|---|---|---|---|---|---|
| water | <2 | <2 | <2 | <2 | <2 | <2 | <2 | <2 | <2 | <2 |
| sediment | <2 | <2 | <2 | <2 | <2 | <2 | <2 | <2 | <2 | <2 |

*3.5.3. Ecological Risk Assessment on Organochlorine Pesticide.* No unified pollution assessment standard has been developed for the risk assessment of organic pollution. Long et al. [20] proposed the use of ERL and ERM to evaluate the risk degree of organic pollution in surface sediments. If pollutant concentration was smaller than ERL, negative ecological effects were rarely produced; if pollutant concentration was between ERL and ERM, negative ecological effects were occasionally produced; if pollutant concentration was greater than ERL, negative ecological effects were often produced. Based on the risk assessment value of surface sediments, the ecological risk of OCPs in surface sediments in the Lanzhou section of Yellow River was assessed (Table 4). The OCP contents at all sampling points were less than ERL, indicating that OCPs exhibited low risk in surface sediments in the Lanzhou section of Yellow River.

**Table 4** Effect range low (ERL) and effect range median (ERM) guideline values for OCPs in sediment

| | ERL (ng/g) | ERM (ng/g) | Sites number | | |
|---|---|---|---|---|---|
| | | | <ERL | ERL-ERM | >ERM |
| DDT | 1 | 7 | 10 | 0 | 0 |
| DDD | 2 | 20 | 10 | 0 | 0 |
| DDE | 2.2 | 27 | 10 | 0 | 0 |
| DDTs | 1.58 | 46.1 | 10 | 0 | 0 |

*3.5.4. Correlation between Organochlorine Pesticide and TOC in Surface Sediments.* Fig. 5 shows the correlation between organochlorine pesticide content and TOC in surface sediments. Organochlorine pesticide content was significantly correlated with TOC in surface sediments ($p = 0.05$); this result indicated that TOC was an important factor affecting organochlorine pesticide distribution in surface sediments. TOC corresponded to organic matter content in sediments. Organic matter content influences organochlorine pesticide in two mechanisms. On the one hand, organochlorine pesticide strongly affects the concentration of organic matter because of its low water solubility and high fat solubility; on the other hand, organic matter indicated high microbial content, which elicit biodegradation effects on organochlorine pesticide[21].

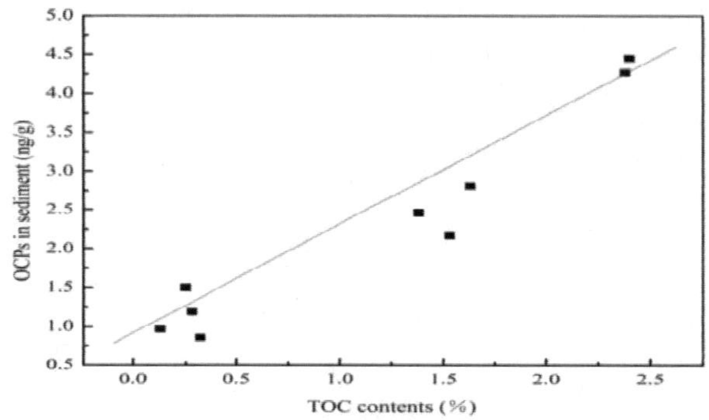


**Fig.5** Relation between concentrations of total OCPs and TOC contents in sediment in the Yellow River of Lanzhou

## 4. Conclusions

In this study, organochlorine pesticide was analysed in surface water, surface sediments and suspended particles in the Lanzhou section of Yellow River. Based on the results, our conclusions are listed as follows: 1) Organochlorine pesticidecontent in surface water, surface sediments and suspended particles ranged from 28.63 ng/L to 123.2 ng/L, from 0.86 ng/g to 4.51 ng/g and from 23.29 ng/g to 126.14 ng/g with averages of 72.13 ng/L, 2.52 ng/g and 65.76 ng/g, respectively; HCH and DDT contents were high. 2)α-HCH and γ-HCH were mainly present in surface sediments and suspended particles, respectively. The ratios of β-HCH to γ-HCH at most sampling points indicated that HCHs in the Lanzhou section of Yellow River were mainly residues of long-term organochlorine pesticide degradation and lindane pesticide application. 3) DDTs in the Lanzhou section of Yellow River were mainly from early residues or long-term weathering residues after pesticides were applied; DDTs in surface sediments were mainly produced by aerobic biodegradation. 4) Organochlorine pesticide content in surface water in the Lanzhou section of Yellow River was moderate compared with that in other rivers in our country and in other countries; organochlorine pesticide content in surface sediments was low, indicating an overall low ecological risk. 5) Organochlorine pesticide content in surface sediments was closely correlated with TOC. TOC was an important factor affecting organochlorine pesticide distribution in surface sediments.

## Acknowledgment

This study was supported by the National Natural Science Foundation of China (21667017), the Provincial Natural Science Foundation of Gansu (1506RJZA102), the College Students Training Project for Creative and Entrepreneurship of China(201510731005).

## References

[1] Gao H, Bao W Y, Zhang S G. Research on pollution chemistry and ecotoxicology in high sand content rivers. The Yellow River Water Conservancy Publisher, Zhengzhou, in Chinese, 2001.
[2] Chen J, He D, Zhang N. Characteristics of and human influences on nitrogen contamination in Yellow River system, China, Environmental Monitoring and Assessment. 2004, 93: 125-138.
[3] Szlinder-Richert J, Barska I, Mazerski J. Organochlorine pesticides in fish from the southern Baltic Sea: levels, bioaccumulation features and temporal trends during the 1995-2006 period, Marine Pollution Bulletin. 2008, 56: 927-940.
[4] Zhang L F, Dong L, Shi S X. Organochlorine pesticides contamination in surface soils from two pesticide factories in Southeast China, Chemosphere 2009, 77: 628-633.
[5] Wang F, Jiang X, Bian Y R. Organochlorine pesticides in soils under different land usage in the Taihu Lake region, Journal of Environmental Sciences-China, 2007, 19: 584-590.
[6] Zhang L F, Dong L, Ren L J. Concentration and source identification of polycyclic aromatic hydrocarbons and phthalic acid esters in the surface water of the Yangtze River Delta, Journal of

Environmental Sciences-China, 2012, 24: 335-342.
[7] Montone R C, Taniguchi S, Weber R R. Polychlorinated biphenyls in marine sediments of Admiralty Bay, King George Island, Antarctica, Marine Pollution Bulletin, 2001, 42: 611-614.
[8] Hu L M, Zhang G, Zheng B H. Occurrence and distribution of organochlorine pesticides (OCPs) in surface sediments of the Bohai Sea, China, Chemosphere 2009, 77(5): 663-672.
[9] Li H P, Li G C, Jen J F. Determination of organochlorine pesticides in water using microwave assisted headspace solid-phase microextraction and gas chromatography, Journal of Chromatography A, 2003, 1012: 129-137.
[9] Na T, Fang Z. The status of pesticide residues in the drinking water sources in Meiliangwan Bay, Taihu Lake of China, Environmental Monitoring and Assessment, 2006, 123: 351-370.
[11] Vagi M C, Petsas A S. Determination of organochlorine pesticides in marine sediments samples using ultrasonic solvent extraction followed by GC/ECD, Desalination 2007, 210: 146-156.
[12] Willett K L, Ulrich E M, Hites R A. Differential toxicity and environmental fates of hexachlorocyclohexane isomers, Environmental Science and Technology. 1998, 32: 2197-2207.
[13] Carvalho P N, Rodrigues P N, Basto M C, Vasconcelos M T. Organochlorine pesticides levels in Portuguese coastal areas, Chemosphere 2009, 75: 595-600.
[14] Yang H, Xue B, Yu P. Residues and enantiomeric profiling of organochlorine pesticides in sediments from Yueqing Bay and Sanmen Bay, East China Sea, Chemosphere 2010, 80: 652-659.
[15] Walker K. Factors influencing the distribution of lindane and other hexachlorocyclohexanes in the environment, Environmental Science and Technology. 1999, 33: 4373-4378.
[16] Willett K L, Ulrich E M, Hites R A. Differential toxicity and environmental fates of hexachlorocyclohexane isomers, Environmental Science and Technology. 1998, 32: 2197-2207.
[17] Mohn W M. Microbial reductive dechlorination, Fems Microbiology Reviews 1992, 56：482-507.
[18] Tan L, He M, Men B, Lin C. Distribution and sources of organochlorine pesticides in water and sediments from Daliao River estuary of Liaodong Bay, Bohai Sea (China), Estuarine, Coastal and Shelf Science. 2009, 84: 119-127.
[19] Mohammed A, Peterman P, Echols K. Polychlorinated biphenyls (PCBs) and organochlorine pesticides (OCPs) in harbor sediments from Sea Lots, Port-of-Spain, Trinidad and Tobago, Marine Pollution Bulletin 2011, 62: 1324-1332.
[20] Long E R, MacDonald D, S. L. Smith S L, Calder F D. Incidence of adverse biological effects within ranges of chemical concentrations in marine and estuarine sediments, Journal of Environmental Management. 1995, 19: 81-97.
[21] Lee K T, Tanabe S, Koh C K H. Distribution of organochlorine pesticides in sediments from Kyeonggi Bay and nearby areas, Korea, Environmental Pollution 2001, 114: 207-213.

IOP Conf. Series: Earth and Environmental Science **61** (2017) 012017 doi:10.1088/1755-1315/61/1/012017

# Study on exploitation and utilization of geothermal resources in Henan Province

**You-li FENG[1 a], Ning MAO[1], Li-jing YU[1], Cong WANG[1], Meng-tao WANG[1]**

School of Resources and Environment, Henan Polytechnic University, P.R.China
fengyouli66@163.com

**Abstract.** Geothermal is a kind of mineral resources with valuable comprehensive. It has many functions and wide use. In this paper, the present situation and existing problems of geothermal development and utilization in Henan Province have been discussed. On this basis, countermeasures including geothermal management and technical aspects have been put forward for rational, efficient exploitation and utilization of geothermal resources in Henan Province.

## 1. The distribution of geothermal resources in Henan Province

Geothermal resources in Henan Province can be divided into two types: one is low geothermal anomalies. Thermal reservoirs buried shallow with high temperature. It mostly in the form of springs exposed to the surface. The other belongs to normal geothermal warming. Thermal reservoirs are buried deep and must be exploited by drilling. [1]

From the geographical distribution, the development and utilization of geothermal resources are wider in Henan Province (table 1). Many cities and counties have developed and utilized it. The water temperature is 40 ~ 50 °C, and some can reach 60 °C or higher. The distribution of hot paddy field: the hot paddy field with high temperature is mainly distributed in the foothill where the thickness of the sediments is less, and the hot paddy fields with lower temperature are mostly distributed in the plain areas where the thickness of the sediments is larger.

## 2. Current situations of exploitation and utilization of geothermal resources in Henan Province

The development and utilization of geothermal in Henan Province can be divided into two stages: First, in the 1980s, the main development and utilization were low geothermal anomalies, and more performed as hot springs, developing recuperation and tourism. Second, since the 1980s, the development and utilization are mainly about the normal geothermal hot water [2].

**Table 1 Distribuition of geothermal resources in Henan Province**

| Area of geothermal resources | Area of thermal storage/km$^2$ | Types of geothermal resources | Depth of well /m | Temperature/°C |
|---|---|---|---|---|
| Plain area in North Henan | 5680.06 | subsiding basin | 1000-1300 | 40-70 |
| upwarping region in Western Henan | 5980.51 | Upwarping, subsiding | 0-2500 | 40-90 |

| | | | | |
|---|---|---|---|---|
| Middle and South Henan | 23437.56 | subsiding basin | 1000-2673 | 42-63 |
| Plain area in Eastern Henan | 6952.57 | subsiding basin | 1000-1600 | 40-60 |

According to incomplete statistics the exploitation of geothermal resources in Henan Province is $4419.28 \times 10^4$ $m^3$ / a till September 2016. The pore-type geothermal water accounted for 84.76% with temperature above 40 °C, the geothermal water accounted for 22.17%. The exploration type of geothermal resources is direct supply and ejectment. It's for the bath, city water supply, breeding, mineral water production, heating, etc.; their proportions were 56.44%, 38.13%, 1.64%, 0.26%, 0.24% respectively, and other uses is 3.29%(Fig.1).

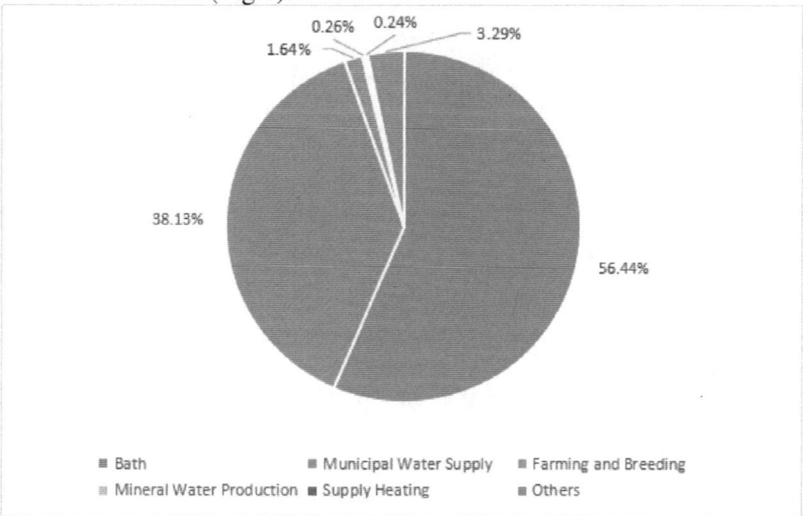


**Fig1 The distribution of geothermal usage in Henan Province**

The development of geothermal resources in Henen is mainly concentrated in the Huanghuaihai Plain including Zhengzhou, Kaifeng, Xinxiang, Anyang, Puyang, Zhoukou, Xuchang, Luohe and Luoyang as the main cities, followed by Luohe- north Zhoukou, Xinxiang-Puyang. In addition, the rest are sporadic mining[3]. It is almost in the blank stage in the development and utilization of dry and hot rock in Henan. Dry-hot rock is a renewable special clean geothermal resources as one of geothermal resources. Experiments in developed countries have shown that the most viable technique for exploiting dry-hot rocks is the EGS (Enhanced Geothermal System), which is not limited by the natural geothermal field because the whole process is carried out in a closed system, and the key technology of heat recovery is to form heat exchange system in impermeable dry and hot rock mass.

## 3. Technologies of geothermal resources development and utilization in Henan Province

Nowadays, according to the characteristics of geothermal resources in Henan Province, geothermal wells is main geothermal resources for the development of drilling technology.

There are four main types in geothermal drilling: 1) Geological exploration wells used to get depth and fault structure of the stratigraphic section in the exploration area. 2) After a series of surveys and analyzes, exploration and mining wells are assumed that the exploration area has the conditions for the formation of underground heat reservoirs, but there are some important data to be identified. Under these circumstances, the exploration wells are arranged. 3) Exploitation wells. When the geothermal field is found and has been controlled in its scope, drilling for the purpose of mining geothermal resources; 4) Recharge wells. With the development of the geothermal field, the water level of the production layer will gradually decline. It needs recharge wells to recharge the water to the hot reservoir, in order to maintain the energy and water column of heat reservoir, and make geothermal fields stably product for a long term[6].

However, comparing with the international, geothermal resources development technology is less backward in Henan. The enhanced geothermal system (EGS) studied by geothermal research institutions in the United States has become the most promising geothermal technology. EGS is different from the traditional method of exploiting the geothermal resources in the permeable rock stratum by manually creating underground heat reservoirs and collecting energy from dry and hot rock layers which don't have economical value. Therefore, Chinese geothermal development should be deeply progress to carry out Chinese "Enhanced Geothermal System (EGS)" development and utilization of pre-research and preparation.

## 4. Problems of geothermal exploitation and utilization in Henan Province

### *4.1. Management issues*

(1) Geothermal authority is not clear. Government and many related departments do not really implement a unified management. Management is very confusing. Chaotic and random mining and disorder are very common. Local government disturb and harass construction.

(2) Lack of exploration funds. Insufficient funds in the pre-exploration of geothermal resources leading to geothermal geological information is not detail, incomplete and geothermal geological foundation work is poor.

(3) Comprehensive utilization level is low. Geothermal resources is rich in Henan, but the utilization is single and disperse. There is almost zero geothermal field comprehensive utilization and cascade utilization, resulting in waste of resources. There are 1/4 of the land in Henan has geothermal resources and nearly half are used for bathing, public heating is less than 1%. Its energy resources utilization structure is irrational resulting in thermal pollution and damage of groundwater environment.

(4) Heat stored value evaluation are not inaccuracy. The government has insufficient funds in the pre-exploration of geothermal resources, which lead to the data released currently are mostly based on private survey data, coupled with the estimates by experts. In fact, no one knows the true geothermal stored value, which affected the province's geothermal energy development and optimization and adjustment of energy structure.

### *4.2. Technical aspects*

(1) The development of geothermal resources in Henan Province is mainly concentrated in the 1000 ~ 1200 m thermal reservoir in Cenozoic Era, because of loose layer, low technical content, low difficult in construction and low investment. The base rock area has high investment and risk, high technical level and equipment requirements, so most units do not want to develop bedrock and deep geothermal resources. This has led to in the bedrock area geothermal development in a low exploitation stage in Henan[7].

(2) It is single that the means of geothermal resources exploration technology. The pre-exploration accuracy of blank area is low. Geothermal geology and geothermal drilling technology develop are slow. Especially in complex formation, it has low drilling efficiency, high cost and high accident rate.

(3) It is uneven that the level of geothermal drilling technology. The team anti-risk ability is low. Construction equipment and personnel are scattered. So they can't unite to share information, and lower prices with each other. The internal construction technology standards and processes are various. Many advanced formation crafts and technologies of wells can't be timely promoted to use and communicated, resulting in geothermal resources exploration and development can't break through.

## 5. Countermeasures of geothermal exploitation and utilization in Henan Province

Aiming at the present situation and existing problems of Henan geothermal resources development and utilization, the following countermeasures are put forward.

(1) To establish geothermal management, unify management of geothermal resources, and be responsible for the province's geothermal energy resources, scientific management and planning and

confirming geothermal properties, specifying geothermal exploration market. To develop "Henan geothermal resources exploration and development management approach" As soon as possible, "Henan shallow geothermal energy exploration and development management approach", and develop relevant preferential policies and measures to encourage new energy exploration and development. At the same time, specify geothermal resources exploration and development team, and implement the construction team access system to avoid the quality of wells formation, the quality problems and series problems of geological environment caused by it, like groundwater pollution and land subsidence.

(2) Set up specialized geothermal exploration team, improve the quality of the corresponding staff, integrate geothermal exploration equipment, personnel, and unify management and schedule in order to provide hardware support for the exploration and development of geothermal resources to achieve a major breakthrough.

(3) Improve the approval and supervision mechanism, increase the investigation and punishment. Government must take strong measures to prohibit strictly "random, arbitrary, arbitrary charges" phenomenon continues to spread in order to develop this new energy resources scientifically and rationally,.

(4) It is relatively simple for geothermal development technology, and it is blank or low exploitation stage in the dry hot rock area geothermal development in Henan Province. Therefore, we should actively introduce foreign advanced technology EGS, the technology has become the most promising geothermal technology.

(5) Develop "Henan geothermal energy development and utilization planning." As for the region where geothermal development and utilization of a higher degree should be designated as key protected areas. Encourage exploration and development of 1600 ~ 3000m geothermal energy resources. It should be classified as general protected areas, and can be moderately developed for the region where geothermal development and utilization of the general level. Other cities should be designated as encouraging development zone combined with their own conditions, developing and utilizing of Neogene and Paleogene geothermal energy resources.

(6) Widen the industrial chain to create geothermal brand. Geothermal is also a tourist resource. In some areas with rich geothermal resources and a certain geological phenomenon, construct "geothermal geology parks" or "spa health care resort parks". Use geothermal brands to promote tourism, real estate, health care and other industries by leaps and bounds.

(7) Increase core technology researches of exploration and development of the geothermal energy resources to solve the technical problems of geothermal exploitation in the re-emergence and avoid the land subsidence caused by decline of geothermal water level. We should be scientifically plan and rationally develop. Especially in the exploration of ultra-deep geothermal resources and drilling technologies, we must achieve breakthroughs and independent innovation.

## 6. Conclusion

Geothermal resources is a renewable clean energy. It can be applied to people's lives, thereby effectively reduce the consumption of civilian resources and pollution, ease the pressure on lack of energy and supply, and make a great contribution for the national energy-saving emission reduction policy. The development and utilization of geothermal resources Henan Province are wider. Many cities and counties are developing and utilizing it, mainly concentrated in Zhengzhou, Kaifeng, Xinxiang and other major cities. The development technology of geothermal resources in Henan Province is mainly geothermal drilling wells, but in the process of geothermal development, there are many problems, most of which mainly in the aspects of management and technical. Put forward the corresponding suggestions and solutions to these two problems. In the 21st century with rapid economic development, it is believed that the development and utilization of geothermal resources in Henan Province will be more and more extensive and the efficiency will be higher and higher.

## References

[1] Baoyu Wang. Henan Geothermal Resources Revelopment and Utilization of the Status and

Prospects [J], Geology and Prospecting, 1999.35 (1): 47-48.
[2] Qingli Wang, Guan Hua, et al. Geological Resources Development and Utilization of Henan Province and Countermeasures [J], Geological Prospecting, 2003.18 (3): 200-202.
[3] J.H Wang. Analysis of Geothermal Resources Development and Protection in Henan Province [J], Groundwater, 2009.31 (4): 45-47.
[4] Jialing Zhu, etc.A Review of Geothermal Energy Resources,Development,and Applications in China:Current Status and Prospects[J],Energy, 2015, 93:466-483.
[5] Hannes Hofmann,etc.A Hybrid Discrete/Finite Element Mdeling Study of Complex Hydraulic Fracture Development for Enhanced Geothermal Systems(EGS) in Granitic Basements[J],Geothermics, 2016,64:362-381.
[6] Zongjun Gao, Hong Cao, et al. Geothermal Water Resources Development and Environmental Protection [J], Groundwater, 2009.31 (1): 78-83.
[7] Yubei Lu, Gubin Zhang, Ying Chen. Research on Present Situation and Problems of Geothermal Resources Development and Utilization in Henan Province [J]. Exploration Engineering (Rock & Soil Drilling and Tunneling), 2010.37 (10): 35-39.
[8] Murad Abuaisha,Benjamin loret,David Eaton.Enhanced Geothermal Systems(EGS):Hydraulic Fracturing in a Thermo-poroelastic Framework[J],Journal of Petroleum Science and Engineering, 2016,146:1179-1191.
[9] Joe Lovenitti,etc.The Basin and Range Dixie Valley Geothermal Wellfieled,Nevada,USA-A Test Bed for Developing an Enhanced Geothermal System Exploration Favorability Methodology[J],Geothermics, 2016,63195-209.
[10] Wenjiong Cao,Wenbo Huang,Fangming Jiang.A Novel Thermal-hydraulic-mechanical Model for the Enhanced Geothermal System Heat Extraction[J],International Journal of Heat and Mass Transfer, 2016,100:661-671.

# Analysis of electric vehicle extended range misalignment based on rigid-flexible dynamics

**Xiaowei Xu, Mingliang Lv, Zibo Chen, Wei Ji, Ruiceng Gao**

School of Automobile and Traffic Engineering, Wuhan University of Science and Technology, Wuhan 430081, China
xxw15@163.com

**Abstract**. The safety of the extended range electric vehicle is seriously affected by the misalignment fault. Therefore, this paper analyzed the electric vehicle extended range misalignment based on rigid-flexible dynamics. Through comprehensively applied the hybrid modeling of rigid-flexible and the method of fault diagnosis of machinery and equipment comprehensively, it established a extender hybrid rigid flexible mechanical model by means of the software ADAMS and ANSYS. By setting the relevant parameters to simulate the misalignment of shafting, the failure phenomenon, the spectrum analysis and the evolution rules were analyzed. It concluded that 0.5th and 1 harmonics are considered as the characteristic parameters of misalignment diagnostics for electric vehicle extended range.

## 1. Introduction

Owing to the driving range of Electric Vehicles, the extended range plays an important role in the electric vehicles development. At this stage, It has been recognized as important electric vehicle products by many car companies from the prospect of the markets advance[1,2]. However, because of high degree of electrification and continuous improvement of the power system, the program of electric vehicles appear various troubles in operation[3,4]. Among them, the shaft misalignment fault of extended range device accounts for much more than 60% in mechanical fault. Meanwhile, the shaft misalignment will lead to the system radial and axial vibration and affect the system stability [5].

Nevertheless, in analysis of the traditional multi-rigid body dynamics, the simulation results can not reflect the actual crankshaft working condition. The reason maybe not considered the increase components as a continuous elastic body. In this paper, taking a diesel generator as the research subject, on the basis of virtual prototyping software ADAMS and finite element analysis software ANSYS, it established the extender hybrid rigid flexible mechanical model[6]. By setting the relevant parameters to simulate the misalignment of shafting, it conducted frequency analysis and researched the failure phenomena and evolution rules of misalignment faults. At last, it extracted the characteristic parameters of misalignment diagnostics for electric vehicle extended range.

## 2. Rigid-flexible hybrid dynamic model

Firstly, it imported the three dimensional CATIA model into the mechanical system simulation software MSC.ADAMS. Secondly, it regarded the key component as linear elastic body by means of the finite element method and combined with other rigid body model through nonlinear connection with boundary Interaction. Then, it defined the material properties of each part and the connection relationship of the kinematic pairs. Finally, it obtained the multi-body dynamic rigid body model by applying external excitation and driving as shown in figure 1.

**Fig.1** Rigid-flexible hybrid dynamic model

## 3. Simulation analysis

### *3.1 Normal state*

The diagnostic characteristic parameters is extracted by comparing and analyzing the normal state and fault condition based on the rigid-flexible hybrid dynamic model. The angular velocity of the crankshaft or the center angle of the flywheel is used to observe the engine transient speed fluctuation, abnormal movement or stagnation phenomenon.

Setting the simulation model in full load and 1500 rpm steady state with normal operation, it collected the X-axis torque signal of shaft coupling and extracted the spectrum signal as shown in figure 2.

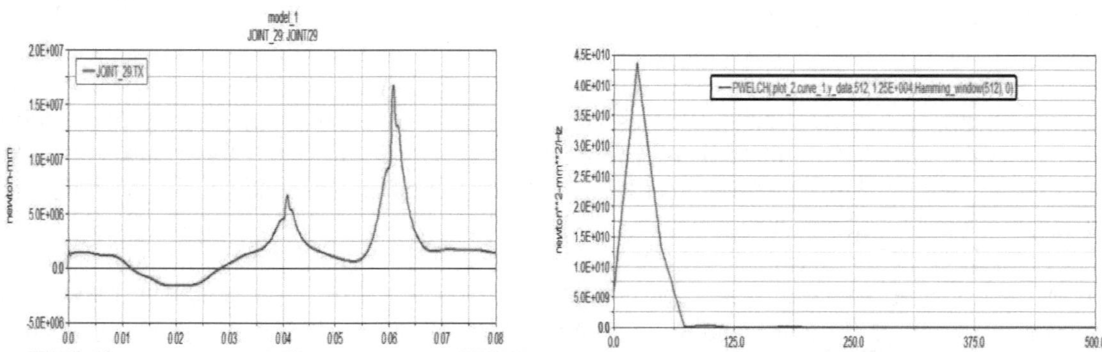


(a) (b)

**Fig.2** Torque signal in time domain and frequency domain with normal state

### *3.2 Misalignment fault in stable speed*

It is obviously that the dynamic response is related to the degree of shaft misalignment. Assuming that the shaft misalignment is 1 mm or 3 mm, the misalignment force is applied in the middle part of the coupling and the shaft is in full load and 1500rpm steady state. The shaft coupling X-axis torque signal of time domain and frequency domain are shown in figure 3 and figure 4.

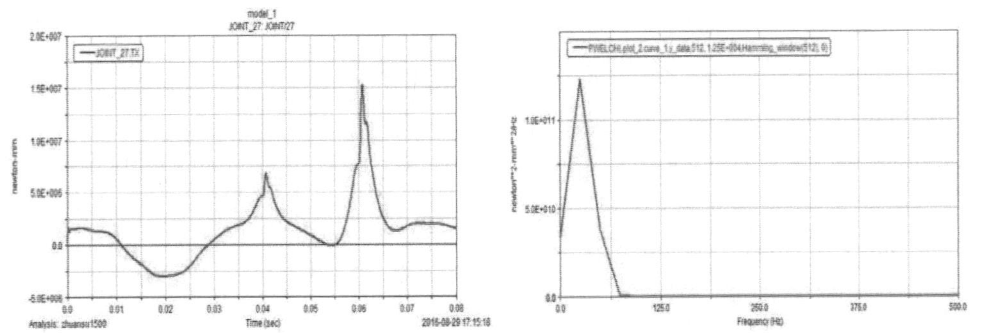

(a) (b)

**Fig. 3** Torque signal in time domain and frequency domain with 1 mm misalignment

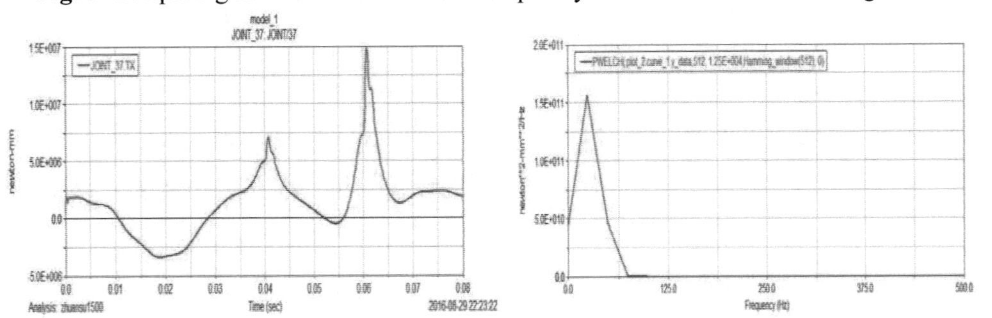

(a) (b)

**Fig. 4** Torque signal in time domain and frequency domain with 3 mm misalignment

In order to research the failure phenomena and evolution rules of misalignment faults, it extracted the spectrum value in some specific frequency.

**Table 1** Spectrum value of X-axis torque signal in 1500 rpm

| Frequency(Hz) | 0 | 24.4141 | 48.8281 | 73.2422 | 97.6563 |
|---|---|---|---|---|---|
| Normal state | 5.9687E+6 | 4.3641E+10 | 1.3532E+10 | 1.5962E+8 | 3.2071E+8 |
| 1mm misalignment | 3.4068E+10 | 1.2276E+11 | 3.8119E+10 | 8.8527E+8 | 3.3344E+8 |
| 3mm misalignment | 4.3540E+10 | 1.5609E+11 | 4.6095E+10 | 8.8995E+8 | 3.7247E+8 |

As shown in figure 3, figure 4 and table 1, viewed from the above three conditions, the shaft coupling X-axis torque signal spectrum value in the frequency of 24.4141Hz appeared a peak value. It means that the power spectral density showed the first times frequency characteristic. With the degree of misalignment fault increased, the spectrum value in different frequency are all improved greatly and the growth rate becomes larger, especially in the first times and the second times frequency characteristic. Compared with the normal condition, the spectrum value of the first times is increased by 181.30% with 1 mm misalignment, 257.67% with 3 mm misalignment. It is because that the power spectral density reflects the power distribution of the torque signal in the frequency domain.When a misalignment fault occurs, the shaft motion is changed, the movement is no longer smooth, the shaft vibration and X-axis torque increased. It resulted the first times frequency characteristic becoming much more obvious.

According to the analysis above, the first times frequency characteristic maybe used as the characteristic parameters of misalignment diagnostics in stable speed. In order to further analyze the universality of the conclusion, it conducted the simulation analysis in 1000 rpm and 2000rpm with different misalignment degree.

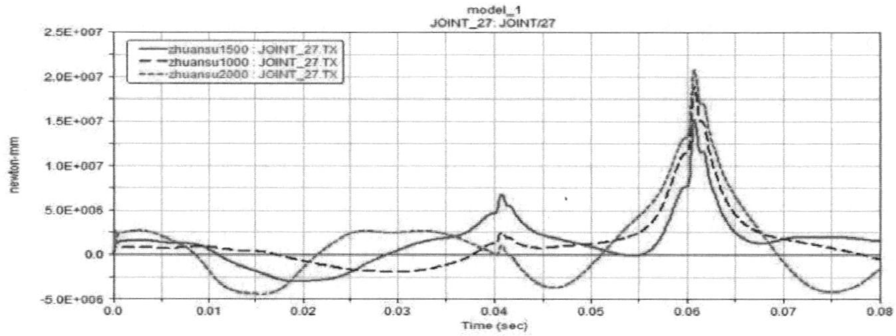


**Fig.5** Torque signal in time domain with 1mm misalignment

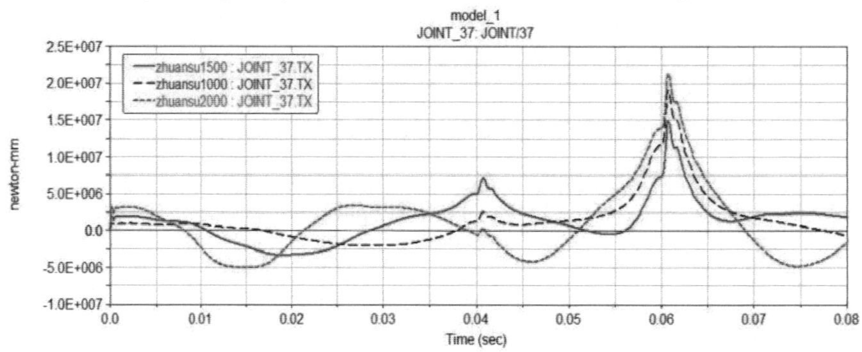


**Fig.6** Torque signal in time domain with 3mm misalignment

**Table 2** Spectrum value of X-axis torque signal in different speed

| Speed (RPM) | 1000 | 1500 | 2000 |
|---|---|---|---|
| Normal state | 1.2593E+10 | 4.3641E+10 | 6.7964E+10 |
| 1mm misalignment | 3.7517E+10 | 1.2276E+11 | 1.7818E+11 |
| 3mm misalignment | 4.5041E+10 | 1.5609E+11 | 2.4314E+11 |

As can be seen from figure 5, figure 6 and table 2, the amplitude of X-axis torque signal becomes larger with the increase of the rotational speed. At the same time, the first times frequency characteristic has much more significant in the power spectral density. Take the example of 1 1 mm misalignment, confronted with the 1000 rpm, the spectrum value of the first times is increased by 227.21% with 1500 rpm and 374.93% with 2000 rpm. It can be seen that the effect of vibration caused by misalignment is also enhanced with the increase of rotational speed. The performance in the time domain is the amplitude of the waveform up and down and the increase in performance in the frequency domain is a multiplier amplitude of the sharp increase.

*3.3 Misalignment fault in variable speed*

In order to study the influence of rotational speed with shaft misalignment, it analyzed the start-up process of a continuous linear change from 0 rpm to 2000 rpm in 2.4 s. Take the normal condition, 1 mm misalignment minor fault and 3 mm misalignment serious failure as the analysis object. The power spectral density of shaft coupling X-axis torque signal are shown in figure 7, figure8 and figure 9.

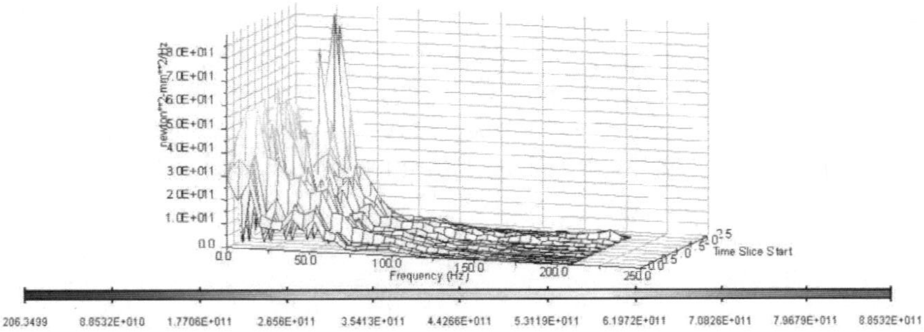


**Fig.7** Spectral density of torque signal with normal state

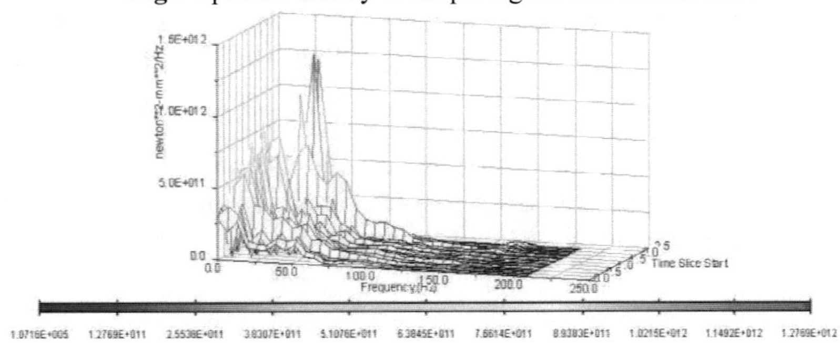


**Fig.8** Spectral density of torque signal with 1 mm misalignment

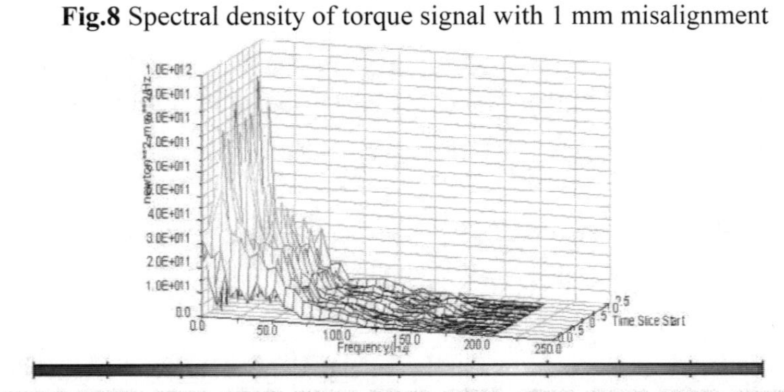


**Fig.9** Spectral density of torque signal with 3 mm misalignment

From figure 7, figure 8 and figure 9, when the rotational speed is changed from 0 to 2000rpm, the amplitude of the torque signal spectral density are also changed and distributed in different frequency range.Firstly, in normal and 1 mm minor misalignment fault conditions, the dominant frequency of power spectral density is 0.5th times and first times in the low or medium speed, and only 1 times in high frequency range.Secondly, under the condition of 3 mm misalignment fault, the dominant frequency is mainly 0.5th times, supplemented by the frequency of first times.

The engine failure will often lead to the occurrence of other faults, such as rubbing fault, rotor loose. The deeper the failure degree, the greater the other faults possibility. In normal and 1 mm minor misalignment fault conditions, the shaft needs to overcome strong inertial and damping forces. There is a slight collision between the main shaft and the bearing, resulting in the rubbing fault of 0.5th times spectral density. As the speed increasing, the effect of rubbing fault is reduced and the first times spectral density starts to appear as a frequency multiplication characteristic. However, in the case of the 3 mm misalignment serious fault, it may cause the change of the shaft physical state and the damage to the electric

IOP Conf. Series: Earth and Environmental Science **61** (2017) 012018 doi:10.1088/1755-1315/61/1/012018

vehicle extended range is larger. Therefore, the friction caused by the rubbing fault plays a significant role and the dominant frequency of power spectral density is 0.5th times.

In summary, the misalignment diagnostic characteristic parameters is 0.5th times in starting speed and first times in stable speed. If the 0.5th times spectrum value in stable speed is still large, it may have occurred in a large amount of misalignment and need for timely treatment.

## 4. Conclusion

The paper established a extender hybrid rigid flexible mechanical model by means of the software ADAMS and ANSYS. By setting the relevant parameters to simulate the misalignment of shafting, it conducted frequency analysis and researched the failure phenomena and evolution rules of misalignment faults. The following conclusions can be drawn.

Firstly, for misalignment fault in stable speed, the first times frequency characteristic maybe used as the diagnostics characteristic parameters. Secondly, for misalignment fault in variable speed, the misalignment diagnostic characteristic parameters is 0.5th times in starting speed and first times in stable speed. If the 0.5th times spectrum value in stable speed is still large, there maybe happened a serious misalignment.

**Acknowledgments**

This work was supported by the National Natural Science Foundation of China (NSFC) (Grant No. 51505345), the Science and technology research project of Education Department of Hubei Province(Grant No. Q20151105) and the Hubei Key Laboratory of Power System Design and Test for Electrical Vehicle(Grant No. HBUASEV2015F005).

## References

[1] E.D. Tate, et al. Electrification the From Conventional Hybrid, to Plug-in Hybrids,to Extended Range Electric Vehicles, SAE paper 2008-01-0458

[2] Fu Hui. Volt: General Motors turned around the world, Detroit landing on moon [J].light vehicles, 2008 (5): p.73-75

[3] Qu Xiaodong, Wang Qingnian, et al. A Study on the Control Strategy for APU in a Extended-range EV [J]. Automotive engineering, 2013,35 (9): p.763-768

[4] Nicola B, Silverio B. PM motors an extended operation transactions on industrial application[C]. Salient-rotor (flux-weakening) for synchronous, 2010,36 (4): p.1118-1125

[5] Ma Hui, Li Huanjun, Liu Yang, Wen Bangchun. Review and prospect for research of coupling faults in rotor systems [J]. Journal of vibration and shock, 2012,31 (17):p.1-9

[6] Du Canyi, Ding Kang, Yang Zhijian. Feature extraction of engine misfire fault based on finite element and multi-body dynamics simulation [J]. vibration and shock, 2012,31 (9): p.18-23

[7] Hong Jiazhen, Liu Zhuyong. Modeling methods of rigid-flexible coupling dynamics[J]. Journal of Shanghai Jiao Tong University, 2008,42 (11): p.1923-1925

[8] Chen Liang, Song Xigeng, et al. Dynamic strength and shock response analysis of a diesel crankshaft based on finite element and multi-body dynamics[J]. Journal of vibration and shock, 2008, 27 (11): p.186-189

IOP Conf. Series: Earth and Environmental Science **61** (2017) 012019 doi:10.1088/1755-1315/61/1/012019

# Research the migration pattern of oil-gas interface in gas cap expansion reservoir

**Heng Song[1], Xuelin Wu[1], Qiangsheng Hao[2] and Dongdong Hou[3]**

1. Research Institute of Petroleum Exploration & Development, Beijing, 100083, china

2. China National Oil and Gas Exploration and Development Corporation, Beijing, 100034, china

3. CNPC-AktobeMunaiGas, 312 Infantry Division Avenue, Building 3, Aktobe, 030006, Kazakhstan

**Abstract.** Gas is continuously driven into the oil rim with the decrease of the oil rim pressure during the process of gas-cap expansion. In order to study the morphological changes in the process of oil-gas interfacial migration, the one-dimensional displacement device is used to analyze the oil-gas interface at different displacement angle and displacement velocity. The results show that the migration of outer boundary is faster than the inner boundary of the oil and gas interface in the process of oil-gas interfacial migration, and there are big differences in migration pattern of oil-gas interface at different displacement angle and displacement velocity. In order to quantitative character the interface morphology with different the displacement angle and displacement velocity, the new factor is used to describe the oil and gas interface morphology and the value of the factor can be calculated using the displacement angle and displacement velocity. The relationship between oil-gas interface form factor and displacement velocity is logarithmically determined for different displacement angles, which can provide guidance for the rational exploitation of gas-cap expansion energy with gas cap reservoir to achieve the aim of improving the overall development effect of gas cap reservoir

## 1. Introduction

With the development of oil and gas fields and the growing demand for natural gas and oil in the world economy, the development of gas cap reservoirs has been paid more and more attention. Gas cap and oil rim are under the same pressure system, any one of the pressure changes will cause the other side of the pressure changes, but also lead to oil-gas interface migration. In order to make full use of natural energy, the conventional development method of gas cap reservoir is to develop the gas cap after exhausting the oil rim [1-2]. During the process of gas cap expansion, the oil-gas interface is continually migration with the decrease of the oil rim pressure. Due to the influence of migration velocity and displacement angel, the oil-gas interface will change its shape, so some of oil production wells which at the border of the interface will be shut due to the gas breakthrough into the production well [3]. It is necessary to carry out the one-dimensional displacement experiment by physical simulation, and to study the law of oil and gas interfacial migration in different displacement angle and displacement velocity, which is helpful to make reasonable use of gas cap expansion energy and improve the overall development effect of gas cap reservoir [4-6].

## 2. One Dimensional Visualization Physical Experiment

One-dimensional visualization experimental device is used to study the migration law of the fluid

IOP Conf. Series: Earth and Environmental Science **61** (2017) 012019 doi:10.1088/1755-1315/61/1/012019

interface during the process reservoir development. The device is 350mm long, the device ends for the round face and the diameter is 27mm. The two end faces each have a 3 mm diameter orifice, which may serve as an inlet end and an outlet end (Figure 1).

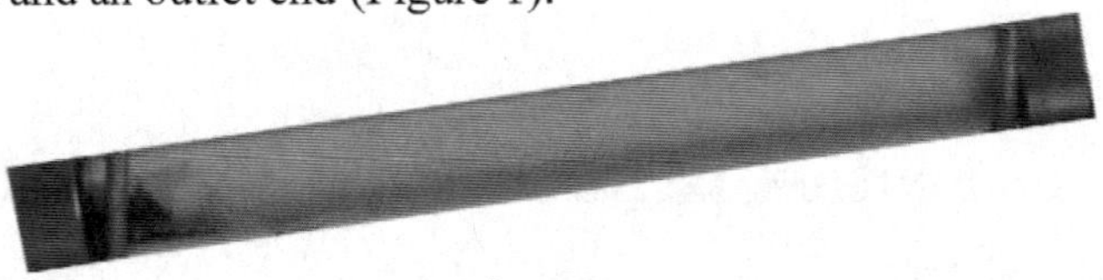

**Figure 1.** One-dimensional visualization experimental device

The device body is made of organic glass, using the glass beads as a porous medium which with a higher degree of visibility. At the same time, in order to make the experimental results easy to observation, the choice of the fluid color has a strong contrasts, the gas sample is nitrogen gas. The oil sample is kerosene which dyed with red ink. Also equipped with nitrogen cylinders, pressure gauges and gas metering device, which composed of the entire experiment system (Figure 2).

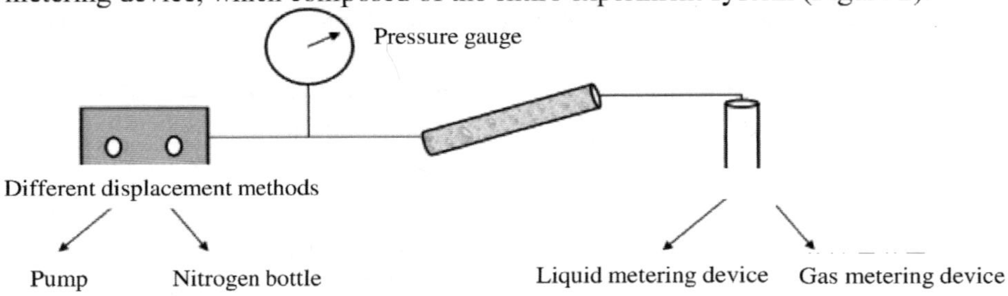


**Figure 2.** The experimental flow chart

## 3. GOC Migration Morphology

In order to study the migration pattern of oil-gas interface at different displacement velocity and dip angle, there are five sets of displacement velocity and four sets of strata dip angle. Since the displacement velocity is affected by the injected nitrogen volume, the designed displacement velocity is replaced by the gas injection rate. The injection rate is 0.01mL/min, 0.05mL/min, 0.1mL/min, 0.5mL/min, 2.5mL/min. And the corresponding actual displacement velocity is 0.0016m / d ~ 0.16m / d, the strata dip angle is 5°, 10°, 15°, 20°.

Taking the formation dip angle of 10° as an example, the oil and gas migration patterns at different gas injection rates are compared. When the gas injection rate is 2.5mL/min, gas flooding fingering serious, so the oil and gas distribution is more dispersive, there is no obvious oil-gas interface. At this displacement velocity, the recovery degree of the oil rim is only 2.8%. When the gas injection velocity is 0.5mL/min, the migration velocity of the oil-gas interface is obviously slowed down compared with the injection velocity of 2.5mL/min, the gas fingering phenomenon is suppressed and the recovery degree of the oil-rim is increased to 7.5%. When the gas injection velocity is 0.05mL / min, the oil-gas interface migration speed is slower, and the recovery degree of the oil-rim is 14.5% when the gas cap breakthrough into oil well (Figure 3, Table 1).

The two characteristics of oil-gas interface migration can be found by displacement test. One is that there is a big difference in vertical migration distance between the outer boundary and inner boundary of oil and gas interface. And the migration distance of the outer boundary is greater than the inner boundary of the oil-gas interface. Also when the oil rim recovery is similar, it is obvious from the displacement test that the difference in the vertical migration distance is greater when the injection rate increased.

IOP Conf. Series: Earth and Environmental Science **61** (2017) 012019 doi:10.1088/1755-1315/61/1/012019

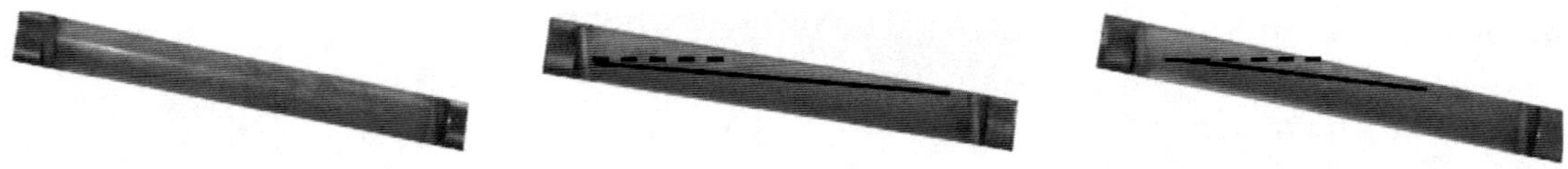

Displacement velocity 2.5mL/min, oil recovery 10.3% | Displacement velocity 0.5mL/min, oil recovery 7.5% | Displacement velocity 0.05mL/min, oil recovery 8.4%

**Figure 3.** oil-gas interfacial at different displacement velocity

**Table 1.** The recovery degree of oil rim at different displacement velocity

| **Displacement velocity (mL/min)** | 0.05 | 0.5 | 2.5 |
|---|---|---|---|
| **Oil rim recovery (%)** | 14.5 | 7.5 | 2.3 |

## 4. Quantitative Characterization of oil - gas Interface Morphology

In order to quantitatively characterize the variation of interface morphology with different displacement velocity and different dip angles, the new factor can be defined as the ratio of the outer boundary migration distance to inner boundary migration distance (Equation 1, Figure 4).

$$\delta = \frac{y1}{y2} \tag{1}$$

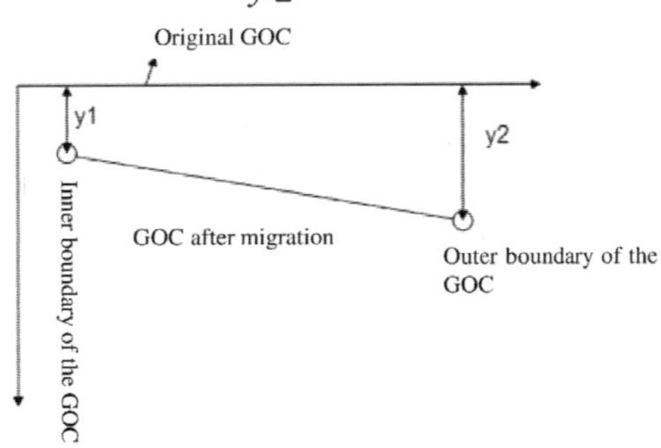


**Figure 4.** The illustration of the oil -gas interface morphology factor

When the strata dip angel is 10°, According to the definition of oil and gas interfacial morphology factors, the factor at different displacement velocity in the one-dimensional displacement experiment is calculated and plotted into the plane rectangular coordinate system (Figure 5). The displacement velocity in the figure has been converted to the actual transport velocity.

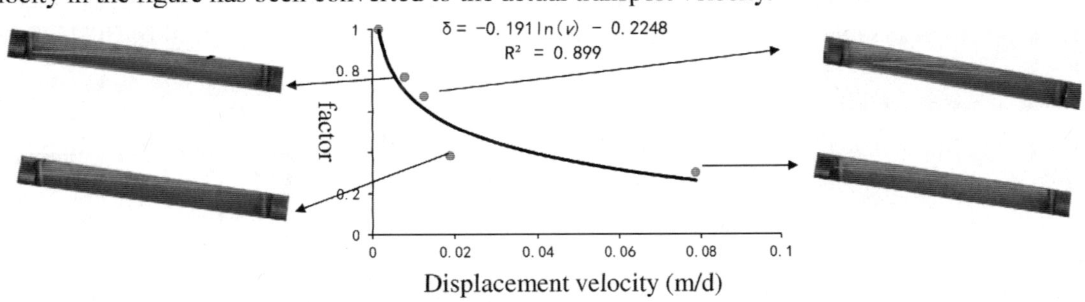


**Figure 5.** Morphological factors of different displacement velocity (strata dip angle 10°)

From the relationship curve, it can be seen that with the increase of displacement velocity, the interfacial morphology factor decreases, which indicates that the difference of the outer boundary migration distance and inner boundary migration distance is bigger, and the oil-gas interface is more unstable. When the dip angle of formation is 10°, the mathematical equation can be acquired

(Equation 2).

$$\delta = -0.191\ln(v)\,0.2248 \tag{2}$$

Where, δ is the interface morphology factor; v is the displacement velocity;

On the basis of pre-research, the interface morphology factor can be calculated with different gas injection rate and displacement angel. The gas injection rate includes 0.01 mL/min, 0.05 mL/min, 0.1 mL/min, 0.5 mL/ min, 2.5 mL/ min. The angle of includes 5°, 15°, 20° (Figure 6).

It can be seen that when the displacement velocity is the same, the larger the dip angle is, the larger the factor is, the greater the dip angle is, and oil and gas interface is more stable, and the difference between the velocity of GOC outer boundary and GOC inner boundary is getting smaller. When the dip angle is the same, with the increase of the displacement velocity, the oil and gas interface morphology factor decreases, which indicate that the difference of the velocity of GOC outer boundary and inner boundary is larger and the oil and gas interface becomes more unstable.

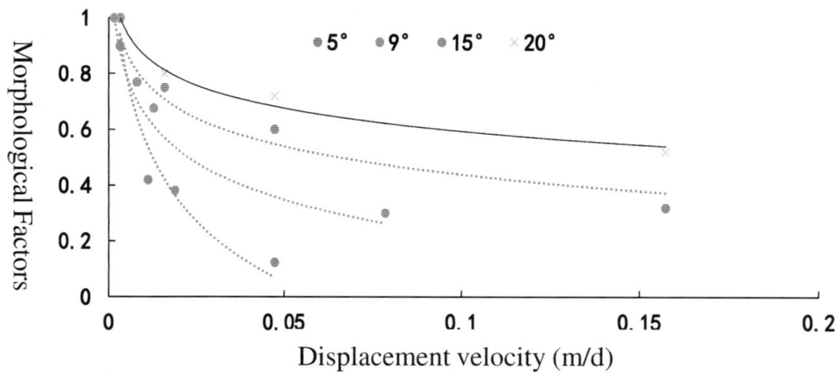


**Figure 6.** The relationship between morphological factors and displacement velocity in different strata dip angle

In the same way, the relationship between oil-gas interface morphological factors and displacement velocity can be obtained by fitting in different strata dip angle including 5°, 15°, 20°.

When the dip angle is 5°, the equation is:

$$\delta = -0.321\ln(v) - 0.9092 \tag{3}$$

When the dip angle is 15°, the equation is:

$$\delta = -0.1451\ln(v) + 0.105 \tag{4}$$

When the dip angle is 20°, the equation is:

$$\delta = -0.118\ln(v) + 0.3212 \tag{5}$$

Based on the above equations, it can be seen that the relationship between oil-gas interface morphological factors and displacement velocity is in logarithm. It can be expressed by the following equation.

$$\delta = \mathrm{a}\ln(v) + b \tag{6}$$

The coefficients a and b of the equation can be regressed, and the relationship between the coefficient and the dip angle is obtained (Figure 7, Equation 7 and Equation 8).

$$\mathrm{a} = -0.0116\alpha^2 + 0.3337\alpha - 2.2873 \tag{7}$$

$$\mathrm{b} = -0.0025\alpha^2 + 0.0673\alpha - 0.5953 \tag{8}$$

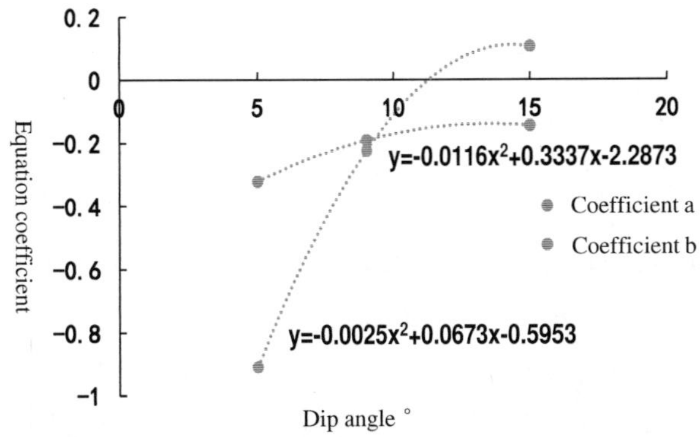


**Figure 7.** The relationship between the coefficients a, b and the strata dip angle

Therefore, for the actual reservoir, the formation dip and displacement velocity are known, so using the above method can calculate the interface morphology factor. It can be used to judge whether the oil-gas interface morphology is reasonable under different development conditions. If unreasonable, it can be adjusted accordingly.

## 5. Conclusion

The one-dimensional displacement test was used to clarify the change law of oil-gas interfacial migration during the development of gas-cap expansion energy flooding, and to quantitatively characterize the relationship between oil-gas interface morphological factors and displacement velocity Also we know that there is a logarithmic relationship between oil-gas interface morphological factors and displacement velocity.

Based on the same displacement velocity, with the increase of the displacement dip angle, the morphological factors of oil-gas interfacial increase, indicating that the higher the dip angle of the formation, the more stable the oil-gas interface. When the displacement dip angle is the same, with the increase of the displacement velocity, the morphological factors of oil-gas interfacial decrease, and the oil-gas interface becomes more unstable. It is helpful to rationally utilize the expansion energy of gas cap and improve the overall development of oil reservoirs with gas cap by studying the changes of oil-gas interface morphology under different dip angle and displacement velocity.

## References

[1] Liu Jia, Cheng Linsong, Fan Zifei, et al. Experimental Studies Oil and Gas Coordinated Development Mechanism of Oil Rim Reservoirs [J]. Journal of Southwest Petroleum University (Science &Technology Edition), 2015, 37(5):99-105.

[2] Liu Shunsheng, Fang Jiazhong, Li Xiaoshan, et al. Feasibility of Using Gas Cap Energy in Janajol Reservoir [J]. XinJiang Petroleum Gelogy, 2003, 24(6):594-597.

[3] Zhang Jianmin, Xu Ming. Study on Recovery Feature of Condensate Gas Reservoir in Wellblock Shuang6, Special Oil and Gas Reservoir, 2003,10(4):41-52.

[4] Fan Zifei, Cheng Linsong, Song Heng, et al. Fluid Interface Migration Mechanism for Concurrent Production of Gas Cap and Oil Rim of Gas Cap Reservoir, Petroleum Exploration and Development, 2015,42(5):624-631.

[5] Yu Lian, zhaolun, chenli, ea al. Reserch on Simultaneous Development of Oil and Gas for Reservoir with Condensate Gas-Cap, XinJiang Oil & Gas(Edition of Natural Science), 2012,8(1):66-69.

[6] Zhang Angang, Fan Zifei, Song Heng, et al. Stability Condition of Oil-Gas Contact of Oil-Rim Reservoir, Geological Science and Technology Information, 2016,35(1):114-118.

IOP Conf. Series: Earth and Environmental Science **61** (2017) 012020 doi:10.1088/1755-1315/61/1/012020

# Analysis of uplift bearing capacity of expanded pile

**Li Bin [1 2 3 4], Zhuge Aijun [2 3 4]**

1. China Institute of Water Resources and Hydropower Research, Beijing, 100048, China
2.Tianjin Port Engineering Institute Ltd. of CCCC First Harbor Engineering Company Ltd., Tianjin, 30222 China
3. Key laboratory of port geotechnical engineering, ministry of communications, PRC Tianjin, 30222 China
4. Key laboratory of port geotechnical engineering of Tianjin, Tianjin, 30222 China
Lee_binbin@163.com

**Abstract**. The research on the uplift bearing capacity of expanded pile is less, which lead to develop slowly of mechanism of expanded pile. The finite element software is used to simulate the bearing capacity of expanded pile, and the friction between the pile and soil is simulated by surface contact. The simulations show the ultimate bearing capacity and the distribution of side friction of the expanded pile. Along with the displacement moves bigger, the side friction of upper expanded pile is fully used. When the displacement moves to a special level, the side friction of whole pile is fully used. If the load is bigger than the 25000kPa, the expanded pile will be destroyed.

## 1. Introduction

The expanded pile is increasing the pile diameter at the pile end of the traditional equal section pile. This method can effectively improve the ultimate bearing capacity of the pile end and the side friction of the pile body, so as to improve the bearing capacity of the whole pile. In recent years, the relevant domestic scholars have done lots of research on theoretical and model test, such as the bearing capacity, pile end resistance, pile side resistance, putting forward a series of important conclusions.

The application of expansion pile in engineering construction has been quite common, but most of the researchers have carried on the theoretical research on the expansion pile. There is little research on the mechanism of the bearing capacity and the side friction resistance of the pile. A small number of scholars are researching on the numerical simulation method of stress characteristics of the expansion pile.

In this paper, the finite element software is used to analyse the pulling capacity of the expanded pile. The distribution of side friction is calculated. The results have some reference significance for designer and constructer of the expanded pile.

## 2. Project Background

In order to facilitate the simulation of finite element software, the author made a pile in the Daxing Experimental Base of Beijing, experimented with pile diameter D = 1.0m, pile height h = 3.5m, the pile material is the reinforced concrete, pile around with the conventional clay.

## 3. Simulation of pulling Capacity of Pile

The Mohr - Coulomb model is used to simulate the soil stress, Mohr-Coulomb yield surface is an irregular hexagonal cross-sectioned pyramidal surface in the principal stress space, and is projected into an equiangular equilateral plane in the π plane. There is sharp angle on the yield surface, as shown in Figure 1.

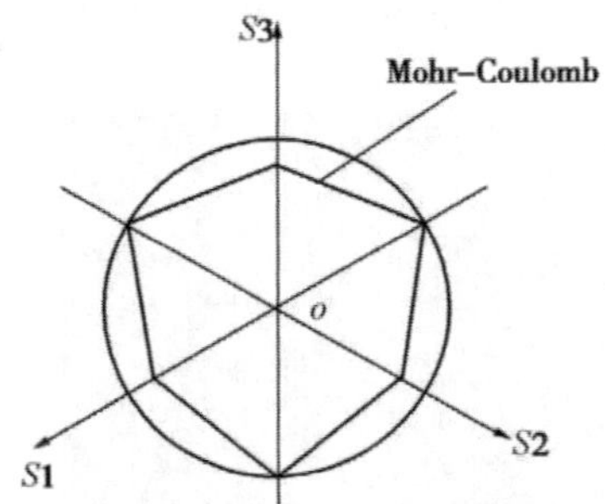


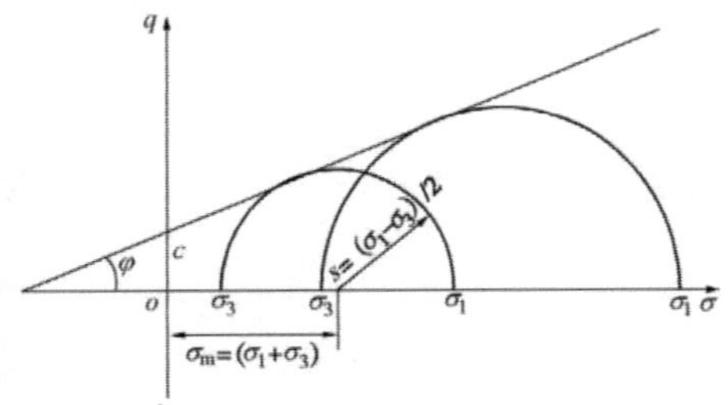


Fig.1 Yield surface on π plane Fig.2 Mohr-Coulomb model

Mohr-Coulomb theories consider that: the shear stress of a yield surface is against by the material cohesion and the friction, the expression is:

$$\tau = c + f\sigma = c + \sigma \tan\varphi \qquad (1)$$

$\tau$、$\sigma$ are the normal stress and shear stress of the plane, $c$ is the cohesion, $f$ is the friction coefficient ,and $\varphi$ is the friction angle.

It can be deduced that when the stress Mohr circle is tangent to the Coulomb criterion line, the rock mass reaches the limit state, and the principal stress should meet the following conditions:

$$\sigma_1 = \frac{2c \cdot \cos\varphi}{1-\sin\varphi} + \sigma_3 \frac{1+\sin\varphi}{1-\sin\varphi} \qquad (2)$$

From the above, it can be seen that the Mohr-Coulomb criterion does not consider the influence of the intermediate principal stress on the strength of rock and soil, but the destruction of material is usually affected by the intermediate principal stress, because of this influence is small, the Mohr-Coulomb criterion is still accurate enough to show the stress level. The result of simulation is very close to the experimental data.

The pile is linear elasticity before the stress reaches the failure strength of material, so it is a linear elastic material in the model. The soil is elastic-plastic material, which is assumed to obey the Mohr-Coulomb criterion. The material parameters are shown in Table 1.

Table 1 parameters table of model

| Name | Elastic modulus Pa | Poisson's ratio | Cohesive force Pa | Friction angle ° | Dilatancy angle ° | Density Kg/$m^3$ |
|---|---|---|---|---|---|---|
| concrete | 3e10 | 0.2 | —— | —— | —— | 2500 |
| Clay | 4e7 | 0.35 | 15000 | 30 | 0 | 1800 |

The elastic models of pile and soil are very different. Under the load, there will be relative sliding between pile and soil. In order to ensure the rationality of the calculation results, the coulomb friction contact is used to simulate the contact between the pile and soil. There is a small sliding model in the software and the friction coefficient is 0.3. Then the contact pairs between the pile and soil are established, and the master-slave contact algorithm is adopted.

The selection of the master and slave faces must follow the principle that the meshing of master surface is finer than the slave surface, even the material of master surface is harder, but the mesh density of two faces must be similar. According to the above principle, the surface of the pile surface is the master face and the soil surface is the slave face.

Coulomb's law is used to calculate the ultimate shear stress in ABAQUS:

$$\tau = \mu p \qquad (3)$$

IOP Conf. Series: Earth and Environmental Science **61** (2017) 012020 doi:10.1088/1755-1315/61/1/012020

$\mu$ is the coefficient of friction, $p$ is the stress between the two contact surfaces. When the shear stress between the two contact surfaces reaches the ultimate shear stress, the contact surface will produce the relatively sliding. Because there is the non-ideal frictional contact between the pile and soil, the concept of "elastic slip deformation" is introduced in the finite element calculation. When the surfaces are bonded together, a small amount of relative slip deformation is allowed. As shown in Figure 3, a small slip between the contact surfaces is allowed when the slip displacement is less than $\omega_s$.

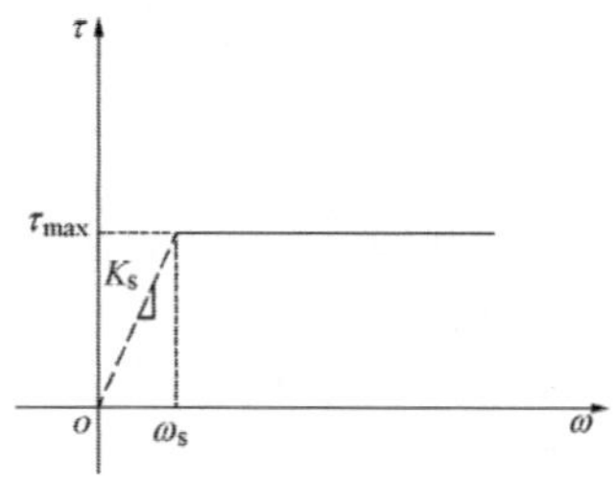

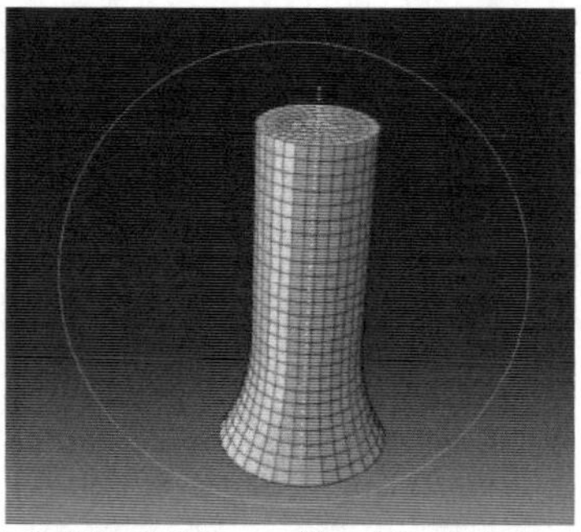

Fig.3 Coulomb friction model Fig.4 Pile-soil grid model

The results of the numerical analysis are closely related to the mesh shape and mesh density. If the results need more accuracy, the grid density must be higher. Therefore, the model should be as far as possible to ensure sufficient accuracy without excessive element, the denser of the mesh close to the pile, the sparser of the mesh away from the pile. Figure 4 shows the pile-soil grid model.

## 4. Result

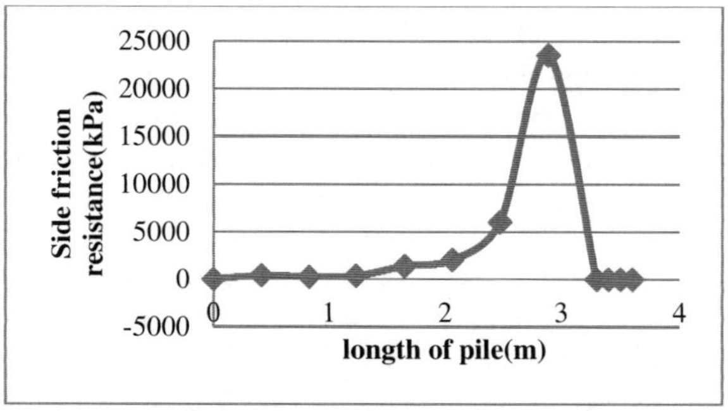


Fig.5 side frictional resistance of pile

Through the simulation of the friction between pile and soil, it can be seen that along with the length of the pile, the friction resistance is increasing continuously. The expansion head of the expanded pile can bear more loads. If the load is more than the 25000kPa, the expanded pile will be destroyed.

Fig.6 strain and stress of pile

Figure 6 shows the stress chart and strain chart of pile and soil. The maximum stress of the pile occurs at the lower part of the pile, and the stress concentration area is also in the lower part of the pile, even at the part of the expansion head. Because the displacement boundary conditions were applied at the top of the pile, the whole pile body is the largest displacement area, and the maximum displacement area of the soil occurs in the upper part of the soil. If the soil was damaged, the pile will be failure.

## 5. Conclusions

The development and application of the finite element software ABAQUS makes it possible to simulate the uplift capacity of the expanded pile.

In this analysis, when the displacement of the expanded pile is very small, the side friction bears the uplift load; when displacement moves to the specific level, the expansion head of the expanded pile bears more load; if the displacement goes on increasing, the pile will be destroyed, the friction between pile and soil will disappear.

## Acknowledgements

This research is financially supported by the natural science foundation of Tianjin No.16JCQNJC06700 and the Open Research Fund of Tianjin Key Laboratory of Soft Soil Characteristics and Engineering Environment No. 2013SCEEKL03 and the Tianjin enterprises postdoctoral innovation projects preferred funding.

## Reference

[1] Pile bearing capacity[M]. Springer Berlin Heidelberg, 2014.
[2] Meyerhof G G. Bearing capacity and settlement of pile foundations (11th Terzaghi Lecture)[J]. Journal of the Geotechnical Engineering Division, 1976, 102(9):195-228.
[3] Gong W. THEORETICAL STUDY ON SELF-BALANCED LOADING TEST FOR PILE BEARING CAPACITY[J]. Industrial Construction, 2002.
[4] Meyerhof G G, Mathur S K, Valsangkar A J. The bearing capacity of rigid piles and pile groups under inclined loa. [J]. Canadian Geotechnical Journal, 2011, 18(2):297-300.
[5] Momeni E, Nazir R, Armaghani D J, et al. Prediction of pile bearing capacity using a hybrid genetic algorithm-based ANN[J]. Measurement, 2014, 57(11):122-131.

# Cherry recognition in natural environment based on the vision of picking robot

**Qirong Zhang** [1, a)], **Shanxiong Chen** [2, b)], **Tingzhong Yu** [1, c)] **and Yan Wang** [1, d)]

1Information Engineering, Guizhou University of Engineering Science, Bijie 551700, China
2 Schools of Computer and Information Science, Southwest University, Chongqing 400715, China
a)Corresponding author: zqr_hk@163.com
b)214282779@qq.com
c)ytz_668@126.com
d)604788055@qq.com

**Abstract**. In order to realize the automatic recognition of cherry in the natural environment, this paper designed a robot vision system recognition method. The first step of this method is to pre-process the cherry image by median filtering. The second step is to identify the colour of the cherry through the 0.9R-G colour difference formula, and then use the Otsu algorithm for threshold segmentation. The third step is to remove noise by using the area threshold. The fourth step is to remove the holes in the cherry image by morphological closed and open operation. The fifth step is to obtain the centroid and contour of cherry by using the smallest external rectangular and the Hough transform. Through this recognition process, we can successfully identify 96% of the cherry without blocking and adhesion.

## 1. Introductions

As cherry retail prices soaring, cherry is planting positively in many places. However, with the increase of cherry planting area, cherry price competition will intensify. Because of cherry fruit small, cherry picking is difficult. Due to picking more time-consuming, this factor may lead to higher cost of cherry picking. Therefore, the use of cherry picking robot has become a trend. It can not only reduce the cost but also can solve the problem that the large area cherry needs to be picked in time in the mature season. Robot vision is the key function of cherry picking robot. It is the main factor to determine the success of cherry picking. There are many researches on pattern recognition [1-9]. In this paper, the recognition of cherry is a key problem in robot vision [10-11].

## 2. Image analysis

In the real environment, the cherry background is mainly the branches, leaves, cherry handle, the sky.

## 3. Cherry recognition methods

### *3.1. Image pre-processing*

Cherry image will produce noise in the process of shooting, shooting techniques and equipment conditions. In this paper, 5 kinds of filtering are used to remove noise, which are block filtering, mean filtering, Gauss filtering, median filtering and bilateral filtering. Their experimental results are shown in Figure 1.

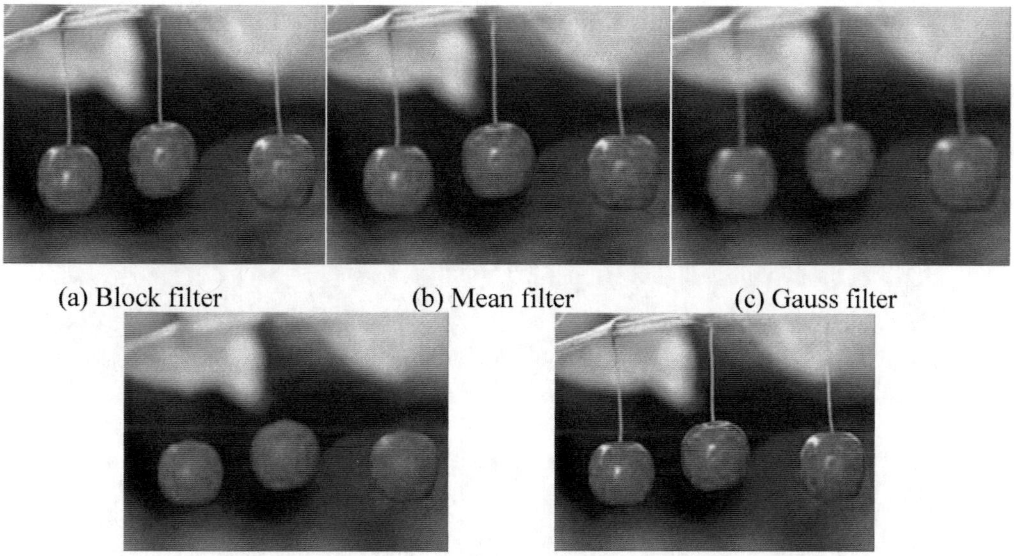

(a) Block filter (b) Mean filter (c) Gauss filter

(d) Median filter (e) Bilateral filtering

**Figure 1**. Pretreatment of cherry image

Figure 1 shows that median filtering is the best. Median filtering not only removes the cherry stalk but also removes the influence of illumination on the surface of a cherry. It removes the noise effectively.

*3.2. Color feature extraction*

The color of ripe cherries varies greatly with the branches, leaves, cherry stems, and the sky. Therefore, color feature extraction is very effective. There are a number of commonly used color models. In this paper, RGB model is used to identify the color image [12]. The cherry image is decomposed into 3 components: red, green and blue, as shown in figure 2.

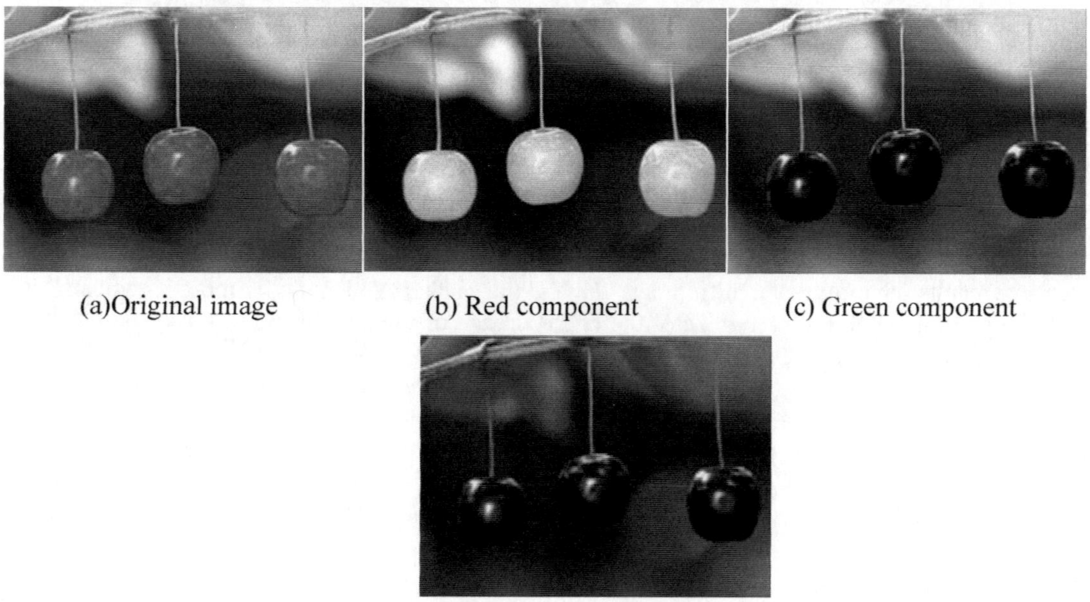

(a)Original image (b) Red component (c) Green component

(d) Blue component

**Figure 2**. The RGB component image of the cherry

Figure 2 shows that it is difficult to identify the color characteristics of cherry using only a single color component. The color of cherry image is produced by different proportion of red component, green component and blue component, so the relationship of the three components is studied. R-G, R-B, G-B, R/G, R/B, 2R-G-B, R-G/2-B/2, 0.9R-G, 0.9R-B were used to process the image of cherry, and the results were shown in figure 3.

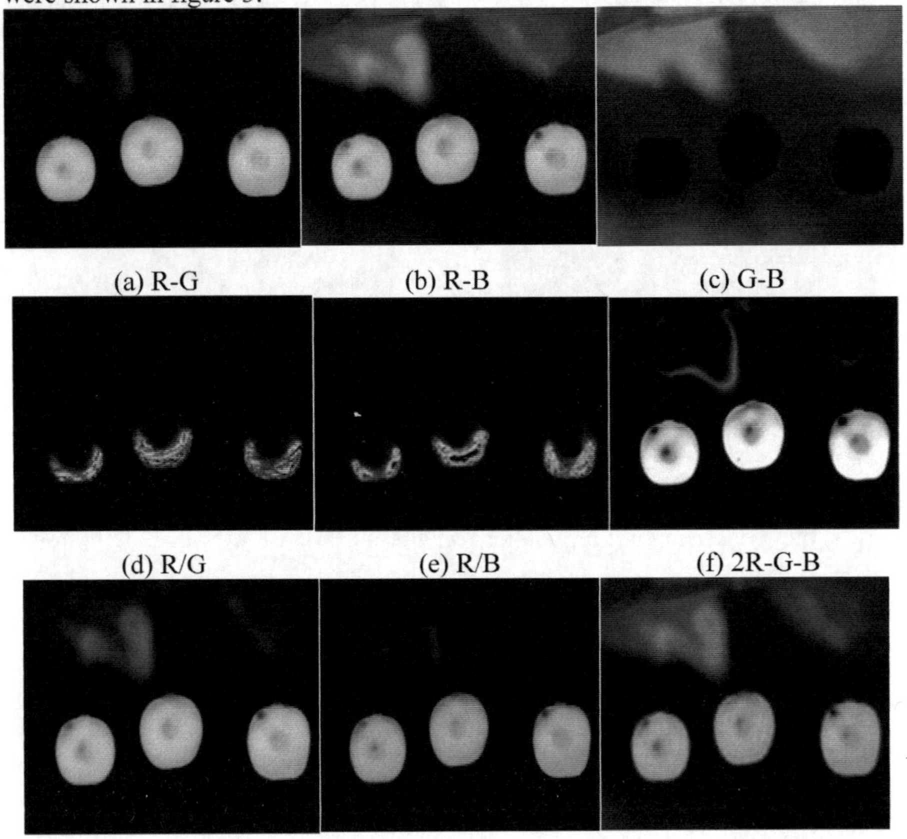

(a) R-G (b) R-B (c) G-B

(d) R/G (e) R/B (f) 2R-G-B

(g) R-G/2-B/2 (h) 0.9R-G (i) 0.9R-B

**Figure 3**. Contrast diagram of nine kinds of color recognition methods

Figure 3 shows that G-B, R/G, and R/B methods are not accurate to identify the cherry shape. So these three kinds of color recognition methods are not successful. The 2R-G-B has produced obvious holes in the treatment of cherry images. Therefore, 2R-G-B is not a better method. R-B, R-G/2-B/2, 0.9R-B and R-G have basically the same effect to identify the image of the cherry, but the background processing results have a greater difference, that is, the background noise is more obvious. Finally, 0.9R-G is the best method to recognize of the cherry image or the processing of background noise, so the 0.9R-G as a method to identify the cherry. After the above color recognition, Otsu algorithm will be used to image segmentation of cherry [13]. If the image segmentation process is found to be a large amount of computation using the Otsu algorithm, we can use the evolution formula of the Otsu algorithm. $P$ is the threshold for segmenting foreground and background, $t_0$ is the probability of the distribution of the foreground, $h_0$ is the average gray value of the foreground, $t_1$ is the probability of the distribution of the background, $h_1$ is the average gray value of the background. The average gray value of the whole image is $h = t_0h_0 + t_1h_1$. The variance formula $\omega^2 = t_0t_1(h_0 - h_1)^2$ is established [14]. The experimental results are shown in the figure 4.

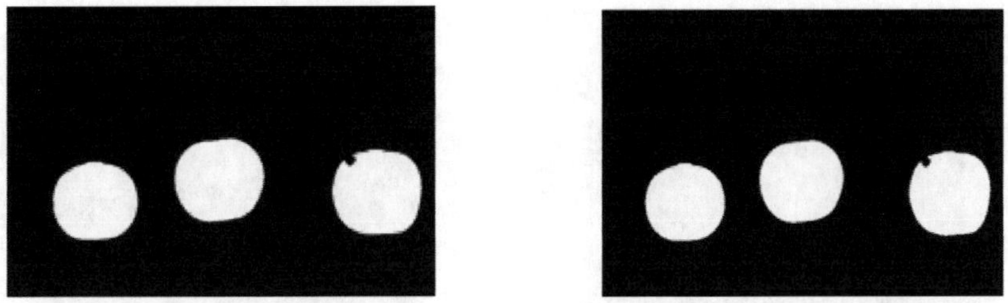

(a) Otsu algorithm (b) The equivalent formula of Otsu algorithm

**Figure 4**. Otsu algorithm and Otsu algorithm of the equivalent formula to deal with cherry image

The experimental results show that the two algorithms have the same effect as shown in figure 4.

### *3.3. Area threshold removal noise*

There are some holes and some noise in the image by the above process. So the area threshold is used to remove noise. We first find out the areas of different cherry in the image, and then find out the largest area of cherry. Finally, half of the largest cherry area is used as threshold to segment the image. Treatment results are shown in figure 5.

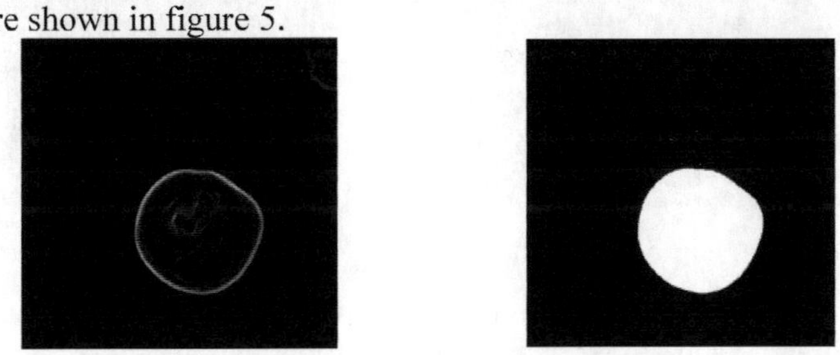

(a) Cherry image with noise inside (b) Area threshold removal noise

**Figure 5.** Contrast figure of area threshold removal noise

Figure 5 shows that there are no holes in the image of the cherry. The noise is removed successfully.

### *3.4. Morphological operation*

In general, there are holes and burrs in the image of cherry. These holes are filled by morphological close operation and open operation. The opening operation is first corrosion and then expansion treatment [15]. Image processing results are shown in figure 6.

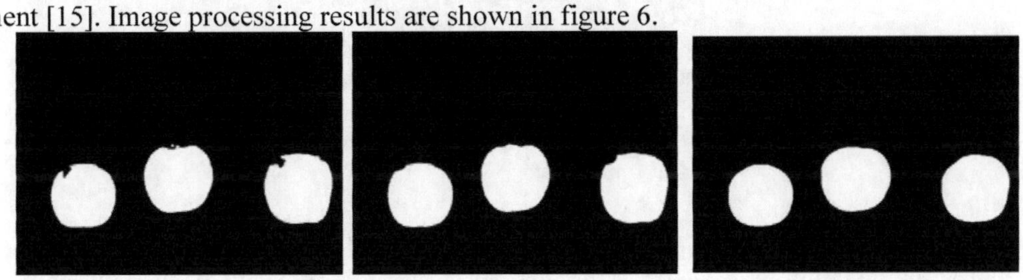

(a) Cherry image with holes (b) Closed operation (c) Open operation

**Figure 6**. Morphological closed and open operation for cherry image processing

### *3.5. Using Hough Circle Transform to Extract Contour*

Using canny operator for edge detection, cherry image processing effect is better. The experiment is shown in figure 7.

IOP Conf. Series: Earth and Environmental Science **61** (2017) 012021 doi:10.1088/1755-1315/61/1/012021

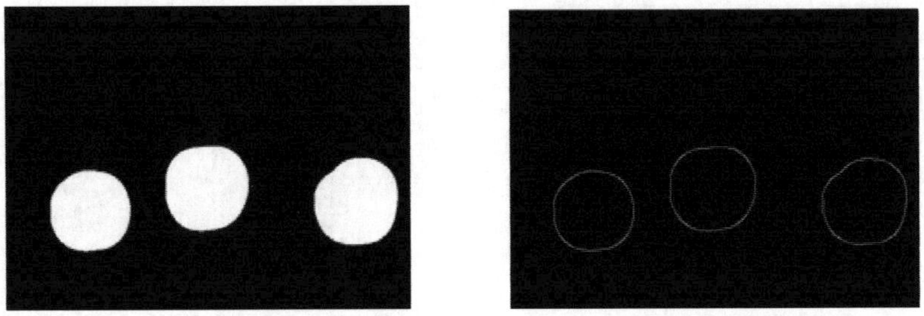

(a) Original (b)Canny operator

**Figure 7**. Cherry image processing using canny operator

The processed cherry image is drawn with a minimum external rectangle, as shown in figure 8.

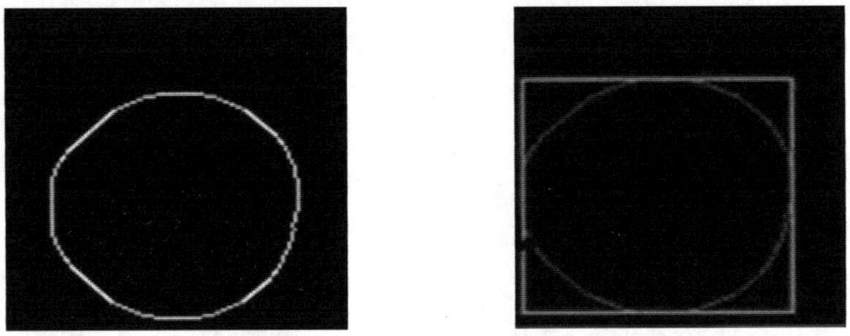

(a) The original contour of a cherry (b) Minimum external rectangle

**Figure 8**. Draws up minimum external rectangle for cherry outline

Hough is more used to circle and ellipse. Through the minimum external rectangle of cherry image, we get the radius of Hough transform. This reduces the amount of computation of Hough transform and avoids the retrieval of multiple centroid and contour in the same cherry. Experimental results are shown in figure 9.

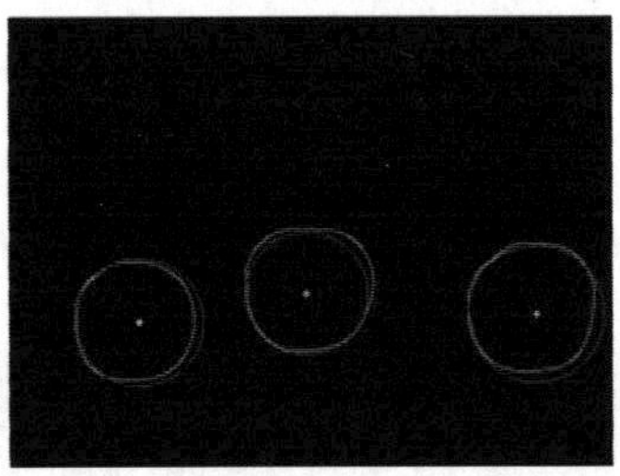

**Figure 9**. Hough transform of the cherry image

## 4. Conclusions

It has great market application value for cherry picking robot. The most important part of the cherry picking robot is the recognition of cherry. This method is to use 0.9R-G for identify the colour, and then use the classical Otsu algorithm for threshold segmentation. If the Otsu algorithm is directly used to deal with a large amount of data, we can also use the equivalent formula of Otsu algorithm. When there are still noises and holes in the image, we can use the area to remove the noise and morphology to remove holes. Small holes can be directly used to open operation, without first for closed operation. Recognition of the key is to draw the centroid and contour of the cherry. We first use the canny operator to extract the outline of the cherry. However, the direct Hough transform will lead to recognition is not successful or long recognition time. So we can first find the minimum enclosing rectangle of contour of the cherry, and the Hough transform can successfully find the centroid and contour of all the cherries in the image.

**References**

[1] Zhang Q R and He Z S 2010 The two-dimensional bidirectional inverse laplacian margin maximum criterion for face recognition *Journal of Computational Information Systems* 6 1101-09

[2] Zhang Q R and He Z S 2010 Dual-tree quaternion wavelets for face recognition *Journal of Computational Information Systems* 6 1151-60.

[3] Zhang Q R and He Z S 2010 Two-dimensional locality discriminant preserving projections for face recognition *Advanced Materials Research* 121-122 391-398

[4] Zhang Q R, Gu J N and Zhang M F 2013 Inverse constrained maximum variance mapping for face recognition *Applied Mechanics and Materials* 462-463 452-457

[5] Zhang Q R 2014 Two-dimensional parameter principal component analysis for face recognition *Advanced Materials Research* 971-973 1838-42.

[6] Zhang Q R and He Z S 2010 The two-dimensionalcombined bidirectional laplacian inverse fisher discriminate analysis method for face recognition *Journal of Information and Computational Science* 7 2919-26

[7] Zhang Q R, Zhang J.Y and Wang Y 2016 The improved relative entropy for face recognition *MATEC Web of Conferences* 63 04006

[8] Zhang Q R and Liu X 2014 Grey incidence measurement method based on dynamic time warping distance and its application *Journal of Chongqing university of posts and telecommunications natural science edition* 26 700-704

[9] Zhang Q R, Liu X and Yan Y 2015 Grey incidence clustering method based on dynamic time warping distance *Journal of Chongqing university of posts and telecommunications natural science edition* 27 563-568

[10] Zhang Q R, Hu R H andYu T Z 2016 Design and implementation of cherry picking robot vision recognition system based on face recognition *MATEC Web of Conference* 63 05017.

[11] Zhang Q R, Peng P and Jin Y M 2016 Cherry picking robot vision recognition system based on openCV *MATEC Web of Conference* 63 01005.

[12] Wang H, Mao W H, Liu G, Hu X A and Li S J 2012 Recognition and location of Apple working robot based on visual combination *Transactions of the Chinese Society for Agricultural Machinery* 43 165-168

[13] Otsu N 1979 A threshold selection method from gray-level histogram I*EEE Trans Syst Man Cybern* 9 62-66.

[14] Lv J D, Zhao D A, Ji W 2014 Target tracking and identifying method for apple picking robot *Transactions of the Chinese Society for Agricultural Machinery* 45 65-72.

[15] Shi X. Q and Cheng X W 2013 Research on target extraction of apple picking robot vision system *Journal of Agricultural Mechanization Research* 10 46-59.

# Numerical simulation of migration behavior of uranium ore dust particles in the human respiratory tract

**Yong-jun Ye [1, 2], An-song Yin [1], Zhi Li [1], Bo Lei[1] , De-xin Ding [1, 2]**

1 Key Discipline Laboratory for National Defense for Biotechnology in Uranium Mining and Hydrometallurgy, University of South China, Hengyang, Hunan 421001, China
2 School of Environmental Protection and Safety Engineering, University of South China, Hengyang, Hunan 421001, China
E-mail: yongjunye@163.com

**Abstract**. There is a certain concentration of radioactive dust particles in the air of workplace of underground uranium mines. Some small diameter particles will pass through the masks and enter the respiratory tract which will cause radiation damage to the human body. In order to study deposition regularity of uranium dust in the human respiratory tract, in this paper, we firstly use the RNG turbulence model to simulate the gas flow field in the human respiratory tract $Z_0$ ~ $Z_3$ level under different respiratory intensity. Then we use DPM discrete phase model to simulate the concentration, particle size distribution, deposition rate and deposition share of uranium dust particles after being filtered through the masks in the human respiratory tract $Z_0$ to $Z_3$ bronchus. According to the simulation results, we have got the following conclusions: the particles' number concentration of uranium dust after being filtered through the mask in the human respiratory tract basically decreases with the increasing of particle size under different respiratory intensities on the environment of uranium mine. In addition, the intensity of respiration and the mass concentration of particles have an important influence on the deposition rate and the deposition of particles in the respiratory tract.

## 1. Introduction

Mine dust comes from mining blasting, drilling and transporting activities under the pit. There are a lot of radioactive nuclides such as radium, uranium, radon daughters and so on. If these radioactive aerosols go into the human body through breathing, the human body would suffer internal radiation damage, which greatly endangers body health. Studies have indicated that particulate matter of aerodynamic equivalent diameter less than 10 microns (PM10) can be directly inhaled into the upper respiratory tract. Aerodynamic equivalent diameter less than 2.5 microns fine particulate matter(PM2.5)can enter the alveolar. The pollution of fine particulate matter is closely related to the health effects of respiration and the occurrence of cardiovascular diseases [1,2].when the concentration of PM2.5 was increased by 10 $\mu g/m^3$, the incidence of total morbidity, cardiovascular disease and respiratory disease increased by 4%, 6% and 8%, respectively[3-5]. With the decrease of the particle size, the toxicity increased [6].

With the rapid development of computer technology, computational fluid dynamics CFD technology has obtained the wide use. CFD numerical simulation technology has the advantages of fastness, accurateness and low cost. It can simulate many experiments in more complex conditions. It broadens the scope of experiment research greatly. Many scholars at home and abroad have applied the CFD technology into the particle simulation experiment of respiratory tract. Cheng et al.[7-9], according to the human true respiratory tract size, established a three-dimensional model of the upper respiratory tract and carried out the relevant experimental study. Zhang et al.[10] pointed that Weibel established a symmetrical three level bronchial model A and later many scholars have used this model to carry out a large number of numerical simulation studies. Luo and Liu [11] used the LRN κ-ε model to simulate the traffic status of the particles in the human respiratory tract, studied the influence of stokes number and Reynolds on deposition rate and deposition distribution in the respiratory tract. Although there are a number of examples about the scholars using CFD to simulate the particle movement of respiratory tract, the simulation study about the uranium mine radioactive aerosol particles in the respiratory tract has not been reported widely. Therefore, we used the CFD numerical simulation technology to simulate uranium ore dust particles size distribution and deposition rate in the different positions of human respiratory tract in different respiratory intensities after being filtered through the mask, and investigated the causes of particle concentration, particle size distribution, deposition rate distribution and deposition ratio in human respiratory tract. As a result, it will provide some referential information for preventing the diseases of respiratory tract.

## 2. Numerical models and method

### *2.1. Respiratory geometric model size*

The depth that particles in different sizes go into the human respiratory system is different. Generally, particles whose size is 5 to 10 μm can be removed by a physical mechanism in the throat. Particles whose size is less than 5 μm can get through the block of the throat and go into the human bronchus. Particles whose size is less than 2.5 μm will go into the deepest part of the lungs with respiratory airflow [12]. Contrasted with previous studies, the mass concentration of uranium dust distribute in less than 3 μm size range after filtering through the mask. These tiny particles easily go through the mouth, nose and throat to human bronchial and alveolar. For this reason, this paper uses the Weibel three-dimensional bronchial model that does not include the human nose and throat to research on particulate deposition situation in $Z_0$ to $Z_3$ bronchial [13]. The bronchial tubes are connected by a circular tube. The respiratory model and its partition are shown in figure 1. The length, diameter and curvature of the bronchial tubes at all levels are shown in table 1.

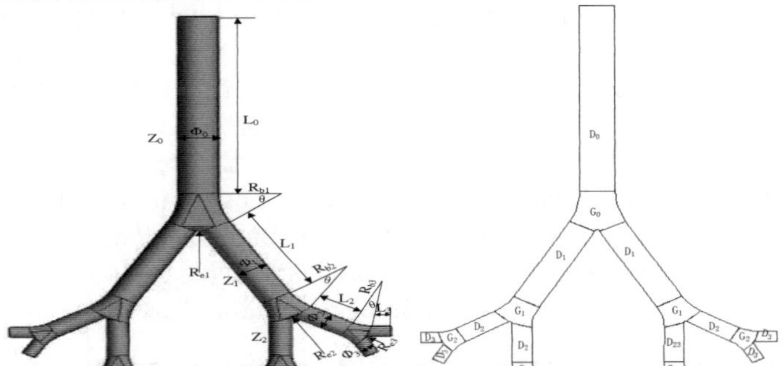

**Figure 1.** Weibel three-dimensional bronchial model A and partition number

**Table 1.** Geometric parameters of Weibel three-dimensional bronchial model A .

| Tracheal bronchus | Diameter D(mm) | Length L(mm) | Radius of curvature of bronchus $R_{bi}$(mm) | Radius of curvature of carina $R_{ei}$(mm) | Bronchial semiangle θ° |
|---|---|---|---|---|---|
| $Z_0$ | $\Phi_0$=15.4 | $L_0$=102.6 | — | — | 30 |
| $Z_1$ | $\Phi_1$=12.2 | $L_1$=47.6 | $R_{b1}$=32.94 | $R_{e1}$=1.22 | 30 |
| $Z_2$ | $\Phi_2$=8.30 | $L_2$=19.0 | $R_{b2}$=22.41 | $R_{e2}$=0.83 | 30 |
| $Z_3$ | $\Phi_3$=5.60 | $L_3$=7.6 | $R_{b3}$=15.12 | $R_{e3}$=0.56 | 30 |

### *2.2. Before the simulation data acquisition*

We place the APS 3321 in planned sampling position and connect backup battery power and notebook computers to make them in the ready state. The mask is put in the APS 3321 air inlet and open the switch for pumping, We observe the data of APS 3321 and wait for the concentration of quality is stable or not occur significant fluctuations. We start sampling for 300 seconds and write the data into computer. The above test is repeated at each sampling point. The monitoring data of APS particle size spectra instrument are put into ORIGIN software for processing. We have got the particle size distribution of uranium dust after being filtered through the mask. As shown in figure 2.

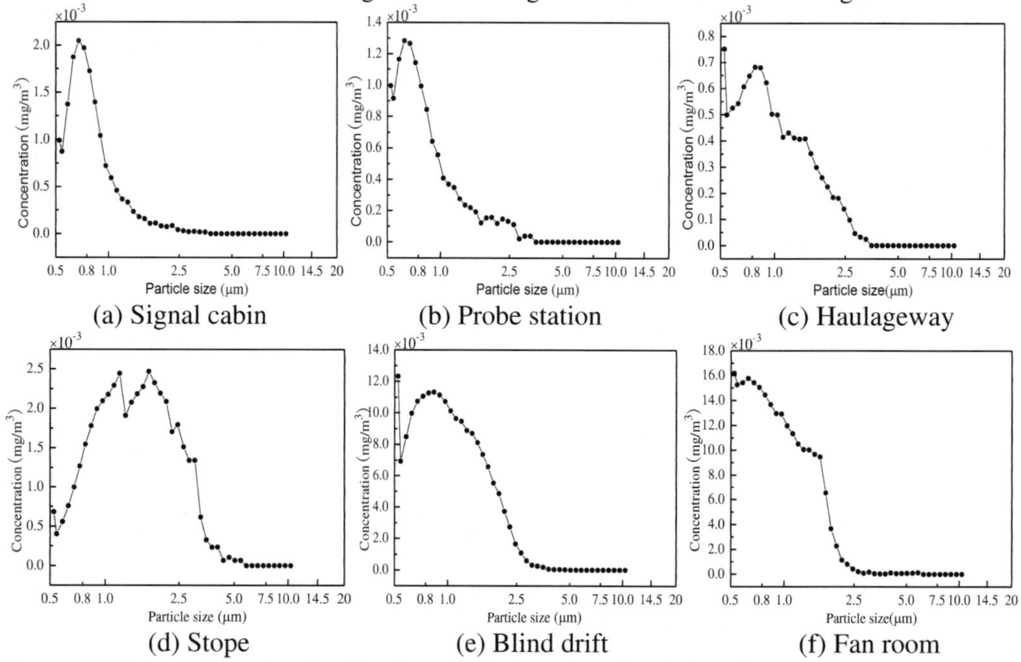


**Figure 2.** The particle size distribution of uranium dustafter being filtered through the mask.

The body's respiratory rate approximate a sine curve. The human body respiration intensity are 30 L/min, 60 L/min and 90 L/min respectively in the process of mild, moderate and high intensity exercise[14]. In this paper, the respiratory intensity is 30 L/min, 60 L/min and 90 L/min by using RNG turbulence model. The flow field distribution in the human respiratory tract was studied under different labor intensity. The mass concentration of PM10 is taken in simulation. The mass flow rate of uranium dust with respiration into the human body is calculated by combining the respiration rate of human body in mild, moderate and high intensity labor. The results are shown in table 2.

**Table 2.** Mass flow rate of uranium dust in human respiratory tract under different respiration intensity (kg/s).

| | Place | respiration intensity（L/min） | | |
|---|---|---|---|---|
| | | 30 | 60 | 90 |
| Gauze mask | Signal cabin | 8.50E-12 | 1.70E-11 | 2.56E-11 |
| | Probe station | 6.50E-12 | 1.30E-11 | 1.95E-11 |
| | Haulageway | 1.30E-11 | 2.60E-11 | 3.90E-10 |
| | Stope | 5.20E-11 | 1.04E-10 | 1.56E-10 |
| | Blind drift | 9.30E-11 | 1.86E-10 | 2.80E-10 |
| | Fan room | 4.80E-11 | 9.60E-11 | 1.44E-10 |

Uranium dust being filtered through the mask is distributed in less than 5μm diameter particle size range.The particle size distribution can be expressed by Rosin-Rammler distribution function.The R-R distribution equation of uranium mineral dust is obtained by statistical calculation ofthe experimental data of uranium mine dust filtered by masks. The results are shown in table 3.

**Table 3.**Uranium dust R - R distribution table after filtered by the masks.

| | Place | $\bar{d}$ (μm) | n | R-square |
|---|---|---|---|---|
| Gauze mask | Signal cabin | 0.880 | 4.695 | 0.995 |
| | Probe station | 1.139 | 2.320 | 0.987 |
| | Haulageway | 2.333 | 1.602 | 0.989 |
| | Stope | 1.729 | 2.142 | 0.996 |
| | Blind drift | 1.172 | 2.425 | 0.993 |
| | Fan room | 1.248 | 2.589 | 0.995 |

## 3. Results and discussion

### *3.1. The distribution characteristics of the flow field*

The flow characteristics of airflow in the respiratory system are closely related to uranium dust deposition in human respiratory system. Therefore, this paper uses RNG turbulence model to simulate the flow field distribution cloud images in the respiratory tract. The results are shown in figure 3.

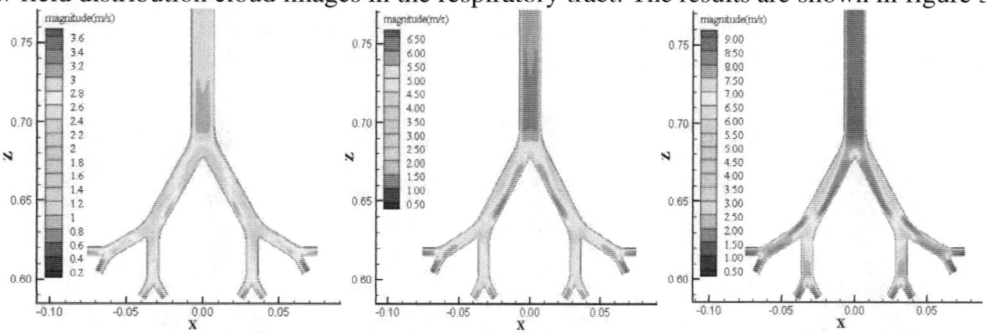

(a) 30 L/min (b) 60 L/min (c) 90 L/min

**Figure 3.** Flow field distribution cloud images under different respiratory intensity in human respiratory tract.

It can be seen from the figure 3 that the airway wall has obvious viscous effects to air, so the air velocity exhibits an obvious parabola distribution. The intermediate speed is larger than the two sides. Air flow produces a small range of reflux at the bottom of the $G_0$, $G_1$, $G_2$ section. High velocity fluid is distributed in the inner side of the center line of the $D_1$ section. The lateral fluid velocity is relatively low. The $D_1$ and $D_2$ sections are further reduced. Air does not appear significant high or low speed separation phenomenon.

### *3.2. Concentration and particle size distribution*

In this paper, we simulate the mass concentration and particle size distribution of uranium dust under the respiratory intensity of 30 L/min, 60 L/min and 90 L/min in the human respiratory system. For the various experimental results, we take the results in signal cabin as an example and display the concentration distribution of uranium dust in the human body. As is shown in figure 4.

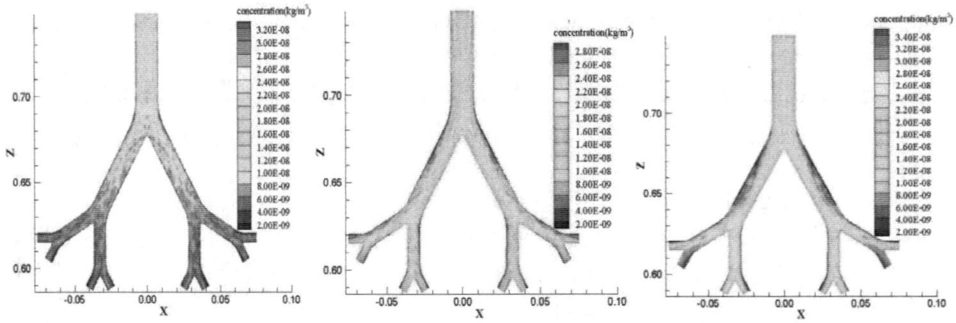


(a) 30 L/min (b) 60 L/min (c) 90 L/min

**Figure 4.** Mass concentration distribution of uranium dust on Y=0 cross section.

It can be seen from the figure4, uranium dust concentration is higher where air velocity is high. Due to the particle's motion trajectory is mainly affected by the impact of air and its trajectory is similar to the trajectories of fluid. When the flow of high-speed airflow is in the respiratory tract, it will take a large number of particles.

### *3.3. The deposition rate of uranium dust particles in the respiratory tract*

In this paper, we have simulated the deposition rate of uranium dust mine after filtering through the mask in different respiratory intensity. Deposition rate distribution diagram of uranium dust in the human respiratory tract is obtained. Because there is a lot of data, we will display the data of signal cabin, as is shown in figure 5.

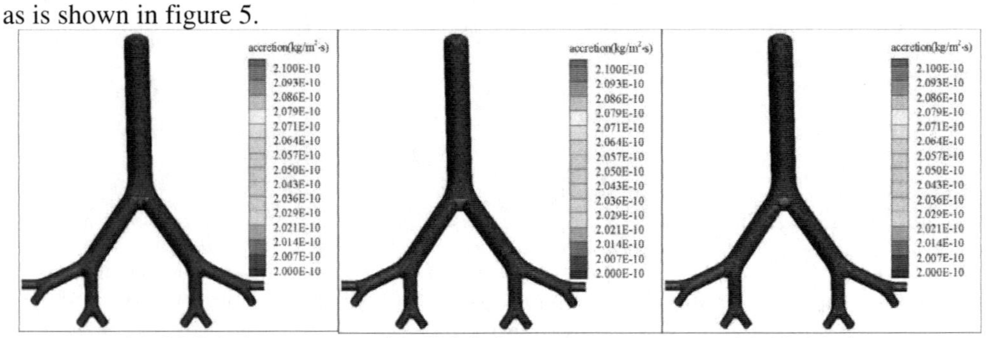


(a) 30 L/min (b) 60 L/min (c) 90 L/min

**Figure 5.** Uranium dust deposition rate in the human respiratory tract.

Through the analysis of figure 5, the deposition rate is bigger in $D_0$ entrance position where uranium dust concentration is higher. With the continuous adsorption of respiratory tract mucous membrane to uranium dust, the deeper is, the lower uranium dust concentration of the respiratory tract is, the lower the particle deposition rate on respiratory tract mucous membrane is. Therefore, mass concentration is an important influential factor of particle deposition rate in the respiratory tract.

The deposition rate is an important parameter of movement to study particle movement in the human upper respiratory tract. Therefore, airway wall is divided in detail on the basis of figure 1 in this article. The deposition rate is counted in the corresponding region. The deposition rate curve of uranium dust is obtained in Weibel 3D bronchial mode under the different respiratory intensity in uranium underground environment after filtering through the mask, As is shown in figure 6.

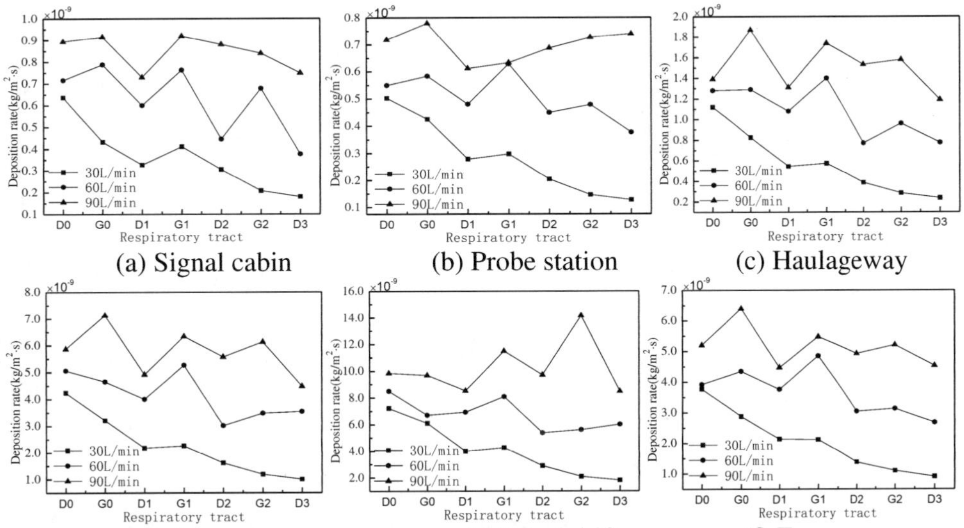


(a) Signal cabin (b) Probe station (c) Haulageway

(d) Stope (e) Blind drift (f) Fan room

**Figure 6.** Deposition rate of uranium dust after being filteredthrough the mask in the human respiratory tract

It can be seen that uranium dust deposition rate was significantly higher than the surrounding straight pipe section in bifurcation segment $G_1$ from figure 6,due to airway narrowed sharply, airflow that coerced particle impact the bottom of bifurcation section. A large number of particles were adhered to the airway mucosa under the effect of inertia collision. Therefore, the inertia impact is one of the main factors affecting of the particles' deposition rate in the cross section of the respiratory tract. In addition, through the analysis of deposition rate data under the three kinds of different respiratory intensity(mild, moderate and severe), it was found that the deposition rate of particles in respiratory tract increases with the increase of respiration intensity. This is because when the intensity of respiration increased, the particle mass increased in the respiratory tract of the human body within the unit time. Therefore, the rate of respiration is one of the main factors affecting the particles' deposition rate in the cross section of the respiratory tract.

## 4. Conclusions

In this study, the particle number concentration of uranium dust after being filtered through the mask in human body $Z_0$~$Z_3$ respiratory tract basically decreases with the increasing of particle size under the different respiratory intensity on the environment of uranium mine. Particle number is mainly distributed in the small size range. The smaller the particle is, the deeper the human respiratory system goes, which is consistent with the results of this study.

Through the analysis of the deposition rate data of light, medium and heavy three kinds of respiration intensity, the quality of the particles concentration, inertia collision and the rate of respiration are the key factors affecting of the particles' deposition rate in the respiratory tract. Pathogenesis and treatment of respiratory diseases has a great relationship with the deposition rate of uranium dust in the mucous membrane of the respiratory tract. Through the study of deposition rate distribution of uranium ore dust in respiratory tract, it will help to locate the most serious and most easily diseased parts.

There is a relationship between the deposition share of uranium dust in the human body $Z_0$ ~ $Z_3$ bronchus and the respiration intensity of the human body. The greater the respiratory intensity of the human body is and the greater the respiratory rate is, the smaller the particle deposition become in the human body $Z_0$ ~ $Z_3$ bronchus under the same conditions. As the air velocity increases, the residence time of particles becomes shorter in respiratory tract, the probability of entering the deep respiratory tract increases.

IOP Conf. Series: Earth and Environmental Science **61** (2017) 012022 doi:10.1088/1755-1315/61/1/012022

**Acknowledgements**
This work was financially supported by Chinese National Natural Science Fund Projects (No.11105069) and (No.11575080), Science and technology project of key technology for prevention and control of major accident in safety production supported by State Administration of Work Safety Supervision (hunan-0015-2015AQ), the Doctoral Scientific Research Foundation of University of South China (2015XQD38).

**References**
[1] Zhou, Y., Su, W. C.,Cheng, Y. S. 2008. Fiber deposition in the tracheobronchialregion: Deposition equations. *Inhalation Toxicology*, 20, 1191–1198.
[2] Heenan, A. F., Matida, E., Pollard, A., Finlay, W. H. 2003. Experimental measurements and computational modeling of the flow field in an idealized humanoropharynx. *Experiments in Fluids*, 35, 70–84.
[3] Pope, C.A., Burnett, R.T., Thun, M.J., Calle, E.E., Krewski, D., Ito, K., Thurston, D.H. 2002. Lung cancer, cardiopulmonary mortality, and long term exposure to fine particulate air pollution.*Journal of the American Medical Association* , 287(9), 1132–1141.
[4] Dai, H.X., Song, W.M., Gao, X.2004.Correlation analysis of atmospheric PM10, PM2.pollution and daily mortality of residents in Shanghai A city.*Institute of health*, 33(3),293-297.
[5] Jiang, X.2000. New research report confirms the relationship between atmospheric particulate matter and health.*China Environmental Science*, 20(6),560.
[6] Laden, F., Neas, L.M., Dockery, D.W., Schwartz, J. 2000. Association of fine particulate matter from different sources with daily mortality in six US cities.*Environmental Health Perspectives*, 108(10),941–947.
[7] Cheng, Y. S., Zhou, Y., Chen, B. T. 1999. Particle deposition in a cast of human oralairways. *Aerosol Science and Technology*, 31(4), 286–300.
[8] Huang, J.H., Zhang, L.Z. 2011. Numerical simulation of micro-particle deposition in a realistic human upper respiratory tract model during transient breathing cycle.*Particuology,*9 (4),423-431.
[9] Yeh, H. C., Schum, G. M. 1980. Models of human lung airways and their application to inhaled particle deposition. *Bulletin of Mathematical Biology*, 42(3), 461–480.
[10] Zhang, Z., Kleinstreuer, C., Kim, C.S. 2001. Effects of curved inlet tubes on air flow and particle deposition in bifurcating lung models. *Journal of Biomechanics*, 34(5),659-69.
[11] Luo, H.Y., Liu, Y. 2009. Particle deposition in a CT-scanned human lung airway.*Journal of Biomechanics,* 42(12),1869-76.
[12] Kameda, Y., Shirai, J., Komai, T., Nakanishi, J., Masunaga, S. 2005.Atmospheric polycyclic aromatic hydrocarbons: size distribution, estimation of their risk and their depositions to the human respiratory tract. *Science of the Total Environment*, 340(1), 71-80.
[13] Weibel E R. 1965. Morphometry of Human Lung.*Anesthesiology*, 26(3), 367.
[14] Chen, X.L., Zhong,W.Q., Jin, B.Y. 2011.Numerical simulation of the movement in the respiratory tract and deposition of particulate pollutants.*Journal of Southeast University*,41(2),393-399.

IOP Conf. Series: Earth and Environmental Science **61** (2017) 012023 doi:10.1088/1755-1315/61/1/012023

# Application and study of 3D visualization model on nuclear power plant construction

**Depeng Guo[1], Chenghao Pu[1], Maolin Li[1], Wenguang Zhang[1], Jing Zhao[2] and Laura Alejandra Arlia[3]**

[1] Nuclear and Radiation Centre, Beijing 100082, China
[2] Tsinghua University, Beijing 100084, China
[3] Autoridad Regulatoria Nuclear, Buenos Aires, Argentina
corresponding author: Jing Zhao, zhao_jing@mail.tsinghua.edu.cn

**Abstract**.3-D visualization model is an advanced tool which addresses a digital expression to the project entity and functional characteristics by integrating the relative engineering data. Construction units lead this technique in the equipment installation activities of the first unit of AP1000 Nuclear Power Plant Project in the world. 3-D technique proved its advantages in construction scheme preparation, second design, collision simulation, construction logic, etc. Through the introduction of the main characteristics and superiority of this technique, the solution and suggestions for some project insufficiencies have been proposed. This will represent a benefit for its application in the future.

## 1. Forward

Large commercial Nuclear Power Plant (NPP) construction is a complex engineering system, which is characterized by strong investment, long cycle period, complicated interfaces, overcrowded people and crossover operation, among others. It means that there are lots of difficulties and complexities in the construction management, such us collision check, construction simulation, crossover operation, the second detail design, etc. 3-D visualization model has been drew step by step in AP1000 (Advanced Passive) NPP Project construction management, showing that some practical difficulties had been solved by the new technique. However, some inefficiencies like insufficient planning and the application range and depth, appeared during the application. It proves that it will be necessary to gather and sum up experience and lessons, analyze the problems and make a more effective use in the future.

## 2. Summary introduction of technique

3-D visual model technique, based on the three-dimensional digital technology, can provide a digital expression of the project entity and functional characteristics by integrating the relative engineering data. It can solve the consistency and global shared problems between the distributed, heterogeneous engineering data-base.

Four levels can be summarized in 3-D visual model technique, including visual simulation, performance simulation, virtual simulation and virtual reality. For detail, the visual simulation (also known as the 3-D visualization) presents the inner-architecture, shape, material, color, lighting and environment in the form of 3-D automatic navigation. The performance simulation can show the traditional scientific computing process and result as an intuitive way of 3-D information model which also can carry out a test of the simulation, including the material structure calculation, acoustic testing, thermal performance and physical environment factors such as sunlight and wind pressure. The virtual building and virtual reality can represent the process of real building construction, including all the physical properties such as gravity, friction, atmosphere and its change, in order to minimize the difference between the construction design effect, process and target.

## 3. Application on Nuclear Power Project

### *3.1. Overall objective*

Revolving around the project operation, the overall policy of introducing the 3-D visual model technique into the nuclear power engineering is to solve the key technology and difficulties in the project, make sure that the technology works in all fields and ensure the risk control management works in the whole construction process. In order to complete the above target, during the construction phase, the 3-D model of all main subjects should be created. It will be optimized with the combined data from the electric, piping, large-scale equipment, ventilation and valve teams, etc. The strict compliance between the 3-D visual model and the drawing should be confirmed. According to the needs of construction, the difficulties and mile-stone activities can be showed in the 3-D visual model after the intensive study and technical clarification by each subject team. Once this step is completed, the 2-D detail drawing used for the construction on-site can be issued.

### *3.2. Model Building*

Solid modeling technique describes the internal geometry of the object with the stereoscopic thinking instead of plane thinking. The spatial object is first abstracted as a series of adjacent 3-D geometry, and then assembled by another method[1]. The main modeling methods include scanning, boundary representation, dimensional geometry, mixed representation and others.

In addition to the construction of 3-D visual model, the most important aspects of 3-D visualization model building is to construct the model base with geological, building, plant and equipment parameters and to build the information database[2]. Due to the complexity of the model, the modeling process is tedious and should be repeated by stretching, lofting, geometry Boolean calculation and arraying. The information model, in this way established, can be combined with the virtual simulation technology and achieve the overall planning of the construction site and formulating emergency plans. In general, the parameters should be modeled separately. For example, REVIT software is more suitable for building 3-D visual information model of the plant, but attention should be paid to the detailed models of structures and installed equipment.

### *3.3. Construction General Layout Management*

There is no reference project for the construction of China AP1000 NPP due to the fact that it is the first unit of its kind in the world. As the most important content of the construction management, construction general layout management should use the 3-D visualization model. The construction site planning, design and layout can be optimized by inputting the time and space arrangement information based on the related machine tool and the NPP structure and equipment. This information includes the layout of 3200t large-scale crawler crane, 500t crawler crane, elevators, concrete hoist, box change and other equipment.

### *3.4. Solution Proposal and Processing Simulation*

Regarding to the operations, such as large-scale hoisting, which are too difficult to control but easy to cause accidents, the 3-D visualization model technology could be used to prepare the large-scale structural module hoisting program, to simulate the operation process and calibrate the safety measures (shown in Figure 1). It can greatly reduce and effectively control the occurrence of safety incidents and improve the construction safety management level. In addition, the 3-D visualization model technology could also be used to simulate the construction process of installation of important

IOP Conf. Series: Earth and Environmental Science **61** (2017) 012023 doi:10.1088/1755-1315/61/1/012023

equipment in nuclear island and high-precision embedded parts. The 3-D visualization makes the preparation, demonstration, review and clarification of the construction program more intuitive.

Figure1. Simulation of the large-scale lifting in a NPP project

### *3.5. Technical Clarification by Visualization*

Construction units should manage the involved staff to understand the relevant drawings and technical documents and integrate the original design information through an intuitive, complete and highly integrated project design information platform [3] in order to obtain the useful management information. This approach can shorten the technical preparation time for the construction team and improve the efficiency and accuracy of drawing technical documentation. Practice has proved that technical clarification by visualization is particularly useful for low-level education workers.

### *3.6. Construction Logic Research*

Modular construction is one of the technical characteristics of the AP1000 NPP construction. Civil engineering and installation are overlapping at the beginning of the project. Two teams should first plan their respective construction activities in the same area of the nuclear island [4] and then rearrange them after careful discussion. The building can be capped after all large-scale equipment has been installed and most of the installation work has been done so it could be said that the construction logic is much more complex. For example, in a system installation process, engineering and technical personnel with the help of 3-D visualization technique can intuitively find that several pieces of pipe need to be installed after the installed module; otherwise it will hinder the module in place. This clearly reflects the three-dimensional technique is an appropriate means of construction logic superiority.

## 4. Suggestion for improvement

Through the preliminary application of 3-D visualization model system in AP1000 NPP project, the construction unit solves a large number of practical engineering problems; speeds up the development and utilization of enterprise information resources; makes the design information resources to the best configuration and construction of enterprise management and operation with higher efficiency and lower costs. However, 3-D visualization model system application could be improved in the following ways:

- Technique application scope can be widening. Using system management methods could extend the range of application of 3-D visualization model system to schedule management, quality management, safety management, procurement management and so on. Furthermore, it can develop the function of automatic retrieval and statistical chart generation. 3-D visualization model system can provide the best means of information management for the construction materials, equipment preparation and engineering statistics.
- The promotion of the technique application. Through business management process analysis and reorganization, the 3-D visualization model should be upgraded from a single technology application

to the engineering system management application [5]. It will raise the value of 3-D visualization model.

- Attention to the accuracy of database. As China AP1000 NPP Project is the first one in the world, the design has to be modified at the same time with the construction activity. Accompanied with the engineering drawings frequent modification, the original model information has to be updated and supplemented. This means that a certain technical force should be prepared to guarantee the accurate data base is available on time or it will easily result in decision making biases.
- Enhance the share of information. The construction of NPP needs the cooperation of the operating units, general constructors, civil construction units, installation units, supervision units, equipment manufacturers, design units and other construction enterprises. The 3-D visual model technique will play a more significant role if all the participating units have used the same information platform.

In short, the application of this new technology should be explored, developed and extended step by step together with the NPP project in the future. The user should make feedback experience summary in order to improve and represent the potential advantages of the 3-D visualization model technology.

**References**

[1] Zhong Denghua, Liu Donghai. Theory and Application of Engineering Visualization Aided Design [M]. Beijing: China Water Power Press, 2004:11-12.
[2] Zhu Xiaochao. The Research and Establishment Of 3d Information Model of Hydraulic and Hydro-Power Engineering[D]. North China Electric Power University. 2014.
[3] LE Yun MA Ji-wei. Development and Application of Construction Project Information Portal[J]. JOURNAL OF TONGJI UNIVERSITY(NATURAL SCIENCE. 2005, 33(4).
[4] Ma Yuanhua. AP1000 nuclear island construction technique [D]. Proceedings of 2010 Nuclear Power Plant construction New Technique Exchange Meeting. 2010.
[5] Zhu Qing. Full Three-Dimensional GIS and Its Key Roles in Smart City[J]. Journal of Geo-Information Science. 2014, (2).

IOP Conf. Series: Earth and Environmental Science **61** (2017) 012024 doi:10.1088/1755-1315/61/1/012024

# Genesis analysis of karst water trichloroethylene pollution in the east of a city

**Peng Jiang[1,2], Zhenmin Ma[1,2,*] and Ming Wen[1,2]**
[1] School of resources and environment, University of Jinan, Jinan 250022, China
[2] Research center of groundwater numerical simulation and pollution control engineering in Shandong province, Jinan 250022, China
*corresponding author's e-mail: stu_mazm@ujn.edu.cn

**Abstract**. To understand the situation of Karst water Trichloroethylene (TCE) pollution in the east of city, Karst water samples sampled at 43 monitoring sites were analyzed. Result shows that relevance ratio of TCE is 100%, over the standard rate of 97.67% and the maximum value is 73.64μg/L, as 14.73 times as the standard value(5μg/L). Causes of groundwater TCE pollution were analyzed. Result shows that indiscriminate discharge of waste water and poor groundwater vulnerability are the main reasons. And based on the reasons, the pollution ways of TCE were found out.

## 1.Introduction

TCE is mainly used to clean metal, extract solvent, chemical materials, dry cleaning, and so on. In recent years, TCE is widely used to electrical and mechanical industry, clean metal parts and electronic components[1-3]. With the widely application of TCE, groundwater TCE pollution problem has become more serious. In groundwater pollution investigation of America, the relevance ratio of TCE was the highest, as much as 36%[4,5]. In 1970s, TCE was found in groundwater and in the latest investigation, groundwater TCE pollution was more serious[6].TCE was found in groundwater of Germany, Britain and Italy[7-9]. In China, TCE was found in groundwater, too. In groundwater organic contamination investigation of Taihu watershed, at 36 monitoring sites, the relevance ratio was 30.6% and serious pollution area fit into the industrial distribution[10].

Karst water samples sampled at 43 monitoring sites will be analyzed to find out the characteristic of groundwater TCE pollution. Thinking about pollution sources of TCE, hydrogeological condition and pollution ways of TCE, causes of groundwater TCE pollution will be analyzed.

## 2.Materials and methods

### *2.1 Study area*

Study area is located in the east suburb of a city, between the Taishan mountain area to the south and alluvial plain of Yellow River to the north, low-lying north south high and low-lying west east high. From south to north, geomorphic type is low hills denudation zone, piedmont clinoplain and alluvial plain of Yellow River. The climate belongs to warm temperate continental monsoon climate, with four different seasons, the rainfall in summer is more than that in winter. Average annual precipitation is 660.3mm. The mainly rainfall focus on July to September.

The distribution of the Quaternary unconsolidated strata is uneven in study area and Quaternary unconsolidated strata is mainly distributed in piedmont clinoplain and alluvial plain of Yellow River, thickness increasing gradually from south to north. Quaternary unconsolidated strata is divided into two layers. The upper layer is brown loam layer mixed with gravel and dinas, even and widely distributed, having good permeability. The under layer is purple clay, dense structure, having poor permeability, which is regarded as water-resisting layer. Ordovician limestone is widely distributed in study area. Ordovician limestone is exposed in the south of study area and becomes concealed to the north. The Carbonifernus, Permian strata and Magmatic rock are distributed in the north of study area, which are regarded as confining boundary because of its poor permeability. Groundwater system mainly include the karst fissure water system and the loose Quaternary pore water system. Shallow pore water mainly lies in brown loam layer, mainly recharged by atmospheric precipitation, which is the main source of agricultural production. Karst water lies in Ordovician limestone, mainly recharged by atmospheric precipitation in uncover limestone area and also recharged by shallow pore water, which is main water aquifer.

Study area is heavy industrial district. Under extensive Mode of development, groundwater over-exploitation has made groundwater level continual lowering, forming landing funnel of groundwater. Waste water produced in the progress of enterprise production discharges into surface water without wastewater treatment, which makes surface water contaminative. Polluted surface water infiltrating into underground can contaminate groundwater.

*2.2 Sampling and analysis*

Thinking about the hydrogeological conditions and the distribution characteristics of different plants in study area, 43 sampling sites are set in study area and the layer of sampling is Ordovician limestone(Figure 1).

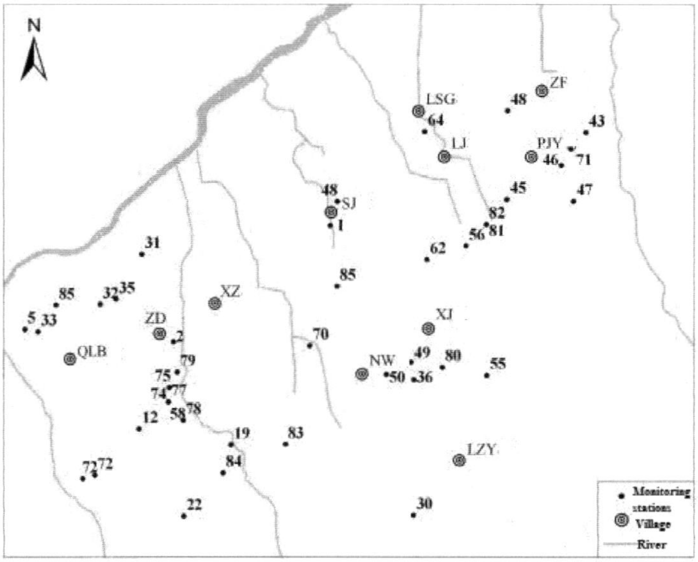


Fig. 1 The location of 43 sampling sites.

Because of the concentration of TCE in groundwater is low, mostly in ppb level. Based on National Standard Method and the U.S. Environment Protection Agency (EPA) method, the concentration of TCE in groundwater are analyzed by GC/MS (gas chromatography/mass spectrum) method.

*2.3 Methods*

*2.3.1 Groundwater pollution characteristics.* Groundwater pollution characteristics is analyzed based on the map of the distribution of contaminant. The map of distribution can show the shape of pollution, migration tendency, the maximum and minimum of contaminant and the area between a concentration value and another concentration value. On the basis of the concentration of TCE at

IOP Conf. Series: Earth and Environmental Science **61** (2017) 012024 doi:10.1088/1755-1315/61/1/012024

sampling sites, firstly, sampling data is dispersed using Kring in mapgis. Secondly, the map of the distribution of TCE in groundwater is drawn through Grid model in mapgis.

*2.3.2 Genesis analysis of groundwater TCE pollution.* Genesis analysis of groundwater pollution is the process of determining pollution sources, finding out pollution way and analyzing controlling factors. On the basis of determining pollution sources, understanding the permeability of surface oil and the thickness of unsaturated zone is necessary, which determines that contaminants go into underground easily or difficultly. Determining the recharge and discharge correlation between shallow groundwater and surface water and determining whether the recharge and discharge correlation between shallow groundwater and deep groundwater are important. Water is the carrier of contaminants migration. Knowing moving path of all kinds of water in the vertical direction is useful to know the pollution way.

## 3.Results and discussion

### 3.1 Groundwater pollution characteristic of TCE

The concentration of TCE at 43 monitoring sites were reported in table 1. From (table 1)we could see that TCE were checked out at all monitoring sites. Among them, the maximum was 73.64μg/L and the minimum was 2.24μg/L. The mean value was 53.37μg/L. Taking EPA drinking water standard(5.00μg/L) as a standard, except monitoring site of 30(2.24μg/L), the concentration of TCE at other sites all exceeded the standard. The rate of exceeding the standard was 97.67% and the maximum was 14.73 times as much as the standard.

Table 1 the concentration of TCE at monitoring sites (μg/L).

| N | C | N | C | N | C | N | C | N | C |
|---|---|---|---|---|---|---|---|---|---|
| 1 | 23.53 | 35 | 39.48 | 71 | 29.66 | 48 | 22.61 | 85 | 43.44 |
| 2 | 44.67 | 36 | 66.54 | 72 | 65.42 | 49 | 69.48 | 72 | 65.43 |
| 5 | 42.11 | 64 | 23.88 | 58 | 70.88 | 50 | 68.87 | 48 | 23.54 |
| 12 | 70.67 | 85 | 43.33 | 73 | 71.22 | 56 | 72.55 | | |
| 19 | 73.64 | 43 | 30.11 | 74 | 70.26 | 62 | 72.62 | | |
| 22 | 66.76 | 44 | 30.11 | 75 | 70.34 | 70 | 72.61 | | |
| 30 | 2.24 | 45 | 29.54 | 77 | 70.22 | 81 | 73.10 | | |
| 31 | 41.50 | 46 | 28.56 | 78 | 71.35 | 82 | 73.11 | | |
| 32 | 41.07 | 47 | 27.43 | 79 | 70.33 | 83 | 73.36 | | |
| 33 | 42.32 | 55 | 65.56 | 80 | 68.76 | 84 | 72.32 | | |

N: the number of monitoring sites; C: the concentration of TCE at monitoring site.

The map of the distribution of groundwater TCE pollution was drawn by Mapgis, as shown in (Figure 2). It can be seen from (Figure 2) that high concentration of TCE distribute in the middle of study area, around the monitoring sites of 36, 50, 80 and 83, whose shape was dumbbell-shaped. Centering on the area of high concentration, TCE spread around with the concentration of TCE diminished gradually. The area of low concentration distributed in the south of study area.

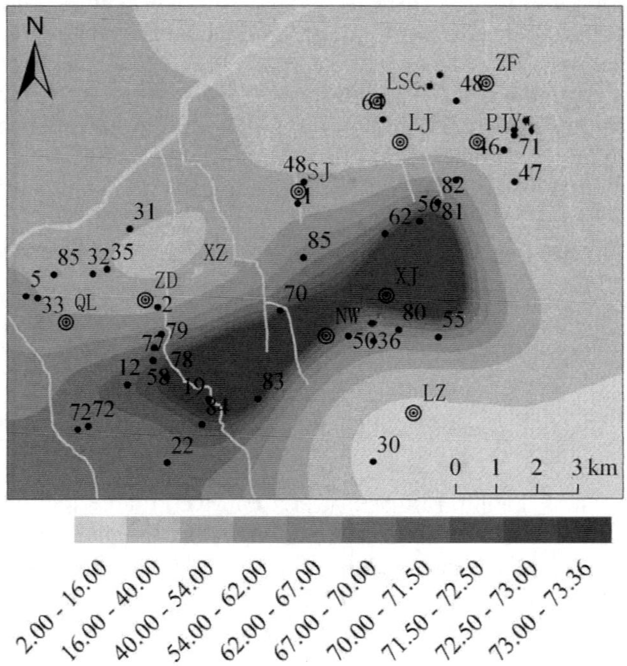


Fig. 2 The distribution of TCE in groundwater.

*3.2 Genesis analysis of groundwater TCE pollution*

*3.2.1 Groundwater TCE pollution sources.* The investigation result shows that there are three pollution sources, s-1, s-2 and s-3 in study area, which are shown in Figure 3. It can be seen from (Figure 3) that the three pollution sources are a zonal distribution. The distribution of pollution sources is similar to the distribution of high concentration of TCE in groundwater and the monitoring sites of checking out high concentration of TCE. They are near the pollution sources.

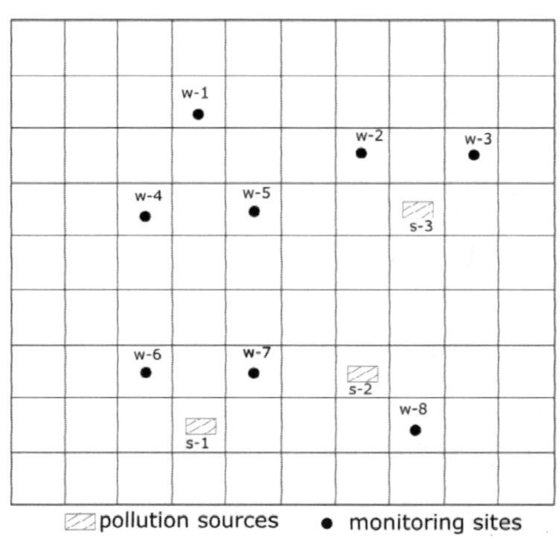


Fig. 3 The distribution of pollution sources.

The three pollution sources are three sewage discharge outlets, there are some blow-off lines near the sewage discharge outlets, including open channel and concealed conduit, and some industries discharge waster water through blow-off lines. Thinking about the use of TCE, ten industries are selected, which are mechanical and electrical industries. TCE is used for cleaning metal parts and electronic components in the production process of the ten industries. So, in waster water discharged by the ten industries, TCE is checked out. Among the ten industries, there are three industries discharging waster water to s-1, three industries discharging waster water to s-2 and four industries discharging waster water to s-3.

*3.2.2 Poor antifouling property of site aquifer.* The buried features of groundwater can decide that contaminants go into underground easily or difficultly. In study area, the lithology of aeration zone is brown loam. The thickness of aeration zone is less than 10m and the hydraulic conductivity of aeration zone is $1.3\times10^{-3}$cm/s. Good permeability and short migration path result in contaminants going into underground easily. So, it is easy for shallow pore water to be contaminated. There is purple clay layer existing between shallow pore water and Karst water. The thickness of purple clay layer is between 30m and 40m. It has poor permeability, whose permeability is less than $1.3\times10^{-6}$cm/s. Because of poor permeability and wide distribution, this layer is regarded as water-resisting layer. And, under the natural state, the water table of Karst water is higher than shallow pore water. So, even if shallow pore water is contaminated, Karst water isn't affected. However, over-exploitation of Karst water makes water table of Karst water decline, forming groundwater depression cone and the water table of Karst is lower than shallow pore . And pumping well casing has been broken because of lacking of repair for long years. For the two reasons given above, shallow pore water supplies Karst water. If shallow pore water is contaminated, Karst water is also contaminated. Therefore, antifouling property of Karst water is poor.

*3.2.3 Groundwater pollution way of TCE.* Water is the carrier of the migration of contaminants. So, research on groundwater pollution ways of TCE, recharge, runoff and discharge of groundwater need to be found out. Types of groundwater in the study area include shallow pore water and Karst water. Shallow pore water is recharged by precipitation, surface water and the lateral flow. The ways of shallow pore water discharge are artificial discharge, lateral flow discharge and recharging Karst water. Karst water is recharged by the lateral flow and shallow pore water. The ways of Karst water discharge are artificial discharge and lateral flow discharge. It can be seen from Figure 2 that the concentration of TCE in groundwater is very low in the south of study area. So, TCE doesn't come from the lateral flow, but comes from vertical infiltration of surface pollution sources.

Waster water produced by industries discharged into channels through blow-off line without waster water treatment. Because of channel bottom without seepage prevention, waster water including TCE permeated into aeration zone. Due to good permeability aeration zone, waster water permeated into shallow pore water further to contaminate shallow pore water. When shallow pore water recharged Karst water along the damaged wall of a well, TCE also went into Karst water to contaminate Karst water. In addition, density of TCE is bigger than water and TCE is easy volatile. Insoluble TCE gathered at the bottom of the Karst aquifer and contaminated Karst water by volatilization.

## 4.Conclusions

The rate of exceeding the standard is 97.67% and the maximum is 14.73 times as much as the standard. The shape of high concentration of TCE is dumbbell-shaped. Centering on the area of high concentration, TCE spread around with the concentration of TCE diminished gradually.

The investigation result shows that there are three pollution sources in study area, waste water outfalls of ten mechanical and electrical industries.

Antifouling property of groundwater is poor. TCE passes through aeration zone and shallow pore water to contaminate Karst water.

## Acknowledgment

The author firstly would like to express their appreciation to the anonymous reviewers. This research was supported by Natural Science Foundation of Shandong Province (No.ZR2014DM011) and Shandong provincial key research project (No.2015GSF117025).

IOP Conf. Series: Earth and Environmental Science **61** (2017) 012024 doi:10.1088/1755-1315/61/1/012024

## References


[1] Yao Lin. Production and Consumption Status of Trichloroethylene and its Prospective Market[J]. Guangzhou Chemical Industry, 2013, 41(7): 42-44.

[2] Grajales-Mesa SJ and Malina G. Screening reactive materials for a permeable barrier to treat TCE-contaminated groundwater: laboratory studies[J]. Environmental Earth Sciences, 2016, 75(9):772.

[3] Elad Dafny.TCE longevity in the vadose zone and loading to the groundwater—The case of episodic NAPL releases from near-surface source[J].Environmental Technology & Innovation, 2017:128–140.

[4] Moran M J, Zogorski J S, Squillace P J. Chlorinated solvents in groundwater of the United States[J]. Environmental science and technology, 2007, 41(1): 74-81.

[5] Lu Jie, Li Menghong, Pan Jiafen. Control and remediation of groundwater environment contaminated by organic chlorinated hydrocarbons[J]. Journal of Shandong University of Technology( Natural Science Edition), 2008, 22(4): 27-30.

[6] Hirata T, Nakasugi O, Yoshioka M, et al. Groundwater pollution by volatile Organochlorines in Japan and related phenomena in the subsurface environment[J]. Water Science and Technology, 1992. 25(11): 9~16.

[7] Cavallaro, A., Corradi, C, De Felice G., and Grassi, P. Underground water pollution in Milan and the Province by industrial chlorinated organic compounds, in J.F. de L. G. Solben (eds.), Effects of Land Use on Fresh Waters, Ellis Horwood, Chichester, 1986:68–84.

[8] Burston M W, Nazari M M, Bishop P K, et al. Pollution of Groundwater in the Coventry Region (Uk) by Chlorinated Hydrocarbon Solvents[J]. Journal of Hydrology, 1993, 149(1): 137-161.

[9] Liu Mingzhu. Study on Numerical Simulation of Migration and Transformation of Chlorinated Hydrocarbon in Shallow Groundwater[D]. Beijing: China University of Geosciences, 2002.

[10] Chen Honghan, He Jiangtao, Liu Fei, et al. Organic contamination characteristics of shallow groundwater in a study area of the Taihu Lake basin, Jiangsu, China[J]. GEOLOGICAL BULLETIN OF CHINA, 2005, 66(1-2):137-148.

IOP Conf. Series: Earth and Environmental Science **61** (2017) 012025 doi:10.1088/1755-1315/61/1/012025

# Landscape structure planning and the urban forest in polycentric city regions

**A J Simson**

Professor of Landscape Architecture + Urban Forestry, Leeds Beckett University, Broadcasting Place, Woodhouse Lane, Leeds LS2 9EN, UK
e: a.simson@leedsbeckett.ac.uk

**Abstract.** The World is continuing to urbanise at an increasing and some say alarming rate, and although urbanism is not uniform in all countries, without a doubt the 21$^{st}$ century is the century of the Polycentric City Region. By the year 2007, for the first time in history, the world hosted more urban dwellers than rural, and in order to deal with this urban expansion in an environmentally acceptable way, the concept of the "sustainable compact city" was advocated. There is now an increasing canon of research however that suggests that such cities may not be quite as sustainable as they are claimed to be. As a consequence, the concept of "urban green infrastructure", which includes the concept of urban forestry, is being incorporated into new thinking on the landscape structure planning of expanding cities and city regions to ensure that they provide an acceptable quality of life for their inhabitants. The environmental, economic, social, health, well-being and cultural benefits that emanate from such an approach to promoting resilient landscape structure planning are considerable. Such an approach to landscape structure planning is well-able to repair the beneficial relationship that people once had with their landscapes, a relationship that has arguably suffered as our scientific and economic cultures have tended to gain the upper hand in the post-industrial times in which we live. Human beings have had a long, deep, cultural relationship with trees, woodlands and the landscape – a relationship which transcends national cultures. The use of the term "landscape" does not refer to the rather shallow modern concept of 'the landscape as a view", but to the more fundamental concept of "landscape as the composition of our world". Thus it refers to both urban, peri-urban and rural areas, and the urban forest is the prime spatial articulator of a landscape structure plan. Although the words "forest" and "forestry" are now generally understood to be connected with trees, the words have arguably been derived from the Latin word "foris", meaning "out of doors" or "unenclosed land" (Porteous, 1928 p34). Thus urban forestry could be described as the "urban out of doors". This presentation will consider the benefits that can be achieved from developing a viable urban forest structure as the backbone of a polycentric city region landscape structure plan. It will focus upon the Leeds Polycentric City Region in the UK and its emerging *Leeds City Region Green and Blue Infrastructure Strategy 2017 - 2036* as a case study.

## 1 Introduction

The aim of this paper is to discuss the merits of re-establishing the concept of Landscape Structure Planning as a creative means of planning, designing, implementing and managing a viable approach to the expansion of our towns and cities into Poly-Centric City Regions. The World is continuing to urbanise at an increasing, and some would say alarming, rate. According to the United Nations (http://www.un.org/esa), only 14% of the world's human inhabitants lived in urban areas at the turn of the 20th century. As late as 1960, only one-third of the World's population resided in urban areas, and it wasn't until 2007 that, for the first time in history, the World hosted more urban dwellers than rural.

IOP Conf. Series: Earth and Environmental Science **61** (2017) 012025 doi:10.1088/1755-1315/61/1/012025

By 2025, this will have risen to two-thirds, and it has been predicted that 60% of the urban areas that will exist in 2030 have yet to be built (EEA, 2016). Much thought will need to be given to exactly what sort of urban futures we should be pursuing. The widespread growth in many countries of car-dependant suburban neighbourhoods has brought unsustainable energy waste and pollution levels, as well as lengthy commuting between homes and workplaces. That said, the success of such suburban housing reveals a growing public dis-satisfaction with the air pollution and hard landscapes of the large city centres, and a desire to live in a greener and more peaceful environment, and for a better contact with nature and natural processes.

## 2 The rationale

Our towns and cities will continue to be subject to constant change, and no urban area is likely to be immune from the forces that bring this about. Indeed, as the 21$^{st}$ century progresses, it is likely that this pace of change will accelerate considerably, and places that were once prosperous may slip into decay and physically, commercially or visually decline, whilst other areas that are currently deemed to be poor or run-down may well benefit from regeneration or revival. The reasons for this state of affairs are many and varied of course, and paradoxically can have as much to do with the "image" of an area – either real, imagined or invented – as with physical re or de-generation. (Shaw & Robinson, 1998). Such urban areas will be different however - urbanism is not uniform in all countries. Although there is still pressure to continue aiming for and developing "sustainable compact cities", there is a growing canon of research that suggests that there might be a finite size for such cities, and that if this is exceeded, they are perhaps not quite as sustainable as they might be, particularly in terms of human health and well-being (NEUMAN, 2005). It might be more viable therefore to plan for and include a number of such cities into a Polycentric City Region.

Change is inevitable, but shaping that change has to be founded upon positive and creative ideas, so as to secure better outcomes for our urban futures. Polycentric City Regions are complex environments. They exhibit both common and unique features that can influence their success or failure, aspects such as the state of the local economy, city identity, social cohesion and safety, green, blue and grey infrastructures and the subsequent health and well-being of their residents. There is no such things as a static polycentric region – they are either on the way up or on the way down.

For polycentric cities to succeed and flourish in the 21$^{st}$ century, they need to be able to attract and retain the best, brightest and most creative of the up-and-coming generations. Thus, there is no creativity in "business as usual", and this presentation will consider what contributions urban forestry can make in promoting the public realm of the city region - the usable and special spaces that can be accessed by as many people as possible. Such spaces are the lifeblood of cities, and the contributions that the macro and micro urban forest make in articulating these spaces is increasingly being recognised. We are now able to engage in "evidence-based design", defined as an approach to design that emphasises that we now have the credible evidence required to support the promotion of urban forestry as one of the critical key elements of the successful polycentric city region of the future (Simson & Ostoić, 2016)

## 3 Case Study: The Leeds City Region

The Leeds City Region sits at the centre of a current UK Government initiative called the "Northern Powerhouse". It comprises 10 District Local Government Authorities, covers an area of over 5000 km², has a population of over 3 million people, a £63 billion economic output and is the largest Local Enterprise Partnership [LEP] outside London. Like many towns and cities in the UK however, the City Region deindustrialised long ago, and although it currently has a diverse and service-orientated economy and a larger manufacturing base than many contemporary UK city regions, it badly needs to attract and retain modern knowledge-intensive business services for its future economic success (Cox, et al, 2016)

Thus the Leeds City Region [LCR] has developed a Green and Blue Infrastructure Strategy 2017-2036, which includes an Urban Forestry Plan, that is wholly integrated with the LCR's Strategic Economic Plan 2017-2036, and which together will deliver their collective vision of "good growth", which combines improvements in economic productivity and output, with social inclusion and a quality environment. Such long-term planning will require a series of larger-than-local Landscape Structure

Plans to graphically articulate the vision, and establish the need, the scale, the role, the purpose and the broad locations of the components of the Green and Blue Strategy Plan.

The provable benefits of such a green and blue infrastructure extend far beyond the intrinsic pleasures of an attractive environment. They include supporting good mental and physical health by improving the experiential quality of commuting, and tackling obesity and diabetes by increasing quality footpaths and cycleways; reducing the frequency and severity of flooding along the City Region rivers; bringing diverse and multi-ethnic communities together; regenerating areas of need; aiding biodiversity and providing a home for wildlife; acting on climate change and enriching human lives. Fundamentally, the Green and Blue Infrastructure Strategy will support a strong and resilient economy. It enables and adds value to new developments, attracts tourism and investment, supports businesses, jobs and training and ensures the resilience of the City Region's assets and its integrated infrastructure. Such a Landscape Structure Plan has the acumen to meld the urban "critical infrastructure" (ie. energy, food, water, transport, etc) with the emerging "urban green infrastructure" (ie. ecological security, multi-functional green spaces, resilience, etc.) into a viable "Integrated Infrastructure", the vital backbone of a Landscape Structure Plan. This will help to deliver its vision for "good growth", which combines improvements in economic activity and output with social inclusion and a quality environment for all.

## 4 Conclusions

It is known that the pace of urbanisation across the world will accelerate as Polycentric City Regions respond to change – changes in their population, in their economies, in their ethnic compositions, climate change and in their people's expectations and demands of the places they inhabit (Brotchie et al, 1995). The planning and design of the urban landscapes of the late 20th and early 21st centuries has all too often not created the quality liveable places that their designers claimed they would. Thus urban design, landscape design and urban green infrastructures are at a crossroads in many countries, as the concept of globalism increases apace, post-industrialisation takes hold and the quality of city centre urban life continues to decline as a result. Thus many cities continue to expand, often with standardised, low-quality developments, which encourages those citizens who have the available resources and opportunities to seek safer, greener, more pleasant edge–of-town surroundings to inhabit, where there is an appreciably higher canopy cover of trees. Urban resilience is increasingly compromised as a result.

Shaping these changes so as to secure better outcomes for urban futures requires us to be confident and brave, and to seek to improve the ways in which we invest in our landscape structure planning, including our multi-functional green infrastructure and urban forestry, to support resilient new settlements and regenerate existing communities.

## References

[1] Cox, E, Raikes, L & Carella, L, (2016) *The State of the North 2016 – building northern resilience in an era of global uncertainty*. Institute for Public Policy Research pp. 26-28.
[2] Brotchie, J, Batty, M, Blakeley, E, Hall, P & Newton, P. (1995) Cities in Competition, Melbourne. Longman.
[3] European Environmental Agency (2016) Secretariat of the Convention on Biological Diversity.
[4] Neuman, M, (2005) The Compact City Fallacy, in Journal of Planning Education Research, Vol. 25, pp.11-26.
[5] Porteous, A, (1928) The Forest in Folk Law and Mythology. New York. Macmillan.
[6] Shaw, K, and Robinson, F, (1998) Learning from Experience? Town Planning Review Vol. 69, No. 1.
[7] Simson, A & Ostoić, S, (2016) Landscape Urbanism and the Building of Sustainable Futures, in Building Sustainable Futures, ed. by Dastbaz, M, Strange, I & Selkowitz, S. Chapter 11, pp.247-269. New York. Springer.

IOP Conf. Series: Earth and Environmental Science **61** (2017) 012026 doi:10.1088/1755-1315/61/1/012026

# Experiment and simulation studies on SPEEK PEM with different sulfonation degrees

**Rujie Wang* and Ming Li**

Department of Environmental Science & Engineering, North China Electric Power University, Baoding, China
rujiewang@gmail.com

**Abstract**. Effects of degrees of sulfonation (DS) on the cluster aggregation, proton conductivity and mechanical properties of sulfonated poly ether ether ketone (SPEEK) proton exchange membranes (PEMs) were investigated by experiment and simulation studies. SPEEK materials with different DS and the corresponding PEMs had been prepared by sulfonation and solution casting. The water uptake, swelling ratio, proton conductivity and mechanical properties of SPEEK PEMs were greatly affected by DS. And the hydrophilic cluster aggregation in SPEEK of different DS was revealed by molecular simulation. The relationship between structure and performance of SPEEK membrane provides theoretical guidance for the preparation of high performance proton exchange membranes.

## 1. Introduction

Because of high efficiency and zero emission, proton exchange membrane fuel cell (PEMFC) is believed to be alternative energy conversion devices in mobile vehicles and power generations [1]. Proton exchange membrane (PEM) acts as a key component of PEMFC, which will directly dominate the performance and lifetime of the PEMFC [2].

Compared with other commonly used PEM materials in PEMFC, sulfonated poly ether ether ketone (SPEEK), a kind of non-fluorinated material, possesses good chemical stability and thermal stability, low cost, and high mechanical stability. However, as is the case with most non-fluorinated material, SPEEK PEMs are limited by poor hydrophilic cluster morphology and insufficient proton conductivity, resulting from low hydrophilic/hydrophobic difference and rigidity of the polymer backbone [3]. It's well known that the increase of degrees of sulfonation (DS) would improve the hydrophilic cluster morphology and the proton conductivity as well. Nevertheless, serious deterioration in swelling and mechanical properties will be yielded by high DS. This study aim to investigate the effects of DS on the cluster aggregation, proton conductivity and mechanical properties of SPEEK PEMs by experiment and simulation studies.

## 2. Experiment and simulation detail

### *2.1. Preparation of SPEEK material and SPEEK membranes*

According to our previous work [4], sufolnation preparation of SPEEK from PEEK was carried out from Poly(ether ether ketone) (PEEK) powder (Degussa, VESTAKEEP 4000 P: density 1.30 g $cm^{-3}$). SPEEK material with different DS was obtained by tuning the reaction time from 0.3 h to 1.5 h at 50 °C. Solution casting method was applied to fabricate SPEEK membranes.

*2.2. Simulation*

SPEEK chain composing of 10 monomers was adopted. The DS values of prepared SPEEK samples were determined by titration and adopted in molecular simulation. Due to chain rigidity and stiffness of SPEEK, initial density of 0.4 g $cm^{-3}$ is set for the periodic cell to avoid ring spearing and catenation [5]. Table 1 exhibits the composition of each periodic cell of SPEEK with different DS. Annealing strategy used in this study was shown in figure 1.

Table 1. Construction of periodic cells of SPEEK with different DS.

| DS | Number of SPEEK chain | Number of $H_2O$ | Number of $H_3O^+$ |
|---|---|---|---|
| 30% | 5 | 30 | 15 |
| 40% | 5 | 40 | 20 |
| 50% | 5 | 50 | 25 |
| 60% | 5 | 60 | 30 |
| 70% | 5 | 70 | 35 |
| 80% | 5 | 80 | 40 |

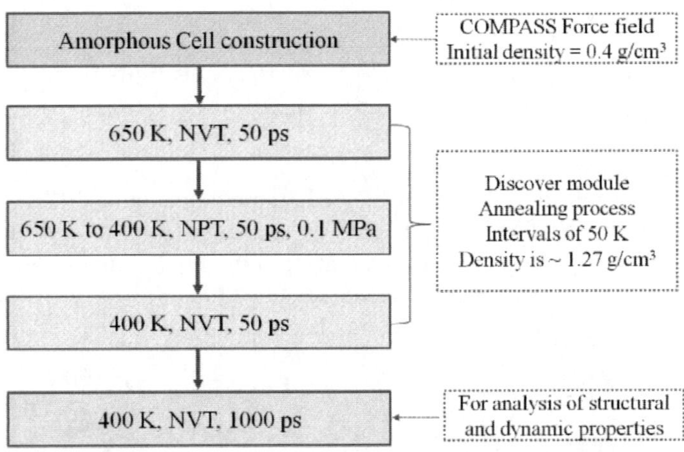


Figure 1. Simulation strategy used in this study.

## 3. Results and discussion

*3.1. Properties of SPEEK membranes*

The DS values of SPEEK mentioned in this research were according to NaCl titration results. As the sulfonation time increased from 0.3 h to 1.5 h, the DS values of SPEEK samples varied from 20% to 70%. This proves that the DS of SPEEK polymers could be easily controlled by sulfonation time.

Figure 2 shows the water uptake, swelling ratio, proton conductivity and XRD spectra of SPEEK membranes with different DS. Enhanced proton conductivity was observed with increasing DS. However, SPEEK sample with a DS of 54% exhibited excessive swelling (as high as 40%) at 80 °C, indicating a decline in dimensional stability of the SPEEK membranes with higher concentrations of sulfonic acid groups. Moreover, as reported in our previous publication [4], tensile strength and modulus of SPEEK samples decreased with increasing DS. When the DS of SPEEK increase from 35% to 54%, tensile strength and modulus of SPEEK membranes gradually decrease from 33.20 MPa and 1156.89 MPa to 23.79 MPa and 564.24 MPa.

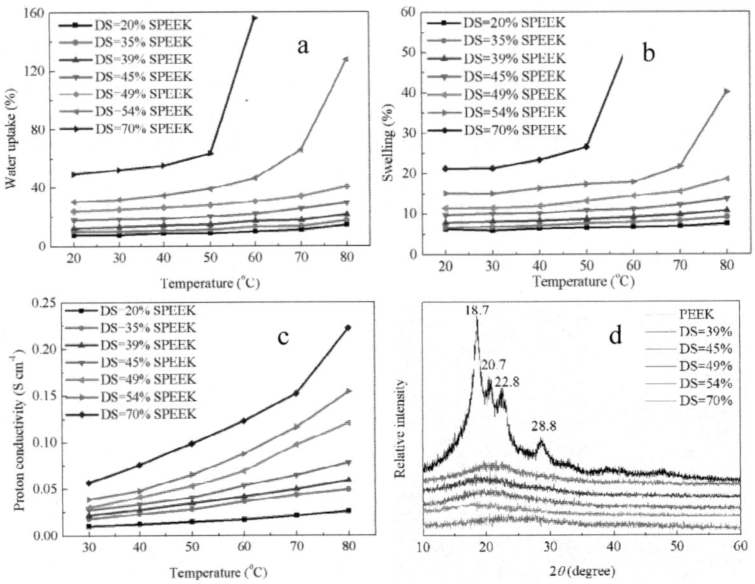


Figure 2. Water uptake (a), swelling ratio (b), proton conductivity (c) and XRD spectra (d) of SPEEK membranes with different DS.

### *3.2. Molecular simulation results*

To explore hydrophilic cluster morphology of SPEEK membranes with different DS, MD simulation was performed. The dispersion of $H_2O$ and $H_3O^+$ was investigated by molecular dynamic simulation. The result was shown in figure 3. With increasing DS, $H_2O$ content increases proportionally. Consequently, the hydrophilic clusters, composed of $H_2O$ and $H_3O^+$, become expanded and connected, resulting in wide and connected proton conducting channels.

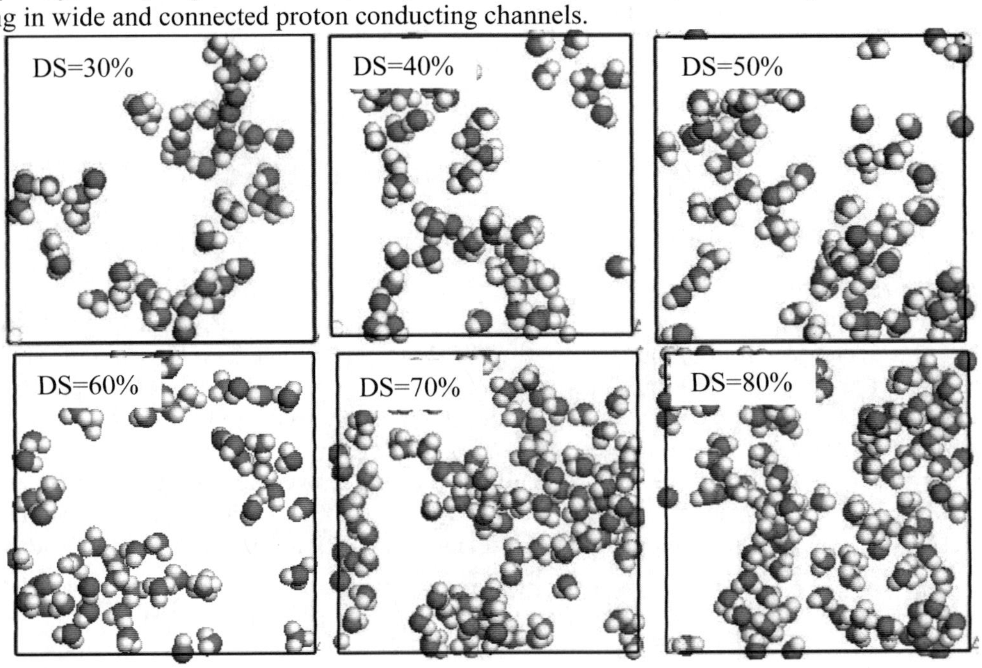


Figure 3. Final snapshot of $H_2O$ and $H_3O^+$ in SPEEK matrix. Red, oxygen in $H_2O$ and $H_3O^+$; white, hydrogen in $H_2O$ and $H_3O^+$; other atoms were set as invisible.

Mean square displacement (MSD) analysis at 300 K was carried out to study the effects of such cluster morphology on the mobility of $H_3O^+$ according to equation (1) [6],

$$MSD = \langle [r_i(t) - r_i(0)]^2 \rangle \quad (1)$$

where $r_i(0)$ and $r_i(t)$ denote the coordinates of atom i at specific time t=0 and t=t respectively, and the bracket represents the ensemble average. Figure 4 reveals the mean square displacements of $H_3O^+$ in SPEEK matrix at 300 K. As shown in figure 4, with the increasing of DS, the MSD of $H_3O^+$ at 300 K was improved gradually. With enhanced $H_3O^+$ mobility, a higher proton transfer is expected, which is consistent with the result of proton conductivity in figure 2c.

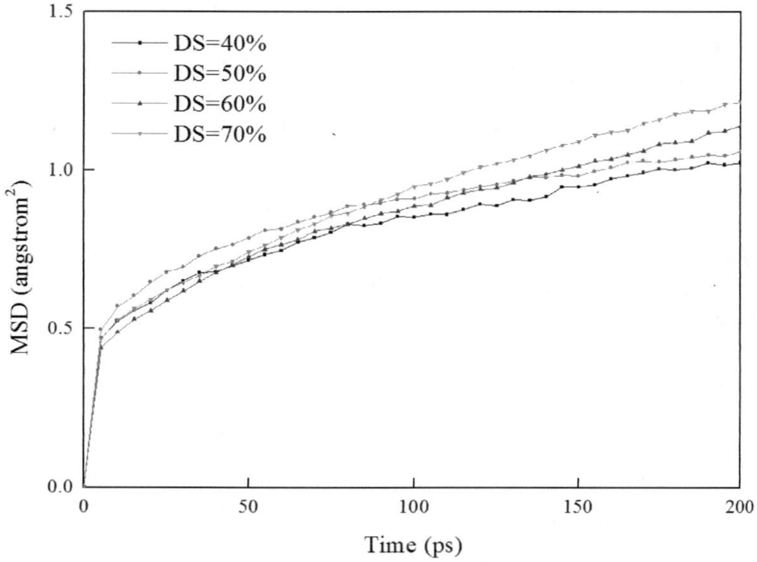


Figure 4. Mean square displacements of $H_3O^+$ in SPEEK matrix at 300 K.

## 4. Conclusion

Enhanced water uptake and proton conductivity of SPEEK membranes are observed with increasing DS. However, SPEEK sample with a DS of 54% exhibits excessive swelling (as high as 40% at 80 °C) and poor mechanical properties, indicating a decline in dimensional stability and mechanical stability. Furthermore, MD simulation was conducted to explore the cluster morphology of SPEEK membranes with different DS. With increasing DS, the hydrophilic clusters are developed to be wide and connected proton conducting channels, and $H_3O^+$ mobility is enhanced as well. Therefore, a higher proton transfer is expected, which is consistent with the result of proton conductivity.

## Acknowledgments

The authors thank the financial support of the Fundamental Research Funds for the Central Universities (Grant nos. 2016MS110 and 2016MS111).

## References

[1] Mader J A and Benicewicz B C 2010 *Macromolecules* **43** 6706–15
[2] Park M J and Kim S Y 2013 *J. Polym. Sci.* B*: Polym. Phys.* **51** 481–493
[3] Park C H, Lee C H, Sohn J Y, Park H B, Guiver M D and Lee Y M 2010 *J. Phys. Chem.* B **114** 12036–45
[4] Wang R, Wu X, Yan X, He G and Hu Z 2015 *J. Membr. Sci.* **479** 46–54
[5] Bahlakeh G, Nikazar M, Hafezi M J, Dashtimoghadam E and Hasani-Sadrabadi M M 2012 *Int. J. Hydrog. Energy* **37** 10256–64
[6] Zhang W, Yue P L and Gao P 2011 *Langmuir* **27** 9520–7

IOP Conf. Series: Earth and Environmental Science **61** (2017) 012027 doi:10.1088/1755-1315/61/1/012027

# Study on the problem-oriented collaborative strategy mechanism for local construction of northeast countryside

**Sun Chaoyang**[1], **Sui Xin***[2(Correspondence author)] and **Li Mo**[3]

[1]Dalian Nationalities University, No. 18, Liaohe West Road, Dalian Economic & Technical Development Zone, China, 116600
[2]City Institute of Dalian University of Technology, No.31, Tieshan West Road, Dalian Economic & Technical Development Zone, China, 116600
[3]City Institute of Dalian University of Technology, No.31, Tieshan West Road, Dalian Economic & Technical Development Zone, China, 116600
Correspondence author: Sui Xin suixin@dlnu.edu.cn

**Abstract.** Under the background of starting the new round of the revitalization of the northeast economy in China, there is a pressing need to set up new idea of the development of the innovative urban - rural integration. Regarding the form of the collaborative construction in the practice of rural construction in recent years as cutting point, through the internal order analysis, seeking problems, classification and analysis of the basic investigation data of typical villages in the countryside of the Northeast and oriented by the problems, the paper presents the applicable and feasible construction strategies suitable for the development of countryside step by step, so as to establish the collaborative mechanism of "up-down collaboration, internal-external coupling, extensive participation, rooting in rural areas and perpetual development", which aims at providing the rule, method and idea with universal significance for the overall revitalization of the northeast countryside.

## 1. Research background and research status at home and abroad

### *1.1 Research background*

In recent years, the national strategy has paid close attention to the issue of "Northeast Revitalization" and "Rural Development". The uneven distribution of resource increases the urban and rural differences, together with the continuous decline in northeast economic development in recent years, rural development has been constrained. Northeast countryside is an important part of Chinese traditional culture and regional diversity. It has a relatively independent social organization and complete system. How to be based on the northeastern regionalism, activate the dilemma faced by the northeast countryside, study the rural activation mechanism with universal significance, characteristics and laws and achieve the goal of inheritance and development of northeast countryside has become an important topic to be focalized.

### *1.2 Externalresearch status*

Foreign theoretical research on rural construction has been started early with mature practice models, such as American model of being driven by urbanization, Canadian model of government becoming the partner of farmers, Japanese model of being driven by intermediaries, German model of being driven by rural industrialization, Swedish model of being promoted by farmer cooperative organizations, Danish model of attaching importance to the development of highly efficient and export-oriented agriculture and achieving remarkable results, Holland model of relying on scientific and technological innovation for the development of high quality and efficient agriculture and creating a miracle, Australian model of increasing labor productivity with the coexistence of extensive and intensive types, and South Korean model of being promoted by the government, etc. [1]

*1.3 Domestic research status*

In the process of rapid promotion of urbanization, the reality of the gradual marginalization and hollowing of rural areas as well as the rapid advancement of the beautiful rural construction around the country has put forward urgent requirements on the study of rural construction. But relevant research is still at the primary stage of fragmentation on the whole, and its main shortcomings include: most are policy propaganda and document interpretation, with less academic nature.For the corresponding research strategy regarding the rural problems, the feasibility and applicability need to be strengthened in both research and practice; staggered and mixed living of all the nations with each other creates rich and diverse cultural patterns in different areas. The research is not comprehensive, and the study and practice on historical evolution of the northeast national settlement and contemporary theories is not very deficient. The research perspectives and the subject position are relatively simple, and lack of analysis from the problem study-oriented perspective and multi-subjective perspective. [2]

## 2. Rural basic research, focusing on existing problems

*2.1 Basic research on the investigation and analysis of rural information data in the northeast minority areas.*

By combining with the basic research topic of rural human settlements of the Ministry of Housing and Urban-Rural Development and collaborating with such ten universities as Tongji University, etc. in China, deep investigation on the northeast countryside, questionnaire, interview, surveying and mapping and video shooting of Manchu, Korean, Mongolian and other ethnic settlement villages is conducted .

(1) Interview with the stay-at-home villagers - questionnaire survey and photographing (family environment andinterviewee).

(2) Photos of interview with key figures (village party secretaries and village heads), public facilities and village environment.

(3) Interview with migrant workers – enterprises withintensive employment.

(4) Interview with leaders of employment enterprises.

(5) Interview with the county-levelleaders in charge of rural construction and interview with relevant functional departments.

By combining with the specialized teaching and social practice and internship, the training on investigation techniques and methods is made to ensure that the information on the questionnaire is accurate.

*2.2 According to the typological method, the rural internal order is analyzed, and the existing focus problems are sought out and induced.*

*2.2.1According to the investigation data, classification research is made, the internal order and main existing problems are sought out.*

Regional attribute, national attribute, geographical attribute (plain and hill), economic attribute (developed andbackward), social attribute, cultural attribute, historical attribute, spatial attribute, resource attribute, traffic attribute, population structure attribute, religious attribute, etc.

*2.2.2 Analysis of the problem*

(1) Regional cultural aphasia

IOP Conf. Series: Earth and Environmental Science **61** (2017) 012027 doi:10.1088/1755-1315/61/1/012027

(2) Environment and ecologicaldegradation
(3) Economic and industrial decline
(4) Morphological convergence of space
(5) Resource and population loss
(6) Endogenous power insufficiency

## 3. Performance of problem-oriented study on the type, role and strategy of local collaborative construction

### *3.1 Types of collaborative construction*

(1) Taking the construction of national or local government as the leading factor and rural independent construction as a supplement, dominated by specific incidents, overall planning and construction is planned, made, managed and implemented.
(2) Dominated by the volunteer group, rural reform movement focusing on social activities and rural education is developed.
(3) Rural construction under the guidance of architects and relevant professional teams.
(4) Public welfare construction activities are carried out by social organizations.
(5) Rural construction model of "art going to the countryside" by artists and cultural scholars.
(6) Example of non-governmental local rural construction initiated by villagers themselves. [3]

### *3.2 Role of collaborative construction*

(1) Rural construction is a complex and systematic project. Collaborative construction is a kind of involvement by all kinds of forces as well as all-round practice of rural culture, industrial development, environmental governance, rural economy and community management, etc.
(2) Demonstration, construction guiding and promotion of follow-up development based on the principle of voluntary participation. In the process of construction, the cooperation power of all the collaborative agents is mobilized, limitations are agreed upon, responsibilities are emphasized, and access rules are prepared.
(3) Collaborative construction has large differences with various ways of intervention and uneven action effect. It is a process of continuous debugging andceaseless update. [4]

### *3.3 Strategy of collaborative construction*
(1) Place shaping of collaborative construction is the thinking foundation leading to organic order.
(2) Wide and sufficient public participation is the fundamental power leading to organic order.
(3) The design strategy with flexible control is the technical means leading to organic order.
(4) The collaboration system of resource sharing is system guarantee leading to organic order.

## 4. Study on the collaborative mechanism of local construction

### *4.1 The system mechanism of "overall planning and authgenic organization derivative"- top-bottom collaboration*
Firstly, as for collaborative construction, it isnecessary to set up the top-most system design from the most basic material design, self-organized construction of villagers from bottom to top and the overall planning of the government from top to bottom, so as to build the top-bottom collaboration mechanism.

### *4.2 The action mechanism of "urban external force driving and rural internal force excitation" - internal-external coupling*
Urban external force driving: external thrust is formed from such resources as urban capital, science and technology policy, etc. -object

Internal order of the village: the activated internal power is continuously fermented. - subject

*4.3 The operating mechanism "dominated by local construction and coordinated by organizations from various circles" - extensive participation*

*4.3.1 Local construction of the "construction camp"*
(1) Historical context inheritance of rural structure and style.
(2) Protection and optimization of the vernacular architecture as well as application of green building technology.
(3) Perfection of the continuous development process towards the direction of integrity and adjustability.
(4) Vernacular integration of local materials, skills, local aesthetics and ecological technology.
(5) Summary and sublimation of folk construction thoughts from the environment of context inheritance.

*4.3.2 Coordination of resource integration design*
(1) Classification, transformation and coordination of different cultural conditions in rural settlements.
(2) Adjustment and upgrade of the dominating and distinctive industries to promote industrial development.
(3) Taking caring as the root, respecting users, and keeping dynamic balance of regional rural values.
(4) Developing and integrating the landscape and ecological agriculture with identification and belongingness and traditional handicraft and culture industry, strengthening the use, and promoting the comprehensive upgrade of ecology, business format and pattern.
(5) Breaking the narrow limits of regions and nations through the participation and driving of international cooperation agencies and volunteers to show international influence.

*4.4 The education mechanism of "independent promotion by the villagers and local cultivation of county sages" - sustainable development*

*4.4.1 Independent promotion by the villagers*
(1) Promotion and autonomy of villagers' subjective consciousness.
(2) Passing on the torch of learning by folk craftsmen.
*4.4.2 Local education of county sages*
(1) Organizing students to participate in the practice of rural construction and reshaping the educational consciousness through ""the cooperative construction society".
(2) Providing a platform for intellectuals and scholars who are committed to rural construction.

*4.5 The development mechanism of "modern science and technology promotion" and "natural ecology integration" – perpetual development*

*4.5.1 "Modern science and technology promotion"*
(1) Instruction of modern technology through the field
(2) Construction of think tank and e-commerce based on "network media", establishment of villagers' community network participation platform
*4.5.2 "Integration of natural ecology"*
(1) Principle of minimal intervention to the natural environment
(2) Principle of harmonious coexistence of society, environment and ecology.

**5. Rural construction experiment of "Collaborative Construction Society"**
Rural cultural buildings in typical ethnic areas are chosen to carry out a series of collaborative work, such as repair, construction, establishment and operation of rural self-organization, etc.During the construction course, the first clue is rural reconstruction and update cases with flexible policy and

professional participation as supplement, and another clue is the practical activities of the rural construction society driven by the private organizations and enterprises keening on rural construction,both of which are taken as experiment means to study the experimental construction and activities developed in the northeast countryside, and inspect the applicability and sustainability of the rural catalysts theory. According to the experiment results, through the return visit on the use of the villagers and the questionnaire survey on satisfaction, comprehensive and dynamic evaluation of rural update projects are carried out. Based on the analysis and evaluation of the experimental data, this paper presents more optimized and reasonable activation mechanism of collaborative construction and self-operation in the villages in Liaoning. The practice way of the actual project here is different from the top-down urbanization construction ways resulting in the rural homogenization at the present stage, aiming at making the policy and capital level flexible, weakening the professional and technical intervention and protecting thegene inheritancein rural area.

The system relationship of up-down collaboration is established under the interaction between the government operating from top to bottom and villages operating from bottom to top; rural collaborative construction mechanism focusing on self-construction of villagers is formed reasonably aided by urban capital, technologies and other resources as the external driving force, through local construction practicefrom bottom to top, attraction of the participation of wide range of rural traditional craftsmen and villagers, enhancement of the villagers' self-recognition and self-confidence on the rural area, exploration of internal factors of rural settlements in the region, nationality, religion, customs, blood ties, culture, history, nature etc. and coupling with urban external force, aiming at improving the quality of human settlement environment and revitalizing the home village in an all-round way. [5]

In summary, refining a kind of mode and theoretical experience with strong innovation and operability will provide theoretical basis for the systematic study on rural construction. Improvement of the rural construction system through scientific means has certain theoretical significance. At the same time, the rural revival plays an important role in enriching the economic structure levels and industrial collaboration. Its innovative development mode will also have a positive interaction with the city. Meanwhile, the "collaborative construction society" way features low involving cost, high efficiency, strong linkage and high adaptability. "Construction Society" work can accumulate teaching experience, and improve the interaction, thus contributing to strengthening specialty construction, training local talents and guiding county sages to be local, with relatively strong practical significance.

**References**

[1] Jorge B and Alvaro C. Towards a new consensus: poverty reduction strategies for Bolivia*[J] International Journal of Development Issues*,2009 (8):18-39

[2] Liang Ying, Zhang Ruide andCaiChengzhi2010Comparison of domestic and foreign rural construction models*Guizhou Agricultural Science*.38(1):237-240

[3] Ge Dandong, Wang Menglu, Wang Zhu and Wang Tao2014 Subject cognition and expression of regionalism of rural settlements *[J]. Human Settlements Forum in West China*29(03): 8-13.

[4] Wang Zhu and Qian Zhenlan 2015 Organic renewal idea and strategy of rural human settlement environment*[J]. Human Settlements Forum in West China*30(02): 15-19.

[5] Yu Kanhua, Liu Jie, Cai Hui, etc. 2016Study on the technical route of rural revival based on humanism orientation*[J]. Urban Development Research*5:43-48.

# Research on the strategies to optimize traditional Korean nationality village residential environment -- Taking the transformation of Chatiao Village in Antu County, Yanbian Korean Nationality Autonomous Prefecture as example

**Sun Chaoyang**[1], **Sui Xin***[2(Correspondence author)], **Li Mo**[3] **and Wang Yongqiang**[4]

[1] Dalian Nationalities University, No. 18, Liaohe West Road, Dalian Economic & Technical Development Zone, China, 116600
[2]Dalian Nationalities University, No. 18, Liaohe West Road, Dalian Economic & Technical Development Zone, China, 116600
[3]City Institute of Dalian University of Technology, No.31, Tieshan West Road, Dalian Economic & Technical Development Zone, China, 116600
[4]Dalian Nationalities University, No. 18, Liaohe West Road, Dalian Economic & Technical Development Zone, China, 116600
Correspondence author: Sui Xin suixin@dlnu.edu.cn

**Abstract.** This research is aimed to make an in-depth research into the strategies and methods to protect and develop the residential environment in the villages and towns with minority group characteristics. In the research on the construction mode and optimization strategy of the residential environment of the original residents in Chatiao Village, Antu County, Korean Nationality Autonomous Prefecture, the contents of architecture and planning were used comprehensively with the philosophy of green design, sociology and economics being combined simultaneously to drive the humanistic and economic development in the minority areas at the same time of providing new employment opportunities and a comfortable residential environment for people, thus realizing the complete development of the characteristic villages in Chinese minority areas.

## 1. Research background and necessity

### *1.1 Analysis to the background of the research on nationality traditional villages*

The Korean nationality is an important part of the minority groups in northern China and is also a typical cross-border nationality. With the modern civilization coming in constantly and due to such reasons as changing in local production and traditional living mode, historic ageing, etc., numerous Korean nationality traditional villages expand disorderly in village space layout, and the infrastructures and public facilities are backward. Under the influence of tourist economy, a lot of houses have been newly built and rebuilt. As a result, new architectures and historic architectures are mixed, and the local style is extremely uncoordinated. At the same time, the natural ecological system is damaged seriously, and the residential environment is faced with a grim situation.

### *1.2 The necessity for the research on the residential environment of Korean nationality villages*

In the process of rapid urbanization, the rural traditional settlement culture, personal connections and social consciousness are subject to the impact of urban culture, and are being integrated and substituted gradually. Such residential environments as rural geographic space environment, social cultural environment and natural ecological environment, etc. are in a disorder and transformational development state. In future 20 years, the urbanization in China will still maintain a rapid development momentum. The rural residential environment is a residential system in rural area, that is, the organic combination of materials and non-materials required by farmers in production and living, which is a dynamic, complicated and huge system. Under such system, the functional conversion and evolution has its inherent laws. Planning and perfecting the infrastructures and public ancillary facilities in a scientific manner, improving the production and living conditions and taking a way of ecological development are very urgent and important work. [1]

## 2. The status quo of the residential environment in Chatiao Village, a traditional Korean nationality village

### *2.1 General introduction to Chatiao Village*

Chatiao Village is located in the northeast of Antu County and the place where the Town Government of Shimen Town is located. It is adjacent to Longjing City in the east and adjoins with such two towns as Laotougou and Tianbaoshan in the west, which connects with Mingyue Town in the north and fronts water with hills on the back. All the hills are luxuriously green. The village integrates with nature. The village construction can date back to early $20^{th}$ century. It is a main village where the population of Changbai Mountain System resides and is also a Korean nationality village which is the largest in scale and the most typical. Its area is around 30 sq. km. The total number of households is 336 and the total population is 1,214, in which the Korean nationality population accounts for 80%. The transportation of Chatiao Village is convenient with rich tourist resources. The traditional industries have certain influential power. The Korean nationality has such humanistic environment advantages as emphasizing education and being open-minded. In 2015, the Party Committee and Government of Antu County determined Shimen Town as a minority group town for key construction in the "$13^{th}$ 5-year Plan" of Antu County. The policy environment is very favorable.

### *2.2 Problems existing at present:*

(1) The planning is lack of systematicness and unable to be geared to the need of development and construction.
(2) The architectural style is lack of unified planning, and there is no landmark architecture.
(3) The public service and infrastructures still need to be further perfected.
(4) The industrial development is lack of effective promotion.
(5) The villagers have no belongingness, and they are lack of cultural self-confidence.

## 3. Strategies to protect and develop the residential environment

### *3.1 The protection of the traditional architectural style and the molding of characteristics*

#### *3.1.1 General layout*

According to the style characteristics of Chatiao Village and the spatial distribution of the environmental elements, the general village style is generalized as such a style characteristic structure as "two hills and two waters, one belt and four zones". The "two hills" are Wuhu Hill and Beijiazi Hill; the "two waters" are Buerhatong River and Zhongping River; the "one belt" is the ethnic and folk custom demonstration belt on both sides of No. 302 national road in Chatiao Village; the "four zones" are folk culture experiencing zone, characteristic commercial service zone, hill and water landscape zone and ecological agriculture sightseeing zone.

IOP Conf. Series: Earth and Environmental Science **61** (2017) 012028 doi:10.1088/1755-1315/61/1/012028

*3.1.2 Zone positioning*

Folk culture experiencing zone - This zone is renovated and repaired on basis of the original architectures with decoration being made mainly by using Korean nationality architectural tones and traditional materials (wood, gray tile, wall brick, etc.). Hotel, shopping store, guesthouse, folk handicraft demonstration center, tourist reception center, etc. are focally built. It is a kernel and key zone planned by Chatiao Village, and mainly consists of soybean paste industrial park, Korean nationality traditional residence demonstration and cultural plaza. The planning generally protects the traditional pattern and style, historic streets and lanes, nodes and historic environmental elements. As a result, the general style is coordinated and orderly with humanistic characteristics being shown. The historic landscape of the original settlement as well as the characteristic industrial landscape of a Korean nationality village is focally showcased. [2]

Characteristic commercial service zone – It covers characteristic food and beverage, guesthouse, happy farmhouse, characteristic commodity transaction market, etc. with nationality characteristics. It mainly satisfies the tourist reception and service function, and guides villagers to repair houses simply to intensify the general appearance style and internal nationality culture elements.

Water and hill landscape zone – It is planned to protect the original terrain and landform, repair the river bank, increase the brightening and leisure atmosphere and showcase the natural appearance of hilly village, river and field in combination with forest ecological tourism.

Ecological agricultural sightseeing zone - On a large area of farmland and hills with a gentle terrain, it is planned to plant rice, vegetables, melons and fruits and economic crops, arrange picking park and process farm products to create an ecological agriculture landscape integrating ecological growing, processing, sightseeing and experiencing in combination with ecological agriculture sightseeing and tourism.

*3.1.3 Focal planning for the kernel zone*

Soybean paste culture industrial park - It is an important folk culture demonstration zone of Chatiao Village, and is a main scenic zone to receive tourists in future. Based on the existing “Korean Soybean Paste Park” and taking the soybean paste production and processing process as the basic chain, a general planning and layout is made to the zone to build cultural market concept.

Korean nationality custom street - Renovation and repair is made on basis of the original architectures on both sides of No. 302 national road with decoration being made mainly by using Korean nationality architectural tones and traditional materials (wood, gray tile, wall brick, etc.). Hotel, shopping store, guesthouse, folk handicraft demonstration center, tourist reception center, etc. are built focally.

Traditional residence protection and renovation - In the planning, classified protection and renovation is made to the architectures in the village according to the historic value, preservation conditions and style characteristics with such measures as original appearance protection, repair, rectification, renovation, removal, etc. being taken. The utilization of courtyard enclosure walls, indoor spaces, courtyard greening layout and internal street and lane planning and design are intensified to mould a characteristic community being “open in large space and close in local space”.

*3.1.4 Environmental rectification*

A comprehensive rectification is made to the external environment of Chatiao Village, including such important nodes as street and lane road surfaces, road greening, courtyard environment, civic pipelines, village and town entrance and exit, etc. Environmental facility furniture is increased. Through environmental beautification and brightening, local flavor and nostalgia is reserved simultaneously, the daily landscape and hometown tune of original settlement is highlighted, and village landscape urbanization and duplication is avoided. For example, for street and lane road surfaces, square paving, construction materials, etc., the principle of using indigenous raw materials is taken by using local wood, stone, red brick, adobe wall, etc. For tree species, the landscapes in four seasons and vision height difference are mainly considered. The local tree species are selected and

IOP Conf. Series: Earth and Environmental Science **61** (2017) 012028 doi:10.1088/1755-1315/61/1/012028

used in an in-situ manner with combination and collocation being strengthened focally to show the landscape characteristics of the Korean nationality villages in Changbai Mountain area. [3]

*3.2 Korean nationality culture protection and inheritance*
Culture is the part that can be perceived most easily in a village as well as an embodiment of the social belief and customs thereof. As the most habitual and real lifestyle of villagers, it embodies the "blood vessel" of villages and towns in a centralized manner, and is also the most important content to be concerned most in the residential environment construction of villages. Through investigation and classification to the history and humanity of Chatiao Village, it is generalized into spiritual belief culture space, cultural education space and daily production and living space. At the same time, pertinent space protection strategies are made, that is, to achieve the purpose of rational protection and utilization of culture space and inheritance of village and town history and culture through such ways as remolding, repair to the original conditions, scene reduction, ritual reproduction, etc. [4]

*3.2.1 Molding a symbolic humanistic landscape*
In 1964, "Antu Man" unearthed on Shimen Hill in Shimen Town is the ancestor creating the heaven and earth in Changbai area that was discovered so far, and is also one of the earliest ancient human beings in northeast China. Shimen Town becomes the town where the human history and culture of Changbai Mountain began. This historic origin is full of cultural imagination. In the cultural furniture, landscapes and sculptures planned for towns and villages, the image of "Antu Man" is to be remolded indifferently in the long cultural corridor to pull the thought of tourists to the prehistoric times long ago.

*3.2.2 Repair to the original conditions*
In the planning, such measures as protection, repair and internal function improvement are taken to protect and utilize the traditional architectures. In the process of repair, the original architectural appearance and the historic information left on the architectures are to be kept as practical as possible, such as material, structure, historic member, etc. They are to be renovated into places for traditional handicraft processing, tourist experiencing activities, etc. to extend the residential function and expand the functions of commerce, tourism and cultural activities.

*3.2.3 Scene reduction*
In order to give play to the demonstration function of cultural space, it is planned to guide villagers to collect clan pedigrees relating to towns and villages as well as tools and implements for production and living activities through historic scene reduction, and open up a folk culture demonstration hall to display them indoors. Simultaneously, the scenes of daily production and living of the residents in towns and villages is demonstrated by way of sculpture, fresco, photo, picture, physical object, etc., such as farming life, wedding ceremony, sacrificial rituals, Korean nationality clothes, the preparation process of such characteristic food as Korean nationality pickled vegetables, glutinous rice cake, etc. [5]

*3.2.4 Ritual reproduction*
Such festivals and ceremonies as Korean nationality traditional wedding ceremony, the rituals for congratulation to elderly people, etc. are set up focally to allow the reproduction of ancient ceremonies and customs. Through the ritual activities, the traditional rituals and customs can be inherited among villagers from generation to generation. On the other hand, tourists are allowed to participate in and experience them to enrich their experience, which is favorable to the dissemination and inheritance of Korean nationality folk culture.

*3.2.5 Culture activity organization*
The masses are organized to compose and rehearse themselves such Korean nationality traditional dances as fan dance, dance of water on the top of head, Xiangmao dance, long drum dance, Korean

nationality folk percussion music, drum dance, etc. In addition, the villagers are actively organized and motivated to participate in physical exercises as well as song and dance performances, so as to strengthen the scale of such cultural bodies as dance team, fitness team, etc. and make their performances become an important item in rural tour.

*3.2.6 Cultural facilities perfection*

Such symbolic public architectures embodying both nationality and local characteristics in a centralized manner are built focally, such as village gate, cultural square, long cultural corridor, etc. The activity spaces and places are provided for villagers and tourists to hold group activities, civilized rituals, theatrical performances, exercises and fitness, etc.

*3.2.7 Cultural product inheritance and innovation*

The development of village and town cultural products mainly surrounds the cultural derivatives with the tourist industry as the dominant part and relating to the village and town culture as well as tourist innovation culture. The types of cultural products are expanded from such aspects as theatrical performance and traditional and tourist handicrafts to enrich the connotations and achieve the purpose of cultural promulgation and influence expansion.

*3.3 Positioning and development of characteristic industries*

*3.3.1 Industrial positioning*

A new type of village and town industrial structure that is harmonious and can make sustainable development is created by taking tourism as the traction and characteristic industrial upgrading as internal kernel, so as to promote population backflow through industrial development. Rich agricultural resources, nationality culture and tourism resources in Shimen Town and such advantages as being close to the downtown area and convenient in transportation are sufficiently used to promote the transformation and upgrade of the traditional agriculture and forestry through developing folk culture tour, ecological agriculture sightseeing and hilly village leisure tour, so as to achieve the purpose of increasing the income of residents, keeping the residents, attracting those going outside to return for business startup and activate the vigor and vitality of villages and towns.

*3.3.2 Industrial layout*

The industrial plan follows the guiding thought of “one kernel and two wings”, and such an industrial spatial structure as “one belt and four stretches” is determined. “One kernel and two wings” means as follows: the development of rural tourism is taken as the kernel, and “the ecological agriculture represented by high-quality organic rice” and “the soybean paste industry with typical cultural characteristics of Korean nationality traditional food culture” are regarded as two wings. “One belt” means taking the commercial street with folk customs as the basic economic belt for commercial service industry. “Four stretches” represents such four functional stretches as Korean nationality folk culture zone, ecological sightseeing and agricultural zone, waterside leisure and sightseeing zone and Wuhu Hill ecological tourist zone. [6]

*3.3.3 Nurturing of characteristic industries*

(1) Stereoscopic ecological agriculture;
(2) Soybean paste cultural industry;
(3) Characteristic tourism.

## 4 Conclusion

In one word, the construction of rural residential environment at the contemporary era is a dynamic and complicated system. Due to the complexity and specialness therein, no uniform example can be followed. As an important integral part of Chinese traditional culture and regional diversity, the Korean nationality villages have a structure system with unique features. By making overall construction and planning for Chatiao Village, Shimen Town and taking a series of measures like improvement of the structure spaces of villages, inheritance of traditional nationality culture, orientation of industrial development, etc., the strategy of overall development and itemized promotion and combination of governmental promotion with market operation is thus formed. In the research course, not only the contents of architecture, planning and ecological design are applied, but the philosophies in anthropology, sociology and economics are also combined to make multi-disciplinary focal study. During the special period of social transformation and the critical stage of building a well-off society comprehensively, emphasizing the protection and revitalization of the residential environment of traditional villages in nationality areas and further exploring the construction mode and optimization strategy to be geared to the regional advantage of Korean nationality and embody the rural residential environment with characteristics has a very important real significance.

**References**

[1] Wu Liangyong 2001 Scientific introduction to residential environment *[M] Beijing: China Architecture & Building Press*
[2] Jin Renhe and Zhang Tai 2006 Research on spatial structure changes in Korean nationality villages *Chinese Landscape Architecture*
[3] Zhang Yulin 2011 Agriculture and rural society in East Asia after the "Modernization" – cases in Japan, Korea and Taiwan and their historical significance *[J] Journal of Nanjing Agricultural University (social science edition)*, 3:1-8
[4] Yu Kanhua, Liu Jie, Cai Hui, etc. 2016 Study on the technical route of rural revival based on humanism orientation *[J]. Urban Development Research* 5:43-48.
[5] Zhang Taixian 2001 Research on external spatial relations for traditional residences in Yanbian, China *IJ. Industrial Science Research*
[6] Peng Song 2004 Nonlinear method – new thinking for research on traditional village space patterns 2004 (4), NOL. 24, No.2.

IOP Conf. Series: Earth and Environmental Science **61** (2017) 012029 doi:10.1088/1755-1315/61/1/012029

# Double-parameter foundation beam modal of overlying thin rock stratum under loose layers grouting mining

**Chen Xianghong[1], Tao Lianjin[1], Liu Honghua[2]**

[1] College of Architecture and Civil Engineering, Beijing University of Technology, Beijing 100124, China
[2] Shenhua geology exploration co., LTD, Beijing 100011, China
07872@bjut.edu.cn

**Abstract.** Strata movement of drilling grouting is distinctly different from that of traditional caving mining, especially for overlying thin rock stratum. Considering the horizontal shear transfer of foundation, a mechanical model of overlying strata movement was established using the theory of Pasternak double-parameter foundation beam. The deflection and internal force equations of rock beam were derived, and the influence mechanisms of grouting material parameter were discussed. The results showed that the thickness of layer grouting was the key influential factor than the foundation modulus coefficient and grout diffusion radius. The comparison with the theory of Winkler elastic foundation beam also showed that double-parameter foundation beam method is more applicable to reinforced strata.

## 1. Introduction

Thin rock stratum can often be found widely scattered in some coalfields in China, and its sudden breakage has significant impact on mining activities, even catastrophic losses, so controlling the roof rock's deformation has remained universally technical challenge facing scientific researchers of mining engineering. The strata movement and its control theory have been introduced and promoted, i.e. pressure arch theory [1, 2], hinged rock block theory [3], voussoir beam theory [4-6], elastic foundation beam theory [7, 8] and plate theory [9]. As to thin rock stratum above coal seam, its related mechanisms have also been studied by researchers, For example, based on the theory of elastic thin plate, Li et al. [10] studied the relationship between the elastic foundation coefficient and the roof deformation; Huang [11] obtained the critical unbroken conditions of main roof using the principle of virtual work. Using numerical simulation, theoretical analysis and in-site measurement, Fang et al [12] investigated the strata movement rules that bedrock thickness and mechanical properties are the key influential elements for the stable structure of "stacked layer of blocks"; Du et al. [13] analysed the deformation and breakage law of semi-arch structural model of immediate roof in Sima Coal Mine of Lu'an; Li et al. [14] analysed the regularity of strata behaviour in mining face under different ratios of rock and loading. The aforementioned investigations are mainly for untreated rock stratum, as to the behaviour of reinforced strata, its correlative theory still need to be discussed.

In this paper we take the grouting technology and site strata in Ordos Coal Field in China. The roof rock's behaviour is outlined based on the mathematical model and analytical solution, in which a Pasternak elastic foundation beam model is proposed for the deformation assessment of rock beam and the transfer matrix method is used for the solution. Finally, the influential effects of grouting parameters are studied.

## 2. Site description

The Stone GeTai Coal Mine lies in Ordos, Inner Mongolia of China, and coal mine 22303 is in the fourth panel of this mine. According to the field drilling investigation and adjacent mining experience, the overlying strata profile revealed: the cover depth of coal seam is about 60-100 m with 10-50 m roof rock mainly composed of fine-grained sandstone. Under such geological conditions, the fracture in roof rock grows quickly during coal mining, in order to improve the mining safety and efficiency, grouting reinforcement measures are proposed to stabilize the overlying strata. The schematic graph of 10 m thick roof rock is illustrated in Figure 1.

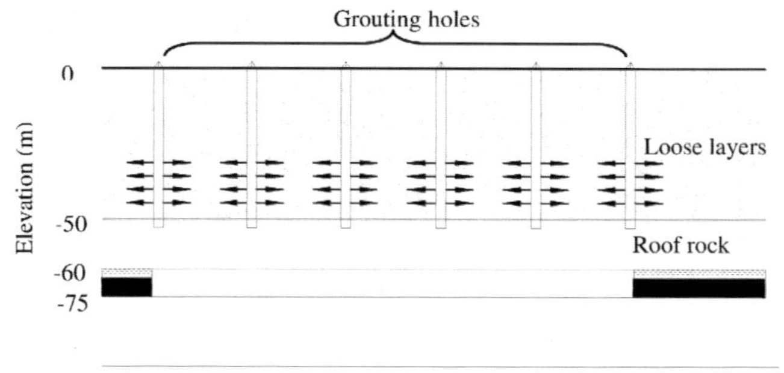


**Figure 1.** Schematic graph of layer grouting

## 3. Pasternak double-parameter beam model

### *3.1. Analytical model and assumptions*

To analyse the mechanism of the roof rock, an Euler-Bernoulli beam is provided, and the overlying grouted loose layers are simulated as Pasternak double-parameter elastic foundation considering the shear transfer mechanism along horizontal, then the solutions for the beam deflection and internal force can then be obtained using the beam differential equation. The simplified analytical model is illustrated in Figure 2.

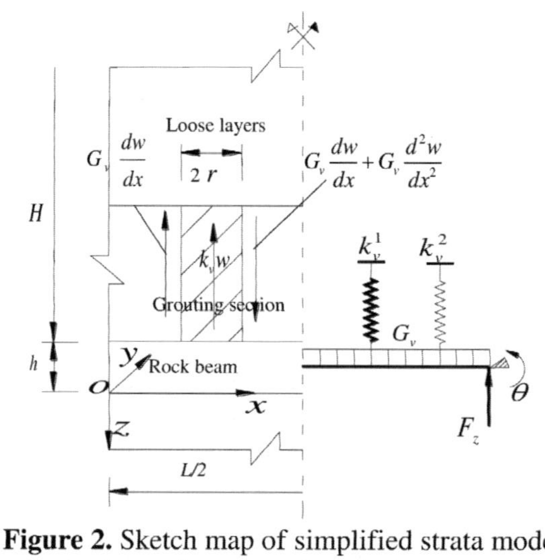


**Figure 2.** Sketch map of simplified strata model

In Cartesian coordinate *xyz*, the beam is assumed to be homogeneous material with thickness *h*, breadth *B* and length *L*; $F_z$ denotes the vertical supporting force at both ends of the rock beam; *H* denotes the depth of loose layers; $k_v$、$G_v$ denote the equivalent foundation reaction coefficient and shear transfer coefficient.

Before analyses, we introduce the following assumptions:

(1) The deformation of rock beam accords with the plane section assumption.

(2) Without considering the influence of other external loadings, the vertical supporting force is:

$$F_z = \frac{1}{2}\sum_{i=1}^{n} \rho_i g H_i \tag{1}$$

where $\rho_i$, $H_i$ are the density and thickness of layer $i$; $g$ is the acceleration due to gravity.

(3) According to Vlazov theory [15], the foundation coefficient $k_v$、$G_v$ can be expressed as:

$$k_v = \sum_{i=1}^{n} E_i \int_0^{H_i} \left[\frac{dh_i(\xi_i)}{dz}\right]^2 d\xi_i \text{, } G_v = \sum_{i=1}^{n} G_i \int_0^{H_i} [h_i(\xi_i)]^2 d\xi_i \tag{2}$$

where $h_i(\xi_i)$ is the vertical displacement distribution function of layer $i$; $\xi_i$ is the local coordinates along the coordinate $z$ direction; $G_i$ is the shearing modulus.

### 3.2. Differential equation and solution

The vertical displacement along the Euler-Bernoulli beam is denoted by $w(x)$, the bending moment $M(x)$ and shearing force $Q(x)$ for unit rock beam element can be expressed as:

$$\left.\begin{aligned} \frac{dQ(x)}{dx} &= p_z(x)B^* \\ \frac{dM(x)}{dx} &= Q(x) \end{aligned}\right\} \tag{3}$$

where $p_z(x) = k_v w(x) - G_v w''(x)$ is the vertical reaction of beam foundation, $B^* = B + \sqrt{G_v/k_v}$ is the equivalent calculation width [16] .

Based on the theory of material mechanics, we get the differential equation:

$$\frac{d^4 w}{dx^4} - \frac{G_v B^*}{EI}\frac{d^2 w}{dx^2} + \frac{k_v B^*}{EI} w = 0 \tag{4}$$

where $E$ denotes the Young's modulus of beam and $I$ is the moment of inertia about $y$-axis.

As to equation (4), the general solution is

$$w(x) = C_1 \sinh(\alpha x)\cos(\beta x) + C_2 \cosh(\alpha x)\cos(\beta x) + C_3 \sinh(\alpha x)\sin(\beta x) + C_4 \cosh(\alpha x)\sin(\beta x) \tag{5}$$

where,

$$\alpha = \sqrt{\sqrt{\frac{k_v B^*}{4EI}} + \frac{G_v B^*}{4EI}} \text{ ; } \beta = \sqrt{\sqrt{\frac{k_v B^*}{4EI}} - \frac{G_v B^*}{4EI}}$$

Then the equivalent internal forces are shown below:

$$\begin{Bmatrix} M(x) \\ V(x) \end{Bmatrix} = \begin{bmatrix} A_1 & A_2 & A_3 & A_4 \\ D_1 & D_2 & D_3 & D_4 \end{bmatrix} \begin{Bmatrix} C_1 \\ C_2 \\ C_3 \\ C_4 \end{Bmatrix} \tag{6}$$

in which,

$$\begin{Bmatrix} A_1 \\ A_2 \\ A_3 \\ A_4 \end{Bmatrix} = -EI \begin{Bmatrix} (\alpha^2 - \beta^2)X_1 - 2\alpha\beta X_4 \\ (\alpha^2 - \beta^2)X_2 - 2\alpha\beta X_3 \\ (\alpha^2 - \beta^2)X_3 + 2\alpha\beta X_2 \\ (\alpha^2 - \beta^2)X_4 + 2\alpha\beta X_1 \end{Bmatrix} ; \begin{Bmatrix} D_1 \\ D_2 \\ D_3 \\ D_4 \end{Bmatrix} = -EI \begin{Bmatrix} (\beta^3 - 3\alpha^2\beta)X_3 + (\alpha^3 - 3\alpha\beta^2)X_2 \\ (\beta^3 - 3\alpha^2\beta)X_4 + (\alpha^3 - 3\alpha\beta^2)X_1 \\ (3\alpha^2\beta - \beta^3)X_1 + (\alpha^3 - 3\alpha\beta^2)X_4 \\ (3\alpha^2\beta - \beta^3)X_2 + (\alpha^3 - 3\alpha\beta^2)X_3 \end{Bmatrix} + G_v B^* \begin{Bmatrix} \alpha X_2 - \beta X_3 \\ \alpha X_1 - \beta X_4 \\ \alpha X_4 + \beta X_1 \\ \alpha X_3 + \beta X_2 \end{Bmatrix}$$

$$X_1 = \sinh(\alpha x)\cos(\beta x); \quad X_2 = \cosh(\alpha x)\cos(\beta x); \quad X_3 = \sinh(\alpha x)\sin(\beta x); \quad X_4 = \cosh(\alpha x)\sin(\beta x)$$

Using the internal force equilibrium and deformation coordination conditions, the beam is divided into m sections according to the grouting influence domains, the mathematical relationship in the interval of unit rock beam $x_i$ and $x_{i+1}$ is

$$\begin{Bmatrix} w(x_{i+1}) \\ \theta(x_{i+1}) \\ M(x_{i+1}) \\ V(x_{i+1}) \end{Bmatrix} = \hat{T}_i \begin{Bmatrix} w(x_i) \\ \theta(x_i) \\ M(x_i) \\ V(x_i) \end{Bmatrix} \tag{7}$$

in which,

$$\hat{T}_i = \begin{bmatrix} X_1 & X_1' & -EIX_1'' & -EIX_1'''+G_vB^*X_1' \\ X_2 & X_2' & -EIX_2'' & -EIX_2'''+G_vB^*X_2' \\ X_3 & X_3' & -EIX_3'' & -EIX_3'''+G_vB^*X_3' \\ X_4 & X_4' & -EIX_4'' & -EIX_4'''+G_vB^*X_4' \end{bmatrix} * \begin{bmatrix} 0 & \dfrac{3\alpha^2-\beta^2-\dfrac{G_vB^*}{EI}}{2\alpha(\alpha^2+\beta^2)} & 0 & \dfrac{-1}{2EI\alpha(\alpha^2+\beta^2)} \\ 1 & 0 & 0 & 0 \\ \dfrac{\beta^2-\alpha^2}{2\alpha\beta} & 0 & \dfrac{1}{2EI\alpha\beta} & 0 \\ 0 & \dfrac{3\beta^2-\alpha^2+\dfrac{G_vB^*}{EI}}{2\beta(\alpha^2+\beta^2)} & 0 & \dfrac{1}{2EI\beta(\alpha^2+\beta^2)} \end{bmatrix}$$

Based on the transfer matrix method, let $S=[w \quad \theta \quad M \quad V]^T$ we get

$$S(x=L)=\prod_{i=1}^{m}\hat{T}_i S(x=0) \tag{8}$$

In order to get the constants of $C_1$-$C_4$, two boundary conditions are proposed

C-1: $\left.\begin{matrix} M(x_0)=M(x_L)=0 \\ V(x_0)=-F_z \\ V(x_L)=F_z \end{matrix}\right\}$; C-2: $\left.\begin{matrix} \theta(x_0)=\theta(x_L)=0 \\ V(x_0)=-F_z \\ V(x_L)=F_z \end{matrix}\right\}$

## 4. Results discussion

To verify the applicability of the proposed model, we define the material parameters as: rock beam thickness is chosen as $h$=10 m, calculation span is $L$=60 m; the overlying loose layer thickness is 50 m. The main material properties are shown in Table 1.

**Table 1** Material properties of foundation

| Properties | Rock stratum | Loose layer | Grouting section |
|---|---|---|---|
| Elastic modulus (Mpa) | 5 000 | 10 | 100 |
| Density (kg/m$^3$) | 2 600 | 1 600 | 1 600 |
| Poisson ratio | 0.2 | 0.3 | 0.25 |

According to grouting technology, the grouting holes spacing is 6 m, diffusion radius is 1.0 m, grouting thickness is 10 m. the calculation results for boundary condition C-1 are shown in Figure 3 whether the shear transfer coefficient is adopted.

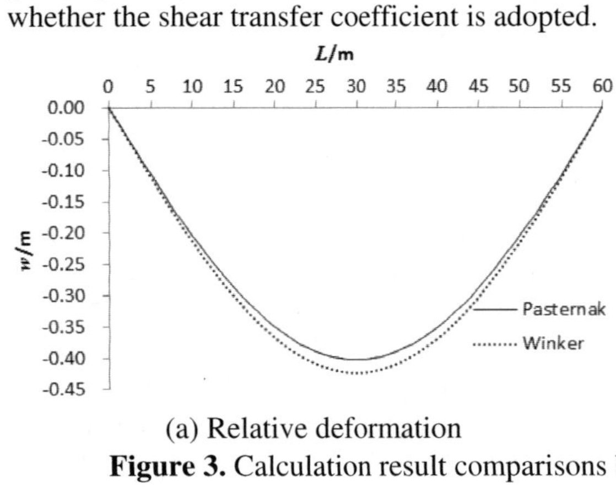


(a) Relative deformation

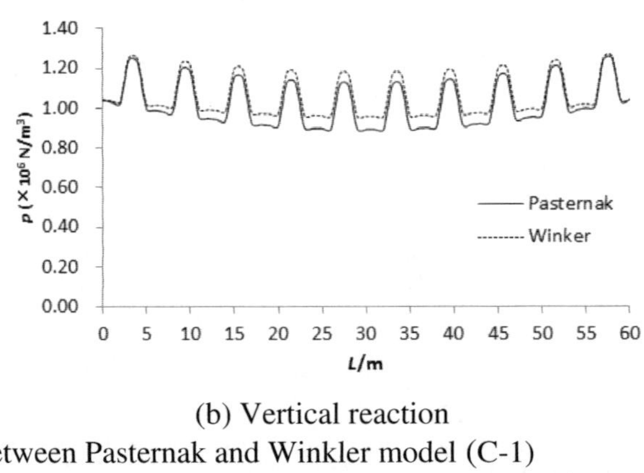


(b) Vertical reaction

**Figure 3.** Calculation result comparisons between Pasternak and Winkler model (C-1)

It can be seen from Figure 3 that, the maximum errors calculated between Pasternak and Winkler method were about 5.0%; and Pasternak double parameter foundation model considering the horizontal shear force is more suitable than Winkler foundation model.

## 5. Parameter analyses

To analyze the effects of drilling grouting on beam deformation, using the Pasternak model proposed above, we introduce the non-dimensional variables $\Delta E$, $\Delta H$ and $\Delta w$:

$$\Delta E = \frac{E_2}{E_1}, \quad \Delta H = \frac{h^*}{h}, \quad \Delta w = \frac{W_2 - W_1}{W_1} \times 100\% \tag{9}$$

where, $E_1$ and $E_2$ denote the elastic modulus of no grouting and grouting region, accordingly $W_1$ and $W_2$ denote the maximum relative displacement of rock beam; $h^*$ is the thickness of layer grouting.

In the next paragraph, we will take $L$=60 m and 120 m for example and investigate the influential mechanisms of grouting material and geometric factors on the deformation of rock beam.

### 5.1. *Elastic modulus ratio* $\Delta E$

Figure 4 shows the non-dimensional displacement $\Delta w$ for different $\Delta E$ in the range of 1-15. It can be seen from Figure 4 that, the non-dimensional displacement $\Delta w$ increases nonlinearly with $\Delta E$ increasing from 1 to 15; and the displacement of rock beam decreases sharply firstly and then slows down. On the other hand, according to the comparisons of two boundary conditions and beam lengths, the results show that, the effects of grouting is more obvious with the increase of beam length especially for boundary condition C-1.

### 5.2. *Poisson's ratio* $\mu$.

Figure 5 shows the effects of Poisson's ratio of grouting material in the range of 0.3-0.05 on the non-dimensional displacement $\Delta w$ . From Figure 5, we conclude that the deformation changes non-obviously with Poisson's ratio.

### 5.3. *Grout diffusion radius r.*

Figure 6 shows the effects of diffusion radius $r$ on the non-dimensional displacement $\Delta w$. As can be seen from Figure 6, the deformation increment $\Delta w$ increase near-linearly by 6.62% (C-1) and 6.48% (C-2) as $L$=60 m, by 9.49% (C-1) and 7.19% (C-2) as $L$=120 m with the diffusion radius.

### 5.4. *Grouting thickness ratio* $\Delta H$.

Figure 7 shows the non-dimensional displacement $\Delta w$ for different $\Delta H$ in the range of 10-50 m. In Figure 7, the non-dimensional displacement $\Delta w$ decreases nonlinearly with the increase of the thickness ratio $\Delta H$ . So, optimal design of grouting parameters is crucial in mine grouting project.

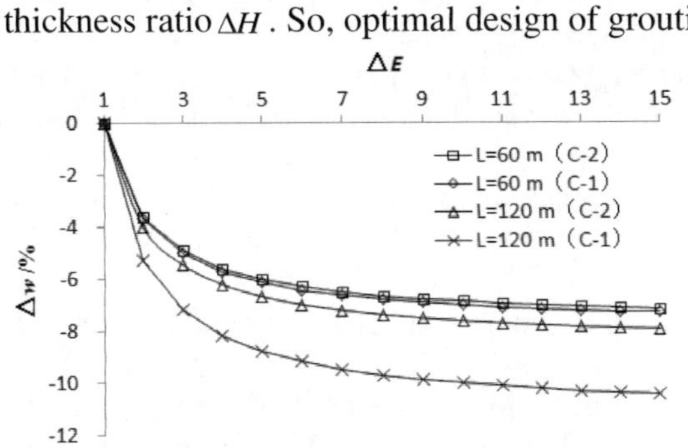


**Figure 4.** Effect of elastic modulus ratio $\Delta E$ on the non-dimensional deformation $\Delta w$

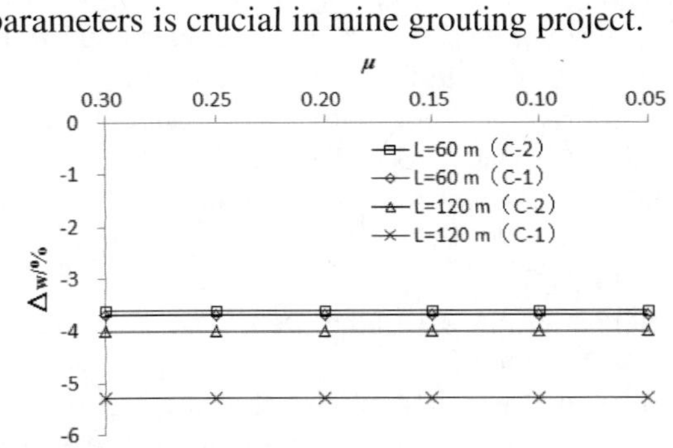


**Figure 5.** Effect of poisson's ratio μ on the non-dimensional deformation $\Delta w$ ( $\Delta E$ =2)

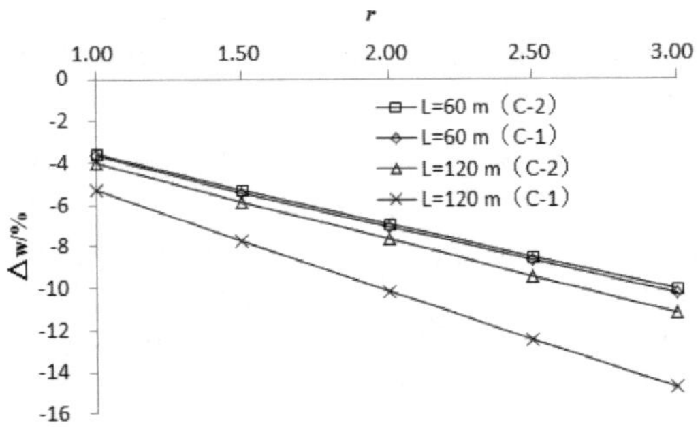


**Figure 6.** Effect of grouting diffusion radius r on the non-dimensional deformation $\Delta w$ ($\Delta E$=2)

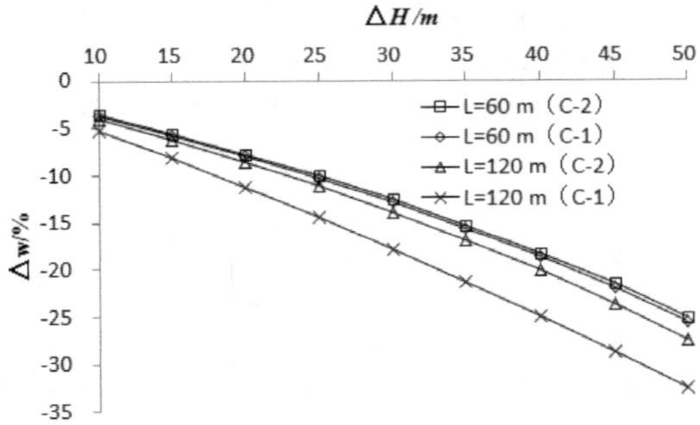


**Figure 7.** Effect of grouting thickness ratio $\Delta H$ on the non-dimensional deformation $\Delta w$ ($\Delta E$=2)

## 6. Conclusions

A simplified mathematical model of the deformation behaviour of thin rock roof during the extraction of coal seam has been proposed that accounts for the compressive stresses created by the overlying strata. Based on Pasternak foundation beam and transfer matrix theory, the rock deformation is investigated and the influence parameters of drilling grouting have been discussed: (1) Pasternak double parameter foundation model is more reasonable to simulate the shear transfer behaviour of reinforced stratum; (2) Different boundary condition assumption has important influence on the deformation characteristics of different calculation length of rock beam; (3) The key factors affecting the characteristics and development regularity of overlying rock beam deformation depends mainly on the elastic modulus, diffusion radius and the thickness of grouting.

## References

[1] Chunlin C 2006 Li. Rock support design based on the concept of pressure arch *International Journal of Rock Mechanics and Mining Sciences* Vol 43 Issue 7 pp 1083-1090

[2] Poulsen B A 2010 Pillar load calculation by pressure arch theory and near field extraction ratio International Journal of Rock Mechanics and Mining Sciences Vol 47 Issue 7 pp 1158-1165

[3] Tianjun He 1998 Gateside supports designed by Wilson theory of hinged rock block Chinese Journal of Rock Mechanics and Engineering Vol 17 Issue 2 pp 173-177

[4] Xiexing Miao and Minggao Qian 1995 Solid structure and model of voussoir beam of face surrounding rock Ground Pressure and Strata Control No3-4 pp 3-12

[5] Sofianos AI 1996 Analysis and design of an underground hard rock voussoir beam roof International Journal of Rock Mechanics and Mining Sciences and Geomechanics Abstracts Vol 33 Issue 2 pp 153-166

[6] Mahdi Shabanimashcool and Charlie C 2015 Analytical approaches for studying the stability of laminated roof strata International Journal of Rock Mechanics & Mining Sciences Vol 79 pp 99–108

[7] Yuehan Wang, Kezhong Deng and Kan Wu 2003 On the dynamic mechanics model of mining subsidence Chinese Journal of Rock Mechanics and Engineering Vol 22 Issue 3 pp 352-357

[8] Jie Chen, Jiping Du, Weisong Zhang, et al 2012 An elastic base beam model of overlying strata movement during coal mining with gangue back-filling Journal of China University of Mining and Technology Vol 41 Issue 1 pp 14-19

[9] Congliang Liu, Zhixiang Tan, Kazhong Deng, et al 2013 Synergistic instability of coal pillar and roof system and filling method based on plate model International Journal of Mining Science and Technology Vol 23 Issue 1 pp 145-149

[10] Meng Li, Jixiong Zhang, Haiqiang Jiang, et al 2014 A thin plate on elastic foundation model of overlying strata for dense solid backfill-mining Journal of China Coal Society Vol 39 Issue 12 pp 2369-2373

[11] Yanli Huang 2012 Ground control theory and application of solid dense backfill in coal mines Beijing: China University of Mining and Technology
[12] Xinqiu Fang, Hanfu Huang, Tao Jin, et al 2008 Movement rules of overlying strata around longwall mining in thin bedrock with thick surface soil Chinese Journal of Rock Mechanics and Engineering Vol 27 Supp 1 pp 2700-2706
[13] Feng Du and Haibo Bai 2013 Mechanical analysis of immediate roof in fully mechanized top coal caving mining with thin bedrock Journal of China Coal Society Vol 38 Issue 8 pp 1331-1337
[14] Fusheng Li, Yong Zhang and Lifeng Xu 2013 Influence of ratio of basement and loadings on mining face rock pressure in thin basement rock thick surface soil Journal of Chin*a Coal Society* Vol 38 Issue 10 pp 1749-1755
[15] Selvadural A P S translated by Wentian Fan, Guanghan He, Shishen Zhang, et al 1984 Elastic Analysis for the Interaction Between Soil and Foundation Beijing: China Railway Publishing House
[16] Yi Huang and Fangshe He 2005 Beam plate and shell on elastic foundation Beijing: Science press

# Yangtze River: the potential ecological risk of heavy metals in sediment from 1996 to 2012

**Xinyu Song, Lina Xu and Yanhui Dai***

Key Laboratory of Marine Environment and Ecology, Ministry of Education and College of Environmental Science and Engineering, Ocean University of China, Qingdao 266100, China
E-mail address: yanhui1437@163.com

**Abstract**. The Yangtze River is a typical river basin in China. As the third largest river in the world and closely related to the life of 4 hundred million people, the Yangtze River has attracted much attention. The heavy metal content of sediments is an important part to evaluate the heavy metal pollution in aquatic ecosystem. In this work, seven kinds of heavy metals including Cu, Hg, Pb, Zn, Cr, Cd and As were selected as the research objects, and the potential ecological risk index method was selected as the analysis method. Based on the results on the investigations of heavy metals in Yangtze River sediments in recent 20 years that we could find until 2016, we conducted a collection and analysis. The results showed that the potential ecological risk of these seven heavy metals in the Yangtze River Basin in the past 20 years had a low level. In addition, the ecological risk factors of Cd and Hg were significantly higher than other heavy metals, which should be paid attention in the future management and control work.

## 1. Introduction

Heavy metals are metals with a density greater than 4.5 g/cm³ and are commonly found in the environments, including beryllium (Be), cadmium (Cd), chromium (Cr), copper (Cu), plumbum (Pb), hydrargyrum (Hg), nickel (Ni), argentum (Ag), thallium (Tl), zinc (Zn), aluminum (Al), barium (Ba), ferrum (Fe), manganese (Mn), molybdenum (Mo), cobalt (Co), titanium (Ti), vanadium (V), stibium (Sb), arsenic (As) and selenium (Se) and other heavy metals and metalloids. They are often considered as important factors causing environmental pollution. Heavy metals have highly toxic, are easy to accumulate and they are also difficult to be degraded [1-4]. Because the heavy metals are readily to be accumulated in various organisms [5-6], and be transferred in the ecosystem through the food chain, their pollution prevention and control is particularly significant.

Aquatic ecosystem is an important type of area in human activity. Not only because of human activities are affected, but also in turn affect the quality of human life and human health. Heavy metals can cause different degrees of water pollution through mining, exhaust emission, sewage irrigation and other channels. The famous Japanese Minamata disease, for example, heavy metal pollution of the water environment has caused serious harm to human health[7].

Sediment is an important food source and habitat of benthic organisms and overlying aquatic organisms [8-9]. So it is an important part of aquatic ecosystems. The sediments have large specific surface area [10] and contain active functional groups [11], which can adsorb heavy metal contaminants from natural water. If we only pay attention to the water quality of the surface water, we cannot control the aquatic ecosystem completely. In addition, studies have shown that the heavy metal pollutants absorbed on the sediments will be released again, thus causing secondary pollution to the aquatic environment and biology [12]. So the harm caused by heavy metal in sediment demands more attention.

## 2. Meterials and methods

### *2.1. Study area and sample collection*
The Yangtze River is the third largest river in the world, with a total length of 6300km. As the basin has a population of 400 million and its economic development is extremely fast, the environmental impact of the Yangtze River is highly concerned. In this paper, we collected the literatures on heavy metals including Cu, Pb, Hg, Zn, Cd, Cr and As in the Yangtze River sediments from 1996 to 2012 and summed up. These 7 elements is mentioned in China prefered controlled pollutant in water, China's marine sediment quality standards (GB 18668-2002), surface water quality standards (GB3838-2002) and hygienic standard for drinking water (GB5749-2006). And the potential ecological risk assessment of heavy metals in different years was analyzed.

### *2.2. Analytical methods*
The potential ecological risk index method is a relatively fast, simple and standard method proposed by the Swedish scientist HAKANSON in 1980 to evaluate the potential ecological risk of heavy metals[13].

Single metal pollution coefficient:

$$C_f^i = \frac{C_s^i}{C_n^i};$$

Potential Ecological Risk Coefficient of Single Metal:

$$E_r^i = T_f^i \cdot C_f^i;$$

Comprehensive potential ecological hazard coefficient:

$$\mathrm{RI} = \sum \mathrm{E}_\mathrm{r}^\mathrm{i}$$

Wherein, $\mathbf{C_s^i}$ represents concentration of heavy metals in sediment, $\mathbf{C_n^i}$ represents background value, $\mathbf{T_f^i}$ represents the toxicity coefficient of heavy metals, which are shown in Table 1. The potential ecological risk levels are shown in Table 2[14].

**Table 1.** The background value and toxicity coefficient of heavy metals.

| | Cu | Hg | Pb | Zn | Cd | Cr | As |
|---|---|---|---|---|---|---|---|
| **Background value[15]** | 30 | 0.2 | 25 | 80 | 0.5 | 60 | 15 |
| **Toxicity coefficient[16]** | 5 | 40 | 5 | 1 | 30 | 2 | 10 |

**Table 2.** The potential ecological risk levels.

| Levels | Potential ecological risk | Code |
|---|---|---|
| **RI≤150** | Low | A |
| **150<RI≤300** | Moderate | B |
| **300<RI≤600** | Considerable | C |
| **600<RI** | Very high | D |

## 3. Results
The potential ecological risk indices (RI) werecalculated to assess pollution by 7 metalsin sediments in the Yangtze River from 1996 to 2012. As shown in Table 3, the RI values of heavy metals in the Yangtze River sediment were basically all lower than150, suggesting that most areas had low ecological risk from heavy metals. Therefore, the RI value in 2007 was ranging from 150 to 300, suggesting that most areas had moderate level from heavy metals in 2007. These results showed that the ecological risk level in the Yangtze River were almost low in the past 20 years.

We also found that from 1996 to 2012, two heavy metals showed obvious ecological risk than other heavy metals. What was shown in figure 2 has proved this. Cd and Hg, especially Hg, showed greater potential ecological risk coefficient than Cu, Pb, Zn, Cr, As. In particular, the potential ecological risk coefficient of Cd was significantly higher than other heavy metals in 2007, and the potential ecological risk coefficient of Hg was higher than others in 2003. We supposed that in these two years the industrial emissions and other anthropogenic pollution led to this situation. Thus, we considered that Cd and Hg as the two elements most in need of attention when we talked about heavy metal pollution issues in sediment.

**Table 3.** Potential ecological risk index of heavy metals in the Yangtze River sediment.

| | 1996 | 2000 | 2001 | 2003 | 2005 | 2006 | 2007 | 2008 | 2009 | 2010 | 2011 | 2012 |
|---|---|---|---|---|---|---|---|---|---|---|---|---|
| **RI** | 13.12/A | 26.63/A | 25.67/A | 102.39/A | 73.49/A | 10.72/A | 235.9/B | 14.57/A | 14.5/A | 30.05/A | 6.2/A | 20.99/A |

**Figure 1.** Comparison of heavy metals in the Yangtze River sediments between 1996 and 2012.

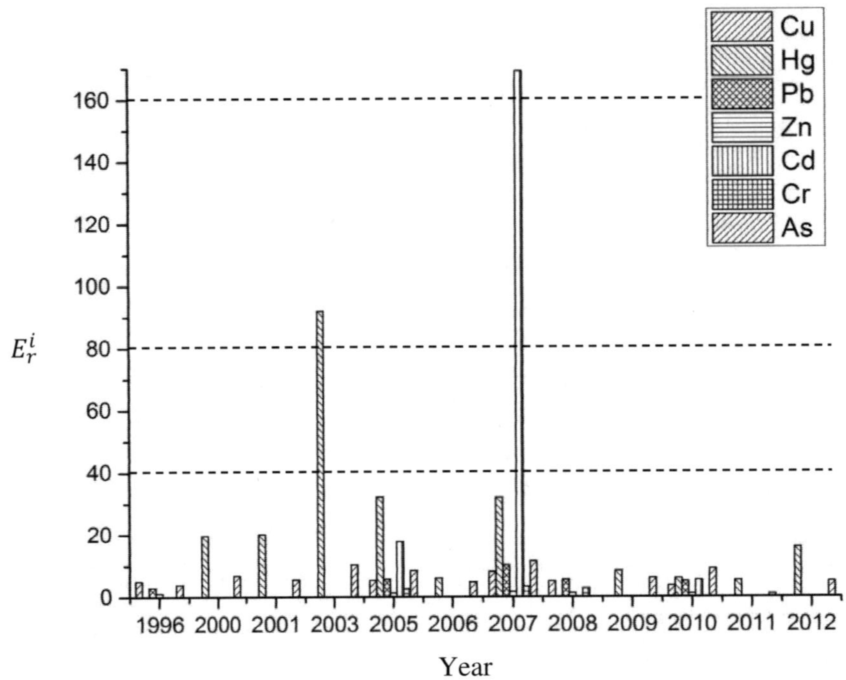


## 4. Conclusion

These results showed that the ecological risk level in the Yangtze River were almost low in the past 20 years. We also considered that Cd and Hg as the two elements most in need of attention when we talked about heavy metal pollution issues in sediment.

**Acknowledgments**

The research was supported by the Ministry of Science and Technology of China (No. 2014FY120603).

**References**

[1] Zhang C, Liu Q and Chen Y 2008 An Review:Toxity of Heavy Metals to Aquatic Organisms*Life Science Instruments*

[2] Ghosh M and Singh S P 2005 A review on phytoremediation of heavy metals and utilization of its byproducts *Applied Ecology & Environmental Research***3(1)**1-18

[3] Asmussen O 1997 Heavy metal concentrations in sediments from the coast of Bahrain*International Journal of Environmental Health Research***7(1)**85-93

[4] Tuncer G, Karakas T, Balkas T I, et al 1998 Land-based sources of pollution along the black sea coast of Turkey: Concentrations and annual loads to the black sea *Marine Pollution Bulletin***36(6)**409-423

[5] Qing L and Zijian W 1996 Research progress on the relationship between heavy metal morphology, biological toxicity and bioavailability *Environmental Science* **1996(1)**89-92

[6] Hongxia Z, Yong Z and Zirong X Research progress on toxicity of heavy metals to aquatic animals *Chinese Journal of Veterinary Medicine* **28(4)**13-18

[7] Jingsheng C and Jiayi Z 1992 Study on Heavy Metals in Water Environment in China *China Environmental Science Press*

[8] J.P.l G.s, Yuduan W and Yiping Y 1984 Chemical oceanography *Ocean Press*

[9] Wurl O and Obbard J P 2004 A review of pollutants in the sea-surface microlayer (SML): a unique habitat for marine organisms*Marine Pollution Bulletin***48(11-12)**1016–1030

[10] Rawlins B G, Turner G, Mounteney I, et al 2010 Estimating specific surface area of fine stream bed sediments from geochemistry*Applied Geochemistry***25(9)**1291-1300

[11] Gibson R N, Barnes M and Atkison R J A 2001 Functional group ecology in softsediment marine benthos: the role of bioturbation*Oceanogr Mar Biol Annu Rev***39** 233-267

[12] Hulscher T E M T, Mol G A J and Lüers F 1992 Release of metals from polluted sediments in a shallow lake: quantifying resuspension*Hydrobiologia***235-236(1)**97-105

[13] Hakanson L 1980 An ecological risk index for aquatic pollution control.a sedimentological approach*Water Research***14(8)**975-1001

[14] Wang Y, Yang Z, Shen Z, et al 2010 Assessment of heavy metals in sediments from a typical catchment of the Yangtze River, China *Environmental Monitoring & Assessment***172(1-4)**407-417

[15] Yi S, Qinglin M, Yunyong S, et al. 2011 Distribution and ecological risk assessment of heavy metals in sediments of the yangtze river estuary and its adjacent sea area *Journal of Zhejiang Ocean University (Natural Science Edition)* **30(2)**107-112

[16] Hui Y, Guoguang C, Hongying L, et al. 2013 Heavy metal pollution and potential ecological risk assessment of major lake sediments in the lower Yangtze River *Earth and Environment* **41(2)**

IOP Conf. Series: Earth and Environmental Science **61** (2017) 012031 doi:10.1088/1755-1315/61/1/012031

# Ethics position towards the exploitation of manganese material in Oenbit Village, East Nusa Tenggara, Indonesia

**Frederikus Fios**

Character Building Development Center, Computer Science Department Bina Nusantara University, Jakarta, Indonesia
fios@binus.ac.id

**Abstract.** Oenbit village is an area that is located in the district of Timor Tengah Utara (TTU), Timor Island, East Nusa Tenggara Province, Indonesia. In Oenbit ongoing a conflict between the economic interests of some parties namely the government, corporation and the local indigenous community. Government of Timor Tengah Utara give legal permission to the Elgari Resources Indonesia (ERI) Company to exploit the mining of Manganese in Oenbit Village which informally is the ancestral land of indigenous peoples Oenbit hereditary called pusuf kelef and Kot-tau niap-tau (king land). Oenbit society has an ethical belief that the ancestral land Oenbit should not be produced by outside parties besides the local community on the orders of the king. Manganese exploitation in Oenbit Village cause problems contradictorily interesting to reflect on the ethical-philosophical. This paper aims to reflect the ethical position against cases of exploitation of manganese in the Oenbit Village with focuses on the local government's decision to issue a permit exploitation and ERI Company exploit Mangan assumed unethical traditional indigenous tribe Oenbit. The study found that the district government and ERI Company has violated the public ethics and society traditional law, especially the rights of local indigenous communities by exploiting manganese material. The method used is the reflection of philosophy with ethical approaches and relevant ethical theories.

## 1. Introduction

Human beings are always doing economical activities to fulfill basic needs in order to survive in life. Unfortunately, the implementation of economic activities done often neglect ethical dimension. Economic activity which does not consider the ethical side will cause problems that harm humans and the natural environment. The primacy of ethical values which are ignored in economic business transactions have a negative impact on humans and the natural environment. This paper is an attempt ethical reflection on the empirical facts manganese mining activities undertaken by PT Energy Resources Indonesia (ERI) in Oenbit Village, East Nusa Tenggara Pprovince that bring ethical controversies on various parties, especially Oenbit indigenous peoples in relation to the environment. This paper is to prove a thesis statement that all forms of exploitation of the natural need to pay attention to the ethical aspects in its implementation so as not to bring ethical issues in human and local cultural environment. In the case of mining Oenbit Mangan, government of Timor Tengah Utara (TTU) and PT ERI assumed violation of ethics and common law of Oenbit society.

The method used philosophy methods particularly ethical reflection to find the meaning of the facts under investigation. Ethical reflection is a technique rational logical reasoning on the basis of

IOP Conf. Series: Earth and Environmental Science **61** (2017) 012031 doi:10.1088/1755-1315/61/1/012031

empirical facts to find ethical meaning to the problem being investigated. Data were collected by field observations and documentation from various sources that are relevant to the research topic.

## 2. Chronology of TTU Regent Permit to PT ERI and Conflict of Oenbit Manganese Exploitation

Oenbit Village located in Timor Tengah Utara District, East Nusa Tenggara Province was inhabited about 21 indigenous groups of Dawanese with customary rights to land are: Ataupah Tribe, Naikofi Tribe, Mantonas Tribe, Haukilo Tribe, Neonbeni Tribe, Taisbenu Tribe, Tanouf Tribe, Amasene Tribe, Noetnana Tribe, Amtasi Tribe, Amkeun Tribe, Naiaki Tribe, Naiboas Tribe, Banusu Tribe, Uskono Tribe, Pakaenoni Tribe, Us Olin Tribe, Taboy Tribe, Neonub/ Saijao Tribe, Aplasi Tribe, and Sonaf Lanasu-Taolin as King of Insana[1]. Oenbit administrative boundary region the eastern with the village of Teba, the northern with Upfaon Village, to the west with the Nunmafo Village, and the south by the Loeram Village. For hundreds of years Oenbit farmers live in communities and develop a pattern of ethical relationship with nature. Natural has spiritual and economics valuable for society of Oenbit. Mr. Amatus Naikofi say that nature is valuable for the local community because it provides benefit spiritually and economically[2]. While Mr. Niko Ataupah and Kornelis Us Olin said the same thing that Oenbit people appreciate ecological unity between human, nature, animals and plants. Oenbit community appreciate living in harmony with nature and preserving nature from being damaged or destroyed by human action that exploits crops such as mining activities. Oenbit community is an example of human ecological support of eco-philosophy (the term of Skolimowski) that claimed human ecological as human beings who live at one with nature[3]. Any mining exploitation definitely caused negative impact to nature and kill all life within it, including humans, animals and plants. Therefore, people of Oenbit reject all forms of exploitation of nature in their land.

Oenbit is a region that contains the natural wealth in the form of manganese. For this reason too many entrepreneurs and corporations have an interest to exploit the manganese in the village of Oenbit. If a company wishes to carry out exploitation or mining Manganese, they necessary legal permits as a condition for mining activities. In the context of Oenbit, PT Elgari Resources Indonesia is companies get permission to exploit manganese of Oenbit. Government of Timor Tengah Utara (TTU) provides mining license in Oenbit to PT Elgari Resources Indonesia issued a decree No. 270 of 2013. In 2008, the Regent Gabriel Manek issued decrees to PT Elgary Resources Indonesia to exploit an area of 1,623 hectares in Oenbit and Loeram Village[4]. In the same year, the Regent Gabriel Manek represents of government signed a memorandum between the Government of TTU and Yayasan Bina Swadaya, Japanese government representatives and community representatives Oenbit to make Banam Neten locations including in rural areas Oenbit area of 110 hectares for planting distances and cashew. Based on the decision letter, then in 2009 PT ERI come to Oenbit communities and told people to dig manganese and must be sold to PT ERI at an initial price of IDR 350/kilograms. These activities occurred until 2010.

In 2010, the regional parliament of Timor Tengah Utara formed a Manganese Special Committee, chaired by Mr. Hendricks Saunoah, SE (Chairman of the Legislative Council of TTU now) to investigate the case of manganese in Oenbit. The findings of the investigation team showed that PT ERI had falsified documents on letter of request to the Regent to invest in mining of Oenbit manganese. The findings are then recommended to the authorities for further action yet taken any action until now. In 2012 without socialization, PT ERI build Stock file Mangan over tribal land in Oenbit Village especially Taesbenu without approach with tribal landowners concerned.

In June 2014, without the permission to the Naikofi Tribe, PT ERI opening a road along 4 kilometers of stock file Mangan to the besin location (on the top of Oenbit mountain) and damaged two traditional rituals (*Bijae Bekuf*) belongs to the Naikofi Tribe thus fined by the Naikofi Tribe IDR 13.000.000 (thirteen millions). Fines is claimed by the PT ERI as a form of submission of tribal land to them to be used, but not so. In October 2014, after becoming aware of the activities of exploitation in besin locations (communal land belongs to the Ataupah, Temuku Ha, Us Taolin Tribe), three representatives of the Ataupah Tribe namely Mr. Nikolas Ataupah, Mr. Anderas Leu Ataupah, and Mr. Markus Tasaeb) reprimand the PT ERI to halt mining activities, but PT ERI argue that their activities are not serious and test tool only. Following up on this, the Head of Oenbit Village bring together people with PT ERI and the Government of Timor Tengah Utara (Mining and Energy). At the time of the dialogue is done, the Government of TTU intimidate people by citing Article of 33, paragraph 1, 2 and 3 that the soil, water, air and controlled by the state so that people speechless and can not argue anything. As if the government is arrogant to Oenbit People average not highly educated[5].

On January 30 2015, after it had been exposed in the media (Kompas Online) Secretary General of the National Student League for Democracy, Emanuel Tabean initiative to give direction to all members of the National Student League Democrat (NSLD) in Kefamenanu City for supporting the struggle of Oenbit People get justice by protesting the exploitation of Manganese of Oenbit. On February 6 2015, NSLD with communities who are members of the People's Alliance Environmental Concern (PAEC), protests to stock file of PT ERI and accepted by Mr. Hendra Castilio as the responsible of PT ERI Management. When it was agreed that the next meeting will be take place on February 23 2015, the deadline for the PT ERI manganese exploitation cease all activities at mine sites as well as agrees to bring the Regent of Timor Tengah Utara and Council of Timor Tengah Utara. But on February 23 2015, the Regent of TTU and the Regional Representatives Council of TTU was not present, so that people Oenbit leaving the location of meeting because they feel cheated by PT ERI. On February 26 2015, the people of Oenbit demonstrate by visiting the office of the Regional Representatives Council of TTU convey the aspirations and accepted by two members of the Regional Representatives Council of TTU namely Mr. Agustinus Tulasi and Mr Heribertus Radja. Both of board members have promised to bring the aspirations of the people Oenbit on the agenda in Legislative Council of TTU and immediately summoned the Regent TTU and PT ERI to resolve this Oenbit case.

On March 16 2015, after receiving an invitation from the Regional Representative Council of TTU, Oenbit people came to the House of TTU Regent and meet the Regent, Chief of Police, and the Chairman of the Regional Representative Council of TTU. Raimundus Fernandez, Regent of TTU has promised to come to Oenbit to see manganese exploitation activities and will investigate the fake names were handed over the land to PT ERI. In accordance with Raimundus promise on March 16, then on March 28 2015, the Regent was accompanied by Vice Regent, Vice Head Police, and Chairman of the Regional Representatives Council of TTU meeting in the office Village Oenbit attended by managers of PT ERI (Mr. Daniel Castilio) with the community of Oenbit demonstrators. During the meeting, strangely, did not disclosed the names of fake transfer their land but instead track the king Insana and the chairmen of the actual rate, because at that time the Naikofi Tribe present flatly refused to hand over their land to PT ERI. On the occasion, the regent said that if until June of 2015 there was no agreement between PT ERI and community of Oenbit, the regent will revoke the mining permit of PT ERI in Oenbit Village.

On April 14 2015, Oenbit society demonstrated by closing the entrance to the mine site. Society closed the entrance with fencing and planting mahogany trees along the road that are above ground of Naikofi Tribe. On April 20 2015, the tribes residing in the territory Insana (15 tribes) to make a statement together to reject the presence of PT ERI submitted to the TTU Regent, the Chairman of the Regional Representatives Council, Chief of Army Region 1618, police chief of TTU, but the letter was not addressed. On June 19 2015, after being suspended 2 times (June 5 and June 12 2015), the community Oenbit coming back to the office of Regent for dialogue but not seriously addressed by the TTU Regent. Feeling the regents do not care about the community of Oenbit are present, the people walk out of the space where dialogue. Until now, the activity of exploitation of manganese in Oenbit discontinued granted. The assets of PT ERI still on the scene and issued a decree for the exploitation of manganese Oenbit had not been canceled. This issue is so silenced without a thoughtful and serious handling of the warring parties.

## 3. Ethical Reflections on the Exploitation of Manganese Material in Oenbit Village

Ethics comes from the word 'ethos' (Greek) which means habit forming characters or character of the individual[6]. Ethics is the study of the merits of human action in social reality together with others. Ethical discipline consists of two (2) part of the general ethics and special ethics. General ethics examines the essence of conscience, freedom and responsibility, values and morals, rights and obligations as well as the ways in which humans live ethically. While special ethics is closely related to the application of ethics in the context of general or specific conditions. From this came the field of applied ethics. According to Beauchamp: "The term 'applied ethics' and its synonym 'practical ethics' Came into use in the 1970s when philosophers and other Academics Began to address the pressing moral problems in society and in professional ethics especially medical ethics and business ethics"[7]. Applied ethics evolved in the decade of the 1970s when the development of the fields of biomedical science and business creates problems for humans in society. This is supported by the occurrence of a variety of complex problems in the world of work or industry. Social movements like the labor movement, the environmental conservation movement, anti-war movement, the movement of gender, democratization movement and others are increasingly being made that drive the need for serious ethical reflection and depth in accordance with the context of the problem that occurred. Ethical reflection is an attempt to do philosophy in practice in accordance with the context of reality that is in empirical reality.

In ethics, there is a general ethical theories include: Ethics of Utilitarianism (Utilitarianism), Ethics of Deontology (Duty Based Ethics), Ethics of Rights (Right Based Ethics), and the Ethics of Virtue (Virtue Ethics). The fourth theory is a fundamental ethical theories related to the paradigm of thought, action, taking a stand and actualization of human behavior in existence of living. Ethical reflection that is developed in this paper applying the theory framework of Virtue Ethics Theory of Classical Greek philosopher, Aristotle as the main reference. The main peculiarities of Virtue Ethics Aristotle noticed dimension of attitude, character and morals of the human person as a moral agent in the actualization of its actions with others entities. According to Plato, the virtuous life is the life associated with behavior patterns that produce goodness and social harmony. "... However, goodness and virtue were Intimately connected with the mode of behavior that produced well-being and harmony"[8]. While Aristotle claimed that the highest aim of every human behavior in life that produces virtue. "... That man aims at a knowledge of the Idea of the Good"[9]. Aristotelian virtue ethics thoughts contained in the main work of the Nichomechean Ethics showing the main and highest goal of human life that seek goodness. "... Aristotle Nichomachean Ethics saying that 'every art and every inquiry, and similarly every action and pursuit, is thought to aim at some good ..."[10]. Virtue Ethics of Aristotle is also named teleological ethics. Thus it can be concluded that human life is an ethical and philosophical meaning of life by basing itself on the qualities and values of virtue directed at acts of goodness in life together. Model of virtuous life is what needs to be comprehended by the human subject as a moral agent in the totality of the existence of life.

Licence Exploitation of Manganese by PT ERI to Regent of TTU assumed to be less scrupulous in examining the document permissions. This is evidenced by the data that PT ERI falsified documents permission to the Regent of TTU without the knowledge of the regents (Documents of 15 Tribes Oenbit). In addition, the regents did not act ethically punish PT ERI who forged permission documents. When people Oenbit demonstrate, there is an impression by TTU Regent and acted indifferent to the aspirations of the people of Oenbit. This indicates that the Regent did not respect the aspirations of indigenous peoples of Oenbit. Regent did not act ethically by ignoring the current society dialogue was conducted in the office of Regent. TTU Regent does not have a good attitude to treat people Oenbit with ethical and respectful manner. TTU Regent ignore customary land rights of Oenbit society. TTU Regent poor decisions in manganese exploitation activities in Oenbit Village.

PT ERI showed quite a lot of unethical include: acting wrongly in the case of document forgery permission to the TTU Regent. PT ERI also not consider Oenbit indigenous communities ethics and destroy the customary ritual of Oenbit communities. PT ERI assumed to perform retrieval manganese Oenbit by way of unlawful and unethical. "Mining activities PT ERI carried out without the consent of land owners in the area of *kot tau-niap tau* and *pusuf kelef* thus assumed to be arbitrary, pilferage and theft of crops belonging to the tribes of native"[11]. This is an unethical action against Oenbit society traditional law.

The act of exploitation of PT Elgari Resources Indonesia conducted by Regent Decree No. 270 of 2013 raises ethical problems and harm the society Oenbit and nature Insana general. Manganese exploitation actions undertaken by PT ERI caused some negative effects, among others: *first*, grazing animals belonging to residents in the location *besin nia'ban* Oenbit reduce economic income communities existing tribes. *Second*, agricultural production Oenbit area disadvantaged communities due to the narrowing of exploration farmland by local farmers in traditional. *Third*, the destruction of places of traditional rituals belonging to residents of Oenbit which is indigenous ancestral tradition passed down through generations. *Fourth*, the destruction of nature for hundreds of large trees have been cuted by PT ERI during the opening of the road to the mine site of manganese. *Fifth*, the presence of PT ERI create horizontal conflict in the Oenbit community because there was a split (disharmony) in the society. *Sixth*, the alleged conspiracy triangle between the Regent, the regional council and PT ERI in the exploitation of Oenbit manganese.

Based on ethical reflection above, it can be concluded that in general the Regent of Timor Tengah Utara and PT ERI commit unethical acts in the exploitation of manganese in Oenbit Village, Timor Tengah Utara, East Nusa Tenggara Province. Government action and PT ERI is not only unethically, but also violated the positive law (falsification of documents permits) and the residents of Oenbit Village customary law (traditional ritual ethics of the local community). Therefore the exploitation of manganese in Oenbit Village can be said to contain ethical and legal contradictions. Manganese exploitation of Oenbit violated three main elements namely: against ethics, against the positive law and against law customary society of Oenbit.

## 4. Conclusion

We have seen together how ethics error, positive law violations and violations against indigenous communities of Oenbit ethics conducted by the TTU Regent, the Regional Representatives Council of TTU and PT ERI. Ethical reflection on the empirical facts manganese mining activities undertaken by PT Energi Resources Indonesia (ERI) in Oenbit Village, East Nusa Tenggara Province brings proven ethical and legal controversy. The thesis statement that all forms of exploitation of the natural need to consider ethical and legal aspects of the implementation is very necessary to be considered so as not to bring ethical and legal problems in humans in the application actions. In the case of mining Manganese of Oenbit, Regent of Timor Tengah Utara and PT ERI proved conclusively in violation of ethics, violate positive law and violated customary law of Oenbit society. Therefore, the authors recommend that the mining of manganese in Oenbit Village need to be stopped because the resulting detrimental to many parties, both in terms of ethics, the human aspect, the essence of the law, as well as nature conservation. If the exploitation of Manganese in Oenbit want to resume needs to be preceded by the analysis of the social and environmental impact comprehensively. Also necessary negotiations and agreements intelligent, ethical and in accordance with the law involving all the parties in a manner that is fair, ethical, legal, democratic and respectful of each other.

## References

[1] Taolin, L. Maksimus 2015 Rejecting Letter of Oenbit Manganese Exploiting A Document (Kefamenanu: Oenbit Tribals) p 2
[2] Fios, Frederikus 2016 Interview Transcript *A Document* (Jakarta: Bina Nusantara University) p 5
[3] Skolimowski, Henryk 2011 World as Sanctuary, The Cosmic Philosophy of Henryk Skolimowski (Arkana: Creative Fire Press)
[4] Taolin, L. Maksimus 2015 Rejecting Letter of Oenbit Manganese Exploiting A Document (Kefamenanu: Oenbit Tribals) p 3
[5] Taolin, L. Maksimus 2015 Rejecting Letter of Oenbit Manganese Exploiting A Document (Kefamenanu: Oenbit Tribals) p 4
[6] Prawironegoro, Darsono 2010 Philosophy of Science (Jakarta: Nusantara Consulting) p 80
[7] Beuchamp, L. Tom 2003 The Nature of Applied Ethics R.G. Frey & Christopher Healt Wellman (ed.) A Companion to Applied Ethics (USA: Blackwell Publishing) p 1
[8] Stumpf, Enoch Samuel 1982 Socrates to Sartre, A History of Philosophy (New York: Mcgraw-Hill) p 64

[9] Stumpf, Enoch Samuel 1982 Socrates to Sartre, A History of Philosophy (New York: Mcgraw-Hill) p 96
[10] Stumpf, Enoch Samuel 1982 Socrates to Sartre, A History of Philosophy (New York: Mcgraw-Hill) p 96
[11] Taolin, L. Maksimus 2015 Rejecting Letter of Oenbit Manganese Exploiting A Document (Kefamenanu: Oenbit Tribals) p 3

IOP Conf. Series: Earth and Environmental Science **61** (2017) 012032 doi:10.1088/1755-1315/61/1/012032

# Research of the leading edge separation vortex characteristics due to the inlet velocity shape

**CHEN Xiao-ming[1], FENG Xiao-qing[1], Hu Xiao-chun [1]**

1. China Huadian Engineering Co., LTD, Beijing 100160, China

**Abstract:** The zero adverse pressure gradient plate is taken as the study object, in order to study the leading edge separation vortex characteristics due to the inlet velocity shape. With the CFD method, we classify the study cases into 6 groups ($h/\delta$ =0.1,0.2,0.5,1.0,1.5,2.0(h for vortex generator(VG) height, $\delta$ of inlet boundary layer thickness)) to study the inflow boundary layer's influence law to the vortex generator characteristics. Calculation results show that, vortex generator leading separation vortex strength is proportional to fluid average kinetic energy within vortex generator height range. Vortex strength shows essentially inverse relationship with the flow boundary layer thickness, while the vortex generator shape resistance also shows the inverse proportion to the flow boundary layer thickness. Namely, heavier flaw boundary layer thickness leads to lower vortex strength and lower vortex generator shape resistance simultaneously.

## 1 Introduction

During wind turbine generation system operating, rotor blades occur surface flow separation and effect the wind turbine power generating efficiency. Vortex generator(VG) is one of the effective flow control equipment for flow separation[1-6]. Its control mechanism is right below: when air flows across the vortex generator, downstream of the VG will generate approximate stream vortex, which can strengthen the momentum exchange between the low energy inside the airfoil surface boundary layer and the high energy outside the boundary layer, so air flow with inverse pressure gradient, VG can postpone the airfoil boundary layer flow separation.

Vortex Generator was firstly proposed by Taylor[7] and initially applied into airplane airfoil. As for the wind turbine, Gyatt[8] did analysis and experimental study on the small two blade horizontal wind turbine(diameter 9.75m) with inverse rotation vortex generator. According to measurement data shown by the wind farm site, wind turbine with VG could increase the output power under the large wind velocity. While in the low wind velocity, wind turbine would generate less power instead. To learn more about the vortex generator technology applied into the wind turbine, the Dutch Delft university wind energy study center carried out the experiment about the Du series wind turbine airfoil lift and drag characteristic by the vortex generator with the different Reynolds number[9]. The result showed that the vortex generator could postpone the stalled attach angle, and improve the lift and drag ratio. Recently, Zhang Lei[10] etc. conducted a numerical simulation on the airfoil with the vortex generator installed. They got a good get a good agreement simulation result with the experiment data, what's more, the result showed that under the condition of the large attach angle, vortex generator was an effective equipment to restrain the boundary flow separation.

The characteristic of the VG includes vortex generator leading edge separation vortex intensity, vortex core area and vortex dissipation etc. When VG is installed on the solid surface, it can be affected by the wall boundary layer, what's more, the fluid flow status differs from the different boundary layer height. So, it is very significant to take a study on the vortex characteristic under the different flow boundary layer thickness. Especially, take the wind turbine blade flow control as example, VG is installed inside the turbulent boundary layer, so the paper studies the vortex generator characteristic under different flow boundary layer thickness.

## 2 Geometry Model and Numeric Method

### *2.1 Geometry Model*

The paper research on the zero pressure gradient plane, and change the boundary layer thickness of inlet turbulent velocity shape(as shown in fig.1) so as to study the vortex generator characteristic by the flow boundary layer thickness. The flow turbulent velocity shape is define as the equation(1):

$$u = \begin{cases} U\left(\dfrac{y}{\delta}\right)^{1/7} & y < \delta \\ U & y \geq \delta \end{cases} \tag{1}$$

Where u is the current velocity, $\delta$ is boundary layer thickness, U is the main stream velocity, and y is wall normal height.

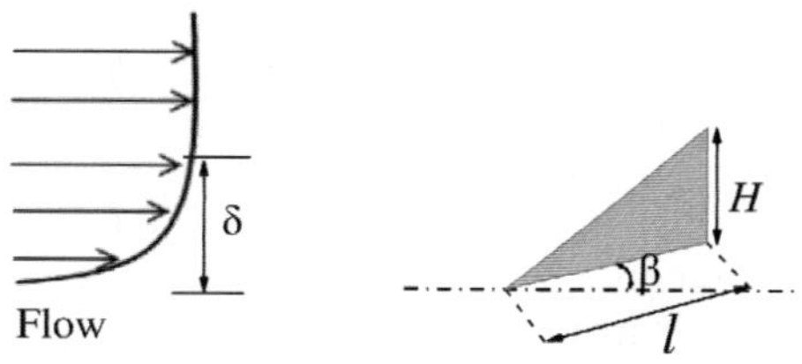


**Fig.1** Flow and schematic of VG

**Table 1 Geometric parameters of VG**

| Install angle $\beta$ (°) | Chord $l$ (mm) | height $H$ (mm) | The proportion of VG Height $H$ and boundary layer thickness $\delta(H/\delta)$ |
|---|---|---|---|
| 20 | 17 | 5 | 0.1, 0.2, 0.5, 1.0, 1.5, 2.0, |

Table 1 shows the VG geometry parameter, VG is wall without thickness. Let flow velocity U 82m/s, and Reynolds number based on the VG height about $3\times10^4$.

### *2.2 Computational Method*

Steady numerical simulation is carried on with fluent software, and finite volume method is used to discretize the control equation, with two order accuracy central difference scheme for space discretion and pressure-velocity coupling SIMPLE algorithm. The full turbulent Spalart-Allmaras(SA) model is selected as turbulent model.

Computation mesh: mesh is generated by ICEM, and the domain is divide with structure mesh. Assigning the VG height and chord direction with 80 grid point, VG downstream with 200 grid point, the total mesh quantity is 5 million. The surface and bottom mesh of VG is shown in fig2.

**Fig.2** Grids

Boundary Condition: for the convenience of monitoring the vortex running situation at the downstream of the VG, set the stream-wise length 144H, while VG is installed in the 9H far from inlet, domain width and height are both 20H. Boundary conditions include velocity inlet, pressure outlet, and symmetry condition at the two side of the computational domain. In addition, the upper side is free slip boundary condition, and both VG and downside is no-slip wall boundary condition.

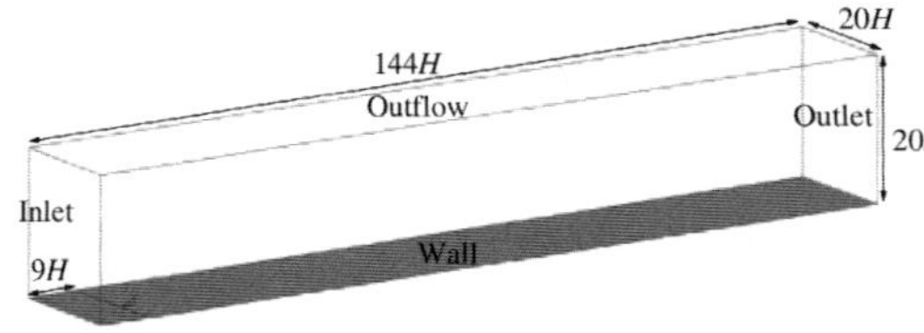


**Fig.3** Computational domain and boundary conditions

## 3 Result

Fig 4 shows the static pressure efficiency curve at the suction surface. We can figure out that the magnitude of Cp increase along with the height of the VG increase no mater the VG installed position along the chord-wise. Especially when H is $1.0\delta$,$1.5\delta$,$2.0\delta$, the magnitude of Cp is close to each other, however, the magnitude of Cp has more difference between H=$0.1\delta$ and H=$0.2\delta$. The magnitude of Cp at the height of $0.5\delta$ is in the middle of the maximum magnitude.

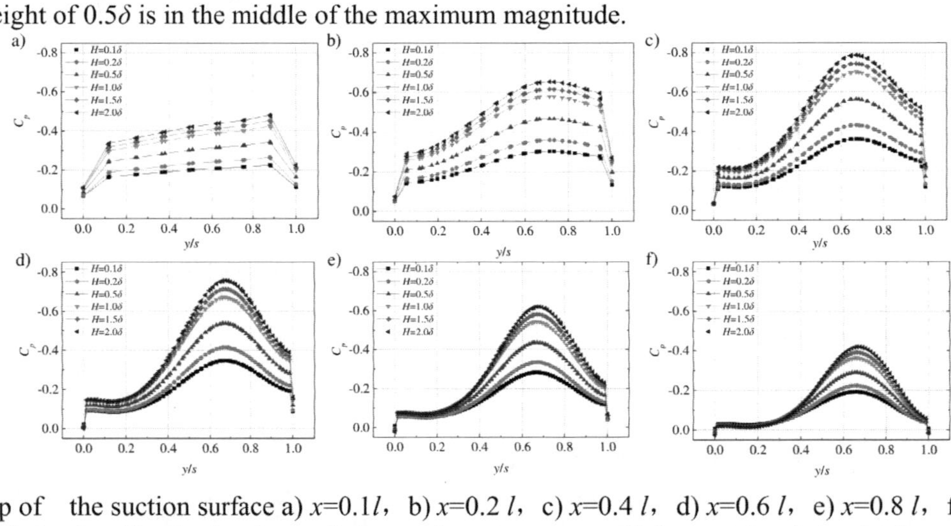


**Fig.4** Cp of the suction surface a) $x$=0.1$l$, b) $x$=0.2 $l$, c) $x$=0.4 $l$, d) $x$=0.6 $l$, e) $x$=0.8 $l$, f) $x$=1.0 $l$

Fig.5 is the distribution law of the static pressure coefficient and axial velocity along the axial-wise. The magnitude of Cp at the vortex core increase first and then decrease, and get the top value at the half of the chord. The variation of axial velocity at the vortex core is consistent with that of Cp, and lower Cp at the vortex core, bigger axial velocity. With the increase of the VG height, Cp at the center of the vortex become higher, and axial velocity become faster, which confirm that the strength of the vortex core will become stronger with the higher height of the VG. Seen from Fig.5 b), axial velocity at the vortex core all increase first and then decrease in the different height of the VG, which illustrate that vortex core become stable and without the vortex core breakout phenomenon et.

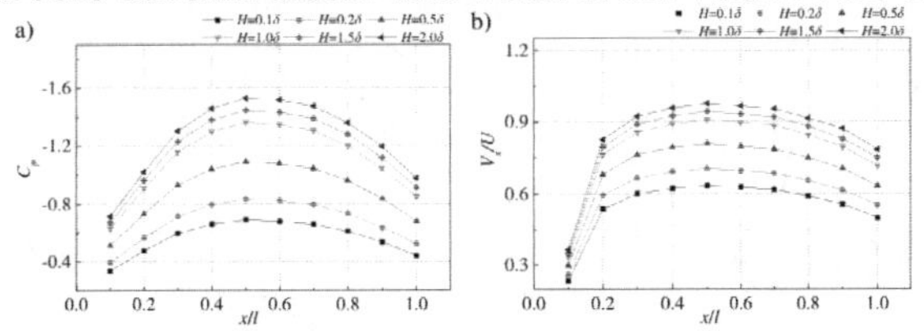

a) Cp, b) Axial velocity

**Fig.5** Cp and axial velocity distribution at the center of vortex

Fig.6 shows the variation law of the vortex core area and circulation in the VG airfoil surface along the stream-wise. As shown in the chart, vortex core area has an approximate distribution along the different height of the VG, only some large difference shows at the trailing edge of the airfoil surface. The increase regularity of vortex core area along the stream-wise distance shows approximately linear change distribution, and vortex core circulation increase as the height of the VG increase, especially at the trailing edge of the VG, the difference of the circulation value become bigger .

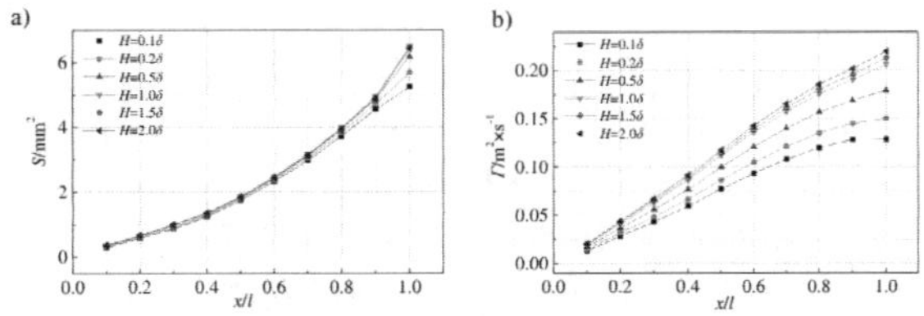

a) Vortex core area, b) Vortex strength

**Fig.6** Vortex core area and vortex strength along with the string

Fig.7 shows the variation law of the vortex core area and vortex circulation at the trailing edge of VG. Seen from Fig.6, vortex core area increase as the flow distance increase, however, vortex circulation inside the VG airfoil surface increase along the stream-wise, and at the downside of VG decrease gradually, namely, the maximum of the circulation exist at the trailing edge of the VG. For the convenience of the comparison, choose the vortex core area and circulation at the trailing edge of the VG. Seen from the below figure, when $H/\delta<1$, vortex core area hit the top, while vortex core area decrease a little when $H/\delta>1$, and vortex core area increase rapidly with the increase of the VG height when $H/\delta<1$. Vortex core circulation is directly proportional to the VG height, namely vortex core strength increase as the height of VG increase. However, vortex core strength is not show the linear change with the VG height. When $H/\delta<1$, the gradient between vortex core strength and VG height is much bigger, but when $H/\delta>1$, the gradient is much smaller.

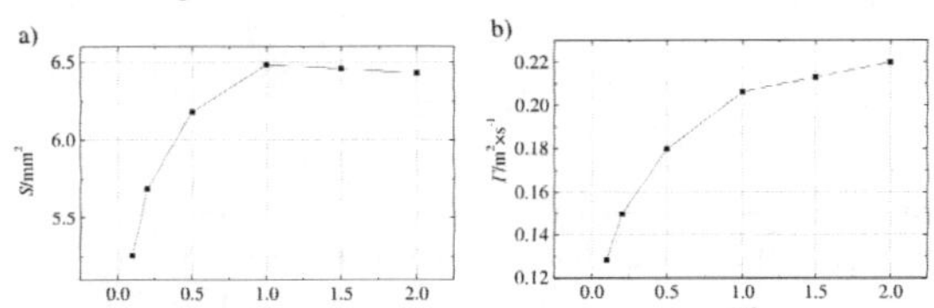

a) Vortex core area, b)Vortex strength

**Fig.7** Vortex core area and vortex strength along with height at trailing edge of VG

Fig.8 a) is the boundary layer velocity shape downside the VG, b) is the fluid average velocity inside the range of VG height. Seen from Fig.8 a), VG is placed in the different boundary layer thickness, the upstream flow velocity shape is different. And the most intuitive knowledge is that with the increase of the boundary layer height, the boundary layer fluid kinetic energy decrease inside the range of the VG height. Seen from b), as the VGs height increase, namely the boundary layer thickness decrease, the fluid average velocity become higher and higher inside the range of the VG height, and the variation law between the fluid velocity and VG height shows the exponent law distribution. When $H/\delta<1$, thee gradient between boundary layer fluid velocity and VG height is a little high. When $H/\delta>1$, gradient become low, and the variation of the velocity show stable.

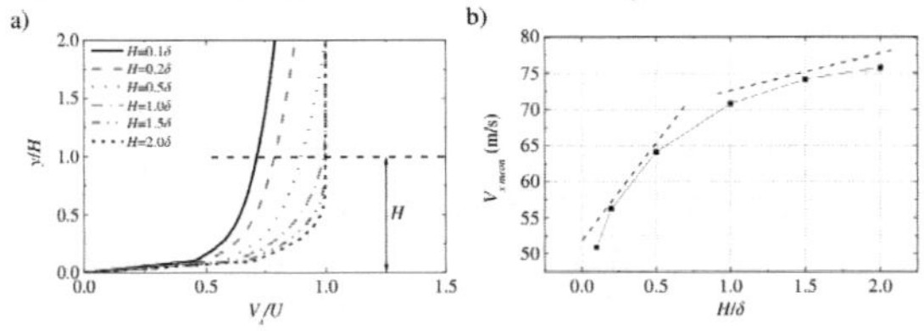


a) The velocity shape, b)Average velocity

**Fig.8** The velocity shape at VG upstream and average velocity within the height of VG

Fig.9 is the variation rule of VG lift coefficient, drag coefficient and lift-drag ratio inside the different boundary layer height. Seen from a), the lift coefficient gradually increases with the VG height increases, especially the maximum lift coefficient is about 0.65 when $H/\delta=2$. Seen from b), the drag coefficient also increases as the height of VG increases and the maximum drag coefficient is 0.25 when $H/\delta=2$. To a great extent, the lift of VG originates in the vortex induced lift from the delft wing. So if the vortex strength of the concentrated vortex induced by the delta wing become stronger, then the VG lift will become stronger. And drag force is pressure drag, namely VG shape resistance. Take Fig.9 a) and Fig.9 b) into comparison, found that the variant law of the lift and drag coefficient along the have consistent distribution. Seen from Fig.7 b) and Fig.8 b), there are consistent variant law among the fluid kinetic energy, VG vortex strength and lift drag coefficient along the different height of VG. So the VG induced vortex strength and shape resistance have close relation with the VG height $H/\delta$ ,and show the direct proportional relation but not linear change, instead the exponent distribution. Namely the effect is less obvious with the bigger VG $H/\delta$. Seen from fig.9 c) the lift-drag ratio hit the top when $H/\delta=0.5$. From a theoretical view of point, when $H/\delta=0.5$, it not only can generate certain vortex strength to conduct flow control, but also the shape resistance is not so big, but whether the ratio of the height and boundary layer thickness is the best or not, the deep analysis should be conducted from the point of the boundary layer fluid kinetic energy.

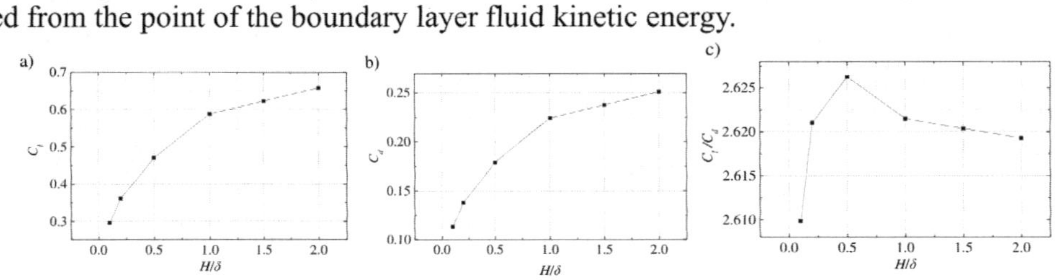


a) $C_l$, b) $C_d$, c) $C_l/C_d$

**Fig.9** $C_l$、 $C_d$ and $C_l/C_d$ of VG

Fig.10 is the distribution of the axis velocity along the normal height at the position of vortex core in the downside of VG. The line in the figure is about the center of the vortex core. Seen from fig.6 a), with the increase of the flow distance, different height of the VG has the different vortex core area and the bigger the vortex core area is, the higher the normal height is. So the vortex core height noted in the figure is the vortex height when $H=1.0\delta$. Firstly, as shown in the figure, as the increase of the flow distance, vortex core height will become higher. Then the velocity shape inside the boundary layer is basically changed with S distribution. Energy conservation can be confirmed by that the increase of fluid kinetic energy inside the vortex core is equal to the decrease of the energy outside of the vortex core. Take comparison with fluid kinetic energy at the different boundary layer height, the boundary layer fluid kinetic energy hit the bottom when H=0.1$\delta$、0.2$\delta$, and both of them have little difference. When H=0.5$\delta$, boundary layer fluid energy have the medium order, namely the value is greater than the energy at H=0.1$\delta$, 0.2$\delta$, simultaneously lower than the energy at H=1.0$\delta$.when H=1.0$\delta$, 1.5$\delta$, 2.0$\delta$, the boundary layer kinetic energy have the approximately same distribution, which means that the boundary layer energy will not increase with the height of the VG increase. While seen from the fig.9, the shape resistance will increase with the increase of the height of the VG. So, the flow control get the best optimal effect at H=1.0$\delta$ for the view of the boundary layer fluid energy, with the much more stronger concentrated vortex strength and less shape resistance.

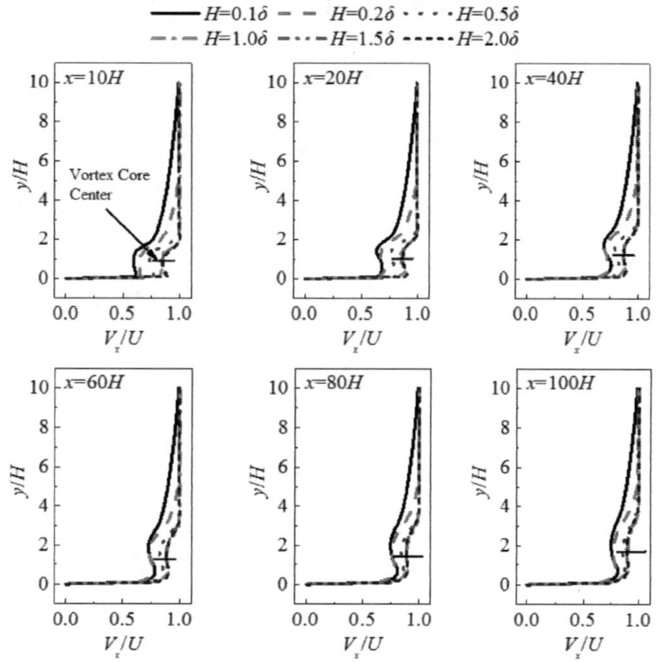


**Fig.10** The velocity shape at VG downstream

**4 Conclusion**

The paper research on the VG vortex characteristic by flow boundary layer height with the numerical method, the main conclusion can be shown below:

(1) The VG leading edge separation vortex strength has a direct proportional relation with the average fluid kinetic energy inside the range of the VG height. Vortex strength and shape resistance induced by VG has inversely proportional relation with the flow boundary layer height. Namely, with the increase of the boundary layer , vortex strength induced by VG become smaller, also the shape resistance.

(2) When $H/\delta$=0.1, 0.2, boundary layer fluid kinetic energy hit the bottom, and both of them have almost not difference. When $H/\delta$=0.5, the fluid energy is at the middle, the VG lift-drag hit the top. When $H/\delta$=1.0, 1.5, 2.0, the boundary layer fluid energy is basically the same, but lower the flow boundary layer height is, bigger the VG shape resistance. So when $H/\delta$=1.0, vortex generator flow control can get the best optimal effect.

**Reference**

[1] Kehro M, Hutcherson S, Liebeck R. Vortex Generators Used to Control Laminar Separation Bubbles[R]. AIAA-1990-0051

[2] Kehro M, Hutcherson S, Blackwelder R F. Vortex Generators Used to Control Laminar Separation Bubbles[J]. Journal of Aircraft, 1993,30(3): 315-319
[3] Nickerso J D. A Study of Vortex Generators at Low Reynolds Numbers[C]. Reno, Nevada: AIAA 24th Aerospace Sciences Meeting, 1986: AIAA-86-0155
[4] Lin J C. Control of Turbulent Boundary-Layer Separation Using Micro-Vortex Generators[R]. AIAA-1999-3404
[5] Johansen J, Soerensen N N, Zahle F, et al. Aerodynamic Accessories[R]. Denmark: Risoe National Laboratory, 2004: Risoe-R-1482
[6] Abdollah.A.Afjeh, Theo G. Keith, Ardeshir Fateh. Predicted Aerodynamic Performance of a Horizontal-Axis Wind Turbine Equipped With Vortex Generators[J]. Journal of Wind Engineering and Industrial Aerodynamics, 1990 (33): 515-529
[7] Taylor H D, et al. Elimination of Diffuser Separation by Vortex Generators[R]. United Aircraft Corporation Report, No R-4012-3,1947
[8] G.W. Gyatt. Development and Testing of Vortex Generators for Small Horizontal Axis Wind Turbines1986[R]. NASA-CR-179514
[9] W.A. Timmer, R.P.J.O.M. van Rooij. Summary of the Delft University Wind Turbine Dedicated Airfoils[R]. AIAA-2003-0352
[10] ZHANG Lei, YANG Ke, XU Jianzhong. Effects on Wind Turbine Airfoils by Vortex Generators[J]. Journal of Engineering Thermophysics , 2010,31(5):749-752. (in Chinese)

IOP Conf. Series: Earth and Environmental Science **61** (2017) 012033 doi:10.1088/1755-1315/61/1/012033

# Emergence of a utopian vision of modernist and futuristic houses and cities in early 20$^{th}$ century

**Nan Ma**

University of Melbourne
nanm@student.unimelb.edu.au

**Abstract**. Throughout the development of literature on urban design theories, utopian thinking has played a crucial role as utopians were among the first designers. Many unrealized utopian projects such as The Radiant City, have presented a research laboratory and positive attempts for all architects, urban designers and theorists. In this essay, a utopian vision following under More's and Jameson's definitions is discussed, examining how the utopian vision of modernist and futuristic houses and cities emerged in the early twentieth century in response to several factors, what urban utopia aimed to represent, and how such version was represented in the built form and the urban landscapes.

## 1. Definition of Utopia: A Brief Overview

Following Ganjavie's article, it is difficult to find a unified definition of utopia.[1] The name of the entire genre is More's utopia where the term 'utopia' can be used to signify any works that are seemingly idyllic and fictional.[2] In general, "utopian has come to be a code word on the left for socialism or communism; while on the right it has become synonymous with 'totalitarianism' or, in effect, with Stalinism".[3] However, when reviewing twentieth century articles in the field of urban design and architecture, describe some shortcomings of More's utopia such as it is failure to correspond to the contemporary society. Therefore, throughout the twentieth century, many architects and urban designers put forward their ideas for utopian cities such as Howard, Wright and Le Corbusier. [4]

## 2. Factors Influencing the Emergence of a Utopian Vision

There are three major factors that motivated the emergence of a utopian vision of modernist/futurist houses and cities, including futurist movements, urbanization and population growth and social background.

Specifically, the publication of The Funding and Manifestos of Futurism in 1909 represents emergence of Futurism. On one hand, this manifesto eulogizes contemporary industrial civilization and scientific techniques totally changed the perception of conventional time and space. On the other hand, this manifesto derogates all the old transitions, supports sweeping all the cultural legacy from ancient Rome and advocates demolition of all museums, libraries and colleges. Marinetti argues that 'the world's magnificence has been enriched by a new beauty: the beauty of speed…' [5] Sant'Elia also articulates his Futurist interpretation of modernity in his manifesto in 1914, claiming that conventional materials are supposed to be replaced by structure of machines and new materials. But Sant'Elia argues that the urban layout should be built corresponding to population-centralization and efficient transportation including underground rails, stereo-crossing road web and moving pavements. [6] Moreover, steel, glass and textile fibre should substitute wood, stone and brick to enable residents to get maximum lightness and elasticity. As the influence of Futurist movements, many of their ideas were

IOP Conf. Series: Earth and Environmental Science **61** (2017) 012033 doi:10.1088/1755-1315/61/1/012033

translated and expressed in architecture and urban designs such as Città Nuova (New City) with machine-like shape created by Antonio Sant'Elia.

In addition, urbanization and increases in population are another substantial factor. During the period of late nineteenth century to early twentieth century, the population of London grew from 900,000 to 4.5 million. The population of Paris increased from 500,000 to 2.5 million over the same period. Berlin's population went from 190,000 to more than 2 million and New York's from 60,000 to 3.4 million. [7] Such fast population growth necessitated reinventions of these cities to respond to the increased number of urban habitants as the cities lost their power to control. Many utopian urban projects at that time emerged such as Le Corbusier's Ville Contemporaine- a city of 3 million inhabitants in 1922 and Wright's Broadacre City.

The third factor driving the formation of utopian vision is the social context. Utopian urban projects are formed gradually and naturally when reality no longer solves new problems happening in cities. Le Corbusier saw around cities stagnation in the countryside, a crisis in the old regional centres, blind forces of chance and profit and class segregation. He proposed a new urban system- The Contemporary City.[7] This new urban design included high-density living places to solve the problem of congested city, such innovation as separation of the pedestrian circulation paths from the roadways to create more efficient urban transport system. [8] Therefore, on the one hand, impulses of utopia provide a critique of existing social order structure; on the other hand, social background or crisis influences the emergence of utopia.[9]

## 3. Aspects Represented by Utopian Thinking

Three major factors motivated the emergence of a utopian vision of modernist/futurist houses and cities, including futurist movements, urbanization and population growth and social background.

If human beings still desire to live in better cities, and for the role of architecture to be sustainable and more just, then utopian approaches to architecture will be used so that planners can begin to figure out new possibilities for human life. [9] Therefore, utopia aims to secure the need to dream, the need for mystery, and sometimes even impulses to create poetry and to construct. Humans are able to realize the world they are living in is in dynamic state of progress rather than constancy through utopian urban projects. With a utopian approach, habitual values are thrown into disarray and utopia's proper and newfound space can be easily entered with an opportunity to educate humans themselves. [9]

Utopia also aims to summon aspiration- to desire more, to dream more and even to dream in new ways. [9] As Satuffer argues, Utopia always presents cities in positive ways and plays an important role in urban planning.[10] However, when cities need to progress, utopia appears to be a viable approach as utopians throughout ages were responding to problems that cities were facing and were able to offer a compelling alternative for their citizens. [11]

In addition, utopia aims to represent ideal types of cities for the future. They are "convenient and attractive intellectual tools that enable each planner to bring together his many innovations in design, and to show them as part of a coherent whole, a total redefinition of the idea of the city". [7] For instance, entering into a new century, Paris's urbanization accelerated continuously under the impetus of Industrial Revolution and a great number of dwellings flock into gathering areas. Governments intended to seek a new urban design solving social problems. The Plan Voisin was a solution for the centre of Paris designed by Le Corbusier including buildings are in a regular grid, the space structured with two new traffic arterial roads through city and skyscrapers rise in the heart.[12] Even though this utopian idea was not accepted by governments, it offered a good beginning and new possibilities for reconstructing Paris.

According to Ganjavie, some attributes of contemporary cities are production of utopian thinking. "The separation of pedestrians from automobiles (Cabet, Garnier, & Hénard), the separation of urban functions, zoning (Garnier), and the generalization of prefabricated systems (Le Corbusier) are just some examples generated by utopian thinkers".[1] In this case, utopia creates new possibilities to improve citizens' living standards and also meets human desires and needs with development of society.

## 4. Epitomisation of a Utopian Vision in Built Environment and Urban Landscape

This section contends that a utopian vision of modernist/futuristic houses and cities somehow is epitomised in built form and urban landscape, but it is not completely embodied due to social, political and other factors.

It is believed that the quality of utopia has been fully expressed in Archizoom's famous project – the No-Stop City – that is an urban presentation of a factory-like building. For Archizoom, this urban project is completely realized in a capitalist city. "The dark images of the No-Stop City have been used as a model to create shock in readers in order to let citizens think about the gaps in a capitalist society".[9]

There are another examples demonstrating that utopian thinking is epitomised in the built form. Mies van der Rohe proposed 'less is more', which means using less materials to show more things in buildings. Bourgeois's House and Terraced House designed by Jacobus Oud and Victor Bourgeois respectively are two of masterpieces in the Deutscher Werkbund exhibition of 1927. Both projects reflect ideals of Werkbund, and in particular, they reveal an aesthetic of function, tectonics, and lightness; the most important point here is that an architecture expressing less is more really is a discourse on beauty.[13] Mies van der Rohe's Tugendhat House in Brno "as an ideal realisation of a new overall conceptualisation not only of architectural space but of living itself" shows weightless living and Mies' design principle of 'less is more' and emphasizes on functional amenities.[14]

More examples still can be found. For example, just as Manifesto of Futurist Architecture 1914 claims that "roofs and underground spaces must be used" [6], Le Corbusier's Five Points of New Architecture proposes free ground design and roof garden, which means ground floor with diverse internal usage is totally released from supporting walls, and a flat roof can be utilized for planting flowers while the whole building would be green. The urban planning of Chandigarh is also based on Le Corbusier's Ville Radieuse.

Nevertheless, Ganjavie believes that the broad conclusion is that all the utopian experiments have generally failed as there is a lack of enthusiasm toward understanding the relationship between utopian thinking and urban design. [9] There is a significant evidence leading readers to believe that the utopian approach to urban design should be seen as an old and dead concept. As Pinder argues, even though utopian thinking has been proclaimed many times throughout history, it always seems like fantasy.[11] Paden also argues that utopia is one human beliefs and this belief can be thought of as neglecting limitations of natural conditions to create ideal (or utopian) cities and societies.[15] What is more, Jacobs in her famous book *The Death and Life of Great American Cities* mentions that all utopian projects are realized arise under particular conditions; in other words, they are under utopian conditions avoiding from reality to create imaginary world.[16] Therefore, utopian projects no matter urban design or architecture, cannot be completely implemented in contemporary cities.

## 5. Conclusion

By referencing and analysing actual designs and utopian theories, this essay argues that a utopian vision of modernist/futuristic houses and cities emerged in the early twentieth century in response to political, economic, social, technological and cultural forces and futurism movement. Generally, a utopian vision aimed to represent the power of contemporary technology, new aesthetic conception, spiritual pursuit and human desires and needs. No-Stop City project, villa designs on Deutscher Werkbund exhibition of 1927 and some other examples given above exemplify utopia can be somehow realized. However, utopianism cannot be fully realized due to natural conditions and reality of contemporary cities. Just as Ganjavie argues, with the use of a utopia approach in urban planning and architecture, "our habitual values are thrown into disarray, and we enter utopia's proper and newfound space with an opportunity to educate ourselves". [9] The way that human beings did for approaching utopia essentially is a process of self-improvement.

Nevertheless, for urban design or planning projects working toward the betterment of human conditions, urban designers or planners alike may need to go beyond the aspire for pure spiritual life and keep away from out-of-tune talks in the name of culture, politics and economy. For a soul that dwells on unmeaningful fantasy and searches for utopian wishes, those refined temperament things, more often than not, would go the wrong way though they sometimes did generate certain relief and success. In this

IOP Conf. Series: Earth and Environmental Science **61** (2017) 012033 doi:10.1088/1755-1315/61/1/012033

sense, urban designers/planners cannot neglect conditions of the real society that may depend on factors beyond the spiritual ballast and aspiration.

## References


[1] Ganjavie, A 2012, 'Role of Utopia for Design of Future Cities: Utopia in Urban Planning Literature', *Studies in Literature and Language,* Vol. 5, No. 3, 2012, pp. 10-19.

[2] Black, J, Conolly, L, Flint, K, Grundy, I, Liuzza, R, McGann, J, Prescott, A, Qualls, B & Waters C 2010, *The Broadview Anthology of British Literature: The Renaissance and the Early Seventeenth Century,* Broadview Press, Canada.

[3] Jameson, F 2004, 'The Politics of Utopia', *New Left Review*, vol. 25, January-February, pp. 35-54.

[4] Picon, A 2008, 'Ville idéale', in M Riot-Sarcey, T Bouchet & A Picon (eds), *Dictionnaire des Utopies,* Larousse, Paris, pp. xiv-296.

[5] Marinetti, FT 1973, 'The Founding and Manifesto of Futurism 1909', in U Appollonio (ed), *Futurist Manifestos,* MFA Publications, New York, pp.19-24.

[6] Sant'Elia, A 1973, 'Manifesto of Futurist Architecture 1914', in U Appollonio (ed), *Futurist Manifestos,* MFA Publications, New York, pp.160-172.

[7] Fishman, R 1996, 'Urban Utopias: Ebenezer Howard and Le Corbusier', in S Campbell & S Fainstein (eds), *Readings in Planning Theory,* Blackwell Press, Cambridge, pp. 19-67.

[8] Anthony, HA 1966, 'Le Corbusier: His Ideas for Cities', *Journal of the American Institute of Planners,* vol. 32, no. 5, pp. 279-288.

[9] Ganjavie, A 2014, 'The Role of Utopian Projects in Urban Design', *Utopian Studies,* Vol. 25, No. 1, pp. 125-149.

[10] Stauffer, M 2002, 'Utopian Reflections, Reflected Utopias: Urban Designs by Archizoom and Superstudio', *AA Files*, vol. 47, Summer, pp.23-36.

[11] Pinder, D 2001, 'Utopian Transfiguration: The Other Spaces of New Babylon', *Architectural Design,* vol. 71, no. 3, pp. 15-19.

[12] Fishman, R 1977, Urban Utopias in the Twentieth Century: Ebenezer Howard, Frank Lloyd Wright, and Le Corbusier, Basic Books, New York.

[13] Dunham, D 2014, 'Beyond the Red Curtain: Less Is More Utopia', *Utopian Studies,* Vol. 25, No. 1, pp. 150-173.

[14] Wagner, M 2012, 'Mies van der Rohe's Tugendhat House: Weightless Living', *Docomomo Journal,* vol. 46, no. 1, pp. 20-25.

[15] Paden, R 2003, 'Marxism, Utopianism, and Modern Urban Planning', *Utopian Studies,* vol. 14, no. 1, pp. 82-111.

[16] Jacobs, J 1992, The Death and Life of Great American Cities, Vintage Press, New York.

IOP Conf. Series: Earth and Environmental Science **61** (2017) 012034 doi:10.1088/1755-1315/61/1/012034

# On public space design for Chinese urban residential area based on integrated architectural physics environment evaluation

**J Y Dong[1,2,3], W Cheng[1] , C P Ma[2] , Y T Tan[2] and L S Xin[2]**
[1] School of Architecture, Harbin Institute of Technology, Harbin 150000, China.
[2] School of Architecture and Design, Changchun Institute of Technology, Changchun 130000, China.
[3]International Asia Art Academy, Banggkok Thonburi University, Bangkok, Thailand.

**Abstract.** The residential public space is an important part in designing the ecological residence, and a proper physics environment of public space is of greater significance to urban residence in China. Actually, the measure to apply computer aided design software into residential design can effectively avoid an inconformity of design intent with actual using condition, and a negative impact on users due to bad architectural physics environment of buildings, etc. The paper largely adopts a design method of analyzing architectural physics environment of residential public space. By analyzing and evaluating various physics environments, a suitability assessment is obtained for residential public space, thereby guiding the space design.

## 1. Introduction

At present, the research on the method of designing a residential public space mainly centers on two aspects: one is the study of the humanized public space, and the other is the application of the physics environment evaluation of the public space. The former focuses on the attributes of public space community and the impact on people's psychology and behavior, while the latter emphasizes digital simulation of physics environment[1]. However, both aspects are closely interacted and integrated with each other without separation. People's outdoor behavior and psychology are largely affected by the physics environment. Thus, only by combining the two aspects can we realize the process design of residential public space and reflect a sustainable concept in the initial design.

Architectural physics environment is a vital part of people's living environment. Every day, people are living in a variety of space environments where stimuli from heat, light, sound and other factors are faced constantly[2]. People's abilities of maintaining the normal physiological and psychological functions and effectively engaging in various kinds of activities depend on their environmental conditions, but people are limited to adjust the spatial environment stimulus. Therefore, it is necessary to adjust and control quantity of stimulus caused by the spatial physics environment, with a view to optimizing the stimulus of environment. Meanwhile, the design plan should also take into account the possible spatial physics environment after construction, and create suitable environment stimulus to provide their users with the best range of environmental stimulus, namely making users feel comfortable[3].

Thanks to the advanced technology, a lot of software can be used to simulate architectural physics environment nowadays. In the design planning, physics environment quality of the public space is the main basis for evaluating ecological feature in the residential area, so the designer should gradually introduce physics environment to the design, with a desire to create a more pleasing space.

## 2. Analysis on architectural physics environments

### *2.1. Research ideas analysis*
The urban space is complicated, variable and uncertain in regard to its physics environment, so is the public space of residential area. Different architectural layouts may exert different impacts on the physics environment. Also, this complexity is far beyond the designer's experience or manual calculation. Fortunately, digital simulation technology specific to the physics environment makes it possible for scientific and accurate construction and sustainable design planning. Wind, light, heat and sound are main environmental factors affecting residential public space. Among the diverse related simulation software, Ecotect Analysis, CFD and Cadna/A are most commonly used[4].

### *2.2. Analysis on architectural physics environment factors*

*2.2.1. Wind environment analysis.* The wind environment of residential area mainly refers to the wind field formed by natural wind when passing through the building group. An improper construction type or layout likely results in a bad wind environment, making pedestrians uncomfortable and impeding the natural ventilation in buildings and pollutant diffusion.

*2.2.2. Thermal environment analysis.* The outdoor thermal environment mainly refers to the impact of solar radiation on the residential public space environment. The amount of solar radiation incident on the ground determines the change of air temperature, and forms a series of thermal environmental indexes. Thus, solar radiation should be focused on in design planning.

*2.2.3. Light environment analysis.* The light environment in residential area does not involve the traditionally direct light, diffuse light and reflected light. Instead, sunlight is the most critical issue in light environment.

*2.2.4. Acoustic environment analysis.* How to avoid and reduce the impact of noise in residence has long been concerned. The acoustic environment researched mainly includes traffic, especially the environmental noise in road.

### *2.3. A combination of integrated architectural physics environment*

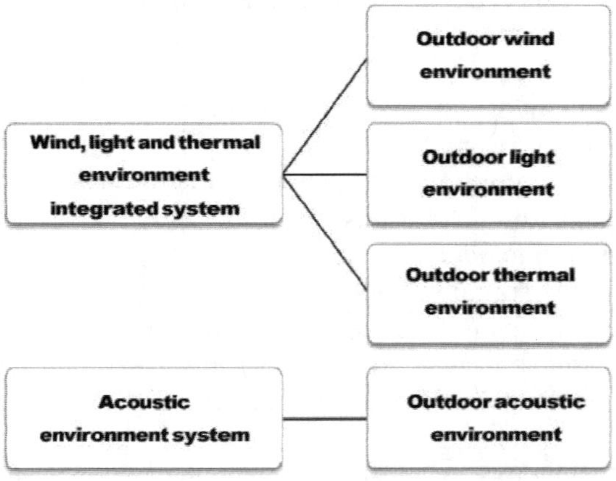


**Figure 1.** Environmental stimulus analysis

IOP Conf. Series: Earth and Environmental Science **61** (2017) 012034 doi:10.1088/1755-1315/61/1/012034

For these 4 kinds of stimuli, thermal environment, wind environment and light environment are more closely related to each other, and they are interrelated in real life and analysis; while acoustic environment is relatively independent (Figure 1). The meteorological data involved in the study are mainly from the *Meteorological Data Set Exclusively for the Analysis of Chinese Architectural Thermal Environment* and the meteorological data downloaded from the Energy Plus website[5].

## 3. Analysis on the architectural physic environment of residential public space—a case study of a residential area in Changchun

### *3.1. Selection of research object*

Located in Yitong River, Changchun City, Jilin province of China, the research object covers a total area of 18.7hm$^2$, and the volume rate of 1.8. The base is divided into 4 groups, centering on the green space. The east of base is the city expressway where a greening protective isolation area is distributed[6]. The public space analyzed largely includes the central green area of community, greening space of 3 groups, and protective greening space and kindergarten at east. In summer, Changchun is extremely hot characterized by sparse cloud and strong solar radiation, and is prone to persistent hot weather; while in winter, Changchun can be rather cold. Therefore, it is quite important for its residents to enjoy a fine architectural physics environment[7].

### *3.2. Outdoor wind environment analysis*

First, a base model is established in Phoenics, and the conditions related to model are set according to the meteorological condition in Changchun, including the wind inlet, wind outlet, wind speed, etc. Then, analysis should be conducted from the most wind direction in winter, summer and all year, calculated by the average wind speed in winter and summer shown in meteorological data. Meanwhile, wind speed is calculated and simulated based on the relationship between wind speed and comfortableness.

Once foreign researchers did a lot of field tests of discomfort resulted from wind in pedestrian areas. As a result, the relation between pedestrian's comfort and wind speed is demonstrated in Table 1. It is set that the CFD analysis reference level is 1.5m.

**Table 1.** Relationship between wind speed and comfort

| Wind speed (m/s) | Human feeling |
|---|---|
| v<5 | Comfortable |
| 5<v<10 | Discomfort, action affected |
| 10<v<15 | Very uncomfortable, the action is seriously affected |
| 15<v<20 | Insufferable |
| v>20 | Danger |

Under the southeast wind in summer, wind environment of base is relatively good, because the entrance, the center green and garage entrance can all form a wind corridor, and each group space can also structure their own gallery. In this instance, wind speed in the front of building is greater than that in the back, which can explain the large wind pressure difference. And this pressure can improve the ventilation inside the building (Figure 2a).

In the case of the prevailing northeast wind all year round, most of the wind environments are favorable except the groups in southwest corner. The upper and lower parts of the central green space are better than the middle in terms of ventilation, and northern and eastern groups enjoy a relatively fine ventilation where wind gallery can be formed (Figure 2b). Based on this, it can well identify the places suitable or unsuitable for activities under the northeast wind.

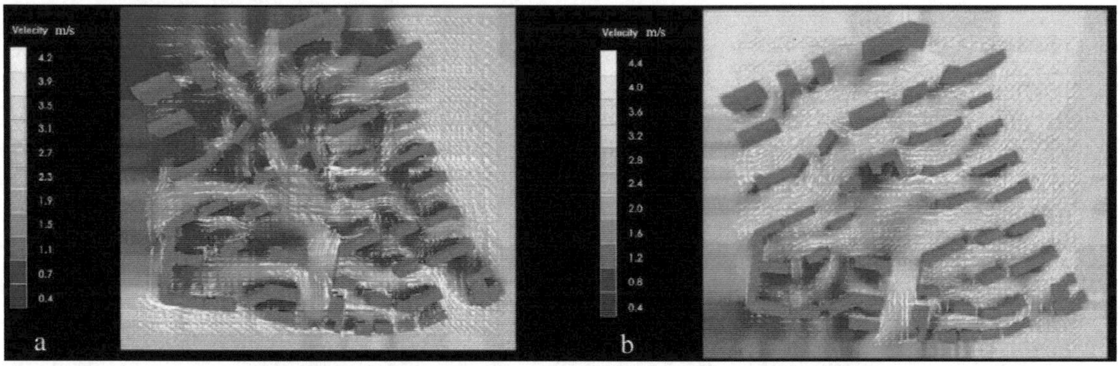


**Figure 2.** Schematic diagram of CFD software simulation

*3.3. Outdoor thermal environment analysis*

Following the Changchun meteorological data downloaded from Energy Plus website, Ecotect model is established. Then, the outdoor thermal environment is analyzed on the basis of this model, but only the solar radiation of base is considered. The base faces a stronger solar radiation in summer, and each group covers a relatively strong solar radiation patch except the central greening space and eastern protective greening area. Therefore, the public activity space should be designed in areas with weak solar radiation (Figure 3a).

The central green area, kindergarten and protective greening space have the largest solar radiation in base throughout the year. An intense radiation belt extends toward south and southwest along the central green land, and all groups have an intense radiation patch. First, according to the annual solar radiation degree, it can determine the scope suitable for the human activities in residential public space, and then further determine a proper position based on the nature of activities; second, it can determine the range suitable for planting, in which heliophilous, largely leafy and deciduous plants are fit for the area with strong radiation, while sciophilous and coniferous evergreen plants are prone to weak radiation (Figure 3b).

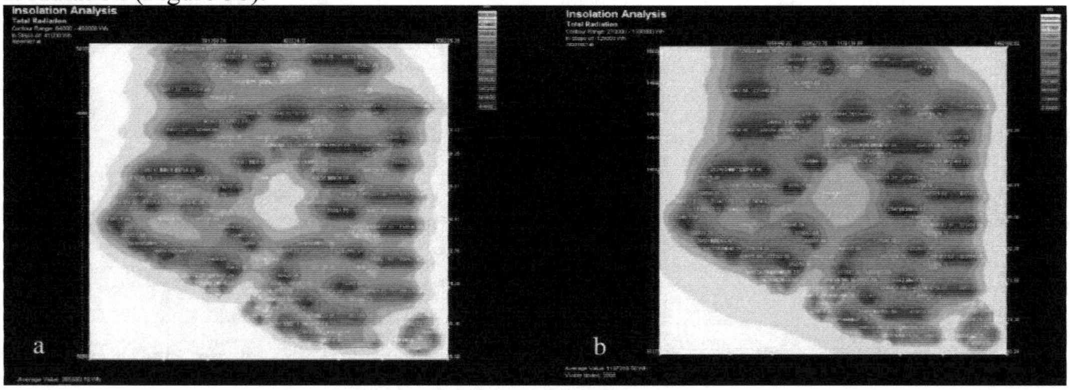


**Figure 3.** Solar radiation pattern

*3.4. Outdoor light environment analysis*

According to the meteorological data, the Ecotect model is established to analyze the outdoor light environment. In cold winter, it is best to cover direct sunlight when having body-building exercise in the morning. In this regard, it should choose the place without shadow above to arrange the site and facilities (Figure 4a).

On the contrary, in hot summer, it is best to avoid direct sunlight when having body-building exercise in the morning. Therefore, it should choose the place with shadow above to arrange the site and facilities (Figure 4b).

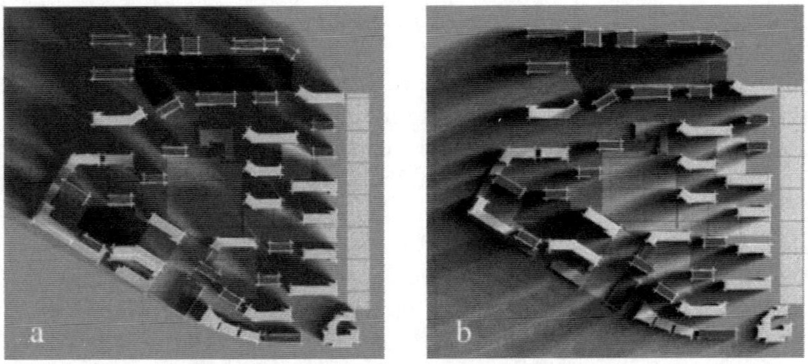


**Figure 4.** Morning sunshine condition

*3.5. A comprehensive analysis of wind, thermal and light*

Through gathering images above of architectural physics environment analysis, a superimposition analysis can be conducted for the previous wind environment, thermal environment and light environment.

According to 5 standards: the best, preferable, ordinary, relatively unfavorable, inappropriate, the public space of base is divided and sorted. By overlay analysis on wind, light and thermal environments in summer and winter, coupled with the conditions of annually maximum wind direction and year-round sunshine radiation, a construction suitability evaluation map of public space is obtained.

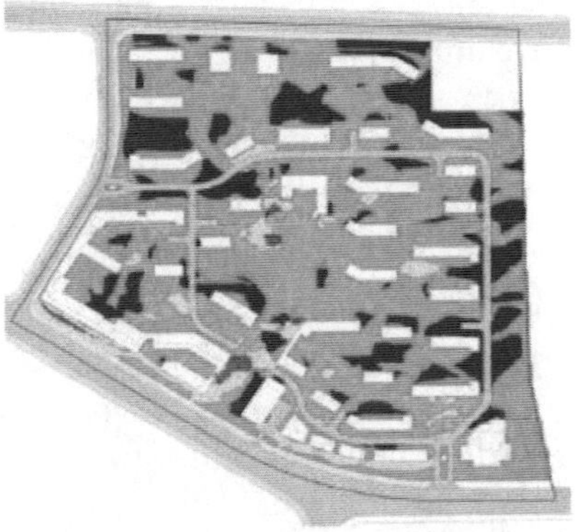

**Figure 5.** Comprehensive analysis of wind, thermal and light

It can be seen that few areas have maintained a good architectural physics environment for a long period during a year, except the west of central green space and the entrance for pedestrians. The west of central green space should be built as the foremost hard open space, used for having morning exercise and entertainment of all ages. Meanwhile, areas for activities in summer and winter should be planned according to their special season situation. In accordance with the suitability degree in the map, each group can establish activity space of group in the yellow and green section, but should strictly limit setting the activity space in the blue section (Figure 5).

*3.6. Sound environment analysis*

The base is located along the urban express way, so the noise will exert a huge impact on future residents. At the same time, the secondary urban road at the south of base and southern Binghe Road can also bring noise to the base to a certain extent. As implied in Cadna/A analysis, the noise situation after the completion of base will be quite bad. In the horizontal distribution, the base noise is fairly serious. Affected by the express way, the east suffers a general noise intensity of over 60db and local noise intensity greater than 75db. For the rest three directions, the street noise is also higher than 60db due to the road noise. In vertical distribution, the first row of buildings adjacent to express way suffers noise of over 75db, and the second and third rows are above 65db noise, far beyond the national standard .Given that the largest source of noise pollution in the base is the express way, the sound insulation screen is provided in the surroundings to reduce its impact on the base. The analysis proves that sound insulation screen of 3 meters in height can achieve a better effect if installed along the express way, the border of northern parking lot and the border of southern kindergarten .

## 4. Conclusion

As residential area is a living place, its design of space environment is particularly important. In addition to fully considering the various influence factors, the design should also cater for the needs of different people, thereby coordinating and unifying the functional, ornamental, and ecological characteristics of residential space environment, and furnishing a comfortable living environment for the residents. Based on thoughts and trials in designing residential area, the paper introduces overall analysis of physics environment into the process of public space design. Furtherly, with Ecotect Analysis, CFD, Cadna/A and other software, the physics environment and space structure are simulated and explained technically. In this way, the land construction suitability of the whole public space gets determined by the final suitability evaluation, which provides a practical guidance for further design.

## Acknowledgments

This work was financially supported by Science and Technology Project Foundation of Ministry of Housing and Urban-Rural Construction of the People's Republic of China (No. 2016-R-2005), Jilin Provincial Social Science Foundation (No.2015-BS-83), Youth Foundation of Changchun Institute of Technology (No.320140006) and Teaching reform project of Changchun Institute of Technology.

## References

[1] Junyan Dong, Hong Jin, Jian Kang. A pilot study of the acoustic environment in residential areas in Harbin, towards the questionnaire design. Journal of Harbin Institute of Technology, 2011(2):319-322.
[2] Junyan Dong, Wen Cheng. Based on the Characteristics of Respondents and the Voice of the Urban Neighborhood Public Space Business Facilities Noise Environment Evaluation Research. Journal of Harbin Institute of Technology, 2011(4):103-109.
[3] Junyan Dong, Hong Jin. The design strategy of green rural housing of Tibetan areas in Yunnan, China. Renewable Energy, Vol.49:63-67.
[4] Junyan Dong, Wen Cheng. Research on Optimized Construction of Sustainable Human Living Environment in Regions where People of a Certain Ethnic Group Live in Compact Communities in China.Renewable Energy and Environmental Sustainability, 2016(1):1-7
[5] Junyan Dong, Hong Jin, Jian Kang, Research of acoustic environment in public space in urban residential community in China, Huazhong Architecure, 2012(2):61-63
[6] Junyan Dong, Hong Jin, Jian Kang, Research on the evaluation of the traffic acoustic environment in public space of residential community based on the characteristics of residents, Architecural Journal. 2013(2):124-129
[7] J.Kang, M.Zhang.Semantic differential analysis of the soundscape in urban open public spaces.Building and Environment.2009(1):265-267

IOP Conf. Series: Earth and Environmental Science **61** (2017) 012035 doi:10.1088/1755-1315/61/1/012035

# A new method for calculating single-point productivity equation coefficient α in heterogeneous gas reservoir

**He Lei[1, 2], Lan Yifei[1, 2], Yuan Yingzhong[*3], Xie Shan[1, 2], You Liangrong[1, 2], Ai Qinglin[1, 2]**

1. National Engineering Laboratory for Low-Permeability Petroleum Exploration and Development, Petro China, Xian 710021, Shanxi, China
2. Research Institute of Exploration and Development, Changqing Oilfield Company, Petro China, Xian 710021, Shanxi, China
3. Chongqing Key Laboratory of Complex Oil & Gas Fields Exploration and Development, Chongqing University of Science & Technology, Shapingba, Chongqing 401331, China

*Corresponding Author, e-mail: yuanyingzh0001@126.com

**Abstract**. Single-point test is mainly based on quantitative statistics of productivity well testing data to obtain average $\alpha$ value of gas fields or blocks, then corresponding single-point productivity equation can be obtained. However, for heterogeneous gas reservoir in Ordos Basin, because of the influences of reservoir physical property and test data, there are relatively big error between calculated average $\alpha$ value of gas field and actual $\alpha$ value of single gas well, which results in application of single-point empirical formula is limited. Through derivation of binomial productivity equation in this paper, it is found that there are good linear relationship between $\alpha$ and reciprocal of the square of pressure $1/p_R^2$. Combined with actual test data of Jingbian gas field, the empirical formula between $\alpha$ and $1/p_R^2$ for 20 gas wells is got through regression analysis. The results indicate that new method greatly improves computational accuracy for absolute open flow of gas wells, which has good application prospect in similar gas field.

## 1. Introduction

Regular productivity well testing include systematic well test, isochronal test, modified isochronal test and single-point test[1-4]. Due to low permeability of lithologic gas reservoir, strong heterogeneity, difference of single well production capacity and large amount of gas wells, regular multipoint production well test are restricted in Ordos Basin. Thus, single-point test is applied in most of gas wells[5-7].

Empirical formula of single-point test is based on large quantity of steady well test data. In another word, single-point productivity equation is derived from average $\alpha$ which is based on reliable productivity equation and corresponding absolute open flow. Generally, the more stable well testing data of gas field, the more representative single-point productivity equation will be[8-10]. However, because of strong heterogeneity, there are big errors between average $\alpha$ which is obtained from a gas block by statistics data and single wells. Considering above problems, in order to accurately forecast gas well productivity, it is essential to find a new method to improve computational accuracy.

## 2. Derivation of basic theory

*2.1 Single-point productivity equation*

Making production of gas wells reach steady state under single working system, single-point test can obtain absolute open flow through substituting reservoir pressure，gas production and corresponding steady flowing bottom pressure into empirical productivity formula.

Binomial productivity equation is defined as follow:

$$P_R^2 - P_{wf}^2 = Aq_g + Bq_g^2 \tag{1}$$

where, the coefficient A expresses viscous resistance and coefficient B expresses inertia resistance along flow path, which can all be got through high pressure physical parameters of reservoir and fluid[11-12]. When $p_{wf}$=0.101MPa(1atm), the maximum potential production capacity is absolute absolute open flow :

$$p_R^2 - (0.101)^2 = Aq_{AOF} + Bq_{AOF}^2 \tag{2}$$

Combination Eq. (1) and Eq. (2), this correlation is given in Eq. (3)

$$\frac{p_R^2 - p_{wf}^2}{p_R^2} = \frac{Aq_g + Bq_g^2}{Aq_{AOF} + Bq_{AOF}^2} \tag{3}$$

where

$$q_D = \frac{q_g}{q_{AOF}} \tag{4}$$

$$p_D = \frac{p_R^2 - p_{wf}^2}{p_R^2} \tag{5}$$

$$\alpha = \frac{A}{A + Bq_{AOF}} \tag{6}$$

Through derivation, single-point productivity equation can be written as:

$$q_{AOF} = \frac{2(1-\alpha)q_g}{\alpha\left[\sqrt{1+4\left(\frac{1-\alpha}{\alpha^2}\right)\left(\frac{p_R^2 - p_{wf}^2}{p_R^2}\right)}-1\right]} \tag{7}$$

*2.2 Analysis of influence factor*

Binomial productivity equation also can be given as[13-15]:

$$p_R^2 - p_{wf}^2 = \frac{1.291\times10^{-3} q_{sc} T \overline{\mu z}}{kh}\left(\ln\frac{0.472 r_e}{r_w} + S'\right) \tag{8}$$

where

$$S' = S + Dq_{sc} \tag{9}$$

$$D = 2.191\times10^{-18}\frac{\beta r_g K}{\overline{\mu} h r_w} \tag{10}$$

$$\beta = 7.644\times10^{10} / k^{1.5} \tag{11}$$

Combination Eq. (8) , Eq. (9) , Eq. (10) and Eq. (11), Eq. (12) can be written as:

$$p_R^2 - p_{wf}^2 = \frac{1.291\times10^{-3} q_{sc} T \overline{\mu z}}{kh}\left(\ln\frac{0.472 r_e}{r_w} + S\right) + \frac{2.828\times10^{-21}\beta \gamma_g \overline{z} T q_{sc}^2}{r_w h^2} \tag{12}$$

where

$$A=\frac{1.291\times10^{-3}T\overline{\mu z}}{kh}(\ln\frac{0.472r_e}{r_w}+S) \quad (13)$$

$$B=\frac{2.828\times10^{-21}\beta\gamma_g\overline{Z}\overline{T}}{r_w h^2} \quad (14)$$

For dry gas reservoir, following equation can be used to calculate production rate

$$q_{sc}=\frac{774.6kh(p_R^2-p_{wf}^2)}{T\overline{\mu z}(\ln\frac{0.472r_e}{r_w}+s')} \quad (15)$$

In fact, wellbore radius, gas viscosity, reservoir temperature, Z-factor and relative gas density is almost the same in the same gas reservoir if ignoring skin effect caused by turbulence flow. Combination Eq.(6), Eq.(13), Eq.(14) and Eq.(15), following simplified formula can be obtained:

$$\alpha=\frac{1}{1+C_1k^{0.5}P_R^2} \quad (16)$$

In low-permeability gas reservoir, permeability is related to formation pressure, especially under the condition of stress sensitive effect[16-17]. According to Eq. (16), there is some correlation between α and permeability as well as reservoir pressure. Actually, coefficient α is an empirical parameter, which is various in different gas wells or blocks along changes of reservoir characteristics[18-19]. For gas wells in Ordos Basin, it is found that relatively to reservoir pressure, the error of permeability k is big and the influence on result is small. Thus，Eq. (17) can be written as

$$\alpha=\frac{1}{1+C_2P_R^2} \quad (17)$$

$P_R$ and $\alpha$ are determined by actual test data of gas field. Empirical formula about reciprocal of pressure square $\frac{1}{P_R^2}$ and $\alpha$ can be obtained through regression analysis of series of $P_R$ and $\alpha$ value for every well.

## 3. Field examples

Based on test data of 20 gas wells in Jingbian gas field, the value of $\alpha$ can be calculated from static pressure($p_R$), $p_{wf}$, $q_{sc}$ , an $q_{AOF}$ . Range of $\alpha$ value in low-permeability and high-permeability area are 0.548~0.853 and 0.218~0.781, with average value of 0.684 and 0.495 respectively (**Table 1**).

Calculation results show that $\alpha$ is variable for each well. Average error in low-permeability and high-permeability area is 13.3% and 41.9%, with range of 3.4%~23.7 and 15.5%~127.1% respectively.

From Eq.(7), it is found that it is relevant between $\alpha$ and reservoir pressure. Through regression analysis of $\alpha$ value and $p_R$ in 20 gas wells, it is shown that as decrease of $1/P_R^2$, value of $\alpha$ will increase, with good linear relation.

Thus, $\alpha$ can be calculated through $1/p_R^2$ which is proposed as follow.

For $\alpha$ in low-permeability area (**Figure 1**):

$$\alpha=1820.3/P^2-1.2645 \quad (18)$$

For $\alpha$ in high-permeability area (**Figure 2**):

$$\alpha=2260.9/P^2-1.8488 \quad (19)$$

**Table 1.** Statistic results of productivity well testing

| Reservoir types | Wells | $q_g$ ($10^4m^3/d$) | $p_{wf}$ (MPa) | $p_R$ (MPa) | $q_{AOF}$ ($10^4m^3/d$) | α Result | α Absolute error(%) |
|---|---|---|---|---|---|---|---|
| Low permeability | X1 | 1.49 | 22.8 | 29.9 | 3.26 | 0.851 | 23.5 |
| | X2 | 2.00 | 28.7 | 31.3 | 8.40 | 0.548 | 18.8 |
| | X3 | 2.00 | 28.3 | 30.7 | 10.77 | 0.768 | 15.2 |
| | X4 | 4.00 | 27.1 | 31.0 | 12.99 | 0.674 | 3.4 |
| | X5 | 4.00 | 28.4 | 31.3 | 15.87 | 0.601 | 8.4 |
| | X6 | 2.50 | 29.5 | 30.9 | 18.58 | 0.585 | 11.3 |
| | X7 | 3.00 | 27.4 | 29.0 | 24.36 | 0.853 | 23.7 |
| | X8 | 2.80 | 28.7 | 30.3 | 19.69 | 0.677 | 3.8 |
| | X9 | 3.50 | 28.2 | 31.2 | 13.01 | 0.563 | 15.6 |
| | X10 | 4.50 | 29.0 | 30.3 | 40.13 | 0.717 | 9.2 |
| | Average | 2.98 | 27.8 | 30.6 | 16.71 | 0.684 | 13.3 |
| High permeability | X11 | 3.01 | 29.5 | 31.1 | 23.15 | 0.781 | 36.5 |
| | X12 | 4.00 | 30.0 | 30.8 | 34.98 | 0.430 | 15.5 |
| | X13 | 8.01 | 26.7 | 29.5 | 36.46 | 0.761 | 34.8 |
| | X14 | 5.97 | 31.0 | 32.1 | 37.50 | 0.332 | 49.5 |
| | X15 | 8.00 | 28.3 | 30.6 | 43.00 | 0.715 | 30.7 |
| | X16 | 15.03 | 30.8 | 32.6 | 58.19 | 0.218 | 127.1 |
| | X17 | 9.85 | 29.3 | 30.5 | 65.17 | 0.419 | 18.3 |
| | X18 | 9.99 | 31.2 | 32.0 | 78.32 | 0.309 | 60.6 |
| | X19 | 4.60 | 30.5 | 31.8 | 27.93 | 0.385 | 28.8 |
| | X20 | 4.10 | 29.3 | 30.0 | 55.71 | 0.597 | 16.9 |
| | Average | 7.26 | 29.6 | 31.1 | 46.04 | 0.495 | 41.9 |

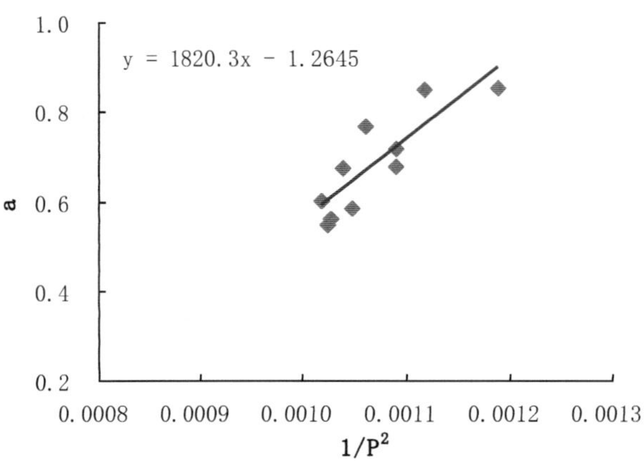


**Figure 1**. $\alpha$ vs. $1/p_R{}^2$ in low-permeability area

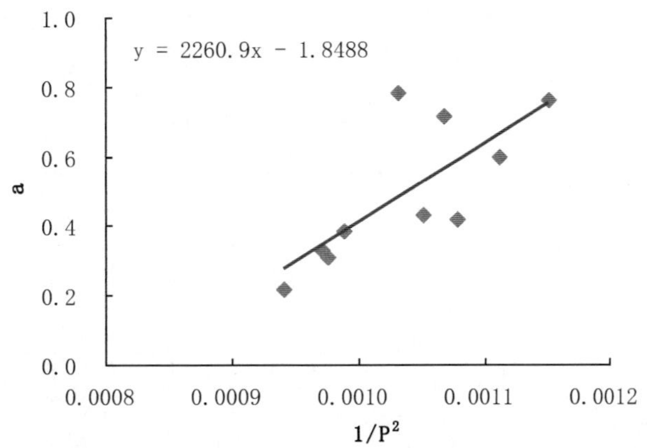


**Figure 2.** $\alpha$ vs. $1/p_R{}^2$ in high-permeability area

The error of new method is obviously smaller than traditional method, where error decline from 13.3% to 7.0% and 41.9% to 18.2 % in low and high permeability area respectively(**Table 2, Figure 3 and Figure 4**). Contrasting with traditional method, there are high accuracy for calculating absolute open flow of gas well with new method, which has good application prospects in similar gas field.

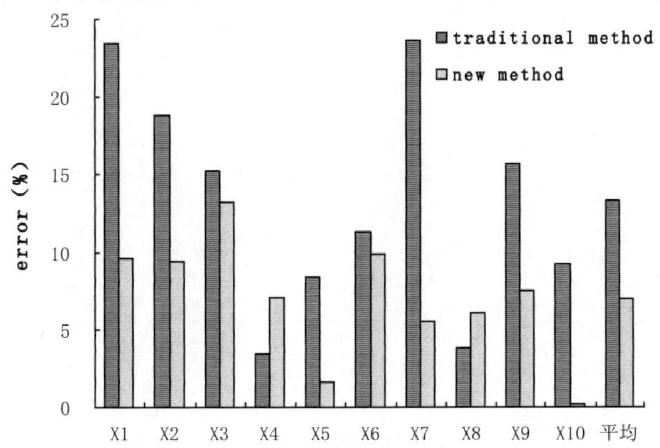


**Figure 3**. Bar graph for error of $\alpha$ in low-permeability area

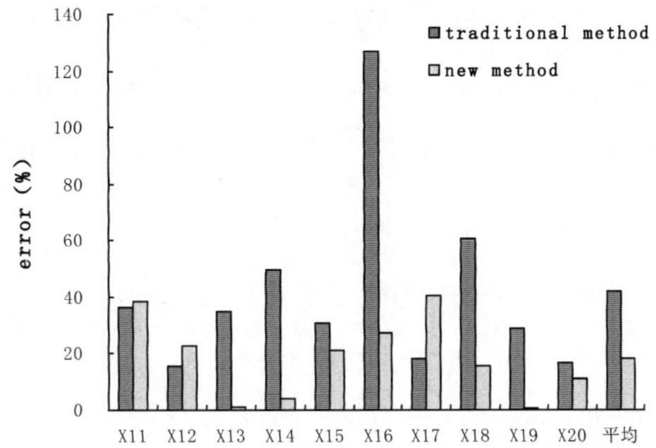


**Figure 4.** Bar graph for error of $\alpha$ in high-permeability area

**Table 2.** Contrast results of traditional and new methods

| Reservoir types | Wells | Traditional method (α) | | New method （α） | |
|---|---|---|---|---|---|
| | | Results | Absolute error(%) | Results | Absolute error(%) |
| Low permeability | X1 | 0.851 | 23.5 | 0.769 | 9.6 |
| | X2 | 0.548 | 18.8 | 0.599 | 9.4 |
| | X3 | 0.768 | 15.2 | 0.666 | 13.2 |
| | X4 | 0.674 | 3.4 | 0.626 | 7.1 |
| | X5 | 0.601 | 8.4 | 0.591 | 1.6 |
| | X6 | 0.585 | 11.3 | 0.643 | 9.9 |
| | X7 | 0.853 | 23.7 | 0.900 | 5.5 |
| | X8 | 0.677 | 3.8 | 0.718 | 6.1 |
| | X9 | 0.563 | 15.6 | 0.605 | 7.5 |
| | X10 | 0.717 | 9.2 | 0.718 | 0.2 |
| | Average | 0.684 | 13.3 | 0.684 | 7.0 |
| High permeability | X11 | 0.781 | 36.5 | 0.482 | 38.3 |
| | X12 | 0.430 | 15.5 | 0.527 | 22.7 |
| | X13 | 0.761 | 34.8 | 0.755 | 0.8 |
| | X14 | 0.332 | 49.5 | 0.345 | 4.1 |
| | X15 | 0.715 | 30.7 | 0.564 | 21.2 |
| | X16 | 0.218 | 127.1 | 0.278 | 27.1 |
| | X17 | 0.419 | 18.3 | 0.589 | 40.4 |
| | X18 | 0.309 | 60.6 | 0.358 | 15.8 |
| | X19 | 0.385 | 28.8 | 0.387 | 0.5 |
| | X20 | 0.597 | 16.9 | 0.663 | 11.1 |
| | Average | 0.495 | 41.9 | 0.495 | 18.2 |

## 4. Conclusions

(1) Because of influences of reservoir property and systematic well testing data, value of $\alpha$ in traditional single-point productivity equation is different from actual α value of every gas well, which results in error generation of absolute open flow.

(2) There are some correlation between α value and formation pressure. Emprical formula calculating $\alpha$ value in high and low permeability area in Jingbian gas field are established.

(3) Contrasting with traditional method, calculation error of new method is obviously smaller, which declines from 13.3% to 7.0% in low permeability area, from 41.9% to 18.2 % in high permeability area respectively. New method can effectively improve calculation accuracy of α value and absolute open flow.

## Nomenclature

$q_g$——wellhead production rate of gas well， $10^4m^3/d$；

$q_{AOF}$——absolute open flow ， $10^4m^3/d$；

k——permeability， mD；

h——thickness of gas reservoir， m；

$p_R$——reservoir pressure， MPa；

$p_{wf}$——flow bottom hole pressure， MPa；

T——reservoir temperature， K；

$\bar{\mu}$——gas viscosity， mPa·s；

$\bar{Z}$——gas deviation factor;
$r_e$——gas drainage radius, m;
$r_w$——well radius, m;
s'——apparent skin factor;
s——skin factor;
D——non-darcy factor;
$\beta$——velocity factor describing influence of turbulent flow in porous media, $m^{-1}$;
$\gamma_g$——gas relative density;
$C_1$——constant;
$C_2$——constant.

**References**

[1] Wang Cai, Li Zhi-Pin, Lai Feng-Peng 2014 A novel binomial deliverability equation for fractured gas well considering non-Darcy effects. *Journal of Natural Gas Science and Engineering* **20(9)** 27-37.
[2] Peng Chaoyang 2010 Discussion on equation application of deliverability with different form for gas well. *Natural Gas Geo-science* **21(1)** 172-174.
[3] Li Zuyou, Yang Xiaobi, Luo Dongming 2008 Binominal deliverability equation of high pressure gas well. *Special Oil and gas Reservoirs* **15(3)** 62-64.
[4] Yang Chengbo, Guo Jianchun, Yang Jian, et al 2014 Modification of horizontal well binomial deliverability equation of low permeability gas reservoir. *Journal of Southwest Petroleum University (Science & Technology Edition)* **36(4)** 123-130.
[5] Lu Yiqun 2008 Application of one point gas testing method in Northeast of Sichuan area. *Well Testing* **17(5)** 34-36.
[6] Han Huiling, Jiang Jianfang, YangYufeng, et al 2008 Research on application of one-point method about rapid gas production testing in shanbei gas fields. *Well Testing* **17(5)** 17-19.
[7] Hu Jianguo, Zhang Zonglin, Zhang Zhenwen 2008 A new method on processing the data from one-point deliverability test in gas fields. *Natural Gas Industry* **28(2)** 111-113.
[8] Hu Junkun, Li Xiaoping, Xiao Qiang, et al 2013 A new method of using dynamic data to determine deliverability equation of gas well. *Natural Gas Geoscience* **24(5)** 1027-1031.
[9] Qiao Zhiguo, Shuai Jianjun, Dong Haifeng, et al 2014 Deliverability equation with dynamic one-point method for Changxing formation, Yuanba gas field. *Natural Gas Technology and Economy* **8(5)** 25-27.
[10] Deng Hui, Feng Xi, Yang Xuefeng, et al 2014 A new method of predicting gas wells deliverability in Longgang reef gas reservoir. *Natural Gas Geosience* **25(9)** 1451-1454.
[11] Yuan Yingzhong, ZhangLiehui, WangJian, et al 2009 A binomial deliverability equation for horizontal gas wells in formationswith nonlinear seepage flow features. *Oil & Gas Geology* **30(1)** 122-126.
[12] Liu Yongliang, Yuan Yingzhong, Deng Lijing, et al 2016 A new method for binomial deliverability equation of horizontal gas well. *Natural Gas Geoscience* **27(2)** 371-376.
[13] Tabatabaei M, Zhu D 2010 Generalized inflow performance relationships for horizontal gas wells. *Journal of Natural Gas Science and Engineering* **2(2-3)** 132-142.
[14] BorsiI, Fusi L, Rosso F, et al 2011 A well deliverability model for non-Darcian flow in geothermal reservoirs. *Computers & Geosciences* **37(10)** 1555-1561.
[15] Kalantariasl A, Farhadi I, Nasriani H R 2013 On the accuracy of dimensionless inflow performance relationships for gas wells. *SPE paper 164602.*
[16] Zhang Hongxue, Liu Weiqun, Zhu Li 2014 A permeability model for changes during shale gas production. *Electronic Journal of Geotechnical Engineering* **19.X** 3847-3860 *Available at ejge.com.*

[17] Hai Tang, Hongyuan Li, Dongliang Lv 2014 Relationship between stress sensitivity and productivity rate in low permeability reservoir. *Electronic Journal of Geotechnical Engineering* **19.V** 6619-6625 *Available at ejge.com.*
[18] Feng Xi, Zhong Fuxun, Wang Hao, et al 2005 Modified single point method to evaluate productivity of gas wells with big production for feixianguan group gas reservoirs in northeast Sichuan. *Natural Gas Industry* **25(supplement A)** 107-109.
[19] Sun Zhidao, Hu Yongle, Fang Xisheng, et al 2011 Applicable scope of the single-point test method for the evaluation of gas well deliverability. *Natural Gas Industry* **31(11)** 63-65.

IOP Conf. Series: Earth and Environmental Science **61** (2017) 012036 doi:10.1088/1755-1315/61/1/012036

# Lightweight Ceramsite based on sewage sludge made by Muffle furnace

**Bi-fan HE**

Chongqing City Management College, Chongqing 401331, P. R. China
Email: 104942562@qq.com

**Abstract:** It took SS (sewage sludge), PFA (pulverized fuel ash) and BS (building spoil) as raw materials to carry out experiments. An orthogonal experimental design was used to arrange experiments on the mixture ratio and processing parameters to find out a better ratio of green pellets and the right parameters of firing process. In accordance with the right ratio and processing parameters, experiments were carried out by MF (Muffle furnace) to produce lightweight aggregates. The characteristics of products such as 1h water absorption rate, packing density, particle strength were tested. According to the results of orthogonal experiment and range analysis, the optimized parameters suitable for different products requirements were selected. The repeatability of the optimized parameters is proved by the repeated experiments of the same batch, and the corresponding product performance is stable.

## 1. Introduction

In recent years, with the improvement of China's sewage treatment capacity, more and more urban sewage treatment plants have been established. With the improvement of sewage treatment rate, the output of sewage sludge is increasing day by day. Located in the middle and upper reaches of the Yangtze River, Chongqing is the political, economic and cultural center of the Three Gorges Reservoir. According to the 2015 Chongqing environmental quality bulletin learned: the city's sewage treatment scale was 3 million 780 thousand tons per-day. According to incomplete statistics, the annual sewage sludge (moisture content 80%) production in Chongqing was more than 1 million tons. How to properly dispose of sewage sludge has been urgent.

Generally, conventional sludge disposal technologies include landfill, incineration, agricultural and sea dumping [1-4]. The high cost of the traditional way of disposal, easy to create secondary pollution, not a sustainable environmental protection strategy. Therefore, the utilization of sludge will be the ultimate way of sludge disposal. It has been reported that the domestic and foreign scholars have made the research on the manufacture of ceramsite by using sewage sludge as the main raw material [5-9]. Chongqing is located in hilly areas; building spoil (mineral composition similar to clay) can be used as the auxiliary material for the production of sludge ceramsite. The output of fly ash from coal-fired power plant (PFA) is stable and easy to obtain, and it can be used as the skeleton material of ceramic production. Therefore, combined with the actual situation in Chongqing, we choose SS, PFA, BS as raw materials to manufacturing ceramsite.

## 2. Experiment

### *2.1 Materials and characters*

It took SS, PFA and BS as raw materials to carry out experiments. Especially, SS came from a municipal sewage treatment plant of Chongqing city. PFA came from a coal power plant of

Chongqing city, and BS came from a stone factory. The characteristics of raw materials were shown in Table 1.

**Table 1**. Chemical analysis of raw materials (mass percentage)

| Category | $SiO_2$ | $Fe_2O_3$ | $Al_2O_3$ | CaO | MgO | $Na_2O$ | $K_2O$ | $SO_3$ |
|---|---|---|---|---|---|---|---|---|
| SS | 32.1 | 9.23 | 11.2 | 3.68 | 2.2 | 0.75 | 1.7 | 2.03 |
| PFA(II) | 44.91 | 16.1 | 24.26 | 3.82 | 0.87 | 0.85 | 1.46 | - |
| BS | 59.84 | 4.59 | 12.1 | 1.33 | 1.46 | 1.7 | 1.97 | 0.23 |

*2.2 Experimental equipment*

A MF was the key experimental equipment with model SX12-2.5. The heating range of the equipment is from room temperature to 1300 degrees Celsius. The auxiliary equipment included mixer, granulating equipment, constant temperature drying box and particle strength tester.

*2.3 Experimental process*

The raw materials aforementioned were all sieved to the particle diameter ranging from 0.10 mm to 0.20 mm. It took the selected raw materials to make pellets with some certain mixture ratio. Experimentally, it needs moisture content of mixture to be about 20% to make green pellets keep in spherical shape. It also needs to dry green pellets before the sintered process to avoid aggregates breaking due to the extravasations of free water in the pellets when firing them. After the pretreatment process, it was time for feeding pellets into the MF for preheating at some certain temperature for several minutes. When the preheat process finished, it should set a certain temperature of the MF for sintering the pretreated pellets. When the MF reached the presetting temperature, the sintering process finished. But, it did not mean the pellets turned out to be ceramsite at that moment. Usually, it needs to hold the final temperature a few minutes to keep pellets burning completely for bloating them. Finally, it's time for turning off the MF to let aggregates cool naturally. Then, it could carry out some physicochemical characteristic tests of the ceramsite.

*2.4 Experimental method*

In accordance with a theory proposed by Rely and Wilson that suitable chemical composition of clay for sintering ceramsite, some researchers had drawn a conclusion that chemical composition of raw materials on burning high-quality ceramsite, the key chemical composition of raw materials should follow this mass percentage $SiO_2$ 48%～79%, $Al_2O_3$ 8%～25%, $\Sigma(CaO+MgO+Fe_2O_3+ Na_2O+K_2O)$ 8%～24%. It could help us to establish raw material mixture ratio. According to the results of chemical composition analysis of raw materials, the chemical composition of BS could meet the elemental composition range on producing high-quality ceramsite. Experimentally, it tried producing ceramsite just by BS through the MF. But, the results showed that products made by BS appeared light gray color, surface cracks and packing density was greater than $1.2g/cm^3$. The experimental results showed it could not produce high quality ceramsite by BS separately. The key chemical composition of PFA and SS was not within the scope of making ceramics. Experiments on sintering ceramsite by PFA and SS alone were carried out. The results showed it could not make lightweight ceramsite by PFA and SS alone.

An orthogonal experimental design method was used to arrange experiments. According to the preliminary experiments, it chose the mixture ratio; preheat temperature, sintered temperature and retention time were key factors of the orthogonal experiment table. In accordance with the DPSS statistic software, an orthogonal table such as $L_{25}$ $(_5{}^6)$ which include five levels of six factors and twenty-five process experimental table was selected to arrange experiments. The experimental factor and levels were detailed in table 2.

**Table 2**. Level of experimental factors

| Level of factor | 1 | 2 | 3 | 4 | 5 |
|---|---|---|---|---|---|
| Mass percentage of SS-A/(%) | 10 | 15 | 20 | 25 | 30 |
| Mass percentage of PFA-B/(%) | 30 | 25 | 20 | 15 | 10 |
| Preheat temperature-C/(C °) | 200 | 250 | 300 | 350 | 400 |
| Sintered temperature-D/(C °) | 1100 | 1120 | 1140 | 1160 | 1180 |
| retention time-E/(min) | 1 | 2 | 3 | 4 | 5 |

*2.5 Experimental results*

The key characters of ceramsite including particle strength, packing density and one hour water absorption rate were selected to identify quality of ceramsite. The experimental process and results were shown in table 3.

**Table 3**. Experimental process and results

| Num. | A/% | B/% | C/ C° | D/ C° | E/min | Packing density/kg·m$^{-3}$ | Particle strength/N | 1h water absorption/% |
|---|---|---|---|---|---|---|---|---|
| 1 | 10 | 30 | 200 | 1100 | 1 | 731 | 520 | 8.12 |
| 2 | 10 | 25 | 250 | 1120 | 2 | 722 | 741 | 8.31 |
| 3 | 10 | 20 | 300 | 1140 | 3 | 778 | 467 | 13.20 |
| 4 | 10 | 15 | 350 | 1160 | 4 | 713 | 740 | 6.90 |
| 5 | 10 | 10 | 400 | 1180 | 5 | 650 | 747 | 8.95 |
| 6 | 15 | 30 | 250 | 1140 | 4 | 665 | 732 | 9.51 |
| 7 | 15 | 25 | 300 | 1160 | 5 | 666 | 640 | 8.13 |
| 8 | 15 | 20 | 350 | 1180 | 1 | 630 | 678 | 9.99 |
| 9 | 15 | 15 | 400 | 1100 | 2 | 688 | 622 | 9.96 |
| 10 | 15 | 10 | 200 | 1120 | 3 | 679 | 656 | 11.33 |
| 11 | 20 | 30 | 300 | 1180 | 2 | 634 | 608 | 12.00 |
| 12 | 20 | 25 | 350 | 1100 | 3 | 656 | 542 | 11.88 |
| 13 | 20 | 20 | 400 | 1120 | 4 | 623 | 666 | 11.08 |
| 14 | 20 | 15 | 200 | 1140 | 5 | 600 | 712 | 9.89 |
| 15 | 20 | 10 | 250 | 1160 | 1 | 620 | 618 | 10.02 |
| 16 | 25 | 30 | 350 | 1120 | 5 | 611 | 602 | 11.38 |
| 17 | 25 | 25 | 400 | 1140 | 1 | 600 | 636 | 11.33 |
| 18 | 25 | 20 | 200 | 1160 | 2 | 620 | 701 | 12.36 |
| 19 | 25 | 15 | 250 | 1180 | 3 | 558 | 603 | 12.11 |
| 20 | 25 | 10 | 300 | 1100 | 4 | 617 | 555 | 12.79 |
| 21 | 30 | 30 | 400 | 1160 | 3 | 589 | 503 | 12.98 |
| 22 | 30 | 25 | 200 | 1180 | 4 | 576 | 413 | 14.33 |
| 23 | 30 | 20 | 250 | 1100 | 5 | 608 | 489 | 14.13 |
| 24 | 30 | 15 | 300 | 1120 | 1 | 578 | 630 | 13.08 |
| 25 | 30 | 10 | 350 | 1140 | 2 | 540 | 641 | 12.22 |

## 3. Discussion

*3.1 Statistical analysis*

The range analysis of key factors affecting the particle strength, packing density and 1h water absorption rate of ceramsite was shown in table 4.

**Table 4**. Range analysis of key factors

| Num. | Packing density /(kg·m$^{-3}$) | | | | | Particle strength /(N) | | | | | 1h water absorption/(%) | | | | |
|---|---|---|---|---|---|---|---|---|---|---|---|---|---|---|---|
| | A | B | C | D | E | A | B | C | D | E | A | B | C | D | E |
| $\overline{K1}$ | 719 | 624 | 645 | 645 | 639 | 643 | 620 | 617 | 617 | 618 | 9.10 | 11.30 | 10.90 | 10.90 | 11.03 |
| $\overline{k2}$ | 666 | 628 | 641 | 638 | 634 | 666 | 623 | 621 | 622 | 622 | 9.78 | 11.15 | 11.03 | 11.11 | 11.16 |
| $\overline{k3}$ | 627 | 637 | 635 | 630 | 642 | 629 | 615 | 616 | 617 | 633 | 10.97 | 11.15 | 11.24 | 11.28 | 10.83 |
| $\overline{k4}$ | 601 | 643 | 624 | 634 | 631 | 619 | 627 | 624 | 642 | 630 | 11.99 | 10.88 | 11.20 | 10.70 | 10.89 |

| | | | | | | | | | | | | | | | |
|---|---|---|---|---|---|---|---|---|---|---|---|---|---|---|---|
| $\overline{k5}$ | 578 | 650 | 630 | 627 | 626 | 535 | 632 | 636 | 624 | 617 | 13.35 | 10.58 | 10.92 | 11.11 | 11.27 |
| R | 141 | 26 | 21 | 18 | 16 | 131 | 17 | 20 | 25 | 16 | 4.25 | 0.72 | 0.34 | 0.58 | 0.44 |
| Optimized | A5 | B1 | C4 | D5 | E5 | A2 | B5 | C5 | D4 | E3 | A1 | B5 | C1 | D4 | E3 |

Remarks: $\overline{Ki}$ is the average result. R is the range value.

It is well known that typical characters of ceramsite were hard, lightweight and low water absorption rate. According to the range analysis of key factors, it made it clear that the project A5B1C4D5E5 was suitable for achieving lower packing density; project A2B5C5D4E3 was suitable for acquiring higher particle strength and project A1B5C1D4E3 fit for getting lower water absorption rate. How to determine the best baking parameters depends on the performance requirements of ceramsite products.

Meanwhile, the repeatability of the optimal scheme needs further study. In this paper, further experiments according to the optimized projects were carried out to verify combined effects about the key factors on the products. Under the same experimental conditions, the experimental scheme of each group was repeated 3 times.

*3.2 Experimental repeatability*

*3.2.1 Project one (A5B1C4D5E5).* Experiments with project one was carried out in MF. Through these experiments, several ceramsites were randomly selected to take scanning electron micrographs. As fig.1 (I) showed, the internal structure of cross section and appearance reveals the reason of ceramsite lightweight. It's easy to find out the porous structure is the main cause leading to ceramsite lightweight. The porous structure is an important characteristic of ceramsite products [10]. In these pore structures, a part of micropores are interconnected. The penetration of micropore is the main factor that results in the increase of water absorption [11]. According to fig.1(I), we can see some larger black holes in the micrograph. This is due to the loss of the larger particles during the firing process.

It is generally believed that the gas produced in the process of sintering ceramsite could make the porous structure [5,12]. The honeycomb structure of uniform distribution is due to the composition of raw materials to produce gas overflow generated at high temperature. The gas composition of the raw material mainly includes the evaporation of the water in the pellet, the gas produced by the combustion of carbon particles and the decomposition of salts. The ceramsite appearance was brown. Some ceramsite surface is similar to enamel material. The surface touch feeling was not smooth, some products appeared chapped cracks. The particle strength, packing density and 1h water absorption rate of project one were tested. The results were showed in table 5.

**Table 5**. Range analysis of key factors

| Num. | Packing density | Particle strength | 1h water absorption |
|---|---|---|---|
| 1 | 620/(kg·m$^{-3}$) | 566/(N) | 11.7/(%) |
| 2 | 611/(kg·m$^{-3}$) | 601/(N) | 9.9/(%) |
| 3 | 628/(kg·m$^{-3}$) | 593/(N) | 10.6/(%) |
| Relative change index * | 2.74% | 5.97% | 16.77% |

* The formula for calculating the relative change index: range/ average value.

The table 5 showed: (i) the relative change index of packing density is very small; (ii) the relative change index of 1h water absorption rate is relatively high. The experimental results show that the repeatability of the experimental results is normal, and the control of bulk density is more successful.

*3.2.2 Project two (A2B5C5D4E3).* Same as the previous experimental method, experiments with project two was carried out in MF. As shown in fig.1 (II), the internal structure and appearance of ceramsite reveals the reason of ceramsite high strength. The surface of the shell is brown ceramic support structure of ceramsite high particle strength. Acicular mullite is widely distributed in the glass phase, the glass phase plays a supporting, strengthening and enhancing the skeleton effect, and the ceramsite has high strength [13].

The particle strength, packing density and 1h water absorption rate of project two were tested. The results were showed in table 6.

**Table 6**. Experimental results

| Num. | Packing density | Particle strength | 1h water absorption |
|---|---|---|---|
| 1 | 651/(kg·m$^{-3}$) | 623/(N) | 12.1/(%) |
| 2 | 667/(kg·m$^{-3}$) | 655/(N) | 10.3/(%) |
| 3 | 639/(kg·m$^{-3}$) | 672/(N) | 11.0/(%) |
| Relative change index | 4.29% | 7.54% | 16.07% |

The table 6 showed: (i) the relative change index of particle strength is small; (ii) the relative change index of 1h water absorption rate is relatively high. The variation of particle strength is mainly due to the randomness of test samples. The experimental results showed that the repeatability of the experimental results is normal, and the control of particle strength is good.

*3.2.3 Project three (A1B5C1D4E3).* Same as the previous experimental method, experiments with project three was carried out in MF. As shown in fig.1 (III), the dense outer enamel structure of ceramsite reveals the reason of lower water absorption. Quartz and silicate minerals (kyanite) have a dense structure, widely distributed in the ceramic surface, which is a very important factor of ceramsite low water absorption rate [13].

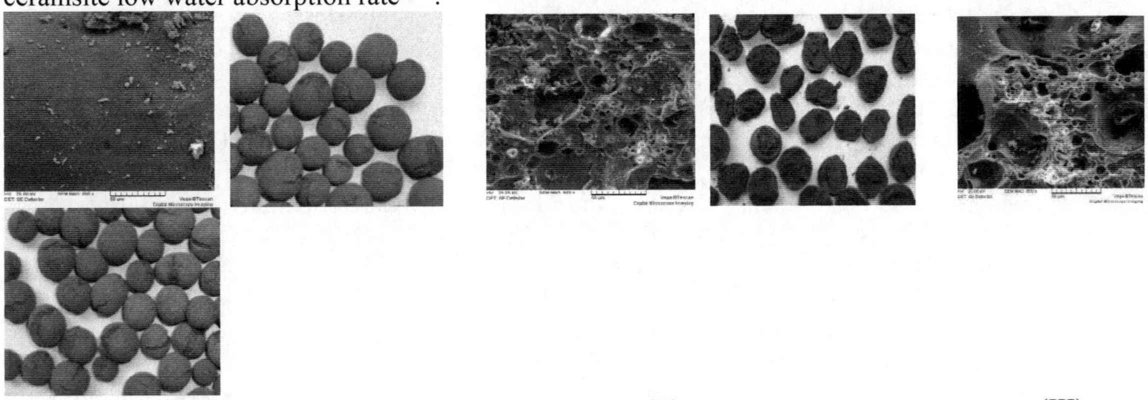

(I) (II) (III)

**Figure 1**. SEM and appearance of ceramsite

The particle strength, packing density and 1h water absorption rate of project three were tested. The results were showed in table 7.

**Table 7.** Experimental results

| Num. | Packing density | Particle strength | 1h water absorption |
|---|---|---|---|
| 1 | 640/(kg·m$^{-3}$) | 590/(N) | 8.9/(%) |
| 2 | 626/(kg·m$^{-3}$) | 633/(N) | 9.4/(%) |
| 3 | 661/(kg·m$^{-3}$) | 650/(N) | 9.0/(%) |
| Relative change index | 5.45% | 9.61% | 5.49% |

The table 7 showed: (i) the relative change index of 1h water absorption rate is small; (ii) the relative change index of particle strength is relatively high. The variation of particle strength is mainly due to the randomness of selected samples. The experimental results showed that the repeatability of the experimental results is normal, and the control of particle strength is well.

## 4. Conclusion

The main conclusions of this study are as follows:

(i). In order to achieve a low packing density, the experimental parameters should meet: mass percentage SS/PFA/BS=30%/30%/40%, preheat temperature 350C °, sintered temperature 1180 C ° and retention time 5 min.

(ii). In order to obtain high strength ceramsite, the experimental parameters should meet: mass percentage SS/PFA/BS=15%/10%/75%, preheat temperature 400 C °, sintered temperature 1160 C ° and retention time 3 min.

(iii). In order to obtain low water absorption ceramsite, the experimental parameters should meet: mass percentage SS/PFA/BS=10%/10%/80%, the preheat temperature 200 C °, sintered temperature 1160 C ° and retention time 3 min.

**References**

[1] HE Jun, LI Feng, LI Yong, etal. Geotechnical Characteristics of Sewage Sludge for Final Cover of Landfill [J]. Chinese Journal of Underground Space and Engineering, 2014(S1):1612-1616.

[2] XU Xin-yang, CHEN Xi, LIU Qing, etal. Research on the Optimization of Sludge Composting Process for Municipal Waste water Treatment Plant [J]. Journal of Northeastern University(Natural Science), 2012, (09) :1340-1343+1348..

[3] QIU Ji-hua, CHENG Ji-dong. The characteristics of sewage sludge during incineration process and its chemical kinetics analysis [J]. Journal of Huazhong University of Science and Technology (Natural Science Edition), 2005, 33(8):25-26.

[4] LU Shu-hong. Seminar on urban sewage sludge discharge technology [J]. China Water & Wastewater, 1988(6):37.

[5] QI Fei, ZHANG Chang-sen, CHEN Jing-hua. Study on Preparation of porous ceramsite by municipal sludge and coal gangue [J]. Journal of Ceramics, 2015(1):58-63.

[6] CHENG Hai-xiang, ZHOU Hai, LI Jian-hui, etal. Experimental study on manufacturing haydite by sintering municipal sludge and its incineration ash[J]. Chinese Journal of Environmental Engineering, 2016(2):845-850.

[7] QV Lie, WANG Yuan, YANG Jiu-jun, etal. Chinese Journal of Environmental Engineering[J]. BULLETIN OF THE CHINESE CERAMIC SOCIETY, 2016, 35(3):970-974.

[8] González-Corrochano B, Alonso-Azcárate J, Rodríguez L, et al. Valorization of washing aggregate sludge and sewage sludge for lightweight aggregates production[J]. Construction & Building Materials, 2016, 116:252-262.

[9] You JH, Hee YJ. The Soil Aggregate Manufacturing Method Which Uses the Sewage Sludge and the Soil Aggregate Which It Uses: http://kns.cnki.net/kns/detail/detail.aspx?FileName=KR100854036 (B1)&DbName=SOPD2010.

[10] WANG Le-le, YANG Ding-yi, AI Yi-mou, etal. Preparation and void structure of sludge ceramsite [J]. Concrete, 2016(1):103-107.

[11] LU Hao. Preparation and sewage treatment Application of ceramsite filter material made from sluge and steel slag [D]. Anhui University of Technology, 2014.

[12] PEI Hui-fang, ZHANG Chang-sen, CHEN Jing-hua. Preparation of Porous Ceramsite with Sludge and Gangue [J]. China Ceramics, 2015(3):72-77.

[13] QU Lie, WANG Yuan, YANG Jiu-jun, etal. Research on the sintering expansion mechanism and preparation of hign-strength ceramisite with municipal sludge and saline soil [J]. New Building Materials 2016, 43(2):47-51.

IOP Conf. Series: Earth and Environmental Science **61** (2017) 012037 doi:10.1088/1755-1315/61/1/012037

# Study on crack development of concrete beams in bending reinforced with FRP bars

**M X Pan**[1,a], **X S Xu**[1,b]

[1] School of Civil Engineering and Architecture, University of Jinan, Jinan 250022, China

[a] meixia_pan@163.com, [b] xinsheng_xu@163.com

**Abstract**. The comparatively high tensile strength and low elastic modulus of fiber reinforced polymer (FRP) bars result in deflection and crack width requirements controlling the design of concrete beams reinforced with FRP bars. Consequently, when it comes to the design of such members, the deflection and crack width are first calculated at serviceability limit state, and then are checked at ultimate limit state. In this paper, five concrete beams reinforced with different FRP reinforcement ratios were tested and the crack development and pattern as well as the effect of reinforcement ratio on crack width and spacing were analyzed. The experimental results were compared with the calculation results using design guidelines recommended by Japan JSCE code, American ACI440.1R-03 code and China GB50608-2010 code. Several design suggestions were proposed for bending concrete beams reinforced with FRP bars.

## 1. Introduction

FRP bars are linear elastic materials with no discernible yield point, and the relatively low elastic modulus, high tensile strength and quite different bond characteristics from steel bars all lead to greater deflection and crack width. Researches have found that serviceability requirements should be as critical as the ultimate limit states in the design of concrete beams reinforced with FRP bars, especially when considering the deflection and crack width requirements [1-2]. In this paper, CFRP and GFRP reinforced concrete beams were tested in bending so as to investigate the flexural behavior in terms of crack pattern and deflection behavior. According to experimental results, several crack-width calculation formulas recommended by design codes home and abroad were verified and compared to experimental results. These studies are discussed hereafter.

## 2. Experimental program

Tests were conducted on five specimen beams subjected to two-point concentrated static loads. Three beams were 180mm in width, 250mm in depth and 1900mm in length, which were numbered CB-1, CB-2 and CB-3. And the remaining two were numbered GB-1 and GB-2 with a cross-section size of 150×200mm and a length of 1300mm. Two steel bars with a diameter of 16mm were adopted as upper longitudinal reinforcement for specimen CB-1 and CB-2 and their lower longitudinal reinforcement were two CFRP bars and three CFRP bars of 9.5mm diameter, respectively. As for CB-3, the upper and lower longitudinal reinforcement were both two CFRP bars. Steel stirrups of 12 mm diameter were used at 100 mm spacing along the tested length for these three beams. The reinforcement details of specimen beam GB-1 and GB-2 were the same as that of CB-1 and CB-2, the only difference was that the upper steel bars were 12mm diameter and the lower reinforcement were GFRP bars, instead.

Moreover, the diameter of steel stirrups was 10mm with the same spacing of 100mm. The material properties of FRP rods adopted in this test were shown in table 1.

**Table 1.** Mechanical properties of FRP bars.

| FRP type | Diameter (mm) | Ultimate strength (MPa) | Yield strength (MPa) | Elastic modulus (GPa) | Ultimate elongation rate (%) |
|---|---|---|---|---|---|
| CFRP | 9.5 | 1779 | 1245 | 136 | 1.65 |
| GFRP | 9.5 | 993 | 695 | 72 | 2.31 |

## 3. Test results

### *3.1. Crack behaviour*

In this section, the crack patterns of CFRP and GFRP reinforced concrete beams under ultimate load were shown in figure 1. The crack patterns of concrete beams reinforced with FRP bars were similar to that of normal RC beams. As can be seen in figure 1, cracks distributed uniformly and densely. Compared with CFRP reinforced concrete beams, GFRP strengthened concrete beams scattered more symmetrical and denser cracks. Based on statistical analysis, the crack spacing of concrete beams reinforced with CFRP and GFRP bars was 88mm and 75mm, respectively.

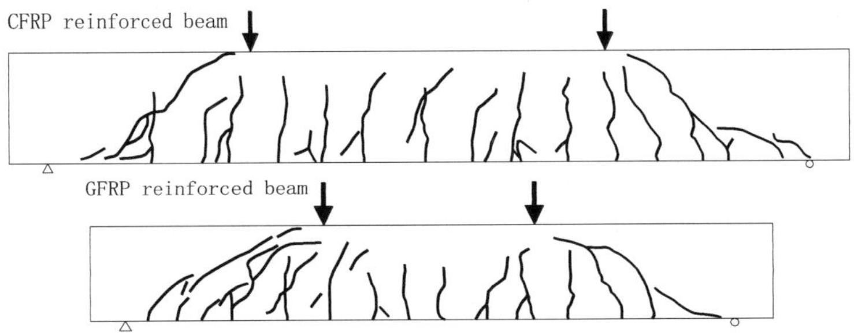


**Figure 1.** Crack patterns at ultimate state.

### *3.2. The effect of reinforcement ratio on crack width and spacing*

An initial vertical crack along the depth of beam appeared in pure bending section and then, with the increasing of applied load, inclined cracks occurred outside pure bending section owing to the increasing of the shear stress, which narrowed the crack spacing. The experiment found that the decrease of crack spacing was slowing down and tended to be stable when the applied load up to 35 percent of ultimate load. The crack spacing and crack width at different load levels of five test beams were listed in table 2, from which we could know that the crack spacing decreased with the increase of reinforcement ratio.

**Table 2.** Crack spacing and width of test beams.

| Beam No. | Sectional dimension (mm) | Reinforcement ratio (%) | Cracks number | Crack spacing (mm) | Crack width (50KN) (mm) | Crack width (90KN) (mm) | Crack width ($35\%P_u$) (mm) |
|---|---|---|---|---|---|---|---|
| GB-1 | 150×200 | 0.51 | 15 | 79 | 0.96 | 1.8 | 0.75 |
| GB-2 | 150×200 | 0.76 | 14 | 71 | 0.66 | 1.17 | 0.46 |
| CB-1 | 180×250 | 0.35 | 16 | 93 | 0.56 | 0.96 | 0.60 |
| CB-2 | 180×250 | 0.52 | 18 | 80 | 0.37 | 0.72 | 0.69 |
| CB-3 | 180×250 | 0.34 | 15 | 90 | 0.62 | 1.02 | 0.57 |

As illustrated in table 2, reinforcement ratio had an evidently influence on crack width. A lager reinforcement ratio meant a greater surface area and better bond strength, which could bear a lager crack stress so as to protect the beam from cracking early. Consequently, the crack width decreased as the reinforcement ratio increased.

## 4. Calculation models for crack width

The crack with of five specimens was calculated and analyzed using the formula suggested by the following guidelines and compared to the test results, and the calculation results of different codes on crack width were discussed.

### *4.1. Crack-width calculation models*

*4.1.1. JSCE (1997)* [3]. Calculation formula of maximum crack width given by *Recommendation for Design and Construction of Concrete Structure Using Continuous Fiber Reinforcing Materials* is as follows:

$$w = k\left[4c + 0.7(c_{\mathrm{f}} - \varphi)\right]\left(\frac{\sigma_{\mathrm{f}}}{E_{\mathrm{f}}} + \varepsilon'_{\mathrm{csd}}\right) \quad (1)$$

where $k$ is a constant that is related to the bond property and reinforcement, usually takes from 1.0 to 1.3; $c$ is the concrete cover depth, mm; $c_{\mathrm{f}}$ is the centroid distance of FRP bar, mm; $\varphi$ is the diameter of FRP bar, mm; $\sigma_{\mathrm{f}}$ is the stress of FRP bars under design load, MPa; $E_{\mathrm{f}}$ is the elastic modulus of FRP bar, MPa; $\varepsilon'_{\mathrm{csd}}$ is compressive strain and can be taken as $150\times10^{-6}$.

*4.1.2. ACI-440.1R-03* [4]. America ACI 440.1R-03 specifications recommend the following formula for the calculation of crack width:

$$w = \frac{2.2}{E_{\mathrm{f}}}\beta k_{\mathrm{b}} f_{\mathrm{f}} \sqrt[3]{d_{\mathrm{c}} A} \quad (2)$$

in which $E_{\mathrm{f}}$ is modulus of elasticity of FRP, MPa; $\beta$ is the ratio of the distance from the neutral axis to extreme tension fiber to the distance from the neutral axis to the center of the tensile reinforcement; $k_{\mathrm{b}}$ is a coefficient that accounts for the degree of bond between FRP bar and surrounding concrete and can be taken as 1.2; $f_{\mathrm{f}}$ is stress in the FRP bar in tension, MPa; $d_{\mathrm{c}}$ is concrete cover depth measured from extreme tension fiber to center of bar, mm; $A$ is the effective tension area of concrete, mm.

*4.1.3. China GB50608-2010* [5]. *Technical Code for Infrastructure Application of FRP Composites* of China suggests that maximum crack width should be computed as shown in Equation (3):

$$\omega_{\max} = 2.1\psi\frac{\sigma_{\mathrm{fk}}}{E_{\mathrm{f}}}\left(1.9c + 0.08\frac{d_{\mathrm{eq}}}{\rho_{\mathrm{te}}}\right) \quad (3)$$

with specific parameters in the formula referring to GB 50608-2010.

### *4.2. Comparison between experimental and calculation results*

Based on available data, above three formulas were used to predict crack width of five specimens and the calculation results were compared to test results, as shown in figure 2. From three charts we can see that JSCE equation overestimates crack width when applied to CFRP bar and underestimates crack width when applied to GFRP bar. As for China GB50608 equation, calculation results of GFRP reinforced member agree well with experimental results, but it overestimates crack width when applied to CFRP bar. While crack width values calculated by ACI formula match best with experimental results, indicating that this recommend model can be used to predict crack width of FRP reinforced concrete member.

## 5. Conclusions

This study was carried out by means of static testing and focused on the crack development and crack-width calculation. According to experimental observations and analysis of tests results, the following conclusions can be summarized.

It was evident that good bond capability between CFRP or GFRP bar and surrounding concrete allows concrete member to work efficiently. GFRP strengthened concrete beams scattered more symmetrical and denser cracks as compared to that of CFRP reinforced beams.

A relationship between reinforcement ratio and crack spacing and width was developed. Results showed that as the reinforcement ratio increased, the crack spacing and width decreased. Crack widths in FRP reinforced members were proved to be larger than those in normal steel reinforced members.

ACI 440.1R-03 code can provide relatively accurate prediction on crack width of FRP reinforced members.

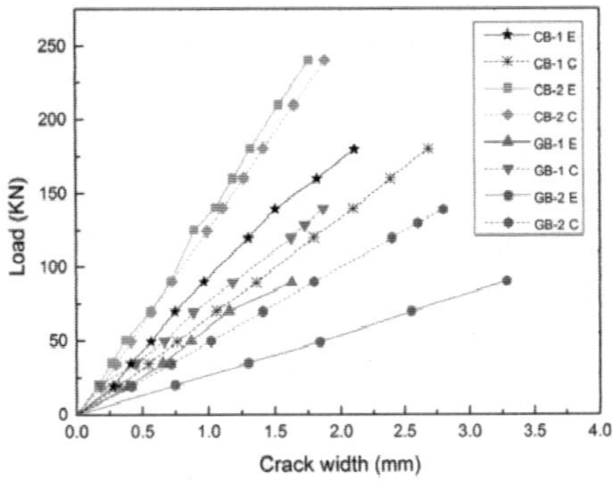


(a) JSCE.

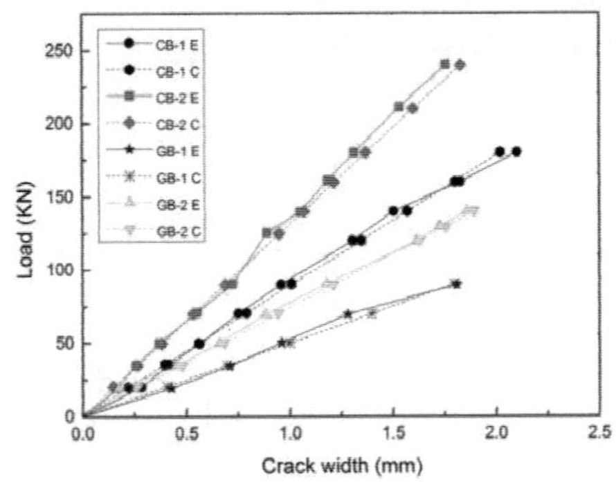


(b) ACI440.

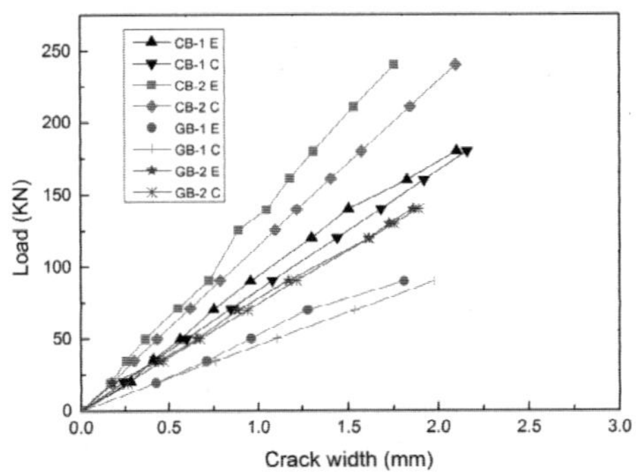


(c) GB50608.

**Figure 2.** Comparison between experimental and calculation results.
Note: "E" represents experimental results, and "C" represents calculation results.

**Acknowledgements**
The author would like to acknowledge the financial support of the Natural Science Foundation of Shandong Province of China No.ZR2014EEM036.

**References**

[1] Almusallam T H 1997 Analytical prediction of flexural behavior of concrete beams reinforced by FRP bars *Journal of Composite Materials* **vol 31** pp 640-57
[2] Masmoudi R, Thériault M and Benmokrane B 1998 Flexural behavior of concrete beams reinforced with deformed fiber reinforced plastic reinforcing rods *ACI Structural Journal* **vol 95** pp 665-76
[3] JSCE 1997 Recommendation for design and construction of concrete structure using continuous

fiber reinforcing materials *Japan: Research Committee on Continuous Fiber Reinforcing Materials*
[4] ACI Committee 440 2002 Guide for the design and construction of concrete reinforced with FRP bars *Detroit: American Concrete Institute*
[5] GB50608-2010 2011 Technical code for infrastructure application of FRP composites *China: China Planning Press*

IOP Conf. Series: Earth and Environmental Science **61** (2017) 012038 doi:10.1088/1755-1315/61/1/012038

# Research on deformation characteristics and design method of concrete beams reinforced with GFRP bars

**M X Pan**[1,a]**, X S Xu**[1,b]

[1] School of Civil Engineering and Architecture, University of Jinan, Jinan 250022, China

[a] meixia_pan@163.com, [b] xinsheng_xu@163.com

**Abstract**. Two different glass fiber reinforced polymer (GFRP) bars were produced to reinforce concrete beams, including plain GFRP bars and ribbed surface GFRP bars. The flexural property of GFRP reinforced concrete beams was studied by testing four specimen beams under two point vertical loads. The crack development and stain distribution of beam cross section during loading were recorded, and the deformation characteristics and failure mode were also analyzed in this experiment. According to the test results, the bond property of plain GFRP bars was too poor to serve as reinforcement bars, while the ribbed surface GFRP bars showed good bond performance and worked well with concrete before final failure. The test beam reinforced with ribbed surface GFRP bars exhibited a uniform crack distribution, and the strain agreed with the assumption of plane cross section. Moreover, calculation method for bending capacity of concrete beam strengthened with GFRP bars was deducted based on this experiment.

## 1. Introduction

The high tensile strength, light weight and corrosion resistance of fiber reinforced polymer (FRP) bars make it an alternative reinforcement to strengthen concrete structure exposed in severe condition, such as salt environment and marine engineering [1-3]. FRP bars reinforced concrete structures possess relatively lighter weight and better durability compared to normal reinforced concrete structures. At present, the most adopted FRP bars in concrete structure are carbon FRP bars, glass FRP bars and aramid FRP bars.

A total of four specimen beams reinforced with GFRP bars were tested under two point vertical loads to investigate strain distribution of beam section, then a design method for FRP reinforced concrete structure members in bending was proposed according to test results, which will be of great significance to provide a theoretical basis for the design of concrete structures reinforced with FRP bars and promote their application in practical engineering.

## 2. Specimen details

Two groups of specimens were tested in total, group A and group B. Group A contained two beams (No. PGB-1 and PGB-2) with a 180×250 mm rectangular cross section and a total length of 1900 mm, and group B comprised two beams (No. SGB-1 and SGB-2) with a 150×200 mm rectangular cross section and a total length of 1300 mm. The upper longitudinal bars of two specimens in group A were both two steel bars with a 14mm diameter, and the lower longitudinal bars were two plain GFRP bars and three plain GFRP bars, respectively. Similarly, the upper longitudinal bars of two beams in group B were two steel bars with a 12mm diameter, and the lower longitudinal bars were two ribbed surfaced

GFRP bars and three ribbed surface GFRP bars, respectively. The stirrups of two groups were steel bars with a 10mm diameter and 100mm space.

According to test results of basic mechanical properties of raw materials, the compressive strength of cubic concrete specimen was measured to be 31.5 MPa, and the yield strength of HPB235 and HRB335 used in the experimental program was 308 MPa and 380 MPa, respectively.

## 3. Test results and discussion

### *3.1. Failure mode*

As shown in figure 1 and figure 2, there were few cracks on the side surface of specimen PGB-1 and PGB-2, and bond failure and slip occurred between the plain GFRP bars and the surrounding concrete with the GFRP bars pulled out while collapsing, indicating that plain GFRP bars were not able to strengthen concrete structure owing to their poor bond capacity. As for specimen SGB-1 and SGB-2, it could be judged from considerable quantity and uniform distribution of cracks that the member possessed good energy dissipation capacity. Eventually, specimen SGB-1 presented a fracture failure mode due to diagonal cracks traversing deeply into the compression zone making it fail to resist the shear force, while SGB-2 failed as a result of the crushing of the confined concrete on the top of the beam inside the pure bending zone.

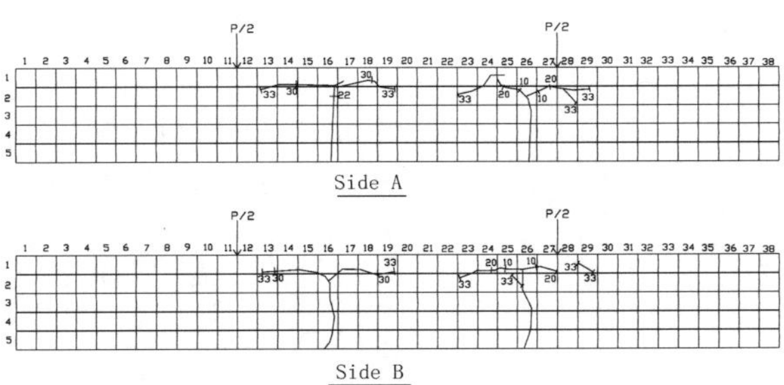


**Figure 1.** Crack patterns of PGB-1.

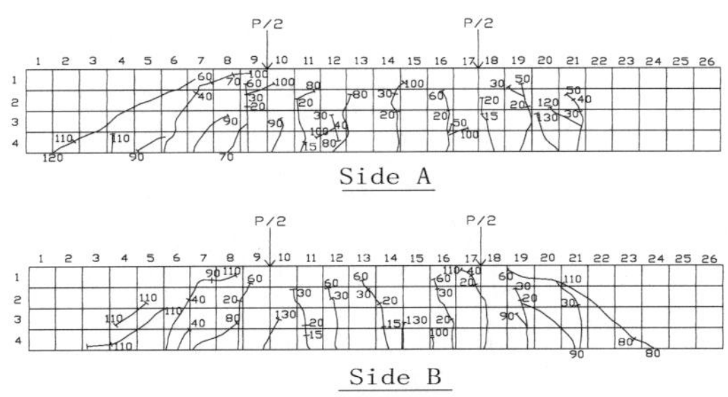


**Figure 2.** Crack patterns of SGB-2.

### *3.2. Cross section strain distribution*

In this section, cross section strain distribution of GFRP reinforced concrete beam under different load was measured and illustrated in figure 3 for specimen SGB-1 and in figure 4 for SGB-2. It can be figured out that the strain distribution meets with the plane section assumption.

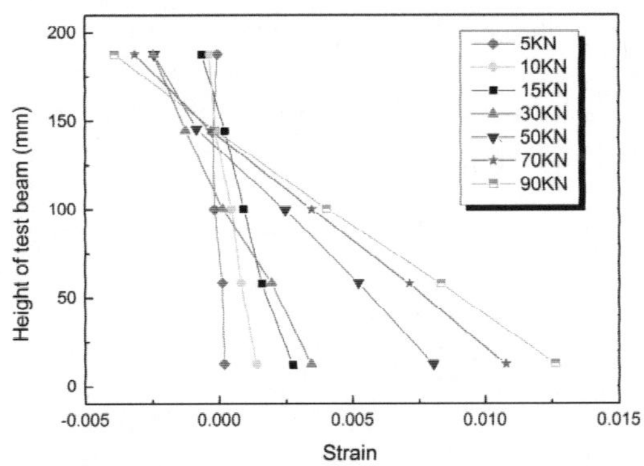


**Figure 3.** Strain distribution of SGB-1.

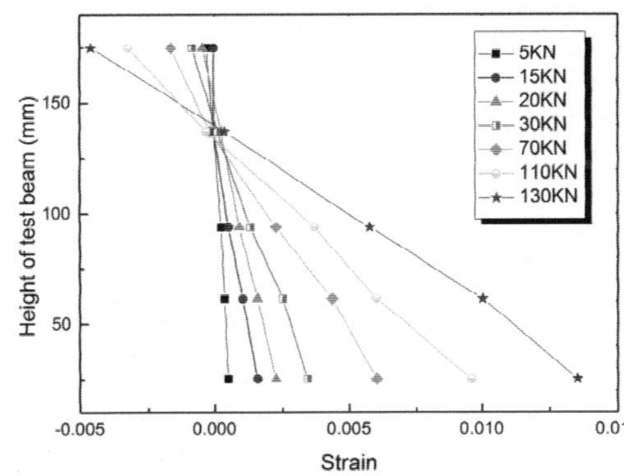


**Figure 4.** Strain distribution of SGB-2.

## 4. Calculation method for flexural capacity

*Reinforced Concrete Structure Design Code* can be referred to while calculating bending capacity of concrete beams reinforced with FRP bars. There are four essential assumptions as follows: ①Ignore the role of concrete in tension after crack appearing. ②Stress-strain relationship of FRP bars is linear elastic [4]. ③Ultimate compressive strain of concrete is 0.0033 [5]. ④There is no slip between FRP bars and the surrounding concrete before ultimate limit states.

Based on test results, FRP reinforced concrete beams were more likely to fail in two modes: One is tensile fracture of FRP bars anterior to concrete crushing, which is called failure mode I. This failure is unsafe for the design of concrete structure. Another is called failure mode II, and this failure is caused by the crushing of concrete in compression zone, and at this time, the strain of FRP bars doesn't reach or exceed the permissible strain.

The balance relative depth of compression zone of FRP reinforced concrete beam can be calculated by the following Equation (1):

$$\xi_b = \frac{0.8}{1 + \frac{f_{fu}}{0.0033E_f}} \tag{1}$$

Where, $f_{fu}$ is design strength of FRP bars, and $E_f$ is elastic modulus of FRP bars, in MPa.

When the relative depth of compression zone $\xi_f$ satisfies with the condition $\xi_f \le \xi_b$, the FRP reinforced beam will fail in failure mode I. Then $\xi_f$ can be calculated as:

$$\xi_f = \frac{f_{fy}A_f}{\alpha_1 f_c b h_0} \tag{2}$$

Where, $f_{fy}$ is design strength of FRP bars, $A_f$ is area of FRP bars in tension ($mm^2$), $\alpha_1$ is a coefficient which can refer to *Reinforced Concrete Structure Design Code* in China, and $h_0$ is effective depth of section.

When $\xi_f$ meets the condition $\xi_f \ge \xi_b$, the FRP reinforced beam will fail in failure mode II. Then, $\xi_f$ is given by Equation (3):

$$\alpha_1 f_c b \xi_f h_0 = \sigma_f A_f - f_y' A_s' \tag{3}$$

Where, $\sigma_f = \varepsilon_f E_f = 0.0033\frac{0.8 - \xi_f}{\xi_f}E_f$, $\sigma_s' = \varepsilon_s' E_s' = 0.0033\frac{\xi_f - 0.8a_s'(h_0)^{-1}}{\xi_f}E_s'$, $\sigma_f$ is stress of FRP bars in tension, and $\sigma_s'$ is stress of steel bars in compression.

Hence, the Equation (3) can be written as the following form,

$$\alpha_1 f_c b \xi_f^2 h_0 + 0.0033(E_f A_f + E_s' A_s')\xi_f - 0.0033 \times 0.8(E_f A_f + \frac{a_s'}{h_0}E_s' A_s') = 0 \tag{4}$$

The relative depth of compression zone $\xi_f$ can be calculated from Equation (4). Finally, the ultimate bending capacity can be calculated by the formula given in Equation (5):

$$M_u = f_c \xi_f (1 - 0.5\xi_f) b h_0^2 + \sigma_s' A_s' (h_0 - a_s') \tag{5}$$

**Table 1.** Comparison between test and calculation results.

| Number of test beam | SGB-1 | SGB-2 | PGB-1 | PGB-2 |
|---|---|---|---|---|
| ①Experimental ultimate load(KN) | 100 | 130 | 33 | 58 |
| ②Experimental bending capacity (KN·m) | 20 | 26 | 8.3 | 14.5 |
| ③Calculated bending capacity (KN·m) | 18.7 | 23.5 | 22.3 | 29.4 |
| ②/③ | 1.07 | 1.11 | 0.37 | 0.49 |

Bending capacity of test beams calculated by proposed equation in this research were compared with experimental values as illustrated in table 1. Close agreement was found between values calculated by the formula proposed in this paper and the experimental results.

## 5. Conclusions

The results of two-point bending tests, which were conducted on beams reinforced with CFRP and steel bars, were presented in this paper. The following conclusions can be summarized:

(1) The plain GFRP bars possess poor bond performance with surrounding concrete, which results in their unable to reinforce concrete structure.

(2) The ribbed surface GFRP bars showed good bond capacity and worked well with concrete before final failure, and the strain distribution of beam cross-section satisfied with the plane section assumption.

(3) FRP reinforced concrete beams presented brittle characteristics while collapsing under loads. The deflection and crack were much larger than that of reinforced concrete beams.

(4) The failure mode II is recommended when designing concrete beam reinforced with FRP bars. Steel bars can be placed in the compression zone and FRP bars can be placed in the tension zone of concrete beam in order to improve the ductility of structure.

## Acknowledgements

The author would like to acknowledge the financial support of the Natural Science Foundation of Shandong Province of China No.ZR2014EEM036.

## References

[1] Rafi M M, Nadjai A and Ali F 2007 Analytical Modeling of Concrete Beams Reinforced with Carbon FRP Bars *Journal of Composite Materials* **vol 41** pp 2657-73

[2] Toutanji H, Deng Y 2003 Deflection and crack-width prediction of concrete beams reinforced with glass FRP rods *Construction and Building Materials* **vol 17** pp 69-74

[3] Abdalla H A 2002 Evaluation of deflection in concrete members reinforced with fibre reinforced polymer (FRP) bars *Composite Structures* **vol 56** pp 63-71

[4] Xu X S 2007 Research on mechanical property of FRP bars and flexural property of concrete beams with FRP bars *Ph.D. thesis* Tianjin Universty

[5] GB 50010-2010 Code for design of concrete structures *S*

# Study on the hydraulic characteristics of side inlet/outlet by physical model test

**KONG Bo[1], YE Fei[2], HU Qiu-yue[2], Zhang Jing[3]**

[1]Xi'an University of Technology,
[2] North China University of Water Conservancy and Electric Power, Zhengzhou 450011, China
[3]Yellow River Conservancy Commission of the Ministry of Water Resources
corresponding author: Ye Fei, yefei@ncwu.edu.cn, North China University of Water Conservancy and Electric Power, Zhengzhou 450011, China

**Abstract**. The hydraulic characteristics at the side inlet/outlet of pumped storage plants were studied by physical model test. The gravity similarity rule was adopted and head loss coefficients under pumped and power conditions were given. The flow distribution under both conditions was studied. Scheme of changing the separation pier section area proportion for minimizing velocity uneven coefficient was brought forward and the cause of test error was researched. Vortex evaluation and observation were studied under the pumped condition at normal and dead reservoir water levels.

## 1. Introduction

The pumped storage plant has a history of more than 100 years. At present, more and more electric power was needed, so the modulation of electric hump and valley became more important than ever before. Due to its electric modulation function, the pumped storage plant can bring huge benefits for economic development. The inlet/outlets of pumped storage plants, according to their pattern, can be classified as side pipe inlet/outlets and vertical pipe inlet/outlets [1]. In china, side pipe inlet/outlets are applied more comprehensively. In this paper, the hydraulic characteristics of side inlet/outlets including flow distribution, head loss, vortexes, etc. were researched under pumped and power conditions by physical model test.

Xi-Longchi pumped storage plant lies in Xinzhou, Shanxi province of China, Machine capacity is 1200MW. Downstream reservoir normal water level is 837.0m, dead water level is 798.0m. Design flow head is 624.0m, flow quantity is 54.18 $m^3/s$ under the power condition and 46.76 $m^3/s$ under the pumped condition.

## 2. Model design and making

### 2.1. Model design

In order to simulate the prototype exactly, six related variables including Froude number (Fr), Reynold number(Re), Weber number(We), circulation intensity parameter (Nr), relative submergence depth(s/d) and geometrical boundary condition should be taken into account. However, all these variables can not be satisfied at the same time because of mutual exclusion among them[1]. Considering that the flow in the inlet/outlet is controlled mainly by gravity, the gravity similarity rule was adopted for the physical model design. The normal model was used for the similarity of flow distribution, submergence depth(s/d) and geometrical boundary condition.

Considering the model material and proving ground, some scales are as follows: model scale, $\lambda_L=39.17$; velocity scale, $\lambda_V=\lambda_L^{0.5}=6.26$; flow scale, $\lambda_Q=\lambda_L^{2.5}=9600.65$; roughness scale, $\lambda_n=\lambda_L^{1/6}=1.84$。

In the prototype, the inhibiting effect on the generation of vortex and circulation by viscosity and surface tension can be ignored because of great values of Re and We. However, in the physical model, the scale effects must be considered [2]. In order to minimize the scale effects, the Re and We should above a critical value in mode making. At present, the common critical values are: (1) Suggested by Amphlett [3], the model Re should be satisfied as $\mathrm{Re}=Q/\upsilon s>3\times10^4$, where $Q$ is the flow quantity, $\upsilon$ is the kinematical viscosity coefficient, $s$ is the submergence depth counted from the orifice center. (2) Suggested by Jain [4], the model We should be satisfied as $We=\rho V^2 d/\sigma\geq120$, where $V$ is the orifice average velocity, $\rho$ is the fluid density, $d$ is the orifice height, $\sigma$ is the surface tension coefficient. The Re and We of the flow in the downstream reservoir inlet/outlet are shown in Tab.1. At different reservoir water levels, both the Re and We were less than the critical values. In order to minimize scale effects, the method of multiplying $Q$ by 2~3 times was adopted for vortex observation in the physical model test[5]. In this paper, $Q$ was multiplied by 2.2 times and both the Re and We were above the critical values at dead reservoir water level.

Tab.1 The Re and We of flow in the model

| Items / Conditions | Prototype values | | | | Model values | |
|---|---|---|---|---|---|---|
| | $Q$ (m$^3$/s) | $V$ (m/s) | $d$ (m) | $s$ (m) | Re ($10^4$) | $We$ |
| Design flow quantity | 46.76 | 0.533 | 6.50 | 47.68 (normal water level) | 0.51 | 26.67 |
| | | | | 7.68 (dead water level) | 3.14 | |
| 2.2 times Design flow quantity | 102.87 | 1.172 | 6.50 | 7.68 (dead water level) | 6.91 | 129.52 |

Annotation: water temperature 15°C, kinematical viscosity coefficient $v=1.139\times10^{-6}m^2/s$, surface tension coefficient $\sigma=0.0735N/m$。

## 2.2. Model making

The physical model included downstream reservoir, inlet/outlet and part of the tail water tunnel. The inlet/outlets were made of organic glass. The downstream reservoir model was built on a platform(11m×10m) 1.5m height from the ground. Under the power condition, the water provided by the water supply tower flowed into the downstream reservoir via the flowmeters, the tail water tunnel and the inlet/outlets ordinally. Under pumped condition, the water provided by the water supply pool flowed into the downstream reservoir, via the inlet/outlet and the tail water tunnel, finally arrived the stabilized pool. The model plane figure is shown in Fig.1.

Under the pumped condition, Q was measured by the right-angled triangle water measuring weir made according to the Japanese Industry Standard(JIS) and the combined error was ±1.4%. Under the power condition, Q was measured by the hole plate flowmeter with a combined error of ±2%. The reservoir water level and weir head were measured by water level measuring needle. The velocity distribution at the inlet/outlet was measured by LS-410 direct-reading propeller meter [6].

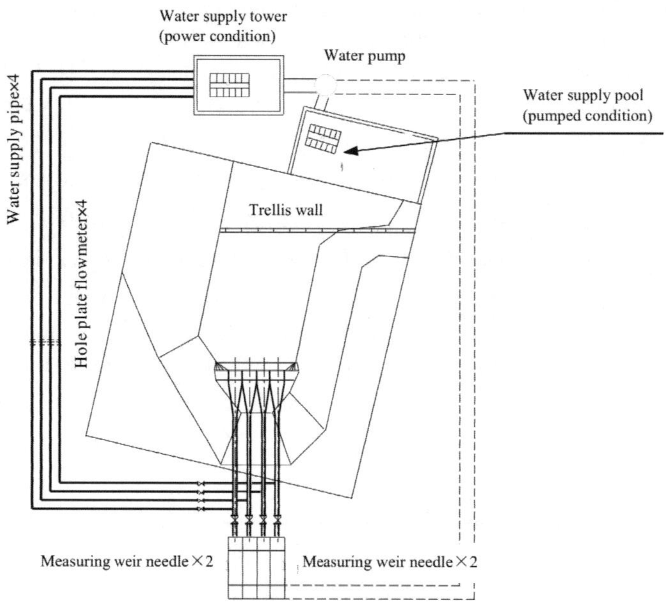


Fig.1 The model ichnography

## 3. Results and discussion

### *3.1. Head loss*

The head loss of the inlet/outlet, mainly local loss, is very important for hydraulic design and flow condition evaluation. The side inlet/outlet of the downstream reservoir was composed of anti-swirl segment, rectification segment, diffusion segment and transition segment as shown in Fig.2.

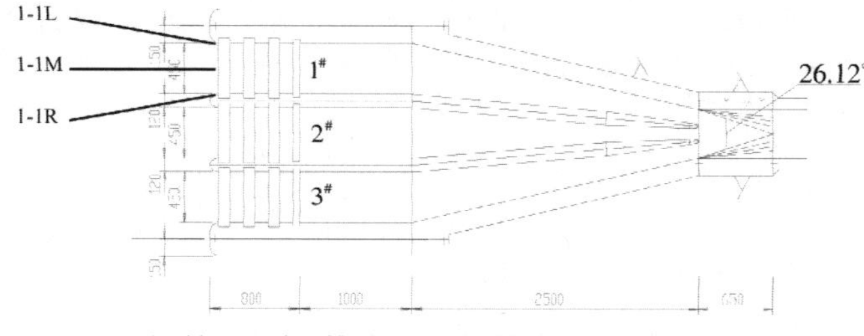


Fig.2 The inlet/outlet ichnography

The head loss was acquired by measuring the reservoir water level ($\nabla_0$), the flow quantity and the piezometer tube level ($\nabla_i$) at the connective section ($i-i$) of transition segment and tail water tunnel[7]. According to the Bernoulli Equation, the formulas for head loss under the pumped and power conditions are shown as follows:

The pumped condition: $h_{0-i} = \nabla_0 - \nabla_i - \alpha v^2 / 2g$ (1)

The power condition: $h_{0-i} = \nabla_i - \nabla_o + \alpha v^2 / 2g$ (2)

The head loss coefficient: $\xi = 2gh_f / \alpha v^2$ (3)

Where $h_f$ is the head loss( $f$ denotes $0-i$ or $i-0$), $\xi$ is the head loss coefficient, $v$ is the average flow velocity of the tail water tunnel, and $\alpha$ is the kinetic energy correction coefficient.

In the prototype, the flow was in square section of flow resistance, so the head loss coefficient was constant and independent of Re and Q. In the model test, the flow was considered to be in square section of flow resistance in the condition that all the head loss coefficients calculated with different Q were basically the same. In this mode test, Q=31.4m$^3$/s～125.4m$^3$/s and Re=4.8×10$^4$～1.9×10$^5$ under the pumped condition, and the corresponding values under the power condition were 25.4m$^3$/s～54.2m$^3$/s and 3.9×10$^4$～8.3×10$^4$. As a result, the head loss coefficients under the pumped and power conditions were 0.23 and 0.33, respectively. The head loss coefficient under the power condition was bigger because the diffusion at transverse and vertical direction led to the flow instability.

### 3.2. Velocity distribution

The flow velocity of each branch orifice(1#, 2# and 3#) was the average value of six measuring points distributed equidistantly along the perpendicular bisector of the section at the trashrack as shown in Fig.2(1-1M). In order to research the change of velocity in transverse direction, the flow velocity of additional perpendicular lines were measured (1-1L , 1-1R).

（1） Pumped condition

In the condition that multi-inlets(inlet1, 2, 3, 4) worked together, the velocity distribution of each inlet was basically the same as that when it worked alone. The flow velocity of three branch orifices of inlet1 under different conditions is shown in Fig.3.

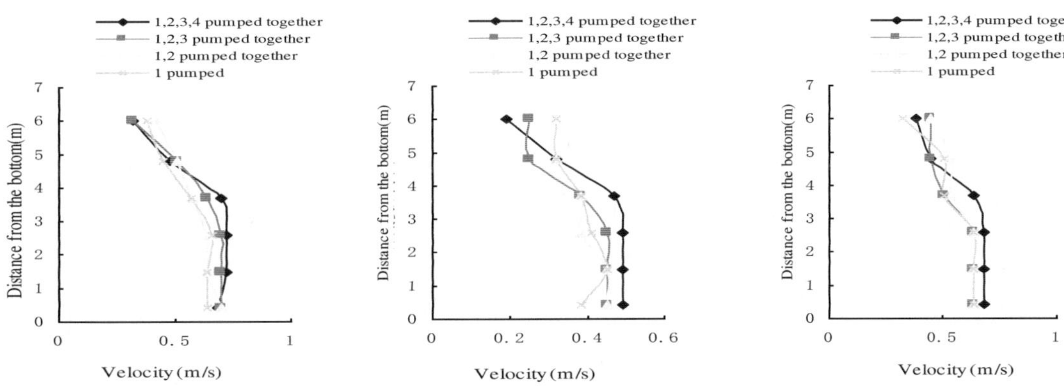


Fig.3 Velocity distribution of inlet1 under different conditions(water level 798.0m, Q=46.76 m$^3$/s)

The velocity distribution at different reservoir water levels is shown in Fig.4. The velocity distribution is basically not influenced by the reservoir water level.

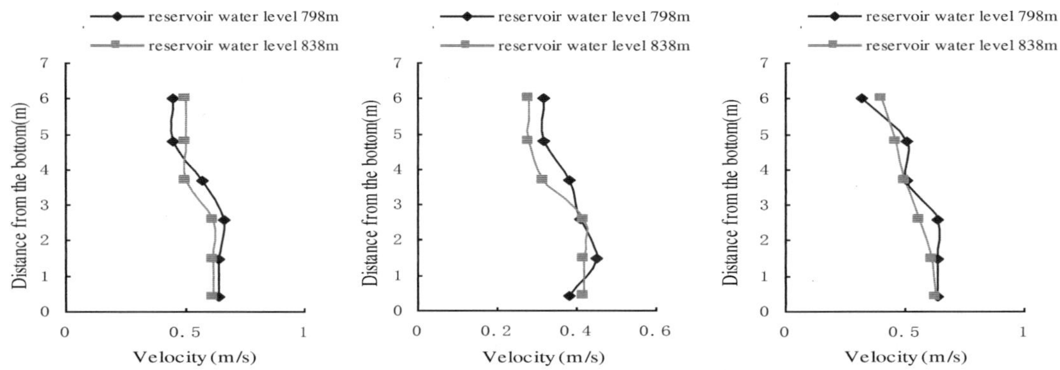


Fig.4 Velocity distribution of inlet1 under different reservoir water levels(Q=46.76 $m^3/s$)

Under different pumped conditions, the average velocity of each branch orifice ranged from 0.42m/s to 0.67m/s, the maximum value was 0.83m/s and the velocity uneven coefficient(the ration of the maximum velocity and average velocity measured at the trashrack) range was 1.21~1.41. For each inlet, the average velocity of the middle orifice (2#) was smaller than that of the two others (1#,3#). The reason is that the section (at the tail of the separation pier) area proportion among the three orifices is 0.35:0.30:0.35. In addition, the press on the flow of the middle orifice by that of the two side orifices at the section also leads to the uneven distribution. This problem can be solved by changing the section area proportion among the three orifices.

Tab.2 The inlet velocity distribution (water level 798.0m, Q=46.76 $m^3/s$)

| Inlets | Measured point | 1 | 2 | 3 | 4 | 5 | 6 | Average velocity (m/s) |
|---|---|---|---|---|---|---|---|---|
| | Distance from the bottom | 0.4 (m) | 1.5 (m) | 2.6 (m) | 3.7 (m) | 4.8 (m) | 6.0 (m) | |
| | Branch orifice | velocity (m/s) | velocity (m/s) | velocity (m/s) | velocity (m/s) | velocity (m/s) | velocity (m/s) | |
| 1 | 1# | 0.76 | 0.76 | 0.76 | 0.70 | 0.48 | 0.32 | 0.63 |
| | 2# | 0.51 | 0.51 | 0.51 | 0.51 | 0.32 | 0.19 | 0.42 |
| | 3# | 0.70 | 0.70 | 0.70 | 0.64 | 0.45 | 0.38 | 0.59 |
| 2 | 1# | 0.76 | 0.76 | 0.76 | 0.64 | 0.45 | 0.38 | 0.63 |
| | 2# | 0.51 | 0.51 | 0.51 | 0.48 | 0.35 | 0.19 | 0.42 |
| | 3# | 0.76 | 0.76 | 0.76 | 0.57 | 0.45 | 0.32 | 0.61 |
| 3 | 1# | 0.64 | 0.70 | 0.70 | 0.57 | 0.45 | 0.41 | 0.58 |
| | 2# | 0.51 | 0.51 | 0.51 | 0.45 | 0.25 | 0.19 | 0.40 |
| | 3# | 0.70 | 0.70 | 0.70 | 0.57 | 0.45 | 0.32 | 0.57 |
| 4 | 1# | 0.83 | 0.76 | 0.76 | 0.70 | 0.45 | 0.38 | 0.65 |
| | 2# | 0.51 | 0.57 | 0.57 | 0.45 | 0.38 | 0.22 | 0.45 |
| | 3# | 0.83 | 0.83 | 0.83 | 0.7 | 0.57 | 0.25 | 0.67 |

（2） power condition

In the condition that multi-outlets(outlet1, 2, 3, 4) worked together, the velocity distribution of each outlet was basically the same as that when it worked alone. The flow velocity of three branch orifices of outlet1 under different conditions is shown in Fig.5.

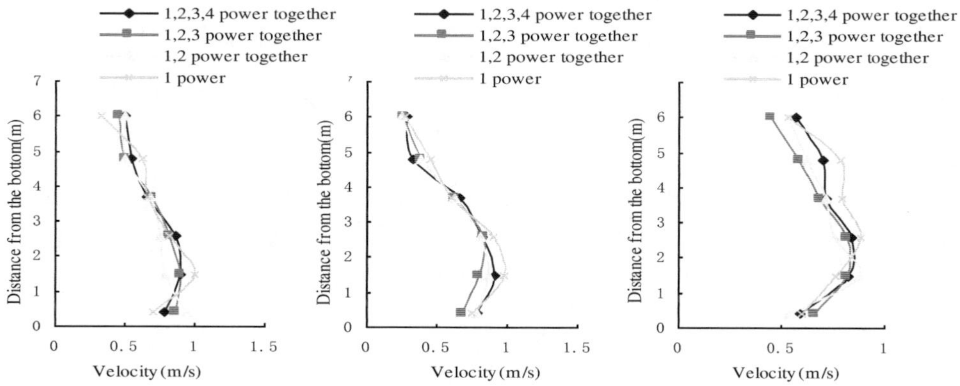


Fig.5 Velocity distribution of outlet1 under different conditions(water level 798.0m, Q=54.18 $m^3/s$)

The velocity distribution of outlet1 at different reservoir water levels is shown in Fig.6. The average flow velocity of each branch orifice at normal water level was less than that at dead water level because of more static water pressure.

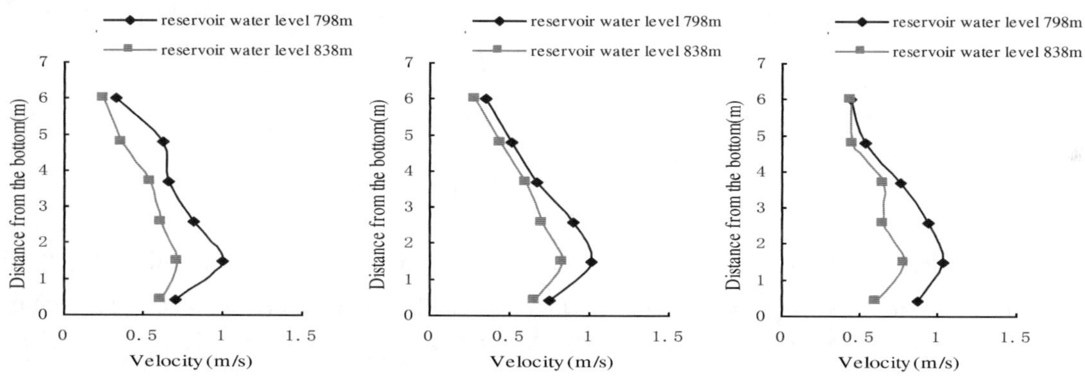


Fig.6 Velocity distribution of outlet1 under different water levels(Q=54.18 $m^3/s$)

Under different power conditions, the orifice average velocity ranged from 0.57m/s to 0.73m/s, the maximum value was 1.05m/s and the velocity uneven coefficient range was 1.30~1.58. Compared with the values(1.21~1.41) under the pumped condition, the uneven coefficient was larger because the outflow diffusion at transverse and vertical direction led to flow instability. For each outlet, the average velocity of the middle orifice(2#) was also smaller than that of the two others(1#,3#). This problem can be solved by changing the section area proportion among the three orifices.

The velocity distribution of outlet1 is shown in Fig.7. For its branch orifice 1#, the average velocity acquired from the measurement along 2-1L, 2-1M and 2-1R were 0.73 m/s, 0.81 m/s and 0.64 m/s, it indicated that the main flow deflected to the left side(facing the upstream direction). For the branch orifice 3#, the corresponding values were 0.62 m/s, 0.85 m/s and 0.78 m/s, therefore, the main flow deflected to the right side. However, the average velocity of branch orifice 2# measured along different lines were basically the same and no deflection was found. Therefore, in order to minimize test errors, the velocity of three perpendicular lines should be measured for calculating the average velocity of branch orifice 1# and 3#.

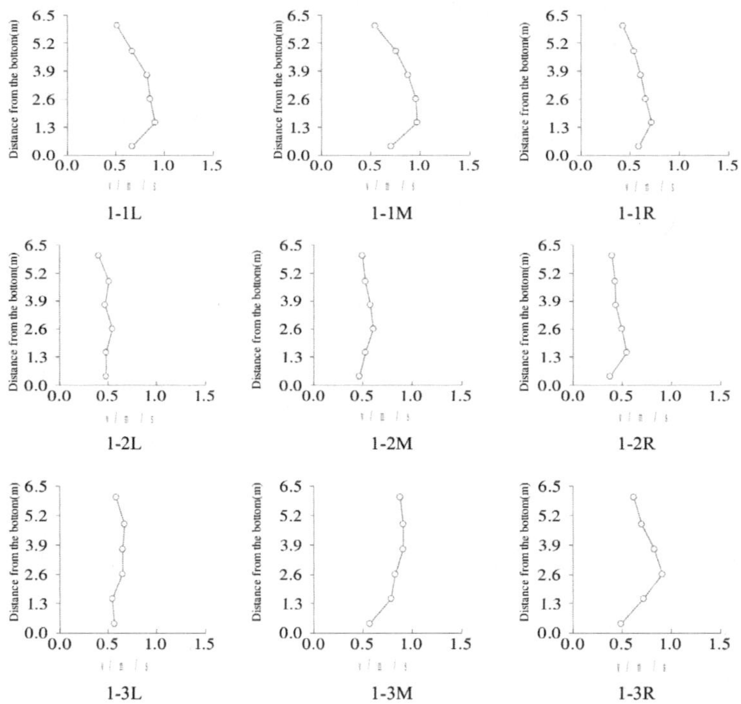


Fig.7 Velocity distribution of outlet1(water level 798.0m, Q= 54.18 $m^3/s$)

### 3.3. *Vortexes*

According to the observation data of 29 prototype inlets, Gordon *et al.* made the conclusion [8] that the generation of vortexes related to the velocity, inlet dimension and submergence depth. The critical submergence depth without vortexes can be described as:

$$S_c = CVd^{1/2} \quad (4)$$

where $S_c$ is the critical submergence depth calculated from the top of the inlet, $d$ is the inlet height, $V$ is the flow velocity of tail water tunnel, $C$ is the coefficient, ($C = 0.55$ for symmetry inflow, and $C = 0.73$ for dissymmetry inflow). According to the prototype data ($V = 0.53m/s$, $d = 6.5m$), the $S_c$ calculated was $0.94m$, lower than the submergence depth ($4.334m$) at dead water level.

According to the data of 13 prototype inlets, Pennino *et al.* draw the conclusion [9] that the inlet Fr without vortexes should be expressed as:

$$Fr = V / \sqrt{gs} < 0.23 \quad (5)$$

where $s$ is the submergence depth calculated from the centre of the inlet, $V$ is the flow velocity, $g$ is the acceleration of gravity. There are almost no vortexes generation if $Fr = 0.06 < 0.23$ and $s/d > 0.5$, where $d$ is the inlet height. Tab.3 shows the vortex evaluation of the inlet according to the prototype data.

Tab.3 The vortex evaluation of the inlet

| Flow quantity Q($m^3$/s) | Flow velocity V(m/s) | Orifice height d(m) | Reservoir water level ▽(m) | Submergence depth s(m) | $F_r$ | $s/d$ | Harmful vortexes (Y/N) |
|---|---|---|---|---|---|---|---|
| 46.76 | 0.53 | 6.50 | Normal water level 838.0 | 47.584 | 0.03 < 0.23 | 7.3 > 0.5 | N |
| | | | Dead water level 798.0 | 7.584 | 0.06 < 0.23 | 1.2 > 0.5 | N |

In the model test, in order to minimize scale effects, Q was multiplied by 2.2 times for vortex observation. When the water level changed from the normal water level to the dead water level, no harmful vortexes (air suction vortexes or other adsorption vortexes) generated.

## 4. Conclusions

In this paper, the flow characteristics of side inlet/outlets were studied by physical model test. The head loss coefficients under the pumped and power conditions were 0.23 and 0.33, respectively. The value under the power condition was larger because the flow diffusion at transverse and vertical direction led to the flow instability.

In the condition that multi-inlet/outlets worked together, the velocity distribution of each inlet/outlet was basically the same as that when it worked alone. The flow distribution was basically not influenced by the reservoir water level under the pump condition, however, under the power condition, The average flow velocity of each branch orifice at normal water level was less than that at dead water level because of more static water pressure. For each inlet/outlet, the average velocity of the middle orifice was smaller than that of the two others, and this problem can be solved by changing the section area proportion among the three orifices. Under the power condition, the main flow of the left (or right) branch orifice deflected from the center to the left (or right) side, in order to minimize test errors, the velocity of three perpendicular lines should be measured for calculating the average velocity of the two side branch orifices.

Under the pumped condition, when the water level changed from the normal level to the dead level, no harmful vortexes generated.

## References

[1] Lu You-mei, Pan Jia-zheng. the Pumped Storage Plant [M]. Bei Jing: Water conservancy and Electric power publisher, 1992, Modeling Hydraulic Structures
[2] Hecker, G.E, S cale effects in vortexes simultion, Symposium on Scale Effects in, Yangtse Rive water and electricity science academe, 1985.10, 219-237.
[3] Amphlett, M.B, et al. Similarity of Free-Vortex at Horizontal Intake[J]. Journal of Hydraulic Research, 1978, 16(2): 95-105
[4] Jain, A.K, et al. Vortex Formation at Vertical Pipe Intake[J]. Journal of the Hydraulics Division, 1978, 104(10):1429-1445
[5] Hecker, G.E. Model-prototype comparison of free surface vortices[J]. Journal of the Hydraulics Division, 1981, 107(10):1243-1259
[6] Tianjin University, Beijing Water conservancy and Electric power Comporation. Report of the up reservoir inlet/outlet of Xilongchi power station[R]. 2002
[7] H.K.Versteeg, W. Malalasekera. An introduction to computational fluid dynamics[M]. New York: Longman Scientific & Technical, 1995
[8] Gordon J L, Wortexes at vertical intakes[J]. Water Power, 1970, (4): 137-138
[9] M.Syamlal, T.J.Brien. Computer Simulation of Bubbles in a Fluidized Bed[J]. AICHE Symp Series, 1989, 85(1): 22-31

IOP Conf. Series: Earth and Environmental Science **61** (2017) 012040 doi:10.1088/1755-1315/61/1/012040

# Investigation of the unsteady motion and formation mechanism of intake vortexes

**YE Fei[1], ZHANG Jing[2], Kong Bo[3]**

[1]North China University of Water Conservancy and Electric Power, Zhengzhou 450011, China
[2]Yellow River Conservancy Commission of the Ministry of Water Resources
[3]XI'AN University of Technology
Corresponding author: YE Fei, yefei@ncwu.edu.cn, North China University of Water Conservancy and Electric Power, Zhengzhou 450011, China

**Abstract**. The unsteady motion and formation mechanism of the intake vortexes are studied by numerical simulations, covering the flow distribution, vortex form, influencing factors, and others. Based on the physical model test, the large eddy simulation (LES) is used in the 3-D simulation of the intake flow. The occurrence of intake vortex presents a coherent characteristic of "dormant period" and "active period", which is approximately consistent with that of the surface fluctuation. According to the turbulent cascade theory, the process of energy balance of the intake vortex is studied. The simulation results compare favorably with that from experimental correlations.

## 1. Introduction

The intake vortexes, with a high occurrence in the intake of a hydraulic structure, can be described roughly in terms of two forms: steady vortexes, whose movement elements keep time-invariant, and unsteady vortexes, whose corresponding elements change frequently with time, rendering the characteristics of frequent variation in size and location, short duration and random occurrence, etc. However, in most current researches these unsteady attributes are always ignored (to be considered as steady vortexes) for simple analysis of the spiral velocity, which leads to a great loss of vortex-motion information[1]. In fact, the unsteady characteristics of intake-vortex motion are very important for hydraulic design and operation. For example, under the inflow conditions, different kinds of vortexes including surface-concavity vortexes, solid-entrained vortexes, air-suction vortexes and so on, may generate with great unsteady attributes to create problems of reduced flow quantity, noises, cavitation, vibration or even destruction of the trashracks, etc.[2] According to the correlation results from prototype observations or physical model tests, many domestic hydraulic projects as MAN Wan, LONG Yang-xia, HUANG Tan-kou, SHI Tou-he, ZI Ping-pu, etc., have a record of harmful intake vortexes, which also can be found in projects in Japan, American, Russia, Canada, Britain, India and so on[3]. Ideally, one would like to perform a physical model test to predict the unsteady vortex motion in the prototype, however, model similarity about vortex motion via the gravity law has always been discussed and no convincible results have been proposed recently.

For these years, such flows with unsteady vortexes present severe physical and numerical modeling challenges, particularly in engineering use, such as the works did by Gao et al.[4], who studied the flow characteristics of surface vortexes in four different intakes by physical model test, to achieve results that under the pumped conditions, random funnel-shaped surface vortexes generate and increase obviously with the flow quantity at the dead water level. Huang et al.[5] studied the effects of the anti-vortex beam on the generation of surface vortex at different working conditions from normal water level to dead water level, indicating that under no anti-vortex beam conditions, more surface vortexes generate and some may even develop to harmful air-suction vortexes with increasing flow quantity. Currently, however, under the limitations of measuring means, most studies today about the unsteady characteristics of intake vortexes can commonly give the average locations of vortex generation and approximate range of vortex motion. Further studies about the unsteady vortex motion and internal flow structure are hampered by the absence of the whole-field-measurement implementation.

So far most studies about the formation mechanism of intake vortexes are always achieved by means of influencing-factor analysis, as results from Zielinski et al[6] and Yildirm et al[7], who studied the effects of viscosity and surface tension on the generation of intake vortexes. Quick et al[8] studied the formation mechanism of intake vortexes by physical model test, indicating that the vortex formation significantly rests with the flow conditions, Froude number, and boundary conditions. In summary, the main influencing factors can be described as follows: flow conditions (velocity, flow quantity, water level, etc.), boundary conditions (structure form and dimension, wall roughness, etc.) and fluid conditions (viscosity, surface tension, etc.). The main influencing parameters include Froude number, Reynolds number, Weber number, circulation parameter and submerse depth. Apparently, accurate flow-field information of the intake is very important for the investigation of vortex generation. During the process of vortex generation and development, energy transmission happens continuously between different vortexes, so the viscosity effects must be taken into account in the analysis of vortex-formation mechanism. According to the Attenuation theory, Schlichtingh et al.[9] acquired the tangential velocity of a vortex based on the fact that a vortex can not maintain itself without more energy supply because of viscosity-dissipation effects. The Turbulent Cascade theory indicates that, the large scale vortexes acquire energy from the mean flow and then transport it to other smaller-scale vortexes, which will finally be dissipated by kinetic viscosity in the micro-scale vortexes. Based on this theory, the author researched the vortex formation mechanism via the energy-balance analysis, to show that the coherent attributes in the process of vortex generation and development are closely associated with the turbulence characteristic parameters[10].

Nowadays, further study about the flow mechanism and internal flow structure is very important for the optimization design of hydraulic structures. Direct numerical simulation(DNS) can present the whole information of the flow field. However, limited by the computational techniques, DNS of the large range of scales spanned by the ensuing dynamics is impractical at high Reynolds numbers, rendering its application to practical flows with complex boundary conditions infeasible[11]. For this reason, nowadays large eddy simulation (LES) is pursued as an alternative, in which the large scale eddies are calculated by the Navier-stokes equation, and the effects of small scale eddies are implemented via an appropriate sub-grid scale(SGS) model[12]. In the complex wall-bounded flow simulations, LES performs far excellent than RANS by giving detail information of turbulent fluctuation at acceptable computational cost[13]. Therefore, in this article the hydraulic characteristics of intake vortexes are studied by LES method, including vortex forms, motion behavior and formation mechanism.

## 2. Numerical model

### 2.1. Control equations

The LES method is used in this paper, in which the large scale eddies are calculated by the Navier-stokes equation, and the effects of small scale eddies are implemented via an appropriate sub-grid scale(SGS) model. The control equations are acquired from the N-S equation of incompressible fluid via the Gauss filter[14] as follows:

$$\frac{\partial \bar{u}_i}{\partial x_i} = 0 \qquad (1)$$

$$\frac{\partial \overline{u_i}}{\partial t}+\frac{\partial\left(\overline{u_i}\,\overline{u_j}\right)}{\partial x_j}=-\frac{1}{\rho}\frac{\partial \overline{p}}{\partial x_i}+\frac{\partial}{\partial x_j}\left[\left(\nu+\nu_t\right)\left(\frac{\partial \overline{u_i}}{\partial x_j}+\frac{\partial \overline{u_j}}{\partial x_i}\right)\right] \tag{2}$$

Where $u_i$ is the velocity, $\nu$ is the coefficient of kinematic viscosity, $\rho$ is the density, $p$ is the pressure, $\mu_t$ is the coefficient of sub-grid eddy viscosity specified using the Smagorinsky-Lilly model as follows:

$$\mu_t=\left(fC_S\Delta\right)\left|\overline{S}\right| \tag{3}$$

Where $C_s$ is the model constant, $C_s=0.1$, $\left|\overline{S}\right|=\sqrt{2\overline{S}_{ij}\overline{S}_{ij}}$ , $\Delta=\sqrt[3]{\Delta x\Delta y\Delta z}$ , $\Delta x$ , $\Delta y$ , $\Delta z$ are the mesh dimensions at $x$ , $y$ and $z$ directions respectively, $f$ is the grid aeolotropic parameters specified as follows[15]:

$$f=\cos h\sqrt{\frac{4(\ln^2 a_1-\ln a_1\ln a_2+\ln^2 a_2)}{27}} \tag{4}$$

Where $a_1=\Delta_1/\Delta_3$, $a_2=\Delta_2/\Delta_3$.

The free surface is an interface of water and air, at which both the dynamic and kinematic conditions should be satisfied as follows[16]:

$$\frac{\partial h}{\partial t}=w^s-u^s\frac{\partial h}{\partial x}-v^s\frac{\partial h}{\partial y} \tag{5}$$

Where $u^s$, $v^s$, $w^s$ are the flow velocity at longitudinal, transverse and verical directions, the local pressure is used at the surface, $p_s=p_a$.

The finite volume method is employed to discretize the control equations with the Admas-Bashforth scheme for temporal differentiation, the second-order upwind scheme is used for advection term and the second-order central difference scheme is used for diffusion term. The PISO algorithm is applied for the pressure-velocity coupling[17].

### 2.2. Boundary conditions

The upstream cross section $30d$ from the intake is the inflow boundary condition given by the average flow velocity with hydrostatic pressure distribution, $\partial p/\partial x=0$ . The Sommerfeld open boundary is used as the outflow boundary condition, $\partial\Phi/\partial t+U_{con}\left(\partial\Phi/\partial x\right)=0$, where $\Phi$ is either physical quantity of $u$ and $\rho$, $U_{con}$ is the weighted average value of the longitudinal velocity along the abscissa axis. The free surface is treated as a pressure boundary given by the local pressure.

Based on the no-slip boundary condition, the particle velocities in all directions need to be zero at the wall. However, this treatment is accurate only when fairly fine meshes are used to resolve the bottom boundary layer. Alternatively, the wall function[18] is introduced to relate the boundary variables to the flow region as follows:

$$\tau_w=\begin{cases}\dfrac{2\mu\left|\overline{u}_p\right|}{\Delta y} & \left|\overline{u}_p\right|\le\dfrac{\mu A^{2/(1-B)}}{2\rho\Delta y}\\[2ex] \rho\left[\dfrac{1-B}{2}A^{(1+B)/(1-B)}\left(\dfrac{\mu}{\rho\Delta y}\right)^{(1+B)}+\dfrac{1+B}{A}\left(\dfrac{\mu}{\rho\Delta y}\right)^{B}\left|\overline{u}_p\right|\right]^{2/(1+B)} & \left|\overline{u}_p\right|>\dfrac{\mu A^{2/(1-B)}}{2\rho\Delta y}\end{cases} \tag{6}$$

Where the $\tau_w$ is the wall shear stress, $\rho$ is the density, $\mu$ the dynamic viscosity coefficient, $\overline{u}_p$ is the sub-layer velocity, $\Delta y$ is the mesh dimension at $y$ direction, $A$ and $B$ are constants, $A=8.3$, $B=1/7$.

IOP Conf. Series: Earth and Environmental Science **61** (2017) 012040 doi:10.1088/1755-1315/61/1/012040

The calculation domain is shown in Fig. 1, where $x \times y \times z = 30d \times 5d \times 10d$, $d$ is the width of the intake. The hexahedral structure mesh is used in the whole simulation field using an adaptive-mesh refinement strategy at the near-wall region. The dimensionless time step is 0.005.

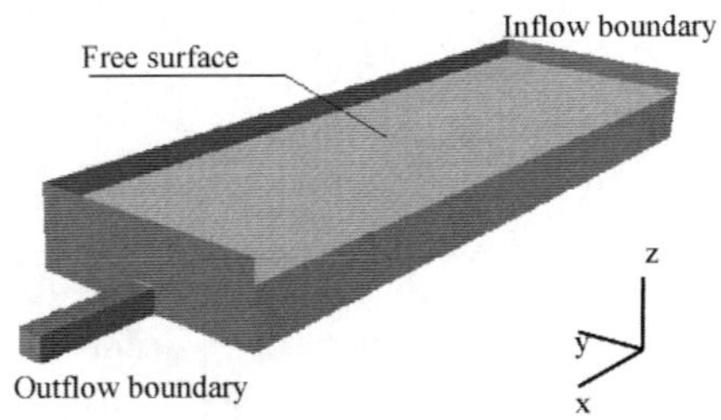


Fig.1 Simulation region

## 3. Results and discussion

Three different working conditions with relative turbulent intensity from 0.05 to 0.22 are studied in this paper as shown in Tab.1.

Tab.1 Working conditions

| Conditions | Relative submerse depth of the intake $s/d$ | Relative turbulent intensity $\sqrt{\overline{u'^2}}/U$ | Turbulent Reynolds number $\mathrm{Re} = \sqrt{\overline{u'^2}}\,R/\nu$ ($10^3$) |
|---|---|---|---|
| 1 | 1.5 | 0.05 | 0.82 |
| 2 | 1.5 | 0.15 | 2.45 |
| 3 | 1.5 | 0.22 | 3.66 |

Annotation: where $s$ is the submerse depth above the intake center, $d$ is the intake diameter, $\sqrt{\overline{u'^2}}$ is the average turbulent intensity, $U$ is the average inflow velocity, $R$ is the intake hydraulic radius, $\nu$ is the kinematic viscosity, $\nu = 0.919 \times 10^{-6} m^2/s$.

Fig.2 presents the instantaneous flow streamlines near the intake, which shows the process of vortex generation and development. First, in the region relatively far from the breast wall (>1.8m), due to the even-distributed incoming flow under symmetric boundary conditions, only a minority of flow separations and unsteady vortexes can be found. Second, in the region at the side-wall corner, more concavity vortexes generate because the flow streamlines curve with the wall boundary to create more small surface circulations. Third, in the region near the top of the intake, caused by the uneven contraction of the flow to the intake, more surface circulations generate leading to a high probability of vortex generation. Mostly, these vortexes can not maintain themselves without continuing provided energy and will so turn to be destructed by other flows. However, at given submerse depths, some vortexes may be combined with others to reinforce to a new larger-scale vortex during the process of floating to the intake. Generally, these large-scale vortexes happen at the top of the intake always induce other new associated vortexes to present a characteristic of paired generation.

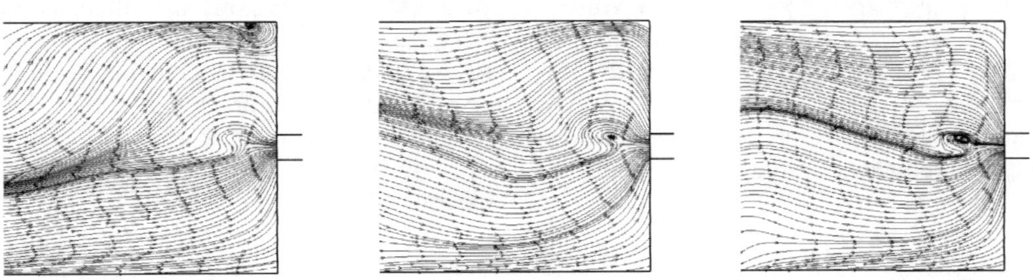

(*a*) $t = 400$ (*b*) $t = 500$ (*c*) $t = 600$

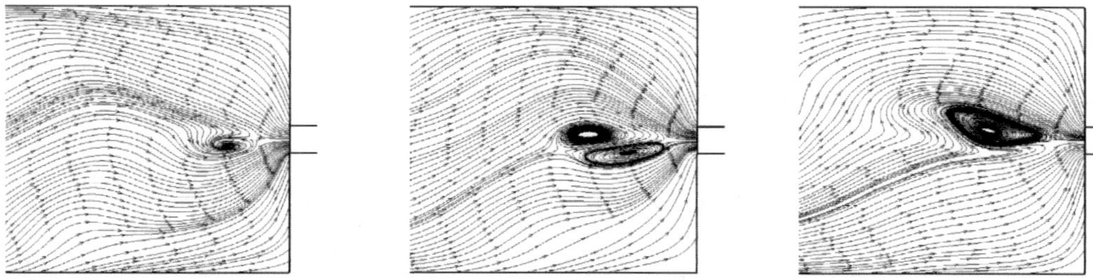

(*d*) $t = 700$ (*e*) $t = 800$ (*f*) $t = 900$

Fig.2 Vortex forms and development ( $s/d = 1.5$ , $\mathrm{Re} = 2.56\times10^5$ , $I = 0.05$ )

Simulation results shows that the region relatively far from the intake (>1.8m), low-intensity surface-concavity vortexes may generate with shallow low-level-rotation tails. Generally these vortexes are not persistent, which may easily be destructed or divided by other flows to present a short-duration characteristic. In the region near the intake, the through-type air-suction vortex may generate with an air-core in the center to be sucked into the intake. These vortexes are mostly developed from the surface-concavity vortexes when they float to the intake and can persist for a while with increasing rotational velocity.

Fig.3 shows the LES results of turbulent intensity under working condition 3, which compare favorably with the results from experimental correlations.

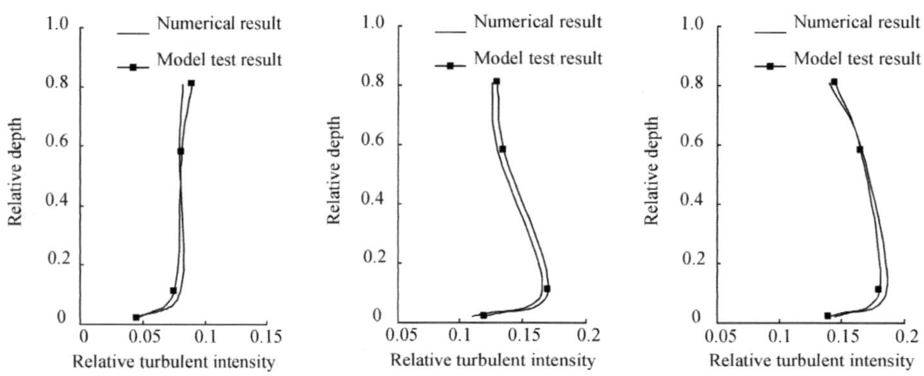


(a) Longitudinal distribution (b) Transverse distribution (c) Vertical distribution

Fig.3 Comparison of the turbulence intensity between the simulation results and model test results (Working condition 3, s/d=1.5, Re=$2.56\times10^5$, I=0.22, Y/B=0)

## 4. Conclusions

In this article, the unsteady motion and formation mechanism of intake vortexes is studied by numerical simulations. First, the vortex forms and flow characteristics are studied. In the region relatively far from the breast wall (>1.8m), only a few low-intensity surface-concavity vortexes can be found with shallow low-level-rotation tails. Generally, these vortexes are not persistent, which may easily be destructed or divided by other flows to present a short-duration characteristic. In the region at the side-wall corner, more concavity vortexes generate because the flow streamlines curve with the wall boundary to create more small surface circulations. In the region near the top of the intake, caused by the uneven contraction of the flow, more surface circulations generate leading to a high probability of vortex generation. At given submerse depths, some vortexes may be combined with others to reinforce to a new large-scale through-type vortex with an air-core in the center. These vortexes are mostly developed from the surface-concavity vortexes when they float to the intake and can persist for a while with increasing rotational velocity.

**References**

[1] CHEN Yun-liang, WU Chao and YE Mao et al. Hydraulic characteristics of vertical vortex at hydraulic intakes[J]. Journal of Hydrodynamics, Ser. B, 2007,19(2):143-149.

[2] DANG YUAN-yuan. Research of the vortex characteristics at dissymmetrical boundary intake[D], Master's degree thesis, Nanjing: Nanjing Hydraulic Research Institute, 2008(in Chinese).

[3] YE Fei. Unsteady vortex motion and formation mechanism[D], Ph.D. Thesis, Tianjin: Tianjin University, 2010(in Chinese).

[4] GAO XUE-ping, ZHANG Xiao-xian and LI Chang-liang et al. The hydraulic characteristics at vertical pipe intake of Xilongchi pumped storage plant[J]. Journal of Hydroelectric Engineering, 2002, 76(1): 52-60.

[5] HUANG Zhi-min, HE Xiao-hui and ZHU Hong-hua et al. Model test research of the lower reservoir inlet/outlet for Guangzhou pumped storage plant[J]. Water Resources and Power, 2005, 23(1):4-7.

[6] ZIELINSKI P. B. and VILLEMONTE J. R. Effect of viscosity on vortex-orifice flow[J]. Journal of the Hydraulics Division, ASCE, 1968, 94(3): 745-752.

[7] YILDRIM N. and JAIN S. C. Surface tension effect on profile of free vortex[J]. Journal of the Hydraulics Division, ASCE, 1981, 107(1): 132-136.

[8] QUICK M. C. Efficiency of air-Entraining Vortex Formation at Water Intake[J]. Journal of the hydraulics division, ASCE, 1970, 96(7):1403-1416.

[9] SCHLICHTING H. Boundary layer theory [M]. New York, N.Y: 7th Ed., McGraw-Hill Book Co., 1979.

[10] DU Min. Formation mechanisms and scale effects of hydraulic intake vortices[D]. Ph.D. Thesis, Tianjin: Tianjin University, 2008(in Chinese).

[11] J. Xu. Benchmarks on tera-scalable models for DNS of turbulent channel flow [J]. Parallel Computing, 2007, 33(12):780-794.

[12] YE Fei, GAO Xue-ping. Numerical simulations of the hydraulic characteristics of side inlet/outlets[J]. Journal of hydrodynamics, Ser. B, 2011, 23(1): 48-54.

[13] YOSHIHIDE T., AKASHI M. and SHUZO M. et al. Comparison of various revised k-ε models and LES applied to flow around a high-rise building model with 1:1:2 shape placed within the surface boundary layer [J]. Journal of Wind Engineering and Industrial Aerodynamics, 2008, 96(4):389-411.

[14] J. W. DEARDORFF. A numerical study of three-dimension turbulent channel flow at large Reynolds numbers[J]. Journal of Fluid Mechanics, 1970, 41 (2): 453-480.

[15] SCOTTI A., MENEVEAU C. and LILLY DK. Generalized Smagorinsky model for anisotropic grids[J]. Physics of Fluids A, 1993, 5(9): 2360-2380.

[16] C. W. Li, L. H. Yu. Hybrid LES/RANS modelling of free surface flow through vegetation[J]. Computers & Fluids, 2010, 39(9): 1722-1732.

[17] CHENG Zeng, C. W. Li. A hybrid RANS-LES model for combining flows in open-channel T-junctions[J]. Journal of Hydrodynamics, Ser. B, 2010, supplement 1, 22(5):154-159.

[18] ALEXANDER O., HANS W. Numerical simulation(DNS and LES) of manipulated turbulent boundary layer flow over a surface-mounted fence[J]. European Journal of Mechanics-B/Fluids, 2000, 19(5): 765-788.

IOP Conf. Series: Earth and Environmental Science **61** (2017) 012041 doi:10.1088/1755-1315/61/1/012041

# Study of the effects of 1-MCP to blueberry under cold storage

**Shenchen Tao[1*], Huailiang Chu[2*], Xiaomin Chen[1*], Huwei Yuan[1*], Lingling Qiu[1], Liang Zhao[1], Daoliang Yan[1,a], Bingsong Zheng[1,b]**

[1] State Key Laboratory of Subtropical Silviculture, Zhejiang A & F University, Linan, Hangzhou 311300, P. R. China
[2] Hengfeng Forestry Bureau, Jiangxi, P. R. China
*These authors contributed equally to this work.
E-mail: [a] liangsie2000@163.com, [b] bszheng@zafu.edu.cn;

**Abstract.** Blueberry is one of the thinnest exocarp fruits in the world, which is difficult to keep fresh due to the special structure of its skin. 1-Methlcyclopropene (1-MCP) is able to combine with ethylene(ETH) receptor. In this study we investigated the effect of 1-MCP on rotting rate, weight loss ratio, soluble sugar content, titratable acid content, antioxidant enzyme activities and malondialdehyde (MDA) content in blueberry (Vaccinium corymbosun 'O`Neal' and 'North Road') under cold storage. 1-MCP reduced the rotting rate, weight loss ratio and MDA content, while keeping high-leveled stability in antioxidant enzyme activities, soluble sugar content and titratable acid content. These results showed the role of 1-MCP in alleviating the negative effects of blueberry and suggested that 1-MCP could be used as a preservative for keeping thin exocarp fruit in fresh.

## 1. Introduction

Blueberry (*Vaccinium corymbosum*), a member of Ericaceae family, is a well-known fruit worldwide and mainly distributed in moist areas with acid soil. The fruit of blueberry is small but rich in vitamins, minerals, carbohydrates and anthocyanins, and it is regarded as one of the five healthy fruits by the Food and Agriculture Organization of the United Nations due to rich nutrients[1]. There are more than 100 varieties cultivated in China, of which "Bluetta", "O`Neal", "Misty" and "Northland" are the main cultivars in Zhejiang Province in the eastern part of China. While blueberry is a kind of berry, it is hard to store for a long time because of the thin exocarp and juice.

1-Methlcyclopropene is a kind of active chemicals, which exists in the stage of gas under normal temperature. 1-MCP can combine with the ethylene receptor specificity, which effectively prevents the generation of endogenous ethylene and the inducing effect of exogenous ethylene [2]. The study found that, 1-MCP can reduce the happening of brown rot in 'Los Angeles' plum [3]; improve the storage quality of loquat during the ripening; affect the active oxygen metabolism related enzymes activity in melon [4].

Many studies on biological, physiological and biochemical, and molecular biological characteristics of blueberry [5-7] have been carried out, but little knows about the effects of 1-MCP on the blueberry under cold storage. Thus, the objective of this study was to investigate whether 1-MCP could alleviate the effect of fruit senescence of blueberry and improve the ability of storage of blueberry.

## 2. Materials and methods

*2.1 Materials*
In early June 2016, blueberry fruits “O`Neal” and “North Land” were collected from the ‘Mu shan nong” blueberry base. Before being transported to the lab, all fruits were needed to be precool at 10°C. After being transported to the lab, divided the blueberry into several retail-used transparent plastic boxes (about 125g per box). Each type of blueberry contained 4 groups (see the following table), and then every group needs to be put into the seal plastic box with same volume. After finishing that, the seal boxes were all placed in the refrigerator at 4°C (with 60% of humidity), at Zhejiang A & F University (N 30°23′, E 119°72′), China. After 48 hours, transfer the blueberry out of the seal box to the refrigerator.

| Group | T1 | T2 | T3 | T4 |
|---|---|---|---|---|
| O`Neal | Control | 10ul/L 1-MCP | 20ul/L 1-MCP | 30ul/L 1-MCP |
| North Land | Control | 10ul/L 1-MCP | 20ul/L 1-MCP | 30ul/L 1-MCP |

Every 5 days, a box of blueberry was used for the determination of soluble sugar content, titratable acid content, soluble solid, enzyme activities and the MDA content analysis.

*2.2 Methods*

*2.2.1 Determination of rotting rate.* About 125g of fresh blueberry were carried out for the detection of the rotting rate. Every 5 days, pick out and record the number of broken and mildew blueberry.
Rotting rate = (0 day`s number - other day`s number) / 0 day`s number *100%

*2.2.2 Determination of weight loss ratio.* About 125g of fresh blueberry were carried out for the detection of the weight loss ratio. Record the weight of the blueberry on the $0^{th}$, $5^{th}$, $10^{th}$, $15^{th}$, $20^{th}$, $25^{th}$, $30^{th}$ day, and the weight loss would be the 0 day`s weight minus the other day`s weight.
Weight loss ratio = ( 0 day`s weight - other day`s weight ) / 0 day`s weight *100%

*2.2.3 Determination of soluble solid.* We used the saccharimeter (HR108ATC) to measure the soluble solid content [8]. 10g of fresh blueberry was grounded at 4°C with an ice-chilled pestle and grinding fully. After filtration with gauze, use the pipette take 0.1ml of the liquid to the saccharimeter and read the soluble solid content.

*2.2.4 Determination of MDA contents and anti-oxidant enzymes activities.* For the MDA content, SOD, CAT and POD activities, 5g of fresh blueberry was ground at 4°C with an ice-chilled pestle and mortar in 10 ml of 50mM phosphate buffer (pH7.8) containing 1% polyvinylpyrrolidone (PVP). The homogenate was centrifuged at 8000rpm at 4°C for 15min and the supernatants were collected for measuring MDA content and enzyme activities according to Ying[9].

*2.2.5 Determination of total soluble sugar content.* The total soluble sugar content was determined using anthrone colorimetry [10]. Briefly, 2g of smashed fresh fruit was placed in a large tube containing 15ml pure water, and increased to a volume of 100ml after 30min in a boiling water bath. Then, 1ml solution was mixed with 5ml anthrone reagent, and measured at 620nm after 10min in a boiling water bath.

*2.2.6 Determination of titratable acid content.* For the titratable acid content, we used the phenolphthalein acid-base neutralization titration, in this research the concentration of NaOH was 0.1%. 5g of fresh blueberry was ground at 4°C with an ice-chilled pestle and mortar in 5ml of water. The homogenate was centrifuged at 8000 rpm at 4°C for 15min and the supernatants were collected for measuring titratable acid content.

Titratable acid content = 0.1% * Volume of NaOH / Volume of extracting solution (0.5ml)

## 3. Statistical analysis

The graphs and analysis of variance were performed using Excel and SPSS Institute Inc. (Version 19.0) software (USA).

## 4. Results

*4.1 Rotting rate of blueberry during 1-MCP treatments under cold storage*

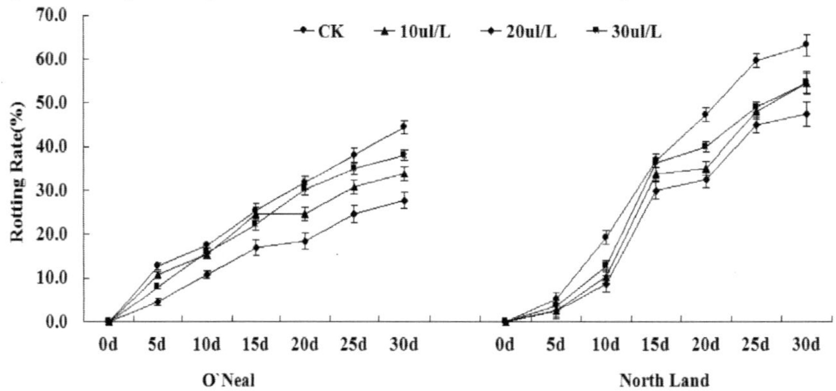


**Fig 1.** Changes in the Rotting rate of blueberry

Rotting rate is one of the most intuitive indicator of the fruits` quality (Fig.1). Rotting rate was increasing from 0d to 30d, and "O`Neal" has a lower rotting rate than the "North Land". At 30d, the rotting rate of the "O`Neal" control was risen to 44.44%, while 1-MCP treated groups (T2-T4) were at 33.85%, 27.69% and 38.10%, and has a 10.59%,16.75% and 6.34% lower than control, respectively; The rotting rate of "North Land" control were at 63.16%, while 1-MCP treated groups (T2-T4) were at 54.55%, 47.5% and 54.55%, with 8.61%, 15.66% and 8.61% lower than control, respectively.

*4.2 Weight loss ratio of blueberry during 1-MCP treatments under cold storage*

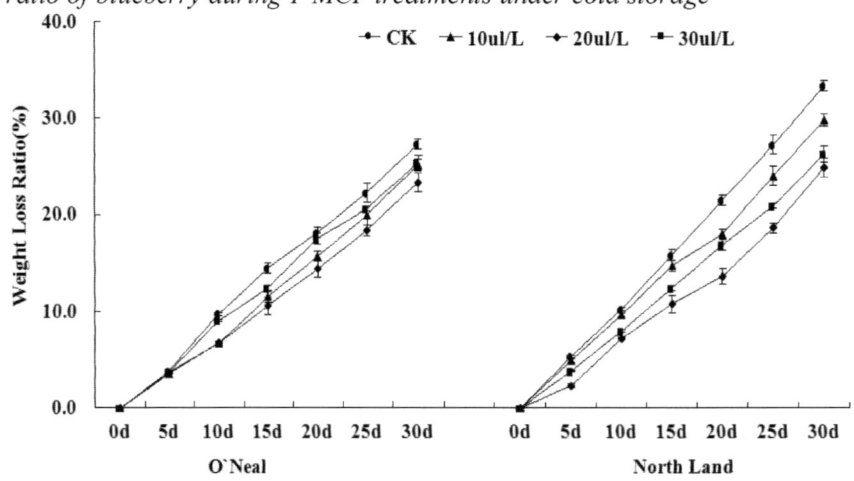


**Fig 2.** Changes in the Weight loss ratio of blueberry

After the harvest, the fruits will gradually lose the weight, because of the water loss and dry matter loss caused by respiration (Fig.2). The weight loss ratio was increasing from 0d-30d, and "O`Neal" has a lower weight loss ratio than "North Land". At 30d in "O`Neal" the control of the weight loss ratio increasing by 27.31% comparing to the 0d, while the 1-MCP treated groups (T2-T4) increased by 25.17%, 23.36% and 25.40%, comparing to 0d respectively; And in "North Land", at 30d the weight loss of control has increasing by 33.36%, while the 1-MCP treated groups (T2-T4) were increased by 29.81%, 24.90% and 26.31%, and has 3.55%, 8.46% and 7.05% lower than control respectively.

*4.3 Soluble solid, soluble sugar and titratable acid content of blueberry during 1-MCP treatments under cold storage*

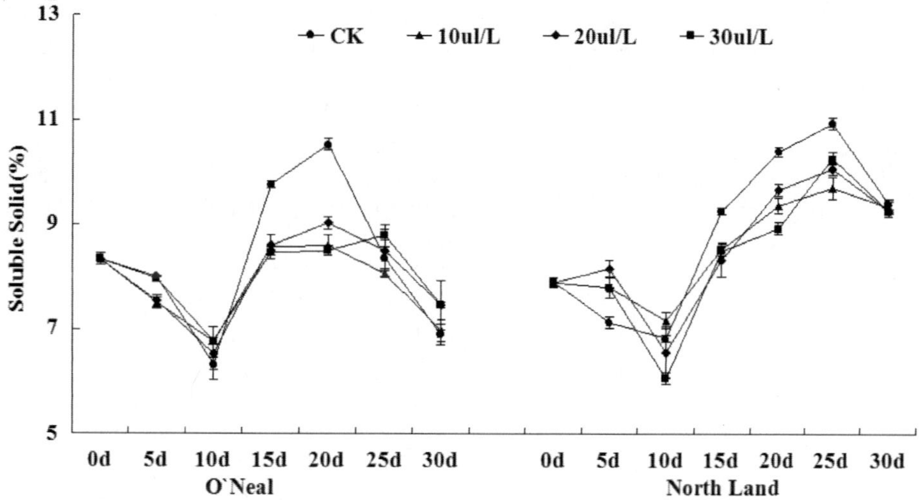


**Fig 3A.** Changes in the soluble solid content of blueberry

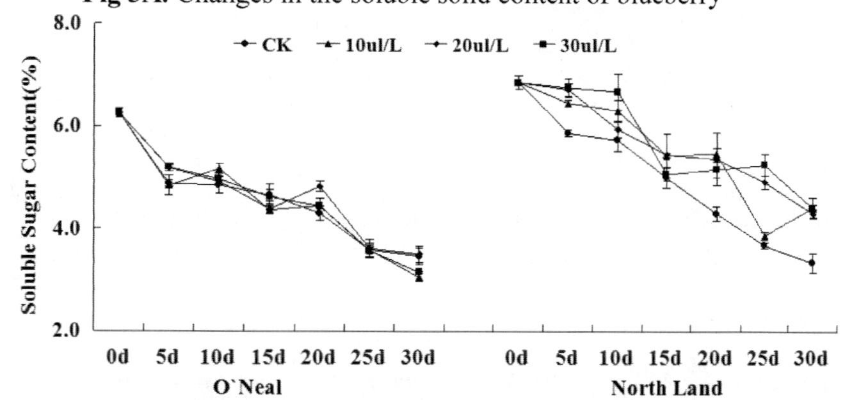


**Fig 3B.** Changes in the soluble sugar content of blueberry

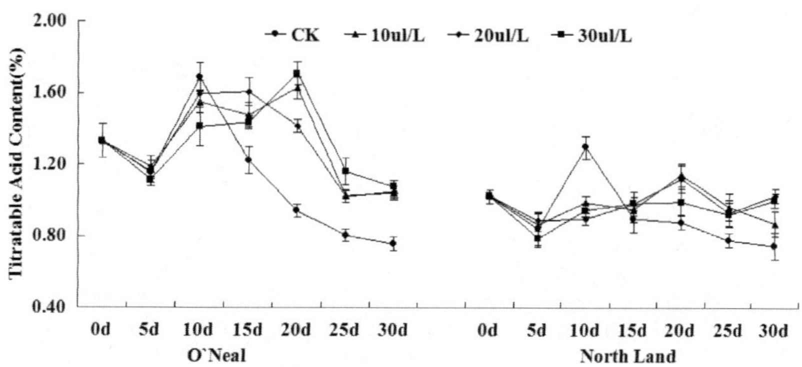


**Fig 3C.** Changes in the titratable acid content of blueberry

In order to investigate the internal quality of blueberry, we have measured the soluble solid, soluble sugar and titratable acid content (Fig.3A-Fig.3C). From 0d to 30d, soluble solid content was decreased and decreased again after the increased (Fig.3). Comparing to the control, 1-MCP treated groups(T2-T4) were in a more stable variation both in "O`Neal" and "North Land", while "O`Neal" has a lower variation than "North Land" during the storage.

Soluble sugar content was in a decline both in "O`Neal" and "North Land" (Fig.3B). At 30d, the soluble sugar content in "O`Neal" were decreasing by 2.79%, 3.20%, 2.75% and 3.10% comparing to the 0d respectively; While in "North Land", the soluble sugar content decreased by 3.51%, 2.44%, 2.54% and 2.44% comparing to the 0d respectively; Comparing to the "O`Neal", "North Land" has been more effected by the 1-MCP treatments.

Titratable acid content is a way to judge the quality stability of blueberry. Fig.3C showed that the titratable acid content decreased with the increase of storage days; While 1-MCP treatment groups has a more gently variation than the control both in "O`Neal" and "North Land"; In the meantime, the titratable acid content in "O`Neal" were higher than that in "North Land".

*4.4 MDA content of blueberry during 1-MCP treatments under cold storage*

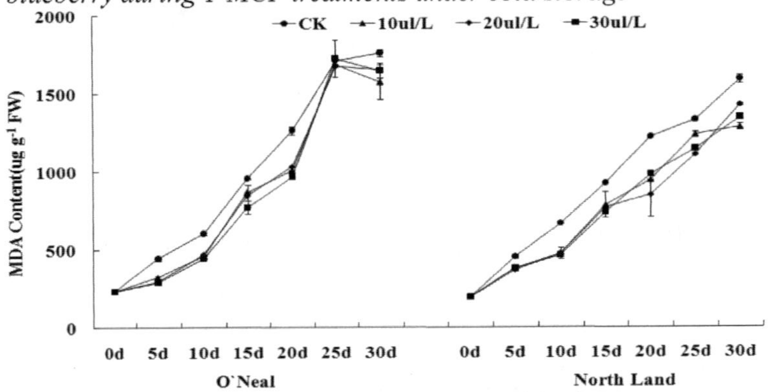


**Fig 4.** Changes in the MDA content of blueberry

The MDA levels under different treatments were analyzed (Fig 4). MDA increased gradually with lasting of days. The treatments (T1-T4) in MDA was as follows: 433.1%, 324.5%, 334.4% and 307.6% from 0d to 20d in "O`Neal" respectively. While in "North Land" from od to 30d the MDA content of treatments (T1-T4) was as follows: 718.3%, 560.5%, 634.2% and 591.8%, respectively. Under 1-MCP treatments the amounts of MDA in blueberry were lower than that of control.

*4.5 Antioxidant enzyme activities of blueberry during 1-MCP treatments under cold storage*

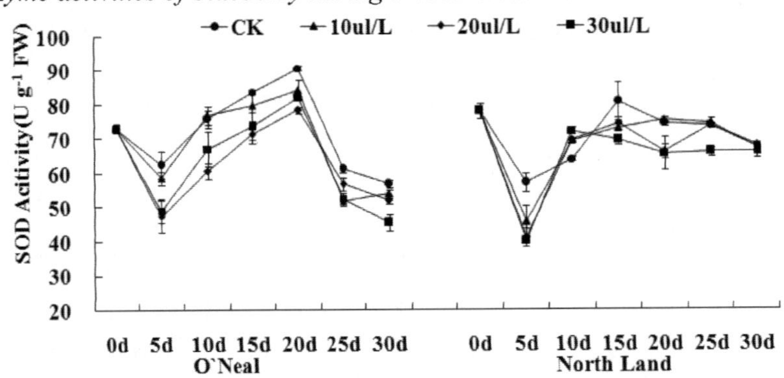


**Fig 5A.** Changes in the SOD activities of blueberry

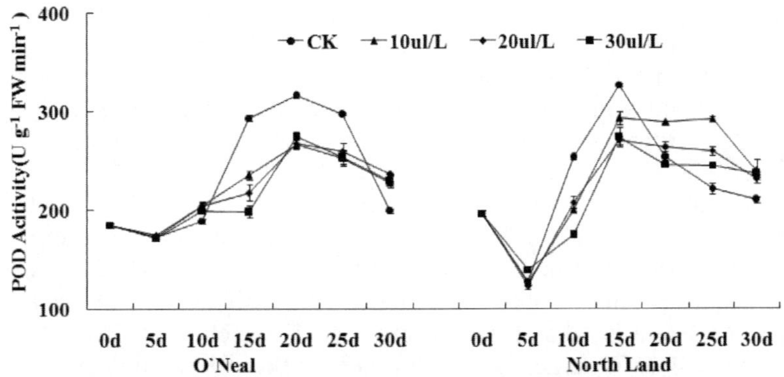


**Fig 5B.** Changes in the POD activities of blueberry

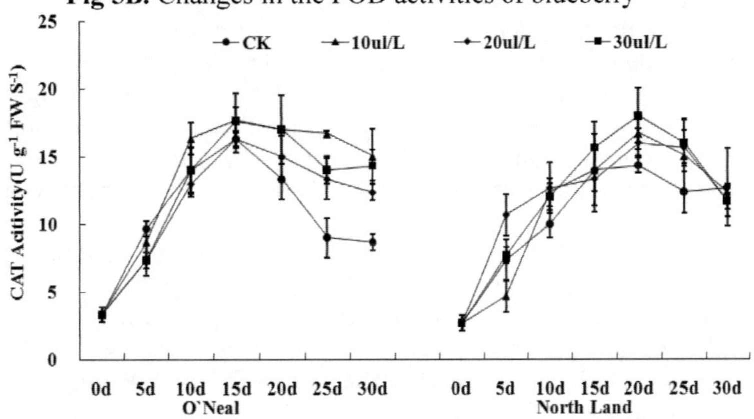


**Fig 5C.** Changes in the CAT activities of blueberry

The antioxidant enzyme activities in the blueberry are shown in Fig.5a -Fig.5c. The SOD activities of blueberry decreased and decreased after the increased (Fig.5a). At the early stages of the treatments (0d–5d), there were in a decline in SOD activities, while at 20d, the SOD activities decreased by 7.01%, 13.58% and 9.41% in "O`Neal" (T2-T4) comparing to the control respectively; While in "North Land", under the 1-MCP treatments (T2-T4) the SOD activities were lower than the control.

Under the 1-MCP treatments, the blueberry has a more stable variation of POD activities both in "O`Neal" and "North Land". In "O`Neal", under different 1-MCP treated (T2-T4), the POD activities were decreased by 19.70%, 15.47% and 14.84% in T2, decreased by 25.76%, 15.11% and 12.52% in T3, and decreased by 32.39%, 13.18% and 14.58% in T3, comparing to the control respectively at 15d, 20d and 25d. And at 30d, under 1-MCP treatments T1-T3 were have a 14.53%, 18.72% and 15.36% higher in POD activities than control; While in "North Land", T1-T3 were decreased by 21.05% and 10.37%, 17.98% and 17.18%, 30.92% and 16.16% at 10d and 15d, comparing to the control respectively; And at 25d and 30d T1-T3 increased by 31.83% and 14.02%, 17.29% and 10.32%, 10.53% and 12.70%, comparing to the control, respectively.

The CAT activities of blueberry decreased after the increased (Fig.5C). And under the 1-MCP treatments (T2-T4) the CAT activities were higher than control both in "O`Neal" and "North Land"; In "O`Neal", under the 1-MCP treatments (T2-T4) the CAT activities increased by 27.5%, 12.5% and 27.5% at 20d, increased by 85.19%, 48.15% and 55.56% at 25d and increased by 73.08%, 42.31% and 65.39% at 30d, comparing to the control respectively; In "North Land", under the 1-MCP treatments (T2-T4) the CAT activities increased by 16.28%, 11.63% and 25.58% at 20d, increased by 21.62%, 27.03% and 29.73% at 25d, comparing to the control, respectively.

## 5. Discussion

Blueberry is one of the most nutrition fruits in the world, while it has a very short period of storage time caused by the effects of the bacteria and mechanical damages [11]. Rotting rate and weight loss ratio are considered as the important indicator, showing the different quality of blueberry [12]. In this study, we can see that: different varieties of blueberry have different quality of shelf-stable; "O`Neal" was better than the "North Land" both in rotting rate and weight loss ratio, while 1-MCP can help the blueberry keep in a good stage, especially when it comes to the concentration of 20μl/L. 1-MCP reduced the water loss, dry matter loss and decay of blueberry, this is congruence with other studies on apple, waxberry and pear [13-14].

Soluble solid content, soluble sugar content and titratable acid content are the important inclusions of blueberry. In the fruits, soluble solid content will be increasing during the mature period, while decreasing when it has started to decay [8]. In this study, soluble solid content decreased from 0d to 10d, that was mainly because of the cold storage, while soluble solid content increased at 10d and decreased at 20d in "O`Neal" and at 25d in "North Land", which shows that "North Land" stays well during the storage and 1-MCP largely help the blueberry keep in a good state. In this study, soluble sugar content and titratable acid content were changing during the storage, while 1-MCP can keep the soluble sugar content and titratable acid content in a more steady state.

The increasing of MDA content was symptoms of damage and deterioration [15]. In this study, MDA content increased during the storage. However, the 1-MCP treatment significantly reduced the MDA content. To protect against oxidative damage, plants produce a variety of antioxidant enzymes, and lipid- and water-soluble molecules to scavenge ROS [16]. In this study, the SOD, CAT and POD activities of blueberry were induced by the 1-MCP, which was in agreement with the results reported in apple and pear [13-14].

As one of the ethylene inhibitor, 1-MCP play an important role in controlling the ripening and senescence of the non-climacteric fruits [17]. Other study also shows that 1-MCP can reduced the gas exchange, decreased the inside oxygen content and delaying senescence of the blueberry [18]. In this study, 1-MCP could keep the blueberry in a good stage in rotting rate, weight loss, soluble substances, MDA content and antioxidant enzyme activities, while different varieties of blueberry showed different during the storage.

## References

[1] Kader F and Rovel B 1996 Fractionation and identification of the phenolic compounds of highbush blueberries (*Vaccinium corymbosum* L.) *J. Food Chemistry* 55(1): 35-40.

[2] Ma Yongqiang and Shi Zhizhong 2006 The effects of 1-MCP on aging and quality of oil beans during the storage *J. Modern food science and technology* 22(4): 8-12

[3] Menniti A M, Donati I and Gregori R 2006 Responses of 1-MCP application in plums stored under air and controlled atmospheres *J. Postharvest Biology and Technology* 39(3): 243-246.

[4] Cai C, Chen KS and Xu WP 2006 Effect of 1-MCP on postharvest quality of loquat fruit *J. Postharvest Biology and Technology* 40(2): 155-162.

[5] Fang ZX, Hu J, Jiang B, Fang J, He Y, Lu W and Zheng BS 2013 Research progress on blueberry (*Vaccinium* spp.) *Journal of Zhejiang A & F University* 30: 599-606

[6] Hyang LE and Sun CH 2013 Influence of Temperature during Transport on Shelf-life Quality of Highbush Blueberries (*Vaccinium corymbosum* L. cvs. Bluetta, Duke) *J. Hort Environ. Biotechnology* 54:128-133

[7] Hyungmin R, Duk JY, Su JK and Hee JL 2012 Limitation Factors for Photosynthesis in 'Bluecrop' Highbush Blueberry (*Vaccinium corymbosum*) Leaves in Response to Moderate Water Stress *J. Plant Biology* 55:450-457

[8] Li J, Jiang Al and Hu WZ 2012 Research of respiratory metabolism and color change on postharvest blueberry *J. Modern gardening* 19:3-4

[9] Ying YQ, Yue YL and Huang XH 2013 Salicylic acid induces physiological and biochemical changes in three Red bayberry (*Myric rubra*) genotypes under water stress *J. Plant Growth Regul* 71:1-9

[10] Wen CF, Dong AW and Li GZ 2005 Detemination of Total sugar and Reducing Sugar in Viola Philippicassp Munda W. Becker by Anthron Colorimetry *J. Modern food Science and Technology* 3:122-130

[11] Zhou Q, Ma C and Cheng SC 2014 Changes in antioxidative metabolism accompanying pitting development in stored blueberry fruit *J. Postharvest Biology and Technology* 88: 88-95
[12] Gonzalez A, Valenzuela S, and Lizardi, MJ 2009 Effect of chitosan coating in preventing deterioration and preserving the quality of fresh-cut papaya'Maradol' *J. Journal of the Science of Food and Agriculture* 89(1): 15-23
[13] Sun XS and Wang WH 2001 Effect of 1-MCP on Cold Storage of Dangshansuli Pears *J. Storage & Process* 6:14-17
[14] Sun XS and Wang WH 2003 Effect of 1-MCP on Physiology of Apples after Harvest *J. Journal of Fruit Science* 20:12-17
[15] Feng Z, Guo A and Feng Z 2003 Amelioration of chilling stress by triadimefon in cucumber seedlings *J. Plant Growth Regul* 39:277-283
[16] Halliwell B and Gutteridge JMC 1999 Free radicals in biology and medicine *Oxford University Press. New York, NY*
[17] Ji SJ, Zhou Q and Ma C 2014 The effects of 1-MCP on blueberry shelf quality change under room temperature *J. Food Science* 35(2): 322-327
[18] Yang SF and Wang JJ 2013 The effects of chitosan coating on blueberry fruit quality in 'rabbit eyes' under normal temperature storage *J. Storage and Process* 6: 14-19

# Study on property and stability mechanism of LAB-AEO-4 system

**Kaifei Song,Jijiang Ge,Yang Wang,Guicai Zhang,Ping Jiang**

School of Petroleum Engineering, China University of Petroleum, Qingdao 266580

**Abstract.** The behaviors of binary blending systems of fatty alcohol polyoxyethylene ether (AEO-4) blended with the laurel amide betaine (LAB) was investigated at 80°C,the results indicated that the optimal ratio of the mixed system of LAB-AEO-4 was 5:2.The stability mechanism of LAB-AEO-4 system was analyzed from three aspects of dynamic surface tension,gas permeation rate and surface rheology.The results showed that the tension of mixed system was easier to achieve balance,the constant of gas permeation rate of the mixed system decreased by about 7% and the elastic modulus and dilational modulus increased by about 2 times compared with the single LAB system.

## 1. Introduction

Anionic and non-ionic surfactant is most widely used in tertiary oil recovery[1],the alkylbenzene sulfonate has been used in tertiary oil recovery of Daqing field and has achieved good economic benefits[2].For high temperature and high salinity reservoirs such as Zhongyuan Oilfield,the temperature reaches 85°C,the concentration of salinity more than $5\times10^4$mg/L,the concentrations of calcium ion and magnesium ion more than $0.1\times10^4$mg/L;the temperature of Tahe oilfield also more than 90°C,the concentration of salinity is more than $1.7\times10^4$~$2.1\times10^4$mg/L[3-4].Anionic surfactant may occur saltout phenomenon under high salinity;nonionic surfactant due to its low cloud point may precipitate under high temperature.Therefore,these two surfactants are not suitable for the high temperature and high salinity reservoirs.It is very important to study the heat-resistant and salt-tolerant surfactant system for the high temperature and high salinity reservoir[5].

Betaine is one of the most common amphoteric surfactants which has been used in the tertiary oil recovery,and with a good heat-resistant property;fatty alcohol polyoxyethylene ether nonionic surfactant (AEO-4) is the fastest growing nonionic surfactant and with a good anti-salt performance,but with poor heat-resistant and foaming property[6-7].Therefore,study of the properties of LAB-AEO-4 system can make up their weakness,which has an actual value for the exploitation of the high temperature and high salinity reservoir[8].

## 2. Experimental materials

Laurel amide betaine (LAB,35wt% aqueous solution);fatty alcohol polyoxyethylene ether（AEO-4,35% aqueous solution）;NaCl (99wt%) and $CaCl_2$ (98wt%),all materials were obtained from China National Pharmaceutical Group Corporation.

DSA100 interfacial rheometer (German KRUSS company);501A super thermostatic bath(Shanghai Precision Instruments Co.,Ltd.);waring blender Agitator;Precision balance （0.0001g）; Constant temperature water bath;Cylinders (100mL、1000mL);Stopwatch and etal.

IOP Conf. Series: Earth and Environmental Science **61** (2017) 012042 doi:10.1088/1755-1315/61/1/012042

## 3. Results and discussion

### *3.1 Study on the foam properties of LAB-AEO-4 system*

The surfactant solution was prepared with simulated formation,the concentration was 1g/L,waring blender agitator was used to produce bubble,the speed and time were 3000r/min and 60s respectively,then recorded the change of bubble volume as a function of time,the concentration of LAB was 1g/L,and the concentration of AEO-4 was changed.Fig.1 shows the optimal proportion of the system is 5:2 at 80°C.

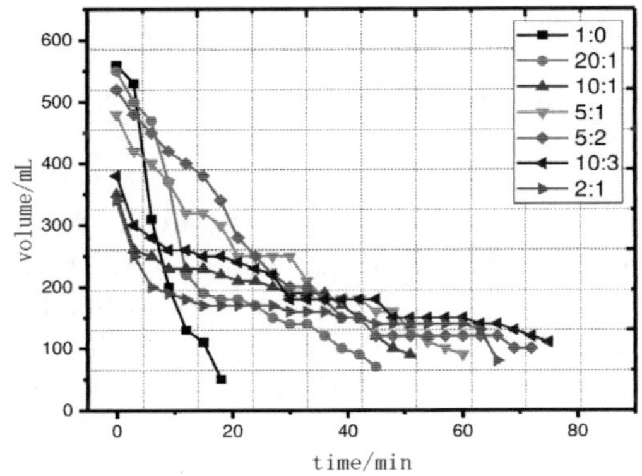


Fig.1 Bubble curve of mixed system of LAB-AEO-4 at 80°C

### *3.2 Dynamic surface tension*

Dynamic surface tension experiments of LAB system and LAB-AEO-4 system have been done at 25°C with the DSA100 interfacial rheometer obtained German KRUSS company,and the concentration of LAB was 1g/L,Fig.2 shows the result.

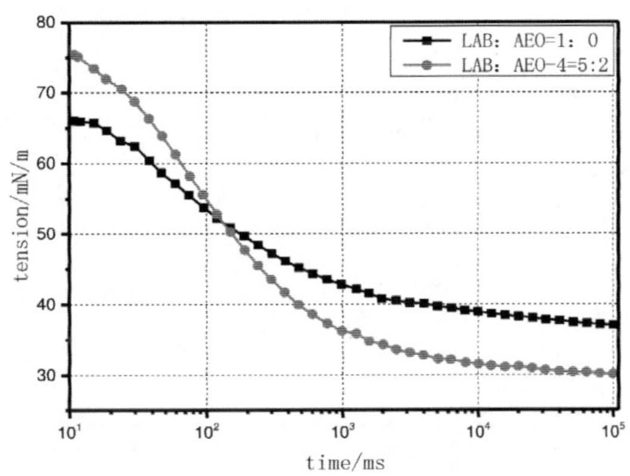


Fig.2 Curve of the dynamic surface tension

The dynamic surface tension curve can be divided into four zones:the induced area,the rapid subsidence area,the over balanced region and the equilibrium region[9].Using equation (1) to fit and analyze the first three zones,then obtained the Curve of the dynamic surface tension fitting,and calculate the dynamic surface tension decrease rate n、t、t* and $t_{1/2}$,which were used to measure dynamic interfacial adsorption process.

$$\gamma_t=\gamma_m+(\gamma_0-\gamma_m)/[1+(t/t^*)n] \quad (1)$$

IOP Conf. Series: Earth and Environmental Science **61** (2017) 012042 doi:10.1088/1755-1315/61/1/012042

$$R_{1/2}=\gamma_0-\gamma/2\ t_{1/2} \qquad (2)$$

$\gamma_m$:surface tension under excessive equilibrium;$\gamma_t$:dynamic surface tension at any moment;$\gamma_0$:initial surface tension;t*:constant;n:no unit constant;$R_{1/2}$:the decreasing rate of dynamic surface tension at $t_{1/2}$.

The dynamic adsorption index were calculated as shown in Table 1.The result shows that the mixed components of LAB and AEO-4 do not effectively accelerate the decline of the surface tension,while the surface tension of single LAB system decreases slightly.The possible reason is that the interaction of the two surfactant molecules increases the molecular mass of the complex,thereby reducing the diffusion coefficient of the surfactants.Therefore,the adsorption of the new interface is decreased,and the volume of foam is lower than that of the single surfactant.

**Table 1** the dynamic surface adsorption index of the LAB-AEO-4 system

| m(LAB):m(AEO-4) | n | t* | $R_{1/2}$ |
|---|---|---|---|
| 1:0 | 1.2 | 46.8 | 46 |
| 5:2 | 0.6 | 75.8 | 39 |

*3.3 Gas permeability*

As everyone knows that the pressure in the small bubbles is higher than that in the large bubbles due to the existence of capillary pressure.Therefore,the gas from the high-pressure bubbles through the liquid film diffuses into the low-pressure bubbles,which results that the small bubbles become smaller,the large bubbles become larger,and bubble burst at the end.In order to further learn the stability mechanism of the mixed system, the gas permeation rate of LAB system and LAB-AEO-4 mixed system were studied.Curve $r_2$~t was drawed according to equation（3）,Fig.3 shows that the curve slope of LAB system and LAB-AEO-4 system were -0.00764 and -0.00752 respectively.Surface tension of LAB system is 37mN/m,and the surface tension of LAB-AEO-4 mixed system is 30mN/m,by calculating the gas permeability constant of LAB system and LAB-AEO-4 system are 0.014cm/s and 0.013cm/s respectively.The experimental results show that the gas permeability of LAB system is stronger than that of LAB-AEO-4 system system,the reason may be that the liquid film of LAB-AEO-4 system become more compact due to the presence of AEO-4.

$$r^2=r_0^2-3kt\gamma/p \quad (3)$$

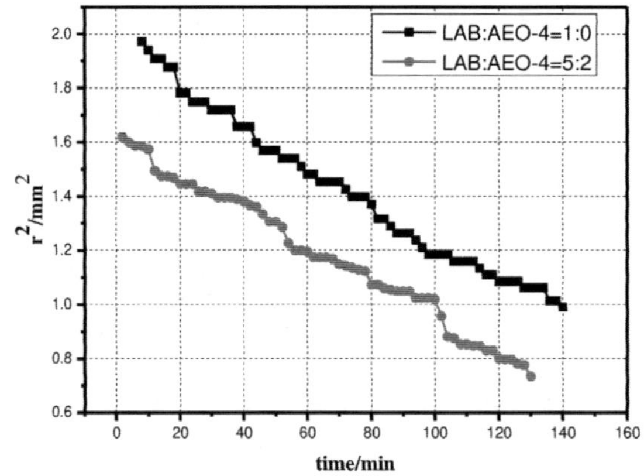


Fig.3 Curve of bubble radius change with time

*3.3.3 Surface dilational rheology*

Study on the surface dilational rheology of LAB-AEO-4 system and LAB system at 25°C,Fig.4 indicates that when the interface deformation rate is relative low, there is enough time to repair interfacial tension gradient through the relaxation process which caused by interfacial area change,so the dilational modulus slows down with the deformation speed decreases.When the dilational frequency is smaller than the characteristic frequency,the double logarithm curve of the expansion frequency and dilational modulus is almost linear,the change range of curve slope is different for different relaxation process.The results show that the slope of the line is less than 0.5,which indicates that the main relaxation mechanism is the diffusion relaxation process between the interface and the bulk phase[10].One of the key factor that affects the stability of foams is the film strength which mainly depends on the interaction between molecules that adsorb on the film surface,the elastic modulus is determined mainly by the interfacial molecular interaction,the interaction is greater,the greater the elastic modulus.

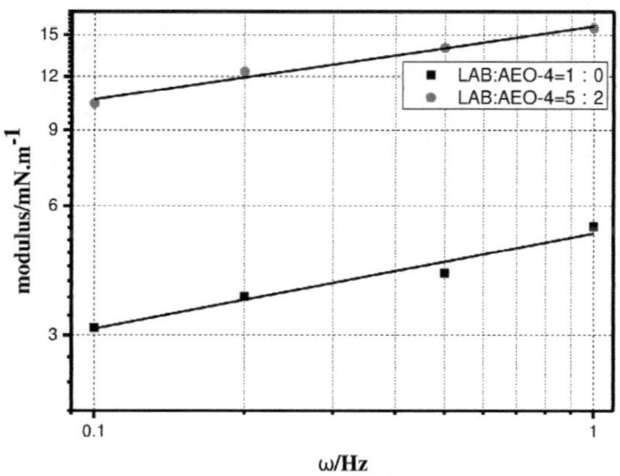


Fig.4 Effect of frequency on dilational modulus of different systems

## 4. Conclusion

The optimal mixed ratio of LAB-AEO-4 system was 5:2 at 80°C,the foam volume was 500mL,and the stable time was 30min.The tension of mixed system is more easy to achieve balance,the gas permeation rate constant of the mixed system decreased by about 7% and the elastic modulus and dilational modulus increased by about 2 times compared with that of the single LAB system.

## Reference

[1] Zhao Guoxi,Zhu Buyao.Principles of Surfactant Action [M].Beijing:China Light Industry Press，2003.
[2] Cui Zhenggang,Liu Shixia,He Jianglian.A STUDY ON THE STATIC ADSORPTION OF HEAVY ALKYLBENZENE SULFONATE ON RESERVOIR SAND OF DAQING OIL FIELD[J].Oilfield Chemistry,2000,17(4):359~363.
[3] Li Liyong,Zhou Zhong,Cui Zhenggang,Wang Jianxin.,Study of Aliphatic Alcohol with Thermal Stability and Salt Resistance[J].Advances in Fine Petrochemicals,2008,9(1):4~7.
[4] Tian Jiang,Yang Zhe,Fang Jicha,et al.Optimization and performance evaluation of high efficiency foamingagent for high-temperature and high-salinity harsh reservoir of Tahe Oilfield[J].Science & Technology Review,2015,33(4):66~71.
[5] Chen Taoping,Zhang Zhiqi.Displacement mode of ultralow interfacial tension system in heterogeneous low permeability oil layers[J].Journal of Daqing Petroleum Institute,2009，33(5):55~59+120~121.
[6] Zhang Fan,Zhang Qun,Zhou Chaohui,Zhu Youyi,Liu Chunde,Kong Ruilei. Characterization and Property of a Heat-Resistant and Salt-Tolerant Betaine Surfactant [J].Oilfield Chemistry ， 2011,28(4):427~430.
[7] Chen Xirong,Huang Fengxing.Research Progress of Temperature-Resistant and Salt-Tolerant Surfactants for Enhance Oil Recovery [J].Petrochemical Technology,2010,39(12)：1307~1312.
[8] Zhao Guozhu,Jin Rengao,Li Shengtao,et al.Performance and application of betain type surfactant TCJ-5 in higher temperature and higher salinity reservoir[J].Journal of Northeast Petroleum University,2015,39(3):111~117+7~8.
[9] D. Kawale,A.T.van Nimwegen,L.M.Portela,et al.The relation between the dynamic surface tension

and the foaming behaviour in a sparger setup [J].Colloids and Surfaces A: Physicochemical and Engineering Aspects,481（2015）328~336.

[10] Nikolai D.Denkov,Slavka Tcholakova， Konstantin Golemanov， et al.The role of surfactant type and bubble surface mobility in foam rheology [J].Soft Matter,2009,5,3389~3408|3389.

IOP Conf. Series: Earth and Environmental Science **61** (2017) 012043 doi:10.1088/1755-1315/61/1/012043

# Synthesis and multi-response research of a highly selective fluorescent chemosensor for $Zn^{2+}$

**Ming Li* and Rujie Wang**

Department of Environmental Science & Engineering, North China Electric Power University, Baoding, China
ming2999@126.com

**Abstract**. A new 1,2,3-triazole receptor bearing quinoline groups have been prepared *via* "click" reaction. The compound shows significant fluorescent response to $Zn^{2+}$ over other cations in $CH_3CN$ solvent. Besides, electrochemical properties including cyclic voltammetry (CV) and differential-pulse voltammetry (DPV) tests are operated to reveal the complex also has good electrochemical response toward $Zn^{2+}$.

## 1. Introduction

Zinc and its derivatives play an essential role in many biological processes in human-beings, and it is the second most abundant transition metal ion in human body after iron [1]. Many diseases, like Alzheimer's disease, and epilepsy are associated with the disorder of $Zn^{2+}$ metabolism [2]. Therefore, measurement of $Zn^{2+}$ is important in the field of biology.

Sensitive chemsensors have attracted more and more attention in recent years because of their potential applications in analytical chemistry, biochemistry, environmental monitoring and convenience of use [3]. Among numerous synthetic methods, Cu(I)-catalyzed "click" reaction has provided an easy strategy for the linkage of 1,2,3-triazole which has good coordination abilities with metal cations to various functionalities. Besides, quinoline groups are rich in optical properties that can be detected easily in fluorescent tests. In this research, we report a new chemosensor toward $Zn^{2+}$ containing both triazole and quinolone groups using ferrocene as the structural matrix.

## 2. Experiment

Synthesis of 8-(prop-2-ynyloxy) quinoline (1). 8.7g (60 mmol) 8-hydroxyquinoline was reacted with 8.5g (70mmol) 3-bromopropyne in 100 mL acetone with 30.0g (220mmol) $K_2CO_3$ as a catalyst. The mixture was stirred in ice-salt bath for three days. When the reaction finished, the solvent was removed under vacuum, and the crude product was washed with petroleum ether for several times to remove the unreacted 8-hydroxyquinoline. 6.5g liquid compound 1 with a deep yellow colour was obtained in 46% yield.

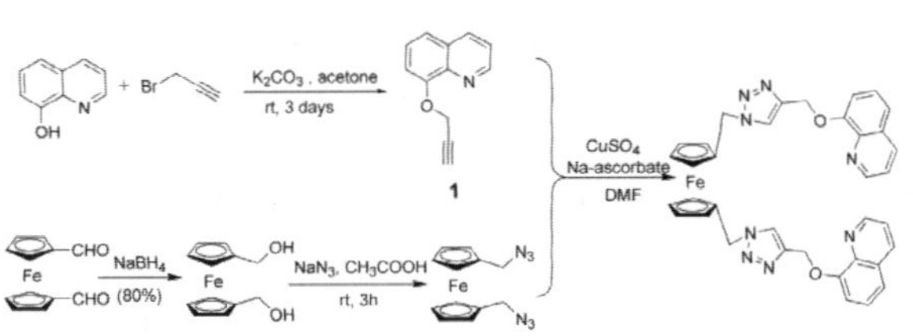


**Figure 1.** Synthesis route of the title compounds

Synthesis of 1,1'-ferrocenedimethanol (2). The reactant 1,1'-ferrocenedicarboxaldehyde was synthesized by literature method [4]. 1.55g (6.4mmol) 1,1'-ferrocenedicarboxaldehyde was dissolved in 25 mL methanol stirring for 10 minutes at 0 °C. Then, 0.54g (14.1mmol) $NaBH_4$ was added into the mixture in batches for another 2 hours of stirring at room temperature. After the reaction, the solvent was removed under vacuum, the solid obtained was extracted with ethyl acetate. The organic layer was evaporated and subjected to the purification by silica-gel column chromatography to get 1.1g yellow needle-like crystals with a yield in 71% (m. p.=105-107 °C).

Synthesis of 1,1'-diazidomethylferrocene (3). 2.4g $NaN_3$ was added into 40 mL acetic acid solution of 0.3g (1.58mmol) 1,1'-ferrocenedimethanol, the mixture was stirred for 3 hours at room temperature. After the reaction, the mixture was diluted with ethyl acetate, and washed with $NaHCO_3$ saturated aqueous solution. Then, the organic layer was evaporated and the crude product was purified with silica-gel column chromatography to give the compound 3 as yellow syrup.

Synthesis of the title compound L1. A mixture of 8-(prop-2-ynyloxy) quinoline (293mg, 1.6mmol) and 1,1'-diazidomethylferrocene (190 mg, 0.64 mmol) was dissolved in 10mL DMF. Then, 64mg (0.26 mmol) $CuSO_4$ and 51 mg (0.26 mmol) sodium ascorbate were added into the above solution as a catalyst. The mixture was stirred for 8hours under nitrogen atmosphere. After the reaction, NaCl saturated aqueous solution was added into the resultant mixture, lots of yellow flocculent precipitate appeared immediately. The mixture was filtered and washed with water for several times to harvest the yellow solid L1 (386mg) with a high yiled in 91% (m.p.=93-96 °C). $^1$HNMR ($CDCl_3$) δ(ppm): 8.90(d, *J*=2.8Hz, 2H), 8.11(d, *J* =8.0Hz, 2H), 7.67(s, 2H), 7.36-7.43(m, 6H), 7.33(d, *J*=7.2Hz, 2H), 5.53(s, 4H), 5.16(s, 4H), 4.15(d, *J* =14.4Hz, 8H). IR (KBr, $cm^{-1}$): 1632 $cm^{-1}$, 1569 $cm^{-1}$, 1500 $cm^{-1}$, 1468 $cm^{-1}$, 1260 $cm^{-1}$, 1050 $cm^{-1}$.

## 3. Results and discussion

In this work, we have synthesized a novel ferrocene-based multi-response chemosensor (**L1**) for $Zn^{2+}$ detection using the "click" reaction method. The synthesis route is shown in Figure 1. To obtain an insight into the sensing properties of **L1** towards metal ions, the UV-visible absorption spectra are examined with different cations such as $Li^+$, $Na^+$, $Mg^{2+}$, $Ba^{2+}$, $Hg^{2+}$, $Cr^{3+}$, $Zn^{2+}$, $Ni^{2+}$, $Co^{2+}$, and $Ag^+$ in $CH_3CN$ (see Figure 2). There is a broad band at 300nm in the UV-vis absorption spectrum of **L1** in $CH_3CN$. Upon the addition of $Zn^{2+}$ to 2 equivalents, the absorption band gradually decreases until disappears, and a new absorption band appears at 312nm, which could be due to the coordination of **L1** and $Zn^{2+}$. Furthermore, the isosbestic point at 305nm implies the undoubted conversion of free **L1** to a zinc complex.

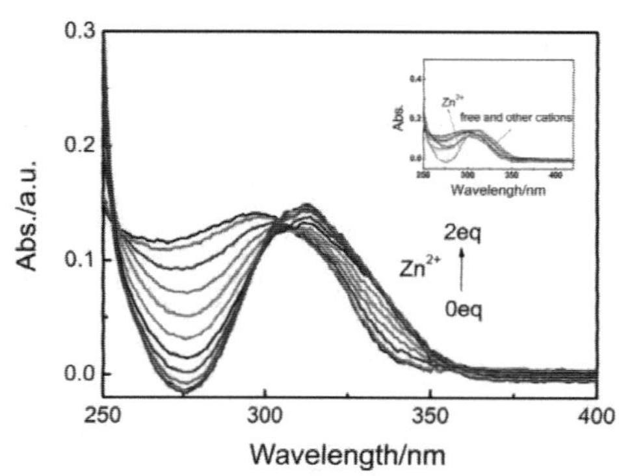


**Figure 2.** UV-vis spectra of **L1**(2×$10^{-5}$M) upon addition of 0-2 equiv. of $Zn^{2+}$ in $CH_3CN$ solution. Inset: UV-vis spectra of **L1** upon addition of 2 equiv. of various of cations in $CH_3CN$ solution.

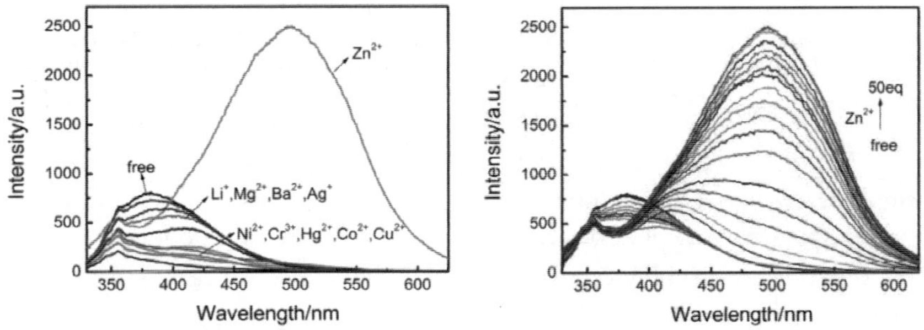


**Figure 3.** Emission spectra ($\lambda_{ex}$=320nm) of **L1** ($2\times10^{-5}$M) upon addition of 50 equiv. of various of cations in $CH_3CN$ solution (left); Changes in emission spectra of **L1** ($2\times10^{-5}$M) upon addition of 0-50 equiv. of $Zn^{2+}$ in $CH_3CN$ solution (right).

The emission titration experiments (see Figure 3) reveals that **L1** shows distinct fluorescent sensing to $Zn^{2+}$, which is consisted with the UV-vis results. When **L1** is exposed in 50 equiv. of various of cations in $CH_3CN$ solution, only **L1**-$Zn^{2+}$ couples shows a great increasing fluorescent emission which is from the new emission band at 496nm. By contrast, **L1**-other cation couples quench their emissions at 381nm which may be ascribed to the quenching effect of a photo-induced electron transfer process from quinolone to triazole [5]. In fact, the emission band of **L1** is also quenched by the addition of $Zn^{2+}$at the first stage as the other **L1**-cation couples, when the large amount of $Zn^{2+}$ added, the new emission band with a distinct red shift appears. It may be due to the formation of rigid structure caused by the chelation from the neighbouring triazoles to $Zn^{2+}$.

Ferrocene-contained compounds often have good electrochemical properties. Cyclic voltammetry (CV) and differential-pulse voltammetry (DPV) titrations are operated for several cations to **L1** (see Figure 4). The electrochemical data are listed in Table 1. The oxidation potential shifts 79mV positively with the reduction potential shifts 47mV positively in the CV curves when $Zn^{2+}$ is added (68mV for DPV test), which could be attributed to the difficulty of ferrocene-oxidation, a result of coordination between **L1** and $Zn^{2+}$.

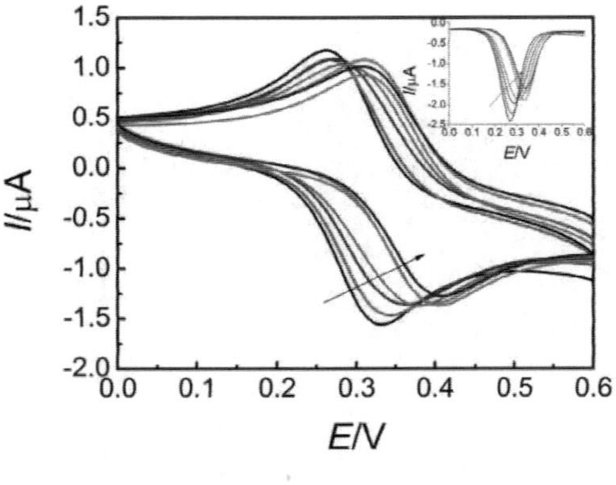


**Figure 4.** Cyclic voltammetry (CV) and differential-pulse voltammetry (DPV) tests for the $Zn^{2+}$ titration to **L1** ($2\times10^{-4}$M).

**Table 1.** Cyclic voltammetry (CV) and differential-pulse voltammetry (DPV) data for the titration of several cations to **L1**.

| *Compound* | *E/V(CV)^aEpa* | *E/V(CV)Epc* | $E_{1/2}$ | $\Delta E_{1/2}$ | *E/V(DPV)^b* |
|---|---|---|---|---|---|
| **L2** | 0.332 | 0.261 | 0.297 | | 0.272 |
| +$Hg^{2+}$ | 0.366 | -- | -- | -- | 0.368 |
| +$Cu^{2+}$ | 0.389 | 0.299 | 0.344 | 0.047 | 0.320 |
| +$Zn^{2+}$ | 0.411 | 0.308 | 0.360 | 0.063 | 0.340 |
| +$Cr^{3+}$ | 0.331 | 0.264 | 0.298 | 0.001 | 0.276 |

| | | | | | |
|---|---|---|---|---|---|
| $+Ni^{2+}$ | 0.382 | 0.292 | 0.337 | 0.040 | 0.312 |

[a] CV, $E_{1/2}$ and $\Delta E$ vs. $Ag^+/Ag$, the $E_{1/2}$ of unsubstituted ferrocene is 0.037 V vs. $Ag^+/Ag$.
[b] DPV, $E_{dpv}$ vs. $Ag^+/Ag$, 50 ms pulse width.

## 4. Conclusion

A novel ferrocene-based multi-response chemosensor **L1** is synthesized successfully. UV-vis spectra, fluorescent emission spectra, and electrochemical tests reveal the compound shows high sensitivity and selectivity toward $Zn^{2+}$. Designing based on the present studies could be further modified to develop more systems for ion probes.

## Acknowledgments

The authors thank the financial support of the Fundamental Research Funds for the Central Universities (Grant nos. 2016MS111 and 2016MS110).

## References

[1] Fraker P J and King L E 2004 *Annu. Rev. Nutr.* **24** 277-298
[2] Walker C F and Black R E 2004 *Annu. Rev. Nutr.* **24** 255-275
[3] Liu M P, Liu T, Li Y, Xu H, Zheng B Z, Wang D M, Du J, and Xiao D 2015 *Talanta* **143** 442-449
[4] Balavoine G G A, Doisneau G, Fillebeen-Khan T 1991 *J. Organomet. Chem.* **412** 381-382
[5] Cao Q Y, Wang Z C, Li M and Liu J H 2013 *Tetrahedron lett.* **54** 3933-3936

# Knowledge spillover and high-tech industry cluster: A literature review

**Shen Qian**

Faculty of Economics, Inner Mongolia University of Finance and Economics, Hohhot 010030, China

**Abstract**. Knowledge spillover is one of the incentive forces for companies using more expenditure on R&D to shape cluster. This paper summarizes and reviews academic research literature from three aspects, of which contain the connotation of knowledge spillover in high-tech industry cluster, conduction mode of knowledge spillover effect in high-tech industry and explain factors of knowledge spillover effect. This paper devotes by assembling anterior divergent tributaries of works to assist in illuminating on this prosperous research area, based on previous research, and finally it notes that future research needs to pay attention to several aspects.

## 1. Introduction

High-tech industry cluster is the most important organization in current economic and social development. Knowledge spillover that is one of the main drivers of formed agglomeration benefits is to explain one of the most vital concepts of enterprise cluster. This paper mainly utilizes Internet academic resources revolving both of the knowledge spillover effect and high-tech industry cluster. It adopts key bibliographic method and reviews the relative databases, such as Wiley Online Library, Springer Link, Engineering Village and Web of Science.

## 2. Connotation of knowledge spillover in high-tech industry cluster

### *2.1. Knowledge spillover meaning*

Knowledge spillover is a transaction of concept, method and theory among individuals and public. Knowledge belongs to a public product with non-competence and externality. Polanyi (1997) mentions that knowledge can be divided into two categories, explicit knowledge and tacit knowledge. Knowledge has three external features as public product. The first one is uncertainty and because knowledge exists in the real world, however, the owner of the knowledge or intellectual has limitations. Second, knowledge obtained from different people in diverse location and on heterogeneity time point holds complement with space and time. Third, because any subject have access to knowledge through mastering, creating and applying. These characters interpret the necessity of knowledge spillover occurrence. Knowledge spillover is a value of intellectual property that is through information exchange and do not or less compensate for the creators. Knowledge spillover is a process through which different subjects interact and communicate and it broadcasts with unconscious during this process.

### *2.2. Knowledge spillover characteristics of high-tech industry cluster*

In knowledge executive economics, knowledge spillover that is in non-rival market has technological advancements through modernization which generally derives from expertness within a high-tech industry. The relationship between knowledge spillover and industry cluster is mutually reinforcing and cumulative cyclic causality. Knowledge spillover prompts industry spatial agglomeration, and spatial concentration improves the economic exchange of thought and consciousness.

Feldman and Audretsch (1999) has put forward that spatial concentration declines the possibility of the scientific discoveries and the scientific commercialization and promotes growth of cluster development and innovation output.

As one of the important drivers of high-tech industrial cluster, Chyi, Lai and Liu (2012) has pointed out that knowledge spillover reduces the high technology cost of industrial innovation to a large extent and promotes the strengthening of the high-tech industry cluster. Knowledge spillover of high-tech industrial agglomeration is an organization of high-tech industry knowledge especially tacit knowledge.

Chang, Wang and Liu (2016) proposed that knowledge spillover spreads, transfers and accesses between different organizations and dissemination in the same area. Tacit knowledge determines the high-tech industries tendency of cluster growth on the grounds that high-tech industries with high innovation potential. Relatively speaking, areas owning substantial stock of knowledge have more knowledge absorbing abilities. High-tech industry cluster is much more prone to form innovation clusters, networks and accelerate diffusion and dissemination of knowledge. Thus it will improve innovation ability and realize industry structure upgrading. Therefore, high-tech industry cluster and knowledge spill interaction and strengthening will accelerate the growth of high-tech industry.

## 3. Conduction mode of knowledge spillover effect in high-tech industry

### *3.1. Origin of knowledge spillover*

Jinji and Mizoguchi (2016) discusses that knowledge spillover always happens in adjacent area. So what is the root of knowledge spillover? There are three respective sources to illustrate the knowledge spillover phenomenon. Knowledge of potential difference refers to a gap between disparate subjects at a fixed time point. This gap causes the knowledge spillover. Although all stock knowledge is freely available and space and time are limitless, the information of specific knowledge is still constrained. There must be knowledge differences in any two individuals for the reason that individual has general differences in capacities, opportunities and conditions, even if they have the same level and stock of knowledge. Knowledge can be obtained through personal channels (including work, social relationships and everyday life). Only some people those who have different knowledge stocks manage to turn knowledge into new inventions or scarce resources, and they develop relatively into skilled workers, scientists, engineers, and so on. These technical person who enter into diverse industries and sectors inaugurate knowledge application and technology innovation. Hence the knowledge of potential difference between sectors and industries is created.

Endogenous economic growth theory concludes that technological innovation is the endogenous variable of economic growth. Superficiality of knowledge spillover and increasing returns scale of knowledge accumulation are two important factors for economic growth. As a productive factor acting on the economic growth, knowledge spillover can cause increasing marginal product of capital. We may assume that other factors are the same and then research and development efforts of the body of economy and innovation differences determine in large extent the gap of economic growth. Economic performance will be improved if industries have access to obtain knowledge without cost on the basis of technological innovation. Therefore, knowledge spillover is the critical ingredient to economic prosperity.

Romer (1994) has said that technical knowledge inputted into production process has significant different with physical capital, such as machinery and equipment inputted into the same process.

IOP Conf. Series: Earth and Environmental Science **61** (2017) 012044 doi:10.1088/1755-1315/61/1/012044

Technical knowledge has the public characteristics of non-competition and non-ranked, but it is not a pure public goods. Because of the non-competitive of technical knowledge, it's hard to avoid spillover effects. Moreover, technology innovation of the knowledge spillover should not be ignored.

### *3.2. Stage process of knowledge spillover in high-tech industry agglomeration*

Knowledge life cycle theory can be used to throw light upon the concept of knowledge generation. Hervas-Oliver, Lleo and Cervello (2017) has put forward that knowledge generation mainly refers to a notion that new ideas and innovations are propagated between people and people, people and enterprises or interactions with external information sources in high-tech industry. Knowledge validation makes reference that publication of new knowledge always goes through a process of accepting, reviewing, testing and approving from other agencies of high-tech industry cluster. After such a long period, it is a previous phase that new knowledge turns into collective knowledge. Subsequently, knowledge synthesis happens. It substantially interpolates a procedure that a transform from new knowledge occurring in industry clusters to integrated knowledge. And this change principally relays on knowledge sharing, transferring, educating and training. Knowledge attenuation of updating refers to knowledge clusters gradually within the curing and storing. Part of it reflecting aging tendency has not adapted to environment changes both inside and outside the high-tech industry organization. Other part gets integrated, updated and added value by spillover and absorptive.

### *3.3. Knowledge spillover approaches in high-tech industry cluster*

Knowledge spillover of high-tech industry proceeds in a variety means. We can divide knowledge spillover into several different ways according to knowledge overflow channel. Liu and Buck (2007) indicates that many scholars classify talent flow that is actors and vectors of knowledge in industry cluster as important ways of knowledge spillover. Relative studies encounter that most manager, engineers and other technical personnel stay in the original region except scientists who live both in the same sector and area or in the same department of different area. Knowledge spillover caused by knowledge and expertise flows in high-tech industry has two kinds. One comes from equivalent talent flow in industry cluster, and the other originates from different regional high-tech industrial agglomeration. China economist, Zhang (2017) discusses that constant innovation pressure and motivation of high-tech industry hail from itself faster upgrading of high returns and product features.

Gathering high-tech enterprises tend to descend low risk of cooperation on research and development, resulting in knowledge spillover and re-engineering. The production-study-research cooperation between researchers in research and development will enable enterprises, universities and research institutes exchange heterogeneous knowledge, based on the research and development of the process or products. General corporate researchers, scientists and research institutions in the university and engineering harbor twin goals that thinking technology subjects will seek to lower costs and high output. In high-tech industry, this decision itself will accommodates high-tech industry agglomeration district effect. Start-up companies will make full use of high-tech industry agglomeration district institutional spillover of knowledge and experience. It certainly finish knowledge creation process in generating new ideas and skills within distinctive groups.

## 4. Explain factors of knowledge spillover effect

### *4.1. Environmental factors*

Lauren, Fielding, Smith and Louis (2016) emphasizes that environmental factors mainly includes government polices on environment and cultural construction of clusters. In order to overcome failure of knowledge overflow, government takes up policies, laws and regulations by means of intervention in the economy. Also, environmental factors conclude customs, cluster cultures and social capital, path dependence formed by them. These factors are such as cultural blessedness, cognitive trap talk. Policies, rules, customs, and culture not only can reduce the opportunism of clusters in knowledge spillover, but also can guide the behavior of enterprises. From two aspects of incentive and restriction, environmental factors implement knowledge spillover in the cluster.

### *4.2. Structural factors*

Puškárová and Piribauer (2016) gives priority to structural influence factors that they incorporate knowledge, structure of cluster, absorption capacity, intellectual properties, and they are associated with specific clusters of industry characteristics. High-tech industries possessing rapid technological transform, short product life cycle and product and process innovation performance characteristics, compared with traditional companies, confront more risks. In the process of market competition and speedy operation in science and technology, any one company cannot continue have forever advantages in technical areas. Every step of the innovation of contemporary high-tech enterprises are required to to be supported by external knowledge sources. Providing the basis for technological innovation in high-tech industries, the knowledge spillover effect of industrial clusters make innovation, technology application and the proliferation more efficient and faster.

### *4.3. Behavior factors*

Kuruppuarachchi and Premachandra (2016) highlights behavioral factors which is an appropriate strategic behavior that high-potential enterprises prevent or promote their own knowledge spillover. The main methods are internal (cooperation, vertical integration, asset investment). There is a certain risk for enterprises of knowledge spillover in the reality, especially among competing enterprises. Innovation of competition will make rent dissipation. Imitators using cost advantage to implement price competition may be allowed to get below-average funding return on innovation productivity. Enterprise integration is most effective on external governance countermeasures.

## References

[1] Chang, Ching-Fu, and Ping Wang, and Jin-Tan Liu, 2016, Knowledge spillovers, human capital and productivity, *Journal of Macroeconomics* 47, Part B, 214-232.

[2] Chyi, Yih-Luan, and Yee-Man Lai, and Wen-Hsien Liu, 2012, Knowledge spillovers and firm performance in the high-technology industrial cluster, *Research Policy* 41, 556-564.

[3] Feldman, Maryann P., and David B. Audretsch, 1999, Innovation in cities:: Science-based diversity, specialization and localized competition, *European Economic Review* 43, 409-429.

[4] Hervas-Oliver, Jose-Luis, and María Lleo, and Roberto Cervello, 2017, The dynamics of cluster entrepreneurship: Knowledge legacy from parents or agglomeration effects? The case of the Castellon ceramic tile district, *Research Policy* 46, 73-92.

[5] Jinji, Naoto, and Yoshihiro Mizoguchi, 2016, Rules of origin and technology spillovers from foreign direct investment under international duopoly, *Japan and the World Economy* 40, 47-60.

[6] Kuruppuarachchi, Duminda, and I. M. Premachandra, 2016, Information spillover dynamics of the energy futures market sector: A novel common factor approach, *Energy Economics* 57, 277-294.

[7] Lauren, Nita, Kelly S. Fielding, Liam Smith, and Winnifred R. Louis, 2016, You did, so you can and you will: Self-efficacy as a mediator of spillover from easy to more difficult pro-environmental behaviour, *Journal of Environmental Psychology* 48, 191-199.

[8] Liu, Xiaohui, and Trevor Buck, 2007, Innovation performance and channels for international technology spillovers: Evidence from Chinese high-tech industries, *Research Policy* 36, 355-366.

[9] Polanyi, Michael, 1997, Chapter 7 - The Tacit Dimension, in Laurence Prusak, ed.: (Butterworth-Heinemann).

[10] Puškárová, Paula, and Philipp Piribauer, 2016, The impact of knowledge spillovers on total factor productivity revisited: New evidence from selected European capital regions, *Economic Systems* 40, 335-344.
[11] Romer, Paul, 1994, New goods, old theory, and the welfare costs of trade restrictions, *Journal of Development Economics* 43, 5-38.
[12] Zhang, Lin, 2017, The knowledge spillover effects of FDI on the productivity and efficiency of research activities in China, *China Economic Review* 42, 1-14.

IOP Conf. Series: Earth and Environmental Science **61** (2017) 012045 doi:10.1088/1755-1315/61/1/012045

# Comparative study on the optimization strategies of the human settlement environment of the rural settlements in Asia

**Sui Xin**[1], **Sun Chaoyang***[2(Correspondence author)], **Li Mo**[3]

[1] Dalian Nationalities University, No. 18, Liaohe West Road, Dalian Economic & Technical Development Zone, China, 116600
[2]Dalian Nationalities University, No. 18, Liaohe West Road, Dalian Economic & Technical Development Zone, China, 116600
[3]City Institute of Dalian University of Technology, No.31, Tieshan West Road, Dalian Economic & Technical Development Zone, China, 116600
Correspondence author: Sun Chaoyang s13591198896@126.com

**Abstract.** There is a similar cultural background between Japan, South Korea and China. The construction of Japanese village has gone through three stages including conception, improvement and promotion, and that of Korean countryside has also experienced three stages of start-up, transformation and development. In the rural construction in both Japan and South Korea, government orientation has sped up the modernization of rural areas in the short term, but has gone against the law of development and the will of farmers in the long run. The key to rural construction is to improve the initiative and enthusiasm of farmers and to stimulate their sense of ownership. During the process of promoting the development of rural economy, we shall also pay attention to the improvement of farmers' spirituality.

From the experience of rural construction in Asia, the development of rural areas cannot be separated from the support and creativity of the villagers, and it is also the process of synchronous coordinated transformation of social culture instead of the single industry adjustment and space construction. The social benefits brought from the modern transformation and cultural activities of the villagers are beneficial to promote the sustainable development of modernization of agriculture and rural areas.

## 1. The present situation and optimization strategies of the human settlement environment of rural settlement in China

### *1.1 Revolution and improvement: The practice of rural construction in the 20s and 30s*

The opening and development of Rural Anthropology and Sociology stays in the 20s and 30s of the twentieth Century when is also the peak of rural construction practice. During this period, rural construction focused on the issues like the quality of farmers, land structure and agricultural development.

Some scholars think the rural construction movement in the early Republic of China to be an improvement movement essentially with the purpose of saving the nation and transforming society, which is the subsidiary product of the revolution. Based on China's national conditions, the movement has provided the new idea for rural construction, and combined intellectuals with farmers, technology promotion and quality education with agricultural economic development. It has also advocated the

IOP Conf. Series: Earth and Environmental Science **61** (2017) 012045 doi:10.1088/1755-1315/61/1/012045

participation of the whole people and the in-depth study of the countryside, and the complex development exploration based on the local is still of great significance to the current rural construction. In the research on village autonomy of Wanxi in Henan in 30s, some scholars have pointed out that the strong or weak change between the state and the local has come into being owing to the difference existing in the local management structure and social conditions of state power construction. Wanxi autonomy has represented the spontaneous action of local elites and shown the endogenous endemicity. In the case of limited capacity of the country, the active village elites can effectively promote rural stability and development, and the locality do not run counter to the development of the nation.

*1.2 Exploration: The practice of rural construction in the early period of reform and opening up*

Since 80s, China has reflected on the rural construction in the early stage of the founding of the People's Republic of China, gradually changing the control of local government behavior on the local and positively promoting the development of independent innovation in rural areas. All countrysides have begun to explore the suitable development model, and many of rural areas have accelerated the development of collective economy and township enterprises after developing non-agricultural industries with the help of resources carrying local characteristics. In that way, a number of distinctive 'Star Village' like Nanjie Village, Wan Feng Village and Huaxi Village have sprung up, and these economic-oriented models have stimulated the initiative of the countryside and presented the a positive role in promoting local development. However, the industrial development model based on economy is more likely to cause serious damage to the natural ecological environment and traditional style. At the same time, in most of the rural areas where the resources and manpower are insufficient in China, agriculture is still the foundation of our country, and such kind of non-agricultural development is difficult to be promoted in other rural areas.

*1.3 Transformation: The practice of rural construction in the late 1990s*

In the face of the development pattern of 'thousands of similar villages' and excessive urbanization in varied countryside, construction strategies of 'beautiful countryside' and 'one village one style' have been put forward, and the characteristics of the village shall be based on the natural environment and social human culture. On account of the similarity of regional environment and social environment, there is a great difference and excessive pursuit of uniqueness existing in the adjacent villages. So, it is prone to appear the vague positioning divorced from reality or to produce the special symbol that is not in accordance with reality like painting the walls and making up the local historical stories. In addition, many scholars have studied the development and construction of new rural area, and there are also non-governmental organizations involved in the practical construction in the countrysides. All of the practices have been affected by both the participation of external forces and the spontaneity of internal force. From the experience of the construction of these villages, it can be found that the rural construction is gradually transformed from a single external force into a stage affected by main internal forces and auxiliary external forces, and the rural internal force plays an important role in rural development.

## 2. The present situation and optimization strategies of the human settlement environment of rural settlement in the Republic of Korea

South Korea had been a colonial country that was invaded and enslaved for a long time, centering on the development of agricultural economy with agricultural productivity lagging behind and the national economy underdeveloped. For the sake of revitalizing the economy and changing the long-term poverty and backwardness, Korean government had adopted the strategy to give priority to the development of industry and neglected the recovery and development of agriculture and rural economy after liberation. Apart from that rural economy had impeded the development of the overall economy, people in the rural areas had still been suffering from the tough living environment and the facilities on the aspects of education, health and cultural life had lagged behind. Moreover, the income of peasant was low, rural young labor had been flowing into the city, farmers were discontent to the government and there were increasing criticism on the government's current policy. Under the circumstance, the South Korean government had set about adjusting the policy of economic development so as to adapt to the changes of economic and social development, stabilize rural economy and accumulate the energy for prosperity.

Since 70s, the urbanization population in Korea has increased to 90% from 41% at a high rate of speed, with low starting point of urban development and urbanization in the initial stage. In spite of the heavy burden of the government at this stage, it had started to focus on rural development to improve agricultural production and farmers living environment and enhance their enthusiasm. With the improvement of urbanization level, the government had gotten increasing pressure and failed to afford the huge expenditure. So, it had commenced to change the form and launched the vigorous new village movement.

From 1970 to 1980, Korean new village movement was the government-oriented rural development model. And the major turning point emerged in December of 1980 that the movement was transformed to be community-driven from government-oriented model, and this transformation had achieved completely after 1990s. And the optimization strategy of promoting the rural human settlements is mainly applied for by the countryside after 2005, and the government will screen before giving support.

The main features of the new village movement in Korea: first of all, the government led has maximized the enthusiasm of the villagers; secondly, the material civilization construction has promoted the construction of ethical spirit; thirdly, the government-oriented top-down movement with mass participation has come into being; fourthly, it has made an obvious influence on the process of the rural modernization; finally, the roles of varied agricultural cooperative organizations are prominent.

South Korea's new village movement has picked up a number of successful cases. The first case is about the improvement of rural income, namely planning green rice for crust of cooked rice. The second and third case are separately about water sightseeing tour by making use of the village resources and the development of celebration and sightseeing activities. The fourth is to mobilize all the people involved in, and the village plants the wild herbs with good sales, which can be strengthened. Next one is about case of artistic creation and leading, in which artists live in the village, so that it can play the leading role of the artists in the local development. The sixth is to make use of the abandoned school to reform, change the abandoned school into the ecological experience school as an educational base where the students in the cities can regularly come to experience the rural living environment, and work together with the education sector. The last case is about the cultural exchange case, establishing the club like dance and musical instruments. Although these cases have been invested a lot, the achievements made are not very exciting.

Something in common of optimization strategies of the human settlement environment of rural settlement between Korea and China is about the governments' strong desire to revitalize the countryside, with great differences existing in the operation of both sides. South Korea pays more attention to mobilize the enthusiasm of the village including the guidance of rural competition to apply for project support, and then the government screen and provide the support. In other words, the model village is engaged in, and the government provides the training and support. If it is made well, the government will provide additional funding and the project will be tracked and maintained after achieving success, instead of implementing what is thought up. There are not only a sound system and mechanism as well as considerable capital investment for rural support, and it also pays attention to the concept of rural development and operation of technology.

South Korea's rural reconstruction has achieved some success, but it is difficult to see the possibility of reversing the loss of population and recession in spite of the more successful cases. It may continue, but it appears difficult to prosper the past and turn back to flourishing population and the sound social structure, which is maybe the cost of modernization. We shall see if we can find our way to avoid this situation.

## 3. The present situation and optimization strategies of the human settlement environment of rural settlement in Japan

Japan is a country with a shortage of natural resources. In 1930s, a large number of young labors in the rural areas had flowed into the city so that the countryside was on the verge of collapse. In order to improve the crisis in rural areas, Japan had adopted legislation to protect the development of agriculture and farmers in rural economy, and gradually carried out the rural construction plan step by step. After the continuing policy funding of the government from 50s to 70s, rural infrastructure had been improved, the modernization of agriculture and the adjustment of industrial structure had been realized, and the production enthusiasm of farmers had been improved with their living standards and income lifting so as to reduce the external flow of the rural population. In the decades of experience, Japan attaches great importance to the combination of traditional cultural resources and rural development. By the end of 70s, it had focused on the characteristics and advantages of rural areas and put forward the idea of "one village, one style" which had promoted the development of rural local economy based on the comprehensive development of its own resources so as to obtain the vitality for development. For example, Mishima Revitalization Plan had advocated the villagers' participation in the process of 'life craft movement' which was to promote the villagers to learn the traditional culture with a series of cultural activities and cohere to the relationship of villagers in the activities. Thought the rural culture carrier, it had promoted the uniqueness and continuity of rural development in rural areas to establish agricultural technology research and promotion institutions, like agricultural selection and the private training institutions to promote farmers' education and modern skills. After construction, Japan's rural life has undergone tremendous changes, and the differences of income, infrastructure and living conditions between the city and the rural areas have basically eliminated. Japan has formed the standard procedure and the method for the historical block protection and the investigation, and these historical blocks are mostly rural communities, reflecting the historical development and evolution of the rural construction in Japan. Wu yun and others have made a research on the methods of investigating and studying the landscape features of Japanese historical and cultural blocks, and brought up that block protection is the main way to learn. During the process of building investigation, it has paid attention to the qualitative and quantitative combination first of all. Through extensive collection of block building color, material, volume, entrance and other features and with the help of the quantitative data like color comparison, material and volume statistics, the visual pictures can be integrated with the quantitative analysis data, and then the overall architectural characteristics of the block are summarized and reflected by the larger sample number statistics and induction.

## 4. The present situation and optimization strategies of the human settlement environment of rural settlement in India

The situation of India is very similar to that of China, with a large proportion of agriculture, a large population, the traditional agriculture having a long history, relatively backward rural economy and duality of urban and rural development. Therefore, the experience of rural construction in India has more practical significance for China. The concept of social stratification and social differentiation in India are very serious and it is difficult for the poor to obtain the necessary educational resources and learning plans. So during the process of strengthening the synchronous development of agriculture, science and technology and education, India attaches great importance to the construction of rural culture. It spares extra efforts on constructing the public cultural facilities such as libraries where seminars, technical training courses and local cultural activities can be held, promotes the "public scientific movement", breaks the hierarchy discrimination in education, enhances the universal participation and lifts the cohesion of rural areas, so as to boost the rural economy. When dealing with the slums in Mumbai, the government's strong eradication and expelling did not get results, and they just hastily winded up the matter by transferring them to the place which was poorer. This lesson had changed the attitude of strong intervention of the government, to study and solve the problem of the poor from a more relaxed level.

## 5. Summary

The way in which China's rural construction attached the importance to the development of farmers in the early stage and the synchronous development between rural agriculture and people's cultivation can be used for reference. In the early period of reform and opening up, the progress had stepped into the one-sided situation owing to its development attitude towards seeking celerity and goodness. Meanwhile, it has also neglected the differences of the rural development in the process of modernization. After reflection and exploration, the participatory endogenous development model which centers on the villagers to accelerate the development of internal power has been gradually explored, and the single pattern will be avoided to apply in the construction by making an adjustment in accordance with specific projects and the actual situation in rural areas.

The status quo of Japan, South Korea and India is closer to that of China with fewer lands for more people as well as resource shortage. In the process of developing the rural areas in these countries, their importance of historical culture, the cultivation and mobilization of villagers' autonomy, the experience of promoting the rural development with low investment, and the combination of qualitative and quantitative research methods can be referred in the construction of rural areas in China.

From the perspective of international human development, the development of a country or region should be based on its own initiative, the positive interaction between endogenous power and foreign aid is necessary, and the enhanced ability of residents shall be paid attention to obtain the fundamental and sustainable motivation for development. The process shall be participated by the public after empowerment, and internal and external forces shall realize communication in the implementation of the actual project and the positive role of local knowledge shall be played. In the pursuit of economic and social development, we should pay attention to the all-round development of human beings, the sustainability of natural ecology and the potential power of culture. Only when the construction of the project is based on the demand from the inside, can the development be actually effective.

## References

[1] Lin Tao, Research on Zhebei Countryside Agglomeration and Its Evolution Pattern of Settlement Space, Zhejiang University, 2012.
[2] Wang Tao, Wang Zhu, Chen Shizhao, Interpretation Architecture and Culture of Rural Settlement Regional Culture from the Perspective of Villagers' Subjective Cognition, 2013, 12
[3] Wang Zhu, Wang Tao, The Journal of Subjective Cognition and Regional Characteristics of Rural Settlement Expression Jixi Human Settlement, 2014.
[4] Li Henan, Research on Regional Distribution and Regularity of Morphological Change of Rural Settlements in Ancient China, Tianjin University, 2006.

IOP Conf. Series: Earth and Environmental Science **61** (2017) 012046 doi:10.1088/1755-1315/61/1/012046

# Evaluation of anti-fatigue and immunomodulating effects of quercetin in strenuous exercise mice

**Wei-qiang Zhang**

Teaching and Research Department of P.E. Central South University, Changsha City, Hunan Province, 410083, China.

**Abstract.** The purpose of the present study was to investigate the anti-fatigue and immunomodulating effects of quercetin in strenuous exercise mice. Mice were given orally either corn oil or quercetin (20, 40 and 60 mg/kg body weight suspended in corn oil) by gavage once a day for 28 day. All mice were sacrificed after rotarod test and the major biochemical parameters were analyzed in serum and liver. The results indicated that quercetin possessed anti-fatigue effects by prolonging retention times, decreasing levels of blood lactate and serum urea nitrogen, and increasing levels of blood glucose, tissue glycogen and serum glucagon. Furthermore, quercetin could improve the immune function of fatigue mice by decreasing tumor necrosis factor-α levels, and elevated interleukin-10 levels. Quercetin possessed anti-fatigue effects may be related to its immunomodulating effects.

## 1. Introduction

Flavonoids, a large group of polyphenolic derivatives of benzo-c-pyrone, are widely distributed in the plant community and have a variety of pharmacological and biological activities. According to the different structure, flavonoids can be classified as flavones, flavonols, flavanones, isoflavones, chalcone, dihydrochalcone, orange ketones, flavans, flavanols, and so on [1]. Quercetin belongs to flavonols group, which consists of 3 rings and 5 hydroxyl groups, and exists as a variety of glycosides or in aglycone form [2]. Quercetin is widely found in the flowers, leaves and fruits of plants. Many food sources, including leeks, tomatoes, broccoli, apples, green tea, black tea, black grapes, blueberries and buckwheat, are rich in quercetin [3]. In addition, there are more than 100 kinds of Chinese medicinal plants containing quercetin, such as *Dendranthema morifolium* (Ramat.)Tzvel, *Plantago asiatica* L, *Gynostemma pentaphyllum* (Thunb.) Makino, *Forsythia suspensa* (Thunb.) Vah, *Houttuynia cordata* Thunb, *Phyllanthus emblica* Linn. etc. Modern pharmacological studies have confirmed that quercetin has multiple biological activities, including anti-oxidation, anti-cancer, anti-thrombosis, anti-diabetes, anti-ulcer, anti-inflammatory, anti-allergy, anti-virus, anti-ageing, anti-apoptosis, cardioprotection, cataract prevention, and neuroprotective [4-5]. Nevertheless, few studies have examined the effects of quercetin on physical fatigue. Thence, the purpose of the current study was to investigate the anti-fatigue and immunomodulating effects of quercetin in exhaustive exercise mice using rotarod test, which will provide a new experimental basis for quercetin as a nutritional supplement.

## 2. Materials and methods

Please follow these instructions as carefully as possible so all articles within a conference have the same style to the title page. This paragraph follows a section title so it should not be indented.

### *2.1. Chemicals and reagents*

Quercetin (> 98% purity) was obtained from Yiji Biomedical Co., Ltd (Shanghai, China). The assay kits for the detection of blood glucose, blood lactate (BLA), and serum urea nitrogen (SUN) were obtained from Jiancheng Biological Reagents Co. (Nanjing, China). The assay kit for the detection of serum glucagon was obtained from Meilian Biological Technology Co., Ltd (Shanghai, China). The assay kits for the detection of tissue glycogen were obtained from Leigen Biotechnology Co., Ltd. (Beijing, China). The assay kits for the detection of interleukin (IL)-10 and tumor necrosis factor-α (TNF-α) were obtained from Yuanye Bioengineering Co., Ltd (Shanghai, China). All other chemicals and reagents (analytical grade) were obtained from commercial channels.

### *2.2. Animals and treatment*

Male Kun-Ming mice, weighing approximately 20 ± 2 g, were obtained from the Drug Safety Evaluation Research Center of Hunan Province (Changsha, China). The animals were fed with standard rodent feed and had free access to drinking water, and housed in a feeding room maintained at 23 ± 2 °C with a 12–12-h light–dark cycle. The experimental animals were processed in strict accordance with the Measures for the Administration of Laboratory Animals in Central South University (Changsha, China). The approval certificate of this experiment was obtained from the Ethics Committee of Central South University. After 7 days of adaptive feeding, the mice were randomly distributed into the following four groups (10 mice per group): control group (C), quercetin treatment at low dose group (LQu), quercetin treatment at medium dose group (MQu), and quercetin treatment at high dose group (HQu). The quercetin treatment groups were administered with various doses of quercetin (20, 40 and 60 mg/kg body weight suspended in 1.0 mL of corn oil). Meanwhile, the C group was administered with 1.0 mL of corn oil. The mice were treated using gavage by oral administration, once a day for 28 days.

### *2.3. Rotarod test*

The mice of all groups were subjected to adapt to the automated rotarod instrument (ZH-YLS-4C, Anhui Zhenghua Biological Instrument Equipment Co., Ltd, Huaibei, China) for two days at a speed of 18 m/min for 10 min before the rotarod test. On the last day of the experiment, 30 min after the last administration, the mice in all the groups were made to carry out the rotarod test at a speed of 36 m/min until exhausted. Exhaustion is defined as the mouse falling from the rotarod, and the retention times of mice on the rotarod were recorded.

### *2.4. Biochemical analysis*

After the rotarod test, the mice were sacrificed by decapitation under anaesthesia (ether). Blood was collected for the BLA and glucose analysis, and then serum was separated by centrifugation (3,000 × g, 25 °C, 10 min) for the SUN, glucagon, TNF-α, and IL-10 analysis. After the blood was collected, the liver and gastrocnemius muscle were immediately resected and washed in ice-cold saline solution. The samples were then frozen in liquid nitrogen and kept at -80 °C for the tissue glycogen analysis. The above biochemical indicators were analyzed by using commercial kits and were carried out according to the manufacturer's recommended method.

### *2.5. Statistical analysis*

Means ± SD was use for data express. Statistical analysis was performed using data analysis software (SPSS 17.0) by one-way analysis of variance (ANOVA) and then Dunnett's test. $P<0.05$ considered to be significant.

## 3. Results and discussion

### *3.1. Effects of quercetin on retention times of mice*

Exercise endurance is the most direct and objective expression of fatigue resistance, and the degree of fatigue can be determined by the length of retention times [6]. As shown in Figure 1, the retention times in the LQu, MQu and HQu groups were significantly longer than those of the C group (P<0.05). These results indicated that quercetin could enhance exercise endurance and possessed anti-fatigue effects.

*3.2. Effects of quercetin on BLA and SUN of mice*

BLA is a carbohydrate glycolytic product under anaerobic conditions, and it is well known that strenuous exercise increases the energy consumption in the body, and therefore enhances the production of lactate. Increased lactate levels further reduce pH value, which can induce a variety of biochemical and physiological side effects, leading to fatigue [7]. Urea nitrogen is a metabolite of protein. During strenuous exercise, protein and amino acid catabolism would be involved in energy supply when the body can't get enough energy through the sugar and fat metabolism, and produce urea nitrogen. There is an inverse correlation between the SUN level in vivo and the exercise tolerance [8]. Therefore, BLA and SUN are important blood biochemical parameter associated with fatigue. As shown in Figure 2, the BLA and SUN levels in the LQu, MQu and HQu groups were significantly lower than those of the C group (P<0.05). These results indicated that quercetin could effectively delay the production of BLA or remove the accumulation of BLA, reduce protein metabolism by decreasing SUN levels, and ultimately alleviates fatigue.

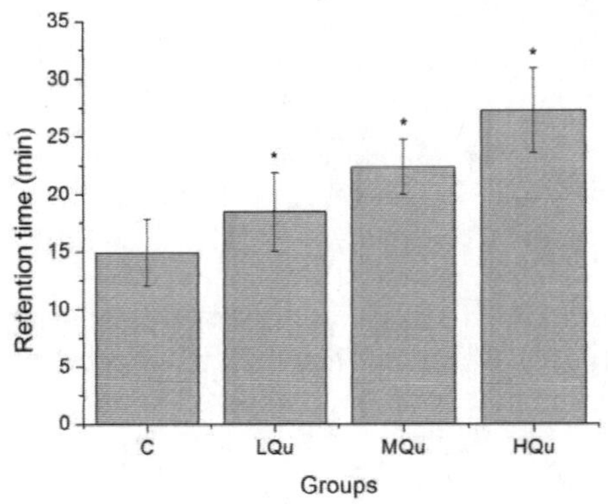


**Figure 1.** Effects of quercetin on retention times of mice. Data were expressed as means ± SD. *P<0.05, compared with C group.

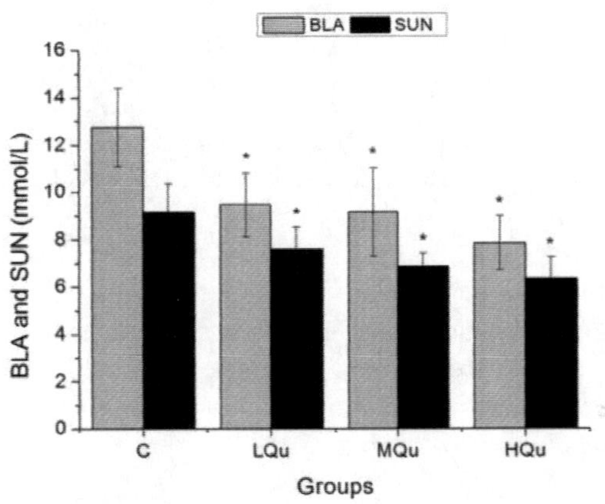


**Figure 2.** Effects of quercetin on BLA and SUN of mice. Data were expressed as means ± SD. *P<0.05, compared with C group.

*3.3. Effects of quercetin on blood glucose and tissue glycogen of mice*

Sugar oxidation has the advantages of low oxygen consumption and large output power, which is the main source of energy required for the anaerobic exercise. It is important to keep blood glucose values within the physiological range, because it is expeditiously available and can be directly oxidized to supply ATP in the blood [9]. During strenuous exercise, blood glucose is quickly depleted, if not promptly supplied, resulting in hypoglycemia, which could inhibit the central nervous system function, thereby reducing exercise endurance. So, blood glucose levels can be used to indicate the rate and extent of fatigue development [10]. Besides, energy for exercise is originally derived from the decomposition of glycogen, and glycogen stored in the liver and muscles as an energy reserve and balancing blood glucose values, which is another important factor related to exercise endurance [11]. As shown in Figure 3, the levels of blood glucose and liver glycogen in the LQu, MQu and HQu groups, as well as muscle glycogen levels in the MQu and HQu group were significantly higher than those of the C group (P<0.05). These results indicated that quercetin made mice resistant to fatigue by means of maintaining blood glucose values within the physiological range and increasing the reserve of liver and muscle glycogen.

### *3.4. Effects of quercetin on serum glucagon of mice*

Glucagon is a peptide hormone, secreted from the islet α-cells. Glucagon plays an important role in glucose homeostasis by regulating liver glucose output in both normo- and hypoglycemic conditions [12]. Glucago can activate the liver phosphorylase, thereby promoting the rapid decomposition of glycogen into glucose. In addition, glucago can rapidly promote gluconeogenesis, including lactic acid, glycerol and sugar amino acid non-sugar substances into glucose, so as to maintain the glucose homeostasis [13]. As shown in Figure 4, the serum glucagon level in the LQu, MQu and HQu groups were significantly higher than those of the C group (P<0.05). These results indicated that quercetin could regulate the secretion of glucagon, which could rapidly promote gluconeogenesis, saving the consumption of liver glycogen during strenuous exercise. This might be another mechanism of its anti-fatigue effects.

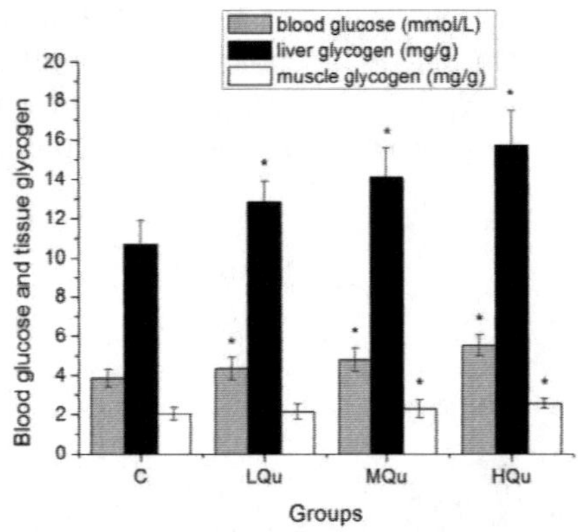


**Figure 3.** Effects of quercetin on blood glucose and tissue glycogen of mice. Data were expressed as means ± SD. *P<0.05, compared with C group.

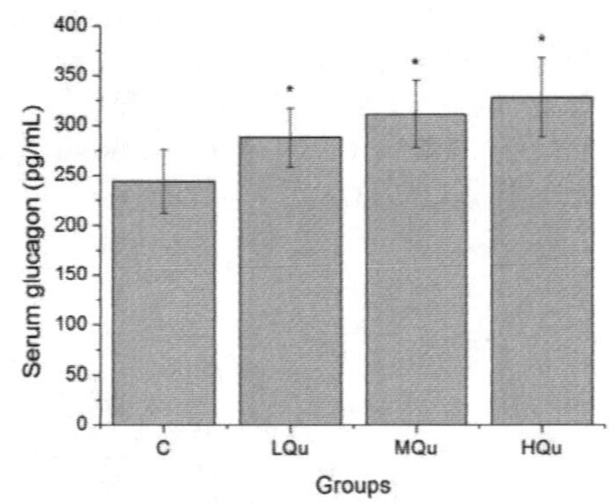


**Figure 4.** Effects of quercetin on serum glucagon of mice. Data were expressed as means ± SD. *P<0.05, compared with C group.

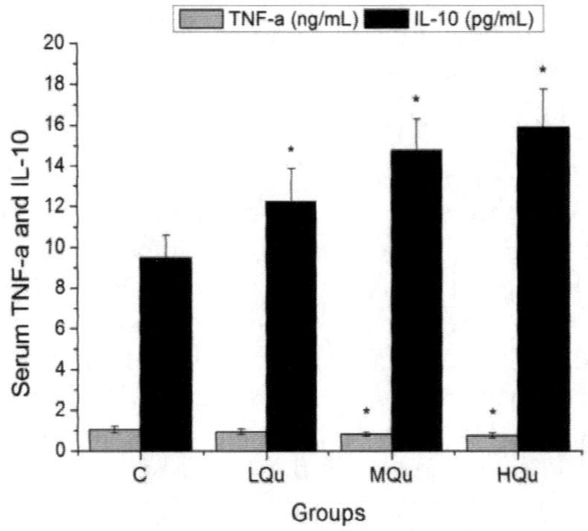


**Figure 5.** Effects of quercetin on serum TNF-α and IL-10 of mice. Data were expressed as means ± SD. *P<0.05, compared with C group.

### *3.5. Effects of quercetin on serum TNF-α and IL-10 of mice*

Fatigue is also related to production of inflammatory cytokines, immune dysfunction, and cytokine imbalances [14]. Several studies have indicated that the causative role of proinflammatory cytokines in fatigue severity, evidenced by elevated production of inflammatory cytokines including TNF-a, IL-1, and IL-6 [15] .TNF-α generates ROS at the mitochondrial inner membrane, which may easily result in the progressive destruction of the mtDNA [14]. IL-10 is an important immunoregulatory cytokine that is produced by numerous cell types. It is an important inhibitor of many aspects of the inflammatory response [16]. As shown in Figure 5, the serum TNF-α levels in the MQu and HQu groups were significantly lower than those of the C group ($P<0.05$). The serum IL-10 levels in the LQu, MQu and HQu groups were significantly higher than those of the C group ($p<0.05$). These results indicated that quercetin could improve the immune function of fatigue mice by decreasing pro-inflammatory cytokines (TNF-α) levels, and elevated anti-inflammatory cytokines (IL-10) levels. quercetin possessed anti-fatigue effects may be related to its immunomodulating effects.

## 4. Conclusions

In conclusion, the present study indicated that quercetin possessed anti-fatigue effects by prolonging retention times, decreasing levels of BLA and SUN, and increasing levels of blood glucose, tissue glycogen and serum glucagon. Furthermore, quercetin could improve the immune function of fatigue mice by decreasing pro-inflammatory cytokines (TNF-α) levels, and elevated anti-inflammatory cytokines (IL-10) levels. Quercetin possessed anti-fatigue effects may be related to its immunomodulating effects. Further work is also needed to clarify the detailed molecular mechanism for the anti-fatigue and immunomodulating effects of quercetin.

## References

[1] Jain P, Jain S, Pareek A and Sharma S 2013 *Bull. Pharm. Res.* **3** 90
[2] Ciolino HP, Daschner PJ and Yeh GC 1999 *Biochem. J.* **340** 715
[3] Neuhouser ML 2004 *Nutr. Cancer.* **50** 3
[4] Lin C, Yu Y, Zhao HG, Yang A, Yan H and Cui Y 2012 *Radiother. Oncol.* **104** 395
[5] Chondrogianni N, Kapeta S, Chinou I, Vassilatou K, Papassideri I and Gonos ES 2010 Exp. Gerontol. **45** 763
[6] Ding JF, Li YY, Xu JJ, Su XR, Gao X and Yue FP 2011 *Food Hydrocoll.* **25** 1350
[7] Yan FW and Hao HT 2013 *J. Food* Agric. Environ. **11** 581
[8] Miao FS, Wu D and Ni G 2016 *Afr. J. Tradit. Complem.* **13** 52
[9] Zhao XN, Liang JL, Chen HB, Liang YE, Guo HZ, Su ZR, Li YC, Zeng HF and Zhang XJ 2015 *Nutrients.* **7** 8657
[10] Wang JJ, Shieh MJ, Kuo SL, Lee CL and Pan TM 2006 *Appl. Microbiol. Biotechnol.* **70** 247
[11] Ikeuchi M, Koyama T, Takahashi J and Yazawa K 2006 *Biol. Pharm Bull.* **29** 2106
[12] Kawamori D, Kurpad AJ, Hu J, Liew CW, Shih JL, Ford EL, Herrera PL, Polonsky KS, McGuinness OP and Kulkarni RN 2009 *Cell Metab.* **9** 350
[13] Chen WD and Yang WL 2008 *J. Xi'An Physi. Edu. Univ. J.* **25** 89
[14] Muluye RA, Bian Y, Wang L, Alemu PN, Cui H, Peng X and Li S 2016 Front. Pharmacol. **7** 328
[15] Nam SY, Kim HM and Jeong HJ 2016 *Biomed. Pharmacother.* **84** 51
[16] Lee BR, Lee JH and An HJ 2012 *Molecules.* **17** 13253

IOP Conf. Series: Earth and Environmental Science **61** (2017) 012047 doi:10.1088/1755-1315/61/1/012047

# The evaluation of the development mode of electric energy and air pollution control in Beijing based on the IPAC-SGM model

**Erdong Zhao, Chaoran Guo, Liwei Liu, Sichen Dai and Shangqi Li**

School of Economics and Management, North China Electric Power University, Beijing, China;
guochaoran609@163.com

**Abstract.** In recent years, China broke out a large-scale of fog and haze, particularly Beijing. Energy production and consumption of fossil fuel combustion emissions is the main source of environmental pollution and haze, and it is most prominent in the power industry. In this paper, we evaluate the relationship between Beijing power structure and the prevention and control of atmospheric pollution by Integrated Policy Assessment Model for China – Second Generation Model (IPAC-SGM). This paper explores the propulsion effect of the new energy industry on Beijing's air pollution prevention and control by simulating the change of development of electric energy in Beijing under three scenarios which are benchmark scenario, general policy scenario and reinforced policy scenario.

## 1. Introduction

With the rapid development of China's economy and society, environmental security issues have become increasingly prominent. The air pollution problem has been widely concerned. Beijing air pollution situation is grim, especially the haze problem. In 2015, there were 186 days that air quality reached the standards, accounting for 51% of the year.Heavy pollution had appeared for 46 days, accounting for 13%. The annual average of PM 2.5 of Beijing was concentration of 80.6 micrograms / cubic meter, which was 1.3 times more than the value of national standard, 10 times more than the maximum annual average of the number which World Health Organization recommended. Air pollution brought a significant negative impact on climate, environment, human health and economic development. It is imperative to further strengthen the haze control and improve the air pollution situation.

For a long time, Beijing local power support and heat supply depend on four coal-fired thermal power plants. These plants only have small units with low energy efficiency and serious pollution problems. In recent years, the government has introduced policies to regulate the production of power industry and reduce pollution emissions from electricity production. Beijing will fully shut down coal-fired units and reduce 9.2 million tons' coal burning and put southeast, southwest, northeast, northwest four gas thermoelectric centers into operation in 2017. When coal-fired power plants complete gas transformation, there will no longer be a new built power plant. At the same time, Beijing also significantly improves clean energy consumption. It is of great theoretical and practical significance to carry out the evaluation of economic model and policy research of air pollution to prevent and control the air pollution and adjust energy structure.

## 2. Construction and implementation of IPAC-SGM model

IPAC-SGM model is a sub-model of IPAC model (Integrated Policy Assessment Model for China). It is based on computable generality model (CGE) Theory, and joins various economic activities of the association and impact of the medium or long-term energy and environment analysis and various types of energy and environmental policies for economic evaluation and analysis.

### *2.1. The objective function of IPAC-SGM*

In **IPAC-SGM** model, economic goal is taken as the main objective function, and environmental target as the constraint condition policies as the scenarios. The minimum input cost of power sources is set as the objective function. The objective function is shown in Equation (1):

$$f(X) = C_1 + C_2 + C_3 + C_4 + C_5 \qquad (1)$$

$C_1$ represents electricity energy supply costs which mainly refer to the fuel cost of thermal power generation and nuclear power generation:

$$C_1 = \sum_{j=1}^{n} (x_{j,t} \times b_{j,t} \times p_{j,t}) \times \frac{1}{(1+i)^t} \qquad (2)$$

In equation(2), $x$ —the power generation of the jth generating unit in the year t; $i$ —Discount rate; $b$ —The fuel consumption rate per unit of the jth generating unit in the year t; $p$ —The first t year j class power unit fuel unit price.

$C_2$ represents new energy input costs which mainly refer to the new energy power generation of fixed capital investment and operation & maintenance costs. For the easy calculation, we take a certain percentage of fixed investment as operation & maintenance costs:

$$C_2 = \sum_{j=1}^{n} (\frac{x_{j,t} - x_{j,t-1}}{H_{j,t}} \times M_{j,t}) \times \frac{1}{(1+i)^t} \qquad (3)$$

In equation (3), $H$ —annual utilization hours of j type of generating units in t year; $M$ — specific investments of j type of generating units in t year.

$C_3$ represents the marginal cost of traditional energy technology innovation.

$$C_3 = \int_0^{G_1} MC(T_k) dG \qquad (4)$$

In equation (4), $MC$ —the marginal cost of $T_k$ technology for traditional energy, the values of k take 1, 2, 3, 4; $T_1$ —reform of traditional coal combustion methods to reduce carbon dioxide emissions; $T_2$ —$CO_2$ capture and storage technology; $T_3$ —$CO_2$ capture and utilization technology; $T_4$ —$CO_2$ Capture, Utilization and Storage technology.

$C_4$ represents environmental management costs calculated by the loss of environmental damage:

$$C_4 = \sum_{k=1}^{n_w} \sum_{j=1}^{n} (x_{j,t} \times b_{j,t} \times s_{j,k,t} \times v_{j,k,t}) \times \frac{1}{(1+i)^t} \qquad (5)$$

In equation (5), $n_w$ —species number of environmental pollutants; $k$ —the type of pollutants, including $SO_2$, $NO_x$ and soot; $s_{j,k,t}$ —the emission rate of k type pollutants from j type of generating units in t year; $v_{j,k,t}$ —the environmental value of k type pollutants from j type of generating units in t year. $C_5$ represents purchased electricity cost which is the product of the incoming input and the electricity price:

$$C_5 = \sum_{j=1}^{n} e_{j,t} \times p_{j,t} \qquad (6)$$

In equation (6), $e_{j,t}$ —purchased power of j class of electricity in t year; $p_{j,t}$ —purchased power price of j class of electricity in t year.

### 2.2. Restrictions

(1) Power supply and demand balance constraints. The yearly sum of all types of energy power generation and purchased power is not less than the total amount of electricity demand in Beijing:

$$\sum_{j=1}^{n}(x_{j,t}\times(1-\varepsilon_{j,t})\times(1-\delta_{j,t})+\sum_{j=1}^{n}e_{j,t}\geq E_t \quad (7)$$

In equation (7), $E_t$—electricity demand of year t; $\varepsilon_{j,t}$—the electricity consumption rate of j type of generators in year t; $\delta_{j,t}$—line loss rate for j type of generators in year t.

(2) Technology evolution constraints. Annual activity amount of energy technology is not greater than the product of allowable maximum installed capacity and available factor:

$$C_{j,t}\leq C_{j,t\,\max}\times\pi_{j,t} \quad (8)$$

In equation (8), $c_{j,t}$—annual activity amount of energy technology for the jth class of energy technology in year t; $C_{j,t\,\max}$—the maximum installed capacity of the jth class of generator sets in year t; $\pi_{j,t}$—annual utilization coefficient of the jth class of generator sets in year t.

(3)Environmental Constraints. Carbon emissions and pollutant emissions from all kinds of generating units must be allowed within the maximum emissions:

$$\sum_{i=1}^{m}x_{j,t}\times b_{j,t}\times \mathrm{E}_{j,t}\leq \mathrm{E}_{\max,t} \quad (9)$$

In equation (9), $E$—carbon emissions and pollutant emissions (respectively carbon emissions, $SO_2$, $NO_x$, and soot emissions) coefficient of the jth generator in year t; $E_{\max,t}$—the maximum carbon emissions allowed by all generating units in year t.

### 2.3. Evaluation index

(1) Low carbon power structure indicators

Low-carbon power structure refers to the installed proportion of non-fossil energy power generation .

$$\eta_1=\frac{G_1}{G_2}\times 100\% \quad (10)$$

In equation (10), $\eta_1$—low carbon power structure indicators; $G_1$—installed capacity of non - fossil energy generating units; $G_2$—total installed capacity of all generating units.

(2) Electricity industry carbon productivity indicators.

Electricity industry carbon productivity refers to the power output generated by carbon emissions of electricity industry. It can directly reflect the efficiency of carbon resources.

$$\eta_2=\frac{P}{T} \quad (11)$$

In equation (11), $\eta_2$—electricity industry carbon productivity index; $P$—total power generation; $T$—carbon emissions from power industry.

(3) Energy efficiency technical indicators

Energy efficiency technical indicators refer to the average energy consumption value of unit production (expressed by standard coal amount). .

$$\lambda_1=\frac{L}{G} \quad (12)$$

In equation (11), $\lambda_1$—energy efficiency technical indicators; $L$—total coal r consumption of power industry; $G$—total power generation of power industry.

(4) Emission indicators

Emission indicators measure carbon emissions and pollutant emissions.

$$\mu_i = \frac{y_i}{G} \tag{13}$$

In equation (13), $\mu_i$ —emission indicators; $y_i$ —total emissions from power industry; $G$ —total power generation.

## 3. Scenario setting and model calculation

### *3.1. Scenario setting*

This article sets three scenarios to evaluate the relationship between Beijing power structure and the prevention and control of atmospheric pollution:

**Scenario 1: Baseline scenario.** The Beijing municipal government continues to follow existing policies and adjusts only to certain circumstances.

**Scenario 2: General policy scenario.** The Beijing government seeks energy improvement policies and measures to gradually improve the status of energy use. In 2020, total coal consumption in Beijing will be controlled below 10 million tons and the proportion of clean energy will be more than 7%. By 2030, total coal consumption in Beijing will be controlled below 6 million tons, and the proportion of renewable energy will be more than 10%.

**Scenario 3: reinforced policy-driven scenario.** The Beijing Municipal Government in energy development policy system focuses on the concept of low-carbon environmental protection. Beijingwill effort to reduce coal. By 2020, total coal consumption will be controlled below7 million tons, proportion of clean energy will be more than 10%; by 2030, total coal consumption will be below 3.5 million tons, proportion of renewable energy will be 18% or more.

### *3.2. Model calculation results and analysis*

*3.2.1. Power and energy production in different situations.* In general policy scenario and reinforced policy scenario, proportions of thermal power structure decrease rapidly, accounting for only 90.9% and 89.6%. In reinforced policy scenario, solar power is encouraged by policies, it grows rapidly from 0.8% in 2016 to 3.69%. Proportion of other clean energy power generation increases gradually. Total proportion exceeds 10% of total power generation. In general policy scenario, new energy power generation also has a certain degree of development, mainly by limiting thermal power development and promoting new energy power generation technology.

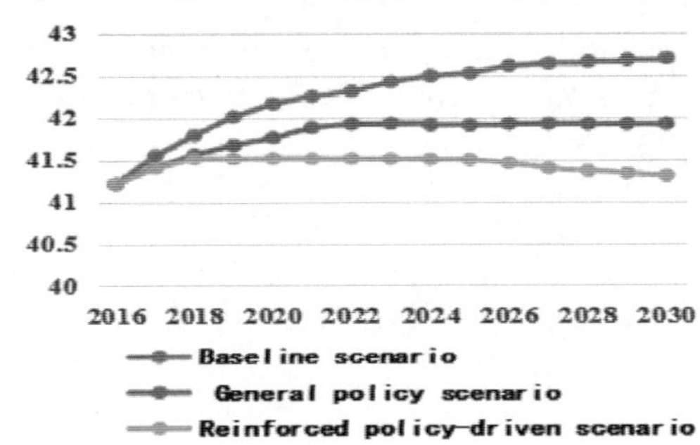


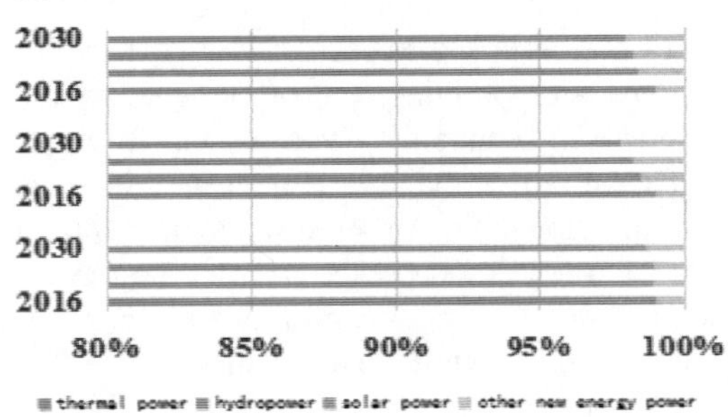


*3.2.2. Comprehensive evaluation of low - carbon economy development of electric energy*

(1) Low carbon power structure indicators

**Table 1.** Structural indicators of low-carbon power supply in Beijing from 2016 to 2020.

| | 2016 | 2020 | 2025 | 2030 |
|---|---|---|---|---|
| Benchmark scenario | 0.032 | 0.044 | 0.0522 | 0.0615 |
| General policy scenario | 0.032 | 0.0641 | 0.0740 | 0.0905 |
| Reinforced policy scenario | 0.032 | 0.0712 | 0.0836 | 0.1039 |

IOP Conf. Series: Earth and Environmental Science **61** (2017) 012047 doi:10.1088/1755-1315/61/1/012047

It can be seen that proportions of non-fossil energy power generation in general policy scenario and reinforced policy scenario are increasing to around 10% by 2030, while proportion of non-fossil energy power generation structures in benchmark scenario increases only 6.15%. Therefore, some constraints and incentive policies can optimize power structure, reduce pollutant emissions and achieve low-carbon development.

(2) Electricity industry carbon productivity indicators

**Table 2.** Carbon productivity indicators of the Beijing electric power industry from 2016 to 2020.

| | 2016 | 2020 | 2025 | 2030 |
|---|---|---|---|---|
| Benchmark scenario | 10.52 | 10.66 | 10.75 | 10.86 |
| General policy scenario | 10.52 | 10.89 | 11.01 | 11.2 |
| Reinforced policy scenario | 10.52 | 10.97 | 11.11 | 11.37 |

From the data in table 2, carbon productivity indicators grow fast in general policy scenario and reinforced policy scenario, which indicates that level of low-carbon development of power industry is high. In particular, in reinforced policy scenario, carbon productivity index in 2020 is close to 11kwh / kg, reaching 11.37kwh / kg in 2030, nearly more1kwh / kg than level in 2016. It indicates that policies of promoting low-carbon development in power industry bring good effects.

(3) Energy efficiency technical indicators

**Table 3.** Technical indicators of energy efficiency in Beijing's power industry from 2016 to 2020.

| | 2016 | 2020 | 2025 | 2030 |
|---|---|---|---|---|
| Benchmark scenario | 370.26 | 364.03 | 360.71 | 357.05 |
| General policy scenario | 370.26 | 356.31 | 352.91 | 345.92 |
| Reinforced policy scenario | 370.26 | 353.77 | 348.81 | 341.13 |

In reinforced policy scenario, energy efficiency technical indicator drops rapidly, from 370.26g / kWh in 2016 down to 341g / kWh in 2030. In this scenario, the Beijing government takes a series of measures, such as shutting down small power generation enterprises, restricting large-scale thermal power plant to reduce energy consumption and emissions, and investing in research and development of new energy power generation technology, promote the development of energy technology effectively.

(4) Emission indicators

Emission indicators response to the effect of power industry's energy-saving emission reduction and low-carbon development. In general policy scenario, the Beijing government requires power generation enterprises to realize low-carbon development, and introduce market mechanism such as carbon emissions trading to promote energy-saving and emission reduction. By 2030, carbon dioxide emissions will reduce to 739.88g/kwh. In reinforced policy scenario, the data will be less 37 g/kwh than that of 2016. At the same time, in general policy scenario and reinforced policy scenario, proportions of pollutant emissions are showing downward trends. It mainly due to the strong constraints on thermal power and the the development of carbon emissions trading. The results show that emission reduction technology and the promotion of new energy power generation technology can reduce carbon emissions to a great extent.

**Table 4.** Beijing's power industry emissions of three scenarios from 2016 to 2020.

| | | 2016 | 2020 | 2025 | 2030 |
|---|---|---|---|---|---|
| Benchmark scenario | $\mu_1$ | 773.21 | 760.16 | 756.28 | 749.62 |
| | $\mu_2$ | 3.01 | 2.98 | 2.95 | 2.93 |
| | $\mu_3$ | 2.7 | 2.65 | 2.63 | 2.6 |
| General policy scenario | $\mu_1$ | 773.21 | 748.09 | 742.85 | 739.88 |
| | $\mu_2$ | 3.01 | 2.94 | 2.93 | 2.91 |
| | $\mu_3$ | 2.7 | 2.6 | 2.57 | 2.52 |
| Reinforced policy scenario | $\mu_1$ | 773.21 | 745.21 | 740.69 | 736.11 |
| | $\mu_2$ | 3.01 | 2.94 | 2.92 | 2.9 |
| | $\mu_3$ | 2.7 | 2.57 | 2.54 | 2.48 |

## 4. Conclusion

According to the calculation results of the model substitution data, the demand for electric energy in Beijing is increasing under three scenarios, but the change of consumption of standard coal is different. In reinforced policy scenario, coal consumption in power industry will reach peak around 2018, then decline after a period of moderate. Solar power will grow rapidly by encourage policies, from 0.8% in 2016 to 3.69%. The proportion of other new energy power generation has gradually increased. The economic evaluation indicators in three scenarios are in the best position, which indicates Beijing Municipal Government's efficient impact on varieties of electric power policies. For instance, it forces to abate coal-fired dramatically, takes financial subsidies to encourage construction of clean energy facilities, develops carbon trading, emissions trading and other market mechanisms to improve structure of electric power, sets up a series of low-carbon energy laboratories to develop more efficient using techniques of new energy etc.. All of the policies mentioned above have positive functions and great impacts on the development of electric power and prevention of air pollution in Beijing.

**Acknowledgement**

This paper is a phased research objective based on the research project of Beijing Social Sciences Foundation " The evaluation of the development mode of electric energy and air pollution control of Beijing based on the IPAC-SGM model" ( project number:15JGB092) and the research project of the Natural Science Foundation of China "The modal of conducting effect posed by carbon allowance auctioning mechanism on carbon emission abatement of China's electric power generation sector " (project number:7167030492) .

## References

[1] Wu Y, Lou S, Lu S. A Model for Power System Interconnection Planning Under Low-Carbon Economy With CO Emission Constraints [J]. Sustainable Energy, IEEE Transactions on, 2011, 2(3): 205-214.

[2] Bi G B, Song W et al. Does environmental regulation affect energy efficiency in China's thermal power generation? Empirical evidence from a slacks-based DEA model [J]. Energy Policy, 2013.

[3] Sahar Shafiei, Ruhul A. Salim. Non-renewable and renewable energy consumption and CO2emissions in OECD countries: A comparative analysis [J]. Energy Policy, 2013.

[4] Jizun Li, Kejun Jiang. Analysis of High Oil Price Based on IPAC - SGM Model and Policy Suggestions [J]. China's prices, 2007(4): 29-31.

[5] Akorede M F, Hizam H, Ab Kadir M Z A, et al. Mitigating the anthropogenic global warming in the electric power industry [J]. Renewable and Sustainable Energy Reviews, 2012, 16(5): 2747-2761.

[6] Golombek R, Kittelsen S A C, Haddeland I. Climate change: impacts on electricity markets in

Western Europe [J]. Climatic change, 2012, 113(2): 357-370.
[7] Chen Q, Kang C, Ming H, et al. Assessing the low-carbon effects of inter-regional energy delivery in China's electricity sector [J]. Renewable and Sustainable Energy Reviews, 2014, 32: 671-683.
[8] Bin Li. Comprehensive Evaluation Model of Low Carbon Energy in Beijing and Energy Strategy of 2030 [D]. Beijing University of Technology, 2014.

# The effect of drying degree on pulverulent rate and strength of lignite

**Xiaonan Zhao[1,2a], Qihai Guo[1,2b], Li Ding[1,2c], and Daohong Wu[1,2d]**

1.Pyrolysis and Quality Engineering Technology Research Center of Low Rank Coal and Organic Waste in Beijing city, Beijing 102200
2.Beijing Shenwu Environment & Energy Technology Co., Ltd. Beijing 102200
[a]Zhaonan354@163.com, [b]guoqihai@163.com, [c]dingli@163.com, [d]wudaohong@163.com

**Abstract**. In order to obtain the drying characteristics of lignite, the effects of drying temperature and drying time on the pulverulent rate and strength of *Tianji* lignite were researched. The experimental results showed that lignite appeared obvious shrinkage, cracking and powdering in the process of drying. The fall strength and compressive strength of dry lignite decreased with the increase of dehydration rate. With the increase of drying time, the damage degree of pore structure of lignite increased, and part of the pore structure was blocked. With the increase of drying temperature, the damage degree of coal pore structure increased, and the pulverulent rate increased obviously, and fall strength and compressive strength decreased significantly. In order to improve the dehydration rate, fall strength and compressive strength, and reduce pulverulent rate, the drying time should be less than 4 hours, and drying temperature should be about 200°C.

## 1. Introduction

International geologists predicted that lignite would be one of the major energy in the future, accounting for about 40% of the world's total coal reserves. Lignite resources in China was also very rich, lignite reserves accounted for about 13% of the country's proven reserves of resources. In future, how to make use of lignite resources scientifically, reasonably and fully will be of great significance to the sustainable development of coal industry [1-3].

As the lowest degree of coal, lignite has characteristics of high moisture content, high ash, high volatile, rich oxygen-containing groups, strong chemical reactivity, and low calorific capacity. It is difficult to separate, reserve and for direct use. The reuse of lignite after upgrading is an effective way of reducing energy consumption and improving lignite utilization efficiency. At present, there are three kinds of lignite upgrading technology: dry separation upgrading, drying upgrading and pyrolysis upgrading. Of these three, drying upgrading is the main means of lignite quality upgrading [4-6].

High moisture is one of the most important characteristics of lignite, and it is also one of the most important parameters of its products properties. Therefore, dehydration is the key of the lignite upgrading. After dehydration treatment, most water in lignite was removed, and the calorific value increased significantly. However, due to the loose structure and the rich internal capillary of the lignite, the strength was significantly decreased in dehydration process, resulting lignite powdering, the large amount of dust in the tar, the low intensity of the semi-coke and so on in the subsequent pyrolysis. Therefore, it is of great significance to study the effect of drying degree on the pulverulent rate and strength of lignite, and to select suitable drying conditions.

## 2. Experimental

The independent research and development fixed bed drying device whose processing capacity was 30kg was used in the experiment. The test procedures were as follows: firstly, analyzed the moisture content, crushing strength and compressive strength of coal samples according to the national standard sampling. Then, loaded the coal sample into the drying device after being crushed, screened and weighed, measured the height of the material layer, purged with nitrogen, set the drying temperature and the drying time, turned off the power when it was reached the setting drying temperature and setting holding time. Last, When the temperature dropped to below 50°C, opened the door, removed the hopper, measured height of the material layer and weight, screened with 10mm sieve, recorded the weight of undersized compost, and analysis of dry coal moisture, crushing strength, compressive strength and pulverulent rate.

*Tianji* lignite was used in the experiment, after being crushed and sieved, 10-50mm grade was selected as the test coal sample. The effects of drying time and drying temperature on the characteristics of lignite were investigated. The analysis results of coal samples were shown in Table 1.

**Table 1. Analysis results of coal samples**

| Total moisture | Industrial analysis wt/% | | | | Elemental analysis wt/% | | | |
|---|---|---|---|---|---|---|---|---|
| $M_t$/% | $M_{ad}$ | $A_d$ | $V_{daf}$ | $FC_d$ | $S_{t,d}$ | $C_{daf}$ | $H_{daf}$ | $N_{daf}$ |
| 28.84 | 8.33 | 10.56 | 38.62 | 54.90 | 0.17 | 74.99 | 4.24 | 0.67 |

## 3. Results and discussions

### *3.1 The influence of drying time on properties of dry coal*

Drying time is one of the most important factors which affect the drying effect, energy consumption and processing capacity. Under the condition of appropriate drying temperature, on the one hand, if the drying time is too short, it will cause low the dehydration rate of lignite and high water content in the lignite, which will reduce the oil production rate and gas production rate, on the other hand, if the drying time is too long, it will cause a lot of energy waste, reduce the amount of treatment.

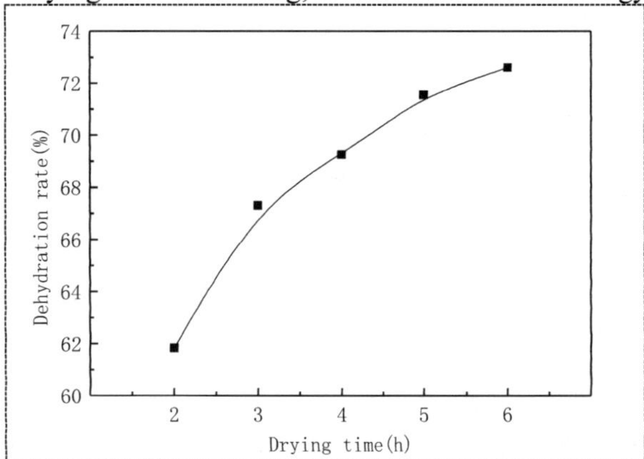


**Figure 1. Dehydration rate of coal samples under different drying time**

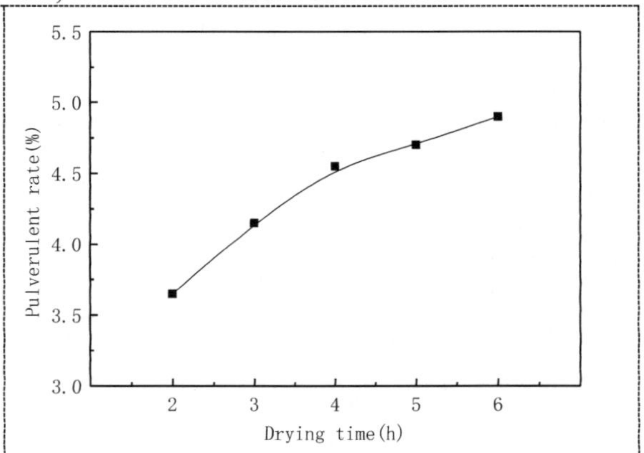


**Figure 2. Pulverulent rate of coal samples under different drying time**

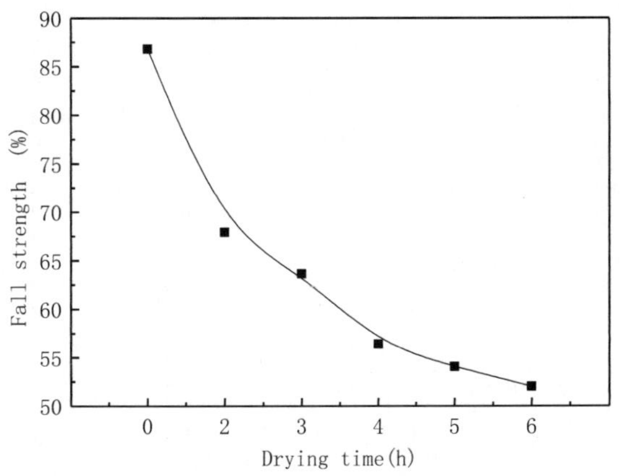


**Figure 3. Fall strength of coal samples under different drying time**

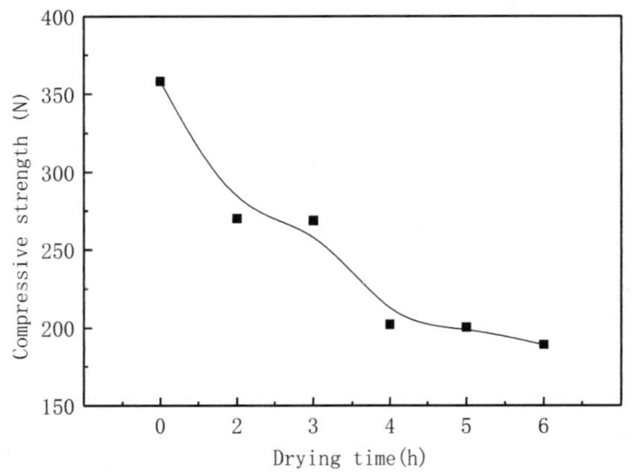


**Figure 4. Compressive strength of coal samples under different drying time**

The relationship between drying time and dehydration rate was shown in Figure.1. With the drying time increase, dehydration rate was increased significantly, however, the slope of the curve decreased gradually. The dehydration rate was 67.31% when the drying time was 3h, and the growth of dehydration rate began to decrease with the drying time sequentially increase. This was because when the lignite was dried a certain degree, it would enter the deceleration drying stage, and the slope of the water yield began to decrease. with the increase of drying time.

The relationship between different drying time and pulverulent rate was shown in Figure 2. With the increase of drying time, pulverulent rate increased significantly, however, the slope of the curve decreased gradually. Pulverulent rate was 4.55% when the drying time was 4h, and the growth of pulverulent rate began to decrease with the drying time sequentially increase.

Effect of drying time on fall strength and compressive strength were shown in Figure 3 and Figure 4 respectively. With the increase of drying time, both fall strength and compressive strength were decreased obviously; however, descend range declined significantly after the drying time was more than 4h. In the range of 0-4h, fall strength and compressive strength decreased fast with the increase of drying time, and more than 4h, fall strength and compressive strength decreased slowly with the increase of drying time.

By investigating the effects of drying time on dehydration rate, pulverulent rate, fall strength and compressive strength, in a relatively short period of drying time, surface structure of lignite was damaged, big hole collapsed firstly, swelling and cracking phenomenon were appeared, specific surface area and porosity were increased, and broken into smaller aperture structure. With the increase of drying time, the damage degree of lignite pore structure was deepened, and the cross reaction between the hole and the hole resulted in the blockage of the pore structure. In order to improve the dehydration rate, fall strength and compressive strength, and reduce the pulverulent rate, the drying time should be less than 4h.

### *3.2 The influence of drying temperature on properties of dry coal*

Drying temperature has a very important effect on the dehydration of lignite. Increasing the drying temperature can improve the heat transfer and mass transfer in the drying process, thus shortening the drying time and improving the drying efficiency. However, the high drying temperature will not only increase the energy consumption, reduce energy utilization, but also make coal pyrolyze, reduce the tar yield and gas yield. Therefore, it is of great significance to study the effect of drying temperature. The effect of drying temperature on the dehydration rate, pulverulent rate, fall strength and compressive strength are shown in Figure 5-8.

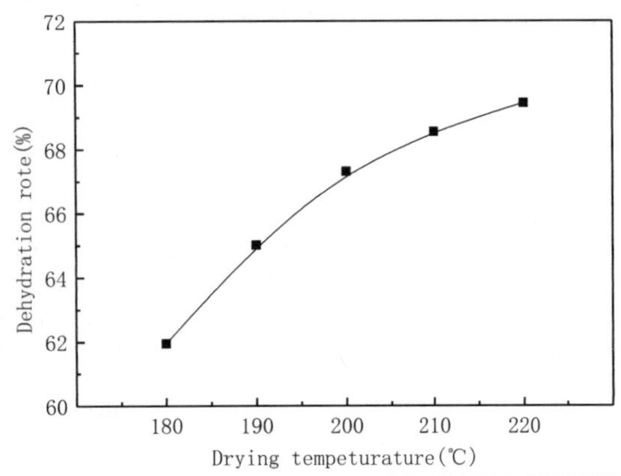


**Figure.5 Dehydration rate of coal samples under different drying temperatures**

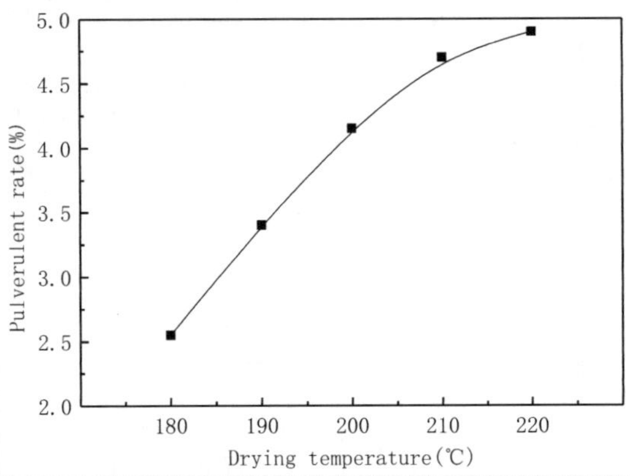


**Figure.6 Pulverulent rate of coal samples under different drying temperatures**

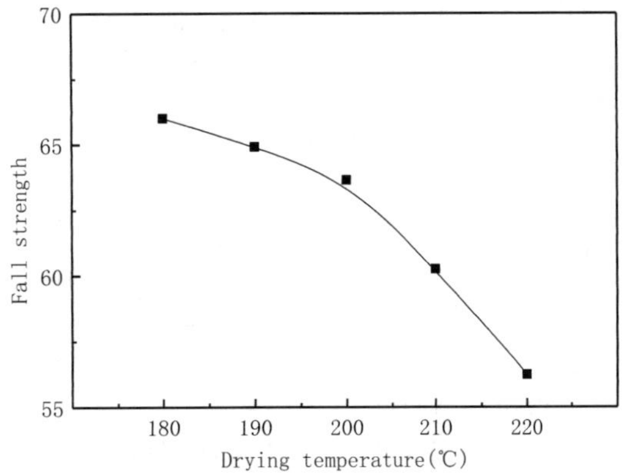


**Figure.7 Fall strength of coal samples under different drying temperatures**

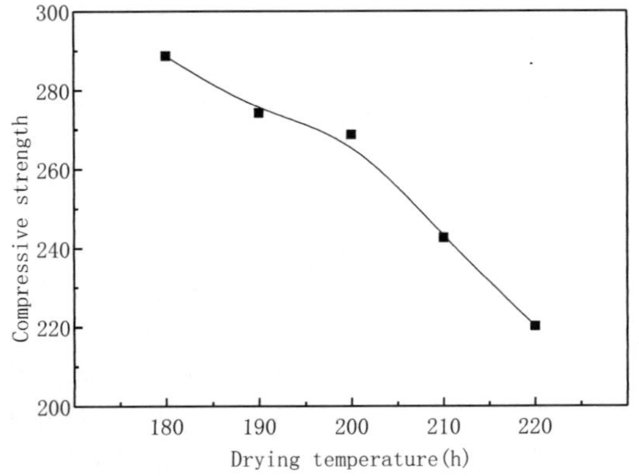


**Figure.8 Compressive strength of coal samples under different drying temperatures**

As shown in Figure 5, dehydration rate increased with the increase of temperature in the drying process, and the damage degree of the coal pore structure increased, which promoted the removal of water in the pores. When the drying temperature was below 200°C, the dehydration rate increased obviously with the increase of drying temperature. When the drying temperature was higher than 200°C, the dehydration rate decreased with the increase of drying temperature.

As shown in Figure 6, pulverulent rate increased with the increase of drying temperature. When the drying temperature was below 200°C, pulverulent rate increased obviously with the increase of drying temperature. When the drying temperature was higher than 200°C, pulverulent rate decreased with the increase of drying temperature.

As shown in Figure 7 and Figure 8, During the drying process, both fall strength and compressive strength of dry coal increased with the dry temperature increase. When the drying temperature was lower than 200°C, both fall strength and compressive strength decreased slowly with drying temperature increase. While when the drying temperature was higher than 200°C, both fall strength and compressive strength decreased fast with drying temperature increase.

Therefore, in the drying process of *Tianji* lignite, when the drying temperature was higher than 200°C, the influence of drying temperature on improving the drying effect was little, pulverulent rate increased, fall strength and compressive strength decreased significantly. Therefore, the optimum drying temperature in the process of *Tianji* lignite drying should be at about 200°C.

IOP Conf. Series: Earth and Environmental Science **61** (2017) 012048 doi:10.1088/1755-1315/61/1/012048

**References**

[1] Yang X Y et.al Increasing oil technology status and trends of low rank coal pyrolysis[J], GuangZhou Chemical Industry and Technology, 2012(13)

[2] Han Y B et.al. Structural characteristics of low-rank coal and its pyrolysis technology development[J], Bulletin of Chinese Academy of Sciences, 2013(6)

[3] Ma L Z et.al Experiment on Zhaotong brown coal pyrogenation[J] ,Journal of Kunming University of Science and Technology, 2005(3)

[4] Zeng C et.al. Effects of pretreatment in steam on the pyrolysis behavior of LoyYang brown coal[J] ,Energy Fuels, 2006(1)

[5] Zeng C et.al. Effects of dewatering on the pyrolysis and gasification reactivity of Victorian brown coa[J] ,Energy Fuels, 2007(2)

[6] Zeng C et.al. Effects of thermal pretreatment in helium on the pyrolysis behaviour of Loy Yang brown coal[J] ,Fuel, 2005(13)

# Research on the strategy of low-carbon urban planning based on residents' living and consumption

**Caixia Hu**

Tianjin Radio and TV University, No1,Yingshui Road, Nankai District, Tianjin, China, 300191
Email: 504670589@qq.com

**Abstract.** Low-carbon urban planning is a key means of lessening the impact of global climate crises, and human activities is the main cause of climate change. Therefore, from the three perspectives of carbon source of human activities, carbon emission and carbon capture, in combination with the theme of spatial, traffic and industrial, this paper puts forward the strategic measures of low-carbon urban planning. These measures provide guidelines for the construction of green low-carbon city to achieve the sustainable development.

## 1. Introduction

Climate is an important factor of the survival of mankind. In recent years, the world has encountered the most extreme weather for more than one hundred years. The bad weather is that we cannot imagine. The fifth assessment report of Intergovernmental Panel on Climate Change (IPCC) made it clear that the average surface temperature of the earth has risen by 0.85 degree since 1880. Since the 20th century, the earth's temperature has risen by 0.89 degree. Over the past century, the sea level has risen by 19 cm[1], which mainly causes by ice melting and sea water swell, while the key to the temperature rise is the increase of carbon emissions. The city is the main cause of carbon emissions, from which the solution of greenhouse effect can be worked out.

In recent years, many cities have introduced a variety of policies to deal with climate warming. Such as "green city", "ecocity", "low-carbon city" are all out of the climate changes, these ideas are helpful to finally reduce carbon emissions. Hence "low carbon" should be taken considerations as the core element. The key measure to cope with the development of low-carbon city is urban planning, which is an action guide of low-carbon city construction.

The report of Intergovernmental Panel on Climate Change (IPCC) in 2013 shows that the possibility of global warming for human activities is 95%[1]. According to the incomplete statistics, the greenhouse gas emissions of the major cities in the world accounted for 80%, while the energy consumption reached up to 75%. This major cities are the focus of high-carbon emissions and high energy consumption. From the analysis of energy use, the city energy is mainly used for industry, transportation, residence and so on; from the root analysis of carbon emissions, city is the concentration of construction, industry, population, transportation and logistics. In the main channel of carbon emissions, industrial emissions ratio is about 29%, transport emissions ratio is 33%, building emissions ratio is 39%. These energy consumption and carbon emissions are mainly used for home life. Therefore, this paper analyzes the factors and measures of low-carbon urban planning from the perspective of residents' living and consumption, which is the key issue of constructing low - carbon city.

## 2. The concept of low-carbon city planning

The concept of "low carbon" comes from the word "low carbon economy" put forward by the British government in the Energy White Paper in 2003. Since then, with the climate change and the increasing awareness of the residents' living environment, "low-carbon economy", "Low-carbon society" and "low-carbon city" have gradually become research hotspot of national experts. At present, the concept of "low-carbon city" has no unified definition international, experts unscrambled it as follows: Low-carbon city is a model of urban construction and social development helpful to reduce carbon emissions through economic development model, the concept of consumption and lifestyle changes, and the premise of ensuring the improving quality of life[2]; Through zero-carbon and low carbon research and its wide application in urban development, Low carbon city saves and intensively uses energy to reduce carbon emissions[3]; Low-carbon city has taken urban space as the carrier to develop low-carbon economy, implement green transportation and construction, change the concept of consumer consumption and innovate low-carbon technology, so as to achieve maximum reduction of greenhouse gas emissions[4]. On the basis of the current research, this study suggests that low-carbon city practices the combination of low-carbon economy and low-carbon society to implement low-carbon production, living and consumption way, so as to form an ecological synthesis catering for residents' life and promoting harmonious and sustainable development of the environment.

## 3. The core element

The establishment of low-carbon city is a systematic project involving all aspects of city works, which calls for the policy support from the government and change the way of residents' life and consumption. Low-carbon city planning includes three aspects including spatial planning, transportation planning and industrial planning. From the perspective of urban planning, the realization of low-carbon city needs three parties working together: reducing the carbon source (the source of carbon reduction), cutting down carbon emissions (the process of carbon reduction) and strengthen the carbon capture (the results of carbon reduction). On the basis of 'the three parties' principle mentioned above, the core elements of low-carbon urban planning is analyzed from the perspective of residents' living and consumption patterns. (see table 1).

**Table 1.**Core elements of low - carbon urban planning based on residents' living and consumption

| Content | Effect | Elements |
|---|---|---|
| Spatial planning | The source of carbon reduction | Using land |
| | The process of carbon reduction | Controlling urban density |
| | The result of carbon reduction | Building ecological unit |
| Traffic planning | The source of carbon reduction | Using low (zero) carbon vehicles |
| | The process of carbon reduction | Developing public transport |
| | The result of carbon reduction | Establishing the Green Railway |
| Industrial planning | The source of carbon reduction | Using low-carbon new energy |
| | The process of carbon reduction | Upgrading traditional industries |
| | The result of carbon reduction | Curing pollution and recycling waste |

## 4. Low-carbon urban planning strategy

### *4.1. Low-carbon urban spatial planning*

*4.1.1. Improving the mixed use of land.* Urban land use planning is a process of rationally arranging and coordinating the nature of the land in each block of the city. At present, in the process of urban land planning, in order to pursue the beautiful view and short-term benefits , the government neglects the actual land use efficiency and the convenient life of the residents, which causes: the degree of mixed land use is low, the urban land use structure is irrational, the proportion of land use is poor, land

IOP Conf. Series: Earth and Environmental Science **61** (2017) 012049 doi:10.1088/1755-1315/61/1/012049

planning function of the suburbs and the rare area of population is single, etc.. These phenomena leads to the increase of unnecessary traffic trips and the waste of land and construction resources in the city and increase of the carbon emissions as well. In the process of land use planning, we should pay attention to the proportion of various types of land use balance and increase the mix use of land in the construction area of the cities. In addition, we should take a notice to the scale of various land blocks, the implementation of compact space use, so that urban residents could meet their demands of life, work, and entertainment in the lowest carbon way.

*4.1.2. Controlling urban density.* According to the survey, the characteristics of modern city was pie-shaped. The residents are separated from workplace and residence. Due to the unhealthy phenomena of urban real estate the school district housing and land property market, the urban regional development is uneven, thus increasing the environment, traffic and resource pressure of the central part of the city and downtown district. We can solve these problems by controlling the city density and reducing its scale.

(1) The reasonable control of construction land is the core of urban scale growth management. The government controls the excessive growth of real estate and makes effective use of land space. It should also consider the layout of the construction land to prevent the erosion of the carbon sink space, but also take into account the differences in regional development. The area which has higher output efficiency and less impact of ecological environment should be rationally distributed.

(2) In the area of community and urban regional planning, the shape of "compact city" and "multi-center network city" should be promoted, and the effective mix of urban land should be encouraged. The root of residents' separation from workplace and residence is caused by uneven housing price gap and educational resources. By narrowing the gap between the two, the government encourages the development of suburban cities and small cities to form a number of central urban areas that can relieve the commercial centers and property pressures of large cities, which solves the problem of "weekend family", "long-distance families". By reducing carbon emissions, the quality of residents' life has been improved.

*4.1.3. Building ecological unit.* Urban planning is designed to draw policy of the "zero-carbon residential" issued by the British government, The sun and wind community of Danish Beder, the low-carbon technological innovations in UAE Masdar, and the eco-city in Tianjin, China[5], the low (zero) carbon community planning and construction should pay attention to:

(1) Construct the pollution-free and low-pollution energy use of the communities ;and build wind power communities and solar energy communities by local resources;

(2) Build public areas of waste recycling and energy recycling in the communities, following the principle of zero carbon, zero waste, etc..

(3) Create green space system in cities based on its ecological and landscape feature. Combined with the urban natural resources and spatial structure, the construction of green space system gives priority to mountain, water, field and forest, supplemented by the green space in the inner part of the city. It is built between the traffic corridor and river system surrounded by the green belt.

### *4.2. Urban public traffic planning*

*4.2.1. Encouraging the use of low (zero) carbon vehicles.* Vehicle exhaust emissions is an important component of urban carbon emissions.

(1) To encourage the purchase of electric energy private cars, the government introduces preferential policies, such as car concessions, driving concessions and others to encourage residents to buy electric energy vehicles.

(2) To encourage residents to travel by bike, the government create a good environment for them, constructing urban bicycle road separated from pedestrian and motor vehicle tracks, setting up a

special bike road in the city overpass and underground passageways, establishing bicycle parking lots with modern management system to prevent stealing and keep the environment clean.

*4.2.2. Vigorously developing public transport.* Public transport mainly refers to the bus tram, subway, light rail, bicycles, taxis and sharing small cars.

(1) Improve urban bus system. The city government can build a wide coverage of the public transport network, design reasonable bus station, provide convenience for residents' travel, designate bus lanes to strengthen the bus operation management shorten the residents' travel time, and reduce bus fares according to the residents' bus cards they hold.

(2) Speed up the construction of rail transit. Because the subway and light rail bear the characteristics of large volume, high speed and low carbon emission, the urban residents prefer to these. The urban government rationally plan urban rail transit with the development of urban overall planning and urban land development.

(3) Innovate sharing of electric cars and bicycles. Firstly, increase the sharing of bicycle usage. Many cities have now put up bike-share program to facilitate the use of residents. By increasing the type and number of bicycles, the city does not ask for returning the rent bikes, simplifying the steps of using the bicycles, in order to encourage citizens to travel by bicycles; secondly, share the innovation of small cars. The idea of sharing the power of private cars has long been proposed , but has not been put into practice. In recent years due to the deterioration of urban environment and fog invasion, the government has increased capital and technology inputs to achieve the sharing of private cars.

(4) Control the amount of private cars. From the perspective of government policy, the city should establish a series of measures on implementing the vehicle tail number limit system, controlling the car brand index, strengthening and regulating the public area parking management system and establishing fees management system of the city's public parking and so on.

*4.2.3. Establish a green wedge green space system for carbon sink absorption.* Carbon sinks refers to the amount of $CO_2$ absorbed and stored by plants. The green space between the bus roads can form a public transport green wedge corridor in conjunction with water systems, wetlands and so on. Besides, it can maximize carbon capture, absorb and reduce the atmospheric $CO_2$ composition in order to promote the development of ecological units evenly.

### *4.3. Urban industry planning*

*4.3.1. Innovating low-carbon new energy.* The city can increase the cultivation of new industries, expand the connotation of urban industry. The city can cultivate low-carbon emerging industries, relying on existing resources, and independently developing new energy and new materials, vigorously developing solar energy, bio-energy, urban waste power generation, green logistics and other industries, so as to improve the proportion of the tertiary industry and light industry.

*4.3.2. Adjusting and optimizing the industrial structure.* Industry is the pillar of urban economic development, the proportion of industrial structure adjustment, optimization and upgrading is a fundamental guarantee to change the way of urban economic growth and achieve urban low-carbon development. Most of the city's economic development depends on the first and second industries, but the tertiary industry development of low energy consumption and low carbon emissions is not complete. Therefore, the adjustment of urban heavy and light industry ratio and the optimization of industrial structure have become an important way to promote the development of low-carbon city. The city can start from the following aspects:

(1) Increase funding and technical support. The city government should gradually improve related laws and regulations of the low-carbon , strengthen the dissemination and education of low-carbon ideas and promotion; through investment, finance , taxation and other means, encourage, guide

and support the transformation of high energy consumption enterprises and the creation of new energy companies, provide funds and talents for low-carbon R & D projects;

(2) Increase the market share of low-carbon products. The government can improve the inherent power of low-carbon production enterprises, strengthen the understanding and use of low-carbon products in order to promote the formation of low-carbon industrial structure.

*4.3.3. Vigorously improving the efficiency of resources, energy efficiency and recycling waste.* The development of renewable resources industry and environmental protection industry is not only conducive to saving resources, energy, but also reduces the greenhouse gas emissions and waste dust on the environment pollution. According to the development needs of the process of urban modernization, the city combines the changes in the national energy consumption structure with the improvement of urban fuel supply, expanding oil and gas consumption, and maximizing the level of gasification and high-quality fuel supply in all types of cities. The city should focus on research and development of advanced power generation technologies such as clean coal, nuclear energy, solar energy, wind energy, advanced energy-saving technologies, carbon capture and storage, and renewable energy.

## 5. Conclusion

Urban planning determines the long-term and sustainable development of urban development, the construction of the city's physical environment requires a lot of resources, and have a great impact on the residents' life and consumption. The realization of low-carbon city needs common efforts through technology innovation, advocacy of healthy lifestyles and building a better green earth home.

**Acknowledgement**

Research on the innovation of the "welfare plan" training mode for the staff training in Tianjin(VE1054)

## References

[1] http://news.xinhuanet.com/politics/2014-11/03/c_127172537.htm
[2] Liu Zhilin &Dai Yixin 2009 *Development Research* Low-carbon City Concept and International *Experience Urban* J (vol 6) pp 1-7
[3] Crawford Jenny French Will 2008 A Low-carbon Future: Spatial Planning's Role in Enhancing Technological Innovation in the Built Environment *Energy Policy* J (vol 12) pp 4575-4579
[4] Shobhakar Dhakal 2009 Use and Carbon Emissions from Cities in China and Policy Implications Energy Policy *Urban Energy* J pp 4208-4219
[5] Ron Nyren 2008. Zero-carbon Cities.*Urban Land* J (vol 6) pp 56-63.

# Annual changes and seasonal variations of Cr in Jiaozhou Bay

**Dongfang Yang[1, 2, a], Haixia Li[1], Longlei Zhang[1], Jiangmin Li[1] and Nan Nan[1]**

[1] Accountancy Shool, Xijing University, Xian 710123, China;
[2]North China Sea Environmental Monitoring Center, SOA, Qingdao 266033, China
[a]dfyang_dfyang@126.com

**Abstract.** Many bays have been polluted by Cr due to the rapid increasing of industry, hence understanding the temporal change trends and seasonal patterns is essential to pollution control and environmental remediation. Jiaozhou Bay is a semi-closed bay located in Shandong Province, China. This paper analyzed the annual changes and seasonal variations of Cr in Jiaozhou Bay in 1979-1983. Results showed that for annual changes, Cr contents were showing decreasing trend. For seasonal variations, Cr contents were in order of spring > summer > autumn. In generally, the pollution level of Cr contents in the early stage of China's Reform and Opening-up was still low. These results were meaningful as basic information for pollution control and environmental remediation in this Bay.

## 1 Introduction

Cr is a widely used heavy metal in various industries, and many bays have been polluted by Cr due to the rapid increasing of industry since marine is the sink of pollutants [1-4]. Jiaozhou Bay is a semi-closed bay located in Shandong Province, eastern China, and has been polluted by various pollutants including Cr due to the rapid increasing of industry along with China's Reform and Opening-up[5-8]. Hence, understanding the temporal change trends and seasonal patterns is essential to pollution control and environmental remediation. This paper analyzed the annual changes and seasonal variations of Cr in Jiaozhou Bay in 1979-1983, and provided basic information for pollution control and environmental remediation in this Bay.

## 2 Materials and method

Jiaozhou Bay (35°55′-36°18′ N, 120°04′-120°23′ E) is located in the south of Shandong Peninsula, eastern China. The area, bay mouth width and average water depth are 390 km$^2$, 2.5 km and 7.0 m, respectively (Fig. 1). This bay is surrounded by cities of Qingdao, Jiaozhou and Jiaonan in the east, north and south, respectively. The bay mouth is located in the south of the bay, and is connected with the Yellow Sea. There are more than ten inflow rivers such as Loushan River, Licun River and Haibo River, all of which are seasonal rivers [9-10].

The investigation on Cr in surface waters in Jiaozhou Bay was conducted by North China Sea Environmental Monitoring Center. The investigation times were in May and August 1979, April and August 1981, April, June, July and October 1982, and May, September and October 1983, respectively [1-6], and the sampling sites were showed in Fig. 1. The investigation and measurement of Cr were following by National Specification for Marine Monitoring [11]. The contents of Cr were listed in Table 1.

Fig.1 Geographic location and sampling sites of Jiaozhou Bay

Table 1 Cr contents in different months during 1979-1983

| Month | 1979 | 1981 | 1982 | 1983 |
|---|---|---|---|---|
| April | | 0.48-32.32 | 0.81-2.11 | |
| May | 0.20-112.30 | | | 0.13-3.96 |
| June | | | 0.33-9.76 | |
| July | | | 1.02-2.42 | |
| August | 0.10-1.40 | 0.18-1.85 | | |
| September | | | | 0.70-3.78 |
| October | | | 0.24-1.35 | 0.44-4.17 |

**3 Annual changes of Cr contents**

In according to investigation data on Cr contents in different months during 1979-1983, the annual changes of both highest and lowest values of Cr contents were showed in Fig. 2. The Cr contents in Jiaozhou Bay in 1979, 1981, 1982 and 1983 were ranging from 0.1-112.3 μg $L^{-1}$, 0.18-32.32 μg $L^{-1}$, 0.24-9.76 μg $L^{-1}$ and 0.13-4.17 μg $L^{-1}$, respectively. For the highest values, the annual change trend were significant decreasing during 1979-1983, yet for the lowest values were showing increasing trend. In according to the guideline of Cr in National Standard of China for Seawater Quality (GB3097-1997) (Table 2), water quality in 1979 was Grade III yet for other years were Grade I. In generally, the pollution level of Cr in Jiaozhou Bay was still slight in the early stage of China's Reform and Opening-up.

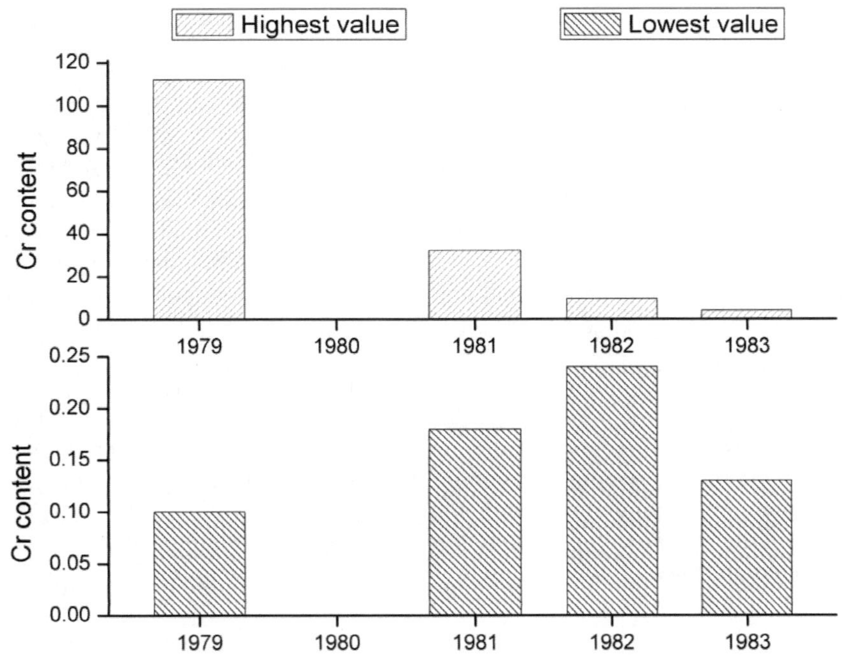


Fig.2 Annual changes of highest and lowest values of Cr contents /μg $L^{-1}$

Table 2 The guideline of Cr in National Standard of China for Seawater Quality (GB3097-1997)

| Grade | I | II | III |
|---|---|---|---|
| Guideline /μg $L^{-1}$ | 50 | 100 | 200 |

**4 Seasonal variations of Cr contents**

In according to the climate in study area, April, May and June are belong to spring, July, August and September are belong to summer, and October, November and December are belong to autumn, respectively. Cr contents in Jiaozhou Bay in different seasons during 1979-1983 were listed in Table 3. For seasonal variations, Cr contents were in order of spring > summer > autumn. In according to the guideline of Cr in National Standard of China for Seawater Quality (GB3097-1997) (Table 2), water quality in spring 1979 was Grade III yet for other seasons during research years were all Grade I (Table 4). In generally, the seasonal variations of Cr contents in Jiaozhou Bay were significant. For temporal change trend of Cr contents in different seasons, the highest values of Cr contents in spring were showing decreasing trend, while the highest values of Cr contents in summer and autumn were showing increasing trend.

Table 3 Cr contents in Jiaozhou Bay in different seasons during 1979-1983 /μg $L^{-1}$

| Season | Spring | Summer | Autumn |
|---|---|---|---|
| 1979 | 0.20-112.30 | 0.10-0.40 | |
| 1981 | 0.48-32.32 | 0.18-1.85 | |
| 1982 | 0.33-9.76 | 1.02-2.42 | 0.24-1.35 |
| 1983 | | 0.70-3.78 | 0.44-4.17 |

Table 4 Assessment results of water quality based on Cr in Jiaozhou Bay in different seasons during 1979-1983

| Season | Spring | Summer | Autumn |
|---|---|---|---|
| 1979 | I, II, III | I | |
| 1981 | I | I | |
| 1982 | I | I | I |

| | | | |
|---|---|---|---|
| 1983 | I | I | I |

## 5 Conclusions

This paper analyzed the annual changes and seasonal variations of Cr in Jiaozhou Bay through 1979 to 1983. The Cr contents in Jiaozhou Bay in 1979, 1981, 1982 and 1983 were ranging from 0.1-112.3 μg $L^{-1}$, 0.18-32.32 μg $L^{-1}$, 0.24-9.76 μg $L^{-1}$ and 0.13-4.17 μg $L^{-1}$, respectively. For the highest values, the annual change trend was significant decreasing during 1979-1983. For seasonal variations, Cr contents were in order of spring > summer > autumn. For temporal change trend of Cr contents in different seasons, the highest values of Cr contents in spring were showing decreasing trend, while the highest values of Cr contents in summer and autumn were showing increasing trend. In general, the pollution level of Cr contents in the early stage of China's Reform and Opening-up was still low.

## Acknowledgment

This research was sponsored by the China National Natural Science Foundations (31560107) , Doctoral Degree Construction Library of Guizhou Nationalities University, Education Ministry's New Century Excellent Talents Supporting Plan (NCET-12-0659) and (31500394), Research Projects of Guizhou Nationalities University ([2014]02), Research Projects of Guizhou Province Ministry of Education (KY [2014] 266), Research Projects of Guizhou Province Ministry of Science and Technology (LH [2014] 7376).

## References

[1] Yang DF and Miao ZQ: Marine Bay Ecology (I): Beijing, Ocean Precess, (2010), p. 1-320. (in Chinese)
[2] Yang DF and Gao ZH: Marine Bay Ecology (II): Beijing, Ocean Precess, (2010), p. 1-330. (in Chinese)
[3] Yang DF, Gao ZH, Sun JY, et al.: Coastal Enginerring, Vol. 27 (2008), p. 48- 53. (in Chinese)
[4] Yang DF, Chen Y, Wang H, et al.: Coastal Engineering, Vol. 29 (2010), p. 73-82. (in Chinese)
[5] Yang DF, Chen Y, Liu CX, et al.: Coastal Engineering, Vol. 32(2013), p. 68-78. (in Chinese)
[6] Yang DF, Wang FY, He HZ, et al.: Applied Mechanics and Materials,Vol. 675-677 (2014), p. 329-331.
[7] Chen Y, Yu QH, Li TJ, et al.: Applied Mechanics and Materials, Vol.644-650 (2014), p. 5333-5335.
[8] Yang DF, Zhu SX, Wang FY, et al.: 2014 IEEE workshop on advanced research and technology industry applications. Part D, Vol. (2014), p. 1018-1020.
[9] Yang DF, Chen Y, Gao ZH, et al.: Chinese Journal of Oceanology and Limnology, Vol. 23 (2005), pp. 72-90. (in Chinese with English Abstract)
[10] Yang DF, Wang F, Gao ZH, et al.: Marine Science, Vol. 28(2004), p. 71-74. (in Chinese with English Abstract)
[11] State Ocean Administration. The specification for marine monitoring: Beijing, Ocean Precess, (1991). (in Chinese)

# Pollution level and source of Hg in Jiaozhou Bay 1987

**Dongfang Yang[1,2,4,a], Danfeng Yang[3], Sixi Zhu[1,2], Ming Wang [1,2], Jianxun Chai [1,2]**

[1]Research Center for Karst Wetland Ecology, Guizhou Minzu University, Guizhou Guiyang, Guizhou Guiyang, China;
[2]College of Chemistry and Environmental Science, Guizhou Minzu University, Shanghai, 550025, China;
[3]College of Information Science and Engineering, Fudan University, Shanghai, 200433, China;
[4]North China Sea Environmental Monitoring Center, SOA, Qingdao 266033, China.
[a] dfyang_dfyang@126.com

**Abstract.** Hg pollution in marine bays has been one of the most critical environmental issues, and it is necessary to understand the distribution, pollution level and sources of Hg. Based on the investigation data on Hg in Jiaozhou Bay in May, July and November 1987, this paper analyzed the content, pollution level, and sources of Hg. Results showed that Hg contents in surface waters in Jiaozhou Bay in May, July and November 1987 were 0.150—0.264 μg $L^{-1}$, 0.088—1.104 μg $L^{-1}$ and 0.007—0.088 μg $L^{-1}$, respectively. The two major Hg sources in Jiaozhou Bay were river flow and marine current, whose source strengths were 0.264—1.104 μg $L^{-1}$ and 0.088—0.376 μg $L^{-1}$, respectively. The source strength of river flow could be worse than Grade IV, while the source strength of marine current was Grade III and IV, indicated that the pollution level of Hg in land river had reached a relative high level in 1987. These findings provided basis for the research and pollution control countermeasures.

## 1 Introduction

Hg is one of the heavy metal widely used in industry, agriculture and everyday life. A large amount of Hg-containing wastes were generated and discharged to rivers and to marine bays so that ocean became the sink of various pollutants. However, Hg is also one of the critical heavy metal due to the high toxicity, therefore it is necessary to understanding the distribution, pollution level and sources of Hg.

Jiaozhou Bay is a semi-closed bay located in Shandong Province, eastern China, and has been polluted by various pollutants including Hg after China's Reform and Opening-up [1-6]. Based on the investigation data on Hg in Jiaozhou Bay in May, July and November 1987, the aim of this paper was to analyze the content, pollution level, and sources of Hg, and to provide basis for the research and pollution control countermeasures.

## 2 Study area and data collection

*Study area.* Jiaozhou Bay is located in the south of Shandong Province, eastern China (35°55′—36°18′ N, 120°04′—120°23′ E), with the total area and average water depth of 446 $km^2$ and 7 m, respectively. This bay is a semi-closed bay connected to the Yellow Sea in the south, with a bay mouth with of 3 km. There are a dozen of rivers including Dagu River, Haibo Rriver, Licun Rriver, and Loushan Rriver etc., all of which are seasonal rivers [7—8].

*Data collection.* The investigation on Hg in surface waters in Jiaozhou Bay was carried on in May, July and November 1987 in six investigation sites (i.e., 2031, 2032, 2033, 2034, 2035, 2047) (Fig. 1). Hg in waters was sampled and monitored follow by National Specification for Marine Monitoring [9].

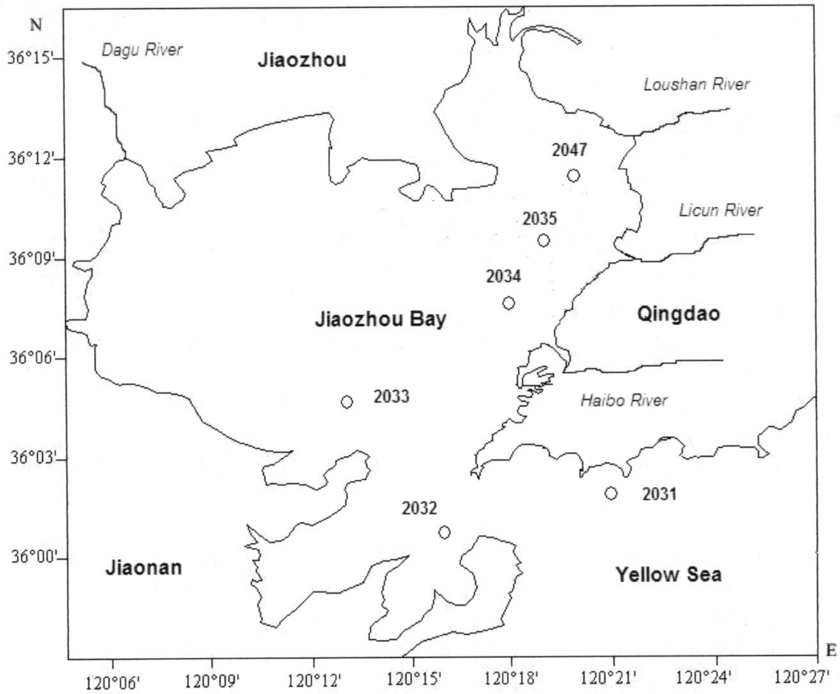


Fig.1 Geographic location and sampling sites of Jiaozhou Bay

## 3 Results and discussion

*Content and pollution of Hg.* Hg contents in surface waters in Jiaozhou Bay in May, July and November 1987 were 0.150—0.264 μg $L^{-1}$, 0.088—1.104 μg $L^{-1}$ and 0.007—0.088 μg $L^{-1}$, respectively (Table 1). In generally, Hg' contents were ranging from 0.007—1.104 μg $L^{-1}$ in 1987, and Hg's pollution level were showing significant spatial-temporal variations (Table 1) in according to National Standard of China for Seawater Quality (GB3097-1997) (Table 2),

Table 1 Content and pollution level of Hg in Jiaozhou bay in 1987

| Month | May | July | November |
|---|---|---|---|
| Content /μg $L^{-1}$ | 0.150—0.264 | 0.088—1.104 | 0.007—0.088 |
| Grade | II, III and IV | II,III and IV, worse than IV | I, II |

Table 2 Guidelines for Hg in National Standard of China for Seawater Quality (GB3097—1997)

| Grade | I | II | III and IV |
|---|---|---|---|
| Content /μg $L^{-1}$ | 0.05 | 0.20 | 0.50 |

In May 1987, Hg contents in Site 2034, Site 2033 and Site 2032 in coastal waters in the east and south of the bay were ranging from 0.241—0.264 μg $L^{-1}$, and were belong to Grade III and IV, while in other Sites were ranging from 0.05—0.20 μg $L^{-1}$, and were relative low and belong to Grade II. In

July 1987, the highest Hg content was in Site 2035 in the east of the bay (1.104 μg $L^{-1}$), and was much more high than Grade III and IV, indicating that the pollution level in this location was worse than Grade IV. Meanwhile, Hg contents was as high as 0.376 μg $L^{-1}$ in the open waters, and the pollution level in this location was also worse than Grade IV. However, Hg contents in other Sites were ranging from 0.05—0.20 μg $L^{-1}$, and were relative low and belong to Grade II. In November 1987, the highest Hg content was in the open water (0.088 μg $L^{-1}$) and was belong to Grade II, while Hg contents in other locations were lower than 0.05 μg $L^{-1}$) and was belong to Grade I. In generally, Hg's contents were relative high in coastal waters in the estuaries of the major inflow rivers and in the open waters from spatial aspect, while Hg's pollution levels were relative high in summer yet relative low in autumn form temporal aspect.

*Distribution and source of Hg.* The horizontal distribution of Hg contents in marine bays was mainly determined by the spatial variation of source input and water exchange, and was important evidence to identify and source of Hg. In May 1987, the high Hg contents were in the estuary of the major inflow river in the northeast of the bay, and Hg contents were forming a series of parallel lines which were decreasing from the bay to the bay mouthflow direction of the major rivers (Fig. 2a). Hence, it could be defined that river was the major source of Hg in May 1987. In July 1987, the high Hg content was in Site 2035 in the estuary of Licun River in the coastal waters in the northeast of the bay, and Hg contents were forming a series of semi-circles which were decreasing from the northeast of the bay to center of the bay along with the flow direction of Licun River (Fig. 2b). Hence, it could be defined that river was the major source of Hg in July 1987. In November 1987, high value (0.088 μg $L^{-1}$) was in Site 2031 in the open waters, and Hg contents were forming a series of parallel lines which were decreasing from the open waters to the bay (0.007 μg $L^{-1}$) along with the flow direction of marine current (Fig. 2c). However, Hg contents were very low in the inside of the bay (0.007—0.014 μg $L^{-1}$). Hence, it could be defined that marine current was the major source of Hg in November 1987.

In generally, the two major Hg sources in Jiaozhou Bay were river flow and marine current, whose source strengths were 0.264—1.104 μg $L^{-1}$ and 0.088—0.376 μg $L^{-1}$, respectively. In according to National Standard of China for Seawater Quality (GB3097-1997) (Table 2), it could be found that the source strength of river flow could be worse than Grade IV, while the source strength of marine current was Grade III and IV. Hence, it could also be found that the pollution level of Hg in land river had reached a relative high level in 1987.

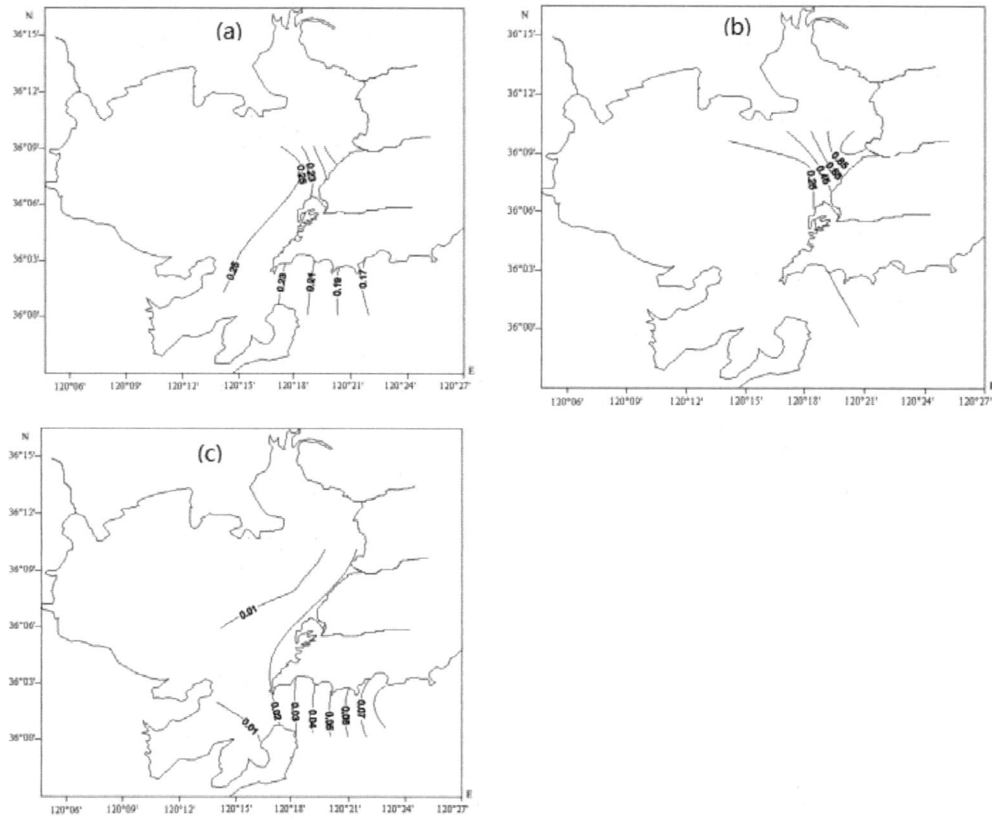


Fig. 2 Horizontal distribution of Hg in in Jiaozhou Bay in a) May, b) July and c) November 1987/μg $L^{-1}$

## 4 Conclusion

Hg contents in surface waters in Jiaozhou Bay in May, July and November 1987 were 0.150—0.264 μg $L^{-1}$, 0.088—1.104 μg $L^{-1}$ and 0.007—0.088 μg $L^{-1}$, and were showing significant spatial-temporal variations. Hg's contents were relatively high in coastal water in the estuaries of the major inflow rivers and in the open water from spatial aspect, while Hg's pollution levels were relatively high in summer yet relatively low in autumn form temporal aspect.

The two major Hg sources in Jiaozhou Bay were river flow and marine current, whose source strengths were 0.264—1.104 μg $L^{-1}$ and 0.088—0.376 μg $L^{-1}$, respectively. The source strength of river flow could be worse than Grade IV, while the source strength of marine current was Grade III and IV, indicated that the pollution level of Hg in land river had reached a relative high level in 1987.

## Acknowledgment

This research was sponsored by the China National Natural Science Foundation (31560107) , Doctoral Degree Construction Library of Guizhou Nationalities University, Education Ministry's New Century Excellent Talents Supporting Plan (NCET-12-0659) and Research Projects of Guizhou Nationalities University ([2014]02), Research Projects of Guizhou Province Ministry of Education (KY [2014] 266), Research Projects of Guizhou Province Ministry of Science and Technology (LH [2014] 7376).

## References

[1] CHEN Y, GAO ZH, QU YH, et al.: Chinese Journal of Oceanology Limnology, Vol. 25(2007), p.

455-458. (in Chinese)
[2] YANG DF, CAO HR, GAO ZH, et al.: Ocean Environmental Science, Vol. 27 (2008), p. 37-39.
[3] YANG DF, WANG LL, GAO ZH, et al.: Ocean Environmental Science, Vol. 28 (2009), p. 501-505 .
[4] CHEN Y, ZHANG YJ, GUO JH, et al.: Ocean Development and Management, Vol. 30 (2013), p. 81-83.
[5] YANG DF, SUN PY, JU L, et al: Coastal Engineering, Vol. 32(2013), p.65-76.
[6] YANG DF, XU ZJ QU YF, et al.:Coastal Engineering, Vol. 33(2014), p. 67-78.
[7] CHEN Y, QU YF, PEI RL, et al.:Advanced Materials Research, Vol.955-959 (2014), p. 2491-2495.
[8] YANG DF, ZHU SX, WANG FY, et al.:Advanced Materials Research, Vol.955-959(2014), p. 2496-2500.
[9] YANG DF, ZHU SX, WANG FY, et al.: Applied Mechanics and Materials, Vol.556-562 (2014), p. 633-636.
[10] YANG DF, WANG FY, HE HZ, et al.: Advanced Materials Research, Vols.955-959(2014),p. 1443-1447.
[11] YANG DF, GENG X, QU YF, et al.: Ocean Development and Management, Vol. 31 (2014), p. 112-118. (in Chinese)
[12] YANG D F, CHEN Y, GAO Z H, et al.: Chinese Journal of Oceanology and Limnology, Vol. 23(2005),p. 72-90.
[13] Yang DF, Wang FY, Gao ZH, et al. Marine Science, Vol. 28 (2004), p. 71-74. (in Chinese)
[14] China's State Oceanic Administration: The specification for marine monitoring (Ocean Press, Beijiang 1991), p.205-282. (in Chinese)

IOP Conf. Series: Earth and Environmental Science **61** (2017) 012052 doi:10.1088/1755-1315/61/1/012052

# Spatial-temporal changes of Cr sources in Jiaozhou Bay

**Dongfang Yang[1, 2, 3, a], Fengyou Wang [1, 2, b,c], Sixi Zhu[1, 2], Chunhua Su[1, 2] and Zhikang Wang [1, 2]**

[1]Research Center for Karst Wetland Ecology, Guizhou Minzu University, Guiyang 550025, China;
[2]College of Chemistry and Environmental Science, Guizhou Minzu University, Guiyang 550025, China;
[3]North China Sea Environmental Monitoring Center, SOA, Qingdao 266033, China.
[a]dfyang_dfyang@126.com, [b]corresponding author, [c]fywang@yahoo.com.cn.

**Abstract.** Cr pollution in marine bay is one of the environmental issues in many countries and regions, and the understanding of the spatial-temporal variation of Cr sources in marine bay is essential to pollution control. Based on investigation on Cr in surface waters, this paper analyzed distributions and spatial-temporal variations of Cr sources in Jiaozhou Bay, eastern China during 1979-1983. Results showed that for spatial variations, the strong Cr sources were in estuaries of the major inflow rivers (i.e. Haibo River, Licun River and Loushan River), and the source strengths were 4.17-112.30 $\mu g\ L^{-1}$. For temporal variations, the source strengths were changing from moderate stage (1979-1981) to slight stage (1982-1983),whose source strengths were 32.32-112.30 $\mu g\ L^{-1}$ and 4.17-19.76 $\mu g\ L^{-1}$, respectively. In general, the spatial-temporal variations of Cr sources in Jiaozhou Bay were significant, which should be taken into account for the countermeasures for pollution control.

## 1 Introduction

Marine is the sink of pollutants, and a lot of marine bay has been polluted due to the rapid increase of industry and population [1-4]. Cr is a widely used heavy metal in various industries, and Cr pollution in marine bay is one of the environmental issues in many countries and regions. Hence, understanding the spatial-temporal variation of Cr sources in marine bays are essential to pollution control.

Jiaozhou Bay is a semi-closed bay located in Shandong Province, eastern China, and has been polluted by various pollutants including Cr due to the rapid increasing of industry along with China's Reform and Opening-up [5-8]. In order to better understand the spatial-temporal variation of Cr sources, this paper analyzed distributions and spatial-temporal variations of Cr sources in Jiaozhou Bay, eastern China during 1979-1983. The results provided basic information for pollution control and environmental remediation in this Bay.

## 2 Materials and method

Jiaozhou Bay (35°55′-36°18′ N, 120°04′-120°23′ E) is located in the south of Shandong Peninsula, eastern China. The area, bay mouth width and average water depth and average water depth are 390 $km^2$, 2.5 km and 7.0 m, respectively (Fig. 1). This bay is surrounding by cities of Qingdao, Jiaozhou and Jiaonan in the east, north and south, respectively. The bay mouth is located in the south of the bay, and is connected with the Yellow Sea. There are more than ten inflow rivers such as Loushan River, Licun River and Haibo River, all of which are seasonal rivers [9-10].

The investigation on Cr in surface waters in Jiaozhou Bay was conducted by North China Sea

Environmental Monitoring Center. The investigation times were in May and August 1979, April and August 1981, June1982, and May, September and October 1983, respectively [1-8], and the sampling sites were showed in Fig. 1. The investigation and measurement of Cr were following by National Specification for Marine Monitoring [11].

Fig.1 Geographic location and sampling sites of Jiaozhou Bay

**3 Spatial variations of Cr source**

In May 1979, high Cr contents were appearing the estuaries of Loushan River and Licun River, and the contour lines of Cr contents were a series of parallel lines that decreasing from the northeast of the bay (112.30 μg $L^{-1}$) to the bay mouth in the south (0.20 μg $L^{-1}$). In August 1979, high Cr contents were appearing the estuaries of Haibo River, and the contour lines of Cr contents were a series of parallel lines that decreasing from the east of the bay (1.40 μg $L^{-1}$) to the west of the bay (0.10 μg $L^{-1}$).

In April 1981, high Cr contents were appearing the estuaries of Haibo River, and the contour lines of Cr contents were a series of semi-concentric circles that decreasing from the east of the bay (32.32 μg $L^{-1}$) to the center of the bay (0.48 μg $L^{-1}$). Meanwhile, high Cr contents were appearing the estuaries of Loushan River, and the contour lines of Cr contents were a series of semi-concentric circles that decreasing from the northeast of the bay (25.40 μg $L^{-1}$) to waters around (0.88 μg $L^{-1}$). In August 1981, high Cr contents were appearing the estuaries of Haibo River, and the contour lines of Cr contents were a series of semi-concentric circles that decreasing from the east (1.85 μg $L^{-1}$) to the center of the bay (0.70 μg $L^{-1}$).

In June 1982, high Cr contents were appearing the estuaries of Loushan River, and the contour lines of Cr contents were a series of semi-concentric circles that decreasing from the northeast of the bay (9.76 μg $L^{-1}$) to the bay mouth (0.33 μg $L^{-1}$).

In May 1983, high Cr contents were appearing the estuaries of Loushan River, and the contour lines of Cr contents were a series of semi-concentric circles that decreasing from the northeast of the bay (3.96 μg $L^{-1}$) to the bay mouth (0.13 μg $L^{-1}$). In September 1983, high Cr contents were appearing the estuaries of Loushan River and Licun River, and the contour lines of Cr contents were a series of semi-concentric circles that decreasing from the northeast of the bay (3.78 μg $L^{-1}$) to the bay mouth (0.70 μg $L^{-1}$). In October 1983, high Cr contents were appearing the estuaries of Loushan River and

Licun River, and the contour lines of Cr contents were a series of semi-concentric circles that decreasing from the northeast of the bay (4.17 μg $L^{-1}$) to the bay mouth (1.44 μg $L^{-1}$).

In generally, the spatial distributions of Cr contents were indicating the spatial variations of Cr sources. In according to the spatial distributions of Cr contents, it could be defined that high value regions were appearing in the estuaries of the inflow rivers, particularly in Haibo River, Loushan River and Licun River, and the source strengths were ranging from 4.17-112.30 μg $L^{-1}$.

## 4 Temporal variations of Cr source

In according to the spatial distributions of Cr contents in different years, the temporal variations of Cr cource were also significant. In 1979, high Cr contents were appearing the estuaries of Haibo River, Loushan River and Licun River, and the source strength could be as high as 112.30 μg $L^{-1}$. In 1981, high Cr contents were appearing the estuaries of Haibo River and Loushan River, and the source strength was 32.32 μg $L^{-1}$. In 1982, high Cr contents were appearing the estuaries of Loushan River, and the source strength was 9.76 μg $L^{-1}$. In 1983, high Cr contents were appearing the estuaries of Loushan River and Licun River, and the source strength was 4.17 μg $L^{-1}$.

For temporal variations, the source strengths were changing from two stages of moderate to slight. The first stage was 1979-1981,whose source strengths were 32.32-112.30 μg $L^{-1}$. The second stage was 1982-1983,whose source strengths were 4.17-19.76 μg $L^{-1}$. In generally, the temporal variations of Cr sources in Jiaozhou Bay were also significant, which should be taken into account the countermeasures for pollution control.

## 5 Conclusion

This paper analyzed distributions and spatial-temporal variations of Cr sources in Jiaozhou Bay, eastern China during 1979-1983. The spatial variation of Cr sources was significant that high value regions were appearing in the estuaries of the inflow rivers, particularly in Haibo River, Loushan River and Licun River, and the source strengths were ranging from 4.17-112.30 μg $L^{-1}$. For temporal variations, the source strengths were changing from moderate stage to slight. The first stage was 1979-1981,whose source strengths were 32.32-112.30 μg $L^{-1}$. The second stage was 1982-1983,whose source strengths were 4.17-19.76 μg $L^{-1}$. The spatial-temporal variations of Cr sources in Jiaozhou Bay were significant, which should be taken into account the countermeasures for pollution control.

## Acknowledgment

This research was sponsored by the China National Natural Science Foundation (31560107), Education Ministry's New Century Excellent Talents Supporting Plan (NCET-12-0659), Doctoral Degree Construction Library of Guizhou Nationalities University and (31500394), Research Projects of Guizhou Nationalities University ([2014]02), Research Projects of Guizhou Province Ministry of Education (KY [2014] 266), Research Projects of Guizhou Province Ministry of Science and Technology (LH [2014] 7376).

## References

[1] Yang DF and Miao ZQ: Marine Bay Ecology (I): Beijing, Ocean Precess, (2010), p. 1-320. (in Chinese)
[2] Yang DF and Gao ZH: Marine Bay Ecology (II): Beijing, Ocean Precess, (2010), p. 1-330. (in Chinese)
[3] Yang DF, Gao ZH, Sun JY, et al.: Coastal Enginerring, Vol. 27 (2008), p. 48- 53. (in Chinese)
[4] Yang DF, Chen Y, Wang H, et al.: Coastal Engineering, Vol. 29 (2010), p. 73-82. (in Chinese)
[5] Yang DF, Chen Y, Liu CX, et al.: Coastal Engineering, Vol. 32(2013), p. 68-78. (in Chinese)
[6] Yang DF, Wang FY, He HZ, et al.: Applied Mechanics and Materials,Vol. 675-677 (2014), p. 329-331.
[7] Chen Y, Yu QH, Li TJ, et al.: Applied Mechanics and Materials, Vol.644-650 (2014), p. 5333-5335.

[8] Yang DF, Zhu SX, Wang FY, et al.: 2014 IEEE workshop on advanced research and technology industry applications. Part D, Vol. (2014), p. 1018-1020.
[9] Yang DF, Chen Y, Gao ZH, et al.: Chinese Journal of Oceanology and Limnology, Vol. 23 (2005), pp. 72-90. (in Chinese with English Abstract)
[10] Yang DF, Wang F, Gao ZH, et al.: Marine Science, Vol. 28(2004), p. 71-74. (in Chinese with English Abstract)
[11] State Ocean Administration. The specification for marine monitoring: Beijing, Ocean Precess, (1991), p. 205-282. (in Chinese)

IOP Conf. Series: Earth and Environmental Science **61** (2017) 012053 doi:10.1088/1755-1315/61/1/012053

# Evaluation of soil nutrients about typical economic forest lands of low hilly areas in eastern part of Zhejiang Province

**Jijiang Chen[1], Fangchun Lu[2a], Yefeng Zou[1], Zhaowei Shen[2], Gang Li[2] and Xijiong Zhou[1]**

1. The Water-Control Bureau of Yuyao City, Zhejiang, China
2. Department of Environmental Engineering, Zhejiang Guangchuan Engineering Consulting Co., Ltd., Zhejiang, China

**Abstract.** Through sampling of seven economic forest lands in low hilly areas of Zhejiang province the samples were tested, including soil organic matter, total nitrogen, available phosphorus and available potassium. The characteristics of soil nutrients in different economic forest lands were analysed by statistical method and the comprehensive evaluation of soil nutrient quality was carried out. The results showed that there were differences in soil nutrient contents in different economic forest lands. The overall status of soil nutrients in the economic forest land is in the "good" level; Among them, the content of soil organic matter and total nitrogen in the economic forest land were relatively abundant, However, the content of available potassium in soil was not enough, which was the main limiting factor for the development of economic forest land.

## 1. Introduction

Soil is an important part of terrestrial ecosystem, which is one of the main conditions for human survival [1]. The most important role of the soil is to provide nutrients and water for plants, and the change of land use has a variety of effects on soil nutrients. It is important to study the characteristics of soil nutrients in different land use patterns in order to understand the maintenance of soil fertility [2]. Soil fertility is the comprehensive performance of various basic properties of soil, how to quantitatively characterize soil fertility and its evaluation has become one of the hot spots of scientists in the world [3]. At present, there were many researches on the comprehensive evaluation of soil quality, such as the evaluation of forest soil quality [4], urban green land soil quality [5], the nursery land soil quality [6], agricultural land soil quality [7], etc. The evaluation methods mainly involved clustering analysis, factor weighting, Nemero comprehensive index method [8-10]. In this study, the low hilly area in southeast of Zhejiang was selected as the study area. Soil samples were collected in the main economic forest lands, the soil organic matter, soil total nitrogen, soil available phosphorus and soil potassium nutrient index were tested to analysis of soil nutrient characteristics. Comprehensive evaluation of soil nutrient level was carried out by using the improved Nemero comprehensive index method. The study can provide reference for the management of land in the low hilly area of Zhejiang province.

## 2. Materials and Methods

In this study, according to the land use types of low hilly area, seven representative economic forest land distributed in four towns (including Myrica rubra, Phyllostachys heterocycla, Camellia sinensis, Acer palmatum Thunb, Cerasus yedoensis, Cerasus pseudocerasus and Grassland), totally eighteen plots were study. In each plot, Sampling according to S-type principle, three-point sampling for mixing (sampling depth 0-20cm). Then, the mixed soil sample into the bag, and brought back to the laboratory to dry, pick root, grinding and sieving standby. The mineral nutrients which can be absorbed directly or transformed by the plant roots are called soil nutrients, which are mainly derived from soil minerals, soil organic matter and fertilization. It is generally believed that the soil organic matter content, total nitrogen, available phosphorus and available potassium in soil are important reflection of soil nutrient status. Therefore, this study is mainly aimed at these four indicators. Preliminary analysis of the test data in the Excel, the relevant statistical data analysis using SPSS software, the required graphics using Origin drawing software. In this study, the soil nutrient classification standard in the second soil survey in China was adopted. At the same time, combined with the actual situation in the low hilly area, the soil organic matter, soil total nitrogen, soil available phosphorus, soil available potassium was divided into four levels according to table 1.

**Table 1.** Gradation standard of soil nutrient index

| Soil nutrient index | $X_{imin}$ | $X_{imid}$ | $X_{imax}$ |
|---|---|---|---|
| Soil organic matter /g·kg$^{-1}$ | 5 | 10 | 20 |
| Soil total nitrogen /g·kg$^{-1}$ | 0.5 | 0.8 | 1.2 |
| Soil available phosphorus /mg·kg$^{-1}$ | 5 | 10 | 20 |
| Soil available potassium /mg·kg$^{-1}$ | 50 | 90 | 150 |

In this paper, the evaluation methods of different kinds of soil nutrients were the improved Nemero comprehensive index method [11]. The calculation method of single soil nutrient index is as follows:

$$P_i = \begin{cases} X_i / X_{imin} & (X_i \leqslant X_{imin}) \\ 1 + (X_i - X_{imin})/(X_{imid} - X_{imin}) & (X_{imin} < X_i < X_{imid}) \\ 2 + (X_i - X_{imid})/(X_{imax} - X_{imid}) & (X_{imid} < X_i < X_{imax}) \\ 3 & (X_i > X_{imax}) \end{cases}$$

Where $P_i$ is single soil nutrient index. $X_i$ is the value of soil nutrient test value. $X_{min}$, $X_{mid}$, $X_{max}$ are the corresponding value of the gradation standards, respectively. $i$ is evaluation index number ($i$ = 1,2,3,4). Comprehensive index of soil nutrients was used to evaluate soil nutrient status of certain economic forest land. the calculation method is:

$$Q = \sqrt{\frac{(P_i)^2 + (P_{imin})^2}{2} \cdot \frac{(n-1)}{n}}$$

Where $Q$ is a certain kind of soil nutrient index. $\bar{Pi}$ is the average value of soil nutrient index. $P_{imin}$ is the minimum value of soil nutrient index. $n$ is soil nutrient index number ($n$ = 1,2,3,4). In this formula, using the $P_{imin}$ instead of the $P_{imax}$ in the Nemero formula. And to increase the correction ($n$ - 1) /$n$. On the one hand, the focal point is made to stand out impact of the worst soil nutrient factors on soil nutrient quality. And this reflected rules of minimum biological growth factor. On the other hand, the more factors, the greater the correction value, the higher the credibility. Table 2 showed the gradation standard of soil nutrient comprehensive index.

**Table 2.** Gradation standard of soil nutrient comprehensive index

| Gradation | I | II | III | IV |
|---|---|---|---|---|
| Range of $Q$ | $\geq 2$ | 1.5 - 2.0 | 1.0 - 1.5 | $< 1.0$ |
| Quality review | excellent | good | qualified | bad |

## 3. Results and Analysis

### *3.1 Soil organic mater*

Soil organic matter refers to the carbonaceous organic matter in soil. Including a variety of animal and plant residues, microorganisms and various organic compounds which decomposed and compounded by them. Soil organic matter is not only the source of mineral nutrition and organic nutrition, but also the energy source of heterotrophic microorganisms in soil. At the same time, it is also an important factor in the formation of soil structure. The content of soil organic matter, to some extent, reflects the fertility of the soil. Figure 1a reflected the characteristics of soil organic matter in different economic forest land. The characteristics of soil organic matter of seven typical economic forest land was the characteristics of soil organic matter of seven typical economic forest land was Acer palmatum Thunb (72.57 g·kg$^{-1}$) > Cerasus yedoensis (68.79 g·kg$^{-1}$) > Camellia sinensis (58.81 g·kg$^{-1}$) > Myrica rubra (55.72 g·kg$^{-1}$) > Phyllostachys heterocycla (55.55 g·kg$^{-1}$) > Cerasus pseudocerasus (47.46 g·kg$^{-1}$) > Grassland (37.15 g·kg$^{-1}$). The average soil organic matter content of Acer palmatum Thunb was 1.95 times that of Grassland. There were also differences in soil organic matter in different regions. For example, soil organic matter content (62.60 g·kg$^{-1}$) of Myrica rubra forest land in Zhangting town was more than that (52.28 g·kg$^{-1}$) in Mazhu town.

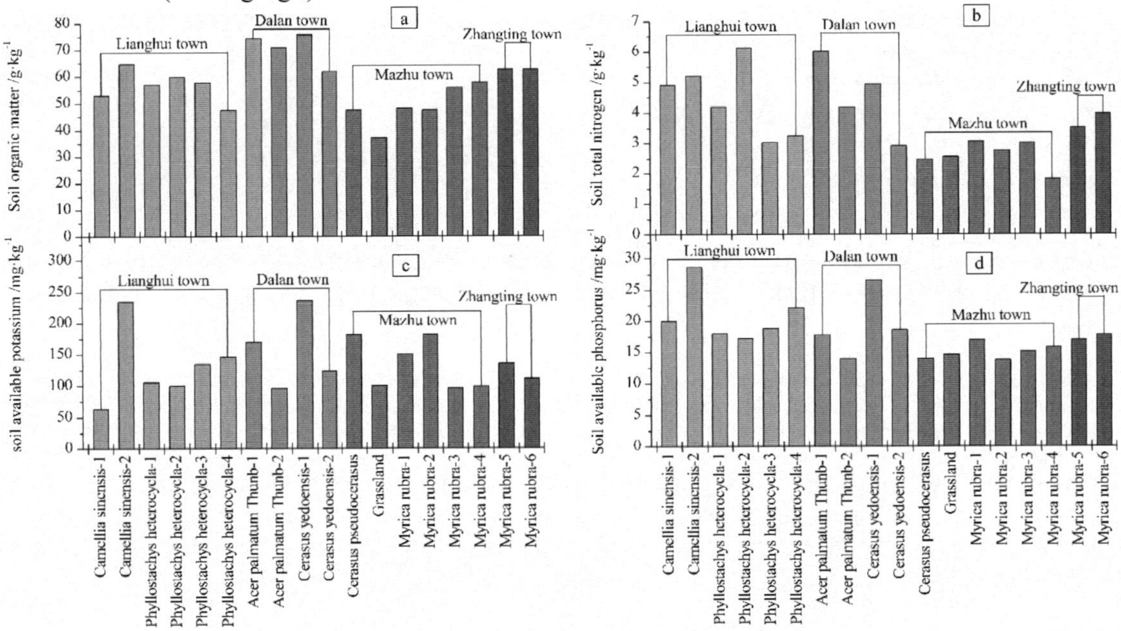


**Figure 1.** Soil nutrients content of different economic forest land

The difference of soil organic matter content in different economic forest land was caused by vegetation type, human disturbance, litter composition and decomposition. because of artificial interference less, Acer palmatum Thunb forest land, Cerasus yedoensis forest land and Camellia sinensis land had more litter and residues of animal and plant. decomposition of humus improved organic matter content in soil. in contrast, Phyllostachys heterocycla forest land and Myrica rubra forest land had lower soil organic matter content because of the artificial removal of understory vegetation and litter.

### *3.2 Soil total nitrogen*

Most of the nitrogen in the soil exists in organic form, and the organic nitrogen is transformed into inorganic nitrogen for crop absorption and utilization under the condition of cultivation and so on. Organic and inorganic nitrogen in soil come to be known as soil total nitrogen. As can be seen from figure 1b, the total nitrogen content of soil was also affected by the type of economic forest, there are obvious differences. The content of total nitrogen in different economic forest land was Acer palmatum Thunb (5.10 g·kg$^{-1}$) > Camellia sinensis (5.06 g·kg$^{-1}$) > Phyllostachys heterocycla (4.15 g·kg$^{-1}$) > Cerasus yedoensis (3.93 g·kg$^{-1}$) > Myrica rubra (3.02 g·kg$^{-1}$) > Grassland (2.56 g·kg$^{-1}$) > Cerasus pseudocerasus (2.46 g·kg$^{-1}$). The soil total nitrogen content of Acer palmatum Thunb forest land was 2.08 times that of Cerasus pseudocerasus forest land. There were also differences in soil total nitrogen content of different regions. For example, soil total nitrogen content (3.74 g·kg$^{-1}$) of Myrica rubra forest land in Zhangting town was more than that (2.65 g·kg$^{-1}$) in Mazhu town. The content of total nitrogen in soil was related to the source of nitrogen and the absorption and utilization of plants. The main sources of soil total nitrogen form soil humus, animal and plant and microbial residues, organic fertilizer. Its absorption was closely related to plant type. Bamboo shoots absorb a large amount of soil inorganic nitrogen, but because of artificial fertilization, the total nitrogen content in soil was not the lowest. Grassland due to not external source of nitrogen, the total nitrogen content in soil was lowest. Because of large litter nitrogen return, soil total nitrogen content of Acer palmatum Thunb forest land and Camellia sinensis forest land were higher.

### *3.3 Soil available potassium*

According to the potassium existing form and potassium uptake by crops, it can be divided into water soluble potassium, exchangeable potassium and stationary potassium in clay minerals. The first two potassium can be absorbed and utilized by the crop, them were referred to as available potassium. As can be seen from figure 1c, the relationship of soil available potassium in different economic forest land was Cerasus pseudocerasus (181.63 $mg \cdot kg^{-1}$) > Cerasus yedoensis (180.33 $mg \cdot kg^{-1}$) > Camellia sinensis (148.86 $mg \cdot kg^{-1}$) > Acer palmatum Thunb (132.84 $mg \cdot kg^{-1}$) > Myrica rubra (128.77 $mg \cdot kg^{-1}$) > Phyllostachys heterocycla (121.83 $mg \cdot kg^{-1}$) > Grassland (100.60 $mg \cdot kg^{-1}$). The average soil available potassium content of Cerasus pseudocerasus was 1.81 times that of Grassland. There were some differences in soil t available potassium content of different regions. For example, soil available potassium content (123.33 $mg \cdot kg^{-1}$) of Myrica rubra forest land in Zhangting town was less than that (131.49 $g \cdot kg^{-1}$) in Mazhu town. There are many factors affecting soil available potassium, such as fertilization, matrix, climate and land use. Due to the limited conditions, the analysis of the above-mentioned factors is not enough, such as vegetation coverage, cropping system and the distribution of residential areas, these factors impact of soil available potassium content, is worthy of further study.

### *3.4 Soil available phosphorus*

The soil available phosphorus is the phosphorus in soil which can be absorbed by plants. In the agricultural production, the soil available phosphorus content was used to guide the application of phosphate fertilizer. The available phosphorus content in soil is the main factor to determine the effect of phosphate fertilizer and its effect. Figure 1d showed the obvious difference of soil available phosphorus in different economic forest land. The content of soil available phosphorus in different economic forest land showed Camellia sinensis (24.33 $g \cdot kg^{-1}$) > Cerasus yedoensis (22.60 $g \cdot kg^{-1}$) > Phyllostachys heterocycla (19.04 $g \cdot kg^{-1}$) > Myrica rubra (16.08 $g \cdot kg^{-1}$) > Acer palmatum Thunb (15.88 $g \cdot kg^{-1}$) > Grassland (14.56 $g \cdot kg^{-1}$) > Cerasus pseudocerasus (13.95 $g \cdot kg^{-1}$). The content of available phosphorus in soil was also different in different economic forest land. The available phosphorus content in Camellia sinensis land was 1.74 times that of soil available phosphorus in Cerasus pseudocerasus forest land. The content of soil available phosphorus in soil was mainly related to the amount of deciduous twigs. The release rate of phosphorus in soil was faster than that of the other nutrient element [12]. At the same time, the content of phosphorus in soil was also affected by fertilization and plant growth and absorption. High content of available phosphorus in the soil of economic forest land has an important influence on eutrophication, which is the main source of non-point source pollution. Therefore, in the control of non-point source pollution, the soil phosphorus output of economic forest land may be more worthy of attention.

### *3.5 Soil nutrients quality evaluation*

According to the calculation of soil nutrient quality index (In table 3), The comprehensive index of soil nutrient quality in different economic forest lands varied between 1.541-2.250. The results of soil quality evaluation were "good" and "excellent", The average condition was "good" (average was 1.978). The average comprehensive index of soil nutrient quality from high to low was Cerasus yedoensis (2.144) > Phyllostachys heterocycla (2.021) > Cerasus pseudocerasus (1.974) > Acer palmatum Thunb (1.965) > Myrica rubra (1.952) > Camellia sinensis (1.896) > Grassland (1.822). It can be concluded from the analysis of the comprehensive index that the comprehensive status of soil nutrients in different economic forest lands in Yuyao is at the middle level, so it is necessary to improve the soil fertility. As can be seen from table 3, the quality of soil nutrients in economic forest land is generally limited by the content of available potassium. Among them, some Camellia sinensis land was very obvious.

**Table 3.** Soil quality index and comprehensive index of different economic forest land

| Regions | Land use types | Soil organic matter | Soil total nitrogen | Soil available phosphorus | Soil available potassium | Soil nutrient quality index |
|---|---|---|---|---|---|---|
| Lianghui town | Camellia sinensis-1 | 3.000 | 3.000 | 3.000 | 1.333 | 1.541 |
| | Camellia sinensis-2 | 3.000 | 3.000 | 3.000 | 3.000 | 2.250 |
| | Phyllostachys heterocycla-1 | 3.000 | 3.000 | 2.802 | 2.269 | 1.898 |
| | Phyllostachys heterocycla-2 | 3.000 | 3.000 | 2.720 | 2.170 | 1.846 |
| | Phyllostachys heterocycla-3 | 3.000 | 3.000 | 2.883 | 2.749 | 2.122 |
| | Phyllostachys heterocycla-4 | 3.000 | 3.000 | 3.000 | 2.933 | 2.219 |
| Dalan town | Acer palmatum Thunb-1 | 3.000 | 3.000 | 2.781 | 3.000 | 2.148 |
| | Acer palmatum Thunb-2 | 3.000 | 3.000 | 2.395 | 2.097 | 1.781 |
| | Cerasus yedoensis-1 | 3.000 | 3.000 | 3.000 | 3.000 | 2.250 |
| | Cerasus yedoensis-2 | 3.000 | 3.000 | 2.863 | 2.570 | 2.038 |
| Mazhu towm | Cerasus pseudocerasus | 3.000 | 3.000 | 2.395 | 3.000 | 1.974 |
| | Grassland | 3.000 | 3.000 | 2.456 | 2.177 | 1.822 |
| | Myrica rubra-1 | 3.000 | 3.000 | 2.700 | 2.997 | 2.111 |
| | Myrica rubra-2 | 3.000 | 3.000 | 2.374 | 3.000 | 1.965 |
| Zhangting town | Myrica rubra-3 | 3.000 | 3.000 | 2.517 | 2.096 | 1.793 |
| | Myrica rubra-4 | 3.000 | 3.000 | 2.578 | 2.149 | 1.822 |
| | Myrica rubra-5 | 3.000 | 3.000 | 2.700 | 2.759 | 2.088 |
| | Myrica rubra-6 | 3.000 | 3.000 | 2.781 | 2.352 | 1.932 |

## 4. Conclusion

There were significant differences in soil nutrient content in different economic forest land. The difference of soil organic matter content was the smallest between seven economic forest lands. The difference of soil available potassium and soil total nitrogen was larger. The soil nutrient status of Cerasus yedoensis forest land and Phyllostachys heterocycla forest land were better. Grassland and Camellia sinensis land soil nutrient was low. In addition, the soil nutrient status of Camellia sinensis land was greatly affected by available potassium.Through the comprehensive evaluation of the Nemero index method, The soil nutrient status of Cerasus yedoensis forest land and Phyllostachys heterocycla forest land belonged to the "excellent" level. The soil nutrient status of other economic forest lands belonged to the "good" level. In the study area, the soil nutrient status of the typical economic forest land was in the "good" level, and the soil nutrient quality was generally limited by the available potassium content.

## Acknowledgements

This work is supported by the Hydraulic Technological Program of Zhejiang Province, The People's Republic of China (Under Grant No. RB1609).

## References

[1] Xiong Y. Discussion on Soil Ecosystem. Soil. 1983, (4):121-125. (in Chinese)
[2] Xiao Y, Zhang YG, Zhang XQ, et al. Review on the Influence of Land Use Changes on Soil Fertility. World Forestry Research, 2007, 20(1):6-9. (in Chinese)

[3] Liu YS, Chen BM. The Study Frame Work of Land Use/Cover Change Based on Sustainable Development in China. Geographical Research, 2002, 5(3):324-330. (in Chinese)
[4] Chen ZX, Wang RS, Song S, et al. Forest Soil Physical and Chemical Properties of Coniferin Xiaoxing'anling. Forest Investigation Design, 2010, (4):73-75. (in Chinese)
[5] Wu ZF, Deng NR, Liu P, et al. Diagnosis and Integrated Evaluation on Soil Fertility of Urban Garden Land — A Case Study of Changhong Nursery of Guangzhou City. Soil and Environmental Sciences, 2000, 9(4):287-289. (in Chinese)
[6] Wu ZF, Wen Y, Z J. Evaluation of Soil Quality of Changhong Nursery in Guangzhou. Chinese Landscape Architecture. 2001, (5):70-71. (in Chinese)
[7] Tian Y, Zhuang Y, Cao Y, et al. Effects of Land Uses on Soil Physical and Chemical Properties in Hilly Areas of Wuling Mountain. Research of Soil and Water Conservation. 2012, 19(6):41-44. (in Chinese)
[8] Luo DQ, Bai J, Xie DT. Evaluation Index and Method of Soil Fertility. Soil and Environmental Sciences. 2002, 11(2):202-205. (in Chinese)
[9] Wang JG, Yang LZ, Shan YH. Application of Fuzzy Mathematics to Soil Quality Evaluation. Acta Pedologica Sinica. 2001,38(2):176-183. (in Chinese)
[10] Chen LM, Gui LG, Lv JL, et al. Evaluation on Soil Fertility Quality of Newly Cultivated Light Sierozem Under Different Fertilization with Methods of Principal Component and Cluster Analyses. Soil. 2008, 40(6):971-975. (in Chinese)
[11] Wang F, Huang M, Sun XH, et al. Evaluation of Soil Nutrients for Different Forest Types in Xing'an Mountains Forest Area. Bulletin of Soil and Water Conservation. 2013, 33(1):182-187. (in Chinese)
[12] Cai HX. Short-Term Effects of Clear-Cutting, Prescribed Burning and Site Preparation on Soil Properties of Chinese Fir Plantation, and Contribution of Harvest Residues to Soil Nutrients. Central South University of Forestry and Technology, Changsha, 2008. (in Chinese)

# A method for extracting elbow feature based on machine vision

**Bing Wu[1], Yajie Chen[2] and Jianmi Wang[1]**

[1]School of Mechanical Engineering, Jiangsu University of Science and Technology, Zhenjiang, China
[2]The 704 Research Institute of China Shipbuilding Industry, Shanghai, China
Corresponding author, Email: chunxiana@126.com

**Abstract.** With the continuous development of computer technology, machine vision has become a hot research topic. And the machine vision is introduced into the field of mechanical processing, the detection of the surface defects of the machined objects. The application of machine vision to industrial production has become more and more popular.
The research content of this paper will have broad prospects and development space in the field of machine vision. It has laid a foundation for solving the problems of manual processing, and plays a very important role in the development of universal CNC system with vision.

## 1. Introduction

This paper mainly introduces the application of machine vision in the feature extraction of elbow. Taking elbow characteristics as an example, put forward the theory and the improved method of image feature extraction, and gives the design scheme of the whole system of feature extraction. This paper mainly includes three aspects: Firstly, design of the overall system scheme based on machine vision; Secondly, propose an improved algorithm based on Canny operator, by comparing the detection results of several commonly used image detection operators, based on the OpenCV; Thirdly, take sample of the outside of the elbow by MATLAB, and use the least square method for curve fitting. Then calculate the curve function and the radius of characteristic parameters of the elbow.

## 2. The design of the elbow feature extraction system

A typical machine vision system includes a light source, an optical system, an image acquisition system, a digital image processing, an intelligent decision making module and a mechanical control execution module. The system includes four parts: vision sensor, high speed image acquisition system, special image acquisition system and PC computer. First of all, the target is converted into an image signal by a vision sensor CCD camera; Next, the digital signal is transmitted to the special image processing system, according to the information of the pixel distribution, brightness and colour; And then, perform various operations to extract the features of the target, judge the result according to the present tolerance and other conditions. Finally, according to the results of the judge to control the equipment movement[2].

The main principle of this system: The image capture system work site (several CCD) and image acquisition system (image acquisition card) to collect the image data of elbow, then the data are sent to PC in the program calls for processing by the image acquisition card, image geometric correction, image enhancement, image smoothing, image segmentation, edge detection and other operations on the work piece prepared using VS algorithm, obtain the characteristic parameters of elbow by curve fitting using MATLAB least squares method[3].

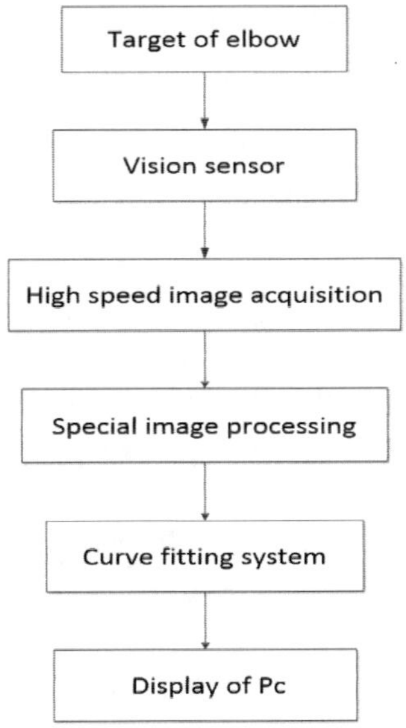


**Figure 1.** The machine vision system

## 3. The optimization algorithm of edge detection based on Canny

The edge is the most significant part of the local brightness of the image. The edge mainly exists in the target and target, target and background, region and region (including different colours). For a work piece, the main feature is the angle of the edge contour. Therefore, edge detection is very important for object recognition, which is an important part of image processing. Poggio has such a definition that the edge detection is the measurement, detection and location of the gray level of the image. The essence of edge detection is to use some algorithm to extract the line between image and background. In general, work piece edge angle can be divided into four types: line-line type, line-arc, arc-arc and arc-line type. When the line is arc tangent, the cross point between the two is the cut off point, the connection point between the convex and concave arc is the inflection point. Detection of target edge angle is mostly based on the change of edge curvature, the discrete curvature is estimated by the change rate of tangent direction, and the vertex with high curvature is considered as the vertex. [4]Now, the widely used method to extract edge feature parameters is the edge detection algorithm based on curvature (edge detection algorithm).

The commonly used operators include Roberts operator, Sobel operator, Laplacian operator, etc. Each operator has its advantages and disadvantages: The Roberts operator has high positioning accuracy, good effect in horizontal and vertical direction, but it is too sensitive to noise; The Sobel operator can produce better results, it has a smoothing effect on noise and can provide the edge direction information accurately, but it also increases the amount of calculation, and will detect the false edge, even the position of image is not accurate; The advantages of the Laplacian operator is that it does not have false edges and always used for known edge pixels to determine the image in the dark or light, but it is sensitive to noise. The edge detection algorithm uses the change rate of tangent direction to estimate the discrete curvature and the vertex with high curvature is considered as the vertex. But the main defect of this algorithm is that the detection method based on edge curvature can not judge the tangent point and inflection point of this kind of common feature points in a work piece, due to its smaller curvature. When the target is rotated toward the edge of a different direction, it will be error detection because of the digital image in discrete curvature measuring is not stable. Compared with the traditional image edge detection methods, this paper uses Canny operator to detect edges[5]. The canny operator first proposed three continuous criteria of edge detection: optimal detection result, optimal location and low repetition response. Its basic idea is to use the Gauss function Gauss to smooth the image, and then determine the edge points by the maximum value of the first order differential. The zero crossing of the two order differential not only corresponds to the maximum of the first derivative, but also corresponds to the minimum value of the first derivative. In other words, the image gray change and slow change point corresponds to two derivative zero crossing point. Consequently, the Canny operator may introduce pseudo edge points

Chalk up $f(x,y)\cdot G_\alpha(x,y)$ after filtering by Gauss function, and α of them are the corresponding scale factors. The gradient vector model $M_\alpha$ and the direction $A_\alpha$:

$$M_\alpha = \|f(x,y)\cdot\nabla G_\alpha(x,y)\| \quad (1)$$

$$A_\alpha = \frac{f(x,y)\cdot\nabla G_\alpha(x,y)}{\|f(x,y)\cdot\nabla G_\alpha(x,y)\|} \quad (2)$$

The edge of the image is a point in the $A_\alpha$ direction that makes the $M_\alpha$ get the local maximum.

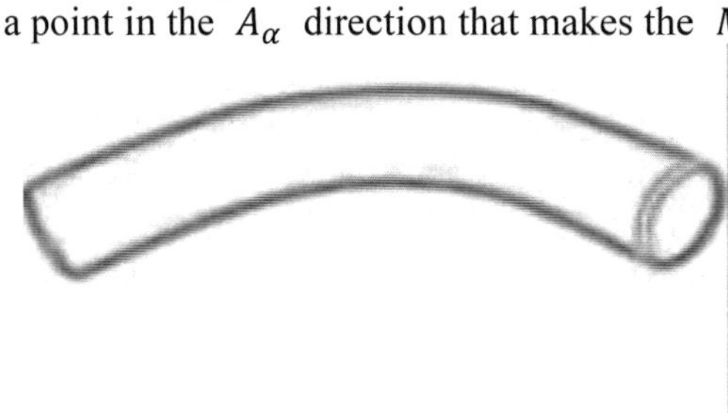

**Figure 2.** The edge detection based on Canny

## 4. The curve fitting of elbow

In the experiment and statistical research, it is often necessary to estimate the function of the $y = f(x)$ approximation function $p(x)$ from a set of measurement data of $(x_i, y_i)(i = 0,1,2,\dots,n)$[6]. Interpolation is a way to deal with such problems, but it has obvious flaws. First of all, it is necessary to approximate all the known points of the function $p(x)$ which is equivalent to the total error, because of the inevitable errors in the observed experimental data and even larger errors. In addition, if still use polynomial interpolation, it will inevitably get a higher number of polynomials because the experimental data are often too much. Not only the calculation is complex, but also the convergence and stability of $p(x)$ are difficult to be guaranteed, so the approximation effect is poor. In order to overcome these shortcomings, often adopts the method of curve fitting for data processing. The curve fitting is to find out the general rules from the data sets of $(x_i, y_i)(i = 0,1,2,\dots,n)$, and construct a curve which can reflect this rule of $p(x)$. It does not require the curve of $p(x)$ across each data point of $(x_i, y_i)(i = 0,1,2,\dots,n)$, but the curve can be as close as possible to all data points, that is, all errors of $\delta_i = p(x_i) - y_i$ $(i = 0,1,2,\dots,n)$ achieve the minimum according to a certain standard[7].

Generally use the following three standards to measure the size of the error:

(1) Select a certain $p(x)$ to minimize the sum of absolute deviations.

$$p(x) = a_0\varphi_0(x) + a_1\varphi_1(x) + \cdots + a_m\varphi_m(x), (m < n)$$

$$F(a_0, a_1, \dots, a_m) = \sum_{i=0}^{n}(\sum_{j=0}^{m} a_j\varphi_j(x) - y_i)^2$$

$$(a_0^*, a_1^*, \dots, a_m^*)$$

$$\frac{\partial F}{\partial a_k} = 0, (k = 0,1,2,\dots,m)$$

(2)Select a certain $p(x)$to minimize the maximum absolute deviations.

$$\max_{0\le i\le n}|\delta_i| = \max_{0\le i\le n}|p(x_i) - y_i| \to min$$

(3)Select a certain $p(x)$ to minimize the sum of squares of deviations.

$$\sum_{i=0}^{n} \delta_i^2 = \sum_{i=0}^{n} [p(x_i) - y_i]^2 \rightarrow min$$

Because there is no absolute value of 2-norm and the calculation is more convenient, it usually adopts 2-norm square as a measure of the total error. We called it is least squares curve fitting which Can make 2-norm minimum square.

In fact, it is a vector of coefficients which is a polynomial when using the least square method to fit the curve. In MATLAB, using the function of *polyfit* to obtain the least squares fitting polynomial coefficient, and according to the given function to calculate the polynomial point approximation by the function of *polyval*. The function of the *polyfit* call format[8]:

$[P, S] = polyfit(X, Y, m)$.

According to the sampling point *X* and the value of sampling point function of *Y*, Produces a *m* degree polynomial of *P* and its error vector of *S* at the sampling point. Among them, *X* and *Y* are the vectors of the same length, *P* is a vector with the magnitude of *m+1*, the elements of *P* are the coefficients of the polynomial. In this paper, use MATALAB to sample the inner side of the elbow image and use the least square fitting method to fit the image , because of the bending radius determined by the radius of curvature.

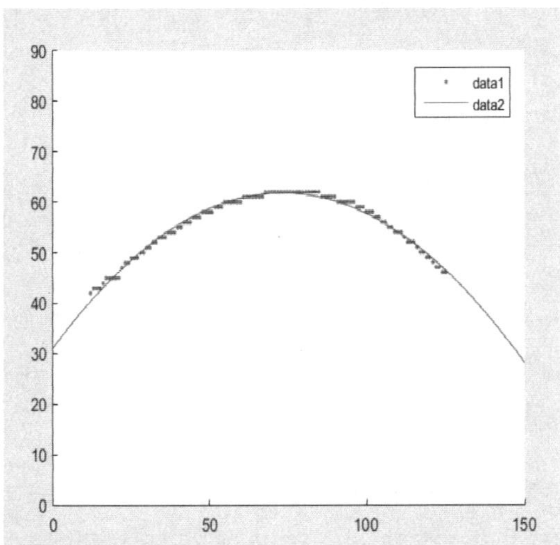


**Figure 3.** Least squares curve fitting

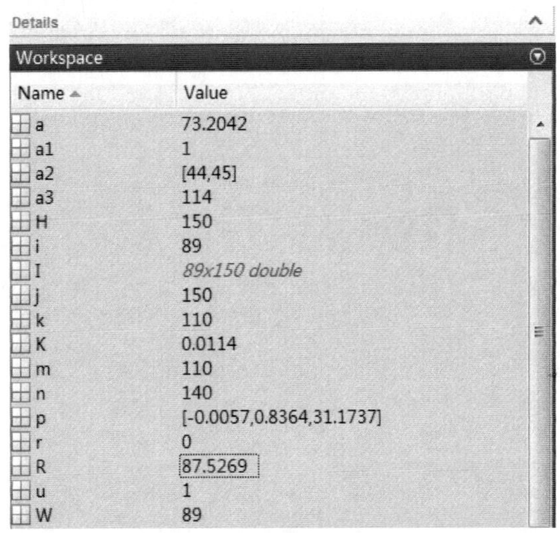


**Figure 4.** Polynomial coefficient and bending radius

## 5. Conclusion

The two time polynomial coefficient is -0.0057, 0.8364, 31.1737, $p(x) = -0.0057x^2 + 0.8364x + 311737$, so the radius of curvature is 87.5269.

## References

[1] Lin Yan, Yang He. Thin-walled tube precision bending process and FEM simulation. In: Yang Huayong, Li Wei, *Proceedings of the 4thInternational Conference on Frontiers of Design and Manufacturing*. Beijing: international Academic Publishers, 2000: 305~308.

[2] Zhang yan. Design of machine vision system based on DSP[J]. *Modern electronic technology*, 2005, 11

[3] Zhang Lei, Luo Xiang, Shi Jinfe. A review of the strategies and methods of machine vision real time problems [J]. *Manufacturing automation*, 2006, 28 (6): 28-32.

[4] Huang Jianling, Chen Bozheng .*An edge detection algorithm based on Canny[J].Computer Simulation*,2010.4:252-254

[5] CHEN Jie， WANG Zhen-hua,DOU Ii-hua． Scale adaptive Canny edge detection method[J]. *Opto-Electronic Engineering*，2008，35(2):79-84.

[6] Cheng Yumin. Research progress and review of moving least squares [J]. *computer aided engineering*, 2009 (02)
[7] Li Ying, Lin Hongsheng.The least squares fitting curve based on relative error [J]. *Journal of Shenyang Normal University: Natural Science Edition*, 2012, 30 (3):338 342
[8] Liu Zhiping,Shi Linying.The principle of least square method and MATLAB implementation[J]. *Science and Technology of West China*.2008.6:33-34

IOP Conf. Series: Earth and Environmental Science **61** (2017) 012055 doi:10.1088/1755-1315/61/1/012055

# An improved adaptive weighting function method for State Estimation in Power Systems with VSC-MTDC

**Kun Zhao[1], Xiaonan Yang[1], Yansheng Lang[1], Xuri Song[1], Minkun Wang[2], Yadi Luo[1], Lingyun Wu[2], Peng Liu[1]**

[1] China Electric Power Research Institute, Beijing 100192, China
[2] State Grid Southwest Dispatching and Control Center, Chengdu 610000, China
Corresponding author e-mail: yangxiaonan@epri.sgcc.com.cn

**Abstract.** This paper presents an effective approach for state estimation in power systems that include multi-terminal voltage source converter based high voltage direct current (VSC-MTDC), called improved adaptive weighting function method. The proposed approach is simplified in which the VSC-MTDC system is solved followed by the AC system. Because the new state estimation method only changes the weight and keeps the matrix dimension unchanged. Accurate and fast convergence of AC/DC system can be realized by adaptive weight function method. This method also provides the technical support for the simulation analysis and accurate regulation of AC/DC system. Both the oretical analysis and numerical tests verify practicability, validity and convergence of new method.

## 1. Introduction

Voltage source converter, as a new transmission technology to support flexible DC power transmission, has drawn much more attentions in electric power system. It is now considered to be a feasible solution for connecting offshore wind farms due to its significant advantages over both current source converter (CSC) HVDC and traditional AC system [1] and the first project which is 400 MW, ±150kV two terminal VSC HVDC system for wind farms connection has been commissioned in Germany in 2010 [2]. The concept of "European Offshore Super grid" [3] has been proposed to build a VSC based subsea meshed DC grid which is used to connect the renewable offshore wind farms to the AC networks. Apart from line commutated converter (LCC), the VSC technology makes the extension to multi-terminal HVDC connections relatively easy. Quite a few papers have been published on VSC-MTDC steady state modeling for load flow [4, 5].

State estimation plays important role in Energy Management System (EMS). With rapid development of HVDC technologies, development of practical methods for AC/DC hybrid power system state estimation is an emergent demand. However, most papers have been published on AC systems, and only a few studies have so far been carried out on AC/DC systems state estimation which includes a generalized VSC MTDC model. In [6, 7], the state estimation was proposed for the two-terminal VSC-HVDC system and the MTDC system was not considered. Glover and Sheikoleslami [8] emphasized on the DC network state estimation without including the AC/DC interconnection system. Leita da Silva et al. [9] separated an AC/MTDC interconnected system into three subsystems, namely AC system, AC/DC interface and DC system. Jagatheesan and Duraiswamy [10] extended the fast decoupled state estimation technique originally applied on the AC system to the AC/DC system and proposed a sequential solution scheme to solve P-Q-DC iteratively. The sequential method is simple and can be easily appended to existing AC fast decoupled state estimators (FDSE). Zhang boming [11] presents a new state estimation algorithm, called improved sequential method, which is based on the polar formulation.

Since VSC-MTDC has significant advantages in power systems and state estimation plays an important role in EMS, this paper presents a new state estimation algorithm, called improved adaptive weighting function method. This method decouples the AC system and the DC system by mathematic method according to the information matrices formed by the alternating solution method. And the residual error information of AC/DC system interface is fully utilized to realize accurate and fast convergence of AC/DC system by adaptive weight function. Tests on IEEE14 test system have been done and the results verify that this proposed method is efficient and valid.

## 2. WLS and FDSE state estimation algorithm

In recent years, with the expansion of the scale of power system and the development of power market, the reliability and accuracy of on-line safety and stability analysis, automatic generation control (AGC) and automatic voltage control (AVC) are becoming higher and higher. Real-time analysis and control of data sources, state estimation has become an indispensable part of the smart grid dispatch control system. Usually assume that the network parameters and wiring status is accurate to each node voltage amplitude and phase angle for the state estimation vector X, to obtain the measured value Z, for the measurement error v, the measurement equation can be written as:

$$z = h(x) + v \tag{1}$$

For a given set of measurement vectors, the estimator is obtained using weighted least square (WLS) criteria by minimizing the objective function:

$$J(x) = [z - h(x)]^T R^{-1}[z - h(x)] \tag{2}$$

The iterative correction formula of the basic weighted least squares method for minimizing the state estimation is obtained.

$$\Delta\hat{x}^k = [H^T(\hat{x}^k)R^{-1}H(\hat{x}^k)]^{-1}H^T(\hat{x}^k)R^{-1}[z - h(\hat{x}^k)] \tag{3}$$

$$\hat{x}^{k+1} = \hat{x}^k + \Delta\hat{x}^k = \hat{x}^k + [H^T(\hat{x}^k)R^{-1}H(\hat{x}^k)]^{-1}H^T(\hat{x}^k)R^{-1}[z - h(\hat{x}^k)] \tag{4}$$

According to the above iterative correction, until convergence, this time close to the minimum value of the objective function.

Since the basic weighted least squares estimation method needs a large amount of memory and a long computation time, a fast decoupled state estimation (FDSE) method is usually adopted. This is a well-recognized model of utility performance on high voltage networks. The quantity measurement, the state quantity and the measurement equation are decomposed into active and reactive power, and the hypothesis is introduced in the Jacobian matrix H:

$$\frac{\partial h_a}{\partial V} \approx 0 \quad \frac{\partial h_r}{\partial \theta} \approx 0 \tag{5}$$

The modified equation of the state estimation of the fast decoupled method can be obtained:

$$\begin{aligned} A\Delta\theta^{(l)} &= a^{(l)} \\ B\Delta V^{(l)} &= b^{(l)} \end{aligned} \tag{6}$$

Where A is the symmetry matrix of active constants, B is the symmetrical matrix of reactive power, a is the active free vector, and b is the reactive free vector.

IOP Conf. Series: Earth and Environmental Science **61** (2017) 012055 doi:10.1088/1755-1315/61/1/012055

$$
\begin{aligned}
A &= V_0^4\left[(-B_a)^T R_a^{-1}(-B_a)\right] \\
B &= V_0^2\left[(-B_r)^T R_r^{-1}(-B_r)\right] \\
a^{(l)} &= V_0^2\left[(-B_a)^T R_a^{-1}(z_a - h_a)V^{(l)},\theta^{(l)}\right] \\
b^{(l)} &= V_0^2\left[(-B_r)^T R_r^{-1}(z_r - h_r)V^{(l)},\theta^{(l)}\right]
\end{aligned}
\tag{7}
$$

The state estimation algorithm is only applicable to the AC system. When the MTDC system is included in the power system, the system state cannot be completely described by the voltage and phase angle calculated by the AC state estimation. After the flexible DC system is put into operation, there may be measurement quality problems. The original state estimation model does not consider the flexible DC system model, the lack of straight state system state estimation function cannot be achieved straight system and AC system mixed state estimation calculation, is not suitable for flexible DC transmission system, it is necessary to the original state estimation Methods to supplement, expansion and innovation.

## 3. VSC-MTDC Model

The steady-state model of the VSC-MTDC is shown as follows:

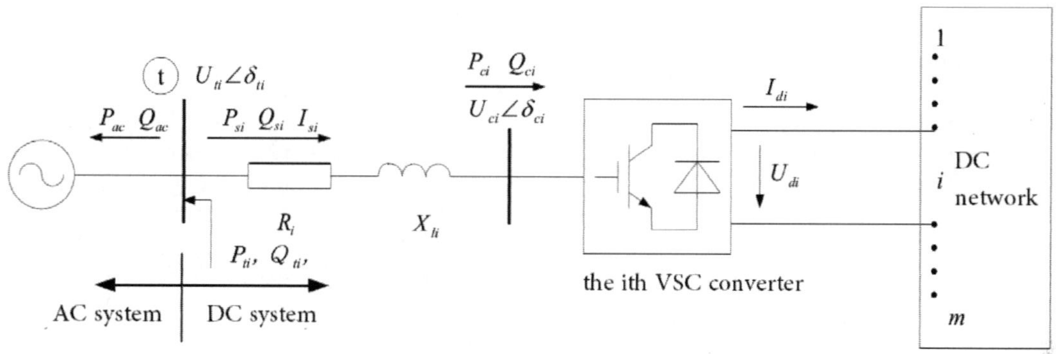


**Figure 1.** Schematic diagram of single -phase VSC –MTDC.

Because of the control characteristics of the flexible DC system, any converter can be used as a rectifier or inverter. The converters in the mathematical form are the same. The subscripts I represent the variables associated with the first inverter and its connected nodes in the HVDC system. Set the converter input voltage to $\dot{U}_{ci} = U_{ci}\angle\delta_{ci}$, from the AC side the node voltage which is connected to inverter is $\dot{U}_{ti} = U_{ti}\angle\delta_{ti}$, The equivalent impedance for the rheological and commutation bridges is $R_i + X_{li}$, The inverter input power is $P_{si} + Q_{si}$.

Since $R_i << X_{li}$, ignoring the branch resistance, the converter input power is:

$$P_{si} = \frac{U_{ti}U_{ci}}{X_{li}} sin\delta_i \tag{8}$$

$$Q_{si} = \frac{U_{ti}}{X_{li}}(U_{ti} - U_{ci}cos\delta_i) \tag{9}$$

The relationship between the converter input voltage $U_{ci}$ and DC voltage is:

$$U_{ci} = \mu_d M_i U_{di} / \sqrt{2} \tag{10}$$

Where: $\delta_i = \delta_{si} - \delta_{ci}$, $M_i$ is the modulation degree, $\mu_d$ is the DC voltage utilization.

As the impedance of the converter bridge forward, so the active power $P_{ci}$ into the converter bridge is equal to the active power output $P_{di}$, $Q_{ci}$ is the reactive power into the converter bridge, $I_{di}$ is the DC current.

$$P_{di} = U_{di} I_{di} \tag{11}$$

$$P_{di} = P_{ci} = \frac{U_{ti} U_{ci}}{X_{li}} sin\delta_i \tag{12}$$

$$Q_{ci} = \frac{-U_{ci}}{X_{li}} (U_{ci} - U_{ti} cos\delta_i) \tag{13}$$

DC system network equation is as follows:

$$I_{di} = U_{di} G \tag{14}$$

Where: G is network conductance.

## 3. Proposed method for state estimation

### *3.1. Dimension invariant function*

The state estimation of flexible DC transmission usually adopts the WLS method. There are a total of 11 variables appeared: $U_{di}, U_{ci}, U_{ti}, M_i, \delta_i, I_{di}, P_{si}, Q_{si}, P_{ci}, Q_{ci}, P_{di}$. The power flow equations of all the AC nodes in the VSC-MTDC system can be expressed as follows:

$$\Delta P_{ai} = P_{ai} - P_{ai}(U_{ai}, \delta_{ai}) \tag{15}$$

$$\Delta Q_{ai} = Q_{ai} - Q_{ai}(U_{ai}, \delta_{ai}) \tag{16}$$

The power flow equations of all the DC nodes in the VSC-MTDC system can be expressed as follows:

$$\Delta P_{ti} = P_{ti} - \sum f_{P_{aci}}(U_{aci}, \delta_{aci}, U_{ti}, \delta_{ti}) - f_{P_{si}}(U_{ti}, \delta_{ti}, U_{di}, \delta_{ci}, \mathrm{M}) = 0 \tag{17}$$

$$\Delta Q_{ti} = Q_{ti} - \sum f_{Q_{aci}}(U_{aci}, \delta_{aci} U_{ti}, \delta_{ti}) - f_{Q_{si}}(U_{ti}, \delta_{ti}, U_{di}, \delta_{ci}, \mathrm{M}) = 0 \tag{18}$$

In addition to the above pseudo-measurement equations, the actual measurement equations in DC system and AC/DC interface system are shown as follows:

$$\Delta d_1 = \Delta P_{si} = P_{si} - f_{Psi}(U_{ti}, \delta_{ti}, U_{di}, \delta_{ci}, \mathrm{M}) = 0 \tag{19}$$

$$\Delta d_2 = \Delta Q_{si} = Q_{si} - f_{Qsi}(U_{ti}, \delta_{ti}, U_{di}, \delta_{ci}, \mathrm{M}) = 0 \tag{20}$$

$$\Delta d_3 = \Delta P_{di} = U_{di} I_{di} - f_{Pci}(U_{ti}, \delta_{ti}, U_{dI}, \delta_{ci}, M_i) = 0 \tag{21}$$

$$\Delta d_4 = \Delta I_{di} = I_{di} - U_{di} G = 0 \tag{22}$$

$$\Delta d_5 = U_{ti}^m - U_{ti}^t = 0 \tag{23}$$

$$\Delta d_6 = U_{di}^m - U_{di}^t = 0 \tag{24}$$

$$\Delta d_7 = \delta_i^m - \delta_i^t = 0 \tag{25}$$

$$\Delta d_8 = M_i^m - M_i^t = 0 \tag{26}$$

$$\Delta d_9 = I_{di}^m - I_{di}^t = 0 \tag{27}$$

Where $U_{ti}^m$ is measurement value, $U_{ti}^t$ is true value.

The VSC HVDC converter can control its active and reactive power injection independently in the system

1) $U_{dc}$, $Q$ control: The DC bus voltage keeps constant through adjusting the active power injection of the converter. The reactive power is scheduled to a set point and the $U_{dc}$, $Q$ control constraint is

$$\begin{aligned} 0 &= U_{di} - U_{di}^{ord} + \eta_1 \\ 0 &= Q_{si} - Q_{si}^{ord} + \eta_2 \end{aligned} \tag{28}$$

2) $U_{dc}$, $U_{ac}$ control: The DC bus voltage and AC bus voltage keeps constant and the $U_{dc}$, $U_{ac}$ control constraint is

$$\begin{aligned} 0 &= U_{di} - U_{di}^{ord} + \eta_3 \\ 0 &= U_{ti} - U_{ti}^{ord} + \eta_4 \end{aligned} \tag{29}$$

3) $P$, $Q$ control: The active power and the reactive power is constant and the $P$, $Q$ control constraint is

$$\begin{aligned} 0 &= P_{si} - P_{si}^{ord} + \eta_5 \\ 0 &= Q_{si} - Q_{si}^{ord} + \eta_6 \end{aligned} \tag{30}$$

4) $P$, $U_{ac}$ control: The active power and AC bus voltage is constant and the $P$, $U_{ac}$ control constraint is

$$\begin{aligned} 0 &= P_{si} - P_{si}^{ord} + \eta_7 \\ 0 &= U_{ti} - U_{ti}^{ord} + \eta_8 \end{aligned} \tag{31}$$

According to each control mode, there are two known quantities are determined, so the corresponding state of state estimation mathematical model also needs to be adjusted. The result shows that the Jacobian matrix and corresponding factor table under different control modes will change in dimension, and because of the switch of control mode, the state estimation mathematical model is changed due to the different states' adding and exiting.

In order to avoid the above-mentioned situation, the state estimation of the VSC-MTDC system needs the algorithm of high precision, high speed and low memory. Therefore, without changing the number of the state, give greater weight to the known state. According to the above-mentioned different control methods, in order to guarantee the fast convergence of the state estimation iterations, we give a large weight $R_{\max}$ to the known quantity measurement in the weighted least squares estimation.

The estimate of known quantities is obtained using WLS criteria by minimizing the objective function:

$$J(x) = [\Delta z - H(x_0)\Delta x]^T R_{\max}^{-1} [\Delta z - H(x_0)\Delta x] \tag{32}$$

Where, the H dimension keeps constant. It is benefit for program development.

### *3.2. Adaptive weight function*

According to equation (19-27), there are only three unknown variables between AC system and DC system by using sequential method. They are $P_{si}$ ,$Q_{si}$,$U_t$.

Control by a single inverter control is analyzed as follows:

1) $U_{dc}$, $Q$ control: AC and DC systems are iterated alternately. The difference $\Delta P_{si}, \Delta U_t$ between the AC and DC results of the two systems are the convergence signs of the state estimation of the VSC-MTDC system.

2) $U_{dc}$, $U_{ac}$ control: $\Delta P_{si}, \Delta Q_{si}$ are the convergence signs of the state estimation

3) $P$, $Q$ control: $\Delta U_t$ is the convergence sign of the state estimation

4) $P$, $U_{ac}$ control: $\Delta Q_{si}$ is the convergence sign of the state estimation

In order to improve accuracy and fast convergence of AC/DC system, adaptive weight function method is proposed. Because most of the known measured values are accurate in the actual power system, they are given a bigger weight than the state quantity $x_0^{(0)}$ measured in the first state estimation calculation. For the AC system iterative calculation and the DC system iteration calculation, after the convergence of the two independently, the state quantity $x_{ac}$ calculated from the AC system and the state quantity $x_{dc}$ calculated from the DC system are obtained.

The absolute value of the difference between the two and the initial state measurement value is calculated:

$$\begin{aligned} \Delta c_{ad} &= \left| x_{ac} - x_{dc} \right| \\ \Delta c_{d0} &= \left| x_{dc} - x_0^{(0)} \right| \\ \Delta c_{0a} &= \left| x_0^{(0)} - x_{ac} \right| \end{aligned} \tag{33}$$

The above three differences from small to large $\Delta c_1 < \Delta c_2 < \Delta c_3$, and set the adaptive comparison threshold $\Delta c_\varepsilon$ compared, there will be the following four cases:

$$\begin{aligned} &\Delta c_\varepsilon < \Delta c_1 < \Delta c_2 < \Delta c_3 \quad ① \\ &\Delta c_1 < \Delta c_\varepsilon < \Delta c_2 < \Delta c_3 \quad ② \\ &\Delta c_1 < \Delta c_2 < \Delta c_\varepsilon < \Delta c_3 \quad ③ \\ &\Delta c_1 < \Delta c_2 < \Delta c_3 < \Delta c_\varepsilon \quad ④ \end{aligned} \tag{34}$$

The next state $x_0^{(l+1)}$ after calculating is different in each case:

$$x_0^{(l+1)} = \begin{cases} \dfrac{x_{ac} + x_{dc} + x_0^{(l)}}{3} & ①、③、④ \\ \dfrac{x_{c1}^1 + x_{c2}^2}{2} & ② \end{cases} \tag{35}$$

Where $x_{c1}^1$、$x_{c2}^2$ is related with $\Delta c_1$, $x_0^{(l+1)}$ is measurement during the next iteration, $l \geq 0$, $x_0^{(0)}$ is initial measurement.

Next, put this average measurement $x_0^{(l+1)}$ into Iterative calculation in AC system and DC system separately. And add bigger weight to this average measurement $x_0^{(l+1)}$:

$$J(x) = [\Delta z - H(x_0^{(l+1)})\Delta x]^T R_{\max}^{-1} [\Delta z - H(x_0^{(l+1)})\Delta x] \tag{36}$$

Keep iterative calculate until satisfied the equation:

$$\Delta c_1 < \Delta c_2 < \Delta c_3 < \varepsilon \tag{37}$$

Where $\varepsilon$ is the VSC-MTDC system initial measurement.
Compared with traditional state estimation, it can only be applied to AC system, and it cannot provide good data for VSC-MTDC system. The sequential state estimation calculation method of the AC and DC system considering the VSC-MTDC can solve the integrated state estimation and calculation problem of the AC system and the multi-terminal flexible DC transmission system hybrid system, and is used for realizing the VSC-MTDC AC and DC system Mixed state estimation calculation, to provide real-time accurate AC-DC system operating status. The method of calculating Jacobian matrix state - of - the - art state estimation based on variable weight is proposed, and the mathematic model of DC system is simplified. AC / DC system is proposed to solve AC / DC system alternately, and the residual error information of AC / DC system interface is fully utilized to realize accurate and fast convergence of AC / DC system by adaptive weight function. And the calculated result provides the technical support for the simulation analysis and accurate regulation of AC / DC system.

## 4. Case Study

In order to test the approach proposed in this paper, IEEE 14 bus system is used for simulation. Figure 2 is the modified IEEE14 node AC-DC hybrid system, where VSC1, VSC2 and VSC3 are assumed to operate at the node 2, 5, 4 orderly. Improved adaptive weighting function method and sequential method are applied to solve the system. All the Equivalent reactance is set to XL= 0.15, R = 0.006 and the resistance of DC circuit is Rd= 0.03, VSC1 is in $U_{dc}$, Q control mode ($U_{d1}$, ref = 2.0000, $Q_{s1}$, ref = 0.2099), VSC2 is in P, Q control mode ($P_{s2}$, ref = -0.3601, $Q_{s2}$, ref = 0.0187); VSC3 is in P, Q control mode ($P_{s3}$, ref = - 0.8663, $Q_{s3}$, ref =- 0.0864).

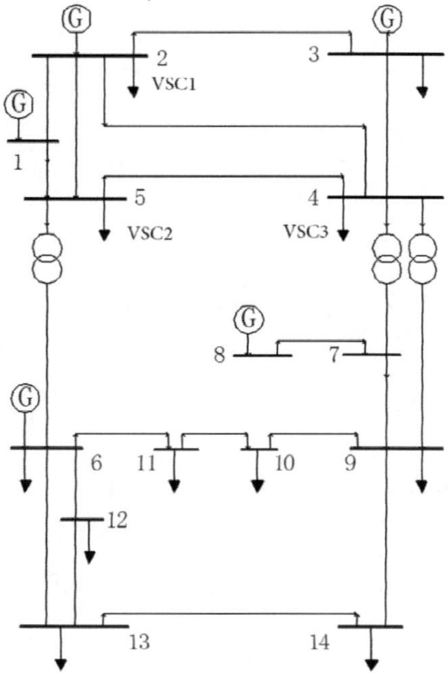


**Figure 2.** Modified IEEE -14 -bus AC -DC system

**Table 1.** State estimation results of AC system.

| Method | Improved method | | Sequential method | |
|---|---|---|---|---|
| Node No. | $U_{ti}$ | $\delta_{ti}$ | $U_{ti}$ | $\delta_{ti}$ |
| 1 | 1.0600 | 0.000 | 1.0600 | 0.000 |
| 2 | 1.0288 | -6.048 | 1.0289 | -6.049 |
| 3 | 1.0116 | -15.129 | 1.0120 | -15.129 |
| 4 | 1.0406 | -14.129 | 1.0406 | -14.122 |
| 5 | 1.0113 | -4.033 | 1.0113 | -4.034 |
| 6 | 1.0702 | -12.159 | 1.0702 | -12.159 |
| 7 | 1.0669 | -15.704 | 1.0666 | -15.696 |
| 8 | 1.0923 | -15.699 | 1.0919 | -15.691 |
| 9 | 1.0543 | -16.546 | 1.0541 | -16.539 |
| 10 | 1.0479 | -16.043 | 1.0476 | -16.037 |
| 11 | 1.0538 | -14.242 | 1.0538 | -14.240 |
| 12 | 1.0555 | -13.259 | 1.0556 | -13.262 |
| 13 | 1.0490 | -13.596 | 1.0491 | -13.597 |
| 14 | 1.0324 | -16.241 | 1.0323 | -16.236 |

**Table 2.** State estimation results of DC system

| DC node No. | method | $P_s(MW)$ | $Q_s(MVAr)$ | $\delta(°)$ | $M$ | $U_d$ | $I_d$ |
|---|---|---|---|---|---|---|---|
| VSC1 | Improved | 124.6501 | 20.9900 | 10.3324 | 0.8220 | 2.0000 | 0.6197 |
| | Sequential | 124.6502 | 20.9900 | 10.3323 | 0.8220 | 2.0000 | 0.6197 |
| VSC2 | Improved | -36.0099 | 1.8700 | -3.0340 | 0.8293 | 1.9919 | -0.1805 |
| | Sequential | -36.0100 | 1.8700 | -3.0340 | 0.8293 | 1.9920 | -0.1808 |
| VSC3 | Improved | -86.6303 | -8.6400 | -6.7075 | 0.8743 | 1.9893 | -0.4392 |
| | Sequential | -86.6300 | -8.6400 | -6.7073 | 0.8743 | 1.9893 | -0.4389 |

Table 1 and Table 2 show the state estimation results of AC/DC system by using different state estimators. Apparently the results of two algorithms are very similar and an inspection of the results indicates that the improved adaptive weighting function method can converge to acceptable state with the required tolerance. And from Table 2, it can be seen that the VSC1 and VSC2 and VSC3 output voltage phase angles are lagging behind the AC bus voltage phase angle of the corresponding node, so that VSC1 absorbs the active power from the AC grid, VSC2, VSC3 to the AC power grid into the active power.

**Table 3.** Comparison of performances between improved and sequential algorithms

| Method | Improved method | Sequential method |
|---|---|---|
| No. of iterations | 5 | 11 |
| Iteration time/s | 0.07285 | 0.1186 |

From table 3, it is noted that the sequential estimator needed more iterations than the improved estimator. It's because in the derivation procedure of the improved method we have assumed Jacobian matrix unchanged in (32). It is also observed that less iterations and less solution time for the improved adaptive weighting function method as compared to the sequential method is because the weights of the same state quantities in different systems are considered.

## 5. Conclusion

This paper presents a novel and efficient approach which called improved adaptive weighting function method is suitable for the state estimation in the VSC-MTDC system. The new state estimation method keeps the matrix dimension unchanged through changing the weight and the progress of incorporation adaptive weight function into the state estimation has been described in detail. Moreover, it is easy to realize the proposed method through extending an existing AC sequential method. Simulation results in modified IEEE 14 bus system demonstrate the effectiveness of the proposed state estimation algorithm.

**Acknowledgments**
This work was financially supported by Beijing Municipal Science & Technology Commission Project (Z161100004816025) and State Grid Corporation of science and technology project fund.

**References**

[1] ABB Group, It's time to connect – Technical description of HVDC Light technology, [Online].
[2] ABB Grid connection of offshore wind farms – BorWin1, [Online].
[3] D. Van Hertem and M. Ghandhari, "Multi-terminal VSC HVDC for theEuropean Supergrid: Obstacles," Renew. Sustain. Energy Rev.,vol. 14,no. 9, pp. 3256-3163, Dec. 2010.
[4] J. Beerten, S. Cole, and R. Belmans, "Generalized Steady-State VSCMTDC Model for Sequential AC/DC Power Flow Algorithms," IEEETans. Power Syst., vol. 27, no. 2, pp. 821-829, May 2012.
[5] Xiao-Ping Zhang, "Multiterminal Voltage-Sourced Converter-BasedHVDC Models for Power Flow Analysis," IEEE Tans. Power Syst., vol.19, no. 4, pp. 1877-1884, Nov., 2004.
[6] Antonio de la Villa Jaén, Eonrique Acha and Antonio Gómez Expósito,"Voltage Source Converter Modeling for Power System StateEstimation: STATCOM and VSC-HVDC," IEEE Tans. Power Syst.,vol. 23, no. 4, pp. 1552-1559, Nov., 2008.
[7] Guoqiang Sun, Yuyan Li, Zhinong Wei and Fang Ye, "State estimationof power system with VSC-HVDC," Electric Power AutomationEquipment, vol. 30, No. 9, pp.6-12, Sep., 2010.
[8] J. D. Glover and M. Sheikoleslami, "State estimation of interconnectedHVDC/AC systems," IEEE Trans. Power Apparatus and Systems, vol.102, pp. 1805–1810, June 1983.
[9] A. M. Leita da Silva, R. B. Prada, and D. M. Falcao, "State estimationfor integrated multi-terminal DC/AC systems," IEEE Trans. Power Apparatus and Systems, vol. 104, pp. 2349–2355, Sept. 1985.
[10] R. Jegatheesan and K. Duraiswamy, "AC/multi-terminal DC powersystem state estimation—A sequential approach," Electric Machinesand Power Systems, vol. 12, pp. 27–42, 1987.
[11] Qifeng Ding, T. S. Chung and Boming Zhang, "An improved sequential method for AC/MTDC power system state estimation," in IEEE Transactions on Power Systems, vol. 16, no. 3, pp. 506-512, Aug 2001.

# Energy efficiency measures in China: A three-stage DEA analysis

**Yu Cai[1,2,b], Siqin Xiong[1,3,a], and Xiaoming Ma[1,3, a]**

[1]School of Environment and Energy, Peking University Shenzhen Graduate School, Shenzhen 518055, China.
[2]Shenzhen Development and Reform Commission, C3073 Citizen's Center, Shenzhen 518035, China
[3]College of Environmental Sciences and Engineering, Peking University, Beijing 100871, China.
[a)]Corresponding author: xiongsq95@163.com
[a)]Corresponding author: xmma@pku.edu.cn
[b)] 2304649750@qq.com

**Abstract.** This paper measures energy efficiency of 30 regions in China during 2010-2014 by using the three-stage data envelopment analysis (DEA)model. The results indict that environmental factors and random error both have significant impacts on energy efficiency. After eliminating these influences, the results present that the energy efficiency in developed regions is almost higher than that in undeveloped or resource-rich regions and low scale technical efficiency is the main constraining factor in inefficient regions. Based on the efficiency characteristics, this paper divides all regions into four types and provide differential energy strategies.

## 1. Introduction

Since China has become the biggest emitter, the pressure to reduce carbon emissions is severe than ever. However, for China is still accelerating the process of industrialization and urbanization, energy is the material basis for development. Therefore, it is of practical significance to explore an optimal path for China's low-carbon development. Considering China is a vast territory with big differences in economic scale, resource endowment, urbanization level and the degree of openness among different regions, to achieve the national reduction targets, it is necessary to begin by elucidating the characteristics of regional carbon emissions, such as using the form of efficiency index.

Current researches on the energy efficiency including two categories: one is single factor index represented by carbon intensity, which is defined as the energy consumption divided by the economic output [1]. Although single factor index has the advantages of simple calculation and easy interpretability, these indicators simply measure the proportional relationship between energy and economic output without considering the substitution effects among energy, capital, labor force and other factors [2]. Another measure is the total-factor energy efficiency method [3], including the parametric method and non-parametric method. Data envelopment analysis (DEA) [4], which can assess the relative efficiency of a set of decision-making units (DMUs) with multiple inputs and outputs, has gained increasingly popularity in energy efficiency measurements.

At present, many scholars pay great attention to China's regional efficiency by using traditional DEA method or extending methods, such as Wang H et al [5], and Zhang et al [6]. Although they have made huge progress towards understanding China's energy efficiency, they fail to fully consider the effects of environmental heterogeneity and stochastic factors. Stochastic frontier analysis (SFA) is a parametric method which can provide a specific function form to void heterogeneity for DMUs. Following the pioneering work of Fried [10] who proposed the three-stage DEA approach combining DEA and SFA, several studies have applied this model to evaluate efficiency considering environmental heterogeneity. Such as Yang et al [7] and Li et al [8]. All these researches validated that environmental factors and statistical error have impact on efficiency.

The paper is organized as follows: Section Two describes the data. Section 3 presents the empirical analysis. Section 4 gives our conclusions and policy implications.

## 2. Data

Based on data availability, this paper takes the 30 provinces of China as the sample set over the period 2010–2014(Hong Kong, Macao, Tibet, and Chinese Taiwan are not included） with capital stock, labor force, energy and carbon emissions as inputs and the GDP of each province as output. Capital stock data are estimated using the perpetual inventory method [9]. Labor force, represented by the number of employed persons by the end of a year, are collected from the China Statistical Yearbooks. The energy data are from The Energy Statistical Yearbook and different energy types are converted into standard coal. Carbon emissions are calculated based on the carbon-emission coefficients provided by IPCC (2006). GDP data are obtained from the China Statistical Yearbook and are converted into standard prices using a price index (2010.100).

## 3. Empirical application and results

### *3.1 Results of Traditional DEA in the First Stage*

From Table1, (1) without considering the influence of environmental factors and random error, the mean comprehensive technical efficiency(CTE) is 0.668, with the mean of pure technical efficiency (TE)and scale efficiency(SE) are 0.899 and 0.756, respectively. (2) The results show a large variation in efficiency values across regions. Beijing lies on the efficiency frontier during the studied period while other provinces somewhat further away from optimal efficiency. (3) CTE presents a divergent trend, with the difference between the maximum and the minimum efficiency value between 30 regions widened from 0.466 in 2010 to 0.648 in 2014. (4) The CTE of most regions present downward trends year by year. The average efficiency values fell from 0.720 in 2010 to 0.580 in 2014, which is not in line with the actual development situation of our country. (5) The average value of SE is higher than TE in most regions, meaning that carbon inefficiency in most provinces emerges from pure technical inefficiency. Contrarily, the economic scale is much different between regions, therefore, it is necessary to strip environment factors and random error to obtain more reliable results.3.2 Results of SFA Regression Analysis in the Second Stage

From Table1, the coefficients on economy structure are all significantly positive, suggesting that this is an unfavourable factor. This is consistent with the viewpoints that the secondary industries are the most intensive energy-consumption part of the economy. The same result is applied to the energy structure. The increase of the proportion of primary energy counts against reductions in input slacks. Urbanization rate and foreign trade are found to have different coefficients for different variables.

**TABLE 1.** The regression results of the second stage.

| Variable | Capital stock | Labour force | Energy | Carbon |
|---|---|---|---|---|
| Constant | -37294.77*** | 87.49*** | -13013.84*** | -52356.01*** |
| Economy structure | 50193.50*** | 566.12*** | 9966.44*** | 53529.68*** |
| Energy structure | 23988.94*** | 2986.99*** | 18126.21*** | 46500.23*** |
| Urbanization rate | 17383.40*** | -2500.62*** | 7327.88*** | 34260.45*** |
| Foreign trade | -23821.07*** | 237.25*** | -7954.20*** | -35755.53*** |
| Sigma Squared | 394247820.00*** | 1825160.10*** | 33758624.00*** | 347714400.00*** |
| Gamma | 0.90* | 0.93* | 0.90* | 0.94* |
| Log likelihood function | 1535.58 | 1110.12 | 1349.48 | 1501.31 |
| LR test of the one-sided | 195.76*** | 233.53*** | 197.05*** | 238.60*** |

*3.3 Results Following an Adjustment of Input Variables in the Third Stage*

From table2, (1) pure technical efficiency is improved in most regions, but comprehensive technical efficiency changes little, mainly due to the decline of scale efficiency in these regions. Comparatively speaking, scale efficiency has much space for improvement in most regions by continuing expansion of the input scale. (2) The differences of efficiency between regions are significant. Jiangsu and Guangdong, perform efficiently, as both scores of comprehensive technical efficiencies are 1. The scores of the other 28 provinces range from 0.137 to 0.877, with Qinghai ranking last among the inefficient provinces. (3) The efficiency does not show the downtrend as the first stage during studied period, which conforms to China's actual conditions.(4) However, the variation between maximum and minimum of the efficiency scores expanded, which shows that some bad performances have been "reward", relying on their advantages of energy resources or the foreign trade level

**Table 2**Energy efficiency results in the first stage and the third stage.

| | **2010** | | **2011** | | **2012** | | **2013** | | **2014** | | **Average** | |
|---|---|---|---|---|---|---|---|---|---|---|---|---|
| | First | Third | First | Third | First | Third | First | Third | First | Third | First | Third |
| Beijing | 1 | 0.67 | 1 | 0.69 | 1 | 0.72 | 1 | 0.75 | 1 | 0.77 | 1 | 0.725 |
| Tianjin | 0.95 | 0.57 | 0.80 | 0.54 | 0.78 | 0.54 | 0.71 | 0.51 | 0.60 | 0.43 | 0.77 | 0.520 |
| Hebei | 0.62 | 0.78 | 0.63 | 0.78 | 0.63 | 0.78 | 0.61 | 0.76 | 0.55 | 0.72 | 0.61 | 0.769 |
| Shanxi | 0.53 | 0.59 | 0.53 | 0.6 | 0.52 | 0.60 | 0.48 | 0.55 | 0.41 | 0.49 | 0.49 | 0.570 |
| Inner | 0.53 | 0.72 | 0.62 | 0.70 | 0.61 | 0.68 | 0.53 | 0.63 | 0.47 | 0.53 | 0.55 | 0.656 |
| Liaoning | 0.63 | 0.81 | 0.6 | 0.78 | 0.60 | 0.78 | 0.57 | 0.77 | 0.51 | 0.72 | 0.58 | 0.778 |
| Jilin | 0.54 | 0.55 | 0.47 | 0.56 | 0.47 | 0.56 | 0.44 | 0.56 | 0.39 | 0.48 | 0.46 | 0.546 |
| Heilongjiang | 0.71 | 0.55 | 0.71 | 0.58 | 0.70 | 0.59 | 0.64 | 0.57 | 0.56 | 0.51 | 0.66 | 0.563 |
| Shanghai | 1 | 0.79 | 1 | 0.75 | 1 | 0.74 | 1 | 0.76 | 0.93 | 0.78 | 0.98 | 0.768 |
| Jiangsu | 0.80 | 1 | 0.72 | 1 | 0.68 | 1 | 0.65 | 1 | 1 | 1 | 0.77 | 1 |
| Zhejiang | 0.80 | 0.85 | 0.79 | 0.86 | 0.79 | 0.86 | 0.8 | 0.87 | 0.78 | 0.88 | 0.79 | 0.867 |
| Anhui | 0.75 | 0.67 | 0.77 | 0.69 | 0.76 | 0.72 | 0.72 | 0.7 | 0.61 | 0.63 | 0.72 | 0.684 |
| Fujian | 0.79 | 0.69 | 0.77 | 0.68 | 0.75 | 0.71 | 0.70 | 0.67 | 0.60 | 0.61 | 0.72 | 0.676 |
| Jiangxi | 0.72 | 0.59 | 0.68 | 0.63 | 0.68 | 0.64 | 0.66 | 0.63 | 0.60 | 0.59 | 0.67 | 0.620 |
| Shandong | 0.68 | 0.94 | 0.64 | 0.85 | 0.65 | 0.86 | 0.64 | 0.87 | 0.59 | 0.83 | 0.64 | 0.876 |
| Henan | 0.56 | 0.78 | 0.54 | 0.74 | 0.54 | 0.75 | 0.50 | 0.73 | 0.44 | 0.59 | 0.52 | 0.720 |
| Hubei | 0.72 | 0.72 | 0.72 | 0.74 | 0.73 | 0.76 | 0.69 | 0.73 | 0.58 | 0.65 | 0.69 | 0.722 |
| Hunan | 0.7 | 0.74 | 0.70 | 0.75 | 0.71 | 0.81 | 0.67 | 0.83 | 0.58 | 0.69 | 0.67 | 0.77 |
| Guangdong | 1 | 1 | 1 | 1 | 1 | 1 | 0.98 | 1 | 0.93 | 1 | 0.98 | 1 |
| Guangxi | 0.59 | 0.55 | 0.53 | 0.58 | 0.50 | 0.58 | 0.45 | 0.58 | 0.37 | 0.43 | 0.49 | 0.549 |
| Hainan | 0.78 | 0.14 | 0.76 | 0.16 | 0.74 | 0.17 | 0.67 | 0.17 | 0.58 | 0.17 | 0.70 | 0.167 |
| Chongqing | 0.63 | 0.51 | 0.63 | 0.53 | 0.64 | 0.53 | 0.60 | 0.48 | 0.48 | 0.41 | 0.60 | 0.497 |
| Sichuan | 0.64 | 0.70 | 0.66 | 0.73 | 0.67 | 0.73 | 0.64 | 0.68 | 0.53 | 0.59 | 0.63 | 0.688 |
| Guizhou | 0.85 | 0.35 | 0.82 | 0.37 | 0.80 | 0.40 | 0.71 | 0.41 | 0.56 | 0.38 | 0.75 | 0.386 |
| Yunnan | 0.57 | 0.43 | 0.56 | 0.44 | 0.55 | 0.46 | 0.35 | 0.31 | 0.41 | 0.39 | 0.48 | 0.409 |
| Shanxi | 0.57 | 0.59 | 0.57 | 0.58 | 0.56 | 0.6 | 0.52 | 0.59 | 0.44 | 0.51 | 0.53 | 0.577 |
| Gansu | 0.74 | 0.34 | 0.74 | 0.36 | 0.72 | 0.38 | 0.68 | 0.38 | 0.56 | 0.36 | 0.69 | 0.368 |

| | | | | | | | | | | | | |
|---|---|---|---|---|---|---|---|---|---|---|---|---|
| Qinghai | 0.65 | 0.13 | 0.61 | 0.13 | 0.55 | 0.14 | 0.46 | 0.14 | 0.35 | 0.12 | 0.53 | 0.136 |
| Ningxia | 0.55 | 0.15 | 0.55 | 0.17 | 0.52 | 0.18 | 0.48 | 0.18 | 0.39 | 0.16 | 0.50 | 0.172 |
| Xinjiang | 0.88 | 0.41 | 0.83 | 0.43 | 0.75 | 0.44 | 0.65 | 0.44 | 0.50 | 0.41 | 0.72 | 0.428 |
| Average | 0.72 | 0.61 | 0.70 | 0.61 | 0.69 | 0.62 | 0.64 | 0.61 | 0.58 | 0.56 | 0.66 | 0.607 |
| Minimum | 0.53 | 0.13 | 0.47 | 0.13 | 0.47 | 0.14 | 0.35 | 0.14 | 0.35 | 0.12 | 0.46 | 0.137 |

**Table 3** Decomposition results of energy efficiency.

| | **2010** | **2010** | **2011** | **2011** | **2012** | **2012** | **2013** | **2013** | **2014** | **2014** | **Average** | **Average** |
|---|---|---|---|---|---|---|---|---|---|---|---|---|
| CTE | 0.720 | 0.615 | 0.702 | 0.617 | 0.691 | 0.627 | 0.644 | 0.613 | 0.580 | 0.566 | 0.668 | 0.607 |
| TE | 0.790 | 0.970 | 0.784 | 0.962 | 0.780 | 0.964 | 0.739 | 0.962 | 0.688 | 0.933 | 0.756 | 0.958 |
| SE | 0.920 | 0.632 | 0.905 | 0.642 | 0.897 | 0.651 | 0.885 | 0.637 | 0.861 | 0.609 | 0.894 | 0.634 |

*3.4 Decomposition Results of Efficiency*

When considering the implications to formulate regional emission reduction policies, many scholars divided China based on the geographic distribution. But we believe specific characteristics would be ignored and the corresponding strategies would be far from convincing. Thus, taking the efficiency value of 0.9 as a critical point, we divide China into four areas based on their carbon efficiency characteristics. First type is "high- high ", that is TE and SE are all above 0.9. These regions have a relative optimal performance and can serve as the example for other regions. Thus, steady improvement of the energy efficiency in these regions should be guaranteed to improve the overall efficiency. The second type is "high-low ", namely high TE but low SE, which should pay attention to improve regional investment attractiveness and expand product scale, taping the energy savings potential of high-energy consumption sectors. A third type of "low-high" is, that TE below 0.9 while SE above 0.9. These technically inefficient regions can cooperate with adjacent technically efficient regions to actively learn advanced energy saving technologies, then update equipment, adopt advanced process, and strengthen energy management. The fourth type is "low-low", whereby TE and SE are both under 0.9. It is difficult for these regions to improve the TE and SE simultaneously, thus we suggest to emphasis on expanding the scale of production to lay a basis for technological progress. Furthermore, governments should formulate relevant supporting policies to promote the efficient flow of capital, technology, and talent, breaking barriers to improve energy efficiency.

**Table 4** The division of 30 regions

| **Type** | **Region** |
|---|---|
| high-high | Beijing, Inner Mongolia, Liaoning, Shanghai, Jiangsu, Zhejiang, Anhui, Hubei, Hunan, Guangdong |
| high-low | Shanxi, Jilin, Jiangxi, Chongqing, Guizhou, Shaanxi, Gansu, Qinghai, Ningxia and Xinjiang |
| low-high | Shanxi, Jilin, Jiangxi, Chongqing, Guizhou, Shaanxi, Gansu, Qinghai, Ningxia and Xinjiang |
| low-low | Tianjin, Heilongjiang, Guangxi, Hainan and Yunnan |

## 4. Conclusion

This study has applied the three-stage DEA model to measure the energy efficiency of 30 provinces in China between 2010 and 2014. The results indicate that:(1) When using the SFA method in the second stage, environmental factors and statistical error both have significant impacts on energy efficiency. (2) There are significant regional differences of energy efficiency in China. In general, energy efficiency in developed regions is higher than undeveloped regions or resources endowment regions, indicating that the developed economy, technology, and human resources in former regions greatly promote the improvement of efficiency, while the latter regions always characterized high energy consuming industries as the pillar industry. (3) The average of TE is higher than SE, namely that SE is the primary cause constraining energy efficiency and has great space for improvement in inefficient regions. (4) Based on the efficiency characteristics, we provide differentiated energy strategies to different types. Steadily improve the energy efficiency of "high-high" regions and update equipment, adopt advanced technology in "high-low" regions. The "high-low" regions should accelerate the transfer of intensive growth by attracting investment and expanding product scale. The "low-low" regions should expand the economy scale and introduce energy-saving technologies.

**References**

[1] M.G. Patterson, What is energy efficiency? Concepts, indicators and methodological issues. Energy Policy 1996;24: 377–390.
[2] K.H. Ghali, El-sakka, Energy use and output growth in Canada: a multivariate cointegration analysis. Energy Economics 2004; 26: 225–238.
[3] Hu JL, Wang SC. Total-factor energy efficiency of regions in China. Energy Policy 2006;34(17):3206–17.
[4] Charnes A, Cooper WW, Rhodes E. Measuring the efficiency of decision making units. European Journal of Operational Research 1979; 2(6):429-444.
[5] Wang H, Zhou P, Zhou DQ. Scenario-based energy efficiency and productivity in China: a non-radial directional distance function analysis. Energy Econ 2013; 40:795–803.
[6] Zhang N, Kong F, Yu Y. Measuring ecological total-factor energy efficiency incorporating regional heterogeneities in China. Ecol Ind 2015; 51:165–72.
[7] Yang, H., Pollitt, M., Incorporating both undesirable outputs and uncontrollable variables into DEA: the performance of Chinese coal-fired power plants. European Journal of Operational Research, 2007;197(3):1095-1105.
[8] Li, K., Lin, B., Impact of energy conservation policies on the green productivity in China's manufacturing sector: evidence from a three-stage DEA model. Appl. Energy2016; 168:351-363.
[9] Zhang J, Wu G, Zhang J. Estimation of China's provincial capital stock (1952–2004) with applications. Journal of Chinese Economic and Business Studies, 2008; 6(2):177-196.

# Post-grouting bored pile technology

**A R Zheng**

CCCC Tianjin Port Engineering Institute Ltd. ,Key Laboratory of Geotechnical Engineering, Ministry of Communications, PRC, Key Laboratory of Geotechnical Engineering, Tianjin , Tianjin 300222, China
airong717@163.com

**Abstract.** Post-grouting is an effective technology to modify the shortcomings of thick bottom slime and shaft mudcake for the slurry bored pile. Construction procedure, parameter selection and strengthening mechanism of post-grouting bored pile had been introduced in this paper. Development of study on physical and mechanical properties of surrounding soil post-grouted was summarized. The bearing capacity behaviour and deformation properties of the pile were also analyzed. It will provide some advice for the research and application of post-grouting bored pile.

## 1. Introduction

In1864, the United Kingdom firstly used the shaft cement grouting in Allihn Puri Bei mine to shut off water. From that time on, the grouting technology has been widely used in the foundation reinforcement and the treatment of bad geological phenomena. In 1961, Venezuela Maracaibo bridge foundation engineering used the grouting pipe to grout the pile foundation. This was the first report of the grouting technology's application in pile foundation reinforcement project. Because of the continual application of grouting technology in the construction of grouting piles, there are relatively perfect post-grouting bored pile construction technologies, including the pile-end grouting, the pile-surrounding grouting and the pile-end and pile-surrounding association grouting, are gradually formed[1, 2]. In addition, deeper and deeper researches on the optimization of post-grouting bored piles' construction technology, the reinforcement mechanism, the bearing capacity and deformation characteristics are done.

## 2. Construction technology of post-grouting bored pile

### *2.1. Construction process for post-grouting technology*

The post-grouting bored pile technology means that after the borehole grouting piles are formed, a certain amount of cement serous fluid will be injected to the pile-surrounding mud cake, the pile-bottom sediment and the soil layer around the pile under certain pressure through the grouting pipe embedded in the pile body so that the bearing capacity of the pile foundation can be improved. When the pile-end and pile-surrounding association grouting way is used, in order to prevent the serous fluid from emerging along the pile-surrounding weak plane during the pile-end grouting process, the pile-surrounding grouting shall be used firstly for plugging before the conduction of the pile-end grouting.

Similarly, when the multilayer pile-surrounding grouting way is adopted, operations shall be done from the top to down layer after layer.

### *2.2. Selection of grouting parameters*

*2.2.1. Grouting quantity.* The grouting quantity is determined by the pile diameter, the pile length, the characteristics of soil layer around the pile and the bearing capacity's improvement requirements. LIU Jinli, etc[2] puts forward the following formula:

Pile-end grouting quantity:

$$G_P = \pi\left(h \cdot t \cdot d + \xi \cdot n_0 \cdot d^3\right) \times 1000 \tag{1}$$

Pile-surrounding grouting quantity:

$$G_s = \pi\left[t \cdot (L-h) + \xi \cdot m_0 \cdot n_0 \cdot d^3\right] \times 1000 \tag{2}$$

In the formula, d means the pile diameter; L means the pile length; h means the upward returning height of the serous fluid along the pile surroundings during the pile-bottom grouting operations; t means the thickness of the serous fluid wrapped in pile surroundings and its value is related to properties of soil around the pile and the construction quality of the grouting pile. The value range is generally between 5mm and 20mm. $n_0$ is the natural porosity of the pile-bottom soil or the pile-surrounding soil. $\xi$ is the grouting rate and its value for gravels and medium sand is between 0.3 and 0.7 while that is between 0.2 and 0.3 for clayey soil, floury soil and sandy silt. $m_0$ is the number of cross sections of the pile-surrounding grouting petals. According to the point distribution situations of the longitudinal wave, the value of $m_0$ is 1/4 of the number of the grouting points.

*2.2.2. Grouting pressure.* Grouting pressure refers to the biggest pressure to be used when the earth's surface does not rise and the foundation piles are not greatly lifted. The grouting pressure value is related to the density, strength and initial stress of the soil layer, the borehole depth, the grouting location, the grouting sequence and other factors. Some of the factors are difficult to predict. Therefore, they shall be determined by the on-site pile testing. The grouting pressure of soil layers with good groutability is under 4MPa generally while that of the soil layer with bad groutability is between 4MPa and 8MPa.

*2.2.3. Concentration of the serous fluid.* Serous fluid with different concentration may have different behavioral characteristics. Thin paste (with the water-cement ratio of about 0.8:1) has strong fluidity which is suitable for permeating and strengthening pile-surrounding and pile-bottom loose soils. Serous fluid of medium concentration (with the water-cement ratio of about 0.6:1) mainly plays the role of filling, densification and compaction. Thick slurry (with the water-cement ratio of about 0.4:1) is used for dehydration of cement serous fluid already injected [3]. Thus, more than one grouting can be done and the cement serous fluid can be injected from thin to thick one level after another. In addition, the method of corresponding grouting pressure can also be used to enhance the grouting effect.

Engineering practice shows that if the thick slurry with the water-cement ratio of 0.5~0.55 is used; better grouting effects can be achieved.

### *2.3. Applicable soil layer of grouting*

The serous fluid has different diffusion areas in different soil layers. Therefore, the selection of the grouting soil layer has a significant influence on the improvement range of the bearing capacity of the post-grouting bored pile. The coarser the soil particles are, the larger permeation range of the serous fluid will be. If the pile-end supporting layer belongs to the thin gravel layer, the diffusion radius of the grouting can reach more than 3.5 times of the pile diameter [3]. In soil layers with small soil

IOP Conf. Series: Earth and Environmental Science 61 (2017) 012057 doi:10.1088/1755-1315/61/1/012057

particles, the serous fluid is not distributed continuously and it is not uniformly filled. Most of them are distributed with thin stratified structures. In addition, their thickness varies greatly and it is relatively hard to mix the slurry and the soil particles. The contrastive analysis of a large amount of on-site static load pile testing results shows that the increasing range of the pile-bottom grouted pile's bearing capacity will increase when soil particles of the pile-end supporting layer become larger. The gravel layer is the most ideal pile-bottom grouting layer [4].

## 3. Post-grouting reinforcement mechanism

At present, in-depth researches have been done to the pressure grouting theory. The technology of the post-grouting bored pile is used as the grouting technology for the reinforcement of the pile foundation and its reinforcement mechanism is similar with that of the traditional grouting technology.

### *3.1. Mechanism of post-grouting cement serous fluid*

(1) Permeation and solidification.
The serous fluid can be injected to a certain range of the sandy soils and gravelly soils with strong permeability and good groutability under low pressure so as to form the stone body with good structure and high strength, to expand the pile-end bearing area and to enhance the pile-end bearing capacity.

(2) Cementing mud cake.
During the pile-surrounding grouting process, the cement serous fluid will squeeze and split off weak mud cakes around the pile and will cement near the piles again so as to form bamboo-like cement reinforced bodies of high strength and to tighten them near the pile. Generally speaking, its thickness is between 2mm and 20mm [2], which is shown as Figure 1.

(3) Filling.
The serous fluid with large concentration will be used to fill soil pores and to compact the pile-surrounding soils with strength reduced because of stress relief and the pile-bottom loose sediment under certain pressure so as to reduce the porous ratio, to solidify the soil again and to enhance the compressive strength of the soil. They can be shown in Figure 2.

(4) Cleaving reinforced strips.
When the grouting pressure is high, the cement serous fluid can overcome the soil resistance and have a cleaving effect on the soil layer, which will form the mesh stone, have the reinforced function to the soil and reduce the non-uniform sedimentation of the pile group. They can be shown in Figure 3.

Through the in-house model experimental study, GAO Wensheng [4] concludes that if the pile-end serous fluid is injected to the floury soil to form the cement stone body, they are out-of-shape and basically belong to the compaction grouting. If the pile-end serous fluid is injected to gravel layer, it will uniformly permeate, will spread to the gravel layer and will condense with the gravel to form the low-grade concrete so as to largely form a rounded expansion chassis on the pile-bottom supporting layer and this belongs to the permeation grouting.

**Figure 1.** Reinforcement effect of shaft mudcake

**Figure 2.** Seep-in and compaction effect

**Figure 3.** Spliting effect

*3.2. Changes of physical and mechanical properties of the soil after the grouting process*

After the grouting process, the effective stress, the internal friction angle and the cohesive force of the soil mass will be enhanced and the shearing strength of the soil mass will also be increased. Changes of physical and mechanical properties of the soil are closely related to the grouting quantity. Zhang Zhongmiao, etc [5] conduct laboratory experiments on mud-cement mixed soil samples which have been solidified for 30 days. When the cement mixing amount increases from 10% to 40%, the moisture content of the mud soil samples will drop from 253.5% to 63.6%; the porous ratio will drop from 7.52 to 2.25; the compression modulus will be improved from 1.86MPa to 8.96MPa; the cohesive force will be enhanced from 2kPa to18kPa and the internal friction angle will be enlarged from 0.5° to 19°. Through in-house laboratory investigation, MA Hailong [6] concludes the relationship among changes of the muddy clay weight, the permeability coefficient, undrained shear strength, the internal friction angle, the unconfined compressive strength, the coefficient of compressibility and the compression modulus and the cement mixing amount. When the cement mixing amount is 20%, the compression modulus can be enhanced for 40 times. These two experiments all use the homogeneous cement soils formed by intensively mixing the fine-grained soil and the cement serous fluid as the objects of study. There are great differences between it and the heterogeneous cement soil formed in the practical grouting project.

In soil layers with small soil particles, the cement paste is discontinuously distributed and non-uniformly filled. They are mostly distributed with the thin stratified structure and their thickness does not vary greatly, which is generally between 0.2cm and 20cm. In addition, it is forbidden to mix the slurry and the soil particles [7]. The grouting reinforced bodies formed in sandy soils and other soil layers with more coarse-grained soil particles and good permeability tend to have high strength. In laboratory model experiments, GAO Wensheng [4] conducts uniaxial compressive strength test to the pile-end grouting reinforced body in the gravel supporting layer and the average ultimate strength he measured is 11.46MPa, which is equivalent to the concrete with the strength level of C15.

After the grouting process, changes of physical and mechanical parameters of soils also lead to changes of soil yield functions. Under the effect of the external loads, the elastic range of the rock and earth mass significantly becomes larger while the yield range of the plastic shear strain and the yield range of the volumetric strain becomes more geographically dispersed rather than relatively centralized before the grouting. In addition, the overall size becomes relatively small, but the stability and bearing capacity of the rock and earth mass are greatly enhanced [8].

## 4. Deformation characteristics of bearing capacity of post-grouting bored piles

After the grouting process, because of the difference of the grouting soil layer, the grouting types, the length-diameter ratio of the pile and the construction technology, the bearing capacity of the grouting

IOP Conf. Series: Earth and Environmental Science **61** (2017) 012057 doi:10.1088/1755-1315/61/1/012057

pile can be improved for 20% to 150% [4] or more.

For common grouting piles, their ends often display the penetrating shear failure and their load sedimentation curves have obvious inflection points. The post-grouting bored pile strengthens the pile-end loose sediment and the pile-end disturbed soil layers, removes the pile-bottom "cushion" and forms a pile-end reinforced body of large strength. It is generally shown as the local shear failure and the load sedimentation curve tends to be mild. In addition, the sedimentation amount also reduced, which is shown in Figure 4 [9]

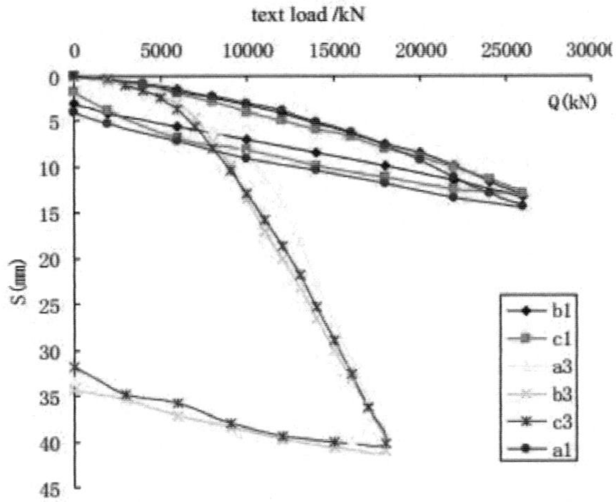


**Figure 4.** Comparison in load-settkenebt curves

Notes: a1, b1 and c1 are non-post-grouting bored piles while a3, b3 and c3 are post-grouting bored piles.

Pile-end and pile-surrounding bearing capacities are all enhanced to different degrees. After the serous fluid of the pile-surrounding grouting cement solidify the mud cake, a hard shell layer tightly fixed on the pile body will be formed. Gao Wenling [4] does a direct shear test relative to the pile body and he measures that the ultimate shearing strength when there is a relative slippage between the pile- surrounding reinforced body and the pile body is 780kPa, which is about 10 times of the pile-surrounding frictional resistance of good foundation soil. In addition, he judges that the shear failure envelope between the pile and the soil will inevitably occur outside the pile-surrounding reinforced body, which means that the grouting has a "Diameter Expansion Effect" on the grouting pile. After the grouting process, the improvement of the compression modulus of the pile-surrounding soils, the reinforcement of the pile-surrounding shear weak surface and the "Diameter Expansion Effect" of the pile all help enhance the pile-surrounding ultimate frictional resistance. The pile-end grouting will solidify the pile-bottom loose sediment. The high strength cement soil at the bottom of the pile will be connected with the pile body and play the "Base Expansion Effect". Besides, the total end resistance will be significantly increased. The seepage consolidation and the cleaving stiffening effect of the post-grouting strengthen soil layers of the pile-bottom sliding surfaces, improving the cohesive force of the soil and its internal friction angle and enhancing the pile-end bearing capacity.

The grouting has an obvious effect on load transmission characteristics of piles. Under the pile top load effect of the non-grouted pile, the axial force will pass down gradually and the lateral resistance will also be played from top to bottom. After the pile-top displacement reaches a certain degree, the resistance will begin to take effect. However, under the pile-end pressure effect, the grouted pile exerts a prestressing force to the pile from bottom to top. In addition, the pile-end soil body and the soils around the pile of some range finish some deformations in advance. This way can make the pile-end

IOP Conf. Series: Earth and Environmental Science **61** (2017) 012057 doi:10.1088/1755-1315/61/1/012057

resistance plays its role from the beginning, which will fully play the strength of the soil and greatly enhance the bearing capacity and reduce the sedimentation amount.

## 5. Conclusion

(1) Deep theories and experiment researches have been done to the post-grouting reinforcement mechanism. The cement serous fluid can enhance the soil strength through permeation, compaction, glue joint and cleaving effect. Under the effect of the load, the elastic range of the soil will be enlarged while the plastic range tends to be relatively dispersed.

(2) The post-grouting brings obvious changes to grouting piles' characteristics. Because of the "Diameter Expansion Effect" and the "Base Expansion Effect" generated by the grouting, the pile-surrounding frictional resistance and the pile-end resistance are greatly increased. After the grouting process, the grouting pile-end resistance is played in advance and the time difference between them is reduced.

## References

[1] Bruce D A 1986 Ground Engingeering 19(4) p 9–15
[2] Liu J L and Zhu J C 1996 Building Science 2 p 13–18
[3] Zhang Z M, Wu S M and Bao F 1999 Chinese Journal of Geotechnical Engineering 21(6) 681–686
[4] Gao W S 1997 Study of Bearing Capacity for Postgrouting Bored Pile (Beijing: China Academy of Building Research Press)
[5] Zhang Z M, Zhang G X, Wu Q Y and Xin G f 2006 Chinese Journal of Geotechnial Engineering 28(6) p 695–699
[6] Ma H L 1995 Journal of Suzhou Institute of Urban Construction and Environmental Protection 8(3) p 34–39
[7] Dai G L, Gong W M, Xue G Y and Tong X D 2006 Rock and Soil Mechanics 27(5) p 849-852 [8] Zhang Y P, Wu S C and Fang Z L 2004 Journal of University of Science and Technology Beijing 26(3) p 240–243
[9] Testing Pile for the Airport Highway Bridge Post-grouting Bored Pile 2007 (Shanxi: Changtong Road and Bridge Engineering Technology co. LTD)

# Voltage on the distribution of the residual chlorine ion of the concrete after the electrochemical dechlorination treatment

**ZHANG Kun [1,2]*, SHAN Wei [1], SUN Wen-bo [3]**

1.Northeast Forestry University, Harbin, China
2.Heilongjiang College of Construction, Harbin, China
3.Harbin Institute of Technology, Harbin, China
zhangkun228@sina.com, shanwei456@163.com,rscswb@hit.edu.cn

**Abstract.** This paper utilizes 60V stable voltage power, saturated calcium hydroxide solution, titanium mesh anode to analyze the influence of the mineral admixtures, applied voltage and electrolyte solution on the residual chloride ions distribution after 14 consecutive days' test of the electrochemical chloride extraction of the cylinder concrete. The results show that after electrochemical chloride extraction, the residual chloride ions content in concrete shows increasing trend from inside to outside of the reinforced interface, with the least at internal layer and the most at external layer and there are 2-3.5 times differences between internal layer and external layer. The difference between the internal layer and external layer increases with the increase of the chloride extraction voltage.

## 1 Introduction

Recent years, because of the corrosion of reinforcement, the concrete structure's performance declines, which causes the service life not to meet design requirements and become disabled. This has been a major problem of science and technology in the civil engineering territory [1-2]. There are several factors which could cause the corrosion of concrete reinforcement. From the current engineering damage statistics, the main reason that causes reinforcement corrosion is sort of chlorine salt invasion [3]. Electrochemical chloride extraction [4] refers to the process of extracting chlorine salt and other harmful components through applied electric filed in the concrete protection layer without damaging it and then passivating the corroded steel surface. It is an efficient and low cost technology for the restoration of reinforced concrete. Since 1990's, developed European and American countries have started spending energy on the research in this field[5], but they mainly focused on the influence of dechloride's effect caused by the external factors such as the electrochemical parameter [6-8]. However, only few concentrate on the influence of the concrete mix proportion on the efficiency of the electrochemical chloride extraction. This paper applies the homemade test units of electrochemical chloride extraction, to analyze the influence of different water-cement ratio on the efficiency of electrochemical chloride extraction according to the test of the electric current through the concrete specimen, the chloride ion discharge, and residual chloride ion content in the same electric field.

## 2 Experiment

### *2.1 Raw material and mixing proportion of concrete*

Using both P. O42.5 cement of Harbin cement co., Yatai Group and II grade fly ash of Harbin No.3 Thermal Power, water requirement ratio is 97%. Anshan iron and steel production of blast furnace slag, its density is 2.86 g/cm3, specific surface area of 390 m2 / kg. Norway's company

IOP Conf. Series: Earth and Environmental Science **61** (2017) 012058 doi:10.1088/1755-1315/61/1/012058

production of Micronesia silica fume, the average particle size of 0.1 mu m, specific surface area of 1.5 x 104 m2 / kg, density: 2.26 g/cm3.Choose the limestone rubble, particle size grading in 5 ~ 25 mm, crushing index was 4.8%, the needle flake content 3%; Fine aggregate used river sand of Songhua river, fineness modulus of Mx = 2.82, belong to medium sand, II district. The Shanghai Huawang company's Mighty 100 high efficiency water reducing agent, recommended dosage 0.7 to 1.2% of the total weight for the gelled material. Ordinary steel reinforcement for 8 mm diameter. Sodium chloride for industrial pure product, NaCl content is more than 96%.

*2.2 Specimens*

Φ100 x 150 mm cylindrical concrete specimen was used, and a Φ 10 mm steel bar was inserted into the specimen along the direction of the axis., the relative position of steel reinforcement and concrete specimen is shown in figure 1.Before concrete formed, steel was made 15mm long and soaked in three ammonia solution to remove the surface rust; After the surface dried, the steel outside of the concrete was covered with surface coating of epoxy resin to prevent this part of the steel being corroded during the concrete curing. Then the processed reinforcement device was fixed well in the special processed mold, to ensure that the coating with epoxy resin reinforced part of the dew on the outside of the concrete, the rest of the concrete axial center location. Concrete specimen was taken out of the mold when it forms after 1 day and then put in a standard curing room for maintenance, wrapped in plastic to prevent the loss of chloride salt in the concrete due to water conservation. To 28 day age, the specimens are retrieved from the room, and then are dealt with electrochemical chloride extraction treatment.

*2.3 Method*

Homemade test apparatus (as shown in Fig. 2) is utilized for the electrochemical chlorine treatment of each concrete. During this experiment, saturated calcium hydroxide solution, 0.1mol/L Na3BO3 solution and tap water are used as the electrolyte respectively. Titanium mesh anode of high corrosion resistance is chosen. The external source is 20V, 40V, and 60V stable AC power. And the rheostat connections are fixed 10 Ω resistance. By measuring the voltage of the fixed resistance connected on the reinforced ends, the current value across the concrete specimens can be converted. Change the electrolyte solution every 2 days. The electrochemical chloride extraction costs 14 days.

After the concrete dechloridation, take out the specimen and remove the residual precipitation of crystal on the surface. Cut the cylinder concrete specimen radially from the reinforcement external diameter into two parts, with the profile shown as Fig. 3. Cut steel trisection along axial direction, from up to down number I, II, III, IV respectively; divide concrete radially into inner, middle, and outer 3 layers. Take concrete samples at the intersection point of layer and layer, which refers to the black point in the figure. Choose mortar instead of coarse aggregate and large granule fine aggregate. Then grind the sample into powder as the fineness of cement and weigh. Because the purpose of this experiment is to measure the related soluble ion concentration, choose the distilled water to soak and dissolve powder sufficiently. A week later, utilize precise ion meter to measure the specimens' chloride ion, sodium ion and potassium ion content, and convert into ion concentration of relevant position

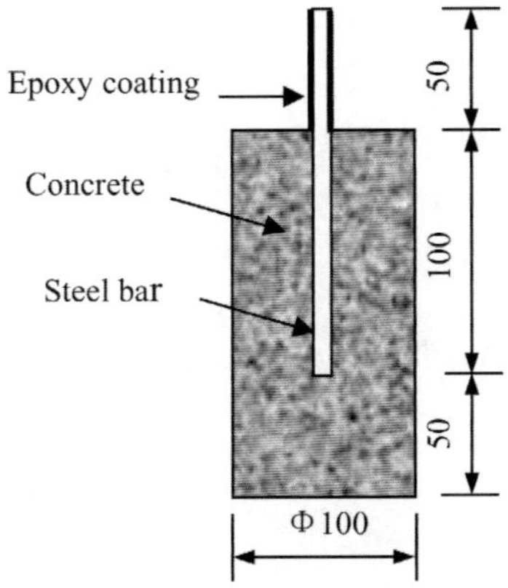


Fig. 1 Immersed steel in concrete specimen

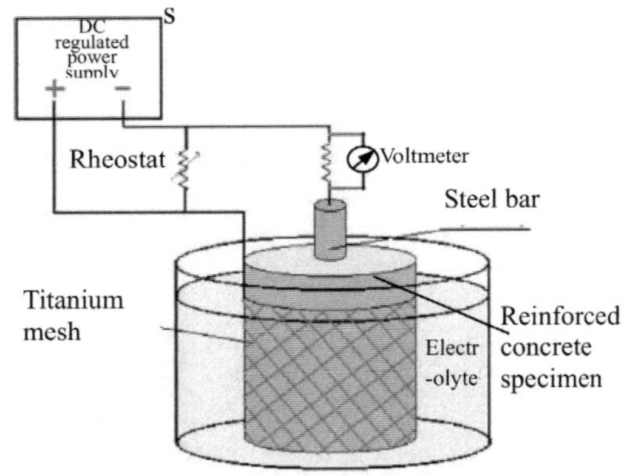


Fig. 2 Experimental ECE set-up.

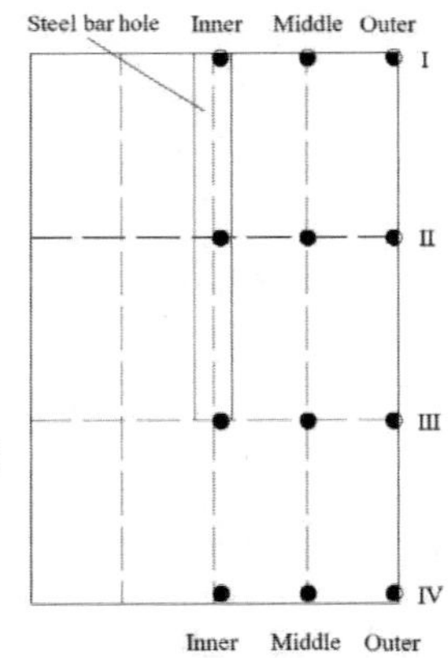


Fig.3 Schematic diagram of the specimen after Cl- has been extracted

## 3 Results and Analysis

*The influence on the residual chloride ion*

Three kinds of specimens are made by changing mix proportion of 0.45, 0.38, 0.32 and using saturated $Ca(OH)_2$ solution as electrolyte solution. The time for dechlorination is 28 days. The influence of voltage on the residual chloride in different layers after dechlorination is shown in figure 4.The influence of voltage on the distribution of residual chloride in vertical direction is shown in figure 5.

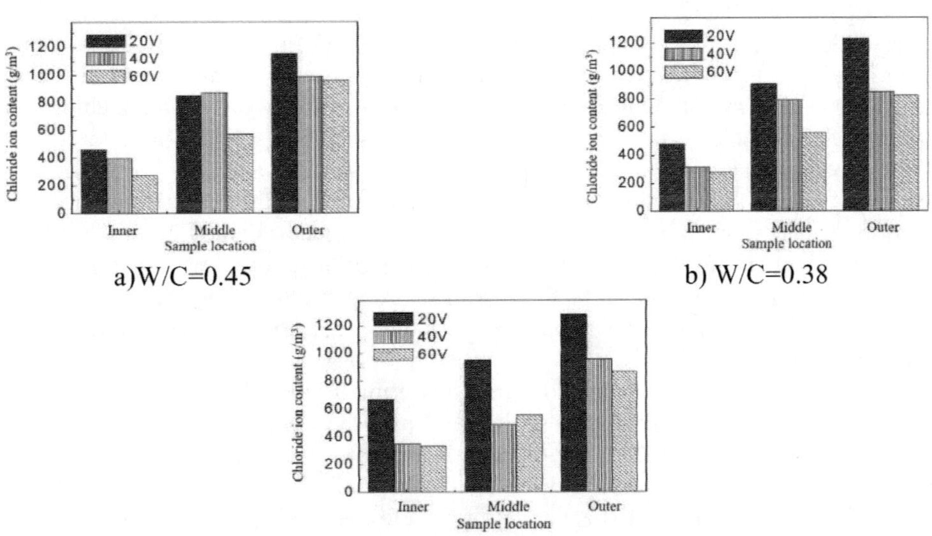


a)W/C=0.45

b) W/C=0.38

c) W/C=0.32

Fig.4 Influence of the potential on the concentration of the $Cl^-$ after electrochemical chloride extraction

IOP Conf. Series: Earth and Environmental Science **61** (2017) 012058 doi:10.1088/1755-1315/61/1/012058

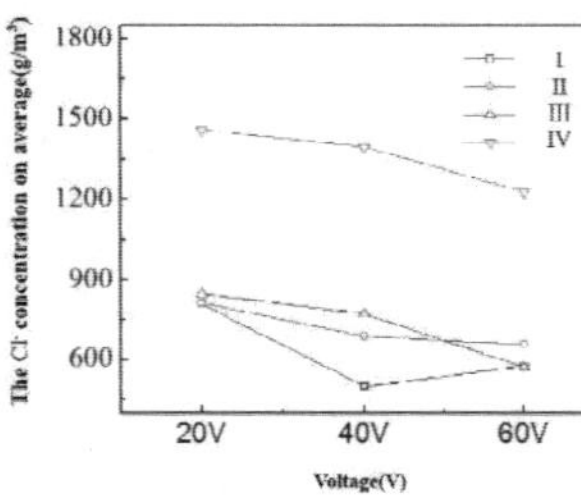


a) W/C=0.45

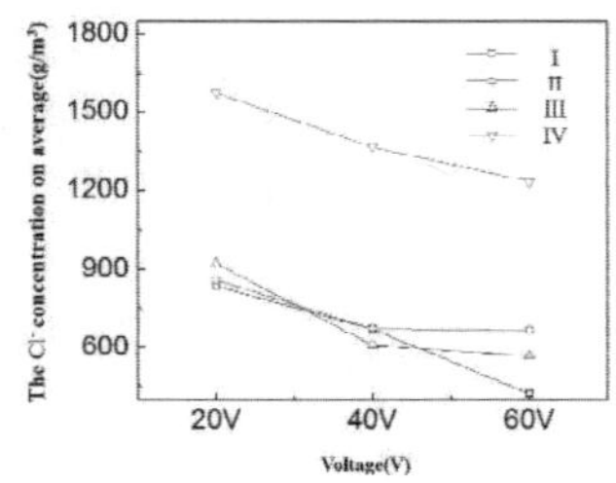


b) W/C=0.38

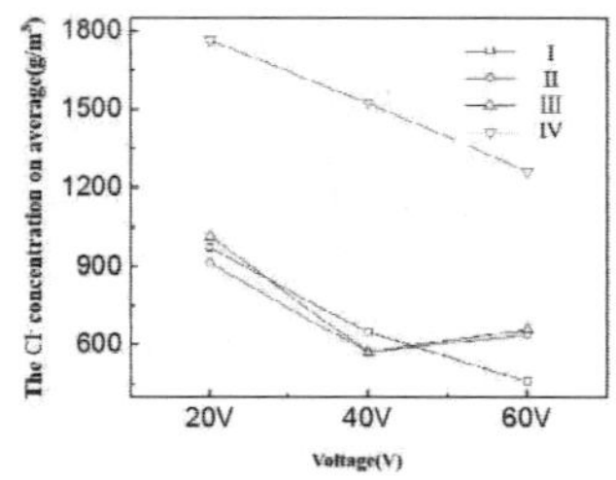


c) W/C=0.32

Fig.5 Influence of the potential on the concentration of the $Cl^-$ in longitude direction after electrochemical chloride extration

Figure 4 compares the influence of different voltage on different water-cement ratio specimens' to analyze the residual chloride ion distribution gradient after dechlorination. From the figure, after the electrochemical chloride extraction process, chloride ion content on the external layer of concrete is obviously higher than the internal layer of concrete. The higher the applied voltage, the lower concentration of the residual chloride on the internal layer is, and the better effect of dechloridation is. This phenomenon on one hand macroscopically reflects the character that the voltage influences the internal chloride ion of concrete moving out to external layer. From figure 4, it is also easy to find out the higher voltage, the lower residual chloride ion content on each layer is, and the better effect of dechloride is. On the contrary, the lower voltage for dechloride, the higher residual chloride ion content on each layer is, the worse effect of dechloride is. For the specimens which have lower water-cement ratio, there is only little difference between dechloride that uses 40V and 60V voltage.

According to figure 4, analyzing the chloride concentration of the internal layer of three kinds of concrete after dechlorination, it is easy to conclude that the higher voltage, the lower residual chloride ion content near the rebar region is. For the specimens which have lower water-cement ratio (W/C are 0.38 and 0.32).The result under 60V voltage is higher that under 40V voltage, but the difference is quite small. However, for the specimens which have W/C as 0.45 ratio, the results are much different among the conditions under three voltages.

Figures 5 compares the residual chloride ion distribution of upper layer and lower layer of concrete after dechlorination under different applied voltages. From the figure, residual chloride ion content is the highest at the IV layer along axial direction, and the lowest content at the I, III layers. This is because the lower layer of the IV layer is farthest away from rebar, and keeps in contact with the vessel during the experiment, for this reason, the tunnel cannot be developed properly which will lead the stack of the chloride ion, and chloride ion content obviously higher at the IV layer of concrete. The chloride ion distribution on the I, II, III layers of every sample are almost the same. For each layer, there is a stronger regularity of the residual chloride ion distribution for different distances from the steel: chloride ion content increases from internal to external in the specimens. Chloride content is almost the same for each layer of concrete. However, sometimes there are some little differences. So it is easy to conclude that the distribution of chloride after dechloride has little relationship with the vertical direction, while it is influenced a lot by the distance from the rebar. So we should consider how to prevent the residual chloride ion at the external layer of concrete from re-corroding internal region of concrete.

According to the data in figure 4, entire dechloride efficiency and the efficiency of dechloride near the rebar of each specimen can be calculated. Kinds of parameters contrast condition are shown in figure

IOP Conf. Series: Earth and Environmental Science **61** (2017) 012058 doi:10.1088/1755-1315/61/1/012058

6 and figure 7. From the figures, three kinds of water-cement ratio are analyzed and compared with each other under the voltage of 20V 40V 60V respectively after 4 weeks' electrochemical chloride treatment. For the specimens, entire efficiency of dechloride lies between 60%~80%, while the efficiency of chloride near the rebar is between 70%~90%. It is obvious that the dechloride efficiency of latter is higher than the former one.

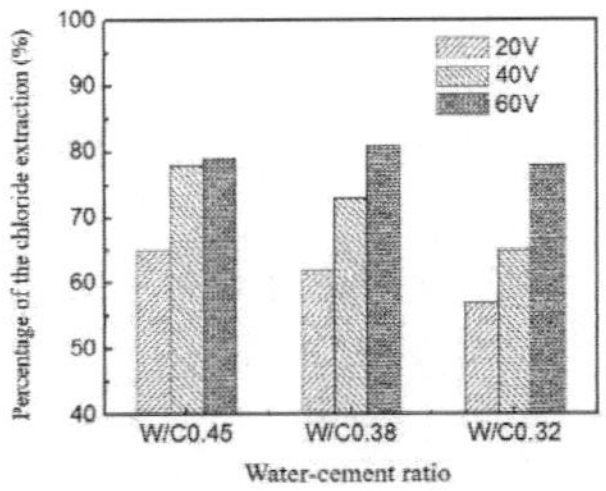


Fig.6 Influence of the potential on the total efficiency of the chloride extraction

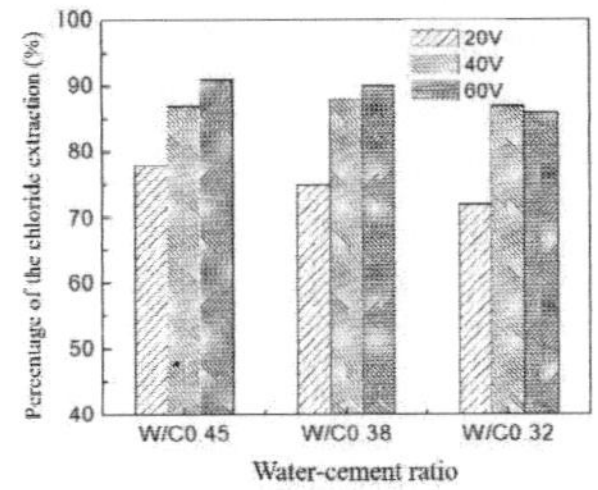


Fig.7 Influence of the potential on the efficiency of the chloride extraction close to the steel bar

For different water-cement ratio specimens, the lager water-cement ratio, the better efficiency of dechloride with the same parameter is. For the specimens with ratio of W/C 0.32 and 0.38, the specimens' entire efficiency of dechloride increase by 10% under the 20V, 40V, 60V voltage condition respectively. But for the specimens with ratio of W/C 0.45, the efficiency of specimens' entire dechloride will increase by 13% under the voltage of 40V than the voltage of 20V, while at the 60V voltage, the efficiency just increases by 1% than the voltage of 40V. According to this, for the concrete which has high water-cement ratio, the eject resistance of chloride in specimens is small. It can fulfill the requirement to dechloride at the 40V voltage. In practical engineering, it's unnecessary to achieve high ratio of ejectment by using higher voltage 60V. If it is necessary to achieve high efficiency of chloride ejectment, the duration of dechloride can be increased.

As shown in figure 7, for the specimens that have high water-cement ratio (W/C=0.45), the efficiency of dechloride of the concrete near the steel is 78%, 87%, 91% under the voltage of 20V, 40V, 60V respectively with the later increase approximately by 10 percentage. But for the lower water-cement ratio specimens (W/C=0.32), the efficiency of dechloride under 40V and 20 V voltage are 86% and 72% respectively with the former increases by 14 percentage, while the efficiency under 60V voltage is almost the same as the efficiency under 60V voltage. Although the voltage is higher, the efficiency of dechloride increases little. According to these experiments, we suggest that in practical engineering for high water-cement ratio specimens, based on practical situation, 40V voltage is utilized at the beginning, and transfer to 20V voltage later. For small water-cement ratio specimens, we suggest not to use the 60V voltage through all, 40V voltage should be considered. And also later it can be transferred into 20V voltage.

Compare the amount of pre-mix salt, chloride content distribution of concrete after dechloride and the content of chloride ion in the electrolyte solution, we can make a conclusion that during the electrochemical chloride extraction process, some parts of chloride ion move out from internal layer to external layer, some parts move to electrolyte solution, and a little of them volatilize into air as chlorine.

## 4 Conclusions

After electrochemical chloride extraction, the residual chloride ions content in concrete shows increasing trend from inside to outside of the reinforced interface, with the least at internal layer and the most at external layer. There are 2-3.5 times differences between internal layer and external layer. And the difference between the internal layer and external layer increases with the increase of the chloride extraction voltage.

## References

[1] Mehta P K, Burrows R W. Building durable structures in the 21st century[J]. Concrete International, 2001, 23(3): 57~63.
[2] Jin Weiliang, Zhao Yuxi. Durability of Concrete Structures [M]. Beijing: Science press, 2002. 1~15.
[3] Broomfield J P. Corrosion of steel in concrete[M]. New York: Chapman and Hall, 1997. 1~60.
[4] Polder R, Vander H J. Electrochemical realkalisation and chloride removal of concrete, state of the art, laboratory and field experience[A]. Proceedings of RILEM conference, rehabilitation of concrete structures[C]. Melbourne: RILEM, 1992, 135~147.
[5] Hansson I L, Hansson C M. Electrochemical extraction of chlorides from concrete part I - a qualitative model of the process[J]. Cement and Concrete Research, 1993, 23(5): 1141~1152.
[6] Yeih W, Chang J J, Hung C C. Selecting an adequate procedure for the electrochemical chloride removal[J]. Cement and Concrete Research, 2006, 36(3): 562~570.
[7] Miguel A C, María J S, Guillem V. Effect of type of anodic arrangements on efficiency of electrochemical chloride removal from concrete[J]. ACI Materials Journal, 2006, 103(4): 243~250
[8] Fajardo G, Escadeillas G, Arliguie G. Electrochemical chloride extraction (ECE) from steel-reinforced concrete specimens contaminated by "artificial" sea-water[J]. Corrosion Science, 2006, 48(1): 110~125.

IOP Conf. Series: Earth and Environmental Science **61** (2017) 012059 doi:10.1088/1755-1315/61/1/012059

# A study on rural low-carbon tourism development in Jiangxi

**XIONG Wen-ping**

Nanchang Normal University, Nanchang Jiangxi, China, 330029
lcshxwp@126.com

**Abstract.** Jiangxi features rich tourism resources in rural areas and favorable eco-environment, and it's worth in-depth exploration how to develop rural tourism while protecting the beautiful rural environment in Jiangxi. Therefore, the new perspective of developing rural low-carbon tourism in Jiangxi was proposed in the paper to revisit rural tourism. Starting with the ecological concerns for rural tourism in Jiangxi, the paper probes into the necessity of rural low-carbon tourism development in the province and the path of rural low-carbon tourism development in the province on the basis of analyzing the principles to follow in developing the tourism, in anticipation of serving as reference for the theories and practices of sustainable development of rural low-carbon tourism in Jiangxi Province.

## 1. Introduction

With fast growth of Chinese economy, people's living standard keeps rising and tourism has become a lifestyle for people to approach nature and release their pressures. And with further industrialization worldwide, consumption of fuels like coal, natural and oil is increasing to lead to continuing lift of worldwide temperatures, exacerbation of the environment and threat to eco-safety. Low carbon is now a global hot topic and rural tourism is also widely discussed these years. During the development of low-carbon economy and modern agriculture, how to development rural tourism has become a common concern for everyone.

Low-carbon tourism advocates diminishing carbon footprints and carbon emissions to minimum while travelling, making it an in-depth manifestation of environment-friendly tourism, an eco-friendly, high-efficiency and sustainable way of tourism product production and consumption, and a lifestyle healthy and civilized oriented to the future; it's also an industry to harvest greater economic, social and environment benefits from tourism by less carbon emissions during tourism development, and to make tourism truly low-carbon, environment-friendly and resource-saving.

Gao Wenling et al. proposed that rural low-carbon tourism is a new-type tourism that is materially based on the natural environment and ultimately oriented to economic growth and integrates the development of rural areas and environmental protection[2]. Ma Dongyue believed that rural low-carbon tourism must be a compound low-carbon tourism pattern that combines the local natural and cultural environments and integrates natural ecology, economic development and social progress in rural areas, so that it should boast functions for production, living and education[3].

Jiangxi features rich tourist resources and beautiful eco-environment, and it's worth in-depth exploration how to develop rural tourism while protecting the beautiful rural environment in Jiangxi. Therefore, the new perspective of developing rural low-carbon tourism in Jiangxi was proposed in the paper to revisit rural tourism. Starting with the ecological concerns for rural tourism in Jiangxi, the paper probes into the necessity of rural low-carbon tourism development in the province and the path of rural low-carbon tourism development in the province on the basis of analyzing the principles to follow in developing the tourism, in anticipation of serving as reference for the theories and practices of sustainable development of rural low-carbon tourism in Jiangxi Province.

## 2. Existing Ecological Concerns for Jiangxi's Rural Tourism

*2.1 Low-carbon infrastructures lag behind and hazards of environmental safety stand out*
First, as tourism operators keep seeking economic benefits but ignore damages to environmental resources, rural tourism lacks effective management mechanism and the eco-environment is easily disrupted; second, some tourists have no clear low-carbon idea of rural travels and insufficient knowledge of the eco-environment, so that they litter around to cause outstanding phenomena of waste pollution in tourist areas; third, due to unreasonable tourism development, the surplus of tourists and vehicles for some tourism destinations has caused undercapacity of rural tourism; fourth, due to weak infrastructures for rural tourism, there're few garbage disposal and sewage disposal systems, causing great damages to the environment. With all these factors, there're clear safety hazards for eco-environment in rural areas.

*2.2 Rural cultural and natural landscapes in Jiangxi have been destroyed*
During rural tourism development, as many tourism operators are local farmers not well educated, they don't have a clear idea of the position and value of rural resources and rural environment in rural tourism development and have no scientific and reasonable planning and management; they blindly seek urbanization by using urbanized buildings and facilities for tourism service, which not only discord with the surrounding environment but the neighboring rural natural views, thus gradually eroding the primitive rural air of rural tourism. Besides, the urban culture that urban tourists take to their rural travels also has certain impact on rural culture; some rural culture has been assimilated by urban culture, affecting the lifestyle of rural residents and the true rural colors. The most essential rural flavor of rural tourism has thus vanished. However, for those living in the city, if rural tourism loses its features different from urban tourism and the local uniqueness, rural tourism will lose its charm and attraction, thus dragging rural tourism development in Jiangxi.

## 3. Necessity of Rural Low-carbon Tourism Development in Jiangxi Province

*3.1 Conducive to protection of rural eco-environment in Jiangxi*
Jiangxi features nice rural eco-environment with the strengths of development rural tourism. The fast development of rural tourism in Jiangxi has contributed greatly to getting the local people rid of poverty and becoming better off. Some rural tourism operators, driven by the want of economic benefits, pursue short-term profits without regard to damages to tourist environment; some tourists don't harbor strong awareness of environmental protection for rural travels; and rural tourism is usually void of efficient management: all these factors cause environmental pollutions to rural tourism. Therefore, rural tourism in Jiangxi must follow the idea of low-carbon development, paying equal attention to rural tourism development and environmental protection. The tourism development must sufficiently and sustainably protect the eco-environment of Jiangxi, guard the brand of rural ecology of Jiangxi and fully exert the rural ecological uniqueness of Jiangxi, so as to avoid turning the rural strengths into weaknesses and fulfill positive cyclic development of rural tourism in the province, ultimately achieving sustainable development of Jiangxi's rural tourism.

*3.2 Conducive to rural tourism development and brand building in Jiangxi*
With the deterioration of the air environment, the idea of low carbon and environmental protection becomes gradually accepted, and low-carbon tourism has thus been accepted by tourist destinations and tourists. Low-carbon tourism is to conduct tourist activities on the basis of preserving the ecological environment; if low-carbon economy is a trend of future economic development, then low-carbon tourism must be the future direction of tourism development. We should stick to rural tourism development with low-carbon idea as it's an idea of sustainable development and foresight. Amid fierce rural tourism competitions, sticking to rural low-carbon tourism development will help

IOP Conf. Series: Earth and Environmental Science **61** (2017) 012059 doi:10.1088/1755-1315/61/1/012059

protect the rural eco-environment, firmly build beautiful villages and continue to exert the ecological strengths of rural areas in Jiangxi, thus making the rural tourism of Jiangxi stand out amid fierce market competition and establish a unique rural tourism brand of Jiangxi, thus winning the approval of tourists and guiding tourists' consumption. Rural low-carbon tourism not only accords with the idea of "basing development on ecological harmony" of Jiangxi Province but meets its anticipation of building a beautiful Jiangxi, and complies with the directions of rural tourism and eco-tourism set by China National Tourism Administration, thus a major trend of tourism development. Also, this allows for new development opportunities for Jiangxi with the advantage of favorable rural ecology.

*3.3 Conducive to rural appearance improvement in Jiangxi*

As a major agricultural province, Jiangxi has a rural population taking up more than half of its total population, and the areas of its rural land cover over half of the total land area of Jiangxi. For these years, Jiangxi has been troubled by a large population and insufficient resources, which doesn't allow it to keep to the development path of "modification after pollution". The development of rural tourism in Jiangxi must be a major factor for New Countryside Construction in Jiangxi Province instead of being the chief cause of damages to rural environment in Jiangxi. Rural tourism is primarily based on a favorable eco-environment, in which tourists make rural production and living, folk customs and cultural relics the tourist attractions. Without the beautiful natural environment, rural tourism competitiveness will be greatly reduced. Low-carbon tourism is a necessary choice for rural tourism to comply with the development of the times and an imperative path for new countryside building in Jiangxi Province. The objective of new countryside building in Jiangxi is to transform the new countryside into one of neat appearance, developed production, high rural culture and well-off living. New countryside building in Jiangxi requires improvements in the eco-environment, living environment and people's life. Rural tourism development will contribute to production growth in the countryside and farmers' affluence, and rural tourism shall stick to protection of eco-environment with low-carbon idea. Therefore, rural low-carbon tourism will be favorable for new countryside building in Jiangxi, making it a trend to accord with the actual situation of the province and the development of the times.

## 4. Principles to be Observed for Rural Low-carbon Tourism Development in Jiangxi

*4.1 The principle of respecting nature and protecting natural environment*

Rural low-carbon tourism development must stress the primitive flavor of rural Jiangxi, guard the natural landscapes there and protect the local eco-environment. The rural environment in Jiangxi should be made the foundation for transforming the production and living views in rural Jiangxi into rural tourism scenery, and attention should be paid to protecting agricultural land and forest land, with building land under control. Rural low-carbon tourism must have countryside images and landscapes in Jiangxi protected. Unlike urban landscapes in Jiangxi, the rural landscapes there are of profound connotations, making a complex of natural landscapes, settlement landscapes and cultural landscapes. The rural images in Jiangxi refer to impressions left on people through material and materialized Jiangxi rural culture like rural buildings and rural environment in the province. The rural images in Jiangxi make a natural atmosphere without artificial making, which contributes to the liberty and vitality sought after by tourists and makes the purpose for tourists to travel there. Therefore, rural low-carbon tourism development in Jiangxi must have natural landscapes and images protected and respect and protect the nature, which should be made the principle of rural low-carbon tourism development there.

*4.2 The principle of creating rural air*

Rural low-carbon tourism development in Jiangxi must avoid urbanization and protect the features of rural tourism, or rather, the rural flavor, as it's the uniqueness of rural tourism. Landscape design for rural tourism in Jiangxi must stress scientific use of local materials and species, which will not only help reduce the amount of traffic and carbon emissions, but harmonize the rural tourism landscapes in

IOP Conf. Series: Earth and Environmental Science **61** (2017) 012059 doi:10.1088/1755-1315/61/1/012059

Jiangxi with the formerly preserved environment. In selecting trees for tourism landscape, those of local characteristics must be chosen to both strengthen the local flavor while maintaining the local ecosystem. Rural low-carbon tourism development in Jiangxi must make full use to local features and create thick rural flavor, with resources consumption reduced, so that rural tourism can be of real rural flavor for achieving sustainable development of rural low-carbon tourism in Jiangxi.

## 5. Paths of Rural Low-carbon Tourism Development in Jiangxi

*5.1 To build low-carbon rural tourism infrastructures to achieve low-carbon conducts during travels*
In order to achieve the goal of low-carbon tourism in rural Jiangxi, low-carbon tourism infrastructures should be first in place. Low-carbon technologies and devices should be applied to achieve tourists' low-carbon travels during the whole process so that they will minimize the emissions of carbon dioxide while travelling. Low-carbon tourism infrastructures in rural Jiangxi include low-carbon traffic facilities, low-carbon environmental health facilities, low-carbon energy facilities, low-carbon entertainment facilities, low-carbon lodging facilities and low-carbon dining facilities. Traffic vehicles of fewer emissions of carbon dioxide should be chosen; public transport means should be the priority; carpool for self-driving travel should be encouraged; hiking and cycling at tourist destinations in rural areas should be promoted; environment-friendly luggage can be carried by tourists, and more favor can be given to environment-friendly hotels, BNBs or camps; hotels should have wider use of energy-saving, water-conserving and emission-reducing technologies and clean energies of solar energy, biological energy and organic energy, and build up green hotels with every effort; hotel rooms may not have bed linens changed every day, with no disposable goods provided; restaurants should not provide disposable tableware and should purchase and take locally seasonable vegetables and fruits[4]. Environment-friendly mobile toilets can be built; eco-friendly trash can should be used; pipelines of distribution of rain and sewage should be promoted; garbage should be classified before treated; low-carbon environmental health facilities should be developed; by using renewable energy technologies like solar energy, wind energy, hydroenergy and marsh gas, new-type low-carbon rural tourism energy supply system can be built; the "Zero Carbon Emission" standard for residence building issued by the UK in December 2006 should be observed to build low-carbon rural tourism lodging and dining facilities, including low-carbon hotels and low-carbon rural tourism resorts; by using low-carbon entertainment facilities, such as fishing, fruit picking, sightseeing boat, bicycles, animal-drawn vehicles and sedans to develop low-carbon sightseeing facilities[5].

*5.2 To strengthen publicities and education about low carbon and environmental protection and boost tourists' awareness and behaviors of low carbon*
Low carbon education will enhance and change tourists' view of environment and their ways of consumption and living; it's workable to make low-carbon tourism a way of publicizing low carbon development philosophy, enhancing all people's awareness and practices of low carbon, environmental protection and energy conservation and fostering low-carbon ways of production and living, so that everyone can take an active part in activities of energy conservation and emission reduction and of environmental protection in order to tackle the common crises of survival, resources and environment that confront the humankind[6]. All media available should be utilized to strengthen publicities so that the idea of rural tourism and low-carbon lifestyle can be deeply rooted in people's hearts. Means of advertising video, exhibition panel and paper pamphlet must be applied to publicities and education to strengthen the awareness and behaviors of low carbon among tourists and local residents, so that low-carbon lifestyle will be made an everyday habit of people and tourists and local residents will try to minimize carbon emissions during travels and living to protect the rural eco-environment. For instance, tourists may choose to rent bicycles or hike at rural scenic spots, not litter around, not use disposable tableware for dining, carry their own bottles instead of having bottled water and try not to take out-of-season fruits and vegetables.

*5.3 To develop rural low carbon tourist products*

Developing rural low-carbon tourist products in Jiangxi must be based on the rural characteristics of the province, fully exert resource superiority of the countryside and integrate rural resources in order to upgrade and alter current landscape resources, seek connotations of rural culture and combine rural folk custom and natural beauties into rural low-carbon tourist products of local characteristics. We should also make full use of low-carbon tourism resources, stick to the idea of low-carbon development, observe the principle of building a resource-saving and environment-friendly society, cooperate with the need of new countryside building, and build up a system of low-carbon tourist products that include cycling, drifting, forest relaxation, geological exploration, hot spring health keeping, rural sightseeing, mountain hiking and outdoor training at tourist destinations [7].

**References**

[1] Cai Meng and Wang Yuming. Low-carbon Economy—A new way of tourism development [J]. Tourism Tribune, 2010(1): 13-16.

[2] Gao Wenling, Shi Shenggao and Xu Li. The Concept of Low-carbon Agriculture and its Value Embodiment [J]. Jiangsu Agricultural Science, 2011, 39(2):13-14.

[3] Ma Dongyue. On Rural Tourism Development in China in the Context of Low-carbon Economy [J]. Journal of Southwest University for Nationalities, 2010, 10(7): 108-110.

[4] Mo Yankai. On Rural Tourism Development in Areas with Ecological Advantages from the Perspective of Low Carbon—An example of Lishui [J]. Ecological Economy, 2011(4).

[5] Chen Chaoqun. On Low-carbon Experiencing Rural Tourism Pattern [J]. Ecological Economy, 2011 (1).

[6] Tang Chengcai, Zhong Linsheng and Cheng Shengkui. On the Connotation of Low-carbon Tourism in China and Strategies of Sustainable Development [J]. Economic Geography, 2011(5).

[7] Yang Hong, Tian Yinhua and Yuan Kaiguo. On Tourism Development in Changzhutan Experimental Zone [J]. China Management Magazine, 2009, (6):81-89.

IOP Conf. Series: Earth and Environmental Science **61** (2017) 012060 doi:10.1088/1755-1315/61/1/012060

# Size effect of concrete column retrofitted by fiber-reinforced polymer (FRP)

**Pengpeng Zhang[1, a], Xinsheng Xu[2, b]**

[1,2]School of Civil Engineering and Architecture, University of Jinan, 336 Nanxinzhuang West Road, Jinan 250022, Shandong, China

[a]pp_chang@126.com, [b]xinsheng_xu@163.com

**Abstract**. At present, the research on the size effect of fiber-reinforced polymer (FRP) reinforced concrete members is not enough, besides the conclusions about size effect are different. It is of great theoretical significance and engineering application value to study the size effect of FRP reinforced concrete members in this field. In this paper, the study status from two aspects containing plain concrete columns and reinforced concrete columns for the retrofitted effect were analyzed and some new problems were put forward in this field. Also the paper mainly discussed the size effect of FRP constraint ratio and column slenderness ratio on FRP retrofitted concrete columns, and the effects of different load forms of FRP reinforced concrete columns under eccentric loading, repeated loading and unloading, and horizontal cyclic loading on the size effect of FRP reinforced concrete columns, and the influence of the single factor change of FRP paste way, type, package type and the concrete strength on the FRP reinforced concrete column. Besides, the changing tendency and regularity of the size effect of FRP strengthened concrete columns under the condition of muti-factors were given.

## 1. Introduction

The concrete structure has become one of the most common and widely used architecture in China, which will still the mainstream of civil engineering construction in a long future period of time. Besides, the geological disasters occur frenquently in China and many buildings are in sub-health status. So the retrofitting of concrete structure has become an important task of the civil engineering industry and a significance research. As a kinds of composite materials, fiber-reinforced ploymer has many advantages, such as light weight, high strength, corrosion resistance, good durability and simply construction. At present, researches on FRP reinforced concrete column have achieved fruitful results, but because of the cost and mechanical restrictions, the scale models were mainly studied by people. The experimental and the simulative data can reflect the actual engineering performance which is still a uncertain problem. The study on the size effect of FRP reinforced column is still relatively limited. The influences of size effect can not be ignored. Besides, a large number of studies have confirmed the existence of size effect. However, the research is not enough, and the conclusions are different.

## 2. Size effect of FRP reinforced plain concrete columns

No size effect. Tests were carried out on carbon fiber-reinforced polymer (CFRP) reinforced concrete columns with different sizes by Akogbe Kokou. M. T. Romuald. [1] The test showed that the CFRP reinforced concrete columns do not have the size effect. At the same time, Elsanadedy H M et al. [2] carried out an experimental study with three kinds of CFRP reinforced concrete columns which were different in size and found that the size effect was not obvious.

Partial size effect. An experiment about 18 concrete circular-columns under axial compressive load was carried out by Fantan Bu et al. [3]. And the test found that there was obvious size effect on the concrete circular-columns. But the ultimate stress and peak strain decrease with the increase of columns' sizes. The study aiming at glass fiber-reinforced polymer (GFRP) reinforced concrete columns with different height, diameter and thickness was carried out by Silva A G et al. [4]. The test showed that there was no obvious size effect on different slenderness ratio columns, but with the increase of diameter, the size effect got clear obvious.

Size effect. Gusheng Tong et al. [5] tested the compressive strength of 27 basalt fiber-reinforced polymer (BFRP) reinforced concrete square-columns with similar geometric of three different sizes. The research showed that the compressive strength has obvious size effect.

## 3. Size effect of FRP reinforced concrete columns

At present, it is found that FRP reinforced concrete columns also have the size effect. Finite element analysis software ANSYS was used to calculate and simulate the size effect of BFRP reinforced concrete columns with similar geometry under axial compression load by Qiulan Wu et al. [6]. According to the date of the above-mentioned references, some results for the ultimate compressive strength data of FRP reinforced concrete columns are shown in Table 1 to Table 3, concrete-columns-size's expression in the table, for circular-column: diameter*height, square-column: length*width*height.

**Table 1** Test results of CFRP reinforced concrete columns in above-mentioned references.

| Date sources | Size (mm) | Number of layers | Average compressive strength (MPa) |
|---|---|---|---|
| Reference [1] | 100*200 | 1 | 64.6 |
| | 200*400 | 2 | 64.9 |
| Reference [2] | 50*100 | 1 | 146.2 |
| | 100*200 | 1 | 94.5 |
| | | 2 | 146.0 |
| | 150*300 | 1 | 76.4 |
| | | 2 | 111.5 |
| | | 3 | 144.2 |
| Reference [3] | 152*304 | 2 | 64.0 |
| | 304*608 | 4 | 66.0 |
| | 51*102 | 1 | 64.0 |
| | 152*304 | 3 | 90.0 |

**Table 2** Test results of GFRP reinforced concrete columns in above-mentioned references.

| Date sources | Size(mm) | Number of layers | Average compressive strength (MPa) |
|---|---|---|---|
| Reference [4] | 150*300 | 2 | 89.5 |
| | 150*450 | 2 | 91.2 |
| | 150*600 | 2 | 85.8 |
| | 150*750 | 2 | 87.6 |
| | 150*750 | 3 | 128.1 |
| | 250*750 | 2 | 55.8 |
| Reference [6] | 150*450 | 3 | 79.6 |

**Table 3** Test results of BFRP reinforced concrete columns in above-mentioned references.

| Date sources | Size(mm) | Number of layers | Average compressive strength (MPa) |
|---|---|---|---|
| Reference [5] | 100*100*300 | 1 | 41.4 |
| | | 2 | 46.9 |
| | 150*150*400 | 1 | 38.7 |
| | | 2 | 41.3 |
| | 200*200*600 | 1 | 36.6 |
| | | 2 | 38.0 |

From above mentioned-tables, it can be concluded that different references on FRP reinforced concrete columns have the different conclusion about size effect. The main factors that influence the size effect of FRP reinforced concrete column are the FRP constraint ratio and the slenderness ratio of columns. There are some conclusions about size effect of FRP reinforced concrete column, and the influence of the component size on the mechanical behavior of FRP reinforced concrete columns is summarized. However, the influences about the size effect of FRP reinforced concrete columns are not summarized. Besides, the following discussions are the influences of FRP constraint ratio and slenderness ratio of column for reinforced concrete columns.

## 4. The influences of FRP constraint ratio

### *4.1 Equal FRP constraint*

Size effect. Some studies showed that test specimen had size effect when their FRP constraint were equal. Simulation analysis on 27 BFRP reinforced concrete columns with the size effect was conducted by Qiulan Wu et al. [6]. There were three kinds of sizes, which were 150*450mm, 200*600mm and 250*750mm, respectively. The number of FRP bonding layer is 3 layers, 4 layers and 5 layers respectively. The study showed that the ultimate load increases with the increase of specimen size, but the strength decreased. The specimens had size effect.

No size effect. However, some studies found that the test specimen which they have the same FRP constraint did not have the size effect. Akogbe Kokou. M. T. Romuald's [1] paper showed that the influences about the size for compressive strength on the CFRP confined concrete cylinder under the axial load. And 24 concrete cylinder test specimen that ratio of height to diameter is 2:1 with different sizes were tested. Small, medium and large diameter sizes of columns were 100mm, 200mm and 300mm, and the number of layers of CFRP corresponding to the test specimen is 1, 2 and 3, respectively. Figure 1 is the relationship between the average value of each specimen size and its compressive strength from the reference [1] and [6]. As can be seen from Figure 1, at the same constraint, the results of the study in reference [1] for FRP reinforced concrete columns do not have the size effect. But the study found the obvious size effect on FRP reinforced concrete columns in reference [6].

### *4.2 Different FRP constraint*

Experimental study on the size effect of BFRP confined concrete square-columns were carried out by Gusheng Tong et al. [5]. The specimen were divided into three kinds of different sizes, a total of nine groups, each group had three test specimen. The size of specimen in the 1-3 group is 100*100*300mm, and the constraint form of BFRP are not pasted, one layer and two layers, respectively. The size of specimen in the 4-6 group is 150*150*450mm. Besides the size of specimen in the 7-9 group is 200*200*600mm. The test showed that the BFRP confined concrete columns had obvious size effect, and with the increase of constrained layer, the size effect got more obvious.

### *4.3 Effect of slenderness ratio on reinforcement*

Silva A G et al. [4] studied the influences of column's height, diameter and GFRP thickness on GFRP reinforced concrete cylinders. The specimen was composed of 18 columns. It was divided into two groups: the first group of columns' diameter is 150mm, and its height is 300mm, 450mm, 600mm and 750mm respectively. The diameter of the second group was 250mm, the height was 750mm. In addition, the two groups of GFRP adhesive layer was not the same. The test showed that there is no size effect for the same thickness of GFRP, the diameter of 150mm, different height of the concrete cylinder.

As shown in the Figure 2, with the increase of diameter, FRP reinforced concrete columns had obvious size effect and its compressive strength was reduced significantly when they have the same ratio of length to diameter and the thickness of GFRP.

The effect of test specimen size, slenderness ratio and FRP reinforced type for the FRP reinforced columns were studied in some other reference. The conclusion was that the slenderness ratio nearly did not have influence on the reinforced effect in the above-mentioned three kinds of columns, no matter it is the CFRP reinforced concrete column or GFRP reinforced concrete columns. The conclusion was that the slenderness ratio nearly did not have influence on the reinforced effect in the above-mentioned three kinds of columns no matter what is the CFRP reinforced concrete columns or GFRP reinforced concrete columns.

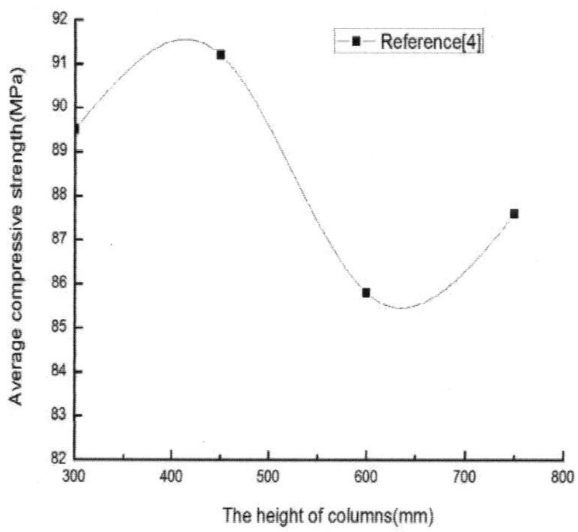


Figure 1 The size effect curve graph of the specimens in some references.

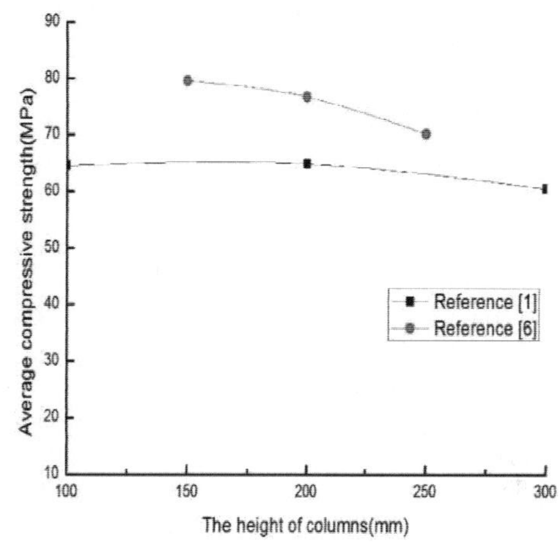


Figure 2 The size effect curve graph of the specimens in References [4].

## 5. Conclusion

(1) Nowadays, most study focused on compressive property with the size of FRP reinforced concrete columns under the axial load. It is necessary to further study on the load forms of eccentric compression, repeated loading and unloading, and horizontal cyclic loading.

(2) The effect of FRP constraint ratio and slenderness ratio of columns is different for the size effect of FRP reinforced concrete columns.

(3) It is necessary to further study the effect of FRP paste type, FRP type, FRP wrapped method and concrete strength grade et al. for the size effect of FRP strengthed concrete columns.

## Acknowledgements

Funding provided by the Natural Science Foundation of Shandong Province of China NO.ZR2014EEM036 and the School of Civil Engineering & Architecture Engineering at University of Jinan are acknowledged.

IOP Conf. Series: Earth and Environmental Science **61** (2017) 012060 doi:10.1088/1755-1315/61/1/012060

**References**

[1] Akogbe Kokou. M. T. Romuald. 2011 *D*. Size effect on axial compressive strength of CFRP confined concrete cylinders. School of Civil Engineering, Dalian University of Technology

[2] Elsanadedy H M, Al-Salloum Y A, Alsayed S H, et al. 2012 *J*. Experimental and numerical investigation of size effects in FRP-wrapped concrete columns. *Construction and Building Materials*, vol **29**, pp56-72

[3] Fangtan B. 2015 D. Experimental study on the size effect of axial compression performance of concrete columns confined with GFRP. School of Civil and Transportation Engineering, Guangdong University of Technology

[4] Silva A G, Rodrigues C C. 2006 *J*. Size and relative stiffness effects on compressive failure of concrete columns wrapped with glass FRP. *Journal of Materials in Civil Engineering*, vol **18(3)**, pp334-42

[5] Gusheng T, Yongsheng L, Qiulan W. 2009 *J*. Size effect on square section column concrete confined with basalt fiber sheets. *Concrete*, vol **21(3)**, pp6-8

[6] Qiulan W, Gusheng T, Yongsheng L, 2011 *J*. Finite element analysis on the compression course of reinforced concrete(RC) quadrate columns confined by BFRP. *Concrete*, vol **34(6)**, pp55-56

IOP Conf. Series: Earth and Environmental Science **61** (2017) 012061 doi:10.1088/1755-1315/61/1/012061

# Experimental research on flexural performance of concrete beams strengthened by hybrid fiber CGFRP sheet

**Long Yuan**[1,a], **Xinsheng Xu**[2,b]

[1,2]School of Civil Engineering and Architecture, University of Jinan, 336 hao Nanxinzhuang West Road, Jinan 250022, Shandong, China
[a]long_yuan@126.com, [b]xinsheng_xu@163.com

**Abstract.** Concrete beams strengthened with Hybrid Fiber carbon/glass fiber strengthened polymer (C/GFRP) sheet can not only improve beam ductility, but reduce the cost of the project effectively. It is plays an important engineering application of meaning. In this paper, three concrete beams are strengthened with different fiber (CFRP, C / GFRP and GFRP) in the normal section and a loading failure test. The stress characteristics, failure form, reasonable reinforcement mode, strain condition, bearing capacity, rigidity and deformation capacity of C / GFRP hybrid fiber strengthened beams were studied. The results showed that the flexural load capability and the ductility of the beams strengthened with hybrid C/GFRP are better than that of normal FRP.

## 1. Introduction

In recent years, the method of strengthening concrete members with high strength carbon fiber is gradually adopted by civil engineer, while the high cost of carbon fiber makes it less widely useful. Previous experiment [1-3] indicate that bearing capacity of using carbon fiber strengthened structure improves effectively, but the cost of material is high, element is poor ductility; The performance of beams with glass fiber is opposite to that of CFRP. This method, which can reduce the cost [4] and improve the comprehensive utilization of the material to improve the Material utilization ratio.

The performance of carbon fiber and glass fiber mixed together is better than that of single material. The high elongation fiber (GFRP) afford the extra load of the low elongation of fiber (CFRP) fracture. Hybrid fiber delay from micro to macro damage process. The average fracture strain of hybrid fiber is improved effectively. The tensile strength of the low elongation fibers in the mixed state is higher (even significantly higher) than the tensile strength of a single fiber. After the low-elongation fibers break, they can work together with the high-elongation fibers in the form of short fibers until all of the fibers are broken. [3]The mechanical properties of hybrid fiber materials in the structure are better than those of the single material stack.

## 2. Experiment program

### *2.1 Specimen details and reinforcement scheme*

Four specimen beams were tested, with a length of 2000mm and the same rectangular cross section of 120x200mm.The concrete strength is C20 (cube compressive strength measured is 23.6MPa).HRP335 is adopted to serve as longitudinal steel bar. HRP300 is adopted to serve as stirrup and top steel bar, yield strength is measured to be 362MPa and 298 MPa respectively. B beam is strengthened. Two layers of carbon fiber is pasted on CB beam, a layer of carbon fiber and two-layer glass fiber is pasted on CGB beam, the 3 layer of glass fiber is pasted on GB beams. Full-length fibers that were 100mm in width were placed in the bottom of four concrete beams and the two ends of fibers should be anchored by U-shape fiber. Reinforcement ways is showed in figure 2. Adhesive tensile strength for seven day is 30 Mpa. The performance index of fiber in table 1.

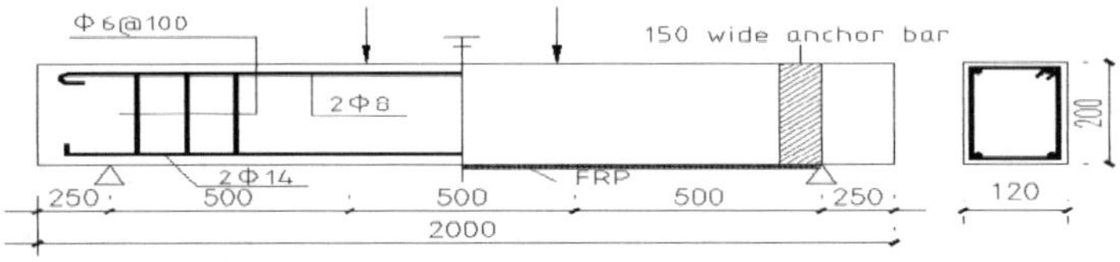


**Figure1**. element size, reinforcing bars and fiber strengthened

*2.2 Experimental facility, experimental scheme and loading system*

**Table 1.**Experiment parameters of fiber properties.

| Fiber type | Tensile strength/MPa | Elastic modulus/GPa | Ultimatetensile rate/% | thickness |
|---|---|---|---|---|
| Carbon fiber | 3450(2100) | 234.5 | 1.5 | 0.13 |
| Glass fiber | (320) | (18) | (3.2) | (0.4) |

Notes, the values in parentheses in the table are measured valued of cured fabric.

Beam of 2 groups under vertical loads at three-dividing point, using a hydraulic jack load, the beams are placed below the hydraulic jack load, two load sensors are arranged between the loading point and the distribution of beam load, imposed by the measurement, the test beam cross is equipped with a dial indicator at the bottom of it, 5 strain gauges should be equipped on the side of the middle part of the beam layout, to measure the strain variation along the beam of strong concerte.

The whole loading process is controlled by load, before the element cracked, load is 5 KN per level, after the element cracking, load is 10 KN per level. When the experimental ended, concrete crushed or fiber fracture, stripping and load do not increase.

## 3. Main test results and analysis

*3.1 The main experimental results*

The measured ultimate load, ultimate displacement and failure mode of all elements are showed in table 2, the direction of strain of mid span of CGB beam is vertical, the strain is showed in figure 2, the bearing capacity of element is showed in figure 3, the load-displacement curve is showed in figure4

**Table 2.**Main characteristic index of the experimental result of beam.

| Beam number | B | CB | CGB | GB |
|---|---|---|---|---|
| Ultimate load | 16.4(18.84) | 34.3 | 25.7 | 20.6 |
| Ultimate displacement | 6.2 | 11.9 | 14.6 | 16.8 |
| Stress and failure mode | Rare damage | failure mode II | failure mode I | failure modes I |

Notes, the ultimate load of the beam in the table is the experimental value (calculated value).

*3.2 Analysis of result*

*3.2.1 Loading feature and failure mode.* Because B beam have a little steel bars, cracks appear in the lower part of the beam B, crack up quickly, concrete of press section is crushed suddenly, it is brittle nature, it belongs to less steel damage, ultimate load value is less than the theoretical value (calculation of proper steel bars). cracks appear in the lower part of the beam CB, cracks up slowly, after cracks through the axis, the cracks on both sides develop toward the loading point, cracks in the middle develop vertically, concrete of compressive zone crushed, beam lose bearing capacity, CFRP sheets in the bottom of beam were close to fracture, at the same time U-shaped anchor fibers became deboned from the surface of beam.

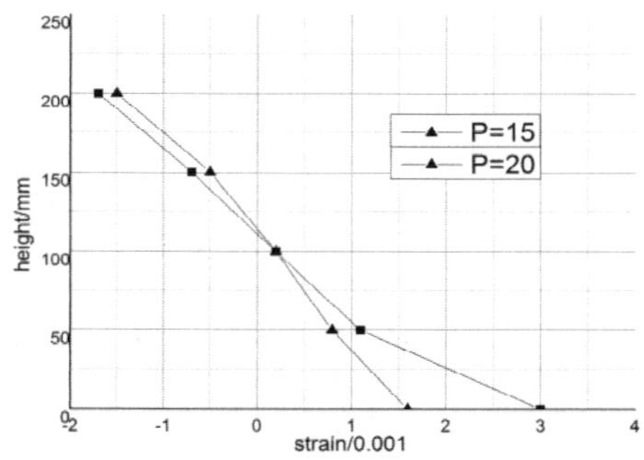


Figure 2.CGB strain.

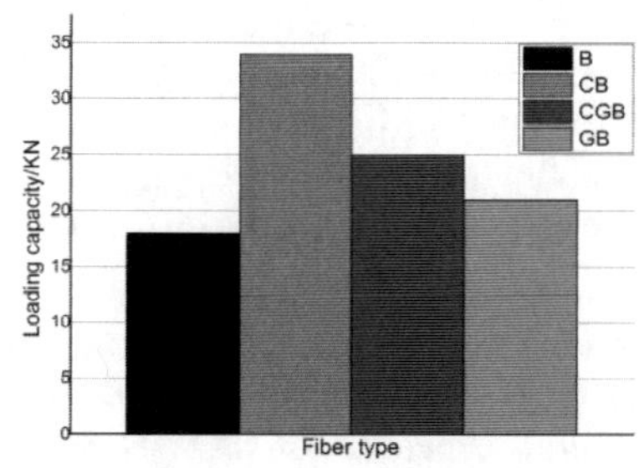


Figure 3. Loading capacity of element.

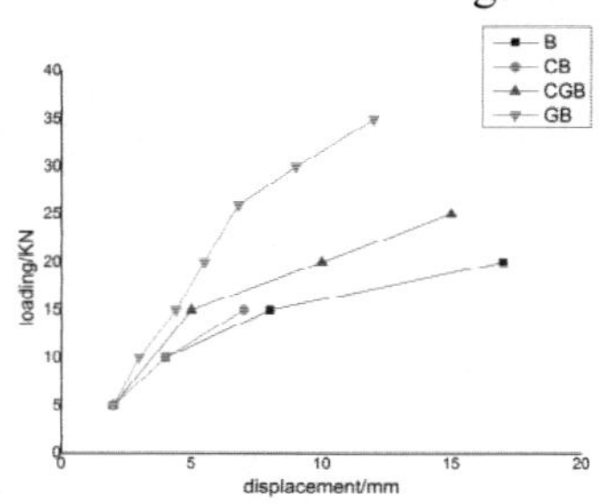


Figure 4. Loading-displacement curve.

CGB cracks are a dense, vertical extension, deformation is obvious, the fibers fabric is broken suddenly inside the anchor fiber of one end and fibers break into strips to became deboned from the bottom of beam with concrete, concrete was crushed and destroyed. The distribution of GB cracks is relatively uniform, the development process of CGB beam and GB beam is similar, the failure process of GB beam belongs to ductile failure.

According to the loading process, the phenomenon of bonding failure does not appear in the whole experiment, even if many layers of fiber are bonded. The anchor fiber avoids the occurrence of bonding failure. From the three reinforcement member failure modes and similar documents, there are two failure modes, failure mode I, the longitudinal reinforcement of element tensile area is yielded, after fiber break, concrete is crushed and damaged; failure modes of II, the concrete has been crushed and fiber cannot be broken.

*3.2.2 Section strain.* The strain of each load is measured in the experiment, the strain of the beam height of CGB under the load of two (P=15KN and P=20KN)is showed in figure three. Section strain is basically in line with the plane section assumption. Good bond capacity was found between fiber and concrete. But cracks appear at the bottom of the beam, with the increase in load, the fiber strain of the bottom of beam increase is more obvious.

*3.2.3 Ultimate loading capacity.* Because of the brittle failure of the contrast beam B, the loading capacity is measured to low, so bearing capacity of proper-strengthened element is compared with the bearing capacity of unstrengthen element. The loading capacity of the reinforcement elements is improved extently, beam CB was the largest, followed by CGB and GB, 82%, 36% and 9%, respectively. From the similar experimental research, [1-3] the loading capacity ratio of beams strengthened with different fibers is not big, and the little of element and the concrete strength distribution of contingency are related, and two fiber tensile strength is different in the experiment. It can be proved that hybrid fiber reinforcement can effectively improve the loading capacity of flexural elements.

*3.2.4 Rigidity and deformation capacity.* The load-displacement curve of the beam is showed in figure 4, the rigidity of the four members did not differ much at the initial stage of loading, indicating that the fiber played no significant role. When the load increases to 10 kN or more, the elements are cracked basically, beam CB showed greater rigidity, while the CGB, GB second. As the load increases, the rigidity and deformation capacity of the three strengthened elements are just opposite to each other, the rigidity of beam GB is the smallest, but the deformation capacity is the largest, and the rigidity and deformation capacity of CGB are larger. it can be concluded that the late rigidity of element strengthened with CGFRP hybrid fiber is smaller than that of element strengthened with CFRP, but larger than GFRP strengthened element, its deformation ability is bigger than CFRP and GFRP. Generally speaking, CGB has better ductility under the condition of high loading capacity.

## 4. Conclusion

(1) Under the premise of ensuring the bond strength and anchoring measures, the failure mode of the fiber reinforcement member can be divided into two types.
(2) The U-shaped anchoring bar can effectively avoid the failure of anchoring.
(3) The cross-sectional strain of fiber-strengthened beams with C / GFRP interlayer is in line with the plane section assumption, between the fiber layers and between the fiber and concrete can be bonded well, they can achieve the same effect of force.
(4) The bearing capacity of low reinforcement flexural element C / GFRP hybrid fiber can be improve significantly.
(5) The later-stage deformation capacity of C / GFRP strengthened element is larger than CFRP, smaller than GFRP, and the beam has better ductility under the condition of high loading capacity.

Through the research in this paper, it is proved that the C/GFRP layer between the form of hybrid fiber strengthened concrete beams to improve the anti-flexural ability is feasible.

**Acknowledgement**
The authors acknowledge the financial support provided by the Natural Science Foundation of Shandong Province of China (No. ZR2014EEM036) and the School of Civil Engineering and Architecture at the University of Jinan.

## References

[1] Xingguo W, Kunjia Z, Dandan Z, etc.2015*J*. Flexural Properties and Influence Factors of Continuous RC Beam Strengthened with Near-surface Mounted Mixing FRP Tendons. *Journal of Traffic and Transportation Engineering*, vol **15(2)**, pp32-41.
[2] Yajun H.2016*D*. The Flexural Capacity of Reinforced Concrete Short Beams Strengthened with Carbon Fiber Reinforced Polymer sheets(CFRP). Zheng zhou, Zhengzhou University.
[3] Ruohong Z, Guangjing X, Jianzhong Y.2002*J*. Present Statuse and Forecast of AppliedResearchintoStrengtheningAxial Compressive Concrete Column with Fiber Composite. *Building Science Research of Sichuan*, vol **28(3)**, pp19-21.
[4] Yaping P, Tiecheng W. 2005*D*. Research on Performance of Concrete Beams, Columns and Joints Strengthened by FRP. Tianjin: Tianjin University School of civil engineering.

IOP Conf. Series: Earth and Environmental Science **61** (2017) 012062 doi:10.1088/1755-1315/61/1/012062

# Study on shear behavior of FRP strengthened concrete beams

**J K Zhao[1,a], X S Xu[2,b]**

[1,2]School of Architecture and Civil Engineering, University of Jinan, Jinan 250022, China

[a]Jiankai_Z@163.com, [b]xinsheng_xu@163.com

**Abstract.** There are many researches on concrete bending beams reinforced with FRP, and the beam section of the bending strength is enhanced, it is necessary to carry out an oblique section bearing capacity review, if the bearing capacity is insufficient, it is also necessary to strengthen, so as to ensure the ductility of the beam and meet the needs of Engineering safety. In this paper, four concrete beams strengthened with different fibers (CFRP, HFRP and GFRP) were used to study the stress characteristics, failure forms, reasonable reinforcement methods, beam strain conditions, bearing capacity, stiffness and deformation capacity of shear behavior of concrete beams strengthened with different fibers and different forms of reinforcement. The experimental results showed that the FRP reinforcement can not only improve the shear bearing capacity and ultimate deformation of beam, but also increase the stiffness of the beam, reduce the bending deformation under the same load beam, and delay the extension of diagonal cracks.

## 1. Shear Test of FRP Strengthened Reinforced Concrete Beams

### *1.1. Introduction*

When FRP material is used to reinforce RC beams, the FRP sheet is attached to the bottom of the beam for bending reinforcement. On the other hand, FRP can be applied to the beam side for shear reinforcement. In many practical projects, there is a shortage of the shear bearing capacity of beam oblique section and the beam need to be strengthened due to the reasons of low strength, insufficient amount of stirrup, use load increase, structural system change. If the bending strength of the beam is enhanced, it is necessary to carry out the recheck of the bearing capacity of the oblique section. If it is insufficient, it also needs to be reinforced so as to ensure the ductility requirement of the beam and meet the engineering safety requirements.

### *1.2. Experimental study on shear behavior of concrete beams reinforced with FRP*

In the experiment, four specimens were tested, all of which had the same shape. Reinforcement with FRP used a total of four sheets and the basic mechanical properties of FRP showed in Table 1. Specimen reinforcement details showed in Figure 1. FRP of C-C1, C-CG have the same U strip width and spacing and the width and spacing of C-C2 U strip is slightly dense.

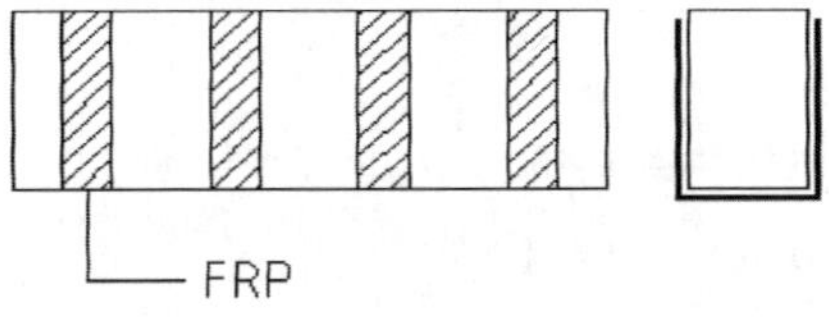


**Figure 1.** Specimen reinforcement details.

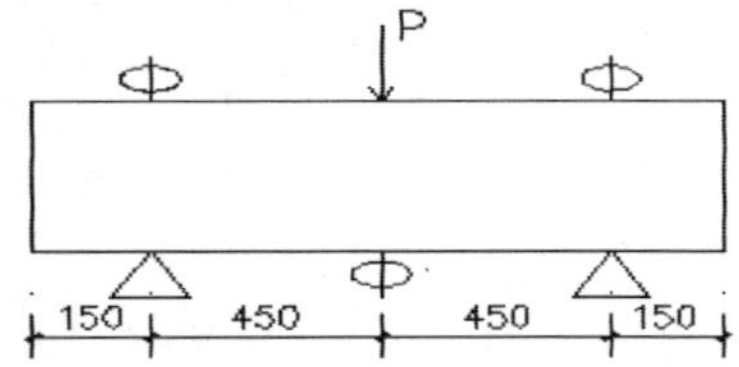


**Figure 2.** Experimental setup and details.

**Table 1.** The basic mechanical properties of FRP.

| No. | Reinforcement ratio | Reinforcement method | Fiber fraction | Remarks |
|---|---|---|---|---|
| C-1 | | Not reinforced | 0 | contrast |
| C-C1 | | 2 layers CFRP sheets(C1), 120 mm in width, 4 | 0.29 | |
| C-C2 | 0.31 | 2 layers CFRP sheets(C2), 100 mm in width, 5 | 0.32 | |
| C-CG | | 1 layers CFRP sheets(C1), 1 layers GFRP sheets(G1),120 mm in width, 4 | 0.70 | |

The method of the counter force frame upside down hydraulic jack was taken in the test and the loading method was single point loading at the top of the beam span controlled by the sensor. Experimental setup and details of test was showed in Figure 2. When the concrete was crushed or the fiber was broken, the test was terminated.

## 2. Test results and analysis

*2.1. Destructive form*

Due to the fact that the C-1 beam was not reinforced, it occurred a typical shear stress failure. The failure process of the three reinforcement members was largely the same. When the critical oblique fracture extended toward Intersection of the loading point and the fiber strip, the U-shaped strip began to peel away from the upper end as the load increasing. Finally, the entire side was completely peeled off, the concrete was crushed at the loading point and the components were destroyed. The reason of the fiber strip stripping wasn't that the bond surface was destroyed, but it occurred shear failure between the concrete protection layer and the core concrete.

*2.2. The load displacement curve of component*

The load displacement curves of the four components measured in the test are plotted as shown in figure 3. The following conclusions can be drawn from the diagram:

1) In addition to C-CG, the ultimate load and displacement of FRP strengthened members are significantly improved compared with those of the corresponding members.

2) The stiffness of FRP strengthened members is obviously higher than that of the contrast members.

3) Reinforcement with relatively narrow and dense form have the best effect.[1]

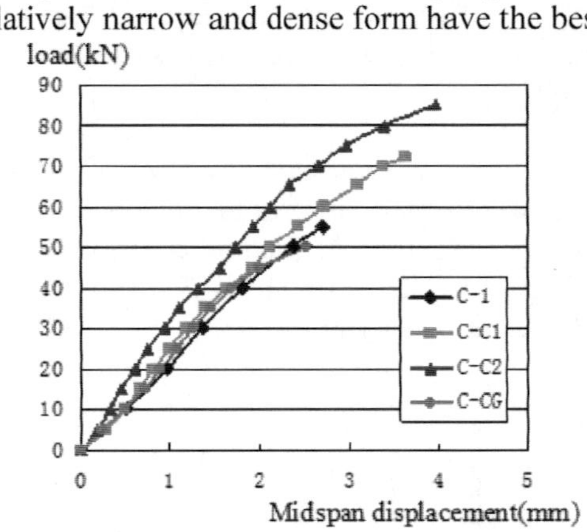


**Figure 3.** Beam lord-span displacement.

*2.3. Bearing capacity*

As shown in Table 2, the shear capacity of the C-C1 beam and the C-C2 beam increased by 31% 55% respectively and the fiber strength and elastic modulus of C2 are both lower than that of C1.So U-shaped fiber strips using narrow and dense paste form to improve the carrying capacity is more effective.[1]

IOP Conf. Series: Earth and Environmental Science **61** (2017) 012062 doi:10.1088/1755-1315/61/1/012062

**Table 2.** Comparison of ultimate load (Pu) and shear capacity (Mu) of beam.

| No. | C-1 | C-C1 | C-C2 | C-CG |
|---|---|---|---|---|
| $P_u(kN)$ | 55.0 | 72.0 | 85.0 | 50.0 |
| $M_u(kN \cdot m)$ | 55.0 | 72.0 | 85.0 | 50.0 |
| Increase ratio | - | 31% | 55% | -9% |

### *2.4. Deformation capacity*

As can be seen from Table 3, although the mechanical property of C2 is lower than that of C1, the limit displacement of C-C2 is larger than that of C-C1. It is obvious that the U shaped fiber strip is more effective in the improvement of the limit displacement by the narrow and dense paste method.

Basted on the above test results, the shear-strengthened FRP strip is most likely to be stripped and damaged without effective anchoring measures. When the reinforcement fails, the fiber bar is intact, and the material fails to play its full role Therefore, reasonable anchoring measures should be taken.[2]

**Table 3.** Comparison of the beam ultimate displacements.

| No. | C-1 | C-C1 | C-C2 | C-CG |
|---|---|---|---|---|
| Ultimate displacement (mm) | 2.70 | 3.63 | 3.97 | 2.52 |
| Increase ratio | - | 34% | 47% | -6.7% |

## 3. Shear capacity analysis of beam oblique section

The ultimate failure sign of FRP shear strengthening of concrete members is that the concrete in the compression zone is crushed. But before this happens, fiber stripe may occur stripping damage or be pulled off and the efficiency of the later is higher than that of the former.That means that the effect of reinforcement is entirely dependent on the effective strain of the fiber bars when the member is damaged.

Failure form I: The strips used for shear reinforcement were stripped and gradually out of work, the critical inclined crack develops rapidly, the concrete in the shear zone is crushed and the member is destroyed.

Failure form II: The fiber strips used for shear reinforcement were pulled off, the critical inclined crack develops rapidly, the concrete in the shear zone is crushed and the member is destroyed.[3]

## 4. Comparison shear capacity of beam between calculated value and experimental value

The test beams C-C1 and C-C2 were calculated and compared with the experimental results. The value of $V_f$ is the shear capacity that FRP can provide when the failure form II happens but the failure form of the two components is failure form I in the experiment, the reaction of calculated values $V_c$ is that the shear capacity of FRP can be provided when fiber stripe occur stripping damage. The ratio of the two components shows the effectiveness of the fiber when the test strip is stripped under the anchoring conditions of the test member. According to the comparison results, it is better to use the relatively narrow and dense paste arrangement to enhance the anti stripping strength of FRP.

## 5. conclusion

The following conclusions can be drawn through the experimental analysis and comparison:

1)The FRP reinforcement can not only improve the shear bearing capacity and ultimate deformation of beam, but also increase the stiffness of the beam, reduce the bending deformation under the same load beam, and delay the extension of diagonal cracks.

2)The ultimate failure sign of FRP shear strengthening of concrete members is that the concrete in the compression zone is crushed. But before this happens, fiber stripe may occur stripping damage or be pulled off and the efficiency of the later is higher than that of the former.That means that the effect of reinforcement is entirely dependent on the effective strain of the fiber bars when the member is damaged.

3)In order to improve the effective strain and shear reinforcement efficiency of FRP, some effective anchoring measures should be taken at the fiber ends.

4)The shear reinforcement effect is relatively good with the form of using relatively narrow and dense fiber paste.[1]

**Acknowledgement**
The authors acknowledge the financial support provided by the Natural Science Foundation of Shandong Province of China (No. ZR2014EEM036) and the School of Civil Engineering and Architecture at the University of Jinan.

**References**

[1] Rizzo A, De Lorenzis L. 2009 Behavior and capacity of RC beams strengthened in shear with NSM FRP reinforcement *Construction and Building Materials* **vol 23** pp 1555-67
[2] Bo C, Qunhui Z, Linjie Z. 2012 Experimental study on a new method to improve the shear capacity of concrete beams *Building Structure* **vol 42** pp 144-47
[3] Teng J G, Lam L, Chen J F. 2004 Shear strengthening of RC beams with FRP composites *Progress in Structural Engineering and Materials* **vol 6** pp 173-84

IOP Conf. Series: Earth and Environmental Science **61** (2017) 012063 doi:10.1088/1755-1315/61/1/012063

# Study on the adverse effects of hydropower development on international shipping

**Changhong Wang**[1]

1.Tianjin Research Institute for Water Transport Engineering, Tianjin 300456,China
45221707@qq.com

**Abstract.** The Lancang - Mekong river is an important international waterway to Southeast Asia and South Asia, which has important strategic significance for promoting regional economic cooperation and safeguarding national economic and security interests. On the Mekong River, the main aim is to develop hydropower resources utilization and shipping. River Hydropower Stations are in Laos and designed by foreign enterprises according to the construction of BOT. In this study, on the basis of a lot of research work and extensive collection of relevant information, and, through in-depth analysis of research, it reveals that the upper Mekong River hydropower development have many adverse effects on international shipping, put forward related suggestions for the healthy and sustainable development of international shipping.

## 1 River profile

The Lancang river flows through China, Burma, Laos, Thailand, Kampuchea and Vietnam, which is the longest river of Southeast Asia. It is the only international rivers in Asia, which originates from China Qinghai province, and its length is 4880km. It is known as "Oriental The Danube". It is called the Lancang River in China, from Burma No. 243 monument after leaving the country, which is called the Mekong river. The Mekong River flows through Burma, Laos, Thailand, Kampuchea and Vietnam to the South China Sea near Ho Chi Minh City. In addition to China and Burma, they are members of the Mekong River Commission [1].The Mekong River from Burma No. 243 tablet, is 2750 km long, which located in Laos and Kampuchea border at the port hole waterfall, a large drop and is unnavigable. In this regard, the waterfall above is the upper Mekong River, a river that is long 1984.6 km. The Mekong River's length is 919.5 km. In recent years, with Laos,Burma, Thailand and other countries for the development of the national economy, the growing importance of the Mekong River hydropower development, multiple cascade hydropower station construction planning.

The construction of power station in Mekong River will provide basic conditions for the Mekong River international waterway improvement. However, the economic and technological development of Laos is relatively backward, and the Mekong River power station was built by multinational companies in the form of BOT investment. The early construction of the Mekong power station was also completed by different design units. Now Laos's construction engineering standards, technical specifications are not perfect, the examination and approval procedures are not standardized, investment and construction companies will often take into account the maximum investment interests, do not pay attention to shipping. Investment construction enterprises and government are lack of effective long-term development planning and coordination mechanism. There will be many adverse effects to the development of the Lancang Mekong River international shipping.

## 2 Overview of Mekong River Hydropower Development

IOP Conf. Series: Earth and Environmental Science **61** (2017) 012063 doi:10.1088/1755-1315/61/1/012063

*2.1 Current situation of hydropower development*

The upper Mekong River is a mountain river, the river drop is large, water yield is abundant,hydropower development is potential. Water resources are the largest natural resources in the Mekong River Basin, which is about 50 million KW, so it is paid more and more attention by the international community. 1957 ~ 1970, generating capacity of up to 2 million 500 thousand KW of the Mekong River tributary hydropower project has been completed. In Thailand, Laos, Kampuchea, Vietnam built a total of 3 million 270 thousand KW power engineering from 1971 to 1980.The countries of Mekong River regard the development of the Mekong River water resources as an important way to get rid of poverty, develop the industry and revitalize the economy, and the development of the Mekong River cascade hydropower stations are formulated by the Mekong River Commission [2]. Once the completion of the cascade power plant,it will have 36950 thousand KW installed capacity, to ensure that the output power of 15730 thousand KW, providing annual power of 146.1 billion degrees. The Mekong River will become one of the most important hydro energy production centers in Asia, which can greatly meet the needs of the economic development of the region.

Water resources is the largest resources in Laos, and there are more than 20 rivers above 200km throughout the process, one of the longest is 1846.8km in Laos, accounting for about 44.4% of the total length of the Mekong River. The Mekong River water resources fall of 484m, more than 60% of the reserves in laos. After years of development, the electric power industry has become one of the industries with the advantages of Laos, is the focus of social and economic development. From 1988 to 1995, foreign investment projects in Laos more than and 10, amounting to $4 billion.

*2.2 On the Mekong River Hydropower Development Planning*

In 1970, the MRC proposed "watershed planning guidance" in 1970, 7 level of water conservancy project in the river river estuary to the Qing Sheng among more than 2400 km. 1989 also proposed amendments to the program, including the 7 and the level of the two comparison program, recommended by the 8 level program. Under the auspices of the United Nations development program and the French government, after more than 2 years of research and exploration, in 1994 proposed a 9 level development program. In 1995, under the auspices of the United Nations Program and the French government, after two years of re investigation of the proposed 12 level development program. At present, the Mekong River planning according to the 12 step development project of [3], in the north, Luang Prabang, Sayaboury, ballee, Sara Kang (Lao river), pamong, bancu (old Thai border), Lastua (diversion, Thakho, Lao river), sombo treng (more than for the lower Mekong River).

In July 1, 2009, Laos Mekong River Hydropower Station Optimization on the ultimate outcome of the meeting held in Vientiane, Laos energy power conference by our Ministry of Mines (DOE) host, Secretary of the Mekong River Commission Office (MRC), Laos Mekong River five project developers and other participants. The meeting informed the final results of the research on the optimization of the Mekong River dry water power station. This paper mainly studies the influence of Laos, the North Luang Prabang station may have on the upper Mekong River upper reaches of Luang Prabang international shipping.

## 3 International shipping situation on the Mekong River

Lancang Mekong international has shipping navigation for over ten years, which is still in a starting period, the freight volume of about 30-40 million tons per year, mainly between Thailand complementary merchandise and agricultural and sideline products, travel in the Thai tourism group, at present, the main international shipping to Jinghong to Thailand kingsone Chinese Hong Kong 345km [4] voyage. Since the formal navigation of international shipping, the rapid development. Cargo cargo transportation from single to diversification transformation, mainly including refrigerated goods and fresh vegetables, groceries, containers, oil products, and the opening of the international shipping water tourism transportation.

The current Sino Burma Laos houayxay 331km channel, from 2002 to 2004 the channel improvement project, the current year navigation 150t class ships, in the flood period is navigable for

200 ~ 300t class ships, and the arrangement of navigation marks; Laos - Luang Prabang channel: 299.5km houayxay without any improvement, but focus on the beach section installed simple navigation in Laos mark, dry season can only 60t level of ship navigation. At present, according to the old Burma and Thailand are jointly prepared to reach a consensus on the Lancang Mekong international shipping development planning, planning on the Mekong River international shipping meet class 500t ship navigation standards, combined with the long-term channel reach 500t level navigation standard.

## 4 Adverse effects of hydropower development on international shipping

On the Mekong River Drainage River power plant, increase the water level upstream, reservoir formation channel, greatly increase the channel depth, in order to improve the channel level; at the same time, the hydropower development of canalized river channel in reservoir area can form a large number of submerged navigation rapids, and has a smooth, navigable flow condition of the current flow of good [5]. Therefore, hydropower development of canalized river channel to enhance the level, improve navigation conditions to provide basic conditions, but must be in the hub of water level between reasonable convergence, simultaneous construction of navigation facilities to meet the standard, general layout, power station, navigation during the construction, lock management should meet the navigation requirements as the premise, otherwise it will produce many adverse effects for international shipping.

### *4.1 Station Mekong mainstream planning scale of the lock standard is not unified*

The Mekong River hydropower development plan 12 cascade development scheme, 8 power plants and old Thai Laos Laos inland river has entered the early stage of construction. These power stations by multinational companies to invest in the construction of BOT, and by the different design units in the country to carry out pre work and design. Navigation facilities are the hub of the North lock, Luang Prabang power station is the 12 cascade development scheme of upstream first, level 2, the lock was grade 500t, 500t grade ship designed ships, according to double lock design, reserved line lock. Some of the lower reaches of the lock according to the 1000t level design, and some locks are designed according to the 500t level, the lock is designed according to a single line. Visible, river power station unified navigation facilities lock scale, standard, through the ability of coordination, will restrict the lock by ability, affecting the sustainable development of ship navigation safety and expedite international shipping.

### *4.2 The North Luang Prabang hub lock scale is small, the ability to meet the requirements of the future development of shipping*

"The MRC" clearly stated in the Mekong River Hydropower "navigation guide": "long term hydropower project with investment and the use of lock, lock and scale must be considered in the next 50 years to meet the development of shipping and traffic demand". The north and Luang Prabang hub lock scale 120 * 12m small, even if the optimization of ship form of single and double line two-way through capacity is only 6 million 30 thousand tons and 12 million 60 thousand tons, it is difficult to meet the requirements of international shipping development in the future, will lead to lock constraints through the ability to form navigation bottleneck.

### *4.3 There is no connection between the water level*

At present, the research of only Laos and Luang Prabang two north of the hub, and has entered the early stage of construction. This is the end of the 5m North hub base level convergence, nearly 30km channel for the dehydration section, in natural state, Luang Prabang hub downstream of 4.43M water level does not converge, nearly 30km channel for dehydration, which is in natural state, can not meet the requirements of navigation depth.

*4.4 North hub layout and lock navigation design needs to be further deepened and optimized*

The North navigable shiplock layout design, the lower lock head and channel connection method is undeserved, can cause adverse flow and sediment deposition, the design of ship in the chamber area can not be used effectively, through the impact of capacity, the chamber and approach channel structure will affect the project cost and economy.

*4.5 Power plant operation mode and lock management mode*

The Mekong River hydropower station is built by overseas enterprises to invest in the construction, often take into account the power generation benefit and business interests, while ignoring the benefits of shipping, such as avionics hub using unified management model, often lead to power plant operation is difficult to fully reflect the demand for shipping, navigation condition and influence through capacity.

**5 Recommendations**

The river development and utilization of resources must be comprehensive and overall planned and utilized. The development of the Mekong River Hydropower should fully embody the principle of comprehensive utilization, realizing the coordinated development of shipping, water and electricity, and the sustainable development of international shipping. In the upper Mekong River hydropower development may have many adverse effects on international shipping advice, (1) to create conditions for the establishment of Lancang Mekong River basin management institutions; (2) the "hanglian committee" should attract the full attention of hydropower development on international shipping, research measures; (3) the "hanglian Committee" should promote the establishment of shipping on the Mekong hydropower, bilateral negotiation mechanism as soon as possible; (4) should adjust hanglian committee functions, rationalize the international shipping management system.

**Reference**

[1]Study on waterway construction and comprehensive utilization of water resources [R]. Ministry of transport Tianjin Institute of water transportation engineering, 2002

[2] Wang Yiliang. Comprehensive development of river resources, accelerate the development process of waterway construction [J], water transport engineering, 2003, (5): 33 ~ 37

[3] Wang Yiliang et al. Channel construction and management policy research, 2001 ~ 2010 transportation policy book [M], press.2001:185-198.

[4] Tianjin Research Institute of water transport engineering, Ministry of communications, Lancang, Yangtze River Delta International Shipping and regional transportation network, [R], 2011

[5] Research Institute of water transport engineering, Ministry of transport, Tianjin, upper Mekong waterway management mechanism and funding channels [R], 2013

IOP Conf. Series: Earth and Environmental Science **61** (2017) 012064 doi:10.1088/1755-1315/61/1/012064

# The optimization of energy storage capacity for distribution networks with the consideration of probability correlation between wind farms based on PSO algorithm

**Bo Wang**[1,2], **Hongxia Li**[1,2], **Xueyuan Cao**[1], **Xiaojun Zhu**[1,2], **and Zhongxue Gan**[1]

1. ENN Science & Technology Ltd, Lang-fang 065001, China
2. State Key Laboratory of Coal-based Low-carbon Energy, ENN, Lang-fang 065001, China
Email: bowangphysics@163.com

**Abstract**. With the rapid development of the energy networks, various forms of renewable energy resources are absorbed into it. Because of the inherent random behaviour of the renewable resources, introducing them into the energy network will destroy the stability of the grids. It is required to use proper energy storages to reduce the uncertain fluctuation from the renewable energy resources. For a concrete model research, this paper presented an explicit method to give suitable capacities of the energy storages in consideration of the economics of the storage, grid losses and the probabilities of the bus voltages violation, for situations of the winds-power generations injected into the power network. Furthermore, the influence of the correlation between the different winds farms on the optimal storage capacity can also be studied by this method.

## 1. Introduction

The distribution network, widely recognized as a better solution for the energy crisis in the world, has been developing rapidly. However, the more renewable energy resources access into the grids, the more instability factors are injected into the distribution network [1]. For example, with the gradual increasing penetration rate of distributed wind generation in distribution networks, the huge impact of the fluctuations generated from the wind power should be considered seriously. One of the solutions is employing the energy storage system to smooth the output of the renewable generations. A great amount of research effort goes into giving the better the storage energy system technologies from different aspects [2-6]. This paper has a concern on how to give an optimal capacity of the energy storage. For this issue, some explorations have been carried out by using probability theory. The research work in reference [7] presented an optimization for determining the capacity of the storage. Energy storage devices are introduced into the power grid system to balance the fluctuation caused by injection power deviation. However, they do not provide the model for how energy storage device balances the fluctuation caused by injection power deviation, and their model can't research the case when there are correlations between the wind farms.

In this paper we use an explicit model to show the modifications of the storage device on the power output probability distribution for the wind generations. Based on the probabilistic modification model, we provide a method for discovering the influence of the correlations between the different wind power generations on the optimization of the storage capacity. In the section 2 we describe the storage probability modifications model for the wind power generations. Then, in the section 3 we present the optimization algorithm model based on the probabilistic power flow. Next, in the section 4 we introduce the flowchart of the method based on the particles swarm optimization (PSO) algorithm model, and then in section 5 we give a concrete case for getting the optimal storage capacity, and research the influence of the correlations between the wind farms on the optimal storage capacity. Finally, some conclusions are made in the section 6.

## 2. The storage probability modified model

In this section, we firstly provide a wind power model, and then describe the method for generating samples of the output power of the correlated wind farms. Finally we introduce a model for generating the samples modified by the access of the storages.

### *2.1. The general wind power model*

Many researches show that wind speed do not follow normal distribution, but can be modelled as Weibull distribution with two parameters. This model is applicable for both short term and long term studies. Weibull distribution is formulated as follows [8]:

$$f(v)=\frac{k}{c}\left(\frac{v}{c}\right)^{k-1}\exp\left[-\left(\frac{v}{c}\right)^{k}\right] \tag{1}$$

Where $v$ is wind speed, $c$ and $k$ are the scale and shape parameters of Weibull distribution, respectively. Output power of wind turbine can be transformed from the wind speed via a nonlinear function formulated in the following equation[9].

$$P_{wind}=\begin{cases}0 & v\le v_{ci}\\ k_1v+k_2 & v_{ci}\le v\le v_r\\ P_r & v_r\le v\le v_{co}\\ 0 & v\ge v_{co}\end{cases} \tag{2}$$

Where $P_r$ is the rated capacity of the wind turbine. $v_{ci}$, $v_{co}$ and $v_r$ , are the cut-in, rated and cut-out speed of the wind turbine, respectively. The parameters $k_1$ and $k_2$ of the equation can be expressed by the others parameters mentioned above as $k_1=P_r(v_r-v_{ci})^{-1}$ and $k_2=-k_1v_{ci}$ In general, the wind farms are considered as PQ buses with a power factor, specially which is 0.95 for the calculations in this paper.

### *2.2. The method for generating the correlated wind speed samples*

In order to research the correlation between the wind farms, we need to obtain the samples of the correlates wind. The general correlation between the normal distributions can be generated by using the orthogonal transformation method. However, wind speeds do not follow normal distribution but rather Weibull distribution. The combination between Cholesky decomposition and Nataf transformation is considered to be better than other methods for generating the correlated samples between the non-normal distributions[10]. Assume that the random vectors in the desired space is $X(x_1,x_2\cdots x_n)$ ,and its correlation matrix with the non-diagonal elements $\rho_{0ij}$ and the diagonal elements $I$ . By equal probability criterion, the standard normal random vector $U(u_1,u_2\cdots u_n)$ is calculated using the following expression:

$$u_i=\Phi^{-1}(F_i(x_i)) \tag{3}$$

Where $\Phi(u_i)$ and $F_i(x_i)$ are the cumulative distribution functions of the normal distribution and any desired distributions. At the same time, $\rho_{0ij}$ which is the non-diagonal elements of correlation matrix $C_U$ of the random vector $U$ can be transformed from the correlation matrix $C_X$ by the empirical expression $\rho_{ij}=T\rho_{0ij}$, which is proposed in [11]. Specially, for two-parameter Weibull distribution, $T$ has the following formulation

$$T=1.063-0.004\rho_{0ij}-0.2\left(\frac{\sigma_i}{\mu_i}+\frac{\sigma_j}{\mu_j}\right)-0.001\rho_{0ij}^{\ 2}+0.337\left(\frac{\sigma_i^2}{\mu_i^2}+\frac{\sigma_j^2}{\mu_j^2}\right)+0.007\rho_{0ij}\left(\frac{\sigma_i}{\mu_i}+\frac{\sigma_j}{\mu_j}\right)-0.007\left(\frac{\sigma_i}{\mu_i}*\frac{\sigma_j}{\mu_j}\right) \tag{4}$$

Since the random vector $U(u_1,u_2\cdots u_n)$ obtained by the above procedure is in the correlated standard normal space, we can get the independent standard normal rand vectors $Z(z_1,z_2\cdots z_n)$ by the equation $Z=B^{-1}U$, where $B$ is a lower triangular matrix from the Cholesky decomposition of correlation matrix $C_U=BB^T$. The above procedure is the Nataf transformation by which the desired random vectors

(RVs) can be transformed from desired input space to independent standard normal space. By the inverse procedure of the Nataf transformation, we can get the desired RVs from the RVs in the standard normal space. The details of the procedure are as follows:
1) Obtaining $C_U$ from using the equation $\rho_{ij} = T\rho_{0ij}$.
2) Calculating the lower triangular matrix $B$ from $C_U = BB^T$.
3) Generating an $n$-dimensional independent standard normally-distributed vector $Z$ whose elements are independent and identically-distributed standard normal random variables.
4) Getting the normal distribution vector $U = BZ$ where the elements of meet the correlation matrix $C_U$.
5) Generating the wind speed vector using the equation $x_i = F_i^{-1}(\Phi(u_i))$ ,where $F_i^{-1}$ is the inverse of the cumulative distribution function of the $i$-th wind speed. Each wind speed RV follows Weibull distribution and follows the given correlation matrix $C_X$ .

*2.3. The probabilistic modification model of the output power by the access of storage*
The reference [12] provides a probabilistic correction method which considered the changes of power injection probability model of node consist of wind power and ESS, based on the combination of the operating characteristic of ESS and mechanism of wind-storage output. By this correction method, for getting the probability distribution of the power output of wind-storage combined systems, the storage system parameters are need as follows: the storage capacity $E_s$, the maximum power output $P_{storage\_max}$, the minimum state of the charge $Soc_{\min}$ , the maximum state of the charge $Soc_{\max}$ . the number of the charging and discharging in a day $n_d$ . The detailed procedure is not introduced here, and can be found in the reference [12].

## 3. The optimization algorithm of the storages by the probabilistic power flow calculation
In this section, we introduce the probabilistic power flow calculation used in this paper and the objective function for the storage optimization process. By means of the probabilistic power flow of the system with the wind power generations, the node voltage cumulative distribution function and the network losses can be obtained, and then the corresponding probability of the system voltage violations can be evaluated. Based on them, the objective function and control variables in the optimal method for the storage capacity are presented.

*3.1. Probabilistic power flow based on the point estimate method and Cornish-fisher expansion*
The Probabilistic power flow is necessary in the process of the storage optimization. The most straightforward method for the Probabilistic power flow is the Monte Carlo simulation [13]. But this technique has a poor efficiency for the many cases. We apply the point estimate method and the Cornish-fisher expansion together to calculate the probabilistic model, which can simple greatly the calculation of cumulative distribution function of state variables [14,15].
The set of the general nonlinear equation can be expressed as $Y = g(X)$ . We use the 2m+1 Point estimate method (PEM) to get the statistical characters of the output random vector $Y$ which can be the voltages of each node in the power network, the network losses for the power flow equation and so on. The basic procedure of the 2m+1 PEM scheme is to use 2m+1 points of uncertain input variables to evaluate the statistical moments of the output variables 2m+1 times. The each input random variable is chosen with three locations and weights which can be written as follows

$$x_{ik} = \mu_i + \xi_{ik}\sigma_i, \quad (k = 1,2,3;\ i = 1,2,...m) \tag{5}$$

Where $\mu_i$ and $\sigma_i$ denote the mean and standard deviation of random variable $x_{ik}$ . $\xi_{ik}$ denotes the location, which is expressed as $\xi_{ik} = \lambda_{i3}/2 + (-1)^{3-k}\sqrt{\lambda_{i4} - (3/4)\lambda_{i3}^2}$ $(k = 1,2)$ $\xi_{i3} = 0$ ,where, $\lambda_{i3}$ and $\lambda_{i4}$ are the third and fourth standardized central moments of $x_{ik}$ known as skewness and kurtosis. Weighting factors related to each location $\xi_{ik}$ are calculated by the following equations:

$$\omega_{ik} = \frac{(-1)^{3-k}}{\xi_{ik}(\xi_{i1} - \xi_{i2})} \quad (k = 1,2) \ , \qquad \omega_{i3} = \frac{1}{m} - \frac{1}{\lambda_{i4} - \lambda_{i3}^2} \tag{6}$$

After calculation above, 2m+1 vectors of input variables $X$ are expressed as follows:

$$X_i(k) = [\mu_1, \mu_2, ..., \mu_3, x_{ik}, \mu_{i+1}, .., \mu_m] \ (k = 1,2; i = 1,2,..m), \ X_{2m+1} = [\mu_1, \mu_2, ..., \mu_m] \tag{7}$$

In the 2m+1 PEM scheme, it requires 2m+1 calculations of function g. Accordingly the *n-th* moments of *Y* can be estimated approximately by the following equation.

$$E(Y_i^n) = \sum_{i=1}^{m} \sum_{k=1}^{2} \omega_{ik} g(X_i(k))^n + \sum_{i=1}^{m} \omega_{i3} g(X_{2m+1})^n \tag{8}$$

*3.2. The cumulative distribution function based on Cornish-fisher expansion*

We can get the cumulative distribution function of state variables by Cornish-Fisher expansion which presents an approximation of a quantile of acumulative distribution function $F(x)$ in terms of the quantile of a normal N(0,1) distribution $\Phi(x)$ and the cumulants of $F(x)$. Using the first five cumulants, the expansion can be formulated as follows[15]:

$$\begin{aligned}\chi(\alpha) \approx \xi(\alpha) + \frac{1}{6}\left(\xi^2(\alpha) - 1\right)\kappa_3 + \frac{1}{24}\left(\xi^3(\alpha) - 3\xi(\alpha)\right)\kappa_4 - \frac{1}{36}\left(2\xi^3(\alpha) - 5\xi(\alpha)\right)(\kappa_3)^2 + \frac{1}{120}\left(\xi^4(\alpha) - 6\xi^2(\alpha) + 3\right)\kappa_5 \\ - \frac{1}{24}\left(\xi^4(\alpha) - 5\xi^2(\alpha) + 2\right)\kappa_3\kappa_4 + \frac{1}{324}\left(12\xi^4(\alpha) - 53\xi^2(\alpha) + 17\right)(\kappa_3)^3\end{aligned} \tag{9}$$

where $\chi(\alpha) = F^{-1}(\alpha)$, $\xi(\alpha) = \Phi^{-1}(\alpha)$ and $\kappa_n$ is the cumulant of order n of the cumulative distribution function $F$ which can be obtained from the above moments calculated by the PEM method in this paper.

*3.3. The objective function for the probabilistic power flow calculation*

By means of the method introduced above, we give the objective function and control variables in the optimal method for the storage capacity as follows [7].

$$F = K_1 \Delta Q + K_2 \Delta P + \lambda \sum_{i}^{n} \left( \frac{P_{i,V} - P_{i,V\,\mathrm{lim}}}{V_{\max} - V_{\min}} \right) \tag{10}$$

Where $\Delta Q$ represents the storage capacity which is the control parameter during the optimal procedure, $\Delta P$ denotes the active losses, $P_{i,V}$ is the probability of the node voltage exceeding limit range from $V_{\min}$ to $V_{\max}$, accordingly $P_{i,V\,\mathrm{lim}}$ is the allowed maximum probability of the node voltage exceeding limit range. $K_1, K_2$ and $\lambda$ are the weight factors respectively for considering the cost of storage, the network losses, and the probability of the violation of the node voltage. The constraints of the objective function are the power flow equations.

## 4. The particles swarm optimizations algorithm and the flowchart of the method

*4.1. The particles swarm optimizations algorithm*

The Particles Swarm Optimization (PSO) is an optimal search approach inspired by the simplified social system such as bird flocking or fish schooling [16, 17]. For the general multivariable optimization search case, the swarm consists of a specified number of particles having their own positions and velocities which can be respectively expressed as follows, $X_i = (x_{i1}, x_{i2}, ..., x_{iM})$ and $V_i = (v_{i1}, v_{i2}, ..., v_{iM})$, where $i = 1, 2, ..., N$, $N$ represents the size of the swarm and $M$ is the dimension of the research space. The fitness value of the particle is related to the objective function. The local best known position is the best solution that achieved by each particle and the global best known position is the best solution among all the achieved solutions during the iteration search process. The main steps for the PSO are summarized below. Firstly PSO also begins with a group of randomly generated solutions, and then updates the position and velocity for each particle in each iteration. For the iteration it can be expressed as follows:

$$\begin{aligned} V_i^{k+1} &= \omega V_i^k + \varphi_1 R_1 (P_{best}^k - X_i^k) + \varphi_2 R_2 (G_{best}^k - X_i^k) \\ X_i^{k+1} &= X_i^k + V_i^{k+1} \end{aligned} \tag{11}$$

where $\omega$ is the inertia weight factor, $\varphi_1$ and $\varphi_2$ are the acceleration factors, $R_1$ and $R_2$ are the random numbers between 0 and 1. $P_{best}^k$ and $G_{best}^k$ are the local best known position, and global best known position respectively in the $k$ iteration, which reflects the cooperation and competition mechanism in PSO algorithm.

It is known that the inertia weight of the particles has a large impart on the efficiency of the search algorithm. Some researches show that selecting large value of $\omega$ at the beginning and gradually decreasing it during the iteration process is a better choice. The inertia weight is given by following equation during the iteration process,

$$\omega = \omega_{\max} - \frac{\omega_{\max} - \omega_{\min}}{iter_{\max}} \times iter \tag{12}$$

Where $\omega_{\max}$ and $\omega_{\min}$ are the maximum and minimum values, $iter$ and $iter_{\max}$ are the current iteration number and the maximum iteration number, respectively. Finally the research process ends when the idea iteration number is achieved or the locations of all particles converge to the same set.

*4.2. The flowchart for the optimal method*

The procedure for the proposed optimum scheme in this paper is described as follows:

1. Reading the initial data: including the data of load flow calculation, the parameters of the wind power generations and the correlation matrix between two wind farms, and the PSO parameters which include particle swarm scale, accelerated coefficient and so on.
2. Randomly generating the location of the each particle which represents the storage capacity value and the velocity of the each particle.
3. Generating the speed samples of the two winds farm, and then getting statistic distributions of the power outputs of the wind power generations.
4. Through the Probabilistic power flow calculation, obtaining the value of the initial individual optimal value $P_{best}$ and the global optimal value $G_{best}$.
5. Judging termination conditions: To stop the algorithm process and output the results when achieving the maximum iteration times. Otherwise continue to update the particles velocities and the locations and to calculate the next probabilistic calculation.

The procedure above can be showed by the figure 1.

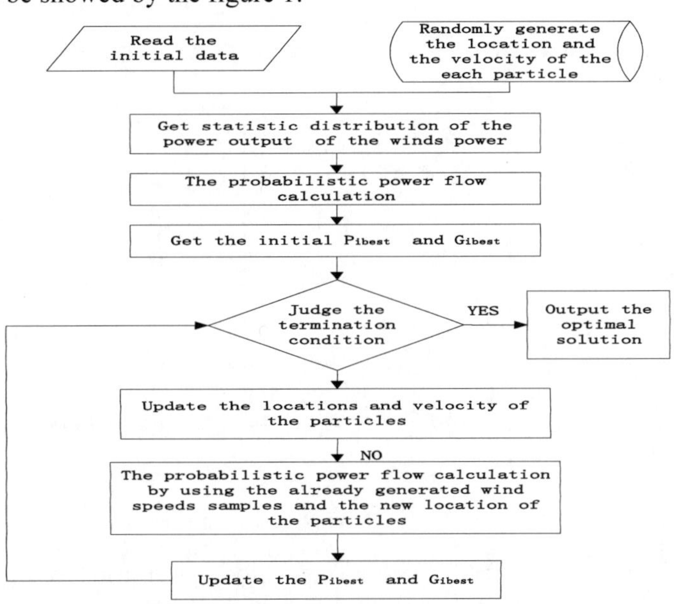


Figure 1. The flowchart for the optimal method

## 5. The study case

The algorithm introduced above is used in the modified IEEE33-bus test system. The topology structure is showed in the figure 2. Twelve wind power generators are added to the node 33 of the system. They are divided into two wind farms A and B. Every wind farm consists of six wind power generators. The scale and shape parameters of Weibull distribution of the wind speed in the wind farm A and B is set as $c_A = 2.8$、$k_A = 5.5$ and $c_B = 2.8$、$k_B = 5.5$, respectively. The rated output power of each wind power generations is 0.6 MW. The cut-in, rated and cut-out speed of the wind turbine are set as $v_{ci} = 3ms^{-1}$, $v_r = 14ms^{-1}$ and $v_{co} = 25ms^{-1}$ respectively. For generating the combined probabilistic distribution of the power output in the 33 node system, the storage parameters need to be set as follows: $P_{storage\_max} = 0.2MW$, $Soc_{min} = 0.1$, $Soc_{max} = 0.8$, $\eta_d = 2$ . By the wind farm and the storage parameters above, we plot the figure 3 which shows the comparison of the power output in the node 33 with the storage and without storage when the correlation between the two wind farms is 0.6. From figure 3 we find that the increasing capacities of the storage can make the power output focused to the average values, which is the basic reason why storage can reduce the fluctuation in the power grid.

The parameters of the original physical IEEE33-bus system can be found in reference [18]. Since the probabilistic power flow calculation need not only the 33 node power output, but also the probabilistic distributions of the loads in all nodes, the normal distribution is used to represent the characteristics of the loads whose standard deviations are given in the table A1 of the Appendix. The parameters for the objective function are set for the particles swarm optimization algorithm as follows: $V_{min} = 0.94$, $V_{max} = 1.06$, $P_{v,lim} = 0.0006$, $K_1 = 10$, $K_2 = 10$, $\lambda = 1$.

The population size for the PSO algorithm is *M*=50, and the iterative times is 200. Through the calculation of certain power flow combined with PSO algorithm, the proper energy storage capacities for the different correlation cases are obtained, which are showed in the table1.

**Table 1.** The optimal storage capacities for different correlations

| Correlation factor | 0.0 | 0.2 | 0.4 | 0.6 | 0.8 | 1.0 |
|---|---|---|---|---|---|---|
| Optimal storage (MW·h) | 5.0 | 5.2 | 5.9 | 6.1 | 6.7 | 7.4 |

From the results in table1, we find that the optimal storage capacity rises with the increase of the correlation between the wind farms.

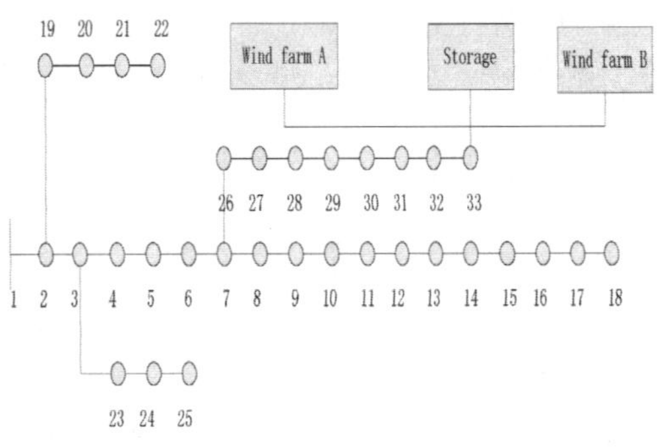


Figure 2. The topology structure for the modified IEEE33-bus grid.

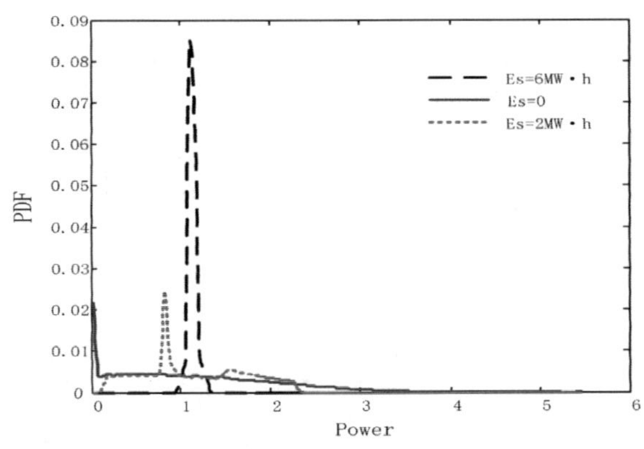


Figure 3. The probabilistic density functions (PDF) for the cases with different storage capacities.

Finally, for showing the effects of the storages on the power grid, the figure 4 and 5 are provided which give the comparisons of between the case with different storage capacities and different correlation factors. From figure 4 and 5, it is found that for the cases with the same correlation factors the active losses and sum of the probability of the node voltage exceeding limit are reduced greatly with the increase of the storage capacities, and that for the cases with the same storage capacities the active power loss and the sum of the probability of the node voltage exceeding limit $P_{sum\_vel}$ is increased with the correlation factors. With the increase of the correlation factors, the storage capacity should be increased in the system for better economy and stability.

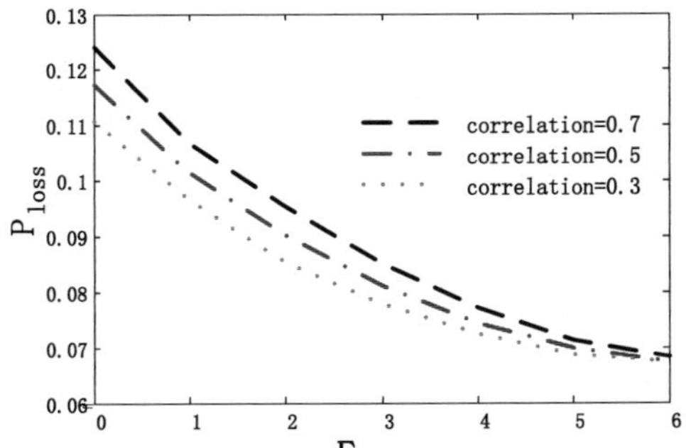


Figure 4. The active power loss changes with the increase of the storage capacities $E_s$ for different correlation cases.

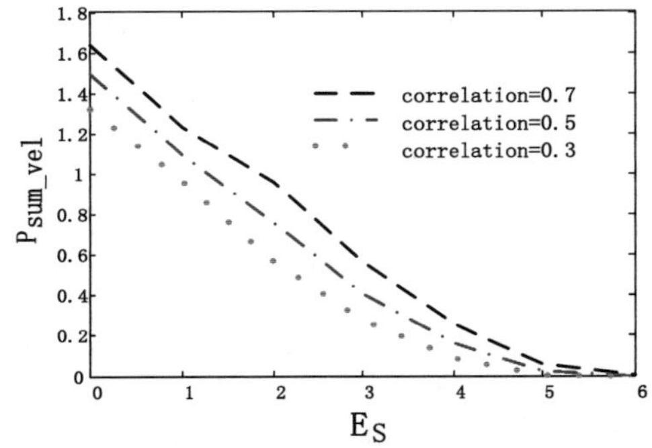


Figure 5. The sums of the probability of the node voltage exceeding limit decreases with the storage capacities $E_s$ for different correlation cases.

## 6. Conclusions

This paper introduced a method for obtaining the optimal storage capacity in the power network with the access of the wind power generations. In this method, the objective function for the optimum algorithm considers the cost of storage, the network losses, and the probability of the violation of the node voltage. The Probabilistic power flow based on the point estimate method and Cornish-fisher expansion is carried out during the optimum process based on PSO, which can reflect effects of the corrections between the wind farms. A concrete example in the IEEE33-bus power grid shows that the application of the storage can reduce fluctuations, and that the correlation between the wind farms has a nonnegligible influence on the optimal capacities of the storage. The results show that we should consider the correlations between the different wind power generations for choosing the proper capacities of the energy storage.

## Acknowledgements

This work is supported by National Basic Research Program of China (973 Program) under the Grants No. 2014CB249203.

## Appendix

**Table A1.** The standard deviation of the load power, where we use the hypothesis that the deviations of the loads active are the same as their reactive power.

| Node number | power deviation | Node number | power deviation |
|---|---|---|---|
| 2 | 0.0522 | 18 | 0.0820 |
| 3 | 0.1650 | 19 | 0.1050 |
| 4 | 0.0870 | 20 | 0.0706 |
| 5 | 0.0750 | 21 | 0.0396 |
| 6 | 0.0637 | 22 | 0.0336 |
| 7 | 0.0225 | 23 | 0.0961 |
| 8 | 0.0421 | 24 | 0.1150 |
| 9 | 0.2650 | 25 | 0.0852 |
| 10 | 0.0760 | 26 | 0.0241 |
| 11 | 0.0690 | 27 | 0.0233 |
| 12 | 0.0950 | 28 | 0.0537 |
| 13 | 0.0238 | 29 | 0.0786 |

| | | | |
|---|---|---|---|
| 14 | 0.0405 | 30 | 0.0378 |
| 15 | 0.0492 | 31 | 0.0512 |
| 16 | 0.2150 | 32 | 0.0595 |
| 17 | 0.0910 | 33 | 0.0923 |

## References

[1] Wang J, Li X Y. 2005 Power system research on distributed generation penetration. *Automation of Electric Power Systems,* **29** : 90-97.

[2] Ghofrani M，Arabali A，Etezadi-Amoli M, et al. 2013 Energy storage application for performance enhancement of wind integration. *Power IEEE Transactions on Power Systems*. **28**(4):4803-4811.

[3] Qiang Fu，Montoya L F，Solanki A，et al. 2012 Microgrid generation capacity design with renewables and energy storage addressing power quality and surety. *IEEE Transactions on Smart Grid*. **3**(4):2019-2027.

[4] Liang Liang, Li J L. 2011 Optimization Configuration for Capacity of Energy Storage System in Large-scale Wind Farm. *High Voltage Engineering*. **37**(4):930-936.

[5] Zhang B H，Zeng J，Mao C X，et al. 2006 Improvement of power quality and stability of wind farms connected to power grid by battery energy storage system. *Power System Technology*. **30**(15):54-58.

[6] Ammar M，Joos G. 2014 A short-term energy storage system for voltage quality improvement in distributed wind power. *IEEE Transactions on Energy Conversion*. **29**(4): 997-1007.

[7] Yao S J, Ma L Y. 2012 Capacity of Energy Storage Based on Probabilistic PowerFlow and the HPSO Algorithm for DS Network with Photovoltaic Power Generation. *10th IET International Conference on ACDC*. 2012 15-19

[8] Garcia A,Torres J L, Prieto E, Francisco A. 1998 Fitting wind speed distributions: a case study. *Solar Energy*. **62** (2): 139-144.

[9] Abbas R, Alireza S, Behnam M I, Mostafa P. 2014 Corrective voltage control scheme considering demand response and stochastic wind power. *IEEE Transactions on Power Systems* **29** (6) :2965-2973.

[10] Chen C, Wu W C, Zhang B M, et al. 2015 Correlated probabilistic load flow using a point estimate method with Nataf transformation. Electrical Power and Energy Systems. *Electrical Power and Energy Systems*. **65**:325-333.

[11] Li H S, Lu Z Z,Yuan X K. 2008 Nataf transformation based point estimated method. *ChinSci Bull*. **53**:627-632.

[12] Yang J H, Ouyang S, Wu Y S, et al. 2016 Stochastic Power Flow of Distribution Network Considering probability model of wind-storage combined systems. *Power System Technology*. **40**:234-241.

[13] Ding M，Li S H，Huang K. 2001 Probabilistic load flow analysis based on Monte-Carlo simulation. *Power System Technology*. **25**(11): 10-14.

[14] Mohammadi M, Shayegani A, Adaminejad H. 2013 A new approach of point estimate method for probabilistic load flow. *Electrical Power and Energy Systems*. **51**:54-60.

[15] Julio U. 2009 Probabilistic load flow with wind production uncertainty using cumulants and Cornish–Fisher expansion. *Electrical Power and Energy Systems*. **31**:474–481.

[16] Yuan X H, Sheng W, Zhang Y C. 2004 The PSO algorithmused in electric power system. *Power System Technology*. **28**(19):14-19.

[17] Kennedy J, Eberhart R C. 1995 Particle Swarm Optimization. *In Proceedings of the 1995 IEEE International Conference On Neural Networks*, Perth, Australia, 1942–1948.

[18] Acharya N, Mahat P, Mithulananthan N. 2006 An analytical approach for DG allocation in primary distribution network. *International Joural of Electrical Power Energy System*, **28**(10):669-678.

# The antioxidant activitives of mango peel among different cultivars

**Yu-Ge Liu, Xiu-Mei Zhang *, Fei-Yue Ma & Qiong Fu**

Key Laboratory of Tropical Fruit Biology, Ministry of Agriculture, South Subtropical Crop Research Institute, Chinese Academy of Tropical Agricultural Science (CATAS), Zhanjiang 524091, Guangdong, China

**Abstract.** In this paper, the contents of total phenol and total flavonoid of 8 mango cultivars were determined. Their antioxidant abilities were also evaluated by 1,1-diphenyl-2-pireyhydrazyl (DPPH) radical scavenging, trolox equivalent antioxidant capacity (TEAC) and ferric reducing antioxidant power (FRAP). Correlations between total phenol, total flavonoid and FRAP as well as TEAC were also analyzed. Results showed that mango peels were rich in natural antioxidant compounds the antioxidant abilities were different among different cultivars. The correlations between total phenol, total flavonoid and FRAP indicated phenolics represent a major part of antioxidant capacity in mango peels. This was also useful in the utilization of mango processing waste.

## 1. Introduction

Mango (Mangifera indica L.) is a widely planted and consumed tropical fruit throughout the world. Besides the fresh fruit, mango can also be processed into juices, nectars, concentrates, jams, jelly powders and so on.

It has been reported that fruits and vegetables contain many antioxidant compounds, such as phenolic compounds, carotenoids, anthocyanins and so on (Naczk and Shahidi, 2006). Among the different parts, fruit peels are rich in polyphenolic compounds, flavonoids, ascorbic acid, and this makes them valuable in making antioxidants.

There has been increasing focus of attention of many researchers searching for potent antioxidants from mango. Different parts of mango, including stem bark, leaves, pulp and peel have been investigated. Results showed that those parts possessed various biomedical activities, including antioxidative, free radical scavenging, anti-inflammatory, and anticancer ((Ajila et al 2007; Rocha Ribeiro et al., 2007; Hernandez et al 2007; Percival et al., 2006; Ling et al, 2009).

Peel is a major by-product of mango processing, and they are discarded as waste. It has been reported that peel was a good source of phytochemicals, such as polyphenols, carotenoids, and it exhibited good antioxidant properties (Ajila et al 2007; Kim et al, 2010). It is commonly considered that the concentration and composition of phenolics is affected by genetic, agronomic and environmental factors (Tomás-Barberán and Espín, 2001). However, the antioxidant activities regarding differences among different mango cultivars have been rarely reported.

In this paper, the contents of total phenol and total flavonoid of 8 mango cultivars were determined. Their antioxidant abilities were also evaluated by 1,1-diphenyl-2-pireyhydrazyl (DPPH) radical scavenging, trolox equivalent antioxidant capacity (TEAC) and ferric reducing antioxidant power (FRAP). Correlations between total phenol, total flavonoid and FRAP as well as TEAC were also analyzed. The results showed that mango peels were rich in natural antioxidant compounds the antioxidant abilities were different among different cultivars, and this was useful in the utilization of mango processing waste.

IOP Conf. Series: Earth and Environmental Science **61** (2017) 012065 doi:10.1088/1755-1315/61/1/012065

## 2. Materials and methods

### *2.1. Materials and regents*

Ripen fruits of 8 mango cultivars in China, namely 'Chenpixiang', 'Boluoxiang', 'Qiumango', 'Zaoshumango', 'Hong kaite', 'Hongmango 6', 'Sijimango 1', and 'Jinhuang' (designed as 1-8 in this reaearch) were collected just from the trees planted in South Subtropical Crop Research Institute. 10 fruits of each cultivar were picked and cleaned with water, then the peels were collected and then ground using a stainless-steel grinder. They were stored in vacuum-packaged polyethylene pouches at -20°C until required for analysis.

Folin-Ciocalteu's (FC) phenol reagent and gallic acid (GA) were purchased from Fluka. The 1,1-diphenyl-2-pireyhydrazyl (DPPH) and 2,2'-amino-di(2-ethyl-benzothiazoline sulphonic acid-6) ammonium salt (ABTS) radical and all other standards were received from Sigma-Aldrich.

### *2.2. Sample extraction*

1 g mango peel of different cultivars was weighed and refluxed with 30 ml of 70% methanol at 60 °C for 2 h under magnetic stirring. The filtrate was separated by centrifugation, and the extraction was repeated for 3 times. All the filtrate was collected and concentrated under reduced pressure at 40 °C with a final volume of 30 ml and the solution used in the following evaluation and detection.

### *2.3. Determination of total phenolic content and total flavonoid in the extracts*

The total phenol content (TPC) was determined using the FC assay described before with some modifications (Du et al., 2014). Typically, 0.025 ml of the extract of different cultivar was introduced into test tubes and followed by the addition of 2.0 ml of FC reagent (diluted 10 times with water in advance) and 5.975 ml of water. The solutions were allowed to stand 5 min at room temperature before the addition 2 ml of sodium carbonate solution (7.5% w/v). After reacting in dark for 30 min at room temperature, the absorbance of the solutions were measured at 760 nm on a UV–vis spectrophotometer (Shimadzu UV-2700, Japan). The calibration curve was prepared using a standard solution of gallic acid. The results were expressed as milligram gallic acid equivalents (GAE)/g dry weight (fresh weight, FW).

Total flavonoid content was determined based on the method described by Kim et al (Kim et al., 2003). One milliliter of extract solution of different cultivar was mixed with 0.3 ml of 5% $NaNO_2$ and 4 ml of distilled water. Then 0.3 ml of $Al(NO)_3$ was added to the mixture followed by adding 2 ml of 1 M NaOH. The solution was immediately diluted to 10 ml using distilled water. The absorbance of the solution was measured at 506 nm and the total flavonoid content was calculated by using a calibration curve of rutin standard and expressed as mg rutin equivalent (QR Equiv)/g FW.

### *2.4. DPPH radical scavenging ability*

The free radical scavenging activity of the extracts was performed by measuring the decrease in absorbance of DPPH solution at 517 nm in the presence of the extracts by the method proposed by Liyana-Pathirana et al (2010) with minor changes. The solution of 0.5 mM was prepared by dissolving DPPH in methanol. For the evaluation of free radical scavenging activity, 3 ml of DPPH was added into 0.5 ml of the extracts with different concentrations. The mixture was then allowed to stand at room temperature for 30 min in dark before the absorbance at 517 nm was read. The control was prepared as above without extract. The antioxidant activity could be expressed as the following equation:

$$\text{Scavenging activity}=\frac{A_0-A_s}{A_0}\times 100\%$$

where $A_0$ and $A_s$ were the absorbance at 517 nm of the control and sample solution, respectively.

### *2.5. Trolox equivalent antioxidant capacity (TEAC) and ferric reducing antioxidant power (FRAP)*

This assay was determined according to the method reported (Benzie et al., 1996). For the determination of FRAP, the extracts (10 μL) were mixed with 1 ml distilled water and 1.8 ml of the FRAP solution. Then the mixture was reacted at 37 °C for 10 min. The absorbance of the reaction solution was recorded at 593 nm. Trolox standard solution was used to perform the calibration curves and the results were expressed as μM trolox/g (fresh weight, FW).

TEAC was calculated according to the ABTS scavenging ability of mango peel extract. This was performed by the procedure described by Re et al (1999). For the scavenging of ABTS, 50 μl of different extracts were added to 4 mL of the solution above. Methanol was used as control. After reaction for 10 min, the absorbance was measured at 734 nm. The free radical scavenging capability was calculated by the equation:

$$\text{ABTS scavenging activity} = (Ac-As)/Ac \times 100\%$$

where Ac and As were the absorbance at 734 nm of the control and sample solution, respectively. Trolox standard solution was used to perform the calibration curves and the results were expressed as μM trolox/g (FW).

## 3. Results and disscussion

### *3.1. Contents of total phenol and total flavonoid*

The contents of total phenol and total flavonoid in each cultivar was summarized in table 1. It can be conclude from the table that the total phenol and to-tal flavonoid ranged from 1.68-13.28 and 0.33-6.08 mg/g, respectively. The cultivar 'Chenpixiang' pos-sess the highest while 'Jinhuang' possessed the low-est contents of total phenol and total flavonoid. This indicated that the bioactive compounds in mango peels varied greatly different mango cultivars. Simi-lar results have been reported in litchi pericarps (Wang et al, 2011).

**Table 1** Total phenol, total flavonoid, FRAP and TEAC of different mango peel extracts

| cultivars | 1 | 2 | 3 | 4 | 5 | 6 | 7 | 8 |
|---|---|---|---|---|---|---|---|---|
| Total phenol (mg/g) | 13.28 | 12.34 | 9.44 | 7.74 | 6.84 | 6.19 | 5.14 | 1.68 |
| Total flavonoid(mg/g) | 6.08 | 5.32 | 4.13 | 4.26 | 1.63 | 2.42 | 1.67 | 0.33 |
| FRAP (μM/g) | 122 | 96 | 82 | 78 | 58 | 42 | 39 | 16 |
| TEAC (μM/g) | 183 | 184 | 108 | 98 | 109 | 65 | 62 | 58 |

### *3.2. Antioxidant activities of mango peel extracts*

Due to its operating simplicity, DPPH is one of the most popular methods employed for the evaluation of antioxidant ability, especially in plant extract. DPPH is a kind of stable organic radical. In the radical form, the molecule of DPPH has an absorbance at 517 nm, which will disappear after the acceptance of an electron or hydrogen radical from an antioxidant in the solution to become a stable diamagnetic molecule (Matthäus, 2002).

The DPPH scavenging ability of the peel extracts of different mango cultivars was given in Figure 1. Different from the results of total phenol and total flavonoid, there was no obvious difference among those values. This may because that the bioactive compounds which could scavenge DPPH might almost be the same in different mango cultivars.

The values of FRAP and TEAC were also given in table 1. It can be seen that FRAP of mango peel ex-tract of different cultivars ranged from 16 to 122 μM/g, and the turn was the same as that of total phenol and total flavonoid. This was because FRAP represented the total antioxidant activity of plants, and it was only connected to the total bioactive compounds. The turn of TEAC was not totally in accordance with that of FRAP, and this may be as-cribed to the same reason of DPPH scavenging ability.

IOP Conf. Series: Earth and Environmental Science **61** (2017) 012065 doi:10.1088/1755-1315/61/1/012065

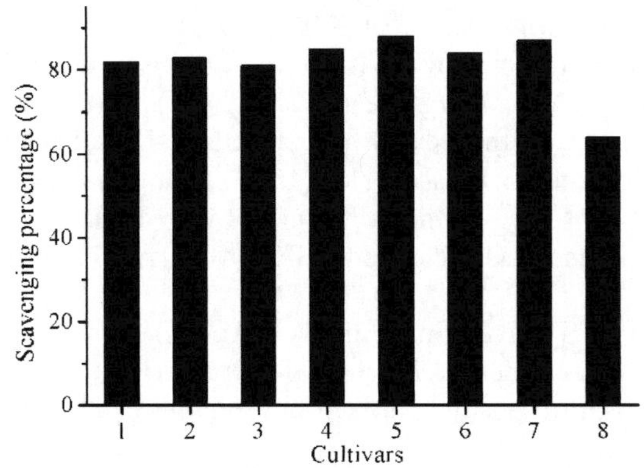


**Figure 1** DPPH scavenging abilities of the peel extracts of different mango cultivars.

### *3.3. Correlations*

Correlations were made in order to determine the contribution of total phenol and total flavonoid to the antioxidant activities of mango peel. The correlation between total phenol and FRAP, TEAC was shown in Figure 2A. It can be seen that the correlation between total phenol and FRAP was almost linear ($r^2$=0.92). The trend was the same of the correlation between total flavonoid and FRAP ($r^2$=0.94). This indicated that phenols and flavonoids represent a major part of antioxidant capacity in mango peel. However, The correlation coefficients between total phenol, total flavonoid and TEAC were lower than those of FRAP ($r^2$=0.69, 0.84, respectively), and this may be due to the fact that TEAC were calculated from the ABTS scavenging ability. Different from FRAP, the ABTS scavenging ability could not represent the total antioxidant capacity of mango peels.

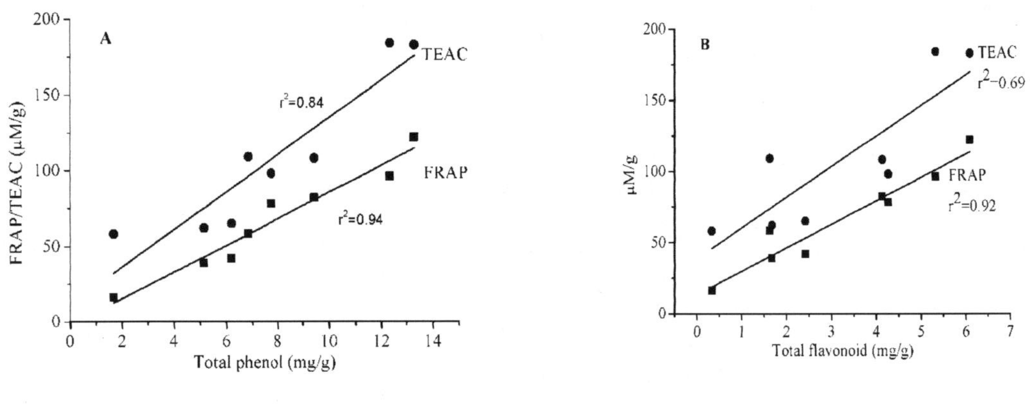


**Figure 2** Correlations of (A) total phenol and FRAP, TEAC, (B) Total flavonoid and FRAP, TEAC.

## 4. Conclusions

The contents of total phenol and total flavonoid of mango peels of 8 different cultivars, namely 'Chenpixiang', 'Boluoxiang', 'Qiumango', 'Zaoshumango', 'Hong kaite', 'Hongmango 6', 'Sijimango 1', and 'Jinhuang' were determined and compared in this research. Their antioxidant abilities were also evaluated by DPPH radical scavenging and FRAP. Results showed that the turns of total phenol and total flavonoid were both 'Chenpixiang'>'Boluoxiang'>'Qiumango'> 'Zaoshumango' >'Hong kaite'> 'Hongmango 6' > 'Sijimango 1' > and 'Jinhuang'. The turn of values of FRAP was the same as that of total phenol and total flavonoid, and the highest and lowest were 122 and 16 μM/g, respectively. This may because FRAP the total antioxidant activity of plants, and it was only connected to the total bioactive compounds. The correlation between total phenol, total flavonoid and FRAP was almost linear, and this was of the same reason above. There was no change among the DPPH radical scavenging abilities of different cultivars, and the turn of TEAC was not totally in accordance with that of FRAP. The reason of this may be that DPPH and TEAC only represented their radical scavenging abilities. Besides, the correlation was not linear. All these indicated that the antioxidant activities mainly came from total phenols and total flavonoids. The research showed that mango peels were rich in natural antioxidant compounds and the antioxidant abilities were different among different cultivars, and phenols represent a major part of antioxidant capacity in mango peels. The results were also helpful in the utilization of mango processing waste.

**Acknowledgements**
The research was supported by the National Natural Science Foundation of Hainan province (No. 20153122) and the Fund on Basic Scientific Research Project of Nonprofit Central Research Institutions (No. 1630062016008).

**References**
[1] Ajila, C. M., S. G. Bhat , & U. J. S. Prasada Rao (2007). Valu-able components of raw and ripe peels from two Indian mango varieties. Food Chem. 102, 1006-1011.
[2] Benzie, I. F. F., & J. J. Strain (1996). The ferric reducing ability of plasma (FRAP) as a measure of "antioxidant power": the FRAP assay. Anal. Biochem. 239, 70-76.
[3] Du, L., Y. Shen, X. Zhang, W. Prinyawiwatkul, & Z. Xu (2014). Antioxidant-rich phytochemicals in miracle berry (Synsepalum dulcificum) and antioxidant activity of its ex-tracts. Food Chem. 153, 279-284.
[4] Gorinstein, S., Martín-Belloso, O., Lojek, A., Soliva-Fortuny, R., Park, Y.S., Caspi, A., I. Libman, & S. Trakh-tenberg (2002). Comparative content of some phytochemi-cals in Spanish apples, peaches and pears. J. Sci. Food Agric. 82, 1166-1170.
[5] Hernandez, P., P. C. Rodriguez, R. Delgado, & H. Walczak (2007). Protective effect of Mangifera indica L. Polyphenols on human T lymphocytes against activation induced cell death. Pharm. Res. 55, 167-173.
[6] Kim, H., J. Y. Moon, H. Kim, D.-S. Lee, M. Cho, H.-K. Choi,Y. S. Kim, A. Mosaddik, & S. Kim Cho (2010). Antioxidant and antiproliferative activities of mango (Mangifera indica L.) flesh and peel. Food Chem. 121, 429-436
[7] Kim, D. O., O. K. Chun, Y. J., Kim, H. Y. Moon, & C. Y. Lee (2003). Quantification of polyphenolics and their anti-oxidant capacity in fresh plums. J. Agric. Food Chem. 51(22), 6509-6515.
[8] Ling, L. T., S. Yap, A. K. Radhakrishnan, T. Subramaniam, H. M. Cheng, & U. D. Palanisamy (2009). Standardised mangi-fera indica extract is an ideal antioxidant. Food Chem. 113, 1154-1159.
[9] Liyana-Pathirana, C. M. & F. Shahidi (2005). Antioxidant activity of commerical soft and hard wheat (Triticum aestivum L.) as affected by gastric pH conditions. J. Agric. Food Chem. 53, 2433-2440.
[10] Matthäus, B. (2002). Antioxidant activity of extracts obtained from residues of different oilseeds. J. Agri. Food Chem. 50, 3444-3452.
[11] Naczk, M. & F. Shahidi (2006). Phenolics in cereals, fruits and vegetables: Occurrence, extraction and analysis. J. Pharm. Bio. Anal. 41, 1523–1542.
[12] Re, R., N. Pellegrini, A. Proteggente, A. Pannala, M. Yang, & C. Rice-Evans (1999). Antioxidant activity applying an im-proved ABTS radical cation decolorization assay. Free Rad. Biol. Med. 26, 1231-1237.
[13] Rocha Ribeiro, S. M., J. H. Queiroz, Ribeiro, Lopes., M. E. L. R. Queiroz, M. F. Campos, & H. M. Pinheiro Sant'ana (2007). Antioxidant in mango (Mangifera indica L.) pulp. Plant Foods for Human Nutri. 62, 13-17.
[14] Tomás-Barberán, F. & J. C. Espín (2001). Phenolic compounds and related enzymes as determinants of quality of fruits and vegetables. J. Sci. Food Agric. 81, 853-876.

IOP Conf. Series: Earth and Environmental Science **61** (2017) 012065 doi:10.1088/1755-1315/61/1/012065

[15] Wang, H.-C., Z.-Q. Hu, Y. Wang, H.-B. Chen, & X.-M. Huang (2011). Phenolic compounds and the antioxidant activities in litchi pericarp: Difference among cultivars. Scientia Hort. 129, 784-789.

IOP Conf. Series: Earth and Environmental Science 61 (2017) 012066 doi:10.1088/1755-1315/61/1/012066

# Applicability research on passive design of residential buildings in hot summer and cold winter zone in China

**Hui Zhang[1], Huihui Wang[1] and Xuan Zhou[1]**

[1]Hubei University of Technology, No.28. Nanli Rd,Wuhan,430068,China
E-mail: 24881430@qq.com,772454268@qq.com,984111425@qq.com

**Abstract**. Passive design has long been a concern as an effective way of building energy efficiency. However, different urban climate characteristics determine the time-effectiveness of passive design. According to the climate characteristics of hot summer and cold winter zone in China, this research chose five cities, Shanghai, Wuhan, Chongqing, Nanjing and Changsha, to analyze their residential building energy consumption and thermal environment conditions. Based on Weather Tool calculation and analysis, the purpose of this research is to put forward the concept of Suitable Degree (SD), namely the applicability of the passive design. In addition, five cities' SD of passive design technology had been analyzed from aspect of ventilation, temperature, solar radiation and envelope, then passive design strategies and methods of five cities' residential building were discussed.

## 1. Introduction

Recently, with the continuous improvement of people's living environment and comfort requirements, residential building energy consumption has increased year by year. As research shows, the national residential building energy consumption accounted for nearly 60% of the total energy consumption of buildings by 2015. For decades, the energy saving of residential buildings in cold regions of China has achieved remarkable results by reducing the ratio of window and wall, strengthening the heat preservation, central heating and optimizing control. In hot summer and cold winter Zone in China, the overall energy utilization efficiency is low and energy consumption of residential building is increasing year by year, due to the special climate characteristics and life characteristics of residents, residential building heating and air conditioning equipment is used locally intermittent.

To further improve residential building thermal environment and reduce building energy consumption in hot summer and cold winter zone, governments had set up a number of standards in China, such as Design Standard for Energy Efficiency of Residential Buildings in Hot Summer and Cold Winter Zone-JGJ134-2010, Design Standard for Residential Buildings of Low Energy Consumption-DB42T559-2013 and so on. Specification and guidance from the energy and thermal aspects of the equipment were also provided. In this regard, the relevant scholars did a lot of related research. On researching the influencing factors of residential buildings, Zhang Xu compared and analyzed the sensitivity of energy consumption of residential buildings in hot summer and cold winter zone, combining with external walls, windows and other energy-saving measures quantitative calculation [1]. Fu Heng used dynamic simulation analysis of the residential building coefficient and natural ventilation on residential building energy consumption impact [2][3][4]. Yu Jinghua researched the best insulation thickness of roof and exterior wall of residential building in hot summer and cold winter zone [5][6]. Research on energy saving potential, Jin Sike used Dest energy consumption to simulate the 65% of residential building energy-saving measures requirements [7]. Yang Zijiang put forward coping strategies in view of the existing residential building energy efficiency in Hubei 50% transformation [8]. Yu Xiaoping proposed the technical route of low energy consumption residential building [9]. Research on passive energy saving in residential buildings, Zhang Hui discussed the technology of passive energy saving of residential buildings in hot summer cold winter regions [10]. For residential buildings, although Ministry of Housing and Construction of Hebei Province, introduced the institute of design standard for energy efficiency of passive low-energy residential buildings, they were mainly used in cold regions, however in hot summer and cold winter zone there were still some deviation.

With the popularization of green building theory and the continuous updating of energy saving technology, low-energy passive residential building has become one of the hot spots at home and abroad. In this research, by using the Weather Tool software analysis of the climate characteristics at different cities in hot summer and cold winter zone and the thermal comfort zone, we provided the ideas for passive energy saving of residential buildings in hot summer and cold winter zone, furthermore combined with analysis of the suitable degree of passive design strategies.

## 2. Influence of Residential buildings' Thermal environment In Hot Summer and Cold Winter zone

For a long time, the Yangtze River basin has been a region of rapid economical development in China, but it is also the unbalanced development of social and economic areas. The differences between economic level and the influence of history and culture, lead to the formation of the region's unique living phenomenon. From the climatic zoning, the vast area of the Yangtze River Basin is located in hot summer and cold winter zone in China, where summer persists high temperature and humidity, the temperature exceeding 34°C, while the temperature is less than 8°C in winter， the spring and autumn season is wet long period of transition, therefore there is a year in the residential heating, cooling, ventilation and desiccant and so on. Affected by climatic characteristics, living habits and economic conditions and other factors, in hot summer and cold winter zone residential building heating and cooling operation are mostly intermittently used for short-cycle. The use of such equipment is compares to the calculation ways of energy efficiency standards. There exists the use of cycles and operating conditions inconsistent, resulting in residential building energy efficiency indicators and the deviation of the actual energy consumption. To a certain extent, also affected the practical effect of residential building energy efficiency.

From the technical application level, in order to achieve the building itself energy saving, passive designing should take advantages of the building's natural ventilation, natural lighting, shading, building envelope thermal insulation, solar energy utilization and other design measures to adjust the building thermal comfort as much as possible. Passive design can provide a good basis for efficient utilization of active technology to achieve the reduction of equipment and the overall energy efficiency of the building. Song Yehao proposed the concept of bio climatic buffer layer in combination with the comfortable climate of human body and put forward the strategy and method of the passive design of ecological architecture from the macro and micro level [11]. This kind of strategy provided a good theoretical foundation for passive energy-saving design of buildings.

From the demand of living environment, the acceptable limit is 28°C in summer and it is 16°C in winter. From the comfort temperature change, the transition season in hot summer and cold winter region provides a good condition for the passive building energy saving. In the process of research, Shanghai, Wuhan, Chongqing, Nanjing, Changsha are chosen for thermal environment comparison. From the enthalpy diagram of different cities, five cities are in the same hot summer and cold winter zone, and results show that the urban climate has different impact on the buildings under the same thermal comfort requirements of residential buildings. (Figure 1)For the thermal comfort of residential buildings and building energy efficiency, although the same is heating, cooling, ventilation and dehumidification requirements, the specific design requirements also reflect a certain degree of difference at different times. Therefore, there are some differences on passive technology in the application of residential buildings between different cities, and it is very important.

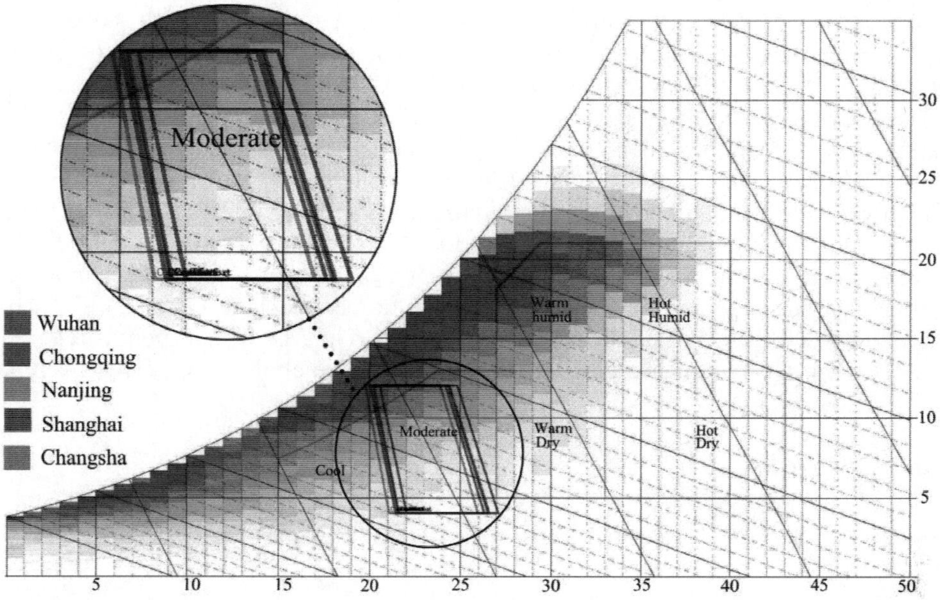


**Figure 1.** Enthalpy wet chart compares of thermal comfort range in five cities

## 3. Applicability passive technical analysis of residential buildings

Combined with Weather Tool calculation and analysis of climate analysis software, based on the passive technology of residential buildings in Shanghai, Wuhan, Chongqing, Nanjing, Changsha, the passive applicability is analysed. In this research, combined with ECOTECT quantitative data analysis results and proposed the concept of SD. It is the percentage improvement of the comfortable period after a certain type of passive technology is used,compared to the comfort period in the absence of any energy-saving technology. SD reflects the application of passive technology in buildings. In other words, the magnitude of the suitable degree reflects the amount of space available for passive technology.

### *3.1. Building ventilation influence*

For residential buildings, building ventilation can be achieved through two ways, natural ventilation and night ventilation.

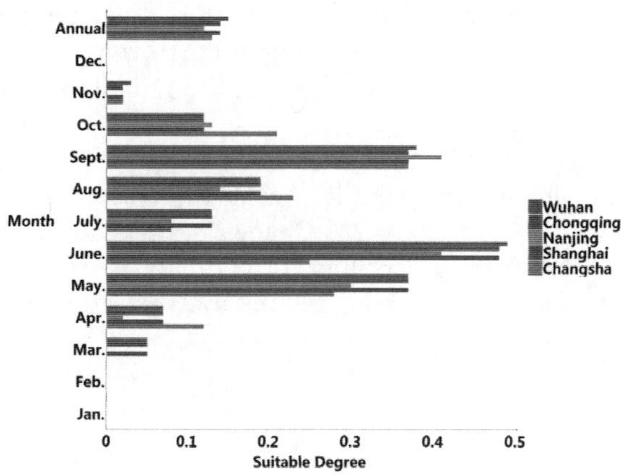


**Figure 2.** Comparison of natural ventilation

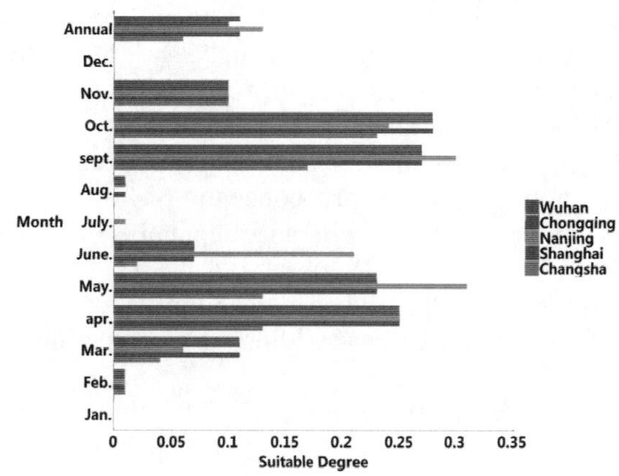


**Figure 3.** Night ventilation

From the comparative results, natural ventilation is an effective passive measures in Shanghai, Wuhan, Chongqing, Nanjing, Changsha, it can be used from April to October, concentrating in May, June and September. The suitable degree of Shanghai and Chongqing in May, June, September reached 0.37, 0.48, 0.37, SD of Wuhan reached 0.37, 0.49, 0.38. SD of Nanjing reached 0.30, 0.41, 0.41. SD of Changsha reached 0.28, 0.25, 0.37. Five cities in August were relatively high in temperature, but the applicability of natural ventilation was still large. Shanghai, Wuhan and Chongqing were 0.19, Nanjing was 0.14 and Changsha was 0.23.

Although the night ventilation has certain requirements on the temperature difference between day and night, It is of great help in the local time for the improvement of building thermal comfort. From the comparative results, night ventilation is concentrated in April, May, September and October. Shanghai, Wuhan, Chongqing four months of suitable degree reached 0.25, 0.23, 0.27, 0.28. The SD of four months reached 0.25, 0.31, 0.30, 0.24 in Nanjing. The SD of four months reached 0.13, 0.13, 0.17, 0.23 in Changsha. In other months, the local area of the application of night ventilation is also relatively high. For example, it reached 0.21 in June of Nanjing.

### 3.2. *Evaporative cooling*

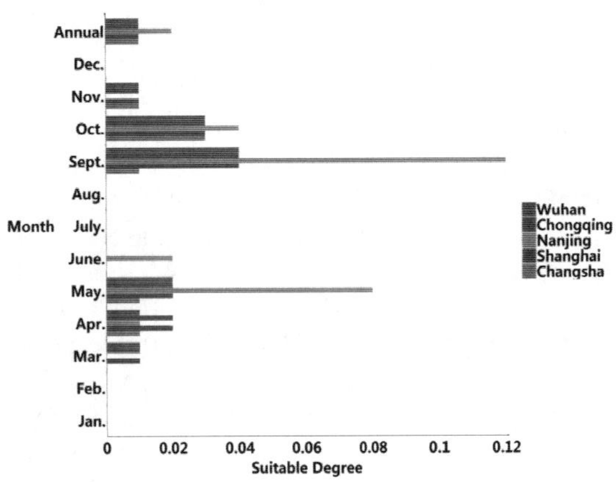


**Figure 4.** Direct evaporative cooling

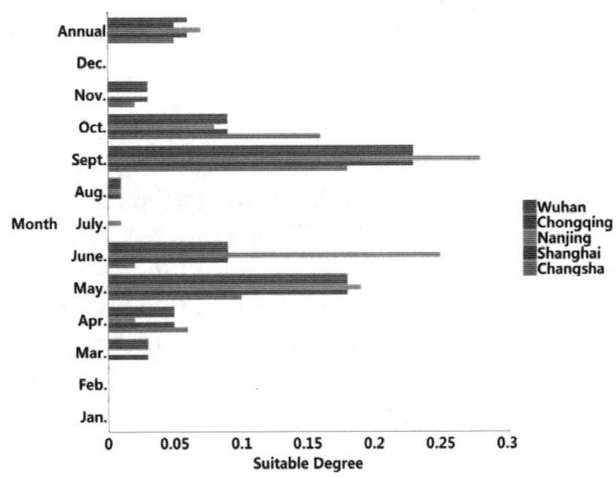


**Figure 5.** Indirect evaporative cooling

Evaporative cooling is a kind of economical and efficient cooling method, which uses the temperature difference between wet and dry bulb in the natural environment and the heat-moisture exchange. Building evaporative cooling can be realized by direct evaporation cooling and indirect evaporative cooling in two ways. From the analysis of the results, direct evaporative cooling effect of five cities in general, in addition to Nanjing in May, September for the degree of 0.08, 0.12, the remaining four cities have not exceeded 0.05.

Compared with direct evaporative, indirect evaporative cooling effect is obvious. Shanghai, Wuhan and Chongqing cities are concentrated in May and September, the SD was 0.18, 0.23. Nanjing concentrated in May, June and September, the SD was 0.19, 0.25, 0.28. Changsha concentrated in May, September and October, and the SD was 0.1, 0.18, 0.16. According to the analysis of SD, it could be seen that indirect evaporative cooling can be used to regulate thermal environmental conditions of residential buildings in some period.

### 3.3. *Solar radiation influence*

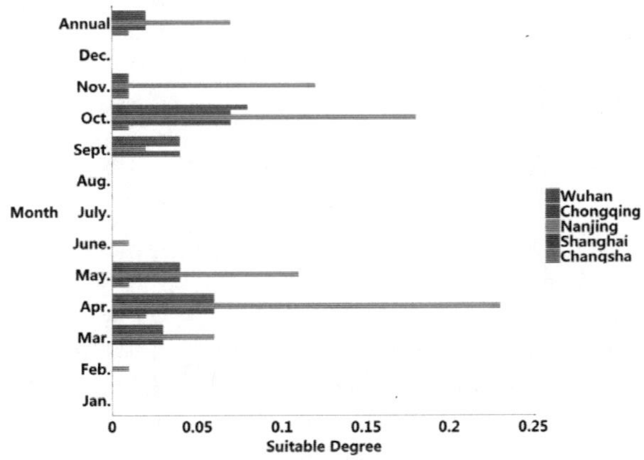


**Figure 6.** Passive solar house

Solar radiation directly affects the light and heat of residential buildings. From the passive use of speaking, the passive solar house is one of the common form. However, the data from the comparative analysis, five cities in June, July and August are not suitable for use. Relatively concentrated in March, April, May, September, October and November, the suitable degree was not high. Shanghai, Wuhan, Chongqing and Changsha suitable degree were below 0.05. However, there were some differences in Nanjing. April, May, October, November suitable degree were 0.23, 0.11, 0.18 and 0.12.

### *3.4. Envelope influence*

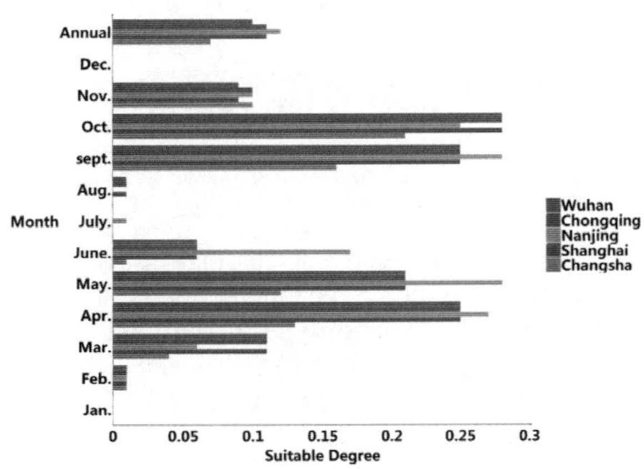


**Figure 7.** High melt materials

In residential buildings, the thermal performance of the envelope directly determines the thermal environment within the building, especially in extreme weather conditions in summer and winter. In the transitional season of hot summer and cold winter zone, the heat storage capacity of the envelope reflects the thermodynamic properties of the building to adjust the indoor thermal conditions. From the comparison results, the five cities in the envelope structure with high fusible materials to adjust the building comfort, more concentrated in April, May, September and October. In these four months, Shanghai, Wuhan, and Chongqing's SD respectively were 0.25, 0.21, 0.25, 0.28. The SD of Nanjing were 0.27, 0.28, 0.28, 0.25. The SD of Changsha were 0.13, 0.12, 0.16, 0.21.

## 4. Discussion

(1)From SD analysis passive technology it could be seen that the size of SD are all within 0.5. Combined with the comparative analysis of related studies on passive design technology, passive technology embodied in many aspects, such as orientation, sunshade, envelope thermal performance, etc. And it usually focused on the building thermal physical demand, such as ventilation, lighting, power consumption, radiation, and so on. From above SD analysis, applicability order for passive technology was ventilation, envelope, evaporative cooling and solar radiation in hot summer and cold winter zone in China.

(2)According to indoor thermal environment changes of residential buildings impacted by annual climate characteristics in hot summer and cold winter zone, it could be roughly divided into five periods, such as heating period, ventilation period, dehumidifying period, cooling dehumidifying period and ventilation period from January To December. From SD analysis of passive design in five cities, it could be seen that application period of passive technology was different in different location. In order to meet the requirements of indoor thermal comfort of residential building, it could be adjusted by heating equipment in heating period, and by refrigeration equipment in cooling dehumidifying period, by ventilation equipment in dehumidifying period. And natural ventilation technology could be used in ventilation period. The application period of five cities could be compared in Figure 8, which combined with SD analysis and China building thermal design of typical meteorological reference data.

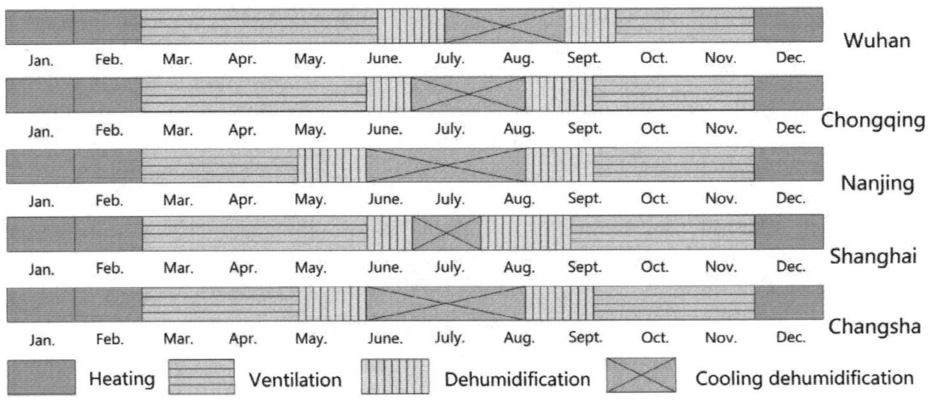


**Figure 8.** Diagram of passive adaptive technology comparative analysis

From Figure 8, it could be seen that deviation of heating period is small in five cities. However, the cycle length was different in wind period, dehumidifying period and cooling dehumidifying period. Cooling dehumidifying period of five cities were all in July, August and September. It could reach 688 hours in Wuhan, 479 hours in Nanjing, 458 hours in Changsha, 465 hours in Chongqing, and 251 hours in Shanghai. It could be affected by the marine climate in sometime in Shanghai, therefore, the cooling dehumidifying period was relatively short and the dehumidifying period was long. From the contrast of ventilation period in Fig.6, it could be seen that cycle length was Shanghai＞Wuhan＞Chongqing＞Nanjing＞Changsha, which directly affected the efficiency of natural ventilation.

(3)From the five cities comparative analysis of passive technology, It could been seen from that natural ventilation is an effective passive way to promote building energy efficiency in hot summer and cold winter zone in China, which could be accounted for over 40% of annual session. And natural ventilation could adjust the internal environment of residential buildings, when the outdoor temperature is less than 30°C, relative humidity is not more than 85% in summer [12]. When the outdoor temperature changes greatly, it could be adjusted by natural ventilation or intermittent ventilation cooling at night. Furthermore, comfort could be regulated by the combination of initiative and passive hybrid ventilation mode during dehumidifying period and cooling dehumidifying period.

(4)As an important factor, solar radiation should be paid attention for affecting the thermal environment of the building in hot-summer and cold-winter zone. Shading is an effective passive technology to regulate thermal radiation effects, and it could effective improve the thermal environment, reduce air conditioning usage cycle, from the Angle of the roof, wall and window shade combined with different cities high angle change.

In addition, it should consider to pay attention to comprehensive utilization of sunshade and natural ventilation, combined with residential building demand. In summer, it could improve comfort by sunshade and natural ventilation during day and by night ventilation in night. In winter, it should consider the use of natural light and heat by passive solar house.

(5) It is a great significance to use reasonable energy storage palisade structure for improving the thermal comfort of indoor environment and realizing building energy efficiency. In part of residential building palisade structure of hot summer and cold winter zone, PCM (Phase Change Materials) could be applied to passive solar houses, combined with ventilation cooling storage in summer night and heating storage in winter [13]. In addition, it could strengthen the natural ventilation with TBW (Trombe wall) in the residential building to improve heating effect in winter [14][15][16].

## 5. Conclusion

The urban climate characteristics and residents living habits directly impact the residential building energy consumption in hot summer and cold winter zone in China. Passive design technology provides better autonomy and flexibility for residential building energy saving, such as natural ventilation, lighting, heat insulation and so on. Appropriate passive technology should have good balance, coordination and unity, to meet the thermal comfort and energy efficiency as the goal, fully combined with systemic quantitative control of regional climate characteristics. Combining Weather Tool data analysis, this paper discussed the passive technology applicability of residential building in different cities in hot summer and cold winter zone in China, and passive energy saving design ideas and methods were provided.

## Acknowledgement

This research was supported by National Natural Science Foundation of China (NO51508169), China Postdoctoral Science Foundation (NO2013M531697) and National Natural Science Foundation of Hubei Provience (NO 2014CFB475).

## References

[1] Zhang X and Li K S 2008 Sensitivity analysis of energy consumption for residential buildings in hot summer and cold winter zone *HV&AC* **38**(7) 10-14

[2] Fu H, Gong Y F and Yu X E 2012 Effects of external wall shading on building's thermal environment and air conditioning energy consumption in hot summer and cold winter area *Architecture Technology* **43**(1) 67-70

[3] Fu H, Gong Y F and Xu J F 2010 Influence of building shape coefficient on energy consumption of residential building in hot summer and cold winter areas *New Building Materials* **37**(01) 44-47

[4] Fu H, Gong Y F and Yu X E 2013 Natural ventilation influencing residential building's thermal environment and energy consumption in hot summer and cold winter area.*Building Energy Efficiency* **41**(08) 21-28

[5] Yu J H and Tian L W 2012 Analysis on optimum insulation thickness of residential roof in hot summer and cold winter zone *Journal of Central South University (Science and Technology)* **43**(04) 1545-50

[6] Yu J H, Yang C Z and Tian L W 2009 Research on the optimum insulation thickness of the external wall for residential buildings in Changsha region *Journal of Hunan University (Natural Sciences)* **36**(09) 16-21

[7] Jin S K and Gong Y F 2011 Energy consumption model and 65% energy conservation measures for typical residential buildings in Nanjing *HV&AC* **41**(02) 94-98

[8] Li B,Yang Z J and Li J F 2009 Investigation on the current situation of existing housing architecture in Hubei province and the energy saving transformation strategies *Journal of Anhui Agricultural Sciences* **37**(11)5243-45

[9] Yu X P, Fu X Z and Liao X F 2008 Technical solutions for low energy consumption residences in subtropical regions *Journal of Chongqing Jianzhu University* **30**(06) 116-119

[10] Zhang H and Zhou X 2015 Analysis of passive energy-saving technology on residential building in hot summer and cold winter zone *Industrial Construction* **2015**(45) 125-128

[11] Song Y H, Wang J L and Zhu N 2013 Pondering over the passive design strategy for native green buildings of China *Architectural Journal* (07) 94-99

[12] Fu X Z and Sun C J 2009 Train of thought and technological system of energy efficiency in residential buildings in Yangtze basin *HV&AC* **39**(10) 69-73

[13] Li N, Yang L and Liu Y F 2014 Analysis of energy saving effect for passive cooling in residential building *Architecture Technology* **45**(08) 743-746

[14] Cheng R, Wang X and Zhang Y P 2012 New concepts and approaches to develop a sustainable building envelope *South Architecture* (1) 86-91

[15] Chen J and Wang H Q 2014 Numerical simulation of solar enhancing natural ventilation in hot summer and cold winter zone *HV&AC* **44**(06) 38-43

[16] Bassiouny R and Koura N S 2008 An analytical and numerical study of solar chimney use for room natural ventilation *Energy and Buildings* **40**(5) 865-873

IOP Conf. Series: Earth and Environmental Science **61** (2017) 012067 doi:10.1088/1755-1315/61/1/012067

# Application of HEC-RAS for flood forecasting in perched river–A case study of hilly region, China

**Pingping Sun [1,3], Shuqian Wang [1], Hong Gan [2,3], Bin Liu [1,2], Ling Jia [3]**

[1] Hebei University of Engineering, Handan 056021, China
[2] State Key Laboratory of Simulation and Regulation of Water Cycle in River Basin, China Institute of Water Resources and Hydropower Research, Beijing 100038, China
[3] China Institute of Water Resources and Hydropower Research, Beijing 100038, China

**Abstract.** Flooding in small and medium rivers are seriously threatening the safety of human beings' life and property. The simulation forecasting of the river flood and bank risk in hilly region has gradually become a hotspot. At present, there are few studies on the simulation of hilly perched river, especially in the case of lacking section flow data. And the method of how to determine the position of the levee breach along the river bank is not much enough. Based on the characteristics of the sections in hilly perched river, an attempt is applied in this paper which establishes the correlation between the flow profile computed by HEC-RAS model and the river bank. A hilly perched river in Lingshi County, Shanxi Province of China, is taken as the study object, the levee breach positions along the bank are simulated under four different design storm. The results show that the flood control standard of upper reach is high, which can withstand the design storm of 100 years. The current standard of lower reach is low, which is the flooding channel with high frequency. As the standard of current channel between the 2rd and the 11th section is low, levee along that channel of the river bank is considered to be heighten and reinforced. The study results can provide some technical support for flood proofing in hilly region and some reference for the reinforcement of river bank.

## 1. Introduction

In recent years, frequent flood disasters caused a great threat to people's life and property in China, the situation affected by this is the most particularly serious especially in 2010, the death toll caused by the flood disaster as high as 2824[1]. In the condition of strong rainfall, the water level of the river and creek in hilly region rises rapidly, which leads to the levee breach easily. A large collection of storm water and flood in the upstream of the high incidence area of mountain torrents, Often make the downstream river facing the serious problem of the levee breach, especially the river or reach relied on the dike to constraint the flow[2]. Such as the perched river, whose river bed or levee above the bank ground, once the river burst its banks will form the new waterlogging area, and then lead to the tragedy of river diversion, flood land, highway interrupted and real pour news. So that the perched river has greatly increased the danger of mountain flood disaster on both sides of the river valley[3]. In view of the above problems and challenges, the river flood simulation and risk prediction in hilly and mountainous regions have gradually become a research hotspot. The ascertain of flow profile is the basis of the analysis of flood routing, and the research on the application of HEC-RAS (Hydrologic Engineering Center - River Analysis System) model to simulate the water surface is very extensive both at home and abroad. Mohsin Jamil et al. constructed the HEC-RAS model in the northern mountain area of Pakistan, and analyzed the dam break risk based on the predicted flow profile[4]; Good results were obtained from the application process of this model in the natural river flow profile calculation both in plain terrain and

IOP Conf. Series: Earth and Environmental Science **61** (2017) 012067 doi:10.1088/1755-1315/61/1/012067

hilly region[5],[6]; Liu Yang compared the HEC-RAS and SOBEK-RURAL model results of water level and velocity in Beijing Huairou District Park River flow, and verified the reliability and rationality of the HEC-RAS application in the natural river flow profile[7][8]; Dai Wenhong compared the simulation results from the HEC-RAS and MIKE11 which about the variation in water level of the Gan River Delta to Nanchang section, and presented a parameter calibration method to improve the simulation precision of the model[9].But the present research shows that the simulation for flood process in hilly perched river based on the HEC-RAS model is relatively less, and the experience and method for the simulation of flood level along the coastline is insufficient. Based on the section characteristics of the hilly perched river, this paper tries to establish the relation between the river shoreline and the flow profile ascertained by the HEC-RAS, and a torrent ditch in Lingshi County of Shanxi Province is taken as a research object, and the levee breach positions along the perched river is analysed under the different scenarios of design storm. The result is expected to provide technical support for flood control and early warning, and provide a reference for the reinforcement of shoreline.

## 2. Study Reach

The Jiaozhong Ditch is located in the territory of Lingshi County of Shanxi Province, southern Jinzhong basin, originated from Zhongshu of Jing Sheng Town in Lingshi County, the three tributaries of the upper reaches flow through the willow village, the Songjiashan village and other places, and eventually flow into the Jingsheng river in Suxi village. The length of the Jiaozhong Ditch is 7.7km, the drainage basin area is 20.5km$^2$ and the slope of the ditch is 19.4‰. In the ground covered with loess, due to frequent flood season rainstorm, strong erosion, there are serious soil erosion, high mountains and steep slopes. Suxi is located in the right of the Jiaozhong Ditch downstream, with the accumulation of long-term sediment and dike reinforcement, the majority residential elevations in Suxi are significantly lower than that of the right bank of the ditch, and formed a huge steep slope. So it is a typical perched river in hilly region, the typical cross section form shown in Figure 1. Affected by the monsoon climate, the region often have "high intensity and large amount" rainstorm occured from July to August[10]. According to the survey, in 1981 the flood caused by the heavy rain broke out of the embankment, submerged most of the Suxi area, the maximum water depth was more than 1m, so that the village residents suffered a huge loss of property. Therefore in order to carry out the corresponding measures for river regulation, it is necessary to understand the flood control capacity and determine the levee breach positions.

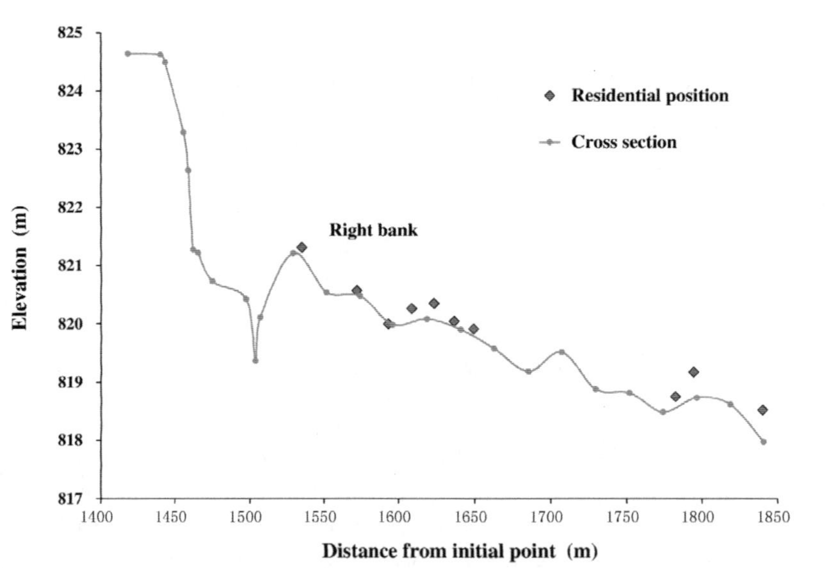


**Figure 1.** Channel pattern of No.6 section in lower reach.

### 3. Geometric and Hydrologic Data

With the assistance of the local water conservancy department, the digitization of the control sections is realized through the recent monitoring data of the river channel. Based on the 30×30m DEM (Digital Elevation Model) data in this study region, river network data is generated with ArcGIS 10.2, and the information such as river central line and catchment area are obtained. After comparing the measured data and the results calculated from ArcGIS 10.2, it shows that the actual river central line is basically consistent, which can be used as the main channel in this study. Combined with 1:20000 high-definition image and site investigation, the work on selection, complement, correction and testing of section control points is completed. The Jiaozhong Ditch flows from north to south. The upper and middle reaches of the selected ditch are in an arc shape, and the cross sections are wide and deep. However the channel in lower reach is straight, the river sections are narrow and shallow. The selected average distance is about 60m each section and 16 cross sections are finally determined. The distribution of river control sections is shown in Figure 2.

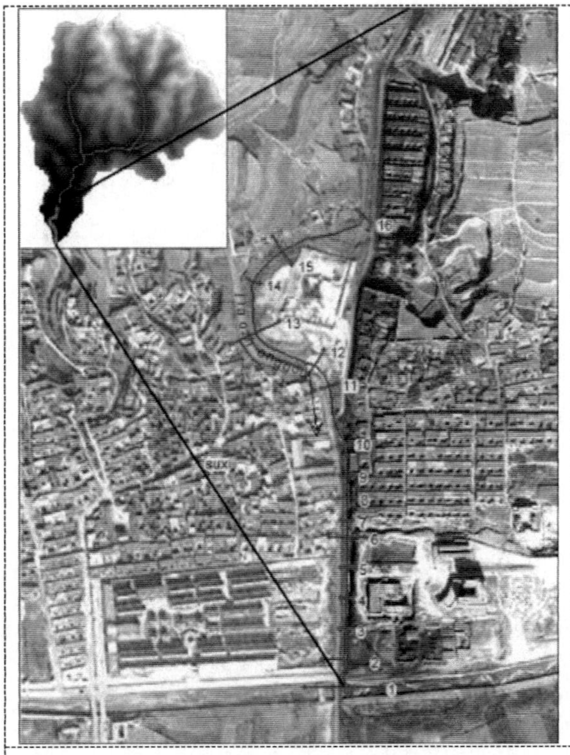

**Figure 2.** Distribution of river control sections.

Hydrological Handbook of Shanxi Province (hereinafter referred to as the Handbook) is regarded as a hydrologic instruction book based on local hydrogeological conditions, and the recommended methods from it have been validated with a certain authority[11][12]. Therefore, the design storm is calculated with the Rainstorm Atlas Method. The design storm duration is selected as 10min, 60min, 6h, 24h and 3D (72h), and the return period is 5 years, 10 years, 20 years, 50 years and 100 years. The mean value, the coefficient of variation (Cv), the ratio of the coefficient of skewness and the coefficient of variation (Cs/Cv), the modulus coefficient (Kp) and the point-surface reduction coefficient ($\eta$) of the fixed-point torrential rain in the study area are determined by the isoline map of rainstorm statistical parameters, and then the results of design storm and its time distribution could be obtained, as shown in Table 1. According to the watershed model method recommended by the handbook, by means of the runoff generation calculation about the net rainfall depth and process, and the flow concentration calculation using the integrated instantaneous unit hydrograph method, the design flood at the same frequency is finally obtained, the flood factors are shown in Table 2. By comparison, the results are basically consistent with the results of the analysis and evaluation of local flood disaster, the peak flow derived by this method can be used as the input data of HEC-RAS model.

**Table 1.** Results of design storm.

| Duration | Mean value (mm) | $C_v$ | $C_s/C_v$ | Precipitation (mm) | | | | |
|---|---|---|---|---|---|---|---|---|
| | | | | 100 years ($H_{1\%}$) | 50 years ($H_{2\%}$) | 20 years ($H_{5\%}$) | 10 years ($H_{10\%}$) | 5 years ($H_{20\%}$) |
| 10min | 12.0 | 0.55 | 3.5 | 31.2 | 27.3 | 22.2 | 18.3 | 14.3 |
| 60min | 30.0 | 0.49 | 3.5 | 65.4 | 57.9 | 47.9 | 40.3 | 32.4 |
| 6h | 44.0 | 0.46 | 3.5 | 111.6 | 99.6 | 83.4 | 70.9 | 58.0 |
| 24h | 61.0 | 0.45 | 3.5 | 143.0 | 127.8 | 107.3 | 91.4 | 74.9 |
| 3d | 77.0 | 0.39 | 3.5 | 169.8 | 153.7 | 131.8 | 114.6 | 96.4 |

**Table 2.** Results of design flood.

| Return period / Flood element | 100 years | 50 years | 20 years | 10 years | 5 years |
|---|---|---|---|---|---|
| Peak discharge ($m^3/s$) | 159 | 132 | 96.4 | 70.2 | 48.1 |
| Flood volume ($m^3$) | 142 | 117 | 86 | 63 | 44 |
| Flood duration (h) | 15 | 15 | 15.5 | 15.5 | 15.5 |
| Peak stage (m) | 824.16 | 823.98 | 823.70 | 823.46 | 823.20 |

## 4. Calibration of HEC-RAS Model

Due to the left bank of the perched river is higher than the right bank about 4m, the right bank residents are frequently affected severely in the historical torrential flood disaster. The main difficulty in the process of calibration is the lack of historical monitoring data in perched river, such as the river flow profile in middle and small floods in recent years. In view of the challenge for ungauged condition, considering the fact that the natural river can be generalized into open channel to simulate the flood process in SWMM (Storm Water Management Model), a method is attempted to apply in this paper that SWMM and HEC-RAS models are be used to determine the levee breach positions and simulate the maximum water depth in a precipitation condition of 5 years return period. Combined with the investigations of the history about torrential flood disaster, site surveys, visiting the nearby villagers and some other ways, the main parameters calibration for HEC-RAS are completed[13],[14]. The perched river drop is finally determined as 13 ‰, and the manning coefficient is determined as 0.03. Simulation results from SWMM and HEC-RAS are shown in Table 2. for the 5 years return period flood, the positions near the 4th and 5th section are both determined to be the levee breach. Basically same results are computed from the two models except a few of cross sections. It can be considered that the model has a certain degree of simulation accuracy and the credibility of flow profile calculation. There are 81.3% of the cross-sectional simulation results with a relative error of less than 10%. In addition, 68.8% simulation results from HEC-RAS are slightly larger than that from SWMM in the maximum water depth of each section, which has a positive effect on river flood warning and levee reinforcement.

**Table 3** Comparison of simulation results between SWMM and HEC-RAS.

| Section number | Levee height | Computed depth (m) | | Relative error (%) |
|---|---|---|---|---|
| | | SWMM | HEC-RAS | |
| 1 | 3.59 | 1.28 | 1.43 | 11.9 |
| 2 | 3.55 | 1.67 | 1.73 | 3.5 |
| 3 | 2.29 | 2.15 | 2.00 | -6.8 |
| 4 | 1.79 | 1.97 | 2.09 | 5.9 |
| 5 | 1.11 | 1.93 | 2.19 | 13.3 |
| 6 | 1.83 | 1.86 | 1.80 | -3.2 |
| 7 | 2.25 | 2.13 | 1.86 | -12.5 |
| 8 | 2.26 | 2.14 | 2.06 | -3.8 |
| 9 | 2.46 | 1.63 | 1.73 | 6.4 |
| 10 | 3.03 | 2.07 | 2.18 | 5.3 |
| 11 | 3.97 | 1.88 | 2.01 | 7.0 |
| 12 | 3.81 | 1.28 | 1.36 | 5.9 |
| 13 | 5.72 | 1.75 | 1.84 | 5.1 |
| 14 | 8.39 | 1.78 | 1.77 | -0.6 |
| 15 | 9.07 | 1.29 | 1.39 | 7.7 |
| 16 | 4.90 | 1.56 | 1.62 | 3.8 |

## 5. Results Analysis

The peak discharges corresponding to the design rainstorm with the return period of 10 years, 20 years, 50 years and 100 years are taken as the precipitation input scenario of the HEC-RAS model, and then the simulation results of the maximum flood stage of each section can be obtained. The simulation results of 16 sections are connected by a smooth curve to establish the correlation between the right bank and the maximum flood level simulated by HEC-RAS, as shown in Fig.3.

The section number of the river section is decreasing from upper reach to lower reach. In this study, the 1st～8th sections are defined as the lower reach section, and the 9th～16th sections are defined as the upper reach section. Obviously, the average elevation of the upper reach channel of the river is much higher than that of the lower reach. The maximum difference in levels between the 15th section and the 5th section is about 15.55m. On the whole, the maximum flood levels from design storm of 100 years in upper reach section are located below the right bank of the ditch, so the levee breach possibility is very less. For lower reach section, with the decrease of the designed storm frequency and the increase of the peak discharge, the positions of the levee breach gradually moves upwards, and the length of submerged channel increases as well as.

Specifically, as for the design storm of 10 years, the higher risk bank of flooding in current situation of the river is between the 3rd～8th section (the 8th section is the initial point); As for the design storm of 20 years, the sections with flooded risk spread to the position near the 3rd～9th section; As for the design storm of 50 years, the current high-risk river bank is continued to be lengthen, levee breach section in upper reach moves to near the 10th section, the lower reach high-risk section moves to near the 3rd section downward for 15m; As for the design storm of 100 years, the levee breach position moves up to about 30m above the 10th section in upper reach, the lower reach's moves downward 25m below the 3rd section. For the status of the river, the flood control standards between the 2nd and 11th section are low, this channel of the perched river should be heightened and reinforced. Considering the right bank residents are more than the left banks, the levee near the 3rd section should be increased by 1.5～2m, and the levee near the 10th section should be increased by 1m.

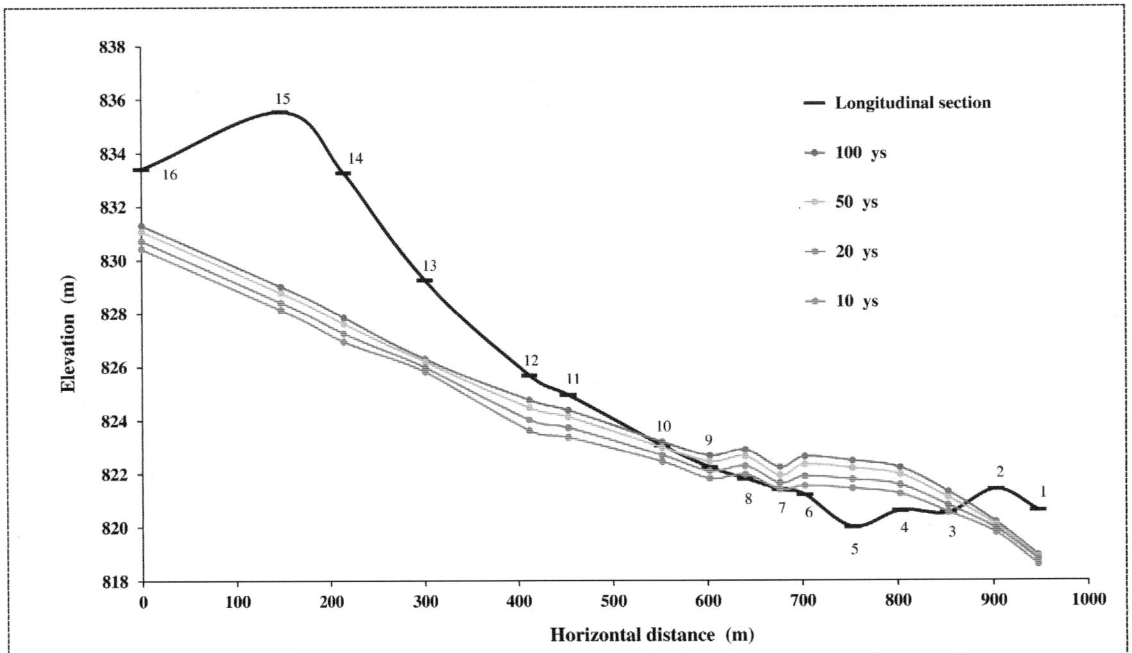


**Figure 3.** Distribution of maximum flood stage in different design storm.

## 6. Conclusion

Parameters calibration for HEC-RAS is completed by comparing the simulation results from SWMM and HEC-RAS in this paper. The flow profiles of 16 sections in different design storm of 4 return periods (10 years, 20 years, 50 years and 100 years) are simulated with HEC-RAS model in perched river, which are established the relationships with the right shore line of the ditch. The distribution of the location of the levee breach positions along the river was presented. On the other hand, the levee breach positions computed from HEC-RAS model are basically the same as the investigation results of the historical floods in this perched river, therefore the simulation results are authentic and credible.

The simulation results show that the flood control standard of the upper reach is high, which can resist the design storm of 100 years. The standard of the lower reach is low, which is the flooding channel with high frequency. The current levee sections from 3rd to 8th are not enough to withstand the design storm of 10 years, the embankment between the 2rd and 11th sections is suggested to build and fill higher. This part of perched river should be focused on inspecting during flood season, which can prevent life and property safety of residents along the river from the flood caused by levee breach. The study results can provide the basis for the suspension of flood warning, the consolidation of river bank and the development of disaster evacuation plan.

## Acknowledgments

The researchers would like to extend their thanks to the Chinese National Natural Science Foundation (No.71573274, No.41401648, No.51279210) and the Chinese National Social Science Fund Project (No. 15ZDB160).

## References

[1] State Flood Control and Drought Relief Headquarters and Ministry of Water Resources of the People's Republic of China 2015 *Chinese Flood and Drought Disaster Bulletin of 2014* (Beijing: China Water & Power Press)

[2] Lexiang Q Wantong W and Shuang L 2005 A review of the study on the problem of *perched river on the ground* in the Yellow River *Yellow River* vol 27 (Zhengzhou: Chinese core journals) pp 1–6

[3] Qianqing X 1999 A Thought on Flood-Control, Disaster Mitigation and Countermeasures in China *Advances in Water Science* vol 10 (Nanjing: American Elsevier) pp 235–241
[4] Mohsin J B Muhammad U and Raheel Q 2013 Landslide dam and subsequent dam-break flood estimation using HEC-RAS model in Northern Pakistan *Natural Hazards* vol 65 pp 241–254
[5] Imbenzi J S Benedict M M and James M R 2014 A review for hydraulic analysis of irrigation canals using HEC-RAS model: A case study of Mwea irrigation scheme, Kenya *Hydrology* vol 2 (American Elsevier) pp 1–5
[6] Thomas A Sherry C Raymond D, et al 2010 The Ohio river community HEC-RAS model *World Environmental & Water Resources Congress* (American Elsevier) pp 1512–1523
[7] Yang L Xiaoying S and Junying W, et al 2008 Calculation for natural river surface in hilly area using HEC-RAS and SOBEK-RURAL software *Beijing Water Affairs* vol 6 (Beijing: Chinese Journal) pp 34–36
[8] Yang L 2011 Comparison of several methods of calculating water surface *Yellow River* vol 33 (Zhengzhou: Chinese core journals) pp 51–53
[9] Wenhong D Song G and Yun Z, et al 2011 Comparison of Manning roughness factor simulation by HEC-RAS and MIKE11 models *Journal of Sediment Research* vol 12 pp 41–45
[10] Council of Lingshi County, Shanxi Province Compiled Compilation 1992 *Lingshi County* ed Fachang C (Beijing: China Social Press)
[11] Jinzhong City Hydrology and Water Resources Survey Branch 2015 *Lingshi County Mountain Flood Disaster Assessment and Evaluation Report*
[12] Shanxi Province Water Resources Department 2011 *Hydrological Computing Manual in Shanxi Province* (Henan: Yellow River Water Resources Publishing House)
[13] Prafulkumar V T Prem L P and Prakash D P 2011 Calibration of HEC-RAS model on prediction of flood for lower Tapi River, India *Journal of Water Resource and Protection* vol 3 (American Elsevier) pp 805–811
[14] Prabeer K P 2013 HEC-RAS model for mannnig's roughness: a case study *Open Journal of Modern Hydrology* vol 3 (American Elsevier) pp 97–101

IOP Conf. Series: Earth and Environmental Science **61** (2017) 012068 doi:10.1088/1755-1315/61/1/012068

# Research on spatial features of streets under the influence of immersion communication technology brought by new media

**Hua-wei Xu, Chen Feng**

Faculty of Architecture and Urban Planning of Chongqing University, Chongqing, China

147455771@qq.com

**Abstract**. The rapid development of new media has exacerbated the complexity of urban street space's information interaction. With the influence of the immersion communication, the streetscape has constructed a special scene like 'media convergence', which has brought a huge challenge for maintaining the urban streetscape order. The Spatial Visual Communication Research Method which should break the limitation of the traditional aesthetic space research, can provide a brand new prospect for this phenomenon research. This study aims to analyze and summarize the communication characteristics of new media and its context, which will be helpful for understanding the social meaning within the order change of the street's spatial and physical environment.

## 1. Introduction

The world is entering a new age permeated by information and network. The evolution of information technology greatly facilitates the development of media. Information and network has become the prevalent technological paradigm in modern society.

New media[1)] is a relative and unceasingly developing concept. For the moment, it mainly refers to the information communication in the support of multi-media technology, mass data and network, which breaks the barriers of time and space. Compared with traditional media, new media is featured by mobility, interactivity and integration. Its development has subverted traditional concept of space and has exerted a profound influence on design concepts and methods. Pure functionalism and aesthetic order no longer work in providing a clear illustration on the dramatic change of street scenes. Thus it's of great necessity to introduce the topic of "new media" into the traditional dimension of city image and to reexamine this phenomenon.

## 2. Research on the Characteristics of Immersion Communication

In the support of new media technology, the modern media offers a brand-new platform for interpersonal contact, bringing about dramatic revolution. "The second media age" [2)] proposed by Mark Poster can no longer accurately define the present spreading mode. New media has formed a dynamic information communication mode, a ubiquitous and unprecedented "immersion communication mode" that operates by unceasingly blending new technology and imaginary space. Media and environment have integrated with each other, forming a new information communication mode that has penetrated into our life[1].

### *2.1. Dynamic communication and Multipoint Information Exchange*

Traditional media mainly adopts linear static scenes as the narration mode. But the overwhelming emergence of new media brings about a revolutionary transformation to the way of information communication; information carriers featured by rigidity, accuracy and authority are undergoing a continuous evolution towards a mode of greater mobility. New media breaks the traditional one-way information communication, the conventional "one-point to multi-point" mode is beginning to disintegrate; the centralized structure has transitioned to a decentralized open-ended one, and continues its way to a multi-point to multi-point reticular communication mode. Audiences can search, sort and customize media information as they want, which marks a transformation from passive receiving to active acquisition, and the communication environment is shifting to a customized one.

### *2.2. Optical Space and Virtual Information Environment*

Paul Virilio proposed the concept of "optical space" long ago, which is a perspective technique in the field of optoelectronics brought by videos, marking a breakthrough of geometric perspective in the general sense[2].New media takes advantage of modern technology including information network, modern digital image and smart mobile terminal, bringing information receivers a special experience with its various features of virtual scene, augmented reality, mixed reality and so on. This accessible experience is characterized by strong infection and is changing the way we receive information.

Virtual reality[3)] contains interactive three-dimensional dynamic vision featured by multi-source information fusion, offering users access to the given environment. Mix reality technology creates new visual environment by combining real and virtual world, where physical and digital objects can exist side by side and make real-time interaction. Augmented reality applies virtual information into a real world, where real environment and virtual objects exist simultaneously through superposition in the same scene or space.

## 3. Street Scenes under the Influence of Immersion Communication

With new media applied in streetscape, the visual system of street scenes is featured by increasing openness, information pluralism and complexity of value orientation. As the traditional street aesthetics is stripped of its mystery, traditional mode of one-way narration is transformed into an interactive narration between men and space, and "Aesthetics of Engagement" [4)] brought up by Arnold Berleant has become the dominant cultural context in street space.

### *3.1. Self-media Communication and Personal Show*

Nicholas Negroponte argues that "in the digital age, I am what I am, rather than a data in the demographic statistics." [3] In the self-media age[5)], the dissemination of information is featured by full participation; individual-centered media has long since walked into everyone's daily life, publicizing personal affairs.

Street artists project thought-provoking images onto the walls, and street scenes become the tool for anti-hegemony. "Traditional order" collapses in front of absurd, game of space and individual ecstasy brings about self-indulgence. As proposed by French sociologist Michel Maffesoi about Le tribalismele and Nomadisme[4]in the digital age, when the mass own rights of "encoding", they are free to take on a flight of imagination in the street scenes like heroes, realizing visual decoding of personal fulfillment in the struggle between graphic signs and spatial ontology.

### *3.2. Decorated Cities and Wanderers' Participation*

Harold Adams Innis and Marshall McLuhan suppose that media is also an environment. Media together with places constructs the way people contact each other, and the rapid mobility of information blurs the limitation of physical places and media places. Since the moment when the equipment of new media is installed in public places, it has become a platform for cultural medium and communication as well as a ceremony in our daily life, fulfilling the tasks of cultural spreading and cultural integration.

The information flow of media breaks through the geographical barriers, thus exotic film signals frequently enter the urban daily life, generating new social scenes. The landscape constructed by new media makes everyplace platforms, which radically deconstructs and subverts the traditional narration of logic and time. In the street campaign, "Europe is just next door", launched by SNCF and TBWA,

IOP Conf. Series: Earth and Environmental Science **61** (2017) 012068 doi:10.1088/1755-1315/61/1/012068

several dimension doors are set in streets of Paris with names of some European cities marked on. These doors will show you real-time exotic scenery shot by remote cameras when open them.

### *3.3. Recombination of Street Space in Screen Window*

New media accelerates the reshaping of new perceptual models. It has become a noticeable trend in the development of modern urban landscape that electronic screens are removed to urban open areas, while constructions in cities perfectly function as screens, redecorating the public space of streets, promoting the generation of new forms and new places of social activities. As Paul Virilio believes, this is a "pervasive architecture-style" [5]; dazzling outer screens, like the windows in "Electronic Gothic" churches, discreetly grasping the morals and thoughts of urban citizens.

Such an experience-based landscape promotes the integration of men with scenes in street space by highlighting the positive perception of humans. For example, on the glass tower of Crown Fountain in Chicago, US, are played 1000 smile faces of Chicagoans. Lummo Blocks from Lummo project tetris onto the giant screens, and the public are free to overturn tetris by their movements in the square. Nina Valkanova, an interactive designer from Barcelona, Span, together with architect MoritzBehrens, works out Smart Citizen Sentiment Dashboard, which provides a tool for urban dwellers to express their emotions[6].

## 4. Trend of Immersion Communication Impacting Street Space

A high-degree interaction is created through the mutual effect between physical environment, digital environment and street campaigns, namely "intermediary interaction" [7] coined by John B. Thompson. This kind of interaction dramatically overturns the traditional urban cognitive structure, obviously bringing tremendous shock to traditional street landscape. On one hand, it breaks the existing barriers, investing new vigor into street space; on the other hand, it aggravates the risks of fragmentation of street space.

### *4.1. Information Integration Fuels the Media Convergence of Street Interface*

With the constant emergence of new technology and new materials, artist forms with more freedom and pluralism featured by patterning, integration, medium, complexity spring up, which deviate from the bondage of mechanics and functionality. Super Flat, Media Architecture, Printed Building and Sculptural Architecture rise to prominence in architectural practices.

Under the influence of interactive spreading brought by new media, streetscape take on a tendency of multifunction and integration, Media Convergence[6)] gradually coming into shape. Everything in the street, diversified scenes including featured landscape, outdoor screens and behavioral pattern are endowed with media attributes of more abundance. New materials and microelectronic technology born with the new age, together with projection facades, window raster animation, display facades and voxel facades influence work together to display the characteristics of street in a more prominent and complicated way.

### *4.2. Street Flowing Scenes Created by Visual Image*

The interactive experience in street scenes accelerates the information communication, and the transient time-space relationship is regarded as the main characteristic of post-modern urban scenes. Inextricable components begin to be movable, traditional structure of street landscape formed by points, lines and surfaces collapses and the surface has transformed from solidity to a flowing data pool. "It looks like we have entered a new metropolitan super space, namely City Of Bits, E-tuopia and The Cyborg Self and the Networked City" [7)].

The high-speed information mobility brought by new media fiercely impacts people's load-carrying limit. Messages get replaced so quickly that they can't even show up in front of people's eyes. Streets have become a flowing network of visual transformation, dotted by electronic footprints and scattered data collection. Manuel Castells points out that the space of flows centered with social practices has gradually replaced traditional space and created "Stream City" [8].

IOP Conf. Series: Earth and Environmental Science **61** (2017) 012068 doi:10.1088/1755-1315/61/1/012068

### *4.3. Collage Breaks the Visual Centralization of Street Space*

Non-linear network structure of new media further aggravates the collapse of traditional elite values. Self-discipline of elite aesthetic culture as well as lofty mask of aesthetics gradually erodes. Aesthetics is no longer the paradigm of value education in the traditional sense, but has sneaked its way to everyone's daily life.

Aesthetic culture in the new media age walks out of pure aesthetics. As media technology develops, aesthetic cultural products get dramatically enriched in their number through bulk copy, and the coding of aesthetic culture gains popularity and simplicity. The mass spreading in urban street marches its way to disintegration, breaking the traditional pattern of being visual centralized and displaying its morphological characteristics of being porous. Street space no longer has intact and clear boundary, scenes with no relations penetrating into each other.

### *4.4. The Third Space Generated by Time-Space Condensation*

New media which breeds information grafting on the global scale generates time-space condensation of street scenes. One image shifted to another, fragmentation is easily found among images arbitrarily shifted, since there is no inner logic in image convergence at all. The "speed of the virtual" breaks through the limitation between the real and virtual world. Virtual space crushes the real space, giving rise to a sense of time-space tense. Increasingly frequent shift of tasks leads to shorter endurance of memory.

Against this background, "the third space" is put forward by Edward W. Soja, who thinks that beyond the real and imaginary space, there exists a discrepant space[9]. The order of events, taking place in the past, at the moment and in the future, becomes blurred. Real time and delayed time get mixed, as supposed by Virilio. People living at the moment can't affirm their identities; they have neither memory about the past, nor the future yet to come. The breakage of time and the prominence of the present strip everything of their inherent links, leaving them pieces in isolation.

## 5. Summaries

Information communication influences people's livelihood, while people's livelihood changes the way of information spreading. The immersion spreading mode bred by the development of new media integrates human power, materials, capital and knowledge together, reshaping the street pattern with its great mobility. With the promotion of new media, the modern urban street space is deviating from the traditional rationalism and the influence of traditional aesthetics of space is eroding, gradually shifting to a sensational form centered with graphs and videos.

Only by responding to the arrival of new media age with a positive mindset, widening horizons and introducing the concept of media into the research on urban space aesthetics can the various issues on visual scenes displayed on urban streets be clearly seen. The author hopes that this research can serve as an introduction in promoting a deeper exploration on the emerging spreading mode that has a profound influence on our daily life.

## Note

1). The concept of "New media"was first brought up by Goldmark in CBS News in 1967. It has quite a few definitions, but in general it refers to all media forms that come up in the support of new technology.

2). He divides mass media age into two time periods, the first media age, when the Internet didn't popularize, and the second media age, when the development of electronic technology and Internet radically transform the communication habits of the mass.

3). It was brought up by Jaron Lanier in the early 1980s and was improved by Kevin Kelly, who extended its definition that it was a simulation system where computers generate interactive virtual environment.

4). Arnold Berleant, an environmental aesthetician supposes that Aesthetics of Engagement breaks the traditional principle of observing from the sidelines, laying more emphasis on sensual experience, which provides a new angle of research for modern aesthetic appreciation on a daily basis.

5). In the traditional sense, media is supposed to have undergone three stages, elite media, mass media and individual media. These three stages represent agricultural age, industrial age and information age respectively. Self-media age marks a new stage of media, when a great number of individuals take full advantage of modern media technology and spread personal information just-in-time.

6). Media Convergence was first brought forth by Professor Paul from University of Massachusetts Institute of Technology, who argued that various media are featured by multifunction and integration.

7). E-topia, City Of Bits and The Cyborg Self and the Networked City, written by Willian J. Mitchell, are considered trilogy inspired by the penetration of information technology in daily life.

## References

[1] Li Q 2013 *The Communication Paradigm of the Third Media Age:Immersive Communication*, (Beijing: Tsinghua University Publishing House)

[2] Virilio P 2004 *The revolutions of speed* (Lu Y C Trans.) (Nanjing: Jiangsu People's Publishing House)

[3] Negroponte N 1997 *Bing Digital* (Hu Y, Fan H Y Trans.) (Haikou: Hainan Publishing House)

[4] Shi T 2015. research on modern aesthetics within the scope of social science *Research on Marxist Aesthetics* **2**

[5] Virilio P 1998 "We may be entering an electronic gothic era" in Architectural Design *Architects in Cyberspace II*, **11-12** p61-65

[6] Behrens M, Valkanova N, gen A F. Schieck, Brumby D P. 2014 *Smart Citizen Sentiment Dashboard: A Case Study Into Media Architectural Interfaces* Proceedings of Pervasive Displays

[7] Thompson J B 2012 *Ideology and modern culture:critical social theory in the eraof mass communication* ( Gao X Trans.) (Nanjing: Yilin Press)

[8] Castor M 2006 *The Rise of Network Society* ( Xia Z J and Wang Z H Trans.) Beijing: Social Sciences Academic Press

[9] Edward W S 2005 *Thirdspace: Journeys to Los Angeles and Other Real-and-Imagined Places* ( Lu Y Trans.) (Shanghai: Shanghai Education Press)

IOP Conf. Series: Earth and Environmental Science **61** (2017) 012069 doi:10.1088/1755-1315/61/1/012069

# Numerical simulation of fracture process of concrete under uniaxial compression via PFC2D

**Jufeng Su[1, 2], Liang Zhang[1], and Yu Zhao[1]**

[1]Shool of Highway, Chang'an University, Xi'an 710064, China.
[2] Email : 181055235@qq.com

**Abstract**. Simulation of cracks and cracking process is one of the difficult problems in the numerical simulation of concrete. In this paper, the discrete element method (DEM) was applied to the numerical simulation of concrete specimen under uniaxial compression and the whole process of crack generation and development were visualized. In order to simulate the three different components of aggregates, mortar and interfaces, the random aggregate model was used in the simulation. The strain-stress curves of concrete under uniaxial compression obtained from the simulation were in good agreement with experiments. These results show that the DEM can be well used to simulate the fracture process of concrete. Therefore, the study lay a foundation for the further study on the cracking mechanism of concrete members.

## 1. Introduction

Concrete was widely used in civil engineering, research on cracking mechanism of concrete was always a hot topic in civil engineering [1,2]. In mesoscopic level, concrete was often viewed as a heterogeneous composite materials composed of coarse aggregate, cement mortar and interfacial zones between the two ingredients [3]. Wittmann [4] thought the mechanical properties of concrete under a certain scale could be explained by a lower scale structure of the concrete, therefore, study of concrete within meso-sacle was helpful to deep understand the macroscopic failure mechanism of concrete and its mechanical characteristics.

Much experimental researches has been conducted on fracture mechanism at the meso-scale in the past [5,6]. In such research, fracture propagation from the interface between mortar and aggregate to the mortar part was observed in compression experiment and the effect of the aggregate on nonlinearity of the macroscopic stress-strain curve of concrete and the failure of concrete are mentioned. However, experiment study was time-consuming and inefficient, therefore, in recent years, as the rapid development of computing power and numerical method, research at the meso-lever from the numerical point of view has begun but has not been conducted far enough yet.

There are many approaches to be used to model the meso-structure and the fracture process of concrete, but most of them fail to describe the behavior of discontinuous zones [7]. Previous numerical simulation work using continuum analysis could well indicate the stress distribution and got a relatively good stress-strain curve. However, fractures cannot be explicitly represented using the continuum approach. Compared with the finite element method, the discrete element method (DEM) has numerous advantages for crack formation [8].

Particle flow code (PFC) provided an embodiment of the discrete element method for the analysis of concrete-mechanics problems. The PFC method was proposed by Cundall [9] in 1979, it use Newton's second law to calculate the motion of particles, so the large deformation of material can be easily capture. Through applying bonds between particles, PFC can simulate the process of material transformation from continuum to dispersion under load. For its unique advantages in simulating granular materials, it has been widely used in geotechnical engineering [10,11] and the failure process of material, especially rocklike materials, within the meso-lever.

In the present work, a random aggregate model of concrete was created first, then the fracture process was simulated by the PFC2d based on the random aggregate model, finally, the numerical simulation results were compared with the experimental results, and the simulation results were discussed.

## 2. PFC method

The research object is discretized into many small rigid disks with quality in PFC2d, the interaction between discs is expressed by the contact model, which can be linear or nonlinear with their corresponding failure criteria. The movement rules are very simple, so particles can be rotated or translated arbitrarily only to obey the Newton's laws of motion [12]. In the PFC model, the constitutive relations of materials are not needed, and all of the macroscopic mechanical properties can only be expressed by defining the interactions between particles. Therefore, the contact type and its parameters selection are the key to success of PFC numerical simulation. There are some built-in contact models in PFC, these models can meet needs of most analysis. Linear Parallel Bond Model is a built-in contact model which was frequently adopted in the simulation of concrete and was verified reliable. Therefore, the LPBM was chosen at the present work to simulate the contact inner concrete.

### *2.1. Linear Parallel Bond Model*

A parallel bond provides the mechanical behavior of a finite-sized piece of cement-like material deposited between the two contacting pieces (similar to the epoxy cementing the glass beads shown in figure 1). The parallel-bond component acts in parallel with the linear component and establishes an elastic interaction between the pieces. The existence of a parallel bond does not preclude the possibility of slip. Parallel bonds can transmit both force and moment between the pieces.

A parallel bond can be envisioned as a set of elastic springs with constant normal and shear stiffnesses, uniformly distributed over a rectangular cross-section lying on the contact plane and centered at the contact point. These springs act in parallel with the springs of the linear component. Relative motion at the contact, occurring after the parallel bond has been created, causes a force and moment to develop within the bond material. This force and moment act on the two contacting pieces and can be related to maximum normal and shear stresses acting within the bond material at the bond periphery. If either of these maximum stresses exceeds its corresponding bond strength, the parallel bond breaks, and the bond material is removed from the model along with its accompanying force, moment, and stiffness.

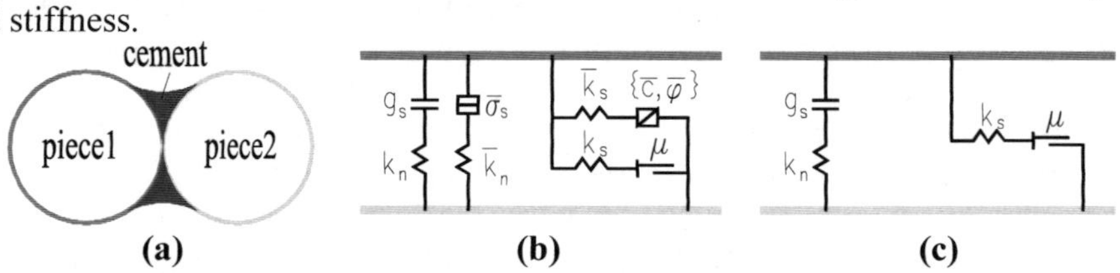


**Figure 1.** Linear parallel bond model: (a) diagram of LPBM; (b) mechanical model of bonded state; (c) mechanical model of unbonded state

The force-displacement law for the linear parallel bond model updates the contact force $F_c$ and moment $M_c$:

$$F_C = F^l + F^d + \bar{F};\ M_C = \bar{M} \tag{1}$$

Where $F_l$ is the linear force, $F_d$ is the dashpot force, $\bar{F}$ is the parallel-bond force, and $\bar{M}$ is the parallel-bond moment.

### *2.2. The method of modelling*

According to the Walraven formula, the size and quantity of coarse aggregate on the two-dimensional specimen were calculated by MATLAB based on the Fuller gradation curve [13]. Then the polygonal aggregates were randomly generated according to the size and quantity and so the geometry of random aggregate model was accomplished. In this study, the size of specimen is 100mm×100mm and the coarse aggregate fraction was 0.25, 0.3 and 0.35 respectively, the maximum and the minimum radius of the coarse aggregate was 25mm and 10mm, respectively.

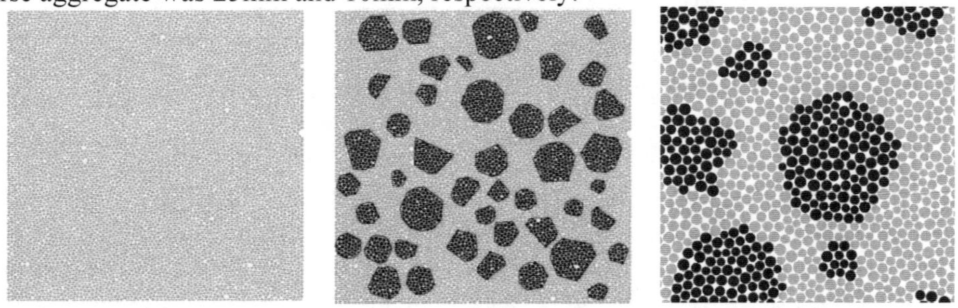

**(a)** **(b)** **(c)**

**Figure 2.** Numerical model: (a) The initial fine particle system; (b) The particle system with coarse aggregate; (c) The details view of model. The dark blue particles denote coarse aggregate, while the light blue particles denote cement.

Particles with radius 0.5~1mm were randomly generated within a rectangular region without gravity, as illustrated in figure 2. The initial generated particles may overlap and produce overlarge contact force in the initial particle system. The velocity of particles accelerated by the overlarge contact force were set to zero per 10 cycles before 2000 cycles to prevent the particles from escaping the rectangular region. To quickly reach the equilibrium state, the initial system was frictionless. In order to further reduce the kinetic energy of the initial particle system, the mass-scale method was adopted to solve until the ratio of the mean static unbalanced force to the mean static contact force was less than $10^{-5}$. Next, the geometry of the random aggregate model created in the previous step was imported to divide the particles into coarse particle group representing coarse aggregate and fine particle group representing mortar (see also figure 2). Based on the groups, the contact can be categorized into three types: coarse particle to coarse particle (CC), fine particle to fine particle (FF) and coarse particle to fine particle (CF).

**Table 1.** Input meso-parameters in the DEM simulations.

| Meso-parameters [Units] | CC | FF | CF |
|---|---|---|---|
| Effective modulus, E, [MPa] | $1.21\times10^{10}$ | $1.21\times10^{9}$ | $1.21\times10^{9}$ |
| Normal-to-shear stiffness ratio, κ, [-] | 1.0 | 1.0 | 1.0 |
| Friction coefficient, μ, [-] | 0.3 | 0.3 | 0.3 |
| Tensile strength,σ, [MPa] | $1.31\times10^{8}$ | $1.31\times10^{7}$ | $2.62\times10^{6}$ |
| Cohesion, c, [MPa] | $2.64\times10^{8}$ | $2.64\times10^{7}$ | $5.28\times10^{6}$ |
| Friction angle, θ, [Degree] | 50 | 30 | 10 |

Eventually, the initial contact type was changed to bond contact and the corresponding parameter values were assigned. It was well knew that the properties of coarse aggregate, cement mortar and interfacial zones between the two ingredients are different, the contacts of CC, FF and CF can used to express this deference by assigning different property values to different contact types. The parameters of the linear parallel bond contact adopted in this study are listed in table1. Generally, the particles and the bonds between the particles respectively have effective modulus and normal-to-shear stiffness ratio. However, in most cases the values of the two could set equal so as to simplify the analysis. Furthermore, the effective modulus and the normal-to-shear stiffness ratio should be converted firstly to normal stiffness ($k_n$) and shear stiffness ($k_s$) as follows,

$$k_n = \frac{E}{L};\ k_s = \frac{k_n}{k^*};\ with\ L = \begin{cases} R^{(1)} + R^{(2)}, & ball - ball \\ R^{(1)}, & ball - facet \end{cases} \tag{2}$$

Where $R^{(1)}$ and $R^{(2)}$ denote the radius of the two contact particles.

### *2.3. Macroscopic Variable Definitions*

Before the beginning of numerical experiment, the wall on both sides of the specimen were removed. The force was load to the specimen by applying a constant velocity to the top and bottom walls, for quasi-static loading, the velocity applied to the wall was limited less than 100με per seconds. The axis strain (i.e., $\varepsilon_1$) is estimated from the boundary movements, which can be defined as follows:

$$\varepsilon_1 = \frac{h_0 - h}{h_0} \tag{3}$$

Where $h_0$ is the initial height and h is the height at current state. The macro-scale axis stress for the meso-scale quantities of contact forces and contact vectors is as follows:

$$\sigma_1 = \frac{1}{2B}\left(\sum_{c\in N_t} f_v^c - \sum_{c\in N_b} f_v^c\right) \tag{4}$$

Where $f_v^c$ is the vertical component of the contact force at contact c within the assembly, B is the width of the assembly, $N_t$ is the total number contacts of particles to top wall and $N_b$ is the total number contacts of particles to bottom wall.

## 3. Results and discussion

The coarse aggregate fraction is an important parameter of the concrete. Therefore, three random aggregate concrete models with different coarse aggregate fractions (0.2, 0.25 and 0.3) were studied in the present work. Loading process was realized by fish language, through the fish programming language, the stress-strain curve can be monitored in real time. When the stress drops to twenty percent of the peak stress, the numerical test stops. The transmission path of contact force and the visualization of cracking are the most important advantages of the PFC method, during the simulation, the visualization of the force chain and the cracking process are shown in figure 4. By formulation 3~4, the macroscopic stress-strain curve can be draw as figure 3.

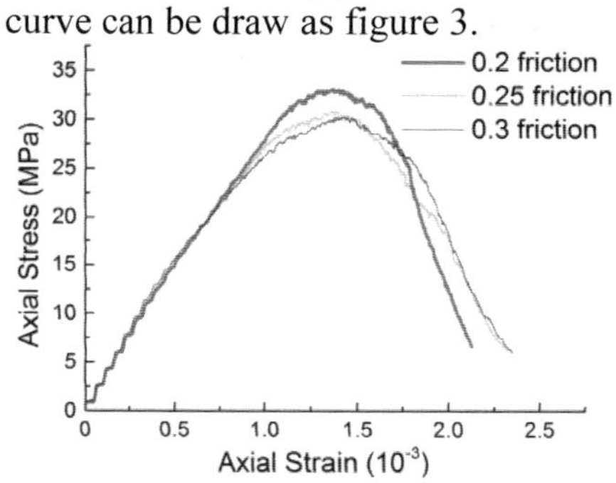


**Figure 3.** Stress-strain curves of different coarse aggregate fraction

As shown in figure 4, at the beginning of the test, the compressive stress level was low and it could be seen that the force chain distribution was uniform, fine and dense. Due to the low strength of interfaces, there were some internal micro cracks on the coarse aggregate surface, however, no macroscopic cracks were visible on the specimen surface. The stress-strain curve was linear, indicating that the specimen was in the state of elastic stress. With the increase of the pressure, the micro cracks on the coarse aggregate surface slowly expanded to the interior of the mortar, then visible cracks begun to form. As the pressure increased, these cracks continued to develop deep into the mortar, finally, the micro cracks in some positions connected with each other, macroscopic cracks were clearly visible. In this stage, the force chain of cracked place disappeared while the uncracked place became stronger, the force chain distribution became not uniform and concentrated to some place. When the macro cracks appeared, the stiffness of the specimen degenerated and the stress state entered the elastic-plastic stage. With the further increase of the pressure, the stress-strain curve arrived the peak and then decreased sharply, cracks developed rapidly and the existing cracks fast connected with each other. Finally, the specimen was divided into small blocks with coarse aggregate as the core, only a few of thick force chain remain, the specimen was completely crushed and the test was over.

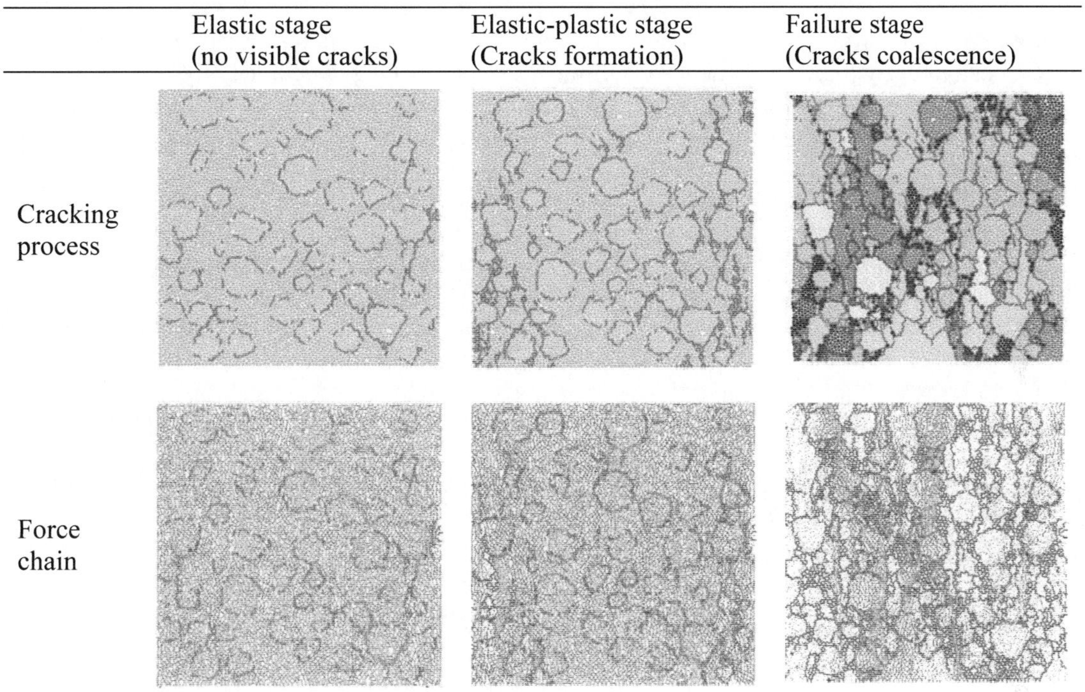


**Figure 4.** Visualization of fracture process

## 4. Conclusion

The PFC method was studied in this work to simulate the fracture process of concrete specimen under compression. The whole process of crack generation and development could clearly observed through the simulation, this is an important advantage of PFC method compared to physical experiment. In order to simulate the concrete specimens of high fidelity, the random aggregate model was used to simulate the specimen geometry, the results show that the method can accurately simulate the mechanical properties of concrete with coarse aggregates, mortar and interfaces and the stress-strain curves obtained from the simulation are in good agreement with physical experiments.

Application of PFC method to concrete is a new way to study the cracking mechanism of concrete structures by numerical simulation. The study on the cracking mechanism of concrete members by using PFC method will be the focus of the next step.

## References

[1] Ooi, E. T., Natarajan, S., Song, C., & Ooi, E. H. (2016). Crack propagation modelling in concrete using the scaled boundary finite element method with hybrid polygon–quadtree meshes. *International Journal of Fracture*, 1-23.

[2] Schlangen, H. E. J. G. (1993). Experimental and numerical analysis of fracture processes in concrete. *Delft University of Technology*, 38.

[3] Nilsen, A. U., & Monteiro, P. J. (1993). Concrete: a three phase material. *Cement and Concrete*

*Research*, 23(1), 147-151.

[4] Wittmann, F. H. (1983). Structure of concrete with respect to crack formation. *Fracture mechanics of concrete*, 43(5), 6.

[5] Kamalibernard, S., Keinde, D., & Bernard, F. (2014). Effect of aggregate type on the concrete matrix/aggregates interface and its influence on the overall mechanical behavior. a numerical study. *Key Engineering Materials*, 617, 14-17.

[6] Gu, X., Hong, L., Wang, Z., & Lin, F. (2013). Experimental study and application of mechanical properties for the interface between cobblestone aggregate and mortar in concrete. *Construction and Building Materials*, 46, 156-166.

[7] Camborde, F., Mariotti, C., & Donzé, F. V. (2000). Numerical study of rock and concrete behaviour by discrete element modelling. *Computers and geotechnics*, 27(4), 225-247.

[8] Wang, Z., Lin, F., & Gu, X. (2008). Numerical simulation of failure process of concrete under compression based on mesoscopic discrete element model. *Tsinghua Science & Technology*, 13, 19-25.

[9] Cundall, P. A., & Hart, R. D. (1992). Numerical modelling of discontinua. *Engineering computations*, 9(2), 101-113.

[10] Potyondy, D. O., & Cundall, P. A. (2004). A bonded-particle model for rock. *International journal of rock mechanics and mining sciences*, 41(8), 1329-1364.

[11] Zhou, S., Zhu, H., Yan, Z., Ju, J. W., & Zhang, L. (2016). A micromechanical study of the breakage mechanism of microcapsules in concrete using PFC2D. *Construction and Building Materials*, 115, 452-463.

[12] Itasca. User's manual for PFC2D version PFC5.0. Minneapolis. USA: Itasca consulting Group, Inc; 2014.

[13] Wriggers, P., & Moftah, S. O. (2006). Mesoscale models for concrete: Homogenisation and damage behaviour. *Finite elements in analysis and design*, 42(7), 623-636.

IOP Conf. Series: Earth and Environmental Science **61** (2017) 012070 doi:10.1088/1755-1315/61/1/012070

# Elastoplastic stress analysis of tunnel considering different principal stress directions

**Shuan-cheng Gu[1,a)], Rong-bin Huang[1,b)], Pei-li Su [1,c)] Xiao Ding[2, d)] and Ang Li[1,e)]**

[1]School of Architecture and Civil Engineering, Xi'an University of Science And Technology, Xi'an 710054, China
[2]School of Civil and Architecture Engineering, Xi`an Technological University, Xi'an 710021, China
[a)]yikaiyizhi@qq.com
[b)] Corresponding author:446413318@qq.com
[c)]1691576328@ qq.com
[d)]yikaiyizhi@163.com
[e)]yikaiyizhi@qq.com

**Abstract**. Based on the unified strength theory, the elastic-plastic stress analysis is carried out under the premise that the first principal stress in two cases of radial stress and circumferential stress, and the formula of the radius of the surrounding rock and the plastic zone is derived. Tunnel surrounding rock has three states which including the elastic state, the elastoplastic state that the maximum principal stress is the radial stress and the elastoplastic state that the maximum principal stress is the circumferential stress. When the elastic-plastic analysis was carried out on the tunnel, the state of the surrounding rock should be judged and the correct formula is chosen to calculate. Analysis results show that the intermediate principal stress is benefit to the rock to give full play to the strength of the potential, so as to improve the stability of surrounding rock of the tunnel and the smaller intermediate principal stress coefficient, the higher sensitivity of the surrounding rock stability to the intermediate stress. When the inside pressure is less than the second critical stress and increasing the hole pressure is beneficial to improving the stability of surrounding rock, while the hole pressure is greater than the second critical stress, the stability of surrounding rock will reduce with the increase of the hole pressure.

## 1 Introduction

The engineering practice usually adopts Mohr-Coulomb strength criterion for tunnel elastic-plastic analysis, the deduced Fenner formula [1], the modified Fenner formula and Kastner formula [2-4] is widely used in tunnel engineering. However, it can not reasonable to consider the influence of the intermediate principal stress on the analysis of surrounding rock by using the Mohr-Coulomb strength criterion, which leads to the analysis result is conservative.

The unified strength theory is not only reasonable considering the effect of intermediate principal stress, but also can degenerate into various corner model between Mohr-Coulomb strength criterion, Mises criterion, Tresca criterion and twin shear strength criterion and between Mohr-Coulomb and the twin shear strength criterion, witch have been widely applied in various engineering fields[5-6]. In this paper, based on the unified strength theory, the formula of plastic zone width under different principal stress directions is derived. The applicable conditions of different formulas are defined at the same time.

## 2 Calculation Model

The radius of circular tunnel is $r_0$, ground stress $P$, internal pressure is $P_i$, radius of plastic zone is $r_p$. In the tunnel of elastoplastic stress analysis, solve the problem of plane strain, the basic assumptions are as follows: (1) Tunnel section equivalent to circular, infinite length; (2) the lateral pressure coefficient is 1; (3) the homogeneous and isotropic rock and the incompressible materials, regardless of physical effects in the process of calculation. Mechanical calculation model shown in Figure 1

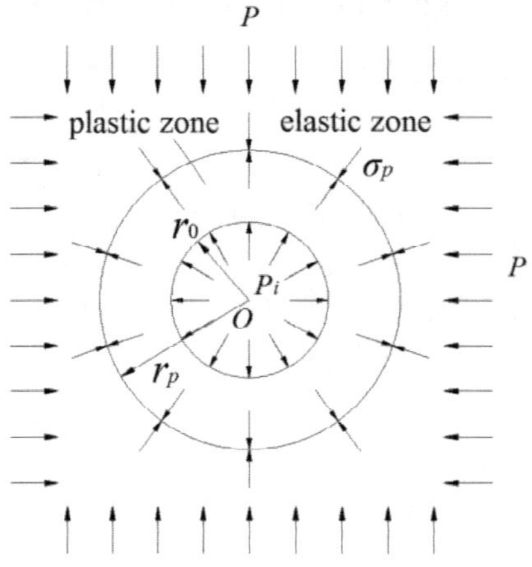


**FIGURE 1.** Mechanical analysis model

## 3 Elastic Plastic Stress Analysis

*Radial Stress is the First Principal Stress*

According to the unified strength theory expression:

$$F=\sigma_r-\frac{1-\sin\varphi}{(1+b)(1+\sin\varphi)}(b\sigma_z+\sigma_\theta)=\frac{2c\cos\varphi}{1+\sin\varphi},\quad \sigma_z\le\frac{\sigma_r+\sigma_\theta}{2}+\frac{\sin\varphi}{2}(\sigma_r-\sigma_\theta) \tag{1a}$$

$$F'=\frac{1}{1+b}(b\sigma_z+\sigma_r)-\frac{1-\sin\varphi}{1+\sin\varphi}\sigma_\theta=\frac{2c\cos\varphi}{1+\sin\varphi},\quad \sigma_z\ge\frac{\sigma_r+\sigma_\theta}{2}+\frac{\sin\varphi}{2}(\sigma_r-\sigma_\theta) \tag{1b}$$

Formula: $c$、$\varphi$ is the cohesion and internal friction angle of the surrounding rock, $b$ is the coefficient of intermediate principal stress.

The equilibrium equation of axisymmetric plane strain mechanics model is

$$\frac{d\sigma_r}{dr}+\frac{\sigma_r-\sigma_\theta}{r}=0 \tag{2}$$

The calculation model can have boundary and elastic-plastic contact surface conditions as follows:

$$\left.\begin{aligned}&(\sigma_r)_{r=r_0}=-P_i\\&(\sigma_r^e)_{r\to\infty}=-P\\&(\sigma_r^e)_{r=r_p}=(\sigma_r^p)_{r=r_p}\\&(\sigma_\theta^e)_{r=r_p}=(\sigma_\theta^p)_{r=r_p}\end{aligned}\right\} \tag{3}$$

Solution as

$$r_p = r_0 \left\{ \frac{\left[(2+b)-(2+3b)\sin\varphi\right](c\cot\varphi+P)}{(2+b-b\sin\varphi)(c\cot\varphi+P_i)} \right\}^{\frac{(2+b)-(2+3b)\sin\varphi}{4\sin\varphi(1+b)}} \quad (4)$$

*Hoop Stress is the First Principal Stress*

When the water pressure or support pressure is too large, the tunnel rock may also yield region. Because the radial compressive stress is larger, the hoop compressive stress is small, and even becomes the tensile stress, so the first principal stress should be $\sigma_\theta$, and the yield criterion expression of surrounding rock becomes

$$F' = \frac{1}{1+b}(b\sigma_z + \sigma_\theta) - \frac{1-\sin\varphi}{1+\sin\varphi}\sigma_r = \frac{2c\cos\varphi}{1+\sin\varphi} \quad (5)$$

Repeat the above analysis process, solution as

$$r_p = r_0 \left\{ \frac{(2+b)(1+\sin\varphi)(c\cot\varphi+P)}{(2+b-b\sin\varphi)(c\cot\varphi+P_i)} \right\}^{-\frac{(2+b)(1+\sin\varphi)}{4(1+b)\sin\varphi}} \quad (6)$$

*Applicable Conditions of the Formula*

Under the action of ground stress, there are three kinds of state of tunnel surrounding rock: elastic state, maximum principal stress under the condition of radial stress, elastic-plastic state and maximum principal stress under the condition of hoop stress. In the tunnel by elastic-plastic analysis before, should first determine the rock in the stress field under the conditions of whether there is plastic failure, and then determine the direction of the maximum principal stress, and then choose the correct formula of tunnel stress analysis.

The critical stress $\sigma_p$ is introduced, witch mean $r_p/r_0 = 1$ when $P_i = \sigma_p$, the surrounding rock will appear plastic failure. Analytical formulas (4) and formula (6) show that the first critical stress

$$\sigma_{p1} = \frac{\left[(2+b)-(2+3b)\sin\varphi\right]P - 2c(1+b)\cos\varphi}{(2+b-b\sin\varphi)} \quad (7)$$

The second critical stress

$$\sigma_{p2} = \frac{(2+b)(1+\sin\varphi)P + 2c(1+b)\cos\varphi}{(2+b-b\sin\varphi)} \quad (8)$$

It can be concluded that the application of formula (4) and (5) to the elastic-plastic stress analysis of tunnel:

(1)when $P_i < \sigma_{p1}$, part of the surrounding rock is in a plastic state, and the first principal stress is radial stress, then the formula (4) is used for elastic-plastic analysis of the tunnel;

(2)When $\sigma_{p1} \le P_i \le \sigma_{p2}$, surrounding rock is in elastic condition completely and stress analysis can be carried out by elastic theory;

(3) when $\sigma_{p2} < P_i$, part of the surrounding rock is in a plastic state, and the first principal stress is hoop stress, then the formula (6) is used for elastic-plastic analysis of the tunnel.

## 4 Example Analysis

Radius of circular tunnel $r_0 = 8.5$ m, buried depth is 180m, original rock stress $P = 3.55$ MPa; surrounding rock cohesion $c = 0.2$ MPa; internal friction angle $\varphi = 30^\circ$, internal pressure $P_i = 0.3$ MPa; intermediate principal stress coefficient $b = 0.25$.

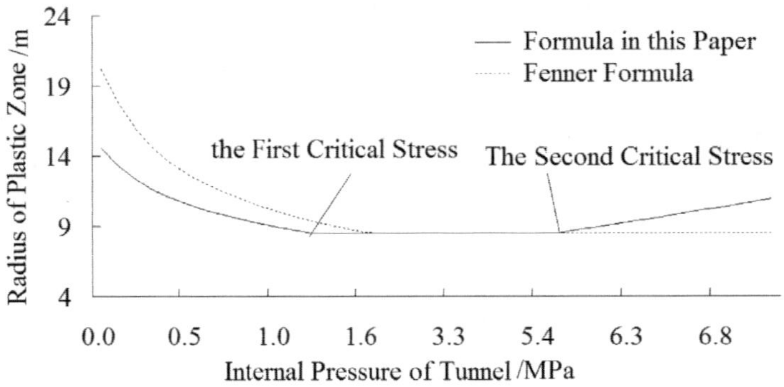


**FIGURE 2.** Radius change curve of plastic zone under different tunnel pressure $P_i$

Figure 2 shows the variation curve of plastic zone radius under different pressure in tunnel. Using the formula (7) and (8) calculation to determine the first critical stress $\sigma_{p1} = 1.26$ MPa, the second critical stress $\sigma_{p2} = 5.84$ MPa. When $P_i < 1.26$ MPa, Due to the stress release caused by tunnel excavation, the surrounding rock is in the elastic-plastic state, and the maximum principal stress is radial stress, so the radius of plastic zone should be calculated by formula (4). With $P_i$ from 0 to 1 $\sigma_{p1}$, the radius of plastic zone decreased from 20.16m to 8.50m. When $1.26\text{MPa} \le P_i \le 5.84\text{MPa}$, surrounding rock is in elastic condition completely and stress analysis can be carried out by elastic theory. When $P_i > 5.84$ MPa, Because of the excessive plastic pressure in the tunnel and the maximum principal stress being hoop stress, the plastic zone radius of the surrounding rock should be calculated by formula (6). Analysis by using the Fenner formula, the first critical stress value is 1.68MPa, compared with the calculation results of (7) 0.42MPa is increased, and the same tunnel pressure under the condition of the radius of the plastic zone is larger, and there is no second critical stress value, which mean that no matter the tunnel pressure increases to any value, the surrounding rock will not appear plastic failure, obviously this conclusion is unreasonable.

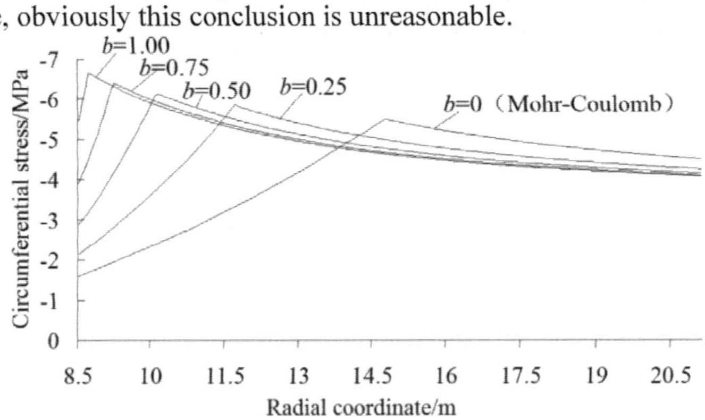


**FIGURE 3.** Radial stress curves under different $b$ values

Figure 3 show the distribution law of circumferential stress when the intermediate principal stress coefficient *b* is taken to be 0, 0.25, 0.5, 0.75 respectively. It can be seen from the figure, with the increase of *b* value, the plastic zone in the same position of the radial compressive stress and circumferential stress are increased, and the stress of the radial coordinate change rate is increased; and in the same place in the elastic region of circumferential stress and radial pressure with the increase of *b* stress decreased; with the increase of *b* value of the plastic zone is obviously reduced, and the elastic-plastic interface circumferential compressive stress increased from 5.49MPa to 6.63MPa, and the radial compressive stress is reduced from 1.60MPa to 0.48MPa, the elastic-plastic interface maximum shear stress increased from 1.95MPa to 3.07MPa, witch shows that rock in the larger shear stress remains stable and does not have the plastic damage.

## 5 Conclusion

(1) The stress solution of the tunnel and the corresponding radius of plastic zone under the condition of radial stress and hoop stress are derived based on unified strength theory. When b=0 and only consider the radial stress as the first principal stress, the theory in this paper is reduced to the Mohr-Coulomb strength theory, and the plastic zone radius formula is reduced to Fenner formula.

(2) With the influence of intermediate principal stress coefficient b increases, the elastic-plastic interface of maximum principal stress and maximum shear stress are increased and the radius of plastic zone is decreased; the surrounding rock is easier to enter the fully elastic state with the maximum principal stress as the radial stress, and the larger the thickness of the surrounding rock is in the elastic state at the same time.

(3) $P_i < \sigma_{p1}$, part of the surrounding rock is in a plastic state, and the first principal stress is radial stress, $\sigma_{p1} \le P_i \le \sigma_{p2}$, surrounding rock is in elastic condition completely, $\sigma_{p2} < P_i$, part of the surrounding rock is in a plastic state, and the first principal stress is hoop stress. In the elastic-plastic analysis of surrounding rock, we should determine the state of the surrounding rock firstly, and then use the corresponding calculation formula for stress analysis.

(4)The above analysis shows that the intermediate principal stress is beneficial to fully exert the strength potential of the surrounding rock, thus improving the stability of tunnel surrounding rock, and the smaller the *b* is, the higher the sensitivity of the surrounding rock stability state to the intermediate stress is.

**Acknowledgments**

This work was financially supported by National Natural Science Foundation of China (Fund number: 51508462) and Shaanxi Science Foundation Research Program (Fund number: 2016JM4014)

## References

[1] Talobre J A. Mecanique des roches [M]. Paris: Dunod,1967.
[2] YU MAO-hong. Unified Strength Theory for Geoma-terials and Its Applications [J]. Chinese Journal of Geotechnical Engineering, 1994, 16(2): 1-10.
[3] CUI Lan, ZHENG Jun-jie. Study of support pressure and surrounding rock deformation of a circular tunnel with an elastoplastic softening model [J]. Rock and Soil Mechanics, 2014,03:717-722+728.
[4] Xu Dongjun, Geng Naiguang.The variation law of rock strength with INCREASE of intermediate principal stress [J]. Acta Mechanica Solida Sinica, 1985, 01: 72-80.
[5] WANG Jun-xiang, JIANG An-nan. An elastoplastic damage constitutive model of rock and its application to tunnel engineering[J]. Rock and Soil Mechanics, 2015, 04:1147-1158.
[6] H Kastner. Statik des tunel und stollenbauess [M]. Berlin:Springer-Verlag, 1962.

IOP Conf. Series: Earth and Environmental Science **61** (2017) 012071 doi:10.1088/1755-1315/61/1/012071

# Research on common methods for evaluating the operation effect of integrated wastewater treatment facilities of iron and steel enterprises

**Xu Bingsheng**

China National Institute of Standardization, No. 4, Zhichun Road, Haidian District, Beijing 100191

dengdaitianqingdeyu@126.com

**Abstract**. Considering the large quantities of wastewater generated from iron and steel enterprises in China, this paper is aimed to research the common methods applied for evaluating the integrated wastewater treatment effect of iron and steel enterprises. Based on survey results on environmental protection performance, technological economy, resource & energy consumption, services and management, an indicator system for evaluating the operation effect of integrated wastewater treatment facilities is set up. By discussing the standards and industrial policies in and out of China, 27 key secondary indicators are further defined on the basis of investigation on main equipment and key processes for wastewater treatment, so as to determine the method for setting key quantitative and qualitative indicators for evaluation indicator system. It is also expected to satisfy the basic requirements of reasonable resource allocation, environmental protection and sustainable economic development, further improve the integrated wastewater treatment effect of iron and steel enterprises, and reduce the emission of hazardous substances and environmental impact.

## 1. Introduction

China is one of the superpowers in iron and steel in the world. According to the statistics, the production of one ton of steel consumes 200-250m$^3$ water and generates a large quantity of wastewater. Statistically, as of December 2016, the member enterprises of China Iron and Steel Industry Association discharged 42 million m$^3$ of wastewater throughout the year, up 5.89% YoY. Among them, COD emission reached 1,050 tons, down 9.31% YoY and up 3.53% MoM; ammonia nitrogen emission reached 134 tons, up 35.66% YoY and up 31.97% MoM; volatile phenol emission reached 0.85 ton; total cyanide emission reached 0.82 ton; suspended-particle emission was up to 600 tons; petroleum emissions were about 19 tons [1]. Currently, most iron and steel enterprises in China have changed their attitude from indifference to water-saving and paying no attention to developing non-traditional water sources to actively using recycling water through water-saving, pollution control and development of new sources[2].

Wastewater from iron and steel enterprises mainly includes wastewater, circulating cooling water, discharge water, desalted water, softened water, strong brine, etc. generated in various production processes, as well as impurities like Calcium, Iron, Manganese, Lead, Zinc, Copper, Arsenic and other heavy metal ions and high-concentration suspended solids[3]. For example, wastewater from a sintering plant mainly contains the wastewater from the wet dust collector and wastewater generated by washing the terrace and conveying belts, while the impurities are mainly the suspended solid of sintered mixtures. Wastewater from a cold rolling plant mainly contains chromate wastewater, acid wastewater, concentrated alkali emulsification wastewater, dilute alkali oily wastewater, etc. The coking wastewater is the most complicated wastewater containing a large quantity of phenol, ammonia, sulfur, cyanide, tar, pyridine and other pollutants, and it is a kind of unmanageable industrial wastewater that can cause severe pollution [4].

All production processes of iron and steel enterprises will generate industrial wastewater. In order to satisfy the production needs and reduce the concentration of wastewater, it is necessary to regularly discharge the pollution in recycling water treatment even though wastewater recycling or treatment facilities are set. The integrated wastewater of iron and steel enterprises refers to the mixed wastewater of wastewater generated in the recycling of untreated wastewater, a small amount of accidental water from certain processes, treated wastewater that meets the emission standard and a small quantity of domestic wastewater. The standard definition of the integrated wastewater of iron and steel enterprises is the wastewater collected and conveyed by the drainage systems within the plant areas of iron and steel enterprises, and drained out through the main drains. Wastewater discharged from production unit (excluding cool rolling wastewater and salt wastewater in high concentration) shall be collected by the drainage system in the plant area and conveyed to the integrated wastewater treatment facility for processing [5]. The integrated wastewater of iron and steel enterprises mainly contains SS, organic/inorganic substances and other impurities and oils. Moreover, it is featured by relatively high conductivity, which significantly affects its recycling.

This paper is aimed to research the common methods applied for evaluating the integrated wastewater treatment effect of iron and steel enterprises. A tertiary evaluation indicator system from such aspects as environment, resources, management, operation and services is set up by investigating and analyzing the relevant policies, regulations, standards as well as main processes and equipment for integrated wastewater treatment facilities of iron and steel enterprises, so as to determine values of evaluation indicators through the data, site investigation and comparison of standards to evaluate the measures taken by iron and steel enterprises for integrated wastewater treatment, thus improving the integrated wastewater treatment effect of iron and steel enterprises and reducing the emission of hazardous substances and environmental impact.

## 2. Research Method

### *2.1. Determination of Indicators*

In this paper, the quantitative and qualitative evaluation indicators are combined to determine the 5 primary indicators of environmental protection performance, technological economy, resource & energy consumption, management & service and conditions of main equipment of the integrated wastewater treatment facilities of iron and steel enterprises based on the full life circle by means of project investigation, questionnaire survey, etc. Several lower level indicators are further set to form an indicator system for evaluating the operation effect of integrated wastewater treatment facilities. The integrated wastewater treatment facilities shall have certain technical innovation and economic benefits, while the evaluation of operation effect of the equipment shall conform to relevant standards and regulations for environmental protection. The equipment shall be safe, reliable and energy-saving, and conform to the basic requirements of reasonable resource allocation, environmental protection and sustainable economic development, provided that the environmental protection requirements are satisfied.

### *2.2. Methods for Determination of Evaluation Values*

(1) The indicator value for environmental protection performance shall be determined as per the existing wastewater discharge standards of China, and the environmental protection evaluation class shall be defined as per the requirements for different regions as stated in such standards.

(2) As for technological economy, resource & energy consumption, and conditions of main equipment, it is necessary to fully investigate the situations of wastewater treatment facilities of the same type, and determine the classification according to the investigation data as follows: Class 1 refers to wastewater treatment facilities with excellent performance, accounting for about 20% of the

total number of wastewater treatment facilities in the industry; Class 2 refers to wastewater treatment facilities with good performance, accounting for 70%; the remaining 10% are those with poor environmental protection performance, which shall be rectified or weeded out.

(3) In terms of management and services, it is required to determine the classification and evaluation values in strict accordance with the requirements of the quality management system, energy management system, environmental management system, as well as the relevant safety management regulations of the enterprises.

## 3. Evaluation Indicator System

After investigating the processes and current situation of application of integrated wastewater treatment facilities of iron and steel enterprises, it is concluded that the indicators in indicator library shall include definitive indicators and general indicators. Specifically, in investigation and indicator selection, the focus is placed on the definitive indicators with significantly impacts on environmental performance or those which shall be strictly adhered to at the request of laws and regulations, indicators covering the limited emission of toxic and hazardous substances, water consumption, energy consumption and power consumption, as well as particular pollutants of the industry, industry access indicators, etc. Moreover, to improve the operation effect of integrated wastewater treatment facilities, importance is attached to investigating general indicators specified in the standards related to the products, mainly including a series of national or industrial standards to be followed by such main equipment as grille cleaner, submersible mixer, oil-water separator, mud scraper, sludge dewatering equipment, air blower, slurry pump, etc.

The method of subjective judgment by experts is applied and factor analysis based on environmental protection, technological economy, resource & energy consumption, management & service and product properties is conducted, so as to build multi-layer indicator architecture for full life cycle of integrated wastewater and finally form an evaluation indicator system containing multiple secondary indicators for purpose of evaluation of operation effect of wastewater treatment facilities of iron and steel enterprises. Specifically, the secondary indicators shall include such key secondary indicators as pH value, SS, COD, ammonia nitrogen, noise emission limit requirement, energy consumption, water consumption, power consumption, hazardous substance content, recycling rate, drug consumption, safe operation, system operation rate and equipment availability, as shown in Figure 1.

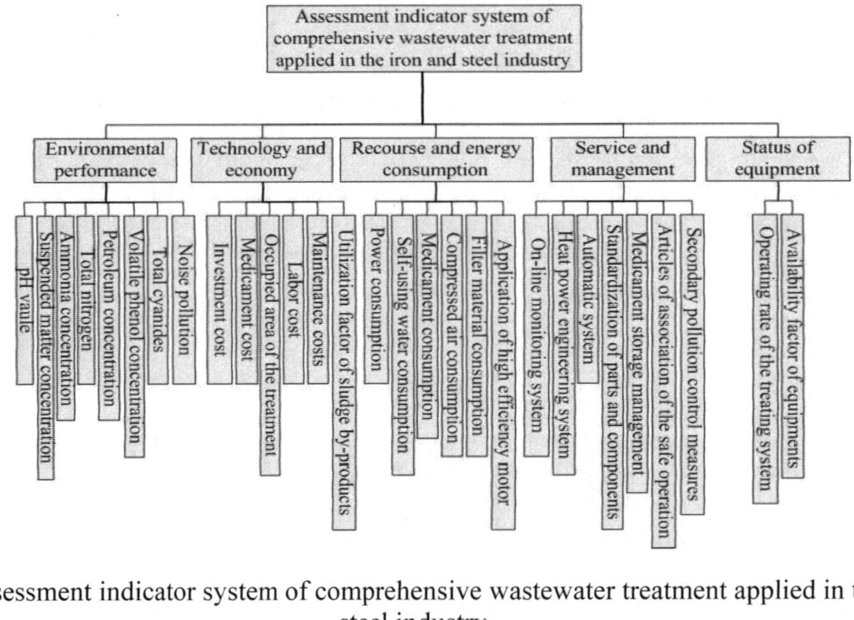


**Figure 1.** Assessment indicator system of comprehensive wastewater treatment applied in the iron and steel industry

IOP Conf. Series: Earth and Environmental Science **61** (2017) 012071 doi:10.1088/1755-1315/61/1/012071

## 4. Analysis on Rationality of Key Indicators for Evaluating the Operation Effect of Wastewater Treatment System of Iron and Steel Enterprises

According to the evaluation indicator system built herein, the focus is placed to investigating and analyzing the rationality of key indicators for evaluating the operation effect in terms of pH value, SS concentration, COD concentration, system operation rate, online monitoring system, etc.

### *4.1. pH Value*

The pH values of wastewater discharged from the integrated wastewater treatment facilities of iron and steel enterprises in China are analyzed, and the statistical results are shown in Figure 2. Among 20 sets of integrated wastewater treatment facilities subject to the investigation and analysis, 18 sets have a pH value larger than 6 but less than 9, accounting for 90%; 0 set has a pH value less than 6, accounting for 0%, and 2 sets have a pH value larger than 9, accounting for 10%. The percent of pass is 90%.

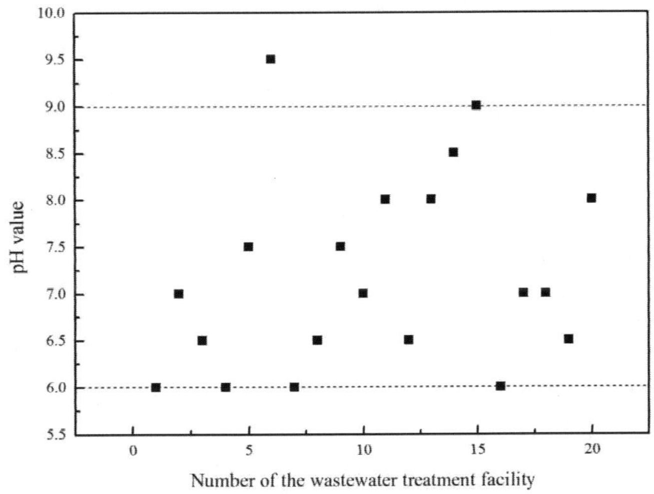


**Figure 2.** Statistical results on pH values of wastewater discharged from the integrated wastewater treatment facilities of iron and steel enterprises

### *4.2. Suspended matter concentration (SS)*

The suspended matter concentration of wastewater discharged from the integrated wastewater treatment facilities of iron and steel enterprises analyzed, and the statistical results are shown in Figure 3. Among 20 sets of integrated wastewater treatment facilities subject to the investigation and analysis, 3 sets have a suspended matter concentration less than 20 mg/L, accounting for 15%; 12 sets has a concentration less than 30 mg/L but larger than 20 mg/L, accounting for 60%, and 5 sets have a concentration larger than 30 mg/L, accounting for 25%. The percent of pass is 75%.

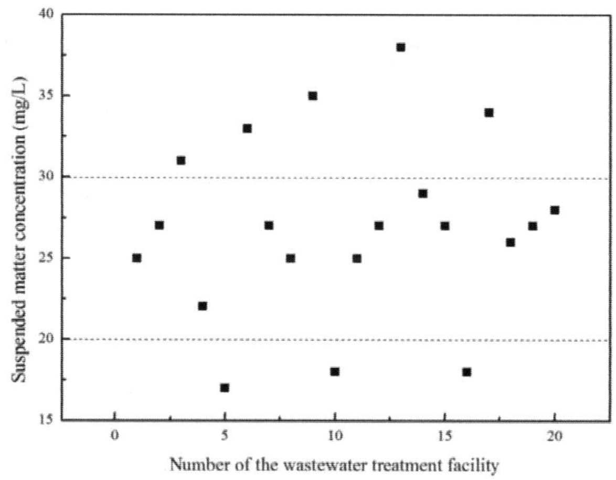


**Figure 3.** Statistical results on SS concentration of wastewater discharged from the integrated wastewater treatment facilities

### *4.3. COD concentration*

The COD concentration of wastewater discharged from the integrated wastewater treatment facilities of iron and steel enterprises analyzed, and the statistical results are shown in Figure 4. Among 20 sets of integrated wastewater treatment facilities subject to the investigation and analysis, 4 sets have a COD concentration less than 30 mg/L, accounting for 20%; 15 sets has a concentration less than 50 mg/L but larger than 30 mg/L, accounting for 75%, and 1 sets have a concentration larger than 50 mg/L, accounting for 5%. The percent of pass is 95%.

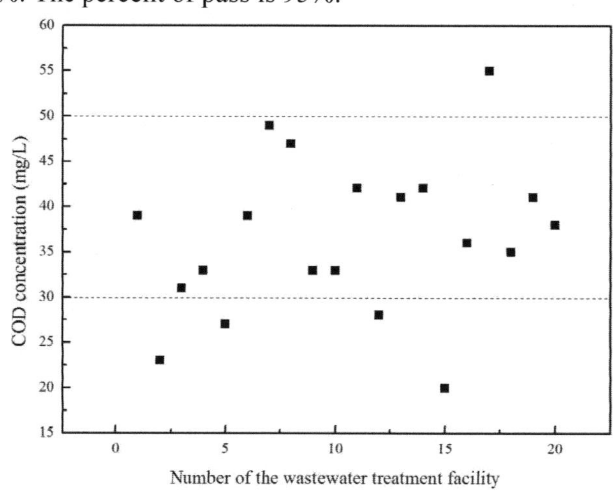


**Figure 4.** Statistical results on COD concentration of wastewater discharged from the integrated wastewater treatment facilities of iron and steel enterprises

### *4.4. Operation Rate of the Treating system*

The system operation rate refers to the percentage of annual normal operation time of the integrated wastewater treatment facility in the total operation time of the main work, and is one of the important factors affecting the quality of wastewater treatment. The improvement of operational rate can effectively control the emission of water pollutants and improve the wastewater treatment efficiency. Frequent system faults and lower operation rate may not only increase the workload of maintenance

IOP Conf. Series: Earth and Environmental Science **61** (2017) 012071 doi:10.1088/1755-1315/61/1/012071

personnel, but also reduce the operation rate and affect the wastewater treatment efficiency. The system reliability and operation rate can be improved by debugging the system for safe, stable and continuous operation.

The operation rate of integrated wastewater treatment facilities of iron and steel enterprises in China is analyzed. Among 20 sets of integrated wastewater treatment facilities subject to the investigation and analysis, the system operation rate reaches 100%.

### *4.5. On-line Monitoring System*

The installation of online monitoring systems for integrated wastewater treatment facilities of iron and steel enterprises in China is analyzed. All of the 20 integrated wastewater treatment facilities subject to the investigation and analysis are equipped with online monitoring systems. The online monitoring system is used to continuously monitor the concentration and total amount of COD, ammonia nitrogen, volatile phenol, oils, phosphorus and other pollutants and particulate matters discharged, and transfer the data to the competent department in real time. The online monitoring system is directly related to the operation conditions of environmental protection device. All wastewater treatment facilities shall be equipped with online monitoring system to satisfy the monitoring requirements proposed by the environmental protection administration and the requirements for comprehensive monitoring of dust removing devices.

## 5. Conclusions

In this paper, integrated wastewater treatment facilities of iron and steel enterprises are analyzed. An indicator system for evaluating the operation effect of integrated wastewater treatment facilities based on five primary indicators of environmental protection performance, technological economy, resource & energy consumption, services and management is set up by investigating the standards and industrial policies in and out of China to further define 27 key secondary indicators (pH value, concentration of suspended matters, COD concentration, system operation rate, online monitoring system, etc.) on the basis of investigation and survey on main equipment and key processes for wastewater treatment, so as to determine the method for setting the key quantitative and qualitative indicators for evaluation indicator system through the rationality evaluation done on the basis of certain indicator values in existing standards. It is expected to build the evaluation system and determine the relevant evaluation methods for the improvement of operation effect of integrated wastewater treatment facilities of iron and steel enterprises in China.

The focus is further placed on the determination of values and weights of the secondary indicators based on the investigation of evaluation indicator system and relevant values, so as to classify and evaluate the integrated wastewater treatment facilities of iron and steel enterprises in China, and provide a standard for the implementation of relevant policy dividend for third-party governance.

## References

[1] 2006 *Iron and steel association information statistics department of China* .
[2] Ma Z and Jiang H 2010 *The Yellow River publishing media group* pp 1-35
[3] Zhou X and Wu G 2013 *Water supply & sewerage* vol 39 pp 36-337
[4] Zhang S and Sun X 2010 *Energy conservation and emission reduction* vol 19 pp 16-17
[5] Wang H and Zhao Y 2007 *The annual meeting of the steel in China* pp 56-59

IOP Conf. Series: Earth and Environmental Science **61** (2017) 012072 doi:10.1088/1755-1315/61/1/012072

# Cloning and bioinformatics analysis of CcPILS gene of Hickory (Carya cathayensis)

**Wenbin Guo[1*], Huwei Yuan[1*], Liuxiao Gao[2*], Haipeng Guo[2], Lingling Qiu[1], Dongbin Xu[1], Daoliang Yan[1, a], Bingsong Zheng[1, b]**

[1] State Key Laboratory of Subtropical Silviculture, Zhejiang A & F University, Linan, Hangzhou 311300, P. R. China; [2] State Key Laboratory of Plant Physiology and Biochemistry, College of Life Sciences, Zhejiang University, Hangzhou 311300, P. R. China

*These authors contributed equally to this work.

E-mail: [a] bszheng@zafu.edu.cn; [b] liangsie2000@163.com

**Abstract.** PILS is a key auxin efflux carrier protein in the auxin signal transduction. A CcPILS gene related to hickory (Carya carthayensis) grafting process was obtained by RACE techniques. The full length of CcPILS gene was1541bp contained a 1263bp length open reading flame (ORF). The CcPILS encoded 294 amino acids with molecular weight of 46 kDa, PI 5.38 and localized at endoplasmic reticulum membrane. The gene contained a central hydrophilic loop separating two hydrophobic domains of about five transmembrane regions each. The gene of CcPILS belonged to Clade III sub-family of PILS and its sequence had high homology with Arabidopsis. Real Time RT-PCR analysis showed that the gene expressions were weakly induced in bud, inflorescence, fruit, leaf and stem, while strongly in root. The expression levels were strongly induced and reached a peak at the third day of grafting in scion and rootstock of hickory, which were 1.45 and 3.45 times higher, respectively, compared to that of control. The results indicated that CcPILS may be involved in regulating the expression of genes related to auxin signal transduction during hickory graft process.

## 1. Introduction

Auxin plays an important role in plant growth and development [1-3]. Polarity transport is an important feature of auxin, and is commonly regulated by the synergy of auxin input carrier AUX/LAX protein family and the output carrier PIN-FORMED (PIN) protein family [4]. Barbez et al. (2004) found that another family of seven-member was also related to intracellular transport of auxin. Although the sequence similarity to PIN is very low (10-18%), it is still named as PILS (PIN-LIKES, PILS) because its topology is similar to that of PIN [5]. PILS can be divided into Clade I, Clade II and Clade III subfamilies. The PILS5 and PILS7 genes included in Clade III subfamily, play an important role in the process of vascular plant evolution [6].

Hickory (Carya cathayensis Sarg.) is an important woody oil plant and economic forest species in Zhejiang Province [7-10]. There are many reports on growth and development of hickory [11-13]. However, the grafting technique and quality improving of hickory is difficult. To understand the physiological and biochemical mechanisms of hickory graft, a series of studies have been carried out by Zheng et al. [15-17]. According to our transcriptomic data [17-18], we found that the expression of PILS gene was markedly induced in the graft process of hickory. In this study, the full-length cDNA of PILS gene was cloned according to the conservative fragment obtained from the transcriptomic data by RACE technique. The functional structure and regulation pattern of CcPILS was analyzed. The

IOP Conf. Series: Earth and Environmental Science **61** (2017) 012072 doi:10.1088/1755-1315/61/1/012072

expression levels of CcPILS were evaluated in the scion and rootstock during the graft process of hickory.

## 2. Materials and methods

*2.1 Materials*

Hickory (Carya cathayensis Sarg.) trees were planted at the breeding base of Zhejiang A&F University, Lin'an, China. The leaves, buds, male inflorescences and fruits of 2-year-old hickory were collected for expression pattern analysis. The samples from rootstock and scion were collected at 0, 3, 7 and 14 days after grafting (DAG). After harvesting, the samples were freeze-clamped quickly at liquid $N_2$ temperature stored at -70°C for qRT-PCR analysis.

*2.2 Methods*

*2.2.1 Rapid amplification of cDNA ends (RACE) of CcPILS.* Total RNA was extracted by modified CTAB method. The first strand of cDNA was synthesized using M-MLV reverse transcriptase. The 5'RACE specific primers (5'-RACE) were designed according to the primers design principle of BDSMARTTM RACE cDNA Amplification Kit. According to conservative sequence of CcPILS， the primers were designed assymRev (5'TAGCTCCTCCCGAATCTGCTGAAG-3') and newpilsR (5'-TCCCGAATCTGCTGAAGAAATCCA-3'). The primers of synRev and newpilsR were used as GST and NGSP1, respectively, for Nested PCR. The 3'RACE specific primers were designed as newpilsF (5'-ATCTCCTCCTTATCATCGTCCCCG-3') and SymFor (5'-AGCACTTCAAGCA ACTGAG GAGGT-3'). The PCR amplification was performed according to the following conditions: 98°C 10s, 55°C10s, 72°C 1.5min, a total of 30 cycles. The PCR products of 5'RACE and 3'RACE were purified by 1.5% agarose gel electrophoresis. The product was ligated with pMD-18 vector and transformed into DH5α. The transformed DH5αwas plated on an LB plate and incubated overnight at 37°C. Monoclonal colonies were picked up for culturing large-scale and the positive clones were identified by PCR and sequenced by Sangon Biotech (Shanghai) Co., Ltd.

*2.2.2 Bioinformatics analysis.* The amino acid sequences of *Arabidopsis thaliana* AtPILS1 (GI:1035997954), AtPILS2 (GI:75169666), AtPILS3 (GI:75169730), AtPILS4 (GI:75169729), AtPILS5 (GI:75205686), AtPILS6 (GI:75181312), AtPILS7 (GI:75171357) and *Daucuscarota* DcPILS3 (GI:1040866677), DcPILS6 (GI:1040913933), DcPILS7 (GI:1040905618) and *Malusdomestica* MdPILS3 (GI:1039893723), MdPILS5 (GI:658038089), MdPILS6 (GI:658061887)were obtained by BLAST searches in NCBI, aligned using ClustalX, and subjected to phylogenetic analysis using the Neighbor-Joining method with MEGA6 using 1,000 bootstraps. The physicochemical properties of the proteins were analyzed using the online tool ProtParam (http://web.expasy.org/protparam/). The hydrophobicity of the proteins was analyzed using the ProtScale program (http://web.expasy.org/protscale/). The amino acid transmembrane region prediction was performed using the TMpred program (http://www.ch.embnet.org/software/TMPRED_form.html). The subcellular location of the gene was predicted using TargetP 1.1 (http://www.cbs.dtu.dk/services/TargetP/). The protein secondary and tertiary structure predictions were performed using the Protein Expert System (http://www.expasy.org/).

*2.2.3 Quantitative RT-PCR analysis.* The total RNA was extracted and then the cDNA was synthesized. Quantitative RT-PCR analysis was carried out by TAKARASYBR Premix ExTaq™ (perfect real time) kit. The RT-PCR primers were designed as symFor (5'-AGCACTTCAAGCAACTGAGGAGGT-3') and symRev (5'-TAGCTCCTCCCG AATCTGCTGAAG-3') according to the sequence of CcPILS. CcACTIN was used as an internal standard. The sequences of primer pairs were (5'-GTGAACGGGAAATTGTC-3') and actin Rev (5'-AGAGATGG CTGGAAGAGG-3'). PCR was performed as follows: 94 °C for 1min, 94°C5 s,

60°C 34 s for 35 cycles. The relative RNA levels for CcPILS were calculated from cycle threshold ($C_T$) values according to the $2^{-\Delta\Delta CT}$method.

## 3. Results

### *3.1 Cloning and sequence analysis of CcPILS Gene*

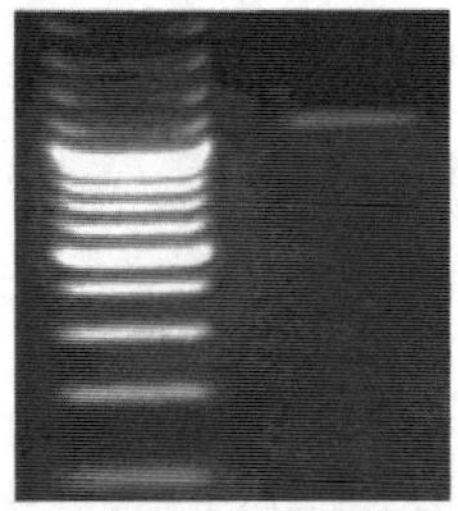

**Fig.1** The PCR results of CcPILS in C.*carthayensis*

Full-length cDNA of CcPILS was cloned from the leaves of hickory by RACE technique with an open reading frame of 1263 bp, which is the same size as expected (Fig. 1). The predicted molecular weight of the protein was 46.22 kDa and the theoretical isoelectric point (PI) was 5.38. The protein was composed of 20 kinds of amino acids, among which the leucine content was the highest (14.0%), and the histidine content is the lowest (0.5%). The CcPILS protein belonged to the labile protein due to its instability index was 41.99. The aliphatic amino acid index of the protein reached 120.64, and the hydrophilic average coefficient was 0.601. ProtScale program predicted the protein hydrophobic region both in the N, C ends of a hydrophobic region, which was separated by the middle of the hydrophilic region. Therefore, the protein was predicated as a hydrophobic and unstable protein. TMpred analysis indicated that CcPILS protein had five transmembrane domains in the N-terminal and C-terminal regions, and the transmembrane region corresponded to the hydrophobic region. Predict protein analysis showed that the protein was localized in the endoplasmic reticulum membrane and belonged to the auxin efflux carrier.

### *3.2 Analysis of CcPILS protein homology and construction of phylogenetic tree*

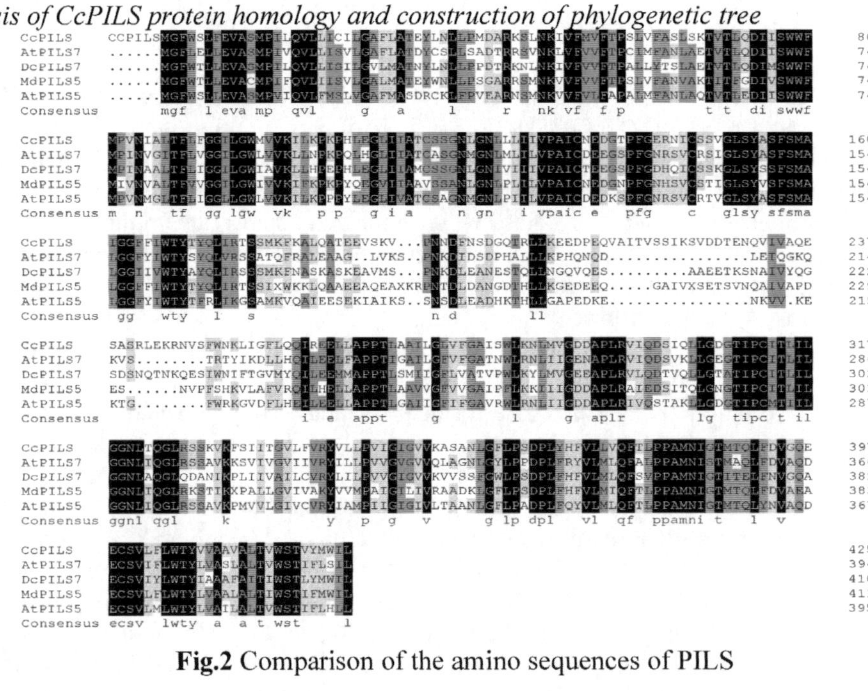

**Fig.2** Comparison of the amino sequences of PILS

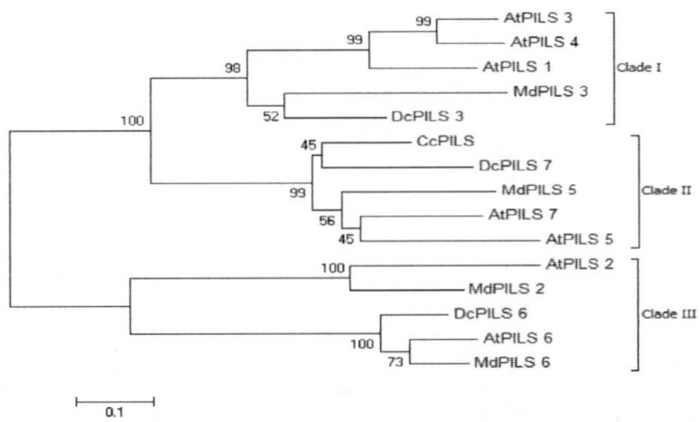


**Fig.3** Phylogenetic tree based on the amino acid of CcPILS and the other PILS genes alignment

To evaluate the homology and the genetic distance of CcPILS protein, we aligned the amino-acid sequence and constructed the phylogenetic tree using the CcPILS, 8 AtPILS, 3 DcPILS and 3 MdPILS. The result showed that the amino acid sequences of CcPILS shared high homology with those of other plants, The consistency was 58.9%, 61.9%, 61.3% and 64.1%, separately (Fig. 2). As shown in Fig.3, CcPILS cluster in the same branch with AtPIL5, AtPIL7, DcPILS7 and MdPILS5.

### *3.3 Spatiotemporal Expression of CcPILS Gene in the graft process*

Expression patterns of CcPILS in different tissues of *C. carthayensis* showed that the relative expression level of CcPILS was the highest in roots, followed by leaves, fruits and male inflorescence (Fig. 4). In addition, the expression levels of CcPILS gene in the graft process of hickory were different at different stages. As shown in the Fig. 5, the relative expression levels of CcPILS were increased and reached the peak at 3 DAG in both rootstock and scion, compared to that of control. The expression level of CcPILS in rootstock was nearly 3 times higher than that in 0 DAG and correspondingly in scion 1.6 times higher. After 7 DAG, its expression level was still induced in rootstock, but inhibited in scion, compared to that of control. The relative expression levels were decreased in both rootstock and scion after 14 DAG, compared to that of control.

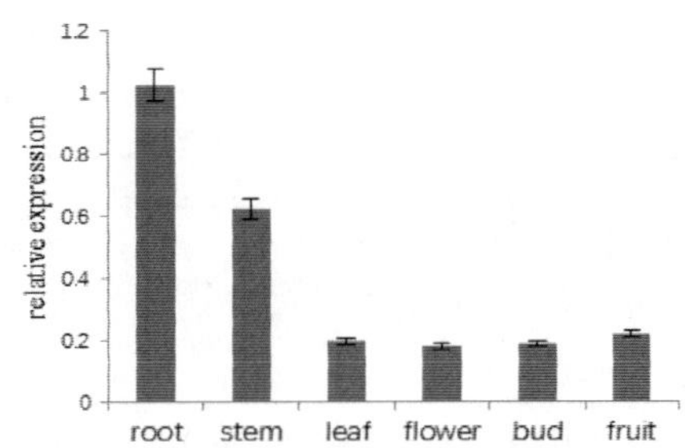


**Fig.4** Expression patterns of CcPILS by real-time PT-PCR analysis in different tissues of *C. carthayensis*

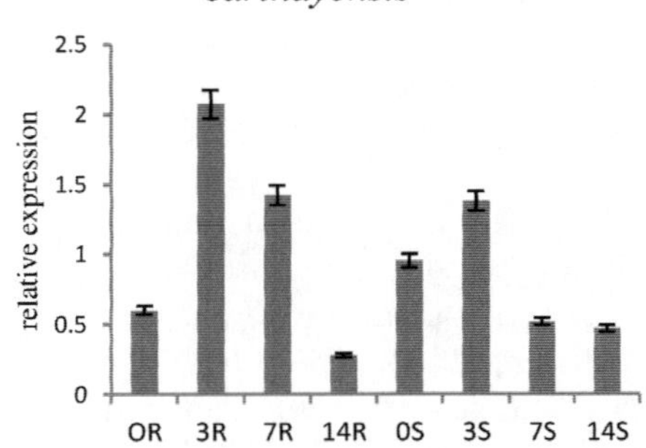


**Fig.5** Expression patterns of CcPILS by real-time PT-PCR analysis during grafting of *C. carthayensis*

## 4. Discussion

Auxin is widely distributed in various organs of higher plants, with a short-distance unidirectional polar transport mode [17]. In this study, the full-length CcPILS gene of hickory was cloned. The results showed that CcPILS had two hydrophobic regions with 3-5 transmembrane domains in each hydrophobic region. There was a hydrophilic ring in the middle of the protein, and the hydrophilic loop was located in the cytoplasm. The intracellular and subcellular locations were similar to PIN5 and PIN8 of Arabidopsis, which were localized in the endoplasmic reticulum and were responsible for the transport of auxin from the cytoplasm to the endoplasmic reticulum, and thus participating in the polar transport of auxin [19-20]. Unlike the PIN family proteins, PILS proteins are also present in lower plants such as algae, suggesting that the PILS family proteins play an important role in biological functions compared to the early appearance of PIN family in evolution [21]. Some research has shown the function of AtPILS5 and AtPILS7 are cellular auxin homeostasis by regulating auxin metabolism [21].

CcPILS protein located in the endoplasmic reticulum and clustered in the same branch with AtPILS5 and AtPILS7, suggesting CcPILS and AtPILS5 and AtPILS7 has a similar function, responsible for auxin from the cytoplasm to the endoplasmic reticulum Transport and thus participate in the auxin-mediated regulation process. Grafting is an important horticultural technique in the asexual reproduction of hickory, and endogenous auxin plays an important role in the process of callus differentiation and formation of vascular bridge [15]. CcPILS, which belongs to Clade III subfamily, may play an important role in this process. CcPILS protein is a high instability index, which belongs to the unstable protein, provides the conditions for rapid regulation of the quantity of CcPILS on the endoplasmic reticulum membrane and the regulation of auxin transport.

The auxin content in 4-6 days to reach the peak in the grafted hickory [15], so auxin efflux carrier gene CcPILS expression was the highest in 3 DAG. Callus between scion and rootstock has formed in 7 DAG [17]. Auxin in the scion efflux slowly to the rootstock, so CcPILS expression in the rootstock was higher than that in the scion. The expression of CcPILS in the scion and rootstock decreased, and researched to the normal level after 14 DAG because of the disappearance of auxin in the grafted hickory.

## 5. Conclusion

CcPILS, a key auxin efflux carrier, encoded 294 amino acids and localized at endoplasmic reticulum membrane. The gene contained a central hydrophilic loop. The CcPILS belonged to Clade III sub-family. Real Time RT-PCR analysis showed that the gene expressions were weakly induced in bud, inflorescence, fruit, leaf and stem, while strongly in root. The expression levels were strongly induced and reached a peak at the third day of grafting in scion and rootstock compared to that of control. CcPILS may be involved in regulating auxin signal transduction during hickory graft process.

## References

[1] Friml J, Vieten A, Sauer M, et al. Efflux-dependent auxin gradients establish the apical–basal axis of Arabidopsis. Nature, 2003, 426(6963):147-53.
[2] Dubrovsky J G, Sauer M, Napsucialymendivil S, et al. Auxin acts as a local morphogenetic trigger to specify lateral root founder cells. Proceedings of the National Academy of Sciences of the United States of America, 2008, 105(25):8790-4.
[3] Zhao Y. Auxin biosynthesis and its role in plant development. Annual Review of Plant Biology, 2010, 61(1):49-64.
[4] Reinhardt D, Pesce E R, Stieger P, et al. Regulation of phyllotaxis by polar auxin transport. Nature, 2003, 426(6964):255-60.
[5] Barbez E, Kubeš M, Rolčík J, et al. A novel putative auxin carrier family regulates intracellular auxin homeostasis in plants. Nature, 2012, 485(7396):119-22.
[6] Feraru E, Vosolsobě S, Feraru M I, et al. Evolution and structural diversification of PILS putative auxin carriers in plants. Frontiers in Plant Science, 2012, 3(4):227.
[7] Huang J, Zhang T, Zhang Q, et al. The mechanism of high contents of oil and oleic acid revealed by transcriptomic and lipidomic analysis during embryogenesis in *Carya cathayensis* Sarg. BMC

genomics 2016, 17:113.

[8] Huang Y, Zhou Q, Huang J, et al. Transcriptional profiling by DDRT-PCR analysis reveals gene expression during seed development in *Carya cathayensis* Sarg. Plant Physiology and Biochemistry 2015, 91:28-35.

[9] Shen C, Xu Y, Huang J, et al. Molecular characterization and expression analysis of the critical floral genes in hickory (*Carya cathayensis* Sarg.). Plant physiology and biochemistry 2014, 83:142-150.

[10] Li J, Zeng Y, Shen D, et al. Development of SSR Markers in Hickory (*Carya cathayensis* Sarg.) and Their Transferability to Other Species of Carya. Current genomics 2014, 15(5):357-379.

[11] Wang Z, Huang J, Huang Y, et al. Deep sequencing of microRNAs from hickory reveals an extensive degradation and 3' end modification. Plant Biotechnology Reports 2013, 8(2):203-209.

[12] Wang Z, Huang J, Huang Y, et al. Discovery and profiling of novel and conserved microRNAs during flower development in *Carya cathayensis* via deep sequencing. Planta 2012, 236(2):613-621.

[13] Wang Z, Huang J, Huang Y, et al. Cloning and Characterization of a Homologue of the FLORICAULA/LEAFY Gene in Hickory (*Carya cathayensis* Sarg). Plant Molecular Biology Reporter 2011, 30(3):794-805.

[14] Jin S, Huang J, Li X, et al. Effects of potassium supply on limitations of photosynthesis by mesophyll diffusion conductance in *Carya cathayensis*. Tree physiology 2011, 31(10):1142-1151.

[15] Zheng B, Chu H, Jin S, et al. cDNA-AFLP analysis of gene expression in hickory (*Carya cathayensis*) during graft process. Tree physiology 2010, 30(2):297-303.

[16] Sima X, Jiang B, Fang J, et al. Identification by deep sequencing and profiling of conserved and novel hickory microRNAs involved in the graft process. Plant Biotechnology Reports 2015, 9(3):115-124.

[17] Qiu L, Jiang B, Fang J, et al. Analysis of transcriptome in hickory (*Carya cathayensis*), and uncover the dynamics in the hormonal signaling pathway during graft process. BMC Genomics, 2016, 17(1):935.

[18] Huang Y, Liu L, Huang J, et al. Use of transcriptome sequencing to understand the pistillate flowering in hickory (*Carya cathayensis*, Sarg.). BMC Genomics, 2013, 14(1):299-299.

[19] Mravec J, Skůpa P, Bailly A, et al. Subcellular homeostasis of phytohormone auxin is mediated by the ER-localized PIN5 transporter. Nature, 2009, 459(7250):1136-40.

[20] Ding Z, Wang B, Moreno I, et al. ER-localized auxin transporter PIN8 regulates auxin homeostasis and male gametophyte development in Arabidopsis. Nature Communications, 2012, 3(1):10-10.

[21] Barbez E. A novel putative auxin carrier family regulates intracellular auxin homeostasis in plants. 2012, 485(7396):119-122.

IOP Conf. Series: Earth and Environmental Science **61** (2017) 012073 doi:10.1088/1755-1315/61/1/012073

# Study on the optimum tilted angle of solar panels in Hainan tropical photovoltaic facility agricultural system

**Jingxuan Wang, Zhiwu Ge*, Xiaoyan Yang, Chunhua Ye, Yanxia Lin**

Hainan normal University, No. 99, Long Kun south road, Haikou, China
Hnnu13@163.com

**Abstract**. Photovoltaic facility agriculture system can effectively alleviate the contradiction between limited land and Photovoltaic power generation. It's flexible to create suitable environment for crop growth, and generate electricity over the same land at the same time. It's necessary to set appropriate solar panel angle to get more solar energy. Through detailed analysis and comparison, we chose the Hay's model as solar radiation model. Based on the official meteorological data got from Haikou Meteorological Bureau, and by comparing the amount of radiation obtained at different tilted angles per month, the optimal placement angle of PV panels at different seasons in Haikou was obtained through calculation, and the optimal placement angle from April to October was also obtained. Through optimized angle and arrangement of solar photovoltaic panels, we can get greater power efficiency.

## 1. Preface

PV facility agriculture system is a system which integrates solar photovoltaic power generation and modern high-tech planting. The system is supported by steel skeleton, and covered by solar photovoltaic panels, which ensures the solar photovoltaic power generation and lighting demands of the crops below at the same time. Reasonable inclination of solar panels can effectively improve the utilization rate of solar energy and increase the power generation efficiency. Hainan province is located in the tropics, the inclination angle of solar panels there is much different from that in mainland.

## 2. Factors affecting the inclination angle

### *2.1.* Continuity

The total solar radiation in a year changes continuously, we can find out the total amount of radiation in each month from the meteorological data, the six consecutive months that have more horizontal radiation are called the summer half year, while the other six months called the winter half year. In the northern hemisphere, most of the summer months range from April to September, winter half year ranges from October to March in next year. The total amount of horizontal radiation in summer half year is always higher than that in winter half year [1].

### *2.2. Uniformity*

In choosing the optimum titled angel, the reception of daily average radiation on the array should be uniform. That is, there should be little difference between summer half year and winter half year, which can avoid the waste of too much radiation in summer while the deficiency in winter[1].

*2.3. Maximum*

In choosing the optimum titled angel, the reception of daily average radiation on the array in winter half year should be maximum[1].

But these three practices are not necessarily applicable to Hainan.

## 3. Calculation of solar radiation

The total solar radiation on inclined plane surface called $H_T$ consists of $H_{bT}$ (direct solar radiation), $H_{dT}$ (diffuse solar radiation from sky) , and $H_{rT}$ (reflection radiation from the earth), that is,

$$H_T = H_{bT} + H_{dT} + H_{rT} \qquad (1)$$

The relationship between $H_{bT}$ and $H_b$ ( direct solar radiation from horizontal plane surface) is [1-4]:

$$H_{bT} = H_b \cdot R_b \qquad (2)$$

$$R_b = \frac{\cos(\varphi-\beta)\cdot\cos\delta\cdot\sin\omega_{sT} + \frac{\pi}{180}\omega_{sT}\cdot\sin(\varphi-\beta)\cdot\sin\delta}{\cos\varphi\cdot\cos\delta\cdot\sin\omega_s + \frac{\pi}{180}\omega_s\cdot\sin\varphi\cdot\sin\delta} \qquad (3)$$

In（3）， $\varphi$ is local latitude， $\beta$ is inclination angle， solar declination can approximately be：

$$\delta = 23.45\sin\left[\frac{360}{365}(284+n)\right] \qquad (4)$$

The solar declination can be approximately expressed as,

$$\delta = 23.45\sin\left[\frac{360}{365}(284+n)\right] \qquad (5)$$

Among them, $n$ is the number of days counted from the New Year's day.

Horizontal sunset angle is[1-4]:

$$\omega_{sT} = \min\left\{\omega_s, \cos^{-1}\left[-\tan(\varphi-\beta)\cdot\tan\delta\right]\right\} \qquad (6)$$

$$H_{dT} = H_d\left[\frac{H-H_d}{H_o}R_b + \frac{1}{2}(1+\cos\beta)\left(1-\frac{H-H_d}{H_o}\right)\right] \qquad (7)$$

Among the above， $H_d$ is diffuse solar radiation on the horizontal plane surface, $H$ is total radiation on the horizontal plane surface, $H_o$ is radiation on the horizontal plane surface outside the atmosphere which can be determined by the formula below[1-4]:

$$H_o = \frac{24}{\pi}I_{SC}\left[1+0.033\frac{360n}{365}\right]\left(\cos\varphi\cdot\cos\delta\cdot\sin\omega_s + \frac{\pi}{180}\omega_s\cdot\sin\varphi\cdot\sin\delta\right) \qquad (8)$$

Among the above, $I_{SC}$ is 1367W/m$^2$ called solar constant. The reflection radiation from the earth is：

$$H_{rT} = \frac{1}{2}\rho H(1 - \cos\beta) \tag{9}$$

Among the above, $\rho$ is called the ground reflectivity, normally $\rho = 0.2$ .

Substituting (2), (7)and(9) into (1),we can get the formula for total solar radiation in the inclined plane surface：

$$H_T = H_b \cdot R_b + H_d\left[\frac{H - H_d}{H_O}R_b + \frac{1}{2}(1+\cos\beta)\left(1 - \frac{H - H_d}{H_O}\right)\right] + \frac{1}{2}\rho H(1 - \cos\beta) \tag{10}$$

If we know the data of the horizontal solar radiation in different months, we can calculate the amount of solar radiation on different slopes.

Table 1 is the amount of sunny days from April to October, 2014 in Haikou, Hainan Province. Table 2 is monthly total radiation from April to October based on the information from Haikou Meteorological Bureau. It can be calculated that the direct solar radiation and diffuse solar radiation in the horizontal plane surface from April to October,2014 in Haikou, Hainan Province according to table 2[5].

Table1. Amounts of Sunny Days from April to October

| Month / Days / Weather | April | May | June | July | August | September | October |
|---|---|---|---|---|---|---|---|
| Sunny | 21 | 19 | 19 | 22 | 18 | 21 | 17 |
| Cloudy/Rainy | 9 | 12 | 11 | 9 | 13 | 9 | 14 |

Table2. Solar Radiation in the Horizontal Plane Surface from April to October

| Month / Radiation | April | May | June | July | August | September | October |
|---|---|---|---|---|---|---|---|
| Total Radiation （$MJ/m^2$） | 423.5 | 540.3 | 508.5 | 628.7 | 526.1 | 448.7 | 381.4 |
| Direct Solar Radiation （$MJ/m^2$） | 251.98 | 281.48 | 273.74 | 379.25 | 259.66 | 266.98 | 177.78 |
| Diffuse Solar Radiation （$MJ/m^2$） | 171.52 | 258.82 | 234.76 | 249.45 | 266.44 | 181.72 | 203.62 |

As for Photovoltaic array, angle $\beta$ normally changes from （$\varphi - 5$） to （$\varphi + 25$） every $5^\circ$, $\delta$ and $\omega_s$ can be calculated by those seven angels. The radiations on inclined plane surface at different angles are shown in Table 4.

Table 3. $\delta$, $\omega_s$ and $H_o$

| n | $\delta$ (degrees) | $\omega_s$ (degrees) | $H_o$ ($W/m^2$) |
|---|---|---|---|
| April 3rd (93) | 4.81 | 91.76 | 41294.60 |
| May 15th (135) | 18.79 | 97.11 | 60244.13 |
| June 15th (166) | 23.31 | 99.02 | 72631.88 |
| July 15th (196) | 21.52 | 98.25 | 83250.54 |
| August 15th (227) | 13.78 | 95.12 | 91477.41 |
| September15th (258) | 2.22 | 90.81 | 94199.37 |
| October15th (288) | -9.6 | 86.47 | 90859.76 |

According to Table 4, in Haikou, the optimum titled angle of Photovoltaic array is 15°in summer half year and it is 25° in winter half year.

## 4. Conclusion

Through detailed analysis and comparison, we choose the Hay's model as the solar radiation model. The model has the advantages of simple expression and high accuracy.Based on meteorological data, and the official data got from Haikou Meteorological Bureau, according to different seasons, the optimal placement angle of PV panels in Haikou is obtained by calculation.

Table 4. Radiation ($MJ/m^2$) on Different Inclined Plane Surfaces in Haikou

| Month \ Degree (°) | 15 | 20 | 25 | 30 | 35 | 40 | 45 |
|---|---|---|---|---|---|---|---|
| April | 427.86 | 424.71 | 419.23 | 411.68 | 402.08 | 390.19 | 376.53 |
| May | 515.99 | 503.28 | 488.35 | 471.20 | 452.15 | 431.21 | 408.72 |
| June | 476.86 | 461.95 | 445.46 | 427.14 | 407.02 | 385.66 | 362.82 |
| July | 590.47 | 572.13 | 551.45 | 528.44 | 502.74 | 475.52 | 446.42 |
| August | 512.43 | 503.03 | 491.05 | 477.59 | 461.59 | 443.53 | 423.69 |
| September | 458.64 | 456.91 | 452.98 | 446.32 | 437.49 | 426.50 | 413.10 |
| October | 404.72 | 408.93 | 410.20 | 409.96 | 407.52 | 403.08 | 399.34 |

Photovoltaic facility agriculture system can effectively alleviate the contradiction between people and land, promoting economic and social sustainable development. It's flexible to create suitable environment for different crops to grow, and generate electricity to meet the agricultural demand. Through optimized arrangement of photovoltaic panels, we can get greater power efficiency. It can also be developed into a kind of sightseeing agriculture. All in all, photovoltaic facility agriculture system has important significance for supply-side structural reform of energy and agricultural industry.

**Acknowledgments**

Thanks for the funding of Hainan province natural science foundation (Grant Number 20165198).

**References**

[1] Yang Jin-Huan, Analysis of optimum tilted angle for fixed photovoltaic arrays, Acta Energiae Solaris Sinica ,1992.vol13,No.1： 86-92.
[2] Shenzheng, Lvjian, Yanghongxing, Jiangying, Investigation on optimum tilt angles of solar radiation absorbing surfaces, Journal of Tianjin Institute of urban construction, Vol.15, No.1, Mar. 2009, 61-75.
[3] Yang Jin-huan, Mao Jia-jun, Chen Zhong-hua, Calculation of solar radiation on variously

oriented tilted surface and optimum tilt angle, Journal of Shanghai Jiaotong University, July 2002, Vol.36 No.7,1032-1036.

[4] Yang Jin-Huan, Further study of optimum tilted angle for fixed photovoltaic arrays — Comments on"Optimum tilted angle for fixed photovoltaic arrays", Acta Energiae Solaris Sinica ,1993, vol.14,No.4：344-347.

[5] Peng Yin-sheng, Wang Ying, Lan Qing ,The climatologically calculation and distributive characteristics of global solar radiation in Hainan Province, Natural science journal of Hainan university, 2007, Vol. 25 , No. 3: 259-264.

IOP Conf. Series: Earth and Environmental Science **61** (2017) 012074 doi:10.1088/1755-1315/61/1/012074

# A typical proficiency testing programmes sample design for electrical and electronic product

**T T Wang, H Zhang, L L Xie and Y Y Wang**

Shanghai Institute of Quality Inspection and Technical Research, No. 900, Jiangyue Road, Minhang District, Shanghai, China

**Abstract.** Creepage distance and clearance testing are the basic testing items in the safety standards for almost all electrical and electronic products. A typical sample group is designed in this paper for the purpose of proficiency testing programmes. The sample group is composed of two kinds of circuit board. The length of the creepage distance of the two circuit boards in pollution degree 2 and 3 are the same but with different paths. This sample group includes three testing points. This sample group is designed beneficial for numerical statistics and avoiding the data complicity in the laboratory. It can be used for effective laboratory monitoring.

## 1 Introduction

Creepage distance is the shortest distance along the surface of a solid insulating material between two conductive parts. Clearance is the shortest distance in air between two conductive parts. A creepage distance cannot be less than the associated clearance so that the shortest creepage distance possible is equal to the required clearance. However, there is no physical relationship, other than this dimensional limitation, between the minimum clearance in air and the minimum acceptable creepage distance [1].

Creepage distance and clearance is two key test items of electrical and electronic products, which is related to the working voltage, insulation type and environment. It is put forward as a safety requirement in kinds of standards [2-7]. If it is too small, the dis-match creepage distance and clearance will cause the equipment breakdown and the electric shock to user. This is a very serious product safety issue and will damage people's life and property safety. Therefore, the test items are always as the focus of the products monitor.

In this paper, a typical proficiency testing programmes sample is designed here. It can be used to control the quality of the testing lab.

## 2 Sample design

The sample is designed as a printed circuit board. The appearance figure of the sample is in figure 1.

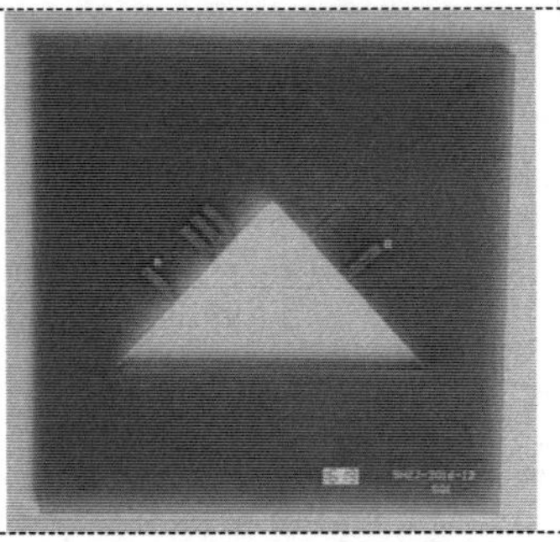

**Figure 1.** The sample for proficiency testing programmes.

A hollow out triangular is in the middle of the sample. Rectangular slot and the conductive metal pieces are designed in the sample. The appearance figure of the sample is in figure 1.

*Design points*

This designing scheme includes two sets of samples made of printed circuit board called sample A and sample B. The production precision is 0.01 mm. The difference between two kinds of samples lies in the vertex angle. The vertex angle of sample A is 75 degrees and sample B is 85 degrees. According to our designing scheme, the length of the creepage distance of two pieces of samples in pollution degree 2 and degree 3 are the same. However, the paths are different.

The most critical part of the creepage distance and clearance testing is how to choose the testing path, which is the key to get the correct results. In this designing scheme, three testing points is included.

- Point one-Groove.

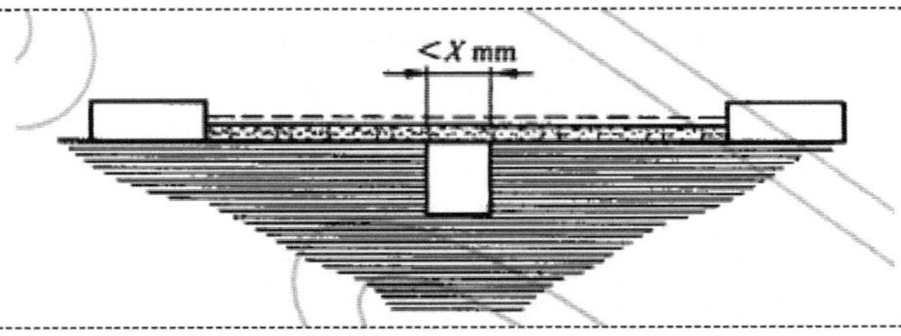


**Figure 2.** Point one-Groove (condition 1).

Condition 1： Path under consideration includes a parallel- or converging-sided groove of any depth with a width less than X mm.

Rule: Creepage distance and clearance are measured directly across the groove as shown.

The dimension X, specified in the examples, has a minimum value depending on the pollution degree as follows.

**Table 1** The relationship between pollution degree and dimension X minimum value

| Pollution degree | Dimension X minimum value |
|---|---|
| 1 | 0.25mm |
| 2 | 1.0mm |
| 3 | 1.5mm |

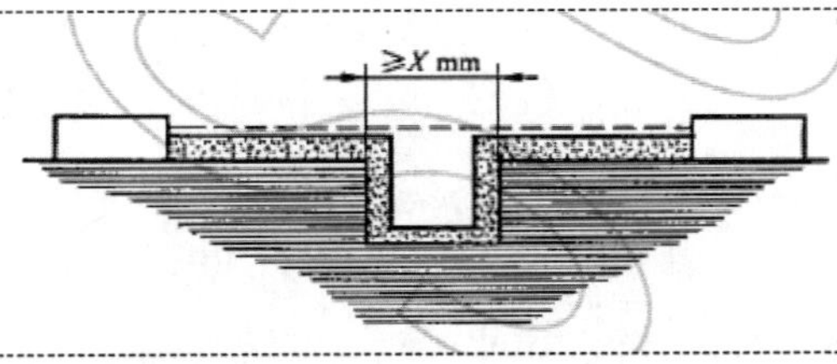


**Figure 3.** Point one-Groove(condition 2).

Condition 2：Path under consideration includes a parallel-sided groove of any depth and equal to or more than X mm.

Rule: Creepage is the "line of sign" distance. Clearance path follows the contour of the groove.

- Point two- V-shaped groove.

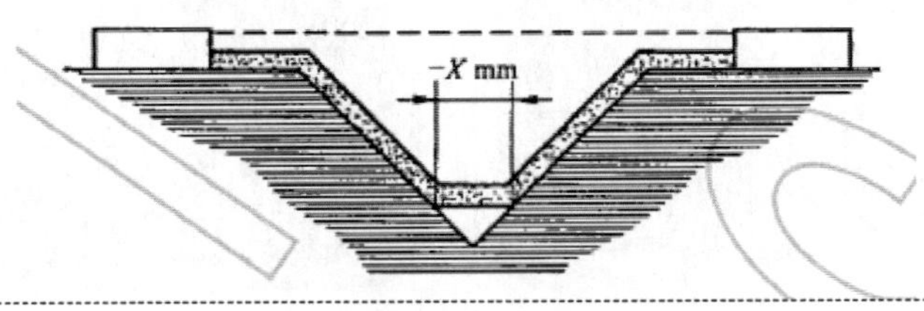


**Figure 4.** Point two- V-shaped groove.

Condition：Path under consideration includes a V-shaped groove with a width greater than X mm.

Rule: Creepage is the "line of sign" distance. Clearance path follows the contour of the groove but "short-circuits" the bottom of the groove by X mm link.

- Point three- Conductive floating part.

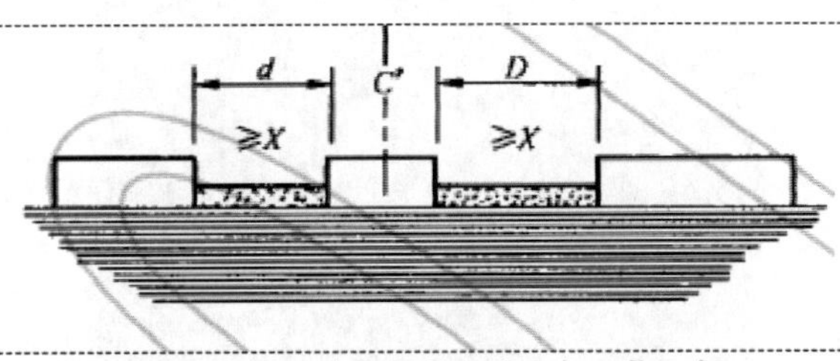


**Figure 5.** Point three-Conductive floating part.

Condition：C is the conductive floating part.

Rule: Clearance is the distance=d+D. Creepage distance is also=d+D.

*Statistical methods*

Using robust statistical methods to evaluate the laboratory's testing result. A calculated value Z is used to represent the satisfaction of the laboratory's test result, which is

$$Z = (x - X)/\sigma \tag{1}$$

Where, x is the testing result of the laboratory. X is the specified value. $\sigma$ is the standard deviation. Here, the specified value is the median value and the standard deviation is the standardized interquartile range (NIQR).

When $[Z] \leq 2$, the result is satisfied. When $2 < [Z] < 3$, the result is suspicious. When $[Z] \geq 3$, the result is not satisfied.

IOP Conf. Series: Earth and Environmental Science **61** (2017) 012074 doi:10.1088/1755-1315/61/1/012074

## 3 The result analysis

### *3.1 The correct path*

In this part, the correct path of the creepage distance and clearance for sample A and sample B in pollution degree 2 and 3 is given here. Draw the path in the red line.

- Sample A (vertex angle is 75 degrees):

The creepage distance and clearance for sample A in pollution degree 2 and 3 are shown in figure 6 to figure 8.

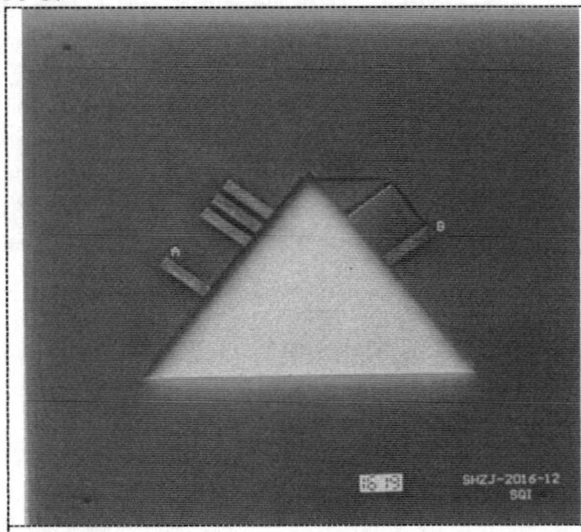

**Figure 6.** The creepage distance for sample A in pollution degree 2.

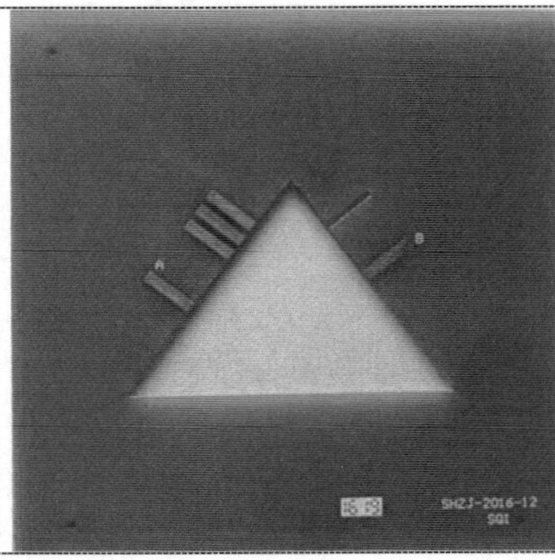

**Figure 7.** The creepage distance for sample A in pollution degree 3.

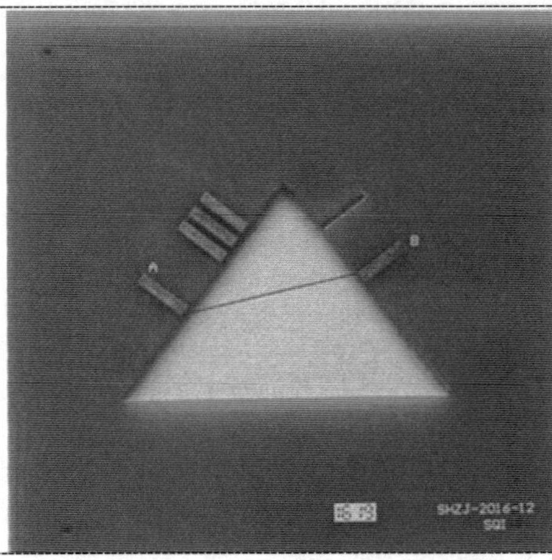

**Figure 8.** The clearance for sample A in pollution degree 2 and 3

- Sample B (vertex angle is 85 degrees):

The creepage distance and clearance for sample B in pollution degree 2 and 3 are shown in figure 9 to figure 12.

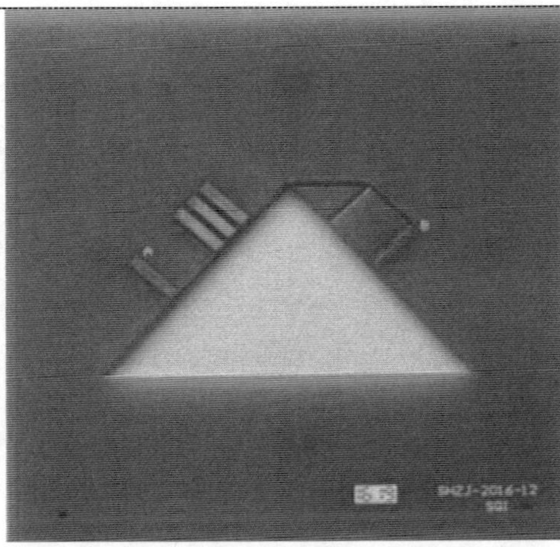

**Figure 9.** The creepage distance for sample B in pollution degree 2.

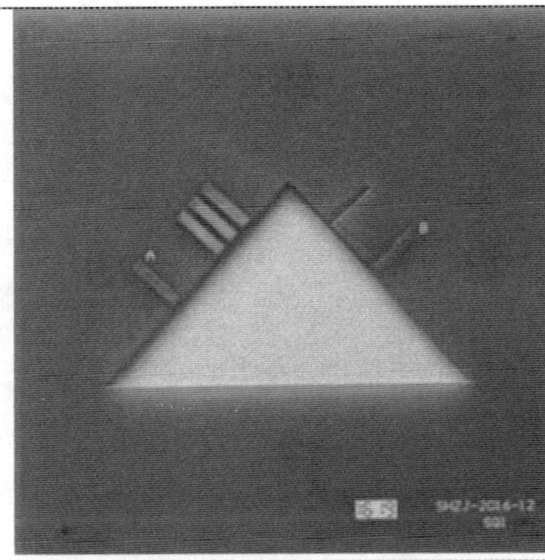

**Figure 10.** The creepage distance for sample B in pollution degree 3.

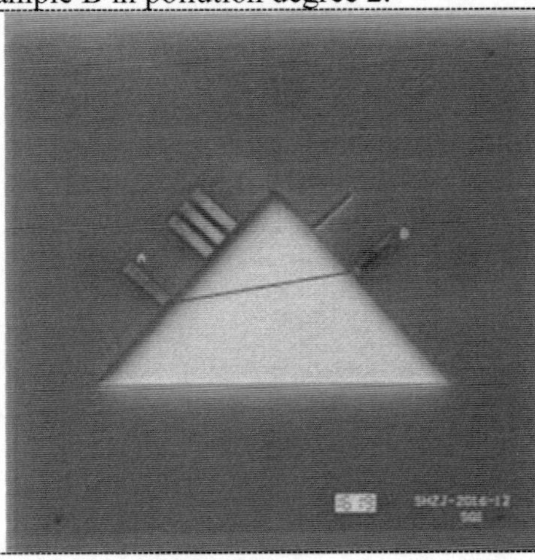

**Figure 11.** The clearance for sample B in pollution degree 2 .

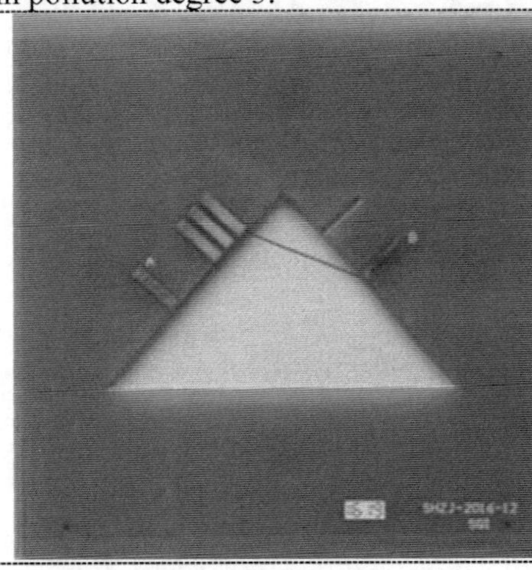

**Figure 12.** The clearance for sample B in pollution degree 3.

### *3.2 Typical error path*

In this part, several typical error paths are presented here.

- Error path one: The groove leads to an error path of the creepage distance in pollution degree 2. Figure 13 is an example for sample A.

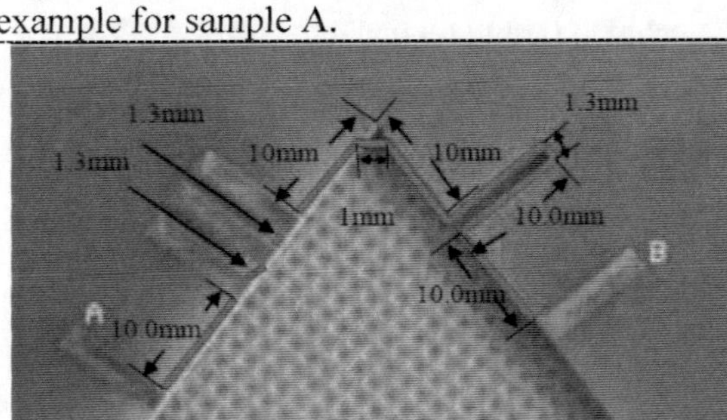


**Figure 13.** The error creepage distance for sample A in pollution degree 2.

IOP Conf. Series: Earth and Environmental Science **61** (2017) 012074 doi:10.1088/1755-1315/61/1/012074

- Error path two: The continuous conduction leads to an error path of the creepage distance or clearance in pollution degree 3. Figure 14 and figure 15 are the examples.

**Figure 14.** The error creepage distance for sample B in pollution degree 3.

**Figure 15.** The error clearance for sample B in pollution degree 3.

## 4 Conclusion

Creepage distance and clearance testing are the basic testing items in the safety standards for almost all electrical and electronic products. A typical sample group is designed in this paper for the purpose of proficiency testing programmes. The sample group is composed of two kinds of circuit board. The length of the creepage distance of the two circuit boards in pollution degree 2 and 3 are the same but with different paths. This sample group includes three testing points. This sample group is designed beneficial for numerical statistics and avoiding the data complicity in the laboratory. It can be used for effective laboratory monitoring.

## Acknowledgments

Authors wishing to acknowledge financial support from the project of the shanghai bureau of quality and technical supervision "The assembly and assessment of the Shanghai energy-saving standard".

## References

[1] IEC 60664-1(ed. 2.0) 2007 *Insulation coordination for equipment within low-voltage systems-Part 1:Priciples,requirements and tests*

[2] IEC 60335-1(ed. 5.2) 2016 Household and similar electrical appliances - Safety - Part 1: General requirements

[3] IEC 60950-1(ed. 2.2) 2013 Information technology equipment - Safety - Part 1: General requirements

[4] IEC 60065(ed.8.0) 2014 Audio, video and similar electronic apparatus - Safety requirements

[5] IEC 61558-1(ed. 2.1) 2009 Safety of power transformers, power supplies, reactors and similar products - Part 1: General requirements and tests

[6] IEC 60598-1(ed. 8.0 ) 2014 Luminaires - Part 1: General requirements and tests

[7] IEC 60884-1 (ed. 3.2) 2013 Plugs and socket-outlets for household and similar purposes - Part 1: General requirements

IOP Conf. Series: Earth and Environmental Science **61** (2017) 012075 doi:10.1088/1755-1315/61/1/012075

# Detection of sulfonamide resistant bacteria and resistance genes in soils

**Nan WU[1], Weiyu ZHANG[1]*, Huifen LIU[2], Xiaobo WANG[2], Fan YANG[1], Ming ZENG[3], Pinpin CHEN[1], and Xiao WANG[1]**

[1] College of Engineering and Technology, Tianjin Agricultural University, Tianjin 300384, China

[2] College of Agronomy and Resources and Environment, Tianjin Agricultural University, Tianjin 300384, China

[3] College of Marine and Environment, Tianjin University of Science and Technology, Tianjin 300457, China

Corresponding author's e-mail address: zhangweiyu@tjau.edu.cn

**Abstract**. Manure application could accelerate the environmental dissemination of antibiotic resistant bacteria (ARB) and antibiotic resistance genes (ARGs) in soils. In this study, the prevalence of sulfonamide resistant bacteria and resistance genes was investigated in agricultural soils to which organic manures had been applied in Tianjin, China. Anti-sulfonamide bacteria were found in the range of $3.29 \times 10^4$ to $1.70 \times 10^5$ CFU/g dry soil, occupying 1.5% to 2.2% of total viable counts. And *sul*I and *sul*II genes were detected in all sampling sites, with relative abundances of $5.69 \times 10^{-5}$ to $6.95 \times 10^{-4}$ and $4.28 \times 10^{-4}$ to $1.25 \times 10^{-3}$ respectively. No significant correlations between cultivable sulfonamide resistant bacteria and *sul* genes were found in this study. While *sul*I showed significant positive correlation with soil organic matter. Overall, the results highlight that soil plays an important role in resistance genes capture as the environmental reservoir.

## 1. Introduction

Manure is a major source of antibiotic pollution in the environment. Since many antibiotics are poorly absorbed by the animals and subsequently are excreted into the environment by urine or feces. The antibiotics tend to persist and accumulate in soils after repeated manure application. Residual antibiotics may exert selection pressure on environmental microorganisms, accelerating the environmental dissemination of antibiotic resistance. Bacteria have been shown to readily share genetic information by horizontal gene transfer (HGT) mechanism driven by mobile genetic elements, permitting the transfer of antibiotic resistance genes (ARGs) among environmental bacteria. Numerous antibiotic resistant bacteria (ARB) and ARGs have been widely detected in soil in the past few decades [1-4]. ARB and ARGs in soils have the potential to pose risks to human health, as susceptible pathogenic bacteria can become resistant by acquiring resistance genes in the environmental media [5]. Therefore, it is of great importance to investigate the behavior of ARB and ARGs in agricultural soils.

Sulfonamides are one of the most commonly used in livestock system. Sulfonamide-resistance genes (*sul* genes) were also observed with high levels of abundances in Tianjin area [6]. In this work, the prevalence of sulfonamide resistance was investigated in agricultural soils to which organic manures had been applied from four districts in Tianjin, China. Culture-dependent method was used to assess the resistance rate of bacteria in soils exposed to different sulfonamide concentrations. The quantitative PCR (qPCR) was used to quantify *sul*I and *sul*II genes. Meanwhile, the relationship between ARGs, ARB and environmental factors was also investigated.

## 2. Materials and Methods

### 2.1. *Soil samples*

Soil samples were collected from agricultural fields to which organic manures had been applied. The samples were taken from Jixian, Xiqing, Wuqing and Dagang Districts (define as JX, XQ, WQ and DG respectively) in Tianjin, China, between March and May 2016. For each site, three subsamples were taken from 0 to 10 cm surface soil and were mixed to form one composite sample. Fresh samples were processed immediately for cultivation of ARB. The other samples were stored at -80°C before DNA extraction and chemical analysis. Weights of wet soils and oven-dry soils were measured to derive soil moistures. Soil pH was determined with a soil to water ratio of 1:2.5. Total carbon (TC) was measured by a by a TOC analyzer (TOC-VCPH, SHIMADZU). Organic matter (OM) was determined by the $K_2Cr_2O_7$ oxidation method. The physicochemical properties of soil samples are summarized in Table 1.

**Table 1.** Physicochemical properties of soil samples

| Samples | Moisture (%) | pH | TC (mg/g) | OM (%) |
|---|---|---|---|---|
| JX | 17.87±0.08 | 7.51±0.22 | 19.59±0.12 | 4.46±0.03 |
| XQ | 21.15±0.10 | 7.38±0.04 | 28.88±2.01 | 3.79±0.40 |
| WQ | 30.13±0.21 | 7.86±0.02 | 26.80±0.28 | 3.12±0.04 |
| DG | 7.67±0.01 | 8.14±0.06 | 25.48±1.26 | 3.75±0.24 |

### 2.2. *Detection of ARB*

Each soil sample was measured in duplicate to determine the numbers of ARB using colony forming units count method described by Chen et al. [5] with minor modifications. Briefly, 5.0 g of wet soil was mixed with 45 mL sterile saline solution (0.85% NaCl) and shaken vigorously on an oscillator at 200 rpm for 20 min. 100 μL of ten-fold serial dilutions for each suspension were spread onto a broth agar plates containing selected concentrations of antibiotics (20 and 50 μg/mL of sulfadiazine). Broth agar plates with no antibiotics was used to determine the total cultivable bacteria numbers.

### 2.3. *Detection of ARGs*

Total DNA was extracted from 0.5 g of soil (fresh weight) by using FastDNA SPIN kit for Soil (MP Blomedicals, LLC., France). The concentration and quality of the extracted DNA was determined by spectrophotometer analysis (NanoDrop ND-1000, Theromo Fisher Scientific). PCR assays were used for broad-scale screening of the presence/absence of *sul* genes, and to get the standards for subsequent qPCR analysis. All PCR assays were conducted in a Peltier Thermal Cycler (Bio-Rad). Primers and annealing temperatures are described in Table 2.

After PCR amplification, gel slices of an agarose gel containing the PCR products were excised, and purified using EasyPure Quick Gel Extraction Kit (TransGen). The purified PCR products were ligated into p-GEM T easy vector (Promega) and then cloned into *Escherichia coli* DH5α (Tiangen). Clones containing target gene inserts were picked and sequenced. If the gene inserts were verified as the object resistance genes using the BLAST alignment tool, clones that had right gene inserts were chosen as the standards for the subsequent qPCR. Plasmids carrying target genes were extracted with Plasmid Kit (TaKaRa). Two target genes (*sul*I and *sul*II) were quantified by qPCR using a SYBR-Green approach. 16S rRNA genes were quantified according to the TaqMan qPCR method [8]. 10-fold serial dilutions of a known copy number of plasmid carrying respective genes were generated to produce the standard curve. The qPCR mixtures (total volume, 20 μL) consisted of 10 μL of SYBR

Premix Ex Taq (TaKaRa), 0.3 μL of each primer (20 μM), 1.6 μL of template DNA and 7.8 μL of dd$H_2O$. The qPCR procedure was conducted using an iCycler IQ5 Thermocycler (Bio-Rad). The protocol was as the following program: 45 s at 95°C, followed by 40 cycles of 10 s at 95°C, 30 s at the annealing temperatures. Product specificity was confirmed by melting curve analysis (55–95°C, 0.5°C per read, 30 s hold). Each reaction ran in triplicate. The PCR efficiencies (96.7–105.1%) were examined to test for inhibition. $R^2$ values were more than 0.99 for all calibration curves.

**Table 2.** Primers and PCR conditions used in this study

| Target genes | Primer | Sequences ( 5′—3′ ) | Amplicon size（bp） | Annealing temp (°C) | Reference |
|---|---|---|---|---|---|
| *sul*I | *sul*I-F | cgcaccggaaacatcgctgcac | 163 | 56 | [7] |
| | *sul*I-R | tgaagttccgccgcaaggctcg | | | |
| *sul*II | *sul*II-F | tccggtggaggccggtatctgg | 191 | 61 | [7] |
| | *sul*II-R | cgggaatgccatctgccttgag | | | |
| 16S rRNA | 1369F | cggtgaatacgttcycgg | 123 | 56 | [8] |
| | 1492R | ggwtaccttgttacgactt | | | |

2.4. *Data analysis*
Averages and standard deviations of all data were determined using Microsoft Excel, 2007. All statistical analyses were performed using SPSS version 19.0, and significant differences were determined by the *t* test. Bivariate correlation analyses were conducted to obtain Spearman's coefficients.

## 3. Results and Discussions

### 3.1. *Occurrence of sulfonamide-resistant bacteria*

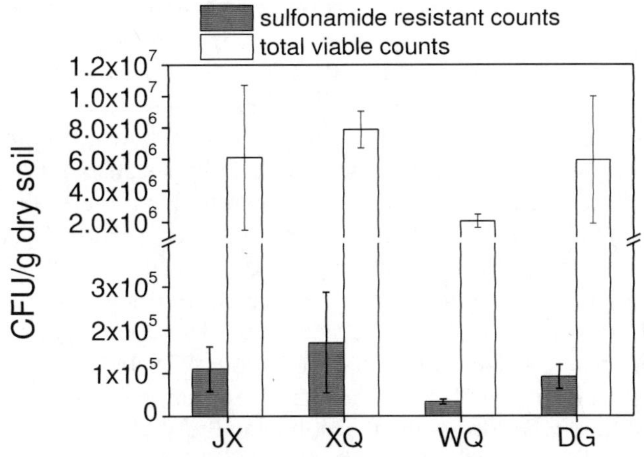


**Figure 1.** Total viable microbial counts and sulfonamide-resistant counts (resistant to 50 μg/mL of sulfadiazine) in soil samples.

As shown in figure 1, total concentrations of cultivable bacterial cells in soils ranged from 2.05 × $10^6$ to 7.86 × $10^6$ CFU/g dry soil. Cultivable live anti-sulfonamide bacteria were observed in all soil samples, ranging from 3.29 × $10^4$ to 1.70 × $10^5$ CFU/g dry soil. The relative abundances of sulfonamide-resistant strains (resistant to 50 μg/mL of sulfadiazine) observed in this study (1.5% to 2.2%) were much lower than those in wastewater-irrigated soils (approximately 10% to 20%) [5].

These resistant bacteria may come directly from the indigenous soil environment. They may also result from the stress of antibiotics in soil which needs to be further studied.

Meanwhile, the sulfonamide resistance rate of bacteria in soils exposed to different sulfonamide concentrations are investigated (figure 2). With the increase of sulfonamide concentration (from 20 to 50 μg/mL of sulfadiazine), the resistance rate of bacteria showed a decreasing trend (4.4%-9.4% at 20 μg/mL, 1.5%-2.2% at 50 μg/mL). Significant difference was observed in resistance rate under different sulfonamide concentrations ($P < 0.01$). But exposed to the same sulfonamide concentration, there is no significant difference in drug resistance of bacteria among different samples.

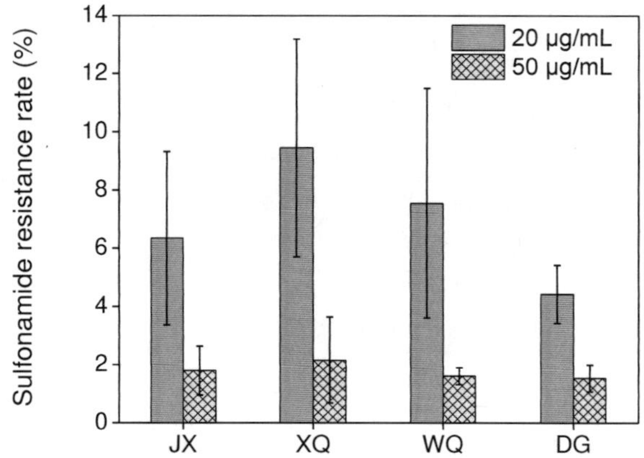


**Figure 2.** The sulfonamide resistance rate of bacteria in soil samples exposed to different sulfonamide concentrations (20 and 50 μg/mL of sulfadiazine).

### 3.2. *Determination of sulfonamide resistance genes*

Absolute gene copy numbers of two *sul* genes (*sul*I and *sul*II) in soil samples were presented in figure 3a. The absolute abundances of *sul*I genes in JX ($1.58 \times 10^7$ copies/g dry soil) and XQ ($1.82 \times 10^7$ copies/g dry soil) were comparable, but significantly higher than those in WQ and DG ($10^6$ copies/g dry soil). The highest abundance of *sul*II genes was occurred in WQ ($1.32 \times 10^8$ copies/g dry soil), approximately 1 order of magnitude higher than those in other three samples. This could be related to the high 16S rRNA gene copies in WQ ($1.58 \times 10^{11}$ copies/g dry soil).

To minimize variance caused by different extraction and analytical efficiencies and differences in background bacterial abundances, the absolute number of all resistance genes were normalized to that of ambient 16S rRNA genes (figure 3b). The same general trends in gene abundance were seen in the normalized data relative to absolute data, since the total number of 16S rRNA gene copies was found to be relatively consistent among different sites at $10^{10}$ copies/g dry soil, except for WQ. The relative abundances (target gene/16S rRNA genes) of *sul* genes showed significant variation over sampling sites, ranging from $10^{-5}$ to $10^{-3}$. In all sampling sites, the relative abundances of *sul*II ($4.28 \times 10^{-4}$ to $1.25 \times 10^{-3}$) were higher than those of *sul*I ($5.69 \times 10^{-5}$ to $6.95 \times 10^{-4}$). This observation is consistent with the results obtained in the aquaculture environment of Tianjin (relative abundances of $10^{-4}$ to $10^{-3}$ for *sul*II, and $10^{-5}$ to $10^{-4}$ for *sul*I) [6]. The total absolute and relative abundances of two *sul* genes (*sul*I and *sul*II) on average in different sites were in the range of $3.70 \times 10^7$ to $1.41 \times 10^8$ copies/g dry soil, and $8.91 \times 10^{-4}$ to $1.95 \times 10^{-3}$ respectively.

In this study, *sul*I and *sul*II genes were detected with high abundance, which were comparable to those in wastewater and reclaimed water irrigated soils in China [4, 5]. The high prevalence of *sul* genes could be related to residual antibiotics in soils introduced by manure application or wastewater irrigation. However, even in the absence of continuous selection of sulfonamides, *sul*I was found to be associated with persistent sulfonamide resistance in microorganisms [6]. Overall, the results highlight that soil plays an important role in resistance genes capture as the environmental reservoir.

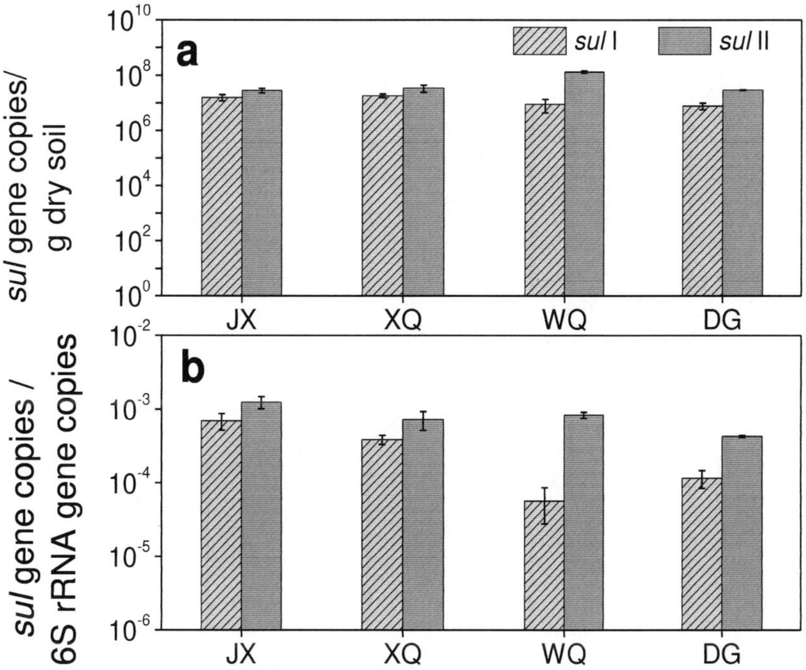


**Figure 3.** Detected levels of *sul* genes (*sul*I and *sul*II) normalized to (a) sample volume (copies per g dry soil) (b) copies of ambient 16S rRNA gene present in soil samples in different sites. Error bars indicate the standard deviation of three replicates.

**Table 3.** Correlation analysis of relative abundances of ARGs, ARB and environmental factors

| | *sul*I | *sul*II | *sul* | pH | TC | OM | sulfonamide resistant bacteria |
|---|---|---|---|---|---|---|---|
| *sul*I | 1.0 | 0.4 | 0.8 | -0.6 | -0.4 | 1.0** | 0.8 |
| *sul*II | 0.4 | 1.0 | 0.8 | -0.4 | -0.4 | 0.4 | 0.0 |
| *sul* | 0.8 | 0.8 | 1.0 | -0.8 | -0.2 | 0.8 | 0.6 |
| pH | -0.6 | -0.4 | -0.8 | 1.0 | -0.4 | -0.6 | -0.8 |
| TC | -0.4 | -0.4 | -0.2 | -0.4 | 1.0 | -0.4 | 0.2 |
| OM | 1.0** | 0.4 | 0.8 | -0.6 | -0.4 | 1.0 | 0.8 |
| sulfonamide resistant bacteria | 0.8 | 0.0 | 0.6 | -0.8 | 0.2 | 0.8 | 1.0 |

** represents statistical significance with $P < 0.01$; *sul* represents the sum of *sul*I and *sul*II genes.

### 3.3. *Relationship between ARGs, ARB and environmental factors*

According to the Spearman's correlation coefficients (Table 3), no significant correlations between cultivable sulfonamide resistant bacteria and *sul* genes were found in this study. This is not surprising given that only a very small fraction of soil microbes were targeted in cultivation. A considerable part of *sul* genes quantified could possibly exist in uncultivable bacteria. In addition, the relative abundance of *sul*I showed significant positive correlation with soil organic matter. This observation was in agreement with previous report [3]. But for *sul*II, no significant positive correlation was found with organic matter. Recent data have shown a significant negative relationship between ARGs and soil pH, since soil pH exerted a strong selection pressure on soil microbes and appeared to have a pervasive effect on the abundance of bacteria [4]. However there is no significant relationship between *sul* genes and pH (r = -0.8, $P > 0.05$) in this study. This observation was in agreement with our previous report [3].

## 4. Conclusions

This study demonstrates the presence of high levels of sulfonamide resistance in agricultural soils in Tianjin area, raising concerns that the overuse of antibiotics in livestock system contributes to the environmental reservoir of resistance, through the potential pathways such as land application of manures. No significant correlations between cultivable sulfonamide resistant bacteria and *sul* genes were found in this study. But *sul*I showed significant positive correlation with soil organic matter. The relationship between ARGs and other environmental factors such as residual antibiotics, needs to be further studied. Overall, the results highlight that soil plays an important role in resistance genes capture as the environmental reservoir.

## Acknowledgments

This research was financially supported by Natural Science Foundation of Tianjin (No. 14JCYBJC43700, 15JCYBJC53700), Scientific Research Foundation for the Returned Overseas Chinese Scholars, National Natural Science Foundation of China (No. 21607114, 51308392), and National Undergraduate Training Programs for Innovation and Entrepreneurship (No. 201610061014).

## References

[1] H. Lin, W.C. Sun, Z.L. Zhang, S.J. Chapman, T.E. Freitag, J.R. Fu, X. Zhang, J.W. Ma, Effects of manure and mineral fertilization strategies on soil antibiotic resistance gene levels and microbial community in a paddy-upland rotation system, Environ. Pollut. 211 (2016) 332-337.

[2] Y.G. Zhu, T.A. Johnson, J.Q. Su, M. Qiao, G.X. Guo, R.D. Stedtfeld, S.A. Hashsham, J.M. Tiedje, Diverse and abundant antibiotic resistance genes in Chinese swine farms, Proc. Natl Acad. Sci. USA. 110 (2013) 3435-3440.

[3] N. Wu, M. Qiao, B. Zhang, W.D. Cheng, Y.G. Zhu, Abundance and diversity of tetracycline resistance genes in soils adjacent to representative swine feedlots in China, Environ. Sci. Technol. 44 (2010) 6933-6939.

[4] F.H. Wang, M. Qiao, Z.E. Lv, G.X. Guo, Y. Jia, Y.H. Su, Y.G. Zhu, Impact of reclaimed water irrigation on antibiotic resistance in public parks, Beijing, China, Environ. Pollut. 184 (2014) 247-253.

[5] C. Chen, J. Li, P. Chen, R. Ding, P. Zhang, X. Li, Occurrence of antibiotics and antibiotic resistances in soils from wastewater irrigation areas in Beijing and Tianjin, China, Environ. Pollut. 193 (2014) 94-101.

[6] P. Gao, D. Mao, Y. Luo, L. Wang, B. Xu, L. Xu, Occurrence of sulfonamide and tetracycline-resistant bacteria and resistance genes in aquaculture environment, Water Res. 46 (2012) 2355-2364.

[7] R.T. Pei, S.C. Kim, K.H. Carlson, A. Pruden, Effect of River Landscape on the sediment concentrations of antibiotics and corresponding antibiotic resistance genes (ARG), Water Res. 40 (2006) 2427-2435.

[8] M.T. Suzuki, L.T. Taylor, E.F. DeLong, Quantitative analysis of small-subunit rRNA genes in mixed microbial populations via 5'-nuclease assays, Appl. Environ. Microbiol. 66 (2000) 4605-4614.

# Development of sustainable ultra-high performance concrete

**Jisong Zhang, Yinghua Zhao**

Institute of Road and Bridge Engineering, Dalian Maritime University, Dalian, 116026, Liaoning, China
zjs_1015@163.com

**Abstract**. To design Ultra-high performance concrete (UHPC) in a sustainable way, this paper investigates the properties of UHPC containing supplementary cementitious materials (SCMs), such as fly ash (FA) and silica fume (SF). The flexural strength, compressive strength, and microstructure of the UHPC are examined. Results indicate that it is possible to design UHPC with very low cement amount. On the basis of 30% FA replacement, the incorporation of 10% and 20% SF shows equivalent or higher mechanical properties compared to the reference samples. The microstructure and pore volume of developed UHPC indicated a high correlation with its compressive strength. Efficiency factor (k-value) is calculated as an indicator to predict the flexural and compressive strength of UHPC with SCMs in terms of their synergistic effects.

## 1. Introduction

Ultra-high Performance Concrete (UHPC) has superior properties compared to normal strength concrete in terms of mechanical property and durability. The benefits of using UHPC in a structure includes reducing concrete amount, which in turn reduce concrete formwork, labor and equipment for erection, and construction time. However, given the recognized benefits, it is surprising that UHPC is not applied universally, which can be attributed to the high cost and high environmental impact per cube meter concrete. On the other hand, the global concrete producing makes up more than 5% of anthropogenic carbon dioxide emissions each year, namely from the production of cement. These issues can be addressed by incorporating supplementary cementitious materials (SCMs) to replace cement in a compressive way [1]. Therefore, great attentions have been paid to produce UHPC with less cost and lower emission while providing the equivalent properties.

For instance, Yu et al. [2] produced UHPC with 620 kg/m$^3$ cement content to obtain a compressive strength of 100 MPa. Ghafari et al. [3] utilized about 950 kg/ m$^3$ cement and 250 kg/ m$^3$ silica fume to produce UHPC. Hassan et al. [4] exhibited mechanical investigation with 657 kg/ m$^3$ cement, 418 kg/ m$^3$ GGBS and 119 kg/ m$^3$ silica fume. Aldahdooh et al. [5] utilized 638 kg/ m$^3$ cement to design UHPC with 120 MPa of compressive strengths.

Although much work has been conducted on this aspect, there still exists plenty of scopes that need to be investigated further [6]. Fly ash, a by-product of the industrial waste, has been proved to be useful in addressing the challenges of minimizing industrial waste and sustainable construction. It is worth mentioned in China, over 620 million tons of fly ash has been generated every year. However, the utilization ratio is still below 50%. Meanwhile, to overcome the deficiency of fly ash's the slowdown in pozzolanic reaction, the incorporation of silica fume is supposed to expedite the pozzolanic reaction at the early age, as the silica fume particles can fill pores between larger particles of cement, sand and other fillers.

The aim of this work is to evaluate the possibility of replacing cement up to 50% of various SCMs through maintaining the equivalent performance compared to reference sample. Emphasis is focused on the synergistic effects and optimum proportion of these materials. For this purpose, the essential properties of UHPC, such as workability, flexural strength, and compressive strength is analyzed. Microstructure characteristics of mixture is examined by scanning electron microscopy (SEM). Barrett–Joyner–Halenda (BJH) method is used to calculate the pore size and distribution. Finally, based on the results, efficiency factor (k-value) is calculated as an indicator of synergistic effects efficiency of these SCMs.

## 2. Materials and experimental methodology

### *2.1. Raw material*

An Ordinary Portland Cement (OPC, P.O 52.5) in accordance with the Chinese Standards GB175-2007 is used to produce HPC. China ISO Standard Sand (in accordance with ISO 679) is used as the aggregate with the fractions of 0-2 mm (D10=320um, D50=930um, D90=1600um). The commercially available superplasticizer (SP, powder, water reducing ratio >30%) is employed to adjust the workability of concrete. The detailed information of raw materials is presented in Table 1.

**Table 1**. Physical properties and chemical composition of raw materials

| Material | Density (kg/m$^3$) | D10 (μm) | D50 (μm) | D90 (μm) | SSA (m$^2$/g) | CaO (%) | $SiO_2$ (%) | $Al_2O_3$ (%) | $Fe_2O_3$ (%) |
|---|---|---|---|---|---|---|---|---|---|
| OPC | 3100 | 2.02 | 14.5 | 44.9 | 0.35 | 61.8 | 20.3 | 5.1 | 3.4 |
| FA | 2300 | 2.80 | 15.9 | 55.6 | 0.33 | 3.26 | 53.5 | 20.6 | 3.18 |
| SF | 2160 | 0.09 | 0.35 | 12.1 | 21.7 | — | 91.2 | 0.41 | 0.32 |

### *2.2. Mixture proportions and specimen preparation*

In this work, the modified A&A theory is used to design the mix proportion of UHPC. The reference UHPC has the highest cement content (875 kg/m$^3$). Afterwards, the volume of FA is held constant by 30% replacement level, the SF in increment of 5%. For example, FA30SF20 represent that 30% cement is replaced by FA and 20% cement by SF. The mix design for each group is presented in Table 2.

**Table 2**. Mixture proportions of designed concrete (kg/m$^3$)

| NO. | OPC | FA | SF | Sand | Water | SP | W/B |
|---|---|---|---|---|---|---|---|
| Reference (C1) | 875 | 0 | 44 | 1273 | 202 | 6.9 | 0.22 |
| FA30SF5 (C2) | 612 | 263 | 44 | 1273 | 202 | 9.5 | 0.22 |
| FA30SF10 (C3) | 568 | 263 | 88 | 1273 | 202 | 10.6 | 0.22 |
| FA30SF15 (C4) | 524 | 263 | 132 | 1273 | 202 | 11.3 | 0.22 |
| FA30SF20 (C5) | 480 | 263 | 176 | 1273 | 202 | 11.9 | 0.22 |

A JJ-5 Cement & Mortar Mixer with two revolving speeds (140 or 285 rpm) in accordance with BS EN196-1 [7] is used to produce mixture. In total, 4 min and 30s is required to produce UHPC in the mixing procedure. After that, the slump flow of the fresh concrete is examined according to BS EN 1015-3 [8]. The amount of superplasticizer is adjusted until achieving a constant flow value between 250 to 280mm at a water to binder ratio (W/B ratio) of 0.22. Finally, the fresh mixture is cast into steel moulds with size of 40×40×160mm and compacted through a vibrating table for 1 min. After 24 hours, the specimen is demoulded and cured under the conditions (20°C±2°C, RH>90%) for the designed ages.

Flexural and compressive strength are examined according to BS EN196-1. At least three samples were put in the test for each group at different ages (7, 28,90 and 365 days). For SEM test, sample is cut into the small cube with the dimension of 1×1×1cm after 90 days. For pore size distribution test, a weight of approximately 1.7-2.5g that extracted from mixture is tested by a ASAP 2020 at 365 days.

## 3. Results and discussion

### *3.1. Flexural strength*

The flexural strength of developed UHPC at 7, 28, 90 and 365 days are presented in Fig.1a. It is apparent that the flexural strength of UHPCs decrease with the increase of FA alone. However, in combination with SF, the flexural strength of UHPCs increase. With the increase of SF, the flexural strength of UHPC increase gradually. The maximum flexural strength is obtained by C5 (FA30SF20), which is followed by C1 and C4. This means that with a total of 50% cement replacement, the flexural strength of C5 is equivalent to C1–marginally higher than that of reference sample. Furthermore, the flexural strength of each group increase with the progression of time, regardless of concrete type. This improvement is especially evident from 90 days to 365days. This can be attribute to the fact that FA's slowdown in early strength development.

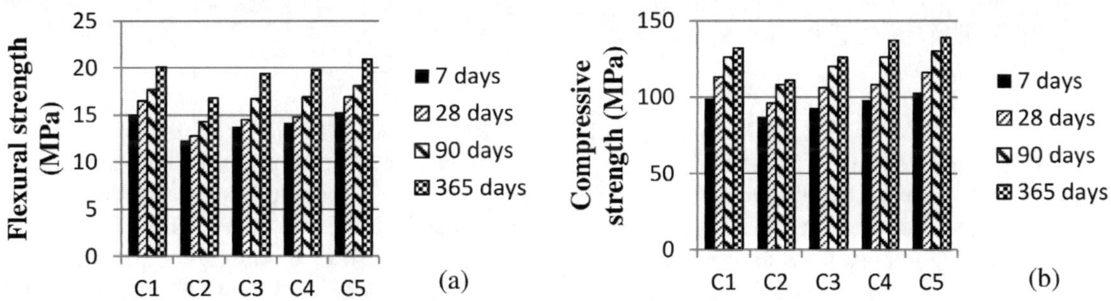


**Figure 1. (a) Flexural strength of UHPC (b) Compressive strength of UHPC**

### *3.2. Compressive strength*

The compressive strength of developed UHPC at 7, 28, 90 and 365 days are presented in Fig.1b. Similar to the tendency of flexural strength, with the increase of SF, the compressive strength of each group increase steadily. At 28 days, the highest compressive strength was obtained by C5, followed by C1 and C4. This fact can be explained by the filler effect of SF, which is responsible for the strength enhancement at early ages. At 365 days, the compressive strength of C2 and C3 approach C1, while compressive strength of both C4 and C5 exceed that of the C1. It is noticed that C5 has cement amount of 480 kg/m$^3$, which means that it is possible to design UHPC with low cement.

It has been reported that the $CaO/SiO_2$ ratio have a strong effect on the strength development and optimal ratio of $CaO/SiO_2$ is about 1.30 [9]. In this study The $CaO/SiO_2$ of C1 to C5 are 2.46, 1.22, 1.02, 0.86 and 0.72, respectively. However, the C2 group fails to have a better compressive strength, although their CaO/SiO2 ratio is most closing to 1.3. As such, it can be concluded that $CaO/SiO_2$ ratio should be used with caution and preferably in conjunction with the fineness and Al/Si or Ca/(Si+Al) ratio as a compressive strength predictor.

To clarify the efficiency of SCMs on compressive strength, the effect of FA+SF can be expressed as an efficiency factor (k-value) [10]. A k-value approaching one means that the addition is equivalent to cement. Based on the experimental results in this work, the k-value for FA30SF5, FA30SF10, FA30SF15 and FA30SF20 are 0.88, 0.95, 1.00 and 1.03, respectively.

### *3.3. Microstructure of UHPC*

The microstructure of concrete can influence the mechanical properties and could be explicated by the SEM observation [11]. Fig 2 shows the microstructure picture of each group observed at 90 days. It can be seen that FA30SF20 have the least capillary pores and the densest matrix compared to other groups, which is consistent with the compressive strength results–they had higher compressive strength. This indicates a positive effect of SF on the microstructure as well as strength enhancement.

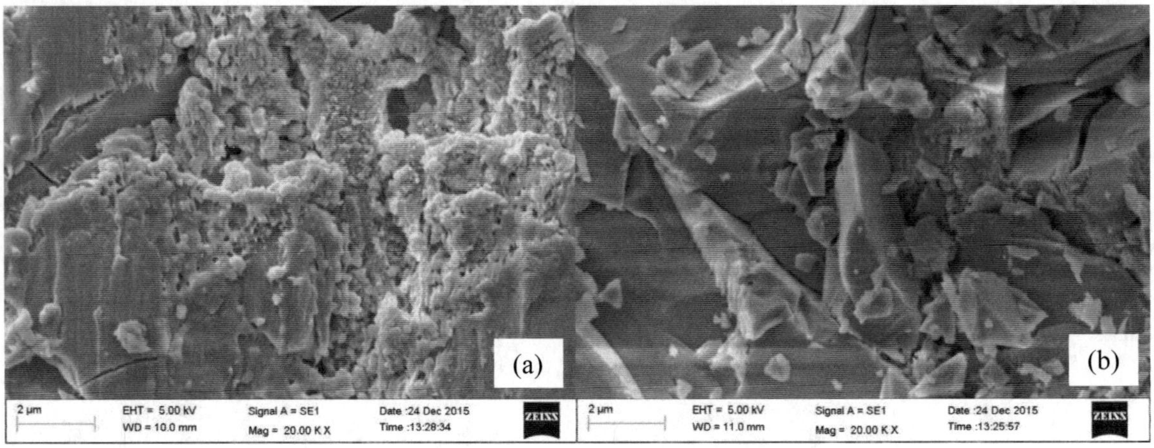


**Figure 2. SEM micrographs of UHPC (a) FA30SF5, (b) FA30SF20**

It is reported that BJH method can be used to quantify the small pores in the range of 2-50nm [12,13]. As an exploratory attempt, the present work employ this method to quantify the pore size and distribution of developed UHPC. The pore size range, total volume and area of the UHPCs are shown in Table 3. The pore volume and area decrease with the increase of compressive strength. When the pore volume and area show the lowest value, the compressive strength obtain the highest value among the UHPCs. Therefore, it can be concluded that the BJH method seems to be an effective way to quantify the pore size and distribution of UHPCs.

**Table 3**. Pore size distribution for designed UHPC.

| No. | >200nm ($10^{-3}$ $cm^3/g$) | 110-200nm ($10^{-3}$ $cm^3/g$) | 40-110nm ($10^{-3}$ $cm^3/g$) | 10-40nm ($10^{-3}$ $cm^3/g$) | <10nm ($10^{-3}$ $cm^3/g$) | Total pore volume ($cm^3/g$) | Total pore area ($m^2/g$) |
|---|---|---|---|---|---|---|---|
| C1 | 0.563 | 0.613 | 1.287 | 1.683 | 0.815 | 0.0050 | 0.969 |
| C2 | 0.758 | 0.985 | 2.136 | 2.274 | 2.659 | 0.0088 | 1.586 |
| C3 | 0.702 | 0.816 | 1.742 | 1.987 | 2.063 | 0.0073 | 1.343 |
| C4 | 0.626 | 0.693 | 1.558 | 1.669 | 1.552 | 0.0061 | 1.159 |
| C5 | 0.537 | 0.572 | 1.241 | 1.533 | 0.762 | 0.0047 | 0.932 |

## 4. Conclusions

This paper evaluates the possibility of designing UHPC with low cement by using SCMs. The flexural strength, compressive strength, and microstructure of the UHPC are examined. From the results, the following conclusions are drawn:

(1) The addition of 30% FA decrease both flexural and compressive strengths compared to the reference sample. However, with the increase of SF, the mechanical properties can be improved gradually. The UHPC made by 30% FA and 20% SF shows the highest flexural and compressive strength in all mixtures. While the UHPC made by 30% FA and 10% SF exhibit equivalent mechanical properties in comparison with the reference sample.

(2) Based on the experimental results, the efficiency factor (k-value), which can be considered as synergic effect efficiency on compressive strength, is calculated. The highest k-value (1.03) was obtained by ternary blend of cement with 30% FA and 20% SF, followed by blend with 30% FA and 10% SF (1.00). It can be concluded that it is possible to produce UHPC with low cement (480 kg/m$^3$).

(3) The microstructure observation and pore volume results indicate a high correlation with mechanical properties. The mixture with densest matrix and least capillary pores shows the highest flexural and compressive properties. Furthermore, the BJH method is validated to be an effective way to quantify the pore size distribution of UHPC.

**Acknowledgments**
The authors gratefully acknowledge the Institute of Road and Bridge Engineering of Dalian Maritime University for funding this research.

**References**

[1] Lothenbach B, Scrivener K, Hooton R 2011 Supplementary cementitious materials. *Cem Concr Res* **41** pp 1244–56.
[2] Yu R, Spiesz P, Brouwers HJH 2015 Development of Ultra-High Performance Fibre Reinforced Concrete (UHPFRC): Towards an efficient utilization of binders and fibres. *Constr Build Mater* **79** pp 273–82.
[3] Ghafari E, Costa H, Julio E 2014 The effect of nanosilica addition on flowability, strength and transport properties of ultra high performance concrete. *Mater Des* **59** pp 1–9.
[4] Hassan A, Jones S, Mahmud G 2012 Experimental test methods to determine the uniaxial tensile and compressive behaviour of ultra-high performance fibre reinforced concrete (UHPFRC). *Constr. Build Mater.* **37** pp 874–82.
[5] Aldahdooh M, Bunnori N, Johari M 2013 Evaluation of ultra-highperformance-fiber reinforced concrete binder content using the response surface method. *Mater Des* **52** pp 957–65.
[6] Jiemger M, Siddique R 2015 Recent advances in understanding the role of supplementary cementing materials in concrete. *Cem Concr Res* **78** pp 71–80.
[7] BS-EN-196-1 2005 Methods of testing cement – Part 1: determination of strength London *British Standards Institution-BSI and CEN European Committee for Standardization*.
[8] BS-EN-1015-3 2007. Methods of test for mortar for masonry – Part 3: Determination of consistence of fresh mortar (by flow table) London. *British Standards Institution BSI and CEN European Committee for Standardization.*
[9] Shi C, Wang D, Wu L 2015 The hydration and microstructure of ultra high-strength concrete with cement-silica fume-slag binder. *Cem Concr Compos* **61** pp 44–52.
[10] BS-EN-206 2013 Concrete-Specification,performance,production and conformity London *British Standards Institution-BSI and CEN European Committee for Standardization*.
[11] Jennings H, Tennis P 1994 Model for the Developing Microstructure in PortlandCement Pastes. *J. Am. Ceram. Soc.***77** pp 3161-72 .
[12] Sing K 2004 Characterization of porous materials: past, present and future. *Colloids Surf A Physicochem Eng Aspects* **241** pp 3–7.
[13] Quercia G, Lazaaro A and Geus J 2013 Characterization of morphology and texture of several amorphous nano-silica particles used in concrete. *Cem Concr Compos* **44** pp 77–92.

# A research on monitoring the canopy temperature variation of rice based on the infrared thermal imaging technique

**Mingchao Gao[1,2], Qiyun Li[1], Jiping Gao[2], Jihong Li [1], Wenzhong Zhang[2]and Yingshan Dong[1]**

[1] Crops Resources Research Institute, Jilin Academy of Agricultural Sciences, Northeast Agricultural Research Center of China, Changchun,136000, China
[2]Rice Research Institute, Key Laboratory of Northeast Rice Biology and Breeding, Ministry of Agriculture, Key Laboratory of Northern Japonica Rice Genetics and Breeding, Ministry of Education, Shenyang Agricultural University, Shenyang, 110866, China
email: gaomingchao99@qq.com, corresponding author: zwzhong@126.com, ysdong@cjaas.com

**Abstract**. The experimental results show that the infrared thermal imaging can observe the overall temperature distribution of canopy in rice much more intuitively.The canopy temperature of different soil water potentials in the field showed a regular change, and the rice canopy reached the highest temperature at 13: 00-15: 00, and the lowest at 2:00-4:00.The range of temperature difference between the crown and the air is gradually increased with the prolongation of the water stress. And the best effect of water stress to observe was at 13: 00-15: 00 through canopy temperature.

## 1. Introduction

The canopy temperature not only can screen and separate out the drought resistance genotype under the same growing environment, but also can be used to diagnose the degree of rice water deficit, and then guide the time and volume of irrigation.

It is becoming a new hotspot of crop quantitative and water saving research that determine the canopy temperature of rice and diagnose the crop water status by infrared thermal imaging and then take the corresponding irrigation measures.

As a convenient technology that applies infrared automatic temperature measurement device to observe the canopy surface temperature of crop, so as to diagnose whether the crop suffered water stress or not, the Infrared Thermal Imaging has already developed maturely [1]. Orta et al. (2004) [2] in Turkey through the study of wheat developed corresponding equation which can be used to qualify crop water stress index Luquet et al. (2004) [3] assessed the limitations of water stress indices using directional thermal infrared (TIR) measurements and 3D simulations. Mahan et al. (2005) [4] determined peanut irrigation threshold by the canopy temperature and its time. Möeller（2007） [5] used infrared thermal image to detect grape canopy water status. Alchanatis et al. (2010) [6] applied infrared thermal image to map the flow of water on cotton leaf. Susan OShaughnessy (2012) [7] established sorghum automatic irrigation model via the study of water stress index. Many experiments showed the infrared thermal imaging technique is feasible to estimate crop water demand.

## 2. Materials and Methods

This experiment took place at the test field of Rice Institute of Shenyang Agriculture University. (N41°8′、E123°38′). Applied terrier 1 variety and established districts in the field by using plastic pool, and the properties of soil in this region was brown. The test established five different water gradient treatments that the field capacity of soil maximum water content 100%(building water courses, CK, corresponding soil water potential is 0), above 80%(corresponding soil water potential 0.025~0.03MP), 60%(corresponding soil water potential 0.045~0.05MP), 40%(S3, corresponding soil water potential 0.055~0.06MP), and 20%(S4, corresponding soil water potential below 0.08MP). At early jointing stage (about 13-14 leaves) began to control the water, soil potential instrument and weighing method combines to be applied in water control. In order to prevent rain from affecting the water stress treatments, the mobile rain-proof sheds were set-up. It took a 24 hours experiment to measure the canopy temperature and air temperature in no wind, sunny day.

Other paragraphs are indented (BodytextIndented style). The convenience and accuracy of Infrared thermal imaging system that can observe the overall temperature distribution of rice canopy temperature much more intuitively, so as to achieve the Synchronous measurement of each part temperature. Earing and full heading time Figure 1. shows after water controlling in earing and full heading time, from the infrared thermal imaging pictures of each test fields that most of the plants in the whole plot came into view under the infrared thermal imaging, and they can freely select the point or space to the canopy Temperature for statistical analysis.

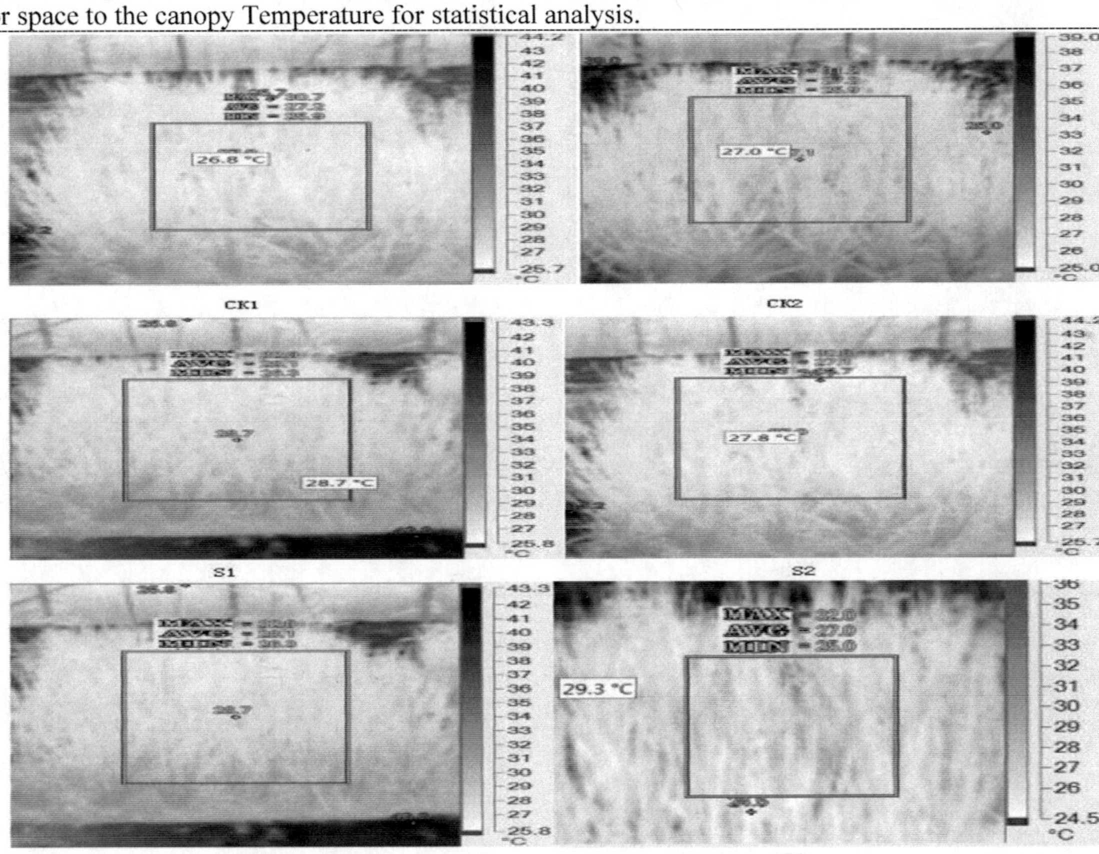


**Figure 1.** The rice canopy infrared imaging figure of different soil moisture in heading - full heading stage at 12:00.

In order to understand the canopy temperature of different rice water potentials, the test chose the infrared thermal imaging in seventh day of earing and full heading time water controlling at 12:00 for 3D-IR drafting. (Figure 2.) In previous studies, many researchers believe that plants become the

IOP Conf. Series: Earth and Environmental Science **61** (2017) 012077 doi:10.1088/1755-1315/61/1/012077

vegetative growth into the reproductive growth stage during this period, and the contradictory of plant and soil water supply conditions is the most outstanding. As the pictures showed that the distribution of canopy temperature under different water treatments was quite different. In the whole water layer CK1 treatment of 3D-IR diagram can be seen, in addition to the four corners of the area affected by the outside, the central area of the district temperature is low, the temperature is evenly divided, and the temperature of the division of the whole area of the plant temperature difference is not significant. From the beginning of CK2 can be seen with the increase in the degree of water stress, 3D-IR diagram of the temperature reaction became more and more intensely. CK2 area due to high soil moisture content, the performance was still relatively uniform, and S1-S3 treatment, it can be seen the temperature distribution was become more and more clutter, the stress treatment area due to low soil moisture that the entire canopy space and the uneven distribution of water within the plant directly leads to the uneven distribution of canopy temperature; S4 treatment is the highest water stress with the lowest moisture in the soil. Although the temperature distribution was more stable than that of the other treatments, the canopy temperature was significantly higher than CK1 treatment. It can be seen from the 3D-IR diagram that the canopy temperature in different soil water potentials is quite different.

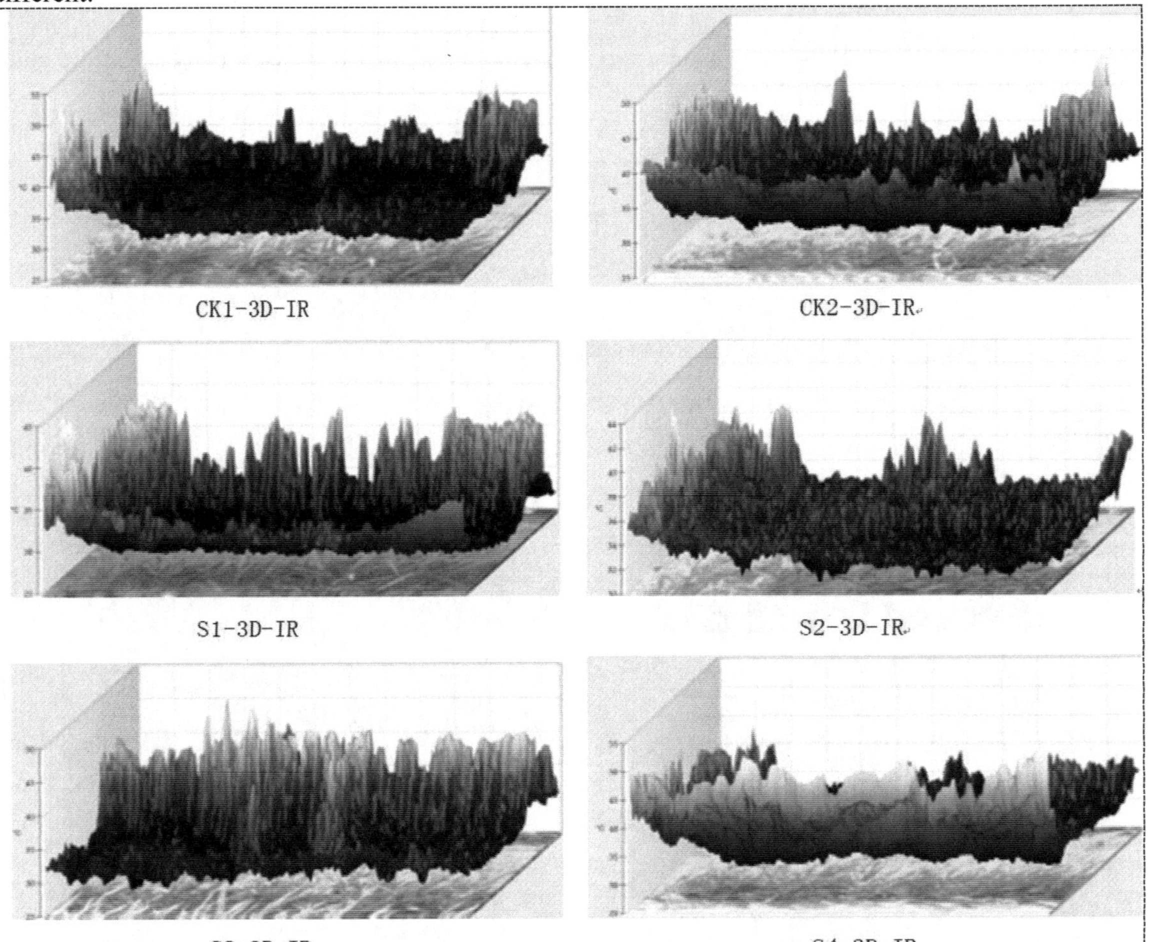


**Figure 2.** The canopy temperature 3D-IR figure of different soil moisture at heading - full heading stage.

Figure 3. and Figure 4. separately showed the diurnal changes of water control at 7th and 14th day of tiller-jointing stage of Kaigeng 1. During the Tillering jointing stage, the canopy temperature changes with the trend of atmospheric temperature. And the canopy temperature of the rice increased gradually with the increase of the atmospheric temperature, and the other canopy temperatures except S4 were peaked at 14: 00-15: 00, which is later than the peak of the temperature. It showed that the change of canopy temperature is generally lagging behind the change of temperature, while the S4 treatment stress is the most serious, which leading to the poor regulation ability of rice itself, and the most close to other temperatures in the daytime. Meanwhile, it also showed that the trend of canopy temperature during daytime was S4> S3> S2> S1> CK2> CK1, and the canopy temperature decreases with the decrease of temperature after 16:00 pm. However, since the rapid decline of solar radiation and the rising of air humidity, that the temperature of the canopy can be reduced gradually, and the difference of the canopy temperature between the treatments is not obvious. After 14 days of water control. The difference in canopy temperature between treatments was more pronounced at10: 00-16: 00, such as S4 and S3 with high water stress, their temperature increased, which were close to the temperature at 7 days. Due to the role of air humidity, the temperature difference at is not obvious among each areas at night 22: 00-4: 00, and the temperature difference is also small, but always lower than the temperature. In the 7 days and 14 days water controlling which measured the temperature of the S4 and S3 treatments were higher than other treatments at night, which may be related to the tiller-tectonic stage, and the canopy temperature is related to the influence of water layer and soil water content at night.

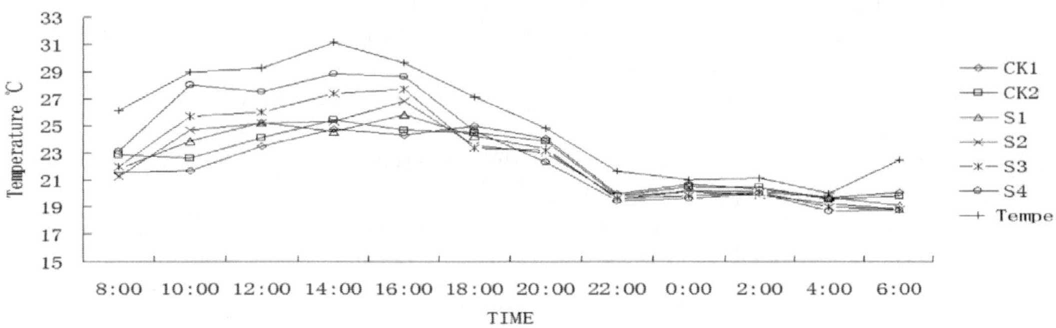


**Figure 3.** The daily changes of the canopy temperature of Kaijing 1 after 7 days under different water treatments at tiller-jointing stage.

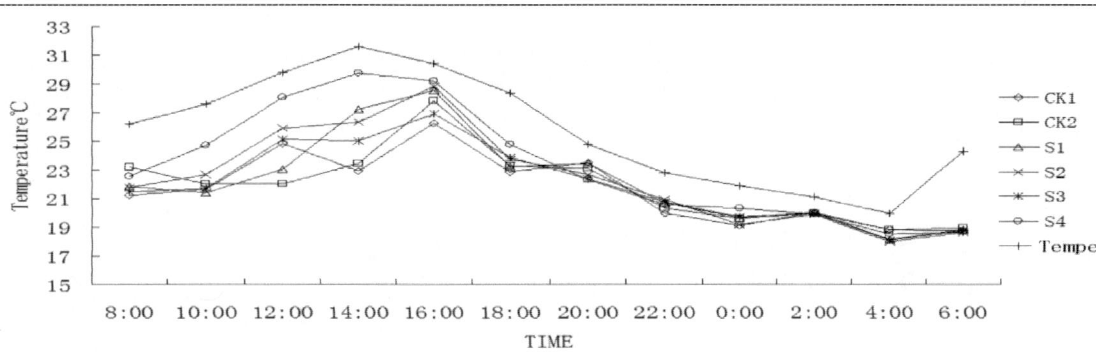


**Figure 4.** The daily changes of the canopy temperature of Kaijing 1 after 14 days under different water treatments at tiller-jointing stag.

## 3. Conclusions

It can observe the status of rice water stress via infrared imaging intuitively. There was a significant difference in canopy temperature between rice under different water treatments, and this difference gradually increased with the prolongation of water stress. In the daytime, the temperature of canopy was higher in rice with higher water stress, which was more obvious than that without stress treatment. The degree of water stress can be observed by the observation of canopy temperature, and the observation effect of different water stress levels is the best at 13: 00-15: 00. At the same time, it can judge whether the rice suffered from water stress or not by the difference of canopy temperature.

**Acknowledgments**
This research was funded by A Doctoral Project of the Academy of Agriculture Sciences in Jilin, The plan LR2015060 about outstanding qualified scientists and technicians of colleges and universities in Liaoning province, The Study on Rice Canopy Temperature Depression and Water Stress Index Construction and Water Saving is the Doctoral Scientific Fund Project of The Ministry of Education of China, 20092103110005.The project on Teachers (Youth)Research Foundation of Shengyang Agricultural University (20131009), and also the Cultivation Plan for Youth Agricultural Science and Technology Innovative Talents of Liaoning Province(2014039).

**References**

[1] Alves I and Pereira L S 2000 Irrigation Science (USA, NY: Springer) 19 p 101-106
[2] Orta A H , Baser I , Sehirali S , Erdem T and Erdem Y 2004 Cereal Research Communications (Hungary: Akademiai ) 32(3) p 363-370
[3] Luquet D , Vidal A , Dauzat J , Begue A , Olioso A and Clouvel P 2004 Remote Sensing of Environment (USA, NY: Elsevier) 90(1) p 53-62
[4] Mahan J R , Burke J J , Wanjura D F and Upchurch D R 2005 Irrigation Science(USA, NY: Springer) 23(4) p145-152
[5] Möller M, Alchanatis V, Cohen Y, et al. 2007 Journal of Experimental Botany,( England Oxford: Oxford Univ Press), 58(4) p 827-838
[6] Alchanatis, V , Cohen Y , Cohen S , et al. 2010 Precision Agriculture (Netherlands: Springer), 11 p 27-41
[7] Susan A, O'Shaughnessy, Steven R. Evett , Paul D Colaizzi and Terry A 2012 Agricultural Water Management (Netherlands: Elsevier) 107 p 122-132

IOP Conf. Series: Earth and Environmental Science **61** (2017) 012078 doi:10.1088/1755-1315/61/1/012078

# Manufacturing technology and application of hemp cigarette paper with dense ash integration

**Yu Yao, Zhan Jian-bo, Wan Hao, Zhang Ying, Li Li-wei, Yu Jiang, Yu Ting-ting, Xie Jiao, Yue Bao-shan**

R&D Centre, China Tobacco Yunnan Industrial Co., Ltd., Kunming, 650231, China
yuyao@ynzy-tobacco.com

**Abstract**. Cigarette paper, as one of the significant materials used for combustion, has special and direct influence on the smoke, also directly influencing the ash appearance of cigarettes before and after combustion. In this paper, full hemp cigarette paper was prepared through creative beating and mixing slurry technology, and the advantages of the preparation process were analyzed. Full hemp cigarette paper was creatively applied to the preparation and verification of slim cigarettes, and the ash integration effect in the process of burning and its influence on whiteness were verified. At the same time, the physical and chemical indexes of cigarette paper were tested and studied, and sensory evaluation was applied to verify the effect of cigarette paper on sensory quality.

## 1. Introduction

Cigarette paper, as an important part of materials applied in cigarettes, not only carries the key image and connotation of products, but also is involved in cigarette burning. Therefore it has special and direct effect on cigarette smoke[1]. On one hand, gas after combustion of the cigarette paper is an important component of cigarette smoke, so the cigarette paper plays an important role for smoking cigarette sensory burning; on the other hand, the inherent quality of cigarette paper will also affect the combustion status of cigarette [2].

Compared with coniferous wood, bamboo and broad-leaved trees and grasses, hemp fiber is quite different in aspects of fiber length, fiber aspect ratio and other parameters. The hemp fiber raw material contains high fiber content, good strength, small cell wall thickness and fiber wall to lumen ratio, and the fiber flexibility is good due to the smooth and soft fiber[3]. Hemp fiber cell cavity leads to fine fibre and appropriate transparency. However, it is easy to form the fiber wire brush, so the paper permeability is relatively low. Therefore, the hemp fiber is rather suitable for making high-grade paper and industrial paper, such as banknote paper, cigarette paper[4].

Slim cigarettes (17.0 mm) market has become one of the fastest growing markets in major markets such as South Korea, Eastern Europe and central Asia. In China, since 2010, the annual growth rate of cigarette smoking has been as high as 40%, which has become a new growth engine in the cigarette sales market. However, the slim cigarette is the circumference of 24.3mm normal cigarette specifications, resulting in combustion in cigarette paper weight percent, compared with the normal cigarette circumference 24.3mm specification 6.53%, slim cigarette with circumference of 17.0mm has significantly increased from 9.44% to 45%, affecting the sensory evaluation. Taste of cigarettes rise, but the harmony evaluation declines. Therefore, it is necessary to study a new type of cigarette paper, which is suitable for special specifications (22.30 mm)[5, 6].

A common problem is that the burning cone of slim cigarettes is longer, which is easy to fall off the combustion cone, resulting in safety risks to consumers. It is difficult to meet the technical requirements of the slim cigarettes, and it is necessary to strengthen the ash content of cigarette paper. In addition, the sharp reduction of the circumference of the cigarette, the aroma of the slim cigarettes decreased a lot more than the normal cigarettes, causing impacts on cigarette paper.

Therefore, it is necessary to combine the general characteristics of cigarette paper and cigarettes, to develop the general package of cigarette paper with dense ash, and to study the method and path of enhancing the contribution of cigarette paper to the flavor of cigarette. According to the slim cigarette paper and general anesthesia, the ash density of cigarette paper was tested at the same time. Moreover, the slim cigarette paper prepared from chemical index, ash effect, sensory quality and other aspects were carried out for the application evaluation.

## 2. Experimental

### *2.1. Cigarette paper composition*
Dense ash cigarette paper consists of general fiber pulp: 55-70%, packing: 25-40%, accelerant: 0.5-3%, ash coating agent 0.2-0.6%, the rest of water.

### *2.2. Preparation of cigarette paper samples*
The preparation of cigarette paper mainly includes the following steps:

(1) Hemp pulp by hydrapulper pulp, pulp concentration control in the range of 3-3.5%, breaking the water into the complete mixture in beating circulation, first through a large taper refiner of hemp pulp fibers were cut off;

(2) Preparation of slurry into two different beating degrees and wet weight of the pulp slurry. The slurry to a fiber cut short by way of beating, and the beating degree of control of 35-65 DEG SR, and the wet weight was about 16-18g. The beating degree control at 80-85 DEG SR, and the wet weight was about 6.5-8.5g;

(3) Prepared slurry is extracted according to the requirements of the quantitative and the air permeability of the paper, respectively;

(4) Filler mixture preparation, selection of liquid calcium carbonate mixed calcium carbonate crystal particle size between 0.15-0.5 m and 6-8 m two material size interval uniform for cigarette paper manufacture;

(5) Mixture of the slurry is prepared. The mixture is copied to the wet paper sheet by a long web paper machine;

(6) The press section of the vacuum roller is transferred for further dewatering;

(7) Transferred from the press section of wet paper into the pre drying section of the paper, the dryer is dried to constant water;

(8) Combustion and ash mixture solution is prepared: accelerant solution prepared by 0.5%-5% solution quality percentage according to requirements; 0.2%-1.5% solution quality percentage requirements ash solution;

(9) Prepared combustion and ash agent mixture into one side coated or double immersion were added to the cigarette paper sizing machine;

(10) Paper plate was rolled into required width.

According to the different components, 4 types of cigarette paper were prepared, which were numbered 1#, 2#, 3#, 4#.

## 3. Results and Discussion

### *3.1. Ash appearance*
As can be seen from Fig. 1, there is obvious difference in the initial stage of combustion common cigarette paper and cigarette paper (4#) under general anesthesia in the combustion process and the effect of ash on the package, and this is possibly due to the initial stage of combustion. The slim cigarette ash characteristics can withstand the weight deviation after the combustion of soot in the combustion inclined; from 1/3 to 1/2 stage, with the gradual accumulation of soot length weight, a

common cigarette paper ash integrity began to appear, and the ash began to crack. The ash density declined seriously, and ash density of the general cigarette paper ensures the full package effect of ash. Common cigarette paper and cigarette paper burning in general anesthesia to the end: the difference shows that common cigarette paper ash density continues to decline, and the ash was grey. The general ash density performance is good, which can appear the whole ash, and the color is snow white, the ash appearance of new cigarette paper was significantly higher than that of ordinary cigarette paper.

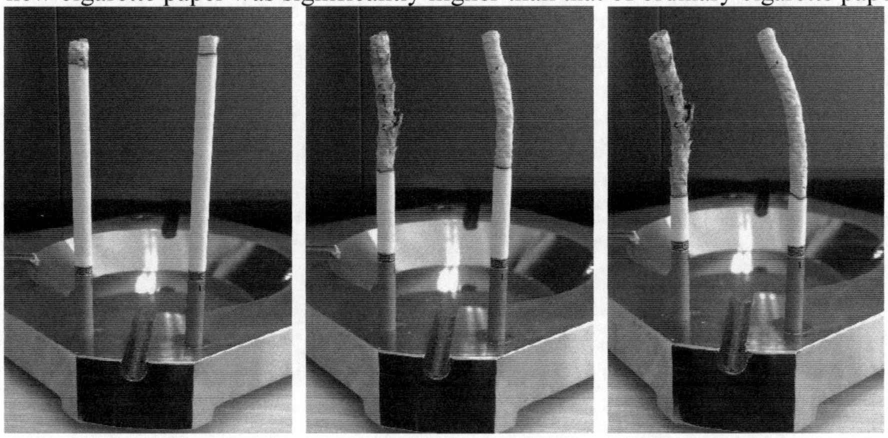

Fig.1 Ash appearance of traditional (left) and hemp (right) cigarette paper at different combustion stages

*3.2. Physical and chemical index detection during combustion process*

Cigarette paper was made into 4 samples, and the cigarette combustion process associated with sensory quality was evaluated during determination of parameters, and the results are shown in Table 1.

Table 1 Physical and chemical testing results of different samples

| Samples | Puff Number (puff/cig) | Tar (mg/cig) | Nicotine (mg/cig) | Humidity (mg/cig) | CO (mg/cig) | Tar correction (mg/cig) | CO correction (mg/cig) |
|---|---|---|---|---|---|---|---|
| 1# | 5.87 | 7.12 | 0.94 | 1.03 | 5.51 | 7.91 | 6.31 |
| 2# | 5.47 | 7.87 | 0.83 | 0.98 | 6.32 | 8.70 | 7.11 |
| 3# | 6.20 | 8.69 | 0.76 | 1.01 | 7.08 | 9.56 | 7.22 |
| 4# | 6.02 | 8.74 | 0.96 | 0.91 | 8.12 | 9.61 | 8.15 |

Where, conditioning and testing atmospheric atmospheric pressure range of 80.5 kPa-81.5kPa, corrected formula: Tar $y = 1.046x + 0.468$, carbon monoxide $y = 0.986x + 0.880$.

As can be seen from Table 1, the relevant physicochemical parameters are in line with the requirements of mainstream smoke technology, but the high value of CO and tar, which may be mainly due to general anesthesia of cigarette paper in the preparation process of the ash results depending on the beating process, calcium carbonate additive quality control and process of comprehensive adjustment, the impact of cigarette paper the air permeability index, thereby affecting the CO value of the mainstream smoke and tar. From the point of view of the ash of cigarette paper, the more dense the cigarette paper, the better the ash content. However, it will also bring the difficulty of raising the air temperature. The problems such as CO and tar rise caused by the low permeability can be improved by the subsequent optimization of the preparation process, which can also be applied to the improvement of the process such as perforating paper.

By increasing the taper of the refiner, beating change cycle, the domestic first batch beating and then mixed into a slurry technology, which is a success as a breakthrough of hemp pulp technical problems. The utility model solves the problems that the pulp fiber is soft and long, and the pipe is easy to be blocked, and the paper defects such as the spot, the transparent point, the eye and the like

IOP Conf. Series: Earth and Environmental Science **61** (2017) 012078 doi:10.1088/1755-1315/61/1/012078

can be caused by the paper cutting. Cigarette paper produced by the present invention can effectively avoid the disadvantages of various ash appearance. Ash calcium carbonate results depend on the beating process, quality control and process of comprehensive adjustment and combination of additives. The ash coating components applies normal dosage, which has no obvious effect. The process could be accomplished with low manufacturing cost, and it does not cause changes in the cigarette sensory. Due to the focus on the ash appearance, such application may lead to the cigarette paper permeability increase limited.

## 4. Conclusions

(1) Applications of compact ash appearance of cigarette paper is feasible for general cigarette products, effectively solving the drawbacks caused by the smell of cigarette paper. To enhance the slim cigarette sensory experience, the quality of cigarette paper could be further improved.

(2) Aiming at the common problems that the burning cone is too long, and it is easy to fall off, the solution of cigarette paper is provided, which further eliminates the potential safety hazard of cigarette smoke.

(3) To obtain a dense balanced consideration of all aspects of cost, quality, ash and other requirements of the package of cigarette paper ash preparation technology, especially the first batch beating and then mixed into a improved technology. Combined application and its supporting process, cost advantages could break the foreign monopoly in the field.

## References

[1] Yu Guolong. An analysis on the effect of relative factors of ash packing performance of cigarette paper[J]. China Pulp & Paper Industry, 2015(6): 37.

[2] Shen J, Li J S, Qian X R, et al. A review on engineering of cellulosic cigarette paper to reduce carbon monoxide delivery of cigarettes[J]. Carbohydrate Polymers, 2014, 101(1): 769.

[3] Zhang Ying, Zhan Jianbo, Yu Zhenhua , et al. Manufacturing Technology and Development Prospects of Aluminum Free Inner Lining paper for Cigarette Packaging [J]. China Pulp & Paper, 2015, 35(11): 58.

[4] Li Guizhen, Wang Pingjun. Research on effect of cigarette paper additives on ash wrapping property of cigarettes[J]. Paper Chemicals, 2011(4): 17.

[5] Shen J X, Sun J, Xiao W Y, et al. The effect of additives on cigarette ash wrapping [J]. China Pulp & Paper Industry, 2012, 33(14): 23.

[6] Cheng Zhangang, Ye Mingqiao, Hu Suxia, et al. Factors Affecting Ash Integration of Burning Cigarette[J]. Tobacco Science & Technology, 2011(2): 9.

# Influence of heating on the weight loss and mineral phase in MSWI ash: LOI of incineration ash

**Shuo Yang**

Rm.1004, No.4, Zhichun Road, Haidian District, Beijing, China
Yangshuo_1984@126.com

**Abstract.** Loss on ignition (LOI) is a very common method for estimating the volatile species in solid sample. Normally, the measurement of LOI can be convenient and accurate, but for municipal solid waste incineration (MSWI) ash, the process may become intricate due to the complexity of the sample. In the incineration ash, there exist various phases, such as mineral, metal, organic and glass. The reaction and transformation of some materials during heating will influence the measurement. 5 incineration ash samples were selected and tested in this study. LOI content was basically measured at high (850°C) and relatively low (440°C) temperatures. The comparison between these two measurements showed a large difference. X-ray diffraction (XRD) and thermal analysis (TG-DTA) were carried out to investigate the mineral changes and weight losses with different ignition temperatures. The mineralogical analysis suggests that the decomposition of hydrate and carbonate phases cannot be neglected for LOI measurement of incineration. A long-time heating under relatively lower temperature (400~450°C) compared with soil sample measurement (≥500°C) was recommended by this study.

## 1. Introduction

Loss on ignition is widely used to evaluate the organic and volatile matter content in solid sample [1, 2]. A lot of work has been done to optimize the procedure and get more accurate LOI result [3], but there is rare research or literature about the LOI measurement on incineration ash sample. Incineration ash is produced from high-temperature treatment (850~1000°C), while the ignition temperature of LOI is normally lower than that, but the LOI of incineration ash still has practical value because it is a gross measure of the relative burnout of the combustor where the incineration ash was produced [4]. It is often to observe something combustible like small pieces of woods or even paper in incineration ash, which indicate the combustion processes are not completely thorough. The main reason is that during incineration, the temperature inside the combustor is not average and some low-temperature spots would preserve those unburned wastes. So an unbiased index is needed to evaluate the incineration process and the condition of the combustor. LOI measurement, as an inexpensive and simple procedure, is very suitable for this purpose.

But the measurement of LOI on incineration ash sample can be intricate because the composition of incineration ash is complex. Depending on the measurement condition, organic, volatile, hydrorate, carbonate and sulphate components in incineration ash can all contribute to the final LOI value [5]. In some cases, the LOI results measured under different circumstances can have big gaps. So it is important to adopt proper measurement condition for different purposes. On the other hand, for the studies on incineration ash, LOI can also be a useful tool. In the field of soil science, there has already been some work about using LOI to estimate certain contents [6]. Because each kind of material has its distinct temperature range of decomposition, referring to the weight losses with temperature we can roughly estimate what and how much of them may exist in the ash sample.

This paper focused on the effect of temperature on the LOI. TG-DTA and XRD were used as the main analytical methods. 5 different ash samples were picked and examined. The temperatures with obvious weight losses were selected by running TG-DTA onto the samples. Then XRD was used to determine the mineral phase changes related to the weight losses. In the end, a temperature-mineral decomposition diagram was made to help decide the optimal LOI measurement conditions.

## 2. Materials and methods

### *2.1 Sample Selection and Preparation*

The MSWI ash samples used in the research were from 3 sources with different generation backgrounds. They were simply referred as No.1 to No.5. The sample No.1 and No.2 were newly produced ash from a real incineration plant. No.1 was air cooled, while No. 2 was water quenched after combustion. Sample No.3 was a 2-year stored ash from another incineration plant. No.4 and No.5 were generated by a pilot incinerator. No.4 was fly ash and No.5 was bottom ash. They were both air cooled after formation.

Before the experiment, big particles were removed from the samples by letting the samples pass through a 4.75 mm mesh. The remains were oven dried (105°C overnight) to remove free water. Then all the samples were homogenized by a mechanic mill. The bulk composition was analyzed by utilizing XRF.

### *2.2 Weight Loss with Temperature*

TG-DTA was used in the experiment to get the relationship between weight loss and temperature. Around 30mg of ash sample was taken for each round. $Al_2O_3$ was used as the inert control. The sample heating chamber was connected with atmosphere directly. The temperature of chamber gradually rose from 25°C to 1400°C by an increment of 10°C per minute.

### *2.3 Mineral Change and Temperature*

With the relationship between weight loss and temperature provided by TG-DTA, several temperatures were selected and let the ash samples exposed to for 2 hours in muffle furnace. Before and after the ignitions, the samples were analyzed by XRD. Through comparison of the XRD profiles, the mineral phase changes could be identified.

## 3. Results and discussion

### *3.1 Characterization of the ash samples*

The bulk chemical composition of the ash samples was presented by Table 1. The 4 bottom ash samples (No.1, 2, 3, 4) were found to be calcium and silicate based materials with various other elements. Besides CaO and $SiO_2$, $Al_2O_3$ and $Fe_2O_3$ were also dominating components. $TiO_2$, $P_2O_5$, MgO, $Na_2O$ and Cl had average contents above 1 wt%, which were considered as major constituents. The others including all the heavy metals were counted as minor (0.1~1 wt %) and trace (<0.1 wt %) constituents. As the only one fly ash sample, No.5 showed a totally different property from the others. It was a calcium based material also had relatively high contents of chloride and zinc. Sample No.4 and No.5 were generated from the same combustion process. The comparison between these two samples revealed that Cl, S, K, Zn and Pb were more concentrated in the fly ash.

**Table 1.** Bulk chemical composition of the incineration ash samples from different sources

| | No.1 | No.2 | No.3 | No.4 | No.5 |
|---|---|---|---|---|---|
| $SiO_2$ (wt.%) | 30.64 | 21.99 | 23.52 | 29.57 | 2.02 |
| $Al_2O_3$ | 10.68 | 12.03 | 11.09 | 10.74 | 0.58 |
| $Fe_2O_3$ | 13.01 | 11.14 | 4.95 | 4.13 | 0.77 |
| CaO | 33.18 | 39.4 | 37.28 | 42.85 | 58.95 |
| $TiO_2$ | 1.3 | 1.41 | 2 | 1.29 | 0.11 |
| MnO | 0.08 | 0.09 | 0.1 | 0.13 | 0.02 |
| $P_2O_5$ | 4.14 | 3.31 | 3.15 | 3.55 | 0.29 |
| MgO | 2.44 | 2.29 | 2.77 | 2.53 | 2.13 |
| $Na_2O$ | 2.08 | 1.75 | 2.53 | 2.11 | 3.94 |
| $K_2O$ | 0.53 | 0.47 | 0.92 | 1.46 | 4.66 |
| Cl | 0.88 | 1.33 | 3.75 | 0.37 | 18.75 |
| S | 0.17 | 0.39 | 0.59 | 0.61 | 2.06 |
| Zn (ppm) | 2707 | 3526 | 2413 | 926 | 7472 |
| Cu | 1912 | 1641 | 1453 | 411 | 391 |
| Pb | 326 | 576 | 1997 | 69 | 2721 |
| Cr | 332 | 332 | 300 | 414 | 64 |
| Ni | 75 | 88 | 84 | 86 | 38 |
| Ba | 1126 | 939 | 1108 | 981 | 28 |
| Sn | 122 | 141 | 126 | 69 | 173 |
| Sr | 424 | 342 | 441 | 362 | 313 |
| As | 2 | 2 | 18 | 0 | 18 |
| Cd | 91 | 95 | 87 | 110 | 258 |

*3.2 LOI measured at 440°C and 850°C*

Table 2 revealed the great influence of temperature on LOI results. The ignition temperatures were chosen as 440°C and 850°C. The heating process lasted for 4 hour to make all possible reactions complete. It was found that water quenched samples (No.2 and No.3) had much higher weight losses than air cooled ones (No.1 and No.4). For sample No.1 and No.2, they were from the same source and had similar chemical composition (see in Table.1). The extra weight loss of water quenched sample must come from the hydrated contents like portlandite losing their structural water [5]. Fly ash sample (No.5) also had a relatively high LOI. Especially under the temperature of 850°C, almost 20% of total weight was lost during ignition. The volatile contents in fly ash must have contributed to the big weight loss. The gaps of LOI results measured under 440°C and 850°C were considerably big, especially for sample No.1 and No.4. The low LOI values of these two bottom ash samples at 440°C indicated very complete burnout of the combustion process, so the possibility that there might remain some combustible or volatile materials could be ruled out. The decomposition of mineral materials like hydrates or carbonates was the main reason for the additional weight loss at 850°C.

**Table 2.** LOI values of the incineration samples measured at 440°C and 850°C

| | No.1 | No.2 | No.3 | No.4 | No.5 |
|---|---|---|---|---|---|
| $SiO_2$ | 30.64 | 21.99 | 23.52 | 29.57 | 2.02 |
| $Al_2O_3$ | 10.68 | 12.03 | 11.09 | 10.74 | 0.58 |

*3.3 TG-DTA analysis*

IOP Conf. Series: Earth and Environmental Science **61** (2017) 012079 doi:10.1088/1755-1315/61/1/012079

There are two curves in the TG-DTA profile (see Figure 1). The blue DTA curve recorded the temperature and heat flow associated with thermal transitions in the sample. The downside peak stood for endothermic process. Despite of the beginning stage and the melting process after 1200°C, there were no very obvious exothermic (upside) peak and only several small endothermic peaks being observed. The flat DTA curve from 400°C~1200°C meant no dramatic physical or chemical reactions like burning had occurred in this range.

The black TG curve stood for the weight change of the sample. Continuous weight losses were observed for all the samples. It is worth to note that the weight loss of water quenched samples (No.2 and No.3) started from the very beginning of the measurement, while the air cooled samples had relatively stable stage until temperature reached 400°C. From 400°C, a cliff of TG curve appeared except for sample No.3.

Besides weight loss, there also existed a small endothermic peak in this area (more obvious for sample No.1). Hydrate materials adsorbing heat and losing their structural water loss could explain the phenomena. From 500°C to 600°C, there was another stable stage except for No.3. From around 600°C, another big weight loss occurred. The drop ended at around 650°C. From 1200°C, vitrification began. The ash samples melt and transformed into glass phase.

For the next step of the research, 450°C and 650°C were selected to determine what mineral phase change contributed to the weight losses. Ignition temperature of 850°C was also carried out to get the highest LOI value before ash sample being vitrificated. For water quenched samples, 400°C was also chosen to investigate the cause of weight loss at relatively low temperature condition.

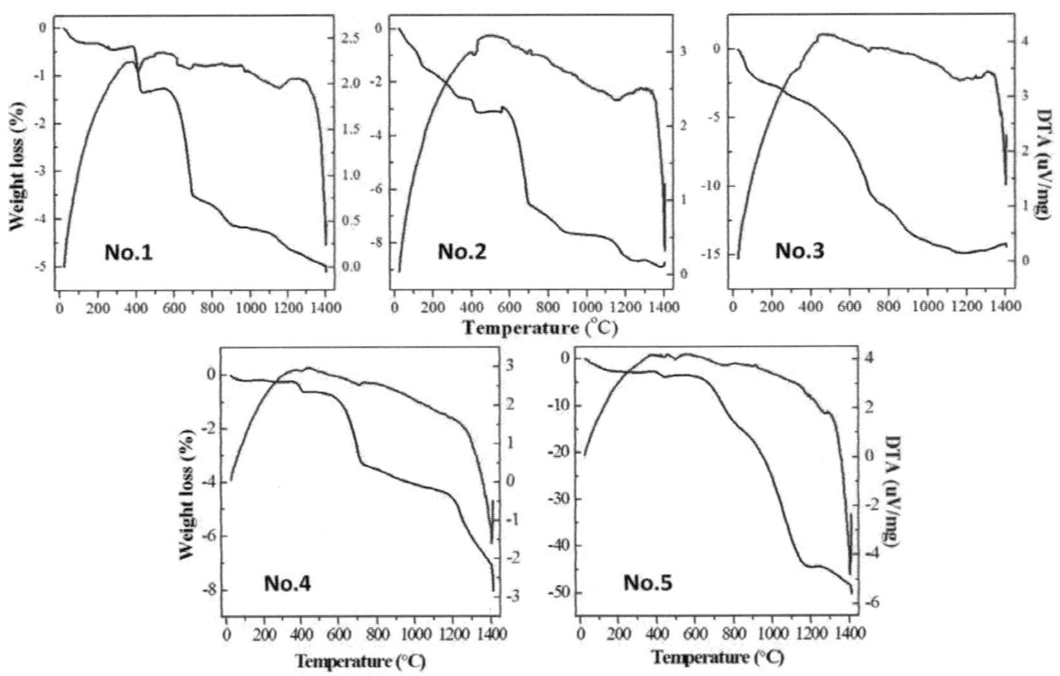


**Figure 1.** The TG-DTA analysis results of 5 incineration ash samples

*3.4 Mineral phase changes with ignition*

Table 3 showed the mineral materials contained in the ash samples. Calcite and quartz were the major mineral phases existing in almost all samples. Gehlenite, portlandite and anorthite also had considerable contents. For water quenched samples, hydrocalumite was found. This mineral was formed during hydration with coexistence of chloride.

IOP Conf. Series: Earth and Environmental Science **61** (2017) 012079 doi:10.1088/1755-1315/61/1/012079

**Table 3.** The thermal constituents in the incineration ash samples

| No.1 | No.2 | No.3 | No.4 | No.5 |
|---|---|---|---|---|
| Calcite ($CaCO_3$) | Calcite | Calcite | Calcite | Calcite |
| Quartz ($SiO_2$) | Quartz | Quartz | Quartz | Portlandite |
| Portlandite ($Ca(OH)_2$) | Portlandite | Halite (NaCl) | Lime (CaO) | Halite |
| Hematitie ($Fe_2O_3$) | Hematitie | Hematitie | Anorthite | Sylvite (KCl) |
| Gehlenite ($Ca_2Al_2SiO_7$) | Gehlenite | Gehlenite | Gehlenite | Arcanite ($K_2SO_4$) |
| Anorthite ($(Ca,Na)(Si,Al)_4O_8$) | Anorthite | Hydrocalimite | | |
| Anhydrite ($CaSO_4$) | Hydrocalimite ($Ca_2Al(OH)_6[Cl_{1-x}(OH)x]\cdot 3(H_2O)$) | | | |

Figure 2 showed the mineral phase change of sample No.1 after being exposed to different ignition temperatures for 2 hours. Through comparison with the original profile, portlandite was found to be decomposed at the ignition temperature of 450°C. The peak of calcite decreased at 650°C and totally disappeared at 850°C. The thermal decomposition of them would cause weight loss as it was showed by TG-DTA profile (Figure 1). Besides the decomposition of certain minerals, there were also some newly formed phases identified by XRD: Mayenite ($Ca_{12}Al_{14}O_{33}$) and larnite ($Ca_2SiO_4$) appeared after the sample being ignited at 650°C and 850°C.

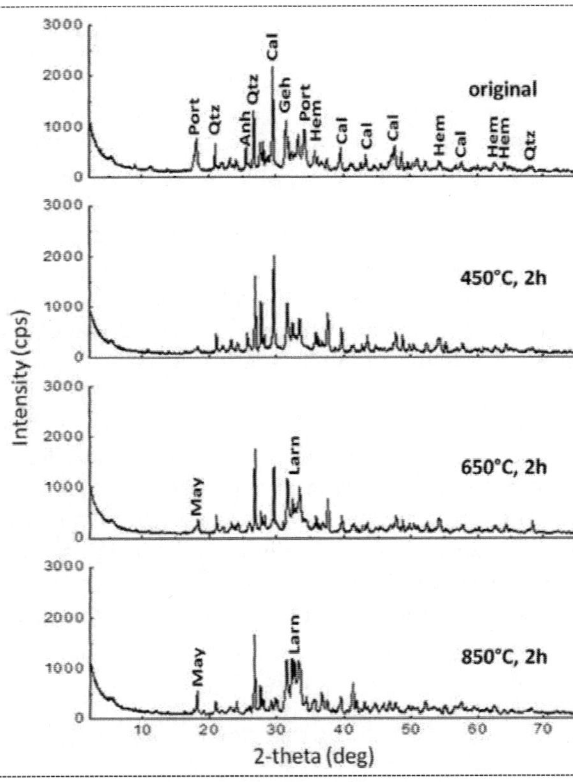


**Figure 2.** XRD profiles of No.1 sample with different treatment conditions. (Port, portlandite; Qtz, quartz; Anh, anjydrite; Cal, calcite; Geh, gehlenite; Hem, hematite; May, mayentite; Larn, larnite).

Same analysis was carried out upon all the other samples. The results were presented by Table 4. Hydrocalumite disappeared under 400°C. Compared with other hydrates, it was less thermal durable and lost structural water at relatively low temperature, which explained the early-stage weight loss of water quenched in Figure 1. After hydrocalumite, portlandite was decomposed at 450~500°C. It also lost structural water and contribute to LOI. Calcite and anhydrite were decomposed after temperature reached 600°C. They would release gases like $CO_2$ or $SO_2$. But not all the reactions would contribute to LOI value. Like lime in sample No.4, which was found disappeared at 650°C, it transformed into another mineral phase and would not be counted in the weight loss.

**Table 4.** Effect of heating temperature (2h) on the mineral phase of the incineration ash samples

| Decomposed minerals | Temperature (°C) | Sample |
|---|---|---|
| Hydrocalumite* | ~400 | No.2, No.3 |
| Portlandite* | 450~500 | No.1, No.2 |
| Calcite* | 600~700 | All the samples |
| Anhydrite* | 600~850 | No.1 |
| Lime | 650 | No.4 |
| **Newly formed minerals** | **Temperature (°C)** | **Sample** |
| Calcium silicate ($CaO{\cdot}SiO_4$) | 850 | No.1, No.2, No.4 |
| Calcium aluminate ($CaO{\cdot}Al_2O_3$) | 850 | No.1, No.2 |
| Lime | 650 | No.5 |

* indicates the decomposition of these minerals would contribute to weight loss.

Table 4 also presents the newly formed minerals in the processes. It needed to be noted that the except for larnite and mayenite, most of the new minerals were calcium -silicate or -aluminate based materials with other ions like chloride or phosphate bound. In No.5, lime was generated through the decomposition of calcite and portlandite. Because there was no much silicate or aluminum contained in this sample, it could preserve rather than turning into other form.

## 4 Conclusions

The LOI value of incineration ash was highly depended on the measurement condition, especially the ignition temperature being used. The decomposition of carbonate and hydrate minerals would become the major source of weight loss if ignition temperature became higher than 600°C. For water quenched ash sample, it might need to be even lower. So ignition temperature of 400°C to 450°C was recommended by this study, and the influence of heating time will be evaluated in the future work.

For mineral phase study, it was found that hydrates' decomposition temperature was around 450°C, but for some thermally unstable materials, the decomposition of them could occur under even lower condition. Calcite would start to decompose when temperature reached 600°C. For anhydrite, its decomposition temperature was found to be 650°C in this research.

## References

[1] M. B. David 1988 Use of loss-on-ignition to assess soil organic carbon in forest soils, Communications in Soil Science and Plant Analysis. **19** pp 1593-1599.
[2] P.J.A. Howard, D.M. Howard 1990 Use of organic carbon and loss-on-ignition to estimate soil organic matter in different soil types and horizons, Biology and Fertility of Soils 9 pp 306-310.

[3] F.J. Larney, L.B. Selinger, A.F. Olson 2005 Influence of loss-on-ignition temperature and heat time on ash content of compost and manure, Communications in soil science and plant analysis 36 pp 2561–2573.
[4] A.J. Chandler, T.T. Eighmy, J. Hartlén. Municipal Solid Waste Incinerator Residues, The international Ash Working Group, pp. 347.
[5] H. B. Sun, M. Nelson, F. Chen, and J. Husch 2009 Soil mineral structural water loss during loss on ignition analysis, Canadian Journal of Soil Science 89 603-610.
[6] O. Heiri, A.F. Lotter, G. Lemcke 2001 Loss on ignition as a method for estimating organic and carbonate content in sediments: reproducibility and comparability of results, Journal of Paleolimnology - J PALEOLIMNOL 25 pp 101-110.

IOP Conf. Series: Earth and Environmental Science **61** (2017) 012080 doi:10.1088/1755-1315/61/1/012080

# Alleviation of cadmium toxicity to Cole (Brassica campestris L. Cruciferae) by exogenous glutathione

**Jun Wang [1], Bin Huang [2], Xin Chen [2], Yi Shi [2, *]**

[1]College of Chemistry, Chemical Engineering and Environmental Engineering, Liaoning Shihua University, Fushun 113001, China
[2]State Key Laboratory of Pollution Ecology and Environmental Engineering, Institute of Applied Ecology, Chinese Academy of Sciences, Shenyang 110164, China
[*]Corresponding author. Tel.: +86 024 83970540; fax: +86 024 83970540. E-mail addresses: shiyi@iae.ac.cn (Yi Shi).

**Abstract.** In this study, we determined the influence of exogenous GSH on cadmium toxicity to cole. GSH addition had beneficial effect on plant development and growth, especially on aboveground biomass and root length. Despite that exogenous GSH insignificantly promoted Cd uptake by the plant, it could decrease of Cd root-to-shoot transport and ameliorate Cd toxicity to the plant. At 6 mg Cd kg-1 soil, GSH addition well countered the Cd-induced significant reduction in CAT activity, but only insignificantly decreased MDA content, suggesting exogenous GSH might indirectly protect plant against oxidative stress via regulating antioxidative enzyme activities. However, at 12 mg Cd kg-1 soil, GSH application insignificantly increased the antioxidant activities but significantly decreased MDA content, indicating external GSH could directly participate in removing radical oxygen species. The results suggest exogenous GSH may have the potential of decreasing Cd accumulation in the edible parts of cultivars and alleviating Cd toxicity.

## 1. Introduction

Inorganic pollution is the main type of soil contamination in China, accounting for over 82% of the sample points exceeding standard value, and the percent of cadmium (Cd) content exceeding standard value ranks first with 7.0%. Cd contamination in soils is becoming a serious problem due to its absorption by plants and transference to human food chain [1]. Cd is a non-essential and toxic metal ion for organisms, and it can induce Cd toxicity in plants such as growth inhibition, chlorosis, and alterations of morphological, physiological, or biochemical properties [2,3]. Studies on different plant species have revealed that the exposure to toxic levels of Cd triggers membrane injury and alteration of antioxidative defense system due to oxidative stress caused by reactive oxygen species (ROS) [4,5]. Correspondingly, plants also have evolved intrinsic mechanisms to regulate ROS levels [6,7]. These mechanisms include enzymatic and non-enzymatic defense systems. As a part of defensive mechanism, antioxidative enzymes, especially superoxide dismutase (SOD; EC 1.15.1.1) and catalase (CAT; EC 1.11.1.6) play an important role in scavenging ROS. SOD is an essential component of antioxidation system in plants as it can dismutate O2•− to H2O2 and O2. CAT is one of the key enzymes for the detoxification of H2O2 via two electron transfer [1]. Therefore, it is imperative to determine the changes in antioxidative enzymes in order to verify the hypothesis that these enzymes can mitigate Cd stress in plants.

Reduced glutathione (γ-Glu-Cys-Gly, GSH), a non-enzymatic antioxidant, is a major low-molecular-weight thiol tripeptide implicated in a variety of life processes and involves in defense system against environmental stresses including heavy metal (HM) [8,9]. GSH can act indirectly as a substrate for some enzymes or phytochelatins (PCs) to alleviate HM toxicity. It is a substrate for glutathione peroxidase (GPX) and glutathione transferase (GST), which are involved in the removal of ROS [10]. GSH is also a precursor for the synthesis of PCs, which are a set of novel heavy metal-binding peptides [11]. This has documented that an increase in intracellular GSH concentration enhanced Cd tolerance and increased Cd accumulation in Brassica campestris [12]. Although GSH indirectly protects plants against HM toxicity by functioning as a substrate, it can also act directly as an antioxidant to detoxify ROS (especially H2O2) associated with biotic and abiotic stress through the ascorbate-glutathione cycle (AsA-GSH cycle) [9]. It plays an important role in maintaining cellular homeostasis as a redox couple since it allows fine-tuning of cellular redox environment under normal conditions and upon the onset of stress. GSH accumulates in response to increased ROS, or to compensate for decrease in the defense capability of other antioxidants and GSH levels are constitutively higher in plants adapted to stress conditions [13,14]. Studies about the effect of intracellular GSH on heavy metal metabolism and toxicity have indicated that GSH can serve to detoxify peroxides, maintain the correct three-dimensional structure of many proteins, synthesize phytochelatins (PCs), mediate enzyme activities, etc [15,16].

Previous studies have revealed that exogenous GSH can trigger a positive effect on antioxidant defense system against abiotic stress. External GSH conferred a high temperature stress tolerance in mung bean (Vigna radiate L.) by increasing chlorophyll content and enzyme activity (e.g. CAT, SOD, GPX) and decreasing MDA and H2O2 content, resulting in better physiological performance [17]. Exogenous application of GSH lessened the adverse effects of Pb on photosynthetic activities and improved cotton's tolerance via reducing lipid peroxidation and antioxidant enzyme activities [18]. GSH application could alleviate the reduction of aboveground biomass, enhance Pb accumulation in shoots and roots, and induce the synthesis of phytochelatins (PCs) [19]. Exogenous GSH appeared to limit the toxicity of herbicide and seemed to overcome the stress status due to enhancing contents of photosynthetic pigments and activities of antioxidant enzymes [20]. However, the mechanism of toxicity alleviation by exogenous GSH in plant under various levels of Cd stress has not been well documented.

Thus, in this study, we chose to investigate the mechanism of toxicity alleviation by exogenous GSH in cole (Brassica campestris L. Cruciferae) exposed to various levels of Cd stress. Cole is a most frequently seen vegetable in the dishes during the hot summer days in China for its low energy but high fiber. We hypothesized that (1) GSH addition will alleviate Cd toxicity, and would be reflected in the negative contribution of malondialdehyde (MDA) and in the positive contribution of plant growth; (2) Extracellular GSH can counteract Cd-induced alterations of certain antioxidant enzyme activities.

## 2. Methods and materials

**Table 1.** Pot experimental design in greenhouse

| **Treatment** | | **Content (mg·kg$^{-1}$ soil)** | | | | |
|---|---|---|---|---|---|---|
| | | **Cd** | **GSH** | **N** | **P** | **K** |
| **Control** | **CK** | 0 | 0 | 150 | 200 | 300 |
| **Group1** | **Cd6** | 6 | 0 | 150 | 200 | 300 |
| | **Cd6+GSH** | 6 | 30 | | | |
| **Group2** | **Cd12** | 12 | 0 | 150 | 200 | 300 |
| | **Cd12+GSH** | 12 | 30 | | | |

The experiment was conducted in Shenyang Station of Experimental Ecology, Chinese Academy of Sciences (41°31′N and 123°41′E), which is located in the North Temperate Zone with a semimoist continental climate. The tested soil is grey meadow soil, and its soil bulk density, pH, total organic carbon (TOC), total N and total P were 1.33 g·cm-3, 7.63, 1.67%, 0.95g·kg-1 and 0.45 g·kg-1, respectively. These soils were passed through a 4-mm sieve. 5.5 kilograms of soils were mixed with CdCl2·5H2O or CdCl2·5H2O accompanied by GSH as listed in Table 1, filled into plastic pots, applied with soluble fertilizers and equilibrated for 2 weeks. Three uniform seedlings of cole (Brassica campestris L. Cruciferae) were transplanted into each pot. Each treatment was replicated five times. No fertilizers were added throughout the cultivation.

After 60 days, the cole was harvested and the experiment was terminated. After being dried with filter papers, the plant samples were dried at 105°C for 10 min, then at 80°C until constant weight, and weighed for dry weight (DW).Dry plant samples were grounded to powders, and digested with a concentration acid mixture of HNO3/HClO4 (3:1, v/v) [21]. The Cd concentration in plant tissues was determined using the AAS method [22].

Lipid peroxidation was estimated by the concentration of malondialdehyde (MDA). MDA content was exacted and estimated with minor modification of the method of Ortega-Villssante et al. (Tieheng et al, 2005). The activities of superoxide dismutase (SOD) and catalase (CAT) were measured as described by Chance and Maehly [23].

The translocation factor (TF) indicated the ability of plants to translocate heavy metals from roots to the leaves [1]. It was calculated as:

TF=the metal concentration in leaves ÷ the metal concentration in roots Eq.

All data obtained from the experiments were performed with SPSS 18.0 and Excel 2007.

**Table 2**. Statistics for plant height, root length, aboveground biomass and underground biomass of tested plants, each with five replicates. Value in tables are presented by mean±SE. Data were compared with single factor analysis of variance. Different lowercase letters in a row mean significant differences (P<0.05).

| **Treatment** | | **Plant height [cm·plant$^{-1}$]** | **Root length [cm·plant$^{-1}$]** | **Aboveground iomass [g·plant$^{-1}$]** | **Underground biomass [g·plant$^{-1}$]** |
|---|---|---|---|---|---|
| **Control** | **CK** | 23.60±2.47 a | 14.20±2.14 a | 16.12±1.36 a | 7.68±0.45 a |
| **Group1** | **Cd6** | 19.20±1.80 c | 13.27±2.14 b | 13.52±0.80 b | 7.28±0.66 b |
| | **Cd6+GSH** | 21.73±1.73 b | 14.73±1.83 a | 15.98±1.27 a | 7.16±0.26 b |
| **Group2** | **Cd12** | 20.33±0.70 c | 13.10±1.71 c | 11.86±1.20 c | 6.97±0.29 b |
| | **Cd12+GSH** | 20.93±1.24 bc | 14.30±2.23 a | 15.56±1.76 a | 7.22±0.24 b |

## 3. Results

As shown in Table 2, GSH application could improve development and growth of cole under Cd stress. For plant height, the supply of GSH led to a significant increase at 6 mg Cd kg-1 soil, and Cd also exerted a profoundly deleterious effect on all the treatments. Furthermore, the application of GSH significantly improved root length and increased aboveground biomass in comparison with the treatments with no GSH supply, and Cd did not have an obvious effect on the treatments with GSH addition. For underground biomass, the application of GSH resulted in a slight increase (without Group 1) in the same Cd treatments, and the biomass of all the treatments was evidently lower than that of the control.

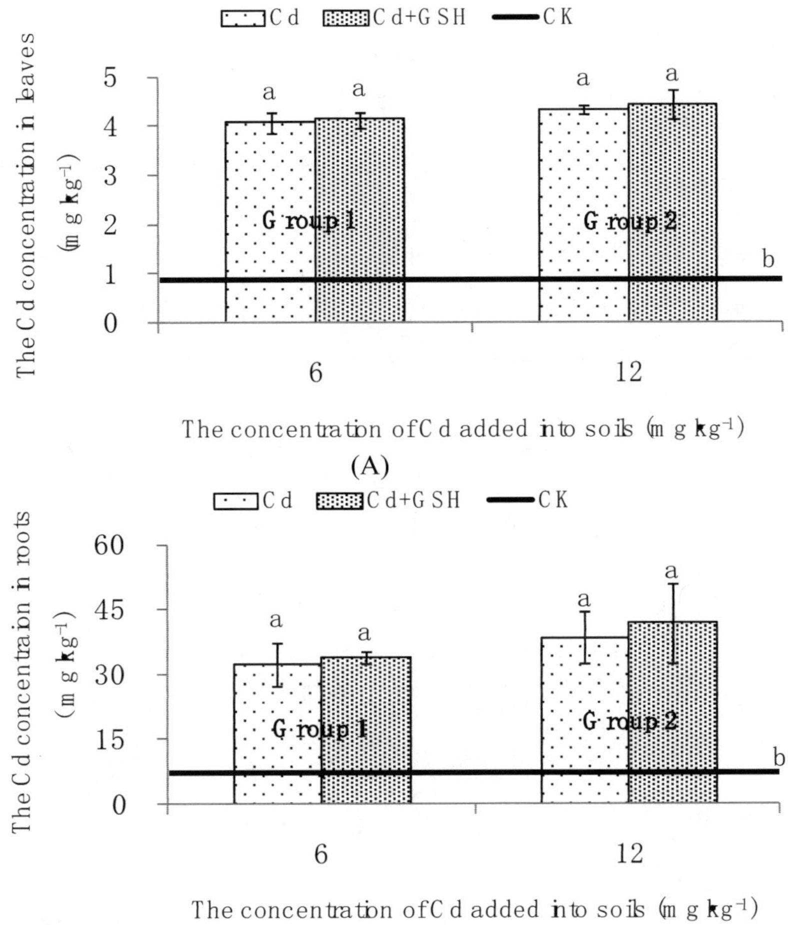


**Figure 1**. Effects of GSH application on Cadmium contents in leaves (A) and roots (B) of cole in soil contaminated with different concentrations of Cd. Error bars represent SE for n=5. Horizontal line represents the control (CK). The different letters indicate a significant difference at $P<0.05$.

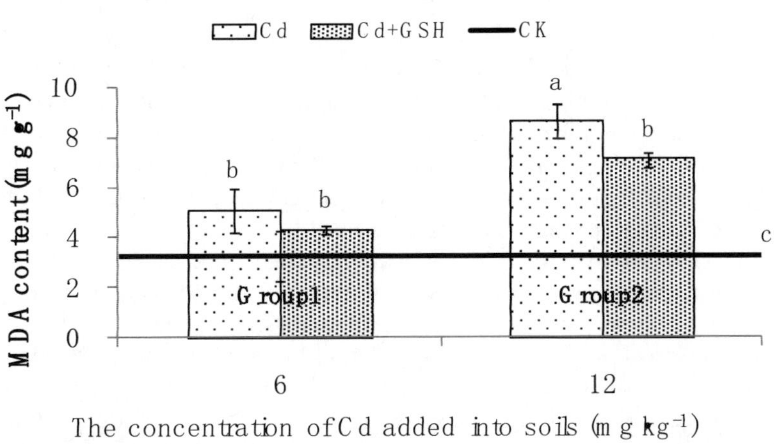


**Figure 2**. Effect of GSH application on malondialdehyde (MDA) content of cole in soil contaminated with different concentrations of Cd. Error bars represent SE for n=5. Horizontal line represents the control (CK). The different letters indicate a significant difference at $P<0.05$.

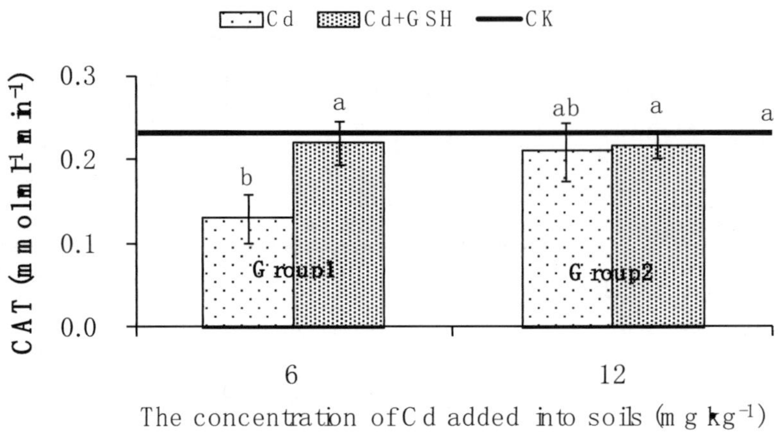


(A)

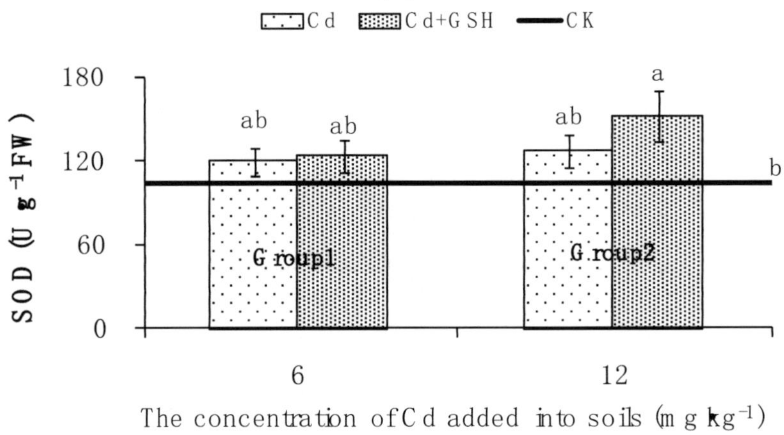


(B)

**Figure 3**. Effect of GSH application on the activities of superoxide dismutase (SOD) (A) and catalase (CAT) (B) of cole in soil contaminated with different concentrations of Cd. Error bars represent SE for n=5. Horizontal line represents the control (CK). The different letters indicate a significant difference at $P<0.05$.

The Cd concentrations in leaves and roots increased with increasing Cd exposure (Figure 1). GSH application promoted Cd accumulation, but the plants exposed to GSH and Cd did not exhibit a significantly higher Cd content than those exposed to Cd only (except for the control). Cd contents of both leaves and roots in all the treatments were obviously higher than those of the control, and increased progressively with the increasing addition of Cd in soil. Cd content in leaves was significantly lower than that of roots in the same treatment. A large increase in Cd contents occurred in leaves and roots at 12 mg Cd kg-1 soil, and similarly after GSH application.

The level of lipid peroxidation in cole leaves was measured in terms of MDA content (Figure 2). MDA contents in the both experimental groups were significantly higher as compared to the control, and increased with the increasing Cd concentration. Under Cd exposure, MDA content increased significantly from 6 mg Cd kg-1 soil to 12 mg Cd kg-1 soil. Under GSH and Cd exposure, MDA content at 6 mg Cd kg-1 soil was significantly lower than that at 12 mg Cd kg-1 soil. GSH application induced detoxification of Cd as evidenced by the decrease in MDA content, although the decrease was only significant when Cd was applied at 12 mg Cd kg-1 soil.

As shown in Figure 3. A, the enzyme activities of superoxide dismutase (SOD) in the both groups were higher than that in the control. Particularly, SOD activity peaked at 12 mg Cd kg-1 soil with GSH

IOP Conf. Series: Earth and Environmental Science **61** (2017) 012080 doi:10.1088/1755-1315/61/1/012080

addition and was significantly higher than that in the control. In the both groups, SOD activities under stress of different levels of Cd with GSH addition were not significantly higher than those without GSH addition.

Catalase (CAT) activities in the control were higher than those in the both groups, and significantly higher than that at 6 mg Cd kg-1 soil without GSH application (Figure 3. B). Under Cd stress without GSH application, CAT activities increased with increasing the level of Cd exposed, but showed no significant difference. Under Cd stress with GSH application, CAT activities did not show any significant difference. At 6 mg Cd kg-1 soil, CAT activity exposed to Cd and GSH exhibited a significant increase than that exposed to Cd.

## 4. Discussion

Reduced glutathione (γ-Glu-Cys-Gly, GSH), a major low-molecular-weight thiol tripeptide, is a nonenzymatic antioxidant. The thiol group of GSH is important in formation of mercaptide bond with metal and for reacting with selected electrophiles [11]. This chemical reactivity makes GSH particularly suitable to serve a broad range of biochemical functions in all organisms. Furthermore, GSH also plays an important role in detoxification processes due to control of H2O2 levels through the AsA-GSH cycle, and functions directly as a free radical scavenger by reacting with $1O_2$, $O_2•-$ and •OH [13]. Because GSH has been documented in enhancing plant tolerance to heavy metals [24], we applied it in all experimental groups as a modifier to alleviate Cd toxicity.

**Table 3.** Statistics for Cd accumulation of tested plants in contaminated soils treated with Cd or Cd and GSH.

| | Treatment | TF |
|---|---|---|
| **Control** | **CK** | 0.124 |
| **Group1** | **Cd6** | 0.131 |
| | **Cd6+GSH** | 0.122 |
| **Group2** | **Cd12** | 0.113 |
| | **Cd12+GSH** | 0.106 |

Being a non-essential and toxic metal ion for organisms, Cd can exhibit detrimental effects on plant growth and yield [25]. In the present study, the statistics for plant height, root length, aboveground biomass and underground biomass were clearly impaired under stress from 6 mg Cd kg-1 soil to 12 mg Cd kg-1 soil in comparison with the control, and GSH addition could mitigate the negative effects of Cd on plant growth and development (Table 2). Several other studies have also documented that Cd can induce visible symptoms of phytotoxicity, such as plant growth inhibition and yield reduction [17,26]. Furthermore, Cd is a major heavy metal threatening food safety due to its high soil-plant mobility and easy accumulation in plant tissues (especially edible parts). Single Cd-treatments significantly increased Cd uptake in roots and leaves compared to the control, GSH addition could slightly enhance Cd accumulation, and the promoting effect of GSH on Cd accumulation in roots was larger than that in leaves (Figure. 1). Translocation factor (TF) is used to demonstrate the ability of Cd translocation from root to shoot. The lower TF with GSH application (Table 3) suggested that GSH could weaken the ability to translocate Cd to leaves. Similarly, several reports have suggested that GSH application can lessen the adverse effects of heavy metals and improve plant's tolerance. Fanrong suggested that exogenous application of GSH in the culture solution obviously alleviated the reduction of plant growth, and caused significant decrease of Cr root-to-shoot transport in the Cr-stressed rice plants [27]. Shin-Ichii N. found that GSH application significantly inhibited root-to-shoot Cd translocation via xylem vessels in oilseed rape. Based on lower TF and higher growth indicators (e.g. plant height, root length and aboveground biomass), GSH has the potential of decreasing Cd accumulation in the edible parts of cultivars to reduce risks associated with Cd toxicity and maintain sustainable crop production [28].

The role of external GSH addition in alleviation of Cd-induced damage was also revealed by its effects on MDA and some antioxidative enzymes. Cd may initiate ROS generation and the process of lipid peroxidation, resulting accumulation of MDA [29], which is a product of cell membrane lipid peroxidation. The change in MDA content indicates the degree of antioxidant damage under abiotic stress. The present results revealed that Cd caused a dramatic increase in MDA content in cole leaves (Figure 2), indicating an occurrence of high oxidative stress due to ROS generation. However, GSH addition reduced MDA accumulation, suggesting that GSH may enhance antioxidant capacity in Cd-stressed plants. Furthermore, GSH addition increased the activities of CAT and SOD. CAT is one of the key enzymes for the detoxification of H2O2 via two electron transfer. SOD is an essential component of antioxidation system in plants as it can dismutate O2•− to H2O2 and O2 (Zhou et al., 2009). In our study, the significant reduction in CAT activity induced by Cd stress (6 mg Cd kg-1 soil) was well countered by the addition of GSH. The results are agreed with Fei C.[30]. Thus, we conclude that in our experiment exogenous GSH might indirectly protect plant against oxidative stress resulted from Cd toxicity via regulating antioxidative enzyme activities.

Another antioxidative defense strategy of GSH is to react with ROS as an antioxidant as evidenced in our study. Under the high Cd stress (12 mg Cd kg-1 soil), the promoting effect of GSH on activity of SOD was larger than on CAT (Figure. 3), indicating that SOD is over expressed in dismutation of O2- to H2O2, resulting in lipid peroxidation due to H2O2 accumulation. More importantly, the application of GSH, which functioned as an intracellular ROS-scavenger through the AsA-GSH cycle [13], resulted in a significant reduction in MDA content under high Cd stress. Our results agree with several previous findings [31]. Thus we speculate that external GSH could directly participate in scavenging ROS as an antioxidant.

## 5. Conclusions

In this study, the influence of exogenous GSH on Cd-induced oxidative stress was investigated. GSH addition could promote development and growth of plants, and ameliorated lipid peroxidation. Exogenous GSH successfully regulated the activities of CAT and SOD, and decreased MDA contents. We speculated that exogenous GSH not only acts indirectly in regulating ROS-scavenging enzyme activity but also acts directly as an antioxidant to remove ROS. The application of additional GSH may provide an effective alternative to alleviate oxidative stress resulted from Cd toxicity.

## Acknowledgments

This work was financially supported by National Science and Technology Infrastructure Program of the Ministry of Science and Technology of P.R. China (2015BAD05B03).

## References

[1] Zhouli L, Xingyuan H, Wei C, Fenghui Y, Kun Y and Dali T 2009 *J Hazard Mater* 169 170-175
[2] Kamrun N, Mirza H, Md Mahabub A, Anisur R., Toshisada S and Masayuki F 2015 *Excotox Environ Saf* 126 245-255
[3] Susana M G, Liliana Beatriz P, Robert A B, Claudia E A, María F I, Eliana P R, Myriam S Z, María D G and María P B 2012 *Environ Exp Bot* 83 33-46
[4] Kelly O, Tony R, Els K, Jaco V and Ann C 2012 *Eviron Exp Bot.* 83 53-61
[5] Zouari M, Ahmed Ben C, Zorrig W, Elloumi N, Rahi M, Delmail D, Rouina Ben B, Labrousse P and Abdallah Ben F 2016 *Ecotox Environ Safe* 128: 100-10
[6] Zhongchun Z and Baosheng Q 2007 *J Environ Sci.* 19 1311-1317
[7] Wojciech P and Zbigniew T 2010 *Ecotox Environ Safe* 73: 1207-1213
[8] Mendoza-Cózatl D, Loza-Tavers H, Hernández-Navarro A and Moreno-Sánchez R 2005 *FEMS Microbiol Rev* 29 653-671.
[9] Pablo M R, Cecilia I M, Facultad J de B, Andrew J R, Ian R B and Mario A 2000 *J Bacteriol* 182 1748-1753
[10] Christine H F and Graham N 2005 *Plant Cell* 17 1866-1875

[11] Sudesh Y 2010 *S AFR J Bot* 76 167-179
[12] Naser A A, Umar S, Altaf A, Muhammad I and Nafees K 2008 *J Plant Interact* 3 189-198
[13] Mohammad Anwar H, Pukclai P, Jaime A, Teixeira da S and Masayuki F 2011 *J Bot* 2012 2-38
[14] May M J and Christopher John L 1993 *Plant Physio* 103 621-627
[15] Christine H F, Frederica L T and Serge D 2001 *Trends Plant Sci* 6 486-492
[16] Stephan C 2006 *J Plant Physiol* 163 319-33
[17] Kamrun N, Mirza H, Md Mahabub A and Masayuki F 2015 *Environ Exp Bot* 112 44-54
[18] Mumtaz K, M. K. Daud, Ali B, Muhammad Jamil K, Azizull ah A, Niza M, Noor M, Zia ur R and Shui Jin Z 2016 *Environ Sci Pollut Res* 23 8431-8440
[19] Haiyun R, Chen C, Xingxing Z, Zhenguo S and Fenqin Z 2016 *J Hazard Mater* 301 304-313
[20] Mamdouh Nemat A and Nemat H 2014 *Pestic Biochem Phys* 112 56-62
[21] Yuebing S, Qixing Z and Chunyan D 2004 *Bioresour Technol* 99 1103-1110
[22] Aldrich M V, Jorge L, Gardea T, Jose Peralta V and Jason G P 2003 *Environ Sci Technol* 37 1859-1864
[23] Harborne J B 1984 Chlorophylls. Chapman and Hall, London.
[24] Piao X, Guangming Z, Danlian H, Liang L, Meihua Z, Cui L, Ningjie L, Zhen W, Chao H and Chen Z 2016 *Ecol Eng* 87 150-156
[25] Haiyan Y, Yongxia Z, Suzhen H, Yongheng Y and Chunsun G 2015 *Environ Sci Pollut Res* 22 2808-2816.
[26] Fernández R, Bertrand A, Reis R, Mourato M P, Martins L L and González A 2013 *J Hazard Mater* 244 555-562
[27] Fanrong Z, Boyin Q, Xiaojian W, Senzhen N and Feibo W 2012 *Biol Trace Elem Res* 148 255-263
[28] Shin-Ichii N, Nobuo S, Toshinori N, Fumiya K, Noriko S I, Sayuri ItoT, Naoki K, Hiroki R, Hiroyuki H, Mitsuo C and Shu F 2013 *J Exp Bot* 64 1073-1081
[29] Shih-Chueh C, Yun-Yang C, Chwan-Yang H and Ching K 2012 *Plant Growth Regul* 66 27–35
[30] Fei C, Fang W, Feibo W, Weihua M, Guoping Z and Meixue Z 2010 *Plant Physiol Bioch* 48 663-672
[31] Fabrizio P, Maria A I, Stefania P and Angelo M 2003 *Plant Physio* 133 829-837

IOP Conf. Series: Earth and Environmental Science **61** (2017) 012081 doi:10.1088/1755-1315/61/1/012081

# Measurement of water content in high water-cut oil-wells based on rayleigh scattering

**MA Weili[1,a],WANG Yansong[2,b], WANG Jianguo[3,c],WANG Yiming[2,d], WANG Jia[2,e], WU Shitong[2,f],WANG Diandong[4,g],GUO Wenyong[5,h], LI Yun[2,i] ,ZHANG Baohui[2,j], LI Zhenlong[2,k],LIU Wensheng[2,l],XU Guanghui[6,m], XU Rui[5,n]**

[1]Huabei Petroleum Tiancheng Industrial Group Co.ltd, Renqiu, 062552, China.
[2]Institute of Oil Production Engineering in Huabei Oilfield, Renqiu, 062552, China.
[3]The Erlian Filiale of Huabei Oilfield Company ,Xilinhote 026000, China.
[4]The First Exploitation Factory Of Huabei Oilfield Company, Renqiu, 062552, China.
[5]The Second Logging Company of CNPC Bohai Drilling Engineering Co.ltd, Renqiu, 062552, China.
[6]Environmental Nonitoring Central Station in Technical Supervision Department of Huabei Oilfield Company ,Renqiu 062552, China.

[a]tc_mwl@petrochina.com.cn [b]cyy_wangys@petrochina.com.cn
[c]hse_wjg@petrochina.com.cn [d]cyy_wangym@petrochina.com.cn
[e]cyy_wj@petrochina.com.cn [f]cyy_wust@petrochina.com.cn
[g]cy1_wddong@petrochina.com.cn [h]476485706 @qq.com
[i]cyy_liy@petrochina.com.cn [j]cyy_zangbh@petrochina.com.cn
[k]cyy_lizhenl@petrochina.com.cn [l]cyy_liuws@petrochina.com.cn
[m]2416370530@qq.com [n]el_xurui@petrochina.com.cn

**Abstract.** According to the lack of traditional method measuring high water-cut production level effectively, this paper discusses the measurement of high water content by Rayleigh scattering. When launching high-frequency electromagnetic waves downward along the axial, the diameter of the oil-in-water emulsion can range from millimeter to centimeter in an oil-water two-phase case. Because the Electromagnetic Wave can be scattered by oil droplets and is related to their quantities and size, we can measure the water content using suitable electromagnetic wave with the right wavelength. Then, the measuring accuracy of Water Content in High Water-cut Oil-wells can be improved and the confluent-liquid umbrella can also be canceled.

## 1 Introduction

Most of the oil fields in our country have came into middle-later developed period and water-cut of oil reservoir is in a high level. In the production, measuring water content of the produced fluid accurately has important significance for the production of oil-wells. At present, measuring production fluid with high water-cut using the traditional method is of bad accuracy and even unable to distinguish. Therefore, the technology measuring the water content of the oil-well production layer accurately is needed when extracting oil at the frontline[1].

Traditionally, we measure water content follow three steps. Putting a metal capacitor into the production layer of production wells firstly. Then, making the oil-water fluid in oil-wells go into the container. Finally, calculating the capacitance. The capacitance is related to the dielectric constant of the mixture, and the dielectric constant of water and oil vary greatly (dielectric constant of water is about 80 while the value of pure oil ranges from 2 to 3). So, we can calculate the water content of oil-water mixture from the capacitance[2]. However, when adopted in an oil-well with high water content, the measurement cannot figure out the available capacitance because the conductive water can lead to the short circuit of two electrodes in the capacitor. Furthermore, the effective capacitance cannot be measured and we can only get the inaccurate data in high water-cut oil-wells. At the same time, the method mentioned is carried out by the closed measurement, which means the measured water content only includes the part entering the container rather than the average water content of the production layer[3]. In order to measure the average water content, an open measurement should be developed.

In this paper, we propose an open measurement based on Rayleigh scattering. Then we can accurately measure the water content with the cancellation of confluent-liquid umbrella in a high water-cut condition.

## 2 Theory of measuring the water content with Rayleigh scattering

When traveling in the air, the acoustic wave and electromagnetic wave can be scattered by some water vapor condensation such as clouds, rain and hail. These scattering bodies can be equivalent to a medium sphere. When the sphere radius a is much smaller than the wavelength$\lambda$, the scattering of sound waves or electromagnetic waves is called Rayleigh scattering. Similarly, in high water-cut oil-wells, the water is used as a continuous phase and the oil droplets play the role of the scattering body.

Generally, the scattering wave is generated because particulate matters in the liquid oscillate and then become the second wave source under the driving of the incident wave. In a homogeneous medium, the wave superposition makes the wave travel in the direction of wave refraction and is offset in other directions because of the interference. While in an inhomogeneous media, the coherence of second wave is destroyed due to the presence of inhomogeneous particles leading to the occurrence of scattering wave. Oil droplets can be considered inhomogeneous medium in high water-cut condition. Hence, the oil content (also the water content) can be obtained by measuring scattering wave. The theory of measuring the oil content with scattering wave and wave characteristics is shown in Figure 1[4].

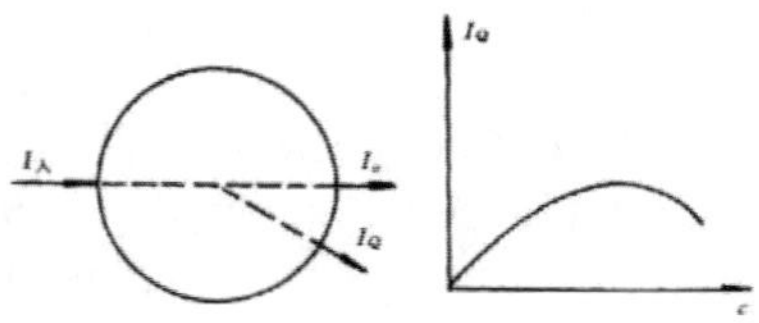

Fig. 1 Theory of measuring the oil content with scattering wave and wave characteristics

As shown in the left of Figure 1, $I_\lambda$, $I_o$ and $I_Q$ represent the intensity of incident wave transmission wave and scattered wave respectively and C means the oil content. According to Figure 1, we can obtain that the scattering intensity is approximately linear with oil content in a certain range. The intensity of the scattered wave decreases when oil content increases to a certain extent because the increase of oil droplets particles with the increase of oil content prevents the scattering of the wave. So, the method above can be applied to the water-cut measurement in a high water-cut condition.

## 3 Mathematical model of measuring water content with Rayleigh scattering

In oil-water two-phase wells, when there are higher water content, the oil exists in the oil-in-water form[5]. Therefore, if water is viewed as the continuous medium, the oil droplets can be viewed as scattering bodies and the quantity of oil content (i.e. water content) can be measured by calculating the strength of Rayleigh scattering.

For an accurate result, the chosen wavelength of the ultrasonic wave or electromagnetic wave should be larger than the diameter of oil droplets. For example, in order to measure oil droplets in millimeter range, the wavelength must be greater than or equal to 10mm. Moreover, assuming that the sound velocity of oil-water two-phase flow is 1000 m/s in a wellbore, oil droplets in millimeter range can be measured with ultrasonic wave, whose frequency should be 100KHz. By the same way, the maximum frequency of the electromagnetic wave is 30GHz[6].

Supposing that a droplet radius is a, the dielectric constant is ε and the dielectric constant of water is ε0. Then the calculating model can be established shown in Figure 2 and the formula of Rayleigh scattering can be represented as

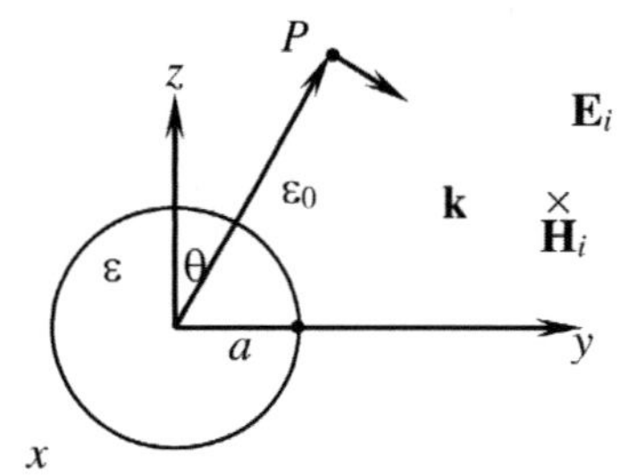


Fig. 2 Mathematical model of Rayleigh scattering of the oil droplets

$$H_{\varphi} = -\frac{E_0(\varepsilon - \varepsilon_0)k^2a^3}{\eta_0(\varepsilon + 2\varepsilon_0)r}\sin\theta e^{-jkr} \quad (1)$$

The meanings of related parameters in the formula are shown below:

$\eta_0$ : Scattering coefficient related to the properties of the medium.

$H_{\varphi}$ : Scattering intensity, Mw/m.

$E_0$ : Wave intensity, Mw/m.

$r$ : The distance between the oil droplets and wave source, m.

$k$ : Coefficient, also a constant.

From the above formula, we can obtain that the Rayleigh scattering is proportional to the three times of the oil droplet diameter and is proportional to the dielectric constant difference between oil and water. For a number of oil droplets, measured results is the superposition of all Rayleigh scattering. So, we can obtain the volume of all oil droplets in the casing pipe statistically. Thus, it is sure that the electromagnetic wave can cover the liquid in cylindrical shaft when launching downward and the water content can be measured accurately without a confluent-liquid device.

## 4 Theoretical experiment and result analysis

### *4.1 Experimental device*

As shown in Figure 3, the test instrument and wave source are placed on central axis of the casing and ultrasonic or electromagnetic waves are launched downward, making oil droplets move along the opposite direction of propagation of the wave. It means that we can get the maximum Rayleigh scattering when the parameter θin formula 1 is set as 90 degree. At the same time, quantities and size of oil droplets can be measured, which correspond to the oil content and water content.

To verify the feasibility of the program quickly, the scalar network analyzer (8756A) produced by HP/Agilent Company in the United States is used to set up a water test instrument with electromagnetic wave. The functional block diagram is shown in Figure 4.

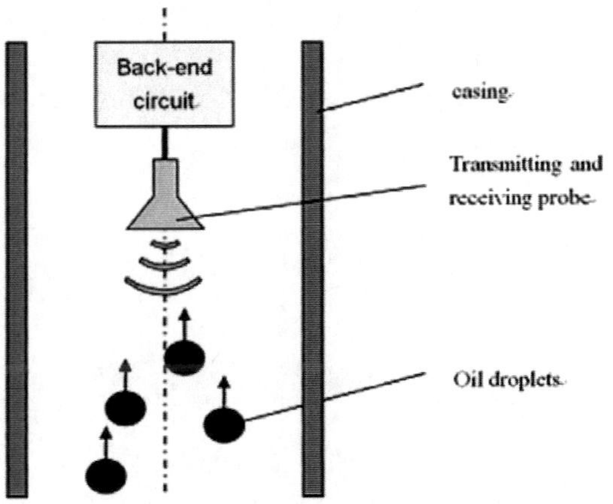


Fig. 3 The testing method and principle

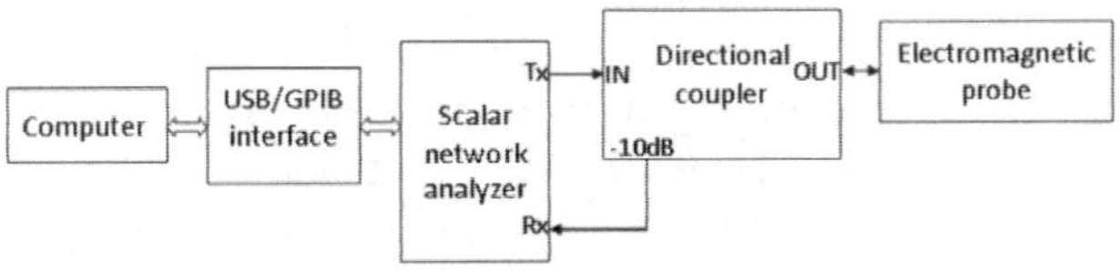


Fig. 4 The functional block diagram

Directional coupler is a kind of microwave device used for signal isolation and separation. The electromagnetic probe is used for the conversion between electromagnetic wave and electrical signal.

*4.2 Experimental method*

With the crude oil water content testing instrument, the theoretical experiment was carried out on the three phase flow experiment device belonging to Institute of oil production engineering in Huabei oilfield. During the experiment, the Electromagnetic probe was stuck into the internal casing and the rest of the device was placed on the ground. The electromagnetic probe was connected with the directional coupler through a coaxial cable. Considering that coaxial cable would attenuate the high-frequency signals, the frequency of electromagnetic waves were set between 250 and 259 MHz ultimately. The three phase flow experiment device was used to allocate different proportions of oil and water mixture, which was sent into the casing. After a lot of experiments, we have got satisfying results.

*4.3 Experimental data analysis*

The Experimental results generated by Matlab are shown as Figure 5.

According to the figure, the Rayleigh scattering intensity is linearly distributed under the irradiation of 250-259MHz electromagnetic wave when the water content is from 80% to 100%. Taking 253MHz wave as an example, the empirical formula describing the connection of scattering intensity and water content can be obtained.

$$Y_{water} = -0.0344\mathrm{H} + 0.0138 \qquad (2)$$

$Y_{water}$ means the water content (%) and $H$ means the scattering intensity (dB).

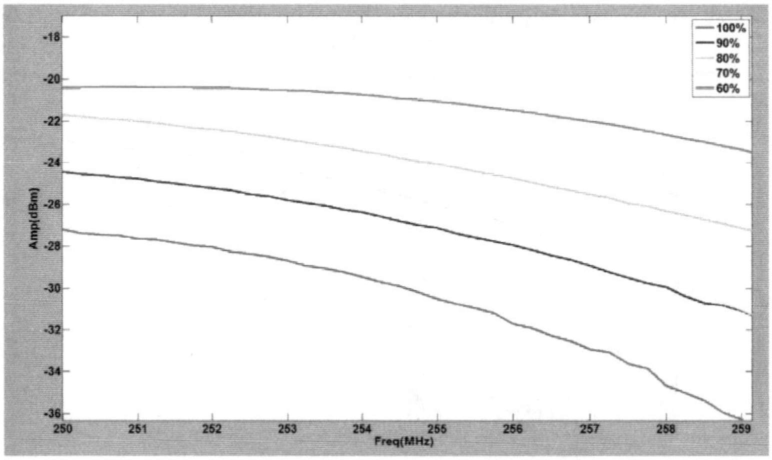


Fig. 5 relationship among the electromagnetic wave,
scattering intensity and water content

It can be seen that curves in the situation where the water content is 60% or 70% do not meet the formula. This is because the increase of liquid droplet in numbers with the increase of oil content can result in the fuzzy boundaries between oil and water. In addition, the scattering effect is unpredictable without a scattering body. So, Rayleigh scattering measurement is suitable for the condition where water content is more than 80%. Similarly, the water used as a scattering body can be adopted to measure the water content with Rayleigh scattering when water content is less than 30%.

## 5 Conclusion

In the oil-water two-phase case where water and oil droplet are viewed as the body and scatterer respectively, water content can be measured through Rayleigh scattering intensity. Thus, theoretical and experimental data show that the method proposed is applicable to oil-wells where the water content is more than 80%.

The method can be used to detect a certain area below the electromagnetic probe. So, the confluent-liquid umbrella can be canceled, which is suitable for high water-cut horizontal wells.

The miniaturization and practicality of the instrument should be achieved as quickly as possible.

## References

[1] Zhou Ximin，Chen Wu: Influencing Factors of Cost Change Rule in Extra-high Water Cut Oilfield, SINO-GLOBAL ENERGY, vol.12(2014), p.59
[2] Wu Xiling: Petroleum Production Logging Principles, Higher Eduction Press (2004), p.221
[3] Wu Hongtao: The measurement model of capacitive water-cut meter in high-water-cut environment, Petroleum Instrument, vol.9(1) (1995), p.31
[4] Zhang Jundong: A Photoelectric Sensor for Measuring Oil Concentration in Water, Domestic technology vol.11(1998), p.39
[5] Chen Jie: Study to Flow Pattern of Oil-water Two-phase Flow, Oil-Gasfield Surface Engineering, Vol.19(2000), p.7
[6] Xie Chufang: Electromagnetic field and electromagnetic wave, Higher Eduction Press (2000), p.164

IOP Conf. Series: Earth and Environmental Science **61** (2017) 012082 doi:10.1088/1755-1315/61/1/012082

# Research on BIM-based building information value chain reengineering

**Zhao Hui, Xie Weishuang**

School of Architecture & Design, Changchun Institute of Technology, 130012, Changchun, China

architectureloving@163.com

**Abstract.** The achievement of value and value-added factor to the building engineering information is accomplished through a chain-flow, that is, building the information value chain. Based on the deconstruction of the information chain on the construction information in the traditional information mode, this paper clarifies the value characteristics and requirements of each stage of the construction project. In order to achieve building information value-added, the paper deconstructs the traditional building information value chain, reengineer the information value chain model on the basis of the theory and techniques of BIM, to build value-added management model and analyse the value of the model.

## 1. Deconstruction of traditional building information value chain

Porter's value chain theory argues that information is a supporting factor in the value-added process, rather than being source of added value. With the further development of building information, information becomes an important production factors in modern society. Through the development and utilization of information resources, it can also create value on its own behalf, promote each phases of the building engineering value chain, so that each phases of the value chain create greater value. The information flow which can create value is the information value chain [1].

### *1.1. The traditional building information value chain model*

The implementation process of building engineering will continue to produce vast quantity of information. Construction information is created, utilised and maintained by numerous project participants. Sharing different information storage and exchange formats which flows back and forth between building participants, accompanied with capital and logistic flow. This produces a fixed quantity according to the generation and change in pattern and then transported to according departments (Figure.1). Due to the limitation in the mode of information transmission, the information transmission process will cause information loss, form an information fault.

*1.1.1. Information value at the decision-making stage.* The decision-making stage primarily defines the overall goal of the project, the stage objectives, quantity of project investment, project functions and other aspects, balancing the relationship between project functionality and cost, majority of which implements non-geometrical information to describe the proposed project, the generated information will affect the design and follow-up work.

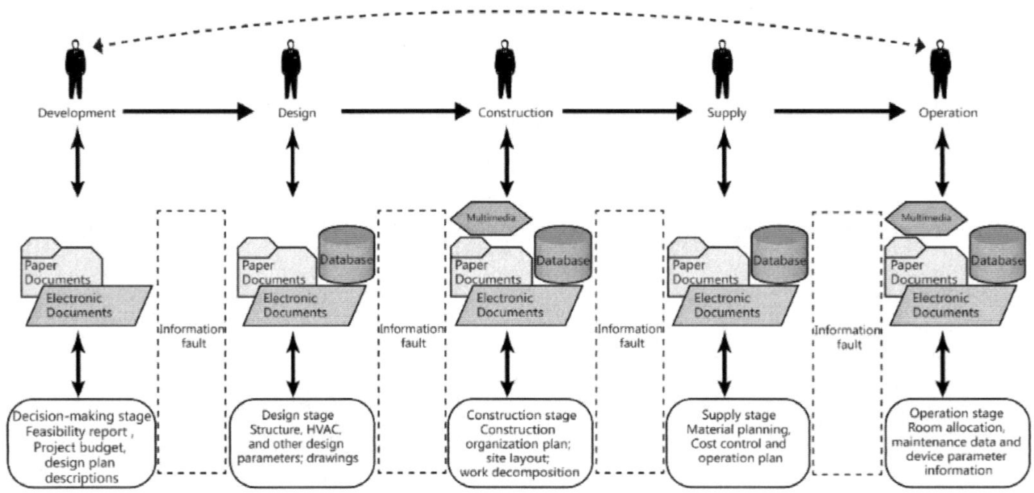


**Figure 1.** Traditional building information value chain model

*1.1.2. Information value at the design stage.* The design stage is to produce a technical representation by converting the owner's construction intents, such as the functional requirements and standards into models that can be implemented. Design work requires multi-professional collaboration, the design will ensure that the intended function of the facility is achieved, maximising the owner's intention.

*1.1.3. Information value at the construction stage.* The construction phase is mainly to establishes the construction plan, building material plan and estimation on the construction cost. The main information comes from the design results, not the direct output of the design information. The information yielded during the construction stage may provide feedback to the design stage, causing alterations in design to optimize the design.

*1.1.4. Information value at the supply stage.* The supply stage is the process of implementing the plan, supporting the transformation of the design information and the construction plan into building entity. This includes more detailed information, such as additional tools, material procurement and equipment allocation.

*1.1.5. Information value at the operation stage.* The operation management is mainly focused on the operation and maintenance of the construction project, guaranteeing sustainable use of project protection. This includes facility operation and maintenance, inclusive of facilities management, equipment operation, building maintenance and so on.

*1.2. The predicament of traditional building information value chain*

Intergrating the above analysis of the construction project information value chain and related literatures [2], problems that arise in project information value chain value-added process are as follows:

*1.2.1. Information cannot be shared effectively.* In the traditional project stage management mode, the responsibility of each stage is based on the objectives of that stage, so that the sharing and use of the project information are phased. Thereby, it is difficult to control and manage the project information between the stages effectively, resulting in the formation of an "information island" whereby affecting the realization of the construction project value chain value-added.

*1.2.2. The loss of information is serious.* The traditional staged management mode lacks of direct communication and contact amongst the construction project participants, lead to the loss of construction project information in the transition phases. In addition, based on the use of both the paper documents and the electronic documents in data transmission, missing information during transcription and data form compatibility restrictions will occur, reducing the efficiency and quality of information transmission.

*1.2.3. The backwardness of information media.* Traditional information media such as paper media, mail, fax, etc., not only increase the cost of communication, but also reduce the efficiency of information transmission. At the same time, most of these traditional means of communication are point-to-point, which cannot guarantee communication and coordination among multiple participants, and it is difficult to achieve knowledge sharing and value-added information.

*1.2.4. Information transmission delay.* There are two main reasons for the delay of information transmission: First, the information media is backward, the speed of information transmission is limited by the geographical distribution of the participants; secondly, the hierarchical organizational structure within each participant is not conducive to the rapid transmission and acquisition of information.

*1.2.5. The information lacks relevance.* Owing to the fact that information creation, exchange and sharing has no unified platform, information on the transition phase thus need to be repeatedly entered. At the same time, due to the application of the parties unable to achieve the association and modification of information, it will also lead to information transmission lag.

## 2. BIM-based building information value chain reengineering

### *2.1. BIM concept and characteristics*

Building Information Modelling (BIM) can effectively assist the information integration, interoperability and collaborative work in the field of construction industry [3], which plays a crucial role in actualising the building lifecycle management (BLM). The term BIM appears to distinguish between the next generation of information technology, the computer-aided design (CAD) and the traditional computer-aided drawing technology. From a narrow level, as first defined by the National BIM Standard-United States® (NBIMS-US™), a BIM "is a digital representation of the physical and functional characteristics of a facility. As such, it serves as a shared intellectual resource for information about a facility, forming a reliable basis for decisions during its life cycle from inception onward [4]."
From a different perspective, BIM has different characteristics. In terms of the model itself, BIM has the following characteristics [5]:

*2.1.1. The objectization of building components.* Building components are the basic digital objects of the BIM model, BIM model can represent the physical and functional characteristics of each component, and has intelligent interactive ability. For example, in the BIM model, the door, window and the wall can be automatically combined between the geometric relations and the functional structure to form a whole.

*2.1.2. The completeness of model information.* In addition to the description of 3D geometric information and topological relation, but also includes a complete description of the project information is included. Such as the object name, structure type, building materials, engineering performance and other design information; the construction process, schedule, cost, quality, labour, machinery, material resources etc. Furthermore, other construction information; engineering safety performance, material durability and, as well as other maintenance information; engineering logic relationship between objects, etc.

*2.1.3. The relevance of model information.* The objects in the information model are identifiable and interrelated, the system can carry out statistics and analysis on the information of the model, develop

IOP Conf. Series: Earth and Environmental Science **61** (2017) 012082 doi:10.1088/1755-1315/61/1/012082

corresponding graphs and documents. If an object in the model changes, all objects associated with it are updated to maintain the integrity and robustness of the model.

*2.1.4. The consistency of model information.* The model information is consistent at different stages of the building life cycle, to ensure that the same information need not be repeated. Moreover, the information model can automatically evolve. The model object at different stages can be simply being modified and extended without the need to re-create, thereby reducing the errors in information inconsistency.

*2.2. BIM-based construction of building information value chain model*

The information value chain model based on BIM is constructed on the basis of its framework, which is processed by the following four points：① The basic framework of the traditional Building Information Value Chain as the overall framework of the model. ②Uses the BIM model, the sub-model center database and the network interaction platform as the support base of information value chain operation. ③ The operations of the construction project participant and its phase work is reorganized according to the information input end, the information flow domain and the value output end of the information value chain. ④ The information input end, the information flow domain and the value output end of the construction engineering information value chain model framework are embodied and visualized (Figure.2).

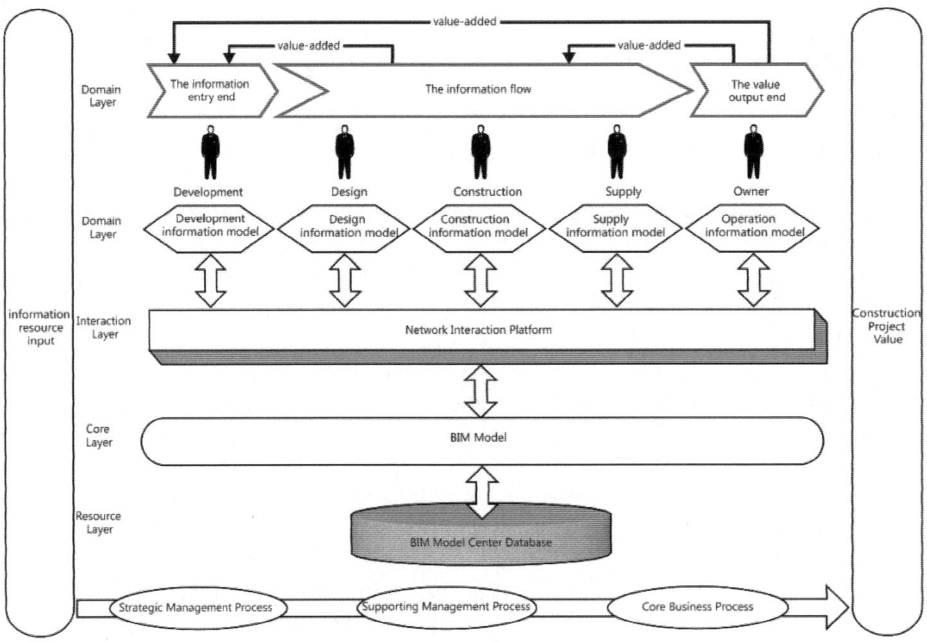


**Figure 2.** BIM-based building information value chain model

The resource layer is the BIM central database, the core layer is a building information model based on IFC and XML standards, created and modified by the design department, other participants in the building supply chain can invoke the building information model [6]. Interaction layer is the network interaction platform, which can rely on similar project information portal or the Autodesk Buzzsaw application software to achieve. The domain layer is different; the designer creates the design information model, such as architecture or structural design system. The construction contractor creates the construction information model, such as the 4D construction management model; the supplier creates the supply information model, in particular the supplier that provides the prefabricated component, and the owner creates an operational information model, such as a property management system.

The advanced feature of the BIM-based building information value chain is that all sides of the building engineering chain can deliver the BIM documents. The BIM document contains all the building information. The information delivery party then transforms the 3D visual building information model into IFC and XML-based BIM documents, and transmits them to the information receiver through the network interactive platform. After the information receiver opens, it also presents a 3D visual information model of the building.

## 3. Value-added management and application value of BIM-based building information value chain

### *3.1. Value-added management of BIM-based building information value chain*

In order to manage the information resources in the building information value chain to an orderly, systematic, repeatable and efficient manner, a value-added management framework for the BIM-based building information value chain is established (Figure.3). According to this framework, the management personnel through the use of the information systems accumulates historical information. Via the mechanism of the four steps; information resources extraction, information processing, information resources management, information resources utilization, the results can be obtained to support management decisions. The conceptual framework can solve the problem of low efficiency of the information flow management from the source. The standardization of information, too, occurs before storage, greatly improving the efficiency of information resources processing. This framework will not only play a reference role in grasping the development of information resources utilization technology, but also plays a guiding role for the development of related special software, thereby promoting the development of special software.

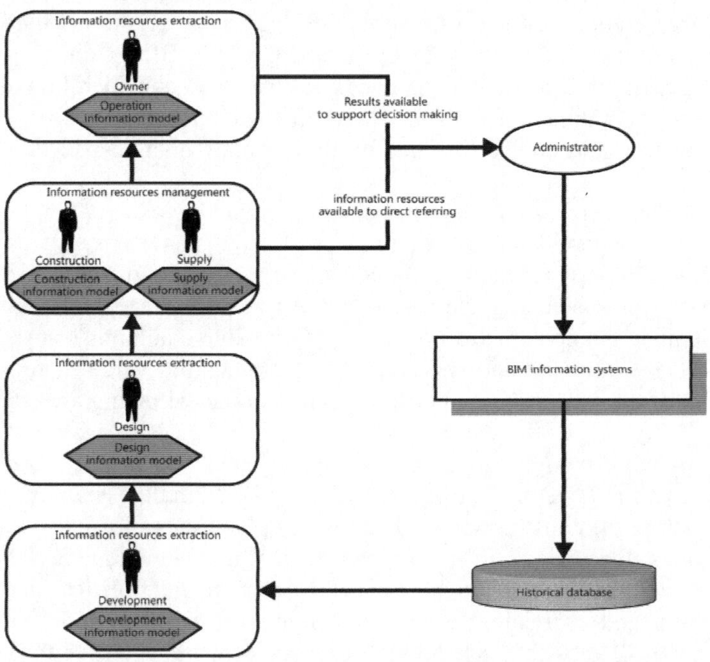


**Figure 3.** Value-added management framework BIM-based building information value chain

*3.1.1. Information resources extraction is carried out in the decision planning stage.* Decision planning stage is the information entry end of the building information value chain. The source of the information resources is the information system based on the BIM technology. According to the analysis on project management in construction, to understand and analyse the main content of the project management and information management, there is a need to carry out the typical user research. This summarizes the essential reusable information resources of the construction enterprise, conducting the first information increment.

*3.1.2. Information resources processing in the design stage.* The design stage is the information flow domain of the building information value chain. These information resources are derived from the BIM technology organizations, which applies the IFC standard for data exchange. Through the analysis of the information resources and the expansion of IFC standards to effectively represent information resources for information understanding and analysis,–ultimately causing the formation of technical solutions for the second value-added information.

*3.1.3. Information resource management is carried out during the construction and supply stages.* This is for the construction and supply stages information flow domain of construction engineering information value chain. The third information value-added involves the following three aspects: ① Information resources. Information resources are expressed in the form of terminology used by the management, such as material procurement records, equipment procurement records, project information, etc. The information resources are combined by collecting the underlying IFC data. ② The Decision-making stage. The decision-making stage is the management link in the construction project, which requires the making of decisions. It is necessary to make use of the appropriate information resources as the basis for decision making according to the appropriate management content and situation. ③ The analysis model. The analysis model is a model of information analysis established under certain decision links in the analysis model.

*3.1.4. Information resource utilization in the operation stage.* The operational phase is the value output end of the building information value chain, combining the characteristics of the BIM for information resources utilization. On the account of the core element of the BIM technology is based on three-dimensional model and is object-oriented, it needs to take into consideration the visualization and intelligence discovery in combination with the three-dimensional model's own characteristics and its core position in the project, thus completing the fourth information value-added.

### *3.2. The application of the BIM-based building information value chain*

The ultimate goal of applying the BIM is to increase productivity, save energy, reduce costs and reduce environmental pollution throughout the design, construction and use of the project. This is consistent with today's advocacy for sustainable development. As a new concept and method, the BIM can overcome the problems in the construction of information value chain, ensuring that the whole life of information is effectively managed and shared [6]. The application value of the BIM-based building information value chain reengineering is mainly reflective in the following aspects:

*3.2.1. Implement associated modifications.* Based on the BIM development software, in the design stage, the subversion of the CAD era sub-graph design form, directly enables the three-dimensional visual design, formulating the construction of the required plans, elevations, sections profile. The BIM-based software provides automatic coordination and modification function, any changes that occur can be automatically coordinated throughout the project, to achieve flat, vertical and section view of the automatic modification and correlate between different profession's modifications. This not only eliminates the coordination errors among the various professions to improve the quality of the design, but also saves a lot manpower, financial resources and time, promising the realization of subsequent goals.

*3.2.2. Information resources extraction is carried out in the decision planning stage.* At present, many applications such as structural analysis, energy analysis, project cost and schedule management are supported by corresponding software. However, the application of the BIM differs in that fact that the BIM is a data carrier that integrates information from all participants. Aforementioned, reducing the need for repetitive input of the same information between the various software and decreasing data errors in the process resulted from human caused. It can improve the reuse rate of information among the participants and at the different stages in the project life cycle, obtaining the necessary information in a timely and effective manner.

*3.2.3. Achieve cooperation of each participant.* The BIM-based information management required each participating party to work together, to attain this BIM-based information management solution needed to provide network collaboration platform for the project participants in different locations. Web-based collaboration platform is employed to achieve cooperation, information sharing, submission, approval, review and use of the various documents, thereby supervises and manages the construction quality, schedule and cost and other management objectives.

*3.2.4. Improve the degree of automation.* Prior to the construction process, the BIM technology can be combined with schedule and cost planning to optimize the construction plan, control the whole construction process, improve the quality and assure the safety of construction. This reduces the rework phenomenon, accomplishing the visualized simulation of the actualized construction process to improve the automation of the entire construction process.

*3.2.5. Support operation and maintenance activities.* The information model based on the BIM integrates the information of the design and construction stage, this information is expanded to form the final information model. Lifecycle information management based on the BIM can maximize the utilization of building information, support the operation and maintenance activities, eliminating the phenomenon that the information cannot be applied in the subsequent stage due to the process fragmentation.

*3.2.6. Realize the data interoperability.* On the grounds that the BIM ideas itself is based on the perspective of the entire life cycle, installing the foundation of the BIM development in various application software, generally to support the corresponding data standards. Therefore, the fundamentals of the BIM information management can smoothly perform the data calling, interoperability and reduce the cost of data interoperability during the process of data storage, exchange and sharing, hence avoiding the phenomenon of "information island" under the traditional circumstance.

## 4. Summary and conclusion

In order to achieve the goal of building engineering and create greater value, future studies should be taken on the appropriate strategies and extensive technical details in constructing the building information value chain and identify the problem. In the research process on the building engineering information value chain reengineering, establishing the BIM-based engineering information value chain model, and proposing the framework of value-added management, based on domestic and foreign construction information technology. Construction of building information value chain is complex system engineering. Therefore, in order to achieve the goal of building engineering and obtain more preeminent significance, future examination should be conducted on the appropriate strategies and extensive technical details in constructing the building information value chain, thus bring about the BIM-based building information value chain for broader application prospects.

## References

[1] Zhang Haitao, Jing Jipeng. Information Value Chain: Connotation, Model, Value-added Mechanism and Strategy[J]. *Information Studies: Theory & Application*， 2009， 32（3）： 16-18.
[2] Ma Shengjun, Xu Youquan. Project Information Management from The Perspective of Life Cycle[J]. Project Management Technology， 2009， 7(2):37 — 40.
[3] Vanlande R, Nicolle C， Cruz C.IFC and building life-cycle management [J]. *Automation in Construction*， 2008， 18(1):70-78.

[4] National BIM Standard-United States™ [EB/OL]. https://web.archive.org/, 2017-02-19.

[5] Zhang Jianping, Cao Ming, Zhang Yang. A 4D Construction Management System Based on IFC Standard and Engineering Information Model[J]. *Engineering Mechanics*, 2005. 22(S1):226-233.

[6] Xu Junqing, Lu Huimin. Information Flow Model for Construction Supply Chain Based on BIM[J]. *Journal of Engineering Management*, 2011, 25（2）: 138 -142.

[7] Umit Isikdag, Jason Underwood. Two design patterns for facilitating building information model-based synchronous collaboration[J]. *Building Information Modeling and Collaborative Working Environments*, 2010, (19): 544-553.

# Preparation and performance evaluation of epoxy-based heat reflective coating for the pavement

**B Hu[1], Y H Liang[1], L Y Guo[1]and T Jiang[1]**

1CCCC First Highway Consultants Co, Ltd, Xi'an, Shaanxi, 700074, China
Corresponding author: 406955182@qq.com

**Abstract**. According to the basic characteristics and composition of heat-reflective coating, combining with the functional requirements of road materials, the experiment selects the epoxy resin with good wear resistance and adhesive force as a film forming material, with TiO2, SiO2 and extinction powder as the main functional filler. The experiment gets a good formula with suitable viscosity, low glossiness and good cooling effect, optimizes by orthogonal experiment. The experiment evaluates the indoor and outdoor cooling effect of heat-reflective coating, and analyses the road performance of the coating. The results shows that the better heat-reflective coating formula included 12% of titanium dioxide, 4% of silica and 4% of extinction powder. When the dosage of coating is 0.8kg/m2, the indoor specimen of heat-reflective coating decrease the temperature of 12 ~ 14°C, and the specimen under solar radiation can reduce the temperature of 7 ~ 9°C. The pavement of heat-reflective coating has good wear resistance, but the road slip resistance partly declines. Therefore, it needs to add the anti-sliding particles to meet the safe driving requirements.

## 1. Introduction

The characteristics of heat absorption and temperature sensitivity can easily lead to pavement rutting, cracks, loose, bleeding and other diseases to asphalt pavement. This has greatly limited the application of asphalt pavement, and the high temperature of asphalt pavement will aggravate the formation of urban heat island effect. At present, the main measures to solve the high temperature disease of asphalt pavement are adjusting the asphalt mixture gradation, improving asphalt grading standards, expanding the application range of modified asphalt, and adding modifier in asphalt mixture. These measures have improved the high temperature performance of asphalt pavement to some extent. But the problem of high temperature does not be solved radically. As a new mode of active cooling, the solar heat-reflective coating has received more and more attention. People did a lot of research on it. Many researchers tried to use Silicone-Acrylate Emulsion, Acrylic Resin and Unsaturated Resin as binder resin to make heat-reflective coating, and applied it to asphalt pavement. But due to the vehicle friction, the coating appears different degrees of damage after a period time of application. Based on this situation, the experiment makes a heat-reflective coating which has good wear-resisting performance and cooling effect. The main content of this paper is about epoxy resin has good wear-resisting and that adhesion as film forming material. Meanwhile, $TiO_2$, as the main reflective material and $SiO_2$ as material improves coating's wear-resisting performance.

## 2. Experimental investigation

### 2.1. Experimental material

**Table1.** Epoxy resin performance parameters

| **Type** | **Epoxy equivalent** | **Viscosity (Mpa▪s)** | **Appearance** |
|---|---|---|---|
| *E51* Epoxy resin | 140 | 600±20 | Transparent liquid |

**Table2.** Curing agent performance parameters

| **Type** | **Amine value (mgKOH/g)** | **Viscosity (Mpa▪s)** | **Appearance** |
|---|---|---|---|
| 107Modified amine | 300±20 | 45±10 | Transparent liquid |

Table 1 shows the performance of E51 modified epoxy parameters. 107 modified amine curing agent. Table 2 shows the performance of parameters.

*2.2. Experimental method*
Low viscosity modified Epoxy Resin as film forming matter, $TiO_2$, $SiO_2$ and Extinction powder is main functional material. Orthogonal test selects out the coating formulation that has suitable viscosity, low glossiness and good cooling effect. The coating is brushed on the rutting plate. The rutting plate is made of 15cm × 15cm × 5cm asphalt mixture of dense gradation (AC). And it tests the indoor cooling effect of the coating by self-made sunlight simulation box. Test outdoor cooling effect of the coating by putting the coating specimens on the sunlight perpendicular incidence open area at the same time. The sliding resistance and wear resistance of epoxy thermal reflective pavement specimens were studied by pendulum friction tester and accelerated polisher.

## 3. Research and analysis

### *3.1. Confection of epoxy heat-reflective coating*

#### *3.1.1. Design of basic formula*

*3.1.1.1. Design orthogonal test.* Filler is main functional material of Heat- reflective Coating. The ratio of filler also has great influence on the performance of Heat-reflective Coating, viscosity and construction time of coating. This test uses orthogonal design to filtrate the mixture ratio of filler quickly, so that we can select out the best ratio of filler which can make the Heat-reflective Coating has better synthetic performance.

L9 $_{(33)}$ orthogonal test, Table 3 shows design orthogonal table. In test, resin dosages are 100 copies. The dosages of the fillers and the resin are in proportion.

**Table3.** L9$_{(33)}$ filler orthogonal design table

| **Factor** | $TiO_2$ | $SiO_2$ | **Extinction powder** |
|---|---|---|---|
| 1 | 8 | 2 | 4 |
| 2 | 10 | 4 | 6 |
| 3 | 12 | 6 | 8 |

Nine tests are carried out according to the design table. Every test repeats three times. Record the values of cooling, viscosity and glossiness of coating after curing, and calculate the average values respectively. Test results are shown in Table 4.

**Table4.** Filler orthogonal test result

| **Test number** | **a** | **b** | **c** | **Filler (filler+resin)** | **Cooling value(°C)** | **Viscosity values(mm/s²)** | **Glossiness values** |
|---|---|---|---|---|---|---|---|
| 1 | 8 | 2 | 4 | 12.3% | 14.3 | 144.4 | 33.4 |

| | | | | | | | |
|---|---|---|---|---|---|---|---|
| 2 | 8 | 4 | 6 | 15.3% | 15.0 | 168.2 | 35.4 |
| 3 | 8 | 6 | 8 | 18.0% | 15.3 | 223.3 | 25.5 |
| 4 | 10 | 2 | 6 | 15.3% | 15.7 | 161.8 | 30.8 |
| 5 | 10 | 4 | 8 | 18.0% | 16.2 | 197.8 | 26.3 |
| 6 | 10 | 6 | 4 | 16.7% | 15.5 | 169.9 | 18.9 |
| 7 | 12 | 2 | 8 | 18.0% | 16.7 | 231.8 | 32.1 |
| 8 | 12 | 4 | 4 | 16.7% | 15.6 | 156.1 | 24.5 |
| 9 | 12 | 6 | 6 | 19.4% | 15.7 | 176.7 | 21 |

*3.1.1.2. Orthogonal design range analysis*

*1) Viscosity range analysis*
Coating's viscosity range analysis and viscosity factors effect are shown in Table 5 and Figure 1.

**Table5.** Viscosity range analysis

| **Factors** | $TiO_2$ | $SiO_2$ | **Extinction powder** |
|---|---|---|---|
| Average 1 | 178.7 | 179.3 | 156.8 |
| Average 2 | 181.5 | 183.0 | 168.9 |
| Average 3 | 187.2 | 195.4 | 217.6 |
| Range | 9.5 | 16.1 | 60.2 |

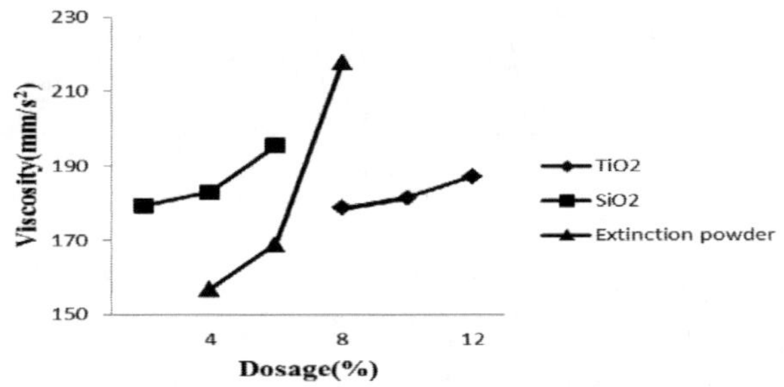


**Figure1.** Viscosity factors effect

From Table 5 and Figure 1, the order of three fillers significant on coating viscosity: Extinction pow-der>$SiO_2$>$TiO_2$.

*2) Value of cooling range analysis*
The coating's cooling value and cooling factors effect are shown in Table 6 and Figure 2.

**Table6.** Cooling value range analysis

| **Factors** | $TiO_2$ | $SiO_2$ | **Extinction powder** |
|---|---|---|---|
| Average 1 | 14.9 | 15.6 | 15.1 |
| Average 2 | 15.8 | 15.6 | 15.5 |
| Average 3 | 16.1 | 15.5 | 16.1 |
| Range | 1.20 | 0.14 | 0.97 |

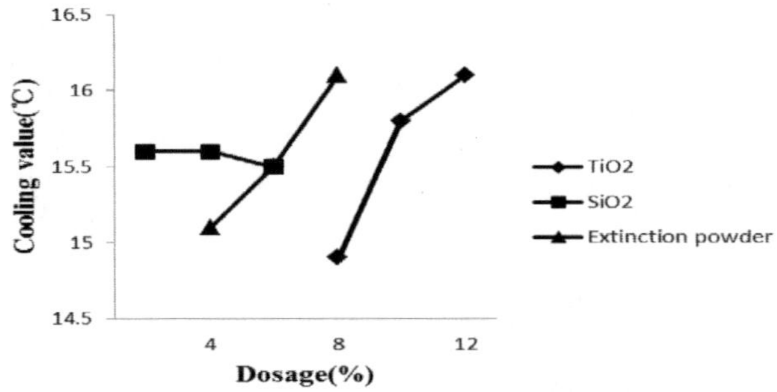


**Figure2.** Cooling factors effect

From Table 6 and Figure 2, the order of three filler influence on coating cooling effect: $TiO_2$>Extinction powder>$SiO_2$.

*3) Glossiness range analysis*

The coating cooling value range analysis and cooling factors effect are shown in Table 7 and Figure 3.

**Table7**. Glossiness range analysis

| **Factors** | ***$TiO_2$*** | ***$SiO_2$*** | **Extinction powder** |
|---|---|---|---|
| Average 1 | 28.2 | 34.4 | 29.8 |
| Average 2 | 26.7 | 27.7 | 21.6 |
| Average 3 | 25.6 | 25.5 | 19.7 |
| Range | 2.6 | 8.9 | 10.1 |

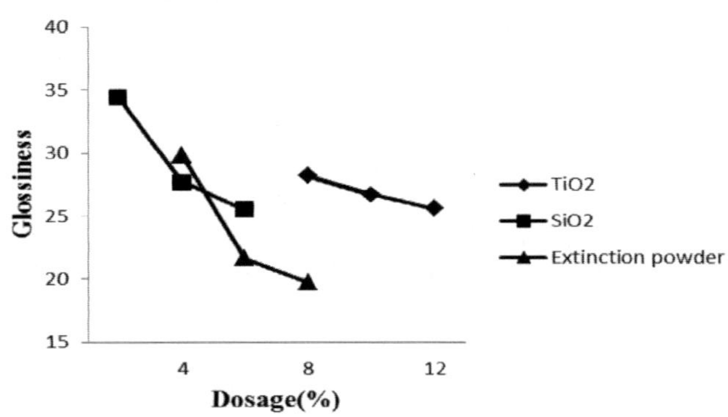


**Figure3.** Glossiness factors effect

From Table 7 and Figure 3, the order of three fillers influence on coating glossiness : Extinction pow-der>$SiO_2$>$TiO_2$.

*3.1.1.3. Orthogonal design optimized formulation and performance parameter.* This test design hopes to get coating with high cooling value, low viscosity and low glossiness. How-ever, the dosage of influencing factors changed will improve some functions of coating while some functions may be decreased. So it needs to take comprehensive balance method to analyze and choose the most optimal proportion. By the range analysis of orthogonal design, it finds that using 12 parts $TiO_2$ will get better cooling effects. While the dosages of $SiO_2$ increase from 2 parts to 4 parts, the glossiness of coating decreases rapidly. It shows great ability of extinction. But when the dosage increases from 4 parts to 6 parts, the glossiness of coating hardly changes. When the dosage of extinction powder is 4 parts, the glossiness of coating has already been lower than 30, completely in full matte state. It meets the needs of driving safety. While the dosage of extinction powder increases from 4 parts to 6 parts, the viscosity of coating increases rapidly. It goes against the painted coating. In conclusion, the most optimized proportion is 100 parts resin, 12 parts $TiO_2$, 4 parts $SiO_2$ and 4 parts extinction powder.

*3.1.2. The confection of color coatings*
Coating's heat-absorptivity will be increased after adding dark pigment. But the cooling effect will decrease. This test chooses the iron oxide red as color pigment. The coating's temperature with different pigment content were compared and tested. And the result shows that white coating is the control group. The cooling effect with different pigment content is calculated. Figure 4 shows the result.

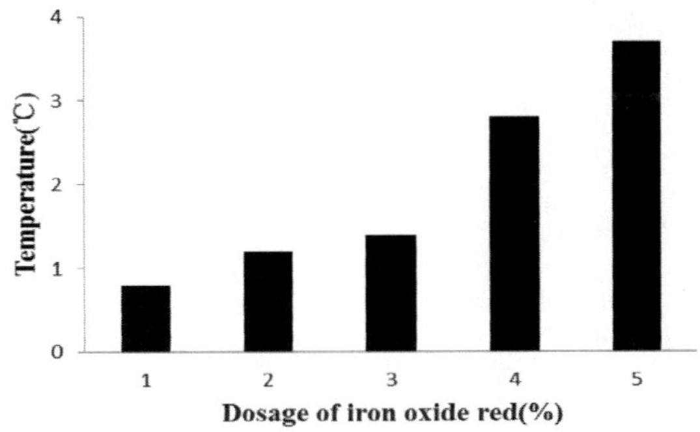


**Figure4.** Influence of color pigment content on cooling effect

From Figure 4, the addition of iron oxide red will increase the temperature of specimens. When the adding amount of iron oxide red is 1%~3%, the cooling effect decreases a little, when over 3%, the temperature difference changes a lot, the cooling effect obviously drops. When the adding amount of iron oxide red reaches to 5%, the temperature difference with white specimens almost reach to 4°C. So the best amount of iron oxide red is less than 3%.

*3.2. The cooling research of epoxy heat-reflective coating*

*3.2.1. The influence on the cooling value of coating dosage*
The surface of AC-13 asphalt concrete specimens are brushed the epoxy reflective coating. Then the specimens are packed well with tin foil. The tin foil plays the role of heat insulation. It will make the test results more representative. The 500W iodine tungsten lampas is used as the simulating indoor solar radiation to test the temperature of specimens with different dosage of coating. The temperature difference between two specimens were calculated when the black specimen's temperature reaches 60°C. Figure 5 shows the results.

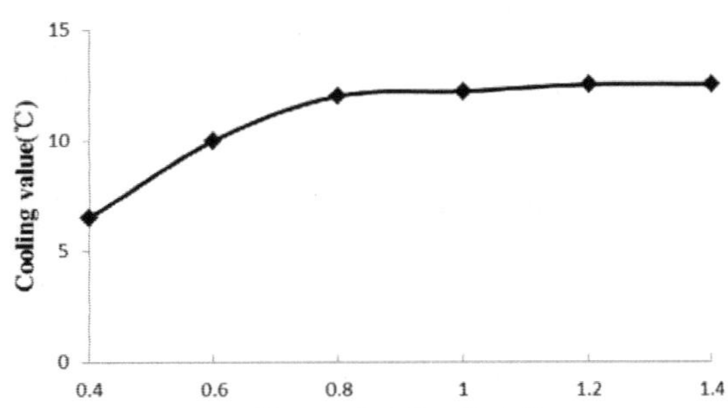


**Figure5.** Influence on the cooling value of coating dosage

From Figure 5, with the increased use of coating, the coating specimens show the tendency of gradual-ly gentle after increasing. And the maximum of cooling value can reach to 12°C. Consider all aspects, the suitable dosage of coating is 0.8kg/m2 for AC asphalt mixture specimens.

*3.2.2. The influence of specimen temperature on the cooling effect*
Take the temperature of the asphalt concrete surface 2 cm depths as the object of study. Calculate the cooling value of specimens under different surface temperature. Figure 6 shows the result.

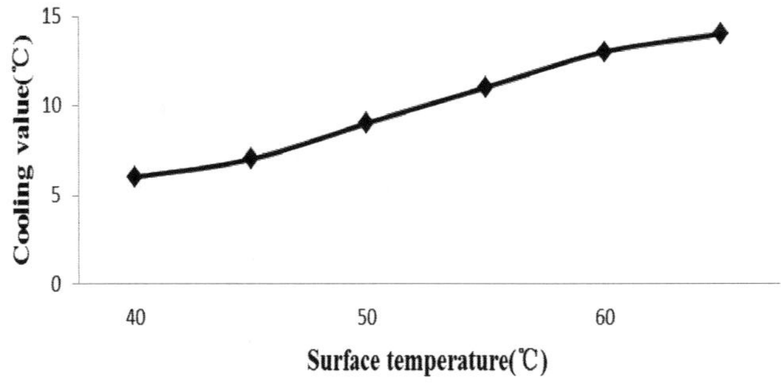


**Figure6.** Contrast of indoor cooling evaluation

From Figure 6, cooling value increases with the specimens temperature increasing. So the heat-reflective coating cooling effect will be more obvious when the pavement temperature is higher.

*3.2.3. The evaluation of the cooling effect under solar radiation*
To further evaluate the cooling effect of heat reflective coating on asphalt concrete in natural environment, two AC-13 asphalt concrete rutting specimens are selected as test specimens. One of them is brushed with nothing. The surface of another is brushed 0.8kg/m2 coating. We put them in the place which is open and also can be exposed to radiation of sunlight. Temperature sensors are embedded in 2 cm away from the surface of specimens. Testing and recording the changes of temperature in a day. The highest temperature that day is 35°C. Figure 7 shows the result.

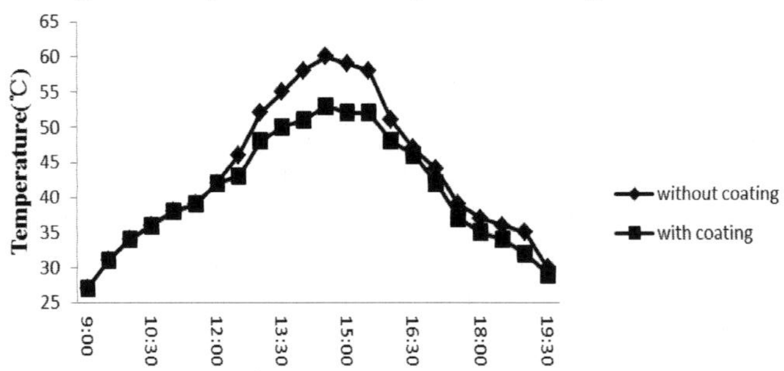


**Figure7.** Change of Asphalt concrete specimens temperature

From Figure 7, two asphalt concrete specimens' temperature increased at the same time. The temperature reached the highest at about 2:30 p.m. At that time, the one's temperature reached to 59.6°Cwithout coating, Another brushed with epoxy heat-reflective coating, its temperature is 51.8°C. The cooling value reaches 7.8°C.

## 4. The wear-resisting research of epoxy heat-reflective coating

The heat-reflective coating on pavement will be repeatedly rolling by vehicle. The wear-resistance is an important performance of heat-reflective coating. But so far, there isn't a unified evaluation criterion for pavement coating's wear-resistance. In this experiment, asphalt mixture accelerated polishing apparatus (developed by Chongqing Transportation research and Design Institute) is used to simulate the friction effect of traffic load on the surface of coating. As shown in Figure 8. Each wheel load is 0.7 MPa, the wheel acceleration run time is 8h.

**Figure8.** Asphalt mixture accelerated polishing apparatus

The experiment tests the epoxy heat-reflective coating's wear-resisting performance before and after put anti-skid. The surface condition before and after the abrasion test is shown in Figure 9.

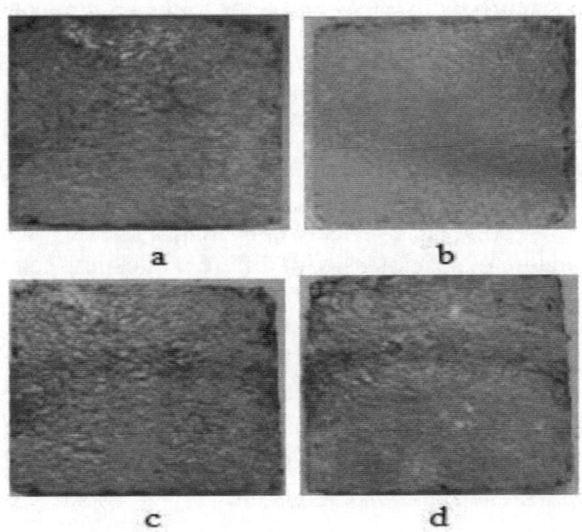


**Figure9.** Surface condition before and after the abrasion

In the Figure 9, a is the heat-reflective coating specimen without anti-skid before abrasion, b is the heat-reflective coating specimen with anti-skid before abrasion, c is the heat-reflective coating specimen without anti-skid after abrasion, d is the heat-reflective coating specimen with anti-skid after abrasion. Figure 9 show that after wheel runs for 8h, no anti-skid heat-reflective coating specimen surface stays well. Only a little coating falls off. A small part of spreading anti-skid pellets reflective coating specimen surface falls off. It is in good condition in general. The result shows that the epoxy reflective coating has better wear-resisting performance.

## 5. The siding resistance research of epoxy heat-reflective coating

The anti-sliding performance of pavement is the key factor affecting road traffic safety. The coating on asphalt pavement will decrease the anti-sliding performance of pavement. So we must research the anti-sliding performance of specimen with heat-reflective coating. AC-13 asphalt concrete slabs are chosen as specimens (30cm×30cm×5cm), which is brushed with different amount of the coating on the surface. And then, we spread a certain amount of anti-sliding on the surface of specimens. We evaluate the anti-sliding performance of asphalt specimens with different amount of coating by pendulum tester, and compare the results of different performance of anti-sliding. Figure 10 shows the result.

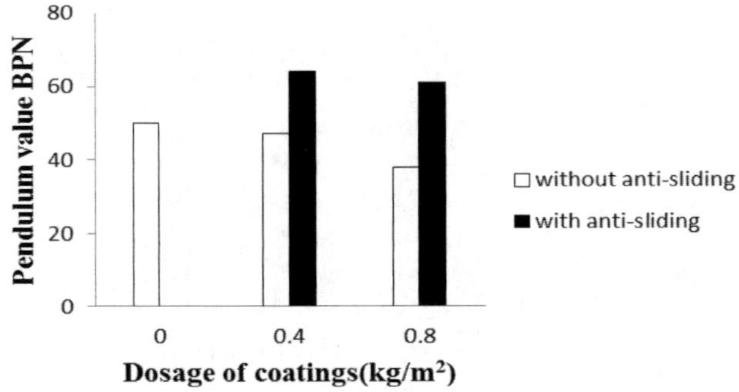


**Figure10.** Pendulum value of different dosage of coatings

From Figure10, anti-sliding performance of asphalt specimen obviously decreased after putting coating. When the dosage of coating is 0.8kg/m2, pendulum value is lower than 40BPN. After spreading anti-sliding on the surface, the pendulum value raises from 38 to 60, increased over 60%. Anti-sliding performance is improved obviously and it meets the requirement of the traffic safety.

## 6. Summary

1. By orthogonal test, the best formula of epoxy heat-reflective coating is 12% $TiO_2$, 4% $SiO_2$ and 4% Extinction powder.
2. White epoxy heat-reflective coating has the best cooling effect. Adding colour coating will affect the cooling effect slightly. Colour coating quantity added is less than 3% for the best.
3. The cooling effect of epoxy heat-reflective coating increases with the thickness through comparing the temperature of specimens with different thickness coating. But when it reaches a certain thickness, the cooling effect tends to be stable. The dosage of coating is 0.8kg/m$^2$.
4. Use solar simulator method measure epoxy heat-reflective coating can reduce the temperature 12~14°C indoor. When the outside temperature is 35°C, the highest cooling value is 7.8°C. The pavement with epoxy heat-reflective coating has good wear-resisting performance.
5. The pavement's wear-resisting performance decreases with the increasing dosage of epoxy heat-reflective coating. The wear-resisting performance of epoxy heat-reflective coating meets the requirement of the traffic safety after spreading anti-sliding particles on the pavement.

## References

[1] Zhao Y Q and Huang D X, "Viscoelastic Behavior of Asphalt Mixtures with Damage Stage," China Journal of Highway and Transport, 21(2008)1, 25-28.
[2] Chen Y R, "Research on Relationship Between the Thermal Characteristic of Underlying Surface and Urban Heat Island," Beijing University of Civil Engineering and Architecture, Beijing, China, 2008.
[3] Tian W Q, Zhou B and Cong L, "High Temperature Stability of Modified Asphalt Mixture and Evaluation Method," Journal of Building Materials, 12(2009)3, 285-287.
[4] Ang J, Yu L M, Wen, et al., "Rutting resistance of asphalt mixtures in the middle course," Journal of Southeast University, 22(2006)2, 270-272.
[5] Song X F, Ling J M and Zhu F H, "The Thermal Reduction property and Effect of Porous As-phalt Pavement," Shanghai Highways, 1(2007)1, 18-20.

[6] Feng D C and Zhang X, "Development of Heat Reflection Coating and Observation of Road Performance," Journal of Highway and Translation Research and Development, 27(2010)10, 17-20.
[7] Wang H, "Asphalt Pavement Light Thermal Reaction Mechanism and Heat-reflection Coating Engineering Research," Harbin Institute of Technology, Harbin, China, 2013.
[8] Zhu X J and Lin A, "Study on Solar Heat Reflection Coating," Equipment Environmental Engineering, 3(2006)2, 29-32.
[9] Cao X J, "Research and Performance Evaluation of Asphalt Pavement unsaturated polyester heat-reflective coating," Chongqing Jiaotong University, Chongqing, China, 2011.

IOP Conf. Series: Earth and Environmental Science **61** (2017) 012084 doi:10.1088/1755-1315/61/1/012084

# Experimental Study on Purification of Low Grade Diatomite

**Liguang Xiao**[1, 2] **and Bo Pang**[1,3]

[1]School of Materials Science and Engineering, Jilin Jianzhu University, Changchun 130118, China

[2] xlg627@163.com

[3]1004625664@qq.com

**Abstract.** This paper presented an innovation for purification of low grade diatomite(DE) by grinding, ultrasonic pretreatment, acid leaching of closed stirring and calcination. The optimum process parameters of DE purification were obtained, the characterizations of original and purified DE were determined by SEM and BET. The results showed that the specific surface area of DE increased from 12.65$m^2/g$ to 23.23$m^2/g$, which increased by 45.54%. SEM analysis revealed that the pore structure of purified DE was dredged highly.

## 1.Introduction

DE is a non-metallic mineral composed of the deposition of unicellular algae, the surface was composed of a large number of regular and orderly micropores, thus DE had a large surface area , strong adsorption properties and high chemical stability, light weight, low density, acid-proof, alkali resistance, high temperature resistant, sound insulation and heat insulation[1-2]. It was widely used in environmental protection, coatings, building materials, functional fillers, filter aid, catalyst carrier, sustained-release agent and other fields[3-4]. DE deposits in China were mostly secondary, tertiary or even lower grade, and the distribution of high grade DE ores was rare[5]. So the DE purification is the key to improving the utilization of DE.

## 2.Experimental process

Purification process was as followed: the original DE→ grinded→ ultrasonic pretreated→ acid leached→ washed→ filtered→ dried→ calcination→ cooled→ pure DE. According to the ratio of liquid to solid of 2.5:1, DE and $H_2SO_4$ of different mass fraction were taken. A certain amount of distilled water was placed in a conical flask and sealed according to the ratio of liquid to solid of 16: 1. After sealing, it was placed in a magnetic stirrer at 60°C for a period of time, the mixture was washed into neutral PH =7, filtered, dried, calcination in a muffle furnace for a certain time, then mixture was cooled to room temperature, so purified DE was obtained.

## 3.Experimental results and analysis

### *3.1. Grind Time*

Fig.5 showed the specific surface area of diatom soil at the same rotation 300 rpm and different grind time, Fig.1-4 was the scanning electron microscope micrograph of DE at the different grind time.

It can be seen from Fig.5, the specific surface area of original DE was 12.65 m2/g, and the specific surface area of DE kept increasing with the increase of grind time, it increased by 3.7%, 6.0% and 7.1% respectively after grinding for 0~1 min, 1~3 min and 3~5 min; The Fig.1-4 SEM has been observed: the total number of diatoms decreased and debris increased when grind time was 0~5min. In order to keep the original porous morphology of diatom, grind time should not be too long, so 1 min was the optimum grind time.

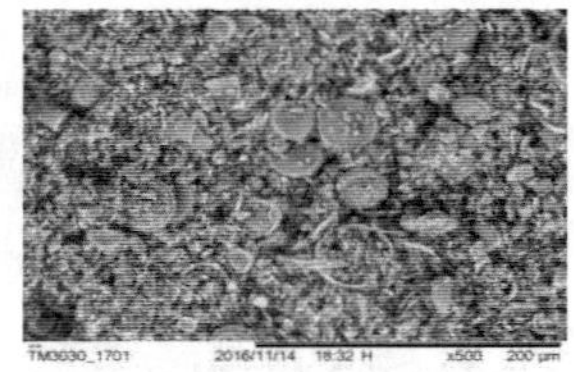

**Fig 1.**SEM of raw soil **Fig 2.**SEM of 1 min grind **Fig 3.**SEM of 3 min grind **Fig 4.**SEM of 5 min grind

### *3.2. Ultrasonic Pretreatment Time*

Figure.6 showed the specific surface area of the diatom soil that grinded for 1 min after ultrasonic pretreatment for different time; Fig.8-11 were SEM at different ultrasonic pretreatment time.

As shown in Figure.6, the specific surface area of DE became larger with the increase of ultrasonic time, specific surface area increased by 17.0%, 17.9%, 18.7% respectively when ultrasonic time was 0~5 min, 5~10 min, 10~15 min; The diatom shells exposed to porous structure when ultrasonic for 5~10 minutes and the number of diatom shells had almost no change. When the ultrasonic time was 15 min, diatom shells were crushed, its integrity was undermined. So the optimal time of ultrasonic pretreatment was 5~10 min. From the point of view of energy conservation, ultrasonic pretreatment time was 5 min in this experiment.

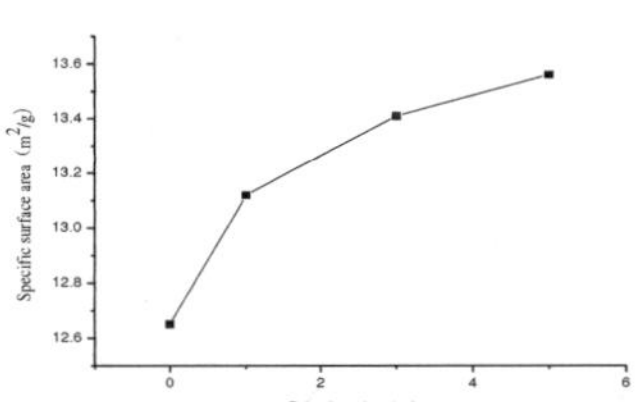


**Fig 5.**Effect of different grind on specific surface area of DE

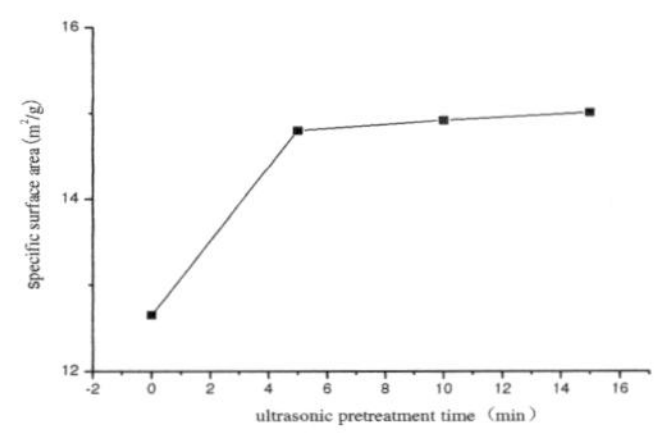


**Fig 6.**Effect of ultrasonic time on specific surface area of DE

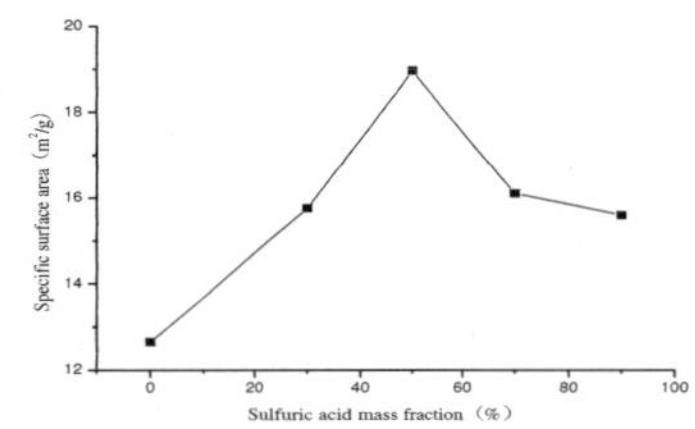


**Fig 7.**Effect of mass fraction of time $H_2SO_4$ on specific surface area of DE

**Fig 8.** SEM of 0 min ultrasonic

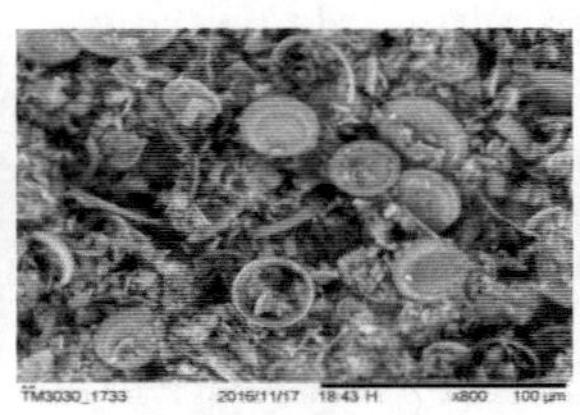

**Fig 9.** SEM of 5 min ultrasonic

**Fig 10.** SEM of 10 min ultrasonic

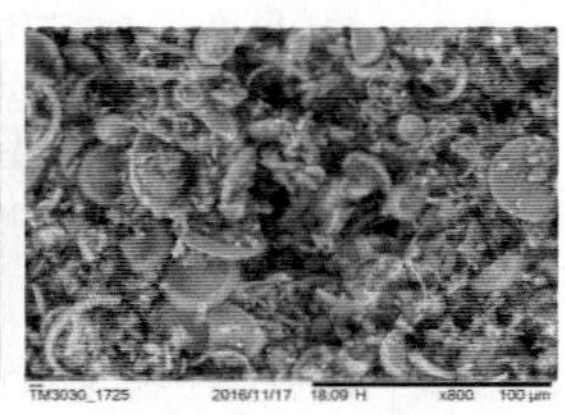

**Fig 11.** SEM of 15 min ultrasonic

### *3.3. $H_2SO_4$ mass fraction*

After grinding for 1 min and ultrasonic for 5 min, the DE was immersed in $H_2SO_4$ for 1h, the temperature of water bath was 60°C, the specific surface area of obtained DE was shown in Fig.7.

From Fig.7, the specific surface area of DE after acid leaching had been improved significantly, which mainly due to $H_2SO_4$ reacted with impurities ($Fe_2O_3$, $Al_2O_3$, MgO, etc.) of DE, generating soluble salts, but the main component of DE ($SiO_2$) did not react with sulfuric acid, and the surface of diatom exposed some new micropores. From the above analysis: 50% sulfuric acid leaching made the specific surface area of DE increase from 14.80 $m^2/g$ to 18.96 $m^2/g$.

**Fig 12.** raw soil

**Fig 13.** 30% $H_2SO_4$

**Fig 14.** 50% $H_2SO_4$

**Fig 15.** 70% $H_2SO_4$

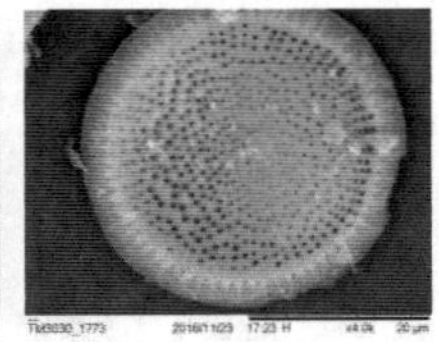
**Fig 16.** 90% $H_2SO_4$

*3.4. $H_2SO_4$ Acid Leaching Time*

After grinding for 1 min and 5 min of ultrasonic, the DE was immersed in 50% $H_2SO_4$ for different time by closed magnetic stirring, the temperature of water bath was 60°C, and the specific surface area of DE was shown in Fig.22. And it can be seen from Fig.22, the specific surface area of DE reached to 18.96 $m^2/g$ after acid leaching for 1h.

**Fig 17.** raw soil

**Fig 18.** $H_2SO_4$ 1h

**Fig 19.** $H_2SO_4$ 2h

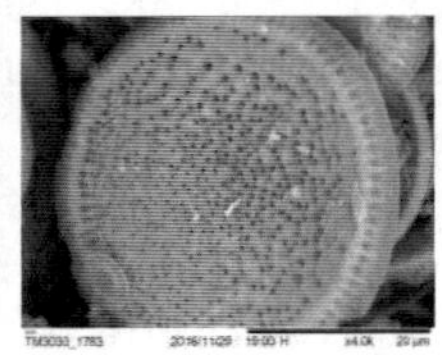
**Fig 20.** $H_2SO_4$ 3h

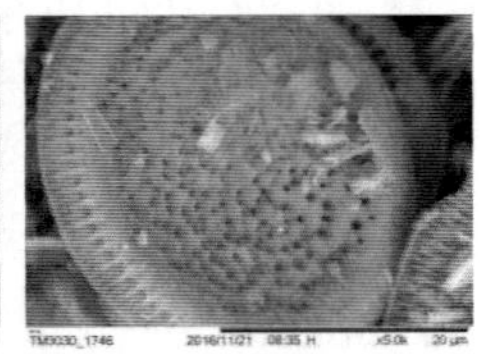
**Fig 21.** $H_2SO_4$4h

*3.5. Calcination Temperature*

Fig.23 was the specific surface area of DE after grinding for 1min, ultrasonic for 5min, 50% $H_2SO_4$ acid leaching for 1h, calcination for 2h at different temperatures. Table 1 was the PH value of calcination of DE at different temperatures.

As can be seen from Fig.23, the specific surface area increased to 21.23$m^2/g$ when the calcination temperature was from 25°C to 600°C, then it decreased with increasing temperature. Therefore, 600°C was the best calcination temperature.

It can be seen from Table 1, from 25°C to 600°C, the surface area of DE increased, the number of inner superficial hydroxyl of hole per square meter increased, the more $H^+$ were ionized, acidity was enhanced, PH value became smaller; When the temperature was risen, the specific surface area decreased, the number of the inner superficial hydroxyl of hole per square meter was lower, the less $H^+$ were ionized, PH value became larger.

**Table 1.** PH value of diatomite in different calcination temperatures

| **Calcination temperature (°C)** | 25 | 400 | 600 | 800 |
|---|---|---|---|---|
| **PH value** | 6 | 5 | 5 | 7 |

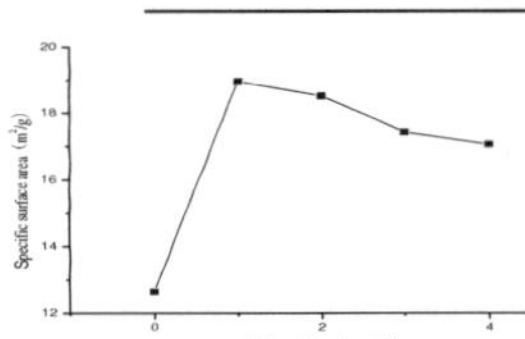

**FIG 22.** Effect of acid leaching time of $H_2SO_4$ on specific surface area of DE

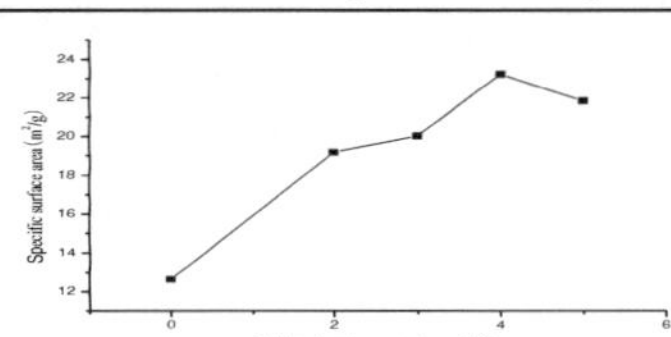

**FIG 23.** Effect of calcination temperature on specific surface area of DE

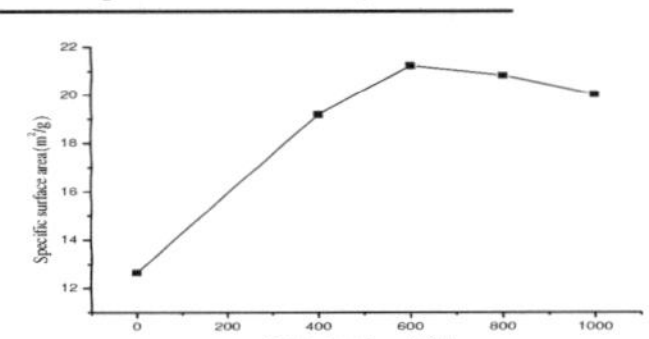

**FIG 24.** Effect of calcination time on specific surface area of DE

**FIG 25.** SEM of raw soil

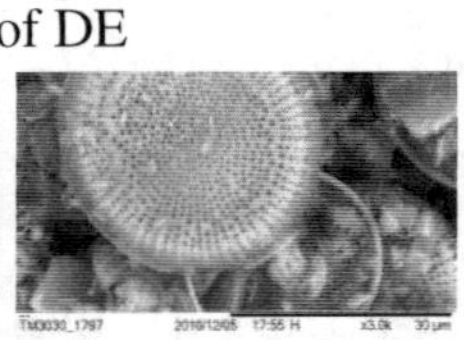
**FIG 26.** SEM of calcination at 400°C

**FIG 27.** SEM of calcination at 600°C

**FIG 28.** SEM of calcination at 800°C

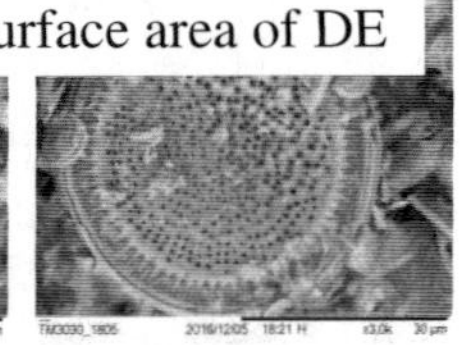
**FIG 29.** SEM of calcination at 1000°C

*3.6. Calcination Time*

IOP Conf. Series: Earth and Environmental Science **61** (2017) 012084 doi:10.1088/1755-1315/61/1/012084

Fig.24 was the specific surface area of DE after grinding for 1min, ultrasonic pre-treatment for 5min, with 50% concentrated sulfuric acid leaching for 1h, 600°C calcination at different time. It can be seen from Fig.24, the specific surface area increased to $23.23m^2/g$ when the calcination time was 4h, so the optimum calcination time was 4h.

According to the pore size distribution of BJH before and after the purification of DE, the DE had a pore size distribution of 2~3nm, and the peak value of pore size distribution of 100~110nm increased. It indicated that the pore structure of purified DE was dredged, the specific surface area and pore volume were increased and the number of small pores and large pores were increased.

## 4.Conclusions

In this experiment, the structure of diatom shell was influenced by acid concentration, acid leaching time, calcination temperature and time, the specific surface area of purified DE increased from $12.65m^2/g$ to $23.23m^2/g$, which increased by 45.54%, the internal pore structure was dredged highly.

## References

[1]Liguang Xiao, Zhuang Zhao. Development status and prospect of diatomite at home and abroad [J]. Journal of Jilin Architectural & Civil Engineering Institute, 2010, 27(2): 27-29.

[2]Khraishem M A M, Al-Degs Y, Mcminn W A M. Remediation of waste water containing heavy metals using raw and modified diatomite [J]. Chemical Engineering, 2004, 99: 177-184.

[3]Larouci M, Sara M, Meddah B. Response surface modeling ofacid activation ofraw diatomite using in sunflower oil bleaching by: Box-Behnken experimental design [J].Journal of Food Science and Technology, 2015, 52(3): 1677-1683.

[4]Penny Crossley. Clarifying matters-world diatomite reviewed [J]. Industrial Minerals, 2000, (3):119-141.

[5]Shuilin Zheng, lijian Wang, Feng Shu, Juntao Chen. Effect of acid leaching and roasting on the properties of diatomite [J]. Chinese Journal of Ceramics, 2006, 34(11): 1382-1386.

# Analysis on design and performance of a solar rotary house

**Xuhong Fan ; Zhaochang Zhang [a)]; Fan Yang; Lilin Cao; Jing Xu; Mingyang Yuan**

Jiangsu university and faculty of civil engineering and mechanics, Zheng Jiang, 212013, China
Corresponding author: 1538719276@qq.com

**Abstract**. A solar rotary house is designed, composed of rotating main structure, fixed cylinder, rotating drive system, solar photovoltaic system and so on, to achieve 360° rotation. Thus it can change the dark and humid situation of the traditional fixed house shade. Its bearing capacity, driving force and safety are analyzed. Rotary driving force and living energy are provided by solar photovoltaic system on roofs and walls. The Phonenics, Ecotect simulation analysis conclude that the rotating house indoor has better natural ventilation effect, more uniform lighting, better the sunshine time compared with traditional houses, becoming a green, energy-saving, comfortable building model.

## 1. Introduction

For traditional housing, the longitudinal depth is extended, housing location is fixed, the sunshine time is relatively short, causing the cold winter and hot summer. Therefore a solar rotary house is designed. Now research on rotating house is few. In China, Genqiang LAN designed "a rotating house"[1], and Qi Xiangchun designed "rotary solar ecological housing"[2]. In foreign countries, Brazil design has 11 layers of rotation of the apartment building[3]. Australia designed a sunny house[4].

## 2. The Deidgn of a Solar Rotary House

### 2.1. *Structure Design*

The solar rotary house is composed of a rotating body structure (1), a fixed inner cylinder (2) and a fixed outer cylinder (3), a pair of thin cylindrical magnetic pole (4), foundation (5), the rotation drive system (6), solar photovoltaic panels (7) and the storage battery (8). The rotary drive system (6) is fixed in an inner cylinder (2) and an outer cylinder (3). Solar photovoltaic panels (7) are mounted on the rotary body structure (1) and a fixed outer cylinder (3). The rotary drive system (6) is composed of engines (15), gears (16) and hinges (17). Electric energy produced by solar photovoltaic panels is stored in the storage battery (8) to provide daily life power and motive power. The shape of floor layer of the rotating body structure (1) is a thin cylinder, whose periphery can be designed as the circular, eight polygon, sixteen polygon. The hinge (17) is fixed on each floor layer of rotating body structure (1). The main structure can be designed with steel structure, shear wall structure. The column (9) and beam (10) can be made of steel, steel and concrete composition of high strength and high performance material. The fixed inner cylinder (2) comprises a solid part and a concave part. The solid part is used as a stair walkway, and the hollow part is used as an elevator shaft and a pipeline well. The engine (15) is fixed on the outer wall of the fixed inner cylinder (2) or on the inner wall of the outer cylinder (3). The gear (16) is fixed on the engine (15). A rotary drive system (6) is installed at every 30°~ 60° between the body structure (1) and the inner cylinder (2) or between the body structure (1) and the

IOP Conf. Series: Earth and Environmental Science **61** (2017) 012085 doi:10.1088/1755-1315/61/1/012085

fixed outer cylinder (3). Figure 1 is a design sketch of a solar rotating house; Figure 2 is a rotary drive system; Figure 3 is a partial view of structure; Figure 4 is a top view of floor; Figure 5 is a pair of thin cylindrical electromagnetic poles; Figure 6 is a circular orbit (14) on the lower electrode (11); Figure 7 is a combination chart of circular orbit (14) and steel ball (15).

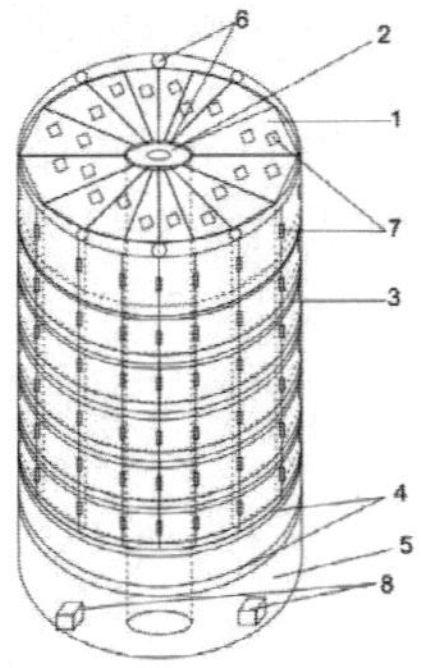


**Fig.1** Design sketch of a solar rotating house

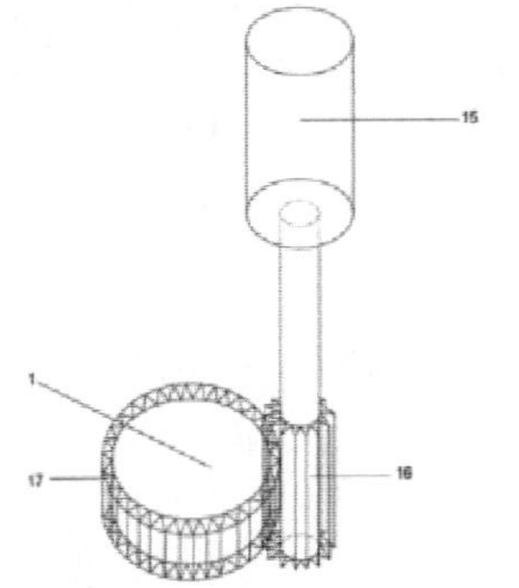


**Fig.2** Rotary drive system（6）

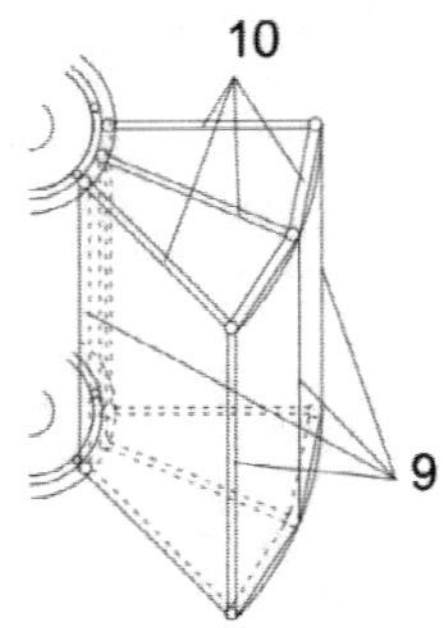


**Fig.3** Partial view of structure (steel or steel concrete composite structure)

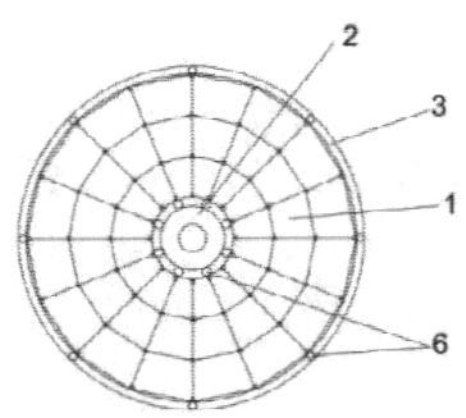


**Fig.4** Top view of floor

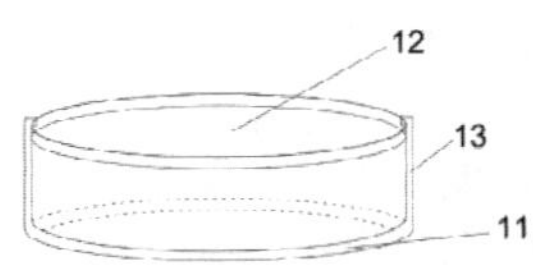


**Fig.5** A pair of thin cylindrical electromagnetic poles

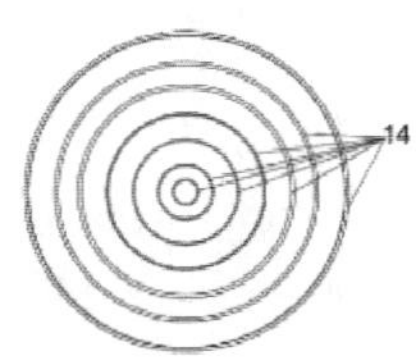


**Fig.6** Circular orbit (14) on the lower electrode (11)

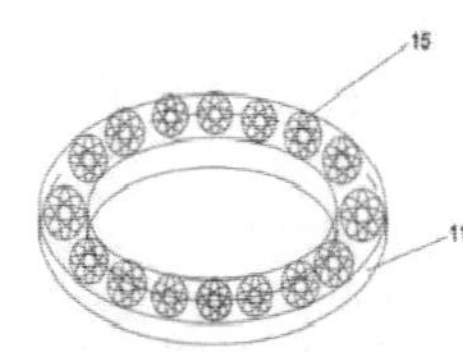


**Fig.7** The combination chart of circular orbit (14) and steel ball (15)

This paper designs a rotary house, whose rotating part is sixteen edge shape. Its circumcircle radius is 12 m, including a 3m fixed part and a 9m rotatable part. The fixed part includes elevator wells, 1m tube well, and the rotatable part includes bedroom and other function rooms. The upper structure is located on a pair of thin cylindrical electromagnets with a radius of 12m and a thickness of 0.15m. The rotary actuator is composed of 32 gears, whose rotary power is provided by four 1000W motors. The solar photovoltaic panels on wall and roof are about 800 m [5].

### 2.2. *Workflow*

When the housing needs to rotate, a rejection reaction controlled by the intelligent control system is formed by the lower electrode (11) and the upper electrode (12) of the thin cylindrical electromagnetic pole (4). The distance between the two electrodes is controlled by fastener (13). Then the intelligent control system of rotary drive system (6) drives the rotary drive system (6) to work. The engine (15) drives the gear (16) to rotate. The gear (16) drives the hinge (17) to rotate. Thus the rotating body structure (1) is driven to rotate. After a certain angle, the rotation can be stopped.

## 3. Key Technologies

### 3.1. *Bearing Capacity of Rotation*

The overall weight of the rotating housing about 6 floors is about 200 t. However, a pair of repulsive force formed by the magnetic induction of the electromagnet can be as high as 600t. Therefore, the separation between superstructure and substructure can be achieved [6].

### 3.2. *Driving Force of Rotation*

IOP Conf. Series: Earth and Environmental Science **61** (2017) 012085 doi:10.1088/1755-1315/61/1/012085

The friction resistance is relatively small, due to the separation of the upper structure and the lower part of the building [7].

### *3.3. Safety of House*

The upper construction is an outer clad steel beam and column composited structure. When a house is not rotating, its structure frame is tube structure; When it rotates, structure frame is the shear wall structure. In addition, a large number of vibration isolation dampers are installed on the structure, which can effectively reduce the damage caused by rotation. So rotating houses are safe and reliable.

### *3.4. Intelligent Control*

The control system is used to realize the intelligent control of building rotation time, rotation angle, elevator and so on.

### *3.5. Energy Saving Measures*

Solar photovoltaic panels are designed on the roof, walls, glass design for Low-E glass to make the housing zero energy consumption.

## 4. Analysis of Main Performance

### *4.1. Analysis of Natural Lighting and Sunshine*

Natural lighting: By the ECOTECT natural lighting: simulation for the rotating housing and the traditional housing, the results are shown in figure 8. The lighting of the rotatable housing along the periphery is uniform, and lighting coefficient above 2% is more than 85% of the area. As for the traditional dormitory, the lighting only near north and south window is good, and the lighting near the east and west is poor.

Sunshine: For the traditional housing, the southern sunshine is long , and the northern is short. For the rotating housing, each function room can receive good sunshine, where is also the biggest advantage for rotating houses than traditional buildings. In addition, a uniform natural light and a longer period of time make the room more comfortable, beneficial to physical and mental health.

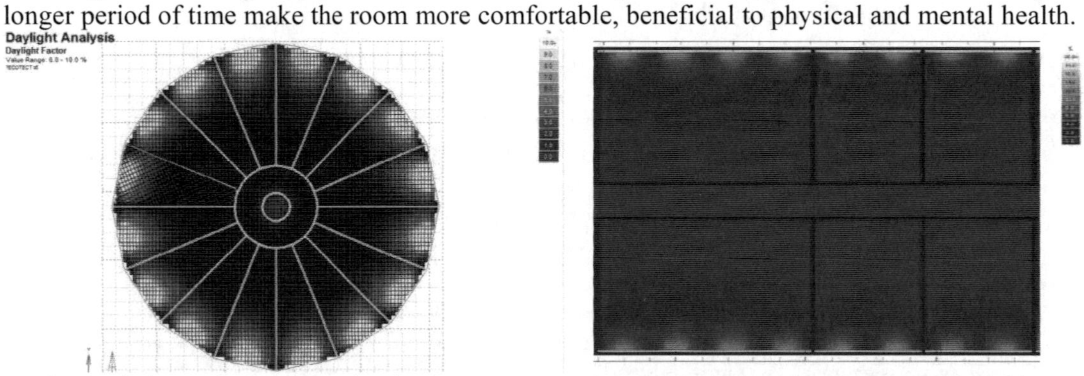


a. The daylight analysis of rotating house b. The daylight analysis of traditional house

**Fig.8** The comparative analysis of daylight

### *4.2. Analysis of Indoor Natural Ventilation*

The PHOENICS simulation analysis of indoor natural ventilation is showed in figure 9. For rotary house, the speed of the main function room is basically controlled in the range of 0.5-2.5m/s, the air age is basically in the range of 50 to 100s, and the air pressure is even, and the airflow direction is smooth. For traditional house, the speed is not uniform, the air age is in the range of 200 to 1000s, and the air pressure is uneven, the direction of the wind is only north and south. Therefore, the indoor comfort for rotary house is better than it for traditional house.

| | Building model | indoor wind velocity cloud chart in summer | Indoor air age cloud chart in summer |
|---|---|---|---|
| Rotary house |  | 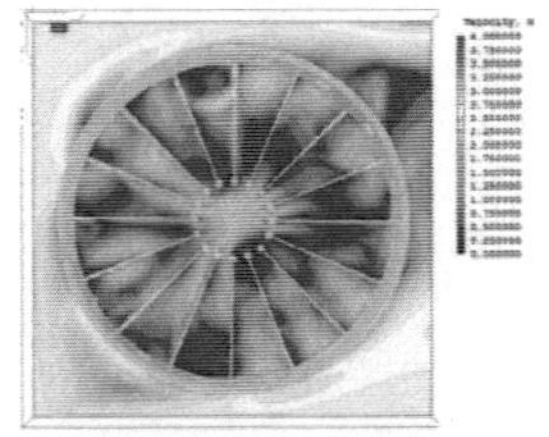 | 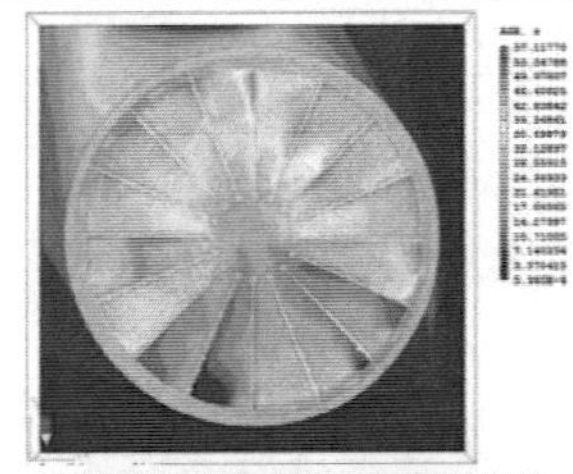 |
| traditional house | 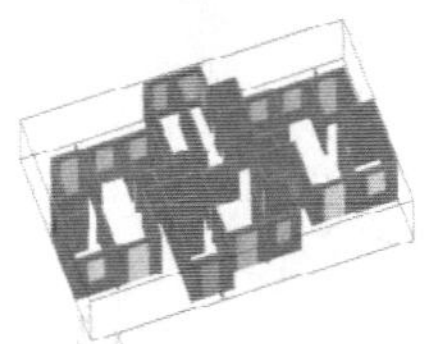 | 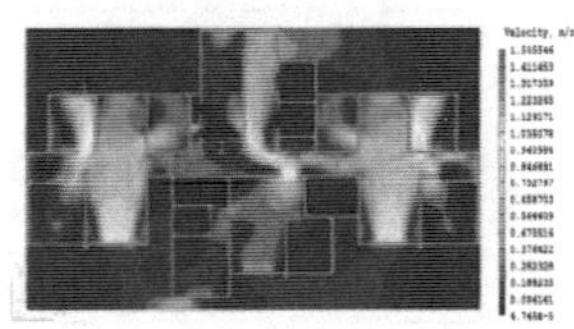 | 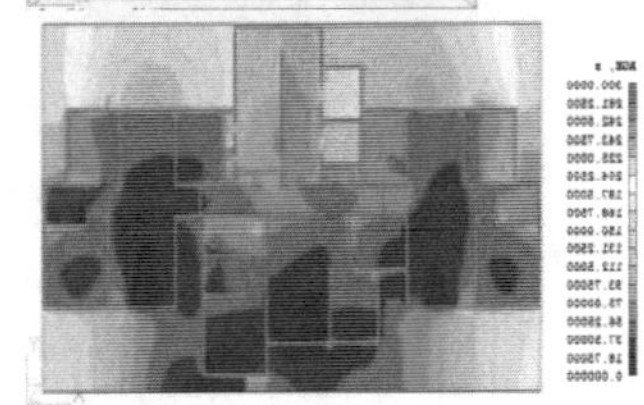 |

**Fig.9** Analysis of natural ventilation in summer

*4.3. Analysis of Energy Saving*

Solar photovoltaic power has the advantages of safe, no noise, no pollution, convenient construction and so on. If the capacity of solar photovoltaic panels is 340 kWp, the total annual generating capacity is 340000 kWh. At the same time, it can save 113 tons of standard coal, greatly reducing emissions of CO2, SO2, dust.

## 5. Conclutions

The designed rotary house based on the solar energy is a green, energy saving, environmental protection, zero energy consumption, intelligent building, but also has the value of artistic appreciation.

**Acknowledgements**

Supported by National Science and Technology Support Project of China (2015BAL02B02), National Natural Science Foundation of China (51408267), Natural Science Foundation of the Jiangsu Higher Education Institutions of China (14KJB560005), Energy saving projects in Colleges and universities in Jiangsu Province (JSSNZH2016301), Graduate Student Practical Innovation Program of 2015 Year in Jiangsu Province (SJLX15_0501).

## References

[1] Qi Chun, rotary Solar Power Eco house: China, CN200520005511.9[P]. April 19, 2006.
[2] Lan Genqiang, Li Xuecai, a rotating house CN201320841968.8 [P].2014 June 4th.
[3] Your house Luan bowl, rotating circle sharing sunshine dream [J], building knowledge, 2005:3, 46.
[4] Peak, the window always to the sun: the rotation of the house came out of [J], the window of science and technology, 2010:2, 20.
[5] Zhai. Junjun Weibei area city construction of renewable energy applications [D], master thesis, Xi'an: Xi'an University of Architecture and Technology, 2014.
[6] Ruan Wei Wen. In the transformation of the old city of the whole house translation, rotation engineering example [J]. building technology, 2008
[7] Zhang Fang. Research and application of dynamic load identification technology [D]. doctoral dissertation, Nanjing: Nanjing University of Aeronautics & Astronautics, 1996.

IOP Conf. Series: Earth and Environmental Science **61** (2017) 012086 doi:10.1088/1755-1315/61/1/012086

# Study on POT model of vehicle load under toll-by-weight mode

**H Y Huang[1,3], J P Zhang[2], W X Xu[1], D Z Huang[1]**

[1]School of Civil Engineering, Guangzhou University, Guangzhou, China
[2]Guangdong Communication Polytechnic, Guangzhou, China
[3]Corresponding author's e-mail: Huanghy@gzhu.edu.cn

**Abstract**. The overload and overweight operations of vehicles in expressway are particularly serious in China recently. Research on the extreme maximum value of the vehicle loads is an urgent problem in the bridge design. On the basis of statistical analysis of the WIM data from Yu-Zhan Expressway under toll-by-weight mode in Guangdong Province, the POT probability model of gross weight of the vehicle was established by the extreme value theory. It can predict the probability distribution of operating vehicle loads in any recurrence period. Results showed that the overweight vehicles were not accidental, and there may be heavier vehicles in the future. It is necessary to improve and modify the current toll-by-weight charging policy.

## 1. Introduction

With the continuous development of economy and society and the automobile industry in China, the growing overload and overweight transportation of operating vehicle produce significant variation. Related research and study had become one of the hot topics. In order to curb the increasingly overload situation and guarantee the traffic safety, toll-by-weight model must be implement. Nowadays, the weight of vehicle got by the Weight in Motion (WIM) technology has been successfully applied in the expressway in our country. Under this weight charging mode and the current Toll-by-Weight policy, whether the overloaded or overweight of vehicle has been effectively controlled? What are the development trends of operating vehicle loads? How can make a more scientific and rational prediction? On these issues, some researchers had made some studies and researches, but in general, the conclusions were very different. In this view, adopted the WIM data of a toll station in Yu-Zhan Expressway, the paper used the extreme value theory to establish the vehicle load model and forecasted the development trend.

## 2. Vehicle load statistics

Yu-Zhan Expressway is an artery highway connecting Chongqing to Zhanjiang with a length of 73 kilometers in Guangdong province, and was opened to traffic in 2005. 307172 WIM data from one vehicle lane at a toll station in Guangdong in 2013 were collected. Through statistical analysis, the numerical characteristics of each wheel vehicle were obtained, as shown in table 1 and figure 1.

Statistical analysis displayed the following. Firstly, 2-wheel vehicles accounted for up to 33.62%, followed by the 4-wheel and the 6-wheel vehicles, accounting for more than 33.57% and 27.01%; 3-wheel and 5-wheel vehicles accounted for less, with less than 5% of the total. This proportion of vehicles reflected the characteristics of the traffic in the post industrial areas. Secondly, the average value of 3~6 wheel vehicles were close to the policy provision limit value of each type, and the maximum value were 2~3 times of the limit value, which reflected the total weight of the operation traffic load of this Expressway were so heavy.

**Table 1.** The numerical characteristics of each type vehicle.

| Type | Number | Proportion (%) | Maximum Value(t) | Average Value(t) | Limit Value(t) | Standard Deviation | Coefficient of Variation |
|---|---|---|---|---|---|---|---|
| 2-wheel | 103275 | 33.62 | 55.5 | 9.71 | 17 | 5.31 | 0.55 |
| 3-wheel | 12874 | 4.19 | 57.9 | 22.96 | 25 | 6.69 | 0.29 |
| 4-wheel | 103117 | 33.57 | 86.5 | 33.99 | 35 | 8.21 | 0.24 |
| 5-wheel | 4931 | 1.61 | 96.8 | 34.47 | 43 | 14.42 | 0.42 |
| 6-wheel | 82975 | 27.01 | 153.4 | 46.66 | 49 | 16.11 | 0.35 |

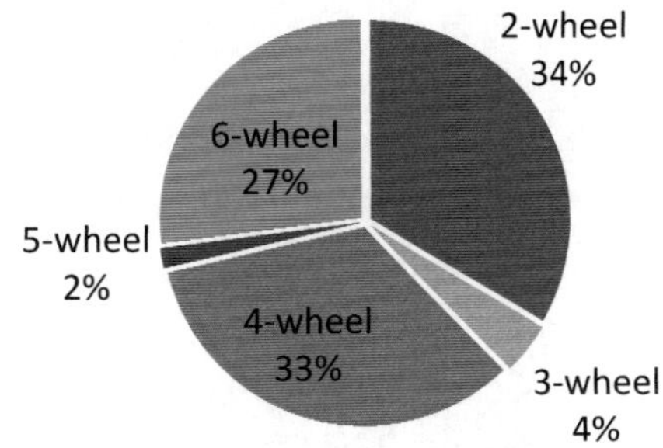


**Figure 1.** Proportion of each vehicle type.

## 3. Principle of POT model

At present, most researchers usually use the methods of hypothesis test, analysis of the parameters of the technical route to simulate the weight distribution of the vehicle load. These trial and error analysis methods have strong subjectivity with exactly simulate the central distribution, but failing to accurately describe the tail distribution. This paper intends to use the POT model based on the extreme value analysis theory to establish the vehicle load model.

F($x$) is an arbitrary distribution function of random variable sequence$\{Xi\}$, which is assumed to have independent identically distributed random variables, F($x$) supports at the top of the point $x^*$, and there is a value of $Xi <\mu< x^*$, called Threshold. $Xi$ is a sequence of random variables, and $\{Xi-\mu\}$ is for out of sequence. $F_\mu(y)$ is defined as the conditional distribution function of the random variable, which can be expressed as in equation (1).

$$F_\mu(y)=P(X-\mu\leq y\,|\,X>\mu)=\frac{F(\mu+y)-F(\mu)}{1-F(\mu)}=\frac{F(x)-F(\mu)}{1-F(\mu)} \tag{1}$$

So

$$F(x)=F_\mu(y)(1-F(\mu))+F(\mu) \tag{2}$$

$X_1, X_2, \ldots, X_n$ is a sample from the same distribution F($x$) population, if the distribution function of the random variable $X$ can be satisfied.

$$G(x;\mu,\sigma,\xi)=1-(1+\xi\frac{x-\mu}{\sigma})^{-\frac{1}{\xi}},x\geq\mu,1+\xi(x-\mu)/\sigma>0 \tag{3}$$

The $X$ is said to obey the General Pareto distribution (GPD distribution). $\mu\in R$ is a function of the position, σ>0 is the scale parameter, and $\xi\in R$ is the shape parameter.

If there are constants like $a_n$ and $b_n$, when reaching $F(x)$ at the upper end of the point, $F_\mu(a_n+b_n)$ have continuous limit distribution , the theorem is established.

$$\lim_{\mu \to x_0} \sup_{0 \le y \le x_0} \left| F_{\mu}(y) - G(y;\mu,\sigma,\xi) \right| = 0 \tag{4}$$

The theorem: for sufficiently large threshold $\mu$, the majority of unknown distribution function F($x$) beyond the volume distribution function $F_\mu(y)$ available GPD distribution. G (y;μ,σ,ξ) approximation is $F_\mu(y) \approx G(y;\mu,\sigma,\xi)$, and the type of substitution as formula (5).

$$F(x) = G(y;\mu,\sigma,\xi)(1-F(\mu)) + F(\mu) \tag{5}$$

After the determination of the $\mu$, you can get a large number of $N_\mu$ than the threshold μ in {$Xi$}, according to the formula (5) with the frequency （1-$N\mu/n$） instead of the F($\mu$), you can get the expression of F($x$).

$$F(x) = F_{\mu}(y)(1-F(\mu)) + F(\mu) = 1 - \frac{N_\mu}{n}\left(1+\frac{\xi}{\sigma}(x-\mu)\right)^{-\frac{1}{\xi}} \tag{6}$$

Through parameter estimation, $\hat{\sigma}$ and $\hat{\xi}$ can be estimated, then equation (7) can be deduced by equation (6).

$$F(x) = 1 - \frac{N_\mu}{n}\left(1+\frac{\hat{\xi}}{\hat{\sigma}}(x-\mu)\right)^{-\frac{1}{\hat{\xi}}} \tag{7}$$

The threshold model was used to estimate the tail distribution of the vehicle load, and the key is the choice of threshold ($\mu$).

## 4. Establishment of a POT model

Using 307172 WIM data of 2013 collected from one lane at a toll station of Yu-Zhan Expressway, according to the weight distribution density frequency diagram of the vehicle load, based on GDP theory, a POT model of gross weight of the vehicle was established with properly fitting the tail probability distribution.

First of all, the threshold was selected by kurtosis method, when the threshold value was 70.5t; the mount of sample and the distribution curve were moderate, which come to the best fitting effect. Then, according to the method of maximum likelihood parameter estimation, a Mat lab program was developed to estimate the GPD distribution parameter by using the GDP distribution to fit the data distribution, and the estimated results are shown in table 2. Finally, the distribution probability function of vehicle weight was obtained as equation (8).

**Table 2.** Results of parameter estimate.

| threshold ($\mu$) | $\hat{\sigma}$ | $\hat{\xi}$ | $N_\mu$ |
|---|---|---|---|
| 70.5 | 5.60754 | 0.028286 | 0.019754 |

$$F(x) = 1 - 0.019752\left[1+\frac{0.028286}{5.60754}(x-70.5)\right]^{-\frac{1}{0.028286}} \tag{8}$$

## 5. The maximum weight of a vehicle prediction

After the probability distribution model of vehicle gross weight is established, if the number of vehicle with a certain weight in a reference period is known, the return level of U(T) over a period of time and the possible maximum value W95 (guaranteed rate is 95%) can be predicted. The return level U(T) means that the weight of more than U (T) of the vehicle will appear 1 times in T years.

Based on above and the vehicle data of Yu-Zhan Expressway, the predicted value of U (T) and the W0.95 (T) in different return period were calculated, and the results were summarized in table 3.

**Table 3.** The predicted return level and maximum gross weight.

| Return Period (year) | Predicted Value (t) | | Return Period (year) | Predicted Value (t) | |
|---|---|---|---|---|---|
| | U(T) | W0.95 | | U(T) | W0.95 |
| 5 | 129.6 | 152.1 | 30 | 143.0 | 166.7 |

| | | | | | |
|---|---|---|---|---|---|
| 10 | 134.7 | 157.7 | 35 | 144.1 | 168.0 |
| 15 | 137.7 | 161.0 | 40 | 145.2 | 169.1 |
| 20 | 139.9 | 163.3 | 45 | 146.1 | 170.1 |
| 25 | 141.6 | 165.2 | 50 | 146.9 | 171.0 |

From the view of predicted results, more than 129.6 tons of heavy vehicles will appear at least once every 5 years, and a maximum of more than 152.1 tons of heavy vehicles will appear once every 5 years, the maximum weight of vehicle will reach 157.7 tons after 10 years. This was far more than the gross weight limit (49 tons) of vehicle on current highway bridge.

## 6. Conclusions

Based on the study of the measured data of the WIM with toll-by-weight model, some conclusions can be obtained.

(1) The average value of 3~6 wheel vehicles were close to the policy provision limit value of each type with the maximum value being 2~3 times of the limit value, which reflected the weight of the operation traffic load in the Yu Zhan Expressway were so heavy.

(2) The POT model of operating vehicle load was established by using the extreme value theory, which can properly describe the gross vehicle weight of the tail distribution characteristics. The predicted results showed that the current toll-by-weight policy had a certain inhibitory effect on overweight vehicles, but can not effectively control the variability of overweight vehicles. With the increase in the return period, there is sufficient reason to believe that more overweight vehicles would appear. Therefore, it is necessary to continue to improve and improve the policy of weighing charges. It is necessary to improve and modify the current weight charging policy in Guangdong province.

## Acknowledgments

These research reported was financially supported by the project of the Science and technology planning of Guangdong Provincial Department of Transportation in 2014(NO.201402022) and the project of National Science Foundation of China(NO.51278134).

## References

[1] Mei G, Qin Q and Lin D J, et al, 2003, Bi-modal probabilistic model of highway and bridge vehicle loads, *Journal of Tsinghua University*, **43**(10), pp 1394-1397

[2] Huang T 2010 Investigation Statistics and Simulation for Random Traffic Loading of Expressway Bridge, Ph.D. thesis, Chang'an University

[3] Wang T, Han W S and Huang P M, 2010 Research status and prospect on traffic loading for highway bridge, Journal of Architecture and Civil Engineering, 4, pp 31-37

[4] Sun J S, Yang C F, Dou Y M, et al, 2009 Study on the Vehicle Load of Bridge Design on Heavy Traffic Highway Basic on Traffic Investigation, .Journal of Hebei University of Technology , (1), pp 91-95

[5] Gong J X , Li W J, Zhao J L, et al, 2010 Research on Probabilistic Model of Highway Bridge Vehicle Loads (2)-Toll-by-weight Area and Overload Controlling Area, Highway traffic science and technology, (7) , pp 56-60

[6] Shi D J 2005 Practical extremum statistical method. (Tianjin: Tianjin Science and Technology Press)

[7] Yu D R, 1992 Highway Bridges of live load and live load model, Central south highway engineering, (3), pp 31-37

[8] Ying T Y, 2009 Bridge vehicle load and its effect analysis based on WIM data, Ph.M. thesis, Tongji University

IOP Conf. Series: Earth and Environmental Science **61** (2017) 012087 doi:10.1088/1755-1315/61/1/012087

# Excavation-caused extra deformation of existing masonry residence in soft soil region

**Y Tang[1,2] and S Franceschelli[1,3]**

[1] Department of Geotechnical Engineering, School of Civil Engineering, Tongji University, Shanghai 200092, China.
[2] Key Laboratory of Geotechnical and Underground Engineering (Tongji University), ministry of Education
[3] Engineering of Building Processes and Systems, curriculum in Historic Building Rehabilitation (Ravenna Campus), University of Bologna, Bologna 85, Italy.
ytang@tongji.edu.cn
stefan.franceschell4@studio.unibo.it

**Abstract.** Growing need for construction of infrastructures and buildings in fast urbanization process creates challenges of interaction between buildings under construction and adjacent existing buildings. This paper presents the mitigation of contradiction between two parties who are involved the interaction using civil engineering techniques. Through the in-depth analysis of the results of monitoring surveys and enhanced accuracy and reliability of surveys, a better understanding of the behavior of deformable buildings is achieved. Combination with the original construction documents, the two parties agree that both of them are responsible for building damages and a better understanding for the rehabilitation of the existing buildings is focused on. Two cases studies are used to demonstrate and describe the importance of better understanding of the behavior of existing buildings and their rehabilitations. The objective of this study is to insight into mechanisms of soil-structure interaction for buildings adjacent to deep excavations, which can result in a damage in existing masonry residence, and to take the optimized measures to make deep excavations safety and economic and adjacent buildings keep good serviceability in urban areas with soft soil conditions.

## 1. Instruction

It is common that some of the existing buildings affected by new construction with excavation nearby are not satisfied with serviceability required. With the high speed development of urbanization in China in the past three decades, the distances between new building constructions and infrastructures and their adjacent existing buildings become smaller and smaller inevitably. In addition, the evaluation of adverse effect caused by adjacent new construction, construction time, budget, and housing privatization do not match the speedy development of urbanization. These results in conflicts between new construction and adjacent existing buildings, especially in soft soil region.

This paper focuses on the elaboration of the interaction between a new construction and adjacent existing buildings from geotechnical engineering and structural engineering, geodesy monitoring, and evaluation methods. Then the effect of new construction on the existing buildings could be described reasonably and quantitatively. The conflicts between the new construction's party and existing buildings' party are mitigated by the reliable evaluations. Then, the two parties would concentrate on rehabilitations or strengthening for the existing buildings if it is necessary rather than arguing with each other. Furthermore, the party of new constructions will take effective measures to compensate the party of existing buildings based on the field monitoring and analysis, as well as strategies for maintenance. Finally, the existing building serviceability is controlled within the allowable condition.

In soft soil region, for instance Shanghai, the conflict is more obvious than other regions due to geological reason. The physical and mechanical properties of typical stratum in Shanghai are listed in table 1.

Two case studies presented in this paper are from Shanghai residences in China. Both are in concrete raft foundation with 2.0 m embedded below ground surface. The residences are masonry structure with concrete ring beams on the floor elevations and concrete columns in the cross positions of brick walls. The floors are of precast hollow slabs, but concrete cast-in-place slabs for kitchens and washrooms. Both had severe cracks on the residence walls. The cracks are also on the ground surface around them. The crack widths on bearing brick walls and self-weighted walls are over limitation (GB/T 50344-2004, DG/TJ 108-79-2008, and DG/TJ 08-804-2005) in diagonal direction, irregular cracks exhibit on the ceramic tiles attached on the decorated layer in kitchen and wash-room, as well as excessive deformations appear on the frame of windows and doors.

**Table 1.** Physical and mechanical properties of typical stratum in Shanghai

| Strata | Name of soils | Thickness | Water content | Bulk density | Void ratio | Permeability coefficient $k_h$ | Permeability coefficient $k_v$ | CU test $C$ | CU test $\varphi$ |
|---|---|---|---|---|---|---|---|---|---|
| | | m | % | kN/m$^3$ | | cm/s | cm/s | kPa | ° |
| ① | Fill | 0.9-2.0 | - | - | - | | | 11 | 17 |
| ② | silty clay | 1.5-2.2 | 35.3 | 18.5 | 1.00 | 5.11E-05 | | 17.0 | 20.8 |
| ③ | Very soft silty | 3.6-4.2 | 39.6 | 18.1 | 1.11 | 1.77E-04 | 2.81E-06 | 11.5 | 22.0 |
| ④ | Very soft clay | 7.2-9.7 | 49.0 | 17.3 | 1.37 | 1.64E-06 | 2.47E-07 | 14.0 | 13.5 |
| ⑤ | Silty clay | 3.4-8.6 | 34.4 | 18.5 | 0.98 | 1.33E-07 | 2.49E-05 | 13.0 | 20.0 |
| ⑥ | Silty clay | 3.2-5.3 | 23.0 | 20.1 | 0.67 | | | 51.0 | 21.0 |
| ⑦ | Sandy with clay | 6.9-7.3 | 31.2 | 18.6 | 0.91 | | | 4.29 | 32.7 |

## 2. Field monitoring

### *2.1 Residence 1*

The residence 1 (photo.1) built in 1995 is located in the center of Shanghai in China (latitude 310 16' 51.58", longitude 1210 25'24.22"). The minimum net distance away from the new construction is 10 m. The maximum depth of excavation of new construction is 9.8 m. The new construction began from December, 2014. The filed monitoring conducted by Tongji University started from January 22, 2016 and ended on April 15, 2016. The measure points are shown in figure 1.

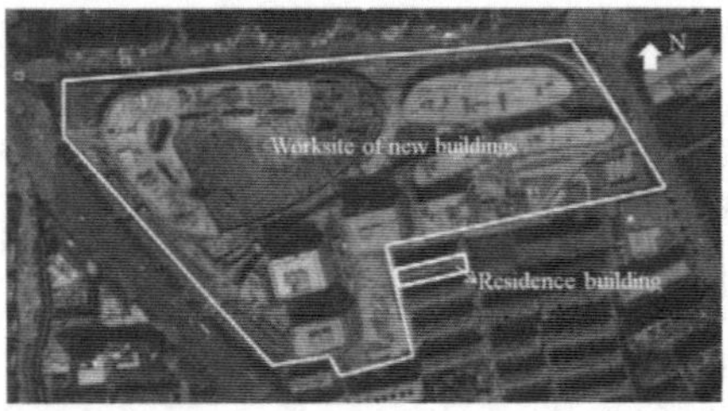


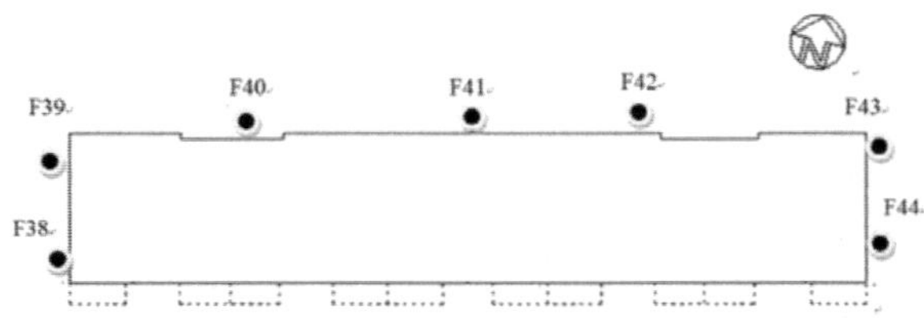


**Photo 1.** Layout of residence 1 (from google earth, July 2016) **Figure 1.** Layout of measure points

The points of the connecting transverse are easy to design when we follow the previous reference point about 50 m in south east of the residence. The measurement points follow the previous monitoring team who conducted settlement measurement from the end of December, 2014 to January 19, 2016. The settlements in different period (figure 2) are listed in table 2.

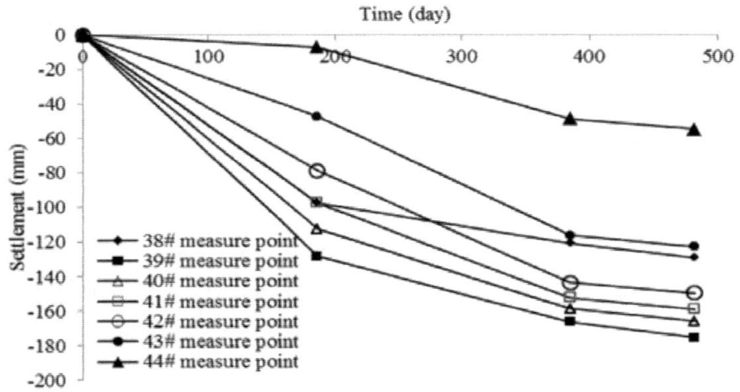


**Figure 2.** Excavation-caused settlement; time 0 denotes May December 20,2014.

**Table 2.** Accumulated settlement in different periods since new construction beginning (mm)

| MP | | F38 | F39 | F40 | F41 | F42 | F43 | F44 |
|---|---|---|---|---|---|---|---|---|
| S1 | P1 | -97.6 | -128.3 | -112.4 | -97.6 | -79.0 | -47.4 | -7.3 |
| | P2 | -120.7 | -166.2 | -158.6 | -152.2 | -143.6 | -116.3 | -48.7 |
| S2 | P3 | -8.4 | -9.3 | -7.5 | -6.7 | -6.0 | -6.5 | -5.9 |
| TS | P4 | -129.1 | -175.5 | -166.1 | -158.9 | -149.6 | -122.8 | -54.6 |

Note: MP denotes Measurement Points; S presents Settlement; S1 and S2 means Settlement from the previous monitoring team and Settlement from Tongji University, respectively; TS for Total Settlement; P stands for measurement period; P1, P2, P3, and P4 denotes the period of Dec.,2014-June 24,2015, Dec. 2014-Jan. 19,2016, Jan.22,2016-April 15,2016, and Dec. 2014- April 15,2016.

It is noted that the Tongji University checked the previous system before monitoring to ensure that the results are reliable. Assume the residence is rigid body despite of moment deformation, combined with settlement data in different periods, we can deduce the initial residence inclination before new construction begins, during of excavation of adjacent new buildings, and after excavation. The analysis results are listed in table 3.

Consider that the inclination of residence 1 is approaching the criteria of 10 ‰ (JGJ 125-99,2004), we calibrated the Total station (RTS112SR5L) made by Germany and monitored 10 lines. Each line had been monitored three times. Each time we read three coordinates. This technical measure can migrate the errors from measurement system and limited data. The residence inclination on April 15, 2016 is shown in figure 3.

**Table 3.** The residence inclination in different periods (‰)

| Position | Initial inclination before new construction beginning | Dec.,2014 - Feb. ,2016 | Feb. 2016 - April 2016 |
|---|---|---|---|
| North west | 1.8 | 7.60 | 1.57 |
| North east | 5.2 | 4.39 | 2.82 |

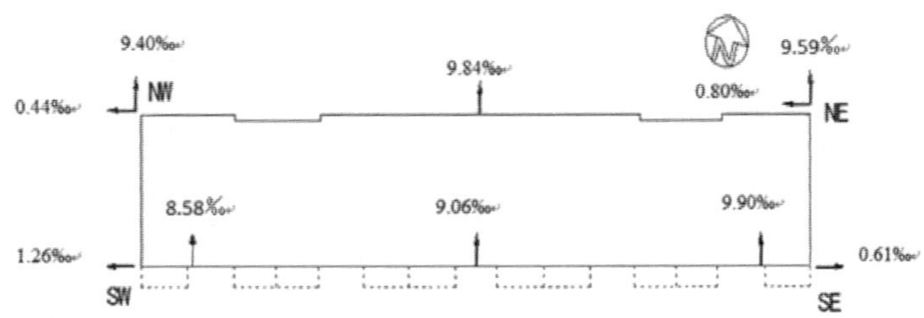


**Figure 3.** Residence inclination in representative positions

## 2.2 *Residence 2*

The residence 2 (photo 2) is located in the south of Shanghai in China (latitude 310 12’ 18.34”, longitude 1210 22’50.37”). They were built in 1993 and 1994. The new construction (tunnel) began from November, 2003. The filed monitoring conducted by Tongji University started from May 24, 2006 and ended on July 28, 2006. The measure points are shown in figure 4. The size of the building and initial condition is listed in table 4. The minimum net distance away from the new construction is 9 m. The maximum depth of excavation of new construction is 6 m. The settlements during and after tunneling construction are shown in figure 5.

**Photo 2.** Layout of residence 2 (from google earth, July 2016)

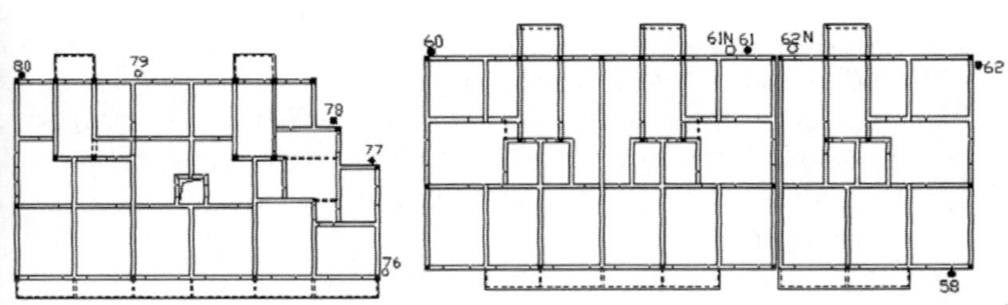


(a) (b)

**Figure 4.** Layout of measurement points: (a) 1# residence; (b) 4# residence

**Table 4.** Size and initial condition of the buildings

| No. of residence | 1# | 2# | 3# | 4# | 5# | 6# |
|---|---|---|---|---|---|---|
| Height of story / m | 2.8 | 2.8 | 2.8 | 2.8 | 2.8 | 2.8 |
| Width / m | 12.3 | 12.3 | 12.3 | 12.3 | 12.3 | 12.3 |
| Length / m | 22.0 | 40.2 | 53.7 | 31.5 | 43.0 | 34.0 |
| Initial condition | Note 1 | Normal | Normal | Normal | Normal | Normal |

Note 1: 60 square meters located on the dark creek processed. Note 2: It includes two parts which are not built at the same time. The settlement joint between two parts is 280 mm width.

IOP Conf. Series: Earth and Environmental Science **61** (2017) 012087 doi:10.1088/1755-1315/61/1/012087

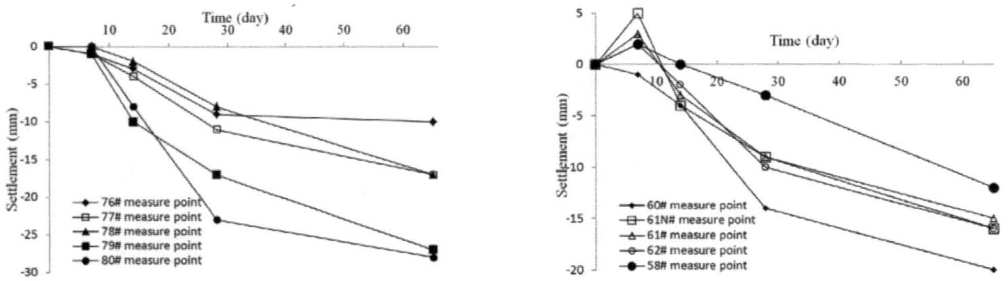


(a) (b)

**Figure 5.** Settlement caused by new construction nearby: (a) 1# of residence 2; (b) 4# of residence 2

Combining the deformation measurements from August to September, 1994, October, 2004 to February, 2006, and May to July,2006 together, the residence inclinations are monitored (Tang, 2010), part of the results is listed in table 5.

**Table 5.** The residence inclination in different periods (‰)

| No. of residence | Before new construction beginning | Duration of new construction[a,b] | July 28,2006 |
|---|---|---|---|
| 1# West | 5.00 | 3.76 | 10.238(11.720) |
| 4# West | 5.00 | 2.66 | 7.705(7.590) |

[a] The values in brackets are from settlement monitoring data. [b] Monitoring duration of new construction from Oct.,2014 to Feb. ,2006

## 3. Result analysis and strategies for rehabilitation

### *3.1 Result Analysis*

Residence 1 and 1# of residence 2 have 5‰ initial inclinations (table 5). Compared with other monitored buildings, they have more increments of inclination. Maximum inclination rates are 2.82‰ and 0.19‰, respectively.

Maximum settlement rates during two months filed monitoring by Tongji University are 0.090 and 0.043 mm/d, respectively. From the slope of settlement rate, they are not stable. Tang and Zhao (2012) found that the soft layer under rafts of residences is not convergent due to constant slope of deformation based on laboratory Triaxial tests. Assume the residence has 10‰ inclination, the maximum allowable inclination (JGJ 125-99), residence 1 has 100 mm differential settlement northward and residence 2 has 210 mm westward. And according to elastic theory, we can deduce the differential stress on the rafts according to equation (1).

$$S = \frac{P}{E} H \qquad (1)$$

Where S is settlement, P is average stress on the raft based on the loads of superstructure and self-weight of foundation, usually take 100 kPa, E is soil elastic modulus, and H is thickness of compressive layers.

Assume the residences have 150 mm average settlements (based on local experiences) after two decades service, then the settlements for residences 1 and 2 after adjacent new constructions are 250 mm and 360 mm, respectively. $S0 = 150$ mm, $S1 = 250$ mm for residence 1, $S1 = 360$ mm for residence 2, $P0 = 100$ kPa.

$S_0 = \frac{P_0}{E} H, S_1 = \frac{P_1}{E} H$, then $\frac{S_0}{S_1} = \frac{150}{250} = \frac{P_0}{P_1} = \frac{100}{P_1}$. Thus, P1 = 240 kPa. Similarly, we can get P1 = 240 kPa for residence 2. Therefore, estimated stress increments due to inclinations are 67 kPa and 140 kPa for residences 1 and 2, respectively. These stress increments will lead to more extra differential settlements and inclinations.

On the basis of the settlement rate, inclination rate vs. initial condition, as well as the deformation caused by different period conducted by scientific analysis, we can better understand the results of monitoring surveys and two parties realize their responsibilities for the present condition of the residences. Then both par-ties will concentrate on how to improve the serviceability of the residences.

We also employed the finite element software to evaluate the bearing capacity and seismic performance of the residences (the process is omitted here). The result is almost satisfied with the requirements (GB 50009-2012, GB 50011-2010, and GB 50007-2011).

### 3.2 *Strategies for Rehabilitation*

There are various measures to avoid existing buildings adjacent new constructions from excessive deformation. For instance, underpin under existing shallow foundations, install diaphragms around existing foundation to isolate existing buildings and new constructions, mitigate excavation deformation itself, and mitigate existing building settlements by effective methods.

## 4. Conclusion

New construction should make sure that adjacent existing building to be in normal service situation by various permanent or temporary measures although it is not difficult to achieve. In soft soil region, multi-story brick masonry buildings with shallow foundations are very sensitive to adjacent new constructions which are with deep foundations accompany with deep excavations. Two case studies show that extra 5~7‰ inclination happens easily if they have initial defects, such as initial inclination. Evaluation of extra deformation and structure safety should be done before the new construction begins. It is recommended to protect existing buildings or mitigate deformation of excavation of new construction if the evaluation exhibits that the existing building is vulnerable.

It is important to understand the existing building behavior to assure field monitoring accuracy, reasonable data process.

## Acknowledgments

Thanks to Professor Zhou Deyi Dr. Feng Jintao who monitored in the field of residence 1 and residence 2, respectively. Thanks to Professor Edward Kujawski who helped to analyze data measured in the field of residence 2. Finally, thanks to the financial support from the National Natural Science Foundation of China (NSFC grant No.50818153 and No. 51278359).

## References

[1] DG/TJ (108-79-2008). 2008. Code of building quality surveying. Shanghai: Shanghai urban construction committee.
[2] DG/TJ (08-804-2005). 2005. Standard of structural inspection and assessment for existing buildings. Shanghai: Shanghai urban construction committee.
[3] GB (50009-2012). 2012. Load code for the design of building structures. Beijing: China architecture & building press.
[4] GB (50011-2010). 2010. Code for seismic design of buildings. Beijing: China architecture & building press.
[5] GB (50007-2011). 2011. Code for design of building foundation. Beijing: China architecture &

building press.
[6] GB/T (50344-2004). 2004. Technical standard for inspection of building structure. Beijing: China architecture & building press.
[7] JGJ (125-99). 2004. Standard of dangerous building appraisal. Beijing: China architecture & building press.
[8] Kerisel J 1975. Old structures in relation to soil conditions. *Geotechnique* 25(3):433-483.
[9] Meli R and Sanchez-Ramirez R 2007. Criteria and experiences on structural rehabilitation of stone masonry buildings. *International journal of architectural heritage* 1(1):3-28.
[10] Tang Y J 2010. Discussion about data mining application in civil engineering deformation measurement analysis. In David G T, Hehua Z and Xiaojun L (eds), *Information technology in geo-engineering; Proceedings of the 1st International Conference (ICITG) Shanghai, Shanghai, 10-12 May 2010*. Delft: IOS Press.
[11] Tang Y J and Zhao WS 2012. Triaxial test research on dynamic settlement of ancient architecture foundation soil under frequent earthquakes. *Journal of Tongji university (natural science)* 40(10):1486-1503.

IOP Conf. Series: Earth and Environmental Science **61** (2017) 012088 doi:10.1088/1755-1315/61/1/012088

# A noise reduction method for Ground Penetrating Radar signal based on wavelet transform and application in tunnel lining

**Jianye Zhu[1], Yiguo Xue[1], Nan Zhang[1], Zhiqiang Li[1], Yufan Tao[1] and Daohong Qiu[1]**

[1]Research Center of Geotechnical and Structural Engineering, Shandong University, Ji'nan 250061, China.
Corresponding author: zjyhappylife@163.com

**Abstract.** One of the major limitations that hinder the use of ground penetrating radar (GPR) in civil engineering is the inevitable noise interference associated with the GPR signal. This study proposes a noise reduction technology for a GPR signal by utilizing the excellent time-frequency localization properties of a wavelet and by analysing the characteristics of the wavelet function and the principles of wavelet noise reduction. A Daubechies wavelet is selected as the wavelet function, and the wavelet decomposition and re-construction is performed on a single channel radar signal. The images of tunnel lining from GPR are compared before and after wavelet transform, and the wavelet transform processing is proved to be capable of separating noise as well as of improving the signal-to-noise ratio. The results show that wavelet transform processing is found useful to improve the recognition of image and the accuracy of radar signal interpretation.

## 1. Introduction

Ground penetrating radar (GPR) is widely used in the quality inspection of tunnel lining. However, due to very complex rock media, radar gets inevitably disturbed by noise and gets cluttered. In early studies, because of the similarities in the propagation form and reflection profiling data of the electromagnetic waves and seismic waves, seismic data processing method was also used to process the data of GPR (Chen and Xiao, 2005)[1]. A thought of applying a wavelet transform in signal de-noising was proposed in 1989 (Larsoneur, 1989; Morlet, 1989) [2]. Based on MRA, a multi-resolution wavelet network was constructed using a compactly supported orthogonal wavelet, where wavelet function and scaling function are applied in wavelet network, and by taking advantages of their complementary properties, the multi-resolution wavelet network was found to have a clear global and local error estimate, lower computational complexity and excellent adaptability (Bahavik, et al. 1993) [3].

The use of an atomic decomposition-based tracking de-noising method was also studied(Chen and Donoho, 1999) [4]. A discrete wavelet transform with an artificial neural network were coupled to propose a method for monitoring the water table (Adamowski and Chan, 2011) [5]. In use of the adaptive threshold method, Joewono studied the use of compression wavelet analysis in retinal recognition technology with low SNR (Joewono, 2015) [6]. The research laid the foundation for the wavelet theory and these wavelet functions and methods have been widely applied and proven in seismic de-noising, image processing and other broad areas. These studies provide a guideline for the use of the wavelet theory in de-noising GPR signal obtained during inspecting a tunnel lining structure.

## 2. The wavelet transform theory and apply on GPR signal processing

### *2.1. Wavelet transform*

The wavelet transform methods include "threshold de-noising" approach (Donoho, 1995) [7]. It accounts that the wavelet coefficients corresponding to the signal contain important information of signal, and are of a small number with large amplitude. The wavelet coefficients corresponding to the noise are uniformly distributed, and are of a large number with small amplitude.

### *2.2. Wavelet transform de-noising threshold method*

The wavelet transform threshold method can be divided into two types, namely, soft threshold and hard threshold (Donoho and Johnstone, 1994; Donoho, 1995) [8]. Hard threshold method is to set a value of 0 for the wavelet coefficients that are smaller than the set threshold value. The values of the wavelet coefficients that are greater than the threshold value are retained without any treatment. On the other hand, while the soft threshold method also sets a value of 0 for the wavelet coefficients that are smaller than the set threshold value, the values of the wavelet coefficients that are greater than the threshold value are set to the new shrunken values obtained by subtracting the threshold value from their original values.

The formula for the hard threshold is given as:

$$\lambda_{hard} = \begin{cases} \chi(t) & |\chi(t)| > \lambda \\ 0 & |\chi(t)| \le \lambda \end{cases} \tag{1}$$

And the formula for the soft threshold is given as:

$$\lambda_{soft} = \begin{cases} (|\chi(t)| - \lambda)\mathrm{sgn}(\chi(t)) & |\chi(t)| > \lambda \\ 0 & |\chi(t)| \le \lambda \end{cases} \tag{2}$$

Where $\chi(t)$ is signal wavelet transfer function; $\lambda$ is selected threshold; sgn represents the sign function, it means that the soft threshold method set the wavelet coefficients , the absolute value of which does not exceed threshold to zero, the other factor reduction processing in accordance with thresholds.

As seen from the above formula (5) (6), since the hard threshold function is a non-continuous function, when high-frequency coefficients are quantized, the coefficients are highly sensitive to the small changes around the threshold value. This leads to a numerical situation which is prone to mutation. Thus, after a hard threshold, the output image is not very smooth, and the image may sometimes consist of ringing, pseudo-Gibbs effects, etc. Nevertheless, the hard threshold method can preserve the local features, such as edges, very well. On the other hand, the soft threshold method may cause blurred edges, but results in a relatively smooth output image.

### *2.3. Determination of wavelet decomposition level*

The actual process of radar signals shows that the wavelet decomposition level is not necessarily better when it is larger. As measured on a single channel radar signal corresponding to the random distribution of white noise de-noising simulation, the original signal sampling points were 512. This paper extracted half the length of it, which means information of the first 256 points. The signal was processed by using a DB4 wavelet. The decomposition level was chosen as 6 and the information about scale change was given for each level. Listed below are the high (low) frequencies of the re-constructed image for decomposition levels 1-6. When s represents the noisy signal; a(i) represents the low-frequency coefficients re-constructed image at the i-scale decomposition; and d(i) represents the high-frequency coefficients of the re-constructed image at the i-scale decomposition, s=a(6)+d(6) +d(5) +d(4) +d(3) +d(2) +d(1). Then, formula (4) was used to reconstruct the desired de-noising image. Figure4 shows that the re-constructed signal basically retains the mutated portion of the original signal; the distortion is small. Figure5 shows that when the scale is enlarged by four times, the effective signal retained little significance. Based on the numerical results, it was found that when the number of layers selected was three. The re-constructed signal resembles the closest to the original signal.

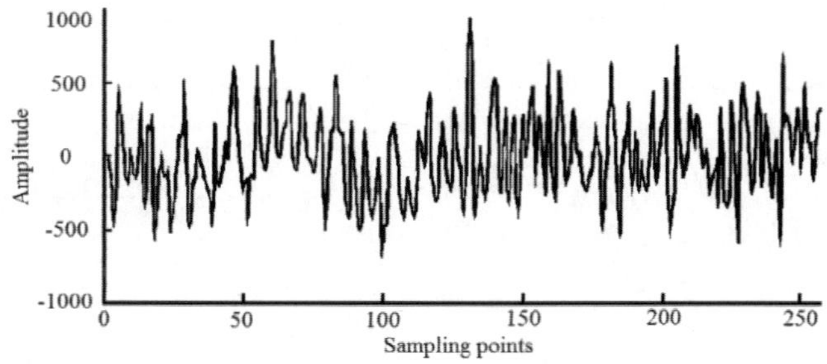


**Figure 1.** Analog random noise

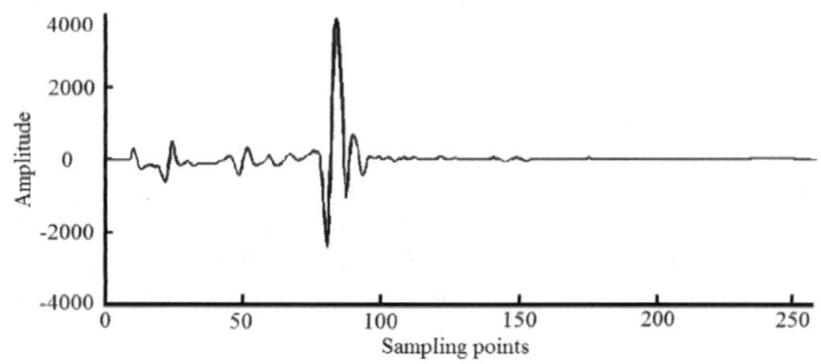


**Figure 2.** One trace GPR signal curve

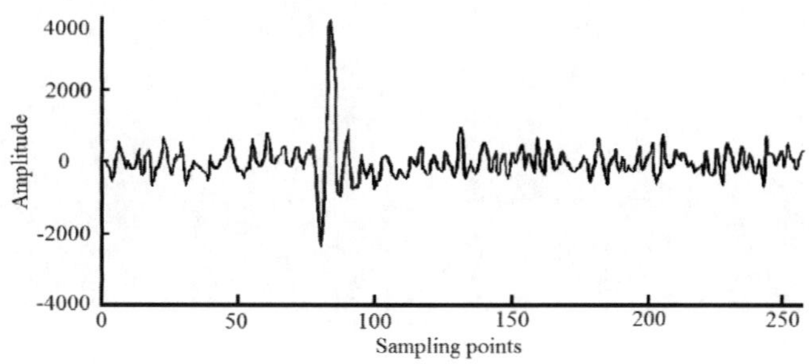


**Figure 3.** Synthesized signal after adding random noise

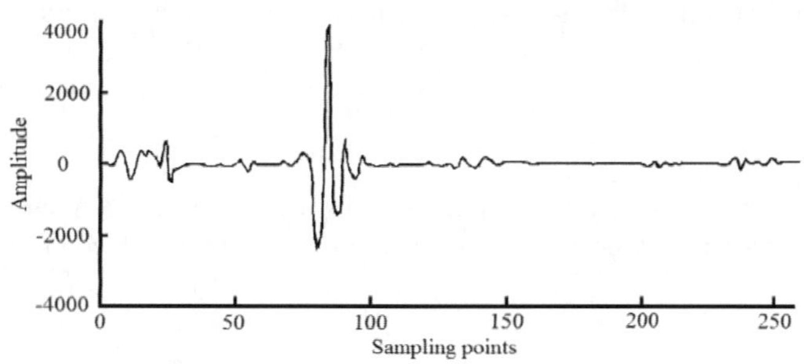


**Figure 4.** Re-constructed signal after de-noise

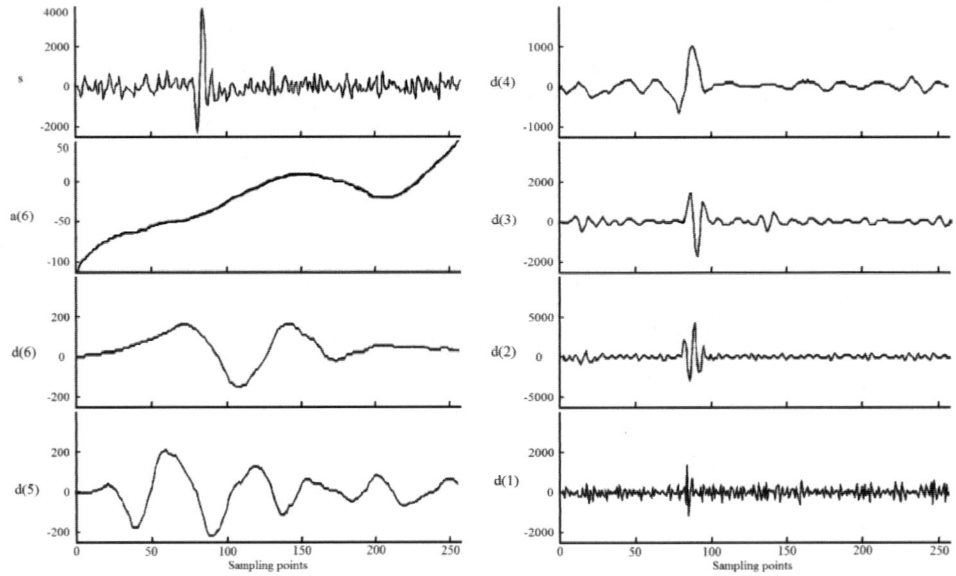


**Figure 5.** One trace signal by DB DWT at level 6

## 3. Engineering application

Yushan tunnel is located in Ji'an city, Jiangxi province, China. The mileage for the survey line at the beginning and the end are ZK190 + 640 and ZK190 + 672 which is located in four surrounding rock area with 0.2 m thick concrete lining with steel arch spaced at 1 m center to center.

**Figure 6.** Data collection during the field quality inspection of Yushan tunnel lining

Quality inspection of the tunnel lining was performed by using GPR equipment of SIR 3000 model produced by US GSSI. Antenna transmit center frequency was 400 MHz and emission rate was 100 KHz. Every signal comprised of 512 sampling points. The data was stored in 16 bit format. The dielectric constant was six. The equipment took continuous measurement for a measuring line along the tunnel vault having a length of 32 m.

A multi-DB4 wavelet decomposition was performed on the measured radar data profiles. The low frequency coefficients and high frequency coefficients re-constructed sections of the signals were obtained.

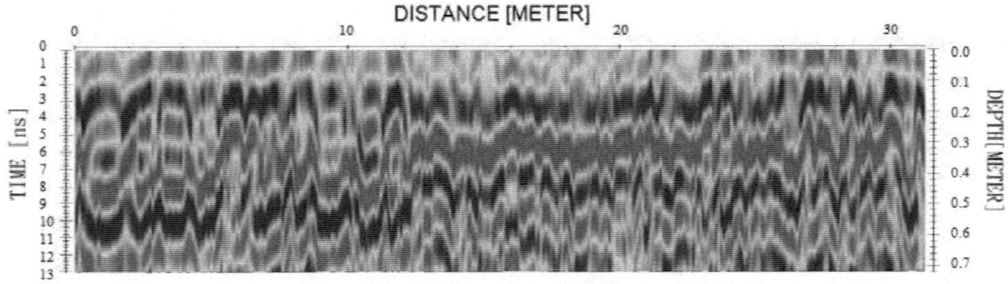


**Figure 7.** Original profile of GPR

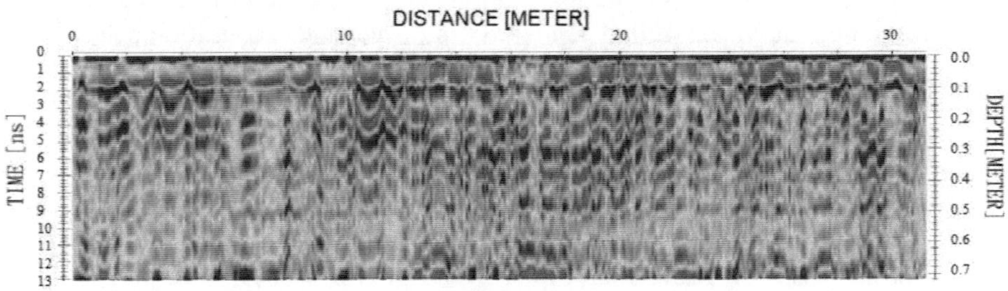


**Figure 8.** Re-constructed profile at level 1 by DB dwt

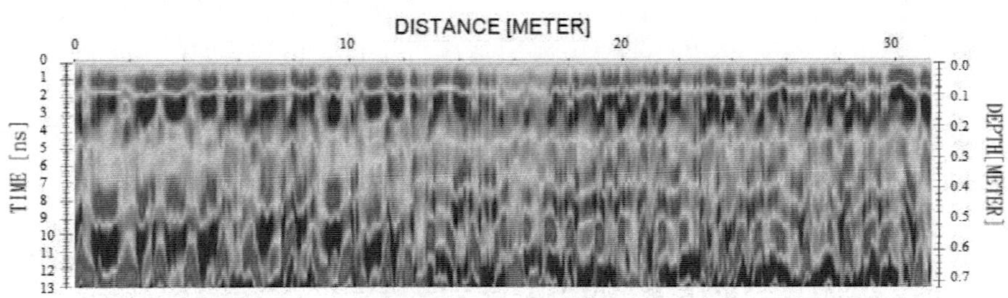


**Figure 9.** Re-constructed profile at level 2 by DB dwt

By comparing Figure 7 to 9, it is found that by multilayer wavelet decomposition, noise was focused on scale 3 profiles, the main signal of steel arch and lining interface remained in scale 1 profile. By combining the profile spectra for several scales, the scope of the band-pass filter is easy to determine and it is also easy to deal with traditional methods. Comparing figure 7 and figure 10, it can be seen that figure.10 more clearly reflect the steel arch and lining interface than figure 7. The number of steel arch is 35 and their intervals are 0.94 m. The lining thickness is 0.2 m and the bond between lining concrete and rock was good without empty area or honeycombing. Thus, the image resolution is higher after wavelet transform processing, and the interpretation can be more accurate.

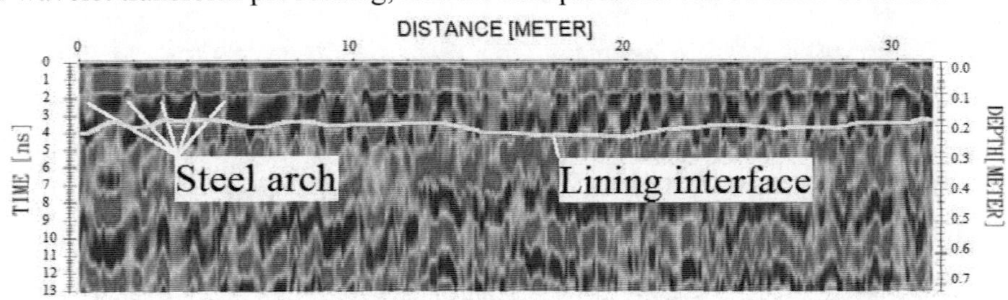


**Figure 10.** Profiles re-constructed by the inverse wavelet transform method

## 4. Conclusion

By using the differences of characteristics in the wavelet domain between the radar signal and the noise, a wavelet transform was applied in ground penetrating radar signal processing. The signal was decomposed into different frequency bands. Depending on the specific circumstances, by selecting a wavelet function and function parameters appropriately, the applied wavelet transform was effective to remove the noise, to implement signal noise separation, and to improve the resolution and the signal-to-noise ratio. Wavelet transform method was used in processing the GPR data obtained during the quality inspection of the lining of Yushan tunnel in China. The results showed that, in the image area, the lining thickness was 0.2 m, steel arch number was 35, lining interface was uniform, the bond between concrete and rock was good and there was no empty area or honeycombing. Compared with the original GPR image, the images processed by wavelet transform clearly reflect the distribution of each reflection surface and steel arch in internal lining. The comparison of the images confirmed that the wavelet transform method works well. The method improves the recognition of various features in the original image and increases the accuracy of the radar signal interpretation. Moreover, the method shows an excellent promise for GPR data processing in future.

## Acknowledgements

The authors wish to thank the anonymous reviewers for their comments. This work was financially supported by a grant from China Natural Science foundation (51379112, 51422904), and the National Program on Key Basic Research Project of China (973 Program) (2013CB036002), and Shandong Natural Science foundation (Grant No. JQ201513). In additional, the authors would like to express appreciation to the reviewers for their valuable comments and suggestions that helped to improve the quality of the paper.

## References

[1] Chen Y.Q., Xiao B.X.: On the status quo and development of ground penetrating radar. *J. Chinese journal of engineering geophysics*, 2(2), 149-155(2005).
[2] Larsonneur, J.L., Morlet J.: Wavelets and Seismic Interpretation. *Wavelets Springer Berlin Heidelberg*, 126-131(1990).
[3] Bahavik R., et al.: Wave-net: A multi-resolution hierarchical neural network with localized learning. *J. AIChE Journal*, 39(1), 57-81(1993).
[4] Chen S.S., Donoho D.L., Saunders M.A.: Atomic decomposition by basis pursuit. *J. SIAM journal on scientific computing*, 20(1), 33-61(1998).
[5] Adamowski, J., Chan, H.F.: A wavelet neural network conjunction model for groundwater level forecasting. *Journal of Hydrology*, 407, 28-40(2011).
[6] Widjaja J.: Noise-robust low-contrast retinal recognition using compression-based joint wavelet transform correlator. *J. Optics & Laser Technology*, 74, 97-102(2015).
[7] Donoho D.L.: Denoising by soft-threshold. *J. IEEE Trans. on Information Theory*, 41(3), 613-627(1995).
[8] Donoho D.L,: Johnstone I M. Ideal spatial adaptation via wavelet shrinkage. *J. Biometrika,* 81, 425-455(1994).

IOP Conf. Series: Earth and Environmental Science **61** (2017) 012089 doi:10.1088/1755-1315/61/1/012089

# Influence of stone powder content on performance of sand stone concrete in freeze-hawing environment

**Mingshi Lu[1], [a]), Dong-mei Zhang[1], [b]) and Jinqiang Wang[1], [c])**

[1]School of Urban and Architectural Engineering, Zaozhuang College, Zaozhuang, Shandong 277160, China.
[a)] 908362598@qq.com
[b)]27374633@qq.com
[c)]1510607928@qq.com

**Abstract**. In this paper, the sand stone concrete was studied on 0%, 2%, 4%, and 6% of stone powder content through experiment. It reports a study on the factors affecting the properties of sand stone concrete include the microstructure, mass loss and compressive strength after freeze-thaw cycles of 0,5,10,15,20,25,30,35,40. The results show that it can improve frost resistance of sand stone concrete with proper amount of stone powder. It can provide experimental basis for good performance of sand stone concrete.

## 1. Introduction

Concrete can resist the environmental media, and long-term maintain its good performance and appearance integrity, and keep the safety and normal use of concrete structures. It is called concrete durability. In cold regions, durability damage is mainly caused by the lack of freeze-thaw resistance, especially in contact with the water and cold environment concrete need high resistance to freezing and thawing. So in cold regions, concrete frost resistance is an important index to measure the durability of concrete [1]. The study on frost resistance of concrete is an important aspect to solve the problem of concrete durability.

Sand stone concrete, also called fine stone concrete [2], is composed of water, cement and gravel. Without natural sand, aggregate gradation of stone concrete is worse than that of ordinary sand aggregate. Due to particles of manufactured aggregate are rough and angular with unsatisfactory grading, it needs more water to prepare concrete. The mixes are of low workability and ease of segregation. But the amount of powder can improve the sand stone concrete workability, increase the density of concrete, and improve the compressive strength. The performance of stone concrete with stone powder content in 4% ~ 8% can be got even better than that of the ordinary concrete. When the powder content is 6%, the comprehensive performance of sand stone concrete is best [3-4]. In this paper, content of dust effect on frost resistance of sand stone concrete is researched. Limestone was disposable crushed by jaw crusher. Powder and gravel were sieved directly. Stone powder accounted of 0%, 2%, 4%, 6% of the concrete was directly prepared without adding natural sand. 0, 10, 15, 20, 25, 30, and 40 times of complete freeze-thaw cycle test of concrete specimens were carried out for 28d ages.

## 2. Test materials

Cement: P.C32.5R, which produced by China United Cement Corporation Lunan Co., Ltd. The performance index of the technology meets national standards and construction requirements (see Table 1).

Aggregates: Both coarse and fine aggregate are produced by siliceous limestone, which have high strength, high hardness and good durability (see Table 2).

Jaw crusher is used in this test, which model is PE60 ×100. By adjusting the size of the crusher outlet, different sand rate are got. The maximum particle size of 15mm have continuous grading gravel, without watering rand screening. The content of stone powder (particle size is <0.15mm, which accounted for about 65% of the total dust <0.075mm) accounted for 2.3%~8.1% of the weight of gravel. Gravel particle size less than 4.75mm fineness modulus is 3.0~3.4, particle size distribution accords with the ordinary concrete sand quality standard and test method of sand (JGJ52-2006 ).The particle size range is 1, coarse sand, sand gradation is qualified. Concrete can be formulated separately.

Water reducing admixture: FDN-C naphthalene series water super reducer, which produced by Yunzhe Concrete Admixture Ltd. It is technical index see Table 3.

Water: tap water.

**Table 1.**Technical index of P.C32.5R cement

| Chemical Constituents (%) | | Setting Time (min) | | Stability | Specific Surface Area ($m^2$/kg) | Flexural Strength (MPa) | | Compressive Strength (MPa) | |
|---|---|---|---|---|---|---|---|---|---|
| MgO | $SO^3$ | Initial | Final | | | 3d | 28d | 3d | 28d |
| 2.1 | 2.73 | 150 | 240 | qualified | 330 | 5.9 | 9.1 | 20.8 | 39.5 |

**Table 2.** Technology index of limestone

| Apparent Density ($kg.m^{-3}$) | Water Absorption (%) | Compressive Strength (MPa) | Mineral Composition (%) | | | |
|---|---|---|---|---|---|---|
| | | | Quartz | Mica | Calcite | Others |
| 2600 | 0.7 | 83 | 7 | 9 | 82 | 2 |

**Table 3.** Parameter of naphthalene series water reducer

| Solid-Containing Content | PH | Surface Tension 1% Solution | $NaSO_4$ | Fluidity of Cement Paste (mm) |
|---|---|---|---|---|
| powder 90% | 7-9 | 70% | 15-20% | 240mm |

## 3. Experimental process and results analysis

The test pieces were divided into 9 groups. Each group have 27 pieces, and the total pieces are 243. Each was respectively marked with number. The 9 groups of specimens were carried out for 0, 5, 10, 15, 20, 25, 30, 35, and 40 freeze-thaw cycles. Before the test, the dry specimens were removed to soak in water for a period of time, and put into the low temperature test box. Set temperature is -15 °C. The specimens were frozen after one day and night, taken out and placed at room temperature to melt a day, then one freeze-thaw cycle was end. After each freeze-thaw cycle, it was necessary to observe the appearance of the specimens, and measure the change of the mass with the increase of freeze-thaw cycles. After completing the corresponding freeze-thaw cycles, the compressive strength of the specimens was measured, and the compressive strength of the specimens was installed between the loading plates of the test machine. The displacement and strain of specimens under pressure, were done by computer dynamic acquisition, record data, and taken the average value. Then specimens were remained into the low temperature test box to next freeze-thaw cycle. This cycle was operated repeatedly. The data was analyzed of and line chart was drawn.

### *3.1. Effects of Stone Power Content on Sand Stone Concrete Microstructure after Freeze-thaw Cycles of 40*

Sand stone concrete specimens with 0%, 2%, 4%, and 6% powder content were performed after 40 freeze-thaw cycles. The sand specimen surface will become rough, not smooth, a tiny hole appears, there will be cement particles dislodged from the specimen. Microstructure performance results are shown in Figure 1.

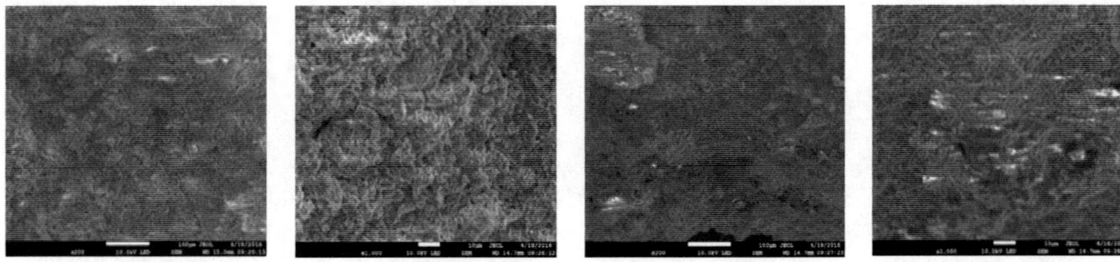

(a) 0% (b) 2% (c) 4% (d) 6%

**Figure 1.** Influences of Stone Power Content on the Microstructure of Concrete.

The following conclusions can be obtained by observing the microstructure of concrete after 40 freeze-thaw cycles:

(1)The freeze-thaw cycle has a great effect on the microstructure of the sandstone concrete.

(2)The amount of stone powder has a certain effect on the damage of the microstructure of the sand concrete.

### *3.2. Effects of Stone Power Content on Sand Stone Concrete Mass Loss after Freeze-thaw Cycles*

Sand stone concrete specimen is dipped in water after a period of time and dried with a damp cloth. Mass of specimen is known as $G_{0i}$. Also dried with a damp cloth after n times of freeze-thaw cycle test, it is called the mass of specimen is $G_{ni}$. The rate of mass loss of the sand stone concrete after freeze-thaw cycle is $\Delta W_n$. The calculation formula is as follows:

$$\Delta W_{ni} = \frac{G_{0i} - G_{ni}}{G_{0i}} \times 100\% \tag{1}$$

Form in: $\Delta W_{ni}$—the mass loss rate of the first i sand concrete specimens after n times freezing and thawing cycles (%).

$G_{0i}$-the mass of the i series sand concrete specimens with 0 times of freeze-thaw cycles (kg).

$G_{ni}$-the mass of the i series sand concrete specimens with n times of freeze-thaw cycles (kg).

$$\Delta W_n = \frac{\sum_{i=1}^{3} \Delta W_{ni}}{3} \times 100\% \tag{2}$$

Form in: $\Delta W_n$—Average mass loss rate of a group of gravel concrete specimens after n times of freeze-thaw cycles.

The variation of mass loss of sandstone concrete specimens with 40 cycles of freeze-thaw cycles under different powder content are shown in Table 4 and Figure 2.

**Table 4.** The mass loss of sandstone concrete with freeze-thaw cycles under different powder content.

| The Number of Freezing-thawing Cycles | Mass Loss of Concrete Specimens under Different Powder Content (%) | | | |
|---|---|---|---|---|
| | 0% | 2% | 4% | 6% |
| 0 | 0 | 0 | 0 | 0 |
| 10 | 1.11 | 0.95 | 1.13 | 1.2 |
| 15 | 1.4 | 1.3 | 1.25 | 1.35 |
| 20 | 2.55 | 2.56 | 2.35 | 2.54 |
| 25 | 2.64 | 2.68 | 2.46 | 2.56 |
| 30 | 3.73 | 3.7 | 3.36 | 3.71 |
| 35 | 3.87 | 3.88 | 3.52 | 4.2 |
| 40 | 4.52 | 4.84 | 4.09 | 4.96 |

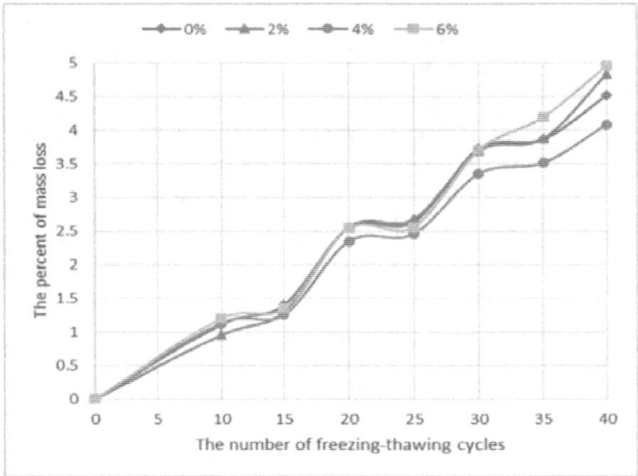


**Figure 2.** Influences of stone powder content on the mass loss of concrete.

The data analysis shows that with the increase of the number of freeze-thaw cycles, the mass of water absorption increased is far less than mass loss of concrete specimens by freeze-thaw cycle erosion. And eventually leading to loss of mass is more and more serious. When the number of freeze-thaw cycle increases to 40 times, sand stone concrete test piece rate of mass loss due to dust incorporation is different. The mass loss of is minimum with 4% dust content. It indicates that proper powder can improve the frost resistance of sand stone concrete.

There are two reasons for the relationship between stone power content and sand stone concrete mass loss after freeze-thaw cycle:

On the one hand, after freeze-thaw cycles the surface of the sand stone concrete specimens will be damaged. The surface mortar layer peeling phenomenon will began to appear, and small particles will fall off. The mass loss of the specimen will be caused. But in the early stage of freeze-thaw cycles, the specimen surface mortar peeling is not very serious, so the mass change little. With the increase of the number of freeze-thaw cycles, peeling test of coarse aggregate and mortar surface is more obvious. Part of the coarse aggregate will fall off, and the specimen surface will form many rugged small pits. So the mass of concrete is obviously reduced, and the mass loss rate increases.

On the other hand, in the freeze-thaw process, both stress of sand stone concrete inside and outside surface with temperature difference and frost heaving pressure surface pore water freezes are beyond bearing of concrete. Sand stone concrete will crack. With the increase of powder content, concrete pores will be filled. It can absorb less water, and improve frost resistance of concrete. Thereby the sand stone concrete specimen mass loss is reduced.

### *3.3. Effects of Stone Power Content on Sand Stone Concrete Compressive Strength after Freeze-thaw Cycles*

After 0, 10, 15, 20, 25, 30, 35, 40 times of freeze-thaw cycle, each group of sand stone concrete specimens with 0%, 2%, 4%, 6% powder content were put on the pressure test machine to get compressive strength test. It obtained the relationship between the content of stone power and compressive strength (see Table 5 and Figure 3) in 0, 10, 15, 20, 25, 30, 35, 40 times of freeze-thaw cycle test.

The frost resistance of concrete is closely related to the degree of compaction. So as to improve the freeze-thaw resistance of concrete, it is necessary to make the inner part of the concrete to form discontinuous pores [5]. The test data shows that stone powder will have an impact on the strength of concrete. With different powder content of concrete subjected to freeze-thaw cycles, the concrete strength decreases continuously. But 4% powder volume of sand stone concrete's compressive strength decrease slowly. The final compressive strength is higher than that of sand stone concrete without mixing powder. Thus it can be seen that the powder with water reducer can improve the compactness of concrete. And the effect of micro aggregate filling can make the inner pores more refined and reduce the porosity of concrete. So it can improve the pore structure of concrete and improve the frost resistance of concrete.

At the same time, the dust particles are dispersed in the pores between the aggregate and the cement particles. And the excess pore water is absorbed to further improve the frost resistance of the concrete. The experimental data also show that with the addition of excessive dust, the hydration products of cement decrease correspondingly, and the compactness of concrete structure reduce. Therefore, under the condition of adding water reducing agent, adding appropriate amount of stone powder into the concrete will improve the frost resistance of concrete.

**Table 5.** The compressive strength of concrete under different powder content.

| The Number of Freezing-thawing Cycles | The Compressive Strength of Concrete under Different Powder Content (%) | | | |
|---|---|---|---|---|
| | 0% | 2% | 4% | 6% |
| 0 | 38.9 | 38.4 | 38.4 | 38.5 |
| 10 | 35.1 | 34.7 | 35.5 | 34.5 |
| 15 | 34.8 | 33.5 | 34.9 | 33.6 |
| 20 | 32.9 | 31.6 | 33.1 | 31.2 |
| 25 | 29.2 | 28.3 | 29.4 | 28.5 |
| 30 | 28.7 | 27.5 | 28.9 | 27 |
| 35 | 27.7 | 27.3 | 28.4 | 27.2 |
| 40 | 25.1 | 24 | 25.4 | 24.3 |

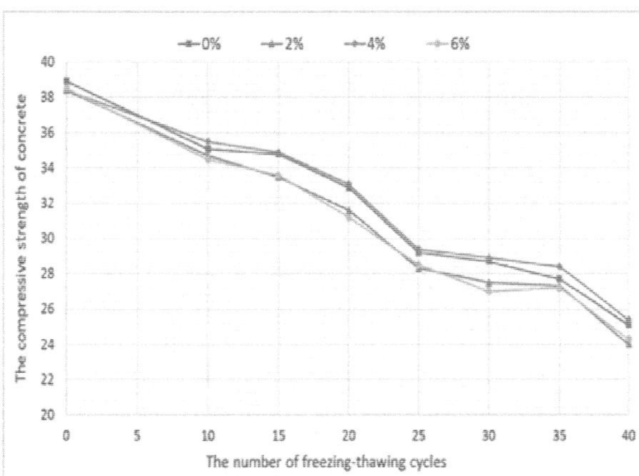


**Figure 3.** Influences of stone powder content on the compressive strength of concrete.

## 4. Conclusions

In this paper, sand stone concrete was studied on 0%, 2%, 4%, and 6% of stone powder content through experiment. It reports a study on the factors affecting the properties of sand stone concrete including the microstructure, mass loss and compressive strength after freeze-thaw cycles of 0,5,10,15,20,25,30,35,40.

(1) The microstructure of sand stone concrete is greatly influenced by freeze-thaw cycle. But proper stone powder content have improving effect of microstructure of concrete.

(2) With the stone powder content mixing into sand stone concrete, the micro aggregate filling effect of the inner pore is refined, the compactness of concrete is improved, and the porosity of concrete is reduced. It can improve the concrete pore structure, reduce powder absorption to pore water, and reduce the presence of pore water. Without affecting the strength of concrete, the frost resistance of concrete is improved.

(3) The results show that it can get best frost resistance of sand stone concrete with 4% amount of stone powder.

IOP Conf. Series: Earth and Environmental Science **61** (2017) 012089 doi:10.1088/1755-1315/61/1/012089

**Acknowledgments**
Project supported by Zaozhuang science and technology development plan (201344).
Author: Mingshi Lu (1976 - ), male, Shandong, Zaozhuang, lecturer, master. Research direction: Research on the performance of high performance concrete.
E-mail: 908362598 @qq.com Tel: 18265277029.

**References**
[1] Lihua Wang, Zhenfeng Zhan, Jia Liu. Comparison study on shape feature of limestone, granite and sandstone manufactured sand using the DIP method [J]. Journal of Hydroelectric Engineering. 2014, 3(1):161-165. (In Chinese).
[2] Li B X,Wang J L,Zhou M K.Effect of limestone fines content in manufactured sand on durability of low -and high-strength concretes[J].Construction and Building Materials,2009,23 (8):2846-2850.
[3] Chaoying Zou, Juan Zhao, Feng Liang. Degradation of mechanical properties of concrete caused by freeze-thaw action Journal of Building by Structures, 2008, 29(1):118-123. (In Chinese).
[4] Jin Weiliang, ZHAO Yuxi. Durability of concrete structure [M].Beijing: Science Press, 2002:70-83. (In Chinese).
[5] Wei Sun, MU Rluo, etal. The effect of chloride salt, freeze-thaw cycling and externally applied load on the performance of the concrete [J].Cement and Concrete Research,2002,32(12):1859-1864. (In Chinese).

# Research on the deformation and failure evolution of sandstone under triaxial compression based on PFC$^{2D}$

**X B Yang, X X Han, T B Zhou and E L Liu**

School of Resource and Safety Engineering, China University of Mining and Technology (Beijing), Beijing 100083, China
hanxxyydy123@126.com

**Abstract:** Through the comparative analysis of the results of the triaxial compression experiments of sandstone and the numerical simulation results of particle flow code PFC$^{2D}$ under the same conditions, the typical simulation curve and the corresponding simulation process were selected to analyze the evolution characteristics of the surface deformation field, the evolution characteristics of the velocity field and displacement field of the deformation localization bands of sandstone under triaxial compression. Research results show that the changes of the velocities and displacements of deformation localization bands corresponds to the change of stress during compression; In the same deformation localization band, the dislocation velocities are always in the same direction, but in the direction vertical to the localization band, the localization band sometimes squeezes and sometimes stretches; At different positions of the same deformation localization band, the dislocation velocities and extrusion velocities are both different at the same time; In the post-peak stage of loading, along the same deformation localization band, the dislocation displacements close to both loaded ends are generally greater than the ones near to the middle position of the specimen, the stretching displacements close to both loaded ends are generally smaller than the ones near to the middle position of the specimen.

## 1. Introduction

Rock is an important and complicated medium material in engineering and its deformation and failure are closely related to various rock geological disasters, so the research of deformation and failure about rock has always been a matter of great concern in the academic and engineering circles [1] and has important scientific and engineering significances [2, 3]. Due to the existence of many cracks and various micro defects in the rock material, damage and localized centralized development of micro cracks will occur under external forces leading to the local deformation and failure of rock [4]. So in most cases, the deformation and failure of rock are triggered by the development of deformation localization [5-9] and it is very important to study the evolution process of deformation localization.

The deformation localization of rock is closely related to the stress state of rock. In the aspect of relationship between the evolution of deformation localization and mechanical properties of loading, Zhang [4] used digital laser speckle method to study the process of the deformation localization of two kinds of soft rocks including mud sandstone and mudstone under uniaxial compression and gained the change laws of local deformation failure of soft rocks under different stress conditions. Song et al. [5, 10] took white light digital speckle correlation method as the mean of experimental observation, studied the deformation evolution of whole loading process and the displacement evolution of deformation concentration areas of rock through uniaxial compression experiments and obtained related evolution laws. Hao and Sun [8] observed synchronously the surface displacement field of rock samples and the load-displacement curve under uniaxial loading through the white light speckle synchronous measurement system which was self built and analyzed the evolution characteristics of surface deformation field. These research results have positive significances for understanding the local deformation evolution of rock under uniaxial compression. However, due to the limitations of the experimental conditions at present, the relationship between the local deformation evolution and the loading stress in the whole process of triaxial compression is rarely reported. Although it is difficult to explore it through experiments, the numerical simulation method provides a new way.

As a powerful discrete element numerical simulation software, PFC$^{2D}$ can better simulate the experimental process of rock under triaxial compression [11, 12]. So this paper simulated the whole process of the triaxial compression of sandstone based on PFC$^{2D}$ and selected a typical simulation curve and the corresponding simulation process to study the evolution characteristics of the surface deformation field, the evolution characteristics of the velocity field and displacement field of the deformation localization bands about sandstone under triaxial compression through the comparative analysis of experimental results and numerical simulation results.

## 2. Experimental and numerical simulation results of the triaxial compression of sandstone under different confining pressures

Electro hydraulic servo testing machine of RLJW-2000 type for the triaxial compression of rock was used to conduct the triaxial compression tests of sandstone samples under the constant confining pressures of 5MPa, 10MPa, 20MPa, 30MPa respectively. Controlled axial strain rate was used in these experiments and the axial strain rate of triaxial compression tests was 0.005mm/s. Based on laboratory experiments, particle flow code PFC$^{2D}$ was used to simulate the whole process of the triaxial compression of sandstone under corresponding different confining pressures. In the simulation, the particles were produced by radius expansion method and the simulation was carried out by parallel bond model. To be the same as the loading mode and loading rate of experiments, the simulation processes were also loaded by controlling axial strain, in which, axial compression was conducted by controlling the speed of the upper and lower walls. While the speed of side walls was controlled by the numerical servo system to maintain a constant confining pressure. The contrast curves of stress-strain between experiments and numerical simulation under different confining pressures were shown in figure1.

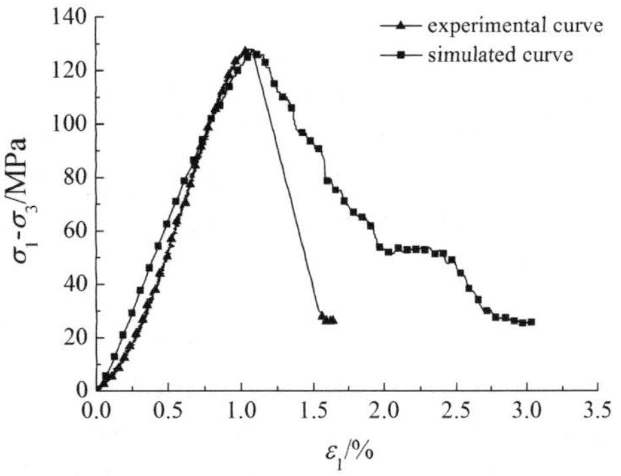


(a)Confining pressure of 5MPa

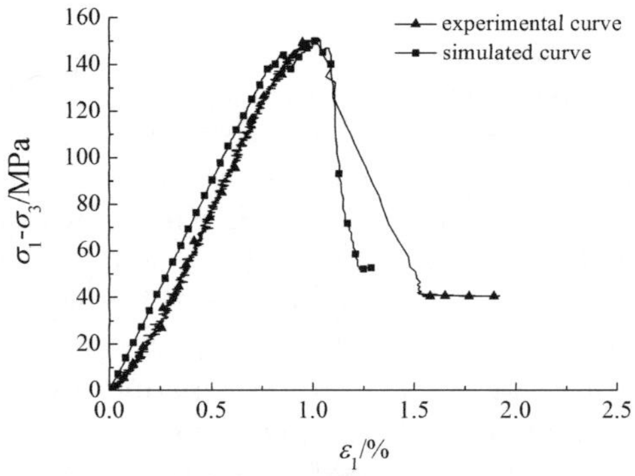


(b) Confining pressure of 10MPa

IOP Conf. Series: Earth and Environmental Science **61** (2017) 012090 doi:10.1088/1755-1315/61/1/012090

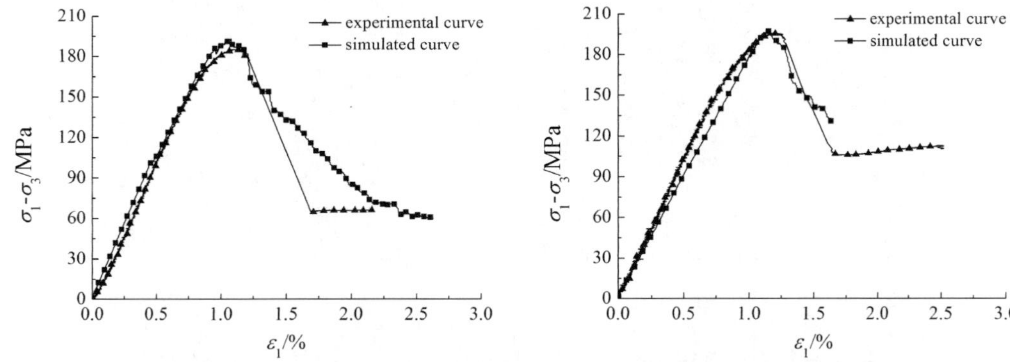


(c) Confining pressure of 20MPa (d)Confining pressure of 30MPa

**Figure 1.** Contrast curves of stress-strain between experiments and simulation under triaxial compression with different confining pressures

As shown in figure 1, under different confining pressures, the stress-strain curves of sandstone in triaxial compression experiments are close to the numerical simulation curves, the triaxial compressive strength between experiments and simulation are basically the same and the final failure form (as shown in figure 3(j)) is shear failure, which is consistent with the actual form of experimental destruction, so the particle flow code $PFC^{2D}$ can be used to simulate the mechanical properties of rock more realistically. Therefore this paper studied the evolution characteristics of the surface deformation field, the evolution characteristics of the velocity field and displacement field of the deformation localization bands of sandstone during deformation and failure under triaxial compression based on the numerical simulation of $PFC^{2D}$.

## 3. Analysis of the evolution characteristics of deformation field

The simulated compression process under the confining pressure of 20MPa was selected to study the evolution laws of the deformation field of sandstone during deformation failure. The stress-strain curve was shown in figure 2 and ten typical identification points were selected on the curve. Then the corresponding failure figures of the specimen about every identification point were shown in figure 3. In figure 3, when cracks occured among particles, the particles would show different colors which indicated bonds among particles had been broken. The particles with the same color indicated that they were well bonded.

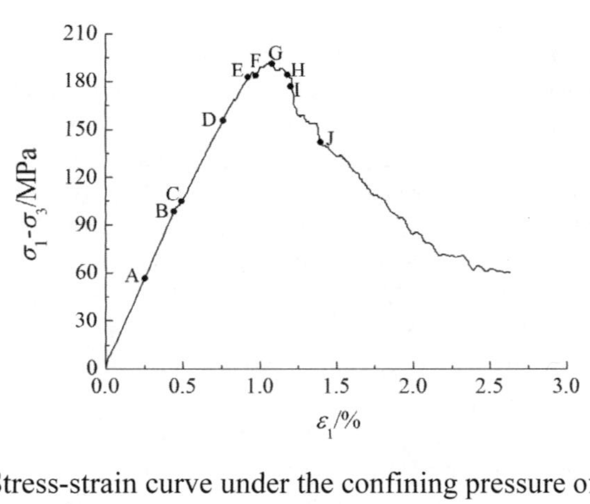


**Figure 2.** Stress-strain curve under the confining pressure of 20MPa

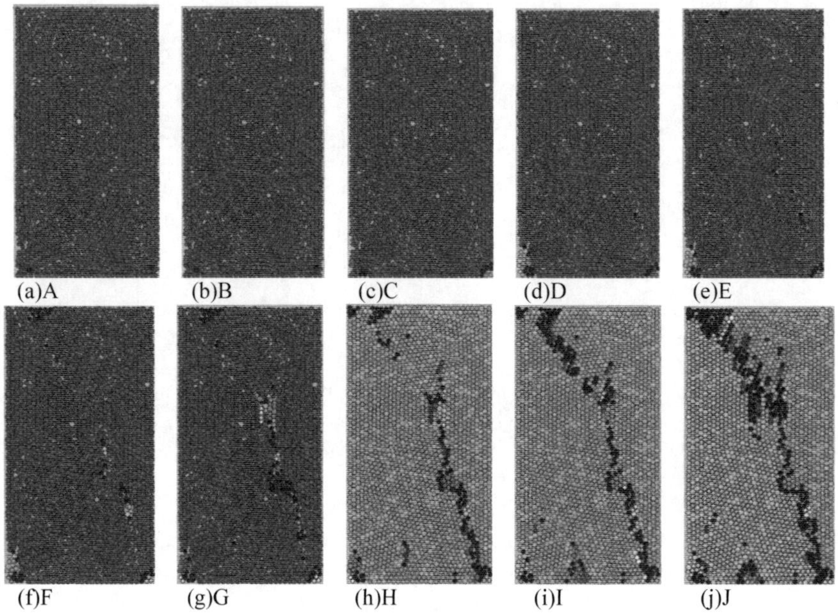

**Figure 3.** Failure figures about different identification points

From figure 2 and figure 3, the identification points A, B, C and D are located in the linear elastic stage. At this stage, the deformation of the specimen is homogeneous. Because of the occurrence of micro damage at the bottom of the specimen, a turn from point B to point C appears on the stress-strain curve. As a result, the stress-strain curve of the specimen don't extend along the original straight line. But due to the extension of cracks at the bottom boundary is slow, the stress-strain curve continues to rise. The identification points E and F are located in the plastic stage, and inhomogeneous deformation occurs. At point E, cracks have appeared in parts of the specimen. As the stress continues to increase, the intersection and penetration of cracks appears in local areas at point F. And at the peak stress of point G, this phenomenon is more obvious. Identification points H, I and J are located in the stage of strain softening, and at this stage, the cracks in the specimen continue to extend, converge and perforate, leading to the final failure of the specimen which is shown at point J. At point J, the colors of left and right part are different, indicating that the bonds between two parts are broken completely, forming shear slip bands.

## 4. Analysis of the evolution characteristics of velocity field in deformation localization bands

In the whole process of rock loading, the velocity and displacement evolution of deformation localization bands are closely related to the stress situation of the specimen. Therefore, it is necessary to study the evolution characteristics of the velocity and displacement of deformation localization bands. Figure 4 is the identification figure of deformation localization bands, in which the deformation localization bands were identified as M, N and the localization bands M and N divided the surface of specimen into two regions: P and Q. In order to explore the formation of deformation localization bands M and N, four sets of symmetry points vertical to localization bands were selected on both sides of the localization bands M and N, namely the points 1-8 in figure 4, in which the distance from each point to the localization band was equal. For the accuracy and reliability of the quantitative study, the velocity values of 7 particles were taken at each point, and the average value was taken as the speed at the point. The horizontal velocities and vertical velocities of each set of symmetry points were decomposed respectively along the orientation parallel and vertical to deformation localization bands, and the differences of the corresponding velocity components between symmetric points were solved as the dislocation velocity and extrusion velocity of this position in the deformation localization bands. In this paper, the dislocation velocities in a clockwise direction along the deformation localization

IOP Conf. Series: Earth and Environmental Science **61** (2017) 012090 doi:10.1088/1755-1315/61/1/012090

bands were regarded as positive, and the extrusion velocities vertical to the deformation localization bands were considered positive. Through processing, the change curves of velocities between different symmetry points were shown in figure 5.

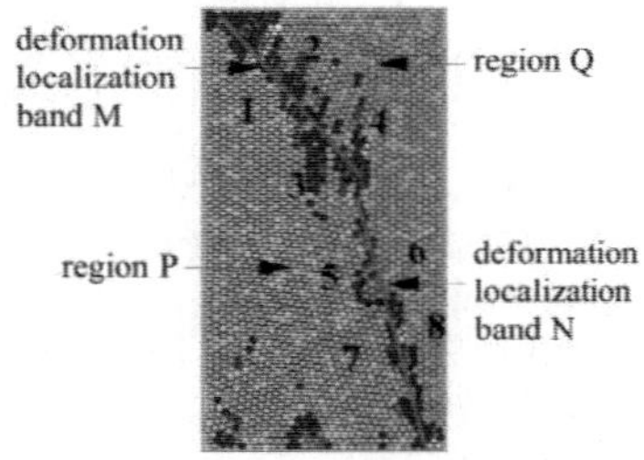


**Figure 4.** Identification figure of deformation localization bands

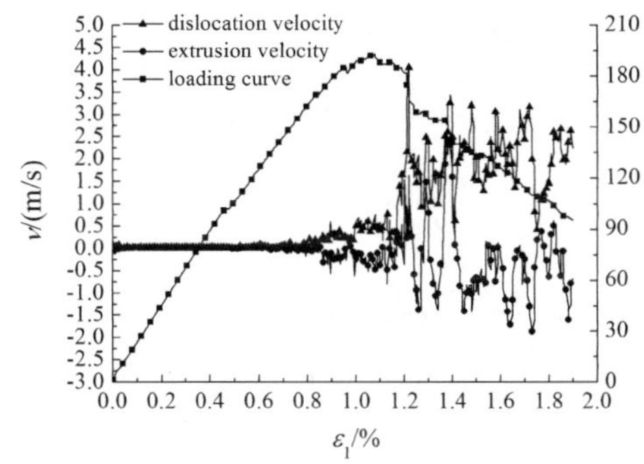


(a)Velocity variation of point 1 relative to point 2

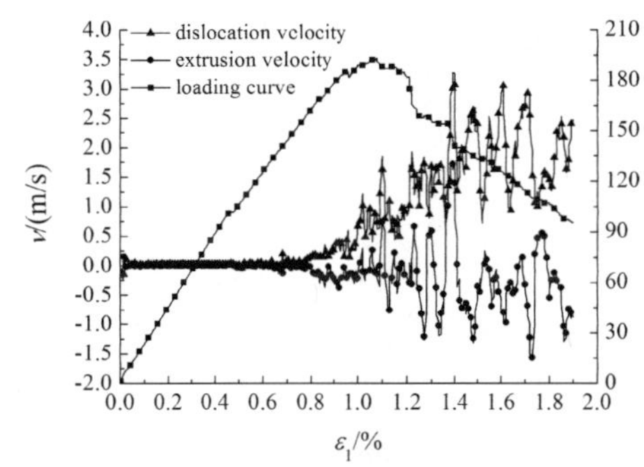


(b)Velocity variation of point 3 relative to point 4

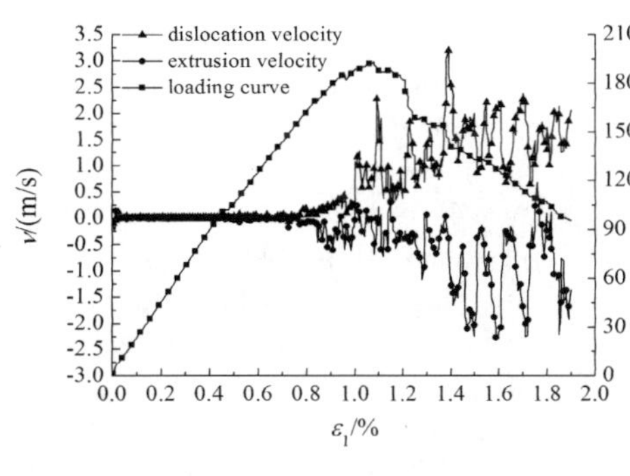


(c)Velocity variation of point 5 relative to point 6

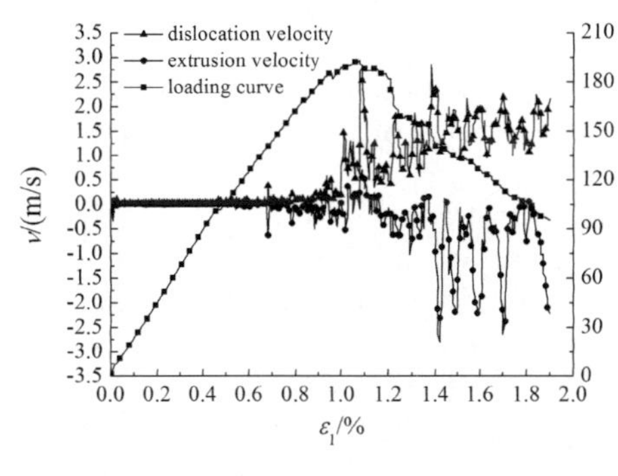


(d)Velocity variation of point 7 relative to point 8

**Figure 5.** Change curves of velocities between different symmetrical points

As shown in figure 5, in the linear elastic stage of stress-strain curve, the dislocation velocities and extrusion velocities between symmetric points of each group are basically zero. In the plastic stage and the post-peak stage, speed gradually present serrated fluctuations. The fluctuation amplitude of the velocities in the plastic stage is small, which indicates that the cracks are slowly expanding. In the whole process of stress loading, the region P and region Q have been moving along the clockwise direction of deformation localization bands. However, the dislocation velocities have not always been increasing, but repeatedly alternating between increase and decrease. In the direction vertical to localization bands, the region P and the region Q sometimes squeeze and sometimes stretch, and the velocities also repeatedly alternate between increase and decrease. But after the cracks are fully connected, the region P and the region Q will show tensile state basically. In figure 5, by comparing

IOP Conf. Series: Earth and Environmental Science **61** (2017) 012090 doi:10.1088/1755-1315/61/1/012090

figure (a), (b), (c) and (d), a conclusion can be gained that at different positions of the same deformation localization band, the velocities are different at the same time, which indicates the expansion rates of cracks are different along the deformation localization band, cracks appear successively and cracks expansion have directionality.

## 5. Analysis of the evolution characteristics of displacement field in deformation localization bands

The analysis method of the evolution characteristics of displacement field in deformation localization bands is the same as the one of velocity field. The horizontal displacements and vertical displacements of each symmetry points were still decomposed respectively along the orientation parallel and vertical to deformation localization bands, and the differences of the corresponding displacement components between symmetric points were solved as the dislocation displacement and extrusion displacement of this position in the deformation localization bands. The dislocation displacements in a clockwise direction along the deformation localization bands were still regarded as positive, and the extrusion displacements vertical to the localization bands were also considered positive. Through processing, the evolution curves of the dislocation and extrusion displacements of the symmetry points on both sides of deformation localization bands were shown in figure 6 and figure 7 respectively.

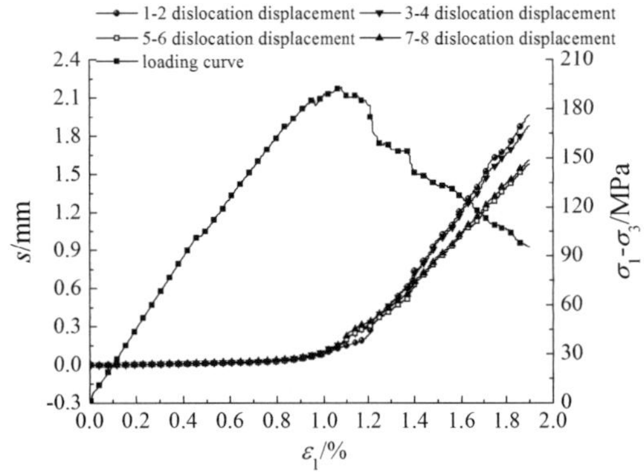


**Figure 6.** Evolution curves of the dislocation displacements of symmetrical points

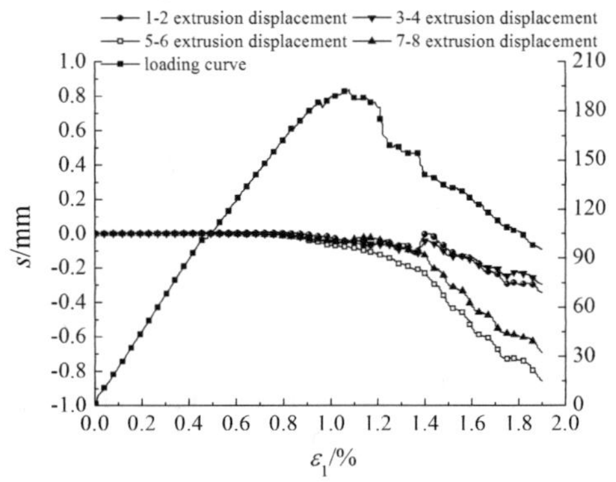


**Figure 7.** Evolution curves of the extrusion displacements of symmetrical points

As shown in figure 6, in the pre-peak stage of stress, the region P and region Q move along the clockwise direction of the deformation localization bands. The dislocation displacements of the symmetric points in each group exhibit linear evolution in the elastic stage, and starts to transite to the nonlinear evolution in the plastic stage. Because of the expansion, intersection and perforation of the cracks before the peak point, and the occurrence of dislocation displacements at the same time, the bearing capacity of the specimen begins to decrease. After the peak point of stress, the dislocation displacements of deformation localization bands increase gradually. And due to the sliding instability of the weak surfaces after peak point, the dislocation displacements increase sharply after the complete rupture between region P and region Q. At different positions of the same deformation localization band, the differences of dislocation displacements are not big. In figure 6, for deformation localization band M, the dislocation displacement between point 1 and point 2 is slightly larger than the one between point 3 and point 4. And for deformation localization band N, the dislocation displacement between point 5 and point 6 is slightly smaller than the one between point 7 and point 8.

In figure 7, region P produces a slight stretching in the direction vertical to the deformation localization bands relative to region Q in the pre-peak stage of stress and the stretching displacements also evolve from the linear evolution of elastic stage to the nonlinear evolution of plastic stage. In the post-peak stage, the stretching displacements of deformation localization bands increase gradually. Like dislocation displacements, the stretching displacements increase sharply after the complete rupture between region P and region Q. In the post-peak stage, the stretching displacements at different positions of the same deformation localization band are different. In figure 7, the stretching displacement between point 1 and point 2 differs little from the one between point 3 and point 4 in the deformation localization band M, but the stretching displacement between point 5 and point 6 is obviously larger than the one between point 7 and point 8 in the deformation localization band N.

Through the comprehensive analysis of figure 6 and figure 7, a conclusion can be obtained that in the post-peak stage of rock loading, along the same deformation localization band, the dislocation displacements close to both loaded ends are generally greater than the ones near to the middle position of the specimen, the stretching displacements close to both loaded ends are generally smaller than the ones near to the middle position of the specimen. However, due to the randomness of damage evolution, abnormal phenomena may occur. In figure 7, for example, the stretching displacement between point 1 and point 2 is first smaller and then slightly larger than the stretching displacement between point 3 and point 4 in the post-peak stage.

## 6. conclusions

The triaxial compression experiments of sandstone under different confining pressures were carried out firstly, and then based on the experiments, numerical simulation was conducted by using $PFC^{2D}$ under the same conditions. Through the comparative analysis of experimental results and numerical simulation results, a typical simulation curve and the corresponding simulation process were selected to study the evolution characteristics of the surface deformation field, the evolution characteristics of the velocity field and displacement field of deformation localization bands about sandstone under triaxial compression and the following conclusions were obtained.

(1) Through the selection of the typical identification points on the stress-strain curve and the comparative analysis with the damage images of corresponding points, the forming process of the deformation localization bands are obtained.

(2) The dislocation velocities and extrusion velocities of deformation localization bands are basically zero in the elastic stage, they have small fluctuations in the plastic stage and have larger fluctuations in the post-peak stage. In the same deformation localization band, the dislocation velocities are always in the same direction, but in the direction vertical to the localization band, the localization band sometimes squeezes and sometimes stretches. And at different positions of the same deformation localization band, the dislocation velocities and extrusion velocities are varied at the same time.

(3) The dislocation displacements and extrusion displacements of deformation localization bands exhibit linear evolution in the elastic stage, and starts to transite to the nonlinear evolution in the plastic stage. In the post-peak stage, the dislocation displacements and extrusion displacements continue to increase, and the values of both displacements increase sharply after the localization bands are completely penetrated.

(4) In the post-peak stage of rock loading, along the same deformation localization band, the dislocation displacements close to both loaded ends are generally greater than the ones near to the middle position of the specimen, the stretching displacements close to both loaded ends are generally smaller than the ones near to the middle position of the specimen.

## Acknowledgements

The authors are grateful to Natural Science Foundation of China (No. 50904071, 51274207) for the financial support to this work.

## References

[1] Zhao Z H and Xie H P 2008 *JOURNAL OF SICHUAN UNIVERSITY (ENGINEERING SCIENCE EDITION)* **40** 26–31
[2] Bieniawski Z T, Denkhaus H G and Vogler U W 1969 *International Journal of Rock Mechanics and Mining Sciences and Geomechanics Abstracts* **6** 323–41
[3] Pan Y S 1999 *Study on Rockburst Initiation and Failure Propagation* (Beijing: Tsinghua University)

[4] Zhang D M 2004 *Theoretical and experimental research on deformation localization and instability of rock* (Chongqing: Chongqing University)
[5] Song Y M, Ma S P, Yang X B and Wang X 2011 *Chinese Journal of Rock Mechanics and Engineering* **30** 170–75
[6] Mogi K 1968 *Bulletin of the Earthquake Research Institute Tokyo Imperial University* **46** 1103–25
[7] Wu L X, Liu S J, Wu Y H and Wu H P 2002 *International Journal of Rock Mechanics and Mining Sciences* **39** 825–31
[8] Hao S W and Sun J 2008 *JOURNAL OF EXPERIMENTAL MECHANICS* **23** 89–95
[9] Wong T F, Baud P and Klein E 2001 *Geophys. Res. Lett.* **28** 2521–24
[10] Song Y M, Jiang Y D, Ma S P, Yang X B and Zhao T B 2012 *Rock and Soil Mechanics* **33** 1352–56
[11] Yao T, Ren W, Que K S and Wang Y D 2012 *Soil Engineering and Foundation* **26** 70–73
[12] Yin C 2014 *Particle DEM Simulation for rock tests of direct tension and compression* (Chengdu: Southwest Jiaotong University)

IOP Conf. Series: Earth and Environmental Science **61** (2017) 012091 doi:10.1088/1755-1315/61/1/012091

# A study on application of squeezed branch pile in clay soil foundation

**Lianzhong Yin** [a*], **Xiaoguang Fan** [a], **Songjiang Wang** [a]
[a] Henan Electric Power Survey and Design Institute, Henan, China
[*]Corresponding author: lianzhong yin, senior engineer, Henan Electric Power Survey and Design Institute, +86 037167162007, yinlianzhong-heny@powerchina.cn.

**Abstract**: Taking a Power Plant Project as an example, the experimental study on clay soil foundation treatment with squeezed branch pile is carried out. The experimental data before and after soil foundation treatment were compared, through the pile capacity test, pile composite foundation soil bearing capacity and static load test, etc.. The theoretical calculated value difference between carrying capacity and the actual test results were obtained to verify the feasibility of Squeezed branch pile in composite foundation program, and get the Squeezed branch pile bearing capacity and other design parameters, which provides a reference for the optimization of engineering piles and the treatment of the thick clay soil foundation.

## 1 Introduction

Squeezed branch pile is a new type of pile developed on the basis of cast-in-place pile [1-2]. It has one or more branches and bearing plates in the pile body. This kind of pile can change the bearing mechanism of original friction pile by sub-bearing force plate and improve the bearing capacity of single pile, reduce the pile diameter and shorten the length of pile, effectively shorten the construction period and reduce material consumption [2]. Many scholars have studied the mechanism and influencing factors. Based on the concrete engineering, this paper has carried on the related experiment research of Squeezed branch pile in clay soil foundation, the effect of foundation treatment is verified by field load test.

## 2 Project Overview

The project is a power plant with installed capacity of 2 × 660MW. **B**ecause the thick clay layer on the upper part of the site cannot meet the requirements of the main building (structure) on the natural foundation, it is proposed to adopt the Squeezed branch pile for foundation treatment. Site soil characteristics are as follows:

layer① plain fill: brown yellow, pale yellow, the main ingredients are silty clay, containing a small amount of calcareous tuberculosis, occasional construction waste. Plastic, layer thickness is 2.40 ~ 3.30 m.

Layer② silty clay: brown yellow, brown, containing a small amount of manganese nodules and calcareous tuberculosis, hard plastic ~ hard, with moderate (low) compression. Thickness of this layer is 6.10～6.80m.

Layer②1 silty clay: Brown, containing manganese oxide and small calcareous tuberculosis, hard plastic, with moderate compression. Thickness of this layer is 1.00～2.80m.

Layer③ clay: brown, containing manganese oxide and small calcareous tuberculosis, hard, with moderate compression. Thickness of this layer is 6.90～7.20m.

Layer③1 clay mixed with calcareous tuberculosis, gravel: Brown, pale yellow , containing manganese oxide and calcareous tuberculosis, hard, with moderate compression. Thickness of this layer is 2.90～4.20m.

Layer④ clay mixed with calcareous tuberculosis: Brown, red brown, mixed 10% to 40% calcareous tuberculosis and gravel, hard, with low compression. The maximum exposure thickness is 10.50m.

The main physical properties and engineering properties of stratum are shown in Table 1.

**Table.1** Index of main physical properties and engineering properties of soil

| Item / Layer | Natural water content ω (%) | unit weight γ (kN/m³) | Natural void ratio e | Liquidity index $I_L$ | Coefficient of compressibility a ($MPa^{-1}$) | Modulus of compressibility Es (MPa) | Cohesion c (kPa) | Angle of internal friction φq φ (°) | SPT value N' (blow) N' (blow) | static point resistance $q_c$ (MPa) | sidewall resistance $f_s$ (kPa) |
|---|---|---|---|---|---|---|---|---|---|---|---|
| ① | 22.4 | 19.5 | 0.722 | 0.24 | 0.22 | 8.7 | 33 | 15.3 | 9 | 2.04 | 108.3 |
| ② | 22.2 | 20.0 | 0.671 | 0.05 | 0.11 | 16.0 | 52 | 17.0 | 28 | 5.24 | 232.9 |
| ②1 | 25.2 | 19.7 | 0.734 | 0.28 | 0.17 | 9.7 | 47 | 16.2 | 20 | 4.06 | 106.8 |
| ③ | 22.4 | 20.2 | 0.667 | -0.06 | 0.08 | 21.7 | 91 | 18.3 | 40 | 5.75 | 346.0 |
| ③1 | 21.3 | 20.1 | 0.654 | -0.07 | 0.09 | 18.0 | 68 | 19.3 | 52 | / | / |
| ④ | 23.1 | 19.8 | 0.705 | 0.16 | 0.20 | 12.3 | 36 | 10.9 | 68 | / | / |

## 3 Test Design

A total of three test pile (No. S1 ~ S3), each pile have two disks and two branches pile ,diameter d = 700mm, disk diameter D = 1400mm. Using the cross branch, branch width is 300mm, branch diameter D = 1400mm. Pile length L = 17.30m, the bearing plate is located at the layer ③ and the top of the layer ④. The layout of test piles is shown in Fig. 1.Main reinforcement of the reinforcing cage test pile are 14 bars 16mmHRB335 rebar, Main reinforcement are all reinforced by reinforcing bars. Spiral stirrups are the 8mm HPB235 rebar, spacing 200mm, within 3 meters near the top of the pile encryption to 100mm. Pile concrete grade using C40, reinforced protective layer thickness is not less than 50mm.

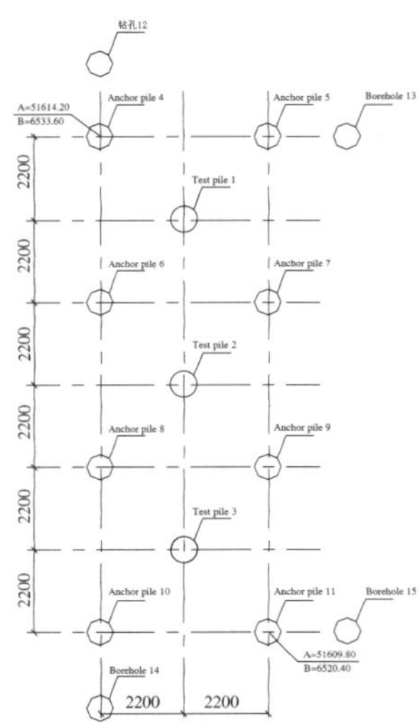


**Fig. 1** layout of test piles

## 4 Test results and analysis

In order to understand the bearing capacity characteristics of squeezed branch pile, the working condition of soil between pile and pile, the vertical compressive static load test of single pile and horizontal static load test of single pile are carried out. Tests were carried out at representative sites of the foundation soil.

### *4.1 Vertical compressive static load test of single pile*

Three test pile Q-s curves is shown in Fig.2 ~ Fig.4.

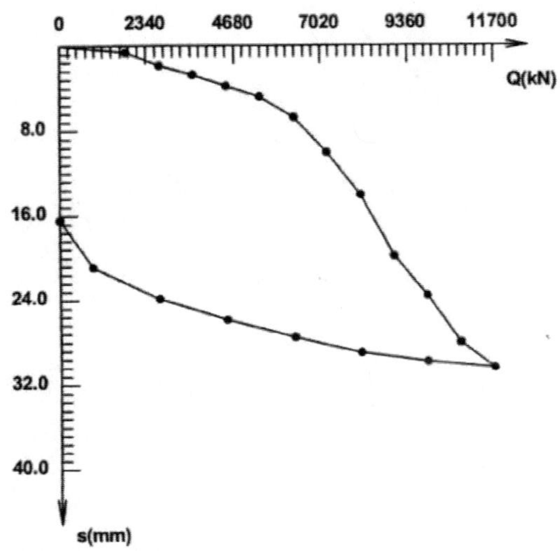


**Fig.2** The Q-s curve of test pile S1

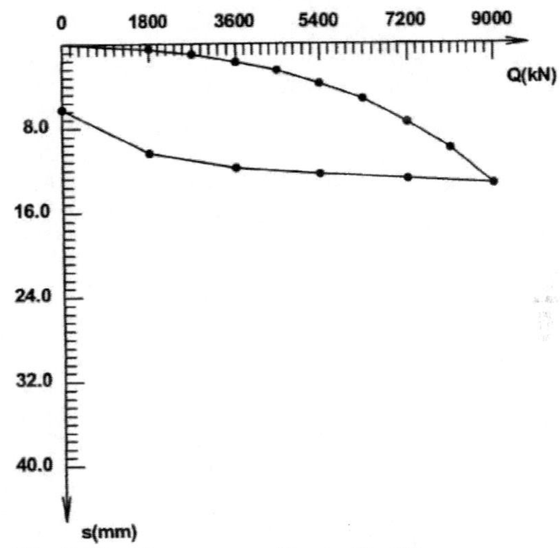


**Fig.3** The Q-s curve of test pile S2

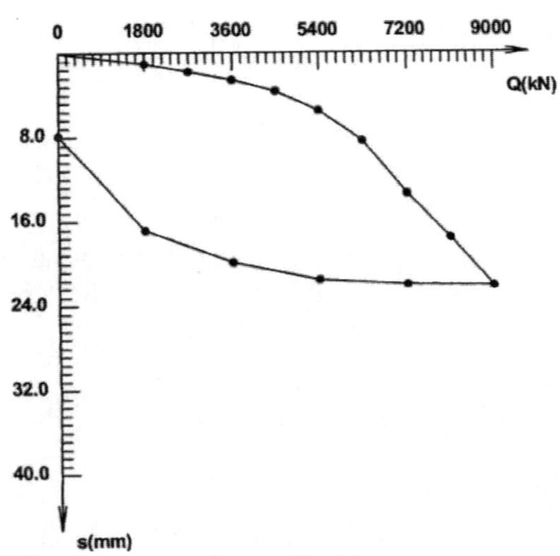


**Fig.4** The Q-s curve of test pile S3

The axial force distribution curves of the three test piles under each load are shown in Fig. 5 ~ Fig.7

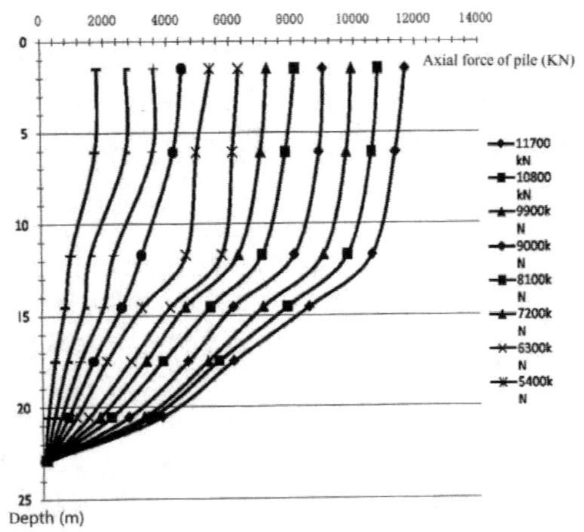


**Fig. 5** The axial force distribution curves of the test pile S1 under each load

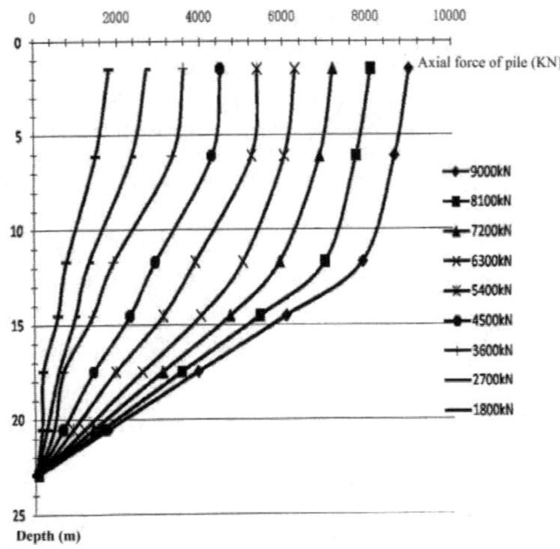


**Fig. 6** The axial force distribution curves of the test pile S2 under each load

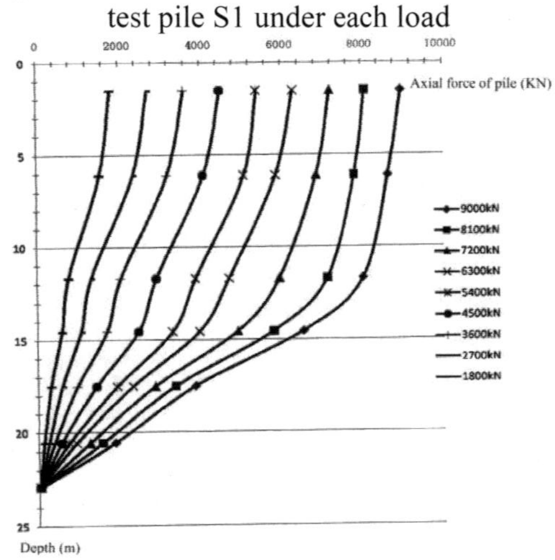


**Fig. 7** The axial force distribution curves of the test pile S3 under each load

According to "Technical code for testing of building foundation piles", comprehensive analysis of pile stress test results and single pile vertical compressive static load test results, vertical bearing capacity values of single pile can be obtained, the results in the following table 2.

**Table.2** the test results of vertical bearing capacity of single pile at base elevation

| Number of test piles | Ultimate bearing capacity of basement elevation(kN) | Statistical value of ultimate bearing capacity(kN) | Characteristic value of bearing capacity (kN) | Explain |
|---|---|---|---|---|
| S1 | 11361 | 9556 | 4778 | The test was carried out at the natural ground, the ultimate bearing capacity of the base elevation measured by the second layer of the stress meter |
| S2 | 8641 | | | |
| S3 | 8667 | | | |

According to the geological data of test pile area, single pile vertical compressive load test results, pile shaft and tip of pile stress test results, overall consideration, ultimate shaft resistance and ultimate tip resistance of the soil around the pile are obtained, the results in the following table 3.

**Table.3** the ultimate shaft resistance and ultimate tip resistance of the soil around the pile

| Layer no. | Layer ① | Layer ② | Layer ②$_1$ | Layer ③ | Layer ③$_1$ | Layer ④ |
|---|---|---|---|---|---|---|
| Standard value of ultimate shaft resistance $q_{si}$(kPa) | 25 | 80 | 82 | 90 | 95 | 105 |
| Standard value of ultimate tip resistance $q_{pk}$(kPa) | / | / | / | 1475 | 1616 | 1955 |

*4.2 Horizontal static load test of single pile*

Horizontal static load test curve of three test pile shown in Fig.8 ~ Fig.10.

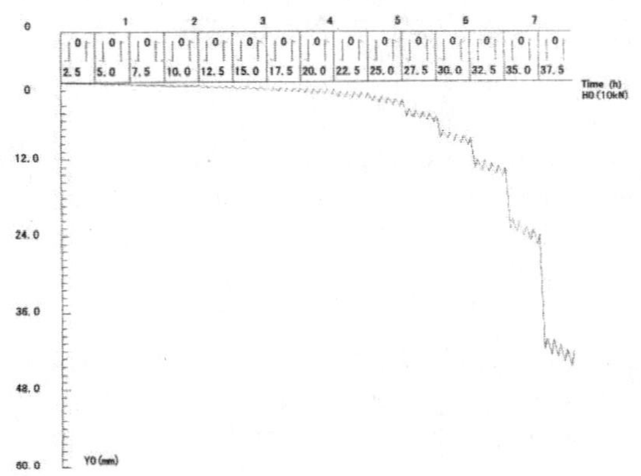


**Fig.8** Horizontal static load test curve of test pile S1

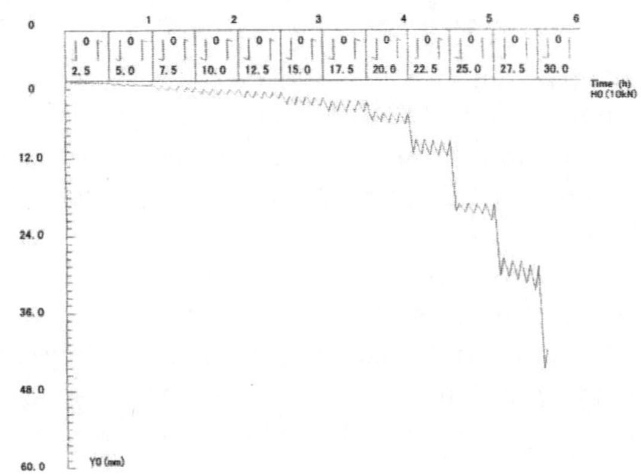


**Fig.9** Horizontal static load test curve of test pile S2

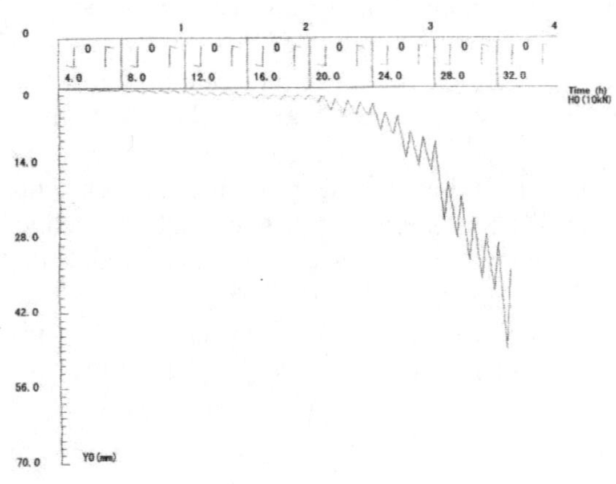


**Fig.10** Horizontal static load test curve of test pile S3

According to "Technical code for testing of building foundation piles", comprehensive analysis of single pile horizontal static load test results, single pile horizontal bearing capacity values can be obtained, the results in the following table 4.

**Table.4** the single pile horizontal static load test results

| Number of test piles | Ultimate bearing capacity (kN) | Critical load (kN) | Statistical value of ultimate bearing capacity (kN) | Statistical value of Critical load (kN) | Explain |
|---|---|---|---|---|---|
| S1 | 325 | 200 | 200 | 150 | The test was carried out at a natural ground level of 2.0 m (corresponding elevation 156.50 m) |
| S2 | 200 | 150 | | | |
| S3 | 200 | 160 | | | |

## 5 Conclusion

In summary, the following conclusions can be drawn:

1) From the test results and engineering practice, it is effective to treat the clay soil foundation with the squeezed branch pile. And the requirements of building (structure) to the foundation can be met.

2) The results of field loading tests show that the squeezed branch pile has obvious strengthening effect to the clay soil layer within the treatment range, the engineering characteristics are improved and the bearing capacity is improved.

3) According to the results of field loading tests, when using standard estimate the bearing capacity of single pile, the selection of parameters should be combined with geotechnical test, in situ test and on-site description, so as to ensure the accuracy of the estimation results.

## References

[1] *Zhang Yan, Chen Pei, Zhao Zhenxin.* Experimental study on squeezed branch pile foundation in soft soil ground. [J].Chinese Journal of Geotechnical Engineering, 2013, S2:994-997.

[2] Lu Chengyuan, Li Hanjie. Experimental study on bearing capacity of the disk-pile group under uplift load. [J]. Journal of Zhejiang University of Technology, 2014, 03:298-301.

[3] Wang Yili, Xu Liangying, Li Biqing, Xu Changjie. Finite element numerical study on the axial bearing behaviors and factors of squeezed branch pile. [J]. China Civil Engineering Journal. 2015, S2:158-162.

[4] Lu Chengyuan, Yan Jianhua. Experimental research on horizontal bearing performance of clustered branch piles with different spacing. [J]. Architecture Technology. 2013, 04:373-377.

[5] Ke Jiang. Analysis of Anti-Pulling Behaviors of Pile with Branch Plates. [J]. Chinese Journal of Underground Space and Engineering. 2013, 01:61-65.

[6] Zhao Zhenxin, Zhou Wei, Zhang Yan, Li Xu, Li Zhanling. Bearing Capacity Calculation and Test of Squeezing Branch Pile in Transmission Line. .[J].Electric Power Survey & Design. 2014, 01:52-56.

IOP Conf. Series: Earth and Environmental Science **61** (2017) 012092 doi:10.1088/1755-1315/61/1/012092

# Research on dynamic collaborative optimization of passenger flow line and waiting resources

**Ying Li[1] , Changfeng Zhu, Zhengkun Zhang**

[1]School of Traffic and Transportation, Lanzhou Jiaotong University, 730070Lanzhou, China

Email: 13919091569@163.com

**Abstract.** In order to improve the resource allocation efficiency of waiting hall, the related factors is discriminated. Based on the discrimination, a dynamic programming model for collaborative optimization about the passenger flowline and waiting halls resource allocation is built by considering the complexity, dynamics and equilibrium of Passenger flow, as well as the cross interference. Besides a progressive optimality algorithm is put forward to solve the model. Finally, a case study is presented to verify the rationality of the model and algorithm. Results show that this model and its optimization algorithm above-mentioned can improve the resource allocation efficiency of waiting hall.

## 1. Introduction

Passenger flow line and waiting resources play an important role in railway passenger station, and it has become a current topic in recent years, therefore, some scholars make lots of research about optimization design of passenger flow line and waiting resources.

Cui et al. studies the passenger organization and analyzes the characteristics about passenger flow line in integrated traffic hub station [1]. Zhou et al. analysis the passenger interaction factors during peak period for urban rail transit station [5]. Cui et al. uses maximum entropyprinciple to research the passenger streamlines about subway transfer station [1]. Congling et al. studies the streamline in metro station by considering the safety strategy during evacuation [3]. Hu et al. evaluates the passenger streamline designation by using micro simulation program.

According to the existing literatures, researches about passenger flow line and waiting resources have got achievements in some degree. However, there is a downside that the passenger flow line and waiting resources are researched respectively, and the optimization results are not comprehensive. Based on this shortcoming, the paper comprehensive studies two questions together from the perspective of dynamic coordination optimization.

## 2. Analysis of Passenger Waiting Characteristics

Suppose that $t_i$ is the departure time of $I_i\,(i=1,2,3,...,n)$, the origin trains and passing by trains is $I=\{I_1,I_2,I_3,...,I_n\}$, the set of waiting resource is $J=\{J_1,J_2,J_3,...,J_m\}$ , $S_j$ is the capacity of waiting resource $J_i$, the set of platforms is $Z=\{Z_1,Z_2,Z_3,...,Z_R\}$, $I_p^f$ is the subset of trains $I_i$ in $p$ period.

Suppose that $t_{ij}^s$ is the beginning time of waiting resource occupied, $t_{ij}^s=t_i-T$ . Then

$$T_{ij}^e=(t_i-\Delta T_i)-(t_i-T) \quad (1)$$

Suppose that $g(t)$ is the density function of passenger arrival. Then

$$g(t)=\begin{cases}\frac{1}{\sqrt{2\pi}\delta t}\exp\frac{-(\ln t-\mu)^2}{2\delta^2} & x\geq 0\\ 0 & x<0\end{cases} \tag{2}$$

$$\int_0^T g(t)dt=1 \tag{3}$$

The statistic distribution of passenger waiting time is shown as Figure 1.

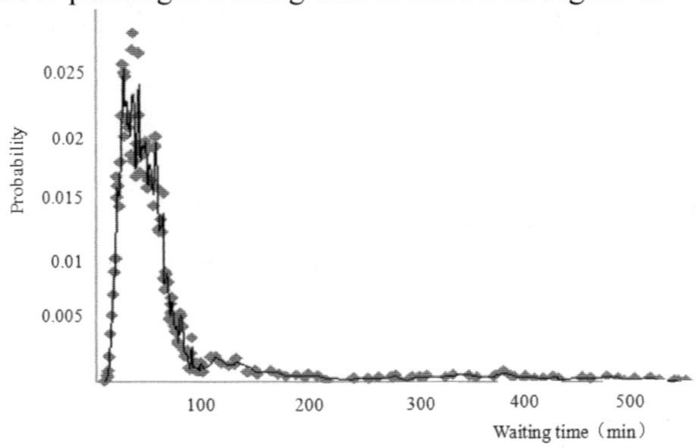


**FIGURE 1.** Statistic Distribution of Passenger Waiting Time

## 3. Establishing Model

Suppose that $S_k=\{s_1,s_2,...,s_k,...s_K\}$ is the state set of waiting resource allocation for the given passenger station, $f_k(s_k)$ is the optimal function value of the *k*-th phase. Then

$$f_k(s_k)=\underset{x_k}{Opt}\{r_k(s_k,x_k)+f_{k+1}(s_{k+1})\} \tag{4}$$

For any two trains $I_a$ and $I_b$, the departure time at the given passenger station is $t_a$ and $t_b$ respectively, the departure time interval of $I_a$ and $I_b$ is $T_{ab}=|t_b\text{-}t_a|$ ,the beginning time of distributing waiting resource is $t_a^s$ and $t_b^s$ respectively, the ending time of distributing waiting resource is $t_a^e, t_b^e$ respectively . Then

$$(t_a^e-t_b^e)(t_a^s-t_b^s)>0 \tag{5}$$

Define 0-1 variable $x_k$ as

$$x_k=\begin{cases}1 & \text{occuping waiting hall for passeenger of train } k\\ 0 & \text{other}\end{cases}$$

Then the minimum waiting resource required is denoted as

$$\min \quad Z=\sum_{k=1}^{n}x_k \tag{6}$$

Suppose that $D=\{D_1,D_2,...,D_k,...D_K\}$ is the phase set of waiting resource allocation for the given passenger station, $I_k^f$ is the trains set to be assigned in the *k*-th phase for the given passenger station. $N_i^k$ is the estimated number of passengers which get on train $I_i$ at the given passenger station. Then

$$N_i^k=\begin{cases}N_i\cdot\int_8^{t-T_k+8}g(t)dt & t_i\geq T_k\\ 0 & t_i<T_k\end{cases} \tag{7}$$

Define decision variation $x_{ij}^k$ as

$$x_{ij}^k=\begin{cases}0 & I_i\notin I_k^f\\ 1 & I_i\notin I_k^f, N_i<S_j^k\end{cases} \tag{8}$$

IOP Conf. Series: Earth and Environmental Science **61** (2017) 012092 doi:10.1088/1755-1315/61/1/012092

Suppose that $N_{ij}^{k}$ is the conversion passenger's number of waiting resource allocation $J_j$ for the given passenger station, $\eta$ is conversion factor which consider the passenger luggage. Then

$$N_{ij}^{k}=\sum_{i=1}^{n}N_{i}^{k}\eta x_{ij}^{k}$$

Suppose that $N_{aver}$ is the number of estimated waiting passengers for any waiting resource allocation $J_j$. Then

$$N_{aver}=\sum_{j=1}^{m}\sum_{i=1}^{n}\sum_{k=1}^{\Gamma}N_{ij}^{k}/m$$

Suppose that $L=\{l_{11},l_{12},...,l_{jr},...l_{mR}\}$ is the set of passenger flow lines for the any given passenger station(See figure 2), $A$ is the entrance of waiting resource allocation $J_j$, $G$ is the platform.

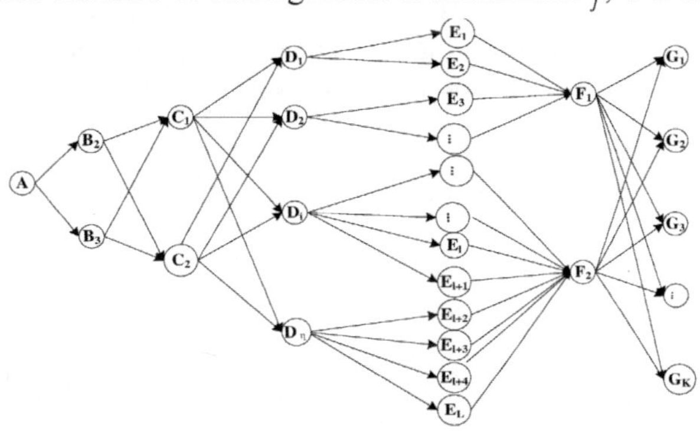


**FIGURE 2.** The Passenger Line in the Any Given Passenger Station

Suppose that $L_{zr}$ is the total size of the passenger flow lines for the any given passenger station. As shown from the figure 2, the total size of the passenger flow lines $L_{zr}$ is proposed as

$$L_{zx}=\sum_{j=1}^{m}\sum_{r=1}^{R}l_{jr}$$

$S_k=\{S_1^k,S_2^k,....,S_j^k,...S_m^k\}$ is the set of state variable for the any given passenger station. $I_k^f\in I$ is the set of departure trains for the any given passenger station .

The state transformation equation of collaborative optimization is proposed as

$$\{S_{ik}\}=\left\{S_{i(k-1)}-\sum_{i=1}^{n_i}N_i^k\eta x_{ij}^k+\sum_{i=1}^{n_{i-1}}N_i^{k\text{-}1}\eta x_{ij}^{k\text{-}1}\right\} \qquad (9)$$

Index of recursion equations of passenger flow line and waiting resources is proposed as

$$\begin{cases} f_k(S_k)=\min\{v_k+f(S_{k-1})\} \\ f_k(S_0)=0 \end{cases} \qquad (10)$$

$$S_{i,k}=d_k(S_{i,k\text{-}1},S_{i,k-2},...,S_{i,k+1-m},x_{ij}^k) \qquad (11)$$

Then, Collaborative optimization objective function is transformed as

$$\min v_k=\{\ (\sum_{i=1}^{n_i}(N_i^k\eta(g_k(S_{k+1},S_k,S_{k-1},...S_{k+1-n}))\text{-}\sum_{j=1}^{m}\sum_{i=1}^{n_i}\sum_{k=1}^{\Gamma}N_{ij}^k/m)^2+(\sum_{r=1}^{R}\sum_{j=1}^{m}\sum_{i=1}^{n_i}N_i^k\eta(g_k(S_{k+1},S_k,S_{k-1},...S_{k+1-n}))\ l_{jr}\mu_{jr}\Big/\sum_{j=1}^{m}\sum_{r=1}^{R}l_{jr})\}$$

## 4. Algorithm Design

Step1: Given any original solution $S_k^0\in X_1, k=2,...,\Gamma$, $S_0=(S_1,S_2^0,...,S_\Gamma^0)$.

Step2: $G_1^1,...G_{\Gamma-1}^1)$ is computed as the following equation

$G_1^1 = \min_{S_2 \in X_2}\{(g_1(S_1,S_2) + \sum_{k=2}^{\Gamma} g_k(S_1,S_2,S_3^0,...S_{k+1}^0)\}$, $G_{\Gamma-1}^1 = \min_{S_{\Gamma-1} \in X_{\Gamma-1}} \{(g_{\Gamma-1}(S_1,S_2^1,...,S_{\Gamma-1}^1,S_\Gamma)$

Step3: $G_1^2,...G_{\Gamma-1}^2)$ is computed as the following equation

$G_1^2 = \min_{S_2 \in X_2} m\{(g_1(S_1,S_2) + \sum_{k=2}^{\Gamma} g_k(S_1,S_2,S_3^1,...,S_{k+1}^1) + \sum_{k=2}^{\Gamma} g_k(S_1,S_2,S_3^1,...,S_{k+1}^1)\}$, $G_{\Gamma-1}^2 = \min_{S_\Gamma \in X_\Gamma} g_{\Gamma-1}(S_1,S_2^2,...,S_{\Gamma-1}^2,S_\Gamma^1)$

If the initial optimal solution $S_k^2(k=2,...,\Gamma,)$ is got, then the total series of solutions for the each phase $S^2$ is also got, namely, $S^2=(S_1,S_2^2,S_3^2,...,S_\Gamma^2)$.

Step4: According to $S^n=(S_1,S_2^n,S_3^n,...,S_\Gamma^n)$, then, $S^{n+1}=(S_1,S_2^{n+1},S_3^{n+1},...,S_\Gamma^{n+1})$ can be got.

## 5. Case Analysis

In this case, suppose that there are four waiting hall. The number of passenger trains and its departure time distribution is presented as figure 3, and the passenger waiting rate for every moment is presented as figure 4. The arrival time and departure time of each train is presented as figure 5.

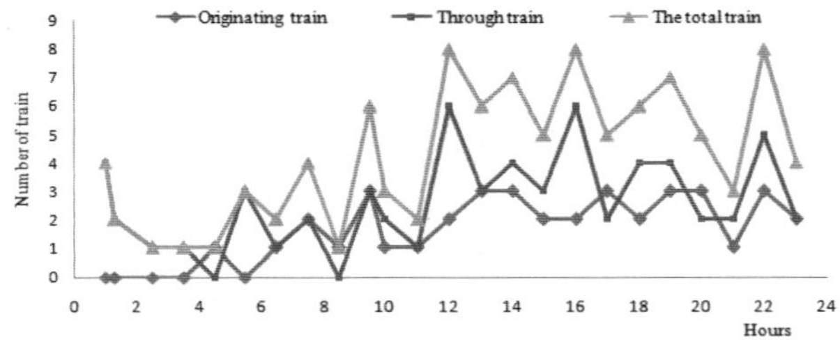


**FIGURE 3.** The Number of Passenger Trains and Its Departure Time Distribution

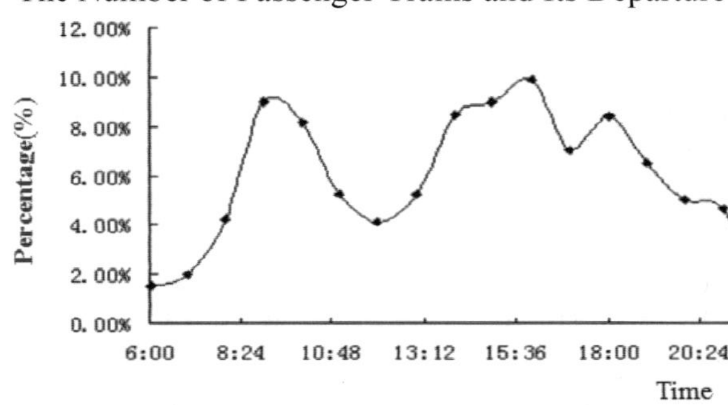


**FIGURE 4.** Passenger Waiting Rate for Every Moment

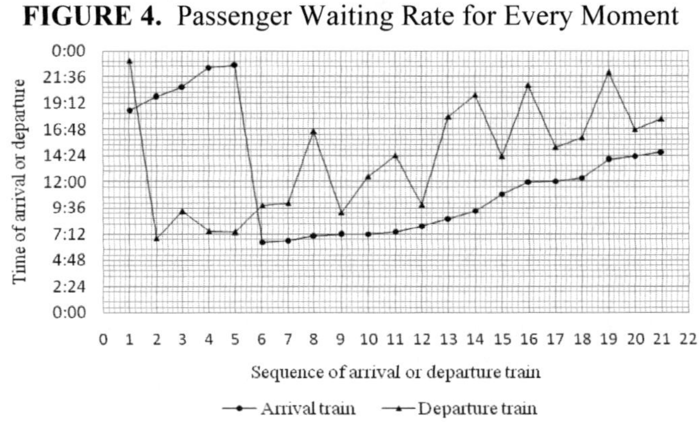


**FIGURE5.** The Arrival Time and Departure Time of Each Train

**Table1.** Optimization Result

| Waiting room | number of waiting passenger in one day | Waiting room | number of waiting passenger in one day |
|---|---|---|---|
| No.1 | 5255 | No.3 | 5902 |
| No.2 | 5637 | No.4 | 5320 |

This optimal result indicates that the model used by the main passenger train station can reach the balance between the waiting room and the shortest passenger line.About mixed stations or stations with sub-end layouts and uninterrupted passenger flow lines, We can use this model in the paper to find a more efficient result.

## 6. Conclusions

For shortage of previous prediction method of emergency supplies, the correlation between properties of emergencies is considered in this paper. Theory of fuzzy rough sets is used to reduce attributes so that redundant attributes of emergencies can be excluded effectively. After reduction of attributes, dimension of attributes can be reduced, thus, the retrieval efficiency can be improved. Furthermore, using the theory of fuzzy rough sets, weight of the main attributes can be distributed, making it more objective to determine the degree of importance of attributes. Thus, the prediction accuracy is improved. Rationality of the method is verified by the experimental results. And it is more efficiently to find out the similar case, and storage space is saved.

## Acknowledgment

This work is supported by the Nature Science Foundation of Gansu Province (No. 1508RJZA061).

## References

[1] Cui Huawei and Jia Junfang, "Research on Streamline Characteristic and Organization of integrated Traffic Terminal for Railway Passenger Transport",*Railway Transport and Economy,* 2007,29(5):26-29.
[2] Qi Kai, "Model of Metro Interchange Station Passenger Line Optimization Based on Maximum Entropy Principle",Logistics Technology, 2011,30(10):87-90.
[3] Shi Congling and Zhong Maohua, "Modeling and Safety Strategy of Passenger Evacuation in a Metro Station in China",Safety Science, 2010,9.
[4] Hu Mingwei and Xie Xiaodong, "Based on Micro Simulation Railway Passenger Passegers Streamline Design Evaluation",Railway Economics Research,2009(3):33-37.
[5] Hou Yanfang and Zhou leishan, "Study on the Influence of Peak Passenger Flow on Urban Rail Transit", Integrated transport, 2010(6):67-70.
[6] Zhu Changfeng and Li Yinzhen, "Collaborative Optimization on Assignment of Passenger Train Stock Based on Given Train Diagram",China Railway Scienc, 2014,35(3):120-128.
[7] Xia Shengli and Yang Hao, "Flow Path Design Theory of Railway Passenger Integrated Transport Hub",China Transportation Review, 2015,37(9):34-40.
[8] Xiao Yifan and Zhu Changfeng, "Railway Train Set Assignment Based on Simulated Annealing Algorithm",Journal of Railway and Engineering, 2012,9(5):108-113.
[9] Anghinolfi Dand Paolucci M,"Freight Transportation in Railway Networks with Automated Terminals",European Journal of Operational Research, 2011,214(3):588-594.
[10] BooS.T, "Dual Detection of Chromosomes and Microtubules by the Chromosomal Passenger Complex Drives Spindle Assembly",Dcvclopmental Cell, 2010,18(6):903-912.

# Earthquake damage of self-built reinforced concrete frame structures in $M_S$ 8.1 Nepal Earthquake

**Peilei Yan and Baitao Sun**

China Key Laboratory of Earthquake Engineering and Engineering Vibration, Institute of Engineering Mechanics, China Earthquake Administration, Harbin 150080 china

Corresponding author: yanpeilei325@163.com

**Abstract.** An earthquake measured Ms 8.1 occurred near the Nepal Pokhara on April 25, 2015(Beijing time). As a member of China to Nepal disaster assessment team, the author conducted thorough investigation about the earthquake damage of all kinds of engineering structure in Nepal. Self-built reinforced concrete frame structure which is widely accepted by the people at all levels in Nepal and is one of the building structure forms which are widely built suffered damage by varying degrees in the strong earthquake and aftershocks. Based on the seismic damage investigation date of the earthquake and its aftershocks, the characteristics of aseismic construction are summed up, the damage characteristics and the reasons of its destruction are analyzed. The main problems in the design and construction of the structure are pointed out. The advices on the existing problem of self-built reinforced concrete frame structures are given.

## 1.Introduction

An earthquake measured $M_s$ 8.1 occurred near the Nepal Pokhara (N28.2 degrees, E84.7 degrees) at 14:11 on April 25, 2015(Beijing time ). The focal depth is 20Km. In Nepal and parts of other countries like India, Bangladesh, China and Pakistan, people could feel clearly. Some counties in Tibet,China like Nielamu County, Tingri County, Geelong County, could feel very strongly.

The author investigated self-built reinforced concrete frame structure deeply in different earthquake affected areas of Nepal. Seismic design ideas weren't considered into the construction of most self-built reinforced concrete frame structure, nor were the standardized construction methods. In the earthquake and strong aftershocks, self-built reinforced concrete frame structure were subjected to various degrees of damage. Particularly in the high intensity region, even collapse happened.

This paper will summarize the structure and disaster characteristics of reinforced concrete frame structure in Nepal. We will also analyse the weak link existing in structure and give some proposals to improve the design and construction. The research can help us understand the seismic capacity of building structure in Nepal preliminary, so that the Chinese government can help Nepal to carry out reconstruction work targeted.

## 2. Seismic Characteristics Of Self-built Reinforced Concrete Frame Structure

Self-built reinforced concrete frame structure are mainly 3-5 layers (Fig. 1).The top of general structure are ready to be followed by half layer of reinforced concrete columns and exposed steel cage(Fig. 2).Storey height is 3.0m.Layout is symmetric regularly.Column space is from 3m to 3.9m.

IOP Conf. Series: Earth and Environmental Science **61** (2017) 012093 doi:10.1088/1755-1315/61/1/012093

The structure use independent column foundation; the size of frame column section is 230mm×230 mm ;longitudinal bar is 6A12 and stirrup is A6@120,without reinforcement of stirrup encryption in frame columns (Figure 3); the size of frame beam section is 300mm×230 mm; the majority of beam section's size is greater than the column section's size,violating the specification requires for a "strong column, weak beam"and component section with minimum size concept design principles of all countries in the world ; floor and roof panels use cast-in-place reinforced concrete floor slab, of which the thickness is 100mm; infilled walls use solid brick (220mm×100mm×50 mm) and mud masonry; the outer wall thickness is 220mm, the inner wall thickness is 100mm;there is no construction rachel measures between the wall and the frame column(Figure 4).

**Fig.1** Overall picture of self-built reinforced concrete frame structure

**Fig.2** Reserved reinforced concrete column and steel cage on the top floor

IOP Conf. Series: Earth and Environmental Science **61** (2017) 012093 doi:10.1088/1755-1315/61/1/012093

**Fig. 3** Reinforcement of frame column

**Fig.4** Structural Rachel measures between infilled walls and frame columns are not set

## 3.Analysis on The Characteristics And Causes Of Seismic Damage Of Self-built Reinforced Concrete Frame Structure

Self-built reinforced concrete frame structure is a better structure relatively in the earthquake affected areas of Nepal. The damage features are: the structures were virtually undamaged in the area of VI degree; slight diagonal cracks and cross cracks occurred in infilled walls in the area of VII degree; a vast of severe diagonal cracks and cross cracks occurred in the infilled walls as well as a small quantity of the stigmas of frame bottom columns was damaged seriously inVIIIdegree;the infilled walls were damaged seriously and the construction collapsed partial ;some frame bottom columns was damaged seriously andthe construction even collapsed in the area of IXdegree.

### *3.1. Damage of Infilled Walls*

In VI zone, there were no obvious damage in infilled walls. In VII zone, the bottom infilled walls run through the inclined cracks or cross cracks (Figure 5). In VIII degree, the bottom infilled walls run through the transparently inclined cracks or cross cracks (Figure 6).In IXdegree, the bottom infilled walls collapsed partially, and were damaged heavily (Figure 7).Using mud masonry in the infilled wall lead to the low level of walls' shear strength. Under the action of reciprocating horizontal earthquake, a large number of shear cracks or cross cracks happened in the infilled walls; at the same time, the infilled walls had external flash, or even collapsed in high intensity area because there were no reinforcement structure filled between the walls and the frame columns.

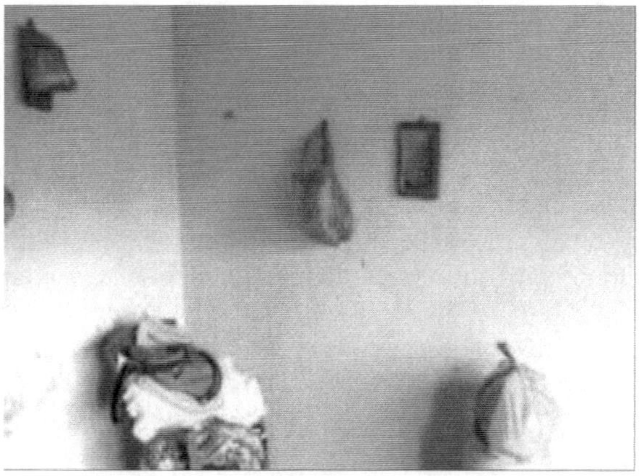

**Fig.5** Destruction of infilled walls in the region of VII degrees

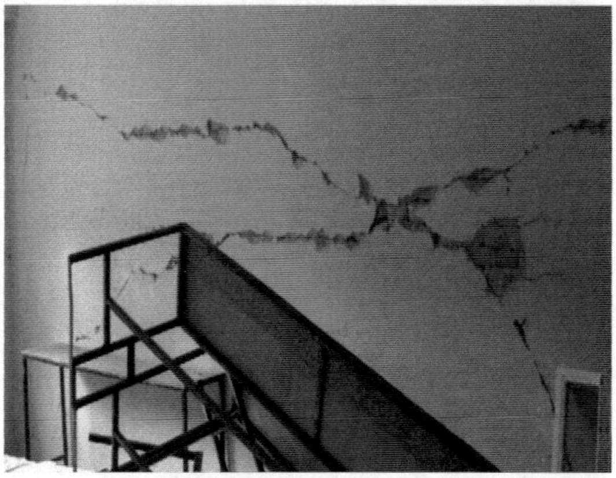

**Fig.6** Destruction of infilled walls in the region of VIII degrees

**Fig.7** Destruction of infilled walls in the region of IX degrees

*3.2. Damage of Frame Column*

In VIII degree area, shear failure occurred at the top of a small quantity of frame bottom columns (Figure 8); In IX degree area, shear failure occurred at the top of some frame bottom columns, shear failure occurred at the middle of a small quantity of frame bottom columns (Figure 9). In self-built reinforced concrete frame structure, the size of beam section is generally greater than that of the column section and the reinforcement of reinforced beams are often stronger than the column, resulting in the formation of "strong beam weak column" mode in the structure [1-3].At the same time, the lack of joint hoop reinforcement near the beam and column node area [4-6] lead to less reinforcement brittle damage on the top of column more easily. At the same time, there is an inherent weak link in the top of the column because most of the frame columns were poured directly to the bottom of the beam during construction. It increased the destruction of the top of the column under the horizontal earthquake reciprocating action, eventually leading to shear failure occurred on the top of the column.

**Fig.8** Destruction of the top of the underlying columns in the region of VIII degrees

**Fig.9** Destruction in the middle of the bottom frame columns

*3.3. Collapse of Self-built Reinforced Concrete Frame Structure*

In IXdegree, the disappearance of bottom layer occurred on parts of self-built reinforced concrete frame structure.The research found that most of the bottom columns suffered severe shear failure in the main shock. In the following 3larger aftershocks,the damage on the top of columns continued to increase until the underlying structure disappeared(Figure 10).

IOP Conf. Series: Earth and Environmental Science **61** (2017) 012093 doi:10.1088/1755-1315/61/1/012093

**Fig.10 Collapse of the self-built reinforced concrete frame structure**

### *3.4. Damage Caused by Collision*

Due to the limited land resources, the density of buildings in Nepal's towns is extremely high and the spacing between adjacent buildings is extremely small. And the design and construction of self-built reinforced concrete frame structure has not been effectively supervised by the government. The above two reasons lead to obvious differences of dynamic performance of adjacent structures, adjacent buildings collide and the structure is damaged in the earthquake. (Figure 11)

Parts of the self-built reinforced concrete structure were built along the road. Structure were close to the road subgrade or connected by rigid connection with the road subgrade. Because the dynamic performance of the structure and subgrade is different, the two collide with each other in the earthquake, resulting in serious structural damage and even collapse. (Figure 12).

**Fig.11** Collision damage to adjacent structures

IOP Conf. Series: Earth and Environmental Science **61** (2017) 012093 doi:10.1088/1755-1315/61/1/012093

**Fig.12** Collision damage between self-built reinforced concrete frame structure and highway roadbed

**4.Some Points of View**

The paper summarize the characteristics of earthquake damage and aseismatic construction of reinforced concrete frame structure built by themselves based on the investigation of 8.1 earthquake in Nepal. The paper indicates the following things:

(1) Self-built reinforced concrete frame structure is one of the housing structures which are widely recognized and built by all sectors of the Nepalese people. Due to the nature of private land ownership and the lack of local government supervision, the ideas of aseismic design has not been considered and the normative construction methods have not been used in self-built reinforced concrete frame structure.

(2) Self-built reinforced concrete frame structure suffered destruction in different degrees in the earthquake and strong aftershock. The structures are virtually undamaged in the area of VI degree and slight diagonal cracks and cross cracks occurred in infilled walls when in the area of VII degree. A vast of severe diagonal cracks and cross cracks occurred in infilled walls as well as a small quantity of the stigmas of frame bottom columns was damaged seriously in the VIIIdegree. The infilled walls were damaged seriously and the construction collapsed partial as well as some of the stigmas of frame bottom columns was damaged seriously and the even the construction collapsed in the area of IXdegree.

(3) The primary design problems existed in Self-built reinforced concrete frame structure in Nepal are followings. The size of beam section is generally greater than that of the column section and the reinforcement of reinforced beams are often stronger than the column resulting in the formation of "strong beam weak column" mode in the structure. Under this mode, destroy occurs in the frame column before frame beam under large earthquake. The lack of joint hoop reinforcement near the beam and column node area leads to shear failure on the top of column more easily.

(4) The primary problems of constructions existed in Self-built reinforced concrete frame structure in Nepal are followings. The concrete was cast to the bottom of the beam when the frame column was constructed which lead to congenital aseismatic defects at the top of the column. Using mud masonry in the infilled wall leads to the low level of walls' shear strength and integrality of structure. Under the action of the earthquake there is the possibility that the wall are pulled off from the main structure, external flashed or collapsed because there were no reinforcement structure filled between the walls and the frame columns.

(5) In view of the existing problems of self-built reinforced concrete frame structure, it is proposed to strengthen the supervision and review of the new reinforced concrete frame structure by the government departments of Nepal.The cross-sectional area of the frame column should be strictly ensured in the structural design, and enough stirrups encryption area should be set up in the beam - column joint area at the same time. We should ensure that the self-built frame structure is a "strong column and weak beam" ductile structure, and the frame column should not be poured directly to the

bottom of the beam during construction.Cement mortar masonry should be used in the infilled wall. And at the same time, enough construction rachel measures should be ensured between the infilled wall and the main structure.

**Acknowledgments**
Yan Li (1982) male, Ph.D., associate professor, mainly engaged in structural seismic and disaster research, Sun Baitao (1961) male, PhD, researcher, mainly engaged in structural seismic and disaster research. Fund project: supported by Scientific Research Fund of Institute of Engineering Mechanics, China Earthquake Administration（Grant No. 2014B11）, the National Natural Science Foundation of China (Grant No. 51508531) and Program for Innovative Research Team in China Earthquake Administration

**References**
[1] Ye Lieping, Qu Zhe, Ma Qianli, Lin Xuchuan, Lu Xinzheng, pan Peng. Talk about how to realize the strong column weak beam yield mechanism from the Wenchuan earthquake frame structure damage [J]. structure: 2008,38 (11): 52-67
[2] Ye Lieping, Ma Qianli, Miu Zhiwei. Study on [J]. engineering mechanics of reinforced concrete frame structure strong column and weak beam design methods:, 2010,27 (12): 102-113
[3] DimitriosG. Lignos, A. M. ASCE, Tsuyoshi Hikino, Yuichi Matsuoka, Masayoshi Nakashima, M. ASCE. Collapse Assessment of Steel Moment Frames Based on E-Defense Full-Scale Shake Table Collapse Tests[J]. Journal of Structural Engineering,2013,139(1):120-132.
[4] Y.C. Sung, T.K. Lin, C.C. Hsiao and M.C. Lai. Pushover Analysis of Reinforced Concrete Frames Considering Shear Failure at Beam-column Joints[J]. Earthquake Engineering and Engineering Vibration,2013, 12(3): 373–383.
[5] M. Fakharifar, M. K. Sharbatdar, Z. Lin, A. Dalvand, A. Sivandi-Pour and G. Chen.Seismic Performance and Global Ductility of RC Frames Rehabilitated with Retrofitted Joints by CFRP Laminates[J]. Earthquake Engineering and Engineering Vibration,2014, 13(1): 59–73.
[6] Hemchandra Chaulagain, Hugo Rodrigues, Enrico Spacone, Ramesh Guragain, Radhakrishna Mallik and Humberto Varum .Response Reduction Factor of Irregular RC Buildings in Kathmandu Valley[J].Earthquake Engineering and Engineering Vibration, 2014,13(3): 455–470.

IOP Conf. Series: Earth and Environmental Science **61** (2017) 012094 doi:10.1088/1755-1315/61/1/012094

# Static and dynamic mechanical characteristic comparison research of v-type insulator string under gale condition

**J C Wang[1], S W Zhu[1, 3], B Peng[2], S B Duan[2] and P Li[1]**

[1] China Electric Power Research Institute, Beijing 100055, China.
[2] State Grid Corporation of China, Beijing 100031, China.
[3] Corresponding author: 15901026934@163.com

**Abstract.** Application of V-type insulator string is most effective method in gale area to reduce wind deflection flashover, and dynamic characteristic of the V-type insulator string under action of pulse wind is an important factor to affect design of the V string. In order to clarify mechanical characteristic of the V-type insulator string under action of dynamic wind load, which is different to traditional rigid straight rod static mechanical method, finite element method is taken to calculate forcing of the V-type insulator string under action of pulse wind. Pulse wind speed time history with space relativity is considered and is converted to wind load time history, establishing non-linear finite element coupling model of insulator string – conductor, calculating mechanical characteristic of V-type insulator string under excitation of pulse wind, analyzing dynamic yield characteristic of the insulator under different angle of the V string, and providing theoretical basis for engineering design through comparison analysis between dynamic calculation result and static calculation result.

## 1. Introduction

Following rapid development of our national economy, power demand of the whole society is increasing. Because the energy bases of our country are mostly located at the west area, energy resource optimization and configuration within the whole country scope can be realized through "Transmitting electricity from western to eastern, mutual power supply between southern and northern, national wide interconnection".

In recent years, several windage yaw flashover accidents occurred in the power transmission line, therefore, the yaw flashover accident is an important safety hidden trouble for the normal operation of the grid, and it's not easy for the line to be reclosed after windage yaw accident, which seriously affects normal operation of the grid and causes great economic loss. Great windage yaw angle of the suspension insulator string is the direct reason for the windage yaw accident. Many newly constructed lines passing through the gale area, in order to reduce windage yaw of the insulator on the straight line tower, the V shaped insulator string is applied to reduce the windage yaw impact of the insulator. The V shaped suspension insulator string has been extensively applied in the domestic and foreign super-high voltage power transmission projects, which has many advantages of reducing size of the tower head, reducing steel consumption quantity, reducing width of the line corridor, saving cost of the line corridor, reducing filth accumulation on the insulator, increasing the self-cleaning capability, and preventing accident expansion during breakage of the insulator string etc. The V shaped insulator string can limit swing of the insulator. Since 2004, several string falling accident of the V shaped insulator string occurred in the power transmission line in our country, for which the reason is as following: the leeward side of the V shaped insulator string is compressed under strong wind, which causes the ball head of the insulator extruding mutually with the socket-clevis eye, and the R shaped pin in the socket-clevis eye being abraded or deformed. All these problems lead the R shaped pin losing limitation function about the ball head, and the conductor falling.

In order to deeply learn the mechanical characteristic of the V shaped insulator string under strong wind, it is necessary to carry out mechanical analysis of the V shaped insulator under pulse action of gale. Forcing of the strain insulator string of the composition insulator in the 220kV line is studied in document[1]. Finite element analysis of the dynamic windage yaw angle of the suspension insulator string is studied in document[2].

This paper used the finite element method to calculate forcing of the V shaped insulator string under action of fluctuating wind. It is firstly simulated that the fluctuating wind speed time-history considering spatial correlation is converted into the wind load time-history, and then the finite element coupling model of the insulator string-power transmission line is established to calculate the stress distribution of the insulator string under excitation of fluctuating wind.

## 2. Calculation of wind load

It generally consists two parts in the time-history curve of wind: one is long period part, of which the value is often longer than 10 min; another is short period part, normally lasting several seconds. Because long period of wind is far greater than natural vibration period of the normal structure, the action of wind on the structure can be taken as static force action. Fluctuating wind is caused by the irregularity of wind, of which the strength is changed over time randomly. The period of fluctuating wind is very short, which will cause vibration of the structure. The horizontal fluctuating wind speed spectrum at tailwind direction applied in our country is Davenport spectrum. The fluctuating wind speed time-history at gale operation condition is obtained through calculation.

Wind speed at any height of the structure $v(z,t)$ may be amount of average wind speed $\bar{v}(z)$ and fluctuating wind speed $v_f(z,t)$, shown as formula (1). Simulation of wind mainly aims at fluctuating wind.

$$v(z,t)=\bar{v}(z)+v_f(z,t) \tag{1}$$

It is extensively acknowledge in wind engineering field that the horizontal fluctuating wind speed spectrum at tailwind direction is Davenport, which is also the wind speed spectrum applied in our national specification. The Davenport assumes that turbulent scale will change along height, and the spectrum density function is shown as formula (2).

$$S_v(\omega)=4K\bar{v}^2(10)\frac{x^2}{\omega\left(1+x^2\right)^{4/3}} \tag{2}$$

In which, $x=600\omega/\pi\bar{v}^2(10)$, $\omega\geq 0$ is frequency (rad/s), $\bar{v}^2(10)$ is average wind speed (m/s) at height 10m away from ground, K is ratio which reflects roughness of ground.

The cosine stage superposition method is applied to simulate the fluctuating wind speed time-history of several points, the fluctuating wind speed at tailwind direction can be simulated by formula (3).

$$\tilde{v}_j(t)=\sum_{m=1}^{j}\sum_{i=1}^{N}\left|H_{jm}(\omega_i)\right|\sqrt{2\Delta\omega}\cos\left(\omega_i t+\theta_{mi}\right) \tag{3}$$

In which, $\tilde{v}_j(t)$ is the fluctuating wind speed of j point at height direction of the high-rise structure, N is isodisperse of the frequency scope in the wind speed spectrum, $\Delta\omega$ is increment of round frequency, $\theta_{mi}$ is the random number distributed in the section of $[0,2\pi]$, $H_{jm}(\omega_i)$ is an element of the lower triangular matrix $[H]$. This matrix is obtained through Cholesky decomposition of the spectrum density function. The spectrum density function matrix $[S]_{j\times j}$ is shown as formula (4).

$$[S(n)]=\begin{bmatrix} S_{11}(n) & \cdots & S_{1j}(n) \\ \cdots & \cdots & \cdots \\ S_{j1}(n) & \cdots & S_{jj}(n) \end{bmatrix} \tag{4}$$

In which, $S_{xx}(n)$ is auto-power spectrum of wind speed at $x$ point, $S_{xy}(n)$ is mutual-power spectrum of wind speed at $x$ point and $y$ point.

Calculation height of reference wind speed takes 10m, and roughness of the ground is considered as B category. Designed basic wind speed is 33m/s（10m high）, and average height of the conductor is calculated as 30m, which is converted to wind speed at this location as 39.34m/s. According to the fluctuating wind spectrum generated from the Davenport spectrum, the fluctuating wind speed time-history curve is shown as Fig.1.

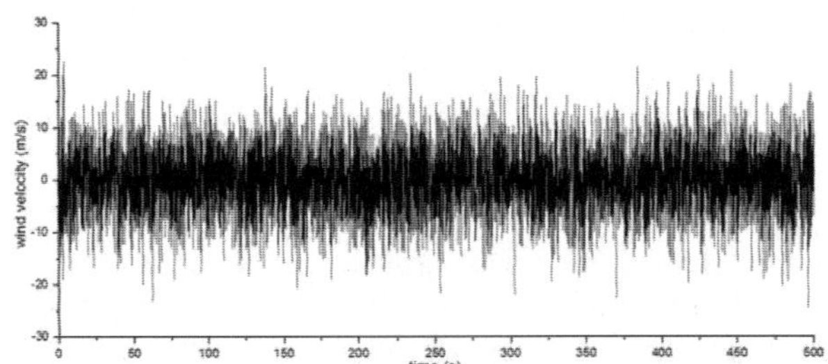


**Figure 1.** Time-history curve of high wind

Wind load of the split conductor is specified in IEC60826 Design code of overhead power transmission line that: wind load of the split conductor equals to the split number multiplying by wind load of the single conductor, and shield effect of the split conductor is neglected.

## 3. Value calculation model

The basic formulas such as balance equation of the suspension chain line and so on, are derived from documents [3-4], and one iterative algorithm is constructed. On such base, documents [5-6] apply the finite element method to derive the rigidity matrix of the suspension chain line unit for analysing the plane and the space suspension rope structure. The basic dynamic movement equation is as following:

$$M\ddot{u} + C\dot{u} + Ku = F \tag{5}$$

In which, M is quality matrix of the structure, K is rigidity matrix, C is damping matrix, $\ddot{u}$ is acceleration vector, $\dot{u}$ is speed vector, u is displacement vector, F is load vector.

The movement equilibrium equation is constant coefficient two order linear ordinary differential equation set, which can be solved theoretically with the standard solution using in the constant differential equation set. However, for the actual dynamic analysis of finite element, two value solution methods are generally applied, i.e., vibration model superposition method and direct integration method.

As the conductor belongs to the flexible structure, the geometrical non-linear issue of great displacement and small deformation shall be considered during analysis process. At same time, initial static balance position of the conductor system is initial condition for analysis of static force and dynamic force, which has a significant impact on the calculation results, therefore, we firstly carried out the static balance analysis of the conductor system. The calculation model is established according to the assembly figure of the insulator string.

We operated the windage yaw analysis of the insulator-fitting-conductor system under the condition of static wind and strong wind respectively, selecting the horizontal span as 500m, the vertical span as 500m and height difference of two spans of the conductor as 0m. We chose the length of whole composite insulator string being 11.0m, and composite insulator being 550kN. The model of the V shaped insulator string is shown as Fig.2. The conductor applied for mechanical calculation was 6×JL/G3A-1000/45-72/7.

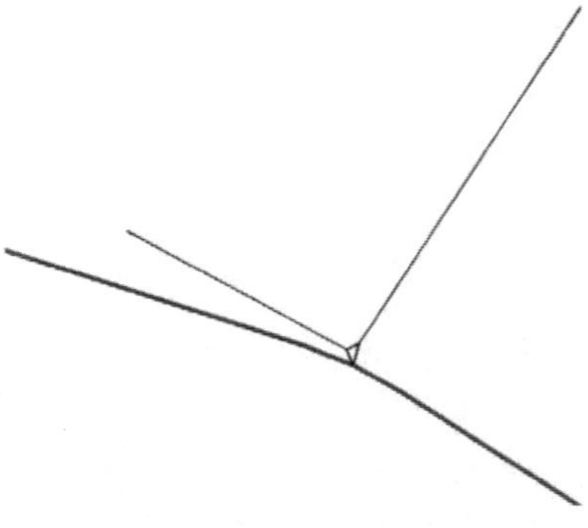

**Figure 2.** Numerical calculation model of V-shaped insulator string

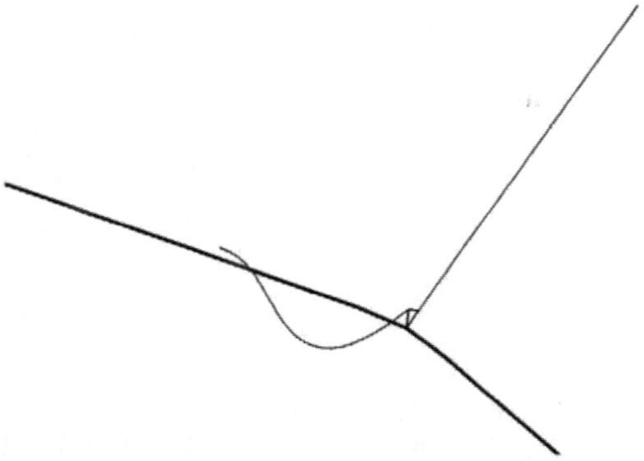

**Figure 3.** Buckling deformation of V-shaped insulator string under high wind

## 4. Static force calculation of V shaped insulator string

Static wind speed applied for calculation is 39.34m/s.

When angle of the V shaped string is 100° and the V shaped string is single connected, forcing of the insulator at the upwind side of the V shaped string under action of calm wind is 148.94kN, and forcing of the single connected insulator at leeward side of the V shaped string is -0.08kN.

When angle of the V shaped string is 110° and the V shaped string is single connected, forcing of the insulator at upwind side of the V shaped string under action of calm wind is 151.27kN and forcing of the single connected insulator at leeward side of the V shaped string is 7.01kN.

When angle of the V shaped string is 120° and the V shaped string is single connected, forcing of the insulator at upwind side of the V shaped string under action of calm wind is 159.03kN and forcing of the single connected insulator at leeward side of the V shaped string is 22.57kN.

When angle under action of calm wind is 100°, the single connected insulator at leeward side is placed at critical status for bending. When angles are 110° and 120°, there is no possibility for bending.

## 5. Mechanical characteristic of insulator string under action of fluctuating wind

Under action of fluctuating wind with the speed being 39.34m/s, the fluctuating wind spectrum was generated according to Davenport spectrum, and the time-history curve of fluctuating wind speed was shown as Fig.1. We calculated the mechanical characteristic of the V shaped string under action of fluctuating wind, when the fluctuating wind caused the insulator at leeward side bending, the deformation figure of the V shaped insulator string is shown as Fig.3.

### *5.1 Calculation Results When Angle of V Shaped String is 100° and V Shaped String is Single Connected*

The force time-history curve of the single connected insulator at upwind side of the V shaped action under action of fluctuating action is shown as Fig.4, maximum value is 256.21kN, which is 1.72 times of static force calculation value.

Force time-history curve of the single connected insulator at leeward side of the V shaped string under action of fluctuating wind is shown as Fig.5, maximum value is 87.35kN. Bending phenomenon occurs in the insulator. Calculation time is 500s, and time step length is 0.1s. During 5000 time steps, the insulator at leeward wind side of the large V shaped string is unloaded at no 2786 time step.

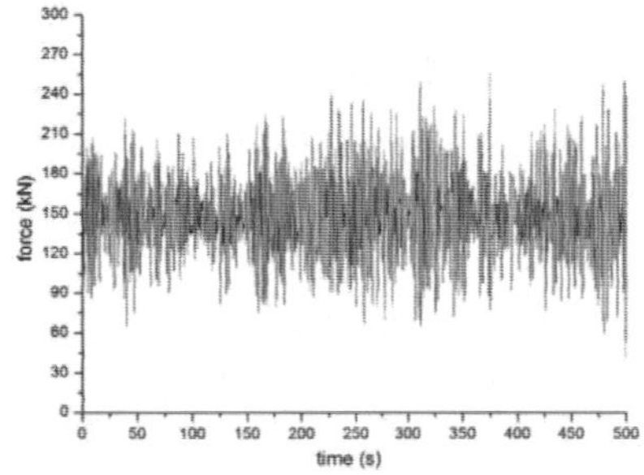


**Figure 4.** Force time-history curve of single-string insulators at windward side of V-string

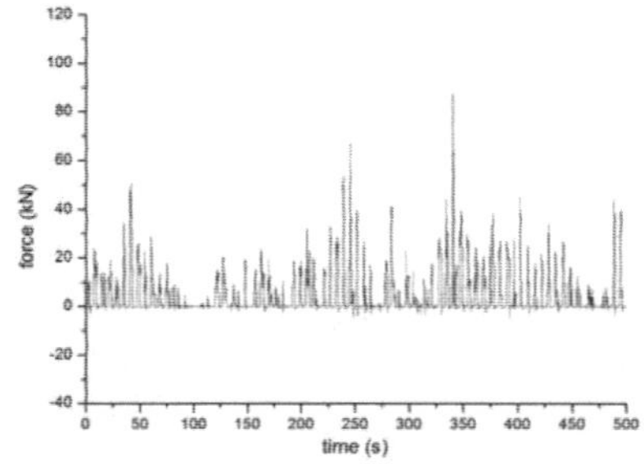


**Figure 5.** Force time-history curve of single-string insulators at leeward side of V-string

### *5.2 Calculation Result When Angle of V Shaped String is 110° and V Shaped String is Single Connected*

The force time-history curve of the single connected insulator at upwind side of the V shaped action under action of fluctuating action is shown as Fig.6, maximum value is 252.81kN, which is 1.67 times of static force calculation value.

Force time-history curve of the single connected insulator at leeward side of the V shaped string under action of fluctuating wind is shown as Fig.7, maximum value is 81.31kN. Bending phenomenon occurs in the insulator. Calculation time is 500s, time step length is 0.1s. During 5000 time steps, the insulator at leeward wind side of the large V shaped string is unloaded at no 1143 time step.

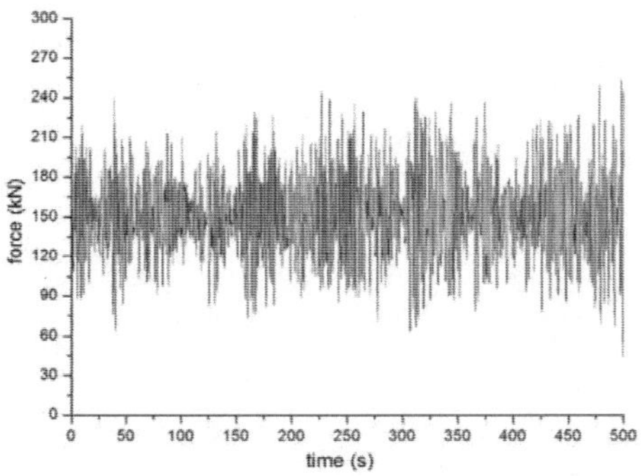


**FIGURE 6.** Force time-history curve of single-string insulators at windward side of V-string

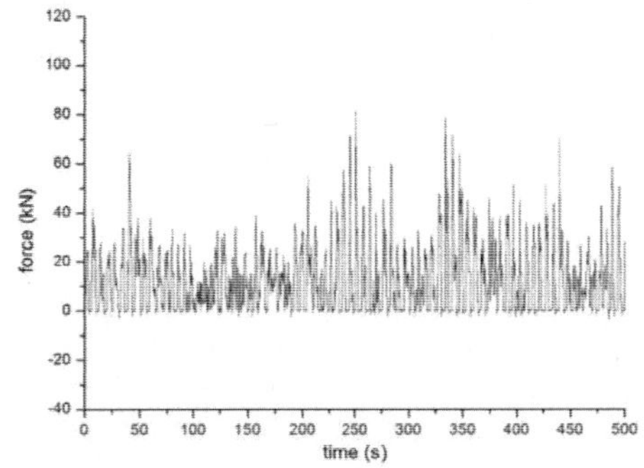


**FIGURE 7.** Force time-history curve of single-string insulators at leeward side of V-string

*5.3 Calculation Result When Angle of V Shaped String is 120° and V Shaped String is Single Connected*

The force time-history curve of the single connected insulator at upwind side of the V shaped action under action of fluctuating action is shown as Fig.8, maximum value is 269.12kN, which is 1.69 times of static force calculation value.

Force time-history curve of the single connected insulator at leeward side of the V shaped string under action of fluctuating wind is shown as Fig.9, maximum value is 98.02kN. Bending phenomenon occurs in the insulator. Calculation time is 500s, time step length is 0.1s. During 5000 time steps, the insulator at leeward wind side of the large V shaped string is unloaded at no 309 time step.

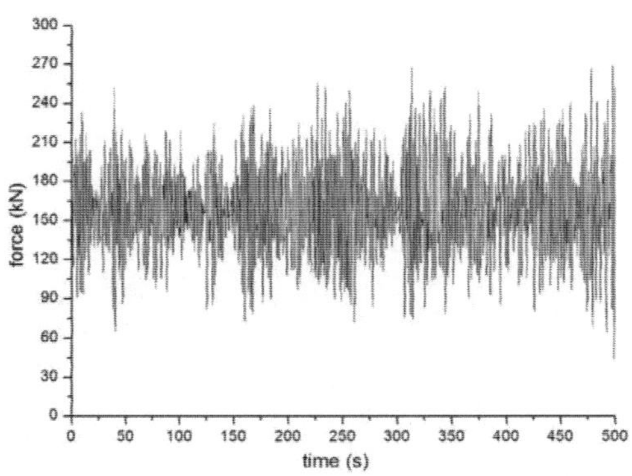


**Figure 8.** Force time-history curve of single-string insulators at windward side of V-string

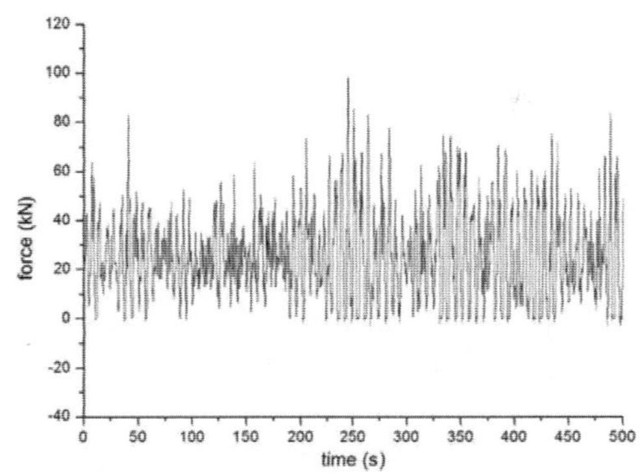


**Figure 9.** Force time-history curve of single-string insulators at leeward side of V-string

## 6. Conclusion

The following conclusions were obtained through calculation results.

Value calculation can effectively simulate forcing process of the V shaped insulator under action of calm wind and fluctuating wind, and obtained the corresponding mechanical characteristic.

With the same angle and same wind speed, the maximum forcing of the single connected insulator at upwind side of the V shaped string under action of fluctuating wind is about 1.8 times forcing under action of calm wind.

Unloading times of the V shaped insulator string at leeward side under action of fluctuating wind can reflect bending condition of the insulator.

The V shaped insulator string can effectively inhibit windage yaw flashover fault under action of general wind load. However, in gale area, dynamic bending of the insulator at leeward wind side of the V shaped insulator string caused by strong fluctuating wind will cause fatigue of the insulator, which shall be considered in the design of V shaped insulator string.

**References**

[1] Pingyuan L and Yi C 2008 *Guangdong Electric Power* 85 15
[2] Dwyi K and Li L 2008 *Electric Power Constrction* 09 5
[3] Terence W, Brien O, Francis A J 1964 *Electric Power Constrction* 90 89
[4] Terence W, Brien O, Francis A J 1976 *Journal of the structure Division Engineering, ASCE* 93 1
[5] Peyrot A H, and Goulois A M 1979 *Computers and Structures* 10 805
[6] Peyrot A, and Osteraas J 1981 *IEEE transactions on power apparatus and systems* 100 3254

IOP Conf. Series: Earth and Environmental Science **61** (2017) 012095 doi:10.1088/1755-1315/61/1/012095

# Research on the shape coefficient of disc insulator strings for UHV transmission lines

**J C Wang[1], S W Zhu[1, 3], B Peng[2], S B Duan[2] and P Li[1]**

[1] China Electric Power Research Institute, Beijing 100055, China.
[2] State Grid Corporation of China, Beijing 100031, China.
[3] Corresponding author: 15901026934@163.com

**Abstract.** Following continuous lengthening of the ultra-high voltage disc shaped insulator strings, wind load plays a more important role in controlling of the disc insulator string. It is necessary to carry out deep study about the shape coefficient of disc insulator strings, learn the concrete wind pressure distribution, and wake flow condition of the insulator string in detail. To this end, we used the SST turbulence model to operate value simulation for the three-dimensional flow field of disc insulator strings. Results showed that value simulation could clearly simulate surface pressure of the insulator and distribution of tail flow; the fitness of calculation results for nonstructural grid and structural grid was very high; the structural grid could control every part of the grid well, but classification was complicated; the structure adaptability of the nonstructural grid was stronger compared with structural grid.

## 1. Foreword

In the ultra-high voltage line which is put into operation at present, length of the insulator string can reach 10m and more. Increment of length can cause reduction of rigidity at lateral resistance side, as a result, control function of wind load about the insulator string is more obvious. The windage yaw and wind vibration phenomenon have been observed for several times in the built-up ultra-high voltage power transmission line.

Wind load of the traditional insulator is generally obtained through approximate calculation, using the product of average wind pressure in 10 min and wind area of the insulator. With the increasing of type, length and connection number for insulator string, its forcing condition becomes more complicated. Therefore, it is necessary to carry out deep research about its shape coefficient, so as to learn concrete distribution of surface wind pressure and tail flow field of the insulator string.

Study of the shape coefficient generally consists of three methods: actual measurement at site, wind tunnel test and simulation of computational fluid dynamic (Computational Fluid Dynamic, CFD). Actual measurement at site is the most direct and effective method, but it also has more limitation. The cost of actual measurement is high and it is limited by meteorologic conditions, what's more, it is hard to obtain data under most adverse operation condition. Because wind tunnel test need a large quantity of model fabrication and repeated test with different operation conditions, test period is long. At the same time, it is difficult to make the Reynolds number similar to the actual situation. In recent years, with the development of computer technology, CFD is characterized with the advantages of low cost, short period, good repeatability and easy control. Congzhen Xie[1] etc took advantage of CFD to study the accumulated filth of the composite insulator and the characteristic of air flow field; Xingliang Jiang[2] etc studied the application of CFD in accumulated filth characteristic of the insulator in the line; Liming Wang[3] etc took advantage of CFD to study accumulated filth characteristic of the pillar insulator and obtained good simulation effect.

## 2. Value analysis

### *2.1 Turbulence model*

Because study about the shape coefficient of the insulator string belongs to the separated bluff body streaming issue generating from gradient of inverse pressure. This paper applied shearing stress conveying k-ω model, which is abbreviated as the SST turbulence model. The SST model was forwarded by Menter[4], taking into consideration transfer of turbulence shearing stress. It can well forecast the starting and quantity for fluid separation under inverse pressure gradient, and simulate flow phenomenon of pure air dynamic with high accuracy. The model applied k-ε model in far wall area, and k-ω model forwarded by Wilcox in wall surface area. We used one blending function for transition between two models. It will fully play advantage of the k-ω model in calculation of high Reynolds number, also it can avoid characteristic that k-ω is sensitive to incoming flow.

### *2.2 Geometrical modeling*

As the insulator string is a composite structure of several insulator strings which are connected at end to end, difference between the shape coefficient of the section model and the shape coefficient of the whole isn't great. Therefore, in order to reduce number of the grid, this paper applied 5 discs of the insulator to build up the ANSYS model. In order to ensure the quality of grids, we carried out polishing treatment at local sharp location, and made its minimum curvature radius above 0.003m. The disc diameter was 0.425m, length was 0.775m and calculation flowing field applied 10m×10m×10m, the insulator string model is located in the middle of the fluid field, and blockage rate is less than 2%.

**Figure 1.** The numerical model of insulator strings

**Figure 2.** The numerical model of insulator strings

### *2.3 Division of grid*

The ICEM fluid grid division tool is applied. In order to investigate application of different grid division way in this subject, the structural grid and nonstructural grid were applied in this paper to divide the fluid field.

Advantage of the k-ω model is that solution expression formula about ω in the viscosity layer is known, ω value of near wall surface must be large enough to obtain correct result[5]. According to requirement of document[6], the dimensionless height y+ value at wall surface is located near 1. For every part of the insulator being different in thickness, in order to obtain the shape coefficient of every part accurately, height of first layer grids of the structural grid and nonstructural grid took 0.00001, taking the 10 layers boundary layer.

For the structural grid, we unified the applicable structural hexahedron grid, and the O grid was applied to directly divide the fluid field. This method could control every part of the grid well, but it had a disadvantage that the division of the grid was complicated. Total number of the structural grid was approximately 3.8 millions, and minimum grid quality was 0.2.

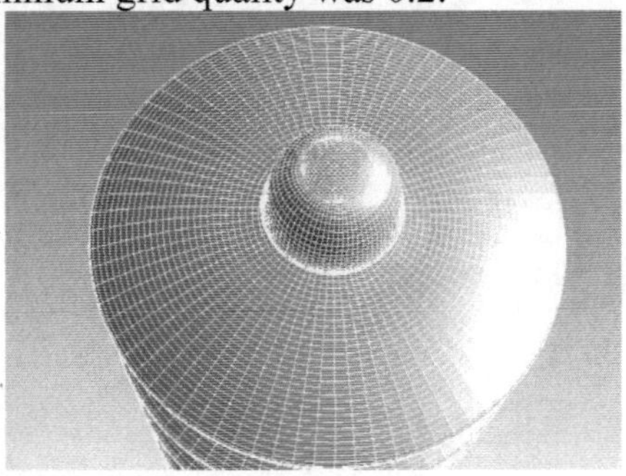

**Figure 3.** Local structure grid on insulator surface

For nonstructural grid, we established one square calculation subdomain within 0.1m scope around the insulator string. The nonstructural tetrahedron grid was applied in the subdomain to fully fit the complicated geometrical boundary, and the blocked structural hexagonal grid was applied out of the subdomain to obtain incoming flow condition accurately. The contact surfaces of two grids applied Interface way to couple. The assembled grid was 5.6 million in total, and minimum grid quality was 0.31. Advantage of the nonstructural grid is that the division is simple and the structural adaptability is strong. While the disadvantage is that the quantity of the grid is more.

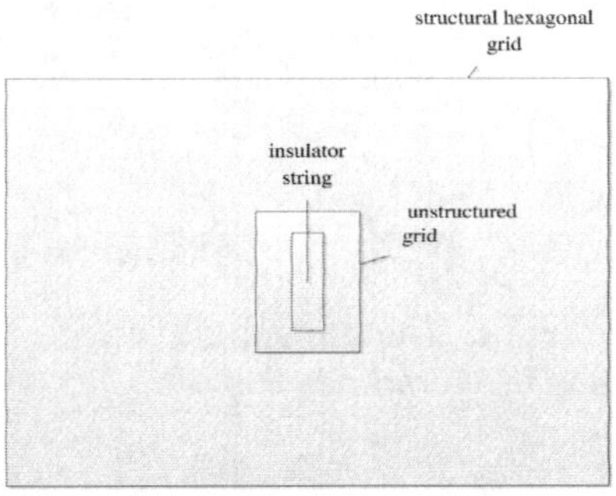


**Figure 4.** Method of unstructured grid

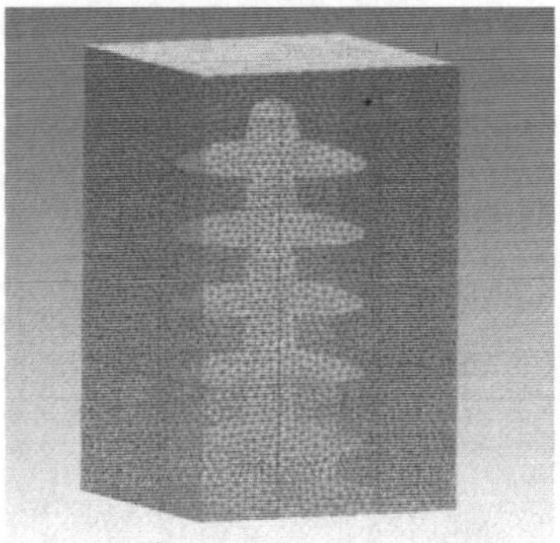

**Figure 5.** Unstructured grid in the subdomain

*2.4 Setting of boundary condition*

The inlet applied boundary condition of speed（Velocity inlet）, with the tangential speed being zero, and there was only vertical speed. The outlet applied the outlet boundary condition, with the relative pressure being zero. The side wall surface applied the free slip (Free slip) wall surface conditions. The upper and lower wall surfaces applied the symmetrical boundary condition (Symmetry), while the column surface applied the no slip (No slip) wall surface boundary condition. We applied the separation type solver with 3D single accuracy, chose the constant density air model which was not compressed, and used the SIMPLE algorithm to disperse the flowing item.

IOP Conf. Series: Earth and Environmental Science **61** (2017) 012095 doi:10.1088/1755-1315/61/1/012095

*2.5 Analysis of calculation result*

It is found that the results of structural grid are similar with the results of nonstructural grid in Fig.6, and in Fig.7, the value simulation can accurately simulate shape coefficient of every part of the insulator body surface. The shape coefficients of the umbrella cap at upwind location, as well as the shape coefficients of connection location between the umbrella cap and the umbrella shirt are maximum. Connection rigidity at this location can be suitably strengthened during design. Fig.9 is the distribution of wake flow around the umbrella cap, while Fig.8 is the distribution at the umbrella shirt. It is also found that the shape of the insulator isn't regular. As a result, the tail flow shows obvious three dimension characteristic. The tail flow speed of the umbrella cap is very high, and the speed gradient is very great, while tail flow of the umbrella shirt is very smooth.

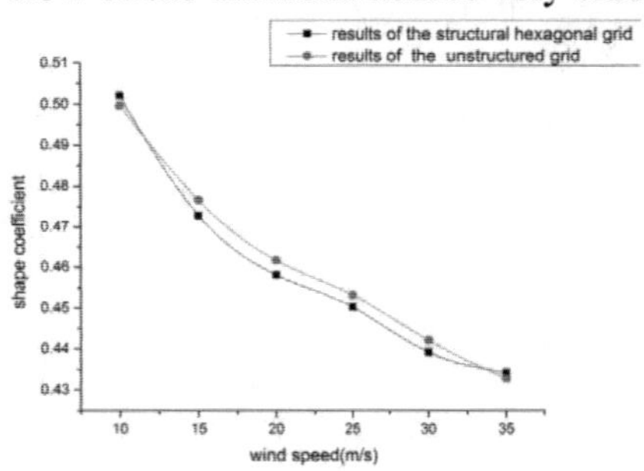


**Figure 6.** Curves of the shape coefficient changing with wind speed

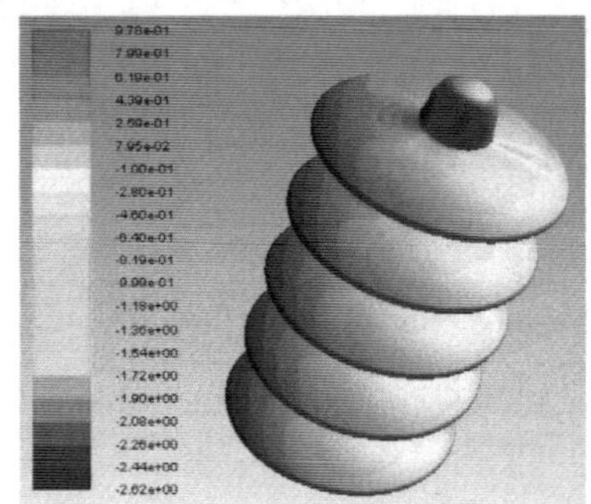

**Figure 7.** Distribution of shape coefficient on the upper surface of the insulator strings

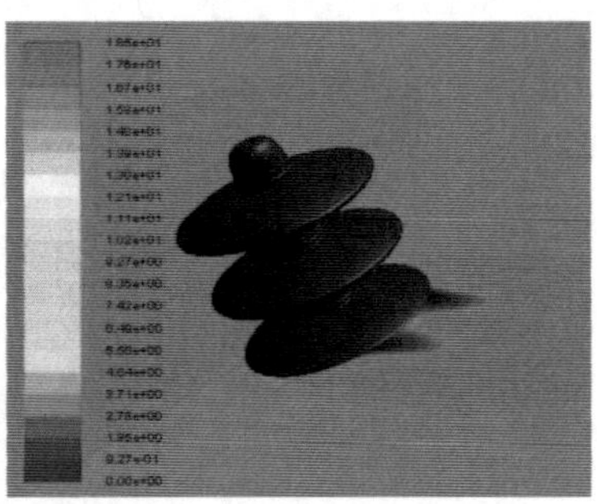

**Figure 8.** The flow field velocity distribution around the umbrella skirt

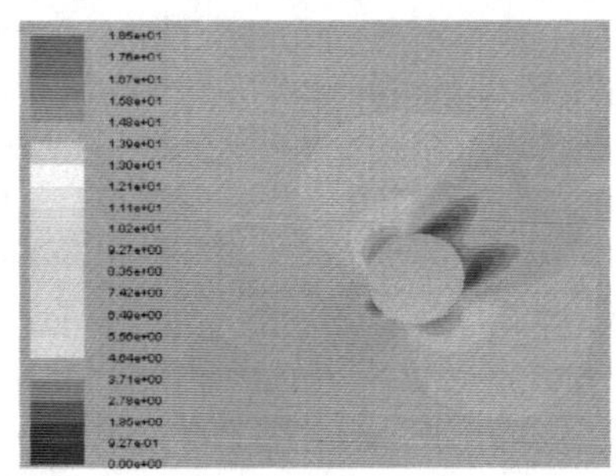

**Figure 9.** The flow field velocity distribution around the umbrella cup

## 3. Conclusion

The following conclusions were obtained through calculation results.

The value simulation could clearly simulate surface pressure of the insulator and distribution of tail flow.

The fitness of calculation results for nonstructural grid and structural grid was very high. The structural grid could control every part of the grid well, but classification was complicated; the structure adaptability of the nonstructural grid was stronger compared with structural grid.

## References

[1] Congzhen X and Xiao Z 2010 *High Voltage Engineering* 36 212

[2] Xingliang J and Haibo L 2010 *High Voltage Engineering* 36 329

[3] Liming W and Ting L 2015 *High Voltage Engineering* 41 329
[4] Menter F R 1994 *AIAA Journal* 32 269
[5] Jinguang Y and Hu W 2014 *Acta Aeronautica et Astronautica Sinica* 35 116
[6] WILCOX D C 1988 *AIAA Journal* 26 1299

IOP Conf. Series: Earth and Environmental Science **61** (2017) 012096 doi:10.1088/1755-1315/61/1/012096

# Nonlinear finite element analysis of PVA fiber reinforced high strength concrete columns under low cyclic loading

**Jun Su[1,2] Qiang Hu[1,3] and Jianping Liu[1,4]**

[1] School of Civil Engineering and Environment, Hubei University of Technology, Wuhan 430068, China
E-mail:[2] Sujun930@163.com, [3] 1178191128@qq.com, [4] 478299036@qq.com

**Abstract.** In this paper, four PVA fiber reinforced super-high-strength concrete columns under the low cyclic reciprocating load were studied by using the finite element analysis software OpenSEES and their hysteretic curves and skeleton curves were studied. The energy dissipation capacity of PVA fiber were analyzed to evaluate the effect of PVA fiber on the seismic performance of concrete columns. The results show that the restoring force curve of the finite element analysis software OpenSEES simulation agrees well with the experimental curve, which can fully reflect the hysteretic behavior of fiber reinforced concrete columns under low cyclic loading. The incorporation of PVA fiber can obviously improve the energy dissipation capacity of ordinary concrete columns.

## 1 Introduction

In this paper, finite element analysis software-Open System for Earthquake Engineering Simulation (OpenSEES) was used to simulate four super-high-strength concrete short columns under axial pressure and low cyclic loading. Through analyzing its energy dissipation capacity, it is proved that the PVA fiber can improve the seismic performance of concrete short columns and it provides effective design method and idea for the structural design compared with the experimental results.

## 2 The basic parameters of the test

The dimensions of the specimens used in the test are shown in Figure 1. It is that column height is 1500mm, the size of cross-sectional is 200mm× 200mm, the specimen shear span is 2.0, longitudinal reinforcement is the 4 $\Phi$ 14 HRB400 grade steel, stirrups is the $\Phi$ 8 HRB400 grade steel and the volume with the hoop rate($\rho_V$) is 2.2% [1]. PVA fiber used Kuraray REC15 × 12 produced by Kuraray Co., Ltd. In the test, all the four specimens were loaded into the failure state and the specimen parameters are shown in Table 1.

**Table 1.**Specimen design parameters

| Specimen Number | Compressive strength $f_{cu}$ / $MPa$ | Splitting tensile strength $f_{t,s}$ / $MPa$ | PVA fiber content $\rho_f$ / % | Test axial compression ratio $n_t$ |
|---|---|---|---|---|
| RC | 103.6 | 5.05 | 0 | 0.35 |
| PFRC-0.17 | 108.1 | 6.16 | 0.17 | 0.35 |
| PFRC-0.33 | 105.7 | 6.86 | 0.33 | 0.35 |
| PFRC-0.50 | 112.1 | 7.41 | 0.5 | 0.35 |

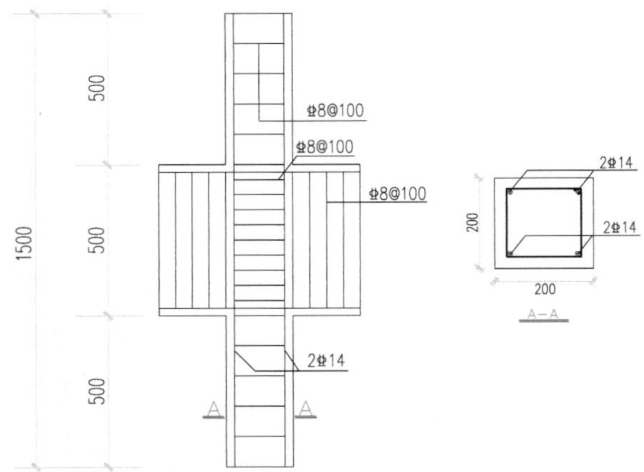


**Figure 1.** Dimensional drawings of test pieces

## 3 Numerical simulation based on OpenSEES

*3.1. Stress-strain Relationship Model of Fiber Reinforced Concrete*
As the number of dividing units is relatively large, the damage constitutive model with relatively simple form is selected under the premise of meeting the requirements of calculation. The constitutive model of each unit of PVA fiber concrete is defined as concrete in order to reflect the brittle performance mainly. A constitutive model of elastic-brittle damage with residual strength is adopted [2]. The corresponding damage evolution equation is expressed as follows:

$$D = \begin{cases} 0 & (\varepsilon \leq \varepsilon_{t_0}) \\ 1 - \mu\frac{\varepsilon_{t_0}}{\varepsilon} & (\varepsilon_{t_0} \leq \varepsilon \leq \varepsilon_{tu}) \\ 1 & (\varepsilon > \varepsilon_{tu}) \end{cases} \quad (1)$$

Where: $\mu$ represents the residual strength coefficient of concrete; $\varepsilon_{t0}$ represents the maximum tensile stress corresponding to the concrete strain; Where $\varepsilon_{tu}$ is the ultimate tensile stress of concrete, $\varepsilon_{tu} = \xi\varepsilon_{t0}$, $\xi$ represents the ultimate strain coefficient of concrete.

PVA fiber using the ideal elastic-plastic stress-strain relationship, namely:

$$\begin{cases} \varepsilon_s = E_s \cdot \varepsilon & (\varepsilon_s \leq \varepsilon_y) \\ \varepsilon_s = f_y & (\varepsilon_s > \varepsilon_y) \end{cases} \quad (2)$$

Where: $E_s$ represents the ideal elastic modulus of concrete; $f_y$ represents the ideal stress of concrete.

*3.2. Constitutive Model of Steel Bar*
The expression of strain explicit function is adopted in this model and its computational efficiency is high while keeping the consistency with the repeated loading test results of steel bars. Therefore, the steel model selected by the paper is isometric stress hardening [3].

*3.3. Comparison of Simulation Results and Test Results*

*3.3.1. Hysteresis Curve.* The hysteresis curve is the load-deformation curve obtained by cyclic loading. It reflects the deformation of structural members in the process of deformation characteristics, stiffness degradation and energy dissipation [4], which is the basis for nonlinear seismic response. The hysteresis curves obtained from the quasi-static tests of four PVA super-high-strength reinforced concrete short columns are compared with the hysteresis curves obtained from the corresponding loading tests, which is shown in Figure 2:

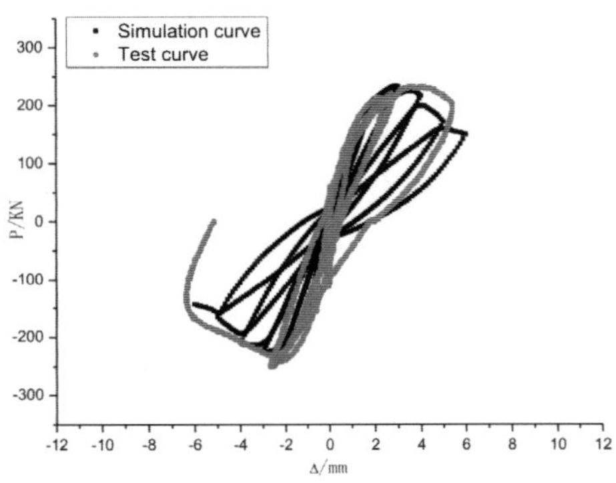


(a) RC column

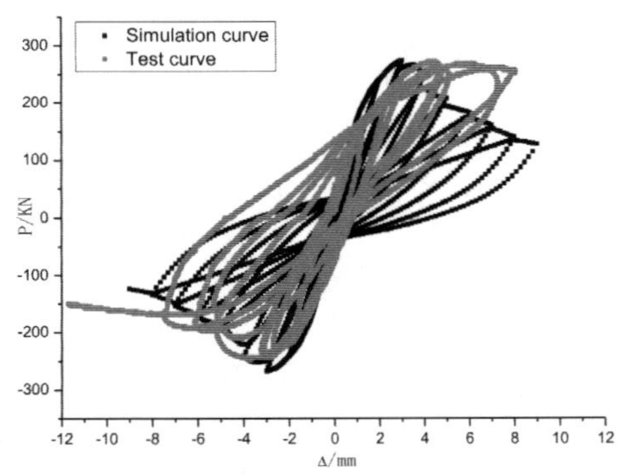


(b) PFRC-0.17 column

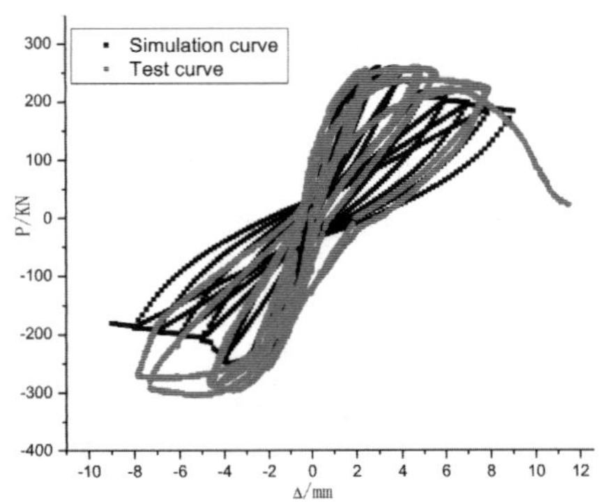


(c) PFRC-0.33 column

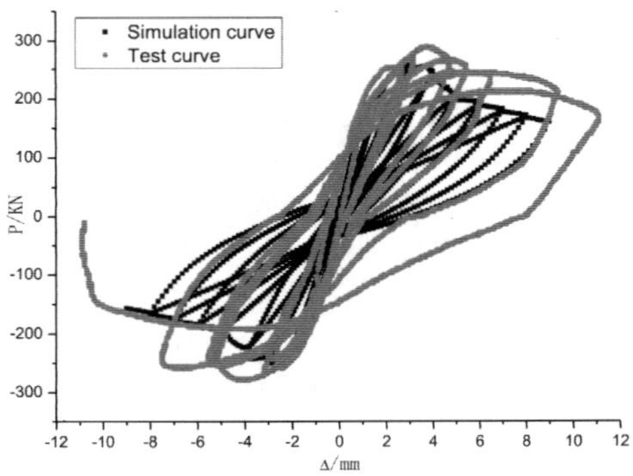


(d) PFRC-0.5 column

**Figure 2.** Hysteresis curve

Figure 2 shows that under the low cyclic loading, the finite element analysis software OpenSEES was used to simulate the super-high-strength concrete (HPC) concrete column with PVA fiber. Compared with the experimental results, the hysteresis curve is represented. At the initial stage of loading, the simulated hysteretic curves of RC columns and PFRC columns coincide with the test curves when the load is not at the peak load, indicating that the components are in the elastic stage.

When the load exceeds the peak load, RC column and PFRC column show different hysteresis characteristics. The hysteresis curve of the RC column becomes narrower, the bearing capacity declines rapidly, and the rigidity degrades. The hysteresis curve of the FRC column becomes fuller, the frequency of the load increases and the sign of the stiffness degradation tends to be slow.

The hysteresis curve of concrete column with PVA fiber content shows: with the increase of fiber content, the fullness of hysteresis curve is better. After peak load, the trend of curve descending is not so obvious. Stiffness degeneration tends to be slow and the plastic property and energy dissipation capacity are also strengthened.

*3.3.2. Skeleton Curve.* The skeleton curve is a curve formed by connecting the peak points of positive and negative loading cycles in turn. The skeleton curves of each specimen were drawn according to the maximum horizontal forces and displacements obtained by the bending test of each loading cycle. When the tensile strength of PVA fiber reinforced concrete columns is yielded, the stiffness and P-Δ curve of the specimens decrease with the increase of load. When the bearing capacity of the specimen is reduced, the specimen undergoes four stages: cracking, yielding, ultimate bearing capacity and failure [5]. For commonly used flexural members, the yielding of the members is the yield strength. The ductility coefficient can be calculated from the yield and the ultimate load and displacement of the member according to the skeleton curve. Skeleton curve reflects the relationship between load-deformation, which is the main basis for seismic-inelastic dynamic reflection of structures [6].

The skeleton curves of the RC and PFRC columns are shown in Figure 3.

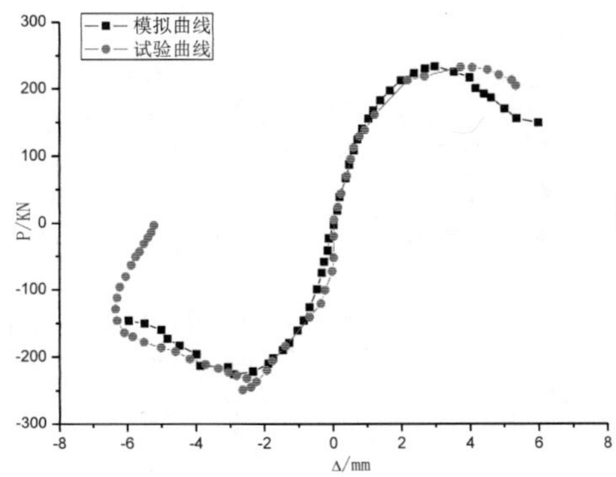


(a) RC column

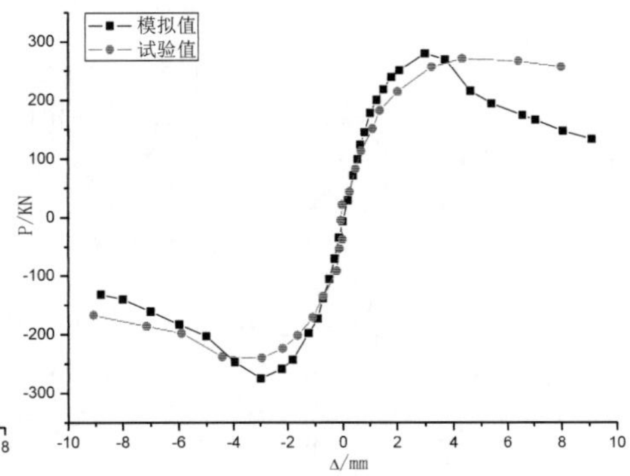


(b) PFRC-0.17 column

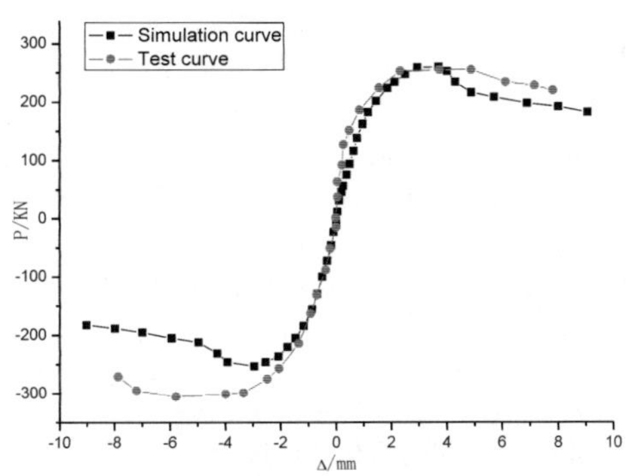


(c) PFRC-0.33 column

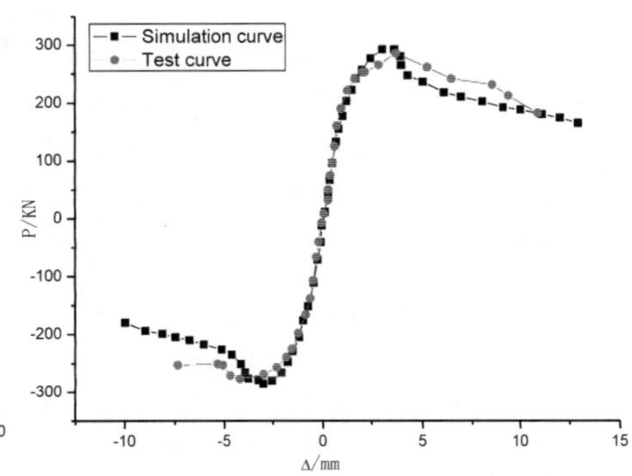


(d) PFRC-0.50 column

**Figure 3.** Skeleton curve

## 4 Energy dissipation analysis

The energy dissipation capacity [1] reflects the energy consumption of building structure under seismic load. The stronger the consumption power is, the better the seismic effect of the structure is. At present, we often use energy consumption ratio, the cumulative energy consumption and the equivalent viscous damping coefficient to represent.

The equivalent viscous damping coefficients of the PVA super-high-strength concrete columns calculated under low cyclic loading are shown in Table 2.

**Table 2.** PVA fiber ultra-high-strength concrete column equivalent viscous damping coefficient

| Component Number | Equivalent Viscous Damping Coefficient $h_{e1}$ | | | Equivalent Viscous Damping Coefficient $h_{e3}$ | | | Equivalent Viscous Damping Coefficient $h_{e5}$ | | |
|---|---|---|---|---|---|---|---|---|---|
| | Test Value | Analog Value | Error | Test Value | Analog Value | Error | Test Value | Analog Value | Error |
| RC | 0.0426 | 0.0467 | 9.62 | 0.0842 | 0.0835 | -0.83 | 0.1785 | 0.1826 | 2.30 |
| PFRC0.17 | 0.0512 | 0.0494 | -3.52 | 0.1096 | 0.1200 | 9.49 | 0.1956 | 0.2130 | 8.90 |
| PFRC0.33 | 0.0585 | 0.0540 | -7.69 | 0.1200 | 0.1255 | 4.58 | 0.2096 | 0.2147 | 2.43 |
| PFRC0.50 | 0.0602 | 0.0576 | -4.32 | 0.1502 | 0.1393 | -7.26 | 0.2131 | 0.2308 | 8.31 |

From the analysis of Table 2:

(1) The error of the equivalent viscous damping coefficient between RC column, PFRC-0.17 column, PFRC-0.33 column and PFRC-0.5 column was 9.62%, 3.52% 7.69% and 4.32% respectively when the first cycle was loaded; The errors were 0.83%, 9.49%, 4.58% and 7.26% at the third cycle; In the fifth cycle, the errors are 2.30%, 8.90%, 2.43% and 8.31% respectively. The results show that the finite element analysis software OpenSEES can accurately analyze the PVA fiber reinforced concrete frame columns quantitatively.

(2) Both the experimental and simulated values show that the equivalent viscous damping coefficient increases with the increase of PVA content under the same conditions. For example, the experimental values of the equivalent viscous damping coefficients of PFRC-0.17, PFRC-0.33 and PFRC-0.50 were 20.2%, 37.3% and 41.3% higher than those of RC column. The simulated values increased by 5.8%, 15.6% and 23.3%. The results show that with the increase of volume fraction of PVA fiber, the energy dissipation capacity of frame column is enhanced, and the deformation ability of the structure is improved effectively.

## 5 Conclusion

Based on OpenSEES, the numerical simulation of PVA fiber reinforced concrete short columns was carried out. The energy dissipation capacity of PVA fiber reinforced concrete columns were analyzed to verify the effect of PVA fiber on the seismic performance of concrete short columns. Compared with the experimental results, the following conclusions can be drawn:

(1) The finite element analysis software OpenSEES can be applied to the numerical simulation of seismic performance of fiber reinforced concrete structures. Compared with the experimental results, it is found that the error values of the hysteresis curve, the skeleton curve and the equivalent viscous damping coefficient calculated by numerical simulation are within the effective range. It shows that the finite element analysis software OpenSEES is able to analyze the seismic performance of PVA fiber reinforced concrete members.

(2) Through experiment and numerical simulation can be found that the incorporation of PVA fiber can improve the seismic capacity of concrete frame columns. Compared with the RC column, the experimental value of the equivalent viscous damping coefficient of the PFRC column increased by 41.5% and the simulated value was 26.4%. The results show that the incorporation of PVA fibers can improve the ultimate elastic-plastic displacement angle of ordinary concrete columns. With the increase of the volume fraction of PVA fibers, the energy dissipation capacity were enhanced.

## References

[1] Jiang Rui. Experimental Study on Seismic Behavior of Super-high-strength Concrete Composite Columns [D]. Dalian University of Technology, 2007.
[2] Tang Chunan, Zhu Wancheng. Damage and Fracture of Concrete [J]. Beijing, 2003.
[3] Mazzoni S, Mckenna F, Scott M H, et al. OpenSees users manual [R]. Berkeley: University of California, 2004.
[4] Guo Zhenhai. *Reinforced concrete theory* [M]. Tsinghua University Press, 2013.
[5] China Academy of Building Research. People's Republic of industry standards: *Building seismic test methods JGJ 101-96* [M]. China Construction Industry Press, 1997.
[6] Liang Xingwen, Kang Li, Jia Jialing, et al. Experimental and analysis on seismic behavior of

partially reinforced concrete columns using fiber reinforced concrete [J]. Engineering Mechanics, 2013, 30 (9): 243-250.

IOP Conf. Series: Earth and Environmental Science **61** (2017) 012097 doi:10.1088/1755-1315/61/1/012097

# Analysis of uplift bearing capacity of pile based on ABAQUS

**Bin Li** [1 2 3 a)], **Jianbao Fu** [4]

1Tianjin Port Engineering Institute Ltd. of CCCC First Harbor Engineering Company Ltd., Tianjin, 30222 China

2 Key laboratory of port geotechnical engineering, ministry of communications, PRC Tianjin, 30222 China

3 Key laboratory of port geotechnical engineering of Tianjin, Tianjin, 30222 China

4 Tianjin Key Laboratory of Soft Soil Characteristics and Engineering Environment

a) Corresponding author： Lee_binbin@163.com

**Abstract**. The uplift bearing capacity of pile in soil foundation is mainly obtained by load test, which is time-consuming and low-efficiency. A small sliding model and the contact pairs in large-scale finite element software ABAQUS are used to simulate the friction between pile and soil, and a good fitting curve of pile foundation bearing capacity is obtained. According to the results of finite element software, the lateral friction of the pile is fully developed, and the stress of the pile gradually reaches the ultimate pullup capacity with the increase of pile displacement. The finite element software ABAQUS can be used to simulate uplift capacity of pile.

## 1. Introduction

In recent years, more and more wind power projects have been constructed; large diameter single pile foundation is the most popular type in many types of wind power foundation. Due to the limitation of not having software in the past few decades, the designers of the project rely on the load test to obtain the Data of uplift capacity of the single pile foundation, but this method is easy to be subject by external factors; another side, the experimental level will also cause a lot of uncertainties, although there are many drawbacks of this method, this method has been used today. There is no better way to solve this problem.

Large-scale finite element software ABAQUS specializes in the calculation of various combinations of materials, complex loading processes and varying contact conditions. The geometrical is a special medium with non-homogeneous and non-linear character, arbitrary geometry and discontinuity. This makes it to be most popular finite element analysis software used. In this paper, the software is used to simulate the pull out bearing capacity of single pile, so that the results can guide the engineering design and construction.

## 2. Project background

The Mohr - Coulomb model was used to simulate the soil stress, Mohr-Coulomb yield surface is an irregular hexagonal cross-sectioned pyramidal surface in the principal stress space, and is projected into an equiangular equilateral plane in the π plane. There is sharp angle on the yield surface, as shown in Figure 1.

IOP Conf. Series: Earth and Environmental Science **61** (2017) 012097 doi:10.1088/1755-1315/61/1/012097

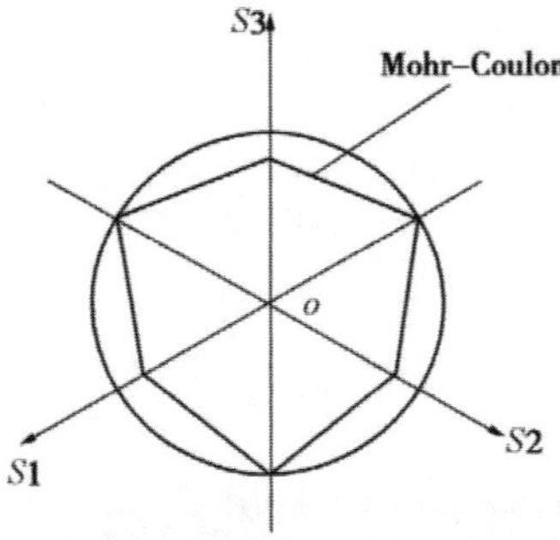


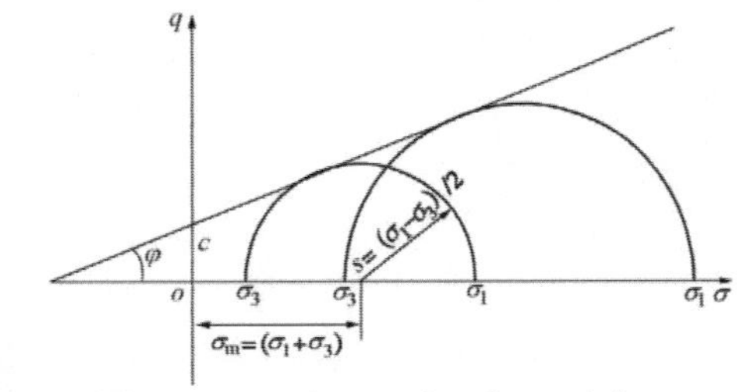


**Figure 1.**Yield surface on π plane **Figure 2.**mohr-coulomb model

Mohr-Coulomb theories consider that: the shear stress of a yield surface is against by the material cohesion and the friction, the expression is:

$$\tau = c + f\sigma = c + \sigma \tan\varphi \qquad (1)$$

$\tau$、$\sigma$ are the normal stress and shear stress of the plane, c is the cohesion, f is the friction coefficient ,and $\varphi$ is the friction angle.

It can be deduced that when the stress Mohr circle is tangent to the Coulomb criterion line, the rock mass reaches the limit state, and the principal stress should meet the following conditions:

$$\sigma_1 = \frac{2c\cdot\cos\varphi}{1-\sin\varphi} + \sigma_3\frac{1+\sin\varphi}{1-\sin\varphi} \qquad (2)$$

From the above, it can be seen that the Mohr-Coulomb criterion does not consider the influence of the intermediate principal stress on the strength of rock and soil, but the destruction of material is usually affected by the intermediate principal stress, because of this influence is small, the Mohr-Coulomb criterion is still accurate enough to show the stress level. The result of simulation is very close to the experimental data.

The pile is linear elasticity before the stress reaches the failure strength of material, so it is a linear elastic material in the model. The soil is elastic-plastic material, which is assumed to obey the Mohr-Coulomb criterion. The material parameters are shown in Table 1.

**Table 1.** parameters table of model

| **Name** | **Elastic modulus Pa** | **Poisson's ratio** | **Cohesive force Pa** | **Friction angle °** | **Dilatancy angle °** | **Density Kg/m$^3$** |
|---|---|---|---|---|---|---|
| concrete | 3e10 | 0.2 | —— | —— | —— | 2500 |
| Clay | 4e7 | 0.35 | 15000 | 30 | 0 | 1800 |

The elastic modulus of pile and soil are very different. Under the load, there will be relative sliding between pile and soil. In order to ensure the rationality of the calculation results, the coulomb friction contact is used to simulate the contact between the pile and soil. There is a small sliding model in the software and the friction coefficient is 0.3. Then the contact pairs between the pile and soil are established, and the master-slave contact algorithm is adopted.

The selection of the master and slave faces must follow the principle that the meshing of master surface is finer than the slave surface, even the material of master surface is harder, but the mesh density of two faces must be similar. According to the above principle, the surface of the pile surface is the master face and the soil surface is the slave face.

Coulomb's law is used to calculate the ultimate shear stress in ABAQUS:

$$\tau = \mu p \tag{3}$$

μ is the coefficient of friction, p is the stress between the two contact surfaces. When the shear stress between the two contact surfaces reaches the ultimate shear stress, the contact surface will produce the relatively sliding. Because of there is the non-ideal frictional contact between the pile and soil, the concept of "elastic slip deformation" is introduced in the finite element calculation. When the surfaces are bonded together, a small amount of relative slip deformation is allowed. As shown in Figure 3, a small slip between the contact surfaces is allowed when the slip displacement is less than ωs.

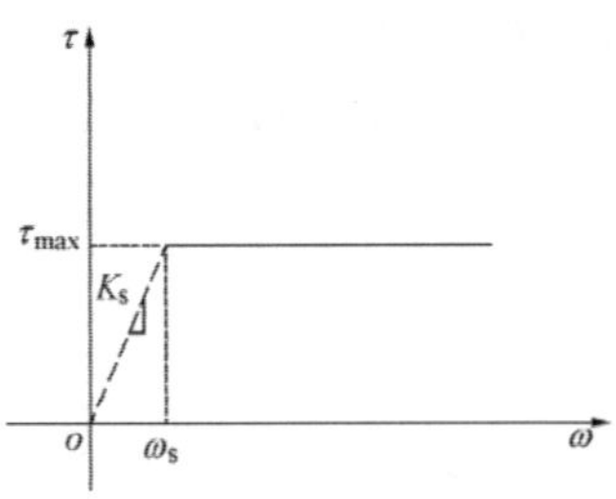


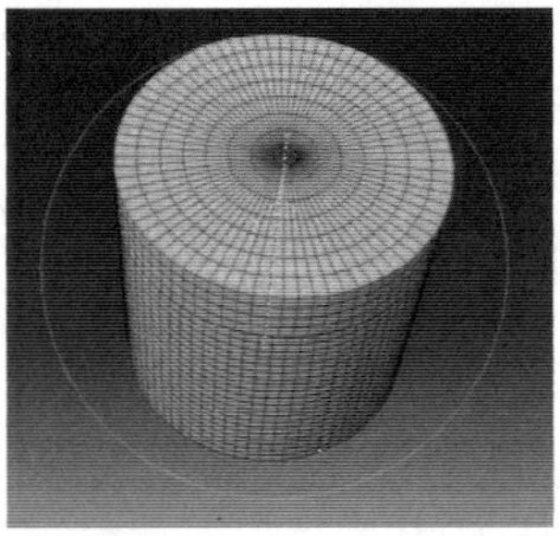

**Figure 3.** Coulomb friction model **Figure 4.** Pile-soil grid model

The results of the numerical analysis are closely related to the mesh shape and mesh density. The more dense of the grid, the higher accuracy of the results, but also increase the time and difficulty of calculation. Therefore, the model should be as far as possible to ensure sufficient accuracy without excessive element, the denser of the mesh close to the pile, the sparser of the mesh away from the pile. Figure 4 shows the pile-soil grid model.

## 3. Result

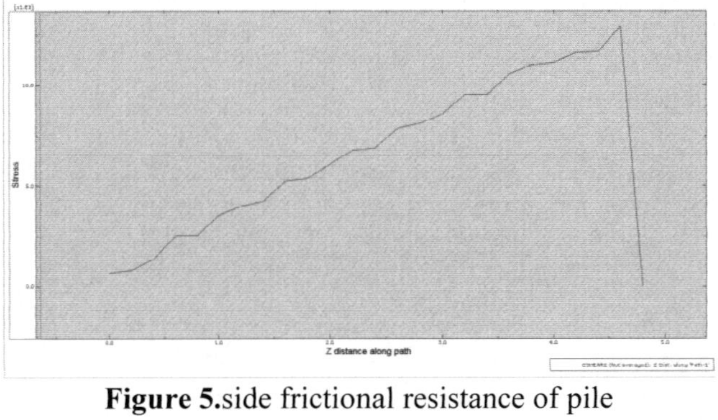


**Figure 5.**side frictional resistance of pile

Through the simulation of the friction between pile and soil, it can be seen that along with the length of the pile, the friction resistance is increasing continuously. This is caused by earth pressure, as the pile pulls out more and more, the side friction increases gradually, reaching a peak and suddenly decreasing, at this time, the pile destroyed.

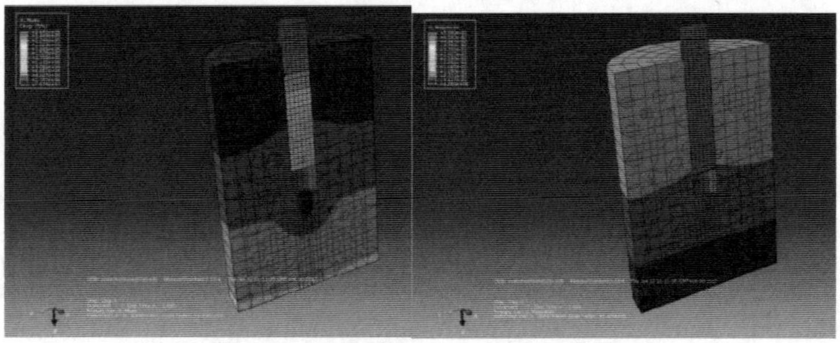

**Figure 6.** Strain and stress of pile

Figure 6 shows the stress chart and strain chart of pile and soil. The maximum stress of the pile occurs at the lower part of the pile, and the stress concentration area is also in the lower part of the pile. Because of the uplift capacity analysis, the whole pile body is the largest displacement area, and the maximum displacement area of the soil occurs in the upper part of the soil.in this research , lots of date of uplift capacity about this pile , because of the limitation of this space ,so this paper will not repeat them.

## 4. Conclusions

The development and application of the finite element software ABAQUS makes it possible to simulate the uplift capacity of the pile. In this analysis, it is feasible to increase the uplift bearing capacity of the pile by increasing the pile length.

**Acknowledgements**

This research is financially supported by the natural science foundation of Tianjin No.16JCQNJC06700 and the Open Research Fund of Tianjin Key Laboratory of Soft Soil Characteristics and Engineering Environment No. 2013SCEEKL03 and the Tianjin enterprises postdoctoral innovation projects preferred funding.

## Reference

[1] Pile bearing capacity[M]. Springer Berlin Heidelberg, 2014.

[2] Meyerhof G G. Bearing capacity and settlement of pile foundations (11th Terzaghi Lecture)[J]. Journal of the Geotechnical Engineering Division, 1976, 102(9):195-228.

[3] Gong W. THEORETICAL STUDY ON SELF-BALANCED LOADING TEST FOR PILE BEARING CAPACITY[J]. Industrial Construction, 2002.

[4] Meyerhof G G, Mathur S K, Valsangkar A J. The bearing capacity of rigid piles and pile groups under inclined loa. [J]. Canadian Geotechnical Journal, 2011, 18(2):297-300.

[5] Momeni E, Nazir R, Armaghani D J, et al. Prediction of pile bearing capacity using a hybrid genetic algorithm-based ANN[J]. Measurement, 2014, 57(11):122-131.

# The Application of FAHP in Decisions of Pavement Maintenance

**Zhaorong Wu**

Tongji Zhejiang College, Jiaxing, Zhejiang 314051, China.
wuzhaorong110@126.com

**Abstract.** In this paper, a method of building the fuzzy complementary judgement matrix and checking consistency is introduced based on the knowledge of the basic theory of FAHP and the procedure to establish the mathematical model corresponded. The scope and the advantages in the problems of multi-objective decisions have also been discussed. The availability of its use in the management system in pavement maintenance is demonstrated by analyzing the optimization for maintenance. Meanwhile, the faulty is also pointed out.

## 1. Introduction

With the rapid growth of the high-grade highway mileages and the traffic volume as well as the increasing severity of the overload problem of transport vehicles, damages to the different extent have occurred on the pavement in succession. At the moment, the highway management department is supposed not only to keep the partially damaged roads, but also to take economic and reasonable maintenance strategies against the trend of reducing the pavement performance. In the process of rapid development of highway transportations, elements that influence the increasing amount of collection volume and calculation volume of the basic data for the pavement maintenance management decisions are gradually increasing and full of fuzziness and uncertainty. To solve this type of problem of multi-objective decisions that man determines the pros and cons of the decision schemes according to various objectives and criteria, analytic hierarchy process is generally applied, which is short of considerations of the fuzziness of the subjective judgment. People's thinking, in which decisions are made, is different from the consistency of the judgment matrix, so it needs to be comprehensively evaluated. Therefore, some academics put forward a method called FAHP, which combines the fuzzy mathematical theories with the analytic hierarchy process and expands the applications to the fuzzy environments which possess many evaluation indexes[1-3].

## 2. The calculation model of FAHP

### *2.1. The establishment of fuzzy complementary matrix*

When every two elements compare with each other in the fuzzy hierarchy analysis, their respective importance in the system can be quantitatively presented as the fuzzy judgment matrix $\mathbf{A}=(a_{ij})_{n\times n}$. If the judgment matrix possesses the following two characteristics, it can be called the fuzzy complementary judgment matrix. 1) $a_{ij}=0.5,\ i=1,2,\ldots,n;$ 2) $a_{ij}+a_{ji}=1,\ j=1,2,\ldots,n.$

To describe any two schemes' relative importance to a criterion, it's very usual to use the two scale methods in Table 1 or Table 2 to mark it quantitatively[4].

**Table 1.** Method of nine marks and its significance

| Scale Value $a_{ij}$ | Definition | Explanation |
|---|---|---|
| 1 | The same important | Two elements are the same important when comparing with each other. |
| 3 | A bit more important | One element is a bit more important than the other one when comparing with each other. |
| 5 | Obviously more important | One element is obviously more important than the other one when comparing with each other. |
| 7 | Much more important | One element is much more important than the other one when comparing with each other. |
| 9 | Extremely important | One element is extremely more important than the other when comparing with each other. |
| 2、4、6、8 | Median | The median of the neighboring comparison above. |
| Reciprocal | Converse comparison | If the result of the comparison between element $a_{ij}$ and element $a_{ji}$ is judgment $r_{ii}$, the result of the comparison between element $a_j$ and element $a_i$ is judgment $r_{ji}=1/r_{ij}$. |

**Table 2.** 0.1~0.9 Scale method and its significance

| Scale Value $a_{ij}$ | Definition | Explanation |
|---|---|---|
| 0.5 | The same important | Two elements are the same important when comparing with each other. |
| 0.6 | A bit more important | One element is a bit more important than the other one when comparing with each other. |
| 0.7 | Obviously more important | One element is obviously more important than the other one when comparing with each other. |
| 0.8 | Much more important | One element is much more important than the other one when comparing with each other. |
| 0.9 | Extremely important | One element is extremely more important than the other when comparing with each other. |
| 0.1, 0.2, 0.3, 0.4 | The converse comparison | If the result of the comparison between element $a_{ij}$ and element $a_{ji}$ is judgment $r_{ii}$, the result of the comparison between element $a_j$ and element $a_i$ is judgment $r_{ji}=1/r_{ij}$. |

The scale value $a_{ii}=0.5$ suggests that two elements share the same importance; while the value $a_{ij} \in [0.1,0.5)$ suggests that element $X_j$ is more important than element $X_i$; And the value $a_{ij} \in [0.5,0.9)$ means that element $X_i$ is more important than element $X_j$.

According to the numerical scales mark above, elements $a_1, a_2, \ldots, a_n$ compare with each other respectively. Then the result is just like the following fuzzy complementary judgment matrix.

$$\mathbf{A} = \begin{bmatrix} a_{11} & a_{12} & \cdots & a_{1n} \\ a_{21} & a_{22} & \cdots & a_{2n} \\ \cdots & \cdots & \cdots & \cdots \\ a_{n1} & a_{n2} & \cdots & a_{nn} \end{bmatrix} \tag{1}$$

*2.2. Determination of the weight of each hierarchy evaluation index*

The main task of the mutual sorting of hierarchies is to define the extent, to which the two elements of hierarchies up and down influence each other. It's usually represented as an index value in the mathematical model. Below are the calculating steps of the index weight[5].

1) The product $\mathbf{M_i}$ of each row is calculated in the judgment matrix: $M_i = \prod_{j=1}^{n} a_{ij}$

2) The nth root $\mathbf{M_i}$ of is calculated: $W_i' = \sqrt[n]{M_i}$

3) The vector is regularized, $W_i = \frac{W_i'}{\sum_{i=1}^{n} W_i'}$ which can make it satisfy the requirement $\sum_{i=1}^{n} W_i' = 1$, then $\mathbf{W}=[W_1, W_2, W_3, W_4, W_5, W_6]$.

4) The greatest characteristic root $\lambda_{max}$ of the judgment matrix is calculated: $\lambda_{max} = \sum_{i=1}^{n} \frac{(AW)_i}{nW_i}$

*2.3. The testing method of consistency of fuzzy complementary judgment matrix*

Whether the weight value from the calculation above is reasonable and whether it can be used to evaluate the rationality of the judgment matrix established and handled by the decision makers should be further compared and judged to examine the consistency. If the judgment matrix obtained is a consistency matrix, it suggests that the qualitative judgments made by the decision makers logically meet the requirement of transitivity. Although in these decisions there are only comparisons between any two of them, the qualitative judgments are consistent and acceptable. To the contrary, they should be corrected.

Therefore, the evaluation indexes $C.I$ and $C.R$ also need to be calculated. It is regulated, when $C.R< 0.10$, that the fuzzy complementary judgment matrix **A** is defined to meet the requirement of consistency and acceptance. The smaller the $C.R$ is, the greater the consistency is. When the value is zero, it means the complete consistency of the judgment matrix[4]. $C.I=\lambda_{max}-n/(n-1)$, $C.R = C.I/R.I$, $R.I$ is the correction value shown in Table 3.

**Table 3.** *R.I* Corrections

| **Order** | 1 | 2 | 3 | 4 | 5 | 6 | 7 | 8 | 9 | 10 | 11 |
|---|---|---|---|---|---|---|---|---|---|---|---|
| ***R.I*** | 0 | 0 | 0.54 | 0.9 | 1.12 | 1.26 | 1.36 | 1.41 | 1.46 | 1.49 | 1.52 |

*2.4. The establishment of fuzzy comprehensive evaluation index matrix*

Members in an experts group figure out the proportion of the risk elements in each hierarchy with analytic hierarchy process and analyze it connecting with the practical situations. Besides, they also evaluate the condition of each risk element in the element level, so they get a judgment matrix **U**. Then the general formula in the reference [6], which is used to obtain the weights of fuzzy complementary judgment matrix, is shown below:

$$U_i = \frac{\sum_{j=1}^{n} a_{ij} + \frac{n}{2} - 1}{n(n-1)}, (i = 1,2, \dots, n) \quad (2)$$

This formula fully includes the advantages and the judging information of the fuzzy consistency judgment matrix. Its amount of calculation is small, which is convenient for the computer programming. Besides, it's very easy to be applied to the practical project.

*2.5. The fuzzy comprehensive evaluation results*

Firstly the evaluation index weight value $U_i$ of each hierarchy is worked out, and then the evaluation index matrix **U** of each hierarchy is further worked out, which is the fuzzy comprehensive evaluation set. At last, the fuzzy comprehensive evaluation results through the formula $\mathbf{R} = \mathbf{U} * \mathbf{W}^T$ are figured out.

**3. The analysis of the application example**

The four alternative schemes mentioned in the pavement maintenance management decisions are regarded as the research objects, which will be optimized and sorted. At the same time, the fuzzy analytic hierarchy process is used to evaluate them comprehensively. The main elements influencing the schemes decision are classified into four categories, which are the pavement performance evaluation, the traffic volume, the road age and the status and function of the road sections in the road network. They are considered as the primary element level.

In general, the top level of the decision-making hierarchy structure is the optimized objective, which is the objective level; all the alternative schemes are in the bottom level, which is the decision level; in the middle are all the elements influencing the decisions, which is the criterion level (it can be multilevel). In this paper, the research objects' corresponding hierarchy structure of program analysis is shown in Table 4.

**Table 4.** Hierarchical chart of program analysis

| **Objective Level** | **Schemes Optimization and Sorting U of Highway Maintenance Management** | | | |
|---|---|---|---|---|
| Criterion level | Evaluation of the pavement performance $U_1$ | Traffic volume $U_2$ | Road age $U_3$ | Road function $U_4$ |
| Decision level | Maintenance scheme 1 | Maintenance scheme 2 | Maintenance scheme 3 | Maintenance scheme 4 |

1) The judgment matrix **A** is determined through scoring after investigating 10 experts with questionnaire survey. According to the qualitative and quantitative empirical analysis of experts, the relative value of the importance of two elements in the same element level is figured out with the method of nine marks, and the weights matrix is obtained, which is shown in Table 5.

**Table 5.** Fuzzy complementary judgment matrix of index weight **U**-$U_i$

| **U** | $U_1$ | $U_2$ | $U_3$ | $U_4$ | $W_i$ | $\lambda_{max}$ | $C.R$ |
|---|---|---|---|---|---|---|---|
| $U_1$ | 1 | 3 | 1 | 1/3 | 0.201 | 4.044 | 0.0161<0.1 |
| $U_2$ | 1/3 | 1 | 1/3 | 1/5 | 0.077 | | |
| $U_3$ | 1 | 3 | 1 | 1/3 | 0.201 | | |
| $U_4$ | 3 | 5 | 3 | 1 | 0.521 | | |

2) The fuzzy judgment matrix $U_i$ of the four alternative maintenance schemes is established respectively and successively according to the order- evaluation of pavement performance, traffic volume, road age, the status and function of the road sections in the road network, which is shown from Table 6 to Table 9. And it's scored with the 0.1~0.9 scale method[4].

**Table 6.** Fuzzy Complementary Judgment Matrix of Pavement Performance index $U_1$

| $U_1$ | $U_{11}$ | $U_{12}$ | $U_{13}$ | $U_{14}$ | $U_i$ |
|---|---|---|---|---|---|
| $U_{11}$ | 0.5 | 0.4 | 0.4 | 0.6 | 0.242 |
| $U_{12}$ | 0.6 | 0.5 | 0.5 | 0.7 | 0.275 |
| $U_{13}$ | 0.6 | 0.5 | 0.5 | 0.7 | 0.275 |
| $U_{14}$ | 0.4 | 0.3 | 0.3 | 0.5 | 0.208 |

**Table 7.** Fuzzy Complementary Judgment Matrix of Traffic Volume Index $U_2$

| $U_2$ | $U_{21}$ | $U_{22}$ | $U_{23}$ | $U_{24}$ | $U_i$ |
|---|---|---|---|---|---|
| $U_{21}$ | 0.5 | 0.3 | 0.7 | 0.6 | 0.258 |
| $U_{22}$ | 0.7 | 0.5 | 0.9 | 0.8 | 0.325 |
| $U_{23}$ | 0.3 | 0.1 | 0.5 | 0.4 | 0.192 |
| $U_{24}$ | 0.4 | 0.2 | 0.6 | 0.5 | 0.225 |

**Table 8.** Fuzzy Complementary Judgment Matrix of Road Age Index $U_3$

| $U_3$ | $U_{31}$ | $U_{32}$ | $U_{33}$ | $U_{34}$ | $U_i$ |
|---|---|---|---|---|---|
| $U_{31}$ | 0.5 | 0.6 | 0.5 | 0.8 | 0.283 |
| $U_{32}$ | 0.4 | 0.5 | 0.4 | 0.7 | 0.250 |
| $U_{33}$ | 0.5 | 0.6 | 0.5 | 0.8 | 0.283 |
| $U_{34}$ | 0.2 | 0.3 | 0.2 | 0.5 | 0.208 |

**Table 9.** Fuzzy Complementary Judgment Matrix of Road Function Index $U_4$

| $U_4$ | $U_{41}$ | $U_{42}$ | $U_{43}$ | $U_{44}$ | $U_i$ |
|---|---|---|---|---|---|
| $U_{41}$ | 0.5 | 0.7 | 0.6 | 0.4 | 0.267 |
| $U_{42}$ | 0.3 | 0.5 | 0.4 | 0.2 | 0.200 |
| $U_{43}$ | 0.4 | 0.6 | 0.5 | 0.3 | 0.233 |
| $U_{44}$ | 0.6 | 0.8 | 0.7 | 0.5 | 0.300 |

According to the four sorting vectors of maintenance schemes which are from every single element above, in the criterion level, the evaluation index weight **U** of schemes optimization sorting criteria of the highway pavement maintenance management can be:

$$U = \begin{bmatrix} 0.242 & 0.275 & 0.275 & 0.208 \\ 0.285 & 0.325 & 0.192 & 0.225 \\ 0.283 & 0.250 & 0.283 & 0.208 \\ 0.267 & 0.200 & 0.233 & 0.300 \end{bmatrix} \tag{3}$$

According the evaluation results of criterion level, the fuzzy comprehensive evaluation results of the objective level can be obtained from the formula $R = U * W$:

$$R = U * W = \begin{bmatrix} 0.242 & 0.275 & 0.275 & 0.208 \\ 0.285 & 0.325 & 0.192 & 0.225 \\ 0.283 & 0.250 & 0.283 & 0.208 \\ 0.267 & 0.200 & 0.233 & 0.300 \end{bmatrix} \begin{bmatrix} 0.201 & 0.077 & 0.201 & 0.521 \end{bmatrix} = \begin{bmatrix} 0.2335 & 0.2327 & 0.2414 & 0.2722 \end{bmatrix} \tag{4}$$

According to the maximum subordination principle[7-8], the optimization sorting of each alternative scheme is just like this: 0.2722(Scheme4)>0.2414(Scheme 3)>0.2335(Scheme 1)>0.2327(Scheme 2). Therefore, Scheme 4 is the optimal choice.

## 4. Conclusion

This paper introduces the mathematical model of FAHP, analyzes its calculation theory and applies this method to the decisions optimization of pavement maintenance management to sort the schemes optimization. And it makes full use of the information in the alternatives, combining the qualitative analysis and quantitative analysis organically to get the best scheme. So it is proved that this method can be generally used in the evaluation of some complicated systems and in the multi-objectives decisions of schemes optimization.

While FAHP has solved this problem- the existing difference between the consistency of matrix and people's thinking, it's still not perfect enough and needs to be improved. This can be specifically divided into the following three aspects:

- If the index weight of various schemes is determined only according to the consistency of the fuzzy complementary judgment matrix when people make analysis and decisions with this method, the result can be less persuasive;
- Because of the quantitative limit of experts taking part in the evaluation and the differences in the individual subjectivity, there are limitations in the data collections to a certain extent, which may also influence the final evaluation results;
- With the increasing hierarchies of the structural layer, the calculation can be more and more complicated. Then the programming software will be needed to solve the problems.

## Acknowledgments

This research was supported by the fund project (Exploration and Research for the Construction Standard of Beautiful Highway) from Anji Highway Bureau (No. 0216008) in Zhejiang China. The author is grateful for the support of the project panel and project team.

## References

[1] Lan Jibin, Xu Yang, Huo Liangan, Liu Jiazhong. *Research on the priorities of fuzzy analytical hierarchy process.* Systems Engineering-Theory & Practice. 2006,09:107-112.

[2] Ji dongchao, Song Bifeng, Yu tianxiang. The method of decision-making based on FAHP and its application. Fire Control and Command Control. 2007,11:38-41.

[3] Li Yong, Hu Xianghong, Qiao Jian. An improved fuzzy AHP method. Journal of Northwest University (Natural Science Edition). 2005,01:11-12-16.

[4] Deng Xue, Li Jiaming, Zeng Haojian, Chen Junyang, Zhao Junfeng. Research on computation methods of AHP wight vector and its applications. Mathematics in Practice and Theory. 2012,07:93-100.

[5] Chen Xin. Application of fuzzy analytic hierarchy process for optimum selection in decision-making. Computer Engineering and Design, 2004,10:1847-1849.
[6] Li Dongchang. Study on the application of modified fuzzy AHP to environmental impact assessment. Shandong Communication Science and Technology, 2011,05:70-72.
[7] Luo Yangqin, Shao Xudong, Hu Boxue. Application of fuzzy analytic hierarchy process for bridge health evaluating. Hunan Communication Science and Technology, 2008,04:97-101.
[8] Liu Pan, Mao Liang. On the affordable housing government risk evaluation based on fuzzy hierarchy analysis. Value Engineering, 2015,03:8-10.

IOP Conf. Series: Earth and Environmental Science **61** (2017) 012099 doi:10.1088/1755-1315/61/1/012099

# The application of dynamic replacement method in deviation rectification of support pile

**Congwen Peng**

School of Urban Construction, Yangtze University, Jingzhou, 434023, China
ypengcw@126.com

**Abstract**. The dynamic replacement method is uppled widely to the soft soil treatment, but is seldom used in retificating pile deviation. By the ways of dynamic replacement method gravel block can form underground, the gravel block can be treated as reaction-froce base, the magnitude of squeeze force needed for retificating pile deviation is related to the scale of gravel block. Take a sheet pile retaining wall for instance, the treatment scheme and paramters calculating method for deviation rectification of support pile was put forward. The project case demonstrate that this method is effective when the layout of gravel column is reasonable to a certain degree, it is a good reference to similar projects.

## 1.Introduction

Dynamic replacement method is mainly used in soft soil reinforcement by squeezing the coarse aggregate, such as gravel, stone, slag, into natural foundation. This kind of method can form different diameters of gravel column underground. The gravel columns (also as gravel pile) combined the roles of composite foundation and vertical drainage channel, and meanwhiles has the effect of soil compaction around the pile body[1-3].

Pile deflection usually happen due to a variety of reasons, unloading and loading methods are commonly used for the deviation rectification of piles[4,5] . The "unloading" means removing soil or constructing stress removal holes for decreasing the earth pressure acting on the pile in the opposite direction of pile deflection, the "loading" means increasing force acting on the pile in the direction of pile deflection by the ways including stacking materials on the ground, anchor pulling or jack pushing . Among them, the jack pushing method is the most effective and most common used way, but this way can not work if there is no counter-force base.Under this condition, how to build a effective counter-force base is a key point. By making full use of the soil compaction effect, dynamic replacement method can bulid counter-force base as well as offer pushing force, In this paper dynamic replacement method is applied to pile deflection rectification.

## 2.Project profile

There is a filling project that plan to fill soil from current elevation of 71.0-73.0 meters to the elevation of 100.0 meters, the filling soil strengthed by dynamic compaction. Because of limited range of land use, this project will finally forming a high artifical slope, so sheet pile rataining wall is used to protect the slope from sliding, the retaining wall structure profile see Fig1, the slope ratio is about 1:1.5, the diameter of support pile is 1.5 meters, the space of pile is 3.0 meters, the shape of pile above ground is square column(also as pile-column) with 1.7 meters side length, the height of retaining wall is 6.0 meters, retaining plate is installed behind square colume,every 5 square colume join together by connecting beam on the top of column.

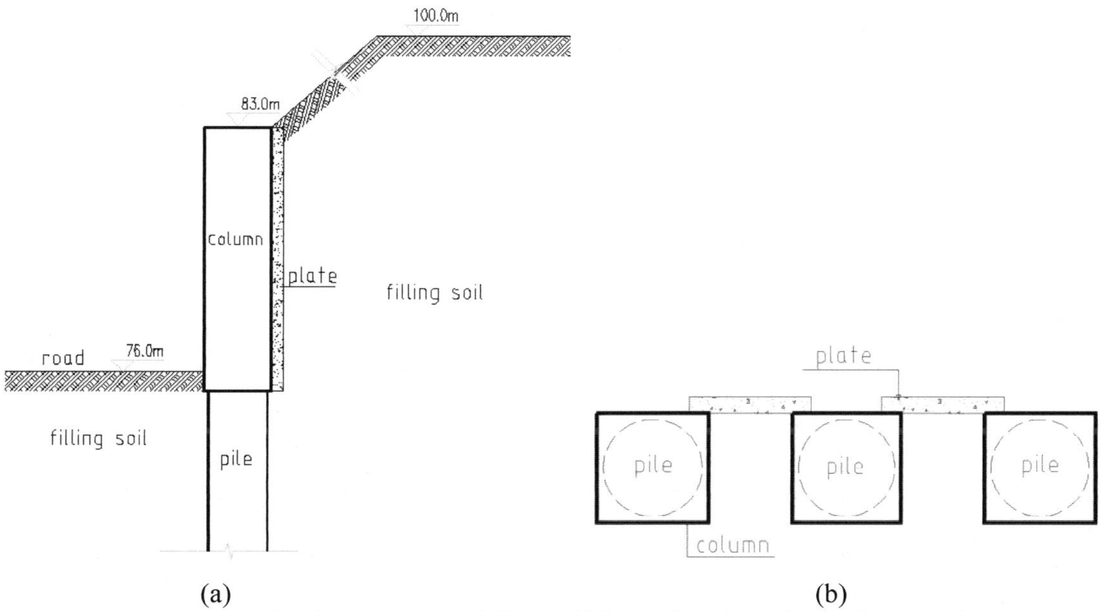


**Figure1.** The drawing of retaining wall.(a)sectional graph, (b)plane graph.

The support piles begin to construct when the elevation of filling surface get to 76.0 meters, after construction of sheet pile retaining wall, the filling working behind wall continue, rainfall occurs during the filling. When the filling soil close to the top of the retaining wall, about 10 square column are found offset in different degrees (see Figure 2). The maxmium offset value of the retaining wall is approximately 20.0-50.0mm, no crack found in the surface of square column. There is apparent soil uplift and lateral fissure phenomenon in the ground close the toe of column, the whole pile body structural is in good condition according to low-strain test.So comprehensive judgment is that the support pile has been offset by earth pressure as a whole.

**Figure2**. The photo of pile deviation.

After detailed investigation, the cause of pile deviation summarized as following: (1)The filling soil was not processed seriously as required. It was relatively loose, resulting in the serious collapse of borehole wall when the bored pile constructing. The high pressure jet grouting method was used for the reinforcement of loose filling, this way lead to form a waterproof curtain along the bottom of retaining wall, which prevented underground water from draining; (2)The filter layer in filling soil was not processed as required after the construction of the retaining wall, resulting that the water in the filling behind the wall cannot be drained in time; (3) Because of the newly-filling soil was relatively loose, the passive resistance in front of support piles was relatively small. All these reasons mentioned above resulted in the excessive earth pressure behind the wall and finally lead to offset of support pile.

## 3. Treatment method

### *3.1 Treatment scheme*

Under the premise of fully taking the safey of the retaining wall into account, the tilted piles should be rectified in the minimize time and expense, and the soil in the passive pressure areas in front of piles should also be strengthen. Only get to this goals can the filling behind the wall be proceeded smoothly.

The basic idea of pile rectification embraces two steps. First, a certain deep of filling soil behind the retaining wall is removed. Second, pushing force on the piles in the opposite direction of piles deflection is applied. The pushing force usually exerted by anchor or jack. According to the site situation, there is no proper anchor point or counter-force base, it is also impossible to build a new temporary counter-force base within time and cost allowed. So these methods of anchor or jack can not be adopted.Through synthetical consideration of treatment requirements and site constructing conditions, dynamic replacement method was applied. Dynamic replacement method can form gravel column underground by squeezing gravels into filling soils, this way can make the soils more compact and exert squeeze pressure at the meantime. It can not only rectify the tilted piles but aslo can strength the filling soils by extrusion force.

The degree of filling compaction relate to the layout and construction process control of gravel column. The first key point is to control the direction of extrusion pressure and make sure the squeeze pressure must pass toward the direction of tilted piles gradually. In general, during the gravel column formation, the squeeze pressure expands to surrounding areas. Therefore, the gravel compaction points should join gravel column together and forming a gravel block, this gravel block is a binding wall that can control the spreading direction of extrusion force. The other key point is to control the magnitude of extrusion force and make sure the tilted pile should not be damaged by the extrusion force. The magnitude of extrusion force is related to confining pressure formed by gravel block. With the above factors taken into consideration, a gravel column layout scheme is proposed as depicted in Fig 3(a). Every circle stand for a gravel column, the number in the circles show the sequence of gravel column constructed. The shadow part is the gravel block planed to form by dynamic replacement method.

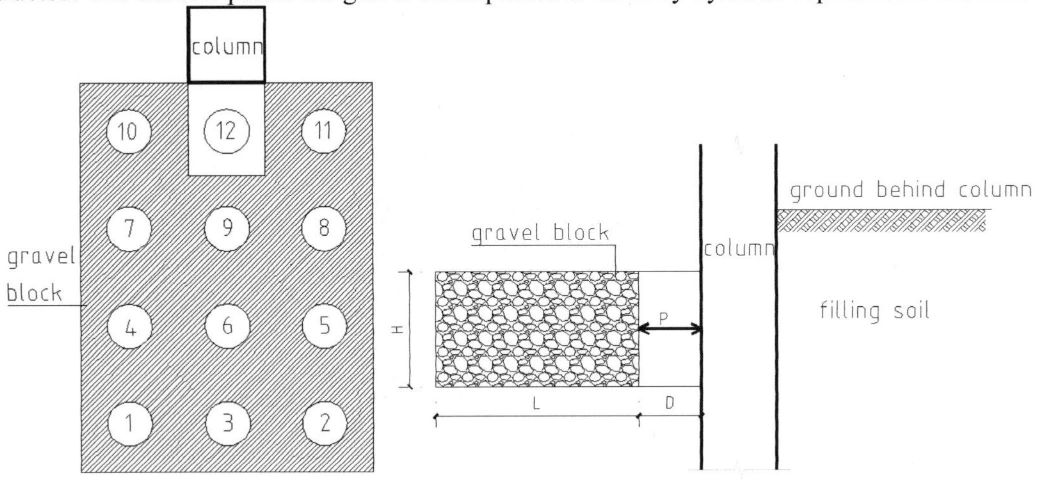


(a) (b)

**Figure 3**. Treatment scheme. (a) the layout of gravel column points, (b)calculation model.

### *3.2 Paramters estimation*

Extrusion pressure,denote as P (see Figure 3b) is a force acting on tilted piles due to gravel compaction effect, the value of extrusion pressure is related to some parameters such as the

IOP Conf. Series: Earth and Environmental Science **61** (2017) 012099 doi:10.1088/1755-1315/61/1/012099

thickness(H) and length(L) of gravel block, the height of soils removed behind walls,the thickness of filling soils.In this paper dynamic reinforcement problem is simplified to a quasi-static problem, and consider the gravel blocks as rigid body. For estimating the required maximum extrusion pressure, the first step is to know the height of removing soil behind the wall according to the actual situation, the thickness of filling soil in front of wall and the thickness of gravel block, the second step is to estimate the required extrusion force according to the formula (1), the third step is to estimate the length of gravel block according to equation (3) under the premise of equation (2).

$$E_p h_p = E_a h_a + Ph \tag{1}$$

$$P \leq 0.25\beta_c f_c b h_0 \tag{2}$$

$$P \leq LH\gamma_g \tan\varphi_s + 2LHE\tan\varphi_g + Hf_a \tag{3}$$

Where: $E_a, E_p$ is active and passive earth pressure respectively; $h_a$ is the distance between active earth pressure and the bottom of the pile column; $h$ is the distance between the resultant force of extrusion pressure and pile column bottom; $r_g, \varphi_g$ are the unit weight and internal friction angle of gravel block respectively; $f_a$ is the bearing capacity of the reinforced soil; $E$ is the earth pressure acting on the side wall of grave block; $D$ is the diameter of gravel column; $\beta_c$ is concrete strength influence coefficient; $b$ is cross section width of pile-column; $h_0$ is section effective height, $h_0 = 1.6r$, $r$ is the radius of support pile.

In this project, the filling soil behind the wall is discharged to the elevation of 76.0 meters, and the soil in front of the wall is filled to the elevation of 77.0 meters. A single gravel column diameter is about 2.0 meters, the depth of reinforcement should be large than the thickness of filling soils, $H = 6.0m$, $r_g = 20kN/m^3$, $\varphi_g = 15^\circ$.The contribution of the end part of the gravel block to the reaction force is not taken into account. The calculation result is $L \geq 5.0m$.The actual arrangement is 4 rows of gravel column with 2.0 meters spacing, the total length of the gravel column is 8.0 meters.

### *3.3 Construction method*

Piles deviation rectification working proceed taking each pile as an object. There is 10 deflection piles, labelled from 1#-10#, the construction sequence is 4# - 6# - 8# - 10# - 9# - 5# - 7# - 3# - 1#. The construction steps is as follows: (1)remove the filling soil behind retaining wall to the bottom of pile-column; (2)cut the connecting beam on the top of the retaining wall; (3)level the site and mark the gravel column position; (4)construct the gravel column according to the sequence shown in Figure 3 (a); (5)repeat the above steps for the next tilted pile; (6)after all the rectification work is completed, re-construction the connecting beam on the top of pile-column.

The following control measures are adopted in the construction process:(1)The surrounding exposed pipeline are effectively covered in case of damage before construction; (2)Tamping energy is controlled. Taking into account the needed reinforcement depth of 6 meters, the requirements of tamping energy is between $3000kN \cdot m - 4000kN \cdot m$, the average settlement of ground induced by the last two hit is not greater than 10cm, and actual formed gravel column diameter is not less than 2.2 meters; (3)Steel wire rope is used for orientation in case of pile-column offset to two side in the construction process, the wire rope is fixed to bulldozer; (4)The workers must pay attention to the change of velocity of pile rectification, adjust the tamping energy and the gravel column position slightly according to the change of pile position during construction.

## 4. Conclusions

The deviation piles rectification working are successfully completed after more than half a month. This engineering practice shows that dynamic replacement method can be applied to the pile rectification effectively. For similar projects, only relevant parameters need adjusted. It should be pointed out that there is large vibration during construction and may have influence on the surrounding environment; in addition, parameters calculate assuming the gravel block as rigid body and do not consider the dynamic effect of pressure change, all these things are not in line with actual situation and needs to be further improved in future work.

**References**

[1] Lei S Q and Chen C X 2012 Struction technology on deviation rectification and strengthening of some bridge *Construction Technology*. **41** 88-91

[2] Shi G and Wu X Z 2015 Design and implementation for pile deviations rectifying scheme for pile foundations of piersof a bridge *Bridge Construction.* **45**(1) 97-102

[3] Bai B,Xu H X and Liu H B 2010 Some problems on dynamic replacement method to improve soft soil ground *Chinese Journal of Rock Mechanics and Engineering.* **29**(suppl) 3001-06

[4] Teng K 2008 Evaluation and improvement of formulas for replacement depth under dynamic compaction *Chinese Journal of Geotechnical Engineering.* **30**(7) 994-98

[5] Wang H J,Han W X and Zhao Q H 2004 The application of dynamic consolidation replacement method in the high embankment treatment *Journal of Disaster Prevention and Mitigation Engineering.* **24**(3) 277-82

IOP Conf. Series: Earth and Environmental Science **61** (2017) 012100 doi:10.1088/1755-1315/61/1/012100

# Calculation and analysis of whiplash effect in multi-DOF system

**Gong Chen[1,2,a)], Qunfu Qi[1,b)], Hongmei Yang[1,c)] and Ling Li [1,d)]**

[1]Chengdu Textile College, Chengdu 611731, China
[2]Chengdu University of Technology, Chengdu 610059, China
[a)]Corresponding author: 13880680121@139.com
[b)]cdfzqqf@sina.com
[c)]33205524@qq.com
[d)]57897864@qq.com

**Abstract.** The response of structural vibration depends on the degree of freedomss, mass, stiffness, external effect of the structure. Natural frequency of structure is its inherent property, also it is related to its own factors, but not related to the external factors. Firstly the nature and characteristics of resonance effect and whiplash effect are analyzed. Secondly whiplash effect of multi-degree freedomss system is analyzed, then orthogonality in main models is used to verify the results of calculation. Then the improvements of whiplash effect are proposed, also the improved models are analyzed. Finally the conclusions are made.

## 1 Introduction

Resonance effect and whiplash effect in the structure are usually happened during vibration, and this is very harmful to the structure. We should avoid resonance effect and whiplash effect when the structure is designed, also we must take precaution to avoid the resonance effect and whiplash effect. Not only the contact, but also the difference are existed between resonance effect and whiplash effect. The contact is that the effects may cause strenuous exercise and large displacement in the structure. However, when the external frequency of structure is equal to or close to its natural frequency, resonance effect which cause strenuous exercise and large displacement in the structure may happen. When the mass and stiffness on top of the structure shrink suddenlly, whiplash effect which cause strenuous exercise and large displacement in the structure may happen. The difference is that resonance effect may happen when the structure is forcing to vibrate, but whiplash effect may happen when the structure is forcing to vibrate or free to vibrate.

In our daily life, whiplash effect is a very common phenomenon. When we grip the root of whip and throw it off energetically, it will move accordingly with our hands. The tip of whip will move sharply, and the displacement and movement will increase. Whiplash effect has been explained in Stanrdard for Terminolog in Earthquake Engineering:"In the earthqueke, the top of the elongate projection in tall buildings and other structures will vibrate intensely." Since the mass and stiffness on top of the structure are both little, the first natural frequency is close to the vibration frequency of earth or natural frequency in mian structure, so whiplash effect may happen. That's the nature of whiplash effect. In the earthqueke, whiplash effect will cause damage easily in the parapet, tower building, attic, etc.

## 2 Calculation and analysis of whiplash effect in multi-DOF system

### *2.1 Multi-DOF System Model*

A three-freedoms-system is analyzed in the model 1, and model 1 is the example. Model 1 is showed by Figure 1.In Figure 1, m is the concentrated mass of every frame, the stiffness of beam is supposed to unlimited; k is the lateral stiffness of floor in the frame, also it is meaning of the force required to be applied when the single displacement is taken place; n is the adjustment factor of mass and stiffness.

### *2.2 Calculation of Free Vibration in Multi-DOF System*

When $n$=1, $n$=0.1, $n$=10, free vibrate of frame is analyzed, the conclusions is made in Table 1, Table 2, Table 3, Table 4. In the Table, $\omega_1$ is the first natural frequency of the vibrate system, $\omega_2$ is the second natural frequency of the vibrate system, $\omega_3$ is the third natural frequency of the vibrate system; $Y_1$ is the displacement amplitude of the first floor in the vibrate system, $Y_2$ is the displacement amplitude of the second floor in the vibrate system, $Y_3$ is the displacement amplitude of the third floor in the vibrate system.

### *2.3 Analysis of Whiplash Effect in Multi-DOF System*

Firstly, the conclusion is made in Table 1: natural frequency of multi-degree of freedoms system is related with the adjustment factor of mass and stiffness, the step of vibrate system, mass factor, stiffness factor.

Secondly, we can make conclusions according to the Table 2, Table 3, Table 4:

When $n$=1, mass and stiffness of structure distribute evently along vertical direction, also the displacement amplitude are changing evently.

When $n$=0.1, although the frontal two-step natural frequency distribute evently along vertical direction, the displacement amplitude of third-step natural frequency in the structure system increase significantly, when $Y_3$=-9.923, the absolute value is approximately ten times as much displacement amplitude as the other floors below. It indicates that whiplash may happen when mass and stiffness of multi-degree of freedoms system reduce suddenly in third-step natural frequency.

When $n$=10, mass and stiffness at the top of strcture is ten times as much as the other floors below, it is impossible in the engineering practice, but mass and stiffness at the top of strcture change suddenly. Displacement amplitude on the second floor where mass and stiffness change suddenly increase obviously when it is at second natural frequency, at the same time, $Y_2$=-10.008, the absolute value is approximately ten times as much displacement amplitude as the other floors below. It indicates that the whiplash effect may happen at the junction where changes suddenly, when mass and stiffness at the top of strcture increase suddenly. Otherwise, when it is at third natural frequency, the displacement amplitude on the first floor is different from the displacement amplitude other floors as twenty times as much. The accident when mass and stiffness change suddenly cause displacement amplitude changes obviously.

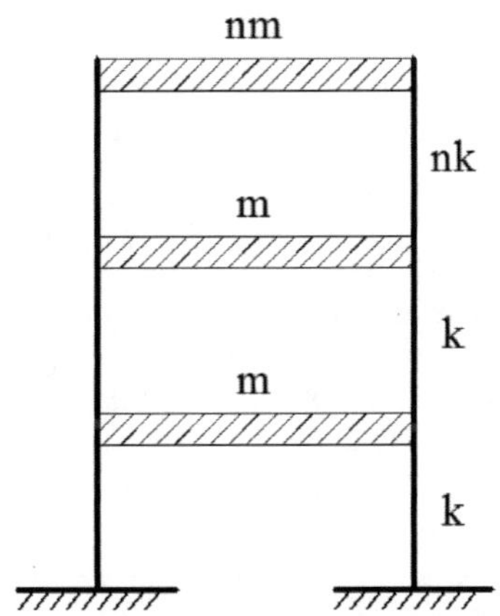


**FIGURE 1** MODEL 1

**Table 1** Relationship between adjustment factor and natural frequency

| $n \backslash \omega$ | $\omega_1$ | $\omega_2$ | $\omega_3$ |
|---|---|---|---|
| $n$=1 | $0.445\sqrt{\frac{k}{m}}$ | $1.802\sqrt{\frac{k}{m}}$ | $1.247\sqrt{\frac{k}{m}}$ |
| $n$=0.1 | $0.586\sqrt{\frac{k}{m}}$ | $1.632\sqrt{\frac{k}{m}}$ | $1.045\sqrt{\frac{k}{m}}$ |
| $n$=10 | $0.207\sqrt{\frac{k}{m}}$ | $3.465\sqrt{\frac{k}{m}}$ | $1.396\sqrt{\frac{k}{m}}$ |

**Table 2** Relationship between displacement amplitude and natural frequency when $n$=1

| $Y \backslash \omega$ | $\omega_1$ | $\omega_2$ | $\omega_3$ |
|---|---|---|---|
| $Y_1$ | 1 | 1 | 1 |
| $Y_2$ | 1.802 | -1.247 | 0.445 |
| $Y_3$ | 2.247 | 0.555 | -0.802 |

**Table 3** Relationship between displacement amplitude and natural frequency when $n$=0.1

| $Y \backslash \omega$ | $\omega_1$ | $\omega_2$ | $\omega_3$ |
|---|---|---|---|
| $Y_1$ | 1 | 1 | 1 |
| $Y_2$ | 1.656 | -0.665 | 0.908 |
| $Y_3$ | 2.524 | 0.399 | -9.923 |

**Table 4** Relationship between displacement amplitude and natural frequency when $n$=10

| $Y \backslash \omega$ | $\omega_1$ | $\omega_2$ | $\omega_3$ |
|---|---|---|---|
| $Y_1$ | 1 | 1 | 1 |
| $Y_2$ | 1.957 | -10.008 | 0.051 |
| $Y_3$ | 2.045 | 0.909 | -0.054 |

*2.4 Verification of Orthogonal Property in Main Model*

It is important that the models are orthogonal when they are vibrating in multi-degree of freedoms system, which is orthogonal of models. The Formulas are showed by Formula 1, Formula 2, Formula 3, Formula 4, Formula 5, Formula 6, Formula 7, Formula 8:

$$Y^{(l)T} M Y^{(k)} = 0 \qquad (1)$$

$$Y = \begin{Bmatrix} Y_1 \\ \vdots \\ Y_n \end{Bmatrix} \qquad (2)$$

$$M = \begin{Bmatrix} m_1 & & \\ & \ddots & \\ & & m_n \end{Bmatrix} \qquad (3)$$

In Formula 1, Formula 2, Formula 3, $Y$ is displacement amplitude vector of multi-degree of freedoms system, where $l \neq k$, $\omega_l \neq \omega_k$. $M$ is mass array of multi-degree of freedoms system. Any displacement amplitude vector and mass array are satisfied with orthogonal property. For example, when $n$=0.1, orthogonal property of main model in multi-degree of freedoms system is verified below:

$$M = \begin{Bmatrix} m & & \\ & m & \\ & & 0.1m \end{Bmatrix} \qquad (4)$$

$$Y^{(\omega_1)T} = (1 \quad 1.656 \quad 2.524) \quad (5)$$

$$Y^{(\omega_3)} = \begin{Bmatrix} 1 \\ 0.908 \\ -9.923 \end{Bmatrix} \qquad (6)$$

$$Y^{(\omega_1)T} M Y^{(\omega_3)} \approx 0 \qquad (7)$$

The orthogonal property of other model can be verified, and be satisfied with orthogonal property. Similarly, the orthogonal property of second model can be deduced below:

$$Y^{(l)T} K Y^{(k)} = 0 \qquad (8)$$

## 3 Improvement measures

To prevent whiplash effect in the structure system from damaging it, the layout of structure should keep regular shape, also the facade stiffness of structure should keep consistency or change uniformly. When $n$=0.1, the facade stiffness of structure is improved to change uniformly along the facade, the improved model is showed by Figure 2. The free vibration of improved frame is analyzed, the conlusions is made to be showed by Table 5 and Table 6. When $n$=0.1, the natural frequency of the original model and improved model is nearly same, but displacement amplitude during floors of improved model is less than displacement amplitude during floors of original model. Otherwise, displacement amplitude of third natural frequency at the top is less than original model. It indicates that whiplash effect of improved model is reducing because of the facade stiffness which has no changing, and which is reducing from top to bottom.

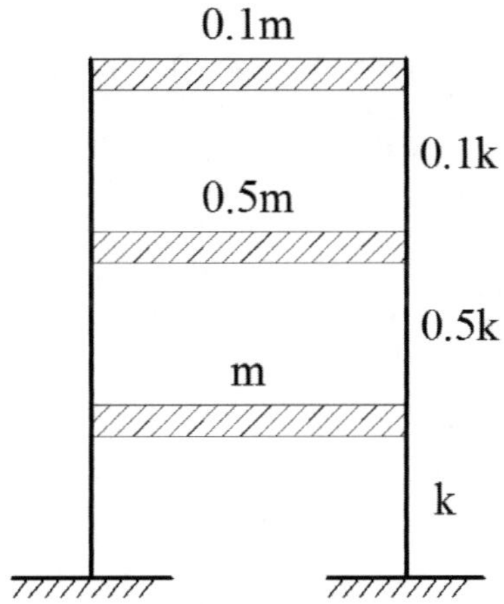


**Figure 2** Model 2

**Table 5** Natural frequency of improved model

| $\omega_1$ | $\omega_2$ | $\omega_3$ |
|---|---|---|
| $0.637\sqrt{\frac{k}{m}}$ | $1.465\sqrt{\frac{k}{m}}$ | $1.071\sqrt{\frac{k}{m}}$ |

**Table 6** Relationship between displacement amplitude and natural frequency in the improved model

| $Y\backslash\omega$ | $\omega_1$ | $\omega_2$ | $\omega_3$ |
|---|---|---|---|
| $Y_1$ | 1 | 1 | 1 |
| $Y_2$ | 2.188 | -1.294 | 0.706 |
| $Y_3$ | 3.684 | 1.128 | -4.812 |

## 4 Conclusion

In summary, in order to prevent whiplash effect from happenning in the structure system, the designers should pay attention to the regular in the plane, facade and vertical cross-section, which would affect seismic performance and economic rationality. The lateral stiffness and bearing capacity should not change suddenly, the unanimous and similar lateral stiffness of structuer system should be preferred. Since earthquake causes the large forcing at the bottom of structure, damping and seismic isolation should be adopted in the structure.

**Acknowledgments**
This research is financially supported by the 2015 Research Project of Chengdu Textile College (Grant No.2015fzlkc06).

**References**
[1] Long Yuqiu, Bao Shihua, Yuan Si. Structural MechanicsI. Beijing, Higher Education Press. 2012: 352-381.
[2] Long Yuqiu, Bao Shihua, Yuan Si. Structural MechanicsII. Beijing, Higher Education Press. 2012: 166-188.
[3] MOHURD. Stanrdard for terminolog in earthquake engineering. Beijing, China Architecture and Building Press. 2011: 25.
[4] Chen Daozheng, Li Aiqun. Seismic analysis and design recommendations of the steel tower on the top . Building Structure. 2004, 34(05): 33-35.
[5] MOHURD. Code for seismic design of buildings. Beijing, China Architecture and Building Press. 2010: 08-09.
[6] Hu Xingfu. Building structures. Beijing, China Architecture and Building Press. 2012: 151.

IOP Conf. Series: Earth and Environmental Science **61** (2017) 012101 doi:10.1088/1755-1315/61/1/012101

# Adjustment method for train disintegration and classification sequence in marshalling station

**Zhenkun Chen [1], Changfeng Zhu[2], Ran Sun[3*]**

[1] School of Traffic and Transportation, Lanzhou Jiaotong University, Lanzhou, China, 730070
[2] School of Traffic and Transportation, Lanzhou Jiaotong University, Lanzhou, China, 730070
[3] School of Traffic and Transportation, Lanzhou Jiaotong University, Lanzhou, China, 730070
*Emsil: czksrr@163.com

**Abstract.** Adjustment of train disintegrating and classifying schemes in the marshalling station is the key of wagon-flow allocating optimization. The train disintegration and classification sequence is subjected to the priority of time departure. Under above-mentioned arrangement, the priority of time departure scheme is difficult to ensure the full-load of each departure train. On the basis of analyzing for interval time of departure train, waiting departure time and the relation among each time parameter, several concepts will be given. Using the method of single machine scheduling problem is to minimize the influence for classification and disintegration sequence adjustment. The necessary conditions of adjustment can be obtained and the influence after adjustment is shown by graphic deduction. Finally, the conception of time pass theory is provided. The rationality of this method has been proved by a case study. This method lays a foundation for adjustment of train classification and disintegration sequences.

## 1.Introduction

The initial disintegration and classification scheme is the foundation of operation scheme, also is precondition for car flow assignment in marshalling station. Foreign scholars' researches on disintegration and classification are as follows: D R Martinelli[1]put forwarded an optimization on railway traffic control operation plan; M A Turnquist[2]set up a machine scheduling model which acquired new scheme by analysis the time parameters; Z Z Avaramovic[3]through analysis the connection of car flow follow get scheme. Domestic scholars' researched are as follows: Li Haodong[4] focused on survey of stage plan for railway marshalling station. Wang Shidong[5] made a mathematical model and algorithm for automatically programming. Xue fen[9-11]'s researches were on stage plan, such kind method established mathematical model and solved this kind of problem by hybrid algorithm.

The concepts of compact continuing classification and the greatest adjustment between adjoin trains were put forwarded by Xue fen, however the influence for whole scheme after an adjustment not be taken into account [12]. Peng Qiyuan combined classification with disintegration process, and focused on after-effect, however not given systematic adjustment approach and condition [13].

According to concepts of compact continuing classification and the greatest adjustment between adjoin trains[12], the marshalling station can be divided into two sub-systems. The operation in each subsystem can be regarded as machine scheduling. Under the condition of one shunting locomotive for disintegration and one for classification, the trains which are not full-load can use partial adjustment method to meet the requirements of railway station. The key part of this thesis is the analysis on the how to guarantee full-load as far as possible and punctuality of departure time of each train.

## 2. Analysis of Operation Systems

Marshalling station is a technical station which includes different types of work. The operation procedure has the same characteristics with machine scheduling. On the basis of analyzing machine scheduling problem, the marshalling station can be divided into two subsystems, one is arrival disintegration subsystem and the other is classification departure subsystem, each subsystem includes three assembly lines. The assembly lines for how to work under horizontal and vertical model can be shown as Fig.1 .

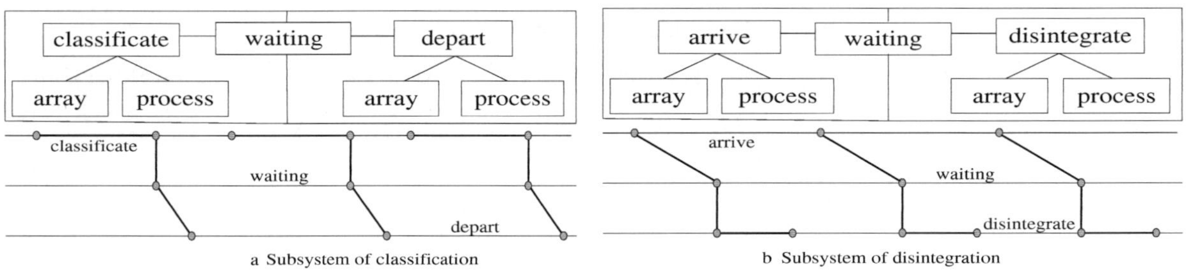


**Figure 1**. Flow operation of subsystem.

## 3. Classification Sequence Adjustment

### *3.1. Parameters Concept*

$m_0$, $n_0$, $B_n$, $J_m$—The number of arrival trains in periods; the number of departure trains in periods; Sequence of classification; Sequence of disintegration.( $m = 1,2,3,\ldots, m_0$, $n = 1,2,3,\ldots, n_0$)

$T_B$, $T_F$, $T_J$, $T_D$—The operation time of classification; The operation time of departure; The operation time of disintegration; The operation time of arrival.

$t_{CF}^{n}$, $t_{D}^{m}$—Departure time ; Arrival time.

$t_{ESJ}^{m}$, $t_{LSJ}^{m}$, $t_{EFJ}^{m}$, $t_{LFJ}^{m}$, $t_{EFJ}^{m'}$, $t_{ESJ}^{m'}$, $t_{LSJ}^{m'}$—The starting time of earliest disintegration; The starting time of latest disintegration; The ending time of earliest disintegration; The ending time of latest disintegration; The ending time of earliest disintegration after adjustment; The starting time of earliest disintegration after adjustment; The starting time of latest disintegration after adjustment.

$t_{ESB}^{n}$, $t_{LSB}^{n}$, $t_{ESB}^{n'}$, $t_{LSB}^{n'}$—The starting time of earliest classification; The starting time of latest classification; The starting time of earliest classification after adjustment; The starting time of latest classification after adjustment.

$T_{DB}^{n(n+1)}$, $T_{DCF}^{n}$, $T_{JCF}^{n(n+1)}$, $T_{DCF}^{n'}$, $T_{JCF}^{n(n+1)'}$—The waiting classification time of departure trains; The waiting departure time; Interval time between adjacent departure trains; The waiting departure time; Interval time between adjacent departure trains after adjustment.

$T_{DJ}^{m(m+1)}$, $T_{DD}^{m}$, $T_{JD}^{m(m+1)}$, $T_{DD}^{m'}$, $T_{JD}^{m(m+1)'}$—The waiting disintegration time of arrival trains; The time of waiting disintegration; Interval time between adjacent arrival trains; The time of waiting disintegration after adjustment; Interval time between adjacent arrival trains after adjustment.

$\Delta t$, $\Delta T$—The change after classification adjustment; The change after disintegration adjustment.

### *3.2. Analysis of Initial Sequence*

First, analyzing the time parameters at the condition of one shunting locomotive is under classification operation. If car flow follow is adequate, that is $T_{JCF}^{n(n+1)} > T_B$ ( $n \in \{n|1, n-1\}$ ) . The waiting classification time $T_{DB}^{n(n+1)}$ will come into being as shown in Fig.2 (a). It will provide enough time for car flow accumulation. If $T_{JCF}^{n(n+1)} \leq T_B$, car flow follow time is short, the trains will be depart compactly and the inspect operation of departure trains will be saturated. The waiting departure time $T_{DCF}^{n}$ will appear as shown in Fig.2 (b).

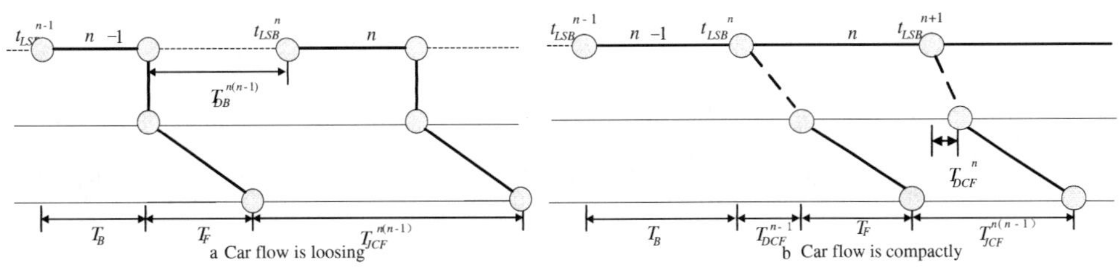

**Figure 2**. Analysis of initial classification sequence.

In order to guarantee full-load and punctuality, $t_{LSB}^{n}$ as an important parameter can get easily according to reference documentation [12] for follow-up adjustment:

$$t_{LSB}^{n_0} = t_{CF}^{n_0} - T_F - T_B \tag{1}$$

The latest classification starting time of other trains are closely linked with tightness degree.

If $T_{JCF}^{n(n+1)} > T_B$, car flow follow is loose, the latest classification starting time will be got by inverse derivation:

$$t_{LSB}^{n} = t_{CF}^{n} - T_F - T_B \tag{2}$$

If $T_{JCF}^{n(n+1)} \leq T_B$, the departure trains are compact classification, the latest classification starting time can be expressed as:

$$t_{LSB}^{n} = t_{LSB}^{n+1} - T_B \tag{3}$$

$$t_{LSB}^{n} = t_F^{n_0} - T_F - (n_0 - n + 1)T_B \tag{4}$$

*3.3 The Process of Classification Sequence Adjustment*

*3.3.1. Classification Sequence Adjustment between Adjacent Trains.* As shown in Fig. 2(b), if the interval of latest classification starting time between adjoint trains can meet $t_{LSB}^{n+1} - t_{LSB}^{n} = T_B$ , it will be named compact continuing classification [12]. If each train can meet the compact continuing classification in periods, that can be defined as section compact continuing classification, that is $t_{JCF}^{nn_0} \leq (n_0 - n)T_B$ . If the discrepancy between $T_{JCF}^{n(n+1)}$ and $T_B$ is large, it will not be compact continuing classification. Summed up above-mentioned, if it meet the condition of $t_{LSB}^{n+1} \leq t_{CF}^{n} - T_F$ , it will not be delayed. The adjustment scheme is as shown in Fig. 3.

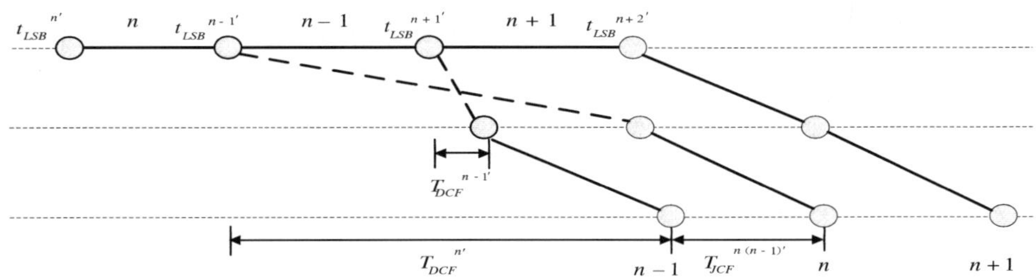


**Figure 3.** Adjustment scheme of classification sequence.

The parameter computation method after adjustment can be referenced to documentation [12]:

$$t_{LSB}^{n-1'} = t_{LSB}^{n-1} + T_B + T_{DCF}^{n-1} - T_{DCF}^{n-1'} - T_B \quad (5)$$

$$t_{LSB}^{n'} = t_{LSB}^{n} + T_B + T_{DCF}^{n} - T_{DCF}^{n'} - T_B = t_{LSB}^{n} + T_{DCF}^{n-1} - T_{DCF}^{n-1'} - 2T_B \quad (6)$$

*3.3.2. The Influence Analysis after Adjustment.* In order to guarantee punctuality of departure time, it has to meet $T_{DCF}^{n-1'} \geq 0 \bigcup T_{DCF}^{n'} \geq 0$. It only needs to prove $T_{DCF}^{n-1'} \geq 0$.

If $T_{DCF}^{n} \geq T_B$, and adjust $t_{LSB}^{n'}$ to $t_{LSB}^{n-1}$, that is $t_{LSB}^{n'} = t_{LSB}^{n-1}$ ,it will ensure $T_{DCF}^{n-1'} \geq 0$ and the delay of departure time will not be occurred . After an adjustment, changes between the latest classification starting time can be defined as $\Delta t$ .When $t_{LSB}^{n'} = t_{LSB}^{n-1}$, that will be $\Delta t = 0$.

The proved method is as shown as follows:

Based on the available conditions: $T_{DCF}^{n-1'} = t_{LSB}^{n} - t_{LSB}^{n'} + T_{DCF}^{n-1} - 2T_B$, $t_{LSB}^{n'} = t_{LSB}^{n-1}$ and $T_{DCF}^{n-1} \geq T_B$.

And $T_{DCF}^{n-1'} = t_{LSB}^{n} - t_{LSB}^{n-1} + T_{DCF}^{n-1} - 2T_B = T_{DCF}^{n-1} - T_B$ will be derived out. Therefore, $T_{DCF}^{n-1'} \geq 0$.

If $T_{DCF}^{n} \leq T_B$, it will lead overall advancement of latest classification starting time. To sum up, the computation method of $\Delta t$ can be derived as shown:

$$\Delta t = t_{LSB}^{n-1} - t_{LSB}^{n'} = T_B - T_{DCF}^{n-1} \quad (7)$$

After classification sequence adjustment, the largest adjustment for $t_{LSB}^{n'}$ is $t_{LSB}^{n-1} - (T_B - T_{DCF}^{n-1})$, under this condition the train will not be delayed.

*3.3.3. The condition of classification sequence adjustment.* Only supported by following conditions the adjustment is effective:

Condition (1): In adjustment, $\rho$ is defined to reflect virtues or defect degree of car flow. If $t_{LSB}^{n'} \geq t_{ESB}^{n}$, that is $\rho = 1$ .If not, $\rho = 0$.

If $T_{DCF}^{n-1} \geq T_B \bigcup t_{LSB}^{n-1} \geq t_{ESB}^{n}$, the adjustment will be optimal between adjoin trains. It just needs to prove $t_{LSB}^{n'} \geq t_{ESB}^{n}$.

Due to $t_{LSB}^{n'} = t_{LSB}^{n-1}$, it just needs to prove $t_{LSB}^{n-1} \geq t_{ESB}^{n}$ .By the analysis of 3.3.2, $T_{DCF}^{n-1'} \geq 0$ will be got.

Condition (2): it needs to judge unfollow-up effect after an adjustment. Under the constraint condition of $t_{LSB}^{n'} \geq t_{LSB}^{n-1}$ [13], it will achieve the aim of full-load.

The stepping trains classification sequence can be adjusted step by step[12].The method can use

$t^{n}_{LSB} - t^{n'}_{LSB} = XT_B$ to express.

## 4. Disintegration Sequence Adjustment

### *4.1. Analysis of Initial Sequence*

Adjusting the disintegration sequence of arrival trains also can guarantee full-load. In order to guarantee the requirement of car flow in classification subsystem, the $t^{m}_{EFJ}$ is an important parameter must be considered. When trains arrived on time, it can provide stable car flow, that is $T^{m(m+1)}_{JD} > T_J$ ( $m \in \{m|1, m-1\}$ ) , as shown in Fig.4 (a). The arrival trains need a period of waiting time for disintegration, it can use $T^{m(m+1)}_{DJ}$ to indicate. The interval time will disappear when trains arrival frequently. At the same time, the disintegrate operation is compact in subsystem. As shown in Fig.4 (b), $T^{m(m+1)}_{DJ}$ will disappear, but the arrival waiting disintegration time $T^{m}_{DD}$ will come with the saturate car flow and busy inspection operations.

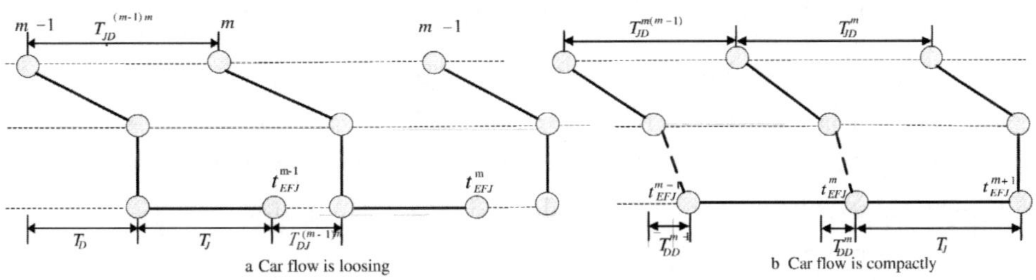

**Figure 4**. Analysis of initial disintegration sequence.

In order to guarantee the demand of car flow, $t^{m}_{EFJ}$ is an important parameter, it can easily get according to tightness degree:

If $T^{m(m+1)}_{JD} > T_J$ , car flow follow is loose, it will be got by extend calculation for time:

$$t^{m}_{EFJ} = t^{m}_{D} + T_D + T_J \tag{8}$$

If $T^{m(m+1)}_{JD} \le T_J$ , car flow is intensive, it will take follow track model:

$$t^{m}_{EFJ} = t^{m_0}_{D} + T_D - (m_0 - m - 1)T_J \tag{9}$$

### *4.2. The Process of Classification Sequence Adjustment*

*4.2.1. Classification Sequence Adjustment between Adjacent Trains.* Only meet the initial condition, $t^{m}_{EFJ} - t^{m-1}_{EFJ} = T_J$, it can adjust the disintegration sequence. The adjustment condition of disintegration sequence will be got by inverse derivation according to arrival time and technical operation time: $t^{m+1}_{EFJ} \ge t^{m}_{D} + T_D + 2T_J$ . The disintegration sequence adjustment scheme is as shown in Fig.5:

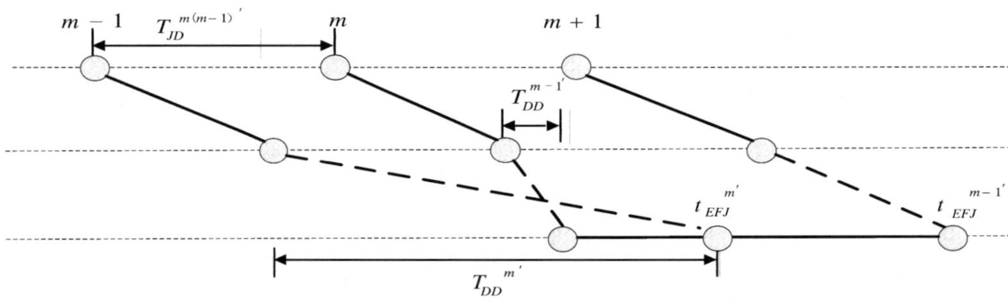


**Figure 5.** Adjustment scheme of disintegration sequence.

The ending time of earliest disintegration after adjustment can be expressed as follow:

$$t_{EFJ}^{m-1'} = t_{EFJ}^{m-1} - T_J - T_{DD}^{m-1} + T_{DD}^{m-1'} + T_J = t_{EFJ}^{m-1} - T_{DD}^{m} + T_{DD}^{m'} + 2T_J \tag{10}$$

$$t_{EFJ}^{m'} = t_{EFJ}^{m} - T_J - T_{DD}^{m} + T_{DD}^{m'} + T_J = t_{EFJ}^{m} - T_{DD}^{m-1} + T_{DD}^{m-1'} - 2T_J \tag{11}$$

*4.2.2. The Influence Analysis after Adjustment.* After initial adjustment, in order to provide effective car flow for classification subsystem, it has to meet $T_{DD}^{m'} \geq 0 \bigcup T_{DD}^{m-1'} \geq 0$.It only needs to prove $T_{DD}^{m'} \geq 0$.

After an adjustment, changes between the ending time of earliest disintegration can be defined as $\Delta t$

If $T_{DD}^{m} > T_J$, the disintegration trains are stable, the greatest adjustment is: $t_{EFJ}^{m-1'} = t_{EFJ}^{m}$.After the greatest adjustment, $\Delta T = 0$. It also meet the adjustment condition $T_{DD}^{m'} \geq 0$.The prove process as shown as flow:

The proved method is as shown as follows:

Based on the available conditions: $t_{DD}^{m'} = t_{EFJ}^{m'} - t_{EFJ}^{m} + t_{DD}^{m}, t_{EFJ}^{m-1'} = t_{EFJ}^{m}$ and $T_{DD}^{m} \geq T_J$.

And $T_{DD}^{m'} = T_{DD}^{m} - T_J$ will be derived out. Therefore, $T_{DD}^{m'} \geq 0$.

If $T_{DD}^{m} \leq T_J$, trains arrive compactly, it will influence other trains disintegration time:

$$\Delta T = t_{EFJ}^{m-1'} - t_{EFJ}^{m} = T_J - T_{DD}^{m} \tag{12}$$

After adjustment, $T_{DD}^{m'} = 0$ when $t_{EFJ}^{m-1'} = t_{EFJ}^{m} + \Delta T$.

*4.2.3. The Condition of Disintegration Sequence Adjustment.* Only supported by following conditions, an adjustment is effective.

Condition (1):In an adjustment, $\rho'$ is defined to reflect defect degree of car flow in disintegration adjustment. If $t_{EFJ}^{m-1'} \leq t_{LFJ}^{m-1}$,that is $\rho' = 1$.Otherwise, $\rho' = 0$.

If $T_{DD}^{m} \geq T_J \bigcup t_{EFJ}^{m} \leq t_{LFJ}^{m-1}$, the adjustment will be optimal between adjoin trains. It just needs to prove $t_{EFJ}^{m-1'} \leq t_{LFJ}^{m-1}$.

By the analysis of 4.2.2, $T_{DD}^{m-1'} \geq 0$ will be got.

Condition (2): By constraint condition of $t_{LSB}^{n'} \geq t_{LSB}^{n-1}$ [13] ,it will guarantee enough car flow for classification.

## 5. A Case Analysis

This case is used to analysis the first shift operating plan in down direction classification and disintegration sequence adjustment. The technical operation time parameters: $T_B = 15$ min , $T_F = 30$ min , $T_J = 15$ min , $T_D = 30$ min .

### *5.1. The Classification Sequence Adjustment*

*5.1.1. The Classification Sequence Adjustment between Adjacent Trains.* This case is used to analysis the first shift operating plan in down direction classification and disintegration sequence. The technical operation time parameters: $T_B = 15$ min , $T_F = 30$ min , $T_J = 15$ min , $T_D = 30$ min .

**Table 1.** Initial classification sequence

| $CF$ | $B_n$ | $t^n_{ESB}$ | $t^n_{LSB}$ | $T^n_{DCF}$(min) | $t^n_{CF}$ |
|---|---|---|---|---|---|
| 40 221 | 5 | 8:13 | 10:35 | 32 | 11:52 |
| 40 217 | 6 | 10:30 | 10:55 | 18 | 11:58 |
| 30 125 | 7 | 10:50 | 11:10 | 20 | 12:15 |
| 30 123 | 8 | 11:09 | 11:25 | 20 | 12:30 |
| 45 307 | 9 | 11:24 | 11:40 | 86 | 13:51 |

Adjustment method is as follow:

Step1: For train No. 40 217, judging whether it meet initial adjustment conditions. Only meet the condition of compact continuing classification, its sequence can be adjusted.

$$\because t^6_F - T_F = 11:28$$

$$\therefore t^7_{LSB} \le t^6_F - T_F$$

Step2: For train No. 40 217, judging whether it meet condition (1). $T^7_{DCF} \ge T_B$ and $t^6_{LSB} \ge t^7_{ESB}$ is clear. So, the train No. 40 217 can meet the condition $\rho = 1$.

Step3: Unfollow-up effect judgment. It must meet condition (2) before adjustment.

Because of $T^{6'}_{DCF} = T^6_{DCF} - T_B = 3$ min and $t^{7'}_{LSB} = t^7_{LSB} + T^6_{DCF} - T^{6'}_{DCF} - 2T_B = 10:55$ ,meet $t^{7'}_{LSB} \ge t^6_{LSB}$ .After adjustment, $\Delta t = 0$.

To sum up, if train No. 40 217 needs more time for waiting car flow, its sequence should be exchange to No.30 125.The result after adjustment is as shown in Table 2:

**Table 2.** Adjustment between track trains.

| $CF$ | $B_n$ | $t^n_{ESB}$ | $t^n_{LSB}$ | $T^n_{DCF}$(min) | $t^n_{CF}$ |
|---|---|---|---|---|---|
| 40 221 | 5 | 8:13 | 10:35 | 32 | 11:52 |
| 30 125 | 6 | 10:30 | 10:55 | 35 | 12:15 |
| 40 217 | 7 | 10:50 | 11:10 | 3 | 11:58 |
| 30 123 | 8 | 11:09 | 11:25 | 20 | 12:30 |

*5.1.2. The Adjustment for Stepping Trains.* Based on adjustment result in table 2, it can continue adjust classification sequence 6 to 8. Due to $T^8_{DCF} < T_B$ , therefore, $\Delta t = t^7_{LSB} - t^{8'}_{LSB} = T_B - T^7_{DCF} = 12$ min .The starting time of latest classification will be advanced 12 min.

**Table 3.** Adjustment among several departure trains.

| $CF$ | $B_n$ | $t^n_{ESB}$ | $t^n_{LSB}$ | $T^n_{DCF}$ (min) | $t^n_{CF}$ |
|---|---|---|---|---|---|
| 30 125 | 6 | 10:30 | 10:43 | 47 | 12:15 |
| 30 123 | 7 | 10:50 | 10:58 | 47 | 12:30 |
| 40 217 | 8 | 11:09 | 11:13 | 0 | 11:58 |

To sum up, it will cause great influences for the whole scheme; therefore, the adjustment should be canceled. And then, only make compact continuing adjustment for No. 40 217.

### *5.2. Disintegration sequence adjustment*

*5.2.1. Disintegration Sequence Adjustment between Adjacent Trains.* If the No. 51 503 needs to disintegrate firstly, and then it can provides car flow for classification subsystem. The partials of initial disintegration sequence are as shown in Table 4.

**Table4.** Initial disintegration sequence.

| $DD$ | $J_m$ | $t^m_D$ | $T^{m(m+1)}_{DJ}$ (min) | $t^m_{ESJ}$ | $t^m_{LFJ}$ |
|---|---|---|---|---|---|
| 41 315 | 6 | 9:47 | 22 | 10:39 | 11:10 |
| 32 137 | 7 | 9:56 | 28 | 10:54 | 11:25 |
| 51 503 | 8 | 10:15 | 24 | 11:09 | 11:40 |
| 41 321 | 9 | 10:39 | 15 | 11:24 | 13:20 |

The adjustment method is similar with classification sequence adjustment. The result after adjustment is as shown in Table 5.

**Table 5.** Adjustment between track trains.

| $DD$ | $J_m$ | $t^m_D$ | $T^{m(m+1)}_{DJ}$ (min) | $t^m_{ESJ}$ | $t^m_{LFJ}$ |
|---|---|---|---|---|---|
| 41 315 | 6 | 9:47 | 22 | 10:39 | 11:10 |
| 51 503 | 7 | 10:15 | 9 | 10:54 | 11:25 |
| 32 137 | 8 | 9:56 | 43 | 11:09 | 11:40 |
| 41 321 | 9 | 10:39 | 15 | 11:24 | 13:20 |

*5.2.2. The Adjustment for Stepping Trains.* If No. 51 501 needs more time for disintegration, adjustment stepping trains for it must be taken into account. Because of $T^7_{DD} \le T_J$, $\Delta T = t^{6'}_{EFJ} - t^7_{EFJ} = T_J - T^7_{DD} = 6$ min. The disintegration time will be delayed 6 min in whole scheme.

**TABLE 6.** Adjustment among several arrival trains.

| $DD$ | $J_m$ | $t^m_D$ | $T^{m(m+1)}_{DJ}$ (min) | $t^m_{ESJ}$ | $t^m_{LFJ}$ |
|---|---|---|---|---|---|
| 51 503 | 6 | 10:15 | 0 | 10:45 | 11:10 |
| 41 315 | 7 | 9:47 | 43 | 11:00 | 11:25 |
| 32 137 | 8 | 9:56 | 79 | 11:15 | 11:40 |

To sum up, it have to cancel the stepping trains disintegration sequence, and only make compact continuing sequence adjustment for No. 51 503.

### *5.3. The conclusion of case*

For the two problems showed in the case, those are as follows, the first one is No. 40 217 needs more time for waiting car flow and the second is No. 51 503 needs emergency disintegrate to provide car flow for classification. Through partial adjustment method can solve those problems effectively. The most important step is judging the adjustment conditions whether it can be meet. The classification and disintegration sequence of stepping trains can be adjusted step by step until achieve the best adjustment scheme. In this case, departure time will be delayed when make stepping trains disintegration sequence adjustment. In practical situation, if adjustment objects are important trains, stepping trains adjustment must be considered. Finally, by eliminating negative effects gradually it can achieve the goal of overall optimum.

## 6. Conclusion

In thesis, the marshalling station is divided into two subsystems according to deferent operation. Each subsystem can be equivalent regarded as machine scheduling operation. Then analysis shows the connection between each parameter of time. Based on the principle of "first inspect first disintegrate, first depart first classificate", it can make partial adjustment. Through this kind of method, the marshalling station will acquire more time for car flow follow, which can achieve the aim of each departure trains is full-load. With help of authentic data in the case, the practical application is proved. The problem for sequence adjustment of disintegration and classification in two subsystems can be solved step by step. It makes foundation for the station stage operating plan.

## Acknowledgement

This work was supported by Chinese National Natural Science Foundation (61364028) and Humanities and Social Sciences of Ministry of Education Project (15XJAZH002).

## References

[1] Martinelli D R,Teng H L. Optimization of railway operation using neural networks [J].Transportation Research Part C,1995, 4(1):33-149.
[2] TURNQUIST M A,DASKIN M S. Queuing models of classification and connection delay in rail yards[J].Transportation Science,1982,16(2):207-230.
[3] AVARAMOVIC Z Z. Method for evaluating the strength of retarding steps on a marshalling yard hump[J].European Journal of Operational Research,1995,85(3):504-514.
[4] Li Haodong,He Shiwei,Wang Baohua,Shen Yongsheng. Survey of stage plan for railway marshalling station[J]. Journal of China Railway Society,2011,33(8):14-15.
[5] Wang Shidong,Zheng Li,Zhang Zhihai,Liu Shucheng. Mathematical model and algorithm for automatically programming the stage plan of marshalling station[J]. China Railway Science, 2008,29(2):121-125.
[6] Zhao Jun,Peng Qiyuan.Integer programming model and algorithm for wagon-flow allocation problem at double-direction marshalling station [J].Journal of China Railway Society,2014,36(9):10-20.
[7] Xue Feng,Wang Ciguang,Zhang Zhanjie.Optimization algorithm for wagon-flow allocation in marshalling station[J]. Journal of Southwest Jiaotong University,2010,45(6): 930-938.
[8] Guo Rui,Guo Jin,Su Yuebin,Ma Liang.Model and approximation algorithm for dynamic wagon-flow allocation based on greedy strategy[J]. Journal of Southwest Jiaotong University,2014,29(4):712-719.
[9] Xue Feng.Collaborative optimization model and algorithm for wagon-flow allocation in railway marshalling station[J].System Engineering-Theory&Practice, 2013,33(11):2930-2936.
[10] Jing Yun,Wang Ciguang.Model and algorithm of dynamic wagon-flow allocating in a marshalling yard under uncertain conditions[J].Journal of China Railway Society,2010,32(4):8-12.
[11] Li Haodong,He Shiwei,Ma Kejun.Research on the influence of dispatching on the stage plan in marshalling station and the plan adjustment[J]. Logistics Technology, 2010,29(1):44-48.
[12] Xue Feng,Wang Ciguang. The adjustment method for train classification sequence in marshalling station[J].Journal of Southwest Jiaotong University,2007,29(4):1-5.
[13] Peng Qiyuan,Zhao Jun. Adjustment method for train break-up and make-up sequence in railway technical station[J].Journal of Southwest Jiaotong University,2009,44(3):385-391.

# Empirical research on the influencing factors of the occupational stress for construction workers

**Xing LV, Xiang WU, Huipeng CI, Qing LIU and Yongzheng YAO**

China University of Geosciences &Beijing
wuxiang@cugb.edu.cn

**Abstract**. Employee's occupational stress and safety performance are highly related, which has been generally recognized by the researchers. We did this research to understand the status of the stress for construction workers, and explore the influence factors of pressure source with characteristics of construction industry. Based on the results of previous studies, we designed questionnaire to collect the influence factors on occupational stressors. The study selected workers from certain construction units at the grass-roots level as sample source. According to the results of the questionnaire, we redesigned the interview outline, and did the semi-structured interviews on workers randomly selected. Finally, we developed a scale which combined the characteristics of construction projects in China. Using SPSS software for factor analysis, reliability analysis, and descriptive statistical analysis, the results show that there are six factors affecting the workers' occupational stress, including The Work Itself, Family-Work, Career Development, Organization Style, Interpersonal Relationship and Role Management Style. The work itself is the main sources of occupational stress. The results can be used by the construction company to provide guidance for workers to control and manage occupational stress.

## 1. The Introduction

Occupational stress has been one of the main problems that bother people's work and life, and attracted many scholars' research. Predecessors' research shows that work stress not only harms the job itself, but also causes great damage to individual in psychology and physiology. [1] China's construction industry has evolved rapidly in recent years, since 2008 the full opening of the market in China. During these years, domestic construction companies are not only facing challenges from advanced construction enterprise, but also experiencing structural, technical and personnel aspects of the deep transformations, which cause more and more workers suffering and experiencing chronic stress. However, a vast number of employees lack of understanding of occupational stress and scientific control methods to pressure, which will create a vicious cycle. [2] And job stress can affect not only individual and organizational level, and even affect the society. [3] Therefore, the effective identification of the occupational stress structure for construction staff is of great significance for the development of construction industry. Over the past few decades, a large number of researchers proposed different concepts and models on occupational stress, as well as some theories, stress scales and questionnaires which are different but overlapping each other. Such as Vagg and Spielberger compiled a work stress questionnaire lists 20 things which lead to occupational stress. [4] Evers, Frese and Cooper developed a second edition occupational stress index to measure the stress source, the result shows that mental health, physical health and job satisfaction affect the job stress. [5] Pan YingXin, Wang Lei, Ren XiangYun and Zeng Yan compiled the Chinese employees work stress questionnaire, found that lack of control, time pressure and insufficient information are three main factors affecting stress.[6] Phillip in the book Stress and Health wrote the main occupational stress source of employees can be divided into the working conditions, role stress, relationship factors, career development, organizational structure and the influence between family and work, etc., he also developed an occupational stress scale including interpersonal relationship, physical conditions and job interest[7]. But, whether this kind of scale's structure is in line with the situation in our country, especially the specific conditions of the occupational stress for construction workers, that still needs further research.

At present, there has no research on the stress of construction project staff, but there have been some researches on the stress of other industry practitioners, which we can use for reference. Xu Fuming found that among primary and secondary school teachers the most pressure events, in turn, are the test, the work itself, occupational prestige and employment pressure [8]. Cooper and Marshal's study found that the occupational stress of white-collar workers are career development, interpersonal relationship, the balance between family and work, role conflict, role ambiguity, organizational structure and atmosphere[9]. The research, Janet and Cary did, shows that the pressure of the civil service mainly comes from the work itself, such as relatively low wages, poor working conditions, or lack of controlling on job and the organization [10]. Chen ZhiXia and Liao JianQiao found that the main factors of occupational stress for knowledge workers are the work task, career development, organizational management, interpersonal relationship and ability, time urgency and the work environment [11]. The mainly purpose of this study is to compile a suitable work stress questionnaire for the workers in construction projects. It is concluded that structure factor of workers' occupational stress, and put forward suggestions according to different factors.

## 2. Research design

### *2.1. Questionnaire design*

The questionnaire designed on the basis of previous relative researches, such as railway locomotive driver [12], knowledge employees from apparel industry [13] and civil aviation practitioners [14], and other workers' occupational stress. We mainly refer to the occupational stress measurement index system which was designed by Cooper and Williams, namely the OSI occupational stress scale. We also refer to the occupational stress scale designed by Bai YuLing. The semi-structured interview outline was designed on the basis of the questionnaire. And then combined with the characteristics of construction project in China, we compiled the occupational stress scale. The factors of this scale include the work itself, management roles, relationships, career development, organization style, conflicts between families and work, six aspects, a total of 20 items, for the form of 1-5 Likert scale. From "strongly disagree" to "strongly agree" meter respectively from 1 to 5.

### *2.2. Data Collection*

Through questionnaire, we collect the structure factors of workers' occupational stress. The sample was selected in a primary construction unit. Before distributing the questionnaire to workers, the investigators explained the aim of this questionnaire and the meaning of each question to the workers. The survey handed out 150 questionnaires and recycle 132. The recovery rate was 88%, excluding eight invalid questionnaires. Finally, we received 124 valid questionnaires, the effective recovery rate were 83%. The basic information of sample is as shown in Table 1.

**Table 1.** Basic information of sample

| variables | gender | | age | | | | Working age | | | | education | | | | matrimony | |
|---|---|---|---|---|---|---|---|---|---|---|---|---|---|---|---|---|
| category | male | female | ≦30 | 31-40 | 41-50 | 51≦ | ≦3 | 3-9 | 10-15 | 16≦ | Primary School or below | Junior High school | Senior High school | Up High school | married | unmarried |
| number | 112 | 12 | 28 | 31 | 42 | 23 | 13 | 40 | 54 | 17 | 31 | 61 | 29 | 3 | 105 | 19 |
| Percentage (%) | 90 | 10 | 23 | 25 | 34 | 18 | 10 | 32 | 44 | 14 | 25 | 49 | 23 | 3 | 85 | 15 |

## 3. Data analyse

### *3.1. Factor analysis*

This research use KMO and Bartlett test to figure out whether the questionnaire is suitable for factor analysis, the results are shown in Table 2.

**Table 2.** KMO and Bartlett test.

| | | |
|---|---|---|
| Kaiser-Meyer-Olkin | | .856 |
| Bartlett sphericity test | chi-squared approximation | 916.275 |
| | Df | 25 |
| | Sig. | .000 |

Test results show that the KMO value between 0.8 and 0.9, Bartlett test showed significant value for.000, which mean that the occupational stress questionnaire is suitable for factor analysis. Occupational stress questionnaire factor analysis uses the statistical analysis software SPSS19.0, and 20 items of work stress questionnaire use principal component analysis method and maximum variance orthogonal rotation method to extract factor, the result of the analysis is shown in Table 3.

**Table 3.** Rotation matrix of occupational stress questionnaire for construction workers.

| Factor | | Factor1 | | | | Factor2 | | | Factor3 | | Factor4 | | | Factor5 | | | Factor6 | | | Factor7 | Factor8 |
|---|---|---|---|---|---|---|---|---|---|---|---|---|---|---|---|---|---|---|---|---|---|
| item | | C3 | C4 | C1 | C2 | C19 | C18 | C20 | C14 | C13 | C16 | C15 | C17 | C9 | C10 | C11 | C7 | C6 | C8 | C5 | C12 |
| component | 1 | 0.867 | 0.817 | 0.799 | 0.784 | 0.199 | 0.087 | -0.259 | -0.142 | 0.159 | 0.15 | -0.278 | -0.059 | -0.244 | 0.179 | -0.012 | -0.065 | -0.217 | 0.109 | 0.226 | 0.291 |
| | 2 | -0.024 | -0.084 | -0.017 | -0.108 | 0.791 | 0.742 | 0.684 | 0.245 | 0.277 | 0.158 | -0.614 | 0.107 | -0.115 | 0.035 | 0.204 | 0.08 | 0.532 | -0.469 | -0.047 | -0.03 |
| | 3 | 0.101 | 0.103 | 0.083 | 0.154 | 0.17 | -0.158 | 0.335 | 0.96 | 0.922 | 0.183 | 0.17 | -0.025 | -0.179 | 0.264 | 0.043 | -0.056 | -0.129 | 0.083 | -0.153 | -0.166 |
| | 4 | -0.065 | -0.065 | -0.108 | 0.137 | -0.04 | -0.224 | 0.061 | 0.324 | 0.2 | 0.844 | 0.779 | 0.625 | 0.126 | -0.108 | 0.476 | 0.134 | 0.012 | 0.157 | 0.107 | -0.109 |
| | 5 | 0.073 | 0.173 | 0.306 | -0.153 | -0.098 | 0.062 | 0.035 | -0.448 | -0.135 | -0.29 | 0.235 | 0.048 | 0.747 | 0.703 | 0.576 | 0.016 | 0.069 | -0.118 | 0.916 | 0.75 |
| | 6 | -0.123 | 0.271 | 0.311 | 0.213 | 0.181 | 0.033 | 0.131 | -0.057 | 0.201 | 0.153 | 0.088 | 0.076 | -0.148 | 0.279 | 0.172 | 0.729 | 0.696 | 0.657 | 0.013 | 0.124 |
| | 7 | 0.101 | -0.198 | 0.061 | 0.098 | 0.221 | -0.312 | -0.241 | -0.14 | 0.019 | 0.119 | -0.024 | 0.055 | 0.031 | 0.119 | -0.124 | 0.324 | -0.067 | 0.112 | 0.784 | -0.024 |
| | 8 | -0.077 | -0.114 | 0.071 | 0.112 | 0.126 | -0.394 | -0.19 | 0.041 | 0.102 | 0.2 | 0.242 | 0.191 | -0.331 | -0.231 | 0.221 | 0.25 | -0.121 | 0.013 | 0.235 | 0.713 |

According to the matrix of occupational stress questionnaire for construction workers, there are totally 8 factors we listed. However, two of them (Factor 7 and Factor 8) only contain a single item respectively, which means they lack of enough evidence to be valid and have to be removed. Therefore, there remain 6 effective factors (dimensions), respectively: The work Itself, Family-Work, Occupation Development, Organization Style, Interpersonal Relationship and Role Management. Thus the rationality of work stress scale dimension is verified. The Work Itself is due to the heavy workload, tight time ,heavy responsibility and poor safety; Family-work is due to lack of communication with families in a long term, work site far from the family or have heavy family burden; Occupation Development refers to workers feel poor working stability; Organization Style refers to some of the workers feel their own rights can't be guaranteed in some unreasonable system; Interpersonal Relationship refers to the relationship between colleagues or leadership is not harmony, or work without supporting and understanding of the leadership; Role Management refers to the uncertainty of their work duties or cannot adjust from different work requirements.

As shown in Table 4, the cumulative explanation of these six factors was 72.495%, namely the extracted six kinds of working pressure factors can summarize 72.495% the information of original variables for workers' job pressure. The construct validity of the questionnaire is ideal. The explanatory rate of the work itself, family-work and occupational development were 21.722%, 16.022% and 14.544% respectively. The explanatory rate of organizational style, interpersonal relationship and role management were 7.256%, 6.944% and 5.967% respectively. This indicates that The Work Itself is the main source of work stress for construction staff. Family-Work and Occupation Development are the most important source of workers occupational stress, While the explain rate of Organizational Style, Interpersonal Relationship and Role Management are relatively low.

**Table 4.** Construction industry staff work stress questionnaire total variance explained.

| Component | Initial Eigenvalues | | | Extraction square and loading | | |
|---|---|---|---|---|---|---|
| | Total | Variance % | Accumulation % | Total | Variance % | Accumulation % |
| 1 | 3.919 | 21.772 | 21.772 | 3.919 | 21.772 | 21.772 |
| 2 | 2.881 | 16.022 | 37.794 | 2.681 | 16.022 | 37.794 |
| 3 | 2.618 | 14.544 | 52.388 | 2.3 18 | 14.544 | 52.388 |
| 4 | 1.306 | 7.256 | 59.594 | 7.256 | 10.033 | 59.594 |
| 5 | 1.250 | 6.944 | 66.538 | 6.944 | 6.944 | 66.538 |
| 6 | 1.074 | 5.967 | 72.495 | 5.967 | 5.967 | 72.495 |

### *3.2. Reliability analysis*

Reliability analysis of occupational stress questionnaire uses SPSS19.0 statistical analysis software. Then we validate the internal consistency of the work stress questionnaire by the 'Cronbach's coefficient. It is generally believed that Cronbach's coefficient reached 0.7; the result is credible [15]. In this study, the reliability coefficients are all above 0.7, as shown in Table 5, so, the questionnaire reliability meets the requirements.

**Table 5.** Reliability statistics of the questionnaire for construction workers' job stress.

| variable name | Cronbach's Alpha | Cronbachs Alpha based on the standard term | term numbers |
|---|---|---|---|
| The work itself | .791 | .753 | 4 |
| Family-work | .727 | .783 | 3 |
| Occupational development | .843 | .761 | 2 |
| organization style | .695 | .723 | 3 |
| interpersonal relationship | .675 | .712 | 3 |
| role management | .698 | .706 | 3 |
| Occupational stress | .731 | .742 | 18 |

### *3.3. Descriptive statistical analysis*

According to the results of the factor analysis for the occupational stress questionnaire, this questionnaire contains 6 dimensions, 18 items, the descriptive statistics of each item and the dimension's score are determined as shown in Table 6.

**Table 6.** Descriptive statistical of workers' occupational stress.

| Factor | Occupational stress questionnaire items | N | Mean value | Standard deviation |
|---|---|---|---|---|
| The work itself | C1 My work is very complex, big workload | 124 | 3.284 | .633 |
| | C2 worry about his safety during working | 124 | 3.651 | .775 |
| | C3 I often work overtime work | 124 | 3.176 | .684 |
| | C4 Work responsibility, afraid of accountability | 124 | 3.053 | .798 |
| | Dimension mean | | 3.291 | |
| Family –work | C18 I felt I bear the burden of family economic income is very heavy | 124 | 3.170 | .789 |
| | C19 Nature of work need to separation, can't take care of family responsibility | 124 | 3.325 | .699 |

| | | | | |
|---|---|---|---|---|
| | C20 Family understanding and support for work is not enough | 124 | 2.926 | .759 |
| | Dimension mean | | 3.207 | |
| Career development | C13 More concern on the development of their future career | 124 | 3.412 | .517 |
| | C14 Work stability is poor, worried that after the end of the project cannot find another project | 124 | 3.856 | .427 |
| | Dimension mean | | 3.634 | |
| Organizational Style | C15 My rights and interests cannot be guaranteed sometimes. | 124 | 3.067 | .726 |
| | C16 organization wage system is not reasonable | 124 | 3.062 | .875 |
| | C17 The organization has not responded to my performance. | 124 | 2.859 | .899 |
| | Dimension mean | | 2.996 | |
| Interpersonal Relationship | C9 Have conflict or unpleasant with colleagues | 124 | 2.983 | .994 |
| | C10 Lack of leadership to support their own | 124 | 2.883 | .950 |
| | C11 Leaders are unwilling or unable to help me with my work. | 124 | 2.536 | .923 |
| | Dimension mean | | 2.801 | |
| Role Management | C6 My work has no clear explanation and explanation. | 124 | 2.693 | 1.112 |
| | C7 Sometimes the leaders have different requirements of the work | 124 | 2.706 | 1.083 |
| | C8 Sometimes being assigned to the different work at the same time | 124 | 2.943 | .966 |
| | Dimension mean | | 2.781 | |
| Occupational Stress | The score of the occupational stress | | 3.118 | |

According to the score standard of occupational stress, score 1-5 in turn represent: do not feel the occupational stress; feel the light occupational stress; feel the moderate occupational stress; feel the high occupational stress; and feel the serious work stress. From the statistical analysis of the results, we can find a total score of 3.118 points of the occupational stress, the overall of the construction project staff feel moderate occupational stress. The professional development gets the highest score, 3.634 points, and the closet to a high degree of occupational stress. The Work itself and Family - Work dimension score, respectively, 3.291 points and 3.207 points, to be higher than the moderate level of occupational stress. Organization style, interpersonal relationship and role management scores are not more than 3 points are 2.996 points, 2.801 points and 2.781 points, slightly lower than the moderate level of occupational stress, which explain these three factors have little effect on construction workers' job stress.

## 4. Discussion

The most important factor of work pressure on construction workers is the pressure from The Work Itself. The construction industry is facing enormous competitive pressure, they must ensure the quality of the project, but also need to improve work efficiency. To reach these purposes, some construction units may require the workers to work overtime. As we all know construction industry is a high-risk industry, so workers have to face the threat of production accidents. Accordingly, the contractor should arrange work and rest time reasonably, do not allow workers to work overtime. At the same time, the construction company should provide safety equipment fully and create a good atmosphere of safety.

The Family-Work dimension is also one of the important structural factors. Construction projects always in remote areas, so the workers have to follow the project and leave far from home. They have not enough time to contact with their families. Besides, most of the workers also face heavy family burdens. These are virtually increasing the occupational stress. The construction company should focus more on workers' family situation. For those workers who have experienced a number of big changes or events and become overstressed, the construction company should transfer these workers to safer post until they calm down.

The Career Development is another important source of occupational stress. It is decided by the nature of the construction industry, each cycle of construction projects is different, and so many construction units often face no project status. The most basic thing is the stability of work can't be guaranteed, which will give the workers a strong psychological burden. In order to avoid the workers' worries of whether they can get a stable job, the construction company needs to keep their project stable, so as to eliminate the menace of the future changes. Construction unit should carry out psychological health education regularly to support the workers take a positive attitude in face of such stress.

Organizational Style, Interpersonal Relationship and Role Management, these three dimensions have little influence on the occupational stress of workers in construction project. But they still need to take seriously. However, some construction companies have not enough attention to the protection of the workers' interest, corporate culture, and humanistic care. These factors might cause occupational stress, too. Construction unit should improve their management level, create a corporate culture environment, and care about the mood of workers to ensure safety in construction.

## 5. Conclusion

This research compiled the occupational stress questionnaire for the workers in China's construction project. Factor analysis using SPSS software, the KMO and Bartlett test shows that the occupational stress questionnaire is suitable for factor analysis. For reliability analysis, the reliability coefficient is above 0.7, which meets the requirement of reliability. Statistical analysis of the results of the occupational stress finds a total score of 3.118 points, represent most of the construction project workers feels moderate occupational stress.

The factors that constitute the occupational stress of workers in the construction project are: Work Itself, Family-Work, Career Development, Organizational Style, Interpersonal Relationship and Role Management. One of the most important factors is the Work Itself, followed by Family-Work and Career Development. Finally, according to different factors, some suggestions can be applied to manage and control occupational stress of workers.

**Acknowledgements**

Funding Project: National Natural Science Foundation of China（51474193）and China university of geosciences (Beijing) college students' innovative entrepreneurial training program（201611415064）

**References**

[1] Shi Lin 2003 Occupational stress research: a review *Psychological Science*.**26**(3):494-497.
[2] Ji Qinghua and Zhang Wen 2011 Primary construction of work stress scale of construction companies for first-line administrators Qufu Normal University.
[3] Viorica P,Raluca I and Mihaela S 2013 Assessment of Work Stress Influence on Work Productivity in Romanian Companies Procedia-Social and Behavioral Sciences. 92 420-425
[4] Vagg P R and Spielberger C D 1998 Occupational Stress Measuring Job Pressure and Organizational Support in the Workplace *Journal of Occupational Health Psychology*.3 294-305
[5] Evers A.,Frese M. and Cooper C L 2000 Revisions and further developments of the Occupational Stress Indicator: LISREL results from four Dutch studies *Journal Of Occupational & Organizational Psychology*. 73 221-240
[6] Pan Yingxin, Wang Lei, Ren Xiangyun and Zeng Yan The Development of a Job Stress Questionnaire for Chinese Employees *Psychological Science*. 29 312-314
[7] Phillip L.Rice 1992 Pressure and health(Beijing: China light industry press) p 149-151.
[8] XU Fuming 2003 A Study on Teachers' Occupational Stress and Burnout Chinese *Journal of Clinical Psychology*. 11 195-197
[9] Cooper C L and Marshal J 1980 Stress amongst nurses In White Collar and Professional

IOP Conf. Series: Earth and Environmental Science **61** (2017) 012102 doi:10.1088/1755-1315/61/1/012102

Stress.Wiley :Chichester

[10] Janet B and Cary C 1995 Job Satisfaction,mental health,and occupational stress among senior civil servants. *Human Relations.* 48 327-338

[11] Chen Zhixia and Liao Jianqiao 2005 Structural components of knowledge - worker's Job stressor and it's relationship with job strain Industrial *Engineering and Management*. 4 26-30

[12] Liu Yv, Wu Zhenye, Zhao Enjiang, et al 2010 Structural Analysis on the Influencing Factors of Locomotive Drivers' Occupational Stress *China Railway Science* . 31 130-133

[13] Bai Yvling 2010 Relationship Between Job Stress, Perceived Organizational Support and Job Burnout Research--Knowledge Workers in Clothing Industry as an Example Beijing:The capital economic and trade university. 25 79-85

[14] Liang Wenjuan and Chen Ming 2009 Research on occupational stress impact factors for workers in China's civil aviation *Journal of China civil aviation flight institute*. 20 13-16

[15] Nunnally J C 1978 Psychometric Theory (2nd Edition) ( New York: McGraw-Hill) p 45 -87

# Test research on influence of admixture agent on engineering characteristics of red sandstone roadbed filling in Dongchang Expressway

**Bo Liang[1,2], Tianxiong Dong[1], Yingjun Mei[1] and Jinqiu Pan[1]**

[1]School of civi engineering, Chongqing Jiaotong university, Chongqing 400074, China.
[2]State Key Laboratory Breeding Base of Mountain Bridge and Tunnel Engineering, Chongqing Jiaotong University, Chongqing 400074, China.
Type the first author's e-mail address is liang_laoshi@126.com.
Type the corresponding author's e-mail address is dongtianxiong_2009@126.com.

**Abstract:** By taking the red sandstone along the Jiangxi Dongchang expressway as the research object, this article takes lime and cement as the admixture agent, and conducts a series of laboratory test investigations on plain soil and improved soil, to research on the influence of the addition proportion on the physical and mechanical properties of red sandstone. According to test results: compared with plain soil, the red sandstones with admixture agents of lime and cement have reduced liquid limit and plastic limit, with reduced plastic index. With the increase of the proportion of lime, the optimal moisture content of the lime-improved soil gradually increases, and the maximum dry density gradually decreases. With the increase of the proportion of cement, the maximum dry density of the cement-improved soil increases, and the optimal moisture content decreases. The CBR improvement effect of cement-improved soil is better than that of the lime-improved soil. The optimal addition proportion of CBR of both the lime-improved soil and the cement-improved soil is 7%. Compared with plain red sandstone, the improvement on compaction functions plays a more significant role on the improvement of the CBR of the improved soil.

## 1 Introduction

Due to the characteristic of water disintegration of red sandstone, the engineering property after disintegration is very poor[1]. Therefore, improper handling of engineering with red sandstone as the roadbed filling will lead to harmful damages such as road surface collapse, cracking and side slope slump; therefore, it is needed to conduct processing to red sandstone before applying it to engineering. By taking the Nanjing-Huaian highway as the engineering background, Zhang Songnan[2] conducts laboratory test research on the expansive soil along with the highway with lime; Yang Jun et al. [3] conduct test research on the strength characteristic of high liquid limit clay after improvement with weathering sand; by utilizing self-made admixture agent composed by curing agent A and common Portland cement, Cheng Yin et al. [4] conduct improvement on the blown filled muck TT of Tianjin Binhai New Area; according to the result, the strength is increased by 2.4 times than that of cement, when the curing agent A is added with the proportion of 10%. By researching on the ion species existing in hydrolysate, Liu Xiaoming et al. [5] conduct research on the measures to restrain the disintegration of the first type of the red sandstone. According to the result, Ca2+ salt can play a preferable role in restraining the disintegration of the red sandstone in a preferable way; certain test is conducted on the resistance to disintegration, and it is worked out that the addition proportion of 7% of lime is the optimal engineering measure for restraining the disintegration of the red sandstone. Liu Daobin et al. [6] conduct improved curing research on the red sandstone near to Dawu-Suizhou highway by adopting lime, cement, coal ash and their combined matching, and mainly conduct test research on the unconfined compressive strength; according to the result, the self-made liquid curing agent has preferable effect. Wang Zhimeng[7] conducts research on the adaptation of the red bed mudstone and the improved soil in high speed railway bed, and they only conduct CBR test on the lime-improved soil with single proportion, and no optimal added quantity can be obtained.

According to document retrievals, current researches are conducted mainly aiming at expansive soil, high liquid limit clay and mucky soil, and little research on red sandstone improved soil has been conducted, especially on intense weathering red sandstone. At the same time, existing researches on red sandstone improved soil are conducted aiming at the improved soil of single proportion, so the optimal added quantity cannot be obtained. Besides, red sandstone has certain regionality; therefore, it is necessary to research the working properties of different addition proportions by selecting the red sandstone with certain representatives. By taking the intense weathering red sandstone adjacent to Dongxiang-Changfu highway in Jiangxi Province as the research object, to conduct a series of CBR tests on red sandstone after improvement with lime and cement, to research on the working performance of red sandstone under different addition proportions, to work out the optimal addition proportion; besides, it analyzes the influence of the admixture agent on the physical characteristic of red sandstone as well as the CBR influencing factors on improved soil, providing certain reference functions to the construction of the engineering.

## 2 Improved soil test scheme design

### *2.1 Physical Property of Plain Soil and Improved Soil*

The oil used for this test is taken from the Dongxiang-Changfu highway in Jiangxi Province with the depth of 3m away from the earth's surface. Crush down the soil with wood grinding and griddle with 20mm sifter, and then mix by adding in lime and cement. The slaked lime is used in this test, with addition proportions of 5%, 7% and 9%, respectively; the Little Nanhai brand 42.5 cement is adopted in this test, with the addition proportions of 3%, 5% and 7%. The admixture agent proportion used in this test is the proportion between the mass of the admixture agent and the mass of the dry soil. Mix the lime and the soil before the test, and spray water to the optimal moisture content, and then enclose material for 24h; the cement is added after the 24h of enclosing material of plain soil after spraying water, which is conducted before the test. According to "Test Method of Soil for Highway Engineering" (E40-2007) [8] and "Test Method of Stabile Materials of Inorganic Binder of Highway Engineering" [9], this test conducts a series of laboratory tests to plain soil and improve soil, to measure the physical property, as shown in Table 1.

**TABLE 1.** Physical Index of Plain Soil and Improved Soil

| Type of the admixture agent | Addition proportion (%) | Liquid limit (%) | Plastic limit (%) | Plastic index IP |
|---|---|---|---|---|
| Plain soil | 0 | 42.96 | 24.617 | 18.343 |
| Lime | 5 | 40.337 | 25.973 | 14.364 |
| | 7 | 39.941 | 27.915 | 12.026 |
| | 9 | 39.104 | 27.583 | 11.521 |
| Cement | 3 | 42.239 | 25.302 | 16.937 |
| | 5 | 41.549 | 25.095 | 16.454 |
| | 7 | 41.835 | 25.996 | 15.839 |

*2.2 Compaction Test Scheme Design*

According to the "Test Method of Soil for Highway Engineering" (E40-2007) [8], this test method conducts compaction test on plain soil and improved soil with cement and lime by adopting the heavy-II compaction test, with the weight of the compaction hammer of 4.5kg, and the drop height of 45cm, and the test tube of 2177cm3; the compaction is conducted in 3 layers, with 98 times for each layer. Crush down the air dried soil with wood grinding, and griddle with 20mm sifter. As for the lime-improved soil, weigh the air dried soil after griddling of 5kg, and sufficiently mix with lime, and then add in water to set moisture content, and then enclose material for 24h, with the lime addition proportions of 5%, 7% and 9%, respectively. As for the cement-improved soil, weigh the air dried soil after griddling, and spray water to the set moisture content, and then enclose material for 24h, and sufficiently mix the soil and the cement as per preset addition proportion before the compaction test, and compete the compaction test within 1 hour. The cement addition proportions are 3%, 5% and7%, respectively. The compaction result is as shown in Table 2.

**TABLE 2.** Result of Plain Soil and Improved Soil Compaction Test

| Admixture agent | Addition proportion (%) | Optimal moisture content (%) | Maximum dry density (g/cm3) |
|---|---|---|---|
| Plain soil | 0 | 15 | 1.803 |
| | 3 | 17.682 | 1.747 |
| Cement | 5 | 17.613 | 1.781 |
| | 7 | 15.771 | 1.792 |
| | 5 | 14.94 | 1.751 |
| Lime | 7 | 16.852 | 1.721 |
| | 9 | 17.96 | 1.701 |

*2.3 CBR Test Scheme Design*

Determine the optimal moisture content and the maximum dry density of the plain soil and the improved soil as per the compaction test. Please refer to Table 2 for the optimal moisture content and the maximum dry density of plain soil, the cement-improved soil and the lime-improved soil. As for plain soil and lime-improved soil, spray water and mix plain soil and the improved soil with lime according to different addition proportions, and then enclose materials in a sealing way for 24h for standby application, and prepare 3 test specimens with the compaction degree of 93% and the compaction degree of 95%, respectively. As for the cement-improved soil, mix the soil as per the optimal moisture content determined by the compaction test, and then enclose materials in a sealing way for 24h. And then, before the preparation of specimens, sufficiently mix the soil after enclosing and the cement of different addition proportions, and finish the preparation of specimens within one hour. And then maintain for 7d and soak for 4d; during the soaking period, the height of the water shall be at least 2.5cm away from the top of the specimens. Please refer to Test Method[8,9] for the specific operation procedures. The CBR test result of plain soil and improved soil are as shown in Table 3.

**TABLE 3.** CBR Test Result of Plain Soil and Improved Soil

| Addition proportion / Compaction degree | Plain soil | Lime | | | Cement | | |
|---|---|---|---|---|---|---|---|
| | 0% | 5% | 7% | 9% | 3% | 5% | 7% |
| 93% | 4.34 | 57.063 | 73.583 | 69.576 | 83.103 | 125.537 | 149.109 |
| 95% | 4.437 | 74.680 | 96.302 | 89.500 | 121.248 | 138.921 | 153.509 |

## 3 Test result analysis

### *3.1 Influence of the Admixture Agent on the Water Ratio Limit*

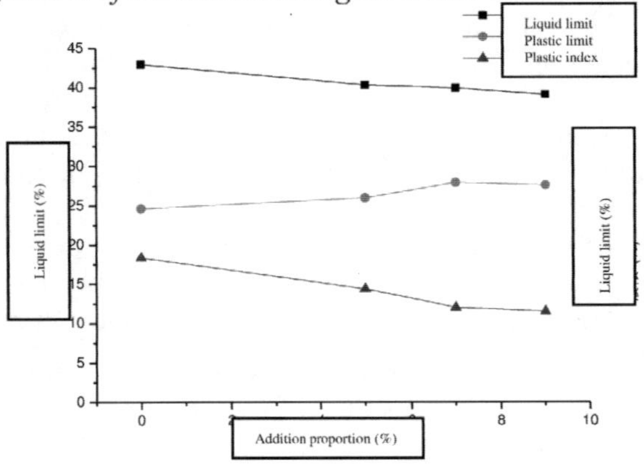


**FIGURE 1.** Influence of Lime Addition proportion on Water Ratio Limit of Red Sandstone

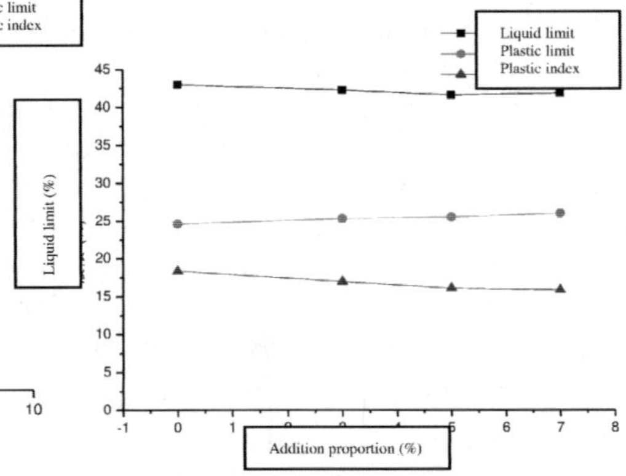


**FIGURE 2.** Influence of Cement Addition proportion on Water Ratio Limit of Red Sandstone

It can be seen from Table 1 that, compared with plain soil, liquid limits of the improved red sandstone after addition of lime and cement decrease, and the plastic limits increase, and the plastic indexes decreases, and there is significant changes on water ratio limit. According to Figure 1, the liquid limit of the red sandstone after improvement with lime gradually decreases with the increase of the admixture agent proportion, and the plastic limit firstly increases and then decreases, and the plastic index gradually decreases. It illustrates that the hydrophilia of the red sandstone after improvement with lime decreases. It can be seen from Figure 2 that, there is no significant change on the plastic limit of the red sandstone after improvement with cement, but the liquid limit is lower than that of the plain soil, appearing in a decline trend, and the plastic limit gradually increases, and the plastic index is in a decline trend.

Generally speaking, the water ratio limits of the red sandstone after improvements with lime and cement change, with improved performance. The performance is improved due to the ion exchange effect between the Ca2+ in the improver and the soil particles, the flocculation and the aggregation effect. On one hand, the ion exchange effect thins the double ionosphere, and weakens the water binding ability, which decreases the hydrophilia[10]. On the other hand, due to the flocculation and the aggregation effect, the small soil particles coagulate into large ones, and the quantity of the small soil particles reduces, which is equivalent to change the proportion of the clay fraction in an artificial way[11]. Therefore, the water ratio limit of the red sandstone after addition of admixture agent is improved.

*3.2 Influence of the Admixture Agent on the Compaction Characteristics*

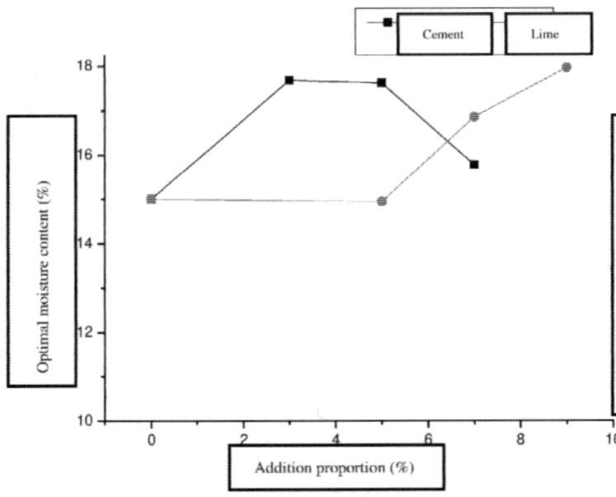


**FIGURE 3.** Influence of the Admixture Agent on the Optimal Moisture content of Red Sandstone

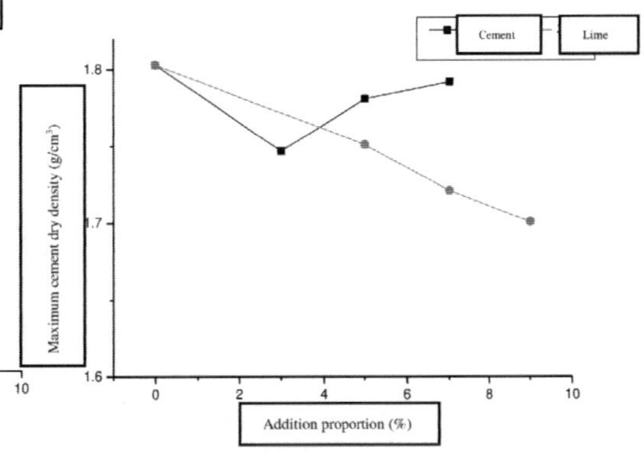


**FIGURE 4.** Influence of the Admixture Agent on the Maximum Dry Density of Red Sandstone

It can be seen from Table 2 that, the maximum dry density of both the lime-improved soil and the cement-improved soil are smaller than that of the plain soil, and with varying patterns. Except for the lime-improved soil with the addition proportion of 5%, the optimal moisture content of the two improved-soils are larger than that of the plain soil, and also with varying patterns.

It can be seen from Figure 2 and Figure 3 that the optimal moisture content of the lime-improved soil gradually increases with the increase of the addition proportion, and is greater than that of the plain soil; the maximum dry density gradually decreases with the increase of the addition proportion of lime, which is consistent with general compaction laws of improved soil. This is mainly related to the mineral compositions of the red sand. After the addition of the lime improver, on one hand, the lime absorbs a part of moisture during the hydration process; on the other hand, the fine particles increase and the specific area increases after the addition of lime due to its large fineness, and it consumes a part of moisture[12] to reduce the friction between soil particles; therefore, the optimal moisture content increases with the increase of the addition proportion of lime.

As for the cement-improved soil, the admixture agent appears different laws on the influence on the compaction characteristic of red sandstone. Different addition proportions of cement lead to different optimal moisture contents and the maximum dry densities; with the increase of the addition proportion, the optimal moisture content of the cement-improved soil gradually decreases, but is larger than that of plain soil, which is different from general acquaintance. The maximum dry density gradually increases with the increase of the addition proportion. It may be caused by the exchange effect between Ca2+ and the positive ion on surface of soil particles after hydration due to the addiction of cement; the Ca2+ replaces the monovalent positive ion, which thins the water film on surface of soil particles and weakens the lubrication action of water on soil particles, to make soil easy for compaction, and increase the dry density.

*3.3 Influence of Admixture Agent on CBR*

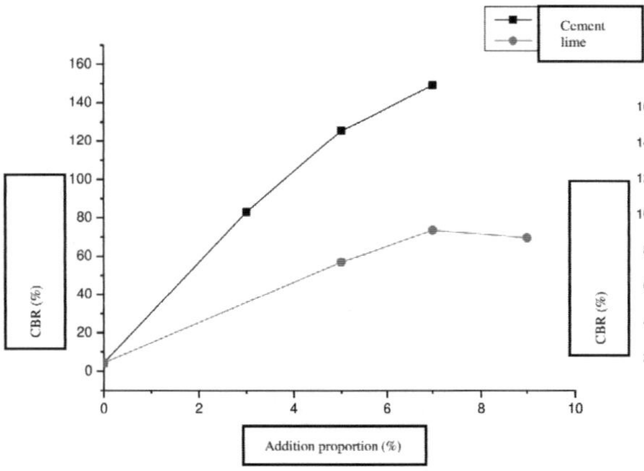


**FIGURE 5.** Influence of Admixture Agent with 93% Compaction on Red Sandstone

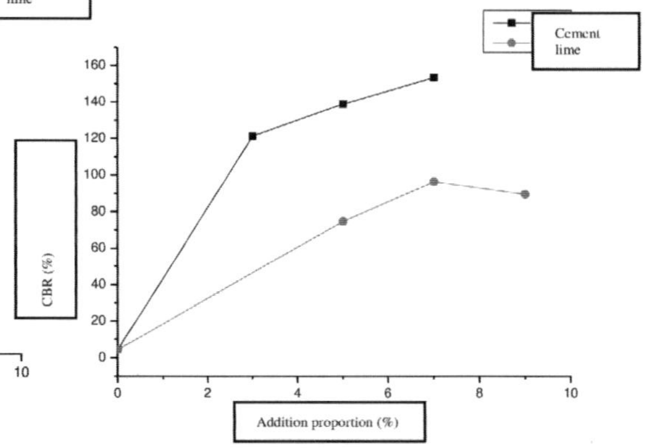


**FIGURE 6.** Influence of Admixture Agent with 95% Compaction on Red Sandstone

Figure 5 and Figure 6 appear the varying patterns of CBR with different types of admixture agents and addition proportions. The CBR of the plain soil is very small, under 5%; Figure 5 and Figure 6 illustrate that the CBR values of the lime-improved soil and the cement-improved soil after improver greatly increase than that of the plain soil. It is caused by the flocculation and the carbonation effect between the soil particles and the improver after the addition of improver. On one hand, due to hydration between lime and cement, it consumes a part of moisture, and it accelerates the evaporation of moisture in soil due to the reaction and heat release of lime, to improve the soil property[13]. On the other hand, the aggregation effect caused by hydration of lime and cement as well as the carbonation effect block the gap between soil clusters, which increases the soundness of soil, and increases soil strength, and increases the CBR of the improved soil[14].

It can be seen from Figure 5 and Figure 6 that, the CBR of lime-improved soil firstly increases and then decreases with the increase of the addition proportion. It illustrates that there is an optimal added quantity for lime; when the addition proportion is increased to certain degree, there will be free lime in soil particles, which decreases the improved soil strength, leading to smaller CBR. The CBR value of the cement-improved soil increases with the addition proportion of cement; in which the CBR value increases in a rapid way when the addition proportion is relatively small; with the increase of the addition proportion, the increase rate slows down. At the same time, it can be seen that the CBR value of the cement-improved soil is larger than that of the lime-improved soil, illustrating that cement has a preferable improvement effect than that of lime.

*3.4 Influence of Compaction Degree on CBR*

In order to reflect the influence of the compaction degree on the CBR of red sandstone, the influencing curve of the influence on the compaction degree on red sandstone is drawn, as shown in Figure 5 and Figure 6. According to Figure 5 and Figure 6, the CBR values of plain soil under the two compaction degrees are almost the same, illustrating that the only increase of the compaction degree fails to have a preferable effect on the plain red sandstone. As for the lime-improved soil, the compaction degree is increased from 93% to 95%, by taking the addition proportion of 5% as the example, the CBR value increases by almost 30%; as for the cement-improved soil, by taking the addition proportion of 3% as the example, the CBR value increases by 50%. It illustrates that, compared with the plain soil, the increase of compaction function on the improved soil has a more significant effect on the increase of the CBR value.

## 4 Conclusion

(1) Lime and cement have varying influence rules on the compaction characteristic of red sandstone.

IOP Conf. Series: Earth and Environmental Science **61** (2017) 012103 doi:10.1088/1755-1315/61/1/012103

The optimal moisture content of the lime-improved soil gradually increases with the increase of the addition proportion, but the dry density gradually decreases; the maximum dry density of the cement-improved soil increases with the increase of the addition proportion, and the optimal moisture content decreases, which is different from general laws on improved soil.

(2) Lime and cement have similar improvement effects on the CBR value of red sandstone, in which cement has better improvement effect than that of lime; it achieves the optimal effect when the addition proportion is 7%; there is also an optimal addition proportion on CBR of the lime-improved soil, and it achieves the optimal value when the addition proportion is 7%; it is necessary to reasonably select the addition proportion according to construction conditions and economic conditions.

(3) As for the red sandstone without improvement, the method of increasing compaction function cannot achieve a significant effect on the increase of the CBR value, and the method of adding admixture agent shall be adopted to improve its DBR value.

**Acknowledgments**

[1]First author:Bo Liang,male,professor, doctoral supervisor, Mainly engaged in tunnel and underground engineering theory and technology research.E-mail address: liang_laoshi@126.com.

[2]Corresponding author:Tianxiong Dong,male, postgraduate student, Mainly engaged in road and geotechnical research. E-mail address: dongtianxiong_2009@126.com.

**References:**

[1] Zhang Jianfeng and Yue Guosheng. Geotechnical Engineering Research on the Wet Stone Masonry Dam Built with Red Sandstone[J].Shanghai Geology, 1989, 30 ( 2): 1 – 6.

[2] Zhang Songnan. Laboratory Test Research on Lime-improved Expansive Soil[D]. Hohai University, 2005, 3.

[3] Yang Jun, Li Xinchun, Tong Lei, et al. Test Research on Strength Characteristic of Weathering Sand-improved High Liquid Limit Red Clay[J]. Journal of Nanjing University of Science and Technology, 2015, 39 (2): 181-187.

[4] Cheng Yin, Li Zhanguo, Deng Xiaoxuan, et al. Test Research on the Curing of the Coastal Chlorine Salty Soil by a New Curing Material [J]. Chinese Journal of Geotechnical Engineering, 2011, 33 (8): 1240-1245.

[5] Liu Xiaoming, Xiong Li and Zhang Liangliang. Test Research on Restraint Measure of Disintegration of Red Sandstone of Class I[J]. Journal of Highway and Communication Technology, 2011, 28 (3): 25-30.

[6] Liu Daobin, Shen Weiguo and Zou Xianjie. Research on Curing Improvement of Red Sandstone Road Base of Class I [J]. The World of Building Materials, 2010, 31 (6): 12-16.

[7] Wang Zhimeng. Adaptation of Red Bed Mudstone and the Improved Soils as Filling of High Speed Railway Bed and the Engineering and Technological Research[D]. Southwest Jiaotong University, 2008, 12.

[8] Ministry of Communications of the People's Republic of China, JTG E40- 2007 Test Method of Soils for Highway Engineering[S]. China Communications Press, 2007 .

[9] Ministry of Communications of the People's Republic of China, JTG E51-2009 Test Method of Stabilizing Materials of Inorganic Binder of Highway Engineering[S]. China Communications Press, 2009.

[10] Ding Honghui and Fu Xuebing. Influence of Addition of Lime on the Compaction Test of High Liquid Limit Soil[J]. Communications Standardization, 2007 (10): 64-66.

[11] Yu Xiangjuan and Wang Yuan. Test Research on Lime-improved Expansive Soil Roadbed Filling[J]. Highway, 2000 (1): 13-16.

[12] Cao Pei, Wang Fang, Yan Lixue, et al. Test Research on the CBR Value of the Lime-improved Cohesive Soil[J]. Chinese Journal of Geotechnical Engineering, 2011, 33 (Supp.1): 298-301.

[13] Wang Yanran and Cao Dingsheng. Test Research on the CBR Value of the Improved Expansive Soil[J]. Chinese Journal of Rock Mechanics and Engineering, 2004, 23 (Supp.1): 4396-4399.

IOP Conf. Series: Earth and Environmental Science **61** (2017) 012103 doi:10.1088/1755-1315/61/1/012103

[14] Xia Mingxiao. Test Research on Strengthening and Improvement of Cohesive Soil Roadbed [D]. Southwest Jiaotong University,2006, 12.

IOP Conf. Series: Earth and Environmental Science **61** (2017) 012104 doi:10.1088/1755-1315/61/1/012104

# Comprehensive benefits analysis of steel structure modular residence based on the entropy evaluation

**Xiaoxiao Zhang[1,a)], Li Wang[1)],Pengming Jiang[1)]**

1Jiangsu University of Science and Technology, ZhenJiang JiangSu CHINA.
a)765937625@qq.com

**Abstract:** Steel structure modular residence is the outstanding residential industrialization. It has many advantages, such as the low whole cost, high resource recovery, a high degree of industrialization. This paper compares the comprehensive benefits of steel structural in modular buildings with prefabricated reinforced concrete residential from economic benefits, environmental benefits, social benefits and technical benefits by the method of entropy evaluation. Finally, it is concluded that the comprehensive benefits of steel structural in modular buildings is better than that of prefabricated reinforced concrete residential. The conclusion of this study will provide certain reference significance to the development of steel structural in modular buildings in China.

## 1. Introductions

At present, the domestic development of the assembly building is mainly a relatively low degree of direct assembly of prefabricated structural system, but modular building structure system is still in its infancy. According to the different structure of the main body, the assembled building can be divided into steel structure and concrete structure. In this paper, steel structural in modular buildings, a collection of steel structure and modular building structure, belongs to the prefabricated building with high degree of prefabrication.

With the entropy method, from the point of economic benefit, environmental benefit, social benefit and benefit perspective, this paper analyzes comprehensive benefit of the above residential structures. Finally, the conclusion is drawn that the comprehensive benefit of steel structure modular residential project is better than that of the assembled reinforced concrete residential project. The research results can provide reference for the further promotion of steel structure modular residential projects.

## 2. Establishment of Comprehensive Benefit Evaluation Index System

Based on studies above some references, the following will be from the economic benefits, environmental benefits, social benefits and technical benefits to compare and analyze the comprehensive benefits of steel structure modular residential projects and prefabricated reinforced concrete residential projects.

### *2.1. Economic Benefits*

About two data of economic benefit analysis based on some cases. Cases are shown in Table1.

**Table 1.** The basic situation of the projects

| | Steel Structure Modular Residential Projects | Prefabricated Reinforced Concrete Residential Projects | The cast-in-place Reinforced Concrete Residential Projects | Remarks |
|---|---|---|---|---|
| **Basic Feature Indexes** | 16 floors above ground; 7064 square meters | 18 floors above ground; 8125 square meters | 20 floors above ground; 9631 square meters; | |
| **Construction Schedule** | 45 days | 120 days | 200 days | |
| **Building Cost** | 2050 yuan / sq. M | 1650 yuan / sq. M | 1200 yuan / sq. M | |

Based on the case study, this paper analyzes the economic benefits of the modular residential projects and the assembled reinforced concrete residential buildings. Among them, the two economic benefits will be compared from the overall cost, operating costs and resource recovery.

*2.1.1 Overall Cost.* The overall cost of the project refers to the sum of project construction cost and the equivalent cost which is transformed from the economic income coming from differences between the effective use area, structure materials and construction schedule. [1] In order to facilitate comparative analysis, here to a place of reinforced concrete residential projects as a benchmark for comparison.

Total cost = project cost - interest savings - rental income - the use of the area to enhance the rate of return the direct economic benefits brought by shortening construction period.

Interest savings per unit area = (the number of days to shorten the duration of each layer × layer) ÷365× per unit area of the project cost ×annual rate; Rental income per unit area = (the number of days per week to shorten the length of the floor × layer)÷30÷rental income per unit area the direct economic benefits brought by the increase of residential area.

Rate of increase in unit area utilization = increase in the use of the area rate × unit area residential price.

According to the above cases and formulas, we can get the overall cost per unit area of modular residential buildings.

**Table2.** Comparison of the overall cost of different residential projects

| | Steel Structure Modular Residential Projects | Prefabricated Reinforced Concrete Residential Projects | The cast-in-place Reinforced Concrete Residential Projects |
|---|---|---|---|
| **Construction Cost per unit area (RMB / square)** | 2050 | 1650 | 1200 |
| **Interest Savings per unit area (RMB / square)** | -51.67 | -16.27 | - |
| **Rental Income per unit area (RMB / square)** | -191.67 | -75 | - |
| **Income raised by unit area Utilization Rate(RMB / square)** | -430.5 | 0 | - |
| **Overall Cost per unit area (RMB / square)** | 1376.16 | 1558.73 | 1200 |

We can find that the overall cost per unit area of the modular residential project of steel structure is slightly lower than or similar to that of the reinforced concrete residential building.

IOP Conf. Series: Earth and Environmental Science **61** (2017) 012104 doi:10.1088/1755-1315/61/1/012104

*2.1.2 Operating costs.* Because of the lack of data on the maintenance cost of the steel structure modular residential and the prefabricated reinforced concrete residential, the experts in the field are invited to estimate the daily maintenance cost and the overhaul cost of the different residential projects according to the parameter estimation method and analogy method. The estimation results are as follows: the modular steel structure housing is 107 RMB / square meters (converted to present value after the maintenance cost per unit area, the same below), prefabricated reinforced concrete residential 81 RMB / square meters.

*2.1.3 Resource recovery* The steel structure and the new type of enclosure materials are the green building materials with high recovery rate. At the end of the life, its resource recovery rate is up to 80%. But reinforced concrete is difficult to achieve resource recovery. At the end of the life, its resource recovery rate is only 10%.

### *2.2. Environmental Benefits*

*2.2.1 Carbon emissions.* The carbon emissions of building life cycle are mainly from the physical and chemical phase, the operation stage and the disposal and recovery stage [3]. In the operation process, the influence of different building materials on the environment is different. In this paper, 6 kinds of main building materials are selected.

**Table3.** $CO_2$ emissions per unit area of major materials [2]

| | **Cement(kg/m²)** | **Wood(kg/m²)** | **Steel products(kg/m²)** | **The total(kg/m²)** |
|---|---|---|---|---|
| **Case1** | 87.6 | 0.3 | 163.5 | 251.4 |
| **Case2** | 182 | 1.6 | 143 | 326.6 |

We find that carbon emissions of steel are much greater than that of cement, but the carbon emissions of the steel structure housing unit area are lower than that of reinforced concrete housing. Because the overall amount of materials to the steel structure housing is lower than concrete, the advantage of carbon emissions per unit is more obvious.

Carbon emissions in operation stage is mainly brought by the requirement for heating and air conditioning and consumption of electricity and gas to the realization of the building itself.

The disposal of carbon emissions in the recovery phase is mainly brought by the removal and disposal of waste. Measured from the whole life cycle, the $CO_2$ emission of reinforced concrete buildings is 1.4 times that of steel structures [2].

### *2.3. Social Benefits*

*2.3.1 The degree of industrialization.* Prefabricated rate refers to the completion of the single building, the proportion of the factory production volume of the total construction volume. According to the case analysis of the steel projects, the current steel structure modular residential prefabricated rate is up to 90% [3]. The prefabricated concrete project of prefabricated rate is only 60% [4].

*2.3.2 The satisfaction of users.* The demand of the customer is mainly reflected in the thermal insulation and sound insulation performance, residential comfort, order purchase, etc. User satisfaction is determined by expert scoring method. This paper invites 6 experts in the field of steel structure modular residential projects to score the user satisfaction of above residential projects.

**Table 4.** Scoring table of user satisfaction

| | Expert 1 | Expert 2 | Expert 3 | Expert 4 | Expert 5 | Expert 6 | Total | Composite Score |
|---|---|---|---|---|---|---|---|---|
| Case1 | 6 | 7 | 5 | 6 | 7 | 6 | 37 | 62 |
| Case2 | 7 | 8 | 6 | 7 | 6 | 7 | 41 | 68 |

### *2.4. Technical Benefits*

The technical benefit analysis of the main body is carried out from the safety of the structure.

IOP Conf. Series: Earth and Environmental Science **61** (2017) 012104 doi:10.1088/1755-1315/61/1/012104

*2.4.1 Structural safety.* Many results of the survey showed good seismic behavior of steel structures. In September 21, 1999, a magnitude 7.6 earthquake occurred in Taiwan, China, which is in the vicinity of Riyuetan Pool, with a magnitude of about 6, and the damage rate of the steel structure is only about 0.6%, and the failure rate of reinforced concrete and brick structure is more than 75% [4].

**Table 5.** Comparison of failure rates of building structures with different structural systems

| Structure Types | RC | Steel Structure | Brick Structure | Wood Structure | SRC | Adobe | Iron | Others |
|---|---|---|---|---|---|---|---|---|
| Failure Rate | 52.5 | 0.6 | 24.1 | 1.5 | 0.2 | 13.1 | 7.9 | 7.2 |

As can be seen from the above table, the damage rate of steel structure and SRC is low, the structural safety is higher. In contrast, the failure rate of reinforced concrete structure and brick structure is high, and the structural safety is low.

*2.5. Establishment of Index System*

Based on the above analysis, the relevant indicators are selected in the following table.

**Table 6.** Comprehensive benefit evaluation index system

| Target Layer | Criterion Layer | Index Layer | Meaning of Indicators |
|---|---|---|---|
| **Comprehensive Benefits(A)** | **Economic Benefits(B1)** | Overall cost | Considering the overall cost of the unit area of the construction period and the room rate |
| | | operating costs | Maintenance cost during operation |
| | | Resource recovery | Resource recovery at the end of life |
| | **Environmental Benefits(B2)** | $CO_2$ emissions | Physical phase carbon emissions |
| | **Social Benefits(B3)** | Industrialization degree | Prefabrication rate |
| | | User satisfaction | User needs satisfaction |
| | **Technical Benefits(B4)** | Main structure safety | Failure rate of building structure |

According to the above established index system and research results, this article sorted out the corresponding index system of the comprehensive benefit of steel structure modular residential and prefabricated reinforced concrete residential with the data shown in the table below.

**Table 7.** Evaluation index data of different residential projects

| | **Case1** | **Case2** |
|---|---|---|
| **Overall Cost** | 1376.16 | 1558.73 |
| **Operating Costs** | 107 | 81 |
| **Resource Recovery** | 80% | 10% |
| **$CO_2$ Emissions** | 251.4 | 326.6 |
| **Prefabrication Rate** | 90% | 60% |
| **User Satisfaction** | 62 | 68 |
| **Failure Rate of Building Structure** | 0.6% | 52.5% |

## 3. Evaluation Method

There are M evaluation objects and N evaluation indexes, $a_{ij}$ is the $J_{th}$ index of the $I_{th}$ evaluation object, all the original data constitute the original matrix $A = (a_{ij})_{m\times n}$. The inverse method is used to forward the reverse index, and the mean value method is used to get the original data $X = (x_{ij})_{m\times n}$.

According to the original data of the research above, we get the matrix X:

$$X = \begin{bmatrix} 1.062 & 0.862 & 1.778 & 1.130 & 1.200 & 0.954 & 1.977 \\ 0.938 & 1.138 & 0.222 & 0.870 & 0.800 & 1.046 & 0.023 \end{bmatrix}$$

The proportion of the target value $P_{ij}$:

$$P_{ij} = \frac{x_{ij}}{\sum_{i=1}^{m} x_{ij}}, (i = 1,2, \ldots\ldots m;\ j = 1,2, \ldots\ldots n) \tag{1}$$

According to the formula (1), the P matrix of modular residential buildings and fabricated reinforced concrete residences.

$$p = \begin{bmatrix} 0.53 & 0.43 & 0.89 & 0.57 & 0.60 & 0.48 & 0.99 \\ 0.47 & 0.57 & 0.11 & 0.43 & 0.40 & 0.52 & 0.01 \end{bmatrix}$$

$$e_j = -k \sum_{i=1}^{m} p_{ij} \ln p_{ij} \qquad k = \frac{1}{\ln m}, 0 \le e_j \le 1. \tag{2}$$

According to the formula (2), $e_j$ can be obtained by the 7 indicators of entropy:

$$e_j = \begin{pmatrix} 0.997 & 0.986 & 0.503 & 0.988 & 0.971 & 0.999 & 0.089 \end{pmatrix}$$

According to $e_j$, $d_j$ can be obtained:

$$d_j = 1 - e_j \tag{3}$$

Coefficient of variation of 8 indexes:

$$d_j = \begin{pmatrix} 0.003 & 0.041 & 0.497 & 0.012 & 0.029 & 0.001 & 0.911 \end{pmatrix}$$

Weight of item $J_{th}$:

$$\omega_j = \frac{d_j}{\sum_{j=1}^{n} d_j} \tag{4}$$

According to the formula (4), entropy weight of 8 indexes are as follows:

**Table 8.** Entropy weight of evaluation indexes

| | B1 | B2 | B3 | B4 | A |
|---|---|---|---|---|---|
| $\omega_1$ | 0.005 | | | | 0.002 |
| $\omega_2$ | 0.027 | | | | 0.009 |
| $\omega_3$ | 0.968 | | | | 0.337 |
| $\omega_4$ | | 1 | | | 0.008 |
| $\omega_5$ | | | 0.956 | | 0.020 |
| $\omega_6$ | | | 0.044 | | 0.001 |
| $\omega_7$ | | | | 1 | 0.623 |

Comprehensive benefit coefficient of the $i_{th}$:

$$v_j = \sum_{j=1}^{n} \omega_j p_{ij} \tag{5}$$

According to the formula (5), comprehensive benefit index of different structural system is as follows.

**Table 9.** Comprehensive benefit index of the housing with two different structural systems

| | B1 | B2 | B3 | B4 | A |
|---|---|---|---|---|---|
| **Case1** | 0.591 | 0.565 | 0.763 | 0.985 | 0.935 |
| **Case2** | 0.409 | 0.435 | 0.237 | 0.015 | 0.065 |

## 4. Conclusions

As can be seen from the table 11, the comprehensive benefits of the modular residential project of steel structure are better than that of the reinforced concrete residential project. Its economic benefits, environmental benefits, social benefits and technical benefits are better than the assembly of reinforced concrete residential projects.

According to the above analysis, the economic benefits of the steel structure modular residential projects are better than that of the assembled steel mixed residential projects. At present, it is important to develop research module technology, and promote the design and construction level of the modular.

## References

[1] Jing Zou, Dewen Liu, Yuanqi Li. “Analysis on the Cost and Comprehensive Economic Performance of High-rise Steel Structure Residence” in *Building Science Research of Sichuan* .2010(36):49-53.

[2] Xiaowei Zhu. “Comparative study on steel structure residence and traditional residence” Zhejiang University, Master's thesis. 2011:32-37.

[3] Zhaoyi Zhang, Xi Zhang, Zhiyan Yang. “Research on the Development Trend of Steel Structure Modular Buildings”. 2014-11-03(08).

[4] Kexin Qu. “Research on Modular Building Structure System of Steel Structure” Tianjin University, Master's thesis .2013:1-13

# Experimental study on mechanical properties of high damping rubber bearing model

**Yale Li[1,2, a], Zhouhong Zong[1,b], Xueyang Huang[3,4,c], Jian Xia[3,4,d] and Lu Liu[1,e]**

[1]School of Civil Engineering, Southeast University, Nanjing 210096, China;
[2]Jincheng college, Nanjing University of Aeronautics and Astronautics, Nanjing 211156, China;
[3]Fujian Academy of Building Research, Fuzhou 350025, China;
[4]Fujian Provincial Key Laboratory of Green Building Technology, Fuzhou 350025, China
E-mail:[a]812044475@qq.com; [b]1600764796@qq.com; [c]656469441@qq.com;[d]811493133@qq.com; [e]657174807@qq.com

**Abstract**. A high damping rubber bearing was designed and manufacturedin order to reduce the seismic response of the bridge structure. The experimental study on its mechanical properties was carried out. The results show: the horizontal equivalent damping ratio is 16.77% when the shear strain is 175%. If the shear strain is larger, the post yield stiffness will be smaller as well as the horizontal equivalent stiffness. However, the equivalent damping ratio and yield force will be larger. If the vertical compressive stress is greater meanwhile the loading frequency is faster, and number of repeated loading is smaller, the equivalent stiffness and the post yield stiffness will be larger as same as the equivalent damping ratio. Destruction didn't occur if the shear strain was no more than allowable shear strain (250%), and the ultimate shear strain was 350%. Compression destruction came when pressure stress was 10MPa. In the case of failure, the bearing was stable, but had large, nonrecoverable vertical and horizontal deformation. The study results can be used as the basis for shaking table test of seismic isolated continuous girder bridge with high damping rubber bearing and the reference for relative seismic isolated bridge design.

## 1. Introduction

Vibration isolation rubber bearings include natural rubber bearing, lead rubber bearing, and high damping rubber (HDR) bearing.The lead rubber bearing has been widely used in the bridge structure at home and abroad due to its good damping effect[1-3]. It is worth noting that environmental pollution is unavoidable in the production of lead core. After experiments, it was found that the lead core which is the source of damping would be fatigue destroyed under low cyclic loading[4]. If the lead rubber bearing fails, pollution will be caused by lead core leakage. Therefore, scholars began to study on other isolation bearings such as HDR bearing. Bhuiyan et al established the elastic viscoplastic rheological model of HDR bearing by the horizontal performance test[5]. Markou et al established a non-linear rheological model to explain the Mullins effect HDR bearing[6]. Monzon et al studied the seismic response of 1:4 steel plate bridge with HDR bearing by shaking table test[7].

HDR bearing is a new type of seismic isolation bearing developed in recent years. It sets the vertical bearing capacity, the level of restoring force and damping performance. Its initial elastic stiffness to resistwind load, is similar with laminated rubber bearings. The damping ratio is greater than 12% under large displacement[8], which is much higher than that of the plate rubber bearing and less affected by temperature[4]. At present, HDR bearing with damping ratio greater than 20%, which is called super high damping rubber (SHDR) bearing, has been developed. SHDR bearings have been applied on Hong Kong-Zhuhai-Macao Bridge[9,10]. It can be predicted that there will be a large number of HDR isolated bridges in the future, so it is of practical significance to study the seismic performance of HDR isolated continuous girder bridges.

## 2. Bearing model design and test conditions

### *2.1. Design of bearing model*

The size of HDR bearing model is smaller than the conventional bearing in bridge engineering, in order to be utilized for the shaking table test of two span continuous girder bridge model. The upper and lower connecting plates size is 24cm × 24cm and the rubber body is a cylinder with a diameter of 15cm. The inner part of the bearing is a laminated structure, which comprises 4 layers of 2mm steel plate and 4 layers of 5mm rubber, shown in figure 1.

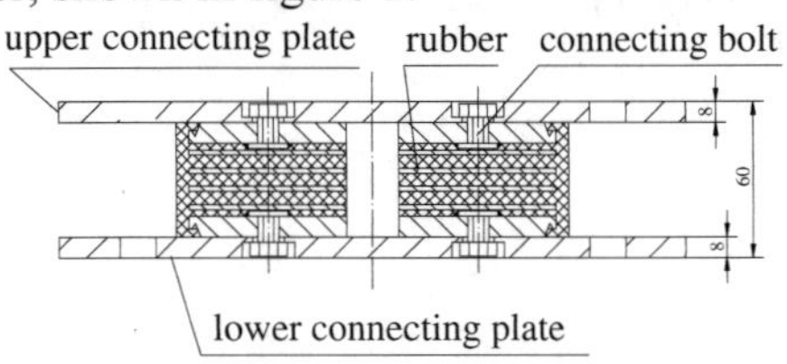


**Figure 1.** HDR bearing structure.

### *2.2. Test condition*

The test was carried out in the laboratory of Hengshui Taiwei New Materials Polytron Technologies Inc. The test equipment is 2000 ton electro-hydraulic servo dynamic pressure shear test machine produced by Ji'nan Sanyue Testing Instruments Co.Ltd, as shown in figure 2.

**Figure 2.** Pressure shear test machine.

## 3. Compression test of bearing model

Bearing model compression test was carried out by using the loading method specified in GB/T 20688.1-2007[11] 6.3.1.3, in accordance with the requirements of the JT/T 842-2012[8] 6.6.1. Loading of 3 cycles was from 0 to maximum design pressure to 0 and only the last cycle was calculated. Compression stiffness can be deduced by Eq.1:

$$K_v = \frac{P_2 - P_1}{Y_2 - Y_1} \tag{1}$$

Where $K_v$=Compression stiffness; $P_1$=26.55kN; $P_2$=106.2kN; $Y_1$, $Y_2$=the corresponding vertical compression displacement of $P_1$ and $P_2$. $K_v$ was calculated as 91.94kN/mm. Therefore, the modified elastic modulus $E_c$ of the bearing model is 103.9MPa.

## 4. Shear test of bearing model

In accordance with the requirements of the JT/T 842-2012[8] 6.6.3, bearing model horizontal shear test was carried out by using the loading method specified in GB/T 20688.1-2007[11] 6.3.2. The vertical stress is 6MPa and the shear strain is 175%. Loading frequency was set to 0.05Hz, in order to ensure the test security. The average equivalent stiffness $K_h$ and equivalent damping ratio of the 2~11th cycle tests were obtained as test results after loading 11 cycles. $K_h$ is 0.924kN/mm and $h_{eq}$ is 16.77%。

## 5. Shear properties correlation testof bearing model

The horizontal shear performance test results were obtained by using certain vertical stress, shear strain, loading frequency and loading times. Performance indicators will be affected if any of the above 4 factorschanges. Thus the shear performance of the bearing model was studied by correlation test.

### *5.1. Shear strain correlation*

The compressive stress was 6MPa and the loading frequency was 0.05Hz. The maximum shear strain were 50%, 100%, 150%, 175%, respectively. The curves of horizontal equivalent stiffness, yield stiffness, yield force and equivalent damping ratio were calculated, as shown in figure 3. The larger the shear strain is, the smaller the equivalent stiffness and yield stiffness are, with greater yield force and equivalent damping force.

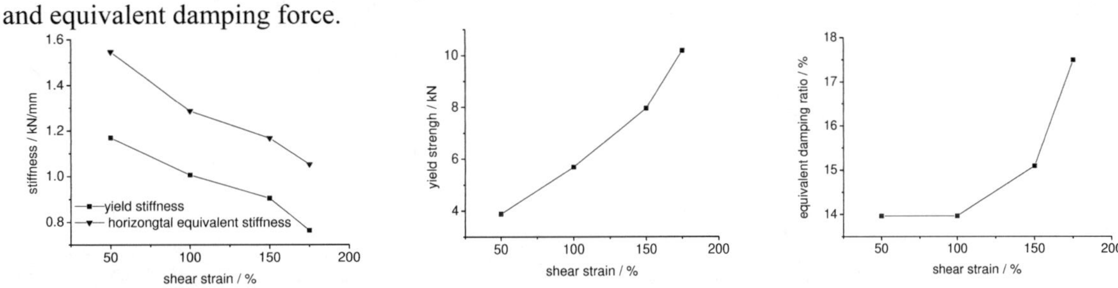


(a) stiffness (b) yield force (c) equivalent damping ratio

**Figure 3.** Horizontal performance indicators under different shear strains.

### *5.2. Vertical compression stress correlation*

The horizontal shear performance was tested with different vertical compression stress which was 2MPa, 4MPa, 6MPa, respectively. The maximum shear strain was 175% and the loading frequency was 0.05Hz. Horizontal shear performance indicators were obtained under the third cycle loading (see table 1). The horizontal equivalent stiffness, the yield stiffness and the equivalent damping ratio will increasing, with less yield force as soon as the vertical compressive stress is greater.

**Table 1.**Horizontal performance indicators under different vertical compression stress.

| Vertical Stress (MPa) | $k_h$ (kN/mm) | $h_{eq}$ (%) | $k_d$ (kN/mm) | $Q_d$ (kN) |
|---|---|---|---|---|
| 2 | 0.935 | 16.834 | 0.594 | 12.016 |
| 4 | 0.981 | 17.253 | 0.673 | 10.86 |
| 6 | 1.053 | 17.494 | 0.764 | 10.184 |

### *5.3. Loading frequency correlation*

The horizontal shear performance was tested with different loading frequency which is 0.02Hz, 0.05Hz, 0.1Hz, respectively. The maximum shear strain was 175% and the vertical compression stress was 6MPa. Horizontal shear performance indicators were obtained under the third cycle loading (see table 2). The horizontal equivalent stiffness, the yield stiffness, the yield force and the equivalent damping ratio are higher as soon as the loading frequency is faster.

**Table 2.**Horizontal performance indicators under differentloading frequency.

| Loading Frequency (Hz) | $k_h$ (kN/mm) | $h_{eq}$ (%) | $k_d$ (kN/mm) | $Q_d$ (kN) |
|---|---|---|---|---|
| 0.02 | 1.017 | 17.366 | 0.744 | 9.595 |

| | | | | |
|---|---|---|---|---|
| 0.05 | 1.053 | 17.494 | 0.764 | 10.184 |
| 0.1 | 1.109 | 18.29 | 0.895 | 10.505 |

### *5.4. Loading times correlation*

The horizontal shear performance was tested with 50 times. The maximum shear strain was 175%. Loading frequency was 0.05Hz and the vertical compression stress was 6MPa. Shear force-shear displacement hysteresis curves of the 1, 3, 5, 10, 20, 30, 50th cycles are plotted in figure 4. The surrounded areas of the hysteresis curves decreased while curves gradually elongated and had the clockwise rotation (see figure 4). The horizontal shear performance changed with the increase of loading times. Horizontal equivalent stiffness $K_h$ and the yield stiffness $K_d$ in figure 5 had a downward trend. The yield strength $Q_d$ decreased with the increase of loading times. The descent speeds of the 1~5th cycles were fast and afterwards the speeds were stable (see figure 6). Horizontal equivalent stiffness $K_h$ and surrounded area of the hysteresis curve $W_d$ were relevant to equivalent damping ratio $h_{eq}$. $W_d$ decreased rapidly of the 1~10th cycles then slowed down. $K_h$ declined faster than $W_d$ in the 10~20th cycles, so $h_{eq}$ increased in the short term. After 20 times, $W_d$ continued to decline and $K_h$ tended to be stable, so $h_{eq}$ decreased but not significant (shown in figure 7).

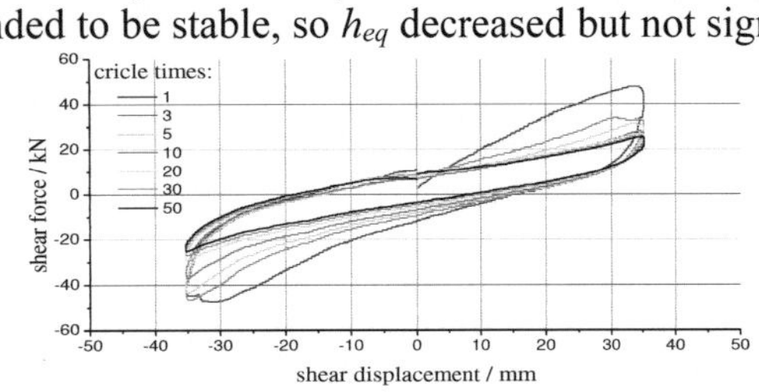


**Figure 4.** Shear force-shear displacement hysteresis curves.

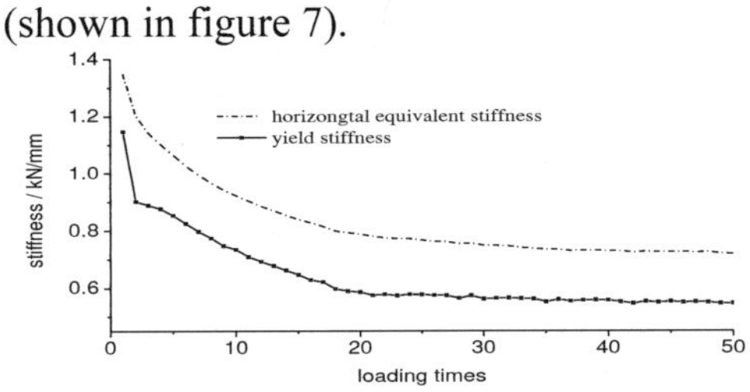


**Figure 5.** Horizontal equivalent stiffness and the yield stiffness.

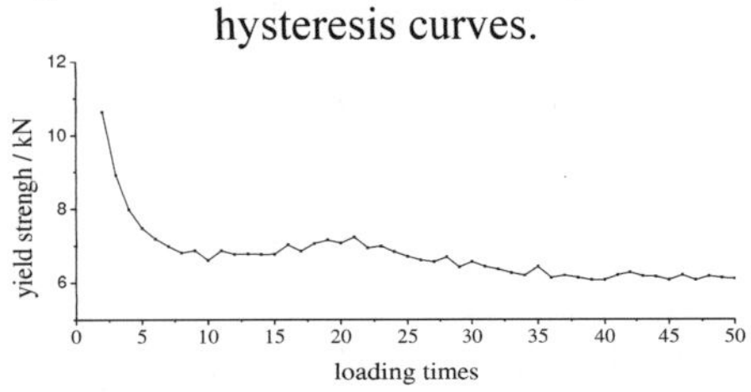


**Figure 6.** Yield strength.

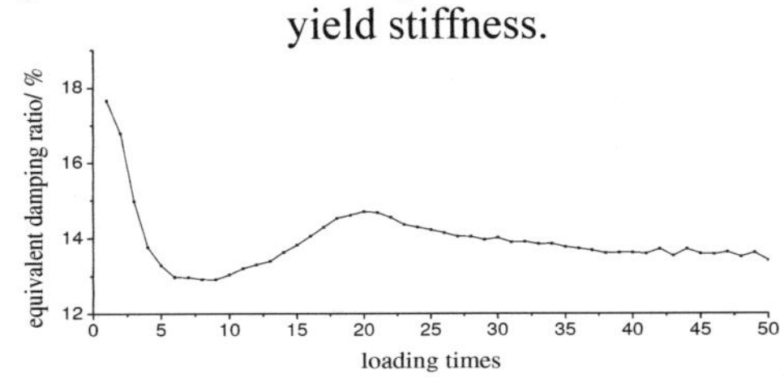


**Figure 7.** Equivalent damping ratio.

## 6. Ultimate shear testof bearing model

### *6.1. Large deformation shear test*

The allowable shear strain was 250%, and the vertical compressive stress was 6MPa. The loading repeated for 3 times with frequency of 0.05Hz. Post test phenomenon is shown in figure 8. The position of the inner steel plate of the bearing model was dislocated, and the surface of the rubber protective layer was marked by steel plates with a small amount of wore surface. But there was no structural damage such as tensile shear failure.

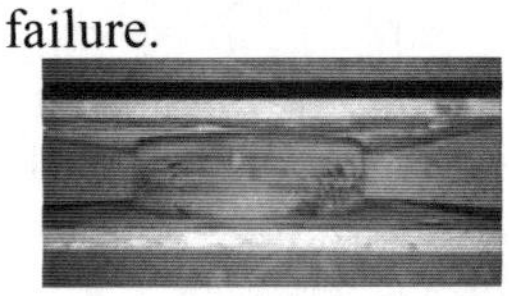

**Figure 8.** Large deformation shear test phenomena.

### *6.2. Ultimate sheartest*

The vertical compressive stress was 6MPa, and the loading frequency was 0.05Hz. Failure occurred when the ultimate shear strain was 350%. Surface wore (see figure 9(a)), 1 of 4 connecting screws for connecting steel plates and rubber body was cut (see figure 9(b)). The rubber body could not recover completely after unloading, and the residual deformation occurred (see figure 9(c)). Measured ultimate shear performance curve is illustrated in figure 10. Failure shear force was 92.3kN. It can be seen from figure 10 that the horizontal stiffness is large when the shear displacement is small, which is suitable for resisting wind load. The horizontal stiffness is small when the shear displacement is large, which means bearing displacement will alleviate bridge seismic response on the condition of the small and medium earthquakes. When the shear displacement is too large, the hardening of the bearing will improve the horizontal stiffness, which is beneficial to controlling the relative displacement between the bridge and the infrastructure.

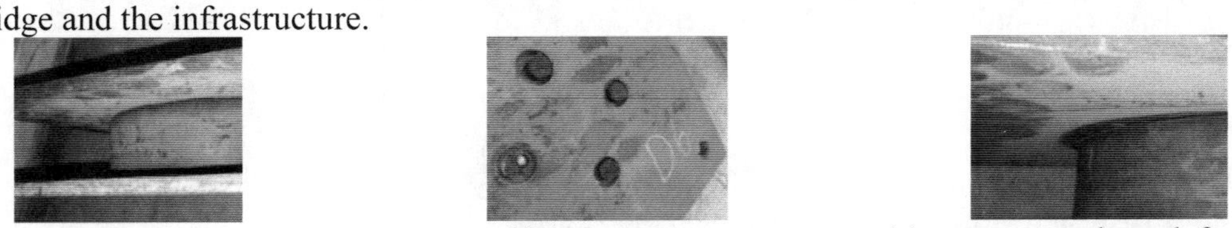

(a) surface wore (b) bolt was cut (c) permanent shear deformation

**Figure 9.** Ultimate shear failure.

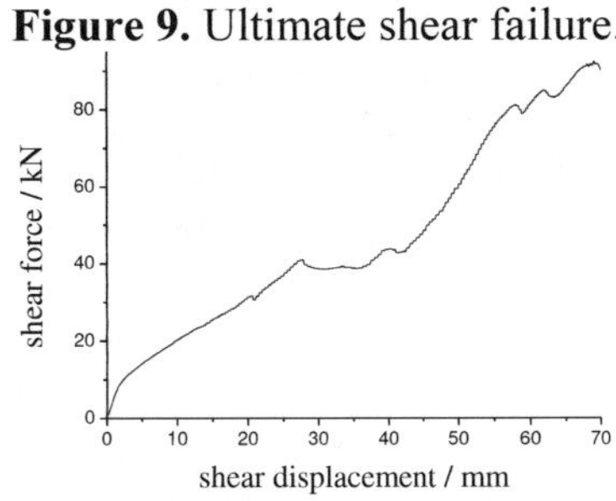


**Figure 10.** Ultimate shear performance curve.

## 7. Ultimate compression test of bearing model

The vertical compression test was carried out to test the ultimate compressive strength. The bearing model ringed to severe bulging under 10MPa compression stress (see figure 11). The horizontal residual deformation of the support was 6mm and vertical residual compression deformation was 3.2mm after unloading. It could be seen that inner rubber was extruded to the original hollow rubber body when the connecting steel plate was invisible, as shown in figure 12.

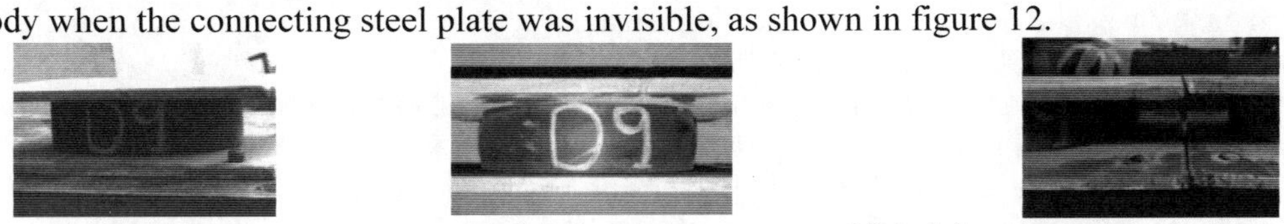

(a)before test (b)after test (c) height comparison before and after test

**Figure 11.** Permanent deformation of rubber.

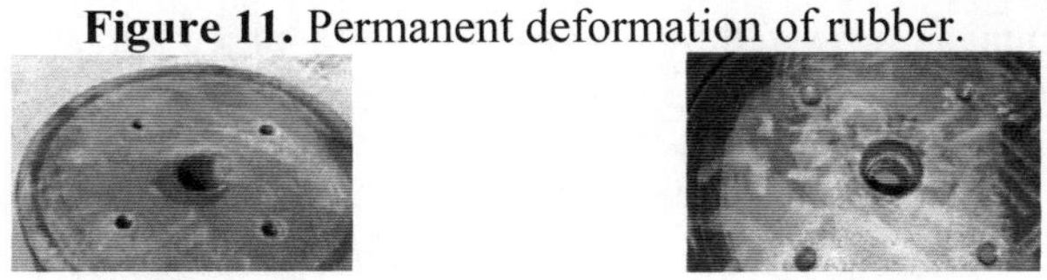

(a)before test (b)after test

**Figure 12.** Partial displacement of rubber.

## 8. Conclusions

In order to test the mechanical properties of the high damping rubber bearing model used in the shaking table test of the continuous beam bridge model, the following conclusions are drawn:

The horizontal equivalent damping ratio is 16.77%, which means the model is HDR bearing.

IOP Conf. Series: Earth and Environmental Science **61** (2017) 012105 doi:10.1088/1755-1315/61/1/012105

The larger the shear strain is, the smaller the equivalent stiffness and yield stiffness are, with greater yield force and equivalent damping force.

The horizontal equivalent stiffness, the yield stiffness and the equivalent damping ratio will increasing, with less yield force as soon as the vertical compressive stress is greater.

The horizontal equivalent stiffness, the yield stiffness, the yield force and the equivalent damping ratio are higher if the loading frequency is faster.

The larger the number of repeated loading is, the smaller the yield force, the horizontal equivalent stiffness and the yield stiffness are. Besides, the equivalent damping ratio also shows a downward trend.

The bearing model will not be destroyed if shear strain is less than 250%. The ultimate shear strain was 350%.

The ultimate compressive stress was 10MPa. When the model was damaged, the bearing was stable though had vertical and circumferential residual compression deformation.

## Acknowledgments

The study was supported by the natural science foundation of China (contract No.51378112), the open foundation of Southwest Jiaotong University of geological disaster prevention and control technology of land transportation (contract No.SWJTU-GGS-2014001), and Jiangsu provincial natural science research project (contract No.15KJD580001). These supports are greatly acknowledged.

## References

[1] B. Yan, X.L Du, Q. Han, et al2014 Application of seismic isolation and mixing device in seismic design of single tower cable stayed bridge *Bridge Construction* **6** pp 101-6.(in Chinese)

[2] D.G. Weng, S. Liu, X.L. Li, et al 2015 Seismic isolation scheme and its engineering application of structure with soft soil foundation *Journal of building structures* **2** pp 41-50. (in Chinese)

[3] O.V.Mkrtychev, G.A.Dzhinchvelashvili, A.A.Bunov 2014 Study of lead rubber bearings operation with varying height buildings at earthquake*Procedia Engineering* **91** pp 48-53.

[4] J.S. Zhuang, Bridge damping, isolation bearings and devices, (China Railway Publishing House, Beijing, 2012), pp.121-123. (in Chinese)

[5] A.R.Bhuiyan, Y.Okui, H.Mitamura, et al 2009 A rheology model of high damping rubber bearings for seismic analysis: Identification of nonlinear viscosity International Journal of Solids and Structures 46(7) pp 1778-92.

[6] A.A.Markou and G.D Manolis 2016 Mechanical models for shear behavior in high damping rubber bearings *Soil Dynamics and Earthquake Engineering* **90** pp 221-6.

[7] E.V.Monzon, I.G.Buckle and A.M.Itani 2016 Seismic performance and response of seismically isolated curved steel I-Girder bridge *Journal of Structural Engineering* **142**(12) pp 04016121.

[8] Ministry of communications of the people's Republic of China 2012 *High damping rubber bearing of Highway Bridge* (JT/T 842-2012). (in Chinese)

[9] H.Y. Yu 2013 Zhuzhou successfully developed the world's largest super high damping rubber bearings *Science and technology Daily* **5**. (in Chinese)

[10] Q.K. Su and H.B. Xie 2016 Survey on the construction of steel bridge of Hong Kong-Zhuhai-Macao Bridge *China Journal of Highway and Transport* **29**(12) pp 1-9. (in Chinese)

[11] China National Standardization Management Committee 2007 *Rubber bearing first part: test method for rubber bearings* (GB/T 20688.1-2007). (in Chinese)

IOP Conf. Series: Earth and Environmental Science **61** (2017) 012106 doi:10.1088/1755-1315/61/1/012106

# Numerical simulation of nanofluids based on power-law fluids with flow and heat transfer

**Lin Li[1, a)], Yongyue Jiang[2] and Aixin Chen[3]**

[1]School of Mathematics and Physics, University of Science and Technology Beijing, Beijing 100083, China.
[2]Beijing Time Technologies Co.,Ltd., Kaituo Road 17, China.
[3]Department of Mathematics and Computer Science, Hengshui University, Hebei 053000, China.
[a)]Corresponding author: exam-19861208@163.com

**Abstract**. In this paper, we investigate the heat transfer of nanofluids based on power-law fluids and movement of nanoparticles with the effect of thermophoresis in a rotating circular groove. The velocity of circular groove rotating is a constant and the temperature on the wall is kept to be zero all the time which is different from the temperature of nanofluids in the initial time. The effects of thermophoresis and Brownian diffusion are considered in temperature and concentration equations, and it is assumed that the thermal conductivity of nanofluids is a function of concentration of nanoparticles. Based on numerical results, it can be found that nanofluids improve the process of heat transfer than base fluids in a rotating circular groove. The enhancement of heat transfer increases as the power law index of base fluids decreases.

## 1. Introduction

Nanofluids which composed of nanoparticles and common fluids has been widely applied in many fields like physics, chemistry and biology, etc., in recent years. As nanofluids have been applied in solving many practical problems, many researchers become interested in studying nanofluids. For example, remedying the soil, removing oily soil and enhancing oil recovery can be done by nanofluids with surfactant micelles and it can improve the effect [1]. Also nanofluids has implication for cooling equipment and inkjet [2]. Actually the enhancement of heat transfer, one of the most important application of nanofluids, is catching more attention. "Nanofluids" is first proposed by Choi [3] who did research on the heat transfer about fluids. He found that putting nanoparticles in fluids could enhance thermal conductivities of base fluids. Since then lots of researches on the process of heat transfer of nanofluids come out and the theme is to find out reasons for the enhancement of thermal conductivity of nanofluids. At the beginning of these researches for the same kind of naoparticles and base fluids, nanoparticle size, concentration, dispersion, nanoparticle structure and microscopic motion are regraded as reasons [4-7]. The different kinds of nanoparticles and base fluids may also enhance heat transfer, e.g., nanofluid with Cu is better than the one with $Al_2O_3$ [5]. In those researches, it is found that experiment data are much higher than results of theoretical models like Maxwell's equation [7]. Absence of microscopic motion and structure-dependent behavior in theoretical models are reasons for this error. In 2006, Buongiorno [16] chose some important reasons of enhancing thermal conductivity to discuss among inertia, Brownian diffusion, thermophoresis, magnus effect, diffusiophoresis, gravity and fluid drainage. He compared these seven physical quantities and concluded that Brownian diffusion and thermophoresis play the most important roles in enhancing the thermal conductivity. Other five physical quantities also influenced the enhancement of thermal conductivity but the influence was too small and could be negligible. Then he proposed a model with the consideration of Brownian discussion and thermophoresis and this model includes the microscopic motion or Brownian diffusion.

Most people will regard the nanofluid as a mixture of nanoparticles and the base fluid. Not like a conventional mixture with solid and fluid, sometimes it behaves more like one kind of common fluids [8]. Based on this assumption, the concentration of nanoparticles is involved in all these new physics parameters such as the new viscosity which is also called the effective viscosity has a definition with $\mu_{eff} = \mu_f /(1-\phi)^{2.5}$, where $\mu_f$ and $\phi$ are the viscosity of the base fluid and concentration of nanoparticles, respectively. In many researches on nanofluids, this one-fluid model performs well and is widely applied to solve many problems especially on boundary layer flow and heat transfer [9-15]. In all these researches, it is assumed that the concentration of nanoparticles is a constant. But according to the model proposed in [16], the nanofluid was considered as a mixture and it has the relative motion between the base fluid and nanoparticles. Not a constant any more, the concentration of nanoparticles is an unknown variable coupled with flow equations and temperature equation. Results of the model are well agreed with some experiments and the model has been applied to solve various physical problems [17-22]. Two modeling ideas are sometimes combined and studied as well [23, 24, 25].

Following the idea proposed by Buongiorno [16], we choose to modify the model in [28] and study the flow and heat transfer of nanofluids based on power-law fluids in a rotating circular groove. In Section 2, the mathematical formulation of the physical problem is introduced. A temporally second order continuous finite element method is proposed in Section 3. Numerical results and discussion are presented in Section 4. Finally, Section 5 concludes the paper.

## 2. Mathematical formulation of the problem

Here the physical model we study is that nanofluids composed of nanofluids (e.g. Cu) and the power-law fluid flow in a rotating circular groove. The rotating angular velocity $\omega$ is $\pi/35$ (see also [28]). Meanwhile, the temperature change of nanofluids is also considered. Also the wall temperature of the circular groove is always zero during the whole process. The corresponding equations are as follows:

$$\nabla \cdot u = 0, \quad (1)$$

$$\rho_f(\partial u/\partial t + (u\cdot\nabla)u) = -\nabla p + \nabla\cdot\sigma_u, \quad (2)$$

$$\rho_f c_f(\partial T/\partial t + u\cdot\nabla T) = \nabla\cdot(k\nabla T) + \rho_p c_p[D_B \nabla C\cdot\nabla T + D_T/T_{in}\nabla T\cdot\nabla T], \quad (3)$$

$$\partial C/\partial t + u\cdot\nabla C = \nabla\cdot[D_B\sigma_C + D_T\nabla T/T_{in}]. \quad (4)$$

Where $\sigma_u$ and $\sigma_C$ represent the viscous part of stress tensor and Brownian diffusion, respectively, and their expressions are given as follows:

$$\sigma_u = \eta(2(\partial u_x/\partial x)^2 + (\partial u_x/\partial y + \partial u_y/\partial x)^2 + 2(\partial u_y/\partial y)^2)^{(m-1)/2}(\nabla u + \nabla u^T), \quad (5)$$

$$\sigma_C = ((\partial C/\partial x)^2 + (\partial C/\partial y)^2)^{(m-1)/2}\nabla C. \quad (6)$$

In addition, here the vector $\vec{x} = (x, y)$ denotes the coordinate system ( $x$ and $y$ indicate the streamwise direction and the normal direction, respectively.). Also $u = (u_x, u_y)$ represents the velocity of the nanofluid, where $u_x$ and $u_y$ denote the velocity components along $x-$ and $y-$axes. $p$, $T$ and $C$ represent the hydrostatic pressure, the temperature and the concentration of nanofluids, respectively. $m$ is the power law index of power law fluids. When $m-1$, the fluid becomes Newtonian fluid. The parameters $\rho_f$, $\rho_p$, $c_f, c_p, \eta$ are the fluid density, nanoparticles density, fluid specific heat, nanoparticles specific heat and relative viscosity coefficient for power law fluids, respectively. As pointed out in [9], here the thermal conductivity $k$ can be written as

$$k = \frac{k_f((k_p + 2k_f) - 2C(k_f - k_p))}{(k_p + 2k_f) + C(k_f - k_p)}, \tag{7}$$

Where $k_f$ and $k_p$ are the thermal conductivity of base fluids and thermal conductivity of nanoparticles, respectively. $D_B$ and $D_T$ denote the Brownian diffusion coefficient and the thermophoresis diffusion coefficient, respectively. The corresponding boundary condition at the edge of the circular groove are

$$u_x = \omega y, \quad u_y = -\omega x, \quad T = T_\omega, \quad \partial C / \partial n = 0. \tag{8}$$

The following dimensionless quantities are introduced:

$$U = u/V, \quad t^* = tV/L, \quad \theta = (T - T_\omega)/(T_{in} - T_\omega), \tag{9}$$

$$p^* = pL^m/(\eta V^m), \quad X = x/L, \quad \phi = C/C_\infty, \tag{10}$$

where $L$ is the characteristic length and $V$ is the characteristic speed. $T_{in}$ is the initial temperature of nanofluids and $C_\infty$ is the characteristic concentration. When substituting these dimensionless quantities into Eqs.(1)-(4) and denoting $u = U$, $x = X$, $p = p^*$, $t = t^*$, we can obtain

$$\nabla \cdot u = 0, \tag{11}$$

$$\mathrm{Re}(\partial u / \partial t + (u \cdot \nabla)u) = -\nabla p + \nabla \cdot \sigma_u', \tag{12}$$

$$\mathrm{Re}\cdot \mathrm{Pr}(\partial \theta / \partial t + u \cdot \nabla \theta) = \nabla \cdot (k/k_f \nabla \theta) + \mathrm{Re}\cdot \mathrm{Pr}[N_B \nabla \phi \cdot \nabla \theta + N_T \nabla \theta \cdot \nabla \theta], \tag{13}$$

$$\mathrm{Re}\cdot Sc(\partial \phi / \partial t + u \cdot \nabla \phi) = \nabla \cdot \sigma_\phi + N_T / N_B \Delta \theta. \tag{14}$$

Some dimensionless parameters in Eqs.(11)-(14) become

$$N_B = \rho_p c_p D_B / (\rho_f c_f VL), \quad N_T = \rho_p c_p D_T (T_{in} - T_\omega) / \rho_f c_f T_{in} VL, \tag{15}$$

$$\mathrm{Re} = \rho_f VL(V/L)^{1-m}/\eta, \quad \mathrm{Pr} = c_f \eta (L/V)^{1-m}/k_f, \tag{16}$$

$$Sc = \eta (L/V)^{1-m} \rho_f D_B, \quad \sigma_\phi = ((\partial \phi/\partial x)^2 + (\partial \phi / \partial y)^2)^{(m-1)/2} \nabla \phi, \tag{17}$$

$$\sigma_u' = (2(\partial u_x/\partial x)^2 + (\partial u_y/\partial x + \partial u_x/\partial y)^2 + 2(\partial u_y/\partial y)^2)^{(m-1)/2}(\nabla u + \nabla u^T). \tag{18}$$

The boundary conditions given in (8) become

$$u_x = \omega y, \quad u_y = -\omega x, \quad \theta = 0, \quad \partial \phi / \partial n = 0. \tag{19}$$

## 3. Finite element scheme

The main aim of this section is to present finite element scheme on Eqs.(11)-(14). Before we do this, some notations used in the subsequent computation are introduced. $\Omega$ denotes the domain of the circular groove in $R^2$. $\Delta t (> 0)$ denotes the time step size and $t^{n+1} = (n+1)\Delta t$. So $u_h^n$, $p_h^n$, $\theta_h^n$, and $\phi_h^n$ represent an approximation of $u$, $p$, $\theta$ and $\phi$ at the time $t^n$. On the one hand, to have a more stable formulation for the pressure $p$ [27], the continuity equation (1) is rewritten as

$$\nabla \cdot u + \delta p = 0, \quad \delta = 10^{-6}. \tag{20}$$

On the other hand, based on the work in [27, 29, 30], the midpoint scheme on the time is used to Eqs.(11)-(14). As a result, the discretized formulas on them become as follows:

$$\int_\Omega ((\nabla \cdot u_h^{n+1/2})q + \delta p_h^{n+1/2} q) dX = 0, \tag{21}$$

$$\int_{\Omega}(\mathrm{Re}(u_{\bar{t}}^{n+1/2}\cdot v+(u_h^{n+1/2}\cdot\nabla u_h^{n+1/2})\cdot v)+p_h^{n+1/2}(\nabla\cdot v)+\sigma_{uh}^{'\,n+1/2}:\nabla v)dX=0, \quad (22)$$

$$\int_{\Omega}(\mathrm{Re}\cdot\mathrm{Pr}(\theta_{\bar{t}}^{n+1}\omega+(u_h^{n+1/2}\cdot\nabla\theta_h^{n+1/2})\omega)+(k_h^{n+1/2}/k_f)\nabla\theta_h^{n+1/2}\cdot\nabla\omega)dX$$

$$-\int_{\Omega}\mathrm{Re}\cdot\mathrm{Pr}(N_B(\nabla\phi_h^{n+1/2}\cdot\nabla\theta_h^{n+1/2})\omega+N_T(\nabla\theta_h^{n+1/2}\cdot\nabla\theta_h^{n+1/2}\omega))dX=0, \quad (23)$$

$$\int_{\Omega}(\mathrm{Re}\cdot Sc(\phi_{\bar{t}}^{n+1}\varphi+(u_h^{n+1/2}\cdot\nabla\phi_h^{n+1/2})\varphi)+\sigma_{\phi,h}^{n+1/2}\cdot\nabla\varphi+N_T/N_B\cdot\nabla\theta_h^{n+1/2}\cdot\nabla\varphi)dX$$

$$-\int_{\Gamma}(N_T/N_B\cdot(\nabla\theta_h^{n+1/2}\cdot n)\varphi)ds=0, \quad (24)$$

where $u_h^{n+1/2}=(u_{x,h}^{n+1/2},u_{y,h}^{n+1/2})$, $p_h^{n+1/2}=(p_h^n+p_h^{n+1})/2$, $u_{\bar{t}}^{n+1/2}=(u_h^{n+1}-u_h^n)/\Delta t$, $u_h^{n+1/2}=(u_h^{n+1}+u_h^n)/2$,

$\theta_{\bar{t}}^{n+1/2}=(\theta_h^{n+1}-\theta_h^n)/\Delta t$, $\theta_h^{n+1/2}=(\theta_h^{n+1}+\theta_h^n)/2$, $\phi_{\bar{t}}^{n+1/2}=(\phi_h^{n+1}-\phi_h^n)/\Delta t$, $\phi_h^{n+1/2}=(\phi_h^{n+1}+\phi_h^n)/2$. In addition, we can obtain

$$k_h^{n+1/2}/k_f=(1-2\phi_h^{n+1/2})k_p+2(1-\phi_h^{n+1/2})k_f/((2+\phi_h^{n+1/2})k_f+(1-\phi_h^{n+1/2})k_p), \quad (25)$$

$$\sigma_{uh}^{'\,n+1/2}=[(2(\partial u_{x,h}^{n+1}/\partial x)^2+(\partial u_{x,h}^{n+1}/\partial y+\partial u_{y,h}^{n+1}/\partial x)^2+2(\partial u_{y,h}^{n+1}/\partial y)^2)^{(m-1)/2}$$

$$+(2(\partial u_{x,h}^{n+1}/\partial x)^2+(\partial u_{x,h}^{n+1}/\partial y+\partial u_{y,h}^{n+1}/\partial x)^2+2(\partial u_{y,h}^{n+1}/\partial y)^2)^{(m-1)/2}](\nabla(u_h^{n+1}+u_h^n)$$
$$+\nabla(u_h^{n+1}+u_h^n)^T)/4, \quad (26)$$

$$\sigma_{\phi,h}^{n+1/2}=[((\partial\phi_h^{n+1}/\partial x)^2+(\partial\phi_h^{n+1}/\partial y)^2)^{(m-1)/2}+((\partial\phi_h^n/\partial x)^2+(\partial\phi_h^n/\partial y)^2)^{(m-1)/2}]\nabla(\phi_h^{n+1}+\phi_h^n)/4. \quad (27)$$

Since Eqs.(21)-(24) are nonlinear implicit equations, a linearization for them should be considered. Here Newtonian method and the fixed point method shown in [27, 31] are used to solve the linear system at each time step. Here $\bar{u}_s=(\bar{u}_{x,s},\bar{u}_{y,s})$, $\bar{p}_s$, $\bar{\theta}_s$ and $\bar{\phi}_s$ represent an approximation of $u_h^{n+1}$, $p_h^{n+1}$, $\theta_h^{n+1}$ and $\phi_h^{n+1}$, respectively. where the initial guess is chosen as $\bar{u}_0=u_h^n=0$, $\bar{p}_0=p_h^n=0$, $\bar{\theta}_0=\theta_h^n=1$ and $\bar{\phi}_0=\phi_h^n=0.01$. Also the parameter $\alpha$, in (30) becomes

$$\alpha=((1-(\bar{\phi}_{s-1}+\phi_h^n))k_p=2(1-(\bar{\phi}_{s-1}+\phi_h^n)/2)k_f)/((2+(\bar{\phi}_{s-1}+\phi_h^n)/2)k_f+(1-(\bar{\phi}_{s-1}+\phi_h^n)/2)k_p), \quad (28)$$

In addition, the above computation are based on the Freefem++ platform [26].

## 4. Numerical results and discussion

The main aim of this section is to present numerical results and discussion. We mainly consider the case: for $m=0.91$, the values of some dimensionless parameters are used as follows:

$$\mathrm{Re}=70.73,\quad \mathrm{Pr}=41.08,\quad Sc=15.80. \quad (29)$$

In addition, $N_B = N_T = 3.29\times10^{-5}$, $k_p = 400W/mK$, $k_f = 0.613W/mK$, $\Delta t = 10^{-5}$ are used for this case.

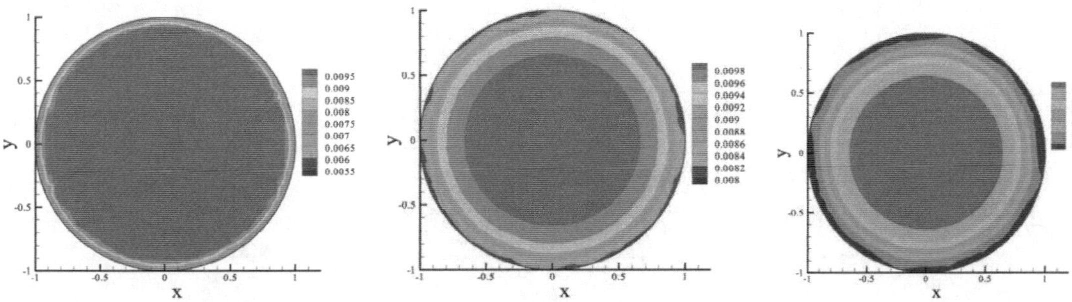


FIGURE 1. The concentration of nanoparticles changes for the nanofluid based on the pseudo-plastic fluid ($m = 0.91$) when $t = 0.04$, 0.3 and 0.8.

Fig.1 shows planar figures about the change of concentration of nanoparticles in the nanofluid based on the pseudo-plastic fluid ($m = 0.91$) for $t = 0.04$, 0.3 and 0.8. Similarly, the change of temperature of nanofluids is shown in Fig.2. As observed in Figs.1-2, we can find that the concentration is not homogeneous close to the boundary of the circular groove. However, the temperature close to the boundary of the groove is still homogeneous when the thermal conductivity $k$ of nanofluids is a function of the concentration of nanoparticles (see (7)). All these facts illustrate that the effect of the concentration on the thermal conductivity of nanofluids is very small for the case.

The changes of the concentration and temperature of the nanofluid based on $m = 0.91$ at $y = 0$ and $x \in [-1,0]$ are shown in Fig.3. When the time increases, nanoparticles go to the boundary from the center of the circular groove and the temperature of the nanofluid decreases. As seen in the left figure of Fig.4, the temperature of nanofluid with three different base fluids ($m = 0.85, m = 0.91, m = 1.0$) at $y = 0$ and $x \in [-1,0]$ when $t = 1$ is shown. Obviously, the dimensionless temperature $\theta$ increases as $m$ increases, and the viscosity of the base fluid still is the main reason to enhance the heat transfer.

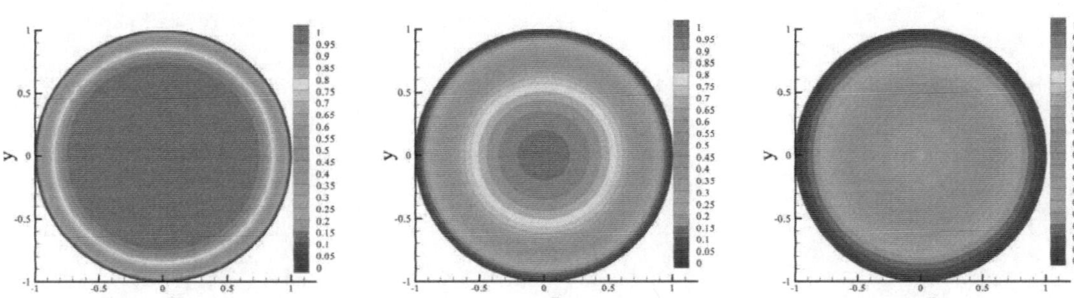


FIGURE 2. The temperature changes for the nanofluid based on the pseudo-plastic fluid ($m = 0.91$) when $t = 0.04$, 0.3 and 0.8.

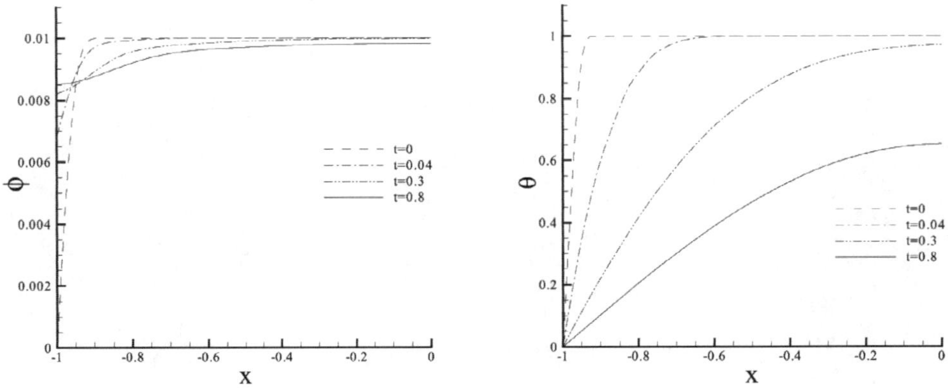


FIGURE 3. The change of physical quantities of nanofluid for pseudo-plastic fluid ($m = 0.91$) with $t$ = 0, 0.04, 0.3 and 0.8. Left: Concentration of nanoparticles; Right: Temperature of nanofluid.

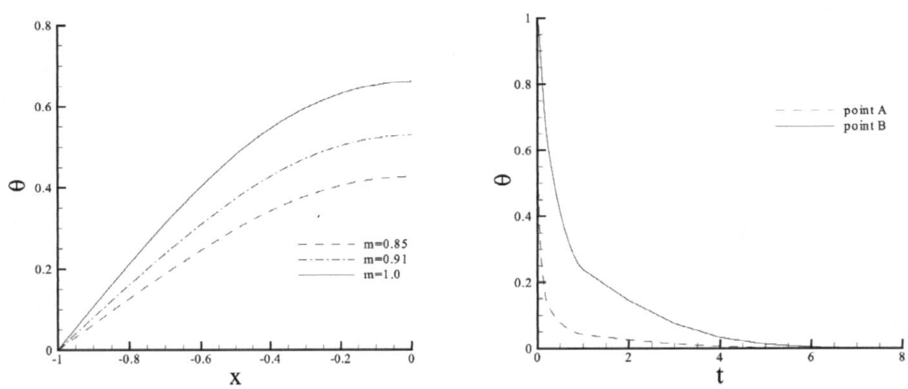


FIGURE 4. The change in temperature for three nanofluids with different base fluids; Right: The change of temperature of nanofluids for the base fluid (i.e. $m = 0.91$).

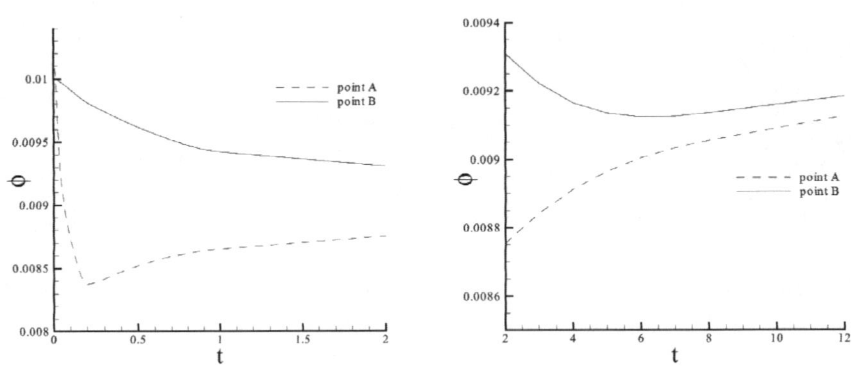


FIGURE 5. The compare on dimensionless concentration $\phi$ between point A and point B. Left: $t$ from 0 to 2; Right: $t$ from 2 to 12.

To show the difference of heat transfer on the concentration, $x = -0.774178$, $y = 0.556429$ and $x = -0.486749$, $y = 0.544328$ are chosen as point A and point B, respectively. According to the domain $\Omega$, point A is much closer to the boundary than point B. The change of the concentration of nanoparticles for point A and point B is shown in Figs.4-5. Obviously, for point A, the concentration decreases first when $t$ goes from 0 to 0.2 and increases later, and its change is greater than the concentration for point B.

## 5. Conclusions

In this paper, the heat transfer of nanofluids based on power-law fluids in a rotating circular groove is investigated numerically with the finite element method. The concentration equation of nanoparticles is coupled with the temperature equation with considering Brownian diffusion and thermophoresis. We have assumed that the thermal conductivity of nanofluids is a function of the concentration of nanoparticles. Some phenomena of nanofluids in this dynamic heat transfer process are discussed (see Section 4), and these results may be helpful to better understand the movement of nanoparticles.

**Acknowledgments**

This work is partially supported by the National Natural Science Foundations of China (Nos. 91430106, 11302024).

## References

[1] D.T. Wasan and A.D. Nikolov, Spreading of nanofluids on solids, Nature, 423 (2003), pp. 156-159.
[2] V.H. Chhasatia, Y. Sun, Interaction of bi-dispersed particles with contact line in an evaporating colloidal drop, Soft Matter, 7 (2011), pp. 10135-10143.
[3] S.U.S. Choi and J.A. Eastman, Enhancing thermal conductivity of fluids with nanparticle, Develpoments and Applocations of Non-Newtonian Flows, ASME MD 231 and FED 66 (1995), pp. 99-105.
[4] S.E.B. Maiga, C.T. Nguyen, N. Galanis and G. Roy, Heat transfer behaviors of nanofluids in a uniformly heated tube, Superlattices and Microstructures, 35 (2004), No. 3-6, pp. 543-557.
[5] J.A. Eastman, S.U.S. Choi, S. Li, W. Yu and L.J. Thompson, Anomalously increased effective thermal conductivities of ethylene glycolbased nanofluids containing copper nanoparticles, Appl. Phys. Lett., 78 (2001), No. 6, pp. 718-720.
[6] B.C. Pak and Y.I. Cho, Hydrodynamic and heat transfer study of dispersed fluids with submicron metallic oxide particles, Exp. Heat Transfer, 11 (1998), pp. 151-170.
[7] X. Wang and X. Xu, Thermal conductivity of nanoparticle-fluid mixture, J. Thermophys. Heat Transfer, 13 (1999), No. 4, pp. 474-480.
[8] Y. Xuan and W. Roetzel, Conception for heat transfer correlation of nanofluids, Int. J. Heat Mass Transfer, 43 (2000), pp. 3701-3707.
[9] N. Bachok, A. Ishak and I. Pop, Stagnation-point flow over a stretching/shrinking sheet in a nanofluid, Nanoscale Research Letters, 6 (2011), pp. 623-632.
[10] S. Dinarvand, R. Hosseini, E. Damangir and I. Pop, Series solutions for steady three-dimensional stagnation point flow of a nanofluid past a circular cylinder with sinusodial radius variation, Meccanica, 48 (2013), No. 3, pp. 643-652.
[11] N.A. Yacob, A. Ishak, I. Pop and K. Vajravelu, Boundary layer flow past a stretching/shrinking surface beneath an external uniform shear flow with a convective surface boundary condition in a nanofluid, Nanoscale Research Letters, 6 (2011), pp. 314-321.
[12] D. Wen and Y. Ding, Effect of particle migration on heat transfer in suspensions of nanoparticles flowing through minichannels, Microfluidics and Nanofluidics, 1 (2005), No. 2, pp. 183-189.
[13] N.M. Arifin, R. Nazar and I. Pop, Non-isobaric marangoni boundary layer flow for Cu nanoparticles in a water based fluid, Meccanica, 46 (2011), pp. 833-843.
[14] J. Niu, C. Fu and W. Tan, Slip-flow and heat transfer of a non-newtonian nanofluid in a microtube, PLoS ONE, 5 (2012), No. 7, pp. 245-255.
[15] K. Vajravelu, K.V. Prasad, J. Lee, C. Lee, I. Pop and R.A.V Gorder, Convective heat transfer in the flow of viscous Ag-water and Cu-water nanofluids over a stretching surface, International Journal of Thermal Science, 50 (2011), pp. 843-851.

[16] J. Buongiorno, Convective transport in nanofluids, Journal of Heat Transfer, 128 (2006), No. 128, pp. 240-250.
[17] W.A. Khan and I. Pop, Boundary-layer flow of a nanofluid past a stretching sheet, International Journal of Heat and Mass Transfer, 53 (2010), pp. 2477-2483.
[18] A.V. Kuznetsov and D.A. Nield, Natural convective boundary-layer flow of a nanofluid past a vertical plate, International Journal of Thermal Science, 49 (2010), pp. 243-247.
[19] P. Rana and R. Bhargava, Finite element simulation of transport phenomena of viscoelastic nanofluid over a stretching sheet with energy dissipation, Journal of Information and Operations Management, 1 (2012), No. 3, pp. 158-161.
[20] S. Nadeem, R. Mehmood and N.S. Akbar, Non-orthogonal stagnation point flow of a nano non-Newtonian fluid towards a stretching surface with heat transfer, International Journal of Heat and Mass Transfer, 57 (2013), pp. 679-689.
[21] P. Rana and R. Bhargava , Flow and heat transfer of a nanofluid over a nonlinearly stretching sheet: A numerical study, Commun. Nonlinear Sci. Numer. Simulat., 17 (2012), pp. 212-226.
[22] R. Kandasamy, I. Muhaimin and R. Mohamad, Thermophoresis and Brownian motion effects on MHD boundary-layer flow of a nanofluid in the presence of the rmalstratification due to solar radiation, International Journal of Mechanical Sciences, 70 (2013), pp. 146-154.
[23] Z. Haddad, E. Abu-Nada, H. F. Oztop, A. Mataoui, Natural convection in nanofluids: Are the thermophoresis and Brownian motion effects significant in nanofluid heat transfer enhancement, International Journal of Thermal Sciences, 57 (2012), pp. 152-162.
[24] H.A. Pakravan and M. Yaghoubi, Combined thermophoresis, Brownian motion and Dufour effects on natural convection of nanofluids, International Journal of Thermal Sciences, 50 (2011), pp. 394-402.
[25] D. Yadav, R. Bhargava, G.S. Agrawal, N. Yadav, J. Lee and M.C. Kim, Thermal instability in a rotating porous layer saturated by a non-Newtonian nanofluid with thermal conductivity and viscosity variation, Microfluid Nanofluid, 16 (2014), pp. 425-440.
[26] F. Hecht, O. Pironneau, A. Le Hyaric and K.. Ohtsuka, FreeFem++ (Version 2.17-1), 2007 (http://www.freefem.org/ff++/ftp/freefem++doc.pdf).
[27] J. Hua, P. Lin, C. Liu and Q. Wang, Energy law preserving $C^0$ finite element schemes for phase field models in two-phase flow computations, J. Comput. Phys, 230 (2011), No. 19, pp. 7115-7131.
[28] L. Zheng, B. Li, P. Lin, X. Zhang, C. Zhang, B. Zhao and T. Wang, Sedimentation and precipitation of nanoparticles in power-law fluids, Microfluidics and Nanofluidics, 15 (2013), No. 1, pp. 11-18.
[29] P. Lin, C. Liu and H. Zhang, An energy law preserving $C^0$ finite element scheme for simulating the kinematic effects in liquid crystal flow dynamics, J. Comput. Phys, 227 (2007), No. 2, pp. 1411-1427.
[30] H. Zhang, A study of the boundary layer on a continuous moving surface in power law fluids. Dissertation, University of Science and Technology Beijing, Beijing, China, 2008.
[31] Y. Jiang, P. Lin, Z. Guo and S. Dong, Numerical Simulation for Moving Contact Line with Continuous Finite Element Schemes, Communications in Computational Physics, accepted.

IOP Conf. Series: Earth and Environmental Science **61** (2017) 012107 doi:10.1088/1755-1315/61/1/012107

# Analysis on the best position and the pile distance of anti-slide pile of reinforced soil slope

**Jinqiu Pan[1], Zhiyong Wang[2], Tianxiong Dong[1]and Bo Liang[1]**

[1]School of civil engineering Chongqing Jiaotong university, Chongqing 400074, China
[2]CHINA RAILWAY SIYUAN SURVEY AND DESIGN GROUP CO., LTD, WUHAN 430063, China
The first author's e-mail address is 1533096492@qq.com.
The corresponding author's e-mail address is 358799558@qq.com.

**Abstract.** Regarding numerical calculation convergence and plastic zone through the slope as the criterion of slope instability, strength reduction method as calculating principle, the stability of soil slope was analyzed by using finite element software ABAQUS, and on the basis of analysis, the best position and pile distance of anti-slide pile of reinforced soil slope were studied. The numerical results showed that with the increasing of distance between anti-slide pile and toe position, safety coefficient raised firstly and then decreased, when the ratio between the horizontal distance of anti-slide to slope toe and horizontal distance of slope is around 0.6, which is the largest safety factor; The pile spacing had certain influence on the soil slope safety factor, with the different pile spacing, the safety factor was different. With the increase of the pile spacing, the safety factor decreased gradually.

## 1.Introduction

As to the studies on the anti-slide pile reinforcement slope, many scholars have done efforts for decades, which mainly focused on the aspects of theoretical analysis, numerical simulation scale model experiments and centrifugal experiments to study the effects of lateral soil movements on sliding pile and the interaction between pile and soil , as well as pile stress and deformation [1-4].Among them, the numerical simulation method is a convenient method to study the slope stability, especially the finite element method and the finite difference method based on the strength reduction method. Ugai[5] and others have found that the strength reduction method is more stable, compared with the traditional limit equilibrium method, the safety factor and the location of the sliding surface are in good agreement. The calculation accuracy and influence factors of the finite element strength reduction method are studied in detail by ZhengYingren[6].Lei Wengjie[7] used finite element method to analyze the pile position of anti-slide piles. Based on the strength reduction method, Deng Jianhui etal[8] proposed a three-dimensional analysis method for slope stability analysis. Based on a water supply project, Liu Zuoqiu etal[9]analyzed the stability and reinforcement of slope by strength reduction method. Zhang Jinghui et al[10]used the strength reduction method to analyze the stability of the landslide before and after reinforcement, and evaluated the reinforcement effect.

However, through the literature search, in author's opinion, the best anti-slide pile position and the most reasonable pile spacing in reinforced soil slope have seldom been studied. Therefore, in this paper the stability of a soil slope were analyzed based on the finite element software ABAQUS by strength reduction method. On the basis of the above analysis, the stability of soil slope reinforced by anti-slide piles were further analyzed, and the most reasonable pile spacing and pile spacing are studied that could supply some reference for the project.

## 2 Theoretical analysis of strength reduction method

*2.1 The principle of strength reduction method*

Safety factor refers to the ratio of actual shear strength that soil slope can provide to resist and shear stress generated by the external load stress, when the constant external load keeps its role in slope and the slope reaches a critical state. Strength reduction refers to a reduction of the shear resistance of the soil under the limit condition, which is equivalent to a discount on shear strength. The implementation process: generally, the shear strength index $c$ and $\varphi$ of soil are firstly reduced which are divided by a reduction factor of *F*, and then the new $c'$, $\varphi'$ value is obtained, a series of reduction is made according to this method. Secondly, input related parameters after reduction into the finite element software and carry out the calculation of reciprocating, until it reaches the end of the computation criterion, at the moment, the factor F is the safety factor of the slope stability.

The reduction formula of $c$ and $\varphi$ is shown below:

$$c' = \frac{c}{F} \quad (1)$$

$$\varphi' = \arctan(\tan\varphi / F) \quad (2)$$

In formula: $c$, $\varphi$, $c'$, $\varphi'$ —cohesion, friction angle, cohesion and friction angle after reduced.

$F$ —reduction factor.

Special attention is paid to that when using of strength reduction method to calculate the stability of the soil slope, the reduction coefficient calculated from the beginning of the F is less than 1 so as to prevent the soil slope at the beginning of calculation from being destroyed and keep the soil in the elastic state. Compared with the traditional limit equilibrium method, the strength reduction method has many advantages[11]:①it does not have to assume the position of the sliding surface;②it can simulate the shape and position of the sliding surface and the dynamic change process;③based on the nonlinear characteristics of soil, the stress-strain relationship of rock and soil is increased.

*2.2. Strength reduction method and instability criterion*

There are a lot of constitutive relations in the finite element method, but the elastic-plastic model is the most used one. The reason is that the calculated results are better than the traditional methods. What's more, in the calculation of rock and soil, we pay more attention to the size of the plastic zone of soil, not the relative displacement. Therefore, the elastic-plastic model was used in this calculation[12],mohr -culomb failure criterion was used as strength criterion, and its function formula[13] was:

$$f(I_1, I_2, I_3) = \frac{1}{2}\left[3(1-\sin\varphi)\sin\theta + \sqrt{3}(3+\sin\varphi)\cos\theta\right]\sqrt{J_2} - I_1\sin\varphi - 3c\cos\varphi = 0 \quad (3)$$

In formula: $I_1, I_2, I_3$—the first, second and third stress invariants.

There are three main judgement standard for judging the failure of the soil slope to the critical state:①whether finite element calculation is convergent;②the displacement changes at the special site; ③whether plastic zone is through. The study of Zhou Yuanfu[14] and Pei Lijian[15] showed that the three judgement standards are consistent, so the first and the third article was used as the judgement standard of instability in this calculation model.

## 3. Example

In this paper, two examples are simulated, and the example one was used to calculate the stability of homogeneous slope. Then, on the basis of the example1, the anti-slide pile was simulated and the most reasonable pile position and the optimal pile spacing are analyzed with the example 2. It was proved that the ABAQUS software based on the strength reduction method was suitable for the stability analysis of soil slope. The specific dimensions of the model involved in this example and the materials used in the numerical experiments were shown in Figure 1 and table 1.

**Table 1.** Nnumerical test parameter.

| Material | Density /(kg/m$^3$) | Adhesion c/(kpa) | Internal Friction Angle φ/(°) | Elastic Modulus *E*/*(Mpa)* | Poisson Ratio υ |
|---|---|---|---|---|---|
| Soil | 18×10$^3$ | 15 | 27 | 90 | 0.25 |
| Anti-slide Pile | 22 | — | — | 32000 | 0.2 |

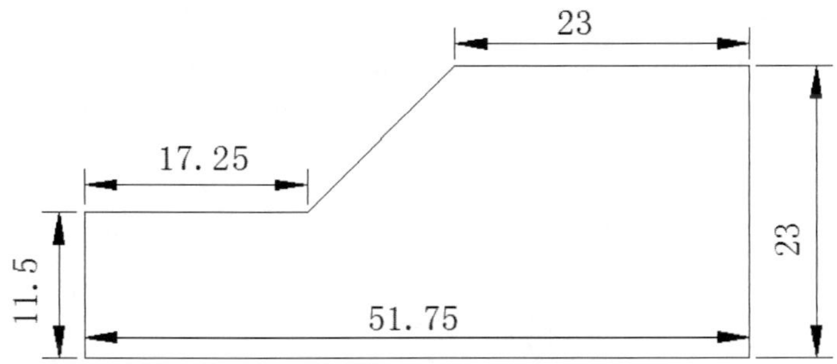


**Figure1.** Dimension of numerical model(unit: m).

*3.1. Example1*

*3.1.1. Parameter selection of model.* Two dimensional model was built by using non associated flow rule, the four node plane strain unit (CPE4)and uniform mesh in this calculation example. The model with mesh was as shown in figure 2.The boundary conditions were that the horizontal displacement was restrained on both sides of slope and horizontal and vertical displacement were restrained in the bottom of slope[15].

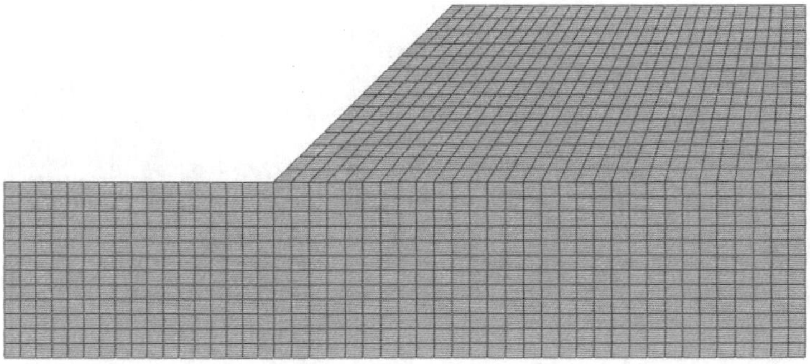

**Figure2.** Mesh model of example.

*3.1.2 Calculation Result Analysis.* According to the judgement standard whether the numerical analysis is convergent and plastic zone is through which decides whether the slope was failure or not, the safety factor was 1.09 after several iterations. The safety factor calculated by traditional slice method is 1.06, which is close to the result calculated by numerical analysis. The result showed that the safety factor obtained by ABAQUS is within the allowable range, and it had a good applicability in slope stability analysis.

*3.2. Example 2*

On the basis of example 1, the reinforcement treatment with anti-slide pile was carried out in example2.In example2, the influence of different pile positions and pile spacing on the safety factor is calculated in order to determine the best pile position and pile spacing.

*3.2.1. Parameter selection of model.* This model was three dimensional calculation model that was built by using the 8 node hexahedron element（C3D8）and uniform mesh with sweep technique. The boundary conditions were that the horizontal displacement on both sides of model, the displacement of three directions in the bottom of the model and the displacement of Y direction in front of and behind the model was restrained.

*3.2.2. Analysis of numerical simulation results.* In this model, the anti-slide piles are buried in different positions of the slope. The position of toe of solpe was as the origin of coordinate and X was defined as horizontal distance from the toe of slope. The maximum horizontal distance D of slope was 11.5m, and the buried position of piles were located at X/D= 0.3, 0.4, 0.5, 0.6 and 0.7.

(1)Analysis of the optimum pile position of anti-slide pile to reinforce soil slope

①Influence of different pile positions on the position and shape of sliding surface

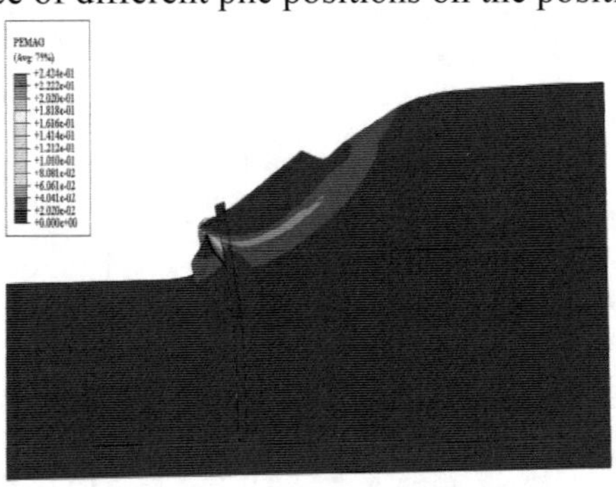

(1) The pile is located at X/D = 0.3

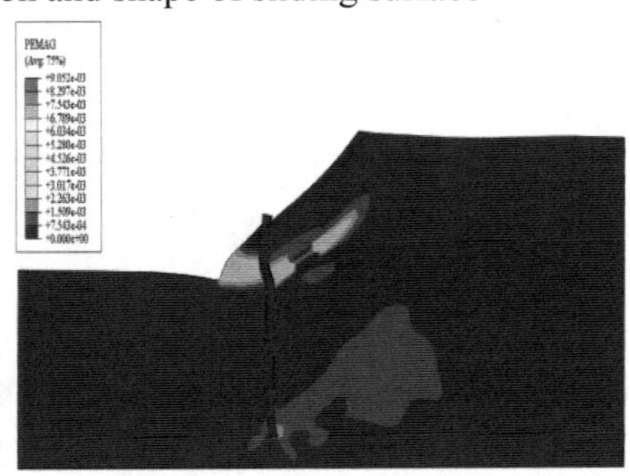

(2) The pile is located at X/D = 0.4

(3) The pile is located at X/D = 0.5

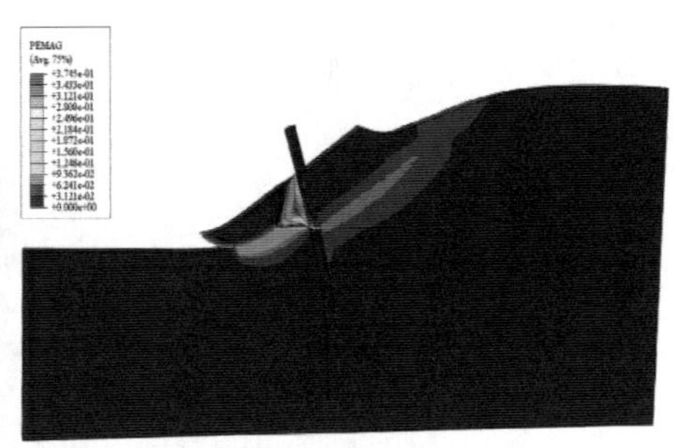

(4) The pile is located at X/D = 0.6

IOP Conf. Series: Earth and Environmental Science **61** (2017) 012107 doi:10.1088/1755-1315/61/1/012107

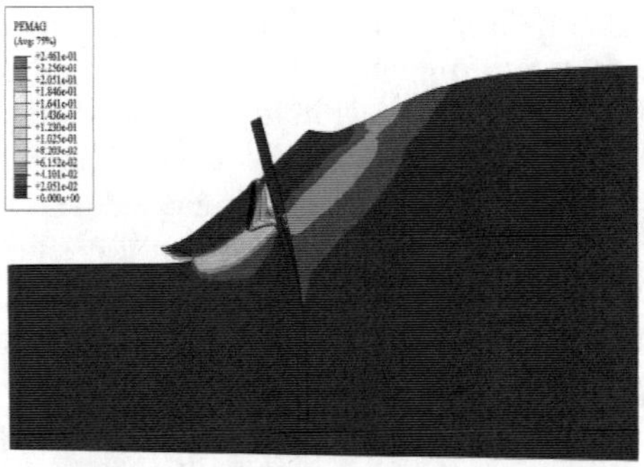

(5) The pile is located at X/D = 0.7

**Figure3.** Position and shape of sliding surface at different positions of pile.

The results showed that the position of the pile produced some influence on the position and shape of the sliding surface. With the position of the pile changing from the toe to the top of the slop, position and shape of the sliding surface changed. The position of the sliding surface became gradually shallow and the shape of sliding surface changed from non-through to the run-through of plastic zone, then changed to the sliding surface crossing over the top of the pile which led to failure before pile.

When the pile was located at X/D = 0.3, the plastic zone of soil slope does not transfix, but the soil slope of pile top cracked and the crack was large. At the same time, the upper soil slid with extrusion; When the pile was located at X/D = 0.3,the anti-slide pile was bent at the position of the sliding surface due to be sheared, Moreover, the plastic zone was the largest in the middle of the slope, which accords with the law that the plastic zone increases from surface to interior of slope; When the pile was located at X/D = 0.5,the plastic zone of the soil slope reached the maximum, and the transfixion phenomenon occurred. At this time, the pile had a serious instability phenomenon, and the reinforcement effect was less than that of X/D = 0.4, but the safety factor was 1.328 and still met the design requirements; When the pile was located at X/D = 0.6, the plastic zone was small. But at the same time, the position of the sliding surface has changed and was damaged before the pile. When the pile is located at X/D = 0.7, the position and shape of the sliding surface is similar to that of X/D = 0.6. However, the plastic zone became larger, and the transfixion phenomenon happened.

② Influence of different pile positions on the safety factor of soil slope

With the calculation model considering the interaction between pile and soil, the anti sliding effect of pile position on the slope safety factor was analyzed. The results were shown in Figure 4 and table 2.The position of pile in table 2was horizontal distance from pile to toe of slope.

**Table 2.** Safety factor of soil slope under different pile positions.

| Pile Position/m | 3.45 | 4.6 | 5.75 | 6.9 | 8.05 |
|---|---|---|---|---|---|
| Safety Factor/$F_s$ | 1.16 | 1.23 | 1.33 | 1.67 | 1.56 |

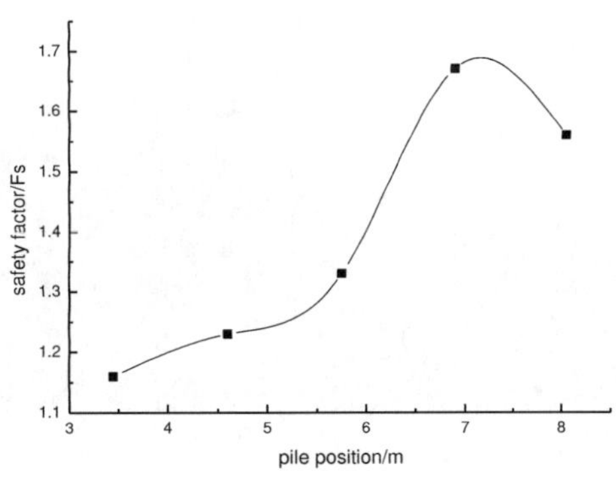


**Figure4.** Relation curve between pile positions and safety factor.

IOP Conf. Series: Earth and Environmental Science **61** (2017) 012107 doi:10.1088/1755-1315/61/1/012107

Figure 4 and table 2 showed that with the changes of pile positions buried in slope, safety factor of soil slope was different, and after the reinforcement with anti-slide pile, safety factor was greater than, 1.06, that of the soil slope unreinforced. The results also showed that with the increase of distance from anti-slide pile to toe of slope, safety factor of soil slope increased first and then decreased. Moreover, the safety factor was largest when the distance reached near 6.9m.

Based on the analysis of the influence of pile spacing on the sliding surface position and safety factor, it could be obtained that the best pile position was When the ratio between distance from pile to toe of slope and horizontal distance of slope was 0.6,the reinforced effect of anti-slide pile was best.

(2)Analysis of the optimum pile spacing of anti-slide pile to reinforce soil slope

In order to study the influence of different pile spacing in soil slope on slope stability, take pile position of 6.9m from toe of slope as an example, the stability of soil slope under 4 times of the pile diameter (4 D), 6 times of the pile diameter (6D) and 8 times of the pile diameter (8 D) was calculated.

①Position and shape of slip surface of soil slope under different pile spacing

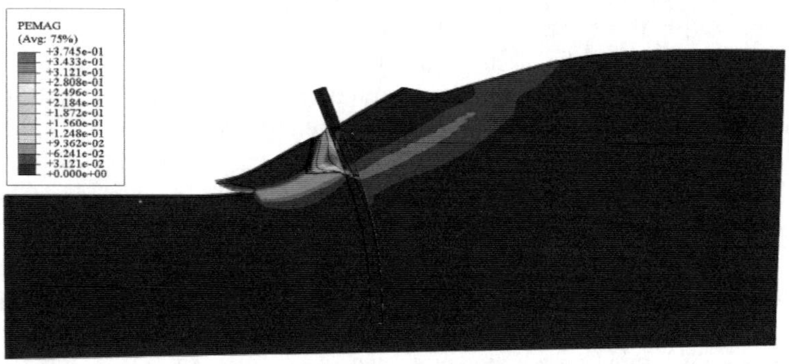


(1) The position and shape of the sliding surface at 4D

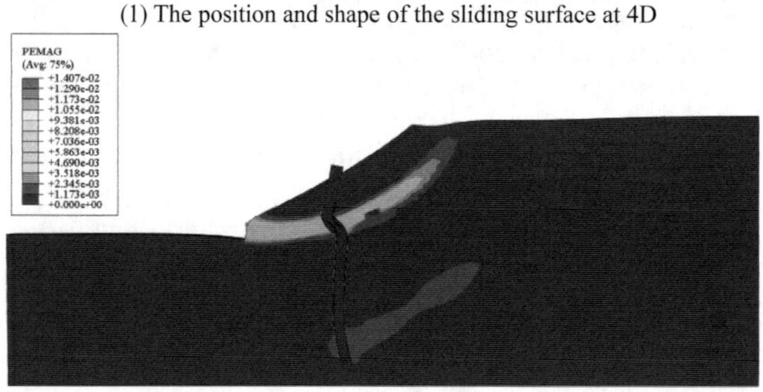


(2) The position and shape of the sliding surface at 6D

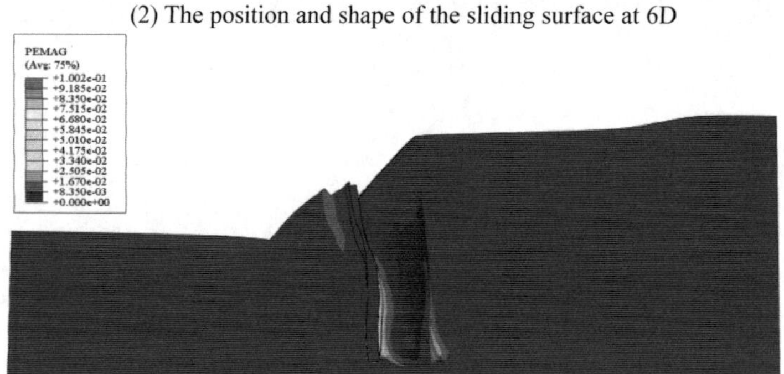


(3) The position and shape of the sliding surface at8D

**Figure5.** Position and shape of slip surface of soil slope under different pile spacing.

As can be seen from Figure 5, the pile spacing had some influence on the position and shape of the sliding surface. With the increase of pile spacing, position of sliding surface was gradually close to the unreinforced soil slope, the reinforcement effect weakened. When the pile spacing was 4 times of the pile diameter, the sliding surface damaged with transfixion in front of the pile and the largest plastic zone was located near the top of the pile. When the pile spacing was 6 times of the pile diameter, the pile was bent, sliding zone was extended to the top of the slope, the reinforcement effect weakened. When the pile spacing was 8 times of the pile diameter, phenomenon of secondary sliding of the slope appeared and phenomenon of soil tympanites appeared at pile bottom. At this time, the reinforcement effect was poor and safety factor was similar to unreinforced soil.

②Safety factor of soil slope under different pile spacing

In order to show the influence of different pile spacing on the safety factor of soil slope, Table 3 and Figure 6 show the safety factor of soil slope under different pile spacing.

**Table 3.** Safety factor of soil slope under different pile spacing.

| **Pile Spacing** | 4D | 6D | 8D |
|---|---|---|---|
| Safety Factor/ $F_s$ | 1.67 | 1.28 | 1.0 |

Note: D is 0.8m.

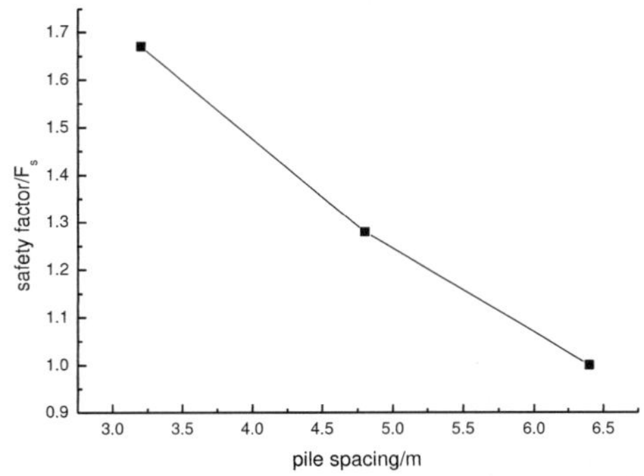


**Figure6.** Curves of pile spacing and safety factor.

As can be seen in table 3 and Figure 6, the safety factor was different with different pile spacing. With the pile spacing increasing, the safety factor decreased and presented linear change. When the pile spacing is 8 times of the pile diameter, the safety factor is 1.0, smaller than unreinforced soil. The reason was that the pile spacing is sparse which leads to the pile functioning badly. At the same time, because of the anti-slide pile, the disturbance to the slope soil aggravated the failure of slope. Considering the sliding surface and safety factor under different pile spacing, the optimal pile spacing of anti-slide pile was near 4 times of the pile diameter.

## 4.Conclusion

(1) ABAQUS software has good applicability for calculation of soil slope stability and its accuracy and applicable scope can meet the requirements.

(2) The stability of slope reinforced by anti-slide pile is greatly improved. Take distance of 3.45m from anti-slide pile to toe of slope as an example, the safety factor is increased by 6.4%.

(3) The pile position of anti-slide pile has some influence on the stability of soil slope. With the increase of distance from anti-slide pile to toe of slope, safety factor of soil slope increased first and then decreased. When the ratio between distance from pile to toe of slope and horizontal distance of slope was 0.6, the safety factor was largest.

(4) The pile spacing had some influence on the safety factor of soil slope. The safety factor was different with different pile spacing, and with the pile spacing increasing, the safety factor decreased. It was recommended optimal pile spacing was 0.4D.

**Acknowledgements**
[1]First author: Pan Jinqiu, male, graduate student of Chongqing Jiaotong University, is mainly engaged in the research on road engineering.
[2]Corresponding author: Wang Zhiyong, male, professor senior engineer of CHINA RAILWAY SIYUAN SURVEY AND DESIGN GROUP CO.,LTD, is mainly engaged in the research on slope stability and karst water system.

**References**

[1] Zhou Depei, XiaoShiguo and XiaXiong 2004 Discussion on rational spacing between adjacent anti-slide piles in some cutting slope project J. Chinese Journal of Geotechnical Engineering.26 132-35
[2] Zhang Jianxun, ChenFuquan and JianHongyu 2004 Numerical analysis of soil arching effects in passive pilesJ. Rock and Soil Mechanics. 25 174-78
[3] Yang Ming, YaoLingkan and WangGuangjun 2008 Study of centrifuge model tests and numerical simulation on soil arching in space of piles J. Rock and Soil Mechanics 29 817-22
[4] Gao Changsheng, Wei Rulong and Chen Shengshui 2009 Centrifugal model tests on deformation of slopes reinforced with piles J. Chinese Journal of Geotechnical Engineering. 31 145-148
[5] UGAI K. 1989 A method of calculation of total factor of safety of slopes by elasto-plastic FEM J. Soils and foundation 29 190-95
[6] Zheng Yingren and Zhao Shangyi 2004 APPLICATION OF STRENGTH REDUCTION FEM IN SOIL AND ROCK SLOPE J.Chinese Journal of Rock Mechanics and Engineering 23 3381-88
[7] Lei Wenjie, Zheng Yingren and Feng Xiating 2006 Analysis of pile location on landslide control J. Rock and Soil Mechanics 27 950-54
[8] Deng Jianhui,Zhang Jiaxiang and MinHong etal 2004 3D stability analysis of landslides based on strength reduction(II):evaluation of reinforcing factor of safety J .Rock and Soil Mechanics. 25 871-76
[9] Liu Zuoqiu, Zhou Cuiying and Dong Liguo 2005 Slope stability and strengthening analysis by strength reduction FEM J. Rock and Soil Mechanics .26 558-61
[10] Zhang Jinghui, Tianbin and TongFuguo etal 2005 Study on Interaction Between Landslide Mass and Anti- Slide Piles by Nonlinear Finite Element Method J. Journal of China Three Gorges Univ. ( Natural Sciences).27 40-51
[11] Liao Gongyun, Huang Xiaoming 2008 Application of ABAQUS finite element software in road engininieering. (Nanjing: Southeast University Press) p 227-30
[12] China Renewable Energy Engineering Institute.SL386-2007 2007 Design code for engineered slopes in water resources and hydropower projects[S]. (Beijing: China Water &Power Press)
[13] Cheng Canyu, Luo Furong and QI Chengzhi etal 2012 Comparative analysis of slope stability by strength reduction method J. Rock and Soil Mechanics. 33 3472-78
[14] Zhou Yuanfu, Deng Jianhu and Cui Yulong etal 2014 instability criterion of three-dimensional slope based on strength reduction method J. Rock and Soil Mechanics. 35(5):1430-37
[15] Pei Li Jian, Qu Benning and Qian Shanguang 2010 Uniformity of slope Instability criteria of strength Reduction with FEM J. Rock and Soil Mechanics.31(10):3337-3341
[16] Fei Kang, Zhang Jianwei 2009 Application of ABAQUS in geotechnical engineering (Beijing: China Water &Power Press) p 393-394

IOP Conf. Series: Earth and Environmental Science **61** (2017) 012108 doi:10.1088/1755-1315/61/1/012108

# Optimal selection on water-supply pipe of building based on analytic hierarchy process

**Tianyun Wei[1, a)], Guiqing Chen[2, b)]**

[1]Department of Road Engineering, Fujian Chuanzheng Communications College, Fuzhou 350007,China.

[2]Department of mechanical engineering, Fujian Chuanzheng Communications College, Fuzhou 350007,China.

[a)]Corresponding author: gq-chen@126.com

[b)]chgq68@163.com

**Abstract:** The main problem of pipes used in water-supply system was analyzed, and the commonly used pipe and their main features were introduced in this paper. The principles that the selection on water-supply pipes should follow were pointed out. Analytic Hierarchy Process (AHP) using 9 scaling was applied to optimize water-supply pipes quantitatively. The optimal water-supply pipes were determined according to the sorting result of comprehensive evaluation index. It could provide the reference to select the reasonable water-supply pipes for the engineers.

## 1 Introduction

Water-supply network has played a key role on people's life and work, which constitutes the main part of urban water-supply system and becomes an integral part of urban infrastructure construction. The construction of water-supply pipe network can reflect a country's economic development level in a certain extent. Pipe is an important part of the water-supply network construction, which costs more than a third of the total cost of pipe network construction, so how to choose pipe in water-supply pipe network engineering construction holds the important status. Pipe is the important restricting factor on the water-supply pipe network. At present, the outstanding problems existing in urban water-supply pipe network is as follows: corrosion is serious, the phenomenon such as pipeline dirt and aged, secondary pollution of water quality is more serious. The operation energy consumption due to leakage is larger, which must be pay enough attention[1]. From what has been discussed above, water-supply pipe must be choose reasonably in order to achieve water pipe network reliable operation, economic and reasonable cost, convenient maintenance, no secondary pollution.

## 2 The main problems existing in the water-supply pipe

In recent years, the development trend of water-supply pipes are diversified with the rapid development of building residential water-supply system, and requirements for the performance of the water-supply pipe is becoming more and more high. At present, urban water-supply network in China still has some problems, such as the leakage of pipe, blasting rate being higher, and pipeline corrosion, aging, and serious dirt, making water produce secondary pollution. The problem cause water quality reduce, water energy consumption increase, seriously affect the economic benefits of water-supply enterprises and water security of users. In view of the city water-supply pipe network construction being not perfect and exiting some problems, in the future city pipe network

IOP Conf. Series: Earth and Environmental Science **61** (2017) 012108 doi:10.1088/1755-1315/61/1/012108

transformation process, the selection of water-supply pipe should be paid more attention to reliable quality, reasonable economic indicators, in order to reduce the operating cost, save investment, ensure water-supply safely and efficiently[2-4].

## 3 Commonly used pipe and its features

In the water distribution system, pipe plays an important role, the factors to consider when choosing pipe have the following several aspects: the first is feed water flow and pressure of water-supply pipe, secondary water quality problems, finally, factors such as pipe installation, and ease of maintenance must be given attention, according to different pipe, the different principle must be followed[5].

### *3.1 Metal Pipe*

When metal pipes such as thin-walled stainless steel pipe and copper pipe were in use process, corrosion phenomena can also appear, which was mainly chloride and sulfide deposit. In the surface oxidation protective film inhibits corrosion, which can maintain and repair for a long time. The copper pipes are influenced easily with the ph of water. If high acidity, corrosion phenomenon will appear. At the same time, high dissolved oxygen and demineralized water, could also result in corrosion of copper, so on these occasions, the brass try not to be used. If the chloride ions in the water content were higher, which will also cause corrosion of austenitic stainless steel pipes, in this kind of circumstance, thin-walled stainless steel pipes is unfavorable choose. Metal pipe connection has a variety of ways, each has advantages and disadvantages. Material, wall thickness and the use occasions must be considered comprehensively to determine pipes. If different metal tube or pipe is connected, the electrochemical corrosion of the pipeline should be pay sufficient attention.

### *3.2 Plastic Pipe*

In general, for the linear expansion coefficient, plastic pipe's was more than metal pipe's, which is also an important difference between plastic pipe and metal pipe. The deformation of plastic pipe occurs in the shaft up generally. The common deformation is bending mainly in the horizontal direction, and the degree of deformation is directly proportional to temperature difference. As a result, the problem must be taken into account in design and construction of installation, the slip joint or angle natural compensation method can be choose. When choosing plastic pipe fittings, adhesive containing benzene class organic solvent should be avoided to choose, so as not to make water quality secondary pollution. The detection qualified tubing must be choose according to the quality supervision and inspection authorities.

### *3.3 Composite Pipe*

The welding aluminum alloy and polyethylene were adhesive by the glue to produce aluminous model multiple tube, whose characteristic was both the advantages of plastic pipe and metal pipe, namely corrosion resistance and resistance to high pressure. According to different uses of hot and cold water, aluminous model multiple tube can be divided into high density polyethylene and crosslinked polyethylene aluminous model multiple tube. Adding a layer of UPVC in the galvanized tube constitutes the plastics-steel multiple tube, which also combines the advantages of steel pipe and plastic steel pipe. The plastics-steel multiple tube commonly was used as a cold water pipe, whose cost is higher. Steel skeleton PE pipe is a new kind of pipeline, which has the advantages of double-sided anti-corrosion pressure. PE pipe has been widely applied in municipal water-supply engineering, but its cost is higher. The plastic coated pipe was produced with polyethylene (PE) resin or non-toxic polypropylene and other organic matter deposited on the inner wall of the steel tube, which is characterized by high strength, water resistance, corrosion resistance, not easy fouling, long service life.

## 4 The principles to be followed when choosing water-supply pipes material

Water-supply pipe plays an important role in supplying the user with clean water. Water quality are greatly influenced by the feed water pipe. Currently, metal pipe, plastic pipe, composite pipe are three

IOP Conf. Series: Earth and Environmental Science **61** (2017) 012108 doi:10.1088/1755-1315/61/1/012108

main water-supply pipeline, each has advantages and disadvantages, and the accurate choice must be based on the principles of weighing the pros and cons to make[6].

*4.1 Clean Sanitation*

Water from the water plant generally go through a long pipeline to reach the user, during this time, chlorine that tap water contains may produce a new physical and chemical reaction with the pipe wall, water quality reaching users was affected, and the inner wall of the pipe was corroded. Therefore, health performance testing qualified pipe must be chosen according to the authority supervision departments of the state.

*4.2 Safe and Reliable*

Water-supply pipe in the process of water distribution and supply must bear a certain amount of water pressure, meet the mechanical strength grade to prevent rupture, and whose sealing must be better to prevent leakage phenomenon. At the same time, the pipe can resist the influence of vibration, water hammer and heat bilges cold shrink, all of these in pipe network design must be considered.

*4.3 Conducive to Environmental Protection*

Feed water pipe can be recycled use and have no pollution to the environment, which has become an important factor when choosing pipe for technical personnel, especially country issued a series of policies to protect earth's environment. Galvanized pipe is easy to rust, which has been basically eliminated. Copper pipe with good anticorrosion performance and pressure performance, has become an important material of high-grade decoration. Most plastic pipes can be recycled, which are now also widely used.

*4.4 Saving Energy*

Any pipe material in the production process and installation process must consume energy. In general, the thin-wall pipe production more efficient than thick wall pipe production. Steel-plastic compound pipe, metal pipe, plastic pipe, their production energy consumption gradually reduce. According to the construction perspective, using besmear brush twisted tooth is more energy-saving than hot melt connection, electric welding and gas welding. If a second pressure was needed in the process of water distribution, the more local head loss of the pipeline, the more energy consumption.

*4.5 Saving Investment*

Under the condition of meeting health and safety, technical and economic analysis must be conducted from the aspects of pipe material price, construction installation cost, to choose appropriate pipes to save the cost of network construction.

## 5 Using AHP to select construction water-supply pipe

In order better to unify the qualitative analysis and quantitative analysis for decision making problem which is influenced by many factors, the analytic hierarchy process (AHP) occurs. AHP was brought up firstly by the T· L· Saaty, which was suitable for problem sorted sequence according to the advantages and disadvantages[7-9]. The influence factors of building water-supply pipe are more, so it is suitable for using AHP to optimize choice.

*5.1 Setting the Structure Model among Levels*

Building water-supply pipe model mainly consists of three layers, namely the target layer, rule layer and solutions. The target layer is the building water-supply pipe. Rule layer according to the survey mainly includes health clean, safety and reliability, and cost, whose importance was based on the opinions of the experts and customers to order. 5 kinds of pipes such as thin-walled copper pipes, thin-walled stainless steel pipes, PVC - U pipes, PP-R pipes, the lining of PVC galvanized pipes were choose as the research object. Building water-supply pipe hierarchical model is shown in Fig.1.

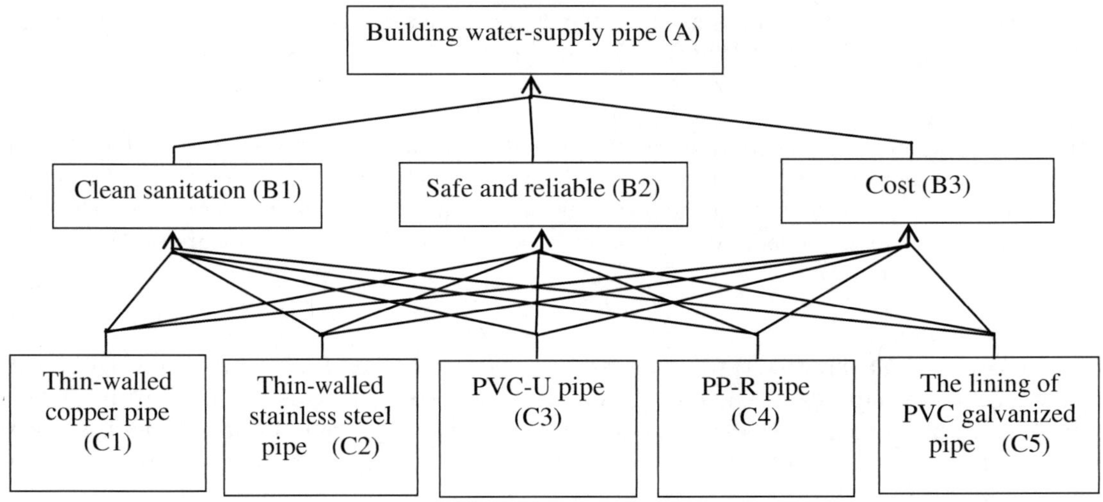


**FIGURE 1.** Building water-supply pipe structure model figure

*5.2 Construction of Judgment Matrix*

According to the experience of the engineers and expert advice, 9 scaling method is used to determine the element size values of judgment matrix , as shown in table 1 to table 4.

**TABLE 1.** Building water-supply pipe decision-making target table

| A | B1 | B2 | B3 |
|---|---|---|---|
| **B1** | 1 | 3 | 5 |
| **B2** | 1/3 | 1 | 3 |
| **B3** | 1/5 | 1/3 | 1 |

**TABLE 2.** Building water-supply pipe clean sanitation indicators table

| B1 | C1 | C2 | C3 | C4 | C5 |
|---|---|---|---|---|---|
| **C1** | 1 | 1 | 1/3 | 1/5 | 1/3 |
| **C2** | 1 | 1 | 1/3 | 1/5 | 1/3 |
| **C3** | 3 | 3 | 1 | 1/3 | 1 |
| **C4** | 5 | 5 | 3 | 1 | 3 |
| **C5** | 3 | 3 | 1 | 1/3 | 1 |

**TABLE 3.** Building water-supply pipe safety and reliability index table

| B2 | C1 | C2 | C3 | C4 | C5 |
|---|---|---|---|---|---|
| **C1** | 1 | 1 | 5 | 5 | 3 |
| **C2** | 1 | 1 | 7 | 7 | 5 |
| **C3** | 1/5 | 1/7 | 1 | 1 | 1/3 |
| **C4** | 1/3 | 1/5 | 3 | 3 | 1 |
| **C5** | 1/3 | 1/5 | 3 | 3 | 1 |

**TABLE 4.** Building water-supply pipe construction cost index table

| B3 | C1 | C2 | C3 | C4 | C5 |
|---|---|---|---|---|---|
| **C1** | 1 | 3 | 5 | 3 | 3 |
| **C2** | 1/3 | 1 | 3 | 3 | 1 |
| **C3** | 1/5 | 1/3 | 1 | 1 | 1/3 |

| | | | | | |
|---|---|---|---|---|---|
| **C4** | 1/3 | 1/3 | 1 | 1 | 1/3 |
| **C5** | 1/3 | 1 | 3 | 3 | 1 |

*5.3 Consistency Check*

The eigenvalue method is used to judge the consistency of matrix:

$$CI = \frac{\lambda_{\max} - n}{n-1} \quad 1)$$

The formula of the *CI* shows consistency index, $\lambda_{max}$ for the largest eigenvalue, *n* shows the number of factors, namely the order number.

The mean random consistency index *RI* can look up to the table 5, the consistency ratio *CR*:

$$CR = \frac{CI}{RI} \quad 2)$$

When *CR*<0.1, hierarchical sorting result consistency can be accepted.

**TABLE 5.** The relations between the mean random consistency index and matrix order

| **Matrix Order** | 1 | 2 | 3 | 4 | 5 | 6 | 7 | 8 | 9 | 10 | 11 | 12 |
|---|---|---|---|---|---|---|---|---|---|---|---|---|
| ***RI*** | 0 | 0 | 0.52 | 0.89 | 1.12 | 1.26 | 1.36 | 1.41 | 1.46 | 1.49 | 1.52 | 1.54 |

Total sorts consistency check:

$$CI_T = \sum_{j=1}^{5} a_j CI_j \quad 3)$$

$$RI_T = \sum_{j=1}^{5} a_j RI_j \quad 4)$$

$$CR_T = \frac{CI_T}{RI_T} \quad 5)$$

*5.4 The Calculation Results and Analysis*

AHP software yaahp v. 9.1 was used to solve the calculation, the main results as shown in table 6. The biggest characteristic value $\lambda_{max}$ was obtained by the software yaahp a. v. 9.1. Consistency index (*CI*) was obtained according to the formula (1), *RI* value can be checked by the table 5, consistency ratio (*CR*) was calculated by the formula (2), total sorts consistency check according to the formula (5), $CR_T$=0.0184<0.1, shows the total sequencing result also has the satisfactory consistency.

**TABLE 6.** Building water-supply pipe weight calculation table

| | **B1** 0.6333 | **B2** 0.2605 | **B3** 0.1062 | **Results Total Sorts** |
|---|---|---|---|---|
| **C1** | 0.0737 | 0.3333 | 0.4285 | 0.3168 |
| **C2** | 0.0737 | 0.4218 | 0.2297 | 0.1809 |
| **C3** | 0.1952 | 0.0566 | 0.0747 | 0.1790 |
| **C4** | 0.4624 | 0.0566 | 0.0868 | 0.1771 |
| **C5** | 0.1952 | 0.1318 | 0.1804 | 0.1463 |
| $\lambda_{max}$ | 5.0555 | 5.0955 | 5.1973 | |
| ***CI*** | 0.0124 | 0.0213 | 0.0440 | |
| ***RI*** | 1.41 | 1.41 | 1.41 | |
| ***CR*** | 0.0088 | 0.0151 | 0.0312 | |

It can be seen from table 6, when considering health clean, PP-R pipe is the best, when considering safety and reliability, thin-walled stainless steel pipe is the best, when considering the cost, PVC - U pipe is the best.

It a word, methods 1 comprehensive weight (0.3168) is the biggest, namely the comprehensive performance of the thin-wall copper pipe is the best, which can be applied to the high standard residential water-supply pipeline installation. The second is the PP-R pipe, whose comprehensive performance is higher, plus the price and the advantages of convenient installation, which is suitable for ordinary residential decoration water-supply pipe.

## 6 Conclusions

(1) It should master the commonly used pipe and its main features, follow the principle of the pipe selection, consider various factors, and do specific analysis, to select the best pipe for exerting its maximum efficiency.

(2) Using 9 scale AHP to optimize building water-supply pipe material, it is concluded that comprehensive weight (0.3168) of thin-wall copper pipe is the biggest, which can be as a priority for building water-supply pipe. Application of AHP can break the limitations of traditional methods and more comprehensively consider various influencing factors, the evaluation results more reliable.

## Acknowledgments

This work is supported by the Educational Commission of Fujian Province of China (Grant No. JA15655、JB12356).

## References

[1] WEI Tian-yun, LIU De-ming. Analysis of water-saving benefit and the engineering application on water-saving appliances. Journal of Anyang Institute of Technology, 2016,15(2):54-56.
[2] WEI Tian-yun. Thinking and suggestion of choice on water-saving appliances. Fujian Construction Science & Technology, 2016,3:73-74+96.
[3] WEI Tian-yun, LIU De-ming. The problems and countermeasures on water saving and energy saving in building water supply system. Journal of Heilongjiang Vocational Institute of Ecological Engineering, 2016,5:32-34.
[4] WEI Tian-yun, LIU De-ming. Optimal selection on water supply pipe based on the fuzzy comprehensive evaluation method. Journal of Anyang Normal University, 2016,2:146-148.
[5] WEI Tian-yun. Application and energy saving analysis of frequency conversion water supply technology in water supply system. Shanxi Architecture, 2016,42(17):114-115.
[6] WEI Tian-yun. Optimal selection on residential life water-supply way based on Analytic Hierarchy Process. Journal of Anyang Institute of Technology, 2016,15(4):55-58.
[7] HONG Ji-hua, SONG Yi-lan. Application research of analytic hierarchy process (AHP) in water Environmental planning. Environmental Science & Technology, 2000(1):33-35.
[8] C.M.Tam, T.K.L.Tong, and G.W.C.Chiu. Comparing non-structural fuzzy decision support system and analytic hierarchy process in decision-making for construction problems. European Journal of Operational Research, 2006,(174): 1317-1324.
[9] WANG Ying-luo. Systems engineering. Beijing: Mechanical industry press, 2001.153-203.

IOP Conf. Series: Earth and Environmental Science **61** (2017) 012109 doi:10.1088/1755-1315/61/1/012109

# Acceleration response research of large span floor under pedestrian load

**Lilin Cao[1], Qingwen Zhuang[1], Lanzhe Chu[2] and Guojun Yu[1]**

[1] Faculty of Civil Engineering and Mechanics, Jiangsu University, Zhenjiang, China
[2] Construction Engineering Quality Supervision Station, Zhenjiang, China
Corresponding author: cll@ujs.edu.cn

**Abstract.** With the progress of technology and the constant improvement of the requirement for the quality of life, people tend to choose light and long- span structure. Due to damping of the long-span floor is small, the fundamental frequency is low, which will produce vertical vibration which exceed a certain limit will affects the security of buildings under the action of pedestrian vibration. Firstly, the large-span floor vertical vibration comfort control standards and design methods were introduced in this paper. Next, the research status of walking load model was summarized. Then, the critical calculation methods of the vibration response were summed up. Finally, the existing problems of comfort control standards, design methods, walking load model and the floor response analysis were discussed and possible research directions were also pointed out.

## 1. Introduction

Large span floor have the characteristics of light weight and low damping, when there are large crowds across the floor, it will produce obvious dynamic response which beyond a certain range that can make users feel uncomfortable or even panic and will significantly reduce the applicability of the structure; Once producing comfort problem on the built floor, the cost of maintenance will be large and the technical problems will be complex. Therefore, large span floor design should not only meet the requirements of strength, deformation, but also take the vibration comfort problem into account adequately. So this paper will introduce the vibration comfort problem from three aspects: 1) the vertical vibration comfort control standards and design methods of large span floor; 2) the research status of walking load model; 3) the critical calculation method of the vibration response.

## 2. Comfort evaluation standard

Floor rigidity and resonance problems are attached great importance by international scholars, standard for structural deformation of finite values are various in different countries, Such as limits the ratio of deflection to span or the smallest depth-span ratio of flexural member section. These standards can only ensure the stiffness of structure and solve the problems of the structure of the static deformation, but not directly ensure the comfort requirements of floor vertical vibration. Researchers solve the problem of large span floor vertical vibration mainly by two ways: 1) Stipulating that floor frequencies surpass a certain threshold; 2) Limiting the response of floor under walking load, namely. The first method has been widely used because it is simple and easy to implement during the design phase, but it cannot directly contain the pedestrian excitation elements; The second approach based on the acceleration is widely applied on vibration comfort evaluation at present stage.

Vibration control permission value was given by *KB* which could be calculate by one of acceleration, velocity and displacement on 1975′s DIN4150 standard [1].

$$KB = \frac{20.2A}{\left[1+(f/f_0)^2\right]^{1/2}} = \frac{0.13Vf}{\left[1+(f/f_0)^2\right]^{1/2}} = \frac{0.8Xf^2}{\left[1+(f/f_0)^2\right]^{1/2}} \tag{1}$$

Where $V$ is the peak speed; $A$ is the peak acceleration; $X$ is the peak displacement ; $f_0$=5.6Hz.

Standards ISO2631-1 [2] adopted right of frequency meter root mean square acceleration($a_{rms}$) as main vibration comfort index. when vibration signal maximum factor was over 9, arms would underestimate the maximum vibration impact, so run root mean square (MTVV) was recommended.

$$a_{rms} = \left[\frac{1}{T}\int_0^T a_w^2(t)dt\right]^{1/2} \tag{2}$$

$$MTVV = \max\left[\frac{1}{\tau}\left(\int_{t_0-t}^{t_0}(a_w(t))^2\right)^{1/2}\right] \tag{3}$$

Where $a_w$(t) is the right of frequency meter acceleration; $T$ is the vibration duration.$T$ is the run average time,(typically take 1); $t_0$ is a transient moment.

Weiliang Jin [3] thought the above comfort standard had certain limitation due to the characteristics of semi-quantitative. In order to achieve quantitative analysis of the vibration, he put out annoyance rate model which calculate the annoyance rate to predict the human subjective response to vibration considering with psychology and relevant experimental research at home and abroad.

## 3. Walking load model

Human walking is actually a process which the left and right leg alternately supports body weight and moving forward, and the process is divided into stance phase and swing phase. Support phase refers to the state that the feet keep in contact with the ground; Swing phase refers to the state that the feet off the ground and swinging forward. Pedestrians walking process can be summed up the repeat of the two phases. Scholars had begun to study pedestrians walking load model and put forward using Fourier series to approximate the continuous walking load in the 1960s [4].

$$F(t)/G = c_0 + \sum_{n=1}^{N} c_n\left[\cos 2nf\pi t\cdot\cos\phi_n + \sin 2nf\pi t\cdot\sin\phi_n\right] = c_0 + \sum_{n=1}^{N} c_n\cos(2nf\pi t - \phi_n) \tag{4}$$

Where $G$ is weight; $C_n$ is the nth order resonance dynamic load factor; $f$ is walking frequency.

When using walking load model we should determine a reasonable calculation of harmonic order N in order to truncate the Fourier series. Ji [5] and Racic [6] pointed out that the determination of Fourier series harmonic order depend on the accuracy of the load curve and the structural vibration response.

Scholars also studied on the key parameters of walking load model. But due to the physical meaning of the phase angle and statistical regularity was not obvious, therefore continuous walking load curve model had certain difficulty in practical application. The methods of dealing with the current specification was only taking the first order dynamic load factor, namely only take the first item of Fourier series to analyze structural response, such as the UK standard BS5400 specification [7] and Canada OHBDC (1991) [8]. Jiansheng Fan[9] put out simple form standard single walking load model on the basis of existing research results, and demonstrated the model to be feasibility and practicability by combining with the measured walking load curves provided by Kerr.

$$f(t) = \sum_{n=1}^{5} A_n \sin\left(\frac{n\pi}{T_e}t\right), t\in[0, T_e] \tag{5}$$

Where $A_n$ is the coefficient; $T_e$ is the cycles of single step load, $T_e$=1/ (0.76$f_s$).

## 4. Theoretical analysis method

The calculation method of vibration response of floor can be roughly classified into the following categories: Fixed on Foot Force Method, Moving on Foot Force Method, Equivalent Number of Synchronization Method, and Response Spectrum Method.

Fixed On Foot Force Method [10,11] supposed on foot force exerting on a fixed position on the floor, the natural frequency of vibration and pedestrian walking pace and the same, in order to inspire the most unfavorable resonance response of the floor, Load applying the role position of the points is maximum floor displacement, but there are the following disadvantages:1) Fixed Single Foot Force Method assumes that the single foot force exerting on the floor of a fixed point, When a large number of crowded through the floor, vibration response is greater than the vibration response of the single person to walk; 2) Fixed Multiplayer Force On Foot Method is equivalent with formed stable flow on the floor to uniform distribution in fixed point load on the walking route. But all pedestrians may not walk with the same frequency and phase. So this method is very conservative in the design phase.

Moving On Foot Force [12] first apply force to simulate the change with time of the step of the point, then carry out dynamics calculation with the ANSYS analysis software etc, finally get the floor response according to the results of calculation. This method is more precise in theory, but difficult to be applied in actual engineering because of the troublesome calculation.

The Equivalent Number of Synchronous Method [10] is simplify to force method for fixed people walking ,suppose there are $N$ pedestrians were distributed evenly on the floor, the length of $Y$ direction is $S$, the length of $X$ direction is $L$, Area is $A$, $\delta$n（$x_i$, $y_i$）is the location of individual of $i_{th}$ of the $n_{th}$ order modal displacement, then get the response of structure $y_n$ (t) according to higher structural dynamics steady-state vibration mode and vibration acceleration response by derivative $y_n$(t) twice.

Response Spectrum Method first standardized load $F$(t) curve ,next loaded $F$(t) to the standard incentive system, then obtained the acceleration response $u''$(t), finally changed frequency and damping ratio of the single degree of freedom system, which can obtain the response spectrum curve. Wei-Liang Jin [13] carried out the maximum beam slab structure vibration acceleration response spectrum based on the vibration equations of the beam with selecting multiple sets of measured single step foot fall curve and adopting method to construct incentive function. Jianguo Nie[14,15] built the standard pedestrian incentive model and get root mean square response spectrum of single span simply supported the pedestrian bridge and the acceleration of 2~4 continuous footbridge according to the vibration differential equation of beam on the basis of walking load model. Chen Juan [16] first got the standard calculation incentive system of floor response under walking load according to the vibration differential equation of rectangle thin plate. Then, the acceleration response of the incentive system under the measured load curve of the 3 motion capture technology was calculated. Finally, the acceleration response spectrum by adjusting the floor frequency was determined.

## 5. Summary

1) Generally, evaluation of the vibration comfort caused by walking load based on the vibration acceleration RMS was in the entire time which may ignore the malaise during the middle period. Comfort evaluation should take the space of three direction acceleration into account; Therefore vibration comfort problem caused by walking load on space and time effect remains to be research.

2) Accurate pedestrian load model is the premise of the response calculation. Using the traditional force plate or a force measuring device for laboratory measurement is different to actual pedestrian load model, and not considering the coupling relationship with the floor vibration. So, the same as biological and medical science, a new experimental method, including graphics processing using a new type of motion tracking (such as three-dimensional gait analyzer) and load inversion technology of indirect measurement were introduced. It's important research direction of load modeling.

3) The acceleration response is an important index of comfort evaluation system. At the same time, it is very important to calculate the acceleration response quickly and effectively because of the frequent changes in the design stage. The response spectrum analysis method only need modal analysis for the structure to get the structural self vibration frequency and generalized mass structure. And then, by using the maximum acceleration response spectrum envelope, we can obtain the approximate value of maximum acceleration response of the structure, which is convenient for engineering application. However, based on the numerical analysis, the following problems are established: ①In the differential equation of vibration, use the beam deflection function to displace lateral displacement. The basic equation is not accurate enough, so the numerical results are inaccurate; ②Instead of the acceleration response of the structure, the first order vibration acceleration response is used, and the result is too small; ③The statistical law and the physical meaning of the

phase angle of the load model function of some scholars are not obvious. The finite element analysis method can eliminate the error of numerical analysis by the beam deflection function instead of lateral displacement reaction, and resolved of the problem of small results caused by using first vibration acceleration response instead of structural acceleration response. Therefore, it is necessary to establish the calculation model of the long-span structure under the standard excitation model and establish the acceleration response spectrum by using the finite element analysis method.

**Acknowledgments**
This study is financially supported by the National Natural Science Foundation of China(Grant Nos.51408267,51508237)and the Natural Science Foundation of the Jiangsu Higher Education Institutions of China(Grant No.14KJB560005)and Energy saving projects in Colleges and universities of Jiangsu Province(Grant No.JSSNZH2016301).

**References**
[1] DIN 4150, Deutsches Institute fur Normung, 1975.
[2] ISO2631-1:1997. Mechanical vibration and shock evaluation of human exposure to whole body vibration. International Standardization Organization, Geneva, Switzerland, 1997.
[3] Song Zhigang, Jin Weiliang. The Annoyance Rate In a Vibrating Building. Structure & Environment Engineering ,2002,29(4),51-58.
[4] Bachmann H. Vibration upgrading of gymnasia, dancehalls and footbridge. Structural Engineer, 1992, (2):118-124.
[5] Ji. Floor vibration induced by dance type loads theory. Structural Engineer, 1994, 72(3/1):37.
[6] Racic V, Pavic A. Stochastic approach to modeling of near periodic jumping loads. Mechanical Systems and Signal Processing, 2010, 24(8):30-37.
[7] British standards association.BS5400.Steel, Concrete and composite bridges.UK: British Standards Association, 1978.
[8] OHBDC. Ontario Highway Bridge Design Code. Ontario Ministry of Transportation, Downsview, Ont, 1991.
[9] Chen Yu. Large span footbridge vibration and vibration reduction research under the walking load. Beijing: Tsinghai university,2007.
[10] Pan Ning, Qian Hongjin, etc. Research on floor vibration response under walking loads for serviceability. Beijing: China Academy of Building Research,2012.
[11] Qi Manyi, Zhang zhiqiang, ect. Stochastic crowd induced vibration response control for large span footbridges and floors. Nanjing: Southeast university,2013.
[12] Wu Junwei, Yan Qiwu.The analysis on comfort degree of long span public building. Changsha Central south university, 2014.
[13] Song Zhigang, Jin Weiliang. Peak acceleration response spectrum of long span floor vibration by pedestrian excitation. Journal of Building Structures, 2004, 04:57-63.
[14] Nie Jianguo, Chen Yu, Fan Jiansheng. RMS acceleration response spectrum method for single-span footbridges under pedestrian load. China Civil Engineering Journal,2010,09:43-51.
[15] Chen Yu, Nie Jianguo. Acceleration response spectrum method for multi-span footbridge vibrations under pedestrian load. China Civil Engineering Journal. 2010,43(9):117-124.
[16] Xue Ruo tia ,Chen Juan, YeTing , etc. Acceleration response spectrum for predicting floor vibration due to occupant walking. Journal of Building Structures,2015,04:133-140.

# The "shape" and "meaning" of the roof arts in Chinese classical architecture

**Xianda Li[1,] and Yu liu[2]**

[1]Tianjin University of Technology， Tianjin 300384， China
421306351@qq.com
[2)]Corresponding author: 2911913920@qq.com

**Abstract:** This paper takes the "roof" in Chinese classical architecture as the research object. The breakthrough point of this paper would be the perspective of design aesthetics. Through the rational and perceptual analysis of the roof art, this paper would reveal that the roof shape has the double artistic features: "beauty of shape" and "beauty of idea". This paper would have a comprehensive analysis for the following aspects: the rational method of roof construction, the emotional feeling of the roof construction and the implied meaning of beauty in the roof construction.

## 1. Introduction

Chinese classical architecture can refer as a "roof art" from the visual and sensory point of view. It is different from the Western classical architecture that pays more attention on the vertical effect of the building. In its shape design, Chinese classical architecture would like to concentrate more on carve and shape of the roof. Among them, the "roof" has the largest body mass in the proportion of the overall building. Beside this, it also has the following features: the most abundant changes, diverse decorative components, the most far-reaching eaves as well as the most diverse type, level and form. The relationship of its volume is enough to cover the body of the whole architecture. Therefore, we can say that the roof of Chinese classical architecture is equal to the Chinese classical architecture itself in its visual image. It has already become the most recognizable component of the Chinese classical architecture. The functional properties, shape modeling and the intrinsic cultural philosophy of the roof are reasons why roof can represent the image of Chinese classical architecture. (Note: The "image" refers to the overall appearance of the Chinese classical architecture and the most intuitive feelings to the audience.) In addition, the form of reverse flow line of the roof is the root for the whole building community to present a beautiful skyline. Therefore, the design of the roof inevitably contains a profound artistic wisdom and creates artistic conception. At the same time, it has the double artistic features: "beauty of shape" and "beauty of idea". At this point, the breakthrough point of this paper would be the perspective of design aesthetics. This paper would have a comprehensive analysis for the following aspects: the rational method of roof construction, the emotional feeling of the roof construction and the implied meaning of beauty of the roof construction. Through the comprehensive analysis, it will reveal that the "roof" in Chinese classical architecture has aesthetic principles and it will interpret its internal "shape" and "meaning".

## 2. The rational method of roof construction

The rational method for the roof of Chinese classical architecture mainly reflected in the following aspects: the scientific feature of construction theory, the propriety of use form, artistry of design technique and order of the Confucian etiquette. Among them, "the scientific feature of construction theory" is starting from the function of the roof to highlight the real function of the roof. "The propriety of use form" refers to the characteristic that the form of the roof should match with the use function. "Artistry of the design technique" is in terms of design techniques. The "order of the Confucian etiquette" refers to the fundamental law that decides the roof level and form. Therefore, the three elements of the rational method of roof construction can conclude as following: scientific functional attributes, artistic style and the cultural attribute of etiquette. The correlation and fusion between these elements becomes the intrinsic motivation of the roof design. The following will have a detailed analysis towards it:

*2.1 The science of roof structure.*

The roof of Chinese classical architecture usually relies on tile. The main reason is to consider the fire protection, waterproof and drainage features of the whole building. The main structure of Chinese classical architecture is made of wood (this paper takes the "Beam lifting structure" as example). Therefore, the main function of roof is for protection of housing and internal frame structure.

In addition, the ridge of Chinese classical architecture uses the curve shape instead of linear type. Let us take the hip roof as example, the central ridge is bent slightly (Note: the "song type" bending larger whereas "Qing type" bending smaller) but vertical ridges at the side show a clear curve shape. (Figure 1) The Purlin beam structure which interior building made by methods of raising and folding eaves makes slope of the roof present multiple segments and derives curved roof. (Figure 2) The efficacy and advantages of the ingenious construction method mainly shows in the rainfall rate and time. According to physical calculation, the curve interface has faster drainage time and it is not easy to cause water phenomenon compared with linear interface. In the arc and angle projections, the roof of Chinese classical architecture is very appropriate.

(Figure 3, figure 4) It has also become an important reason for the curve shape of the roof. In addition, the roof of Chinese classical architecture is out of the house that has a far-reaching characteristic for the eaves. Its main purpose is to prevent the erosion of rainwater and to protect the integrity of the internal pillars and brackets. Therefore, the scientific feature of the roof in Chinese classical architecture presents the rational method.

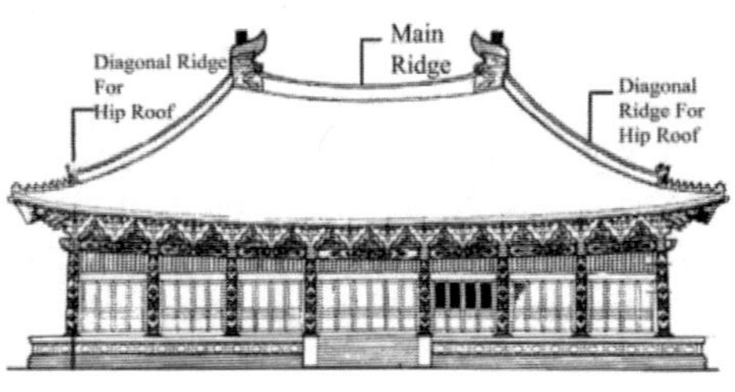


**FIGURE 1.** Roof curve
Source: Author

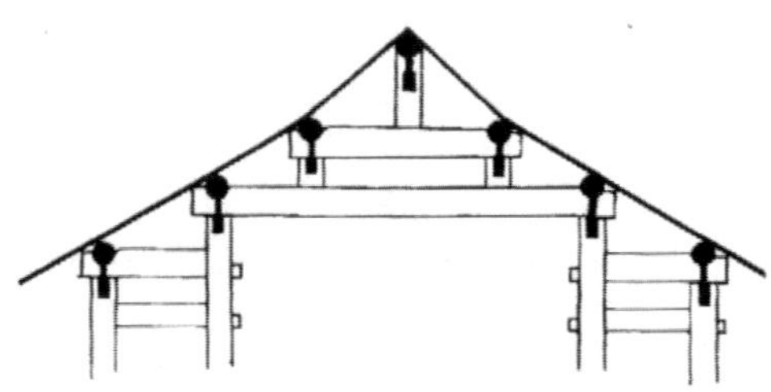

**FIGURE 2.** Line bending
Source: A pictorial history of ancient Chinese architecture, Hou Youbin, Li Wanzhen

*2.2 The use of appropriate roof shapes.*

There are many different kinds of roofs in Chinese classical architecture that mainly contains veranda hall, hanging hill, hard hill, tip save, Xie Shan, etc. Different roof forms are suitable for different functions and ranges. For example, veranda hall and Xie Shan mainly applies to the palace, temples, gardens and other official buildings, hard hill are mostly used in civil buildings, whereas the tip save and Juan Peng are used in the garden pavilions and corridors as well as the sacrificial ritual buildings (such as: qiniandian belongs to the round tip save). In addition, the shape of the roof will accumulate or overlap due to the function and level of the building, such as the building like the attic, veranda hall and Xie Shan. Therefore, the rational method of the roof can be seen from the unity of function and form.

*2.3 The art and advance of roof modeling.*

The roof of Chinese classical architecture has the principle of modern design aesthetics on the base of inheriting the ancient wisdom. That adopts the design method of point, line and surface combination to constitute the streamlined cornices. Each link is placed on top of the roof. Among them, the "point" mainly focuses on Chiwei parts and the decorative cornice and the "line" is reflected in the ridge and the "surface" is composed of four sides of the slope and tile stack. (Figure 5) The essence of this kind of aesthetic principle aims at establishing a complete art form through the stitching and occlusion of the different parts of the "point, line and surface". Then it can produce a harmonious aesthetic effect. Therefore, it can be said that the roof of Chinese classical architecture contains the aesthetic rules of modern art.

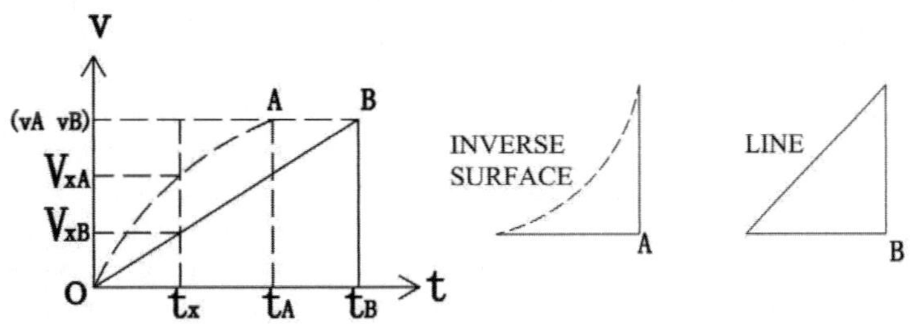


**FIGURE 3.** Mechanical Analysis of curved surface and straight line
(Note： the acceleration of the surface fast at the beginning, so the time would be quick )
Source: author draw

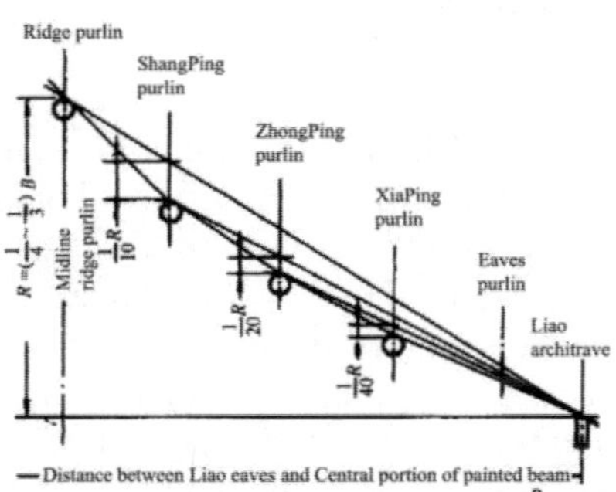


**FIGURE 4.** "Lift fold system" angle, radian chart

Source: A pictorial history of ancient Chinese architecture Hou Youbin, Li Wanzhen

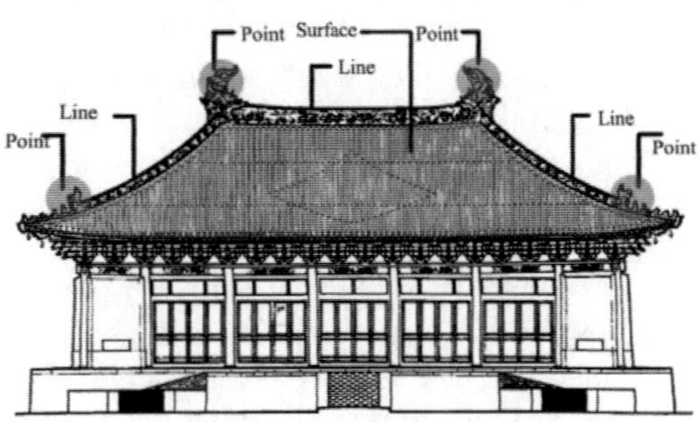


**FIGURE 5.** Roof "point, line, surface" analysis Source: Author draw

*2.4 The order of the roof construction.*

The roof of Chinese classical architecture has the strict hierarchy and institutional requirements instead of random building. It has its own unique building system that needs to in accordance with the contents of the rite. Such as the higher level roof form like veranda hall, eaves and tip save can only serve royal family. At the same time, residential buildings shall not use yellow glazed tiles. Each roof form represents a different level. Such as the level of veranda hall and tip save is higher than that of Xie Shan and hanging hill. The eaves roof form is higher than that of single roof. Therefore, the roof of Chinese classical architecture has the strict order and hierarchy. From the point of view of design, this rank and order subjective stability and strengthen the structure and content of the roof and makes it have the characteristics of "model" and "standardization". It can seek unity on the base of obtaining rich form. Therefore, it also constitutes the cultural factor of the rational method.

In summary, the "beauty of form" in the roof of Chinese classical architecture is a combination of

IOP Conf. Series: Earth and Environmental Science **61** (2017) 012110 doi:10.1088/1755-1315/61/1/012110

functional science, artistic design and cultural etiquette. The "beauty of form" can also bring to strong sensibility that contains rich moral beauty. The following will have a theoretical exposition about it.

## 3. The emotional feeling of the roof construction

The roof of Chinese classical architecture can bring out the "lively" and "vivid" emotional experience through the beautiful curve and neutralization culture on the base of following the principle of scientific, rational, etiquette principles. (Note: "Neutralization" culture refers to the influence and inspiration of Chinese classical culture on architectural art). The deepening and understanding of this kind of emotion mainly comes from the collision and blending between the different cultures and it also comes from perfect integration between modular system and the curve modeling under this system. At this point, the neutralization of culture and the conformity between "regular system" and "natural curve" are important reasons for roof art to convey a sense of emotion.

### *3.1 The cultural neutralization in the roof of Chinese classical architecture.*

From the above discussion on the scientific nature of roof shape, we can conclude that the shape design of ancient roof following the specific program, that is the Confucian "rites". It fully embodies the Confucian culture for the influence of roof shape. The modeling also shows the design principles of equilibrium, symmetry and modulus. From the design of the roof curve can also see that the ancient roof has smart, elegant, streamlined features. Its style is the highlight of Taoist Culture. Therefore, Chinese ancient roof contains both rational "ritual", but also contains the emotional "music". It is under the strict system but keeps its flexibility. Although it is in the method of modular system, it is sill flexible. Therefore, China ancient roof design is generated by fusing multicultural and has obvious cultural characteristics of "Neutralization". The solemn and intelligent features of roof itself can be sensed in the viewer's visual attention to resonate emotional experience with each other.

### *3.2 the perfect integration of the "data" and the change curve.*

The ancient Chinese roof has uniform system. Although there are some differences between the song style and the Qing style roof in the positioning method of the frame, they are still based on fixed data. They take the "bucket mouth" and "material" as a unit. (Figure 6) Therefore, the method of its manufacture is unified. In addition, the ancient roof can also form a graceful curve, so it has the aesthetic feature of "line seeking". It is hard to imagine that the inner structure of the roof contains complex rational principles and complicated data just from perceiving individually from the form. Therefore, the roof in ancient architecture is a combination of "rational method" and "sensibility". It explains the reason for the "curve shape" of the roof. At the same time, it interprets the original implied testimonies through the beautiful shape.

In summary, the emotional feeling of the roof construction in Chinese classical architecture is to resonate emotional experience with each other through the viewer's intuitive feel for the roof of the external form. It makes the audience further comprehend its deep beauty through its own cultural connotations on the base of audience to appreciate the architectural "beauty".

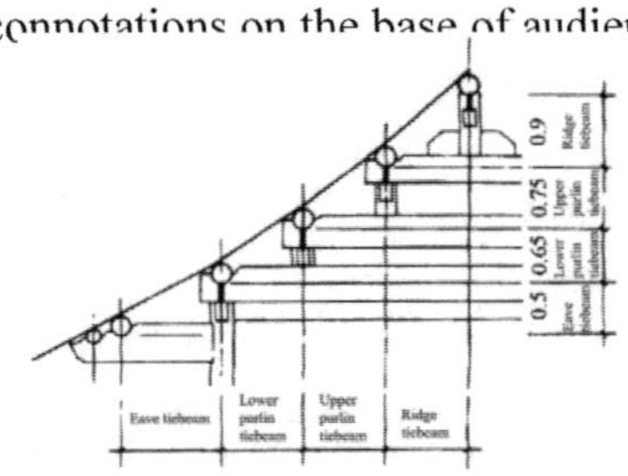


**Figure 6.** Frame location

Source: A pictorial history of ancient

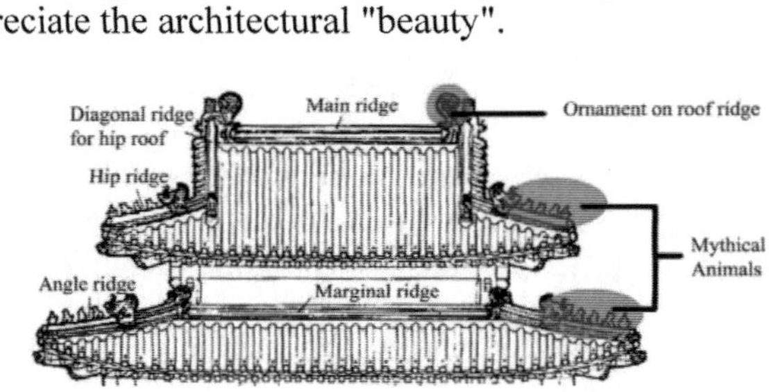


**Figure 7.** Decorative component location map

Source: Author draw Chinese architecture
Hou Youbin, Li Wanzhen

## 4. The implied meaning of beauty in the roof consruction

The meaning of roof in Chinese classical architecture mainly highlight through the implication beauty of decorative component and the symbolic meaning of form itself. Among them, the "decorative component" is a specific materialized form to convey the meaning. It is the dominant media of the "meaning". The form of the "pictographic meaning" is relying on the decorative elements and curve shape. It is the metaphysical experience from the perspective of Philosophy. It is the recessive element of the "meaning". The following will respectively elaborate the implied meaning of the roof art from the aspects of materialized decoration and metaphysical philosophy.

### *4.1 The implied beauty of materialized decoration.*

The roof of Chinese classical architecture has decorated many component and animal on the base of integral streamline modeling. These elements are arranged at both ends of the roof ridge, before the ridge, and at both ends of the ridge. These decorations select animal or a row of "animal" as a component. (Figure 7) It contains a profound moral and functional thinking when it plays a decorative role. The roof of Chinese classical architecture has been adhering to the concept of "fire prevention" in its design. Thus, the designer starts from the function thinking to introduce the Mythical Animals on the roof that has the fire prevention implication. It symbolizes that the animal can protect buildings and controls the fire. For example, the Chiwen at he both ends of the ridge is the animal that can spit water. There are other Pegasus, hippocampus, dragon, Phoenix and suanni, etc. In addition, the roof of the animal also provides with a Xiezhi that represents moral and justice. The use of these elements also subjectively enriches the curve shape of the roof. In addition, from the decorative component itself as a symbol of the role for a deeper extension, we can also see that this symbolic meaning contains "personification" characteristics. To seek shelter, security; pray for wealth, good luck; justice, goodness belong to the scope of human nature. This kind of personification and ideographic characteristics are in accordance with the expression of "borrowing objects and love" in literature and art. Therefore, the decorative component of the roof both contains allegorical and anthropomorphic features. It has become the "Artifacts" media for ancient craftsmen to design roof and express their inner thoughts and circumstances. At this point, it can be said that the decoration of the ancient roof is not random filling. It is based on the premise of functional thinking and desire of human nature to enrich the roof form and content and to highlight the unique design idea of craftsmen through the symbolic animal. At the same time, this idea is consistent with the modern design aesthetics that expresses the design idea in the form of concrete images. Therefore, the decorative components of the roof are through the two attributes to reflect "beauty of meaning".

### *4.2 metaphysical meaning of the roof shape.*

Chinese presents the curve shape of ancient roof that is unique in the history of world architecture (Notes: the architecture history of Japan, South Korea, North Korea have inherited the Chinese architectural system. Therefore, Chinese ancient is the source and fundamental of the world architecture). This curve form deeply expresses the philosophy meaning that Chinese people pursue "refined dust" and "Nirvana rebirth". The Book of Changes has recorded that the things beyond the shape is law and the things beyond the shape is device. The modeling and decoration of the roof curve is a device. The philosophy and meaning that above the shape is law. Chinese people call their living land as the Divine Land from the ancient time. And it can also be called Phoenix land. The ancient people have linked the rise and fall of the Boulevard as well as the emergence of the great man with the Phoenix. The Phoenix's rebirth has also become the longing for the ancients. In the design of roof in Chinese classical architecture, the mainly feature are far-reaching eaves and reverse flow line. This also contributed to the construction of the community presents a beautiful "curved corridor" as if merging with the sky. This flying and elegant shape just as the Phoenix wings that can fly in the world, the shape of the roof of the "metaphysical meaning" is to express the rebirth of immortality and convey the eternal philosophy. Although the wood frame building itself is temporary, the connotation of the roof form of the building is infinite and eternal. The building can be burned, but rebirth after the ashes will become more perfect. At this point, the roof curve shape by the expression of "metaphysical

IOP Conf. Series: Earth and Environmental Science **61** (2017) 012110 doi:10.1088/1755-1315/61/1/012110

meaning" is reborn, perfection and eternity "Daoism" spirit.

## 5. Conclusion

From the above analysis of the rational and emotional analysis of the roof, we can see that the roof design in ancient China is based on rational and testimonies. The goal is beautiful, vivid and elegant. It has the dual attributes of "beauty of shape" and "beauty of meaning". It can through the culture of "Neutralization" and "blend" to get a deep sense of beauty. So far, we can draw a conclusion that roof in Chinese classical architecture is simple but with profound meaning. Through the combination of theory and form, it can reach the unity of shape and meaning. It not only realizes the perfect combination of function and form in the design of "form" and "meaning", at the same time, they can resonate with each other with the design ideas in modern times. It can be seen as the typical work in the combination of shape and meaning.

## Reference

[1] Liu Dunzhen. 1984.3-17[2], *Chinese Architecture History* [M]. Second Edition. Beijing: China Architecture and Building Press,

[2] Li Yunhe. 2005. Chinese Artistic Conception [M]. Tianjin:Tianjin University press, p185-p189,

[3] Li Jie. 2007, *Create a Form (Volume 1)* [M]. Beijing: China Building Industry Press.

[4] Hou Youbin, 2002, Li Wanzhen. *A Pictorial History of Ancient Chinese Architecture* [M]. First edition. Beijing: Chinese Building Industry Press, p184-p200

[5] Wang Hui. 2012, *Architectural aesthetic form and meaning* [M]. First edition. Beijing: China Architecture and Building Press, p118-p121

[6] Zhu Liangzhi. 2013 *Ten lectures on Chinese Art* [M]. First edition. Beijing: Peking University press, p134-p137

[7] *《Book of Change》*

# The analysis on characteristic of pre-stressed reticulated mega-structure with grid sub-structure

**Chunye Wu[1,2,a)] and Zhixiang Yin[1,b)]**

[1]Liaoning Technical University, Fuxin, Liaoning, 123000
[2]Tiandi Science and Technology Co., LTD , Beijing, 100013
[a)] Corresponding author: wcy830302@163.com
[b)] Yin zhixiang@shouhu.com

**Abstract**. The author brings a new structure named pre-stressed mega grid structure which can be applied to super large span architectural structure through combining the advantages of mega grid structure and beam string structure. This new structure has many advantages, such as, light weight, strong rigidity and large span. This article has secondly developed ANSYS software with APDL language for modeling establishment and solving integration, and also edited related programs. Referring to the contractive analysis of dynamic characteristics, static characteristics of pre-stressed mega grid structure and mega grid structure; and by adopting time history method to analyze the seismic response of pre-stressed mega grid structure. One may get the conclusions of strengths in deflection characteristics and mechanical characteristics of pre-stressed mega grid structure. Through analyzing the dynamic characteristics, static characteristics and seismic response of pre-stressed mega grid structure which is across 280m by the method of changing original pre-stress, loss ratio of main structure and the number of grids. It may get conclusions of range of evaluation of all above parameters.

## 1. Introduction

With the development of social economic development and the new requirements of structural span, the long span of mega grid structure has become main research direction of many researchers and exporters. Especially the appearance of haze weather in recent years, the architecture in whole sealing structure which spans across 300m, 500m and 1000m is the mainstream development trend of future city. Although research of long-span spatial structure in China has reached a significant development, it still has its' weakness as lacking research classification and engineering projects in 200m span across structure.

The author brings a new structure named pre-stressed mega grid structure combining the advantages of mega grid structure and beam string structure (see Figure 1). This new structure is that the horizontal truss of mega grid structure is designed as rigid beam-column, and the pre-stress steel cables are added between the junctions of horizontal truss and longitudinal truss of contiguous grids. The anchor strut is the connection between steel cables and truss arches. The truss will be adopted as anchor strut because of large span of steel cable and great pre-stress power. The paper gets the conclusion of the reasonableness for the new structure through contractive analysis of pre-stressed mega grid structure and mega grid structure by usage of software ANSYS, also gets a series of important conclusions which could provide significant reference for related research.

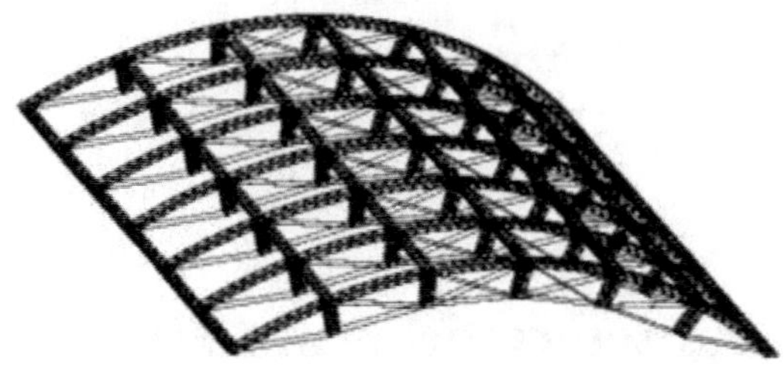

**FIGURE 1** Pre-stress mega grid structure without sub-structure

## 2. Model establishment and secondary development of ansys

### *2.1. Model Establishment*

(1) Set following assumptions in computer for convenience to calculation:

(2) Space hinged node, and neglect the impact of node rigidity,

(3) The member is in elastic state, and the whole structure is in small strain condition,

(4) The effect of loading is only in the node without considering the plastic deformation,

The member of main structure and sub-structure, and the anchor strut will be simulated using the link 8 element, the cable will be simulated using the link 10 element.

Furthermore, in the characteristics analysis of pre-stressed mega grid structure and mega grid structure, the span is 280m, the number of horizontal grid of main structure is 6, and the number of longitudinal truss is 7, the number of horizontal grid of sub-structure is 6, the number of longitudinal grid of sub-structure is 7, the size of small grid of longitudinal truss is 3.5m, the height of the truss is 3.5m, and the span ratio of main structure is 1/5. The type of sub-structure is flat-panel grid structure. The height of flat-panel grid is 1.5m, the grid size of main structure and sub-structure is the same, the height of anchor strut is 10.5m, the steel density of structure is 7800kg/m$^3$, the elasticity modulus is E=2.1×108 kN/m$^2$. The dead loading is 1.0 kN/m$^2$. The living loading is 0.5 kN/m$^2$. The transverse section of the main structure after the full stress optimization will be showed in table 1.

**TABLE 1** Transverse section specifications

| Location of Members | Models of Members(mm) |
|---|---|
| Top Chord And Lower Chord Of Main Structure | Φ219*14 |
| Diagonal Struts Of Main Structure | Φ180*12 |
| Top Chord And Lower Chord Of Sub Structure | 140*10 |
| Diagonal Struts Of Sub Structure | 114*10 |
| Anchor Strut | 102*10 |

### *2.2. Secondary Development*

Author has secondary developed the APDL parameters design language of ANSYS to set the related program for the convenience to calculation, and the program and design language is shown in figure 2, the program can make an integration for structural modeling and calculation.

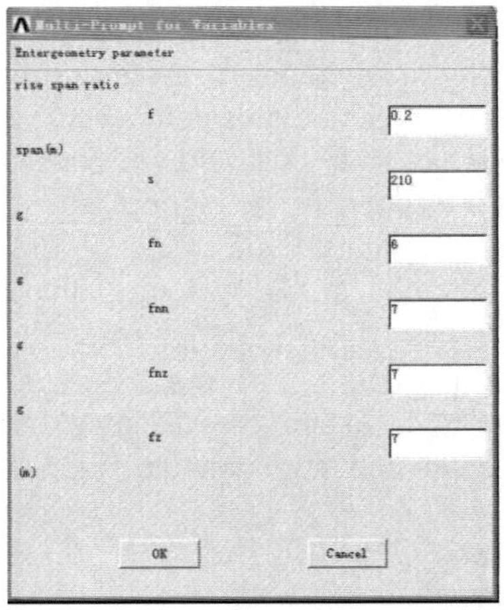


```
multipro, 'start', 7
*cset, 1, 3, ro1, 'Radius (m)', 0.240
*cset, 4, 6, ro2, 'Radius (m)', 0.240
*cset, 7, 9, ro3, 'Radius (m)', 0.240
*cset, 10, 12, ro4, 'Radius (m)', 0.240
*cset, 13, 15, ro5, 'Radius (m)', 0.240
*cset, 16, 18, ro6, 'Radius (m)', 0.240
*cset, 19, 21, ro7, 'Radius (m)', 0.140
*cset, 61, 62, Enter, 'pipe Radius',
Multipro, 'end'
```

**FIGURE 2** Program order and UI

## 3. Static characteristics analysis

*3.1. Static Characteristics Analysis of Pre-stressed Mega Grid Structure and Mega Grid Structure*

It may get a conclusion that maximum displacement has occurred in the middle of span in two structures, of which node displacement of top chord of mega grid structure is 0.14294m, node of lower chord of mega grid structure is 0.13352m, node of top chord of pre-stressed mega grid structure is 0.0864m, node of lower chord of mega grid structure is 0.079807m (see Figure 3). Through contractive analysis of node displacement of same location of main structure and stress of member under the normal working condition, it proves that for pre-stressed mega grid structure, portrait displacement has been decreased for the reason of which loading of main structure and pre-stress has been balanced.

Otherwise, for pre-stressed mega grid structure, the member stress of horizontal truss of main structure is larger than the member stress of longitudinal truss of main structure. This shows that main bearing member in pre-stressed mega grid structure is horizontal truss arch. The compression member is top chord member of main structure, the members in middle of structure is tension member, the stress will become smaller from both ends to middle part; compression member is lower chord member of main structure in pre-stressed mega grid structure, internal force of lower chord member will become smaller from both ends to middle part.

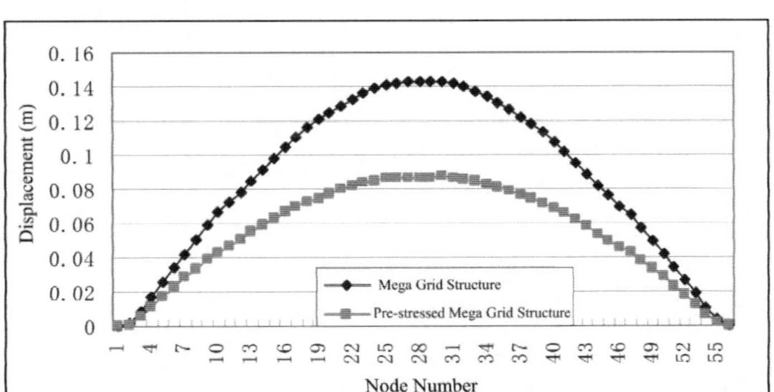


**FIGURE 3** Comparison chart displacement of top chord node of two structures

*3.2. Parameter Analysis of Static Characteristic of Pre-Stressed Mega Grid Structure*

IOP Conf. Series: Earth and Environmental Science **61** (2017) 012111 doi:10.1088/1755-1315/61/1/012111

Through analyzing the impacts of node displacement of same location in pre-stressed mega grid structure which crossed 280m and stress of member by the method of changing original pre-stress, loss ratio of main structure, the differences of number of grids, it may get a conclusions as below:

(1) It has decreasing trend in displacement of node in same location with enlarging pre-stress; this proves that pre-stress and external loading is one pair of opposite acting force.

(2) Displacement of same location has been enlarged with the increasing of loss ratio, and it proves that horizontal main truss is the main compression member in the structure, then loss ratio should be design as 1/5～1/6.

(3) With the decreasing of number of grids in main structure, i.e. enlarged the size of one grid, both member stress and displacements of top and lower node in pre-stressed mega grid structure will be increased. Then number of grids should not be design too few, and the size of it should be set as 35m～45m.

## 4. ANALYSIS OF DYNAMIC CHARACTERISTICS

### *4.1. Contractive Analysis of Dynamic Characteristics in Pre-Stressed Mega Grid Structure and Mega Grid Structure*

It may get a conclusion of that the frequency of the first 6-stage of pre-stressed mega grid structure is bigger than the frequency of the first 6-stage of mega grid structure through the free vibration characteristics contractive analysis with the iterative methods of above two structures (figure 4), basic frequency of pre-stressed mega grid structure and mega grid structure which is 0.9487Hz and 0.62665Hz. There are fewer differences in frequency between 7th~16th stage of two structures, generally speaking, the frequency of pre-stressed mega grid structure is larger than the other one, and the frequency in $17^{th}$ stage starts increasing in pre-stressed mega grid structure. The reason of all above appearance is the rigidity of structure has been strengthened after adding pre-stress steel cable, then the span of pre-stressed mega grid structure has increased as well, so seismic performance of structure with pre-stress is better than the structure which is not with pre-stress.

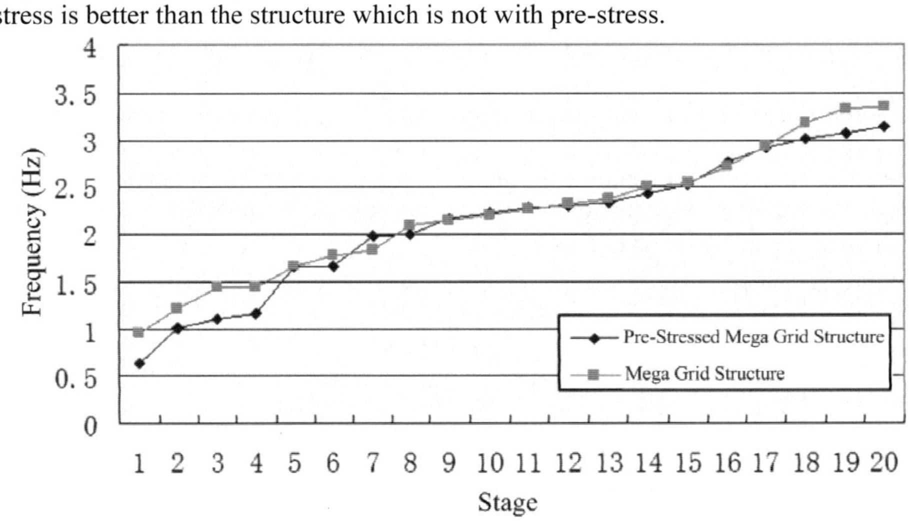


**FIGURE 4** Comparison chart of free vibration frequency in two structures

Moreover, model 1 and 8 show that they vibrated in transverse horizontal direction. Model 2 and 5 shows that they vibrated in transverse horizontal direction with up and down vibration. Model 3, 6, 7, 9 and 10 shows that they vibrate in up and down direction. Model 4 shows that it vibrated anti-symmetric up and down direction. Model 11 shows that it vibrated in vertical direction, and it has a portrait symmetry vibration in horizontal direction with four half waves, and they will keep in same step of each arches, but only the vibration range of middle part is slightly larger than both ends. Model

IOP Conf. Series: Earth and Environmental Science **61** (2017) 012111 doi:10.1088/1755-1315/61/1/012111

12 shows that it vibrated in portrait direction, and it has a portrait symmetry vibration in portrait direction with two half waves, and they will keep in same step of each arches. The less component of longitudinal displacement in two models of those two structures proves that rigidity in vertical direction is the most maximum rigidity after adding in sub structures of flat framework.

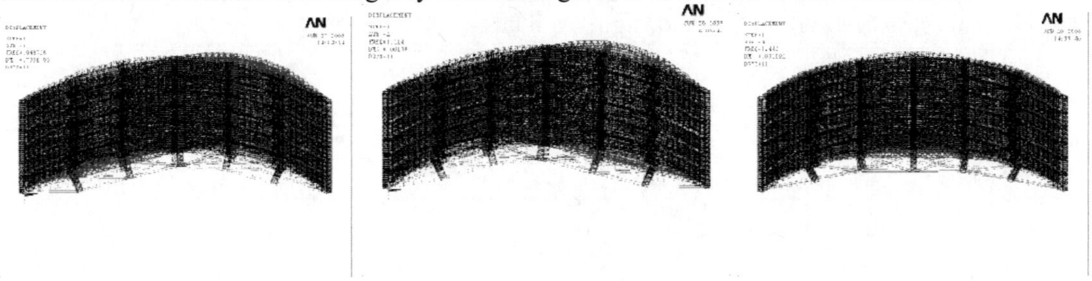

(a)Model 1 （f1=0.9487Hz） （b）Model 2（f2=1.2137Hz） （c）Model 3（f3=1.4345Hz）

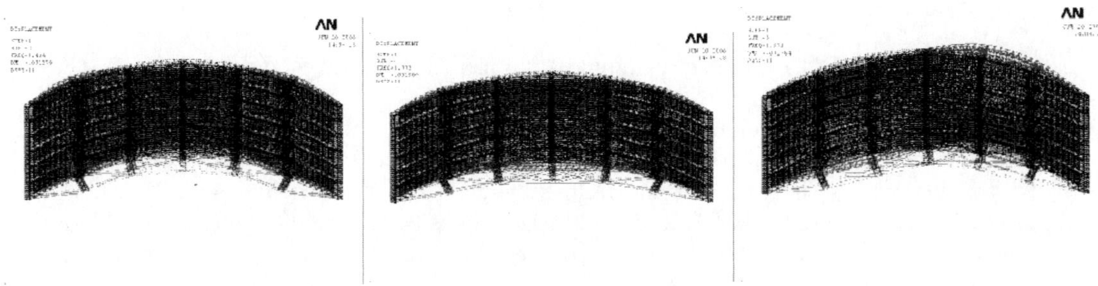

（d）Model 4（f4=1.4421Hz） （e）Model 5（f5=1.6711Hz） （f）Mode 6 (f6=1.7731H)

**FIGURE 5** Vibration chart of pre-stressed mega grid structure

*4.2. Parameter Analysis of Dynamic Characteristic of Pre-Stressed Mega Grid Structure*

Through analyzing the impact to the parameters of dynamic characteristics of pre-stressed mega grid structure which is across 280m by the method of changing original pre-stress, loss ratio of main structure and the number of grids, it may get following conclusions:

(1) Assumed pre-stress as S, then pre-stress will be $S_1$=0.8S； $S_2$=S； $S_3$=1.2S； $S_4$=1.4S，through analyzing above four pre-stress models of calculation, one can conclude(figure 6): By increasing the pre-stress, free vibration of each stage in its structure will be increased as well, this is the reason that the strengthening thr whole rigidity of structure became, the larger of original pre-stress of steel cable would be made.

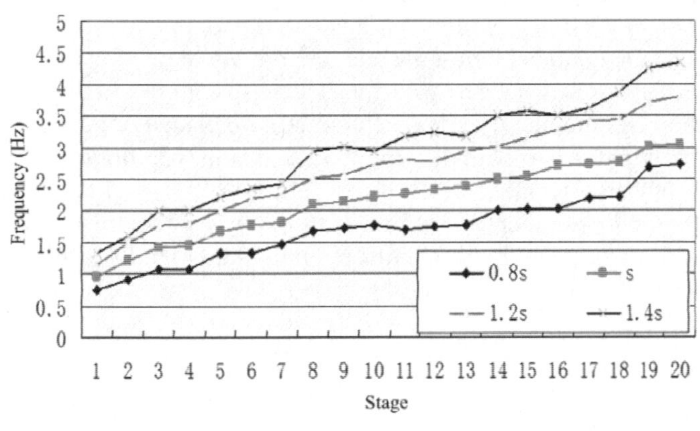


**FIGURE 6** Frequency comparison for different pre-stress models

(2) Trough analyzed above calculation of four pre-stress models which loss ratio of main structure is 1/3、1/4、1/5、1/6 respectively, it can be concluded as (figure 7): with the increase of loss ratio of main structure, frequency of free vibration of each stage in its structure will be decreased, and this is the result of decreasing in rigidity which leaded by increasing of the height of arch in main structure. Then ratio of main structure should not be designed too large.

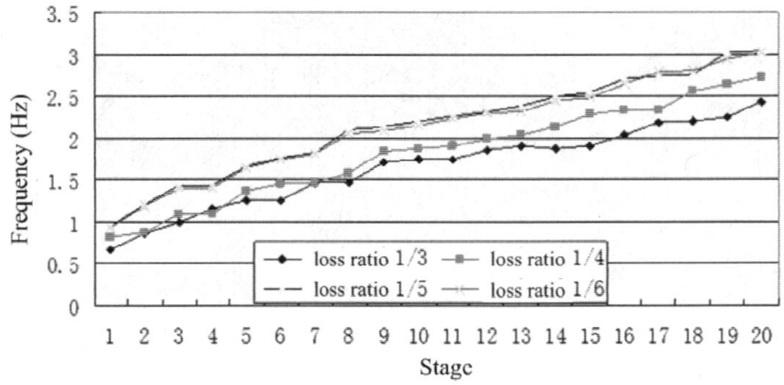


**FIGURE 7** Frequency comparison for different ratio model

(3) Trough analyzing above calculation of four grids which grid number is 4, 5, 6 and 7 respectively (table 7): with the increase of numbers of grid in main structure, frequency of free vibration of each stage in its structure will be increased, basic frequency of 7 grid number and 4 grid number pre-stressed mega grid structure are 0.8348Hz and 1.0436Hz. This is because the numbers of grid has been decreased, then each size of unit grid will become smaller, then the whole rigidity has been decreased.

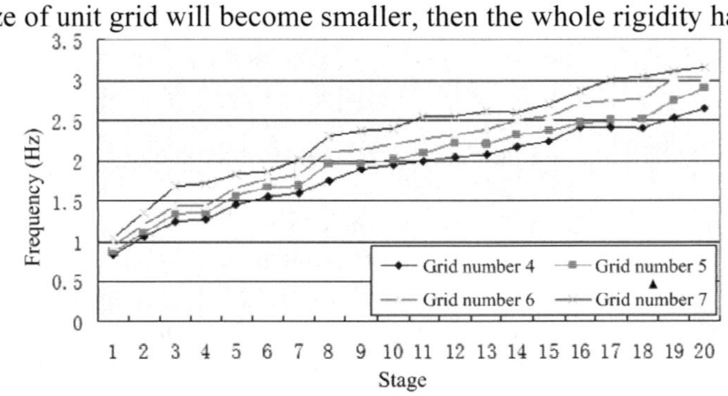


**FIGURE 8** Frequency comparison for different grid number

## 5. Analysis of multi-dimensional seismic response

*5.1. Analysis of Multi-Dimensional Seismic Response of Pre-Stressed Mega Grid Structure*

Through analyzing the response of pre-stressed mega grid structure which is across 280m in Tianjin earthquake by the method of time history (figure 9 and 10): displacement of each node of main truss in pre-stressed mega grid structure, especially the displacement of top chord is mainly in portrait direction (Z direction), but displacement peak of top chord node is slightly larger than the number of lower chord node. Maximum number of top chord node in mid-span is 0.162m, while maximum displacement number in horizontal and vertical direction is -0.093m and 0.062m.

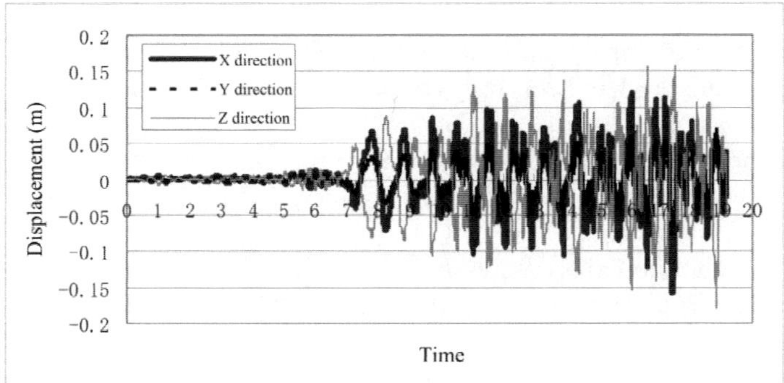


**FIGURE 9** Time history curve of lower chord node of mid-span in x、y、z direction

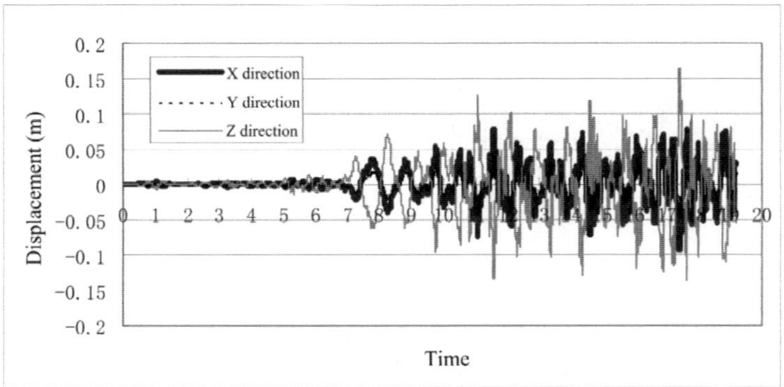


**FIGURE 10** Time history curve of top chord node of mid-span in x、y、z direction

Moreover, both displacement responses in horizontal transverse (X direction) and longitudinal horizontal (Y direction) are shorter than portrait direction (Z direction), this is the reason of sub-structure settlement in main structure, the appearance of sub-structure greatly increased the rigidity of structure in vertical (Y direction) and portrait direction (Z direction).

*5.2. Analysis for Parameters of Seismic Response*

Through analyzing the impact to the parameters of seismic response of pre-stressed mega grid structure which across 280m by the method of changing original pre-stress, loss ratio of main structure and the number of grids, it may get following conclusions:

(1) With the increase of pre-stress, both responses in displacement of top and lower chord of the structures has been decreased, which proves that adding pre-stress will strengthen rigidity.

(2) With the decreasing of loss ratio of main structure, both responses in displacement of top and lower chord of the structures have been decreased. The response of maximum displacement in structure has been raised to -0.23m when loss ratio is 1/3, while the response of maximum displacement in structure has been raised to -0.062m when loss ratio is 1/6. This proves that loss ratio has a great significant influence on structural displacement. Also proves that rigidity of structure will be decreased after increased loss ratio of main structure, and the height of arch in main structure has been raised.

(3) With the increase of number of grid in main structure, both responses of displacement in top and lower chord of the structures will be decreased. This is the reason of decreasing of number of grids in main structure which leads to the size will become larger in each grid, i.e, the longer of each member being made, the whole rigidity in structure will be decreased.

IOP Conf. Series: Earth and Environmental Science **61** (2017) 012111 doi:10.1088/1755-1315/61/1/012111

## 6 Conclusions

In this article, we can get following conclusions through above analysis:

(1) A new long span structure has been brought based on the advantages of pre-stressed mega gird structure and mega gird structure.

(2) Under the conditions of same span and loading effects, the statically characteristics in node displacement and member stress of pre-stressed mega gird structure are better than those characteristics in mega gird structure.

(3) The free vibration characteristics of pre-stressed mega gird structure is better than that mega gird structure, i.e. the whole rigidity of pre-stressed mega gird structure is better than that mega gird structure.

(4) The best design parameters will be determined after contractively analyzed the structure characteristics by changing loss ratio, pre-stress and number of grids, which can provide significances in actual design.

## References

[1] Lu Xilin,Yin Siming. Modern Prestressed Steel Structures[M]. Beijing: China Communications Press,2003.

[2] Luo Yaozhi. Design And Construction of Large-Span Coal Storage Structures[M].Beijing: China Electric Power Press,2007.

[3] GB5107-2003-Code for Design of Steel Structures[S].Beijing: China Building Industry Press,2003.

[4] GB5007-2001- Load Code for the Design of Buildings Structures[S]. .Beijing: China Building Industry Press,2001.

[5] Yin Zhixiang,Wu Chunye. Analysis on the dynamic characteristic of prestressed reticulated mega-structure with grid sub-structure[J].Building Structures,2009,4:95-96.

[6] Wu Chunye, Yin Zhixiang. The Analysis of Multi-Dimensional Seismic Response of pre-stressed mega gird structure [J]. The eighth national conference on modern structural engineering,Tianjin,2008.

[7] He Yongjun, Zhou Xuhong. Super-span reticulated mega-structure[J].Journal of Architectue and Civil Engineering,2005,9:25-27.

IOP Conf. Series: Earth and Environmental Science **61** (2017) 012112 doi:10.1088/1755-1315/61/1/012112

# Analysis of the influence of Nanchang Metro Line 4 on groundwater

**Yingying Lan**

Academy of Hydraulic and Ecology Engineering, Nanchang Institute of Technology, Nanchang, China.
E-mail: lan-yy@163.com

**Abstract**. Nanchang city subway excavation depth and the Quaternary aquifer layers are located at approximately the same depth. After the completion of the subway, that is equivalent to adding a retaining wall in the phreatic aquifer. The metro line 4 influence on groundwater flow field was analyzed based on the groundwater flow field and the dynamic relationship between groundwater and surface water. The result was that the groundwater level would rise in the area of facing groundwater movement, while others decrease. The influence of metro was apparent at the place where hydraulic contact of Gan River with groundwater was good, and vice versa.

## 1. Introduction

Nanchang metro project is the forward-looking engineering of Nanchang development. It is of great significance to ease the situation of urban traffic and promote the economic development of Nanchang. Nanchang City belongs to the Poyang Lake basin, located in the leading edge of Poyang Lake. There are abundant groundwater resources in Nanchang area and its water-bearing stratum is very good. The hydrogeological condition is special for Nanchang City to build metro. Nanchang city metro excavation depth and the Quaternary aquifer layers are located at approximately the same depth. After the completion of the metro, that is equivalent to adding a retaining wall in the phreatic aquifer [1]. This will hinder groundwater runoff and lead to changes in the groundwater flow field [2]. Therefore, that is very important significance to research that whether or not to change the original groundwater flow field and cause the related hydrogeological problems after the operation of Nanchang metro. What problems should be paid attention to under the condition of new urban development and in the exploitation, utilization of groundwater resources in Nanchang City.

## 2. Background

Nanchang City is located in the middle and lower reaches of the Yangtze River, the west coast of Poyang Lake. It has a subtropical humid climate, four distinct seasons. The main surface water bodies are Gan River, Fuhe River, Yudai River, Qingshanhu Lake, Aixihu Lake, Yaohu Lake, etc. The whole terrain is high in the northwest and low in the southeast, Low Mountain hilly, downland and plain are developed, showing the layered geomorphologic characteristics. The tectonic is denudation hillock in the northwest of Gan River, and erosion accumulation plain in the east of the Gan River. The study area is mainly the erosion accumulation plain. The erosion accumulation plain consists of holocene, middle and upper pleistocene alluvium, and the terrain is flat. The Quaternary aquifer consist of middle Pleistocene, upper Pleistocene, Pleistocene sand, sand gravel and gravel. The overlying strata are a relatively aquiclude those are composed of clay, silt soil, and silt clay lens with weak water conductivity. The lower part is the main storage space of groundwater that composed of gravel [3]. The upper Pleistocene is the most widely distributed among the three times of sand gravel layer, followed by the Holocene, and the distribution of Middle Pleistocene is less. The three formed a unified aquifer with closed hydraulic connection, between the three is the relationship of overlapping, and there is little difference in the elevation of the bottom of the junction.

## 3. Analysis and results

### *3.1. Analysis of groundwater effect*

Nanchang metro line 4, as shown in Figure 1: High speed passenger west station→North Avenue of Zhanqian Road→Chaoyangzhou South→Fousheng Road→Hongcheng Road→Hongdu Avenue→Nanjing East Road→Shanghai Road→Guowei Road→Huoju Avenue. Planning route length is 24.6 km, and 19 stations were set up. It is the auxiliary line that connection of East District, Chaoyang area and Hongjiaozhou area, and strengthen the rapid connection of the peripheral chip interval. But, metro line excavation position lies 10 meters below the ground. That would obstruct the hydraulic connection between groundwater and Gan River. This is bound to have a corresponding impact on groundwater flow field [4].

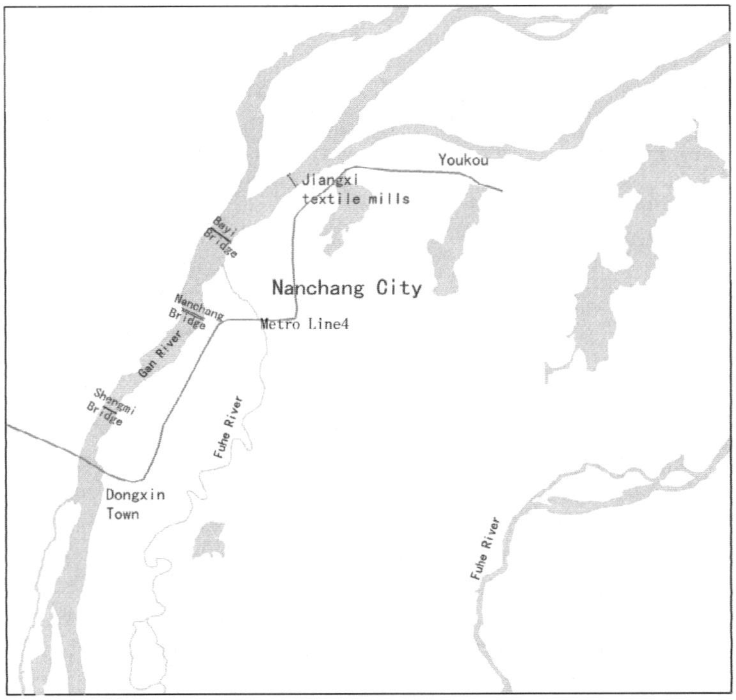


Fig.1 Distribution map of Nanchang Metro Line 4

### *3.2. Effect result*

*3.2.1. Xindong Town to Bayi Bridge.* The general trend of groundwater flow field in Nanchang city is from the round to the center of the groundwater depression cone. Gan River is northwestern boundary of the study area, and that is an important recharge boundary and discharge boundary of groundwater in Nanchang city. In the section from Xindong Town to Bayi Bridge, The bed of Gan River is located over the holocene sand and gravel layer. The main function of Gan River is to discharge of groundwater that closely related to surfer water (Fig.2-a). But it also can recharge groundwater in the

IOP Conf. Series: Earth and Environmental Science **61** (2017) 012112 doi:10.1088/1755-1315/61/1/012112

flood season. The movement direction of groundwater is from east to west, and finally into the Gan River. The upcoming metro line 4 will impede groundwater discharge to Gan River. This will lead to groundwater level rise in the eastern side of the metro line 4 in the section from Dongxin Town to Bayi Bridge. However, the west side of the metro line 4 will be reduced due to runoff reduction.

*3.2.2. Bayi Bridge to Jiangxi textile mill.* The section from Bayi Bridge to Jiangxi textile mill, Gan River cutted the sand loam and clay layer of Pleistocene. The river bed was deposited of sand and gravel of Holocene, so that kept in close contact with aquifer, strong ability to recharge groundwater (Fig.2-b). In this section, the groundwater flow direction that was from west to East, the Gan River recharged groundwater. Metro line 4 would hinder the Gan River recharge to groundwater when it was completed. This will inevitably lead to the rise of groundwater level near the side of Gan River, but the level would reduced on the eastern side of the river due to the reduction of supply flow.

*3.2.3. Jiangxi textile mill to Youkou.* It was affected that Gan River recharged to groundwater, and the recharge was less because there were upper Pleistocene or Holocene sub clay, silty loam in the river bed in the section from Jiangxi textile mill to Youkou. Therefore, the influence of metro line 4 on the groundwater level is relatively weak.

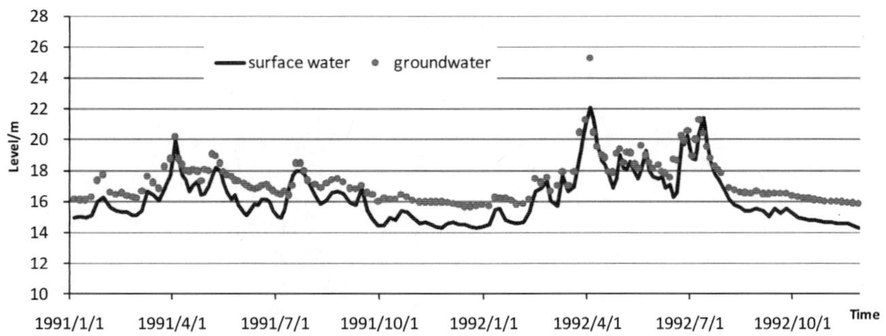


(a) Shengmi Bridge

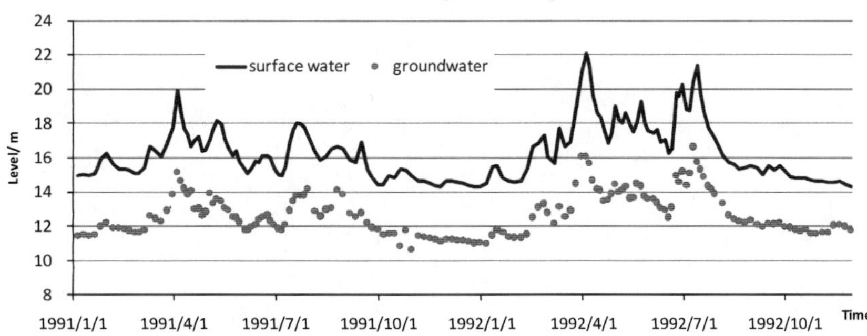


(b) Bayi Bridge

Fig.2 Dynamic relationship between groundwater level and surface water level

## 4. Conclusions

According to the above analysis we can see that the metro project will directly affect the existing situation of Nanchang city groundwater environment because of Nanchang urban construction intensive, numerous rivers, and underground pipeline complex. However, this change is often lagging, if the scope and extent of its impact do not be considered fully, not only does not help the expansion of urban space, but also will cause adverse social and economic consequences. Therefore, the Nanchang

IOP Conf. Series: Earth and Environmental Science **61** (2017) 012112 doi:10.1088/1755-1315/61/1/012112

metro project must take full account of the regional engineering geological and hydrogeological characteristics, scientific, reasonable, carefully carried out.

## Acknowledegments

The work reported in this paper was performed under the Science and Technology Department of Jiangxi Province, project number 20151BBG70008. The authors thank also the experts of Jiangxi Province Survey and Design Institute for their constructive comments, which undoubtedly contributed to the improvement of this manuscript.

## References

[1] M. Bata, “Effect on Buildings of Vibrations Caused by Traffic,” Building Science, 1985,99(1), pp. 1-2.
[2] G R. Watts, “Case Studies of the Effects of Traffic Induced Vibrations on Heritage Buildings,” Berkshire: Transport and Road Research Laboratory, 1987.
[3] Kim Y. Y. and Lee K. K, “Disturbance of groundwater table by metro construction in the Seoul area, Korea,” Geosciences Journal, 2003, 7 (1) pp.37-46.
[4] Jiao J. J., Leung C. M. and Ding G. P., “Changes to the groundwater system, from 1888 to present, in a highly-urbanized coastal area in Hong Kong, China,” Hydrogeology Journal (2008) 16, pp.1527–1539.

IOP Conf. Series: Earth and Environmental Science **61** (2017) 012113 doi:10.1088/1755-1315/61/1/012113

# Analysis of coupling between high-speed railway and common speed railway system in transportation corridor

**Hongchang Zhou[1*], Haijun Li[1], Xiaohong Chen[2] and Changfeng Zhu[1]**

[1] Lanzhou Jiao-Tong University, School of Traffic and Transportation, Gansu province 730070, China.
[2]Xi'an Railway Bureau Xi'an west station, Xi'an 710082, China.
*Corresponding author: 972349233@qq.com

**Abstract.** The high-speed railway and common speed railway subsystems as important components of the railway transportation system, can make railway traffic organization more orderly, when there are a rational division and balance development between them. In order to quantitatively evaluate the coordinate relations between high-speed railway subsystem and common speed railway subsystem, this paper takes the railway transportation corridor from Baoji to Lanzhou as an example. Firstly, using Logit model and grey forecasting model predict the passenger volume, passenger turnover and time value of high-speed railway and common speed railway in the Baoji-Lanzhou corridor. And then, the coupling forecast model of these two subsystems is established. Lastly, the coupling and coupling coordination of these two subsystems using are predicted and analyzed at theatrically level.

## 1. Introduction

At present, predecessors have done a lot of researches which are related the transport internal system and other systems' coupling relationship. ZAMBELLI[1] and CREPIN.A.S[2] introduced the coupling theory to study the coordination degree between subsystems. Shuai Bin et al[3] divided China regional economy into three types through studying coupling relationship between regional economy and the railway development. Zhu Lei[4] by combining coupling degree and coordination degree, building coupled coordination degree model analyzed the coupling and coordination relationship among various traffic modes of the corridor. Li Xin et al[5] used the coupling function to analyze the relationship between the China railway transport situation and opencast coal mine production, but the paper used analytic hierarchy process in the coupled model for weight calibration, which was influenced by subjective factors greatly, and might reduce the accuracy of assessment. Zhang Yaodong [6] took the urbanization and transport systems as the two main indexes, and built the coupling degree model to study traffic modes in Lanzhou and the coupling mechanism of urban development. Yang Li et al[7] took the Xinjiang as an example and built the coupling degree model of the Xinjiang transportation infrastructure and economic to analyze their coupling mechanism, but the paper did not divide the index weights.

Based on the previous experience, this paper constructed a coupling evaluation model of high speed railway and the common speed railway in Baoji-Lanzhou corridor, and does make a brief forecast and analysis on the coupling degree and coupling coordination of two traffic modes in the next ten years. The result has guide significance for planning the two systems' rail transport corridors development.

## 2. Coupling evaluation model

According to characteristics of Baoji-Lanzhou transport corridor, firstly, the coupling index systems of high-speed railway and common speed railway are established in this paper. As shown in Figure 1.

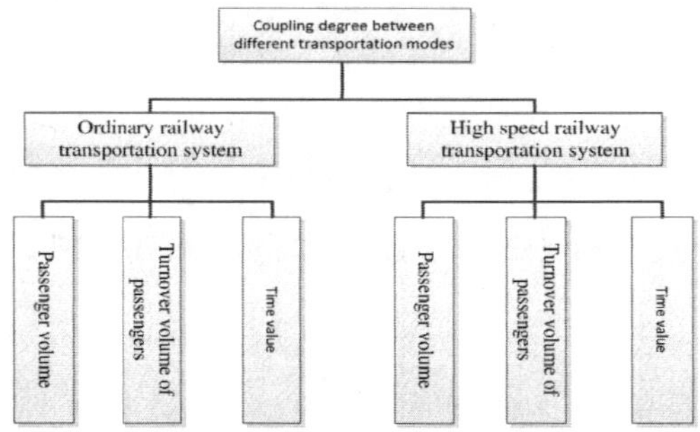


**Figure 1.** Index system of coupling in different transportation systems

The various indicators in subsystems have different dimensions, because they represent different physical meanings. Based on this, these indicators must be dimensionless before the coupling model is established. The linear power function is used in this paper, as shown in equation (1) .

$$u_i = \begin{cases} \dfrac{X - X_{\min}}{X_{\max} - X_{\min}} \\ \dfrac{X_{\max} - X}{X_{\max} - X_{\min}} \end{cases} \quad (i = 1,2,\ldots\ldots n) \tag{1}$$

Where, $U_i$ is the effective contribution of x to the system; $X_{\min}$ is the top value of order parameters; $X_{\max}$ is the lower limit of the order parameters.

In a transport corridor, the different transport modes are located in different and interacting subsystems. This paper uses the linear weighting method to calculate order contribution from subsystem to total system. A specific formula is shown as formula (2).

$$U_i = \left(\prod_{i=1}^{n} u_i\right)^{\frac{1}{n}} = \sum_{i=1}^{n} \lambda_i u_i \tag{2}$$

Where $U_i$ is the total order parameter; $\lambda_i$ represents the weights of the each order parameter.

This paper builds calculation formulas of coupling degree, as shown in equation (3).

$$C = \left[\frac{n(u_1u_2 + u_1u_3 + \ldots\ldots + u_{n-1}u_n)}{(u_1 + u_2 + \ldots\ldots + u_n)}\right]^{\frac{1}{n}} \tag{3}$$

Where, the coupling degree of specific criteria is shown in table 1.

**Table 1.** Coupling grade

| Coupling degree | [0,0.3] | (0.3,0.5] | (0.5,0.8] | (0.8,1] |
|---|---|---|---|---|
| Coupling degree grade | Weak | secondary | preferably | high quality |

Entropy weight method could analyze the coordination of indicators in objective and comprehensive way which is not affected by the influence of subjective factors. This paper uses entropy-right method to determine the index weights [8]. The steps of entropy-right method are shown in the following .

The index of j of entropy $e_j$ :

$$e_j = -k\sum_{i=1}^{n}\frac{u_{ij}}{\sum_{i=1}^{n}u_{ij}} In \frac{u_{ij}}{\sum_{i=1}^{n}u_{ij}} \tag{4}$$

The entropy weight of the $j$ index:

$$w_j = \frac{1-e_j}{\sum_{j=1}^{m}1-e_j} \tag{5}$$

This paper builds coupling and coordinating functions among subsystems to fully reflect the coupling strength among subsystems, and the relationships between the efficacy and collaborative effects.

$$D = (C \times T)^{\frac{1}{2}} \tag{6}$$

$$T = k_1u_1 + k_2u_2 + \ldots\ldots + k_nu_n \tag{7}$$

Where, $D$ represents coupling and coordinating degree; $T$ represents the coordinating index among subsystems.

**Table 2.** Coupling coordination degree

| D | [0,0.3] | (0.3,0.5] | (0.5,0.7] | (0.7,1.0] |
|---|---|---|---|---|
| Degree | Low coordination | Basic coordination | Moderate coordination | Highly coordination |

## 3. Parameter calibration

The time value of common speed railway refers to that the passenger use the extra travel time that the common speed railway mode costs than the high speed railway creating the production value, and it is a negative indicator. On the contrary, high-speed rail will create a positive time value due to the travel time saving. The calculation formula is shown in the formula 8.

$$R_{(t)} = Q_{(t)} \times R \times W_{(O,D)} \times P_{(t)} \tag{8}$$

where, $R_{(t)}$ is the value of travel time; $Q_{(t)}$ is the number of passengers change transport modes; $R$ is the effective utilization coefficient of passenger travel; $W_{O,D}$ is the saved time that passengers change transportation from O to D; $P_{(t)}$ is the unit of time value.

Logit model is a linear equation which considered the major factors as a characteristic varible when people choose travel modes[9]. This paper only studies the relationship between high speed railway and common speed railway in Baoji-Lanzhou corridor. $U_i$ and $U_j$ represent utility functions of high speed railway travel mode and common speed railway travel mode, representively. Therefore, passengers flow share rate in Baoji-Lanzhou corridor can be expressed as:

$$P_{(j)} = expU_j / \left[expU_i + expU_j\right] \tag{9}$$

The main factors affecting passengers travel modes choice in Baoji-Lanzhou transport corridor are safety, quickness, ticket price and convenience. By referencing relevant information, the safety index of common speed railway and high speed railway are (0.95, 1), respectively. The ticket prices of the main cities in Baoji-Lanzhou corridor are set by the market survey. Quickness and convenience index describe the total travel time and waiting time targets from the starting point to the target point. The calculation formula is as shown in the formula 10 and 11.

$$K_i = T_i / P_{(t)} \quad (10)$$

$$F_{(i)} = T_j / P_{(t)} \quad (11)$$

As a result, the utility function of passengers transport mode choice is shown in the formula (12).

$$U_i = (\theta_1 P_i + \theta_2 K_i + \theta_3 F_i) A_i \quad (12)$$

Where, $K_i$ and $F_i$ refer to the utility value of the quickness and convenience index, respectively; $P_i$ refers to the utility value of the ticket price; $A_i$ refers to the utility value of safety factors; $T_i$ refers to the whole time needed to complete the trip, $T_j$ refers to a time sum of access time and waiting time by transport mode j .

Based on the above analysis, the share rate model are solved by using Trans CAD software, the passengers share rates between high speed railway and common speed railway of major cities in Baoji-Lanzhou corridor in 2012 to 2015, can be obtained , and specific values are provided in table 3.

**Table 3.** Share ratio of high-speed railway and ordinary railway

| Section | 2012 | | 2013 | | 2014 | | 2015 | |
|---|---|---|---|---|---|---|---|---|
| | Ordinary | high-speed | Ordinary | high-speed | Ordinary | high-speed | Ordinary | high-speed |
| Baoji--Tianshui | 0.79 | 0.21 | 0.75 | 0.25 | 0.73 | 0.27 | 0.69 | 0.31 |
| Baoji--Dingxi | 0.75 | 0.25 | 0.70 | 0.30 | 0.68 | 0.32 | 0.65 | 0.35 |
| Baoji--Lanzhou | 0.61 | 0.39 | 0.58 | 0.42 | 0.57 | 0.43 | 0.55 | 0.45 |
| Tianshui--Dingxi | 0.40 | 0.60 | 0.42 | 0.58 | 0.44 | 0.56 | 0.45 | 0.55 |
| Tianshui--Lanzhou | 0.28 | 0.72 | 0.30 | 0.71 | 0.31 | 0.69 | 0.32 | 0.68 |
| Dingxi--Lanzhou | 0.65 | 0.35 | 0.68 | 0.32 | 0.71 | 0.29 | 0.73 | 0.27 |

Based on the travel characteristics survey in Baoji-Lanzhou corridor, using Logit model calculates the sharing rate of common speed railway and high-speed railway. The passengers flow of high speed rail can be simulate by the data of the ordinary railway from 2012 to 2015. Grey system theory using the raw data directly added and moving average-weighted sum method to make the resulting sequences show certain regularity, and using typical curve approximate its corresponding curve to predict the systems.

## 4. Data collection and processing

By collecting the main common speed railway passengers flow of Baoji-Lanzhou corridor from 2012 to 2015, using share rate model can simulate the high-speed rail passengers flow in Baoji-Lanzhou corridor. The specific values are provided in table 4.

**Table 4.** Passenger of High speed rail from 2012 to 2015

| Section | 2012 | | 2013 | | 2014 | | 2015 | |
|---|---|---|---|---|---|---|---|---|
| | Ordinary | high-speed | Ordinary | high-speed | Ordinary | high-speed | Ordinary | high-speed |
| Baoji--Tianshui | 43.68 | 11.61 | 42.9 | 14.3 | 43.6 | 16.13 | 44.1 | 19.81 |
| Baoji--Dingxi | 41.37 | 13.79 | 40.1 | 17.19 | 40.7 | 19.18 | 41.7 | 22.45 |
| Baoji--Lanzhou | 33.68 | 21.53 | 33.3 | 24.11 | 34.2 | 25.80 | 35.2 | 28.80 |

| | | | | | | | | |
|---|---|---|---|---|---|---|---|---|
| Tianshui--Dingxi | 22.13 | 33.20 | 24.0 | 33.14 | 26.3 | 33.47 | 28.9 | 35.32 |
| Tianshui--Lanzhou | 15.39 | 39.57 | 17.1 | 40.47 | 18.5 | 41.18 | 20.4 | 43.35 |
| Dingxi--Lanzhou | 35.98 | 19.37 | 38.9 | 18.31 | 42.4 | 17.32 | 46.7 | 17.27 |

Based on these four years passengers flow data, grey prediction model is built to predict the common speed railway and high-speed rail passengers flow and Passenger transport turnover of Baoji-Lanzhou corridor in 2017 to 2026, as shown in table 5 and table 6.

**Table 5.** Passenger forecasts from 2017 to 2026 between Baoji-Lanzhou (ten thousand people)

| Years | 2017 | 2018 | 2019 | 2020 | 2021 | 2022 | 2023 | 2024 | 2025 | 2026 |
|---|---|---|---|---|---|---|---|---|---|---|
| High speed railway passenger flow | 195 | 209 | 225 | 241 | 259 | 278 | 298 | 320 | 344 | 369 |
| Ordinary railway passenger flow | 238 | 250 | 263 | 276 | 290 | 304 | 320 | 336 | 353 | 370 |

**Table 6.** The forecasts of passenger transport volume (ten thousand people/ kilometer)

| Years | 2017 | 2018 | 2019 | 2020 | 2021 | 2022 | 2023 | 2024 | 2025 | 2026 |
|---|---|---|---|---|---|---|---|---|---|---|
| Ordinary railway | 65196 | 68081 | 71094 | 74240 | 77525 | 80956 | 84538 | 99279 | 92185 | 96265 |
| high-speed railway | 61111 | 66110 | 71515 | 77363 | 83689 | 90533 | 97935 | 105945 | 114607 | 123979 |

Through the yearbook data of Gansu and Shanxi provinces' from 2012 to 2015, it could get the GDP of the main cities' and the number of resident population. From equation 8 and 9, and combing grey prediction model, the travel time value of common speed railway and high speed railway in 2017 to 2026 can be predicted, and Specific values are provided in table 7.

**Table 7.** The travel time value in the corridor of Baoji-Lanzhou (ten thousand yuan)

| Years | 2017 | 2018 | 2019 | 2020 | 2021 | 2022 | 2023 | 2024 | 2025 | 2026 |
|---|---|---|---|---|---|---|---|---|---|---|
| Ordinary railway | 26429 | 30367 | 34892 | 40092 | 46066 | 52930 | 60817 | 69879 | 80292 | 92256 |
| High-speed | 22128 | 25370 | 29086 | 33347 | 38232 | 43832 | 50253 | 57615 | 66054 | 75730 |

Dimensionless the index is the main way to resolve the problem that different indicators represent different meanings in coupling system. Using equation (1) quantify the indicators in the railway system. In order to clear the effect degree of each indictor in high speed railway and common speed railway system on the whole railway transport system development, each indictors are given the weights, and they are calculated using the formula (4) and (5), and quantitative results are shown in table 8.

**Table 8.** The weight of order parameter in transport system

| Index | Ordinary railway transportation system | | | High speed railway transportation system | | |
|---|---|---|---|---|---|---|
| | Passenger volume | Turnover volume | Time value | Passenger volume | Turnover volume | Time value |
| $e_j$ | 0.96537 | 0.96772 | 0.91377 | 0.94524 | 0.94691 | 0.87825 |
| $w_j$ | 0.22613 | 0.21078 | 0.56307 | 0.23850 | 0.23123 | 0.53027 |

In order to clearly understand the stability of Baoji-Lanzhou railway system, based on the formula (3) and (6) in coupling degree model, it can get the coupling degree and the coupling coordination degree between common speed railway and high-speed railway from 2017 to 2026, specific values are provided in table 9.

**Table 9.** The coupling of high-speed railway and ordinary railway between 2017 to 2026

| Years | 2017 | 2018 | 2019 | 2020 | 2021 | 2022 | 2023 | 2024 | 2025 | 2026 |
|---|---|---|---|---|---|---|---|---|---|---|
| Coupling | 0.628 | 0.659 | 0.681 | 0.695 | 0.704 | 0.707 | 0.704 | 0.694 | 0.677 | 0.651 |
| Benignity | 0.500 | 0.530 | 0.558 | 0.583 | 0.606 | 0.626 | 0.647 | 0.663 | 0.676 | 0.684 |

This paper makes the coupling and coordination degree line of high speed railway and common speed railway in Baoji-Lanzhou corridor, as shown in Figure 2.

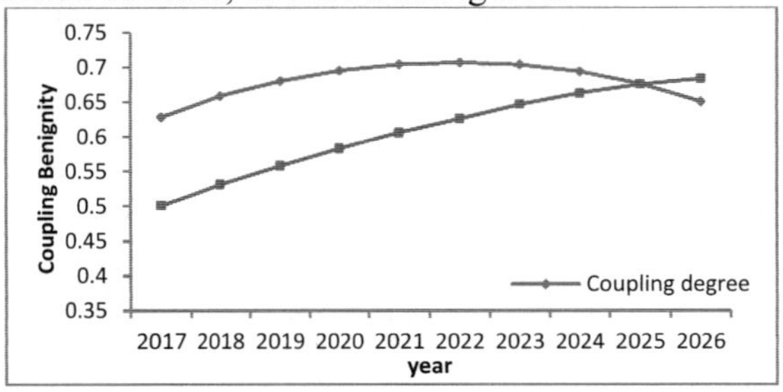


**Figure 2.** Benignity of coupling between high-speed railway and ordinary railway in the corridor of Baoji-Lanzhou

Combining table 1, table 2 and Figure 2, it can be seen that Baoji-Lanzhou high speed railway had better coupling and medium coupling coordination with common speed railway in the early decade, and the coupling degree has a growing trend from 2017 to 2023, this is because the completion of the high-speed rail will induced some passengers flow so as to increase the passenger traffic volume. However, after the years of 2024, with the region's economic growth, passengers will have a deeply recognition of the time value concept, and the high speed railway passengers flow volume will increase dramatically ,and coupling degree of the high-speed rail and common speed railway appeares to be a certain degree of decline. The overall coordination degree will being a high level, and their interaction effects promote railway system to develop in a more healthy and orderly way.

Based on existing references, this paper sets up the coupled coordination degrees forecast model of Baoji-Lanzhou corridor by combining the logit model and Grey forecasting model. Secondly, based on the existing traffic data, it simulates future ten years data, and it analyzes and predicts coupled coordination relationship between two subsystems within 10 years after the Baoji-Lanzhou high speed railway is finished. As can be seen from above, after the Baoji-Lanzhou high speed railway is completed, it will share passengers transportation tasks with common speed railway.

## Acknowledgments

This work is supported by The Research Planning Funds of Humanities and social sciences for the Ministry of Education(14YJA790023，15XJAZH002); National Natural Science Foundation of China (61364028).

## References

[1] ZAMBELLI S 2011. Flexible accelerator economic systems as coupled oscillators. *Journal of Economic Surveys*, pp 608-633.
[2] CREPIN A S, NORBERG J and MALER K 2011. Coupled economic-ecological systems with slow and fast dynamics-modelling and analysis method. *Ecological Economics*, pp 1448-1458.
[3] Shuai B, Zhong S l and LI J 2013. Region division in China for railway transportation development. *Journal of railway science and engineering*, 10(01), pp 108-111.
[4] Zhu L 2014. Integrated transport channel coupling coordination research on the mode of

transportation. *Chang-an university*.
[5] Li X, Shang T and Zhou W 2012. Evaluation of coordination between railway transportation and Open-Pit mining based on efficacy and coupling. *Journal of southwest Jiaotong university*, 47(3), pp 490-494.
[6] Zhang Y D 2013. Coupled coordination evaluation of urbanization and systems in Lanzhou city. *Lanzhou university*.
[7] Yang L, Yang D G, Zhang Y F et al 2009. Grey associative analysis of infrastructure system and economic development coupling in Xinjiang . *Progress in Geography*, 28(3), pp 345-352.
[8] Sun X H 2009. Research on the coordination development between highway traffic and regional economy.*Chang-sha: Institutes of technology of Chang-sha*.
[9] Xie R H, Qiu Z Q, Li Q Y and Wang R H 2006. Application of logit model in estimating the distribution rate of passenger flows on Guangzhou-Shenzhen railway. *China railway science*, 27(05), pp111-115.

# The application of MATLAB-based analytic hierarchy process in Hainan residential quarter function factor evaluation

Dongdong Qiu[1, a)], Peiqi Liu [1]

[1]College of Civil Engineering and Architecture, Wenzhou University, Wenzhou 325035, China.

[a)]Corresponding author: ddqiu@126.com

**Abstract.** Since being designated as an international tourist island, Hainan has become an overwhelmingly favored choice of real estate investment. This paper first constructed Hainan residential quarter function factor index system, then evaluated relevant factors, and finally solved the problem of factor importance ranking. In this specific case, the software MATLAB was used to facilitate AHP calculation. The evaluation results have guiding and referential value to both real estate developers and residential consumers.

## 1. Introduction

As an integral part of modern people's life, house is characterized as an investment with high values, high returns as well as high risks. In recent years, the real estate development boom in Hainan has overwhelmed. Hainan has changed a lot since it was rated as an international tourist island. As a base for leisure and health， Hainan market attracts developers and capital investment. However, now that the real estate development is highly-invested, highly-risky and highly-technological, the failure of the decision-making is sure to cause a huge waste of national and enterprise resources and even social wealth. At present, the function-cost analysis is based on the experiences of researchers which are susceptible to such influencing factors as personal knowledge and judgment. Therefore, it is necessary that the advanced decisive ideas to be introduced, and the residential quarter evalu ation system be established for a better function-cost factor analysis, a more accurate decision and a reduction of property market risks.

## 2. Function factor index system of Hainan residential quarters

Hainan province, located at the north edge of the tropics, is an island with mountains and plateaus surrounded by sea. With a monsoon climate, its average temperature is 22-26 °C and is vulnerable to typhoons. As to a function factor index system, the residential quarter is expected to meet not only the living but also the social and environmental needs. Specific to Hainan, the index system is briefly classified into six parts, of which the living condition is the most basic index while the living taste functions as the index for living improvement. Currently, residents have a stronger awareness of privacy and emphasize the communication with others, so it is necessary to consider the facilities within the community. Hainan is a tourism-oriented province with a good ecological environment where people have naturally higher demands for residential greening, so the environment should be included in the evaluation system as well. Furthermore, a large proportion of urban residents are the elderly and the young who have relatively higher requirements for the infrastructure and life services. Considering all the above, we built Hainan residential quarter function factor index system as shown in Table 1.

**TABLE 1.** Function factor index system for residential quarters in Hainan.

| First level indexes | Second level indexes | First level indexes | Second level indexes |
|---|---|---|---|
| Living conditions | Building quality | Living taste | Indoor decoration |
| | Type design | | Public place decoration |
| | Equipment and facilities | | Humanistic environment |
| Residential environment | Greening | Living services | Shopping conditions |
| | Vegetation growth | | Education and medical service |
| | Landscape | | Fitness and Recreation |
| | Square | | Property Management |
| Infrastructure | Water supply and drainage | Facilities | Security |
| | Power supply | | Management and Monitoring |
| | Gas supply | | Information Network |
| | Road | | |
| | Parking space | | |

## 3. The application of AHP and MATLAB

*3.1 Introduction of AHP*

The Analytic Hierarchy Process (AHP) is a systematic analytic method proposed in the 1970s by Professor A. L. Saaty of the University of Pittsburgh, U.S.A. As a qualitative and quantitative multi-objective technology for decision analysis, it simulates human decision-making and helps solve complex problems. The basic principle is to break the complex problem into several levels, the decision-makers or experts give scores through pair-wise comparison of the respective importance of each level, to construct the judgment matrix with the scores gained, to figure out the feature vectors to ensure the next-level index's contribution, and to finally achieve the importance ranking results. The general steps are as follow:

(1)Establish the hierarchical structure of the model.

(2)Construct the judgment matrix. On the basis of questionnaires, the scaling method of 1-9 and their multiplicative inverses is used to give pair-wise scores for the identification of ranking differences. The specific meaning of the scale is shown in Table 2.

**TABLE 2.** The meaning of the 1-9 scale.

| Scale $a_{ij}$ | Meaning |
|---|---|
| 1 | Factor i and factors j are equally important |
| 3 | Factor i is slightly important than the j factor |
| 5 | Factor i is more important than factor j |
| 7 | Factor i is much more important than factor j |
| 9 | Factor i is absolutely important than factor j |
| 2、4、6、8 | In-between states |
| Multiplicative inverses | If factor j is compared with factor i, the result is $1/a_{ij}$ |

(3) Rank the relative weights and levels of each function index, and have the consistency test. Firstly, the matrix is regulated and the factors in every vertical line are added up. Secondly, regulate again to get the feature vector, calculate the maximum feature root E to get the value of the consistency test CI=(E-n)/(n-1). Finally, have the consistency test with the CR value. If CR=CI/RI$<$ 0.1, the matrix consistency is considered acceptable. If not, the matrix is supposed to be corrected.

(4) Sort an importance order for the general goal factors. Sum up all the multiplication results of the general goal factor weight vectors and respective sub-factor weight vectors. And then, sum up the multiplication results of the general goal factor weight vectors and the random consistency index. The consistency index RI is obtained by the division of the former summation by the latter one. When the RI$\leq$0.1, the ranking consistency is satisfactory.

### *3.2 MATLAB for AHP computing*

MATLAB (Matrix Laboratory) has been selected as the analytical tools in this study in that the software offers a good working platform, a pleasant programming environment, and an easy-to-use programming language. Specifically, the latest version of MATLAB R2011b is used.

There are three computing steps:

(1)AHP programming. Firstly, go to user interface, fill in the index matrix values of various levels, then a comparative matrix for the function factor index system will be naturally formed; calculate for the weight vectors and the maximum feature values at all levels; and lastly have consistency test and judge the acceptability of the matrix.

(2)When in real practice, the software default window is used. In this case, fill AHP program into the Command window and press "Enter" to run the program.

(3) Fill the function factor index system and feature values in the format of [a b c; d e f] and press "Enter" to get the corresponding weight vectors and the maximum feature values.

## 4. AHP analysis of Hainan residential quarter function factors

### *4.1 Operation*

(1)On the basis of the questionnaires and the newly-built function factor index system, we get the function weight table of Hainan residential quarters and construct judgment matrix. (For space limitation, this process is omitted)

(2)Calculate the weight vectors of every index as well as the maximum feature values, and then give the consistency test. The test results are shown in Table 3.

**TABLE 3.** Hainan residential quarter function weight table.

| First level indexes | Weight vectors | Maximum feature values | Matrix consistency | Second level indexes | Weight vectors | Maximum feature values | Matrix consistency |
|---|---|---|---|---|---|---|---|
| Living conditions | 0.4481 | 6.4681 | acceptable | Building quality | 0.6370 | 3.0385 | acceptable |
| | | | | Type design | 0.2583 | | |
| | | | | Equipment and Facilities | 0.1047 | | |
| Living taste | 0.0549 | | | Indoor decoration | 0.7143 | 3 | acceptable |
| | | | | public place decoration | 0.1429 | | |
| | | | | Humanistic environment | 0.1429 | | |
| Residential environment | 0.2181 | | | Greening | 0.2778 | 4.0458 | acceptable |
| | | | | Vegetation growth | 0.3659 | | |
| | | | | Landscape | 0.2326 | | |
| | | | | Square | 0.1238 | | |
| Living services | 0.1088 | | | Shopping conditions | 0.1759 | 4.0658 | acceptable |
| | | | | Education and medical services | 0.2888 | | |
| | | | | Fitness and Recreation | 0.0591 | | |
| | | | | Property Management | 0.4762 | | |
| Infrastructure | 0.1469 | | | Water supply and Drainage | 0.1829 | 5.2916 | acceptable |
| | | | | Power supply | 0.6037 | | |
| | | | | Gas supply | 0.0439 | | |
| | | | | Road | 0.0847 | | |
| | | | | Parking space | 0.0847 | | |
| Facilities | 0.0233 | | | Security | 0.5591 | 3.0536 | acceptable |
| | | | | Management and Monitoring | 0.0887 | | |
| | | | | Information Network | 0.3522 | | |

(3)Rank the general goal factors according to their respective importance. The results are shown in Table 4.

**TABLE 4.** Hainan residential quarter function ranking.

| | **Living conditions** | **Living taste** | **Residential environment** | **Living services** | **Infrastructure** | **Facilities** | **Weight values** | **Ranking** |
|---|---|---|---|---|---|---|---|---|
| | 0.4481 | 0.0549 | 0.2181 | 0.1088 | 0.1469 | 0.0233 | | |
| **Building quality** | 0.6370 | | | | | | 0.2854 | 1 |
| **Type design** | 0.2583 | | | | | | 0.1157 | 4 |
| **Equipment and facilities** | 0.1047 | | | | | | 0.0469 | 10 |
| **Indoor decoration** | | 0.7143 | | | | | 0.0392 | 11 |
| **Public place decoration** | | 0.1429 | | | | | 0.0157 | 16 |
| **Humanistic environment** | | 0.1429 | | | | | 0.0157 | 17 |
| **Greening** | | | 0.2778 | | | | 0.0606 | 7 |
| **Vegetation growth** | | | 0.3659 | | | | 0.0798 | 6 |
| **Landscape** | | | 0.2326 | | | | 0.0507 | 9 |
| **Square** | | | 0.1238 | | | | 0.0270 | 13 |
| **Shopping conditions** | | | | 0.1759 | | | 0.0191 | 15 |
| **Education and Medical services** | | | | 0.2888 | | | 0.0314 | 12 |
| **Fitness and Recreation** | | | | 0.0591 | | | 0.0064 | 21 |
| **Property Management** | | | | 0.4762 | | | 0.0518 | 8 |
| **Water supply and drainage** | | | | | 0.1829 | | 0.0269 | 14 |
| **Power supply** | | | | | 0.6037 | | 0.0887 | 5 |
| **Gas supply** | | | | | 0.0439 | | 0.0065 | 20 |
| **Road** | | | | | 0.0847 | | 0.1244 | 2 |
| **Parking space** | | | | | 0.0847 | | 0.1244 | 3 |
| **Security** | | | | | | 0.5591 | 0.0130 | 18 |
| **Management and Monitoring** | | | | | | 0.0887 | 0.0021 | 22 |
| **Information network** | | | | | | 0.3522 | 0.0082 | 19 |
| $\sum a_j \cdot CI$ | 0.0086 | 0 | 0.0033 | 0.0024 | 0.0107 | 0.0006 | | |
| $\sum a_j \cdot RI$ | 0.2599 | 0 | 0.1963 | 0.0979 | 0.1645 | 0.0135 | | |

(4)Test the ranking consistency.

$$RI = \frac{0.0086 + 0 + 0.0033 + 0.0024 + 0.0107 + 0.0006}{0.2599 + 0 + 0.1963 + 0.0979 + 0.1645 + 0.0135} = 0.035 \le 0.1$$

It can be safely concluded that the consistency is satisfactory.

*4.2 Result Analysis*

The above research shows that the building quality, road, parking space, type design and power supply are the five factors people most concern in Hainan house consumption, among which the building quality is always the key index. Considering the Hainan's humidity, a good building quality is the guarantee for long service life. With the improvement of living standards in Hainan, automobile has become the common means of transportation, so people have special requirements for roads and parking services. Additionally, interior design and power supply are factors currently concerned, since interior design affects sunlight and ventilation while power supply ensures the convenience of life. By contrast, the fitness and recreation facilities as well as the management and monitoring are minor factors. Real estate developers are suggested reducing investments in these two areas while saving money and energy for more important function factors.

## 5. Conclusions and prospects

Residential building is a risky investment. A scientific evaluation of residential function factors can not only help coordinate the relationship between human, nature and society, but also help reduce the resource waste and bring huge economic returns to real estate developers. This study concerns project management, computer programming and software development, and its research results provide guidance and suggestions to both developers and residents in terms of their respective investments. With the speeding construction of Hainan into an international tourist island, residential preferences consequently change, so the function factor index system is supposed to change as well. Particularly, as the aging population increases in Hainan, the "people-orientation" is more likely to become the core value of local community. It is expected that the relationship between nature and society becomes even closer, and the evaluation index system and methods that better match appear in further studies.

## Acknowledgments

The Project Supported by Department of Education of Zhejiang Province (No. Y201635200).

## References

[1] Zeng HuiQun. The Study of Function Configuration about Residential District by AHP [J]. Science Mosaic, 2007,(8):247-249. (in Chinese)

[2] Wang Yi, Wang Qinming. The Foundation of MATLAB and Its Applications in Economics and Management Science [J].Beijing：Mechanical Industry Press，2009.7. (in Chinese)

[3] Dong Shougui. MATLAB-based Analytic Hierarchy Process [J]. Aero Weaponry, 2003,(4):16-18. (in Chinese)

[4] Guo Dongsuo, Cheng Zhengming, Peng Qian. MATLAB-based Analytic Hierarchy Process and Its Application [J]. Management & Technolosy of SME, 2010,(11):269-270. (in Chinese)

[5] Xuan Jianfeng, Liu Peng. Analytical Hierarchy Process Based on MATLAB for Radar Anti-Jamming Performance Evaluation [J]. Radar Science and Technology, 2007,(4):239-241,245. (in Chinese)

IOP Conf. Series: Earth and Environmental Science **61** (2017) 012115 doi:10.1088/1755-1315/61/1/012115

# Study on friction coefficient of soft soil based on particle flow code

**Xiaohong Lei, Zhongwei Zhang**

School of Urban Cconstruction,Yangtze University Jingzhou,Hubei ,434023,China
34973907@qq.com

**Abstract**.There has no uniform method for determining the micro parameters in particle flow code, and the corresponding formulas obtained by each scholar can only be applied to similar situations. In this paper, the relationship between the micro parameters friction coefficient and macro parameters friction angle is established by using the two axis servo compression as the calibration experiment, and the corresponding formula is fitted to solve the difficulties of determining the PFC micro parameters which provide a reference for determination of the micro parameters of soft soil.

## 1. Research status of PFC micro parameters

PFC is a kind of numerical simulation tool based on discrete element, which is often used to simulate the granular particles. When a geotechnical problem is hard to explain from macro perspective, particle flow code provides micro way for analysis. Micro parameters for input are the following: the contact stiffness between particles as kn, ks; parallel contact stiffness between particles as pb_kn, pb_ks; particle friction coefficient as fric; particle density as ρ. For the problem of how the micro parameters corresponds to the properties of the actual simulated specimen, Zhao has studied the influence of micro parameters on the macro characteristics in parallel bond model[1]. Zhou Guoqing has studied the applicability of damping coefficient by simulating Curved cantilever beam and ball falling collision[2]. Zengyuan has carried out biaxial simulation experiments and discussed the effects of particle shape, friction coefficient and porosity on macroscopic properties[3]. The micro parameters in the simulation are obtained by trial and error method. The trial-and-error method is based on the simulation of the biaxial servo compression as the calibration test and achieved by constantly modifying the micro parameters that may affect the stress-strain curve to make stress-strain curve of the simulated compression test the same as the stress-strain curve of the actual experiment[4,5]. The adjustment of micro parameters has relevant rules. Zhang zhongwei has studied the effect of different micro parameter[6]. In recent years, the researchers have tried to obtain the empirical formula for the relationship between microscopic parameters and macroscopic parameters.Zhi jiewang has established empirical formula for 3D porosity and 2D porosity in practice in a new suggestion for determining porosities in DEM study[7].

How to select the micro parameters to make the simulation sample close to the real sample is always a hot issue in the study of particle flow parameters The method of trial and error is the only way to identify micro parameters, by constantly adjusting the micro parameters to make stress-strain curves in calibration test as close as reality. The existing formulas obtained by each scholar can only be applied to similar situations. In this paper, the relationship between the friction coefficient and friction angle is analyzed, which provide a reference for determination of the micro parameters of soft soil.

## 2. Study on the relationship between friction coefficient and soil friction angle

The micro parameters in the PFC are summarized in Table 1. The micro parameters which affects the friction angle are as follows: contact stiffness between the particles as kn, ks; friction coefficient as fric; parallel bond strength as pb_sn, pb_ss. From qualitative point of view, larger soil friction angle corresponds to larger contact stiffness, Stiffness is mainly used to calculate the displacement and force of particles in PFC and is not the main factor that influences the friction angle. To study the relationship between the friction coefficiency and the friction angle, it is necessary to control the parameters other than friction coefficiency.

**Table 1**.Micro parameters.

| Micro para | kn | ks | pb_kn | pb_ks | fric | pb_sn | pb_ss | ρ |
|---|---|---|---|---|---|---|---|---|
| unit | pa | pa | Pa | pa | 1 | pa | pa | Kg/m^-3 |

The number of necessary simulation can be reduced by Orthogonal analysis, which can also facilitates parameter control.Let: $fric^{a}L^{b}R^{c}kn^{d}ks^{e}\rho^{f}T^{g}$ be dimensionless, where L, R are the length and width of the specimen, and T is the time step of the loading,We get:

$$1^{a}L^{b}L^{c}(MT^{-2}L^{-1})^{d}(MT^{-2}L^{-1})^{e}(ML^{-3})^{f}T^{g}=1 \quad (1)$$

M, L, T corresponding index should be equal, so there are:

$$\begin{cases} d+e+f=0 \\ b+c-d-e-3f=0 \\ -2d-2e+g=0 \end{cases} \quad (2)$$

Three independent solutions are:(1,-1,0,0,0,0), (0,0,-1,1,0,0),(0,2,-1,0,1,2). The three dimensionless quantities are: $LR^{-1}$, $kn^{-1}ks$, $R^{2}kn^{-1}\rho T^{-2}$.

Obtained by Buckingham theorem:

$$\phi = F(fric, LR^{-1}, kn^{-1}ks, R^{2}kn^{-}\rho T^{-2}) \quad (3)$$

For the PFC calibration test, $LR^{-}$, $kn^{-1}ks$, $R^{2}kn^{-1}\rho T^{-2}$ are set value, then we can vary value of fric from 1 to 20 and do 20 set of biaxial servo simulation, each confining pressure is taken 1Mpa, 3Mpa, 5Mpa. Draw Mohr stress circle, calculate common tangent, then we can get corresponding tangent friction angle. Drawing of Mohr stress circle is shown in Figure 1.

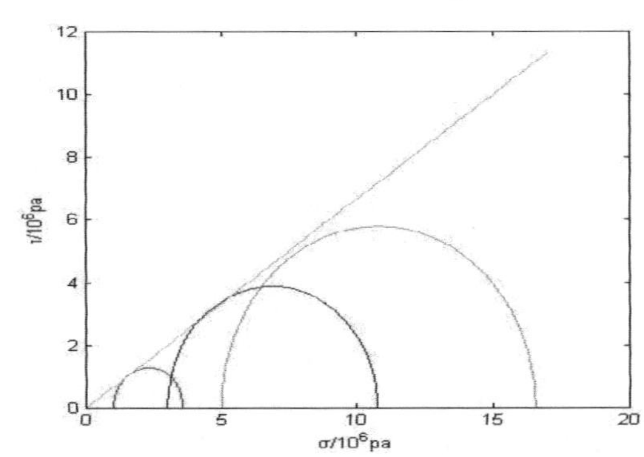


**Figure 1**.Moore stress circle.

In Figure 1 we can see that common tangent pass through the origin, which represent that the simulated soil sample has no cohesive force. The parallel bond in PFC is used to prevent the relative rotation between particles, we set micro parameter relate to parallel bond 0 at beginning for that the soil sample we simulate is silt or sand.

The biaxial servo pressure simulation test is carried out, and the friction angle corresponding to the microscopic parameter fric is obtained as shown in Table 2.

**Table 2**.Fric and corresponding friction angle.

| | | | | | | | | | | |
|---|---|---|---|---|---|---|---|---|---|---|
| fric | 1 | 2 | 3 | 4 | 5 | 6 | 7 | 8 | 9 | 10 |
| fai | 33.66 | 38.448 | 40.74 | 41.49 | 42.06 | 42.15 | 42.43 | 42.92 | 42.68 | 42.108 |
| fric | 11 | 12 | 13 | 14 | 15 | 16 | 17 | 18 | 19 | 20 |
| fai | 42.503 | 42.66 | 42.81 | 42.37 | 43.01 | 42.97 | 42.95 | 42.38 | 42.78 | 43 |

The data points in Table 2 are plotted in the form of scatter plots, as shown in Figure 2.

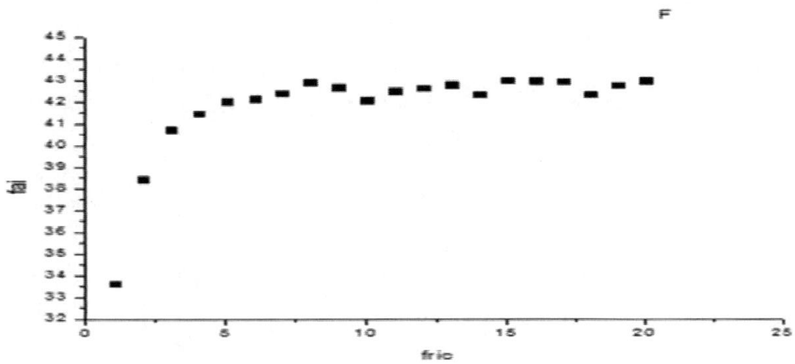


**Figure 2**.Fric and corresponding friction angle scatter diagram.

It can be seen from the scatter diagram that while fric increase from 1 to 20,corresponding friction angle increase, which is consistent with the actual law. When the fric value in the range of [1,8], fric and the friction angle is approximately linear relationship, this trend is in accord with conclusion get by Oda M[9] , In the range of [8,20], relationship is no longer linear so it is not wise to fit all data point with single linear equation. In addition, when the fric is increased to 10, the friction angle is stable at about 42 degrees, and the friction angle of the sand in the actual project is generally in the range of 20 degrees to 40 degrees, most of which is about 30 degrees. It is proved that the selection for the parameters of silty soil is correct. The difference table is used to determine the characteristics of the data distribution. Form analysis of difference table we draw the conclusion that the data can be fitted by the cubic function, In order to make the fitting function smooth at the data points, the three order spline function is used to fit the data. When the value of fric is 0, it is considered that there is no bearing capacity of soil, and a total of 20 data points are fitted to the 19 three order spline function. As shown in formula (4), x represents the microscopic parameter fric, the calculated function value is the corresponding friction angle. The data point calculated by formula and the experimental data points are drawn together in Figure 3, from which we can see that the fitting of the three order spline function is in good agreement.

$$\begin{cases} 28.872 + 3.54x + 1.872x^2 - 0.624x^3 & (1 \le x < 2) \\ 18.888 + 18.516x - 5.616x^2 + 0.624x^3 & (2 \le x < 3) \\ 39.57 - 0.42x + 0.405x^2 - 0.045x^3 & (3 \le x < 4) \\ 33.81 + 3.9x - 0.675x^2 + 0.045x^3 & (4 \le x < 5) \\ 35.91 + 3.6050x - 0.7125x^2 + 0.0475x^3 & (5 \le x < 6) \\ 56.43 - 6.655x + 0.9975x^2 - 0.0475x^3 & (6 \le x < 7) \\ 100.32 - 26.155x + 3.832x^2 - 0.1825x^3 & (7 \le x < 8) \\ -86.56 + 43.92x - 4.927x^2 + 0.1825x^3 & (8 \le x < 9) \\ -126.232 + 57.931x - 6.527x^2 + 0.2418x^3 & (9 \le x < 10) \\ 97.063 - 17.3955x + 1.785x^2 - 0.0595x^3 & (10 \le x < 11) \\ -61.326 + 25.8015x - 2.142x^2 + 0.0595x^3 & (11 \le x < 12) \\ 37.038 + 1.035x - 0.068x^2 + 0.0017x^3 & (12 \le x < 13) \\ -541.15 + 136.18x - 10.53x^2 + 0.27x^3 & (13 \le x < 14) \\ 940.61 - 181.34x + 12.15x^2 - 0.27x^3 & (14 \le x < 15) \\ 26.81 + 3.33x - 0.225x^2 + 0.005x^3 & (15 \le x < 16) \\ 67.77 - 4.35x + 0.255x^2 - 0.005x^3 & (16 \le x < 17) \\ -746.785 + 132.9425x - 7.425x^2 + 0.1375x^3 & (17 \le x < 18) \\ 296.81 - 43.295x + 2.43x^2 - 0.045x^3 & (18 \le x < 19) \\ -320.50 + 54.175x - 2.7x^2 + 0.045x^3 & (19 \le x \le 20) \end{cases} \quad (4)$$

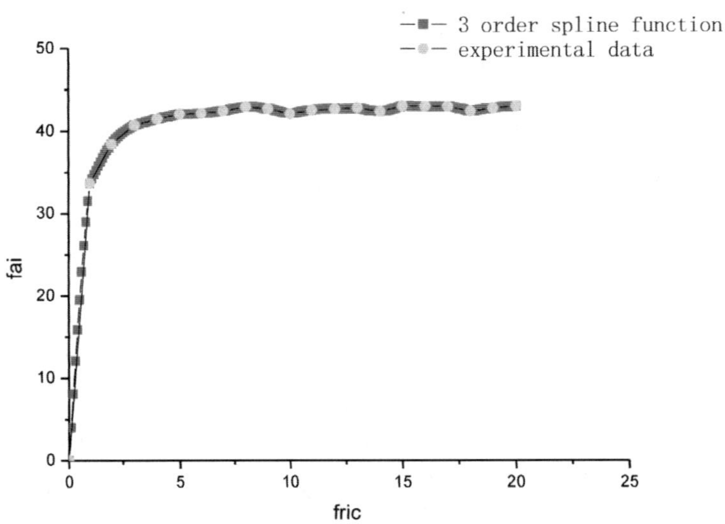


**Figure 3**.Comparison of fitting results.

## 3. Conclusion

In this paper, the method to determine micro parameters is established, and put into use of large area load problem of the foundation. The load limit criterion for simulation of PFC foundation is also provided. The main conclusions are as follow: (1) the relationship between the friction angle and the friction coefficient is not linear relationship.(2)the empirical formula for determining micro parameters is proposed.

**References**

[1] Zhao Guoyan, Dai Bing and Ma Chi 2012 Study on the influence of micro parameters on the macroscopic properties of the parallel bond model. *Chinese Journal of Rock Mechanics and Engineering*.**31** (7) 1-3

[2] Zhou Guoqing, Zhou Jie and Lu Yong 2011 Application of damping parameters in particle flow program (PFC2D) *Journal of China University of Mining and Technology*. **40** (5) 1-4

[3] Zeng yuan and Zhou Jian 2008 Study on the influence of micro parameters on the macroscopic properties of sandy soil *Journal of Underground Space and Engineering*. **4** (3) 1-3

[4] Frank R.Giordanio 2003 A first course in mathematical modeling *tomson learning* **3** 134-235

[5] Zhou Guoqing, Zhou Jie and Lu Yong 2011 Determination of damping parameters in particle flow program *Journal of China University of Mining and Technology*. **40** (5) 2-6

[6] Zhang Zhongwei, Chen Liang 2016 Research on the influence of the simulation of the micro parameters of the double axle compression.PFC2D. *Sichuan Architecture*. **36**(4)1-5

[7] Zhi Jiewang 2014 A new suggestion for determining porosities in DEM study G*eomechanics and Engineering,*.**7**(6) 2-14

[8] Oda M 1972 The mechanics of fabric changes during compressional deformation of sand *Soil and Foundations*.**12**(2)

IOP Conf. Series: Earth and Environmental Science **61** (2017) 012116 doi:10.1088/1755-1315/61/1/012116

# The effect of fly ash on the mechanical properties of cinder lightweight aggregate concrete

**Liguang Xiao[1,2] and Dawei Jiang[1,3]**

[1]School of Materials Science and Engineering, Jilin Jianzhu University, Changchun 130118, China
[2] xlg627@163.com
[3]775008831@qq.com

**Abstract.** The volcanic cinder is a kind of porous rock with good performance and is formed by the cooling of the volcanic eruption. This paper studies the effect of fly ash on the mechanical properties of cinder lightweight aggregate concrete. Results show the mechanical properties of cinder lightweight aggregate concrete is significantly improved with fly ash, and it has a maximum strength value when the dosage is 35%; the early strength of lightweight aggregate concrete is greatly improved, when we extend the grinding time of fly ash, decrease the particle size and increase the activity.

## 1. Introduction

The volcanic cinder is a kind of porous rock which is formed by the cooling of the volcanic eruption, and the porous system is caused by gas expansion. Cinder is a kind of good natural lightweight aggregate. The inner pore for volcanic cinder is like honeycomb. Moreover, the pores are tough, light and pure, mostly used for high quality products. The new wall products will have a broad application prospects which produced by cinder, this product does not only have the advantages of light weight, heat insulation, good sound insulation, simple process and low price, but also meets the demands of building energy saving and reducing structure weight [1-4].

## 2. Materials

(1) Cement: adopt the cement of ordinary Portland cement, and it has the strength grade of 42.5.

(2) The chemical composition of fly ash is shown in Table 1. The microstructure is shown in Fig.1.

**Table 1.**Chemical composition of fly ash (W %)

| The ignition loss | $SiO_2$ | $Al_2O_3$+$TiO_2$ | $Fe_2O_3$ | CaO | MgO |
|---|---|---|---|---|---|
| 2.86 | 62.52 | 22.89 | 4.00 | 0.67 | 1.15 |

(3) The cinder: The volcanic cinder is produced in Jingyu County of Jilin Province, the physical properties are shown in Table 2. The chemical composition is shown in Table 3, and the micro structure is shown in Fig.2.

**Table 2.**The physical properties of the cinder

| Max size | volume weight (kg/m³) | Void fraction (%) | Porosity (%) | Water absorption (%) | Softening coefficient |
|---|---|---|---|---|---|
| 15 | 670 | 52.2 | 49.2 | 13 | 1.0 |

IOP Conf. Series: Earth and Environmental Science **61** (2017) 012116 doi:10.1088/1755-1315/61/1/012116

**Table 3.** The chemical composition of cinder

| $SiO_2$ | $Al_2O_3$ | $Fe_2O_3$ | CaO | MgO | $SO_3$ | $K_2O$ | $Na_2O$ |
|---|---|---|---|---|---|---|---|
| 49.11 | 17.02 | 10.04 | 6.59 | 5.40 | 0.03 | 2.64 | 4.26 |

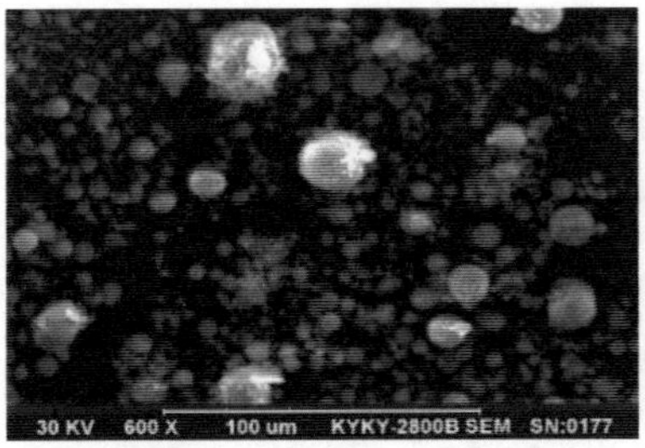

**Figure 1.** SEM images of fly ash

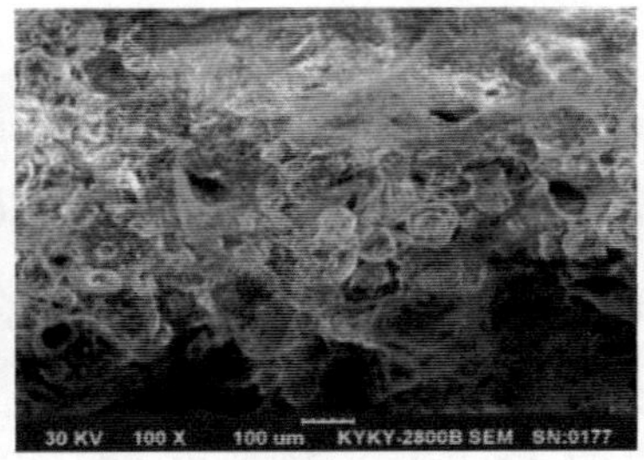

**Figure 2.** SEM images of cinder

**Figure 3.** the SEM images of grinding fly ash by 40 min

## 3. The experimental results and discussion

*3.1. The effect of fly ash replacement on the mechanical properties*

From Fig.5, when we increased the amount of fly ash, it had a maximum strength value when the dosage was 35%; due to the pozzolanic activity of fly ash, and the active ingredient had an Alkali reaction with the $Ca(OH)_2$ which came from hydration of cement, the hydration of calcium, the calcium silicate hydrated calcium aluminate and the hydrated silicon aluminum products were produced in this process, and the bonding ability of the matrix and the structural compactness of the hydration products were improved. Therefore, the strength of this concrete is improved.

*3.2. The fly ash farticle size influence of grinding time change*

**Table 4.** The influence on the particle size of fly ash with grinding time

| Grinding time（min） | Max size (μm) | Average size (μm) | Surface area/volume（$cm^2/cm^3$） |
|---|---|---|---|
| 0 | 58 | 22.42 | 7154 |
| 20 | 42 | 14.15 | 11701 |
| 30 | 44 | 11.77 | 13565 |
| 40 | 36 | 9.72 | 16167 |
| 50 | 44 | 8.69 | 17645 |
| 60 | 24 | 8.16 | 18257 |

The fly ash is composed of vitreous mineral, and its surface is hard and dense, and we can improve the activity of fly ash by grinding, destroy the coating of the vitreous body or crush the particles. This experiment through analyzing different grinding time for particle size of fly ash, the experimental results were shown in Table 4 and Fig.4, and microscopic state was shown in Fig.3.

From Fig.4, the activity of fly ash particles was increased with the grinding time increased and the average particle size decreased. The early strength and the fluidity of cement were improved.

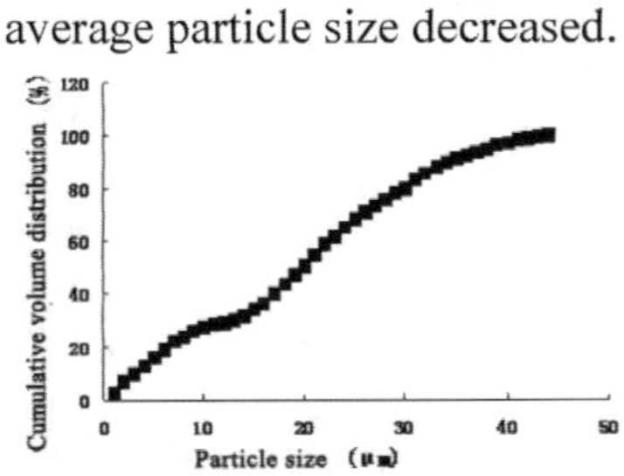

a. Grinding 0min

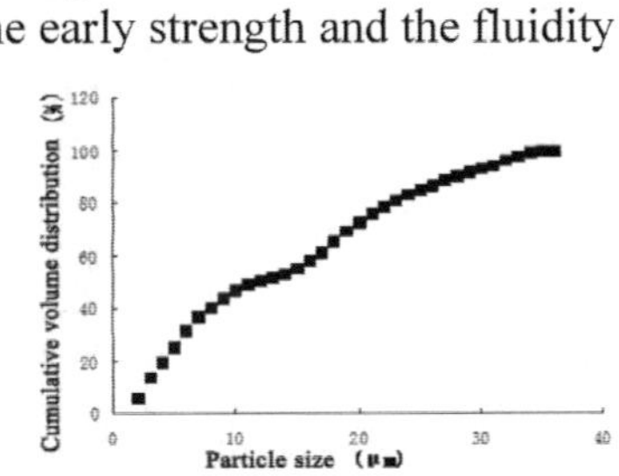

b. Grinding 20min

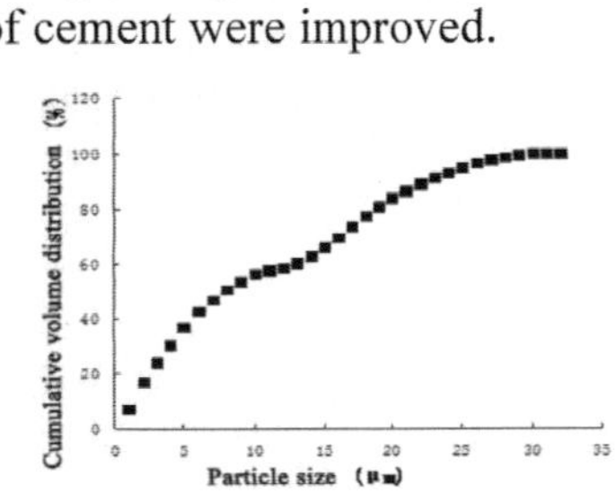

c. Grinding 30min

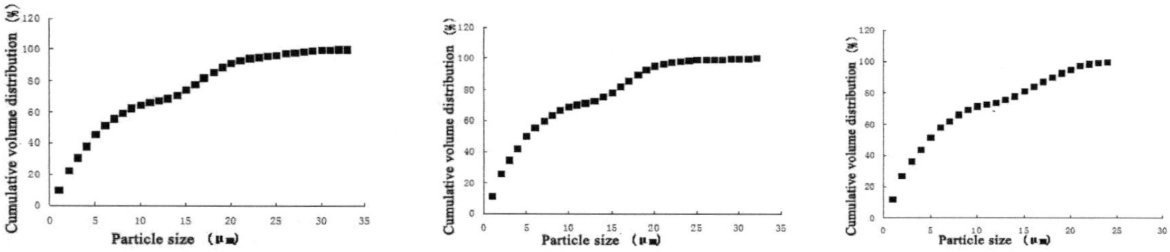


d. Grinding 40min e. Grinding 50min f. Grinding 60min

**Figure 4.**Particle size distribution with different grinding time of fly ash

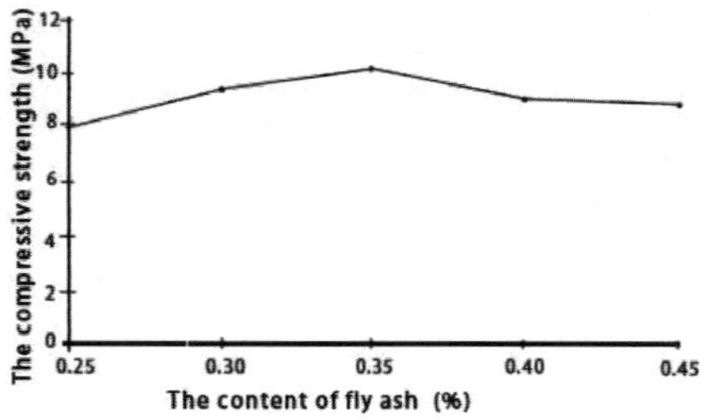


**Figure 5.**The effect of fly ash replacement on the mechanical properties

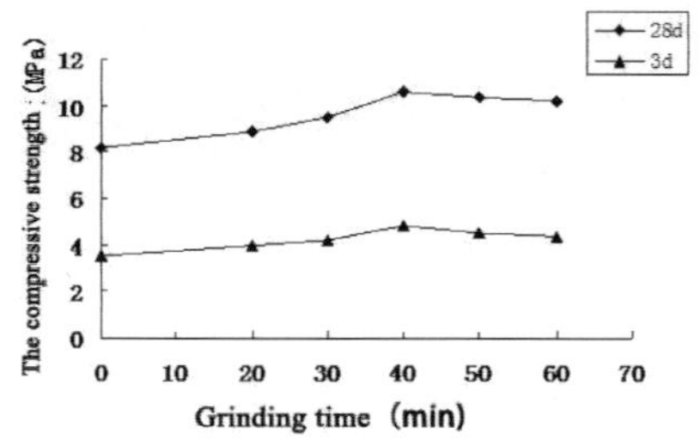


**Figure 6.**The influence of different grinding time of fly ash

*3.3. The influence of different grinding time of fly ash*

The fly ash was grinding with 0min, 30min, 40min, 50min, 60min respectively, and it was added 35%, and we had studied the influence of the different grinding time of fly ash on the mechanical properties of lightweight aggregate concrete, the experimental results was shown in Fig.6, and the interface of cinder aggregate and hydration products scanning electron microscopic diagram was shown in Fig.7.

From Fig.6, the compressive strength of cinder lightweight aggregate concrete was increased with the grinding time increased of fly ash, and the strength reaches the maximum value, when the fly ash was grinding 40min. The main reason: The crystal structure was destroyed after grinding of fly ash particle surface, and the chemical bond was broken in the chemical bond. At this time, the crystal surface was in a metastable state, which is prone to hydration reaction ,and the activity of fly ash was increased; on the other hand, the particle size became smaller after the fly ash was grinding, the specific surface area was increased, the area of the unit mass of fly ash and the hydration reaction was increased, and more hydration products can be generated. It was conducive to the improvement of strength. At the same time, the micro aggregate effect of fly ash was increased, and it makes the structure more compact. And a large number of loose porous carbon particle structure were exposed when we had a too long grinding time, and the water demanded of hydration reaction was increased, and the strength was decreased at the same time. From Fig.7, the hydration products are more abundant in the slag aggregate, and the interfacial transition zone is continuous, which shows that the slag and cement stone matrix can form a solid interface effect. The cinder was fully wrapped by hydration products, and the interface transition zone was continuous, this shows a solid interface could be formed between the cinder and cement stone.

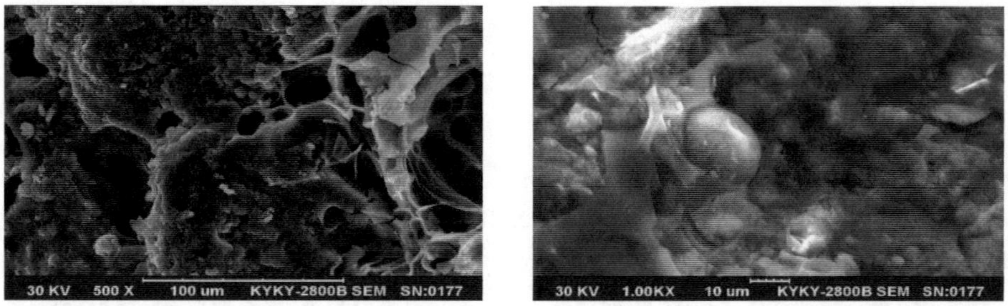


**Figure 7.**The SEM diagram of the interface of cinder aggregate and hydration products

## 4. Conclusion

The fly ash had a great influence on the mechanical properties of cinder lightweight aggregate concrete, and with the fly ash content increased, the compressive strength of cinder lightweight aggregate concrete had been improved obviously, and it had a maximum strength value when the dosage was 35%. The activity was increased and the average particle size was decreased of fly ash with the grinding time increased. All of these had obviously improved the fluidity and the early strength of lightweight aggregate concrete.

## References

[1] Jiangyang Hu, Jun Mao, Hao Zhang, Xuesen Zhe, Cinder as subgrade filling test research, Journal of Shenzhen University Science and Engineering, 2015(11) , pp.625-631.

[2] Jian Chen, Qiang Luo, Mengshi Liu, Duowei Liang, Study on the Seepage Stability of Railway Embankment Filled with Volcanic Cinder Gravels, Railway standard design, 2014(12), pp.14-17.

[3] Xiaodong Zhu, Huisheng Qian, Hongyu Qian, Yuping Wang, The development of the fly ash cinder foam concrete block, Brick & Tile, 2010(5) , pp.43-45.

[4] Bhaskar Desai, Sathyam, Mallikarjunappa, An experimental investigation on some strength properties of light weight blended aggregate concrete, International Journal of Earth Sciences and Engineering, 2014(7), pp.349-355.

IOP Conf. Series: Earth and Environmental Science **61** (2017) 012117 doi:10.1088/1755-1315/61/1/012117

# Experimental study on pore pressure generation mode of saturated remolded loess during dynamic liquefaction

**Jianmin Liu [1,2], TingtingGuo [1,2, a], Guangcheng Zhang [1, 2] and Haiqing Fu[1, 2]**

1Shandong Institute of Earthquake Engineering, Jinan 250021, China
2Earthquake Administration of Shandong Province, Jinan 250014, China
a)Corresponding author: gtt.yy@163.com

**Abstract.**Through dynamic triaxial tests of the saturated remolded loess, the pore water pressure generation characteristics of the samples are analysed during liquefaction processes. Based on dynamic triaxial test results, pore pressure generation characteristics of remolded sample are discussed, the effect of confining pressure on the generation of excess pore water pressure is evaluated, the particularity of pore pressure generation by comparing with the predecessor's research results is pointed out and the normalized pore pressure growth model is given. Finally, fitting curve of pore water pressure growth is drawnby using this model and fitting parameters are given.

## 1. Introduction

Nowadays, our country is still in seismic active period and the loess area is located in strong earthquake activity area [1]. In this area, the seismic disasters such as land slide and subsidence often occur during earthquakes. Under earthquake loads, pore water pressure can increase rapidly and effective stress can decrease, which will lead to soil liquefaction [2]. Therefore, it is important to research pore pressure generation and seismic stability during liquefaction of loess area. Initial liquefaction and cyclic mobility both can cause extensive damage during earthquakes, including sand boils, settlement and tilting of buildings and bridge abutments, collapse of offshore structures, lateral spreading and cracking of slopes, flow failures of earth dams, cracking of pavements, and flotation to the ground surface of buried concrete tanks.

Cyclic loading of saturated granular soils can result in the generation of excess pore water pressure under undrained conditions. In loose soils, the developed excess pore water pressure can be as high as the existing overburdened pressure on soil. This condition corresponds to zero effective stress and is called initial liquefaction of the granular soil [3]. Fitting pore pressure formula is given combining with the influence of dynamic size, dynamic form, vibration frequency and consolidation ratio to saturated loess vibration of pore pressure [4]. Pore pressure growth model of saturated undisturbed loess under dynamic loading is given [5-6].

For pore pressure change regularpattern of saturated undisturbed loess specimen under dynamic loading, many pore pressure growth models have been given, but research result about remolded sample is little. In this paper, loess liquefaction is studied experimentallywith a dynamic triaxial apparatus. Through the vibration liquefaction test of saturated remolded loess, the pore pressure generation characteristics of remolded sample is discussed, the particularity of pore pressure generation by comparing with the predecessor's research, is pointed out, and the normalized pore pressure growth model is given finally.

IOP Conf. Series: Earth and Environmental Science **61** (2017) 012117 doi:10.1088/1755-1315/61/1/012117

## 2. The test instruments and samples preparation

*2.1The Test Instruments*

The experiments were conducted with motor control dynamic triaxia test system manufactured by GDS Co. in the British. This instrument uses digital pressure/volume controller which improve the precision of the pressure to 0.1 kPa and the precision of the volume change to 1 cubic millimeter. All experiences can through the computer GDSLAB software for servo-type control and all the parameters are directly recorded.

*2.2Samples Preparation*

The soil of test are taken from XiAn area, brown yellow and soil evenly, The property indexsare shown in table 1 and granule analysis curves areshown in figure 1：

**TABLE 1**. Major physical parameters of silt

| Specific Gravity/Gs | Natural Moisture Content/% | Wet Density/ g/cm3 | Dry Density/ g/cm3 | Liquid Limit/% | Plastic Limit/% | Plastic Index/Ip |
|---|---|---|---|---|---|---|
| 2.71 | 24.5 | 1.87 | 1.50 | 28.0 | 17.8 | 10.2 |

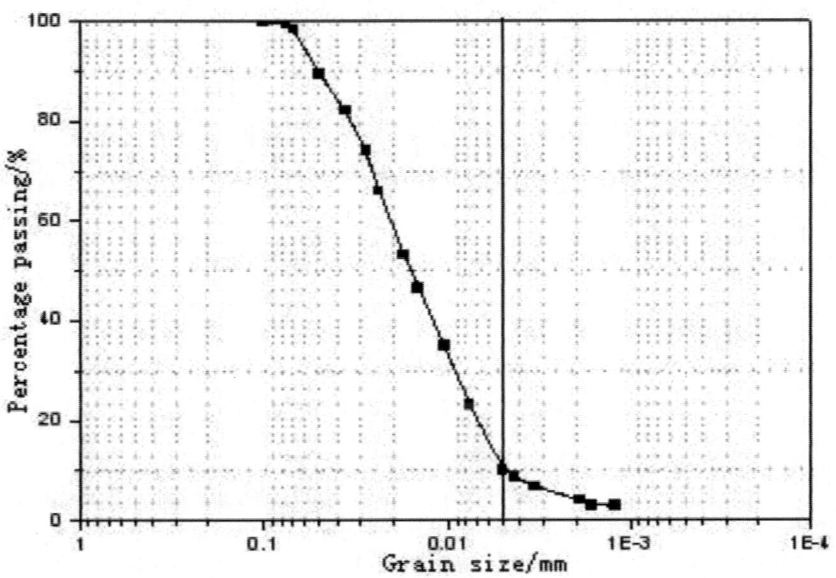


**FIGURE1.** Grain size distribution of tested soil.

The grain size distribution is given in FIGURE1. About 10% by weight of particles of this clay is smaller than 0.005mm, but it almost does not contain any sand. Different researchers around the world have proposed throughout the years a great number of preparation methods of sand samples for laboratory testing. For example, the water sedimentation method is one of the most popular methods for sample preparation (Kuerbis and Vaid, 1988). Wet installed molding, it is used for easy sedimentation of sand soil, silt cohesionless soil, etc； Compaction of the sample is performed by use of a thin rod, it is used for all kinds of clay soil.

During specimen preparation process, first make the soil airing dry and next shake them through the 2 mm standard soil sieve, large particles which cannot through the sieve should be ground scattered by mallets. The clay mineral composition of loess are mainly composed of montmorillonite, and bentonite predominantly montmorillonite clay minerals, so different clay content of the sample by adding bentonite directly to the loess.

The samples used in the experiment are all prepared by dry pack compaction methods which are determined by averagedry density of soil sample in situ test. Samplessize are50 * 100 mm and samples are compacted for five layers; Samples are prepared by special compaction instrument and each layer needs contact well to avoid statified phenomenon.

## 3. The test method

After the completion of the samples preparation, put them in a cylinder for vacuum suction to attain saturation; Then make the samples saturated by back pressure on dynamic triaxial apparatus, and when pore water pressure coefficient is equal or greater than 0.98, drainage consolidation of samples can be done. After the completion of the consolidation,close drain valve and use different dynamic load in test. In FIGURE2, The test can recordthe dynamic loading, dynamic pore pressure and dynamic strain test data.

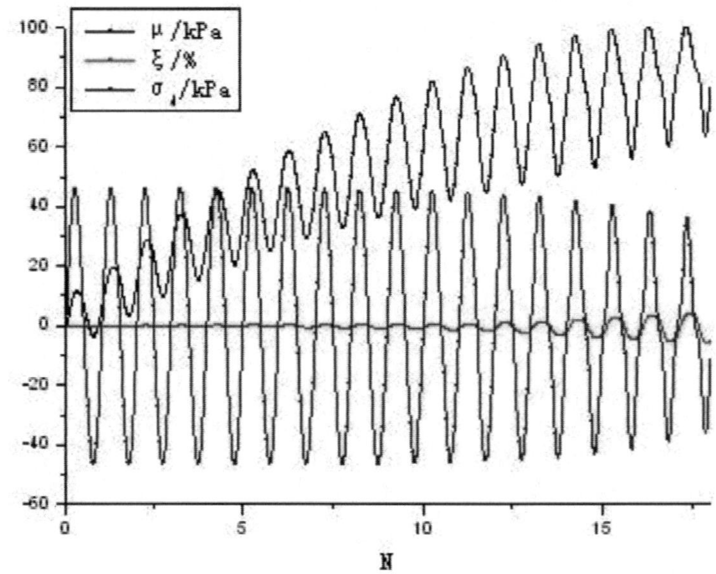


**FIGURE2.** Test record curve under dynamic loading

## 4. Analysis of test results

Through the analysis of pore pressure growth pattern by Jianmin Zhang, pore pressure growth curve is divided into A, B and C three categories (FIGURE3). It is used to simulate the seismic load which producethe pore pressure curve in the front, central and rear of seismic time history. Among them, type B curve area can better simulate hole pressure growth process of sand and loess soilpore water pressure growth curve is more close to type A curve area.

$$U_d = U_f(1 - e^{-\beta \frac{N}{N_f}}) \quad (1)$$

$$U_d = \frac{2}{\pi} U_f \sin^{-1}(\frac{N}{N_f})^{\frac{1}{2\alpha}} \quad (2)$$

$$U_d = U_f[\frac{1}{2}(1 - \cos\pi \frac{N}{N_f})]^b \quad (3)$$

(1) Type A pore water pressure growth curve
(2) Type B pore water pressure growth curve

(3) Type C pore water pressure growth curve

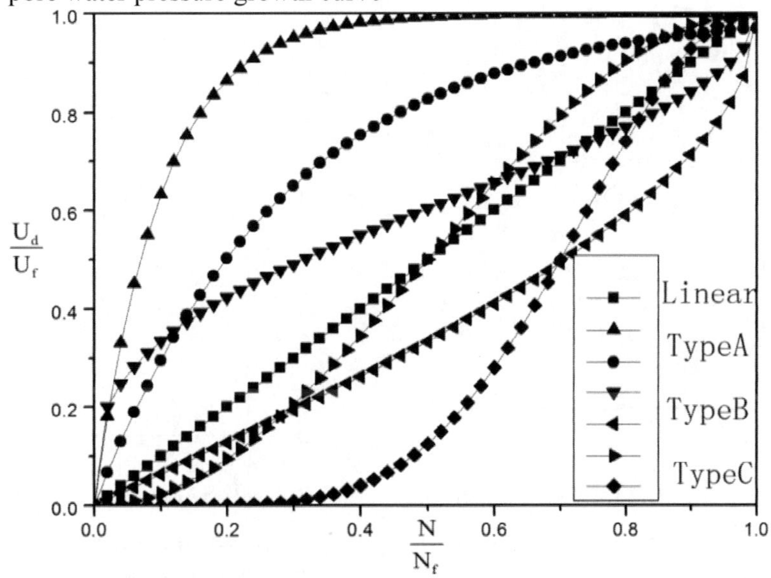


**FIGURE3.** Pore pressure growthRegional division

From pore water pressure growth curve under dynamic loading in figure 2, it can be seen that pore water pressure growthcharacteristics of remolded saturated loess sample is different from sand sample. This is mainly due to the uniquestructural characteristics and different contentof clay in loess.

Figures 4 shows the results of the tests performed on specimens, that is, the excess pore water pressure( μ = measured excess pore water pressure ) is presented versus the number of loading cycles(N), for shear stress levels of 25, 30, 35, 40 and 50 kPa. As can be seen from Figure 4, the excess pore water pressureincreased significantly during the first 10 loading cycles for all shear stress levels, and continued its development at a slower rate with further application of loading cycles.

From figure 4, we also can see that the specimen tested at 25kPa shear stress showed a progressive increase with increasing number of loading cycles. The excess pore water pressure ratio at this stress level was 3% after the first cycle, and ascend to approximately 38% at the end of 50 loading cycles. Similarly, at the biggest applied shear stress of 50kPa, the excess pore water pressure ratio progressively increased from 10% at the end of the first cycle of loading to 100% after 17 loading cycles. At this point the specimen has reached full liquefaction in 17 cycles. It is clear that as the shear stress level increases, more excess pore water pressure is generated for the same number of loading cycles.

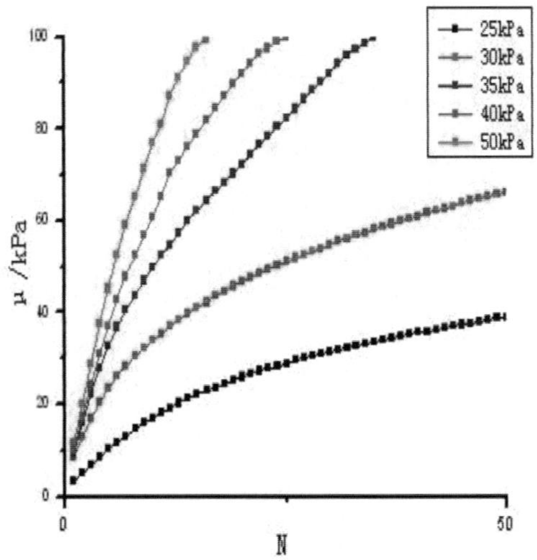


**FIGURE 4.**Excess pore water pressure generation as a function of number of loading cycles

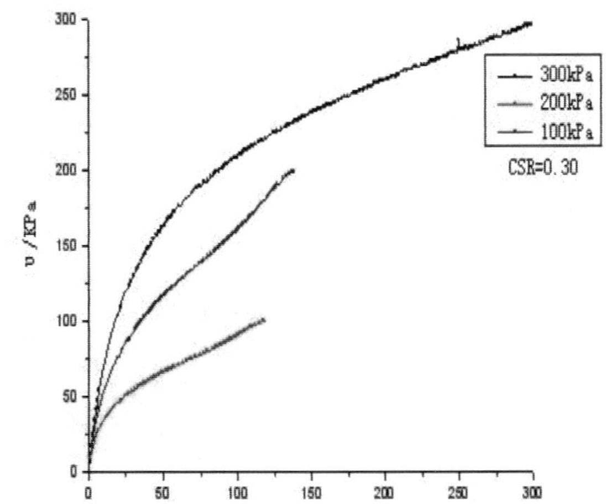


**FIGURE 5.**Comparison of excess pore water pressure measurements from σ 3=100kPa、200kPa and 300kPa specimens at CSR=0.30

Figure 5 displays the comparison at the smallest shear stress level of CSR=0.30, The shapes of the pore water pressure responses of the big and small confining pressure specimens are again similar.The main difference is that the excess pore water pressure in the big confining pressure is about 5% more than it in the small confining pressure. This difference appears to be maintained approximately constant at all loading cycles.Interestingly, the rate of excess pore water pressure generation are similar in the different confining pressure (FIGURE 6).

From figure 6, we can see that pore water pressure growth characteristics of saturated remodedloess samples are different from sand during vibration process. In the earlyvibration days, the pore pressure is growing faster, when the cycles ratio is 0.2, pore pressure ratio has risen to about 0.5; and with vibration continues, pore pressure ratio rising will slow down and reach the final stability, pore water pressure will present a fast-slow-flat trend.DuringThe whole process, the change of pore water pressure will not appear sharply or sudden increase phenomenon, but gradually become stable, This regular pattern is different from saturated loose sand. Based on this, the author put forward pore water pressure growth normalized relations model of saturated reshapingloess samples as following:

$$\frac{U_d}{U_f}=(\frac{N}{N_f})^{\beta}\sin[\frac{\pi}{2}(\frac{N}{N_f})^{\alpha}]$$

Where $U_f$ :maximum pore pressure under liquefaction, $U_d$ :instantaneous pore pressureundervibration, $N_f$ :liquefaction cyclic, $N$ : vibration cyclic, $\beta$ are parameters of test which are determined by dynamic load size, vibration frequency, $\alpha$ the nature of the earthsoil such as mineral composition, grain size distribution and density; $\alpha$ is smaller, the curve shape is more fullness, $\alpha$ is 0.34 in this paper.

Using this model, the fitting curve of pore pore water pressure growth is shown in figure 7:

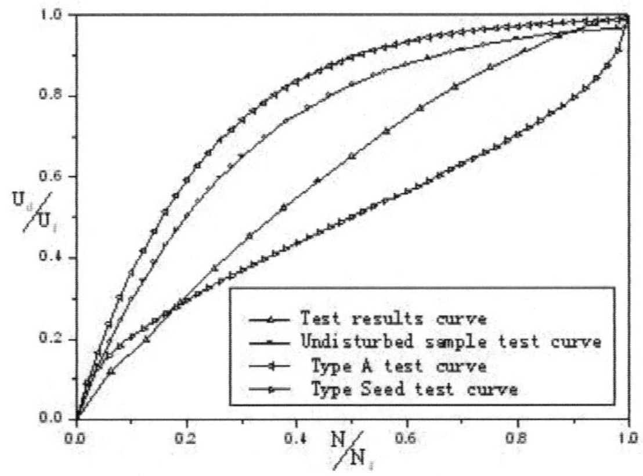


**FIGURE 6.**Different types of pore water pressure growth curve

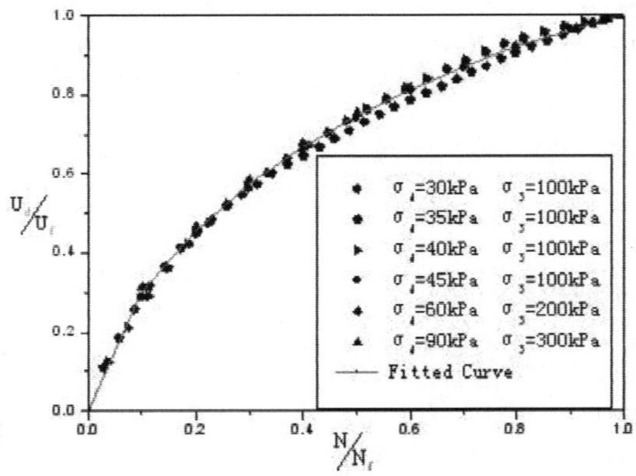


**FIGURE 7.** Fitting resultsof pore water pressure growth curve

## 5. Conclusion

The excess pore water pressure generation insaturated remolded loess is evaluated using stress-controlled and undrained cyclic triaxial tests. Then, the effect of confining pressure on the generation of excess pore water pressure is evaluated by comparing the responses of 100kPa, 200kPa and 300kPa. This comparison shows that, the excess pore water pressure generation in different confining pressure is similar in general. The pore water pressure growth characteristics of saturated remolded loess samples are different from sands'. Through the analysis of data, pore water pressure growth model is established.

## Acknowledgments

[1]Liu Jian-min, Male, Master graduate student, Mainly engaged in the study of soil dynamics, tel: (15963101199).

[2]Shandong Provincial Natural Science Foundation,China（ZR2014DQ022）

Title: Research on Statistical Analysis and Numerical Simulation of "Avoiding Belt" Width of Active Fault

[3]ContractalResearchProjectofEarthquakeAdministrationofShandongProvince,China（13Y86）

[4]Spark ProgramofEarthquakeSciences,China（XH16014）

## References

[1] WangLanmin 2003 *Loess Dynamics*（BeiJing:SeismologicalPress）p282

[2] Ishihara Ket al1990Liquefactioninduced flow slide in the collapsible loess deposit in Soviet Tajik*Soils and Foundations.*473-87

[3] SeedHB 1979 Soil Liquefaction and Cyclic Mobility Evaluation for Level Ground During*Earthquakesournal of the GeotechnicalEngineeringDivision.***2**201-255

[4] SUN Haimei, WANG Lanmin2010 Experimental study of development of strain and pore water pressure duringliquefaction of saturated Lanzhou loess*Rock and Soil Mechanics.***11**3464-3468

[5] SHE Yuexin, LIU Hanlong 2006Study on pore water pressure mode and cyclic softening of saturated compacted loess*Northwestern Seismological Journal.***2**123-128

[6] SHE Yuexin, LIU Hanong, Gao yufeng2002 Study on liquefaction mechanism and pore water pressure mode of saturated orginal loess *Rock and Soil Mechanics***4**395-399

[7] Castro.G, Liquefaction and Cyclic 1975 Mobility of Saturated Sands,*Journal of the Geotechnical Engineering Division.***6**551-569

[8] Dobry.R1985 Liquefaction of Soils During EarthquakesNational Research Council (NRC)*Committee on Earthquake Engineering.*CETS-EE-001

[9] Finn, W. D. L, Lee.K.Wand Martin. G. R1977 An Effective Stress Model for Liquefaction*Journal of the Geotechnical Engineering Division.***6**517-533
[10] Ishihara1996 *SoilBehaviour in Earthquake Geotechnics*（Oxford:Clarendon Press）p350-358

IOP Conf. Series: Earth and Environmental Science **61** (2017) 012118 doi:10.1088/1755-1315/61/1/012118

# Steady heat transfer analysis of 2D plates using the numerical manifold method

**Ji Lu[1], Huihua Zhang[2, 3] and Shangyu Han[2]**

[1] Insitute of Computer Application, China Academy of Engineering Physics, Mianyang, Sichuan, 621000, PR China.
[2] School of Civil Engineering and Architecture, Nanchang Hangkong University, Nanchang, Jiangxi, 330063, PR China.
[3] Corresponding author: hhzhang@nchu.edu.cn

**Abstract.** In the numerical manifold method (NMM), the mathematical cover system can be inconsistent with the physical boundary, which can alleviate the meshing cost to some extent. In the present paper, the NMM is developed to solve two-dimensional steady heat conduction problems with the square mathematical elements. The corresponding formulations are derived and the proposed approach is validated through a typical example.

## 1. Introduction

The numerical manifold method (NMM) was proposed by Shi [1] in the beginning of 1990s for rock engineering problems. Attributing to the two cover systems, i.e., the mathematical cover system and the physical cover system, the NMM is capable of solving both continuous and discontinuous problems in the same framework. One of the major advantages of the NMM is that the mathematical cover system (similar to the concept of "mesh" in the finite element method (FEM)) can be independent of the physical boundaries.

Recent years, the NMM has been further developed and applied to solve various problems, e.g., see [2-16]. In the present paper, the NMM is further extended to study two-dimensional (2D) steady heat transfer problems. To this end, the rest of the paper is organized as follows. Firstly, the governing equations and associated boundary conditions are presented. Secondly, the NMM formulations for 2D heat conduction analysis are addressed; then, a typical numerical example is considered to verify the developed method. Finally, the corresponding conclusions are given.

## 2. Governing equations

The governing equations for steady heat conduction problems is [13]

$$-\nabla \mathbf{q} + Q = 0 \tag{1}$$

where $\nabla$ is the gradient operator. $\mathbf{q}$ is the heat flux vector determined by the Fourier's law as $\mathbf{q} = -k\nabla T$ with $k$ the thermal conductivity for materials and $T$ the temperature. $Q$ is the heat source. The corresponding boundary conditions are

$$T = \overline{T}\ (\mathbf{x} \in \Gamma_T) \tag{2}$$

$$\mathbf{q} \cdot \mathbf{n} = \overline{q}\ \ (\mathbf{x} \in \Gamma_q) \tag{3}$$

where $\Gamma_T$ is the temperature boundary and $\Gamma_q$ is the flux boundary. $\overline{T}$ and $\overline{q}$ are, respectively, the prescribed temperature and flux on corresponding boundary. **n** is the outward unit normal to the domain.

## 3. The NMM for steady heat transfer problems

To solve a physical problem with the NMM, the mathematical cover (MC) system is firstly constructed. Theoretically, the MC formed by mathematical elements can be of any shape and the MC system may be non-conforming to all domain boundaries once it is large enough to cover the considered region. On each MC, a weight function is defined. Next, the physical cover (PC) system is formed by the intersection of MCs and physical domain. On each PC, the cover function is constructed to represent the local physical property. Then, the manifold elements (MEs) are obtained through the shared region of PCs. Accordingly, the NMM approximation on each ME is obtained by pasting the cover functions using the associated weight functions.

For the concerned problem, the temperature in any ME $e$ is expressed as [13]

$$T^h(\mathbf{x}) = \sum_{i=1}^{n_\mathrm{p}} w_i(\mathbf{x}) T_i(\mathbf{x}) \tag{4}$$

where $n_\mathrm{p}$ is the number of total PCs shared by $e$. $w_i(\mathbf{x})$ is the weight function defined on the MC containing the $i$th PC, often borrowed from the shape functions in the FEM. $T_i(\mathbf{x})$ is the cover function for the $i$th PC, and is usually chosen as

$$T_i(\mathbf{x}) = \mathbf{P}(\mathbf{x})\mathbf{a}_i \tag{5}$$

for 2D continuous problems. $\mathbf{P}(x)$ is the polynomial basis being

$$\mathbf{P}(\mathbf{x}) = [1, x, y, ...] \tag{6}$$

and $\mathbf{a}_i$ is the vector of unknowns defined on the $i$th PC.

The NMM discrete equations are derived using the weighted residual method as [13]

$$\mathbf{KT} = \mathbf{F} \tag{7}$$

where **T** is the vector of unknowns; **K** and **F** are, respectively, the global thermal conductivity matrix and equivalent thermal load vector. The contributions of the ME e to **K** and **F** are

$$\mathbf{K}^e = \int_{\Omega^e} \mathbf{B}^\mathrm{T} k \mathbf{B} d\Omega + \lambda \int_{\Gamma_1^e} (w_i \mathbf{P})^\mathrm{T} (w_i \mathbf{P}) d\Gamma \tag{8}$$

$$\mathbf{F}^e = \int_{\Omega^e} w_i \mathbf{P}^\mathrm{T} Q d\Omega + \lambda \int_{\Gamma_1^e} w_i \mathbf{P}^\mathrm{T} \overline{T} d\Gamma - \int_{\Gamma_2^e} w_i \mathbf{P}^\mathrm{T} \overline{q} d\Gamma \tag{9}$$

where the superscript **T** denotes the matrix transpose. $\Omega^e$ is the domain occupied by $e$. $\Gamma_T^e$ and $\Gamma_q^e$ are, respectively, the essential and natural boundary associated with $e$. $\lambda$ is the penalty factor for the enforcement of essential boundary condition. The elements of the matrix **B** are

IOP Conf. Series: Earth and Environmental Science **61** (2017) 012118 doi:10.1088/1755-1315/61/1/012118

$$\mathbf{B}_i = \begin{bmatrix} (w_i\mathbf{P})_{,x} \\ (w_i\mathbf{P})_{,y} \end{bmatrix} \tag{10}$$

To efficiently and accurately calculate the integration in equation (8) and (9), all the MEs are firstly divided into several sub-triangles, and then the 3-point Gaussian quadrature rules are used on each sub-triangle, the corresponding results on which finally adds up to the integration of the ME.

## 4. Numerical example

In this part, to verify the proposed method, a typical 2D heat conduction problem is tested under steady state.

Consider the square domain plotted in figure 1. The dimension of the plate, the Cartesian coordinates system and the boundary conditions are all provide in this figure. The analytical solutions to this problem can be derived as

$$T = 4x - \frac{x^2}{2} \tag{11}$$

When modeling, in view that the MC system can be independent of the physical domain, MCs composed of square mathematical elements are adopted to cover the plate. The discretized domain (when element size side length of square mathematical elements is $h$=0.3) is shown in figure 2, which includes 121 MEs and 144 PCs. Further, for simplicity, the polynomial basis in Eq. (6) is set to be constant. The penalty factor $\lambda$ in equation (8) and (9) is taken as $1.0\times10^{10}k$.

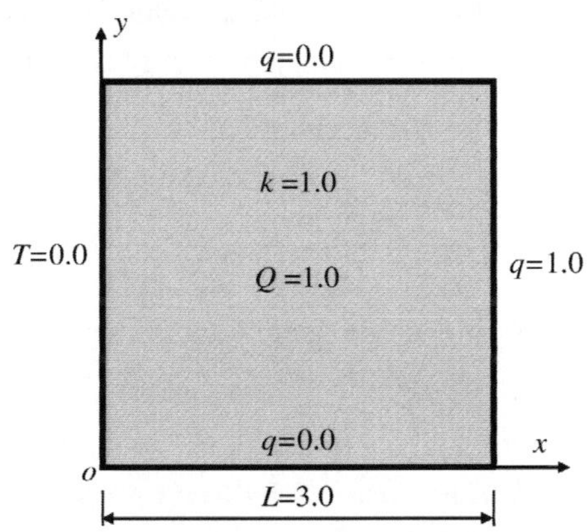


Figure 1. Physical problem.

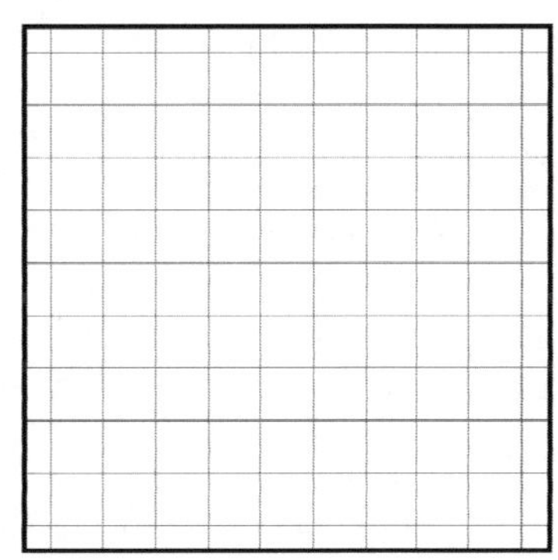

Figure 2. Discretized plate when $h$=0.3.

The temperature contour is displayed in figure 3, which infers that the value of temperature only varies along $x$-direction, as revealed in equation (11). The computed temperatures at several sample points, i.e., (0.5, 1.5), (1.0, 1.5), (1.5, 1.5) … and (3.0, 1.5) by the NMM are listed in table 1, together with the exact solutions from equation (11). It is obvious that the NMM results agree very well with the analytical ones.

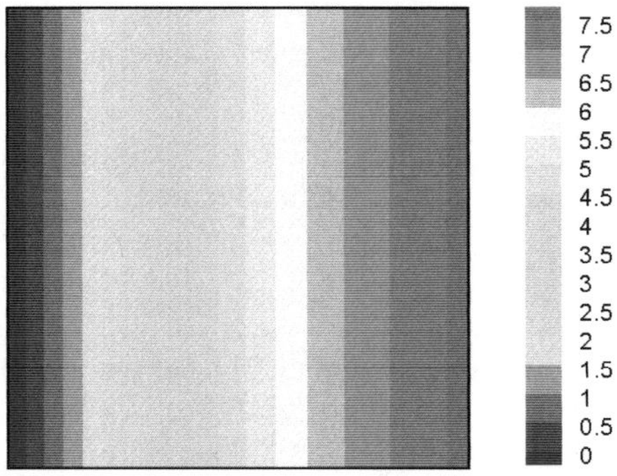


Figure 3. Distribution of temperature fields.

**Table 1.** Computed temperatures at sample points

| Sample point | (0.5, 1.5) | (1.0, 1.5) | (1.5, 1.5) | (2.0, 1.5) | (2.5, 1.5) | (3.0, 1.5) |
|---|---|---|---|---|---|---|
| NMM results | 1.869 | 3.494 | 4.864 | 5.994 | 6.869 | 7.499 |
| Exact solutions | 1.875 | 3.500 | 4.875 | 6.000 | 6.875 | 7.500 |

## 5. Concluding remarks

The numerical manifold method was developed to study 2D steady heat conduction problems. Details about the governing equations, the NMM discrete formulations and the numerical integration schemes are provided. A typical example is analyzed to validate the proposed method. Mathematical covers formed by square elements are adopted for numerical modeling due to the inconsistence of the mathematical cover system and physical boundaries. The great agreement between the NMM solutions and the reference results demonstrates the high accuracy of the present method.

## Acknowledgements

The present work was supported by the National Natural Science Foundation of China (Grant No. 11462014), the Provincial Natural Science Foundation of Jiangxi, China (Grant No. 20151BAB202003) and the Science and Technology Program of Educational Committee of Jiangxi Province of China (Grant Nos. GJJ14526 and GJJ150752).

## References

[1] Shi G H 1991 Manifold method of material analysis *Transaction of 9th Army Conference on Applied Mathematics and Computing* (Minneapolis, Minnesota) p 57-76

[2] Tsay R J, Chiou Y J and Chuang WL 1999 Crack growth prediction by manifold method *J. Eng. Mech-ASCE* **125** 884-90

[3] Chiou YJ, Lee YM and Tsay RJ 2002 Mixed mode fracture propagation by manifold method *Int. J. Fracture* **114** 327-47

[4] Li S C, Li S C and Cheng Y M 2005 Enriched meshless manifold method for two-dimensional crack modeling *Theor. Appl. Fract. Mec.* **44** 234-48

[5] Terada K, Ishii T, Kyoya T and Kishino Y 2007 Finite cover method for progressive failure with cohesive zone fracture in heterogeneous solids and structures *Comput. Mech.* **39** 191-210

[6] Ma G W, An X M, Zhang H H and Li L X 2009 Modeling complex crack problems using the numerical manifold method *Int. J. Fracture* **156** 21-35

[7] Zhang H H, Li L X, An X M and Ma G W 2010 Numerical analysis of 2-D crack propagation problems using the numerical manifold method *Eng. Anal. Bound. Elem.* **34** 41-50

[8] Zhao G F, Zhao J, Zhang H H and Ma G W 2010 A numerical manifold method for plane

micropolar elasticity *Int. J. Comp. Meth.-Sing.* **7** 151-66
[9] Zhang H H and Zhang S Q 2012 Extract of stress intensity factors on honeycomb elements by the numerical manifold method *Finite Elem. Anal. Des.* **59** 55-65
[10] An X M Zhao Z Y, Zhang H H and He L 2013 Modeling bimaterial interface cracks using the numerical manifold method *Eng. Anal. Bound. Elem.* **37** 464-474
[11] Wu Z J and Wong L N Y 2013 Elastic-plastic cracking analysis for brittle-ductile rocks using manifold method *Int. J. Fracture* **180** 71-91
[12] Zhang H H and Ma G W 2014 Fracture modeling of isotropic functionally graded materials by the numerical manifold method *Eng. Anal. Bound. Elem.* **38** 61-71
[13] Zhang H H, Ma G W and Ren F 2014 Implementation of the numerical manifold method for thermo-mechanical fracture of planar solids *Eng. Anal. Bound. Elem.* **44** 45-54
[14] Zheng H, Liu F and Li C G 2015 Primal mixed solution to unconfined seepage flow in porous media with numerical manifold method *Appl. Math. Model.* **39** 794-808
[15] Fan H, Zheng H and He S M 2016 S-R decomposition based numerical manifold method *Comput. Method. Appl. M.* **304** 452-78
[16] Zhang H H Ma G W and Fan L F 2017 Thermal shock analysis of 2D cracked solids using the numerical manifold method and precise time integration *Eng. Anal. Bound. Elem.* **75** 46-56

IOP Conf. Series: Earth and Environmental Science **61** (2017) 012119 doi:10.1088/1755-1315/61/1/012119

# Study on aging of single pile in soft soil foundation

**Tiejun Chen[1] , Haiying Cao[2]**

[1]Hebei Vocational & Technical College of Building Materials, Qinhuangdao Hebei 066004, China.
[2]School of Civil Engineering and Mechanics, Yanshan University, Qinhuangdao Hebei 066004, China.
chen4490@126.com

**Abstract**. In this paper, the single pile under load is analyzed by using the Biot consolidation equation and the modified method of the coupling of the Komala - Huang model. The consolidation of soil is simulated by Biot consolidation theory, and the whole process of the deformation of single pile is presented in the paper by means of the modified Komala - Huang model. Some useful conclusions are obtained by analyzing the variation of the pore pressure and the settlement of the soil under the action of load.

## 1. Introduction

It is obvious that the stress and strain of soil are affected by time, and not only the deformation of cohesive soil, but also of cohesionless sand, is not instantaneous. According to the different stress state, the deformation velocity of cohesive soil is very slow, and finally tends to stop. Others grow, eventually leading to destruction.

The deformation of piled raft foundation on soft soil foundation is a long-term settlement process under the load. It is generally considered that the deformation includes three parts: instantaneous deformation, consolidation deformation and secondary consolidation deformation. With the consolidation of soil and pore water pressure dissipation, the load is transferred to the soil skeleton. The soil skele ton under stress due to the viscous particle surface absorbed water, the rearrangement of particles and the dislocation of skeleton have time effect, the deformation of soil is related to time, so the rheological behavior of soil is considered in the settlement analysis of pile foundation, it is necessary. In particular, when the soft soil layer at the bottom of the pile is relatively soft, the rheological deformation of the soil occupies a significant proportion in the total deformation[1].

## 2. Visco elastic plastic constitutive model of soil

In general, the simplest model can be summarized under the premise of soft soil rheology, which is an ideal model suitable for numerical analysis of soft soil engineering. As shown in Fig.1, the model can describe the elastic deformation at the moment of loading, at the same time, it can simulate the delay phenomenon of soil deformation and the residual deformation after unloading.

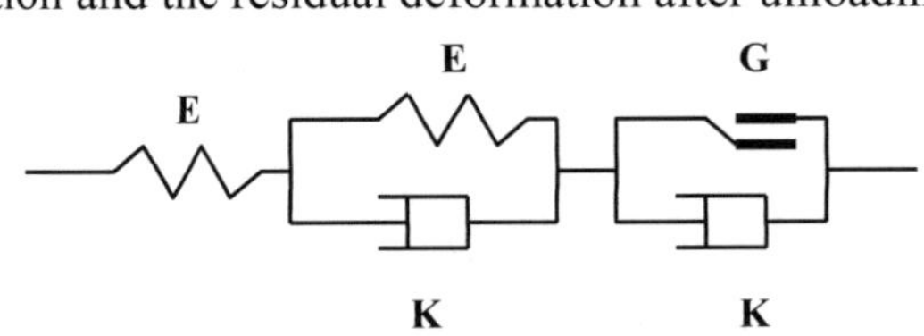


Figure.1 Komala - huang model

The total strain can be seen as consisting of three parts: $\{\varepsilon\}=\{\varepsilon_e\}+\{\varepsilon_{ve}\}+\{\varepsilon_{vp}\}$,the $\{\varepsilon_e\}$ is elastic strain; $\{\varepsilon_{ve}\}$ is viscoelastic strain; $\{\varepsilon_{vp}\}$ Is a viscoplastic strain. See $\{\varepsilon_{ve}\}$ and $\{\varepsilon_{vp}\}$ as initial strain, thus the solution of the viscoelasticity problem will be reduced to the solution of the elastic problem with initial strain.

The physical equation is:

$$\{\sigma\}=[D]\{\varepsilon_e\}=[D](\{\varepsilon\}-\{\varepsilon_{ve}\}-\{\varepsilon_{vp}\}) \tag{1}$$

The geometric equation is:

$$\{\varepsilon\}=[B]\{\delta\} \tag{2}$$

The unit node force is:

$$\{F\}=\int_v[B]^{\mathrm{T}}[D]([B]\{\delta\}-\{\varepsilon_{ve}\}-\{\varepsilon_{vp}\})\mathrm{d}V=[K](\{\delta\}-\{F_{ve}\}-\{F_{vp}\}) \tag{3}$$

$$[K]\{\delta\}=\{F\}+\{F_{ve}\}+\{F_{vp}\} \tag{4}$$

For the strain increment of the viscoplastic rheology, the following calculation method can be used:

$$\Delta\varepsilon=\varepsilon_{vp}\cdot\Delta t \tag{5}$$

$$\dot{\varepsilon}=\gamma<\Phi(F)>\frac{\partial Q}{\partial\sigma} \tag{6}$$

In the formula: $Q=Q(\sigma,\varepsilon_{vp},\kappa)$ is the plastic potential; $\gamma$ is the flow parameter that controls the plastic flow rate; $\Phi(F)$ is the yield function,when $x>0$, $\Phi(F)$ is a positive monotonically increasing function; The symbol < > is a switching function,

when $x>0$, $<\Phi(x)>=\Phi(x)$;

when $x<0$, $<\Phi(x)>=0$;

Using the associated plastic flow law, then there are $Q=F$, Then the formula (6) can be transformed into:

$$\dot{\varepsilon}=\gamma<\Phi(F)>\frac{\partial Q}{\partial\sigma}=\gamma<\Phi>\alpha \tag{7}$$

Where: $\alpha$ is the flow vector, the specific form is as follows:

$$\alpha^{\mathrm{T}}=\frac{\partial F}{\partial\sigma}=\begin{bmatrix}\frac{\partial F}{\partial\sigma_x} & \frac{\partial F}{\partial\sigma_y} & \frac{\partial F}{\partial\sigma_z} & \frac{\partial F}{\partial\tau_{yz}} & \frac{\partial F}{\partial\tau_{zx}} & \frac{\partial F}{\partial\tau_{xyz}}\end{bmatrix}$$

The yield function $F$ uses the Mohr-Coulomb yield criterion, which is widely used in engineering, to compensate for the shortcomings of the simple plastic slider. [2] The Mohr-Coulomb yield criterion is described in detail in many of the teaching materials that introduce the soil constitutive model[3].

## 3. Finite element analysis of consolidation and rheological coupling

The rheological properties of the soil is reflected by the time effect of stress - strain relationship, with the dissipation of pore water pressure, soil consolidation deformation, also produce secondary consolidation deformation, which is the main and secondary consolidation are produced at the same time. In the finite element analysis, the secondary consolidation is regarded as a posterior creep deformation.

The modified Komala - Huang model is used to reflect the visco - elastic - plasticity of the soil, and the model is combined with the consolidation theory to derive the incremental form of the Biot consolidation equation:

$$\begin{bmatrix}\boldsymbol{K} & \boldsymbol{K}' \\ \boldsymbol{K}'^{\mathrm{T}} & \tilde{\boldsymbol{K}}\end{bmatrix}\begin{Bmatrix}\Delta\delta \\ \beta\end{Bmatrix}=\begin{Bmatrix}R-R_t+R_v \\ 0\end{Bmatrix} \tag{8}$$

IOP Conf. Series: Earth and Environmental Science **61** (2017) 012119 doi:10.1088/1755-1315/61/1/012119

Where: $\boldsymbol{K}$ is the stiffness matrix; $\boldsymbol{K}'$ is the coupling matrix; $\tilde{\boldsymbol{K}}$ is the percolation matrix; $\Delta\delta$ is the displacement increment; $\beta$ is the ultra-static pore pressure; $R$ is the equivalent load of the corresponding load of the external load; $R_t$ is the $t-\Delta t$ time Of the load corresponding to the stress balance of the part of the load; $R_v$ is $t-\Delta t$ due to the soil viscosity corresponding to that part of the load.

According to the above principles and methods, a two - dimensional 8 - node Biot consolidation visco - elastoplastic finite element program is compiled,pile-soil contact plane unit selection Goodman unit, the program can consider plane stress, plane strain and axisymmetric problem[4][5].

## 4. Finite element analysis of single pile

### *4.1. Selection of model parameters*

Single pile finite element analysis model selection pile length $l = 16m$, pile diameter $d = 0.6m$, pile elastic modulus $E_p = 30Gpa$ soil compressive modulus $E_s = 4MPa$, according to the Shanghai area statistics, soil elastic modulus and compression The relationship between the modulus is $E = (2 \sim 5)\ E_s$, the elastic modulus of the pile side soil is $E = 8MPa$, the Poisson's ratio of soil is $\nu = 0.4$, the permeability coefficient is $k_x=k_y=1\times10^{-4}$ m • $d^{-1}$, The calculated depth is 14 m, the total depth of the foundation is 30 m, and the calculated width is 50 m. The total load in the analysis was 250 kN, applied in 5 times, 50 kN each time.

### *4.2. Variation of pore water pressure at pile tip*

Figure 2 is a pile at the end of the curve of pore pressure changes with time, it can be seen from the figure at the pile end Mandel effect in the early stage of consolidation, the pore pressure has not decreased, but increased gradually, reached a certain time (200 d) after the fall, which is consistent with the measured results[6].

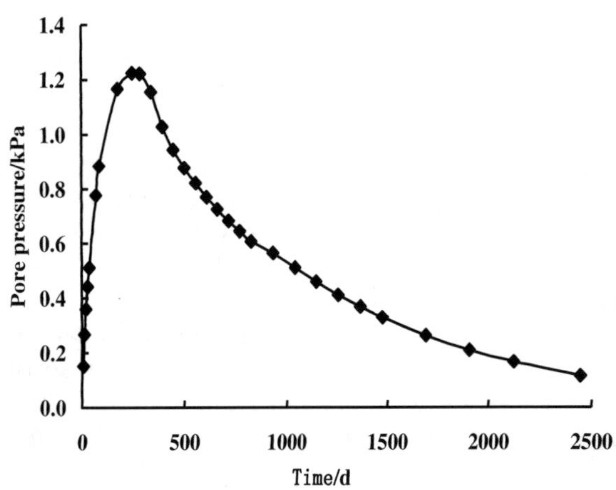


Figure. 2 Variation of pore pressure with time

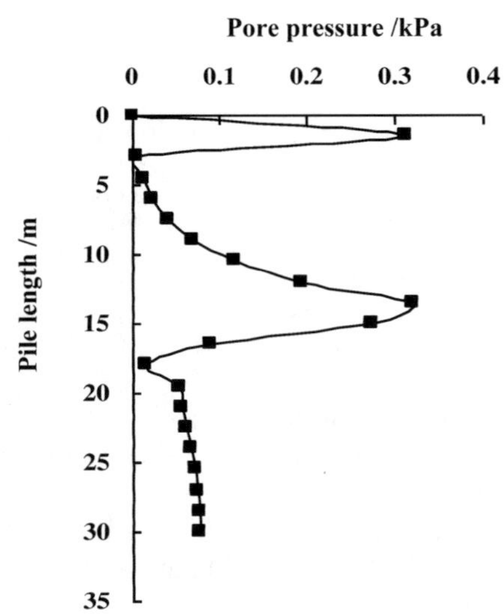


Figure. 3 distribution of pore pressure of pile side and pile tip soil when loading 50kN

Figure 3 shows the application of load 50kN, experienced 30d; Figure 4 shows the application of load 150kN, for 90d; Figure 5 shows the change of excess pore water pressure along the depth of the soil at the pile and pile ends after 150 d of the applied load of 250 kN. It can be seen from the figure that the super-static pore water pressure in the vicinity of the pile and the pile top is relatively large, and the excess pore water pressure in the soil at the pile and pile ends is small. With the increase of the load, the excess pore water pressure Change is not obvious.

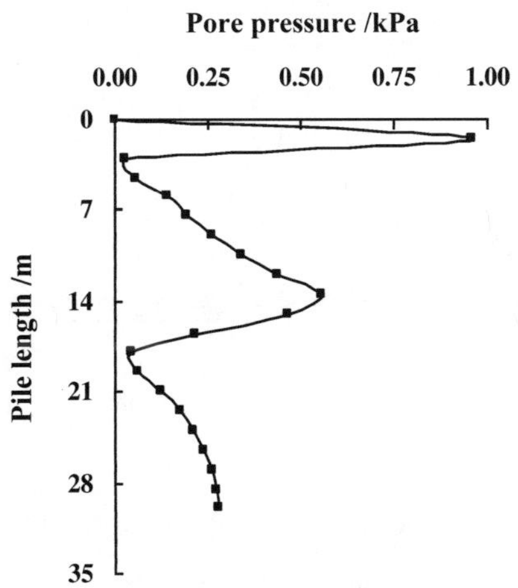


Figure. 4 distribution of pore pressure of pile side and pile tip soil when loading 150kN

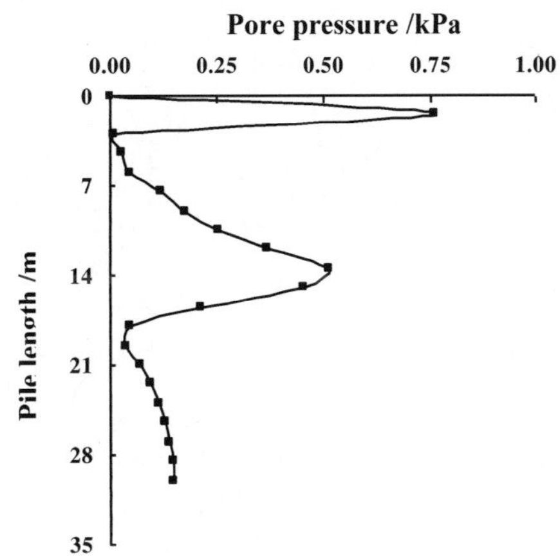


Figure. 5 distribution of pore pressure of pile side and pile tip soil when loading 250kN

*4.3. Interaction between pile and soil*

It can be seen from the variation curves of the stress of the pile soil contact surface element along the depth in Figure 6 that the stress of the pile soil contact surface gradually decreases with the loading time, and the phenomenon of the middle part is smaller, the pile tip and the pile end increase rapidly.

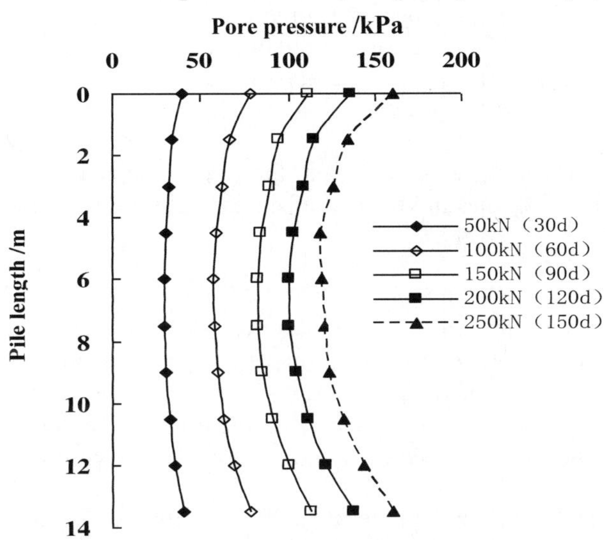


Figure.6 change of stress along the depth of pile interface

Figure 7 and figure 8 are for the changes of pile end soil shear stress along the horizontal direction and the pile end soil stress along the horizontal direction, it can be seen that the stress variation law in the two figures is more consistent. Pile soil at normal stress and shear stress are increased with the increase of pile spacing of horizontal distance decreases rapidly, more straight from the pile after 5 m curves of stress in soil is very small, can be neglected in the analysis of pile group. The influence of the stress at the pile end under load is limited, which is consistent with the conclusion obtained by Randoph et al.

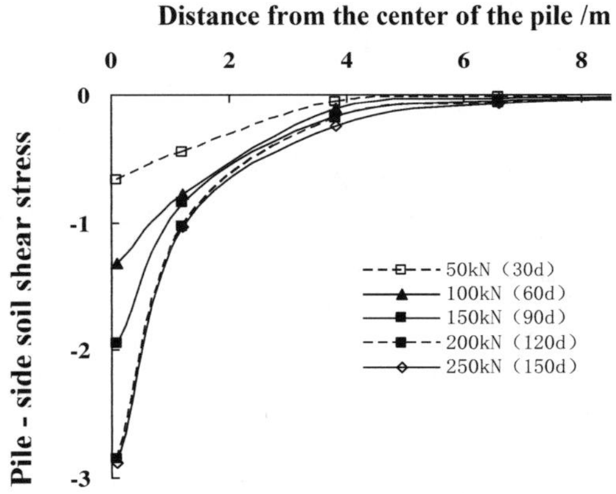


Figure. 7 Variation of $\tau_{zx}$ along the horizontal direction of soil at pile tip

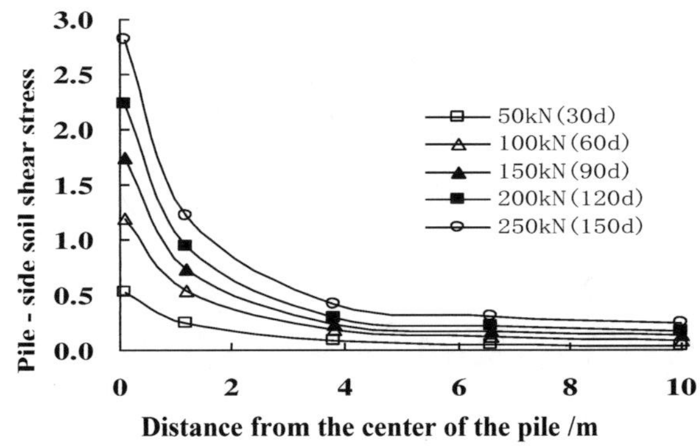


Figure. 8 variation of $\sigma_z$ along the horizontal direction of soil at pile tip

## 5. Conclusion

In this paper, the whole process of settlement and deformation of single pile under load is analyzed by the method of combining the Biot consolidation equation and the modified Komala - Huang model. Based on the simulation of the settlement process of the soil around the pile, the following conclusions can be drawn:

Under the action of load, there is a significant Mendel effect at the pore water pressure at the pile tip. At the early stage of consolidation, the pore pressure did not decrease, but gradually increased, and then began to decline after a certain period of time.

The horizontal displacement of soil caused by pile foundation settlement decreases rapidly with the increase of pile distance. Especially in the vicinity of the pile, the displacement decreases sharply, which shows that the influence of the pile on the surrounding soil displacement is limited, beyond this range, the effect is negligible.

The normal stress and shear stress in the soil at the pile tip decreases rapidly with the increase of the horizontal distance from the pile, and the curve becomes more flat after a certain distance from the pile,the stress in the soil is very small.

## References

[1] Erwin Hinton. Plastic finite element mechanics [M]. Beijing: Weapon Industry Press, 1989
[2] Han Jun. Analysis of vertically loaded piles in elastic plastic foundation [J]. Journal of Changsha Railway Institute, 2013, 18 (2): 6-12
[3] Li Zhenya, Wang Kuihua, et al., et al. Study on the effect of settlement of single pile under the static load of pile side [J]. Chinese Journal of rock mechanics and engineering, 2015 (5): 1022-1030
[4] WU Wen-bing, DOU Bin, WANG Kui-hua, et al.Time effect of single pile settlement of viscoelastic foundation under static load [J]. Journal of Building Science and Engineering, 2012, 29 (3): 73-80
[5] Yao Zhiran. Study on the settlement of pile composite foundation in homogeneous soil [D]. Taiyuan University of Technology, 2014
[6] Luo Yongchun. Study on bearing capacity and settlement of loess foundation treated by cement soil pile composite foundation [D]. LanZhou JiaoTong University, 2013

IOP Conf. Series: Earth and Environmental Science **61** (2017) 012120 doi:10.1088/1755-1315/61/1/012120

# Analysis on pile testing results of post-grouting bored pile

**A R Zheng**

CCCC Tianjin Port Engineering Institute Ltd. ,Key Laboratory of Geotechnical Engineering, Ministry of Communications, PRC, Key Laboratory of Geotechnical Engineering, Tianjin , Tianjin 300222, China
airong717@163.com

**Abstract**. Based on static load test results, the bearing capacity of bored piles with pile-toe and pile-shaft post-grouting has been analyzed. The analysis reveals that: with post-grouting, the interface between pile and surrounding soil are strengthened and the relative sliding displacement in between is reduced; end resistance of pile is enhanced and can be mobilized at earlier stage with smaller sliding displacement. As a result, the performance of bored pile is improved with increased bearing capacity and reduced settlement.

## 1. Introduction

Pile-bottom sediment and pile-surrounding slurry cake are two inherent defects which can affect the pile-end resistance and pile-surrounding frictional force of grouting piles with shaft wall protection by drilling mud [1]. The post-grouting technology is widely used because it has good prevention effects on pile-bottom sediment and pile-surrounding slurry cake. Therefore, more specific researches on bearing capacity and reinforcement mechanism of post-grouting grouting piles are conducted [2~5]. This paper will analyze the vertical bearing capacity of post-grouting piles after the pile-end and pile-surrounding association according to the on-site pile testing.

## 2. Pile testing

More than 18000 piles are arranged for the foundation of grouting pile with shaft wall protection by drilling mud. In order to improve the bearing capacity of pile foundation, the quality of piles will be improved by the way of pile-end and pile-surrounding joint grouting. The pile-end grouting quantity is 1000kg. Different grouting ways will be applied to sites around the pile according to different pile length. A pile-surrounding grouting with the grouting quantity of 600kg is set within the pile length of 30m. If the pile length exceeds 30m, two lines of pile-surrounding grouting will be set and the grouting quantity of each line is 400kg.

The destructive static load tests are done to 66 post-grouting piles and 6 non-grouting piles. The average characteristic value of bearing capacity of some pile types is shown in Table 1. After the grouting process, their vertical bearing capacity shall be respectively improved for 144%, 120% and 97%.

**Table 1.** Eigenvalue of bearing capacity of piles

| Pile type | Stress Type | Bearing stratum | Pile diameter /mm | Valid pile length /m | Characteristic value of bearing capacity of non-grouting pile /kN | Characteristic value of bearing capacity of grouting pile /kN |
|---|---|---|---|---|---|---|
| 1 | Compression resistance | ⑨ Fine sand | 800 | 25 | 2050 | 5000 |
| 2 | Compression resistance | ⑨ Fine sand | 1000 | 25 | 2750 | 6050 |
| 3 | Compression resistance | ⑪ Medium fine sand | 1000 | 27.5 | 3500 | 6900 |

## 3. Calculation results and analysis

### 3.1. *Axial force transfer characteristics*

The length of non-grouting pile TP4 and post-grouting pile TP2 is 27.5m, both the diameter of the piles is 1.0m. During different loading stages, the buried depth of axial forces of cross sections of a pile along the pile is distributed as Figure 1 and Figure 2. The axial force of cross section of a pile will increase when the loads increase and will decrease with the increasing of depth under the load effect of the same level. In addition, the decreasing rate is different on different soil layers and during different loading stages. It represents that the slope of the axial force curve has differences. The slope of the axial force curve reflects the size of the pile-surrounding resistance. The smaller the slope is, the axial force difference between levels will be larger, and in other words, the pile-surrounding frictional resistance will be larger. Comparison of the axial force figure of the piles before and after the grouting shows that when the pile-top loads of the two testing piles are close to the maximum load value, the axial force curve slope of post-grouting piles is smaller than that of the non-grouting pile. This means that the frictional resistance of the pile-surrounding soil after the grouting is higher than that of non-grouted.

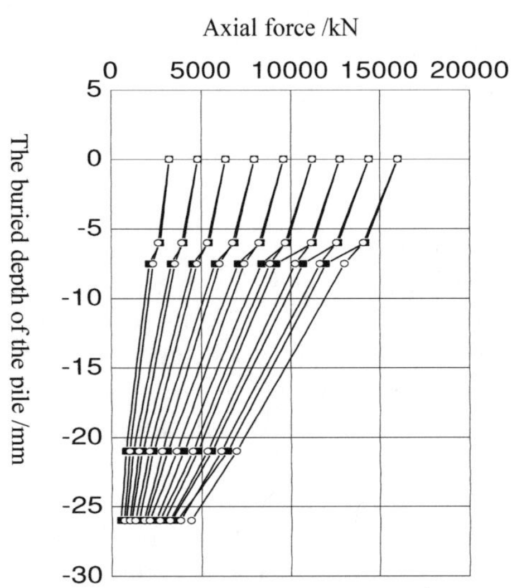


**Figure 1.** Axial force distribution of post-grouted bored pile

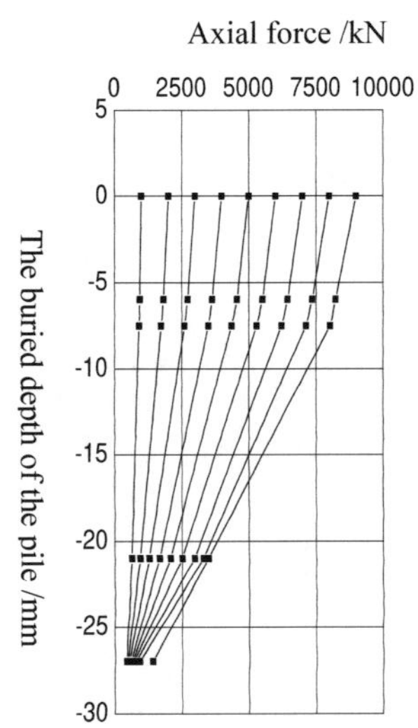


**Figure 2.** Axial force distribution of non-grouting bored pile

### 3.2. *Performance features of pile-end resistance*

The relationship among the pile-surrounding frictional resistance, the load and the displacement of pile tip is shown as Figure 5 and Figure 6. Under small loading effect, the difference of pile-surrounding frictional resistance before and after the grouting becomes small, which presents a linear relation with the load. When it is loaded to 8000kN, the pile-surrounding frictional resistance of non-grouting pile TP4 will decrease along with the increase of test load; the settlement amount will increase suddenly and finally reach the limited lateral frictional resistance. During the loading process of post-grouting pile TP2, the pile-surrounding frictional resistance basically has the linear relation with the pile-top load and the changes of pile-tip displacement relation curves are also mild, failing to reach the peak. This is because the post-grouting greatly improves the pile-surrounding interface conditions and the limited frictional resistance of the pile-surrounding soil, which can finally improve the pile-surrounding frictional resistance.

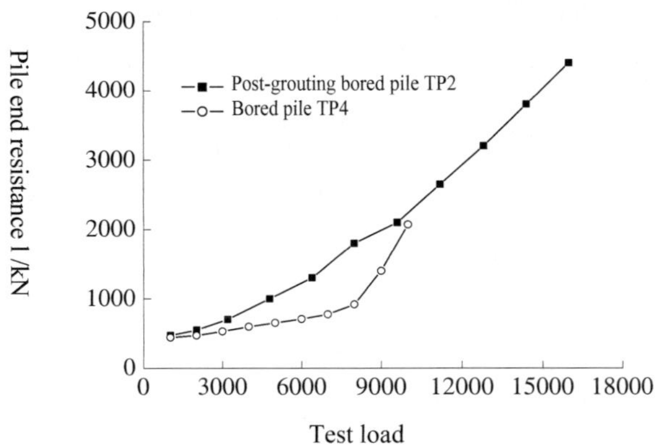


**Figure 3.** Relationship between pile end resistance and test load

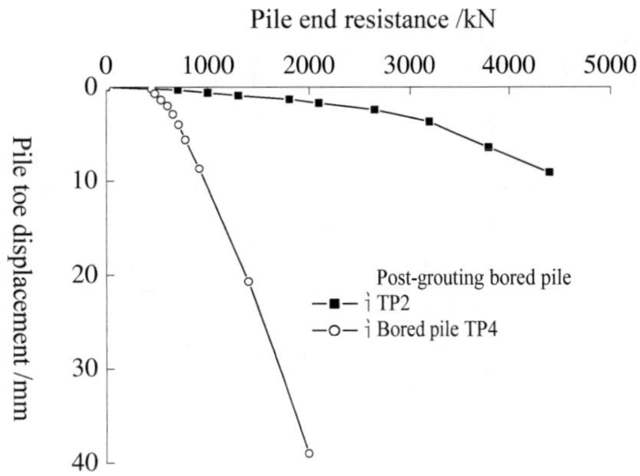


**Figure 4.** Relationship between pile end resistance and pile toe displacement

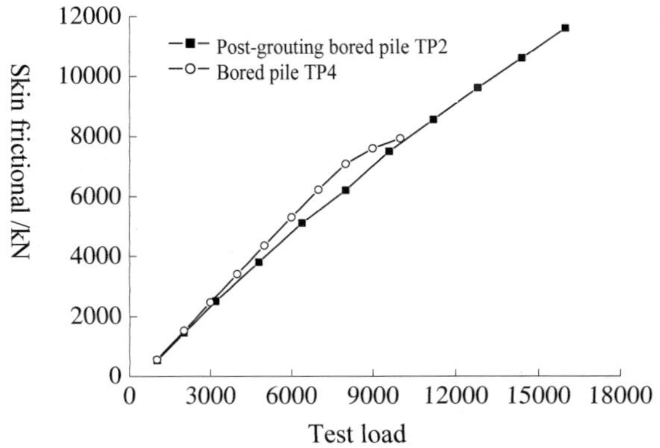


**Figure 5.** Relationship between skin friction and test load

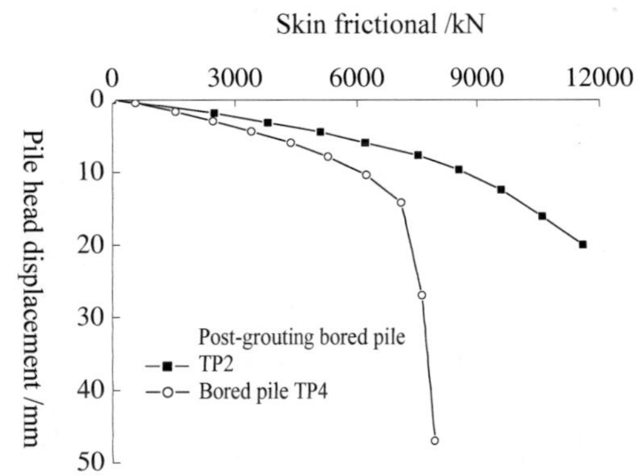


**Figure 6.** Relationship between skin friction and pile head displacement

## 4. Conclusion

(1) Pile testing results show that compared with the grouting pile under the vertical loading effect, the settlement amount of the pile-end and pile-surrounding association grouting pile is significantly reduced while the bearing capacity is greatly improved.

(2) After the grouting process, because of the enhancement of the strength of soil around the pile and the improvement of conditions of the interface between pile and soil, the pile-surrounding frictional resistance increases obviously. It still maintains the linear relationship with load under large pile-top loads. In addition, the pile-soil relative displacement is also reduced.

(3) The reinforcement of sediment and the improvement of strength of the pile-end bearing stratum greatly reduce the pile-end displacement. With only a small pile-end displacement, great tip resistance can be played. It is played in advance, which reduces the difference between that of the pile-surrounding frictional resistance. They can bear the pile-top load better, which will improve the vertical bearing capacity of piles.

**References**

[1] YUAN Zhen, CHEN Jin-jian and WANG Jian-hua 2006 *Rock and Soil Mechanics* 27(8) p 1398–140
[2] D. A. Bruce 1986 *Ground Engineering* 19(4) p 9–15
[3] ZHANG Zhong-miao, WU Shi-ming and BAO Feng 1999 *Chinese Journal of Geotechnical Engineering* 21(6) p 681–686
[4] ZHANG Zhong-miao, BAO Feng and CHEN Yun-min 2000 *Chinese Journal of Geotechnical Engineering* 22(2) p 243–246
[5] HUANG Sheng-gen and GONG Wei-ming 2006 *Chinese Journal of Geotechnical Engineering* 28(1) p 114–117

# Study on strength and deformation characteristics of sandy mudstone based on triaxial unloading test

**Xianfan Yuan**

PowerChina Chengdu Engineering Corporation Limited, Chengdu 610072, China
Corresponding author:xianfanyuan@163.com

**Abstract.** To achieve the strength and deformation characteristics of sandy mudstone under the unloading condition, a triaxial unloading test was performed on a group of sandy mudstone taken from the diversion tunnel of a hydropower station. The results show that the stress-strain curve slope of sandy mudstone increases suddenly at the moment of unloading confining pressure, and there is no significant decline in the whole test. There is usually only a whole shear failure surface in the samples damaged under low initial unloading confining pressure while those damaged under high one develop tensile fracture. The introduction of semilog method helps to solve the problem that it's difficult to get the peak intensity of sandy mudstone directly based on the variation characteristics of the stress-strain curve. From the initial stage of unloading to the damage of the samples, the reduced percentage of sandy mudstone deformation modulus is positively correlated with initial unloading confining pressure. The decrease rate of deformation modulus and increase rate of axial strain of sandy mudstone are both mutate under destructive confining pressure. The research results can provide an useful reference to research on the mechanical properties of similar soft rocks.

## 1. Introduction

With the rapid development of infrastructure construction and the advancement of Western Development in China, "More, longer, bigger and deeper" become the general trend of tunnel engineering. A large number of deep-buried long tunnel projects have been launched. Especially the big long tunnels for water conservancy and hydropower, highway and railway develop unprecedentedly. More and more projects face the problem of constructing underground caverns in the weak rock mass[1]. Research has shown that soft rocks' mechanical properties are significantly different from hard rocks as engineering rock mass, which should be given much attention.

The triaxial test is one of the primary ones for determining geotechnical strength parameters, and the accurate geotechnical strength parameters thus acquired is the basis of solving related geotechnical theories and engineering problems[2,3,4]. To provide construction advice for a hydropower station's diversion tunnel which goes through the sandy mudstone layer, samples of the sandy mudstone are taken specially to do triaxial unloading test, whose strength and deformation characteristics are studied.

## 2. Test conditions and test methods

The samples come from the diversion tunnel of a hydropower station in Western China, with a height of 100mm and a diameter of 50mm. Tests were done on the RMT-150C rock mechanics test system(as shown in Fig.1), and the temperature of the laboratory maintains at 20±1°C in the process of test.

**Figure 1.** RMT-150C rock mechanics test system

The triaxial unloading test methodsadopted are as follows:

Firstly, use stress control mode, exerting $\sigma_1=\sigma_3$ to predetermined value (10, 20, 30MPa) gradually under the condition of hydrostatic pressure with loading rate 0.05 MPa/s.

Secondly, continue to adopt stress control mode, keep $\sigma_3$ constant and gradually increase σ1 to 80% of the ultimate compressive strength under the same confining pressure at the loading rate of 0.05 MPa/s. Make sure the stress level of $\sigma_1$ is more than the uniaxial compressive strength and less than the triaxial compressive strength under corresponding confining pressure.

Lastly, carry on adopting stress control mode and reduce $\sigma_3$ gradually at the rate of 0.05 MPa/s while keeping $\sigma_1$ constant until the sample is damaged.

## 3. Stress-strain curve and failure characteristics of samples

The stress-strain curve of sandy mudstone under different initial unloading confining pressure is in Fig.2.

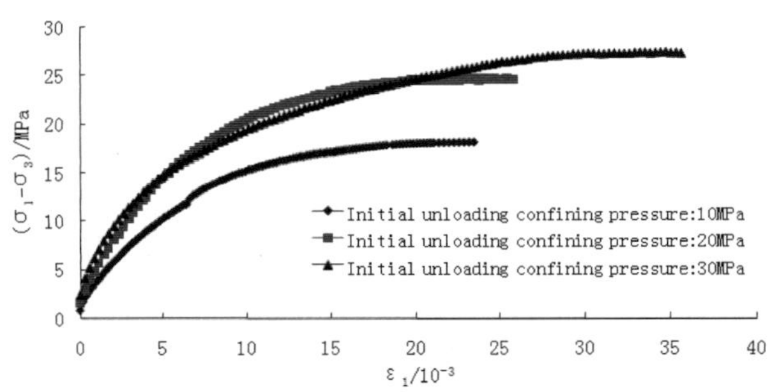


**Figure 2.** Stress-strain curves of sandy mudstone with triaxial unloading test

Fig.2 shows that the stress-strain curve slope of the sample increases suddenly in the instant of unloading confining pressure, which showed that the growth rate of axial strain decreases suddenly at the moment of unloading confining pressure, and this phenomenon is especially evident under lower initial unloading confining pressure. The analysis shows that this is becausethe higher the confining pressure is, the greater the constraints on the lateral expansion of the rock becomes. The sample enters destruction state gradually with the deviator stress ($\sigma_1$-$\sigma_3$) increasing, a small change in deviator stress can cause great axial strain, and there is no distinct decline period on the stress-strain curve throughout the whole test, the sample shows significant strain hardening characteristics.

Fig.3 shows the failure pattern photos and sketch of the crack of sandy mudstone samples under different initial unloading confining pressure.

IOP Conf. Series: Earth and Environmental Science **61** (2017) 012121 doi:10.1088/1755-1315/61/1/012121

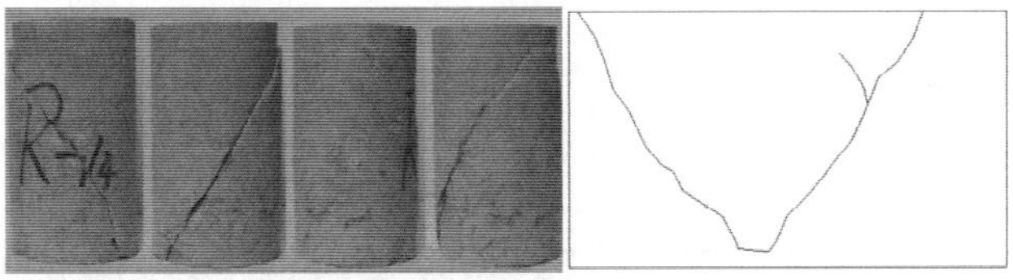

(a) R-14 (Initial unloading confining pressure: 10MPa)

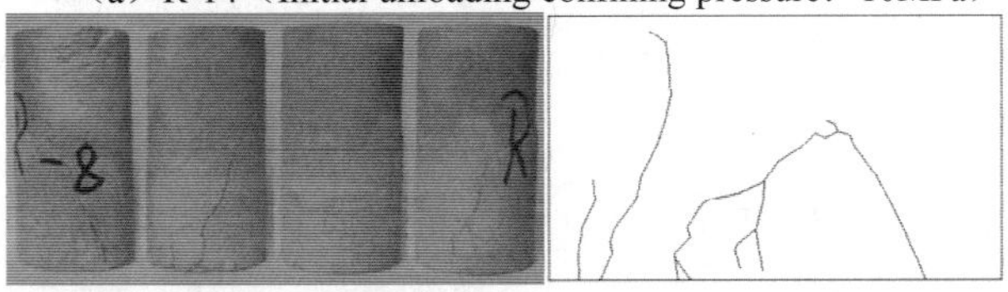

(b) R-8 (Initial unloading confining pressure: 20MPa)

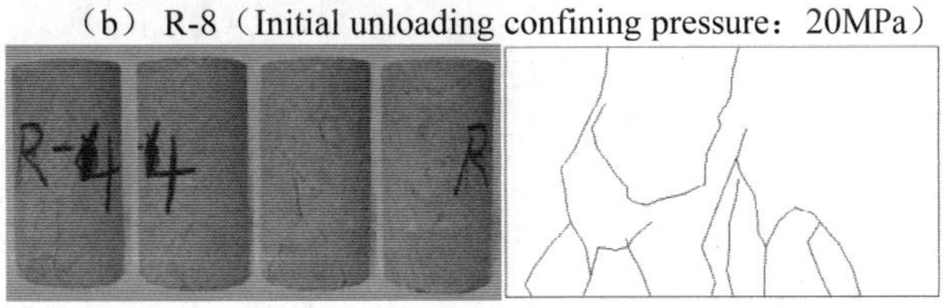

(c) R-4 (Initial unloading confining pressure: 30MPa)

**Figure 3.** Failure pattern photos and sketch of the crack of sandy mudstone samples with triaxial unloading test

As shown in Fig. 3, although the stress-strain curve shows no significant decline, there is obvious damage in the sample taken from the testing machine. The sample remains a relatively complete cylinder form after the destruction and does not look fragmentized like hard rock. Shear failure occurs in the sample under lower initial unloading confining pressure, and there is only a whole shear plane in the sample and no evident tensional fracture. With the increase of initial unloading confining pressure, lateral unloading degree of the sample increases, and tensional fractures produced in the direction of the maximum principal stress increase, with gradually increasing the degree of fragmentation.

## 4. Analysis of strength characteristics

Based on above analysis, it can be seen that after the destruction of the sample, the stress-strain curve shows no significant decline. Therefore, it is impossible to determine the peak intensity of the sample directly based on variation features of the stress-strain curve. Tu et al.[5] proposed that when determining the peak intensity of an enormous outwash deposits based on large-scale triaxial test results, "when there is no distinct peak intensity, it should be the axial stress value when the axial strain is 15%". However, many soft rock test results show that the plastic deformation of soft rock increases as the confining pressure increases. Therefore, it is obviously unreasonable to adopt the stress value at the same axial strain level as the sample's peak intensity under different confining pressure. Meanwhile, You[6] analyses that confining pressure determines the bearing capacity of the sample in the process of plastic deformation, namely it keeps carrying axial stress through friction and its bearing capacity doesn't decrease with the increase of axial deformation. So, when the strain is maximum, it's not right to choose the stress value as the sample's peak intensity either.

To acquire the exact peak intensity of sandy mudstone in the triaxial unloading test, the semilog method which is often used in geo-technique triaxial and shear test data processing is introduced here after many trials. Fig.4 shows the $\sigma_3$-lg$\varepsilon_1$ relation curve of the samples.

IOP Conf. Series: Earth and Environmental Science **61** (2017) 012121 doi:10.1088/1755-1315/61/1/012121

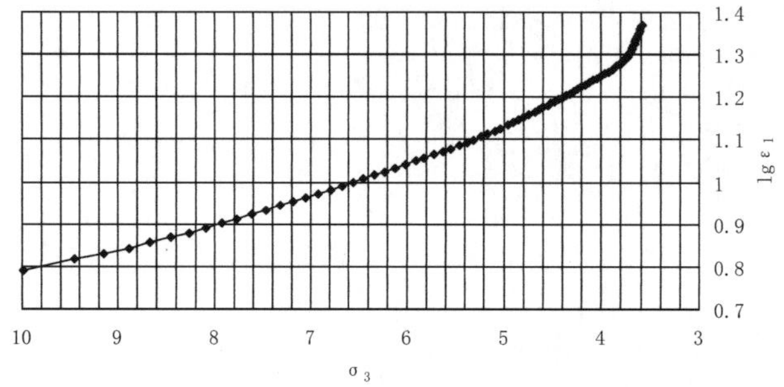


(a) R-14 (Initial unloading confining pressure: 10MPa)

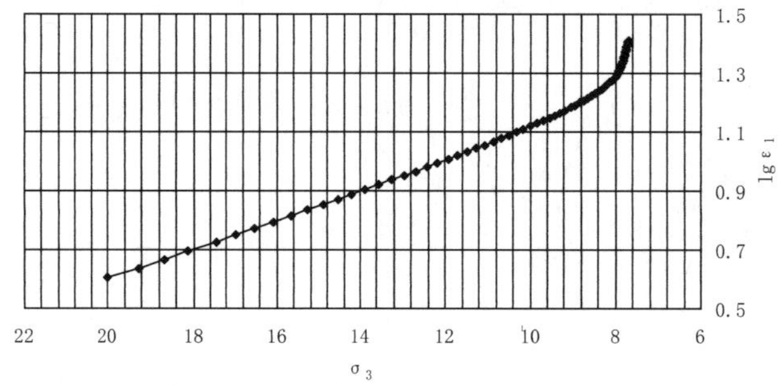


(b) R-8 (Initial unloading confining pressure: 20MPa)

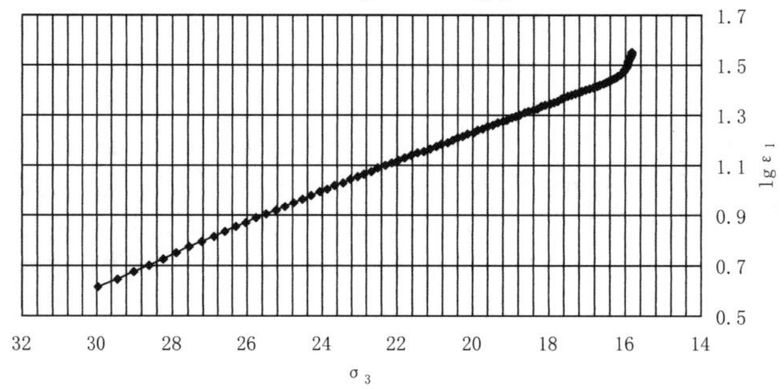


(c) R-4 (Initial unloading confining pressure: 30MPa)

**Figure 4.** $\sigma_3$ -lg$\varepsilon_1$ relation curve of sandy mudstone with triaxial unloading test

As can be seen in Fig. 4, there is an evident inflexion on the $\sigma_3$-lg$\varepsilon_1$ relation curve of sandy mudstone with the triaxial unloading test, which divides the curve into approximately two straight lines.

(1) The stage of slow rise: the sample is not yet destroyed. Before the unloading confining pressure decreases to the certain range of value, lg$\varepsilon_1$ gradually increases at a basically constant rate. This shows that the axial strain of the sample increases at a constant rate during this period, but the sample shows no evident damage.

(2) The stage of quick rise: the shear failure surface of the sample expands quickly. When the confining pressure unloading reaches the certain value, lg$\varepsilon_1$ increases sharply in a short time and the slope of $\sigma_3$-lg$\varepsilon_1$ relation curve increases rapidly, which shows that the sample was already destroyed then. After that, the axial strain of the sample increases quickly as the confining pressure declines.

Based on the above analysis, the demarcation point of the slow rise and quick rise corresponds to is the peak intensity of the sample. As is regulated in the reference[7], the stress value that the inflection corresponds to is the peak intensity of the sample. Table 1 shows the peak intensity of different samples.

**Table 1.** Peak intensity of sandy mudstone with triaxial unloading test

| Initial unloadingconfining pressure[MPa] | Axial strain[MPa] | Destroy confining pressure[MPa] | Peak intensity（$\sigma_1$-$\sigma_3$）[MPa] |
|---|---|---|---|
| 10 | 20 | 3.80 | 16.20 |
| 20 | 30 | 8.26 | 21.74 |
| 30 | 40 | 16.50 | 23.50 |

## 5. Analysis of deformation characteristics

### *5.1 The Relationship Between Deformation Modulus and Confining Pressure*

Fig.5 shows the relationship between deformation modulus and the confining pressure of the samples during unloading. As can be seen in the picture, the deformation modulus of the rock declines gradually as the confining pressure declines during unloading. When the initial unloading confining pressure is relatively low, the variation range of deformation modulus relatively small in the initial period of unloading confining pressure, when the confining pressure decrease to destructive confining pressure, the deformation modulus declines sharply, and an obvious inflection occurs on the curve. When the initial unloading confining pressure is relatively high, the deformation modulus shows an obvious downtrend in the initial period of unloading， and the curve shows obvious non-linear characteristic.

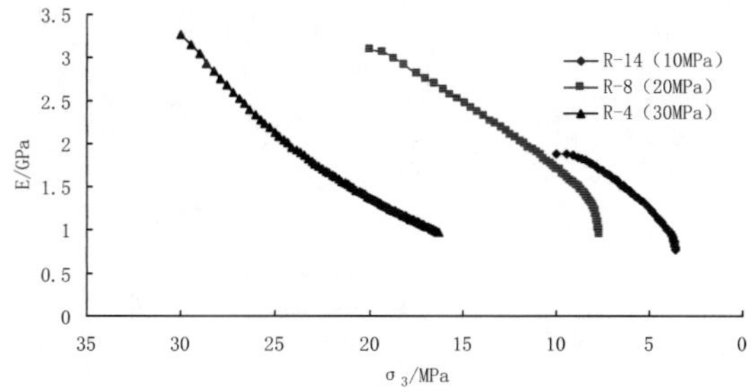


**Figure 5.** Deformation modulus-confining pressure relation curve during unloading

To study the relationship between deformation modulus and initial unloading confining pressure of the samples from the initial stage of unloading to the time when the yield failure occurs, the statistics about the deformation modulus of the samples at initialunloading and destruction stage are shown in Table 2. As can be seen from the table, the higher the initial unloading confining pressure is, the morethe deformation modulus declines during failure of the samples. The relation between the two is as shown in Fig.6. Based on linear fitting analysis:

$$\Delta E=0.9665\sigma_3+39.17 \tag{1}$$

Where $\Delta E$ represents the declining percentage of deformation modulus, $\sigma_3$ represents the initial unloading confining pressure.

As can be seen in Fig.6, with the increase of initial unloading confining pressure, the declining percentage of deformation modulus from initial unloading stage to the time when the sample yield failure occurs increases linearly.

**Table 2.** Variation of Deformation Modulus in Unloading Process

| Initial unloading confining pressure [MPa] | Initial deformation modulus [GPa] | Deformation modulus when destruction [GPa] | Decline percentage of deformation modulus [%] |
|---|---|---|---|
| 10 | 1.88 | 0.94 | 50.00 |

| | | | |
|---|---|---|---|
| 20 | 3.08 | 1.35 | 56.17 |
| 30 | 3.26 | 1.00 | 69.33 |

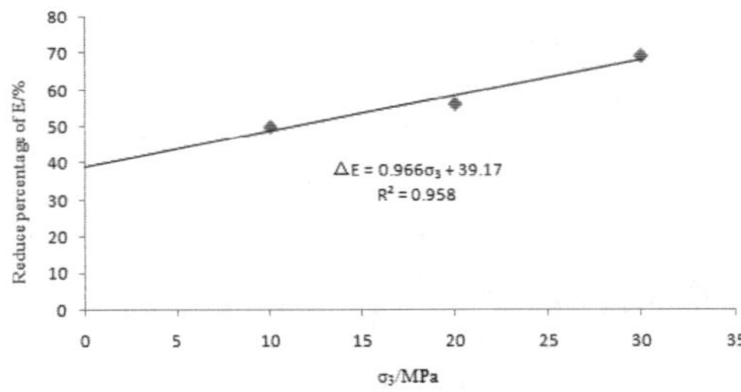


**Figure 6.** Relation curve of declining percentage of deformation modulus and initial unloading confining

*5.2 The Relationship Between Axial Strain and Confining Pressure*

Fig.7 shows the relationship between axial strain and confining pressure during unloading process. The increase rate of the sample's axial strain is relatively small at the initial unloading stage, and the increase rate of axial strain rises sharply as confining pressure decreases to destructive confining pressure, then the strain-confining pressure relation curve is almost horizontal. The higher the initial unloading confining pressure is, the bigger the axial strain of the sample is.

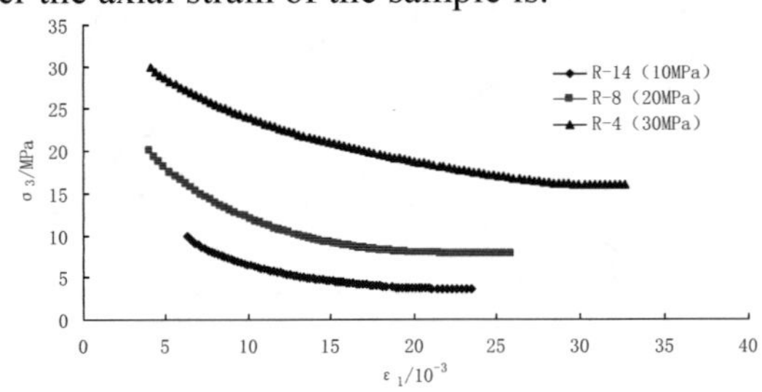


**Figure 7.** Axial strain-confining pressure relation curve

## 6. Conclusion

Through analyzing the test results of sandy mudstone under triaxial unloading test, the conclusions are as follows:

(1)There is no obvious decline in the strain-stress curve of sandy mudstone, which shows evident features of strain hardening. So, it is impossible to acquire the peak intensity of the sample directly based on the variation characteristics of the strain-stress curve.

(2)The sample of sandy mudstone remains relatively complete after destruction.There is usually only a whole shear failure surface in the samples destroyed under lower initial unloadingconfining pressure while samples destroyed in higher initial unloading confining pressure develop tensional fractures.

(3) By introducing semilog method into the data analysis of triaxial unloading test, the peak intensity of the samples is determined conveniently through $\sigma_3$-lg$\varepsilon_1$ relation curve.

(4) From the initial stage of unloading to the destruction of the sample, the declining percentage of sandy mudstone's deformation modulus increases linearly with the rise of initial unloading confining pressure. Meanwhile, both the declining rate of its deformation modulus and the increase rate of axial strain increase suddenly when the confining pressure reduces to destructive confining pressure.

## References

[1] X.F. Yuan. Research on the unloading rheological mechanical properties of sandy mudstone.

Yichang: China Three Gorges University. (2014)
[2] M.C. He, T.H. Xue, Y.F. Peng. A new way of determining mechanical parameters of engineering rock masses. Journal of Rock Mechanics and Engineering, Vol. 20 (2001), pp. 225-229.
[3] Z.X. Zou, H.M. Tang, W. XU, et al.. Fitting for mohr envelope model in triaxial test. Geological Science and Technology Information, Vol. 30 (2011) , pp. 118-122.
[4] H.F. Deng, X.F. Yuan, J.L. Li, et al. Research on failure characteristics and determination method for compressive strength of soft rock in triaxial loading and unloading tests. Rock and Soil Mechanics, Vol. 35 (2014) , pp. 959-964+971.
[5] G.X. Tu, R.Q. Huang, H. Deng, et al. Study on the strength and deformation behavior of a huge outwash deposits based on large-scale triaxial tests. Journal of Mountain Science, Vol. 28 ( 2010) , pp. 147-153.
[6] M.Q. You. Mechanical properties of rock. Beijing: Geology Publishing House. (2007)
[7] Z.Y. Lin. Geotechnical testing and monitoring test manual. Beijing: China Building Industry Press. (2005)

IOP Conf. Series: Earth and Environmental Science **61** (2017) 012122 doi:10.1088/1755-1315/61/1/012122

# Construction method of foam glass thermal insulation material in sloping roof

**Longwei Hu[1] , Fangming Bu[1] , Fenglu Guo[1] and Zimeng Zhang[1]**

[1]Qingdao University of Technology, Qingdao 266520, China
Corresponding author: ttyy8885755@163.com

**Abstract.** Foam glass thermal insulation board has the characteristics of fireproof, waterproof, corrosion resistant, noncombustible, mothproof, non-toxic, non-aging, non-radioactive, high mechanical strength and good dimensional stability. Foam glass thermal insulation material in sloping roof construction method is an effective solution to large angle sloping roof construction operation difficulties.

## 1. Introduction

Sloping roof is mainly for the slope angle greater than 30 ° of the roof, because of its beautiful appearance. In recent years, it is used on more and more buildings. Construction process of these roofs during the construction process emphasizes aesthetics, but also pays more attention to the reliability of its roof construction quality. Foam glass thermal insulation board is a new kind of inorganic thermal insulation material. It has the characteristics of fireproof, waterproof, corrosion resistant, noncombustible, mothproof, non-toxic, non-aging, non-radioactive, high mechanical strength and good dimensional stability. We in the construction of Qingdao Zhengda Haier Pharmaceutical Co., Ltd. bio-pharmaceutical R & D Center project have innovated foam glass insulation materials used in slope roof insulation construction, and explored a practical, quality assurance of construction methods.

Foam glass thermal insulation material in sloping roof construction method can fully utilize its characteristics of light weight and high strength. We can effectively solve the problems that the construction operation and the control of quality are difficult in sloping roof.

## 2. Characteristics of Construction Method

Material innovation: The construction method of the new thermal insulation foam glass board instead of traditional insulation materials used in sloping roof insulation works, is to promote new materials, new crafts, new technologies, new methods in the construction.

The quality control: We analyze the process in the new roof insulation works and find the difficulties we may meet before construction. The problems encountered in sloping roof insulation works should be timely detection, feedback and seek effectively solutions.

Simple and efficient: through the application of new lightweight insulation materials ,we can simplify the construction process and the complexity of the operation of workers, improve the efficiency of roof insulation works, cost savings and significantly shorten the duration.

## 3. Process Principle

The construction method is applicable to the construction of sloping roof foam glass thermal insulation material with slope between 30 ° and 60 °. The construction of sloping roof glass foam insulation material with slope range from 30 ° to 60 ° can also be carried out with reference to this construction method.

The construction method full use of foam glass insulation board fire insulation performance, durability and good weather performance, high strength, non-toxic, non-radioactive, harmless, easy processing and deformation, long life, safety and environmental performance advantages, applied to sloping roof insulation works, in accordance with the optimization of the process of "foam glass insulation board material preparation → grass-roots cleanup →snap the line and line test(deepening layout) → laying foam glass thermal insulation board → roof special treatment → insulation layer acceptance". We can adjust the standard by deepening the insulation plate typesetting, and in a certain area of the bomb line test row before the commencement of large-scale construction.

## 4. Operating Points

Foam glass thermal insulation board should be packed in a special box. The transport process of anti-vibration, moisture-proof measures and handling should be gently prevent mechanical and man-made damage. Foam glass thermal insulation board should accord with the requirements of the admission check and be stacked in a solid, dry room. Before foam glass thermal insulation board is used into the construction site, the acceptance should meet the appearance criteria, and do the re-examination about the density of materials, thermal conductivity and compressive strength.

Before the construction, the water and dust in the primary surface of the base course should be cleaned up. The base course must keep clean without any mortar, floating ash, debris and so on. At the same time, cleaning up the process can not use tools such as spade clean recklessly. It is strictly prohibited undermining the lower structure.

Laying foam glass thermal insulation board:Foam glass thermal insulation board should be directly attached to the grassroots by mortar full stick method (adhesive thickness ≥ 5mm). Slit should be packed between the boards and the gap of plate should be controlled within 2mm. Slit between the boards should be wedge packed by the foam glass thermal insulation. People can not trample on the boards before adhesives curing. Foam glass insulation board in the sides and the node (sink, drain, etc.) of the roof must use special adhesive (or organic binder) to bond with the base course. Adhesives should be applied to the back and the side of the insulation board, the base course.When foam glass insulation board laid, we should according to the line snapped from one side to the other (or from the center to around) to reduce the stampede during the construction process.

The djacent foam glass insulation boards should be staggered stitched. The staggered length should satisfy 1/2 standard board length and ≥ 100mm. Minimum non-standard insulation sheet size should be ≥ 150mm, and should not be set at the edges. Layered laying boards between the upper and lower layers should be staggered with each other, the gap between the boards using the cutting foam glass thermal insulation boards matted stability and inlaid dense.The cutting processing of foam galss thermal insulation usually use portablecutting machine or hacksaw.foam glass insulation board should be promptly applied by the pressure level, to ensure that the insulation board and roof slope to keep in line. The perturbation shall not be percussed during the condensation process. Foam glass paving after 2h, the slope does not meet the requirements of the available special scrubbing plate leveling.

## 5. Quality Control

Master project:1 The thermal insulation properties of foam glass insulation board must meet the relevant specification requirements;2 The laying of the insulation layer in the appearance of the quality of the construction should not have serious flaws;3 The insulation board overlap between the staggered joints, lap length must meet the requirements, the gap between the board filled with mortar dense

General items:1The density of glass foam, thermal conductivity, compressive strength must comply with regulatory requirements;2 When bonding the insulation, we should make it flatten by guiding rule to ensure the slope of the insulation board roof;3 The appearance quality defects of the insulation layer shall be governed by the parties concerned, such as supervision (construction) units, construction units and so on according to its influence on the thermal insulation performance;4 After the completion of construction, the supervision (construction) units and construction units shall check the appearance quality and size deviation, make a good record, and deal with the defects in time according to the technical scheme of construction.

The key points of quality control in key parts and key process:1 Insulation layer and the grass-roots mortar with tile bonding, full compaction, shall not appear empty;2 The flatness and the sensory quality of roof should meet the general quality of plastering requirements;3 The slope and elevation of the roof should meet the design requirements and construction specifications;4Snap the line must be fully reasonable, in accordance with the order from the middle to the surrounding to avoid cross-construction;5 Insulation board should be staggered stitching and the staggered stitch length should be 1/2 standard board length, and ≥ 100mm.

## 6. Safety Measures

Everyone who enters the construction site must wear a helmet.When carrying out high-altitude construction operations, the relevant provisions of JGJ80-91, the current national standard, "Technical Code for Safe Operation of High Building Work" shall be complied with.Slope roofing operations must be tied with a safety rope, fasten seat belts.In the roof operations, to wear flat shoes, high heels and easy to slip shoes is strictly prohibited.Around the roof to be set up Weidang protective measures, protective height should be higher than 1.5m above the operating surface.Rainy weather and high winds above Grade 6 are prohibited.

## 7. Environmental Protection Measures

The use of construction waste refuse closed Road or the use of container hoisting, is strictly prohibited volley run to cause dust.Site-friendly use of directional lighting can be removed lamp cover type, should be used to prevent light pollution.Each time after the completion of the construction should be promptly clean down the hopper, you can re-use of the Treasury in a timely manner to prepare for the next use, can not be recycled back to a place, timely clean-up.

## 8. Conclusion

The whole construction process did not appear serious quality problems and overall construction quality is excellent. We achieved the expected quality and time limit objectives. Practice has proved the feasibility of this new material construction method.

Economic Benefits:At this stage, although the foam glass insulation board compared to XPS, ESP and other traditional insulation material unit price is higher, but its material performance, energy saving and environmental protection, no post-maintenance costs, worthy of promotion and popularization.The use of new materials in the construction process and deepening the layout design, construction speed, short cycle, reducing a large number of rejection, repair and clean-up work, while saving less work days, for follow-up work to provide adequate time limit. Cost savings during the construction are about 8500 yuan.

Each roof to save labor 10 days; a total of 8 roofings, labor costs 200yuan / working days. So artificial saving: $10 \times 8 \times 200 = 16000$ yuan.

Foam glass insulation materials due to higher than the traditional price of insulation materials, construction materials for the use of new materials, the cost increase of about 1,500 yuan.So the actual cost savings: 16000-1500 =14500 yuan

Social benefits: We originality ues this new and efficient insulation materials in sloping roof insulation works and actively promote the idea of building environmental protection and energy conservation. Foam glass insulation materials in the construction of insulation promote the project play a leading role in the demonstration. It is conducive to more environmentally friendly and efficient construction of new materials promotion.

## References

[1] Yan Ruping and Li Jinxia, Application of Foam Glass Thermal Insulation Board in Sloping Roof Project (Architecture Technology, Beijing, 2015), pp. 651–655.

[2] Situ Wenlin, The Foam Glasshouse Facel Insulation System Construction Technology (Guangdong Architecture Civil Engineering, Guangzhou, 2013), pp. 54–56.
[3] Li Sen and Wang Yuejun, Application of Deposed Foam Glass in Roof Insulation Project (Construction Technology, Beijing, 2013), pp. 64–67.
[4] Yu jinyong, Cheng jia and Yu Yangqing, Roofing Foam Glass Insulation Technology in One City of Vanke (Guangdong Architecture Civil Engineering, Guangzhou, 2014), pp. 39–41.
[5] Lu xinhao and Dai Chao, Inverted Roof Insulation Foam Plate Mounting Construction Technology (Architecture Technology, Beijing, 2015), pp. 113–116.

IOP Conf. Series: Earth and Environmental Science **61** (2017) 012123 doi:10.1088/1755-1315/61/1/012123

# Research on construction technology of shaped steel reinforced concrete beam and column

**Shan Zhu[1], Zimeng Zhang[1], Lengjing Liu[1] and Fangming Bu[1]**

[1]Qingdao University of Technology, Qingdao 266520, China
Corresponding author: 290353345@qq.com

**Abstract.** The Shaped Steel Reinforced Concrete structures are increasingly used in large-span or irregularly shaped public buildings. In the actual construction, cross-operation of steel structures and concrete is more frequent. Traditional construction method causes a lot of quality problems easily like installation deviation of beam and column, improper treatment of steel at the joints of beams and columns, and unevenness of concrete surface. If you can not take timely measures to deal with or prevent, it will inevitably have a great impact on the reliability of the structure, and also delay the construction schedule. This paper makes related research about the quality issues, proposes corresponding solutions and provides a complete set of construction technology processes. At last, through engineering case, we analyze the economic and social benefits.

## 1. Introduction

With the development of economy, the number of new buildings which has different functions is increasing day by day. The Shaped Steel Reinforced Concrete structures are increasingly used in large-span or irregularly shaped public buildings. This structure has characteristics of a large stiffness, high strength, light weight and so on. Compared with the prestressed structure, this structure reduces the number of processes. Due to the external role of the concrete wrapped in steel, it overcomes the shortcomings of poor corrosion resistance, poor rust resistance, poor fire performance, and local or total instability.

However, in actual Shaped Steel Reinforced Concrete structures construction, steel and concrete cross jobs more frequently. Especially in the beam-column joints at steel complex cross. If not handled properly it will have a huge impact on the reliability of the structure, and also delay the construction schedule, affecting project quality. Due to the complicated beam and column joints of steel-reinforced concrete structures, steel banding, formwork support and concrete pouring process are different from ordinary reinforced concrete structure, there is still a series of problems in the construction process.

## 2. The key technologies and application principle

Space borrow measurement positioning technology: According to the shaped beam in 3D space to change the form of borrowed measurement method. According to the shape of the beam in the three-dimensional space changes, to develop borrowing measurement method.

Shaped beam cross-pillar node processing technology: In the design of the shaped beam, according to the position of the beam inside the beam column to draw the exact location of the cavities of the special beam wing, to ensure that the reserved hole does not affect the structural force of the special beam. In the factory production of shaped beams, according to the size of the drawings to control the location of the hole to ensure that the bottom of the steel bar in the hoisting accurately through the shaped beam.

Adjusting Perfusion and shaped twisted corbels concrete Pouring Technology: In ensuring the strength of concrete at the same time, adjust the ratio, try to use fine aggregate, timely adjustment of pouring method. First with a large rod diameter vibrating rods on both sides of the beam evenly drainage, and then a small rod diameter vibrating rods tightly vibrating, making the beam column force balance, to ensure the density of concrete pouring and concrete forming aesthetics. In the twist of the heel, control the distance from the beam. Pouring concrete with a small vibrating rod vibrating, to prevent the concrete pouring side of the pressure caused by the phenomenon of mold, to ensure that the concrete molding no hemp.

## 3. Construction process and operation points

### *3.1. Construction Process*

(1) The processing and production of steel members
(2) Space borrow measurement positioning
(3) Steel Column in place installation
(4) Steel Beam place installation
(5) Outsourcing reinforced installation
(6) Columns, beams template installation
(7) Concrete pouring

### *3.2. The Key Points of Space Borrow Measurement Positioning Technology*

First of all, we use CAD to draw the drawings of the beams, columns of the plane line, the spatial line and steel layout.

Use the AutoCAD to establish the user's coordinate system. Through a complete control point database, we can get the coordinates of the points anytime for accurate measurement outside. Use CAD to accurately measure the horizontal distance between the points and the corresponding axis so that the steel structure, steel, template and other professional operations.

On the horizontal plane, use the Total Station to measure the coordinates of each control point to ensure that the coordinates on the screen is less than 3mm.

In the facade, since the fall line point error is greater, measured using a Laser Plumb Instrument.

### *3.3. Installation points of shaped steel columns in place*

This project is a rigid steel reinforced concrete structure, supported by the steel reinforced concrete column, the roof steel reinforced concrete beam irregular arrangement, resulting in the formation of irregular columns of steel reinforced concrete column cap. The steel cap is welded by a three-layer steel plate and a vertical stiffener.

Before lifting the steel plate with steel cap column symmetrical four-sided welding ear plate, with the tower crane hanging to the installed with the steel column alignment, with two 70*400 connection plate and pre-welded to the steel column on the ear plate with bolts to do temporary fixed, in order to ensure the stability of the steel column, adjust the position and elevation, cut off the hanging ear. Steel pillar assembly node.

### *3.4. Shaped beam cross-pillar within the node processing techniques*

In the steel beam diagram of the second deepening design, according to the beam at the end of the column inside the steel bar to draw the upper and lower wing of the steel beam to set the exact location of the blank.

In the lifting process, with the tower crane to the first truncated to the steel cap next to the pre-erected installation platform, and steel cap (legs) to do temporary connection, and find the location, and then lifting the second cut steel beam.

In the installation of steel beams, we need to repeatedly observe and correct the axis, elevation, vertical deviation, until the requirements meet the requirements, before the butt welding. After the steel beam is welded, the vertical elevation of the steel beam is re-examined.

## 4. Outsourcing reinforced installation

beam height <700mm, the beam is not a retractor; beam height ≥ 700mm, the hook will be welded at one end of the steel beam in the web, the other end with the beam barbed strong.

steel girder main reinforcement: beam in the beam at the junction of steel, the beam on both sides of the outermost main bar around the steel column, the outer second row of the main bar through the steel column web reserved hole, the middle of the main bar welded the steel beam is pre-welded on the connecting plate. At the intersection of primary and secondary beams, perforations are made on the web of the secondary girder, and the main girders are passed through the reserved holes.

steel beam with the first tie and then support the beam at the end of the way the template construction. Lashing steel bars before the steel beam in the bottom spot welding I-shaped positioning of steel, I-shaped positioning of steel spacing according to the beam span, generally a beam set three positioning ribs. Beam under the steel bar and steel reinforced with a solid band.

## 5. Column and beam template installation

The beam side mold uses a 15mm thick multilayer board, and the beam bottom mold uses a 15mm thick multilayer board. Beam under the keel of 50mm×100mm square wood, spacing 300mm; main keel for the 100mm×100mm square wood, spacing 600mm. When the beam height <700mm, the beam side mold with 100mm×100mm square as a bracing, spacing 500mm; when the beam height≥700mm, the beam side mold with 100mm×100mm square wood as a bracing, spacing 500mm, and set to pull Bolts, bolts and welded plates welded to steel webs, with a spacing of 500.

first install the beam at the end of the template, according to the design drawings, according to 2/1000 from the arch. Pad with self-made mortar pad, spacing 600, plum-shaped arrangement. And then install the beam side of the template, and 100mm×100mm wooden square braided solid, tighten the pull bolts.

## 6. Innovation in this paper

The space borrowing measurement and positioning technology solves the problem of the construction of profiled steel reinforced concrete beam. This technique ensures accurate transmission of structural forces.

Shaped beam cross-pillar within the node processing technology works on each connected node of the structure to draw steel through the steel flange or web perforation and reinforcement nodes. Factory custom components to ensure the accuracy and dimensional accuracy member opening position, to avoid the on-site repair work.

Adjusting Perfusion and shaped twisted corbels concrete Pouring Technology is reasonable to adjust the concrete ratio, to ensure the density of concrete pouring and forming aesthetics.

## 7. Conclusion

By deepening the design, the amount of steel bars is reduced, and the amount of steel bars is reduced by about 10% before optimizing the design.

In this paper, through the use of steel components of the production and installation methods, welding methods, the installation of reinforced bars, template reinforcement methods and concrete pouring methods and other construction techniques, can improve efficiency and speed up the construction speed. This technology can save the construction machinery classes, steel and labor costs about 350,000 yuan, with a considerable economic benefit.

## References

[1] Chen Xueying and Li Weikui, On Construction Technology Optimization of Steel Reinforced Concrete Column for an Office Building Project (THEORIES&TECHNOLOGIES, Beijing,2013), pp. 114-118.
[2] Zhang Xiuchuan and Ding Quanmin, Construction Technology of Steel Reinforced Concrete Beam - column Joints (Tianjin Construction Science and Technology, Tianjin,2016), pp.13-15.
[3] Xu Feng, Construction Technology for Connected Rigid Structure of Twin Sister Super Tall

Buildings (BUILDING CONSTRUCTION, Shanghai,2009), pp.356-357.
[4] Liu Chunyan, Construction Technology of Steel Reinforced Concrete Beam - column Joints (SHANXI ARCHITECTURE, Shanxi, 2016), pp.89-90.
[5] Ye Feng and Han Qiuhong, Construction Points and Joint Construction of Stiffened Concrete Structure (CONSTRUCTION TECHNOLOGY, Chongqing,2008), pp.72-73.

IOP Conf. Series: Earth and Environmental Science **61** (2017) 012124 doi:10.1088/1755-1315/61/1/012124

# The key quality control technology of main cable erection in long-span suspension bridge construction

**Yongrui Chen[1, a)], Wei Wei[2, b)] and Jie Dai[1, c)]**

[1] School of Highway, Chang'an University, Xi'an 710064, China.
[2]School of Environmental Science and Engineering, Chang'an University, Xi'an 710064, China.
[a)] chenyr@chd.edu.cn
[b)] Corresponding author:weiwei8587@163.com
[c)] counter_dj@163.com

**Abstract.** Main cable is one of the most important structure of suspension Bridges, which bear all the dead and live load from upper structure. Cable erection is one of the most critical process in suspension bridge construction. Key points about strand erection are studied in this paper, including strand traction, lateral movement, section adjustment, placing into saddle , anchoring, line shape adjustment and keeping, and tension control. The technology has helped a long-span suspension bridge in Yunnan Province, China get a ideal finished state.

Main cable is one of the most important structure of suspension Bridges, which bear all the dead load and live load from upper structure. Main cable erection is the key of the suspension bridge construction. Main cables in modern suspension bridges are made from high strength galvanized steel wire. Mature construction technology include prefabricated parallel wire beam method (PPWS method) and air spin method (AS). As advantages such as high erection efficiency, easy control of fabrication in factory and higher precision in site, the PPWS method is deeply adopted in China.

## 1. The goal of main cable erection

### *1.1. Precision requirements of main cable erection*

**Table 1** Precision Requirements of Main Cable Erection

| Subject | | | Allowable deviation |
|---|---|---|---|
| Strand elevation (mm) | Standard strand | Midspan of main span | ±L/20000 |
| | | Attitude difference between upstream and downstream strand | 10 |
| | Normal strand | Interval to normal strand | 0，+5 |
| Anchor span force deviation | | | Comply with design requirements |
| Main cable void fraction （%） | | | ±2 |
| Out of roundness of main cable diameter （%） | | | 2 |

### *1.2. Appearance Requirement*

(1) The strand wires are parallel and straight; there is no drum and overlap in strand.
(2) Straight, no cross, no twist
(3) The galvanized layer are protected in good condition; surface of wire is clean.

## 2. Key control points of main cable erection

### *2.1. Strand traction*

Lift strand by mobile crane and release the strand to necessary length so that anchor head can reach the traction. Lay strand cable on level rollers in front of the cable tray. Adjust the section of strand to standard hexagon starting from the anchor head and fix temporary fixture. Then pull the strand after strand is in position.

**Figure 1** Strand Tracking

Strand traction speed should be controlled in 15 ~ 20 m/min. But when anchor head gets through door frame, the speed should be controlled in 5 m/min. The pulling speed should shift smoothly with start, stop, shift.

Monitor strand traction in real-time to prevent twist, wear away, binder fracture, drum in strand.

When the strand tag reach the bulk cable saddle on the other side, stop strand traction. Take the anchor head away from the traction by mobile crane, and temporarily fixed the other anchor head by counter loop wire rope.

### *2.2. Lateral movement of strand*

Install bond tools on positions 20m on both sides of tower tops and cable saddle. Before installation, the section of strand should be adjusted to standard hexagon. Tighten the bolt on bond tools by several times.

Before lateral movement, connect bond tools with hoist on door frames at tower tops and anchorages, then start hoists to raise whole strand from rollers on cat walk in finite lift distance.Use the lateral move devices on the top of the tower and loose saddle to move strand to the given position. After the whole span of strand leaving rollers, strand can be moved on the saddle with suitable tension.

### *2.3. Section adjustment, placing in saddle and anchoring*

**Figure 2** Section Adjustment of Strand

**Figure 3** Strands in saddle

Set hexagonal fixtures between the two torpedo clip at tag position in order to clamp strand, then remove fiber bundle bands between fixtures and torpedo clips. Adjust strand to quadrilateral section 1m next to hexagonal fixtures and fix quadrilateral fixtures. Starting from quadrilateral fixture, using stalloy comb steel wire while tap strand with rubber hammer, and insert lateral and vertical stalloy in adjusted strand (when necessary, fix the strand temporarily with galvanized sheet fasten or adhensive bonding belt). Strand adjustment should start from mid-span direction to saddle direction.

After strand adjustment complete both at main cable saddle and bulk cable saddle, remove galvanized sheet fasten and adhensive bonding belt. Then place strand into saddle according to direction , order and goove number from design, and remove strand clamp. At the tower top, strand should be placed into saddle from side-span direction to main-span direction, otherwise, at bulk saddle strand should be placed into saddle from anchorage direction to side-span direction. Strand should be placed in main-cable-saddle before bulk-cable-saddle. After strand is placed into saddle, it need to be clamped by chock temporarily to prevent slide. At last, the right steel seperators are installed in saddle according to blue print.

### *2.4. Line shape adjustment*

#### *2.4.1. Line shape repetition survey before erection*

After main tower, abutment pier and anchorage are completed, all related geometric parameters need to be remeasured precisely under stable air temperature before cable erection, including the coordinate of reference points on towers and piers, the perpendicularity of towers, the posture of main saddles and bulk saddles, center line error of saddles. After the erection of catwalks, these parameters should be remeasured again.

#### *2.4.2. Sag adjustment of standard strand*

IOP Conf. Series: Earth and Environmental Science **61** (2017) 012124 doi:10.1088/1755-1315/61/1/012124

For strand, the critical process of line shape adjustment is to adjust the sag about the point at mid-span. Standart strand adjustment base on absolute height, besides on relative sag to standard strand. Strand in mid-span is prior to side-span. The adjustment starts late in night with stable air temperature and weak wind.

Elevation is measured through trigonometric leveling by total station. The theoretical sag and internal force of strand in anchorage span is computed based on dead load, air temperature, real span between saddles, and other parameters measured in-site. If sag of mid-span is match with theoretical requirement, adjust sag in side-span, and temperarily fix strand in bulk saddles. When sag adjustment in all spans is finished, the internal force of strand in anchorage span should be adjusted to theoretical value through jack pushing and screwing the nuts on rod at the end of strand.

After the first round of adjustment, the strand should be surveyed at least 3 days and nights. Everynight, make sure that there are at least 3 valid survey times, and the results of each time is in the range of allowable error. In the end, take the average level at mid-sapn as the control level of standard strand.

### *2.4.3. Sag adjustment of normal strand*
Sag adjustment of normal strands takes standard strand as the reference, using relative sag adjustment method. That follow the same order with the standard strand. Mark method is used during the day, night by sag adjustment method.

Specifically speaking, in night with stable temperature, measure the sag difference to standard strand at control points in every span. The tension or loosen value is computed according to sag difference, span and temperature correction. The sag is adjusted till the relative sag difference is in the range of allowable error according to disign.

For convenient, the normal strand should be pre-evevated about 200mm～300mm higher than design in main span, and 100mm～200mm in side span. Let the strand suspen in the air, then loose strand according to computed value till the normal strand get close to standard strand.

### *2.5. Line shape keeping*
After every 10 strands are erectied, vertical steel shim line shape keeper should be insert into main cable with interval of 160m in order to keep the relative position of strands.

**Figure 4** Steel Shim Insert between Strands

Adjusted strand between line shape keepers should be bind with φ2cm manila rope till subsequent strand is ready for adjustment.

### *2.6. Tension control in anchorage span*
After the strand in side span has been adjusted and fix in bulk saddle, start tension jack to adjust strand tension. Considering the effection of temperature, tension of each strand is compared with design till its difference is in the range of allowable error. The tension is evaluated by jacking force and displacement of anchor slab which is measured by micrometer gauge.

in the process of tensioning, jacks should be feed oil slowly and evently. Make sure oil guage is stable after every level of load, then start the next level. Meanwhile, check whether strand anchor head is tilted or the strand anchor plate is perpendicular with the pull rod. Otherwise, unload the jack and adjust the anchor, then tense the strand again until the tension is in the allowable range.

When jack is working and achor moves to anchor surface, the hexagonal nut should be screw up and locked in time in personnel. The torque value of two nut on rod fixed the same strand should be consistent.

**Figure 5** Tension Adjustment of Starnd

## 3. Conclusion

By tracking survey the construction of a long-span steel box-girder suspension bridge in Yunnan province, China, perfect line shape and ideal inter force are achieved with adoption of the key points to control the erection of main cable in this paper.

**Figure 6** Research in Construction of Suspension bridge

**Figure 7** The Overall View of the Bridge

**Acknowledgement**

This work was supported by Chinese Universities Scientific Foud of Chang'an University (NO. 310821171011, 310829161013, 310821171020). We gratefully acknowledge the invaluable contribution of Pro. Laijun Liu, Lecture Wei Wei, Dr. Xiaoyu Zhang, Dr.Jie Dai in this paper.

## References

[1] Arzoumanidis S G, Bieniek M. Finite element analysis of suspension bridges[J]. computers &Structures, 1985, 21(6): 1237-1253.
[2] Ohshima H, Sato K, Watanabe N. Structural analysis of suspension bridges[J]. Journal ofengineering mechanics, 1984, 110(3): 392-404.
[3] Dumanogluid A, Soyluk K. A stochastic analysis of long span structures subjected to spatiallyvarying ground motions including the site-response effect[J]. Engineering Structures, 2003, 25(10):1301-1310.
[4] Suzumura K, Nakamura S-I. Environmental factors affecting corrosion of galvanized steel wires[J].Journal of materials in civil engineering, 2004, 16(1): 1-7.
[5] Jens Vejlby T, Matthew L. Little belt suspension bridge-corrosion protection of the main

cablesand maintenance of major components [Z][J]. Copenhagen Denmark, 2004, 221.
[6] Montgomery D C. Introduction to statistical quality control[M]. John Wiley & Sons, 2007.
[7] Housner G W, Bergman L A, Caughey T, et al. Structural control: past, present, and future[J].Journal of engineering mechanics, 1997, 123(9): 897-971.
[8] Hayes-Roth F, Waterman D A, Lenat D B. Building expert system[J], 1983.

IOP Conf. Series: Earth and Environmental Science **61** (2017) 012125 doi:10.1088/1755-1315/61/1/012125

# Investigation on unified model of constitutive relations for masonry

**Zhong Huang[1,a) ], Tianquan Liao[2,b)], Bingzhang Huang[3,c)], Bangbiao Huang[4], Zhi Li[2], Bei Zhang[4], Jiayu Pan[4], Weiwei Qi[4], Xiao Li[4] and Junyun Wang[4]**

[1]County Government of LuZhai,Liuzhou 545006, China.545600.
[2]PinXiang Development reform bureau,PinXiang 532699.
[3]Zhong She Fan Huan engineering consulting company,Liuzhou 545006, China.
[4]College of Civil and Architectural Engineering, Guangxi University of Science and Technology, Liuzhou,545006, China.
[a)]Corresponding author:1532375023@qq.com
[b)]Corresponding author:2247123294@qq.com
[c)]Corresponding author:1173997381@qq.com

**Abstract.** In order to establish a unified model of constitutive relations for masonry explain the relationship between energy dissipation and damage energy in the process of uniaxial masonry compression.Based on the application of energy principle and the theory of damage mechanics,a constitutive relation of masonry with parameters is proposed by combining a damage function and the stress-strain curve's characteristics which is under axial compression. The constitutive relation expressions of masonry for different materials can also be determined, only in combination with the peak stress and strain of the axial compression test.To verify this model, the constitutive relation of the perforated shale brick masonry was put forward according to it's compression tests results and the constitutive relation fits well with the experimental data.Besides,not only in the small deflection,it's curve shape is basically in line with these constitutive relations suggested by others, but it can also avoid the defects of these relations to reflect the curve features in the large deformation.

## 1 Introduction

In both seismic regions and non-seismic regions,the masonry structure has a wide range of applications in the multi-storey building and the high-rise buildings.So far, researchers at home and abroad have done a lot of studies on the basic mechanical properties of masonry[1-5], but the constitutive relations researches are relatively few and all these constitutive relations are empirical formulas which have been put forward only based on the experimental results[6-10]. A perfectable constitutive relation model of masonry should have the following characteristics. Firstly,for the different materials of masonry,the constitutive relation can also be proposed by determining the parameters which are related to the masonry compression strength and the peak strain. Secondly, the expression must be terse in form and can show the relationship between stress and strain of masonry under the action of axial compression.Thirdly,the constitutive relation curve must be continuous and smooth in the peak point to ensure the calculation results of convergence and so that it can be applied in the finite element simulation software,and besides,there is only an inflection point in the curve of the decline period.Apart from these features,the constitutive relation suggested not only based on the test results,but also have the theoretical inference.

## 2 Unified model of constitutive

Masonry is a two-phase hybrid material, and it's blocks in the process of blocking and sintering forms some micro cracks and holes.From the macroscopic mechanical properties, the rise of the masonry stress-strain curve shows a stiffness degradation phenomenon ,this suggests that loading process of masonry under the compression is essentially a material degradation process caused by it's internal defect.Therefore, damage mechanics can be used to explain the process of crack extension until failure due to accumulation along masonry damage.

### *2.1 Masonry compression energy dissipation*

The variation of the state or the motion state of objects is a thermodynamic process,and this process is not reversible while the variations accompany an energy dissipation.Due to the masonry damage process under static load changes slowly,so in any short period of time , the masonry can be considered as a state of balance.

Assume that masonry axial compression test is conducted in an adiabatic system with no heat loss,and thus the work done by compressive stress $\Delta W$ is converted into masonry's strain energy $\delta K$ and internal energy $\delta U$ and kinetic energy $\delta E$ .Where the internal energy increased $\delta U$ is essentially the heat masonry accepted $\delta Q$ .According to the first law of thermodynamics, it can be derived equation (1).

$$\Delta W = \delta K + \delta U + \delta E \qquad 1)$$

In the loading process of masonry under axial compression, the kinetic energy $\delta E$ keeps very close to zero for the speed of lower loading.But In practical,the loading process is conducted in a thermostat system,for this reason, the internal energy of masonry is unchanged and masonry have $\delta Q$ of heat released.so in the process of masonry compression, the process of masonry internal damage the same as heat dissipation is an irreversible process.Therefore, the energy dissipation of damage owing to development of cracks can be considered as heat released in the processes of masonry compression.So get equation(2).

$$\Delta W = \delta K + \delta Q \qquad 2)$$

Make a division on the opposite side of the equal sign by divide the volume V of masonry,and get equation(3).

$$W(\varepsilon) = W_Q(\varepsilon) + W_K(\varepsilon) \qquad 3)$$

Where $W(\varepsilon)$ is strain energy density of the elastic system when the strain is $\varepsilon$, $W_Q(\varepsilon)$ is the energy density released as the strain from 0 to $\varepsilon$ ,during the masonry compression damage, $W_K(\varepsilon)$ is strain energy density of the elastoplastic system.

On the basis of above study,a damage variable D caused by external compressive strain can be defined.

$$D = \frac{W_Q(\varepsilon)}{W(\varepsilon)} = 1 - \frac{2\int_0^{\varepsilon} \sigma d\varepsilon}{E\varepsilon^2} \qquad 4)$$

Where E is the initial elastic modulus of masonry compression, $\sigma$ and $\varepsilon$ is stress and strain,respectively.Take the derivative of equation(3) and get (5).

$$\sigma = E\varepsilon(1-D) + \frac{1}{2}E\varepsilon\frac{dD}{d\varepsilon} \qquad 5)$$

Damage variable is a constant for the sick of the irreversible nature of the masonry compression in the unloading process and so $\frac{dD}{d\varepsilon}$ =0. Then,get the masonry damage constitutive relationship expression (6).

$$\sigma = E\varepsilon(1-D) \qquad 6)$$

Expression(6) shows that the masonry Elastic Modulus is $E(1-D)$ result from stiffness deterioration cause by the damage development.

Due to the masonry damage is irreversible, and it's degree increased along with the strain growth.so the damage function D is a monotone increasing function,when D=0 is denote masonry without damage, and D=1 is entirely damaged.According to the characteristics of damage function D,various expressions of damage functions can be used,such as polynomial,piecewise function,as well as probability distribution functions like Weibull distribution function,logarithmic normal distribution ,etc.Thus there are varieties of forms for constitutive relation expression can be constructed.In order to establish a simple expression so that it can be accepted by engineers,expression(7) is chosen as the damage function D.

$$D = 1 - \frac{a}{a + (1-a) \bullet (\frac{\varepsilon}{\varepsilon_0})^b} \qquad 7)$$

Where $a$ and $b$ are parameters determined by the compressive strength of masonry and the peak strain.

Expression(8) can be derived by combining expression(6) and (7).

$$\sigma = \frac{E\varepsilon a}{a + (1-a) \bullet (\frac{\varepsilon}{\varepsilon_0})^b} \qquad 8)$$

Masonry compression stress-strain curve have the following characteristics:① at the origin of stress strain curve,where $\varepsilon$ =0,there are $\sigma$ =0 and $\mathrm{d}\sigma/\mathrm{d}\varepsilon = \mathrm{E}$ .That is,the elastic modulus of the origin is E.②Where $\varepsilon = \varepsilon_0$ ,there are $\sigma = f_m$ and $\mathrm{d}\sigma/\mathrm{d}\varepsilon = 0$ .Obviously, expression(8) meets the feature ① According to the feature ②, the following results can be derived.

$$a = \frac{f_m}{E\varepsilon_0} \text{ , } b = \frac{1}{1-a} \qquad 9)$$

According to equation (8) and (9),the unified constitutive relation model with parameter for masonry-expression(10)-been put forward.Obviously,for different materials of brick masonry,the constitutive relation can be proposed by determine the parameters $a$ and $b$.

$$y = \frac{x}{a + (1-a)x^b} \qquad 10)$$

## 3 Application and validation

### *3.1 Experiment introduction*

Adopting sintered shale porous brick with strength grade of MU10 and ordinary cement mortar with strength grade of M7.5,making 9 groups of specimens.The specimens size and constructing procedure and method of loading,enforcing strictly according to the standards specifications,namely Standard for Test Method of Basic Mechanics Properties of Masonry(GB/T 50129-50129). The height and length and width of each specimen are 710 mm×365 mm×240 mm.After 28-day standard curing at room temperature, the specimens are conducted compression tests using pressure testing machine with pressures value 200-T.Fig.1 presents the test unit for specimens.

In the process of loading, the test data and the phenomena in the loading will be Carefully recorded.

**Fig1.**Test device for specimens compression

*3.2 Experimental results*

According to the standards, Code for Design of Masonry Struction (GB500003-2011), expression(11) was recommended for calculating the average compressive strength for masonry .

$$f_m = k_1 k_2 f_1^{\alpha}(1+0.07f_2) \qquad 11)$$

Where $f_m$ is the average compressive strength of masonry（MPa）, $k_1$ is parameters related to the types of masonry,and for sintered porous brick $k_1$=0.78, $\alpha$ is also parameters related the kind of mortar, usually k=1.0, $\alpha$ is determined by the block materials,here a=0.5, $f_1$ and $f_2$ are the average compressive strength for bricks and mortar（MPa）,respectively.The main test results are shown in table 1.where f is the measured compressive strength of the specimens in the axial compression test, $\varepsilon_0$ is the measured peak strain for masonry.According to the table,it was clear that the measued values were greater than the values of calculating suggested by expression(12).

**Table 1.** The major data of compressive strength

| Number | Compressive Strength | Average Compressive Strength | PeakStrain | Average Strain |
|---|---|---|---|---|
| | $f$/MPa | $f_m$/MPa | $\varepsilon_0 \times 10^{-3}$ | $\varepsilon_0 \times 10^{-3}$ |
| KY--1 | 4.20 | | 2.19 | |
| KY--2 | 4.19 | | 2.26 | |
| KY--3 | 4.13 | | 2.18 | |
| KY--4 | 4.03 | | 2.55 | |
| KY--5 | 4.31 | 3.78 | 2.41 | 2.27 |
| KY--6 | 4.14 | | 2.02 | |
| KA--7 | 4.05 | | 2.23 | |
| KA--8 | 4.11 | | 2.13 | |
| KA--9 | 4.05 | | 2.42 | |

*3.3 Constitutive relation for sintere-shale porous brick*

The elastic value of masonry is determined by formula (12)[9].

$$E = 370 f_m \sqrt{f_m} \qquad 12)$$

The constitutive relation for sintere-shale porous brick-expression(13) -can be put forward as the parameters $a$ and $b$ was determined by combine with the test results of masonry and expression(10).

$$y = \frac{x}{0.6+0.4x^{2.5}} \qquad 13)$$

### *3.4 Validation and analysis*

The test data contains only the data in the process of loading,on account of the insufficient stiffness of the testing machine.As is shown in Fig.2,in the rising step these results fit well with the stress-strain's curve shape ,which is suggested in this paper.In order to further verify the curve of expression(16) in the downward section,A comparison between constitutive relations which were suggested by others and in this paper is shown in Fig.3.In small deformation range of the decline period,the curve of the formula(16) coincides better with other formulas.In addition to this,the formula(16) also avoids the deficiency of the existing constitutive relations.So the compressive constitutive relations of masonry, formula(16),can be used to reflect the stress-strain relationship of masonry under compression.And thus,the unified constitutive relation model with parameters for masonry,expression(14) is also a rational model.

**Table 2.**Expressions of compressive constitutive relations of masonry

| Number | Formulas | References |
|---|---|---|
| 14 | $y=\begin{cases}\dfrac{1.52x-0.279x^2}{1-0.483x+0.724x^2} & x\le 1\\ \dfrac{3.4x-1.13x^2}{1+1.4x-0.13x^2} & x>1\end{cases}$ | [13] |
| 15 | $y=6.4x-5.4x^{1.17}$ | [12] |
| 16 | $y=2.3x-1.555x^2+0.195x^3+0.075x^4-0.015x^5$ | [14] |
| 27 | $y=xe^{1-x}$ | [15] |

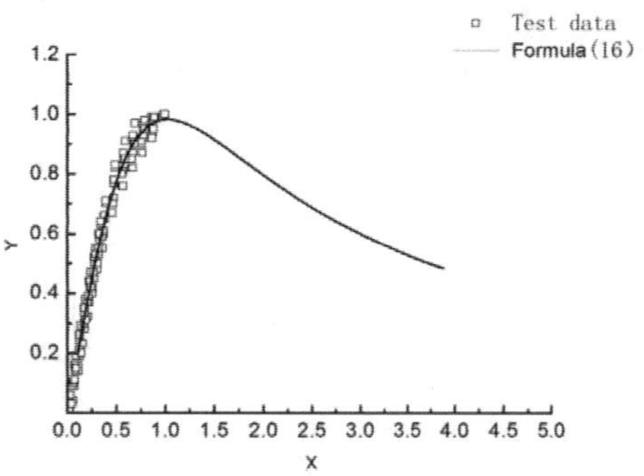


**Fig2.** Comparison between formula(16) and test data.

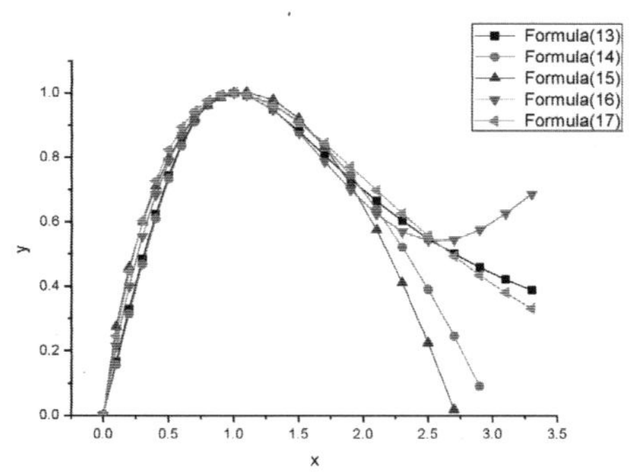


**Fig3.** comparison between formulas of compressive constitutive relations of masonry

## 4 Conclusions

According to the principle of conservation of energy,combining with the damage mechanics theory,established a constitutive relation model of masonry.This model is only related with three parameters-elastic modulus, peak stress and strain.All these parameters can be obtained from masonry axial compression test.According to the axial compression test results of sintered shale porous brick masonry structure,gets the constitutive relationship formula(16).The curve of the formula(16) is agree very well with test data.More importantly,this model avoids the deficiency of the existing constitutive relations.

## References

[1] Bryan D E, Mervyn J K. Compressive behavior of unconfined and confined clay brick masonry [J]. Journal of Structural Engineering, 2004,(130):650-661.

[2] Gregoire Y. Compressive strength of masonry according to Euro code6: a contribution to the study of the influence of shape factors[J]. Masonry International Journal of the British Masonry Sociy,2007,20(2):69-74.

[3] P.Chindaprasirt, K. Pimraksa.A study of fly ash-lime granule unfired brick[J].Powder Technology, 2008,182(1): 33-41.

[4] Dhanasekar R, Ferozkhan M, Dhanasekar M.Behaviour of dry stack concrete masonry blocks under eccentric compression[C].International Brick and Block Masonry Conference, 2008,(13):427-436.

[5] Oti,KinuthiaJ.M.Compressive strength and microstructural analysis of unfired clay masonry bricks[J].Engineering Geology,2009,(109):230-240.

[6] Claudius S,Barbosa,Paulo B.Lourenço.On the compressive strength prediction for concrete masonry prisms[J].Materials and Structures,2010,( 43):331–344.

[7] Oti Kinuthia J.M.Engineering properties of unfired clay masonry bricks[J].Engineering Geology,2009,(1007):130-139.

[8] Gregoire Y.Compressive strength of masonry according to Euro code 6: acontribution to the study of the influence of shape factors[J]. Masonry International Journal of the British Masonry Sociy,2007,20(2)： 69-74.

[9] LIU Gui-qiu,YAN You-qing,SHI Chu-xian.Research on the Unified Model of the Compressive Constitutive Relations of Masonry[J].Journal of Hunan University(Natural Sciences), 2008,4(35),29-32.116-119.(In Chinese)

[10] ZAVALIS R,JONAITIS B.Numerical and experimental analysis of grouted hollow block masonry under compression [J].Engineering Structures and Technologies,2013,5( 2) : 45-53.

[11] ZHU Bo-long.Masonry structure design principle (1st edition) [M].Shanghai: tongji university press,1998.(In Chinese)

[12] B.Powell,H.R.Hodgkinson Determination of stress/strain relationship of brick work,TN 249,B.C.R.A,Stoke on Trent. 1976.

[13] ZHUANG Yi-zhou.The experimental research of model brick masonry mechanical behaviors[J].Building Structure,1997,27(2):22 -25.(In Chinese)

[14] SHI Chu-xian.Masonry structure theory and design (2ed edition) [M].Beijing: China building industry press, 2003,24-78.

[15] NARAINE K,SINHA S. Behavior of brick masonry under cyclic compressive loading [J].Journal of Construction En-gineering and Management,1989,115(2): 1432 -1444.

IOP Conf. Series: Earth and Environmental Science **61** (2017) 012126 doi:10.1088/1755-1315/61/1/012126

# Analysis about equivalent conversion of two dimensions for large cylinder structure

**Guojun Feng[1, 2], Bo Song[1, 2]**

1 School of Civil and Resource Engineering, University of Science and Technology Beijing, Beijing 100083, China.
2 Beijing International Cooperation Base for Science and Technology-Aseismic Research of the Rail Transit Engineering in the Strong Motion Area, Beijing 100083, China.
Corresponding author: fgjshx@126.com

**Abstract**. There was practical engineering of large cylinder structure used in China. Elastic mechanics problems of large cylinder structure quay wall were space problems. Sometimes we needed to transform space problem of large cylinder structure into plane problem of it. Circular section of signal large cylinder structure was converted into rectangular one by using the same bending stiffness and section area for breakwater.

## 1. Introduction

Large cylinder structure is the abbreviation of large diameter thin-walled cylinder structure, and it usually refers to shell structure of over six meters in the diameter, bottomless, no internal partition and thin-walled cylinder. Large cylinder structures mostly were made of reinforced concrete or steel. According to the characteristics of port engineering, large cylinder structure could be roughly divided into two types of bedding and embedded, and the embedded could be divided into shallow and deep buried. The large cylinder structure was shown in Fig. 1.

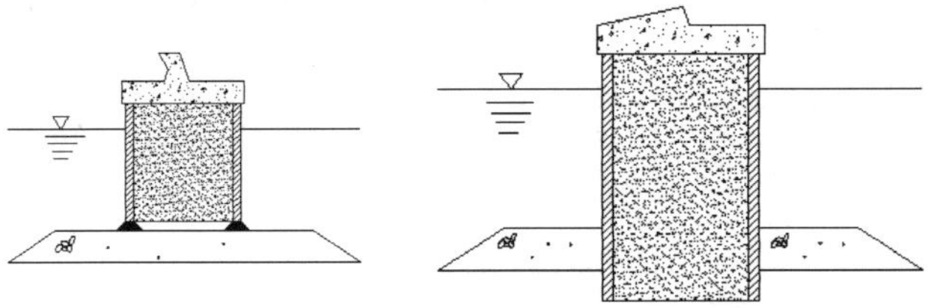

(a) bedding (b) embedded

**FIGURE 1.** Types of large cylinder structure.

Hydraulic structures began to apply large cylinder structure in the 1940s. the structure of new type was at the beginning of 50s in France for the construction of the port at first, and quay wall of large cylinder structure also was used in the Soviet Union, Spain, Britain, Denmark, Canada, Japan and other countries , large cylinder structures were introduced to China in the 1980s.

Many test and engineering practice works of large cylinder structure were done in China. Through the relevant scientific research institutes, universities, design consulting and construction enterprises according to the working mechanism, model and physical model test, design method and construction technology of large cylinder structure were carried out. A lot of research topics had made many remarkable achievements, and practical engineering was used in the project. Advances and Characteristics of large cylinder structure were studied[1], and new design approach of it were analyzed[2]. Some parameters of large cylinder structure in China were shown in the Table 1.

**TABLE 1.** Some parameters of large cylinder structure.

| Type | Parameters | | | Material |
|---|---|---|---|---|
| | diameter | height | thickness | |
| bedding | 7 | 4.75 | 0.6 | reinforced concrete |
| embedded | 12 | 17.1 | 0.28 | reinforced concrete |
| bedding | 16 | 14.4 | 0.3 | reinforced concrete |
| bedding | 8 | | 0.25 | reinforced concrete |
| bedding | 8.36 | 7 | 0.18 | reinforced concrete |
| bedding | 14 | 13/11.2 | 0.3 | reinforced concrete |
| embedded | 15 | 25～26 | 0.28 | reinforced concrete |
| bedding | 15 | 13 | 0.35 | reinforced concrete |
| embedded | 10 | 12 | 0.2 | reinforced concrete |
| bedding | 10 | 12 | 0.2 | reinforced concrete |
| bedding | 14 | 13 | 0.3 | reinforced concrete |
| embedded | 13.5 | 13～34 | 0.014<br>0.012 | steel |
| embedded | 22 | 40～50 | 0.016 | steel |

## 2. Two dimensions theory of equivalent conversion

Elastic mechanics problems of large cylinder structure are space problems. when we will transform space problem of large cylinder structure into plane problem of it, the analysis and calculation of the workload will be reduced, and the results of it also can meet the practical engineering requirements on the accuracy[3] [4]. Diagram of plane stress is shown in Fig. 2.

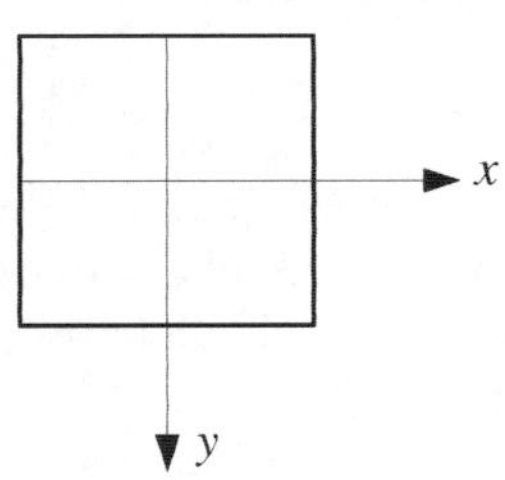


**FIGURE 2.** Diagram of two dimensions.

In the process of elastic mechanics analysis, three equations are established by considering the three aspects of statics, geometry and physics. The relation between the stress component and the physical component in the plane problem can be obtained by the equilibrium differential equation in the plane problem, and the relation of it was shown in formula 1.

IOP Conf. Series: Earth and Environmental Science **61** (2017) 012126 doi:10.1088/1755-1315/61/1/012126

$$\left.\begin{aligned}\frac{\partial\sigma_x}{\partial x}+\frac{\partial\tau_{yx}}{\partial y}+f_x=0\\\frac{\partial\sigma_y}{\partial y}+\frac{\partial\tau_{xy}}{\partial x}+f_y=0\end{aligned}\right\}\tag{1}$$

When the physics of the plane problem were considered, the relationship between the deformation component and the stress component is derived, and the relationship is the physical equation in the plane problem. Because of $\sigma_z=0$, physical equations of plane stress problems are shown in formula 2. because of $\varepsilon_z=0$, physical equations of plane stain problems are shown in formula 3.

$$\left.\begin{aligned}\varepsilon_x&=\frac{1}{E}\left(\sigma_x-\mu\sigma_y\right)\\\varepsilon_y&=\frac{1}{E}\left(\sigma_y-\mu\sigma_x\right)\\\gamma_{xy}&=\frac{2(1+\mu)}{E}\tau_{xy}\end{aligned}\right\}\tag{2}$$

$$\left.\begin{aligned}\varepsilon_x&=\frac{1-\mu^2}{E}\left[\sigma_x-\frac{\mu}{1-\mu}\sigma_y\right]\\\varepsilon_y&=\frac{1-\mu^2}{E}\left[\sigma_y-\frac{\mu}{1-\mu}\sigma_x\right]\\\gamma_{xy}&=\frac{2(1+\mu)}{E}\tau_{xy}\end{aligned}\right\}\tag{3}$$

Where $x$ is the direction of the horizontal axis, $y$ is the direction of the Vertical axis, $\sigma$ is the normal stress, $\tau$ is the shear stress, $f$ is the body force, $\varepsilon$ is the linear strain, $E$ is the Elastic modulus, $\mu$ is the poisson ratio and $\gamma$ is the Shear strain.

## 3. Equivalent conversion for section

According to Technical Standards and Commentaries for Port and Harbour Facilities in Japan[5] and interpretation of the harbor facilities, when the number of large cylinder breakwater is connected with the arc plate, the equivalent width of the wall is calculated according to equivalent conversion in Fig. 3 by formula 4.

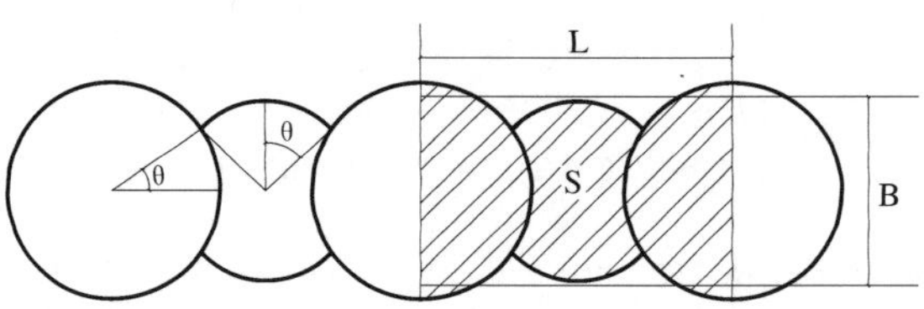


**FIGURE 3.** equivalent conversion for section.

$$B=S/L\tag{4}$$

Where $B$ is the width of conversion wall, $L$ is the length of conversion wall and $S$ is the area of section.

The embedded steel cylinder structure is used on the breakwater. The breakwater consists of a series of steel cylinders, and the steel cylinders are connected through concrete membrane bags. A typical section of breakwater is shown in Fig. 4.

IOP Conf. Series: Earth and Environmental Science **61** (2017) 012126 doi:10.1088/1755-1315/61/1/012126

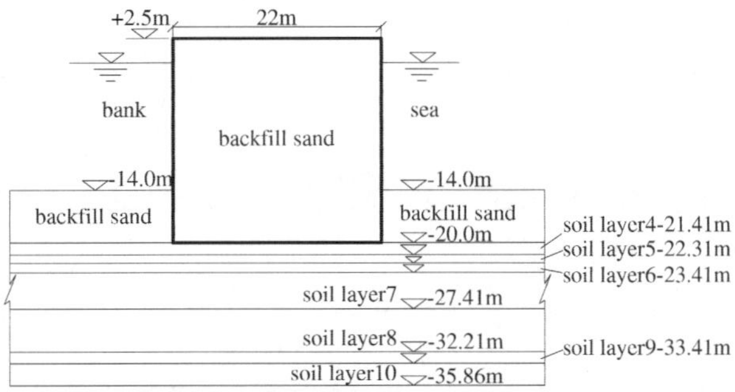


**FIGURE 4.** Diagram of a typical section for breakwater

Circular section is converted into rectangular one (Fig. 5) by using the same bending stiffness and section area of breakwater for signal large cylinder structure. In addition, the width (B) and length (L) of rectangular section are converted using formula (5) and (6). The plane strain model of steel cylinder breakwater can be established according to Fig. 5.

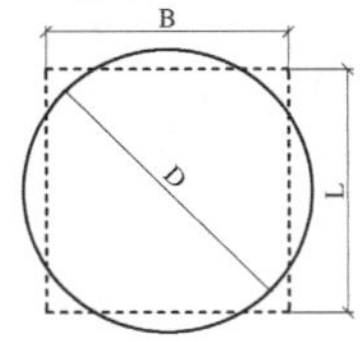


**FIGURE 5.** equivalent conversion for section.

$$A = \frac{\pi D^2}{4} = B \times L \tag{5}$$

$$I = \frac{\pi D^4}{64} = \frac{B^3 \times L}{12} \tag{6}$$

Where $A$ is the section area, $I$ is the inertial moment of section, and $D$ is the diameter of steel cylinder.

## 4. Equivalent conversion for section

Elastic mechanics problems of large cylinder structure are space problems. We may transform space problem of large cylinder structure into plane problem of it. Circular section of signal large cylinder structure is converted into rectangular one by using the same bending stiffness and section area for breakwater.

**Acknowledgments**

This study was supported by the project of Beijing international cooperation base for science and technology-aseismic research of the rail transit engineering in the strong motion area.

## References

[1] ZHANG Zhen-yu, YAO Wen-juan, Advances in the Research of Large Cylinder Shell Structures (Port & Waterway Engineering, 2004, 10(1), 82-90).
[2] FENG Hai-bo, New design approach for wharf structure of large steel cylinder (Port & Waterway Engineering, 2008, 12, 98-104).
[3] Zhilun Xu, Elastic mechanics (Beijing: Higher Education Press, 1984).
[4] Zhilun Xu, Finite element method (Beijing: tsinghua university press, 2003).
[5] MLIT, Technical Standards and Commentaries for Port and Harbour Facilities in Japan (The Overseas Coastal Area Development Institute of Japan, Publisher City, 2009).

IOP Conf. Series: Earth and Environmental Science **61** (2017) 012127 doi:10.1088/1755-1315/61/1/012127

# Evaluation of embankment stability on soft ground considering service loads

**Binbin Xu**[1]

[1] Tianjin Port Engineering Institute Ltd. of CCCC, Tianjin, China; Key Lab. of Geotechnical Engineering of Tianjin, Tianjin, China; Key Lab. of Geotechnical Engineering, Ministry of Communication, Tianjin, China.
xubinbin@tpei.com.cn

**Abstract**. The stability of road embankment on soft ground is always the highlight during construction. Usually the construction is carried out on the reinforced ground and the degree of reinforcement is influenced by the methods and period. In order to evaluate such influence, changing the compression index is a simple way to represent it. Also the service load can be represented by assigning different bulk density of surcharge. Finite element strength reduction method is used to evaluate the stability of embankment at different stages.

## 1. Introduction

In the coastal area, due to the high phreatic line there usually is soft ground with high water content and low bearing capacity. Ground reinforcement is necessary before the construction. The typical characteristics of soft ground are high compressibility, low permeability and low shear strength. The void ratio of soft soil is very large (larger than 1). The compression coefficient is around 1.5~2.0MPa-1 and the modulus of compressibility is less than 4MPa. Because the compressibility of soil is greatly influence by its mineral component, the larger the liquid limit is, the greater the compressibility is. The vertical permeability coefficient is usually smaller than 10-6cm/s. Under the load action, it takes a very long time to reach the high degree of consolidation. The shear strength of soft soil is very small. The drained effective cohesive is smaller than 15kPa and inner frictional angle is smaller than 10.

**Table 1.** Ground reinforcement method

| Method | Principle | Applicability |
|---|---|---|
| Replacement | remove the soft soil and replace by proper soil with high shear strength and low compressibility | all kinds of soft ground |
| Reinforced cushion | install geosynthetics into the ground to improve its stability | all kinds of soft ground |
| Drainage consolidation | install sand well or plastic vertical drains to expel pore water from the ground | deep, saturated soft ground with low plasticity index |
| Dynamic compaction | apply impacting force on the ground by heavy hammer to make the soil | soft soil with high permeability |

| | | |
|---|---|---|
| | denser | |
| Cement mix | rotate soft soil with cement to solidify the ground | all kinds of soft ground |

In recent years, the technics of ground reinforcement have been greatly developed, from the drainage consolidation method to geosynthetics, cement mixed method and vibrating method. The usual ground treatment methods are listed in Table 1. After reinforcement, one of the concerning problems is the stability of the road embankment under different load according to the consolidation effect. Therefore, it is necessary to evaluate the influence of serve load and compressibility of soil on the stability of the road embankment on the soft ground.

## 2. Determination method of stability

In 1955, Bishop[1] proposed another simplified method based on the Swedish method in which the arc-shape slide surface and the equilibrium of the moment are kept while the interaction between two strips is considered. The safety coefficient is represented by the ratio between the resistant shear strength and the actual shear strength on the sliding surface, which is called safety factor of shear stress:

$$F_S = min(F_S^i) = min\left(\frac{\tau_f}{\tau}\right) = \min\left(\frac{c_f + \sigma \tan\varphi_f}{\tau}\right) \quad 1)$$

Although the strength reduction method has been widely used in the stability analysis of the slope, the final safety coefficient is always depended on the evaluation criterion of instability. Usually the numerical convergence is taken as the judge criterion, but the influence factors of the convergence of FEM are multiple and it is difficult to explain whether the divergency is attributed to instability or not. Now there are mainly three judge criterions for the instability of the slope: convergence of calculation, run-through of the plastic range and sudden change of characteristic position.

When using strength reduction method for the slope stability, the calculation convergence is usually used as the criterion[2-6]. If the convergent solution cannot be obtained under the designated convergence judgment, the slope loses its stability. For the run-through of the plastic range, some significant shear bands have been observed during the damage of the slope. Similar regulation can be forecasted. If there is run-through plastic deformation in the slope under certain reduction coefficient, such a coefficient can be regarded as the safety coefficient. For the sudden change of characteristic position, the treatment is building the relationship between the displacement of certain point and the reduction coefficient in the FE calculation. If there is a sudden change in the curve, the slope is in a critical state.

## 3. Calculation results

A FE code called Plaxis3D is used to carry out the strength reduction method. In Plaxis3D, a module called safety analysis can be used for the slope stability analysis. In the calculation, the reduction coefficient is gradually reduced and the total displacement at toe of the slope is also monitored for each reduction coefficient. When the relationship between the reduction coefficient and the total displacement is plotted in the diagram, the safety coefficient is taken as the reduction coefficient if there is no variation even though the total displacement increases. Cam-clay model is used in the calculation. The parameters are listed in Table 2.

**TABLE 2.** Parameters in the Calculation

| | |
|---|---|
| Drainage type | Undrained A |
| Unsaturated bulk density (kN/m$^3$) | 14/16/18 |

IOP Conf. Series: Earth and Environmental Science **61** (2017) 012127 doi:10.1088/1755-1315/61/1/012127

| | |
|---|---|
| Compression index | 0.10/0.12/0.15/0.18 |
| Swelling index | 0.02 |
| Equivalent permeability (cm/s) | $6.0\times10^{-6}$ |

In order to evaluate the influence of service load, the bulk density of surcharge and compression index of soft ground is varying from active and passibe aspects. Fig. 1 gives the settlement of the ground under different bulk densities of surcharge. As can be seen, when the bulk density is 14 kN/m$^3$ the settlement is much smaller than other two densities, while there is no much difference between the settlements when bulk density is 16 and 18. Therefore, it is very useful to decrease the total settlement by reduce the servic load. The settlement at the center of embankment with different compression index is shown in Fig. 2. It is easy to understand that if the soft ground is reinforced under various methods or period the compressibility of final ground might be quite different. It is a feasible way to represent such difference by assign different compression index to the reinforced ground. Obviously, the larger the compression index is, the greater the final settlement is. It can also be noticed that if the compression index is small, the settlement becomes comnvergent soon after the second load and the settlement get stable. However, as the compression index gets greater, namely the soil becomes weaker, the rate of settlement is accelarated.

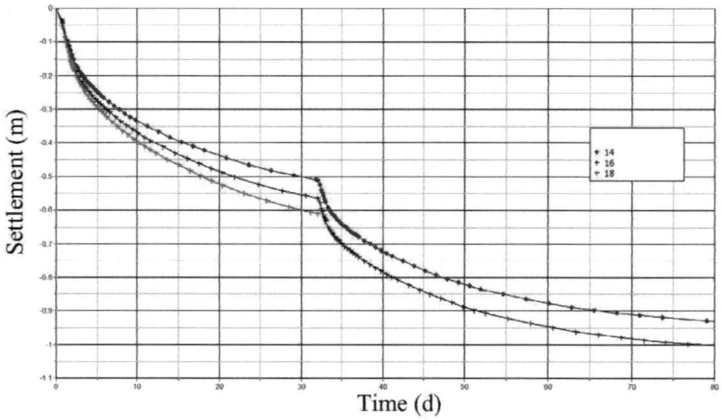


**Figure 1.** Settlement at the center of bottom surface of embankment under different bulk densities.

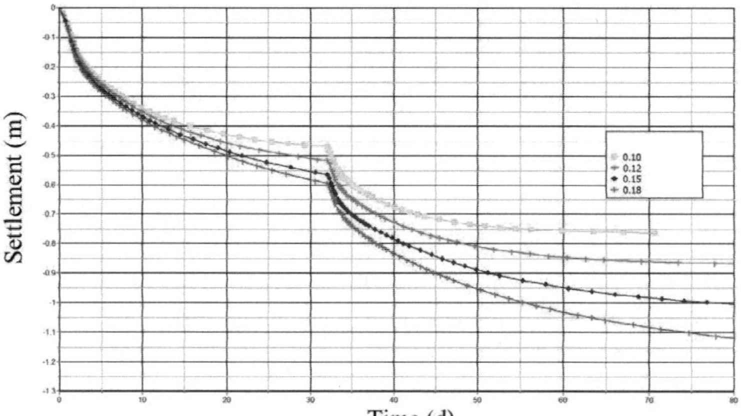


**Figure 2.** Settlement at the center of bottom surface of embankment under different compression index.

IOP Conf. Series: Earth and Environmental Science **61** (2017) 012127 doi:10.1088/1755-1315/61/1/012127

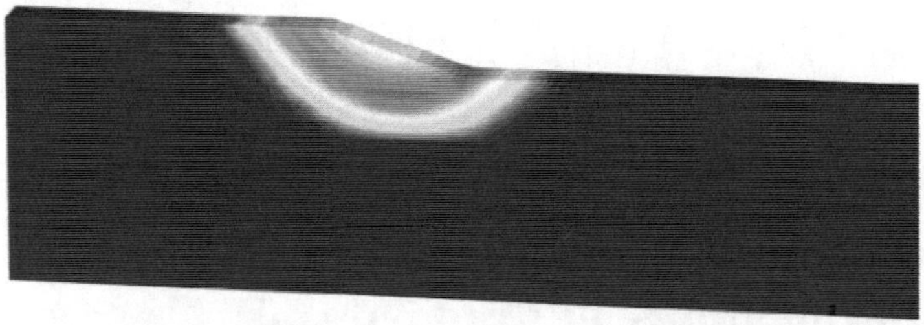

**Figure 3.** Slide surface inside the embankment.

Fig.3 demonstrates the slide surface inside the embankment when the compression index is 0.18. As mentioned above, when the compression index becomes larger, the rate of settlement also gets greater, which results in the slide surface through the embankment. Therefore, there is a risk that the embankment may lose its stability if the reinforced strength is not enough.

The stability coefficient of embankment at different stages is shown in Fig. 4. As can be seen, the coefficient after the first construction is around 1.07 and then 1.76, 1.12, 1.46 followed by first consolidation, second construction and second consolidation. Due to the consolidation, there is a significant improvement for the stability of the embankment.

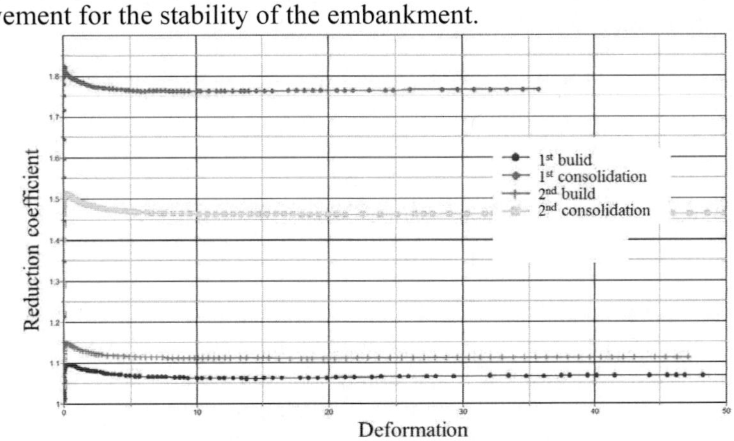


**Figure 4.** Settlement at the center of bottom surface of embankment.

## 4. Conclusions

The influence factors on the road embankment are investigated numerically from both active and positive aspects by varying the bulk density of the surcharge and compression index of soft ground. The conclusions are as follows:

1) When the bulk density is small the settlement is much smaller than other two densities. Therefore, it is very useful to decrease the total settlement by reduce the servic load.

2) When the compression index becomes greater, the final settlement also becomes larger. If the compression index is small, the settlement becomes comnvergent soon after the second load and the settlement get stable. However, as the compression index gets greater, the rate of settlement is accelarated.

3) Enough consolidation time is very useful to increase the stability of the road embankment.

## References

[1] A.W. Bishop, Geotechnique, 5, 7 (1955)
[2] K. Ugai, Soils and Foundations, 29, 190 (1989)
[3] D.V. Grifiths, P.A. Lane, Geotechnique, 49, 387 (1999)
[4] E.M. Dawson, W.H. Roth, Geotechnique, 49, 835, (1999)
[5] J.M. Duncan, Journal of Geotech. Engineering, (1996)
[6] I. Donald, Z.Y. Chen, Canadian Geotechnical Journal, **34**, 853 (1997)

IOP Conf. Series: Earth and Environmental Science **61** (2017) 012128 doi:10.1088/1755-1315/61/1/012128

# Surcharge preloading consolidation analysis of coastal road considering permeability of dredged soil

**Binbin Xu[1] and Wei Si[2]**

[1] Tianjin Port Engineering Institute Ltd. of CCCC, Tianjin, China; Key Lab. of Geotechnical Engineering of Tianjin, Tianjin, China; Key Lab. of Geotechnical Engineering, Ministry of Communication, Tianjin, China.
[2] Tianjin Port Engineering Institute Ltd. of CCCC, Tianjin, China.
xubinbin@tpei.com.cn

**Abstract**. Due to the high plasticity index and low permeability of reclaimed ground composed of dredged soil, the reinforcement period is very long, which makes the merits of vacuum preloading, short reinforcement time, disappear. Surcharge preloading method becomes the most suitable and economic way. The equivalent permeability method is used to simulate the installation of the plastic vertical drain and the influence of different permeabilities, namely the interval space is calculated numerically.

## 1. Introduction

As the expansion of the infrastructure construction, the contradiction between the land resource and the road development becomes more and more intense. One of the effective ways is increasing the area of reclaimed land. However, the difficulty of development of reclaimed land lies in the soft dredged soils, which is pumped over by the slurry pump from the seabed. During the reclamation, the structure of the soft soil is destroyed and the soil particles gradually deposits as fine particles. Therefore, the typical characteristics of reclaimed soil are large plasticity index, high water content, high compressibility and low permeability. Because of poor physical and mechanical property, the reclaimed ground should be reinforced before acting as foundations [1].

For the reinforcement methods of soft ground, usually there are replacement, surcharge preloading, vacuum preloading, composite foundation etc., in which the drainage method is of economic and effective to apply in the large area. Compared with the surcharge preloading method, the merit of the vacuum preloading method is the short period of reinforcement, even though the cost of vacuum preloading is higher. Considering the high plasticity index and low permeability of the reclaimed soil, the reinforcement time of both the vacuum preloading and the surcharge preloading usually lasts at least six months, even two years [2-5]. In light of this, there would be no advantage for using vacuum preloading. In practical engineering, after the reclamation of the ground there will be a long-time placement before it is used as ground foundation and at this case it is very suitable to use the surcharge preloading method to deal with the dredged soil.

## 2. Calculation conditions

### *2.1. Plastic vertical drains*

In order to accelerate the drainage of reclaimed ground, the plastic vertical drains are installed at certain space. One of the difficulties in the calculation of drainage consolidation is the simulation of the vertical drains. The common method is the equivalent transferring method from the axial-symmetric sand well to the plane-strain sand wall based on the equal degree of consolidation at various stages. Viewing from the macro level, the installation of plastic vertical drain improves the permeability of the reclaim ground. Chai et al. [6] proposed a simple method to calculate the drainage problem where the installation of vertical drains can be regarded as the improvement of the vertical permeability of the ground and the transferring equation is shown in Eq. 1.

$$k_{ve} = \left(1 + \frac{2.5l^2}{\mu D_e^2}\frac{k_h}{k_v}\right) \qquad 1)$$

where $l$ means the length of plastic vertical drain, $D_e$ means equivalent diameters of drain space, $k_h$ and $k_v$ are horizontal and vertical permeability coefficients respectively, $k_{ve}$ is the equivalent coefficient of permeability of the improved reclaimed ground. As can be seen, the smaller the space of the plastic vertical drains, the larger the equivalent coefficient of permeability is. In this calculation, the influence of different equivalent permeability, namely different space, on the reclaimed ground is investigated.

### *2.2. Calculation parameters*

Plaxis3D is used in this calculation and the constitutive model is Mohr- coulomb. The parameters are listed in Table 1. In order to evaluate the influence of equivalent permeability, five values are adopted varying from $1.2\times10^{-7}$cm/s to $1.2\times10^{-5}$cm/s.

**TABLE 1. Parameters in the Calculation**

| | |
|---|---|
| Drainage type | Undrained A |
| Unsaturated bulk density (kN/m$^3$) | 16.8 |
| Effective Young's modulus (kPa) | 1800 |
| Poisson's ratio | 0.3 |
| Effective cohesive strength (kPa) | 4.0 |
| Effective internal frictional angle (°) | 5.0 |
| Equivalent permeability (cm/s) | $1.2\times10^{-7}$<br>$4.8\times10^{-7}$<br>$1.2\times10^{-6}$<br>$6.0\times10^{-6}$<br>$1.2\times10^{-5}$ |

### *2.3. Calculation Stages*

The construction process is shown in Fig. 1. As can be seen, the construction of the road is divided into two layers to avoid the instability of embankment. Between the construction of two layers, there is 10-day interval for the consolidation of the ground.

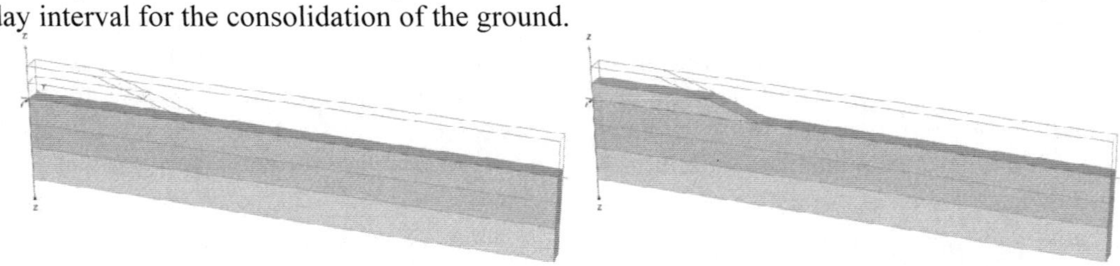

(a) Equilibrium of geostress (b) Filling of 1st layer

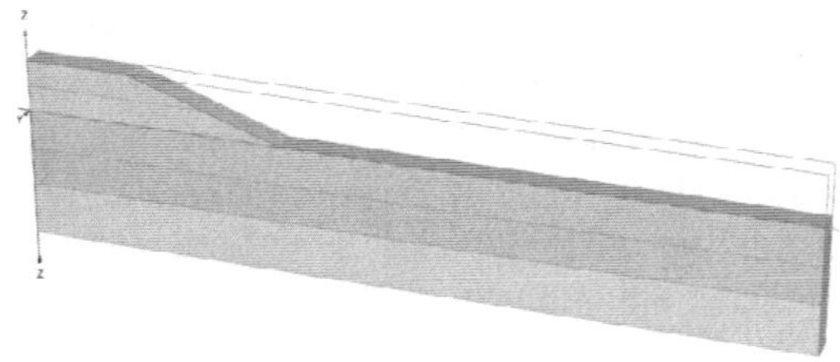

(c) Filling of 2nd layer
**Figure 1.** Construction stages.

## 3. Calculation results

The settlement of the reclaimed ground under the surcharge preloading is shown in Fig. 2. The permeability of improved ground is used as a benchmark and other four permeabilities is demonstrated. Because the road embankment is constructed in two stages, the settlement can be also clearly divided into two stages. Once the surcharge load is applied on the reclaimed ground, there would be a sudden increase of settlement due to the undrained behavior of the reclaimed soil. After then, the consolidation process becomes predominate with assist of the plastic vertical drain. The settlement gradually increase but the rate of the settlement become smaller and smaller. When the permeability coefficient is very small, the consolidation settlements at two stages are also very small. With the increase of permeability coefficient, the final settlement becomes larger and the time reaching the maximum settlement becomes shorter, which means the consolidation stage gets faster and faster. After 80 days, the settlement of improved ground is 0.84m while the final settlement is 1.03m. Namely, if there is no improvement of the permeability of the reclaimed ground the residual settlement would be 0.19m. When the permeability coefficient is enlarged 100 times, the time to reach the final settlement is only 42 days, which is much accelerated. Usually, it is deemed that the permeability of the soft ground can be enlarged 100 times after the installation of the plastic vertical drains.

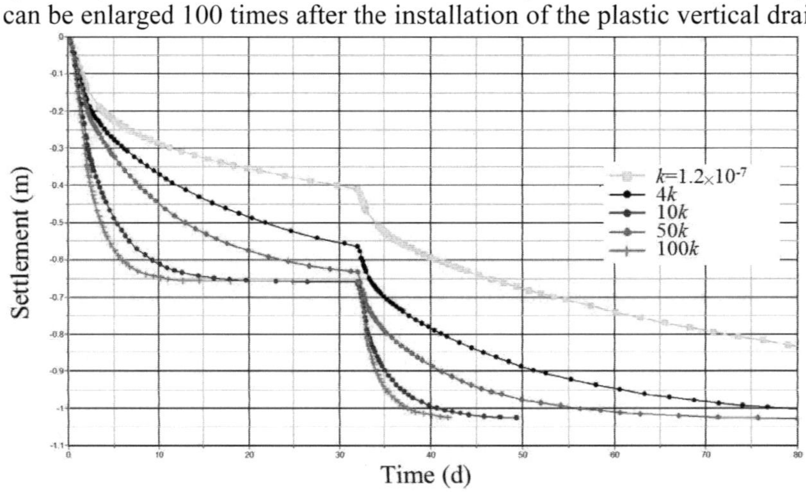


**Figure 2.** Settlement at the center of bottom surface of embankment.

The distribution of settlement in both the reclaimed ground and the road embankment is shown in Fig. 3. As mentioned above, when the layer is applied suddenly, there is a significant increase in the settlement. Along with the consolidation process, the settlement then gradually spreads towards the deep layer of the ground. After the filling of the second layer, the settlement mainly concentrates at the range of first layer and the maximum settlement occurs in the middle of the road embankment.

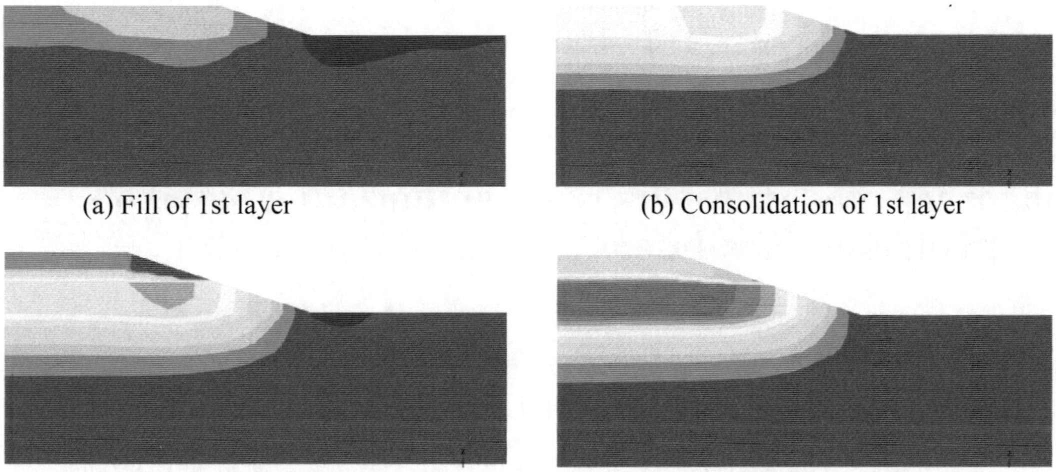

(a) Fill of 1st layer (b) Consolidation of 1st layer

(c) Fill of 2nd layer (d) Consolidation of 2nd layer

**Figure 3.** Distibution of settlement in the ground

## 4. Conclusions

The surcharge preloading method is numerically investigated considering the influence of equivalent permeability coefficient, which provides a way the deal with such problems. The conclusions are as follows:

1) After the embankment load, the consolidation process becomes predominate and the settlement gradually increase, but the rate of the settlement become smaller and smaller.

2) When the permeability increases, the drainage process is significantly accelerated and the risk of instability is reduced.

As the development of the coastal area, land reclamation has been one of the most important ways to solve the land shortage. The reclaimed soil is of high plasticity index, low permeability and complex physical mechanism, which make it always the research highlights in geotechnical engineering. Until now, the drainage consolidation methods such as vacuum preloading, surcharge preloading, electroosmosis preloading or the combination of two or three methods are still the most effective reinforcement methods. Recently, some novel methods involving chemistry methods are also proposed. For example, the water reducing agent is used to expel the pore water in the reclaimed soil as much as possible. However, the cost of such new method is much higher and the applicability range is also very limited. Therefore, the emphasis should be put on the composition of mineral components and microscopic structure of the reclaimed soil to get deep comprehension.

## References

[1] Z. L.Dong, et al.. Rock and Soil Mechanics, 2010(31), 1452-1456.
[2] H. Chen and X.Q. Bao. , Chinese Journal of Ge-otechnical Engineering, 1984(5): 39-47.
[3] H. Chen and F. Lin, Chinese Journal of Geotechnical Engineering, 1987(4):13-21.
[4] C.H. Zhang, Chinese Journal of Geotechnical Engineering, 1990(1): 56-60.
[5] Z.Q. Yu, Y. T. Zhu, Z. F. Yu, China Harbor Engineering,, 2001(1), 26-30.
[6] J. C.Chai, S. L.Shen N.Miura and D. T.Bergado, J. Geotech. and Geoenviron. Engrg. ASCE. 2001(11):965-972.

# Temporary road transport route optimization based on ArcGIS platform conditions

**Zhang Baoliang[1, a] , Li Dahua[1, b], Shao Xianfeng[2, c]**

[1] School of Civil Engineering, Anhui jianzhu University,230601,Heifei,China;
[2] State Grid Anhui Economic Research Institute
[a]1169273902@qq.com; [b]Ldh2006a@163.com; [c]xianfengyihao83@163.com

**Abstract:** Based on the ArcGIS software platform, the method of digital elevation model (DEM) is established by using the irregular triangulation (TIN) of the "three links and one leveling" project of the Guquan converter station. The calculation principle and the realization steps are discussed. In the clear fill area, site formation and earth temporary transport route design applications. Research shows that the simulation of ground simulation through ArcGIS can be more rapid and efficient land leveling, design optimization route.

## 1. Introduction

In the earthwork, in order to meet the construction requirements and ensure the smooth progress of the earthmoving project, it is necessary to design and optimize the temporary transport route of the earth. In this paper, based on the design method of temporary transportation route of large-scale field engineering in the past, this paper uses ArcGIS software's powerful spatial analysis ability and modeling ability to solve the complicated geoscience application problem, and comprehensively studies the " "The characteristics of the original topography and geomorphology before the construction of the project are discussed, and a new method for the design of the temporary transport route of the earth and stone in the large level project is discussed.

## 2. Project overview

± 1100kV Guquan converter station site is located in Xuancheng City, Anhui Province, 8 km southwest of ancient springs, covering 37 hectares as much. The site is dominated by natural terrain, mainly forest land, shrubbery, the original geomorphology complex, potholes more artificial activity is weak, limited to the local valley lowland farmland farming, station area and adjacent non-plant facilities distribution.

## 3. Based on the TIN model to calculate the principle of earthwork

### *3.1 On the TIN model*

TIN is the abbreviation of irregular triangle net, its principle is by a finite number of points in the area of set (i.e., elevation point) division will be interconnected and overlapping network of triangle surface in the area of irregular distribution of the density of measurement point and location determines the shape and size of the triangle, triangle of continuous surface constitutes the digital elevation [1].

Data structure of TIN advantages: to the resolution of the different code number to simulate the surface morphology. Compared with grid data model, TIN model under a particular resolution can use less space and time said more complex ground contour more accurately. Especially local form contains a large number of features such as fault lines, lines, TIN model can better take into account these characteristics [2].

In all possible triangle net, Delaunay (Delaunay triangulation of common used in terrain fitting, so is often used in the production of TIN, [3].

Purpose: use this kind of data structure, can more convenient the surface-to-surface shape is analyzed, such as slope and slope to the extraction of information, fill in excavation volume calculation and its region, contour automatically generated and topography of the three-dimensional display, etc. [4].

*3.2 On DEM and its generation method*

The Digital Elevation Model (DEM) [5], which is a digital representation of terrain surface by finite terrain elevation data, is a digital representation of terrain surface by a set of ordered numerical array Elevation of a solid ground model. DEM is a single-order digital geomorphology model with zero order. Other topographic features such as slope, aspect and rate of slope change can be derived on the basis of DEM [6].

TIN can be generated directly from the contour data, or a raster DEM can be generated directly. The grid DEM can also be interpolated from contour lines to generate TIN. In the past, it has been proved that the efficiency and precision of the TIN model generated by the contour line and the interpolation grid DEM are very good [7].

## 4. Design temporary transportation routes with ArcGIS

The TIN model can be established in ArcMap using the elevation points and contours of the original terrain. The TIN and DEM can be used to generate the raster map, slope and raster map of the area. , You can also ArcSence the original terrain of the target area of the three-dimensional, can make the site more 3D terrain visualization.

*4.1 Use ArcGIS to generate the target area DEM*

(1) First open the ArcMap software, through the tool table Arctoolbox in the conversion tool to Autocad in the original terrain file elevation points, contour files from dwg format converted to Shapefile format.

(2) open the Shapefile format in ArcMap point file elevation, in the left content list to find the elevation point file and right-click to open the property sheet to check whether the Elevation field is empty, if empty it will be deleted; Empty, then the elevation value is valid.

(3) created by TIN Arctoolbox>3D Analyst ArcMap tool in >TIN>, click on the input elements to determine the elevation points for Shapefile format and target area, established the TIN model file, the target area, as shown in figure 1.

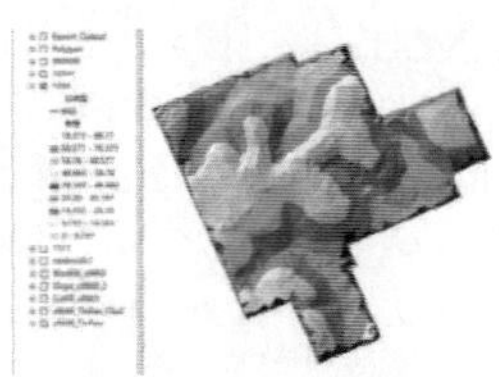

**Fig. 1 TIN model of the target area**

(4) To know the design height of the target area, draw the same area in the drawing editor as the contour of the target area with the design height. Open the ArcGIS Spatial Analyst tool> Surface Analysis> Filling and Filling, and enter the target area grid Grid model and the design elevation plane, click OK, you can fill in the region of the excavation operation and generate the appropriate fill-digging grid diagram, as shown in Figure 2, the blue area represents the excavation area, the red area represents the fill area.

(5) From the Spatial Analyst tool> Surface Analysis> Slopes in ArcMap, enter the raster model for the target area to generate a gradient raster map for the target area, as shown in Figure 3:

**Fig.2 Grid diagram of the target area**

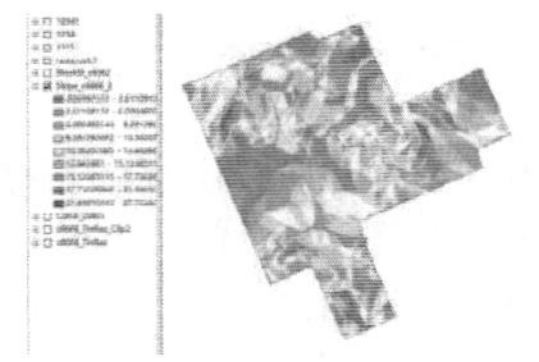

**Fig. 3: Raster map of the target area**

*4.2 The use of ArcGIS raster map for temporary transport route design*

temporary road approach: the original central station and station area on the west side of the village road widening to 6 meters, the excavator rolling compaction, flat road, within the scope of such as weak areas or hollows using large stones, gravel replacement. The central station on the west side of the road and the main road transport for the vegetation cleaning stage, temporary support of road transport were laid in the excavation area and fill area, see Figure 4. Clean up the stage of the road to meet the temporary traffic can be temporary traffic road slope shall not be 15 degrees.

**Fig. 4 Based on the slope of the temporary clean-up grid layout of the grid design**

## 5. Conclusion

ArcGIS software by simulating field topography, make elevation points and contour data visualization, and then according to the different situation of the extraction of the key factor of the constraint conditions, such as slope, slope direction, curvature, generates the corresponding grid diagram, get the geography of intuitive, comprehensive data, combined with the traditional design method on the basis of allows designers to more rapid, accurate and reasonable design of earthwork transportation route and scheme. This method through the successful application in the project, earthwork transportation route for large flat engineering design provides a successful case.

## Acknowledgments

Item number: （1P12001500010681000000）

Unit name: State Grid Anhui Economic Research Institute

Project name: The research and application of key technology-extra large-scale earthwork for±1100kV UHV DC converter station.

## References

[1] Wang Z. Research on three-dimensional visualization model of orebodies based on binomial model [J]. Information Technology, 2011,180 (12): 31-32.

[2] TANG Guo-an, GONG Jian-ya, CHEN Zheng-jiang, CHENG Yan-hui, WANG Zhan-hong.Research on Quantitative Simulation of Terrain Representation Accuracy in Digital Elevation Model [J] .Journal of Surveying and Mapping, 2001,30 (4): 361-365.

[3] Ming Anfa. Research and implementation of 3D terrain visualization based on wavelet terrain simplification technology [D]. Guizhou University, 2006.

[4] CHEN Nan-xiang, LI Dan-dan, LI Yao-hui, LIU Feng-lin.Research and Application of Earthwork Calculation Method Based on TIN Model [J] .Anhui Agricultural Sciences, 2013,41 (26): 10870-10873.

[5] Thompaonja, Belljc, Butlerca. Digital elevation model resoution: effects on terrain attribute calculation and quantitative sail landscape modeling[J].Geoderma, 2001,100(1):1-5.
[6] Li Jin.Digital elevation model of performance and application [J]. Technological innovation and productivity, 2010,196 (5): 77-78.
[7] Liu Rong. Vector topographic data processing and quality control [D]. PLA Information Engineering University, 2001.

IOP Conf. Series: Earth and Environmental Science **61** (2017) 012130 doi:10.1088/1755-1315/61/1/012130

# Experiment study on constraint condition of distributed sensing optical fiber in deformation measure of soil

**Zhijie Sun**[1,a] **and Ziyang Zhao**[2,b]

[1]Key Lab of Highway Construction and Maintenance Technology in Loess Region Ministry of Transport, Taiyuan 030006, China.
[2]Shanxi Transportation Research Institute, Taiyuan 030006, China.
[a)] Corresponding author: 46395114@qq.com
[b)]759802022@qq.com

**Abstract.** Although the optical fiber sensing technology has been successfully applied in the geological and geotechnical engineering field, there is a deformation discrepancy between the sensing optical fiber and the rock and soil to be measured in the actual project, which has a greater impact on the monitoring results. The influences of constraint condition with optical fiber are simulated in test, the deformation coordination between the optical fiber and soil is studied through comparative test analysis. The relationship between optical fiber and load is established, as well as the relationship between actual soiled formation and load.

## 1. Introduction

Fiber-optic sensing technology appeared in the 1970 of the 20th century with the development of optical fiber communication technology and the rapid development of a kind of light as a carrier, fiber optic media, sensing and transmitting external signal of new sensing technology[1-3].Fiber optic sensors have many advantages, such as resistant to electromagnetic interference, waterproof and damp proof, corrosion resistance and long durability and many other advantages, which is suitable for use in complex geotechnical engineering monitoring environment[4, 5].Another notable characteristics of optical fiber sensor is small volume, light weight, ease of installation layout, having little impact on the material properties of the monitoring object and mechanical parameters, being suitable for condition monitoring of rock mass and structure and the test evaluation[6].In addition, the optical fiber sensing technology with fiber optic signal transmission, which has the low propagation loss, easy to implement remote signal transmission. Optical fiber bandwidth is large, through the sensor multiplexing capabilities of multiple sensors can be installed on a single fiber, implement multiple points of monitoring at the same time. It is because of these advantages of the technology of optical fiber sensing, makes its successful application in the geotechnical engineering and geological engineering[7-8].

By more groups of loading test, this paper analyzed the coordination deformation characteristics between the fiber and the soil in different constraint condition, which illustrates the coordination deformation evolution between the fiber and the soil, thus to further establish the coordination deformation mechanism between the optical fiber and the soil.

## 2. Experimental design

### *2.1. Test Device*

IOP Conf. Series: Earth and Environmental Science **61** (2017) 012130 doi:10.1088/1755-1315/61/1/012130

A series of loading tests are designed for the study of coordination deformation characteristics between embedment strain sensing optical fiber and the soil. The specifications of experimental model box and counter force frame are as shown in table 1. The model box used 10 mm thick toughened glass as retaining structure, which can monitor the deformation of the fiber through the visibility glass.

**Table 1.** The specifications of experimental model box and counterforce frame.

| Caption | Length/cm | Width/cm | Height/cm |
|---|---|---|---|
| Model Box | 60 | 30 | 30 |
| Counterforce Frame | 70 | 40 | 55 |

*2.2. Sensor Placement*

*2.2.1. Optical fiber placement.* Place the model box in the counterforce frame and layer the soil into the box. The thickness of each layered soil is 5 cm. Test the density of each compaction soil with ring sampler method. When the soil reaches fiber laying height, laid the optic fiber close to the glass panel and bound on the two ends of the fiber optical fiber, and then fill the upper soil layer by layer, tamping and leveling.

*2.2.2. Displacement meter placement.* Displacement meter is used to monitor the soil subsidence. The displacement meter was buried in the design depth of the soil, the activity side of displacement meter was buried into the same height which the optical fiber was imbed, the fixed end of displacement meter is fixed in lower surface of the thrust frame upper plate.

*2.2.3. Pressure gauge placement.* After the last layer of soil compaction, decorate the pressure gauge in the surface layer of soil,pressure gauge placed on the surface of the jack, the pressure gauge is used for monitoring the load values, the test process is shown in figure 1.

**Figure 1.** Sensor placement.

## 3. The effect of bound condition of optical fiber on coordinate deformation

*3.1. Test Procedure*

This test was divided into two kinds of working condition, the Steel tight-rope optical fiber embedded in 10 cm deep in the soil and was set in level. On the first condition, both ends of the optical fiber were without constraints. On the other condition, both ends of the fiber fixed on the pillar of reaction frame, used to simulate constraints of fibers buried in soil, which are shown in figure 2.

**Figure 2.** Bound condition.

### *3.2. Results Analysis*

The load - deformation curve of optical fiber and the soil on bound condition are shown in figure 3.

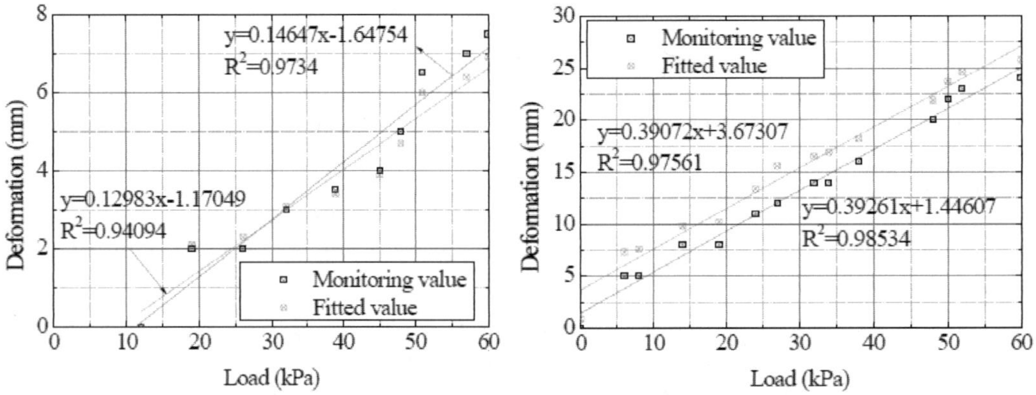


(a)Without constraint (b)Constraint

**Figure 3.** The load - deformation curve of optical fiber and the soil on bound condition.

Figure 3 show that with the overlying load increasing, the deformation of optical fiber and soil increase gradually in different constraint condition. The deformation of fiber and soil on constraint condition is larger than fiber and soil which without constraints. The deformation of Steel tight-rope optical fiber is larger than the Tight polyurethane optical fiber.

Figure 3 (a) shows that on unconstrained condition, the deformation difference is small between fiber and soil within load 50kPa.While the load is over 50kPa, the deformation difference increases gradually, imposed the same fiber optic deformation is greater than the soil deformation under load. In the same load condition, the deformation of fiber is larger than the actual deformation of soil. Figure 3 (b) shows that on constrained condition, the deformation difference are almost consistent with the increase of the applied load, which larger than unconstrained condition.

In unconstrained condition, it found that both ends of fiber optic which not buried in soil at the beginning of the test both had exposed into the soil, with the increase of load, a certain area on both ends of the fiber optic relative slip has occurred with the soil. Even though the across parts of the optical fiber has good coordination with the soil, but due to both ends of the fiber relative slip with the soil causing parts of the optical fiber unloading, which lead to the actual stress of optical fiber decreased.

With the load increasing, the deformation of fiber optic on constraint condition basically stable, may be due to constraints of optical fiber's ends to prevent the coordination deformation between fiber and soil, Which caused the deformation of fiber not to reflect the deformation of soil. So when laying fibers in actual project, the boundary condition of fiber at both ends cannot be simple constraints, it should be intensive studied and be specified specific constraint methods.

## 4. Conclusions

The influences of constraint condition with optical fiber are simulated in test, the deformation coordination between the optical fiber and soil is studied through comparative test analysis. The relationship between optical fiber and load is established, as well as the relationship between actual soiled formation and load.

In actual monitoring, the deformation is obtained by calculation formula of the monitoring stress of optical fiber, so the measured deformation of optical fiber is smaller than the real deformation of the optical fiber in test condition, which need further amendment. When laying fibers in actual project, the boundary condition of fiber at both ends cannot be simple constraints, it should be intensively studied and with specifically constraint methods.

## Acknowledgments

This work is supported by the Shanxi Basic Research Program of China through Grant No. 2015021113.These supports are gratefully acknowledged.

## References

[1] Barnoski M K, Jensen S M, "Fiber waveguides: anovel technique for investigating attenuation characteristics," in *Applied Optics* **15**, 2112-2115 (1976).
[2] Horiguchi T, Shimizu K, Kurashima T, et al, "Development of a distributed sensing technique using Brillouin scattering," in *Journal of Lightwave Technology* **13**,1296-1302 (1995).
[3] Hill K O, Fujii Y, Johnson D C, et al, "Photosensitivity in optical fiber waveguides: application to reflection filter fabrication," in *Applied Physics Letters* **32**, 647-649 (1978).
[4] Liao Yanbiao, "The Promotion of OFS to the Development of Industry," in , *Optoelectronic Technology & Information* **16(5),** 1-6 (2003).
[5] QiangXifu. Sensor.Beijing: China Machine Press, (1999).
[6] She Junkuan, Zhu Honghu, Zhang Chengcheng, et al,. "Experiment study on mechanical properties of interface between sensing optical fiber and sand," in Journal of Engineering Geology 22(5) , 855-860 (2015).
[7] Zhu Honghu, Albert N L, Yin J H, et al, "An optical fiber monitoring system for evaluating the performance of a soil nailed slope," in *Smart Structures and Systems* **9(5)**, 393-410(2012).
[8] Pei Huafu, Yin Jianhua, Zhu Honghu, et al, "In-situ monitoring of displacements and stability evaluation of slope based on fiber Bragg grating sensing technology," in *Chinese Journal of Rock Mechanics and Engineering* **9(8)**, 1576-1583 (2010).

IOP Conf. Series: Earth and Environmental Science **61** (2017) 012131 doi:10.1088/1755-1315/61/1/012131

# Numerical simulation of the low-cyclic loading of new-type assembled integral beam-column joints

**Jiangchuan Wu[1, 2], Jiwen Zhang[2], ,Wanyun Yin[3] and Rencai Jin[3]**

[1] School of Civil Engineering, Southeast University, Nanjing 210096, China.
[2] School of Civil and Ocean Engineering, Huaihai Institute of Technology, Lianyungang 222005, China.
[3] MCC17 Group Co., Ltd. , Maanshan 243000, China.
Corresponding author: wujch@sina.com

**Abstract**. In this paper, the low-cycle cyclic loading test and the elasto-plastic numerical simulation by OpenSEES of assembled integral beam-column joints of the three new prefabricated prestressed concrete beams are carried out. The key technologies such as unit selection, material model selection and prestressing in the numerical simulation of the new assembly-type beam-column joint is studied deeply. The results show that the relative model and parameters of the OpenSEES software can be used to simulate the hysteresis performance of the low-cycle cyclic loading of the new-type assembled integral beam-column joints.

## 1. Introduction

With the development of construction industrialization and green building in China, precast assembled integral structure has been studied and applied frequently. However, at present, China's the level of construction industrialization is very low and precast assembly technology is relatively backward in China. So it is necessary to carry out in-depth study of the precast assembly structure, to promote the application and development of the precast structure in China. China is a multi-seismic country, and most of the city are in the seismic fortification zone. Therefore, the development of precast assembled integral structure must take into account its seismic capacity.

The Precast Prestressed Assembled Structural System (referred to as PPAS system) studied by Southeast University and other universities and research institutions, is based on the introduction of Japanese prestressed assembled structure, and combines the development of China's national conditions. It is a kind of cost - effective building construction technology. The joint reinforcement diagram and beam end structure details are shown in Figure 1 and Figure 2. PPAS system frame beam is Precast prestressed composite beam. The main beam end is embedded channel steel connector, which can not only be used for rapid bolt connection between beam and column, but also be used as a shear part of old and new concrete joint in joint area. Besides, it can also control the location of the plastic hinge, shift the plastic hinge, improve the stress state of the node area, avoid brittle damage. The secondary beam end use short channel steel for the connection between primary and secondary beam, and weld shear plate to increase the shear section. According to the needs, PPAS system precast prestressed beam end use PVC casing release part prestress of steel strand. PPAS system column using Concrete filled steel tube columns or reinforced concrete composite column node plusing short steel tube .The ribbed floor and the beam and column joints are cast on the site with patent templates.

The full name of OpenSEES is Open System for Earthquake Engineering Simulation (open system for seismic engineering simulation). OpenSEES is funded by the National Natural Science Foundation of the United States (NSF), led by Western University and the Pacific Earthquake Engineering Center (PEER), mainly developed at the university of California, Berkeley. It is a more comprehensive and continuous development of the open program software system for seismic response simulation in structural and geotechnical aspects. Since its introduction in 1999, it has been widely used in the Pacific Earthquake Engineering Research Center and some universities and scientific research projects of research institutions in US. It simulated a large number of practical engineering and shaking table test project, including reinforced concrete structures, bridges, geotechnical engineering, which proved that it has a good nonlinear numerical simulation accuracy.

In the OpenSEES program, the component section can be divided into several kinds of fibers and the user can define the location of each fiber, cross-sectional area and constitutive relationship. The program automatically assumes the strain of each fiber based on the flat section assumption and iteratively computes to ensure that the section is subjected to force balance. The fiber model links the microscopic constitutive relation of the material with the macroscopic component response, taking into account the advantages of macroscopic finite element and micro finite element. At the same time, the so-called flexibility method is used to form the stiffness matrix of the beam-column element based on the fiber model. The iterative solution process includes the iterative solution of the structure level, the component level and the section level. This feature makes the flexibility method have a unique advantage in analyzing the nonlinear response of the component.

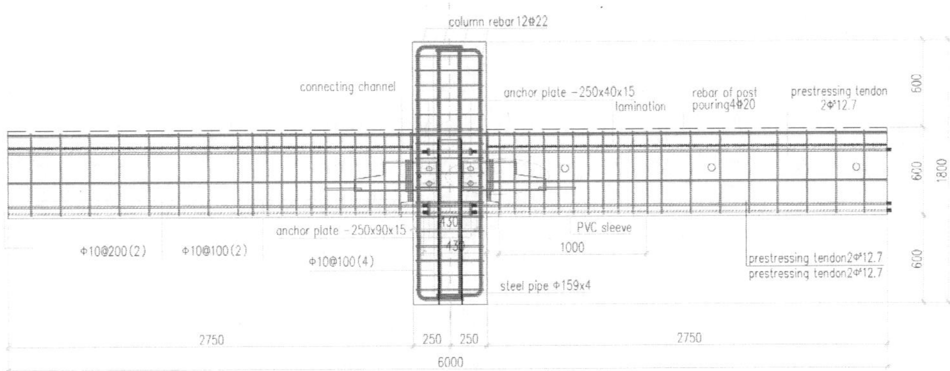


**Figure 1**. Reinforcement of PPAS joints.

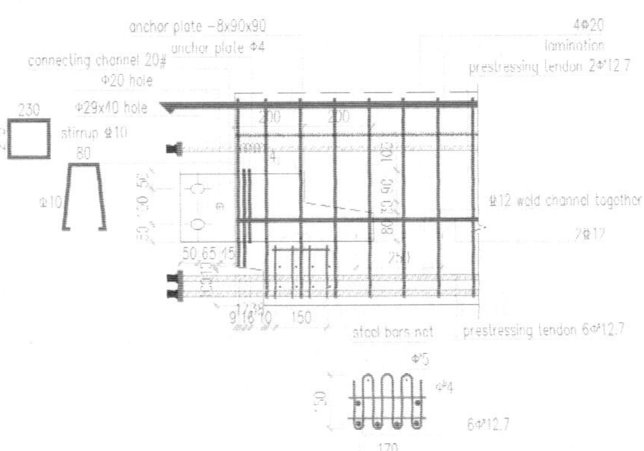


**Figure 2.** Beam end construction detail.

## 2. Specimen design

In order to more truly reflect and restore the stress state of the joint, the test use the full test pieces based on the prototype. And according to the design drawings of actual engineering, they are made the appropriate adjustments, combined with the actual size of the laboratory joint frame. The detailed dimensions and reinforcement of the components are shown in Table 1.

**Table 1.** Size and reinforcement of specimens.

| specimen | Section size (mm×mm) | Concrete strength grade | By the beam / column edge extension length | Longitudinal reinforcement | Stirrup |
|---|---|---|---|---|---|
| precast beam | 250×480 | C50 | 2750mm | Top 2Φ$^s$12.7<br>Bottom6Φ$^s$12.7 | Φ10@100/200(2) |
| Laminated layer | 120 | C40 | 2750mm | 4Φ22 | Φ10@100/200(2) |
| Cast-in-site column | 500×500 | C40 | 600mm | 12Φ22 | Φ10@100(4) |

The precast prestressed beams of the PPAS system are prevented from cracking or crushing due to excessive prestress during the production process. The influence of different prestressed casing release on the joints is considered in the process of making the precast prestressed beams. See Table 2 for specific parameters.

**Table 2.** The change in parameters of different groups of specimens.

| Grouping instruction | Specimens number | Information about Stress release |
|---|---|---|
| Cast-in-site comparative joint | RJ0 | Ordinary reinforced concrete comparative specimens |
| Different information about Stress release | PJ1 | The out side of the lower row of the two stranded end 1m range of the release |
| | PJ2 | The out side of the lower row of the two stranded end 1m range of the release, and the middle of the low rower of the two stranded end 0.6m range of the of the release and the upper row of two steel wire ends 0.6m range of release |

The mechanical propertiy of concrete and steels are shown in Table 3 and Table 4.

**Table 3.** Mechanical property of concrete(/MPa).

| Design strength | fcu | fc | ft | Ec/（×10$^4$） |
|---|---|---|---|---|
| C50 | 63.6 | 41.2 | 3.3 | 3.636 |
| C40 | 38.6 | 25.8 | 2.6 | 3.222 |

**Table 4.** Mechanical properties of steels.

| Reinforcement type | Area/mm$^2$ | fy/MPa | fu/MPa | Es/(×10$^5$)MPa |
|---|---|---|---|---|
| Φ$^s$12.7 | 98.70 | 1851.0 | 2026.0 | 2.06 |
| Φ22 | 380.13 | 524.1 | 669.6 | 2.10 |
| Φ10 | 78.54 | 296.9 | 451.8 | 2.00 |

The low cycle repeated load test of precast prestressed assembled Structural frame Beam and Column is carried out on the joint testing machine in structure laboratory, the four-storey campus of Southeast University. The loading device is shown in Figure 3, and the joint test chart is shown in Figure 4.

During the test, the column top was loaded with a 320t hydraulic jack, and the axial compression ratio of the control column was 0.2 during the loading process and remained unchanged. In the experiment, the two-point reverse loading method was used. The anti-symmetrical low-cycle cyclic loading with two 50t hydraulic jacks was loaded in left and right beam end distance from the column 2.5m. Before the yield of the specimen, the pressure is controlled by three-stage loading, which is 1/3F, 2/3F and F of the yield load. After the yield, the displacement control is taken, and the maximum displacement of the specimen is 0.5 times, The test is terminated when the load drops to 80% of the maximum load. Before yield, each level of load repeated once; after yield, repeated twice.

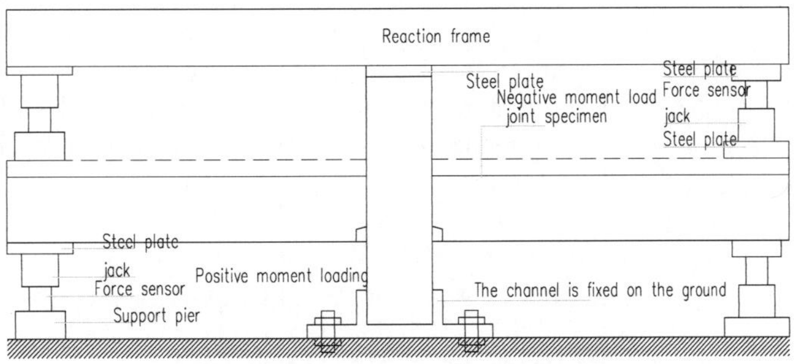


**Figure 3.** Loading equipment.

**Figure 4.** Test joints.

## 3. Result analysis

The nonlinear beam and column model is used as the simulation unit. It is a finite element model and a nonlinear element based on the fiber model method. It allows the stiffness to change along the length of the component. It is a distributed plasticity model. In the solution, the rod end displacement is obtained by the displacement of the node, and then the deformation of the cross section is solved according to the preset displacement interpolation function. Next, the material constitutive model of the section is used to determine the resistance of the section and the tangent stiffness matrix. The resistance of the whole unit and the unit tangent stiffness matrix is calculated according to the Gauss-Legendre integral. The unit model can be widely applied to the static and dynamic nonlinear analysis of the bar structure, and the P-$\Delta$ effect can be fully considered. Prestressed tendons are modeled using truss units. Rigid Link units connect truss units for simulating prestressed steel strand wire to reinforced concrete using fiber units for simulating, reflecting the relationship of deformation and coordination between the two. In this study, the steels in beams and columns is made of steel01 model, and the steel01 model is a bilinear follower strengthening model. The elastic modulus of the strengthening section is 0.01ES. ES is the initial elastic modulus of the steel, The effect of softening is not taken into account. Steel stranded Wires in precast beams use steel02 model, which can consider the prestressing effect by setting the initial stress.

IOP Conf. Series: Earth and Environmental Science **61** (2017) 012131 doi:10.1088/1755-1315/61/1/012131

Constitutive model of concrete uses the concrete02 model, which is the modified Kent-Park model by Scott et al. The model considers the effect of transverse stirrups by changing the peak stress, peak strain and slope of the softening section of the concrete compression skeleton curve, and it can consider the residual strength of the concrete, and the rising and falling segments of the concrete are all straight, and the initial cracking of concrete can be considered. Concrete02 model keep a good balance between simplification and precision, and the reinforcement nonlinear analysis of concrete structures has good accuracy.

The hysteresis curve of each specimen and calculated by OpenSEES software are shown in Figure 6. The shape of the hysteresis curve of cast-in-place reinforced concrete comparative speciment joint RJ0 and assembled crossed joint specimen PJ1 are similar and relatively full. The hysteresis curves of PJ2 is slightly collapsed showing that the number and range of stress release of the steel strand have a significant effect on hysteretic behavior of the component. The numerical simulation results of the OpenSEES are similar to those of the experimental hysteresis curves. The numerical simulation results can predict the unloading stiffness, reloading stiffness and stiffness degradation of each cycle of prefabricated prestressed beam assembly integral frame joints under the reciprocating load. While the simulation results also reflect the shape and pinching effect of the hysteresis curve of the actual specimen, which can better reflect the hysteresis characteristics of the test specimen.

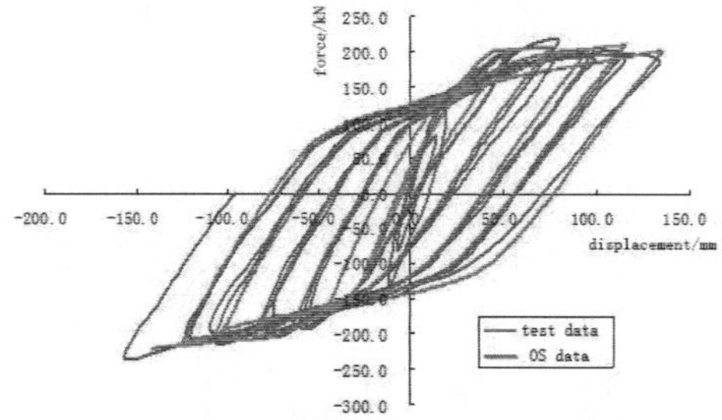

**Figure 5.** Hysteresis curve of PJO.

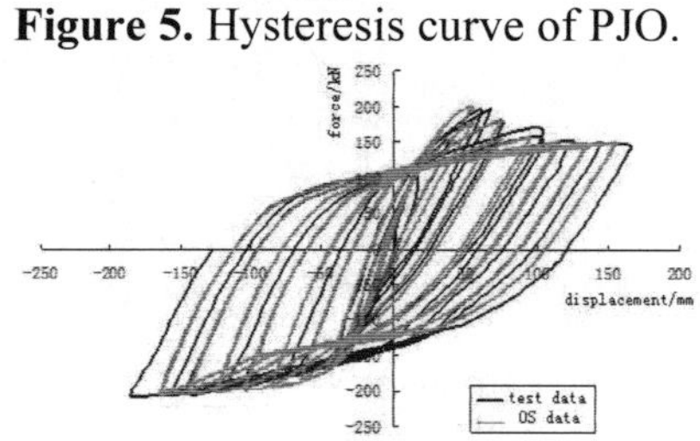

**Figure 6.** Hysteresis curve of PJ1.

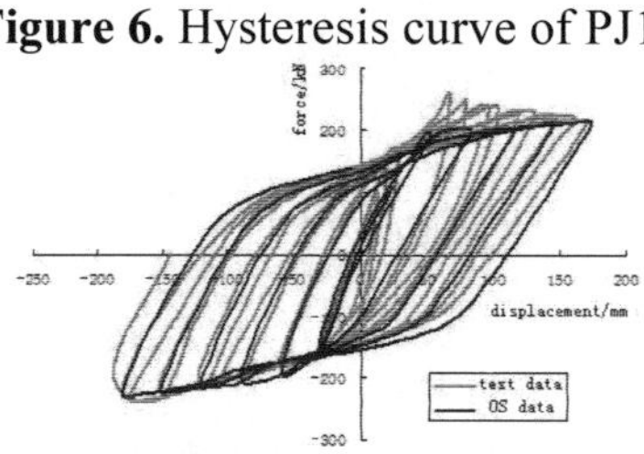

**Figure 7.** Hysteresis curve of PJ2.

## 4. Conclusion

In this paper, the finite element software OpenSEES is used as the analysis platform. The Nonlinear BeamColumn model is used as the simulation unit. The steel01 normal steel bar model, the steel02 prestressed steel strand model and the concrete02 concrete model are selected. Rigid Link units connect truss units for simulating prestressed steel strand wire to reinforced concrete using fiber units for simulating, reflecting the relationship of deformation and coordination between the two. OpenSEES software can simulate the unloading stiffness, reloading stiffness and stiffness degradation of each cycle under the action of low cyclic reversed load. The simulation results also reflect the shape and pinching effect of the hysteresis curve of the actual specimen. which better reflects the hysteresis characteristics of the test specimen.

**Acknowledgments**
MCC "three five" major science and technology special (MCC technology plan [2013]1) funded projects; Lianyungang Civil and Architectural Innovation Research Institute (KH16020); Lianyungang City Science and Technology Project (SH1628); Jiangsu Province Construction System (China) Technology Project (2016ZD34).

**References**

[1] Cai Jianguo, Feng Jian, Wang Zan, et al. Seismic Behavior of Frame Comprised of Precast Prestressed Concrete Components[J]. ACTA SCIENTIARUM NATURALIUM UNIVERSITAIIS SUNYATSENI . 2009 (2): 136-140.
[2] Cai Xiao-ning, MENG Shao-ping, Sun Wei-wei. Experimental Study on the Force Behavior of Self-resetting Prefabricated Frame Edge Node Components[J]. Engineering Mechanics. 2014 (3): 160-167.
[3] Chen Yaogang, Guo Zhengxing, Dong Niancai, et al. Joint Connection Construction Technology of the New Precast Concrete Shear Wall Structure[J]. CONSTRUCTION TECHNOLOGY. 2011 (11): 3-5.
[4] Cheok G S, Lew H S. Performance of Precast Concrete Beam-to-column Connections Subject to Cyclic Loading [J]. PCI Journal. 1991, 36 (3): 56-67.
[5] Dou Zu-rong, Wang Yi, Xue Wei-chen. Seismic Performance of Precast Monolithic Concrete Frame Exterior Connections Under Low Cyclic Loading[J]. Journal of Architecture and Civil Engineering of Building Science and Engineering. 2010 (2): 61-66.
[6] Liang Peixin, Guo Zhengxing. Experimental Study on New Prefabricated Unsymmetrical Hybrid Subassemblages [J]. Construction Technology. 2010 (3): 63-65.
[7] Li zhongxian, Zhang Xuesong, DingYang,Seismic Behavior of Steel Reinforced Concrete Joints with Beam End Plastic Hinges [J]. Journal of South China University of Technology (Natural Science Edition). 2007 (3): 77-82.
[8] Loo Y C, Yao B Z. Static and Repeated Load Tests on Precast Concrete Beam-to-column Connections [J]. PCI Journal. 1995, 40 (2): 106-115.
[9] Wang Dongyan, Li Zhenbao, Lu Haibo, et al.Strengthing Experimental Research of Unbonded Post-tensioned Precast Concrete Structure[J]. Earthquake Resistant Engineering and Retrofitting. 2007: 333-335.
[10] Wang Lili, Zhang Jiwen. Experimental Vestigation on Flexural Behavior of Novative Concrete Frame Joint Assembled with Precast Prestressed Concrete Beam[J]. Industrial Architecture. 2015(8):73-77.
[11] YU Jian-bing, GUO Zheng-xing, GUAN Dong-zhi. Experimental Study on Seismic Behavior of a New Precast Beam-to-column Connection [J]. Construction Technology. 2014 (22): 26-29.

IOP Conf. Series: Earth and Environmental Science **61** (2017) 012132 doi:10.1088/1755-1315/61/1/012132

# Study on the preparation process of cross-linked porous cassava starch

**Xiulian Yin, Qinghong You*, Miaomiao Wan, Xuejuan Zhang, Chunhua Dai**

Huaiyin Institute of Technology, School of Life Science and Food Engineering, Huaian, Jiangsu 223003, China

* youqhong@163.com

**Abstract**. Using cassava starch as raw material, preparation process of porous cross-linked cassava starch was studied. Using TSTP as cross-linking agents, Orthogonal design was applied for the optimization of cross-linked porous starch preparation process. The results showed that the opitmal conditions of cross-linked porous cassava starch were as follows: reaction temperature 45°C, reaction time 20 h, 1% of the amount of the enzyme, the enzyme ratio of 1:5, pH5.50, substrate concentration of 40%.

## 1. Experimental

### *1.1. Materials and chemicals*

Cassava starch, Glucoamylase and Soybean oil were food grades. Sodium trimetaphosphate, Sodium hydroxide,α-amylase and other reagents were all analytical degree.

### *1.2. Prepareation of Cross-linked starch*

#### *1.2.1. Determination of cross-linking degree*

It is difficult to directly determine the cross-linking degree of the cross-linked starch for the cross-linking degree is very low[1-8]. In the research, the sedimentation method was used to measure the cross-linking degree.

0.5g cross-linked starch with 25mL distilled water was put into a100mL beaker. The beaker was put into a water bath of 85°C and was stirred slightly for 2 min. Then the beaker was taken out of the water bath and was cooled to room temperature. The reaction products were put into some centrifuge tubes and were centrifuged at a speed of 4000r/min for 10min. The volume of sedimentation was calculated through the following formula:

$$S=10-V \qquad (1)$$

S-sedimentation, mL;

V-supernatant volume, mL.

### *1.3. Preparation of porous starch*

#### *1.3.1. Optimization of preparationprocess*

Using reaction time (A), starch concentration (B), pH (C), enzyme dosage (D) and reaction temperature (E) as five factors, L16 (45) orthogonal test design was conducted(Tab.1). The optimum technological parameters for preparation of cassava porous starch by enzymatic hydrolysis were

IOP Conf. Series: Earth and Environmental Science **61** (2017) 012132 doi:10.1088/1755-1315/61/1/012132

determined by the index of oil absorption. 2g cross-linked starch was put into the reaction flask with a certain concentration,α-amylase and glucoamylase were dissolved in bufferto certain concentration ,then transfered the enzyme liquid to the starch suspension, stirring for a certain time, After the reaction, adding 0.2mL 4%NaOH to stop the reaction, filtrating. The samples are placed in a drying oven at 45°C to a constant weight[9-10].

Tab.1 Factors and levels of orthogonal experiment design

| levles | factors | | | | |
|---|---|---|---|---|---|
| | A | B | C | D | E |
| | Time/h | Starch breast concentration /% | pH | Enzyme dosage /% | Temperature /°C |
| 1 | 20 | 25 | 4.00 | 1.0 | 40 |
| 2 | 24 | 30 | 4.50 | 1.5 | 45 |
| 3 | 28 | 35 | 5.00 | 2.0 | 50 |
| 4 | 32 | 40 | 5.50 | 2.5 | 55 |

The oil absorption rate was calculated according the following equation:

$$\text{Oil absorption rate S} = \frac{M_2 - M_1}{M_1} * 100\% \tag{2}$$

Where M1 was the dried weight of the starch, M2 was the weight of starch that has absorbed oil.

## 2. Results and discussions

### *2.1. Enzymatic hydrolysis*

#### *2.1.1. Orthogonal test results*

Tab.2 The results of orthogonal test

| NO. | factors | | | | | Oil absorption/% |
|---|---|---|---|---|---|---|
| | A | B | C | D | E | |
| 1 | 1 | 1 | 1 | 1 | 1 | 71.76 |
| 2 | 1 | 2 | 2 | 2 | 2 | 68.06 |
| 3 | 1 | 3 | 3 | 3 | 3 | 72.5 |
| 4 | 1 | 4 | 4 | 4 | 4 | 64.55 |
| 5 | 2 | 1 | 2 | 3 | 4 | 52.86 |
| 6 | 2 | 2 | 1 | 4 | 3 | 54.19 |
| 7 | 2 | 3 | 4 | 1 | 2 | 90 |
| 8 | 2 | 4 | 3 | 2 | 1 | 52.42 |
| 9 | 3 | 1 | 3 | 4 | 2 | 68.62 |

| | | | | | | |
|---|---|---|---|---|---|---|
| 10 | 3 | 2 | 4 | 3 | 1 | 52.86 |
| 11 | 3 | 3 | 1 | 2 | 4 | 66.76 |
| 12 | 3 | 4 | 2 | 1 | 3 | 83.17 |
| 13 | 4 | 1 | 4 | 2 | 3 | 69.26 |
| 14 | 4 | 2 | 3 | 1 | 4 | 55.65 |
| 15 | 4 | 3 | 2 | 4 | 1 | 33.63 |
| 16 | 4 | 4 | 1 | 3 | 2 | 75.71 |
| K1 | 69.218 | 65.625 | 67.105 | 75.145 | 52.668 | |
| K2 | 62.368 | 57.69 | 59.43 | 64.125 | 75.597 | |
| K3 | 67.852 | 65.722 | 62.297 | 63.483 | 69.78 | |
| K4 | 58.563 | 68.962 | 69.168 | 55.247 | 59.955 | |
| R | 10.655 | 11.272 | 9.738 | 19.898 | 22.929 | |

By orthogonal test, we can choose the optimal preparation process for cross-linked cassava porous starch preparation, the order of influencing factors for cross-linked porous starch absorption rate was reaction temperature > enzyme dosage > starch emulsion concentration > reaction time >pH value. Considering the factors that affect the preparation of cross-linked porous starch, the optimum technological conditions were $A_1B_4C_4D_1E_2$, those are temperature 45°C, the enzyme content 1%, the concentration of starch 40%, the reaction time 20h, pH5.50.

## 3. Conclusions

In this study, the preparation process and the optimum conditions of cassava starch were discussed. After cross-linking, enzymatic hydrolysis and other conditions, the oil absorption rate of cassava starch was increased, which indicated that the degree of resistance to paste was improved, and the adsorption ability was improved. The best conditions for cross-linking: reaction temperature 50°C, reaction time 3 h, pH10.00, cross-linking agent dosage of 0.5% cassava starch concentration 40%; the optimal preparation conditions for cross-linked cassava porous starch, reaction temperature 45 °C, reaction time 20 h, enzyme dosage 1%, pH5.50, substrate concentration 40%.

## Acknowledgements

This work was supported by Six Talent Peaks Project of Jiangsu Province (No:XNY-012(2015) and XNY-010(2015)), Independent Innovation Fund of Agricultural Science and Technology of Jiangsu Province (No: CX(15)1009), Huaian Key Research and Development Project (Modern Agriculture) (No:HAN2015003), Colleges and universities in Jiangsu Province Natural Science Fund(No:13KJD550001), Jiangsu province science and technology plan (BY2015051-03), Open project of Jiangsu Provincial Engineering Laboratory for Biomass Conversion , Process Integration (No: JPELBCPL2014004) , The Ministry of science and Technology Spark Program (No: 2013GA690267, 2014GA690218, 2014GA690163, and 2015GA690288)

## References

[1] Wu,Y.,Du,X.,Ge,H.,et al. Preparation of microporous starch by glu-coamylase and ultrasound. Starch (Stärke), 2011(63): 217-225.
[2] Zhang, B., Cui,D., Liu,M., et al.Corn porous starch: Preparation, characterization and adsorption property. International Journal of Biological Macromolecules, 2012(50):250-

256.

[3] Mahsa M., Sara H., Asgar F.. Functional properties of microporous wheat starch produced by α-amylase and sonication. FoodBioscience, 2015 ( 11):79-84

[4] Bai Y., Shi Y.C.. Structure and preparation of octenyl succinic esters of granular starch, microporous starch and soluble maltodextrin. Carbohydrate Polymers, 2011 (83) :520-527

[5] Xu, A., Seib, P. A.. Determination of the level and position of substitution in hydroxypropylated starch by high-resolution 1H-NMR spectroscopy of alphalimit dextrins. Journal of Cereal Science, 1997 (25):17-26.

[6] Bao, J. S., Xing, J., Phillips, D. L., et al.. Physical properties of octenyl succinic anhydride modified rice, wheat, and potato starches. Journal of Agricultural and Food Chemistry, 2003 (51):2283-2287.

[7] Bhosale, R., & Singhal, R. Process optimization for the synthesis of octenylsuccinyl derivative of waxy corn and amaranth starches. Carbohydrate Polymers, 2006 (66): 521-527.

[8] Segura-Campos, M., Chel-Guerrero, L., Betancur-Ancona, D.. Synthesis and partial characterization of octenylsuccinic starch from Phaseolus lunatus. Food Hydrocolloids, 2008 (22):1467-1474.

[9] Song, X. Y., He, G. Q., Ruan, H.,et al. Preparation and properties of octenyl succinic anhydride modified early indica rice starch. Starch/Stärke, 2006 (58): 109-117.

[10] Hung, P. V., Huong, N. T. M., Phi, N. T. L., et aln. Physicochemical characteristics and in vitro digestibility of potato andcassava starches under organic acid and heat-moisture treatments. International Journal of Biological Macromolecules,2017 (95) 299-305

IOP Conf. Series: Earth and Environmental Science **61** (2017) 012133 doi:10.1088/1755-1315/61/1/012133

# Modeling of scattering intensity of spheroid particles with a Gaussian beam

**Hong Tang**[1, 2*] **Wenbin Zheng**[2]

1 Sichuan Normal University, College of Engineering, Chengdu, Sichuan province
2 China Jiliang University, Hangzhou, Zhejiang province

* tangbenben@126.com

**Abstract**. Based on the Generalized Lorenz Mie Theory (GLMT), the scattering intensity of spheroid particles is investigated within the Gaussian beam. The spheroid model is applied to represent the real non-spherical particles, and the scattering intensity of particles is deduced according to the GLMT. On the other hand, the sphere of the same volume for the spheroid is used for calculating the scattering intensity of the spheroid within the Gaussian beam. For a comparison, the scattering intensity of the spheroid with the pane wave is also calculated. Simulation data results indicate that fairly reasonable results of the scattering intensity for the spheroids can be obtained with this model, and it can provide a reliable and efficient approach to research the scattering intensity of the non-spherical particles by the Gaussian beam.

## 1. Introduction

The light scattering of particles is very important in the engineering, Chemical, medicine and other regions. Since the theory of light scattering has been established, some researches have studied the electromagnetic light scattering of particles for the plane wave case [1]. And some common theories and methods have been utilized to analyze this problem. When the plane light wave is incident the particle, the classical Mie theory, the discrete dipole approximation (DDA), the T matrix method, and the finite difference time domain (FDTD) method can be used for sphere and non-sphere particles [2-4]. Nevertheless, none of those methods can be applied to analyze and calculate the scattering of particles for the non-plane waves, i.e., the Gaussian beam incidence.

In this paper, we study the scattering intensity of particles based on the Gaussian beam incidence. Since the computation of light scattering by the non-spherical particles is very important in many measurement areas, great progress has been made in the study of light scattering for the non-spherical particles [5]. Actually, the non-spherical calculations and measurements show significant differences from the sphere particles. Here, we choose the spheroid model as the non-spherical particles to study the scattering intensity of non-spherical particles within the Gaussian beam incidence.

## 2. Theory and method

With the development of the laser technique, the laser are usually used for the measuring the light scattering of particles. Especially, many optical particle sizing techniques rely on the interaction between the particles and the laser sources. So this research on the light scattering by the particles and the interaction between the particles and the incident non-plane beam for the case in which the particles are illuminated by the laser beam has been on growing up [6].

It is well known that the Generalized Lorenz Mie Theory (GLMT) proposed by Gouesbet et al. is a generalization of the Lorenz Mie theory for the Gaussian beam [7]. The GLMT can describe the properties of incident Gaussian beam and the scattered field of the homogeneous spherical particles as well as the interaction between the spheres and the Gaussian beam. Actually, the GLMT is a direct extension of the plane wave Mie theory to the case of Gaussian beam. What is most important is that the GLMT is a rigorous theory that depends on the Maxwell electromagnetic scattering equations [6].

In the GLMT framework, the incident Gaussian beam field including the electric and the magnetic fields can be described by the Bromwich Scalar Potentials (BSP) in the spherical coordinate system (r, θ, φ) [6]. The field components are then found to be:

$$E^i{}_r = \frac{\partial^2 U_{TM}}{\partial r^2} + k^2 U_{TM}$$

$$E^i{}_\theta = \frac{1}{r}\frac{\partial^2 U_{TM}}{\partial r \partial \theta} - \frac{i\omega\mu}{r\sin\theta}\frac{\partial U_{TE}}{\partial \varphi}$$

$$E^i{}_\varphi = \frac{1}{r\sin\theta}\frac{\partial^2 U_{TM}}{\partial r \partial \varphi} + \frac{i\omega\mu}{r}\frac{\partial U_{TE}}{\partial \theta}$$

$$H^i{}_r = \frac{\partial^2 U_{TE}}{\partial r^2} + k^2 U_{TE}$$

$$H^i{}_\theta = \frac{i\omega\varepsilon}{r\sin\theta}\frac{\partial U_{TM}}{\partial \varphi} + \frac{1}{r}\frac{\partial^2 U_{TE}}{\partial r \partial \theta}$$

$$H^i{}_\varphi = \frac{1}{r\sin\theta}\frac{\partial^2 U_{TE}}{\partial r \partial \varphi} - \frac{i\omega\varepsilon}{r}\frac{\partial U_{TM}}{\partial \theta} \tag{1}$$

where $U_{TM}$ and $U_{TE}$ are the transverse magnetic (TM) and transverse electric (TE) BSP, respectively, $i = \sqrt{-1}$, $k$ is the wave number, $\omega$ is the angular frequency of the electromagnetic wave, $\mu$ and $\varepsilon$ are the permeability and the permittivity of the medium, respectively.

When the beam is incident upon a spherical particle with radius $r$ and complex refractive index $M$, the scattered field components in the far field approximation are given as [6]:

$$E^s{}_r = H^s{}_r = 0$$

$$E^s{}_\theta = \frac{iE_0}{kr}\exp(-ikr)S_2$$

$$= \frac{iE_0}{kr}\exp(-ikr)\sum_{n=1}^{\infty}\sum_{m=-n}^{n}\frac{2n+1}{n(n+1)}[a_n g^m_{n,TM}\tau^{|m|}_n(\cos\theta) + imb_n g^m_{n,TE}\pi^{|m|}_n(\cos\theta)]\exp(im\varphi)$$

$$E^s{}_\varphi = \frac{-E_0}{kr}\exp(-ikr)S_1$$

$$= \frac{-E_0}{kr}\exp(-ikr)\sum_{n=1}^{\infty}\sum_{m=-n}^{n}\frac{2n+1}{n(n+1)}[ma_n g^m_{n,TM}\pi^{|m|}_n(\cos\theta) + ib_n g^m_{n,TE}\tau^{|m|}_n(\cos\theta)]\exp(im\varphi)$$

$$H^s{}_\theta = \frac{-H_0}{E_0}E^s{}_\varphi$$

$$H^s{}_\varphi = \frac{H_0}{E_0}E^s{}_\theta \tag{2}$$

where $\tau_n$ and $\pi_n$ are the generalized Legendre functions, $a_n$ and $b_n$ are the scattering coefficient of Mie theory, $S_1$ and $S_2$ are the scattering amplitudes.

The far field scattering intensity can be expressed:

$$I = \lim_{r \to \infty} \frac{\lambda^2 E_0}{4\pi^2 r^2} (|S_1|^2 + |S_2|^2) \tag{3}$$

In order to study the scattering intensity of spheroid particles based on the GLMT, it is still practice to use the sphere model to approximate the scattering properties of spheroid particles. A spheroid particle may be represented by a sphere of the same volume. This method could be used to readily calculate the scattering intensity of spheroids.

## 3. Numerical calculations

In this section, we will discuss the scattering properties of spheroids with different sizes and ratio aspects in the case of the incident Gaussian beam.

Fig.1 shows the scattering intensity of spheroid particles. In this figure, the incident wavelength is 0.5μm, the beam waist $w_0$=1μm, and the relative complex reflective indices of particles is 1.33. We can see that the scattering of spheroid is different from the sphere, and the difference is more obviator with the increasing aspect of spheroid.

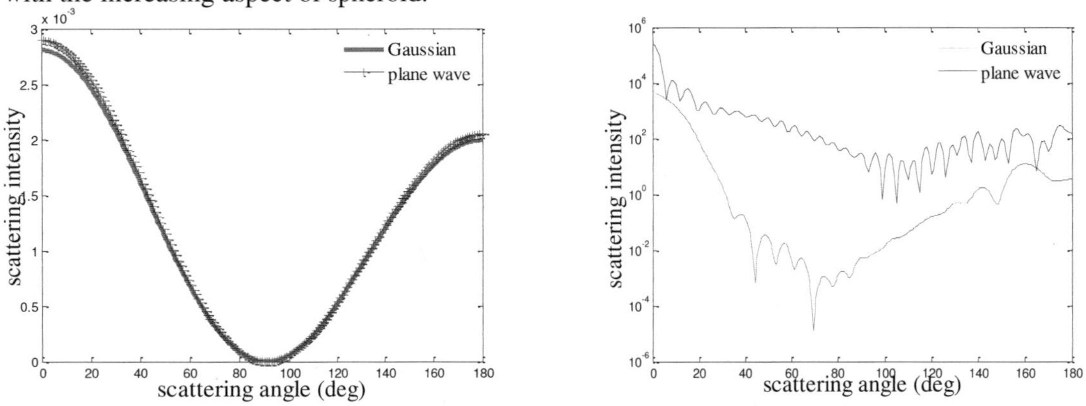


(a) $a$=1μm,aspect= 31 (b) $a$=1μm,aspect= 0.08

Fig.1 Scattering intensity of spheroid with different aspects

Fig.2 shows the scattering intensity of spheroid particles with different sizes. In this figure, the incident wavelength is 0.5μm, the beam waist $w_0$=1μm, and the relative complex reflective indices of particles is 1.33. The aspect of spheroid is invariant and the size of minor axis of the spheroid is increasing. We can see that the scattering of spheroid is also different from the sphere.

## 4. Conclusions

In this paper, the scattering of spheroid particle with in the Gaussian beam based on the GLMT is studied. In the framework of GLMT, the scattering intensity of particles is calculated. In order to research the scattering of spheroid particle with more efficiency, the a sphere of the same volume with the spheroid is used to calculate the scattering intensity of spheroid particles within the Gaussian beam, and then the scattering intensity of spheroid particles is also calculated within the plane wave incidence. In the simulations, the results show that the scattering of spheroid is different from the sphere, and the difference is more obviator with the increasing aspect of spheroid.

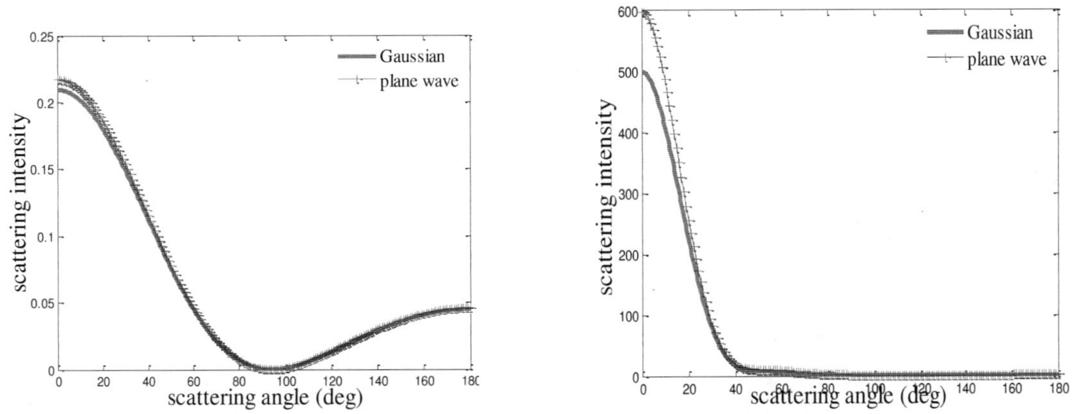


(a) $a$=0.06μm,aspect= 0.2 (b) $a$=2.7μm,aspect= 0.2

Fig.2 Scattering intensity of spheroid with different sizes

**Acknowledgements**

This work was supported by the Zhejiang Province Natural Science Funds (LY15A020003, LY14F020041). The authors are grateful to Gérard Gouesbet from INSA de Rouen, France, for useful discussions on GLMT and some helpful codes.

## References

[1] James A. Lock and Gérard Gouesbet, Rigorous justification of the localized approximation to the beam-shape coefficients in generalized Lorenz–Mie theory. I. On-axis beams, Journal of the Optical Society of America A Vol. 11, Issue 9, pp. 2503-2515 (1994)

[2] Wenbo Sun, Gorden Videen, Qiang Fu, Yongxiang Hu. Scattered-field FDTD and PSTD algorithms with CPML absorbing boundary conditions for light scattering by aerosols. Journal of Quantitative Spectroscopy and Radiative Transfer, Volume 131, December 2013, Pages 166-174

[3] Cui, Z.W. Han, Y.P.; Wang, J.J.; Zhao, W.J. Scattering of Gaussian beam by arbitrarily shaped inhomogeneous particles, Journal of Quantitative Spectroscopy and Radiative Transfer, v 113, n 6, p 480-488, 2012

[4] Zhou. B., S.S.Li, and K.Stamnes (2003), Geometrical-optics code for computing the optical properties of large dielectric spheres, Applied Optics 42 (21), 4295-4306

[5] Haihua Wang, Xianming Sun, Huayong Zhang, Scattering by a spheroidal particle illuminated with a couple of on-axis Gaussian beams, Optics & Laser Technology, Volume 44, Issue 5, July 2012, Pages 1290-1293

[6] G. Gouesbet, J.J. Wang, Y.P. Han, Transformations of spherical beam shape coefficients in generalized Lorenz–Mie theories through rotations of coordinate systems: I. General formulation. Optics Communications, Volume 283, Issue 17, 1 September 2010, Pages 3218-3225

[7] G. Gouesbet, F. Xu, Y.P. Han, Expanded description of electromagnetic arbitrary shaped beams in spheroidal coordinates, for use in light scattering theories: A review. Journal of Quantitative Spectroscopy and Radiative Transfer, Volume 112, Issue 14, September 2011, Pages 2249-2267.

IOP Conf. Series: Earth and Environmental Science 61 (2017) 012134 doi:10.1088/1755-1315/61/1/012134

# Key technologies of laser power transmission for in-flight UAVs recharging

**Z H Cui, W S Hua *, X G Liu, T Guo and Y Yan**

Department of Electronic & Optical Engineering, Mechanical Engineering College, Shijiazhuang, Hebei 050003, China

optic123456@163.com

**Abstract.** UAVs have played important roles in many fields. However, due to the insufficient energy of electric UAVs, the future development has a major obstacle. This problem can be solved by laser power transmission. This paper summarizes the research results at abroad and domestic, introduces the work flow of the whole system, discusses the key technologies and puts forward the prospect.

## 1. Introduction

In 1987, the small UAV "Pointer" was developed by Aero Vironment, and evaluated by the "Jane's Defence Weekly". Although there are some shortcomings, its potential is enormous [1]. The small electric UAVs have become a research hotspot. The main development directions of the UAVs are low cost, light, small, high altitude and long endurance, etc. [2]. It manifests small electric UAVs still be an important direction for the development of UAVs, which also has a higher demand for the energy supply. It depends on increasing the airborne battery capacity or in-flight recharging to extend its flying endurance for small electric UAVs. However, increasing airborne battery capacity will increase the UAVs' weight and reduce the UAVs' effective pay-load. The existing recharging mainly relies on charging after landing, but the UAVs' damage often occurs in the takeoff and landing process. Laser wireless charging can solve this problem.

## 2. Research status

At present, the laser wireless charging system is widely researched in some countries, such as Japan and the United States, etc.

In 1997, Yugami H carried out the ground experiment of laser power transmission [3]. It was observed that the laser affected by atmospheric changed for a long time. The performance of various types of lasers and photovoltaic (PV) panels were compared. This experiment provided references for the subsequent research. In 2006, the laser energy was used to power a robot by Kinki University. The kite and the helicopter powered by laser beam were also done in this experiment [4]. The kite continuously and stably flew more than an hour. It proved that it is possible to apply the laser power transmission technology in recharging UAVs.

In 2011, "Feasibility of Laser Power Transmission to a High-Altitude Unmanned Aerial Vehicle" was issued by the Rand Corporation [5]. According to the report, it was feasible to recharge the UAV by laser beam based on the existing technology. In 2012, it was successful to make the "Stalker" UAV powered with the laser charging system flying more than 48 hours by Lockheed Martin Space Systems Company and LaserMotive.

Chinese also had corresponding studies of the laser power transmission, but mostly in the theoretical research and experimental stage. There is currently no overall feasibility of the experiment. In 2007, a device using light energy was designed by Wang J P [6]. In 2013, the American laser power transmission experiment for in-flight UAVs was introduced by Li X Q. Difficulties and prospects were analyzed for the UAVs powered by laser beam [7].

## 3. The system structure

The overall structure of the UAV laser wireless charging system includes three major parts: ground-based equipment, data link and airborne equipment, as shown in figure 1.

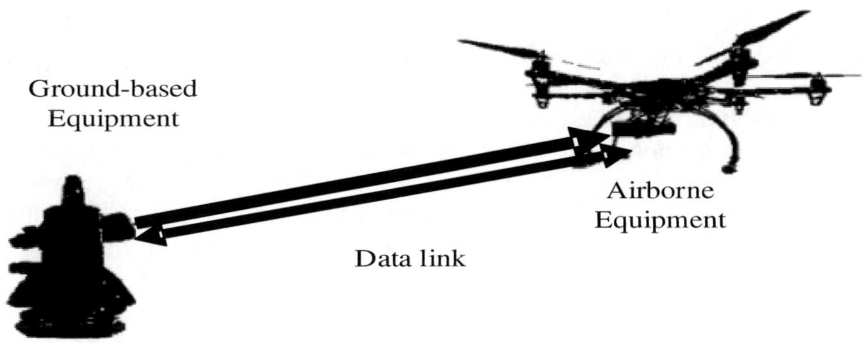


**Figure 1.** The diagram of the laser wireless charging system.

The ground-based equipment includes the power supply circuit for lasers, lasers, the laser cooling system (not required), laser beam shaping elements, optical trackers and sensors, control circuits and the beam direction device. The main work is positioning and aiming at the optical antenna of the airborne equipment, powering the whole system and lasing.

The data link refers to the transmission process of the laser and the communication signal in the atmosphere. The request of data link is unimpeded. It needs to consider the effect of atmosphere for the laser beam energy and beam quality.

The airborne equipment composes optical trackers and sensors, the optical antenna, PV panels, the cooling system for PV panels (not required) and the power management circuit. The main function is sending the position of the UAV or aiming error to the ground-based equipment, and converting the laser energy to the electrical energy for powering equipment.

## 4. The work flow

The work flow includes ten steps, as shown in figure 2. Two optical trackers and sensors of the ground-based equipment and the airborne equipment establish a connection for real-time communication. and transmit the aiming error to the control circuit of the ground-based equipment in the whole charging process. The control circuit makes beam direction device to aim at the optical antenna. Starting the power supply circuit for the laser, the laser beam goes through the beam direction device to the airborne optical antenna, and be received by the PV panels. The laser energy is transferred to the electric energy by the PV panels. The power management circuits distribute electric energy to the battery, other equipment and the motor.

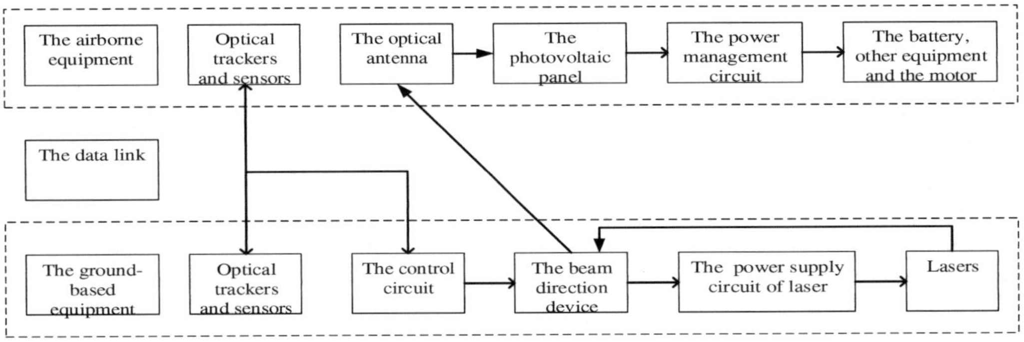


**Figure 2.** The work flow chart of laser power transmission for in-flight UAVs recharging.

## 5. Key technologies and difficulties

The most important evaluation criterion for the system is the overall conversion efficiency of the electro-optical-to-electrical process [8]. The factors that affect the overall conversion efficiency are the tracking and aiming, the electro-optical conversion efficiency of the laser and the output beam quality, the effect of the atmosphere on the laser beam, the conversion efficiency of the PV panel, the maximum power point tracking and the power management circuit.

### *5.1. The tracking and aiming*

The tracking and aiming research is a difficult spot, because the electric UAV is small and flexible, which makes the UAV tracking requires a lot. Aiming error will decrease the using efficiency of laser beam and the conversion efficiency of PV cells. Therefore, it is necessary to design the high precision tracking and aiming system with UAVs as a cooperation target.

### *5.2. The laser*

It is important for improving the overall conversion efficiency that the laser has the high electro-optical conversion efficiency and higher beam quality, where the laser diode has the highest conversion efficiency, reaching 55% ~ 60% [9]. So, it was used in the laser power transmission system. Most of the laser beam have a large divergence angle, and gently it is a class of Gaussian spot. The system requires decreasing the divergence angle to save the laser energy, and designing the structure of array-photovoltaic to match a class of Gaussian spot.

### *5.3. The laser transmission*

Laser transmission in the atmosphere will produce linear and nonlinear effects. Linear and nonlinear effects will decrease the aiming accuracy and the beam quality, which reduces the overall conversion efficiency.

### *5.4. The PV panel*

At present, commercial PV cells have low conversion efficiency, generally 20%. If the beam quality is poor, the conversion efficiency of PV panels will be significantly reduced, and PV panels may even be burned out. According to the beam quality, it is necessary to design a reasonable structure of array-photovoltaic.

### *5.5. The maximum power point tracking algorithm*

There is a nonlinear relationship between the output power and voltage of PV panels. There is also a maximum power point. Due to the uneven illumination, there are many extremes in output characteristic curve of the PV panel. It is necessary to use the maximum power point tracking technology for improving the efficiency of PV panels. But traditional maximum power point tracking methods, such as perturbation observation, incremental conductivity method and open-circuit voltage

IOP Conf. Series: Earth and Environmental Science **61** (2017) 012134 doi:10.1088/1755-1315/61/1/012134

method, are proposed for uniform illumination conditions, and not applicable to the present. Therefore, it is urgent to research a maximum power point tracking technique for uneven illumination.

### *5.6. The power management*

The function of the electric energy management circuit is distributing the electric energy to the storage battery, other equipment and the motor. If the power supplies the equipment or the motor, it only needs to design the step-down rectifier circuit. But charging the battery will need to consider the way of charging and so on. The existing charging methods are constant current charging, constant voltage charging, constant current-constant voltage charging, variable current intermittent charging, fast charging and intelligent charging. A suitable charging mode matching the laser wireless charging system is needed.

## 6. Summary

In this paper, after a short state-of the art review on the laser power transmission, the system structure and charging process model are proposed for small electric UAVs recharging, and the key technologies and problems are presented. Improving energy conversion efficiency from the aspects of laser selection, tracking and aiming control, the array-photovoltaic structure, charging mode and power management is important.

## References

[1] Fan L J, Review of American and Israeli mini-UAVs, *Aeronaut. Sci. Technol.* **2007-03** 13-16.
[2] Zhen Y H, Related Technologies and Development of UAV, *Ordance Ind. Automation* **2009-28(1)** 14-16.
[3] Yugami H, Kanamori Y, Arashi H, Niino M and Moro A, Field experiment of laser energy transmission and laser to electric conversion, *Proceedings of the 32nd Intersociety IEEE* **1997-1** 625-630.
[4] Howell J T, O'Neill M J, Fork R L and Fork R L, Advanced receiver/converter experiments for laser wireless power transmission, Solar Power from Space-SPS'04 2004-567 187.
[5] Mason R Feasibility of Laser Power Transmission to a High-Altitude Unmanned Aerial Vehicle, (Rand Corporation 2011).
[6] Wang J P and Cheng Y, the laser energy supply by utilization of solar energy, *Power Energy* **2007-28(2)** 98-101.
[7] Li X Q. laser power beaming for UAVs, *Laser J.* **2013-34(4)** 18-19.
[8] Liu X, Hua W, Liu X, Factors influencing efficiency of laser wireless power transmission system for micro unmanned aerial vehicles, International Symposium on Optoelectronic Technology & Application Development & Application of High Power Lasers SPIE **2014-9294** 929403-6.
[9] Li Y, Zhang Q, Zhang C. research of UAV laser wireless energy transmission technology, *Wing Missil J.* **2015-01** 58-61

IOP Conf. Series: Earth and Environmental Science **61** (2017) 012135 doi:10.1088/1755-1315/61/1/012135

# A global analysis on water-based fire extinguishing agent

**Shuai WANG**

Tianjin Fire Research Institute, Tianjin, China
wangshuai@tfri.com.cn

**Abstract**. Due to the superiority of the attribute of water, water-based fire extinguishing agent is considered as one of most effectively fire extinguishing agents. NFPA has developed two standards regarding to water-based fire extinguishing agents. ISO technical committee working group is also preparing for developing a standard about the subject fire extinguishing agent. China also has its own national GB standard about water-based standard. This paper aims at to elaborate standard requirements and methods in different technical documents and standards currently available around the world with a view to summarize the main concern in different standards, and trying to find out valuable information for readers in future research and development.

## 1. Need for the study on water-based fire extinguishing agent

Water has been recognized as a fire extinguishing agent in ancient times. This is because of the characteristics of strong heat absorption and cooling ability of ambient environment. At the same time, water could be transformed to large amount of steam after being vaporized, and the steam could prevent air from entering into the burning area, cut down the entering of combustion accelerator, and thus cut down the burning process. But it is also noted that water has some deficiencies in being applied as fire extinguishing agent. Water has high degree of fluidity, hard to stay on the surface of the burning item, and thus it cannot linger on the combusting item for a longer time and cannot cooling down the temperature and will probably lead to vapor explosion if more vapor accumulated within an enclosed space. Water has large value of surface tension, low wetting ability and these factors will reduce the efficiency of the fire extinguishing process. Under such circumstances, more research are conducted about the analysis to add water additives in water and to improve the performance of the fire extinguishing agents.

This paper aims at finding out relevant standards and technical documents about water-based fire extinguishing agents currently available and trying to elaborate the main contents and pinpoint the differences and pros and cons of different documents for later technical reference.

## 2. Standards review

### *2.1. General*

In doing systematic review of fire extinguishing agents, it is found that some western countries have developed relevant standards regarding to the products, and some have already been developed for several editions. China also has national GB standard about water-based fire extinguishing agent, which was developed based on the products available in China's market and China's situations.

### *2.2. Elaboration of documents*

IOP Conf. Series: Earth and Environmental Science **61** (2017) 012135 doi:10.1088/1755-1315/61/1/012135

ISO/TC21/SC6 Technical committee for Foam and powder media and firefighting systems using foam and powder in International Standardization recently has launched standards review about water-based fire extinguishing agents currently available around the world. The research found out five main standards documents in the market

Based on the document table indicated, ISO and EN standards have only standards about chemical surface active agents rather than standards about water additives applied in fire extinguishing, while NFPA has two standards about wetting agents and water additives for fire control and vapor mitigation.

Table 1. Standards review on water-based fire extinguishing agents

| No. | Standard/Document | Standard/Document Name |
|---|---|---|
| 1 | DIN EN 1772:2000 | Surface active agents – Determination of wetting power by immersion (ISO 8022:1990,modified) |
| 2 | ISO 8022 | Surface active agents – Determination of wetting power by immersion |
| 3 | NFPA 18-1995 | Standard on Wetting Agents |
| 4 | NFPA 18A-2011 | Standard on Water Additives for Fire Control and Vapor Mitigation |
| 5 | GB 17835-2008 | Water based extinguishing agent |

### *2.3. NFPA Standards*

#### *2.3.1. NFPA 18 Standard on wetting agents*

This standard has been confirmed and updated in several editions since 1949. Currently, the most recent standard is 2011 edition. This edition has undergone extensive technical and editorial revision. Technical changes include limits for aquatic toxicity for parity and consistency with other product standards.

Regarding to technical performances, NFPA has many detailed requirements and many requirements are evaluated both for concentrations and solutions. About Concentrates pour point, it shall be determined in accordance with ASTM D97. Concentrates also shall meet the requirements of miscibility, separation and impact of low temperature of wetting agent concentrate on surface tension. Wetting agent concentrate shall be between 6 and 9 at 18°C±2.7°C.Viscosity of concentrate shall be measured. A Brookfield viscometer is recommended to use. Wetting agent has a server requirement of toxicity about both concentrate and solution. The toxicity shall be inspected on four levels, for acute oral toxicity, acute dermal toxicity, acute eye irritation and acute dermal irritation. All the test methods shall confirm to EPA OPPTS or their equivalents. Wetting agent concentrate shall also conform to requirement of biodegradability, which shall be in accordance with U.S.EPA OPPTS 835.3110 or equivalent.

Both concentrate and solution shall undergo severe corrosion test, which shall be in accordance with NACE TM 0169 or with ASTM G 1 and ASTM G31. Results shall be less than or equal to the values found in Table 2. Wetting agent and its solutions shall be tested for corrosion with samples of 4130 mild steel, 2024-T3 aluminum, and UNS C27000 yellow brass (65 percent copper, 35 percent zinc). The wetting agent and its solutions shall be tested at the maximum and minimum use concentrations specified by the manufacturer.

Regarding to fire tests, NFPA requires products listed for use on Class A fires shall pass all the fire tests specified regarding to Class A fire extinguishment tests in NFPA 18. The tests include wood crib fire test, deep-seated fire test and wood fibreboard penetration tests. These tests mainly test the penetration and deep seated ability of the wetting agent. NFPA also requires products listed for use on Class B fire shall pass all the fire tests specified regarding to Class B fire extinguishment tests in

NFPA 18. Wetting agent solutions at the concentrations specified by the manufacturer shall be evaluated to and comply with the requirements of UL 711/ULC S508 for Class B fires.

Table 2. Maximum allowable corrosion rates (Adopted from NFPA 18)

| | 2024-T3 Aluminum | | | | 4130 Steel | | | | Brass[a] | AZ31B Magnesium | | | |
|---|---|---|---|---|---|---|---|---|---|---|---|---|---|
| | Total Immersion | | Partial Immersion | | Total Immersion | | Partial Immersion | | Partial Immersion | Total Immersion | | Partial Immersion | |
| Application | 21°C (70°F) | 49°C (120°F) | 21°C (70°F) | 49°C (120°F) | 21°C (70°F) | 49°C (120°F) | 21°C (70°F) | 49°C (120°F) | 49°C (120°F) | 21°C (70°F) | 49°C (120°F) | 21°C (70°F) | 49°C (120°F) |
| Wetting agent concentrates | 2.0 | 2.0 | 5.0 | 5.0 | 5.0 | 5.0 | 5.0 | 5.0 | 5.0 | 5.0[b] | 5.0[b] | 5.0[b] | 5.0[b] |
| Wetting agent solutions | N/A | N/A | N/A | N/A | N/A | N/A | N/A | N/A | N/A | N/A | N/A | N/A | N/A |
| Fixed-wing aircraft | 2.0[c] | 2.0[c] | 2.0[c] | 2.0[b] | 5.0 | 5.0 | 5.0 | 5.0 | 5.0 | — | — | — | — |
| Helicopter with fixed tank | 2.0[c] | 2.0[c] | 2.0[c] | 2.0[b] | 5.0 | 5.0 | 5.0 | 5.0 | 5.0 | 4.0[c] | 4.0[c] | 4.0[c] | 4.0[c] |
| Helicopter with bucket | 2.0 | 2.0 | 2.0 | 2.0 | 5.0 | 5.0 | 5.0 | 5.0 | 5.0 | — | — | — | — |
| Ground application[d] | 2.0 | 2.0 | 2.0 | 2.0 | 5.0 | 5.0 | 5.0 | 5.0 | 5.0 | — | — | — | — |

Note: All values in milli-inches (mil) per year; for SI units, 1 mil = $2.54 \times 10^{-2}$ mm.
[a]Composition of brass is 65 percent copper, 35 percent zinc.
[b]Required only if submitted for use in helicopters equipped with fixed tanks or if the concentrate is contained on board the helicopter.
[c]Intergranular corrosion tests also required; see 4.3.4.2 of NFPA 1150.
[d]Includes fire apparatus, portable pumps, backpacks, and other such devices.

*2.3.2. NFPA 18A Standard on water additives for fire control and vapour mitigation*

Initially, the committee responsible on water additives for fire control and vapor mitigation proposed to combine wetting agents and water additives under on standard. This effort was returned to the committee by Association action in June 2003. As a result, the committee decided to divide this work into two subject areas and standards, retaining and revising NFPA 18 and creating a new standard addressing water additives, NFPA 18A, the first edition of which was issued in 2007. Changes in the 2011 edition include a reorganization of Chapter 5 and 6, clarification of test procedures and criteria, and the removal of secondary (U.S.) units throughout much of the text.

This standard regulates that water additive concentrate shall be mixed only with water. Regarding to technical performances in NFPA 18A, they share with NFPA 18 in many items, i.e. toxicity and environment test (including mammalian toxicity, aquatic toxicity, and biodegradability), concentrate pour point, PH value and viscosity.

This standard highlighted the performance requirement of concentrate stability. Three 19L samples of water additive concentrate from a single production lot shall be stored in sealed containers. The samples shall be designated as Sample 1, Sample 2, and Sample 3. The three samples shall be treated in different conditions and be inspected the status of stability.

The standard also has more requirements about corrosion except for uniform corrosion regulated in NFPA 18. The standard asks for the test of intergranular corrosion test. There shall be no intergranular corrosion on aluminum 2024-T3 when tested if the solution is recommended for application by fixed wing aircraft and magnesium AZ31B for rotary-wing aircraft. The standard also required compatibility with nonmetallic materials. The tests include sample exposure tests, volume test and hardness of test. Regarding to fire test methods, water additive solutions for Class A fuels shall be tested in accordance with all of the following fire test procedure, including wood panel fire test and wood crib fire test. As far as Class B fire test methods are concerned, water additive solutions for Class B fuels shall be tested in accordance with one or more of the test procedures, including spill fire test, pool fire test, three-dimensional fuel fire test, polar solvent fire test, emulsification and fuel in depth fire test.

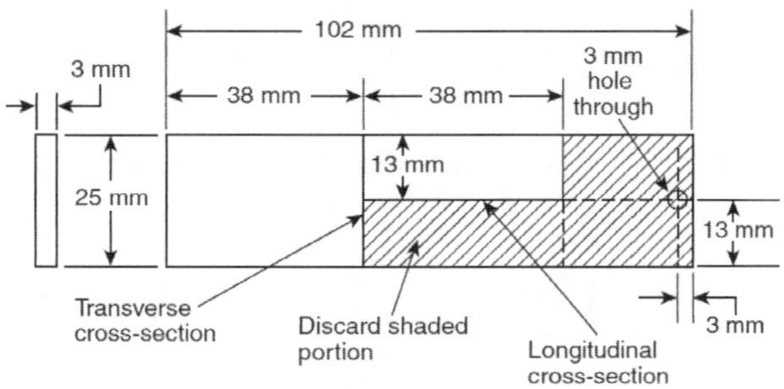


Figure 1.Intergranular Corrosion Test (Adopted from NFPA 18A)

### *2.4. China's GB Standard*

China has its own national GB standard regarding to water-based fire extinguishing agent. GB 17835 has two editions with the current edition published in 2008. Most of the technical performances and test methods comply with GB 15308-2006 Foam extinguishing agent, including pour point, PH value, surface tension and corrosion test.

The GB standard has requirement about toxicity, but with comparatively less sever requirement. The test is to let zebra fish stay in solution mixed with water-based agent and prepared solution for 96 hours. Death rate shall be not more than 50%. Regarding to fire tests, GB standard has performance requirements for Class B and Class A fire tests. About Class B fire test, the agent shall extinguish solvent oil, and 99% acetone fire. As to Class A fire test, the agent shall extinguish wood crib with more than 1A grade.

## 3. Comparative research and limitation of this study

Based on the standards review of current standards available, the study found out that China's GB standard has many obvious differences with peer standards in US. China's standard is developed based on China's foam standard while NFPA standards have more severe requirements and test methods. There are longer specimen treatment process in NFPA standards and more requirements about corrosion, toxicity and wood fibreboard penetration. NFPA standard has many requirements about toxicity, including for mammalian toxicity, concentrate aquatic toxicity, and concentrate biodegradability. As to the status that chemical toxicity and environment problems have been highlighted, it is recommended to research on the possible solution to downgrade the toxicity to human beings and the earth while to realize the fire performance requirements in NFPA standards. ISO technical committee working group responsible for fire extinguishing agents has realized the same challenge and intend to add biodegradability as an important performance in future ISO water-based fire extinguishing standard.

Combining to the global concern and realistic status in China, the research suggest add more severe requirements about biological toxicity and relevant fire test performance in China's future research and standard development projects. Even if this research has accumulated most of standards about water-based extinguishing agent currently available, there may be some more standards and documents around world which need us spend more time to re-search. Due to limitation of time and research funding, this paper only review the same and differences of the content of the subject standards, while further experimental research need to be developed in future.

## References

[1] Standardization Administration of China. GB 15308-2006 *Foam extinguishing agent*

[2] Standardization Administration of China. GB 17835-2008 *Water-based extinguishing agent*

[3] NFPA 18 *Standard on Wetting Agents*, 2011.

[4] NFPA 18A *Standard on Water Additives for Fire Control and Vapor Mitigation*, 2011.

[5] ISO 8022 *Surface active agents — Determination of setting powder by immersion,* 1990.

[6] EN 1772 *Surface active agents — Determination of setting powder by immersion* (ISO 8022:1990 modified), 2000.

[7] Yang Liang et al 2015 Fire extinguishing agents and standardization (Beijing:Scientific and technical documentation press) chapter 1 pp 64–68.

IOP Conf. Series: Earth and Environmental Science **61** (2017) 012136 doi:10.1088/1755-1315/61/1/012136

# Short-term toxicity of polystryrene microplastics on mysid shrimps *Neomysis japonica*

**Mingxin Wang, Xiao Wang, Xianxiang Luo and Hao Zheng***

Key Laboratory of Marine Environment and Ecology, Ministry of Education and College of Environmental Science and Engineering, Ocean University of China, Qingdao 266100, China

E-mail address: zhenghao2013@ouc.edu.cn

**Abstract.** Plastic debris especially microplastics (MPs) have become a global concern for the aggravating pollution in the oceans. In this study, the physico-chemical properties of fluorescently labeled polystyrene (PS) beads and the effects of PS-MPs on the survival of mysid shrimps (*Neomysis japonica*) were investigated. PS-MPs were identified to have spherical shape, uniform size and stable green fluorescence. The results showed that PS beads had little effects on the mortality of shrimps under a short-term (72 h) exposure with concentrations of 50 μg $L^{-1}$ and 500 μg $L^{-1}$. However, PS-MPs had severe short-term toxicity on the survival of mysid shrimps, resulting in 30% mortality especially in a 72 h exposure with the higher concentration of 1000 μg $L^{-1}$. These findings provide new insights into the toxic effects of MPs on marine invertebrates.

## 1. Introduction

Global plastic production increased rapidly annually due to the widespread use in both industrial products and consumer goods, which was estimated to reach up to 322 million tons in 2015 [1]. However, lack of appropriate disposal measures of plastic waste led to excessive accumulation of plastic debris in the ocean ecosystem [2], which further aggravated environmental threats such as smothering the seabed, preventing gas-exchange and threatening functioning of marine ecosystems [2,3]. Moreover, large plastic items can be gradually fractured into microplastics (MPs), which defined as plastic particles less than 5 mm in diameter, because of photodegradation, mechanical breakdown and microbiological decomposition [4]. In recent years, concern is continuously rising on the ubiquitous and persistent contamination of MPs in marine ecosystems, especially their threats to marine biota.

Polystyrene (PS), one of the five main types of plastic items, is a primary component of plastic debris observed in the environment [2,4]. Increasing evidences have supported that PS derived MPs (PS-MPs) could be ingested by diverse organisms (e.g. fish, crab and copepod) [5,6,7,8]. Entanglement and blocking of PS-MPs in the tissues may disrupt physiological processes [6], hamper energy flows [7] and even affect the survival and growth of marine biota [8]. Mysid shrimps *(Neomysis japonica)*, as an ecotoxicological model for environmental sensitivity [9], have been used in evaluating toxicity of heavy metal [10], PAHs [11] and other environmental pollutants. However, few studies have been focused on the effects of different concentrations of MPs on mysid shrimps. The aim of this study was to investigate the physico-chemical properties of PS-MPs and evaluate the effects of different concentrations of MPs on the survival of mysid shrimps.

IOP Conf. Series: Earth and Environmental Science **61** (2017) 012136 doi:10.1088/1755-1315/61/1/012136

## 2. Materials and methods

### *2.1. Preparation and characterization of PS-MPs*

Fluorescently labeled PS-MPs (5μm, 450 nm excitation, 498 nm emission) were purchased from Phosphorex, Inc. (Lot #: 2106B, Hopkinton, USA). PS suspensions (10 mg/mL) were prepared in filtered natural seawater (NSW) (0.45 μm filter membrane, Jinteng, China) and kept at 4 °C in dark. The suspensions were vortexed for 10 seconds prior to ecotoxicological studies.

Primary size and shape of PS-MPs were identified by transmission electron microscopy (TEM, JEM-1011, JEOL, Japan) and scanning electron microscope (SEM, Hatachi S4800, Hatachi, Japan). The fluorescent intensity of PS-MPs was ascertained by fluorescence microscope (DM2500, Leica, Germany). In addition, the zeta-potential of the suspensions was determined using dynamic light scattering (Nano-ZS90, Malvern, UK).

### *2.2. Mysid shrimps maintenance and PS-MPs exposure*

Mysid Shrimps (*Neomysis japonica*) at the age of 27 days were obtained from an aquafarm in western part of Liaoning province. They were maintained in 18 L glass cylinder containing 12 L of filtered NSW (pH 8.14 ± 0.03, salinity 32.5 ± 0.5 ‰, temperature 22.6 ± 0.6 °C) with continuously aeration for 1 day to allow full gut depuration. The short-term toxicity test was performed in glass beakers containing 2 L of filtered NSW. The mysid shrimps were exposed to different concentrations of PS suspensions (0, 50, 500, 1000 μg $L^{-1}$) for 72 h, and were fed with brine shrimp nauplius (40 baits/shrimp) twice a day (08:00-09:00 and 17:00-18:00). Dead mysid shrimps, faeces and excess food were removed at regular intervals and the mortality was recorded every day during the experiment.

### *2.3. Statistical analysis*

The significance of the various parameters was tested by one-way analysis of variance (ANOVA) using Duncan's multiple range test ($P = 0.05$) for different exposure concentrations (i.e. 0, 50, 500, 1000 μg $L^{-1}$). All of the analyses were performed by means of Statistical Product and Service Solutions Software (SPSS, version 20.0).

## 3. Results and discussions

### *3.1. Properties of PS-MPs*

Both the SEM and TEM images confirmed that the pristine diameter of PS-MPs was 5 μm and they had regularly spherical shape (figure 1 a-c). PS microspheres were labeled with stable green fluorescence, which dispersed evenly in the aqueous phase (figure 1 b, d). Additionally, the zeta-potential of PS microspheres was -17.25 ± 2.84 mV.

### *3.2. Effects of PS-MPs on the survival of mysid shrimps*

Regardless of the differences in the exposure time, the toxicity of PS-MPs on the survival of mysid shrimps increased with increasing exposure concentrations, indicating that PS-MPs exhibited significant concentration-dependent toxicity of mortality for the tested mysid shrimps. PS-MPs exposure at the highest concentration (1000 μg $L^{-1}$) caused 30% mortality of the shrimps after waterborne exposure for 72 h ($P < 0.01$), but no similar effect was observed at lower concentrations (≤ 500 μg $L^{-1}$). Additionally, there was also no significant difference between PS-MPs exposure concentrations of 50 μg $L^{-1}$ and 500 μg $L^{-1}$ and control treatment during the whole incubation (figure 2). These results could be attributed to the large amounts of MPs accumulation in stomach, intestine or other tissues of the mysid shrimps [12] at the higher concentrations of PS-MPs exposure, which could interfere feeding, digestion, excretion and other normal physiological processes [2,7], consequently suppressing the growth and further posing threats to the survival of the mysid shrimps.

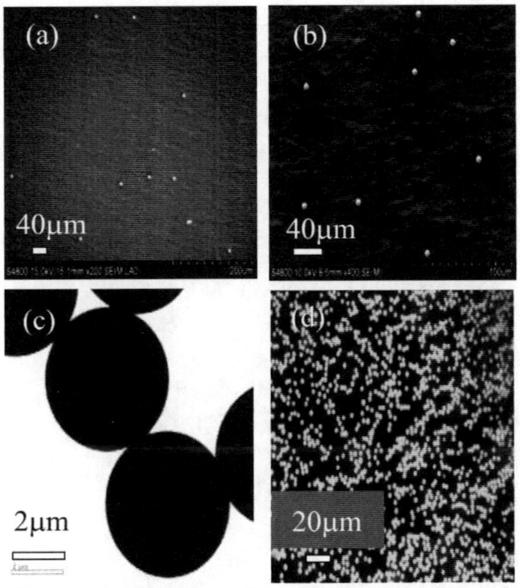


**Figure 1.** Physico-chemical properties of PS-MPs: (a), (b) SEM images; (c) TEM image and (d) optical image taken from fluorescence microscope.

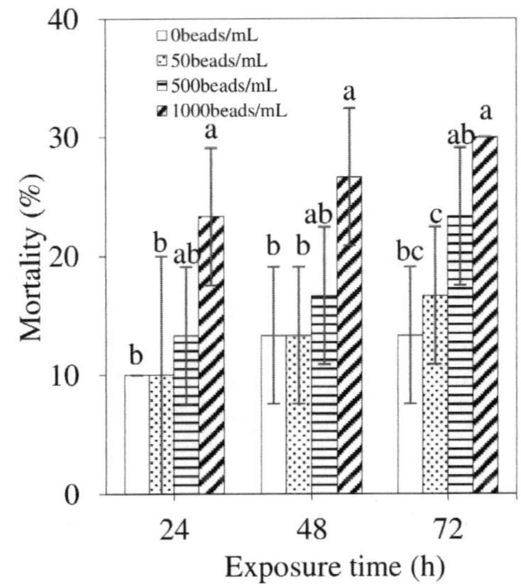


**Figure 2.** Mortality of the mysid shrimps with PS-MPs at concentrations of 0, 50, 500 and 1000 μg $L^{-1}$ during a short-term exposure (72h). Different small and capital letters indicate significant difference between different treatments ($P < 0.05$).

## 4. Conclusion

In this study, the spherical PS-MPs with particle size of 5 μm, were stable and dispersed uniformly in aqueous phase. The effects of PS-MPs on the mortality of mysid shrimps were concentration-dependent. The higher concentration of PS-MPs (1000 μg $L^{-1}$) caused a 30% mortality, but lower concentrations (≤ 500 μg $L^{-1}$) had no obvious adverse effect on the survival of mysid shrimps. These results could contribute to verify threats of MPs to marine biota.

**Acknowledgments**

This study was supported by National Natural Science Foundation of China (41325013, 41406085) and Foundation of Shandong Provincial Key Laboratory of Marine Ecology and Environment and Disaster Prevention and Mitigation (201512).

## References

[1] Europe P 2015 Plastics–the facts 2016 *Occasion of the special presentation of K 2016* Messe Düsseldorf (Brussels: PlasticEurope offices) pp 1-38
[2] Kärrman A, Schönlau C and Engwall M 2016 *Exposure and effects of microplastics on wildlife MTM Research Centre Report* (Sweden: DiVA ) pp 1-39
[3] Gregory R 1996 Plastic ‘scrubbers’ in hand cleansers: a further (and minor) source for marine pollution identified *Mar. Pollut. Bull.* **32** 867-71
[4] Andrady A L 2011 Microplastics in the marine environment *Mar. Pollut. Bull.* **62** 1596-1605
[5] Lu Y, Zhang Y, Deng Y, Jiang W, Zhao Y, Geng J, Ding L and Ren H 2016 Uptake and accumulation of polystyrene microplastics in zebrafish (*Danio rerio*) and toxic effects in liver *Environ Sci Technol.* **50** 4054-60
[6] Lönnstedt O M and Eklöv P 2016 Environmentally relevant concentrations of microplastic particles influence larval fish ecology *Science.* **352** 1213-6

IOP Conf. Series: Earth and Environmental Science **61** (2017) 012136 doi:10.1088/1755-1315/61/1/012136

[7] Watts A J, Urbina M A, Corr S, Lewis C and Galloway T S 2015 Ingestion of plastic microfibers by the crab *Carcinus maenas* and its effect on food consumption and energy balance *Environ Sci Technol.* **49** 14597-604
[8] Cole M, Lindeque P, Fileman E, Halsband C and Galloway T S 2015 The impact of polystyrene microplastics on feeding, function and fecundity in the marine copepod *Calanus helgolandicus Environ Sci Technol.* **49** 1130-7
[9] American Society for Testing and Materials 2012 *Annual book of ASTM standards* vol 11 (IHS West Conshohacken: ASTM international ) p E1463-92
[10] De Figuerêdo L P, Nilin J, Da Silva A Q, Loureiro S and Costa-Lotufo L V 2016 Development of a short-term chronic toxicity test with a tropical mysid *Marine Pollution Bulletin.* **106** 104-8
[11] Poelmans S, Verslycke T, Monteyne E, Noppe H, Verheyden K, Janssen C R and De Brabander H F 2006 Testosterone metabolism in Neomysis integer following exposure to benzo (a) pyrene *Comparative Biochemistry and Physiology Part B: Biochemistry and Molecular Biology.* **144** 405-12
[12] Setälä O, Fleming-Lehtinen V and Lehtiniemi M 2014 Ingestion and transfer of microplastics in the planktonic food web *Environ Pollut.* **185** 77-83

IOP Conf. Series: Earth and Environmental Science **61** (2017) 012137 doi:10.1088/1755-1315/61/1/012137

# The application of microbial combination flooding oil recovery technology in heavy oil reservoir with low temperature

**Yongbin Li[1,1,a], Jing Wang[1],Yanjun Wang[2], Dengfeng Ju[1], Yaxiu Fu[3],Xiaoyang Lei[1], Jizhe Jing[1] and Guiying Liu[2]**

[1]Petroleum Production Engineering Research Institute, Huabei Oilfield Company, Renqiu ,Hebei062552, china
[2]Chemical Industry Branch of Huabei Petroleum Tiancheng Industrial Group Co.Ltd in Hebei, Renqiu ,Hebei062552, china
[3]The Erlian Filiale of Huabei Oilfield Company of petrochina, Xilin Hot026017,china

[a]email:cyy_liyb@petrochina.com.cn

**Abstract**. HuabeiBaolige Oilfield belongs to the common heavy oil reservoirs with low temperature, which were tapped by the conventional waterflooding. The formation temperature of Baolige Oilfield is 38~58°C, and the oil viscosity of reservoir is 13.7~2000mPa•s. Thanks to the high oil-water viscosity ratio and strong heterogeneity, the small waterflooding swept volume and serious water breakthrough are caused by waterflooding fingering, causing that the workable reserve cannot be used efficiently during the oilfield development. According to the characteristic that the environment of the reservoirs is fit for the growth and reproduction of microorganism, the microbial enhanced oil recovery (MEOR) technology is used to improve oilfield development status. On the basis of continuous and further studies of MEOR, the industrialized application of MEOR has been fulfilled. By the continuous and further study, the efficient system of the combination flooding technology with oil displacement microbial fields was formed, and MEOR technologies have been enriched. All the above researches could provide technical ideas for the comprehensive treatment for similar blocks.

## 1. Introduction

In April 2004, the Ba 19, Ba 38, Ba 48 block was put into production. Late on, the Ba 51, Ba 10 block was built continuously, achieving production to succeed. In the process of waterflooding development, contradictions were increasingly apparent. On the one hand, with Ba 19, Ba 38 thin oil reservoirs, because of serious reservoir heterogeneity, large plane and interlayer contradiction and relatively high oil-water viscosity, swept volume of water flooding was small, the water cut rising speed was high, production decline was rapid. On the other hand, with Ba 48, Ba 51 ordinary heavy oil reservoirs, due to their high oil-water viscosity, water injection fingering is serious, water cut rising is fast, still in the low speed and inefficient development. Therefore, for the purpose of the overall improvement of oilfield waterflooding development effect, the overall management technology of waterflooding

[1]Biography: Yongbin Li, born in 1984, male, engineer, graduated from Department of Geochemistry of Yangtze University in 2007, majored in bio-engineering, worked on the tertiary oil recovery research, Petroleum Engineering Research Institute of Huabei Oilfield Branch Company of PetroChina Company Ltd.. Postal address: Petroleum Engineering Research Institute of Huabei Oilfield Company, Renqiu, Heibei Province.

IOP Conf. Series: Earth and Environmental Science **61** (2017) 012137 doi:10.1088/1755-1315/61/1/012137

development to improve low temperature and heavy oil reservoir is researched and developed, which is the based microbial enhanced recovery of low temperature and heavy oil reservoir as the core, with gel displacement to improve the effect of microbial oil recovery. Through the large scale application of MEOR technology, microbial field had been preliminarily established, the cyclic microbial drive had been realized, production decline has been controlled effectively, the oilfield development effect had been improved effectively, solving the need of oil field production, filling the blank of the related technology of Huabei oilfield.

## 2. Basic conditions of reservoir

### 2.1. Geologic characteristics

Baolige oilfield is located in Bayindulan Sag north Sub-sag of Erlian Basin: Ba 19, Ba 38 block is located in BaII tectonic zone, while Ba 48, Ba 51 block is located in BaI tectonic zone (Fig. 1). Ba 48 block is monoclinal structure, while the other three blocks are fault-anticline structure, which more than one internal down faults are developed, making the structure further complicated.

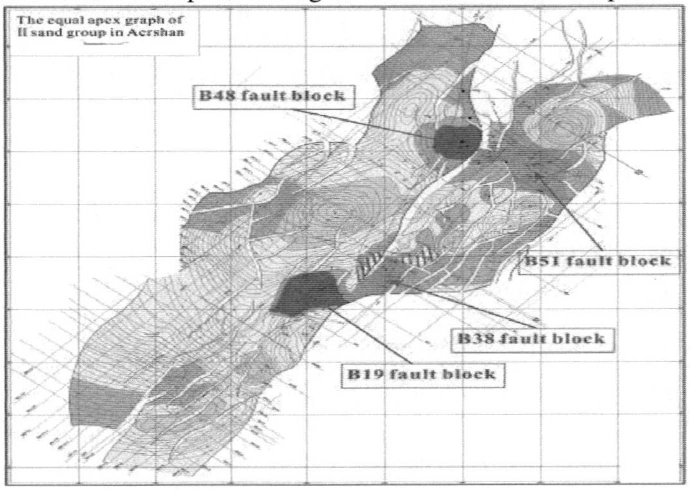


**Fig. 1** Distribution map of Baolige oilfield

The sedimentary environment is fan-delta front subfacies, the provenance is from the southeast, the main developed microfacies is the underwater distributary channel, mouth bar, distal bar, frontal sand sheet, subsea overbank sand, interdistributary bay and frontal fan- delta mud. Ba 19 and Ba 38 block is medium- porosity and low&medium-permeability reservoir, Ba 48 and Ba 51 block is low&medium-porosity and low permeability reservoir. According to formation test data, every block of fields is the normal pressure and temperature system.

### 2.2. Development characteristics

The water drive efficiency of thin oil reservoirs of BaII tectonic zone is high, of which working condition is multidirection-based. The distribution areas of relatively coarse sand conglomerate, conglomeratic sandstone and fine sandstone are better working, while other lithologic distribution area is relatively poor. The producing degree and control degree of waterflooding of major reservoir is higher, and the control degree of waterflooding is up to 68.8%. The sand suction-production condition of oil layer is well, but the interlayer contradiction is prominent. Water cut rising is faster after water breakthrough, water cut rising rate is higher than the forecast. The water cut rising rate of Ba 19 block is high to 18.6% and Ba 38 block is high to 16.4%.

The waterflooding control degree of major reservoir in heavy oil reservoirs of BaI tectonic zone is higher, but the producing degree is low. At the same time, the large longitudinal property differences lead to relatively large interlayer contradiction. Water absorbing capacity of 6.7% of the thickness in the water accepting layer is 59.7%, while liquid producing capacity of 21% of the thickness in fluid producing layer is 73.9%. Ba 48 and Ba 51 block is breakthrough instantly when put into production. After water injection response, the water cut rising is rapid, initial water cut rising rate is as high as 62.4% and the calculated recovery of water drive is only 4.2%.

## 3. MEOR technology research

### *3.1. Research solution*

From the basic study and understanding of the reservoirs, by using the static and dynamic contrastive analysis methods, the influence factors of causing injection- production contradictions are analyzed in-depth. According to the production features showed by every well group of reservoirs, management technical countermeasures is researched divisionally. On the basis of correct understanding of the reservoirs, further for the block and the main influencing factors, through the data query comparison and laboratory experiment verification, the technological measures for oil and water wells method are chose and studied, realizing the technology innovation in management thinking (Fig. 2).

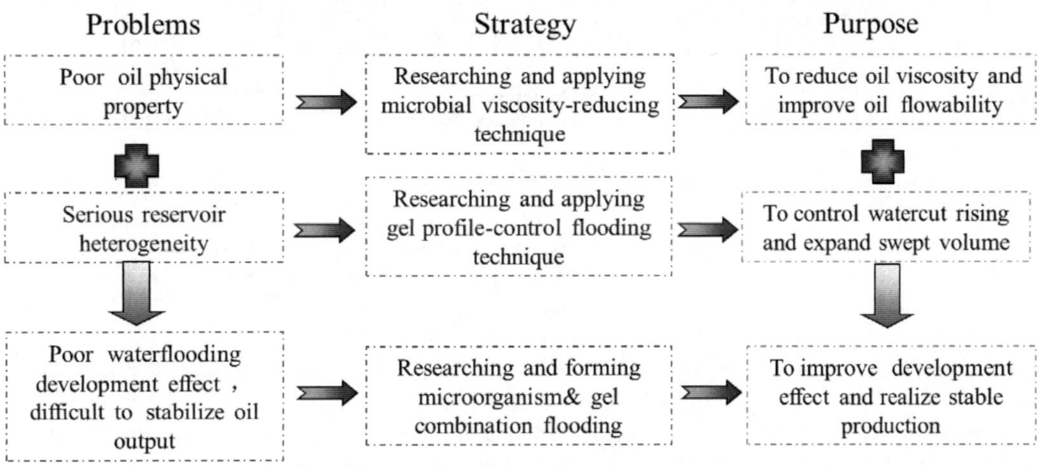


**Fig. 2** Management technical countermeasures chart

In the study of technology, the technological method that lowers crude oil viscosity, controls the oil-water viscosity ratio and expands the formation sweep volume of microorganisms is the key research contents. By learning from mature theory and applying field test experience, combined with the laboratory evaluation experiment study, optimization design of the scheme is explored and completed. At last, the technical pilot field test is carried out step by step, to verify the reliability of technology. On this basis, facing the whole Baolige oilfield, deployed overall, implemented step by step, through the periodical and periodic technology implementation, the driving ability of formation oil is enhanced, the oil-water mobility ratio is improved, the water cut rising rate is controlled, the development speed is improved, realizing the industrial application of microbial recovery technique in Baolige oilfield.

### *3.2. Microbial technical system research*

On the basis of the overall determinate thought, since 2007, through the related institutes cooperation, from the filtration of microbial strains with efficient production, species that can reduce the viscosity of crude oil are separated and domesticated and the evaluation method is constructed. And then the microorganism metabolites are analyzed qualitatively and quantitatively, establishing analysis standards of various metabolic products. The mechanism of MEOR is further understood, and the basic microbial flooding system has been established which is suitable for the Baolige oilfield (Fig. 3).

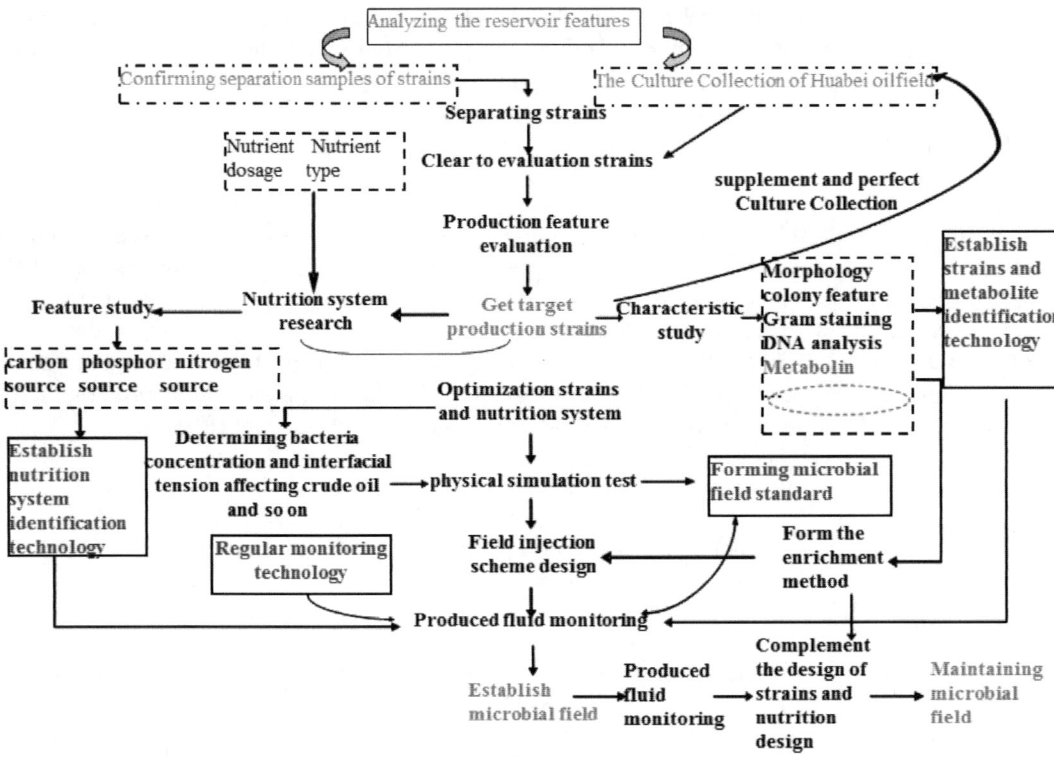


**Fig. 3** Flow chart of strain screening and evaluation

Eventually, 6 cases of microbial strains are determined by microbiological analysis and evaluation, of which emulsifying performance of indoor evaluation can reach more than level 4, average viscosity break rate is above 40% and average decreasing interfacial tension is 80% or more. At the same time, on the aspect of adaptability to environment evaluation, the screened six strains grow well under the condition of the reservoir temperature. Through the puncture test, these strains are confirmed as facultative anaerobe and grow normally in formation environment with dilute oxygen. They do not produce inhibition zone each other and the antagonism compatibility between the strains is well. In order to make full use of advantages of interaction between different strains, strain compounded system is established, whose performance evaluation is excellent and improved oil recovery in indoor physical simulation is above nine percent.

### *3.3. Gel flooding system research*

In microorganism displacement process, because of reservoir heterogeneity, many problems are caused such as serious microbial channeling in the formation, being difficult to build microbial field effectively. Through technical research, under the condition of microorganism, gel formula system and injection technology is formed to satisfy the requirement of deep profile control of different reservoirs, which relieves effectively the reservoir contradictions between plane and vertical, inner layer and interlayer, and gives full play to the function of the microorganism field in the formation, to reach the aim to reduce oil viscosity, enhance the swept volume of injected water and improve the overall development effect of Baolige oilfield.

On the first stage, focusing on reservoir physical property and fluid property of Baolige oilfield, gel profile control technology system based on polymer- organic chromium is built. Considering that the formation temperature, fluid property and dosing water are the important factors influencing the gelling in the gel system, represented by Ba 19 and Ba 51 block, the laboratory experiment of formula system is carried out to determine the basic formula of gel profile control. Compatibility experiments show that the microbial and gel system has nice compatibility. Through physical simulation test of gel profile control under the condition of microbial, oilfield development effect could be improved effectively. On the basis of laboratory experiments, field test is carried on. On the second stage, according to the field application situation, in view of the polymer gelling reaction mechanism and the effects of oxygen bearing in water on the stability of gel, the formulation of gel profile control system is optimized. On the basis of analyzing early profile controlling working practice in-depth, injection technology is perfected and optimized emphasizing from injection concentration, injection rate, injection volume and so on to improve the measures effect effectively. The optimum gel injection timing is determined through indoor physical simulation test. For the technical problems that conventional profile control system cannot realize sewage with fluid, low-temperature promoting agent of polyacrylamide and water soluble phenol-formaldehyde resin is researched and developed independently.

## 4. The field effect analysis

Since 2007, combination flooding technical critical of microbial and gel is carried out comprehensively in Baolige oilfield. Pilot test technology such as single well microbial stimulation, microorganism&gel combination flooding and cyclic flooding is taken on, and put into large-scale application. Since 2007, by measures such as single well stimulation, microbial flooding, gel flooding, natural decline is slowed down 2.3 percent. By saddle point method, the calculated cumulative increasing oil production is 190 000 tons, and the input-output ratio is 1:4.8(Fig. 4), obtaining the good economic benefit and social benefit.

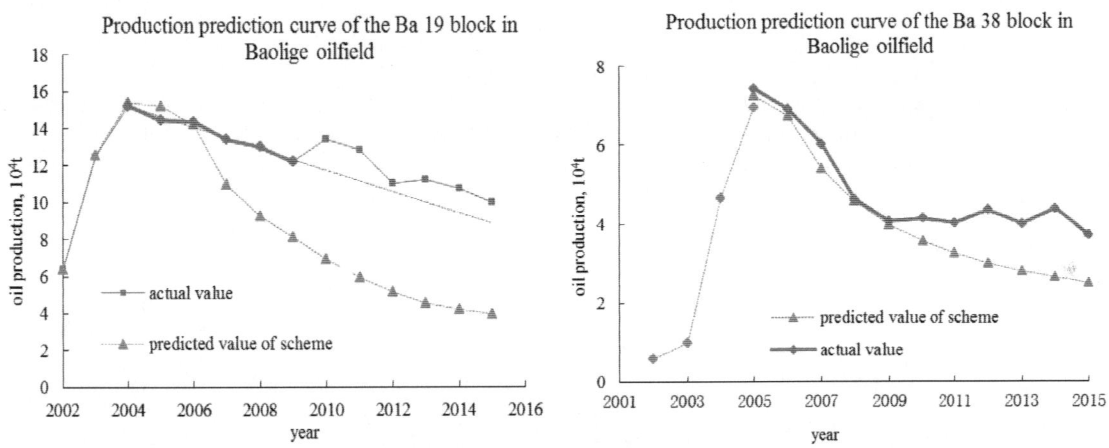


**Fig. 4** Contrast curve of production

## 5. Conclusions

Through application of microorganism&gel combination flooding technology, the development effect of thin oil reservoir of Ba 19 and Ba 38 block is effectively improved, realizing the stable oil production, at the same time it shows that the combination flooding technology thought which set build underground microbial field as the core idea is correct and effective.

Using molecular biology techniques, the system analysis and evaluation of microbial oil recovery technology and monitoring technology after microorganism displacement is established, realizing the relative quantitative analysis of the component analysis, which has a guiding significance for further research of microbial oil recovery technology.

By using the principle of the bioreactor, floor fermentation and loop injection technology of produced liquid from micro flooding are innovated to form, realizing high reinjection of produced fluid.

For eight years, through the deepening research and field application, the system of microbial combination flooding technology is gradually enriched and improved, which has important practical significance on Huabei oilfield and even the similar oil reservoir of China petroleum.

**Acknowledgements**
Key S&T Special Projects of PetroChina Company Ltd. "Key technology research and application for increasing production and stable yields of 8 million tons in Huabei oilfield" (Project number 2014E-3507).

**References**

[1] Chen Yujun, Ju Dengfeng, Song Yiwei, et. 2005 *Profile Control and Oil Displacement by Joint Function of WeakGel and Microbe (Special Oil & Gas Reservoirs vol 1)* chapter 12 pp 84–86.
[2] Gang Wu, Fuping Ren, Jing You, et. 2013 *Luteimonas Huabeiensis sp. nov., isolated from stratum water (International Journal of Systematic and Evolutionary Microbiology vol 9)* chapter 63 pp 3352–3357.
[3] Ke Congyu, Wu Gang, You Jing, et. 2013 *The Large Scale Application Research on Overall Microbial Flooding Technology in Baolige Oilfield (Oilfield Chemistry vol 2)* chapter 30 pp 246–250.
[4] Ke Congyu, Wu Gang, You Jing, et. 2013 *The Growth, Migration and Distribution Law of Microbes During Microbial Displacement Oil (Microbiology China vol 5)* chapter 40 pp 849–856.
[5] Li Yongbin, Yu Lixin, Zhang Lili, et. 2013 *Experimental Study of Microbial Flooding Physical Simulation for Baolige Oilfield (Oil Drilling & Production Technology vol 6)* chapter 35 pp 95–97.
[6] Sun Sangdun. 2014 *Field Practice and Analysis of MEOR in Shengli Oilfield (Journal of Oil and Gas Technology vol 2)* chapter 36 pp 149–152.
[7] Lu Mansheng, Liu Zhicheng, Tang Sihong. 2012 *Experimental Selection Method and Effect Evaluation of Microbial Enhanced Oil Recovery (Drilling & Production Technology vol 2)* chapter 35 pp 90–93.

IOP Conf. Series: Earth and Environmental Science **61** (2017) 012138 doi:10.1088/1755-1315/61/1/012138

# The characteristics of dryland aggregates in mouding

**Minxia Cao[1], Zhantai Wang[2], Heping Zeng[3]**

Kunming University of Science and Technology, Kunming, Yunnan 650500, People's Republic of China.
First author:Minxia Cao(1991 -), female, chongqing people, master degree, students, mainly engaged in soil and water conservation research in Longchuan river basin.
Corresponding author: Heping Zeng, Kunming University of Science and Technology, Kunming 650500, China.

[1]546545805@qq.com (Minxia Cao)
[2]1345137350@qq.com
[3]dabatou@126.com(Heping Zeng)

**Abstract.**Taking purple soil from Mouding city as the research object, the aim of this paper is to discuss the number characteristics of dryland soil aggregates. The results indicated that cumulative amounts of dryland soil aggregate obtaining from artificial screening method were obeyed the log-linear distribution, and fractal dimension of soil was ranged from 2.11 to 2.48. Moreover, the higher aggregates content was contained, the larger aggregates fractal dimension in the soil particles, which was smaller than 0.25 mm. The fractal dimension of dryland aggregates was decreased with the increasing of soil depth.

## 1. Introduction

Pedosphere is one of the most active and the most vitality spheres in the global ecological environment , and is the important place of material cycle and energy conversion in natural environment(Zhao, 200 1). Soil aggregate is the identification and important index of soil fertility, the size and content of soil aggregate is an important physical properties of soil(Zhang and Shao,2000). Soil aggregate is the mai n index of soil quality high or low ,and the ability to resist erosion(Zheng and Liu, 2003.Ning, et al, 20 05). Its characteristic is porous and water stability. The characteristics of specific performance in the soil porosity size are moderate, coexists the pore water and pore gas, and has a proper quantity and proportion. At present, researches of scholars mainly focus on the factors such as different fertilization conditions, different cultivation patterns and different ways of land use and different cultivated fixed number of year produced an effect on soil aggregate composition distribution and stability(Pei, et al, 2015 ).

In order to understand the ecological system of dryland better, linear distribution of soil aggregate and fractal dimension was observed in this study, which selected purple soil from Mouding town as the research object. As an objects mouding purple soil, this paper studied the logarithmic linear distribution and fractal dimension of soil aggregate. Aiming to further understand the dryland ecological system will provide a scientific basis for soil characteristics and ecological environment recovery.

## 2. Materials and methods

IOP Conf. Series: Earth and Environmental Science **61** (2017) 012138 doi:10.1088/1755-1315/61/1/012138

### *2.1. Study area*
The study area is located in Mouding town, Chuxiong yi autonomous prefecture, Yunnan province, China. It is located in 101°19'~101°51'E and 25°09'~25°40'N, and belongs to north subtropical monsoon climate zone. The climate is mild and there is just a little, rainfall in every year. The average rainfall of annual year is 880.1 mm. The elevation of 80% county region is about 1600-1900 m. The average sunshine time of every year is 2117.9 h.

### *2.2. Sample collection*
The soil samples are from the dryland located in Samachang Country of MouDing County, Yunnan Province, China. Then, about 1kg soil is taken in each soil horizon (0~10cm, 10~20cm and 20~30cm) of each sample point. Then, each layer soil of the ten soil sample points are mixed evenly in situ and take a quarter of mixed soil sample in three times. Finally, we got three different soil horizon samples and the soil sample of each soil horizon is about 1kg in each sampling area. Finally, the soil sample is peeled into small pieces less than 12 mm by hand along the natural soil structure. Finally, the soil is air-dried.

### *2.3. Analysis method*
The total nitrogen contents were determined with semimicro Kjeldahl. The total phosphorus were determined by $HClO_4$-$H_2SO_4$. Available P was extracted by ammonium fluoride and hydrochloric acid. Soil pH value was determined by potentiometric (soil-water ratio, 1:2.5). Available N was extracted by alkali solution diffusion method. Soil aggregate analysis was determined with artificial screening method(Bao SD, 2008).

### *2.4. Soil aggregate fractal feature model*
It was generally believed that soil aggregate structure had self-similarity. Turcotte(Turcotte, 1986) had put forword that the calculation method of soil aggregate fractal dimension was by using the relationship between the weight distribution and the average particle size: $(d_i/d_{max})^{3-d}=W(\delta<d_i)/W_0$. Among them, $d_i$ was the two screening average particle size of between $d_i$ and $d_{i+1}$ , $d_{max}$ was average diameter of biggest graded soil particles. $W(\delta<d_i)$ was less than the accumulation of $d_i$ soil grain weight, $W_0$ was each graded soil weight, D was soil aggregate fractal dimension. Finally, respectively by lg $(W(\delta<d_i))/W_0)$, lg $(d_i/d_{max})$ for vertical and horizontal, D was calculated by the regression analysis.

## 3. The results and discussion

### *3.1. The basic physical and chemical properties of soil*
The pH of dryland was approximately 5.37. Total P was 0.40g $kg^{-1}$. Total N was 2.71 g $kg^{-1}$. Effective P was 20.81 mg $kg^{-1}$. Effective N was 831.67 mg $kg^{-1}$.

### *3.2. Soil aggregate fractal characteristics*
Under different soil depth, the results of soil aggregate analysis were showed in table 1. Being carried out Logarithmic transformation on the three soil aggregate composition, it was found that $lg(W(\delta<d_i))/W_0)$ had significant linear relationship with $lg(d_i/d_{max})$ (figure 1). According to the fitting of the logarithmic transformation, it was got the linear equation Lg(y) = aLg(x) + b, and among them, x and y were respectively $lg(W(\delta<d_i))/W_0)$ and $lg(d_i/d_{max})$. a, b, and the determination coefficient $R^2$ of each group of the parameters of the equation was presented in table 2. From table 2, it showed that the decision coefficient of the regression equation was greater than 0.9633, the highest could be reached 0.9981.

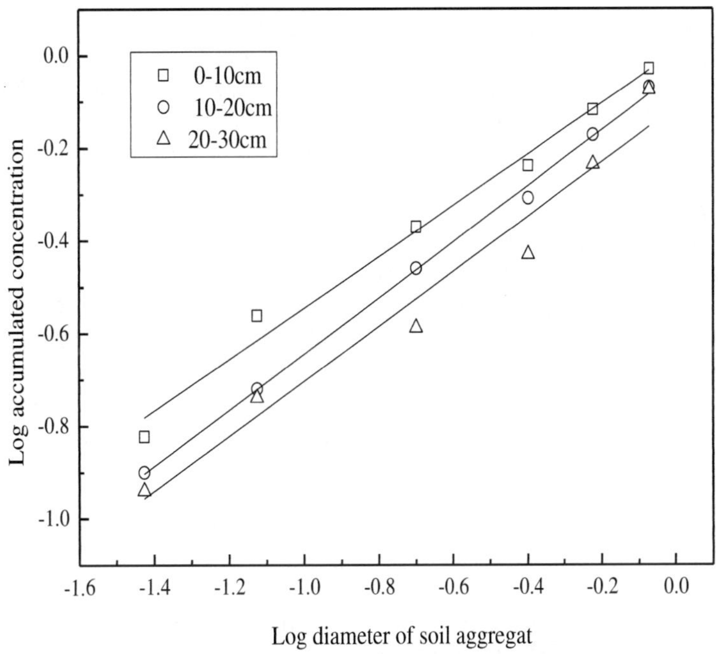


**Fig.1** Log-probability plot of aggregate-size distribution

**Table.1** Composition of dryland aggregates under different soil depth

| Soil depth | Processing method | Composition of dry soil aggregates (%) | | | | | | |
|---|---|---|---|---|---|---|---|---|
| | | 10-7 mm | 7-5 mm | 5-3 mm | 3-1 mm | 1-0.5 mm | 0.5-0.25mm | <0.25mm |
| 0-10cm | Dry | 13.14 | 14.08 | 20.2 | 16.74 | 15.14 | 6.78 | 13.92 |
| | Wet | 0.00 | 20.2 | 16.8 | 13.6 | 15.2 | 18 | 16.2 |
| 10-20cm | Dry | 29.4 | 20.72 | 23.72 | 10.82 | 3.42 | 1.32 | 10.6 |
| | Wet | 0.00 | 15.2 | 12.6 | 18.2 | 27.8 | 11.6 | 14.6 |
| 20-30cm | Dry | 31.66 | 26.84 | 25.24 | 6.6 | 0.98 | 0.66 | 8.02 |
| | Wet | 0.00 | 25.2 | 16.8 | 16 | 14.2 | 12.8 | 15 |

**Table.2** Logarithmic linear distribution characteristics of soil aggregate

| Item | 0-10cm | 10-20cm | 20-30cm |
|---|---|---|---|
| a | 0.5531 | 0.6045 | 0.5917 |
| b | 0.0068 | -0.0041 | -0.1124 |
| R2 | 0.988 | 0.9981 | 0.9633 |

**Table.3** Aggregate fractal dimension

| Ssmples | 0-10cm | 10-20cm | 20-30cm |
|---|---|---|---|
| Fraetal dimension | 2.48 | 2.26 | 2.11 |

Fractal dimension reflected the characteristics of soil aggregate composition, the effects of land use and the way of soil management. From table 3, soil aggregate fractal dimensions were different under different soil depth. The fractal dimension was arranged from 2.11 to 2.48. For 0 to 10 cm layer of the soil, aggregate content of < 0.25 mm soil was the highest (30.12%), the maximum value of fractal dimension was 2.48. For 10 to 20 cm layer of soil, aggregate content of < 0.25 mm soil was 25.2%, the fractal dimension was 2.26. For 20 to 30 cm layer of soil, aggregate content of < 0.25 mm soil was 23.02%, the fractal dimension was 2.11. The study found that the higher the clay content was, the finer quality of soil was, the higher the soil fractal dimension of particle size was(Yang PL et al, 1993). In this study, the aggregate content of < 0.25 mm soil was higher, so the fractal dimension of soil aggregate was bigger. Furthermore, the fractal texture had the similarity trends. The higher of the aggregate content of < 0.25 mm soil was, which showed that worse structural stability of soil was, the bigger of fractal dimension was.

## 4. Summary

(1)Theresearch area, cumulative amounts of dryland soil aggregate obtaining from artificial screening method were obeyed the log-linear distribution.

(2)fractal dimension of soil was ranged between 2.11 to 2.48. Moreover, the higher aggregates content was contained, the larger aggregates fractal dimension in the soil particles, which was smaller than 0.25 mm.

(3)The fractal dimension of dryland aggregates was decreased with the increasing of soil depth, which showed that soil structure stability was better with the increasing of soil depth.

**Acknowledgements**

This work was financially supported by National Natural Science Foundation (NO. 6492-20140037).

## Reference

[1]Zhan X, Shao M. [Soil nitrogen and organic matter losses under water erosion][J]. Chinese Journal of Applied Ecology, 2000, 11(2):231-4(in Chinese).

[2]Zhao QG. Prosrects of siol science the 21st century[J]. Advance in Earth Sciences, 2001.

[3]Zheng ZP, Liu ZX. [Soil quality and its evaluation].[J]. Chinese Journal of Applied Ecology, 2003, 14(1):131-4(in Chinese).

[4]Ning LD, Shi H, Zhou HJ. [Quantitative characteristics of soil aggregates under different vegetations in upper reach of Minjiang River].[J]. Chinese Journal of Applied Ecology, 2005, 16(8):1405-1410(in Chinese).

[5]PEI ZJ, LIANG CH, YIN Y. [Size Hierarchy and Stability of Soil Aggregates in Solar Greenhouse Soils with Different Planted Years].[J]. Chinese Journal of Bulletin of Soil and Water Conservatio,2015, 35(6):70-74(in Chinese).

[6]Chan KY, Heenan DP, OatesA. Soil carbon fract ions and relationship to soil quality under different tillage and stubble management[J]. Soil and Tillage Research,2002,63: 133-139.

[7]Elliott E T. Aggregatest ructure and carbon, nitrogen, and phosphorus in native and cultivated soils[J]. Soil Science Society America Journal,1986,50: 627-633.

[8]Pedect E, Blevins RL. Fractal characterizations of soil aggregation and fragmentation as influenced by tillage treatment[J]. Soil Sci Soc Am J, 1997,61:896-900

[9]Kasper M, Buchan G D, Mentler A, et al. Influence of soil tillage systems on aggregate stability and the distribution of C and N in different aggregate fractions[J]. Soil & Tillage Research, 2009, 105(2):192-199.

[10]Li K, Dou S. Effects of chemical fertilizer combined corn stalk application on the quantity and infrared spectra of humic acids in soil aggregates.[J]. Journal of Jilin Agricultural University, 2009, 31(3):273-278.

[11]Kristova, P/Hopkinson, L/Rutt, et al. Quantitative analyses of powdered multi-minerallic carbonate aggregates using a portable Raman spectrometer[J]. American Mineralogist, 2015, 98(2-3):401-409.

[12]Jia J, Yu D, Zhou W, et al. Variations of soil aggregates and soil organic carbon mineralization across forest types on the northern slope of Changbai Mountain[J]. Acta Ecologica Sinica, 2015, 35(2):1-7.

[13]Zhang Z, Wei C, Xie D, et al. Effects of land use patterns on soil aggregate stability in Sichuan

Basin,China[J]. Particuology, 2008, 6(3):157-166.
[14]Zhou H, Lv Y, Yang Z, et al. Influence of conservation tillage on soil aggregates features in North China Plain[J]. Journal of Integrative Agriculture, 2007, 6(9):1099-1106.
[15]Bissonnais Y L. Aggregate stability and assessment of soil crustability and erodibility: I. Theory and methodology[J]. European Journal of Soil Science, 2016, 67(1):11–21.
[16]Jiang Y, Shao M. Effects of soil structural properties on saturated hydraulic conductivity under different land-use types[J]. Soil Research, 2014, 52(4):340-348.
[17]Nweke I A, Nnabude P C. Aggregate stability of four soils as evaluated by different indices.[J]. Journal of Experimental Biology & Agricultural Sciences, 2015, 3(3):246-252.
[18]Abdollahi L, P Schjønning, Elmholt S, et al. The effects of organic matter application and intensive tillage and traffic on soil structure formation and stability[J]. Soil & Tillage Research, 2014, 136:28-37.
[19]Bao SD. The method of analysis for soil agro-chemistry (3rd edition). [M]. Agriculture press, 2008(in Chinese).
[20]Niewczas J, Witkowska B. Index of soil aggregates stability as linear function value of transition matrix elements[J]. Soil & Tillage Research, 2003, 70(2):121-130.
[21]Turcotte DL. Fractal fragmentation. Geogr Res,1986,91(12):1921-1926.
[22]Yang PL, Luo YP, Shi YC. Fractal feature of soil particle by using mass distribution[J]. Chin Sci Bull,1993,38(20):1896-1899(in Chinese).

IOP Conf. Series: Earth and Environmental Science **61** (2017) 012139 doi:10.1088/1755-1315/61/1/012139

# The application of natural ventilation of residential architecture in the integrated design

**Ji Yao**

Chongqing university, key laboratory of new technology for construction of cities in mountain area(chongqing university), ministry of education,(chongqing,400045,china)

**Abstract:** As one of the important parts in the architecture design, ventilation fully reflects one important factor of energy conservation to some extent. To ensure that decreasing use of energy resources to the maximum extent in architectures and make architectures harmonize with their surroundings, which have become important pursuing gradually for residential architecture, at present, integrated design of residential architecture should not only highlight natural ventilation technique, but also should integrate natural ventilation and main feature of local climate, etc. What's more, designers shall carry out unified analysis on the base of taking all factors into consideration. This point is more important. By this way, architecture technique can be more comprehensive and complete. With the guidance of sustainable development concept, natural ventilation as one of ecological technologies is applied extensively in many architecture designs owing to its economic effect and health benefit.

With the rapid development of society and progress of times, heath issue has attracted people's eyes. To make indoor air fresh, people always open windows regularly to welcome fresh air enter and make the original polluted air go out for four seasons. Meanwhile, health factor is taken into full consideration. When designing the residential architectures, designers need to carry out comprehensive application on the base of detail knowing to natural ventilation designing methods. They use this way to improve indoor environment to the greatest extent and build a good and comfortable living environment for people.

## 1. Important meaning of natural ventilation

At present, owing to health concept and economic effect of natural ventilation, it is applied extensively in the integrated design for a large number of residential architectures. So it can replace the position of air conditioner and other systems to some degree and decrease consumption of all kinds of energy resources to the great extent.

On one hand, natural ventilation of residential architecture enables effectively realizing passive cooling. When humidity and temperature of outdoor environment are lower relatively, natural ventilation can automatically carry out some activity to decrease humidity and lower temperature without using any energy resources. This can provide a good and comfortable living environment. Besides that natural ventilation can reduce usage of all varieties of energy resources and cut down environment pollution, which corresponds to sustainable development concept advocated by the nation principle of complete harmony. on the other hand, natural ventilation of residential architecture brings new wind that is fresh enough to a great extent. This plays an important part in improving health and emotion of people. One of important reasons why current indoor air quality is not good and decreases is that there is no enough new wind and new wind is not fresh enough. The long usage of air conditioner leads to a series of problems. But natural ventilation enables effectively clearing out stale

IOP Conf. Series: Earth and Environmental Science **61** (2017) 012139 doi:10.1088/1755-1315/61/1/012139

air and smell to a great extent. And it benefits to realize harmonious development of people and nature.

## 2. Main principles of natural ventilation

Natural ventilation of residential architecture can be divided into two models complete natural ventilation and machine assisting ventilation according to the source of driving force. Complete natural ventilation has direct relation with density difference between indoor air and outdoor air and wind pressure.

### *2.1. The function of wind pressure*

To a large degree, the forming of wind has direct relation with pressure difference existing in the air. The function of wind pressure is that using mainly pressure difference existing in windward side and leeward side to have ventilation. When wind blows from the left side of architecture, windward side can produce a positive pressure that is 1.5 -1.8 times of the dynamic pressure under normal wind speed with the air promotion. Leeward side produces a negative pressure that is 1.3-1.4 times of dynamic pressure under normal wind speed with strong influence of air flowing around. Under the effects of positive and negative pressure, natural ventilation of wind pressure type that can have air changing comprehensively is formed owing to this. but there is a matter that wind pressure around architecture can be influenced by many factors, for example buildings' geometries, concrete positions, wind speed, surrounding environment and others, as the following picture 1-1.

### *2.2. Heat pressing function*

The forming of heat pressing relates to temperature difference between the indoor temperature and outdoor temperature to a great degree. Somehow, the principle that natural ventilation is under the effect of heat pressing is similar to the principle of hot air rising. House building designers usually set some certain air outlet on the upper side of architectures, which is to exhaust indoor polluted air in time and introduce outdoor fresh air. it realizes effective conversion between indoor air and outdoor air. Meanwhile, owing to the influence of temperature difference, there is some certain density difference; baric gradient also produces towards the direction that is perpendicular to the wall. If indoor temperature is higher obviously than outdoor temperature, pressure above the architecture will be higher than the temperature below the architecture; when there are some certain orifices on the upper part and below part of architecture, air can enter from the lower orifice and flow out from the upper. If outdoor temperature is higher than indoor temperature, the direction of air flowing is totally opposite. So an important conclusion can be drawn that heat pressing has a direct relation with temperature difference of indoor air and outdoor air and height difference of orifice.

· Therefore, when we carry out integrated design to natural ventilation of residential architecture, we can use fully piping shaft, pull air shaft, stair half, atrium and other vertical cavities to adapt and meet the height difference between foot entering and air outlet. Then it can exhaust a large amount of hot air existing in the architecture in time, which makes ventilation effect of architecture become perfect, as the following picture 1-2.

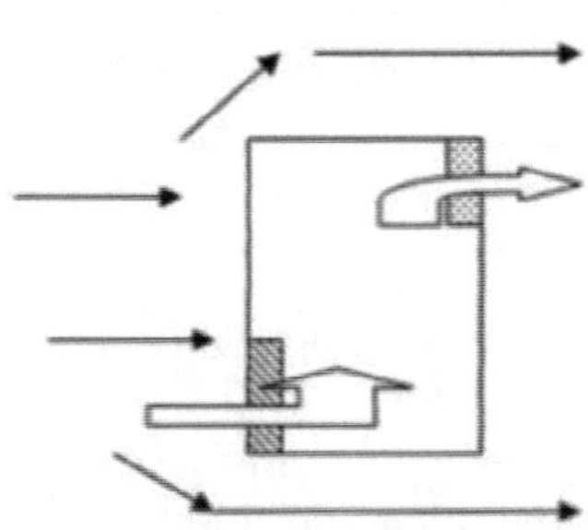

Fig.1 Under the action of wind pressure of natural ventilation

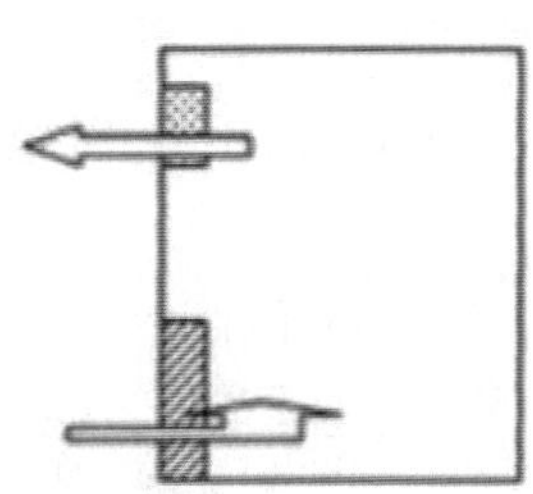

Fig.2. Under the action of thermal natural ventilation

*2.3. Combined action of air pressure and heat pressing*

As a matter of fact, natural ventilation in the residential architecture is the result with combined action of air pressure and heat pressing. However air pressure and heat pressing are influenced by all varieties of factors, e.g. climate, shape of architecture, specific direction of outdoor air and surrounding environment situation and others. and the effects of these two pressures are not just the linear superposition, for example, in the same architecture, if the heat pressing is below the windward side, to ensure that direction of wind pressure effect is consistent, air entering volume should be bigger than the one under the effect of heat pressing. But if wind pressure on the upper part of windward side is much bigger than the heat pressing, correspondingly air can't be exhausted from the upper hole. Air will flow backward.

Meanwhile, outdoor wind speed and wind direction change all the time. in order to ensure good natural ventilation effect, wind pressure usually is not taken into consideration in the calculation, but wind pressure exists all the time. Its influence on natural ventilation exists, so it is possible that alternate using the effect between wind pressure and heat pressing.

*2.4. The effect of machine assisting*

In some big architecture designs relatively, ventilation path usually is bigger, which produce very strong flow resistance. Under this condition, if just depending on the combined effect of wind pressure and heat pressing to realize natural ventilation, it is not real at all. For the cities polluted seriously, like noise and air pollution, if having ventilation directly, it will take the outdoor pollution into indoor room. This is harmful to people' health to a great degree. At this time, use related machine assisting effect to have natural ventilation, which can achieve good ventilation result.

## 3. Integrated design methods and natural ventilation design methods

To some certain degree, integrated design method is a method taking organic combination, mainly including spiraling and integrated platform. Natural ventilation of residential architectures usually can view the whole building as a organic integrated system in the integrated design. Then they can carry out unified process and consideration from the whole period of life. On the base of these, residential architecture can adapt better to local geography condition, climate features and others. it mainly includes building concept, structure concept, energy and environment concepts about houses. Realizing these concepts can't be separate with the interaction of many majors.

However, integrated design method is not perfect. There are very serious problems in these aspects ventilation, getting heat, light selecting, floodlight and other energy resource consumption. So it is necessary to take natural ventilation, light selecting, heat keeping and others into consideration in the design. And it is possible that effectively decrease the consumption of important energy resources.

What's more, in the natural ventilation design of residential architecture, all kinds of assisting measures for decreasing temperature are required to decrease the indoor temperature in the summer. in the process of decreasing indoor temperature, natural ventilation and air conditioner play their roles. This can decrease effectively consumption of related energy resources. So when designers use natural ventilation, they are required that considering fully the basic requirements of local architecture, climate feature and other factors. on the base of these, natural ventilation can play a perfect role.

## 4. The application of natural ventilation in the integrated design

*4.1. Designing natural ventilation under the different climate conditions*

There are many factors influencing natural ventilation of residential architecture, e.g. layout of building, location and others. If designers can have scientific layout to building, it can achieve perfect natural ventilation result. Meanwhile, natural ventilation of residential architecture is influenced greatly by climate condition. So designers need to choose geography location reasonably and scientifically, consider fully space range of climate. Generally, climate condition concludes the following factors.

Firstly, designers should consider chinese real climate region in the design. chinese climate region mainly includes season changing region which refers to that wind direction changes with the season changing. Most frequent wind direction region refers to that wind direction doesn't change within one year, as nei monggol, tibetan plateau and others. There are the most frequent wind direction region, which refers to that wind direction always changes within one year. if the building is located here, designers need to consider wind speed. ( quasi-stationary wind is that wind speed is below 1.5 meters) secondly, take local climate into consideration, local climate is called local circumfluence including water and land wind, forest and plain wind and others, keep building orientation be consistent with the wind direction. Thirdly, fully consider local micro climate. On the base of knowing micro climate, consider building layout, separation distance and the influence of surrounding vegetation on natural ventilation.

### *4.2. Designing natural ventilation scientifically and reasonably according to building features*

Firstly, take these factors including building material, structure, psychology of resident into consideration, then ascertain the storey height of the whole building. Storey is usually divided as stratum, high-rise and multilayer. Height of building decides the number of layer. But in the real design, high-rise mainly includes small high-rise, super high-rise. Generally, there are many factors influencing the number of layer, e.g. real using condition of local land resource, related laws and regulations and others, so designers need to think of above factors and design the number of layer reasonably and scientifically. Somehow, use this point to ensure good natural ventilation result.

Secondly, wind speed spreading in the vertical direction of building will increase with the increasing of number of layer. so when carry out specific building design, think of wind speed.

Thirdly, in the integrated design of natural ventilation, the design of hole setting can influence indoor air circulation and real wind amount. in the most cases, hole opening is bigger, wind amount will be bigger. But designers also need to consider solar radiation, wind speed and building location.

### *4.3. Designing natural ventilation reasonably and scientifically according to the details of building*

Firstly, main methods for designing the size of window

Windows as the detail part of building is the main media of light selecting and ventilation of buildings, it plays an important part in the natural ventilation. in the process of designing size of windows, designers should not only consider the opening hole, but also should choose scientific and reasonable way to open windows, as the window-blinds, sliding window and others. These window opening methods have a good ventilation result.

Secondly, designing the specific shade form

In the integrated design of natural ventilation, designers ought to think of shade results with different forms, choose appropriate shade form according to real situations of buildings.

Conclusions

Ventilation as one of the important parts in the architecture design can give the full representation of energy saving result to a certain degree; decrease the usage of energy resource to the great extent; make it be a part of surrounding environment. Integrated design method is a method taking organic major combination, mainly including spiraling and integrated platform. This paper introduces briefly meaning of natural ventilation. It explains main natural ventilation principles from wind pressure effect, heat pressing effect, combined effect of wind pressure and heat pressing and machine assisting effect. Next, this paper describes that designing natural ventilation under different climate conditions, designing natural ventilation reasonably and scientifically according building' features and details. And introduce species application of natural ventilation in the integrated design.

**Acknowledgements**

The integrated reform and practice of curriculum in succession graphics of major architecture design 106112014cdjzr200022

IOP Conf. Series: Earth and Environmental Science **61** (2017) 012139 doi:10.1088/1755-1315/61/1/012139

**References:**

[1] Jin xi, shen shouyun, xie mingjing. architecture designing strategy based on ecology technology concept [j]. journal of central south university of forestry and technology, 2013, (02).
[2] Leng yanfeng, zhao hui, yuan rubing. study on natural ventilation measures of residential architecture in the area with hot summer and cold winter [j]. construction science, 2011, (02).
[3] Wang jie, wei jia. exploration and practice of integrated design in the green residential area [j]. architecture design managing, 2011, (10).
[4] Tan hongwei, ji liang, pu zhen. natural ventilation concept and efficiency under the condition with changing wind [j]. journal of central south university (science and technology)
[5] Wan xin, su yaxin. the application of natural ventilation technology in the modern architecture [j]. architecture and saving energy, 2007, (09).
[6] Zhou min. study on ventilation design of chinese traditional residential architecture—as old street of nanjing [j]. urban studies, 2015, (12).
[7] Zhang wei, wang chuan. discussion about architecture' integrated design with the concept of sustainable development [j]. journal of shenyang jianzhu university (social science edition), 2009, (03).
[8] Zhang qun, che xiaomin, liu jiaping, hui shankang, wu xiaoxiang. real measure analysis and design strategy of natural ventilation in the sunan region [j]. journal of xi'an university of architecture and technology (science and technology), 2015, (01).

IOP Conf. Series: Earth and Environmental Science **61** (2017) 012140 doi:10.1088/1755-1315/61/1/012140

# Density and hardness of Nd-doped zircon ceramics as nuclear waste forms

**Xun Yuan[1], Xudong Liu[2], Lan Wang[2] and Xirui Lu[2,*]**

[1] Tianfu College of Southwestern University of Finance and Economics, Mianyang 621010, P. R. China

[2] Key Subject Laboratory of National Defense for Radioactive Waste and Environmental Security, Southwest University of Science and Technology, Mianyang 621010, P. R. China

Corresponding author's e-mail address: luxiruimvp116@163.com (X. R. Lu)

**Abstract.** Zircon ceramics were employed as host material for immobilization of element Nd, which was used as the surrogate of trivalent actinide nuclide. A series of $Zr_{2-x}Nd_xSi_2O_{8-0.5x}$ compounds were prepared by solid-state reaction method at 1500 °C for 72 h in air and investigated in respect of their physics performances. These performances were characterized by density and hardness. The results show that the density raises as the value of $x$ increases from 0.02 to 0.20. Moreover, the hardness of the Nd-doped ceramic also increase with enhanced Nd content.

## 1. Introduction

Nuclear energy is one of the most promising energy, which has high energy density than fossil fuels and does not cause air pollution [1-3]. The development and utilization of nuclear energy have brought great economic benefits and social benefits to human beings. However, such nuclear waste from the nuclear industry pose a considerable biological hazard on human beings, and produce a threat to environmental safety at the same time [4, 5]. Thus, safe disposal of these nuclear waste, especial for radioactive nuclides (Uranium, Plutonium, Americium and Thorium), has become more and more important in nuclear industry [6, 7]. In the past, some host materials, like glasses and cements, have been considered for this purpose. Nevertheless, the high leaching rate in glasses and cements limit their further application in the disposal of radioactive waste [8-10]. Moreover, glasses belong to metastable state, which has a tendency to get devitrified at high temperature and pressure [11]. By contrast, composites of mineral analog titanates, named as SYNROC, such as zircon based materials have excellent physical properties, and it is considered as suitable to immobilize nuclear wastes [12].

Zircon (I41/amd, Z = 4) belongs to the island of silicate minerals and tetragonal crystal system, and the theoretic composition consists of 32.8 wt% $SiO_2$ and 32.8 wt% $ZrO_2$. Due to its fantastic properties, such as high thermal decomposition temperature (≥ 1500°C), low thermal expansion coefficient (about $4.1 \times 10^{-6}$ $K^{-1}$ at 25-1400°C), good chemical stability as well as the ability to immobilize actinides in the lattices, zircon has been attracted intensive attention in recent years [13, 14]. Zircon in nature (crystalline rocks) contains a certain amount of radioactive elements, as much as $UO_2$ 5 wt% or $ThO_2$ 15 wt%, and keeps its structure well after millions of years [15]. In the past 10 years, great progress has been achieved in the research of zircon immobilized Pu and other radioactive elements. A certain amount of radionuclide replaced Zr at the Zr site to synthesize ASiO4 ($A^{4+}$=Zr, Hf, Th, Pa, U, Np, Pu, and Am) has been investigated by Keller at el. [16, 17].

The aim of the present work is to synthesize a series of $Zr_{2-x}Nd_xSi_2O_{8-0.5x}$ ($0.02 \leq x \leq 0.20$) ceramic by gradually substituting Nd for Zr in the composition. The effect of Nd substitution on the density as well as the hardness of the compositions was investigated. The density meter and hardness tester were implemented to measure the density and hardness of the each sinered sample induced by substituting Nd, respectively.

## 2. Experiment

Nd has so far gained widespread recognition and become representative substitutes for An (An: actinide) owing to their similar ionic radius and ion valence to some actinide elements. Therefore, in the present work Nd is selected as an alternative to simulate the trivalent actinides, and to substitute for Zr in $ZrSiO_4$ to simulate ceramics solidifying radioactive nuclides like Pu, Am, U and Th.

In the present study, AR grade $ZrO_2$, $CeO_2$ and $SiO_2$ were selected as the starting materials. A series of $Zr_{2-x}Nd_xSi_2O_{8-0.5x}$ ($0.02 \leq x \leq 0.20$) were prepared by solid-state reaction method. Before weighing, all the raw powders were heated at 100°C for 6 h to remove adsorptive water. Then stoichiometric amounts of the powders in appropriate ratios were weighed, and ground in analytically pure ethanol medium. The dried powders were compacted in a pellet form (12 mm diameter and ~2 mm thickness) at a pressure of 10 MPa. The pellets were sintered at 1500°C for 72 h in air atmosphere to fabricate dense bulk ceramic at a heating rate of 5 °C/min. The sintered compounds were taken out of the furnace after being naturally cooled to room temperature. Table 1 presents the contents of starting materials for synthesizing $Zr_{2-x}Nd_xSi_2O_{8-0.5x}$ ($0.02 \leq x \leq 0.20$). Figure 1 shows the T-t curve of the sintering progress.

**Table 1.** Compound design of $Zr_{2-x}Nd_xSi_2O_{8-0.5x}$ ($0.02 \leq x \leq 0.20$)

| Target compounds | Additive amount of raw materials/g | | |
|---|---|---|---|
| | $ZrO_2$ | $Nd_2O_3$ | $SiO_2$ |
| $Zr_{1.98}Nd_{0.02}Si_2O_{7.99}$ | 1.9908 | 0.0274 | 0.9805 |
| $Zr_{1.96}Nd_{0.04}Si_2O_{7.98}$ | 1.9659 | 0.0548 | 0.9781 |
| $Zr_{1.94}Nd_{0.06}Si_2O_{7.97}$ | 1.9411 | 0.0820 | 0.9757 |
| $Zr_{1.92}Nd_{0.08}Si_2O_{7.96}$ | 1.9163 | 0.1090 | 0.9733 |
| $Zr_{1.90}Nd_{0.10}Si_2O_{7.95}$ | 1.8917 | 0.1359 | 0.9709 |
| $Zr_{1.88}Nd_{0.12}Si_2O_{7.94}$ | 1.8671 | 0.1627 | 0.9685 |
| $Zr_{1.86}Nd_{0.14}Si_2O_{7.93}$ | 1.8450 | 0.1896 | 0.9673 |
| $Zr_{1.84}Nd_{0.16}Si_2O_{7.92}$ | 1.8206 | 0.2161 | 0.9649 |
| $Zr_{1.82}Nd_{0.18}Si_2O_{7.91}$ | 1.7963 | 0.2426 | 0.9625 |
| $Zr_{1.80}Nd_{0.20}Si_2O_{7.90}$ | 1.7744 | 0.2692 | 0.9613 |

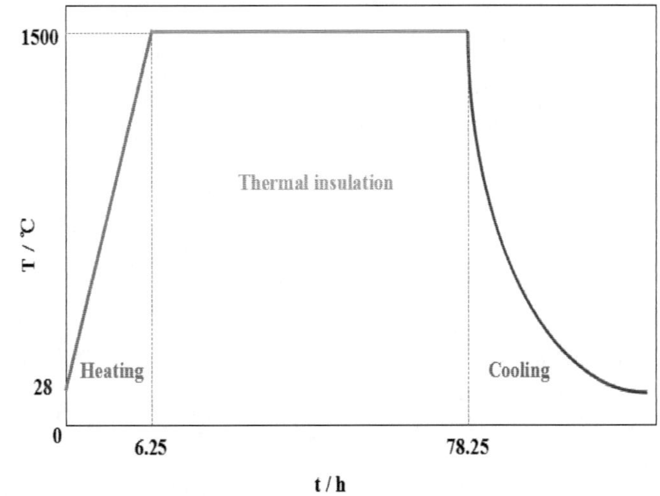


**Figure 1.** The T-t curve of sintering progress.

The density of the each sinered sample was measured by a full automatic density meter (GH-120D) based on the Archimedes principle. As much as possible to improve the accuracy of the solidified body density test values, the density testing were carried out for three times, and then take the average value as the density value of the sample. The micro-hardness test was carried out on Vickers Hardness Tester Hv-1000A with a diamond tip at a test load of 500 g and dwelling time of 15 s. In order to improve the accuracy of hardness test of the solidified samples, the test were randomly selected in 3 different locations, and then take the average value as the sample hardness value.

## 3. Results and discussion

The density and hardness are closely related, which are important indexes to evaluate the performance of ceramic materials. As a highly radioactive waste material, it is better to have a larger density and higher hardness [18]. Table 2 lists the tested density of $Zr_{2-x}Nd_xSi_2O_{8-0.5x}$ $(0.02 \leq x \leq 0.2)$ compounds and Figure 2 shows the relation curve between the densities and $x$ in $Zr_{2-x}Nd_xSi_2O_{8-0.5x}$ $(0.02 \leq x \leq 0.2)$.

Figure 2 shows the relation curve between the densities and $x$ in $Zr_{2-x}Nd_xSi_2O_{8-0.5x}$ $(0.02 \leq x \leq 0.2)$. It could be found that the density value display a gradual increase as the content Nd increase in the solid solution. The highest density achieved as 3.784 g·cm$^{-3}$ for $Zr_{1.80}Nd_{0.20}Si_2O_{7.90}$, while the lowest density is obtained as 3.223 g·cm$^{-3}$ for $Zr_{1.98}Nd_{0.02}Si_2O_{7.99}$. The reason for this is mainly due to the relative atomic weight of the element. The relative atomic weight of Nd (144.242) is greater than that of Zr (91.224) and trivalent simulated nuclide Nd can completely replace Zr in the solidified crystal lattice [19]. So the density of Nd-doped ceramics increases with the amount of simulated nuclide Nd.

**Table 2.** Tested density of $Zr_{2-x}Nd_xSi_2O_{8-0.5x}$ $(0.02 \leq x \leq 0.2)$ compounds.

| Target compounds | Tested density / g·cm$^{-3}$ | | | | | |
|---|---|---|---|---|---|---|
| | 1 | 2 | 3 | 4 | 5 | Average |
| $Zr_{1.98}Nd_{0.02}Si_2O_{7.99}$ | 3.209 | 3.206 | 3.265 | 3.232 | 3.205 | 3.223 |
| $Zr_{1.96}Nd_{0.04}Si_2O_{7.98}$ | 3.300 | 3.313 | 3.332 | 3.286 | 3.215 | 3.289 |
| $Zr_{1.94}Nd_{0.06}Si_2O_{7.97}$ | 3.296 | 3.361 | 3.404 | 3.396 | 3.361 | 3.364 |
| $Zr_{1.92}Nd_{0.08}Si_2O_{7.96}$ | 3.402 | 3.415 | 3.395 | 3.33 | 3.408 | 3.390 |
| $Zr_{1.90}Nd_{0.10}Si_2O_{7.95}$ | 3.434 | 3.368 | 3.381 | 3.379 | 3.425 | 3.397 |

| | | | | | | |
|---|---|---|---|---|---|---|
| **$Zr_{1.88}Nd_{0.12}Si_2O_{7.94}$** | 3.400 | 3.482 | 3.483 | 3.406 | 3.441 | 3.442 |
| **$Zr_{1.86}Nd_{0.14}Si_2O_{7.93}$** | 3.472 | 3.526 | 3.419 | 3.453 | 3.532 | 3.480 |
| **$Zr_{1.84}Nd_{0.16}Si_2O_{7.92}$** | 3.711 | 3.685 | 3.592 | 3.571 | 3.502 | 3.612 |
| **$Zr_{1.82}Nd_{0.18}Si_2O_{7.91}$** | 3.734 | 3.699 | 3.744 | 3.646 | 3.739 | 3.712 |
| **$Zr_{1.80}Nd_{0.20}Si_2O_{7.90}$** | 3.789 | 3.779 | 3.787 | 3.803 | 3.764 | 3.784 |

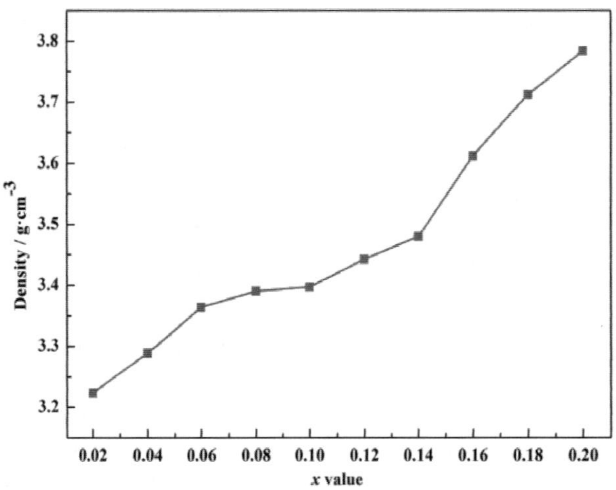


**Figure 2.** Relation curve between the densities and $x$ in $Zr_{2-x}Nd_xSi_2O_{8-0.5x}$ ($0.02 \le x \le 0.2$).

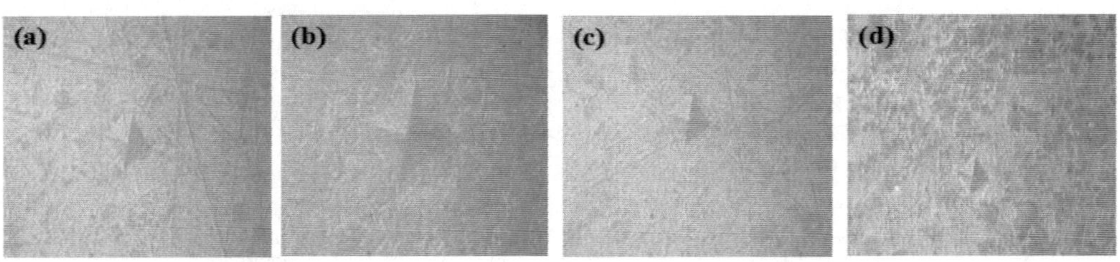


**Figure 3.** Vickers hardness test photos of some samples.

**Table 3.** Vickers hardness of $Zr_{2-x}Nd_xSi_2O_{8-0.5x}$ ($0.02 \le x \le 0.2$) samples.

| **Target compounds** | **Hardness / kg·mm$^{-2}$** | | | |
|---|---|---|---|---|
| | 1 | 2 | 3 | **Average** |
| **$Zr_{1.98}Nd_{0.02}Si_2O_{7.99}$** | 413.03 | 366.052 | 352.355 | 377.146 |
| **$Zr_{1.96}Nd_{0.04}Si_2O_{7.98}$** | 453.237 | 441.261 | 465.171 | 453.223 |
| **$Zr_{1.94}Nd_{0.06}Si_2O_{7.97}$** | 453.73 | 447.216 | 458.27 | 453.072 |
| **$Zr_{1.92}Nd_{0.08}Si_2O_{7.96}$** | 543.438 | 440.531 | 486.665 | 490.211 |
| **$Zr_{1.90}Nd_{0.10}Si_2O_{7.95}$** | 521.59 | 512.187 | 503.317 | 512.365 |
| **$Zr_{1.88}Nd_{0.12}Si_2O_{7.94}$** | 728.826 | 660.721 | 660.085 | 683.211 |
| **$Zr_{1.86}Nd_{0.14}Si_2O_{7.93}$** | 732.336 | 699.195 | 809.206 | 746.912 |
| **$Zr_{1.84}Nd_{0.16}Si_2O_{7.92}$** | 803.128 | 894.93 | 952.408 | 883.489 |
| **$Zr_{1.82}Nd_{0.18}Si_2O_{7.91}$** | 933.39 | 1093.815 | 916.067 | 981.091 |
| **$Zr_{1.80}Nd_{0.20}Si_2O_{7.90}$** | 1145.181 | 1063.015 | 1024.197 | 1077.464 |

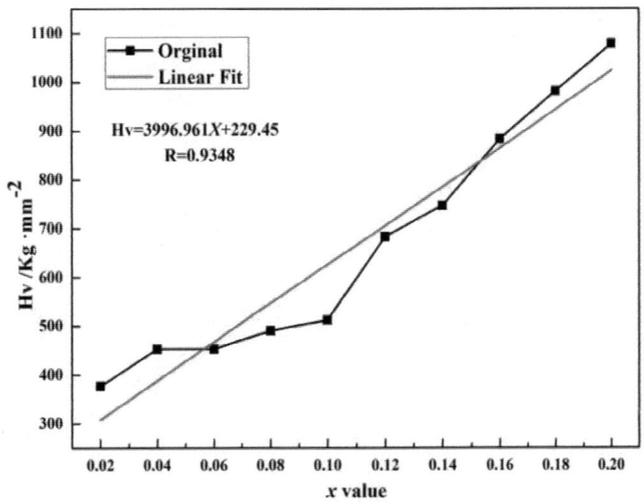


**Figure 4.** Relation curves between $H_v$ and $x$ of $Zr_{2-x}Nd_xSi_2O_{8-0.5x}$ ($0.02 \leq x \leq 0.2$) compounds.

Figure 3 shows part of Vickers hardness test photos, which reveals the process of the testing. Table 3 gives a series of Vickers hardness test results, and Figure 4 displays the relation curves between $H_v$ and $x$ of $Zr_{2-x}Nd_xSi_2O_{8-0.5x}$ ($0.02 \leq x \leq 0.2$) compounds. The results show that the Vickers hardness of $Zr_{2-x}Nd_xSi_2O_{8-0.5x}$ ($0.02 \leq x \leq 0.2$) samples ranges from 377.146 kg·mm$^{-2}$ for $Zr_{1.98}Nd_{0.02}Si_2O_{7.99}$ to 1077.464 kg·mm$^{-2}$ for $Zr_{1.80}Nd_{0.20}Si_2O_{7.90}$. From Figure 4, it can be clearly seen that with the increase in the amount of $Nd_2O_3$, Vickers hardness value appears to be gradually increasing trend. By using the linear equation $y = a + b\,x$, the Vickers hardness can be linear fitted by the $x$ in the sample. The linear relationship is expressed as $H_v = 229.45 + 3996.96\,x$, and the $R^2$ value is greater than 0.8739, indicating that Vickers hardness and the content of Nd directly have a good linear relationship.

## 4. Conclusions

A series of $Zr_{2-x}Nd_xSi_2O_{8-0.5x}$ compounds were prepared by solid-state reaction method at 1500°C for 72 h in air. The density tests show that the density of $Zr_{2-x}Nd_xSi_2O_{8-0.5x}$ ($0.02 \leq x \leq 0.20$) compounds ranges from 3.223 to 3.784 g·cm$^{-3}$, and with increasing of simulated nuclide Nd, the density increased gradually. From the hardness results, it can be found that the hardness of $Zr_{2-x}Nd_xSi_2O_{8-0.5x}$ ($0.02 \leq x \leq 0.20$) compounds extent from 377.146 kg·mm$^{-2}$ to 1077.464 kg·mm$^{-2}$. Vickers hardness of $Zr_{2-x}Nd_xSi_2O_{8-0.5x}$ ($0.02 \leq x \leq 0.20$) compounds increase gradually with the increase of $Nd_2O_3$ content. Vickers hardness and $x$ value has linear fitted as $H_v = 229.45 + 3996.96\,x$, $R^2$ value is greater than 0.87.

## Acknowledgments

This work was supported by the Pre-research Foundation of Fundamental Science on Nuclear Wastes and Environmental Safety Laboratory (No. 15yyhk01).

## References

[1] Zhao X F, Teng Y C, Wu L, Huang Y, Ma J M and Wang G L 2015 *J. Nucl. Mater.* **466** 187.
[2] Graves C, Ebbesen S D, Mogensen M and Lackner K S 2011 *Renew. Sust. Energy Rev.* **15** 1.
[3] Lenzen M 2008 *Energ. Convers. Manag.* **49** 2178.
[4] Binelli A, Pedriali A, Riva C and Parolini M 2012 *Chemosphere.* **86** 906.
[5] Kim H, Park C and Kwon O J 2016 *Energy.* ***115*** 155.
[6] Li L, Luo S and Wang D 1998 *ChemInform.* **80** 250.
[7] Taneja N and Biswal M 2006 *J. Hosp. Infect.* **62** 525.
[8] Rebiscoul D, Lee A V D and Rieutord F 2004 *J. Nucl. Mater.* **326** 9.

[9] Day D E, Wu Z, Ray C S and Hrma P 1998 *J. Non-Cryst. Solids.* **241** 1.
[10] Lange L C, Hills C D and Poole A B 1997 *J. Hazard. Mater.* **52** 193.
[11] Lu X R, Fan L, Shu X Y, Su S J, Ding Y and Yi F C 2015 *Ceram. Int.* **41** 6344.
[12] Ringwood A E, Kesson S E, Ware N G, Hibberson W and Major A 1979 *Nature.* **278** 219.
[13] Tu H, Duan T, Ding Y, Lu X R, Tang Y J and Li Y X 2014 *Mater. Lett.* **131** 171.
[14] Vage J L, Doi K and Mazieres C 2006 *J. Am. Ceram. Soc.* **51** 349.
[15] Krogh T E 1973 *Geochim. Cosmochi. Ac.* **37** 485.
[16] Ferriss E D A, Ewing R C and Becher U 2010 *Am. Mineral.* **95** 229.
[17] Galambos M, Rosskopfva O, Kufcakova J and Rajec P 2011 *J. Radioanal. Nucl. Ch.* **288** 765.
[18] Marinin D V and Brown G N 2000 *Waste Manage.* **20** 545.
[19] Ding Y, Lu X R, Dan H, Shu X Y, Zhang S and Duan T 2015 *Ceram. Int.* **41** 10044.

IOP Conf. Series: Earth and Environmental Science **61** (2017) 012141 doi:10.1088/1755-1315/61/1/012141

# Research on safety protection for live working on ±800kV DC transmission line with AC and DC multi-circuit transmission lines

**Yan Xie[1], Tao Xiang[2] and Mu Li[3]**

[1]School of Electrical and Electronic Engineering，Wuhan Polytechnic University, Wuhan 430048, China
[2]Wuhan electric power supply company of State Grid, Wuhan 430015, China
[3]Wuhan Nari Limited Liability Company of State Grid Electric Power Research Institute, Wuhan 430074, China

**Abstract**. In recent years, in order to save line corridors and reduce the scope of demolition, some new transmission modes for high voltage AC and DC power transmission line are proposed. According to the characteristics of the parallel transmission line, this paper studies the distribution characteristics of the electric field intensity of multi circuit parallel transmission lines and the single circuit transmission line. Comparing and analysing the simulation results of the electric field strength, the safety measures for live working of ±800kV DC transmission line with multi circuit parallel transmission lines is put forward. That is the electric field protective equipment and measures used in the live line work of the ±800kV UHDC single circuit line can meet the requirements of the live operation of the ±800kV UHDC multi-circuit transmission line.

## 1. Introduction

With the rapid development of AC and DC transmission project construction, the transmission voltage level is higher. The number of extra high voltage and ultra-high voltage AC and DC transmission lines are increasing, and the transmission line corridor is becoming more and tenser. Meanwhile, the transmission line construction must also meet the requirements of the environment. [1-2] In recent years, in order to save line corridors and reduce the scope of demolition, some new transmission modes for high voltage AC and DC power transmission line are proposed, such as multi loop, parallel erection and so on. [3-4]

There are four parallel erection of high voltage transmission line in the province, including: two ±800kV extra-HVDC transmission lines, ±500kV double circuit ultra-HVDC transmission line, and 1000kV extra- high-voltage AC transmission lines.

According to the characteristics of the parallel transmission line, this paper studies the influence of the electric field on the human body, and verifies the safety and feasibility of the live line work on the multi circuit transmission line. And compared with the single tower with the voltage level of live working methods, the safety measures for live working on ±800 kV transmission line is put forward.

## 2. Calculation and analysis of electric field

Because of the low frequency electric field of operating transmission line, the electric field can be regarded as an electrostatic field when the finite element method is used to calculate the electric field.

The field of each tower in typical operations is calculated based on multi circuit transmission lines running at the same time. [5, 6]

For the analysis of parallel multiple loop circuit of HVDC transmission lines on the same tower space field intensity level, the simulation calculation model is considering the tower and the DC line operation mode of the electric field, and ignoring the effect of insulator effect on electric field distribution and ion current on DC electric field. The conductor is equivalent to single conductor to be calculated. The tower surface is considered as a conductor plane. The earth is considered as an infinite conductor plane. Finally, the calculated results are compared with the spatial field strength of single circuit.

### *2.1. Calculation parameters*

When the electric field intensity is calculated, the related calculation conditions are as follows. The electric field of 1000kV transmission line is calculated by reverse order, and the maximum operating voltage of the system is 1100kV. The voltage of B is $1100/\sqrt{3} * \sqrt{2}$ kV, and the voltage of A and C were respectively loaded with $1100/\sqrt{3} * \sqrt{2} *\cos(0+120)$kV and $1100/\sqrt{3} * \sqrt{2} *\cos(0-120)$kV. The polarity of the DC transmission line is set to the left as the negative pole and the right as the positive pole, and its voltage is calculated in accordance with the rated operating voltage.

#### *2.1.1. Structure of towers*

Figure 1 shows the tower section model of multi circuit transmission line with same corridor. Figure 2 shows the structure of transmission line tower.

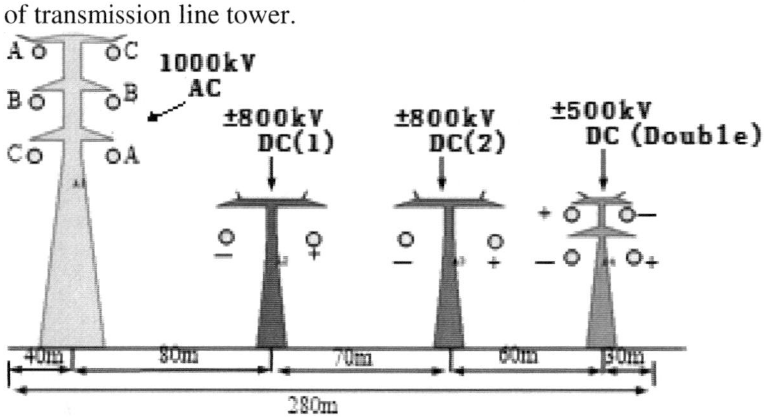


**Figure 1.** Location map of the tower with transmission corridor.

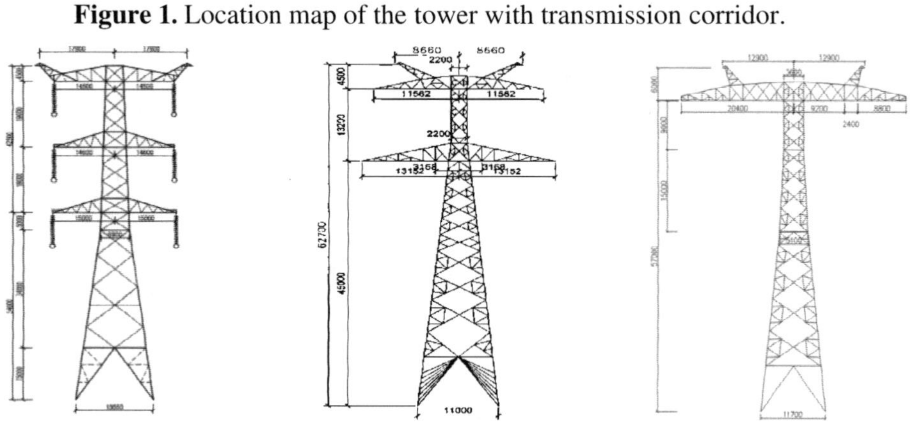

(a) (b) (c)

**Figure 2.** Structure of AC and DC transmission lines tower.

IOP Conf. Series: Earth and Environmental Science **61** (2017) 012141 doi:10.1088/1755-1315/61/1/012141

### *2.1.2. Transmission lines parameters*
Transmission line parameters are set as shown in table 1.

**Table 1.** Transmission lines parameters.

| Line name | 1000kV AC | ±800kV DC(1) | ±800kV DC(2) | ±500kVDC |
|---|---|---|---|---|
| Wire type | 8×ACSR-LGJ630/45 | 6×ACSR-720/50 | 6×JL/G3A-900/40 | 4×ACSR-720/50 |
| Split interval/mm | 400 | 450 | 450 | 450 |
| Sub wire radius /mm | 16.8 | 15.14 | 16.93 | 15.14 |
| Calculating radius/mm | 441.21 | 344.66 | 351.14 | 258.2 |
| Hanging point distance/m | 13 | 11 | 11 | 7 |

### *2.2. Results of calculation*
The electric field distribution on the vertical cross section of the tower and the conductor is shown in Figure 3(a). Figure 3(b) shows the contour map of the electric field strength on the plane.

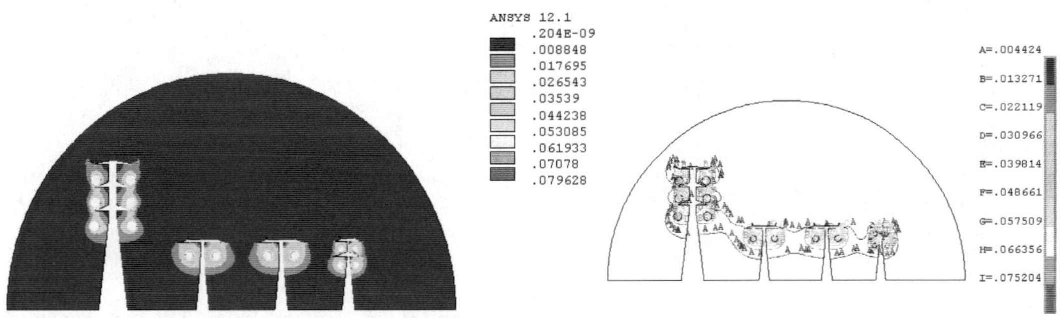

(a) (b)

**Figure 3.** Distribution map of space electric field and its contour

From Figure 3, it shows that the electric field distributions on the right side of ±800 kV DC (1) transmission line, on the left side of ±800 kV DC (2) transmission line and the double direct current transmission line are obviously distorted.

## 3. Safety protection for live working on ±800kV DC high voltage transmission line
Comparing the electric field intensity of ±800 kV DC transmission line key points under multi circuit transmission line parallel erection condition with the electric field intensity of ±800 kV DC single circuit transmission line with the cathead tower, safety protection measures for live working of ±800 kV DC transmission line with AC and DC multi-circuit transmission lines can be obtained.

### *3.1. Analysis of space electric field of multi circuit transmission line tower*
Figure 4 shows the location of the point where the electric field intensity is calculated.

IOP Conf. Series: Earth and Environmental Science **61** (2017) 012141 doi:10.1088/1755-1315/61/1/012141

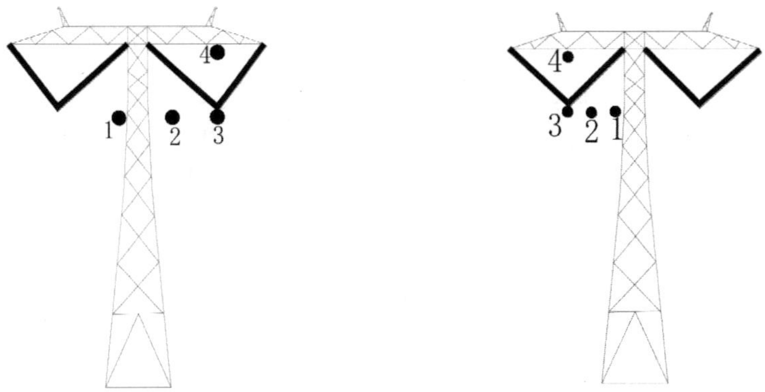


(a) DC (1) transmission line (b) DC (2) transmission line

**Figure 4.** Position distribution diagram of calculation points of ±800 kV DC transmission lines

The electric field distribution of the central section of the DC (1) transmission line tower is shown in Figure 5(a). Figure 5(b) shows the contour map of the electric field strength.

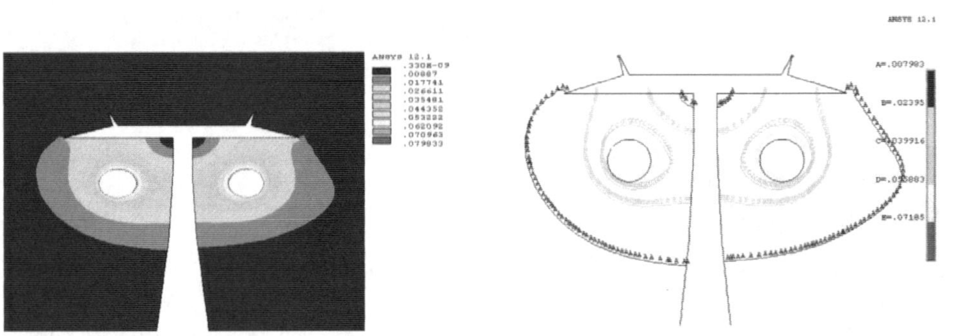

(a) (b)

**Figure 5.** Distribution map of space electric field and its contour of DC (1) transmission line

Table 2 shows the simulation results of the calculation of the electric field strength of the tower of ±800 kV DC (1) transmission line.

**Table 2.** The effective value of the electric field strength of ±800 kV DC (1) transmission line

| Calculation point | electric field intensity (kV/m) | Calculation point | electric field intensity (kV/m) |
|---|---|---|---|
| 1 | 35.0 | 3 | 709.0 |
| 2 | 65.0 | 4 | 22.0 |

The electric field distribution of the central section of the DC (2) transmission line tower is shown in Figure 6(a). Figure 6(b) shows the contour map of the electric field strength.

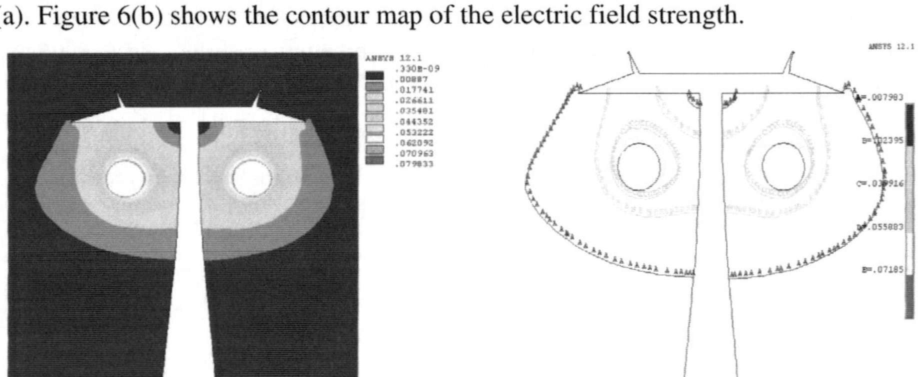

(a) (b)

**Figure 6.** Distribution map of space electric field and its contour of DC (2) transmission line

Table 3 shows the simulation results of the calculation of the electric field strength of the tower of ±800 kV DC (2) transmission line.

**Table 3.** The effective value of the electric field strength of ±800 kV DC (2) transmission line

| Calculation point | electric field intensity (kV/m) | Calculation point | electric field intensity (kV/m) |
|---|---|---|---|
| 1 | 24.0 | 3 | 699.0 |
| 2 | 81.0 | 4 | 11.0 |

As can be seen from the table 2 and table 3, the distribution of the electric field intensity has the following characteristics.

1) The electric field intensity on the surface of the conductor is maximum. Along with the increase of the distance of the conductor, the electric field intensity decays rapidly.

2) The electric field intensity of tower surface with the same height of the conductor, the insulator strings on the cross arm and the lower of the conductor is relatively large because they are close to the conductor.

3) The electric field intensity at the corner intersection with the tower of tower cross arm is relatively small.

### *3.2. Analysis of electric field of single circuit tower*

For the convenience of analysis and comparison, the single circuit tower with ±800 kV DC transmission line is selected as the research object. Figure 7 shows distribution of space electric field and its contour.The electric field distribution of the central section of the line tower is shown in Figure 7(a). Figure 7(b) shows the contour map of the electric field strength.

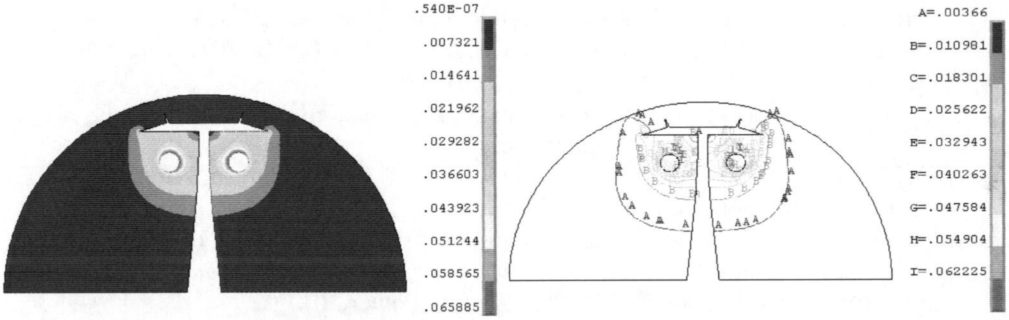


(a) (b)

**Figure 7.** Distribution map of space electric field and its contour

Table 4 shows the simulation results of the calculation of the electric field strength of ±800 kV DC transmission line.

**Table 4.** Electric field intensity of ±800 kV DC tower calculation point

| Calculation point | electric field intensity (kV/m) | Calculation point | electric field intensity (kV/m) |
|---|---|---|---|
| 1 | 12.0 | 3 | 852.0 |
| 2 | 38.0 | 4 | 12.0 |

From the results in Table 4, we can know that the maximum electric field strength of the wire surface. Along with the increase of the distance to the wire, the electric field intensity is quickly fading.

### *3.3. Comparison and analysis of results*

Comparing the electric field strength of the typical operation line tower working position of multi loop parallel charged transmission line and independent single tower corridor, the calculation results show that electric field strength has the following characteristics and rules for ±800 kV UHDC electrification circuit tower.

IOP Conf. Series: Earth and Environmental Science **61** (2017) 012141 doi:10.1088/1755-1315/61/1/012141

The electric field strength of the surface of the conductor is the largest for the two kinds of transmission lines, and the value of the field strength is basically equivalent. The maximum value of electric field strength on the surface of the conductor of the single tower transmission line is slightly larger than that of the electric field strength on the surface of the conductor of the multi loop parallel tower transmission lines.

## 4. Conclusion

By the above analysis, it can be known that the electric field protective equipment and measures used in the live line work of the ±800 kV UHDC single circuit line can meet the requirements of the live working of the ±800kV DC transmission line with AC and DC multi-circuit transmission lines. That is live working personnel (including the ground potential conditions of live working personnel) should be wearing a full set of ±800kV special shielding clothing. Special shielding clothing is made of shielding clothing cloth , the maximum electric field inside it is not more than 15kV/m, and the current flowing through the human body is not more than 50μA. In order to ensure the shielding clothing has high shielding efficiency and ion current blocking ability, the design of shielding clothes uses integral structure, conductive material and flame retardant fibre mesh mask to reduce the surface area exposed.

Because the domestic manufacturers have developed 1000kV AC shielding for live working clothes, the type of shielding clothing used in the research and application of ±800kV live working is the same as the use of 1000kV shielding clothing model. It can meet the relevant technical standards, so it can be directly applied to ±800kV live working.

## References

[1] Research Status and Development Trend of Live Working Key Technology, HU Yi, LIU Kai, PENG Yong, SU Ziming, WU Tian. High Voltage Engineering,2014.07

[2] Live Woeking on EHV/UHV Transmission Lines,HU Yi,LIU Kai,LIU Ting,XIAO Bin,PENG Yong, SU Zing. High Voltage Engineering,2012.08

[3] Safety Protection for Live Working on Transmission Line, HU Tao , HU Yi , LI Jing lu , Long lihong. High Voltage Engineering,2006.05

[4] Safety Hazard and its Prevention for Hot-line Work of 500 kV Transmission Line, XIE Yu-gan. Industrial Technology Innovation,2016.05

[5] Simulation Analysis of Electric Field Protection for Live-Working on 1 000 kV and 500 kV AC Four-Circuit Transmission Lines on One Tower, SU Zi-ming, PENG Yong, LIU Kai, LIU Ting, XIAO Bin,WU Tian, TANG Pan. Electric Power,2014.02

[6] Computation of Body Surface Electric Field during Live Working on 1 000 kV AC Compact Transmission Lines, FANG Yaqi, PENG Yong, SU Ziming, TANG Pan, LIU Kai,WU Tian,WANG Linong. Electric Power,2015.10

# Biological removal effect for odor in medical waste steam treatment process

**Jin Dengchao[1,a], Teng Hongjun[2,b], Bao Zhenbo[1,c] and Li Yang [1,d]**

[1]Engineering and Technology College, Tianjin Agricultural University, Tianjin, China, 300384
[2]Clinic, Tianjin Agricultural University, Tianjin, China, 300384
[a]jindengchao@163.com; [b]hongjun-teng@163.com; [c]baozhenbo@sohu.com; [d]liyang.tjac@gmail.com

**Abstract.** Malodorous gas will be produced in medical waste steam treatment process, which mainly consist of volatile organic compounds. Biological deodorization is a low cost deodorization technology. In this paper, biological deodorization technology in an actual medical waste disposal project was studied, and its removal efficiencies of VOCs were analyzed. The contents of four volatile organic compounds in exhaust gas were determined by gas chromatography and methylene blue spectrophotometry. The average concentration of them were: methyl-sulfhydryl 0.037mg/$m^3$, dimethyl sulfide 0.046mg/$m^3$, methyl disulfide 0.904mg/$m^3$ and hydrogen sulfide 0.204mg/$m^3$. After they were treated by biological deodorization, the average removal rates of the four compounds were methyl-sulfhydryl 93%, dimethyl sulfide 92.6%, methyl disulfide 97.6% and hydrogen sulphide 92.6%.The results showed that the four kinds of VOCs had good deodorization effect, and the national standards for emissions requirements can be met well.

## 1.Introduction

Medical waste steam treatment is one kind of non-incineration technology[1-3], which basic principle is moist heat sterilization[4]. The organic compounds in medical waste will evaporate and cause odor pollution when medical waste is sterilizing by high temperature steam. Therefore, it is necessary to purify the waste gases by removing odor. Biological deodorization is a widely used and low cost technology[5], its basic principle is that microorganisms can absorb and decompose the odor components. In this paper, an actual medical waste treatment engineering was investigated ,which adopted steam treatment technology to dispose medical waste and used patented biological deodorization technology for waste gas purification. According to the national detection standard, the content of four kinds of odor components, which were methyl-sulfhydry, dimethyl sulfide, methyl disulfide and hydrogen sulfide, were analyzed before and after deodorization. The purpose of this research is to verify the deodorization effect of biological deodorization technology.

Medical waste disposal capacity of the project was 5 tons per day, basic operation process of medical waste treatment was described as below: medical waste was put into the sterilizer firstly, after alternate operation of steam injection and vacuum pumping was finished, steam was then injected into sterilizer and the temperature of the medical waste was raised to 134 °C. Keeping at this temperature for 45 minutes to complete the sterilization, medical waste was then vacuum dried. After drying, sterilized medical waste was pulled out from sterilizer and was shredded outside. At last, they were landfilled.

In the process above, odor was mainly produced in four aspects: the first was pre-vacuum stage, the second one was drying stage after sterilization, the third was discharging stage of medical waste from sterilizer, and the last was crushing stage. In the first two stages above, exhaust carrying fetor was discharged outward from sterilizer by vacuum pump, it belonged to control emissions. while, in the next two stage, malodorous gases were directly diffused into the operating environment, it belonged to non- control emissions. In order to treat all odor gas, it needed to collect both control and non-control odor emissions together firstly, then to remove them centrally.

To do this, the discharging door of sterilizer was provided with a collecting cover in this study, exhaust gas collected by cover together with the vacuum pump exhaust were drawn into the biological deodorization system.

Odor deodorizing system in this project was one of patented products manufactured by Tianjin greentech environmental technology limited company. Main equipment of the system included four parts: heat exchanger, deodorization tower, draught fan and chimney. Biological deodorization tower was core equipment which was a cylindrical structure with four layers, each layer was filled with biological packing filler, odor gas was firstly absorbed by porous biological filler and then was degraded by deodorizing organisms adhered on the filler. Heat exchanger could regulated the temperature of the gas entering deodorization tower, ensuring the temperature suitable for biological bacteria. Centrifugal fan supplied power to the gas entering the tower, and took the purified gas out to environment through 15 meter height chimney.

## 2. Detection method and apparatus

### *2.1. Criterion for testing*

(1) GB14554-1993, emission standard of odor pollutants.

(2) GB/T14678-1993, air quality ——determination of sulfuretted hydrogen, methyl-sulfhydryl, dimethyl sulfide and methyl disulfide——gas chromatography.

### *2.2. Detection method*

The method to determine methyl-sulfhydryl, dimethyl sulfide and methyl disulfide were gas chromatography. The method to determine hydrogen sulfide was methylene blue spectrophotometry, which detailed operation based on the reference[6].

### *2.3. Core testing instruments*

(1) Gas chromatography flame photometric detector, manufactured by SHIMADZU, model: GC-2010 Plus-FPD.

(2) Gas chromatography mass spectrometer, manufactured by Agilent, model: 7890A-5975C GC-MS.

(3) Ultraviolet visible spectrophotometer, manufactured by SHIMADZU, model: UV-1800.

## 3. Results and analysis

Figure 1 shows concentration values of methyl-sulfhydryl in exhaust before and after purification for six times, their average values are shown in Figure 2. Figure 3 and Figure 4 indicates correspondingly values for dimethyl sulfide. Six concentration values and their average values of methyl disulfide are presented in Figure 5 and Figure 6. As for as hydrogen sulfide, its correspondingly values are showed in Figure 7 and Figure 8.

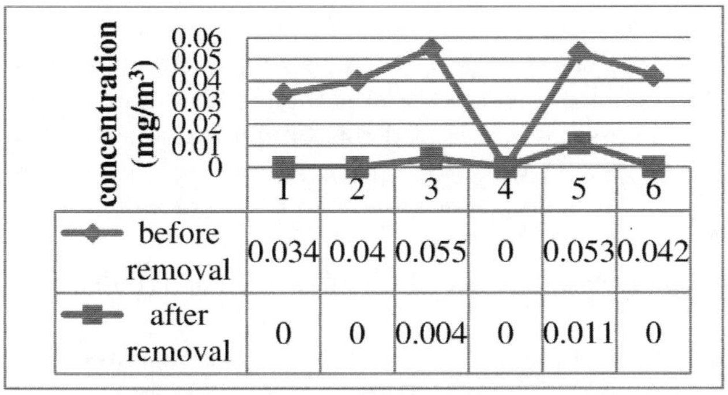


**Figure 1. Methyl-sulfhydryl concentration in waste gas**

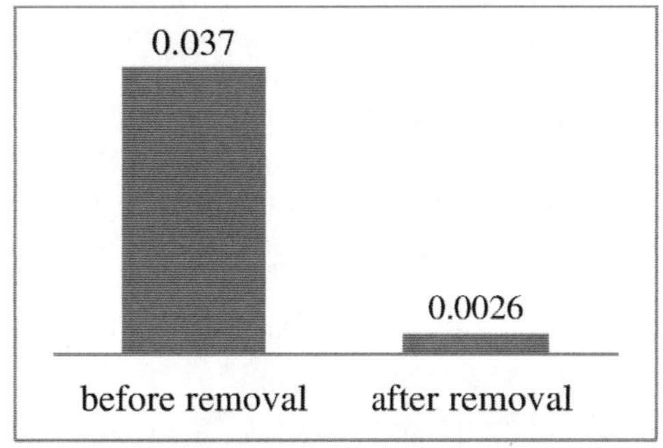


**Figure 2. Methyl-sulfhydryl average concentration in waste gas(mg/m$^3$)**

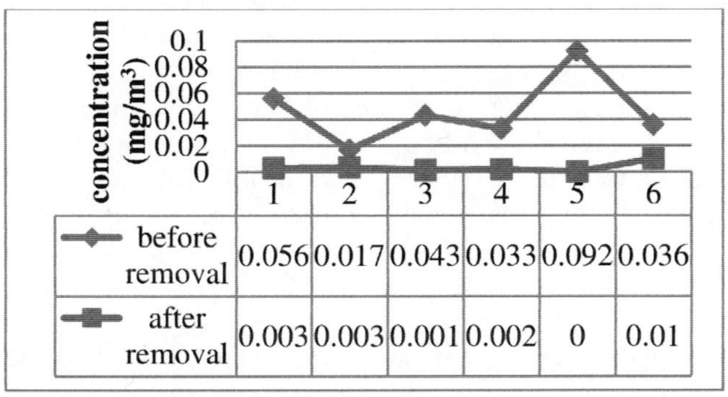


**Figure 3. Dimethyl sulfide concentration in waste gas**

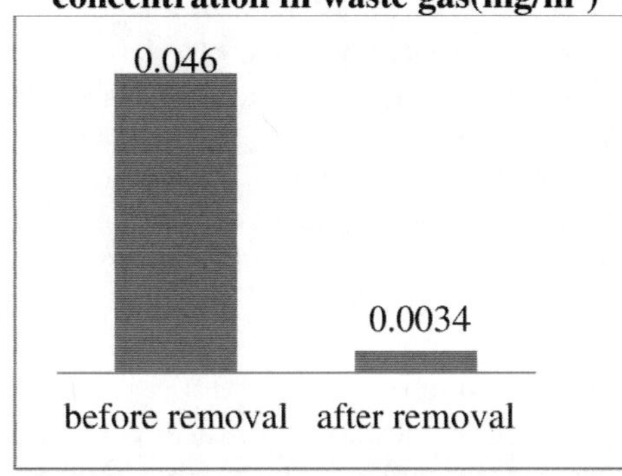


**Figure 4. Dimethyl sulfide average concentration in waste gas(mg/m$^3$)**

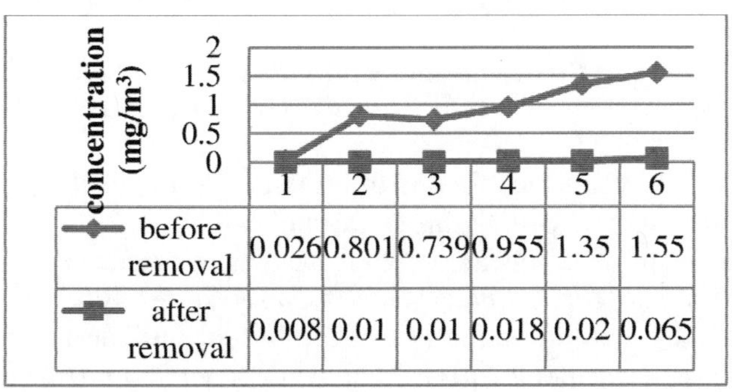


**Figure 5. Methyl disulfide concentration in waste gas**

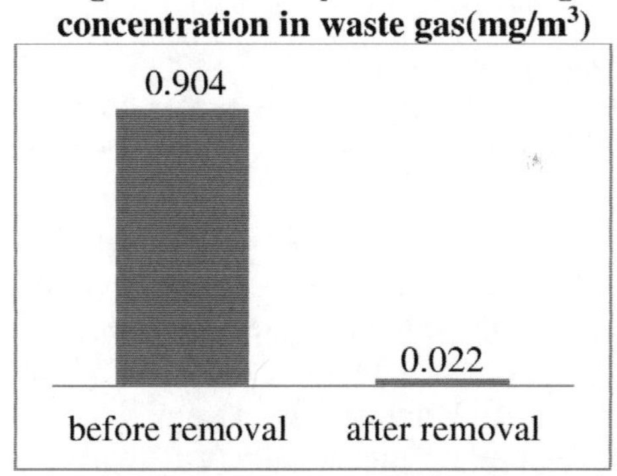


**Figure 6. Methyl disulfide average concentration in waste gas(mg/m$^3$)**

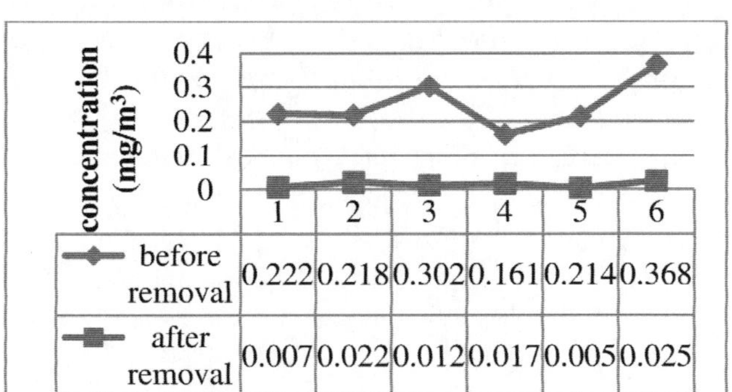


**Figure 7. Hydrogen sulfide concentration in waste gas**

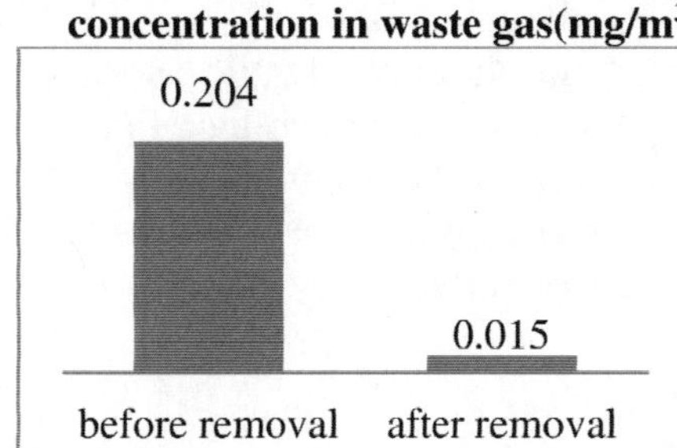


**Figure 8. Hydrogen sulfide average concentration in waste gas(mg/m$^3$)**

IOP Conf. Series: Earth and Environmental Science **61** (2017) 012142 doi:10.1088/1755-1315/61/1/012142

By analysis of above testing result, it could be found:

(1) In the process of medical steam treatment, the content of VOCs varied with treatment stages. For example, as for as hydrogen sulfide, its most content in waste gas was the sixth sampling time, and the lowest content on the fourth time. Other components also had this obvious characteristics. Comparing the sampling time with the state of the process, it was found that the highest concentration of organic always existed in the drying stage, and the lowest concentration time were pre-vacuum and crushing stage.

(2) Among the four VOCs, the most odor components in medical waste gas was methyl disulfide, its average was 0.904mg/m$^3$.

(3) The deodorization efficiency for the four VOCs in this study were: methyl mercaptan 93%, dimethyl sulfide 92.6%, methyl disulfide 97.6% and hydrogen sulfide 92.6%.

(4) Mean concentration of methyl mercaptan after purification is 0.0026 mg/m$^3$, its discharge rate at sampling times was 9.3x10-8~1.1x10-5kg/h. Dimethyl sulfide mean concentration was 0.0034 mg/m$^3$, and discharge rate was 7.7×10-7~1.0x10-5kg/h. Formethyl disulfide mean concentration was 0.022 mg/m$^3$ and discharge rate was 7.6x10-6~6.6x10-5kg/h. As for as hydrogen sulfide, its mean concentration was 0.015 mg/m3 and discharge rate was 4.8x10-6~2.6x10-5kg/h. According the standard GB14554-1993, if the chimney height is 15 meters , the limiting values for emission are as follows: methyl mercaptan0.04 kg/h, dimethyl sulfide 0.33 kg/h, methyl disulfide0.43kg/h and hydrogen sulfide0.33 kg/h. So to speak, deodorization effect of this four VOCs could meet requirements of national standards.

## 4. Conclusion

(1)Odor concentration was different in different stages in medical waste steam treatment process, the maximum amount of odor produced during the drying stage.

(2) Biological deodorization technology could effectively reduce VOCs in medical waste treatment.

## References

[1] Wei Zhao, Ester van der Voet, Gjalt Huppes 2009 Comparative life cycle assessments of incineration and non-incineration treatments for medical waste *Int J Life Cycle Assess* **14**(2)pp 114-121.

[2] Diaz L F, Savage G M, Eggerth L L2005 Alternatives for the treatment and disposal of healthcare wastes in developing countries *Waste Management* **25**pp626-637.

[3] Majid Sartaj.Raheleh Arabgol 2015 Assessment of healthcare waste management practices and associated problems in Isfahan Province (Iran) *J Mater Cycles Waste Manag.* **17**pp99-106.

[4] Jin Dengchao, Li Yang, Bao Zhenbo 2015 kinetic equation of biological indicator for moist heat sterilization in medical waste treatment *Chinese journal of environmental engineering* **9(9)** pp4531-4533.

[5] Guillaume Darracq, Annabelle Couvert, Catherine Couriol 2012 Removal of Hydrophobic Volatile Organic Compounds in an Integrated Process Coupling Absorption and Biodegradation—Selection of an Organic Liquid Phase *Water Air Soil Pollut* **223**pp4969-4997.

[6] China EPA 2003 *Method for analysis of air and waste gas(Fourth Edition)*(Beijing: China Environmental Press)p512-513.

IOP Conf. Series: Earth and Environmental Science **61** (2017) 012143 doi:10.1088/1755-1315/61/1/012143

# Adsorption of bovine serum albumin and urease by biochar

**Wenjing Wang, Lei Chen, Yipeng Zhang, Guocheng Liu***

Key Laboratory of Marine Environment and Ecology, Ministry of Education and College of Environmental Science and Engineering, Ocean University of China, Qingdao 266100, China

*Corresponding author: iceliugc@126.com

**Abstract.** The application of biochar to soil improvement inevitably affects free soil enzymes. However, there is little information on the interaction of soil enzymes with biochar to our knowledge. We thus investigated the adsorption of bovine serum albumin (BSA) and urease onto two biochars from giant reed pyrolyzed at 300 and 600 °C (BCF300 and BCF600). The adsorption amount of BSA and urease on BCF300 and BCF600 was up to 45.6-209 mg/g and 75.3-808 mg/g, respectively, suggesting that the test proteins could be adsorbed onto the biochars effectively. The sorption rate of BSA and urease significantly decreased as the protein concentration increased, suggesting that their adsorption was nonlinear. For the same initial concentration (50 or 200 mg/L), the adsorption amount of BSA on the biochars was lower, only 25.9-60.5% of that of urease. The high specific surface area and hydrophobicity of the biochars may play important roles on the immobilization of the proteins by biochars. These findings will be helpful for better understanding the effects of biochar adding on the soil enzymes.

## 1. Introduction

Biochar is a porous carbonaceous material derived from pyrolysis of biomass (e.g., agricultural and forestry residues, poultry manure and municipal sludge) in an anaerobic or anoxic condition [1]. The properties of biochar including alkalinity, specific surface area, pore structure, elemental composition and functional groups are diversified due to different feedstocks and pyrolysis conditions, especially pyrolysis temperature [2]. Biochar has been described as a soil amendment which can be stored in soil for hundreds or even thousands of years to achieve carbon sequestration [3], and can also improve microbial activity and soil fertility, decrease soil nutrient leaching, reduce bioavailability and phytotoxicity of heavy metals and organic pollutants in soils [4].

Soil enzymes participates in nutrients cycling and energy metabolism in soil, determining the intensities and directions of soil biochemical reaction process (e.g. soil organic matter formation or decomposition, plant productivity and resource supply), which plays important roles in soil ecosystems [5]. Urease, one of soil hydrolases, is involved in purine metabolism and ornithine cycle [6]. Urea, as a widely used nitrogen (N) fertilizer added into soil, can be easily degraded to ammonia via urease catalysis [7]. Application of biochar can adsorb $NH_4^+$, reduce $NH_3$ volatilization and $N_2O$ emission, as well as increase plant uptake of N [8]. Lu et al. reported that the addition of eucalyptus-derived and poultry litter-derived biochar to a cropland area caused different effects on b-glucosidase, β-glucosaminidase, urease and phosphomonoesterase activities [9]. Demisie et al. also reported that 0.5%, 1% and 2.0% oak wood biochar and bamboo biochar significantly improved the urease activity in a degraded red soil. [10]. It is thus clear that biochar will inevitably interact with soil enzymes, but this interaction is overlooked.

The objectives of this study are to: (1) examine the effect of solid to liquid ratio on the sorption rate of BSA by the biochars; (2) study the adsorption kinetics of BSA and urease on the biochars. Bovine serum albumin (BSA) as a model protein due to its high structure stability and relatively clear nature was selected to be a reference adsorbate. This work will provide valuable data about environmental behavior of urease in the biochar-amended soils.

## 2. Materials and methods

### *2.1. Materials*

The biochars were produced from giant reed (*Arundo donax* L.) at 300 and 600 °C as described by Zheng et al. [10], respectively, and sieved through a 0.125 mm sieve for further characterization and experiments. A washing treatment was employed to remove excess minerals from the biochars. Briefly, the ground biochars were repeatedly soaked in 1 M HCl solution 4 times at a ratio of 1:20 (w/v), and then treated with 1 M HCl-HF solution for the same procedure. These biochar samples were thoroughly washed with distilled water several times until their pH values retained stable, and then oven-dried at 80 °C to get the ash-free biochars. The deashed biochars were hereafter named as BCF*X*, where *X* represents temperature. Bovine serum albumin (BSA) (V900933) and urease (U1500) were both purchased from Sigma-Aldrich.

### *2.2. Stability assays of BSA*

BSA stock solution (5000 mg/L) was prepared in a acetate buffer solution (NaAc-HAc, 0.1M, pH 5.0±0.1) containing 200 mg/L sodium azide ($NaN_3$). The experiments were conducted in 8-mL glass vials with Teflon-lined screw caps. The BSA stock solution was diluted to 50 mg/L and 200 mg/L. The samples were kept in dark and shaken at 150 rpm at 25 °C for 120 h and sampled periodically. The vials were centrifuged at 3500 rpm for 10 min, and the absorbance of the supernatant was measured at 280 nm by a UV spectrophotometer (Lambda 35, PerkinElmer, USA) [11]. All the samples were run in triplicates.

### *2.3. Solid to liquid ratio experiment*

The BSA stock solution was diluted to 50-1000 mg/L using the acetate buffer solution. BCF300 (5 mg) or BCF600 (2 mg) was added into 8-mL vials containing 7.5 mL BSA solution with different concentrations (50, 100, 500 and 1000 mg/L). All the vials were shaken at 150 rpm for 24 h at 25 °C. After centrifuging at 3500 rpm for 10 min, the BSA concentrations in the supernatants were determined at 280 nm by Lambda 35. The adsorption rate (%) was calculated as follows:

$$R = \frac{C_0 - C_{24}}{C_0} \times 100\% \quad (1)$$

Where $R$ is adsorption rate (%), and $C_0$ and $C_{24}$ (mg/L) is concentration of BSA at 0 and 24 h in solutions, respectively.

### *2.4. Adsorption kinetics of BSA and urease on the biochars*

The BSA or urease stock solution was prepared and diluted to 50 and 200 mg/L using the acetate buffer solution in 2.2. A 5-mg BCF300 or 2-mg BCF600 were added into 8-mL vials with the BSA or urease solution at two initial concentration (50 and 200 mg/L). All the vials were shaken at 150 rpm at 25 °C, and sampled at different time (8-120 h) and centrifuged at 3500 rpm for 10 min to examine the BSA or urease concentration. The urease concentration was determined by Coomassie brilliant blue method [12]. The adsorption capacity ($Q_t$) was calculated as follows:

$$Q_t = \frac{C_0 - C_t}{m} \times V \quad (2)$$

Where $Q_t$ is adsorption amount (mg/g), $C_0$ and $C_t$ is protein concentration (mg/L) at time 0 and t, respectively, $m$ is biochar amount (g), $V$ is solution volume (L).

### *2.5. Data analysis*

Microsoft office excel (version 2013) was used for data analysis. Significant differences were tested using least significant difference (LSD) ($P < 0.05$) by Statistical Product and Service Solutions (version 20.0).

## 3. Results and discussion

The absorbance values of BSA solution after shaking for different times are shown in Figure 1. Whether low or high concentration of the BSA solutions, there was no significant difference for the absorbance within 120 h, implying that BSA can sustain structure stability in the tested buffer solution for 120 h.

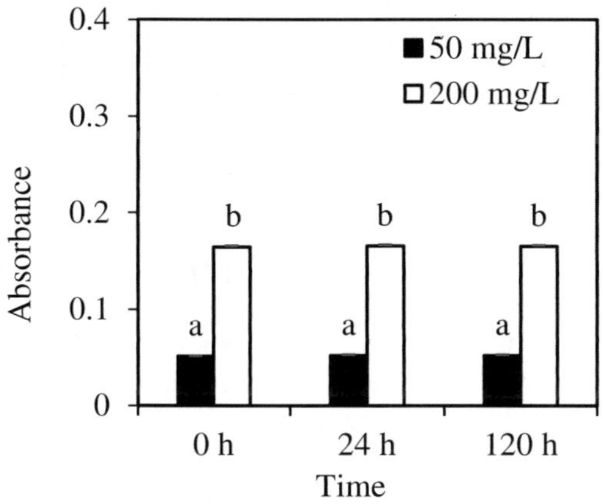


**Figure 1** Absorbance of BSA solution after shaking for different time. For a given BSA concentration, the bars with lowercase letters represent the significant difference of the BSA stability among different times ($n = 3$, $P < 0.05$).

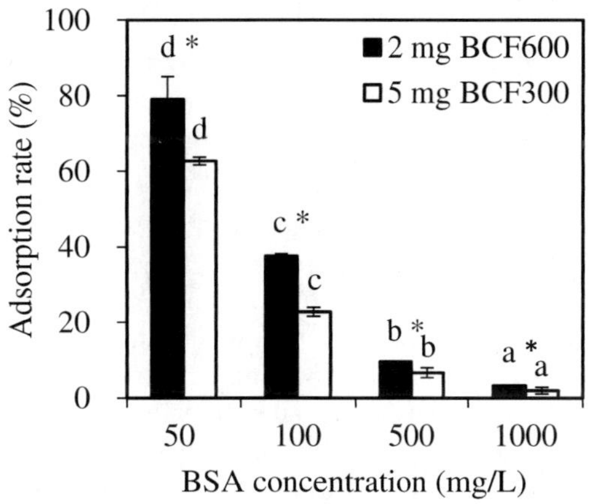


**Figure 2** Adsorption rate of BSA by the biochars. Lowercase letters represent the significant difference of the BSA adsorption on the same biochar among the different BSA concentration. For a given BSA concentration, asterisk represents the significant difference of the adsorption rate of BSA between BCF300 and BCF600 ($n = 3$, $P < 0.05$).

The adsorption rate decreased with increasing BSA concentration for both BCF300 and BCF600 (Figure 2). For instance, the adsorption rates of BCF300 and BCF600 reduced to 9.66% and 6.67% from 79.0% and 62.7% when the BSA concentration increased from 50 mg/L to 1000 mg/L, respectively. This indicated that the adsorption efficiency of BSA by the biochars greatly depended on the adsorbent dose.

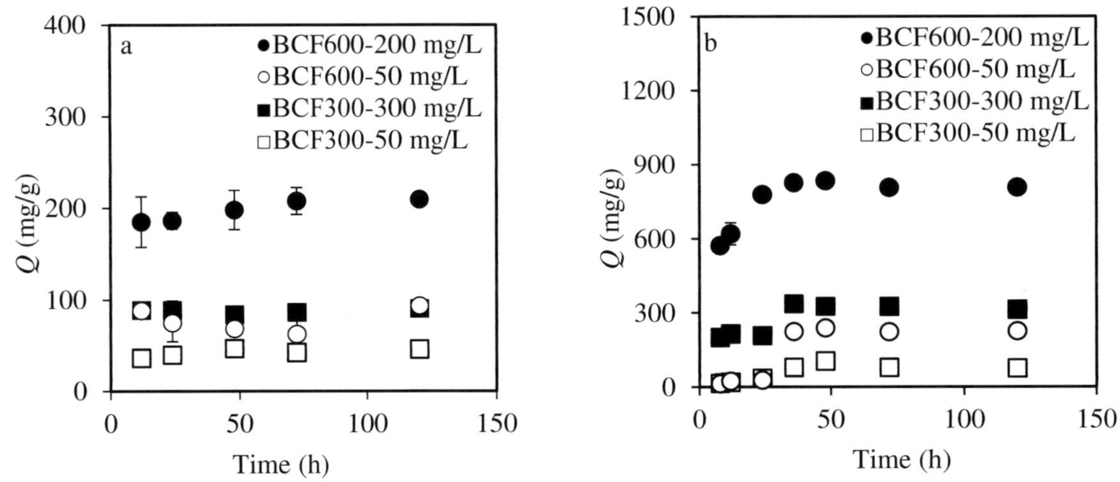


**Figure 3** Adsorption kinetics of (a) BSA and (b) urease on the biochars.

BSA and urease were effectively sorbed by BCF300 and BCF600, and the sorption reached up to equilibrium at 24 h and 36 h (Figure 3), respectively. The adsorption capacities of the two adsorbates were different. For example, when the initial concentration was 200 mg/L, the sorption amount of BSA on BCF300 was 88.5 mg/g, significantly lower than that of urease (337 mg/g), implying that urease was easier to be adsorbed by biochar (Figure 3). That may depend on the differences of structure and properties of the tested protein. Furthermore, the adsorption amount of BSA and urease on BCF600 was both higher than that on BCF300, indicating that BCF600 had stronger affinity than BCF300. In our previous report, H/C ratio of BCF300 and BCF600 was 0.82 and 0.34, respectively, and the specific surface area was 2.72 and 50 $m^2/g$, respectively [9]. Therefore, the greater sorption of BSA and urease on BCF600 is likely related to its higher surface area and hydrophobicity. Hitherto, there is no information about the interaction mechanisms between soil enzymes and biochar to our knowledge. Further investigations therefore should be carried out to explore the mechanisms of protein uptake by biochar.

## 4. Conclusions

The results suggested that BSA could retain structure stability in a acetate buffer solution within 120 h, and be effectively immobilized onto the deashed biochars. The adsorption of BSA and urease on the biochars reached to equilibrium at 24 h and 36 h, respectively. BCF600 exhibited stronger adsorption capacity for both BSA and urease than BCF300. And urease exhibited better affinity onto the biochars than BSA. In addition, the high hydrophobicity and specific surface area may be possible mechanisms for BSA and urease adsorption on the biochars.

## Acknowledgements

This work was financially supported by *Open Foundation of Shandong Provincial Key Laboratory of Marine Ecology and Environment and Disaster Prevention and Mitigation* (201511, 201512) and *Natural Science Foundation of China* (41325013, 41406085, 41503093).

## References

[1] Zheng H, Wang Z, Deng X, Herbert S and Xing B 2013 *Geoderma* ***206*** 32-39
[2] Jeong C Y, Dodla S K and Wang J J 2016 *Chemosphere* **142** 7-12
[3] Lehmann J 2007 *Nature* **447** 143-44
[4] Inyang M and Dickenson E 2015 *Chemosphere* **134** 235-37
[5] Burns R G, DeForest J L, Marxsen J, Sinsabaugh R L, Stromberger M E, Wallenstein M D, Weintraub M N and Zoppini A 2013 *Soil Biol. Biochem.* **58** 216-34
[6] Wang L, Wang S, Deng X, Zhang Y and Xiong C 2013 *ACS Sustainable Chem. Eng.* **2** 433-39

[7] Singh J, Kunhikrishnan A, Bolan N S and Saggar S 2013 *Sci. Total Environ.* **465** 56-63
[8] Luo X, Chen L, Zheng H, Chang J, Wang H, Wang Z and Xing B 2016 *Geoderma* **282** 120-28
[9] Lu H, Li Z, Fu S, Méndez A, Gascó G and Paz-Ferreiro J 2015 *Chemosphere* **119** 212-14
[10] Demisie W, Liu Z and Zhang M 2014 *Catena* **121** 217-19
[11] Zheng H, Wang Z, Deng X, Zhao J, Luo Y, Novak J, Herbert S and Xing B 2013 *Bioresour. technol.* **130** 463-71
[12] Ravindran A, Singh A, Raichur A M, Chandrasekaran N and Mukherjee A 2010 *Colloids and Surfaces* B **76** 33-34
[13] Ke C Y, Wu Y T and Tseng W L 2015 *Biosens. Bioelectron.* **69** 47-48

IOP Conf. Series: Earth and Environmental Science **61** (2017) 012144 doi:10.1088/1755-1315/61/1/012144

# Study on the thermal storage properties of Al-Si-Cu-Mg alloy

**XIE Yi[1], WAN Ke-yang[1], XIE Guo-sheng[2], HU Jia-rui[1], YIN Fu-cheng[3,*], GONG Jing[2]**

1. Hunan Electric Power Corporation Research Institute, Changsha 410007
2. Hunan Xiangdian Testreseach Institute Co. Ltd, Changsha 410007
3. Key Laboratory of Materials Design and Preparation Technology of Hunan Province, Xiangtan University, Xiangtan 411105

Correspondent: Prof. YIN Fu-cheng, fuchengyin@xtu.edu.cn

**Abstract**: Based on thermodynamic calculation technology, the thermodynamic data of six alloys (inc. Al-Si and Al-Si-Cu-Mg systems) was calculated. The microstructure, phase transformation temperature and latent heat of the Al-4%Cu-12%Mg-7%Si alloy and Al-13%Si alloy were also verified by X-ray diffraction(XRD), scanning electron microscopy/Energy-dispersive system (SEM/EDS) and Differential scanning calorimetry (DSC). The results show that The enthalpy change value of Al-Si-Cu-Mg alloy is larger than that of Al-Si alloy from 800 °C to 400 °C, while its phase transformation temperature is lower. In particular, the enthalpy value of Al-4%Cu-12%Mg-7%Si alloy is 85% higher than that of Al-13%Si near eutectic alloy, and its initial temperature of phase transformation is about 74 °C lower. The former has relatively low phase transformation temperature and over-dimensioned latent heat of phase transformation, displaying the excellent thermal storage property. Therefore, the alloy is a good potential solar energy thermal storage material. The results in the paper also indicated that thermodynamic calculation is of value in seeking new potential solar energy thermal storage materials for solar thermal power generation system.

## 1. Introduction

Energy crisis can promote the rapid development of solar energy utilization technology, so that the study on thermal storage materials features great practical significance. Al alloy materials can achieve the high temperature energy storage of solar energy on solar cookers, with the thermal storage temperature of 650-700 °C. Meanwhile, compared with inorganic salt thermal storage materials, the Al alloys have large thermal storage density, high thermal conductivity and fast thermal exchange speed. Therefore, it is a potential candidate for future thermal storage materials [1].

The thermal storage capacity of materials mainly consists of two aspects: the first one is the latent heat with phase transformation and the other is the sensible heat. In recent years, studies on thermal storage properties of Al alloys mainly concentrate on the latent heat with phase transformation. However, sensible heat is also an important part of materials' thermal storage in practical use and lack of deep views in it.

According to the previous studies, Al-Mg, Al-Si, A1-Si-Mg, Al-Si-Cu and other alloys are widely applied in the high temperature phase transformation thermal storage area. The phase transformation temperature range of Al-Mg alloy is relatively narrow [2], A1-Si-Mg alloy has high latent heat of phase transformation, while Al-Si-Cu alloy has long thermal storage life [3]. Nevertheless, there is few study on Al-Si-Cu-Mg alloy, especially on their comprehensive thermal storage property. In this paper, the energy storage property and microstructure of Al-Si-Cu-Mg alloy were studied by thermodynamic calculation, XRD, SEM/EDS and DSC.

IOP Conf. Series: Earth and Environmental Science **61** (2017) 012144 doi:10.1088/1755-1315/61/1/012144

## 2. Phase diagram calculation and experiment methods

Thermodynamic calculation (CALPHAD) is the most mature and widely used phase diagram calculation technology in the world now [4], which constructs proper thermodynamics model, describes the thermodynamic functions of each phase in the system, and calculates the needed phase diagram data (enthalpy) in accordance with the characteristics and data of each phase in duality and multicomponent system (inc. measured phase diagram data, thermodynamic data, crystal structure data, and metastable phase diagram data). However, a series of unknown parameters in thermodynamic function expressions should be optimized and obtained through thermodynamic calculation software, and with the combination of thermodynamic data and phase diagram data. The thermodynamic database adopted in this paper is cost507 [5], and the calculation software is Thermo_Calc [6]. The calculation is based on the following procedure: (1) calculate the enthalpy of the alloy with given components under different temperatures, with the temperature interval of 1 °C, and temperature scope of 400-800 °C; (2) when the calculation is completed, the diagram and data documents of the calculation results are output; (3) find out the enthalpy with 400 °C and 800 °C from the calculated data documents, to get the enthalpy change.

In order to make comparative analysis, Al-Si alloy, which is a binary near eutectic alloy with commonly accepted good thermal storage performance, is especially selected as the control one. Three kinds of Al-Si and Al-Si-Cu-Mg alloy are calculated, respectively, as shown in Table 1. Then, based on the calculation results, an alloy with good thermal storage property is selected from the Al-Si-Cu-Mg. And its thermal storage property is studied via experiment study, by comparing that of Al-13%Si.

**Table 1.** Composition of studied alloys /wt.%

| Sample No. | Si | Cu | Mg | Al |
|---|---|---|---|---|
| A1 | 7 | - | - | Bal. |
| A2 | 13 | - | - | Bal. |
| A3 | 20 | - | - | Bal. |
| B1 | 7 | 4 | 4 | Bal. |
| B2 | 13 | 4 | 4 | Bal. |
| B3 | 7 | 4 | 12 | Bal. |

In the experiment, the primeval alloy ingredients are melted in the well resistance furnace, and then cast to slab through iron pattern, with the alterant of NaCl+KCl+NaF, and mold discharging agent of $Na_2SO_4$:ZnO:$H_2O$=3:12:85.

D/max 2500 VB-type XRD is adopted to analyze the phase of the alloy sample, and the microstructure is observed and analyzed through Philips Sirion200 SEM. DSC analysis is conducted on NETZSCH STA 449C DSC, and powder sample is used to test its latent heat of phase transformation and melting point. The DSC test parameters are as follows: heating temperature scope is 30~700 °C, and heating rate is 20 °C/min; protective medium is nitrogen with 99.999% purity, and current speed is 50 ml/min. Before analysis, high purity aluminum melting point is used to adjust the DSC analysis system.

## 3. Results and discussion

### *3.1. Calculation results*

Fig. 1 shows the relationship between enthalpy and temperature for Al-Si alloys A1-A3, wherein A2 is near eutectic alloy. The horizontal ordinate represents temperature, with the unit of K, while the longitudinal coordinate represents enthalpy value, with the unit of J. From the diagram, we can get the enthalpy value ($H$) with the alloy between 673 K (400 °C) and 1 073 K (800 °C). According to these data, we can calculate the enthalpy value change ($\Delta H$)of the alloy between 400-800 °C. Besides, we can also obtain the phase transformation temperature ($T_m$)area of the alloy, and the data is shown in Table 2.

IOP Conf. Series: Earth and Environmental Science **61** (2017) 012144 doi:10.1088/1755-1315/61/1/012144

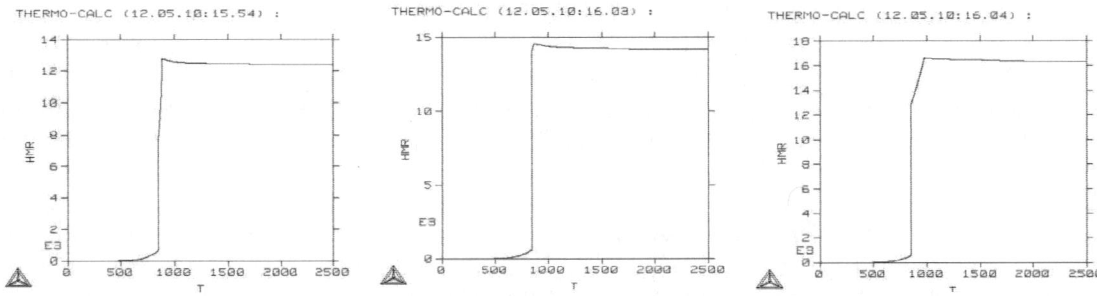


**Figure 1.** Relationship between enthalpy and temperature for Al-Si alloys

According to the previous paper [7], the latent heat of phase transformation for Al-13%Si near eutectic alloy is larger than that of Al-22%Si. As shown in Table 2, with the content of Si rising from 7% to 20%, the enthalpy change value of A1-A3 between 400 and 800 °C increases from 458 J $g^{-1}$ to 605 J $g^{-1}$. On the other hand, the enthalpy change value rises along with the increasing content of silicon, while without much relationship with eutectic structure.

**Table 2.** Enthalpy change of the studied alloys between 400-800 °C

| Alloy No. | $H$ (400 °C) J $mol^{-1}$ | $H$ (800 °C) J $mol^{-1}$ | $\Delta H$ (400-800 °C) J $mol^{-1}$ | $\Delta H$ (400-800 °C) J $g^{-1}$ | $T_m$ °C |
|---|---|---|---|---|---|
| A1 | 109 | 12539 | 12430 | 458 | 578-615 |
| A2 | 102 | 14326 | 14224 | 524 | 578-596 |
| A3 | 93 | 16550 | 16457 | 605 | 578-701 |
| B1 | 6927 | 34360 | 27433 | 963 | 500-574 |
| B2 | 7039 | 32960 | 25921 | 912 | 500-574 |
| B3 | 4318 | 31723 | 27405 | 972 | 504-610 |

The initial temperature of phase transformation of Al-Si alloy is related to its melting phase. It was once reported that the $\alpha$-A1 solid solution firstly changes from solid to liquid in Al-Si binary system, and the phase transformation temperature does not differ too much due to the similar structures for $\alpha$-A1 solid solution with different Si contents [7]. In Table 2, the calculated initial temperature of phase transformation for A1-A3 alloy is the same, which is 578 °C. The results is also pretty close to the measured phase transformation temperature of Al-12.5%Si eutectic alloy, where the temperature is 577.2 °C [8].

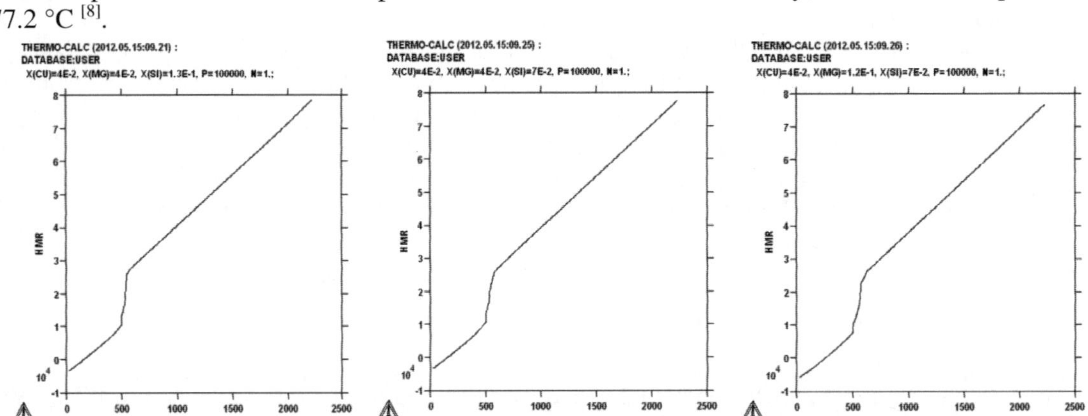


**Figure 2.** Relationship between enthalpy and temperature for Al-Si-Cu-Mg alloys

Fig. 2 shows the relationship between enthalpy and temperature for Al-Si-Cu-Mg alloys B1-B3. Have to state here, the unit of temperature on horizontal coordinate is °C. The corresponding thermodynamic data are also shown in Table 2. The enthalpy change value B3 alloy is the largest (60 J $g^{-1}$), while its phase transformation temperature is slightly higher than that of B1 and B2 alloy.

Compared with Al-Si binary system alloy (Table 2), Al-Si-Cu-Mg alloy has the larger enthalpy value between 400-800 °C, and with lower phase transformation temperature. Taking A2 alloy (Al-

13%Si) as a example, the enthalpy change value of B3 alloy is 85% higher and, its initial temperature of phase transformation is about 80 °C lower. Therefore, from the above calculation results, we can draw the conclusion that the thermal storage performance of Al-Si-Cu-Mg is better than that of Al-Si alloy and the Al-4%Cu-12%Mg-7%Si alloy displays the best thermal storage performance among the studied alloys.

### *3.2. Experiment results*

Based on the thermodynamic calculations of phase diagram, the microstructure and thermal storage performance for B3 alloy (Al-4%Cu-12%Mg-7%Si) are contrasted with commonly used Al-13%Si binary near eutectic alloy.

#### *3.2.1. Microstructure analysis*

SEM structure of Al-13%Si is shown in Fig. 3, which is combined with grey solid solution, grey acicular phase and a few white particles. The length of grey acicular phase is about 25-50 μm. Meanwhile, there are coarse bulk phases in the structure, with the color of grey and the size of about 20 μm. The corresponding EDS results of Al-13%Si are listed in Table 3. Combining the component data and Al-Si binary phase diagram [9], we can confirm that position A is $\alpha$-Al solid solution, position B is primary silicon, and position C is Al-Fe-Si compound, while needle like phase is eutectic silicon. As stated in Section 2, the alloy casting process was conducted by iron casting and so the metal cools slowly, hence acicular eutectic but not evenly distributed lamellar eutectic generates as shown in Fig.3.

**Figure 3.** Microstructure of Al-13%Si alloy

**Table 3.** EDS results of Al-13%Si and Al-4%Cu-12%Mg-7%Si alloys / at. %

| Alloy | Position | Al | Si | Cu | Mg | Fe |
|---|---|---|---|---|---|---|
| Al-13%Si | A | 99.32 | 0.68 | - | - | - |
| | B | 1.89 | 98.11 | - | - | - |
| | C | 64.18 | 26.54 | - | - | 9.28 |
| Al-4%Cu-12%Mg-7%Si | D | 96.64 | 0.59 | 1.02 | 1.75 | - |
| | E | 78.80 | 5.36 | 14.95 | 0.88 | - |
| | F | 51.55 | 19.12 | 6.07 | 23.26 | - |
| | G | 6.36 | 29.20 | 0.65 | 63.79 | - |

The microstructure of B3 (Al-4%Cu-12%Mg-7%Si) alloy is shown in Fig.4. There are grey matrix, white second phase structure, a few grey structures and grey-black bulk structure in the alloy. The corresponding EDS and XRD results are showed in Table 3 and Fig. 4, respectively. Based on the EDS and XRD results, we can confirm that the grey matrix in position D is $\alpha$-Al solid solution, the second phase in position E is $Al_2Cu$, grey structure in position F is $Al_2CuMg$ and eutectic silicon structure, while-grey black position G is $Mg_2Si$ phase. Al-Si eutectic in microstructure does not have characteristic peak in X ray diffraction test, maybe it is because Al-Si eutectic in B3 alloy is less than the detection limit of diffraction measurement.

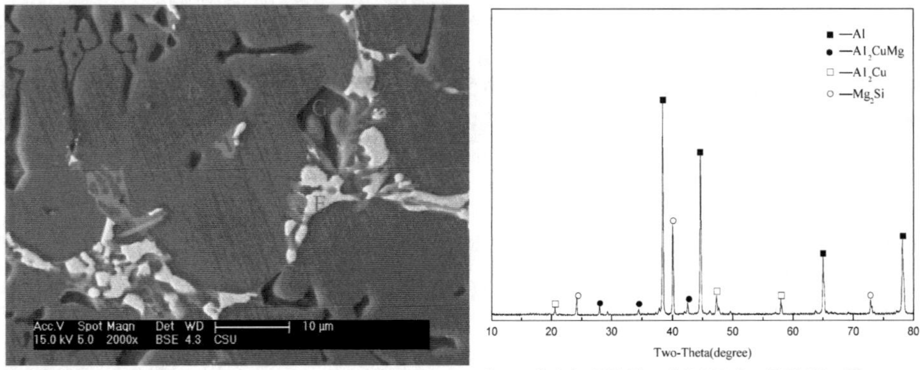


**Figure 4.** Microstructure and XRD results of Al-4%Cu-12%Mg-7%Si alloy

*3.2.2. Analysis for latent heat of phase transformation*

Fig. 5 is the DSC curves of A2 (Al-13%Si) and B3 (Al-4%Cu-12%Mg-7%Si) alloys. For the Al-13%Si system, the alloys experience the melting phase transformation from 577 °C to 638 °C, and its peak temperature is about 590 °C. And the measured initial temperature of phase transformation (577 °C) is pretty close to that by phase diagram calculation (578 °C, as shown in Table 2). In theory, the transformation temperature of Al-Si eutectic is 577 °C [10], which is the same as the measured and calculated initial temperature of phase transformation. The fact illustrates that the calculation data accords with the tested result perfectly and the calculation methods in this paper are reasonable. In addition, many structures will undergo constant temperature melting at the eutectic temperature of 577 °C in Al-Si system, resulting into a large number of heat absorption, as shown in Fig. 5.

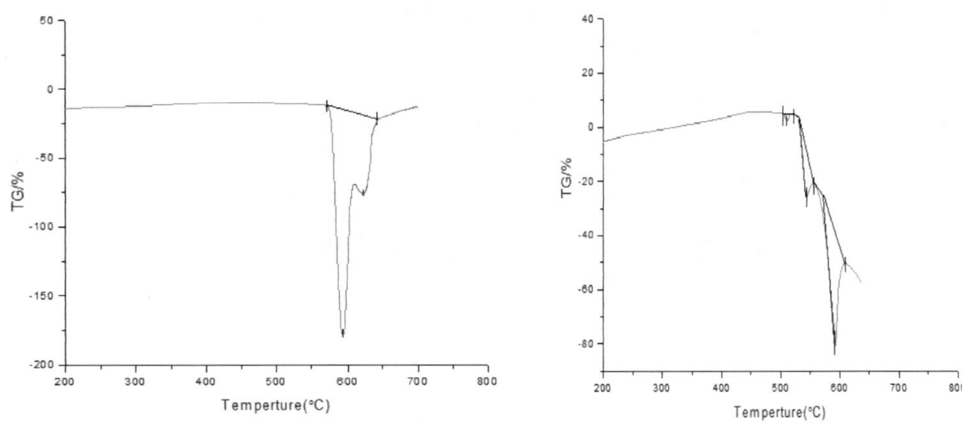


**Figure 5.** DSC curve of Al-13%Si and Al-4%Cu-12%Mg-7%Si alloy

The phase transformation temperature of B3 (Al-4%Cu-12%Mg-7%Si) alloy are in range of 503.5 °C ~610.3 °C, which is very close to the above calculation results (504 °C-610 °C, as shown in Table 2). As displayed in Fig. 5, DSC curve of B3 alloy includes three endothermic peaks, which are 510.5 °C, 543.3 °C and 591.7 °C, respectively. The melting point of $\alpha$-Al+$Al_2Cu$+$Al_2CuMg$ eutectic is 507 °C, and the melting point of $\alpha$-Al+$Al_2Cu$ eutectic is 548 °C [11]. During the phase transformation process, the $Al_2Cu$ phase with low melting point will melt firstly at 510.5 °C, followed by the melting of $Al_2CuMg$ at 543.3 °C, and finally the $Mg_2Si$ and α-Al solid solution experience melting process, resulting into the endothermic peak at 591.7 °C.

The latent heat and temperature of phase transformation for A2 and B3 alloys can be calculated form DSC curves and listed in Table 4. Compared with A2 alloy, the initial and terminal temperatures of phase transformation for B3 alloy have reduced 74 °C and 27 °C, respectively, and its overall phase transformation temperature is lower. In spite of that the latent heat of phase transformation for B3 alloy is slightly lower than that of A1 alloy, it still reaches 425 J $g^{-1}$, which is far higher than the general requirement for metal-matrix phase transformation energy storage materials used in solar thermal power generation system, namely 200 J $g^{-1}$ [12].

**Table 4.** Latent heat of phase transformation for Al-13%Si and Al-4%Cu-12%Mg-7%Si

| Alloy | Latent heat / J $g^{-1}$ | Temperature / °C |
|---|---|---|
| Al-13%Si | 486 | 577.3~637.7 |
| Al-4%Cu-12%Mg-7%Si | 425 | 503.5~610.3 |

## 4. Conclusions

To study the the energy storage property of Al-Si-Cu-Mg alloy, the thermodynamic data of six alloys (inc. Al-Si and Al-Si-Cu-Mg systems) was calculated. Then the microstructure, phase transformation temperature and latent heat of the Al-4%Cu-12%Mg-7%Si alloy and Al-13%Si alloy were also verified by XRD, SEM/EDS and DSC. The following conclusions can be drew:

(1) The enthalpy change value of Al-Si-Cu-Mg alloy is larger than that of Al-Si alloy from 800 °C to 400 °C, while its phase transformation temperature is lower. In particular, the enthalpy value of Al-4%Cu-12%Mg-7%Si alloy is 85% higher than that of Al-13%Si near eutectic alloy, and its initial temperature of phase transformation is about 74 °C lower.

(2) The Al-4%Cu-12%Mg-7%Si alloy have relatively low phase transformation temperature and over-dimensioned latent heat of phase transformation, displaying the excellent thermal storage property. Therefore, the alloy is a good potential solar energy thermal storage material.

(3) In the paper, the calculated results are in complete agreement with the measured ones, indicating that the calculation result is reliable and thermodynamic calculation is of value in seeking new potential solar energy thermal storage materials for solar thermal power generation system.

## References

[1] Huang Zhi-guang, Wu Guang-zhong, Dai Xu-qi, Chen Zheng-rong, Xing Deng-qing. Thermal storage of phase change metal for concentrating solar cooker[J], Tibet Institute of Solar Energy, 1992, 13(3): 271-275.
[2] Zhao Zhi-yuan, Beijing: Knowledge Press, 2005, 594.
[3] Huang Zhi-guang, Mei Shao-hua, Wu Guang-zhong. Techniques for thermal storage of phase change metal[J], New Energy, 1990, 12(8): 11-16.
[4] L. Kaufman, H. Bernstain, Computer calculation of phase diagram, New York: Academic Press,1970.
[5] N. Saunders, I. Ansara, A.T. Dinsdale, M.H. Rand, COST-507: Thermodynamic database for light metal alloys, Luxemburg: European Communities, 1998.
[6] B. Sundman, B. Jansson, and T.O. Anderson, The program for optimization, Calphad-computer coupling of phase diagram & thermochemistry, 1985,9: 153.
[7] Zhang Ren-yuan, Sun Jian-qiang, Ke Xiu-fang, Zhou Xiao-xia, Heat storage properties of Al-Si alloy[J]. Chinese Journal of Materials Research, 2006, 20(2): 156-161.
[8] L. F. Mondolfo, Wang Zhu-tang, Zhang Zhen-lu, Zheng Xuan, Aluminum alloys, structure and propertys[M], Beijing: Metallurgical Industry Press, 1988: 313.
[9] T. B. Massalski, H. O. kamoto, P. R. Subramanian, L. Kalprzak, Binary alloy phase diagrams. Ohio: ASM International, 1990: 183-184.
[10] C. Kammer, Aluminium handbook[M]. Beijing: Chemical Industry Press 2008: 44-47.
[11] Wang Zhu-tang. Heat treatment for 2024 aluminum alloy[J]. Metal World, 2009, 23(2): 43-48.
[12] Gong Dian-qing, Properties of high-temperature phase change thermal storage Al-Si-Cu-Mg-Zn Alloy[D], Wuhan University of Technology, 2008.

IOP Conf. Series: Earth and Environmental Science **61** (2017) 012145 doi:10.1088/1755-1315/61/1/012145

# The experimental verification on the shear bearing capacity of exposed steel column foot

**LIU Xijin**

Infrastructure and Planning Department, Beijing Jiaotong University, Beijing 100044, China
liuxijin2004@163.com

**Abstract.**In terms of the shear bearing capacity of the exposed steel column foot, there are many researches both home and abroad. However, the majority of the researches are limited to the theoretical analysis sector and few of them make the experimental analysis. In accordance with the prototype of an industrial plant in Beijing, this paper designs the experimental model. The experimental model is composed of six steel structural members in two groups, with three members without shear key and three members with shear key. The paper checks the shear bearing capacity of two groups respectively under different axial forces. The experiment shows: The anchor bolt of the exposed steel column foot features relatively large shear bearing capacity which could not be neglected. The results deducted through calculation methods proposed by this paper under two situations match the experimental results in terms of the shear bearing capacity of the steel column foot. Besides, it also proposed suggestions on revising the Code for Design of Steel Structure in the aspect of setting the shear key in the steel column foot.

## 1. Introduction

Steel structure is one of the important construction structures. As its weight is light and the construction is simple, it is extensively used in the large-scale plants, stadiums and the super-high buildings. With the development of the real estate industry, the aseismic design of the building structure is attracting increasingly more attention. The article 8.4.13 of Code for Design of Steel Structures [1] in China regulates: The anchor bolt at the column foot is inappropriate to bear the horizontal reaction force at the bottom of the column foot. Such a horizontal reaction force shall be borne by the force of friction between the baseboard with the concrete (the friction coefficient could take 0.4) or the shear key. In the nations suffering frequent earthquakes, Japan and European and American nations, they all consider the involvement of the steel column foot in the shear bearing. What's more, Code for Design of Steel Structures does not give specific illustration on how to calculate the shear bearing capacity of the exposed steel column foot. The domestic and overseas researchers made a large quantity of theoretical studies on the shear bearing capacity of the exposed steel column foot [2]-[8]. However, experimental verification is in shortage. On the basis of the domestic and overseas theoretical research, this paper comprehensively considers the contribution of the anchor bolt and the shear key to the shear bearing capacity of the steel column foot. It verifies the reliability of the theoretical analysis on the shear bearing capacity, and suggests raising the separation point of the shear key for the exposed steel column foot, reducing the construction difficulty of the exposed steel column foot and providing certain experimental evidences to the design and research of the steel column foot.

## 2.Theoretical evidence

This paper suggests to calculate the shear bearing capacity of the steel column foot in accordance with the literature of the Theoretical Research on the Shear Bearing Capacity of the Exposed Steel Column Foot, i.e.,

### *2.1.The calculation method for the shear bearing capacity of the steel column foot without shear key*

$$\begin{cases} V = \mu N + \beta\ \mu n_1 A_e \sigma + \beta n_2 A_e \tau \\ \sqrt{\sigma^2 + 3\tau^2} \leqslant f_y \end{cases} \tag{1}$$

formula, $n_1$ is the quantity of the anchor bolt at the tensile side; $n_2$ is the quantity of the anchor bolts involved in the shear bearing;$\sigma$ is the positive stress of the single anchor bolt at the tensile side;$\tau$ is the shear stress of the single anchor bolt at the tensile side;$A_e$ is the effective cross-sectional area of the single anchor bolt;$f_y$ is the yield strength of single anchor bolt; N is the axial pressure;$\mu$ is the friction coefficient between the column baseboard with the concrete, taking the value of the American ACI349-85;$\beta$ is the reduction coefficient for considering unevenly distributed loads. It is suggested to take 0.65.

### *2.2.The calculation method for the shear bearing capacity of the steel column foot with shear key*

$$V' \leq \frac{f_c b}{l\varepsilon_0}\left[\frac{4h}{\pi}\times\frac{12\pi^3 h^2\left(\frac{N}{EI}+\frac{2f_c b}{kAG\varepsilon_0 l}\right)+48\pi h^4\frac{2f_c b}{EI\varepsilon_0 l}-3\pi^5}{16\pi^2 h^2\frac{2f_c b}{kAG\varepsilon_0^2 l^2}+128h^4\frac{f_c b}{EI\varepsilon_0^2 l^2}} - \frac{h}{2l\varepsilon_0}\times\left(\frac{12\pi^3 h^2\left(\frac{N}{EI}+\frac{2f_c b}{kAG\varepsilon_0 l}\right)+48\pi h^4\frac{2f_c b}{EI\varepsilon_0 l}-3\pi^5}{16\pi^2 h^2\frac{2f_c b}{kAG\varepsilon_0^2 l^2}+128h^4\frac{f_c b}{EI\varepsilon_0^2 l^2}}\right)^2\right] \tag{2}$$

formula, $f_c$ is the designing value of the compressive strength of teh axle center for the concrete;b is the width of the shear key; $\varepsilon$is the strain of the foundation concrete under the pressure of the shear key; l is the calculation width of the foundation concrete along the forced direction; $\varepsilon_0$ is 0.002; N is the pressure at the top of column;EI is the flexural rigidity of the shear key; KA is the effective shearing area of the shear key;G is the shearing modulus;h is the buried depth of the shear key.

## 3.Experimental verification

To verify the reliability of the above mentioned theory, it designs the two experiential models to study the bearing capacity of the steel column foot with shear key and that of the steel column foot without shear key respectively; it establishes three components for each group to study the shear bearing capacity of the steel column foot under the different axial forces.

### *3.1.Basic assumptions for the experiment*

(1) Under the horizontal force, it neglects the deflection of the steel column;

(2) When the column foot bears horizontal force, the horizontal reaction force generated by the depressed deformation of the concrete at the largest compression side of the eccentric compression baseboard is negligible;

(3) It neglects the bending of the anchor bolt at the tensile side and only considers the tensile-shear function of the anchor bolt at the tensile side.

### *3.2. Model design and manufacturing*

The prototype of the experimental model is the real industry plant engineering in Beijing. Its structure is the portal frame structure; the column foot is the exposed low - position anchor bolt steel column

foot; the height of the storey is 5.4m; one storey in total; the flat area is 60.25 x 40.6m. The model selects the connection part between the steel column foot at the bottom of the prototype structure with the foundation. Due to the limitation of the experimental conditions, the column foot takes the height of 700mm. The dimension of the sectional area is the same with the original structure. The material of the steel column is Q345A. The foundation bolt adopts Q235. The diameter is 28mm. The dimension of the experimental model is shown as figure 1:
The physical experimental models are shown as figure 2, figure 3 and figure 4:

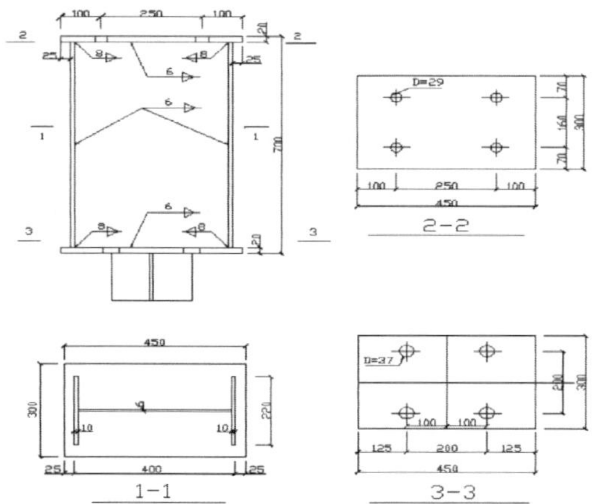


**Figure 1.** experimental model dimension

**Figure 2.** physical steel column

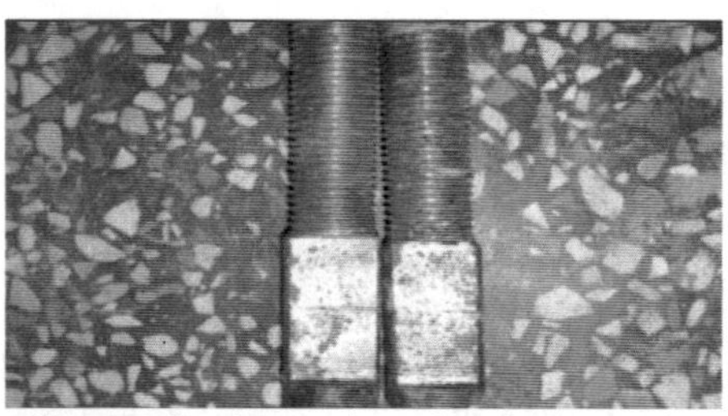

**Figure 3.** physical anchor bolt specimen

**Figure 4.** physical secondary pouring completion

The steel column foot model is manufactured in the steel structure processing factory of China Jingye Engineering Corporation limited. The concrete foundation is manufactured in the new Youth Palace and the components are assembled in the structure laboratory of Beijing University of Civil Engineering and Architecture. The mix proportion of the concrete is shown as table 1.

**Table 1.** The mix proportion of the concrete

| The construction mix proportion of the concrete（kg/m3） | | | | | |
|---|---|---|---|---|---|
| Cement | Coal ash | Sand | Pebble | Water | water reducing agent |
| 350 | 100 | 950（2% moisture content) | 860 | 150 | 5.18 |

There are three blocks of test cubes of 100 mm×100 mm ×100 mm during the pouring process of the concrete foundation, which are maintained with the experimental model foundation under the same conditions. It takes the average value as the material indicator for the concrete, which is shown as table 2.

**Table 2.** Concrete actual test material indicator

| concrete actual test material indicator | |
|---|---|
| Foundation concrete cube Compressive strength (MPa） | secondary pouring concrete cube Compressive strength (MPa)） |

| | |
|---|---|
| 42.00 | 44.28 |

Φ28 anchor bolt adopts Q235. It cuts three pieces of specimen with the length of 50cm to receive the tensile test in the universal testing machine in the material laboratory of Beijing University of Civil Engineering and Architecture. The average value of the yield strength of the anchor bolt is 288 MPa.

### *3.3.Experiment loading device*

The experiment loading device adopts MTS. The experiment is conducted in the static pedestal. The foundation baseboard is fixed in the static pedestal with ground anchor. It places an oil jack in the center of the top of the steel column foot to impose the fixed vertical loading. Meanwhile, it is connected with the model top by the connection accessories through the tensile-compression jack in the reaction wall to impose the low frequency cyclic loading. The loading device of the experiment is shown as the figure:

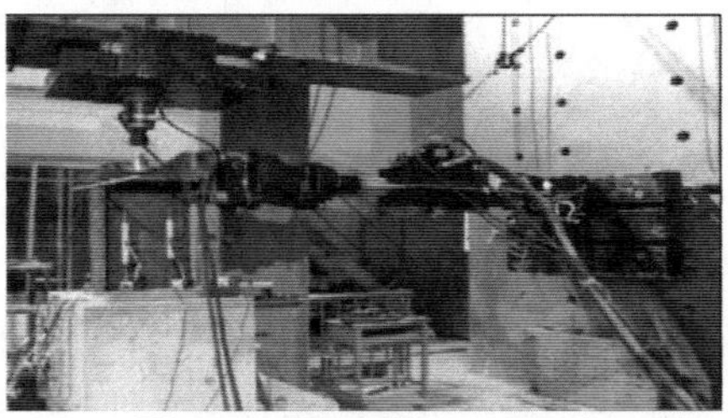

**Figure 5.** MTS connection picture

**Figure 6.** Displacement meter and the strain gage connection picture

The experimental damage is shown as figure 7 and figure 8:

**Figure 7.** The extension and deformation of anchor bolt yield

**Figure 8.** Foundation concrete is cracked and damaged

### *3.4 The comparison between the theoretical analysis value and the experimental results*

**Table 3.** The calculation parameter for the steel column foot without shear key

| Column foot code | The effective cross-sectional area of single anchor bolt Ae(mm) | Anchor bolt yield strength (N/mm) | a (mm) | b (mm) | c (mm) |
|---|---|---|---|---|---|
| GZ-1 | | | | | |
| GZ-2 | 448 | 312 | 225 | 100 | 125 |
| GZ-3 | | | | | |

**Table 4.** The comparison between the theoretical value of the column foot's shear bearing capacity calculated as per formula (1) with the experimental value

| Axial force N (kN) | Columa foot code | $\sigma$ (N/mm2) | $\tau$ (N/mm2) | $n_2 = 2$ Vmax (KN) Calculation value | $\bar{V}_{max}$ (KN) Experimental value | $\frac{V_{max}}{\bar{V}_{max}}$ |
|---|---|---|---|---|---|---|

| | | | | | | |
|---|---|---|---|---|---|---|
| 0 | GZ-1 | 292.72 | 62.34 | 90.86 | 110.11 | 0.83 |
| 166.13 | GZ-2 | 292.62 | 62.58 | 144.15 | 186.87 | 0.77 |
| 158.47 | GZ-3 | 292.58 | 62.64 | 141.66 | 166.32 | 0.85 |

The experimental data of the table shows: Under the conditions of the same section of the steel column and the same anchor bolt configuration, with the increase of N value, the shear bearing capacity of the steel column foot increases while the ratio of the shear bearing capacity of the anchor bolt in the overall shear bearing capacity of the whole steel column foot decreases. When the N value is: 0KN, 166.13KN and 158.47KN, the ratios of the shear bearing capacity of the anchor bolt in the overall shear bearing capacity of the whole steel column are 33%, 20% and 22% respectively. Even if the axial force reaches 166.13KN, the shear bearing capacity of the anchor bolt is still as high as 20%. Therefore, the shear bearing capacity of the anchor bolt is significant for the design.

**Table 5.** The calculation parameter for the steel column foot without shear key

| Columa foot code | The effective cross-sectional area of anchor bolt Ae(mm) | Anchor bolt yield strength fy（N/mm） | Concrete compressive strength fc（Mpa） | a （mm） | b （mm） | c （mm） |
|---|---|---|---|---|---|---|
| GZ-4 | | | | | | |
| GZ-5 | 422 | 288 | 42 | 225 | 100 | 125 |
| GZ-6 | | | | | | |

**Table 6.** The comparison between the theoretical value of the column foot's shear bearing capacity calculated as per formula (2) with the experimental value

| | | | | $n_2 = 2$ | | |
|---|---|---|---|---|---|---|
| Axial force N（kN） | Column foot code | $\sigma$ （N/mm2） | $\tau$ （N/mm2） | Vmax（KN） Calculated value | $\overline{V}_{max}$ （KN） experimental value | $\frac{V_{max}}{\overline{V}_{max}}$ |
| 0 | GZ-4 | 278.59 | 42.16 | 155.86 | 148.29 | 1.05 |
| 89.05 | GZ-5 | 278.18 | 43.05 | 182.08 | 179.38 | 1.02 |
| 131.87 | GZ-6 | 277.98 | 43.47 | 207.68 | 206.75 | 1.00 |

Table 4 and table 6 show: The shear bearing capacity of the steel column foot with shear key is larger than that of the steel column foot without shear key, but the increase is not significant.

### *3.5.Experimental analysis*

*3.5.1.The damage process of the steel column foot without shear key.*As for the three experimental models without shear key, they stay at the elastic stage at the early phase of the loading and the structural displacement and the anchor bolt strain present linear relationship. With the increase of the horizontal force, the GZ-1 column foot anchor bolt yields. The horizontal force at this moment is P=110.11kN. After that, with the further increase of the horizontal thrust, the anchor bolt generates plastic deformation and the residual deformation generates correspondingly. Since the steel material used by the column foot and the anchor bolt have sound plastic capacity, the column foot at last is damaged as the anchor bolt crushes the concrete. The damage mechanism of GZ-2 and GZ-3 is similar to that of GZ-1. The difference is as follows. Due to the function of vertical loading of GZ-2 and GZ-3, the shear force in the column foot would firstly be borne by the force of friction between the column baseboard with the foundation concrete. Besides, the function of the vertical loading increases the rotational stiffness of the steel column foot, the horizontal overturning displacement is relatively small.

When the GZ-2 anchor bolt yields, the horizontal force P=156.57kN; when the GZ-3 anchor bolt yields, its horizontal force P=166.32kN.

*3.5.2.The damage process of the steel column foot with shear key.*As for the three experimental models without shear key, they also stay at the elastic stage at the early phase of the loading and the structural displacement and the anchor bolt strain present linear relationship. As for GZ-4, as the shear key is set up, the shear force at the column foot is jointly borne by the shear key and the anchor bolt. With the further increase of the horizontal force, the anchor bolt of the column foot reaches the yield point. At this moment, the horizontal force is 148.29KN. The anchor bolt yields and the foundation concrete is crushed and damaged. The damage mechanism of GZ-5 and GZ-6 is similar to that of GZ-4. The difference is as follows. Due to the function of vertical loading of GZ-5 and GZ-6, the shear force in the column foot would firstly be borne by the force of friction between the column baseboard with the foundation concrete. Besides, the function of the vertical loading increases the rotational stiffness of the steel column foot, the horizontal overturning displacement is relatively small compared with the GZ-4. When the GZ-5 anchor bolt yields, its horizontal force P=179.38kN; when the GZ-6 anchor bolt yields, its horizontal force P=206.75kN.

*3.6.The location of the counterforce action point of the concrete foundation*

The location of the counterforce action point of the concrete foundation is complicated from theoretical perspective. In the eccentric compression, the counterforce action point location is related to the finite rigidity or the infinite rigidity. According to the research results on the location of counterforce action point of the concrete foundation both home and abroad, it is reasonable to regard the column foot baseboard as the infinite rigidly. However, regarding it as the finite rigidity does not have great differences with the results by regarding it as the infinite rigidity. When the column foot is the infinite rigidity, the compressive resistance figure of the concrete could be the triangle, rectangle and different parabolic waves. Among them, the quadratic parabola is the most advanced on in theory. When it adopts the quadratic parabola, the action point of the foundation compressive resistance joint force is very close to the edge of the biggest compression side of the baseboard of the column foot. The focus of the research in this paper lies in the shear bearing of the column foot. Although the action point of the counterforce is related to the tensile strength of the anchor bolt, it has minor influences on the shear bearing performance. Therefore, it is better to adopt simple and reliable method. Therefore, this research adopts: When the anchor bolt begins to bear the force, the action point of the counterforce is always in the edge of the compressive side of the column web; with the increase of the tensile borne by the anchor bolt, the action point of the counterforce moves outside gradually; when the anchor bolt yields, the action point of the counterforce is in the edge of the baseboard at the largest compressive force side. The above mentioned simplified value taking for the action force location of the counterforce is set for the calculation convenience in the research.

## 4.Conclusion

(1).The anchor bolt of the steel column foot has certain shear bearing capacity. It is suggested to consider the involvement of the anchor bolt in the shear bearing when calculating the shear bearing capacity of the exposed steel column foot.

(2).It is suggested to raise the separation point for the shear key in the exposed steel column foot; the experimental result shows that it is appropriate to calculate the shear bearing capacity of the steel column foot in accordance with the formula (1).

(3).The shear bearing capacity of the steel column foot with shear key is larger than that of the steel column foot without shear key, but the increase is not significant.

(4).As for the steel column foot which shall be equipped with the shear key, the experimental results show that it is appropriate to calculate its shear bearing capacity in accordance with the formula (2).

(5).Under the function of the horizontal shear force, the shear key might tear the concrete foundation. Therefore, it is suggested that the steel column foot with the shear key shall increase the stirrup reinforcement ratio of the foundation short column.

(6).The theoretical analysis and the experimental verification in this paper do not consider the influences of the factor of the beam bracing and column bracing. It is suggested to design the physical model for further study.

## References

[1] Ministry of Housing and Urban-Rural Development of China 2003 *Code for Design of Steel Structures (GB50017-2002) B.* China Architecture & Building Press, p85

[2] Yu Anlin et al. 1992 Research on Shear Behavior of Shaped Steel Columns (I) *J. Industrial Construction*, **5**, p 24-26

[3] Yu Anlin et al. 1992 Study on Shear Behavior of Shaped Steel Columns (II) *J. Industrial Construction*, **6**, p 29-33

[4] Tong Genshu et al. 2004 The Design on the Bearing Capacity of Single Anchor Bolt of the Steel Column Foot *J. Architectural Structure*, **34**, p 10-14

[5] Tong Genshu et al. 2005 The Designing Method for the Anchor Bolt of the Steel Column Foot *J. Progress of the Steel Structure of the Building*, **1**, p 27-36

[6] LI De-zi. 1983 Stress analysis and design of steel column foot anchor *J.Industrial Building*, **10**, p 30-36

[7] Li Weiping et al. 2007 The Theoretical and Experimental Research on the Shear Bearing Capacity of the Exposed Steel Column Foot *J .Steel Structure*. **1**, p45-48

[8] Ueda T, S Kitipornchai, Ling K, 1990 Experimental Investigation of Anchor Bolts under Shear *J. Journal of Structural Engineering*, **4**, p910-924

IOP Conf. Series: Earth and Environmental Science **61** (2017) 012146 doi:10.1088/1755-1315/61/1/012146

# Theoretical research on the shear bearing capacity of exposed steel column foot

**LIU Xijin**

Infrastructure and Planning Department, Beijing Jiaotong University, Beijing 100044, China

liuxijin2004@163.com

**Abstract.**This paper makes an overview on the domestic research on the steel column shear capacity in detail. On such basis, in accordance with the 4th theory of strength and the flexural vibration of straight bar, it makes theoretical analysis on the shear-bearing capacity of the steel column foot with the shear key and without the shear key respectively. It proposes the methods for calculating the shear-bearing capacity of steel column foot under two situations. It suggests that Code for Design of Steel Structures should raise the separation point of exposed steel column foot, which would reduce the construction difficulty of the exposed steel column foot, shorten the construction period, improve the construction quality and provide certain theoretical references for the design and research on the steel column foot.

## 1.Introduction

As parts of shear bearing are different, the exposed steel column foot could be divided into steel column foot with shear key and the steel column foot without shear key. The base shear of steel column foot without shear key is mainly borne by the force of friction and anchor bolt while the base shear of the steel column foot with shear key is mainly borne by the shear key, force of friction and anchor bolt. Code for Design of Steel Structures of China regulates: The anchor bolt of the steel column foot is inappropriate to bear the horizontal reacting force of the steel column foot. Such a horizontal reacting force shall be borne by the force of friction between the baseboard and the concrete (friction coefficient could be 0.4) or borne by the shear key [1]. European and American nations all consider make the anchor bolt be involved in the share-bearing. In terms of the value setting of the strength of extension, European and American nations set the strength designing value of anchor bolt as the corresponding designing value of common bolt or materials of bolt. Japan is the nation of frequent earthquakes and it has much research on the shear-bearing performance of steel column foot. In accordance with the Code for Design of Steel Structures, when the force of friction between the steel column foot with the foundation concrete could not resist the horizontal shear, the anchor bolt could be used for the shear bearing.

Besides, Code for Design of Steel Structures only illustrates how to set the shear key when the horizontal reaction force of the steel column foot is large, but does not give specific explanation on the regulation of shear key calculation.

This paper would make analysis from two aspects on the shear bearing capacity of exposed steel column foot: The shear bearing capacity of steel column foot without shear key and that of steel column foot with shear key; besides, it also suggests the new separation points for the steel column foot in setting the shear key, providing certain references for the staff of engineering designing and related researchers. By doing so, it expects to cancel the shear key in common situations, facilitate

IOP Conf. Series: Earth and Environmental Science **61** (2017) 012146 doi:10.1088/1755-1315/61/1/012146

construction, save labor force, physical resource and financial resources.

## 2.Theoretical research on the shear-bearing capacity of steel column foot without shear key

### *2.1.The damage mode of steel column foot without shear key*

The interactive mechanism between shearing force and tensile force of the anchor is quite complicated. At present, it has not been clarified completely. Research on the simultaneous working of flexural force and shearing force is insufficient. To study the tensile status and shear status of the anchor bolt, it analyzes the damage mode of the column foot firstly. When the strength and stability of plates of the column foot are guaranteed, the rigidity of the base board and the whole column foot is relatively large. The damage mode of the column foot mainly has the following several varieties:

(1) The tensile side anchor bolt yields; without minor sliding, the column foot rigid body rotation could develop continuously;

(2) The dimension for the obtrusive edge is over small, which leads to the cracking of foundation concrete under the pressure of base board;

(3) The pressure bearing strength of the concrete is insufficient. Under the pressure of R, the foundation suffers local pressure bearing damage;

Among the above mentioned 3 failure types, the first damage model is the ideal damage model and the other damage models are the concrete brittle failures which shall be avoided in the engineering design. For the other brittle damage models, it could avoid them by increasing the extra reinforced steel bars and extending the length of anchor bolt.

### *2.2.Analysis on the calculation method for the shear-bearing force of steel column foot without shear key*

The shear-bearing force of the steel column foot without shear key is composed of two parts, including: the force of friction between the base board of column foot with the foundation concrete surface, and the shear bearing force of the anchor force of column foot. In terms of the issue about whether the column foot anchor bolt could participate in the shear bearing or not, many overseas experts have done lots of theoretical analysis and experimental study. It introduces these calculation methods one by one.

*2.2.1.The calculation method proposed by Yu Anlin*[2] As the anchor bolt has sound plastic deformation capability, it shall guarantee the soundness of concrete during the designing process. For the steel column foot featuring appropriate damage model, the horizontal loading of the anchor bolt at its tensile side at the tensile moment is its shear bearing capacity $V_{max}$.Therefore, Yu Anlin et al. consider the beneficial influences of the phenomenon that the force of friction increases as the bending torque increase. It makes the experiment on the 12-frame test specimen for the low, medium and high-position anchor bolt column foot under the loading of horizontal reciprocating force, gets the anti-bending M-θ hysteretic curve and the anti-shear P-δ hysteretic curve and points out the computational formula of anti-bending bearing capacity:

$$M_y = T_y \left(d_t + d_c\right) + Nd_c \tag{1}$$

IFormula: dt — the distance from the anchor bolt at the tensile side to the axis of the column;dc— the distance from the foundation counterforce R to the axis of the column, it could take simplified value. The computational formula of shear bearing capacity:

$$V_{max} = \mu\left(T_y + N\right) + V_b = 0.4\left(T_y + N\right) + \beta n_c A_e \frac{f_y}{\sqrt{3}} \tag{2}$$

$$T_y = n_t A_e f_y \tag{3}$$

formula:N— axial force, the pressure force is positive and the tensile force is negative; $\left(T_y + N\right)$ — the value of this item is zero when the result is the negative value;$T_y$— total tensile force of the anchor bolt at the tensile side;$n_t$— quantity of anchor force at the tensile side;$n_e$— quantity of the anchor bolt at the tensile side;$A_e$— effective cross-sectional area of an anchor bolt;$f_y$— yield strength of the anchor bolt material;0.4— the friction coefficient between the column foot board with the concrete, take it as 0.4;β— adjustment coefficient, reflect the unevenly stressing of low-position anchor bolt and the equivalent conversion of medium and high-position anchor bolt, the low-position anchor bolt takesβ=0.65 and the medium and high-position anchor bolt takesβ=0.4;

The calculation method of professor Yu Anlin et al. tacitly proves that the anchor bolt yield at the tensile side and the anchor bolt yield at the pressure side incur at the same time. However, the simultaneous yield of the anchor bolt at the tensile side and at the pressure side is almost impossible. From the probability perspective, the anchor bolt at the tensile side also bears a half of the horizontal force. Besides, as the research is done under the status of complete elasticity and it declares the column foot failure with the anchor bolt failure at the tensile side, so it is not appropriate to take $f_y$ in the calculation method.

*2.2.2.The calculation method of the American ACI349-85* [(3)(4)] According to the American ACI349-85[3] [4], the anchor bolt designing depends on the anchor bolt yield failure and the designing value of the bearing capacity of anchor bolt is larger than or the same with the equivalent equal force under the common work of tensile force and shear force. That is:

$$CN_v + N_t \leqslant \Phi A_e f_y \tag{4}$$

formula: φ=0.9 is the reduction coefficient of the bearing capacity, C is the shearing coefficient which equals the reciprocal of the product between the friction coefficient with $\sqrt{3}$ ,the value is: When the top surface of the column foot baseboard stays in the same level with the foundation concrete surface, it takes C=1/0.9=1.11 (friction coefficient is 0.52); when the undersurface of the column foot board stays at the same level with the foundation concrete surface, it takes C=1/0.7=1.43 (the friction coefficient is 0.4); when there is the undercourse of cement mortar under the column foot baseboard, it takes C=1/0.55=1.82 (friction coefficient is 0.32). Formula (4) considers the influences of the relative coefficient between the column foot baseboard with the concrete surface on the friction coefficient. It is an appropriate factor, but its theoretical evidence is not strong. It is only an experience formula.

*2.2.3.The calculation formula proposed by professor Li Dezi* [5].Li Dezi takes the friction coefficient between the column foot baseboard with the concrete contact surface as 0.4. The horizontal loading which is larger than the fraction is borne by the anchor bolt. The calculation formula for the tensile-shear of anchor bolt is:

$$N_t \leqslant 1.25\left[N_t^b\right] - 1.6N_v \tag{5}$$

$$N_t \leqslant \left[N_t^b\right] \tag{6}$$

Formula: $\left[N_t^b\right]$— tensile capacity of single anchor bolt under the tensile force;Nt— the tensile force suffered by single anchor bolt under the force of tensile-shear;Nv— the shear force suffered by single anchor bolt under the force of tensile-shear;

The calculation method recommended by professor Li Dezi only considers the shear capacity of single anchor bolt and does not take the friction force of column foot and the overall shear capacity of the anchor bolt at the column foot. Besides, the formula (5) is simply the combination of shear and tensile of the anchor bolt, which is not appropriate according to the 4th theory of strength.

*2.3.The method proposed by this paper*
According to the 4th theory of strength, it is suggested that the shear bearing capacity of the steel column foot without shear key to be calculated as per the formula (7):

$$\begin{cases} V_0 = \mu N + \mu n_1 A_e \sigma + n_2 A_e \tau \\ \sqrt{\sigma^2 + 3\tau^2} \leq f_y \end{cases} \quad (7)$$

formula,$n_1$ is the quantity of the anchor bolt at the tensile side; $n_2$ is the quantity of anchor bolt participated in the shearing;σ is the positive stress of the single anchor bolt at the tensile side; τ is the shearing strength of the single anchor bolt at the tensile side;Ae is the effective cross-sectional area of single anchor bolt; fy is the yield strength of single anchor bolt;N is the axial compressive force;μ is the friction coefficient between the column foot baseboard with the concrete, taking the value as per the American ACI349-85.

When the anchor bolt is placed at the outer side of the steel column, it is shown as figure 1:

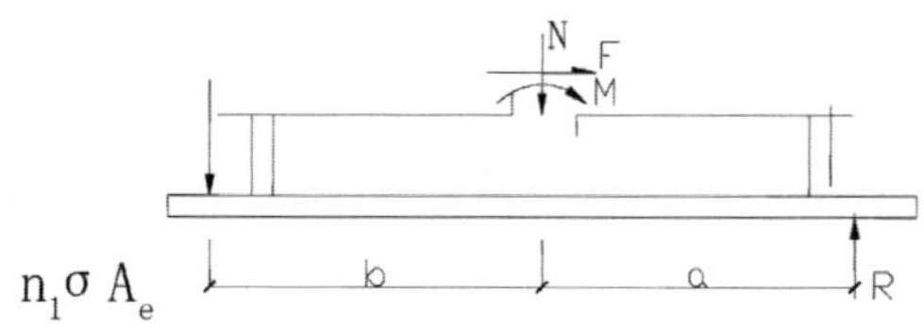


**Figure 1** Sketch for the anchor bolt staying at the outer side of the steel column

$$\sigma = \frac{M_y - Na}{(a+b)n_1 A_e} \quad (8)$$

When the anchor bolt stays at the inner side of the steel column, it assumes the column foot baseboard is completely rigid, which is shown as figure 2:

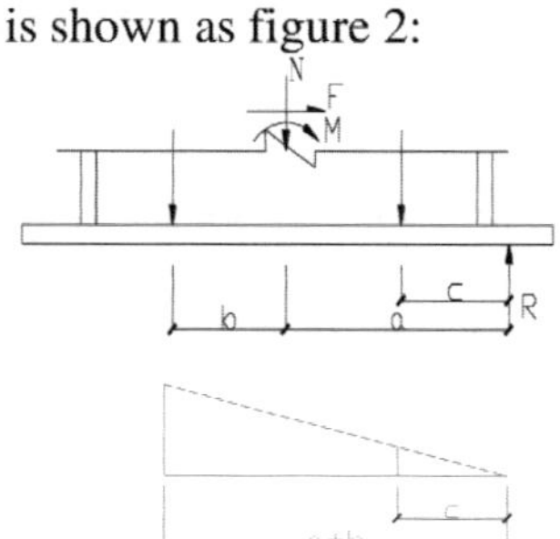


**Figure 2** The sketch when the anchor bolt stays at the inner side of the steel column

$$M_y = (a+b)\, n_1 \sigma A_e + n_3 \sigma' A_e c + Na \quad (9)$$

$$\sigma = \frac{M_y - Na}{(a+b)n_1 A_e + n_3 A_e c^2/(a+b)} \quad (10)$$

formula,$n_3$ is the quantity of anchor bolt under pressure;My is the bending moment of the anchor bolt yield under tensile.

Formula (1) is established in accordance with the tensile-shear yield of the anchor bolt at the tensile side, that is, the column foot declares to be damaged when the anchor bolt at the tensile side yields. As the bending moment suffered by the anchor bolt is relatively small, which contributes little to the anchor bolt yield, this formula neglects the bending of the anchor bolt at the tensile side. Besides, if it considers the influences of the unevenly stress of the anchor bolt at the tensile side and at the pressure side, it could calculate according to the following formula:

$$\begin{cases} V = \mu N + \beta \mu n_1 A_e \sigma + \beta n_2 A_e \tau \\ \sqrt{\sigma^2 + 3\tau^2} \leq f_y \end{cases} \quad (11)$$

formula,β is the reduction factor considering unevenly stress of the anchor bolt. It is suggested to take

0.75 in accordance with the literature [2] and [3] and other significances of symbols are the same with formula (7).

This paper suggests taking the calculation value of formula (11) as the separation point for the exposed column foot in setting the shear key. When the actual column foot shear force is larger than the calculated value in formula (11), it considers to set the shear key; otherwise, it would not set the shear key.

## 3.The Calculation Method for the Shear Bearing Capacity of the Column Foot with Shear Key

*3.1.Set up the damage model for the column foot with shear key*

In the regions where the wind loading is large or the earthquake fortification intensity is high, the horizontal force (shear force) at the column foot bottom is relatively large. The sum of the force of friction between the concrete with the baseboard and the shear bearing capacity of the anchor bolt could not offset the shear force at the column foot. Under such a situation, it needs to set up the shear key under the column foot baseboard. The common types for the shear key include cross steel plate and U-bar. There are reserved holes in the foundation column during the construction. When it installs the column, the shear key is located in the reserved hole. It should use the grouting material to fill the reserved hole and the screed coat. In this way, when the column foot is bearing the horizontal shear force, the shear key and the foundation concrete could work as an integral part. When the column foot is bearing the horizontal shear force, three damage models are possible, including: The damage of the shear key, the tensile failure of the welding seam between the shear key with the column foot baseboard and the foundation concrete is cracked to damage; as the compressive strength of the steel material is much larger than that othe concrete and the strength of the welding seam between the shear key with the column foot baseboard is also much larger than that of the concrete under the prerequisite of appropriate designing and construction, so the damage for the foundation concrete would occur firstly when the column foot bears the horizontal shear force. That is to say, it should take the damage of foundation concrete as the damage control condition when the column foot bears horizontal shear force.

*3.2.The analysis on the calculation method for the column foot shear bearing capacity of the shear key*

*3.2.1.The calculation method proposed by Yang Fengdi*[6]This method only considers the strength of the welding seam between the shear key and the column foot baseboard, failing to consider the strength of the shear key and the foundation concrete.

*3.3.The method proposed by this paper*

The shear bearing capacity of the column foot with shear key is mainly composed of three parts: The force of friction between the column foot baseboard with the top surface of the foundation concrete, the shear bearing capacity of the anchor bolt and the shear bearing capacity of the shear key. Considering the influences of the shear deformation and axial force on the shear key, it takes the shear key as the object for analysis. The working forces on it include the axial force N of the column transmited through the column body, the shear force at the bottom of the column foot and the shear force V' at the column foot baseboard, and the join force of the stress force of concrete F(x), which are shown as figure 3:

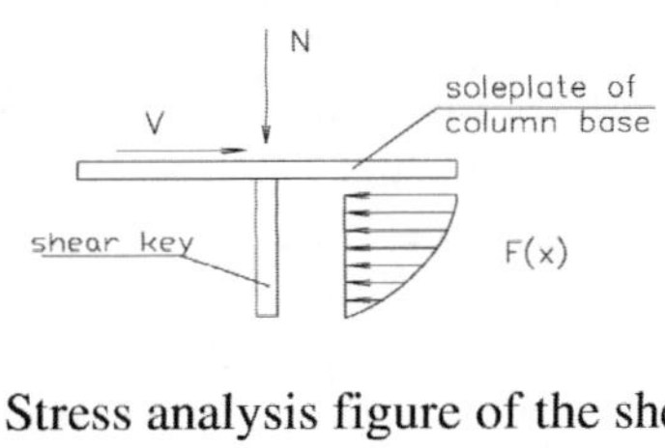


**Figure 3** Stress analysis figure of the shear key

It takes the micro-segment dx as the isolator, the stress force borne by the micro-segment is shown as figure 4.

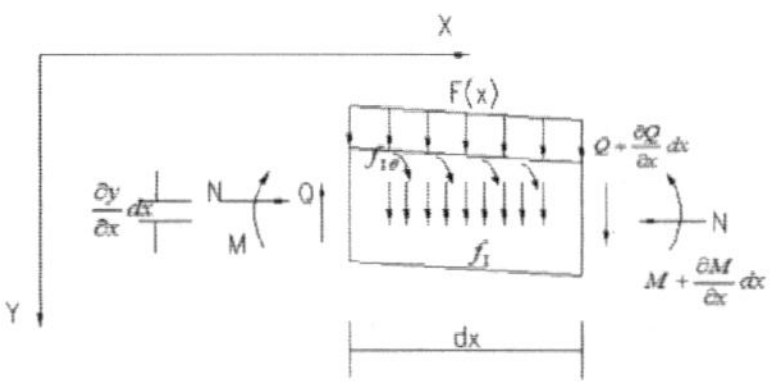


**Figure 4.** The force diagram of the micro-segment isolator

Under the extreme balance status, the join force of the compressive resistance of the shear key borne by the concrete shall offset the shear force of the column foot, which indicates that it should satisfy:

$$\int_0^h F(x)d(x) \geqslant V' \tag{12}$$

$$F(x)=f_c b\left(2\frac{\varepsilon}{\varepsilon_0}-\frac{\varepsilon^2}{\varepsilon_0^2}\right),$$

Among it,$f_c$ is the strength designing value of the concrete axial compressive resistance strength;b is the width of the shear key; is the strain of foundation concrete under the pressure of shear key;l is the calculation width of the foundation concrete along the forced direction; $\varepsilon_0$ is 0.002.

It establishes the equation through the bending vibration of the straight-bar and solve the above mentioned equation in accordance with the method of Galerkin.

The solution is as follows:

$$V' \le \frac{f_c b}{l\varepsilon_0}\left[\frac{4h}{\pi}\times\frac{12\pi^3h^2\left(\frac{N}{EI}+\frac{2f_cb}{kAG\varepsilon_0 l}\right)+48\pi h^4\frac{2f_cb}{EI\varepsilon_0 l}-3\pi^5}{16\pi^2h^2\frac{2f_cb}{kAG\varepsilon_0^2l^2}+128h^4\frac{f_cb}{EI\varepsilon_0^2l^2}}-\frac{h}{2l\varepsilon_0}\times\left(\frac{12\pi^3h^2\left(\frac{N}{EI}+\frac{2f_cb}{kAG\varepsilon_0 l}\right)+48\pi h^4\frac{2f_cb}{EI\varepsilon_0 l}-3\pi^5}{16\pi^2h^2\frac{2f_cb}{kAG\varepsilon_0^2l^2}+128h^4\frac{f_cb}{EI\varepsilon_0^2l^2}}\right)^2\right] \tag{13}$$

In the formula,N is the pressure of column top;EI is the flexural rigidity of the shear key;KA is the effective shearing area of the shear key;G is the shearing modulus; h is the buried depth of the shear key; the meaning of other symbols is the same with the above context.

The above formula is the exclusive shear bearing capacity provided by the shear key. If it includes the force of friction and the shear bearing capacity of the anchor bolt, the shear bearing capacity of the whole column foot should be:

$$V_{total}=V+V' \tag{14}$$

The secondary pouring concrete could not vibrate, which would generate certain influences on the strength of concrete. It suggests to reduce the in the formula(14) (reduction coefficient $\beta'$ is 0.6).

## 4.Conclusion

(1) The exposed column foot without shear key has certain shear bearing capacity and the shear bearing capacity of its anchor bolt is significant; this paper suggests considering the participation of the column foot anchor bolt in the shear bearing and raising the separation point for the shear key of the exposed column foot. That is, it should calculate whether the shear key should be set in accordance with formula (11).

(2) As for the light-steel plant column foot which shall be equipped with the shear key in accordance with the regulation, it is suggested to enhance the shear bearing capacity by increasing the diameter of the anchor bolt at the column foot. By doing so, it could avoid the shear key, reduce the construction difficulty and shorten the construction period.

(3)As for the column foot which shall be equipped with the shear key, the shear force at the column foot is borne jointly by three parts, including the force of friction between the baseboard with the foundation concrete, the anchor bolt of the column foot and the shear key. This paper suggests making the calculation as per formula (14).

(4) The prerequisite for the function of shear key is that the secondary grouting shall ensure the compactness. Only if the compactness of the secondary grouting is guaranteed, it could ensure the foundation concrete and the shear key work as an integral part and the shear key thus could play its due function.

(5) This paper makes detailed analysis on the shear bearing capacity of the exposed column foot from the theoretical perspective. It is suggested to verify the correctness of the theoretical analysis in this paper by establishing the experimental model.

## References

[1] Ministry of Housing and Urban-Rural Development of China 2003 *Code for Design of Steel Structures (GB50017-2002) B.* China Architecture & Building Press, p85

[2] Yu Anlin et al. 1992 Study on Shear Behavior of Shaped Steel Columns (II) *J. Industrial Construction*, **6**, p29-33

[3] Chipp. John G. and Edward R. Haninger, 1983 Design of Headed Anchor Bolts *J. Engineering Journal*, **2**, p57-59

[4] Dewolf John T, Sarisley E F . 1980 Column foot Plates with axial loads, & moments *J. Journal of the Structural Division*, **106**, p63-65

[5] Li Dezi. 1983 The Analysis and Design on the Strain of Column foot Anchor Bolt *Conf. Steel Structure Research Selections Volume*, **2**, p17-19

[6] Yang Fengdi et al. 2005 Analysis on the Design of Light-steel Structure *J. House Column foo*t, **5**, p12-14

IOP Conf. Series: Earth and Environmental Science **61** (2017) 012147 doi:10.1088/1755-1315/61/1/012147

# Biochar from Chinese herb residues as adsorbent for toxic metals removal

**Guocheng Liu[1], Yu Huang[2], Lina Xu[1,*]**

[1]Key Laboratory of Marine Environment and Ecology, Ministry of Education and College of Environmental Science and Engineering, Ocean University of China, Qingdao 266100, China

[2]Gansu Institute of Urban Planning and Design, Lanzhou 730000, China

[*]Corresponding author: xln1984@126.com

**Abstract.** Two biochars were prepared form Chinese herb residues by slow pyrolysis at 300 °C and 600 °C (CHR300 and CHR600) for removing two toxic metal ions ($Pb^{2+}$ and $Cd^{2+}$) from aqueous phase. In this study, both $Pb^{2+}$ and $Cd^{2+}$ were effectively immobilized by CHR300 and CHR600 from water. For an initial concentration of $Pb^{2+}$ and $Cd^{2+}$ ($C_0$ = 10 mg/L), the removal rate by CHR300 and CHR600 were all greater than 90.0% at a solid:liquid ratio of 50 mg biochar in 10 mL solution. For $C_0$ of the two toxic metals was 100 mg/L, the $Pb^{2+}$ removal by CHR600 was significantly stronger than that by CHR300, but there was no significant difference of the removal rate of $Cd^{2+}$ between CHR300 and CHR600. Moreover, the removal rate of $Pb^{2+}$ by CHR300 and CHR600 was both markedly greater than that of $Cd^{2+}$, indicating that the biochars had stronger adsorption favorite for $Pb^{2+}$ than $Cd^{2+}$. The SEM-EDX data of the biochars after the toxic metals sorption drew strong evidences on the $Pb^{2+}$ and $Cd^{2+}$ immobilization by CHR300 and CHR600. The existing of phosphorus (P) and sulphur (S) in CHR300 and CHR600 implied that the heavy metals might be removed by forming Pb/Cd-P and Pb/Cd-S precipitates. These results suggested that the biochars from Chinese herb residues would be likely to be good adsorbents for $Pb^{2+}$ and $Cd^{2+}$ removal in water.

## 1. Introduction

Biochar, as an environmental-friendly and economic material, is developed for reducing or immobilizing contaminants, such as PAHs (e.g., phenanthrene) [1], herbicides and pesticides (e.g., diuron, atrazine, chlorpyrifos and carbofuran) [2], antibiotics (e.g., sulfamethoxazole) [3], and heavy metals (e.g., Pb, Cd, Cu and Zn) [4] in waters or soils. It is well known that the surface oxygen-containing functional groups (e.g., -OH, -COOH and -R-OH), ash components and π electrons (C=C) of carbon fraction of biochar are responsible for the sorption of toxic metals via ion exchange and/or complextion, precipitating action and cation-π interaction. Hence, the biochar production will have a potential market as an engineered sorbent or ameliorant to remove or immobilize some given contaminants in the polluted waters and soils.

In China, the production of Chinese herb residues originated from pharmacy industry is more than 12 million tons every year [5]. The careless disposal or stacking of Chinese herb residues with high moisture content can lead to serious water and soil pollution because that their leachates into environment by runoff and percolating remained a mass of active pharmaceutical ingredients. To minimize the environmental impacts, the production of biochars from this low-value material may be turned into a clean and efficient product for utilizing and reducing environmental and health impacts.

In addition, the biochars prepared from Chinese herb residues have relatively better cost advantage than other biomass (e.g., agricultural straws and woods) due to the easier-collecting and lower-value.

Overall, the slow pyrolysis technology was used to prepare biochars with Chinese herb residues as biomass feedstock to investigate their removal ability of the given heavy metals ($Pb^{2+}$ and $Cd^{2+}$) from waters. Scanning electron microscopy (SEM) equipped with an electron dispersive X-ray analysis (EDX) was employed to observe surface structure of the biochar particles and qualitatively examine the elemental composition of biochar samples before and after $Pb^{2+}$ and $Cd^{2+}$ sorption. This work will provide a high-efficiency biochar adsorbent for $Pb^{2+}$ and $Cd^{2+}$ removal.

## 2. Materials and methods

### *2.1 Materials*

Chinese herb residues (CHRs), the mixture of angelica sinensis, gastrodia elata and other herbs after boiling with drinking water and pouring extract liquid, were collected from some dispensaries of traditional chinese medicine in Qingdao, Shandong province, China. CHRs were dried at 105 °C for 48 h at least, and then smashed into powders (passing through 0.25 mm sieve). 50-g CHRs powders were weighed into a quartz ark and pyrolyzed at 300 °C and 600 °C in a $N_2$ environment for 3 h into biochars [6]. The biochar samples from CHRs were severally grinded and passed through 0.125 mm sieve and stored in a vacuum dryer before the following adsorption experiments, and are hereafter named as CHR300 and CHR600, respectively.

### *2.2 Sorption experiment*

$Pb^{2+}$ and $Cd^{2+}$ stock solution (1 g/L) were prepared using guaranteed reagent $Pb(NO_3)_2$ and $Cd(NO_3)_2$, and were diluted to 10 and 100 mg/L, respectively. The toxic metal solutions contained 0.01 M $NaNO_3$ for maintaining ionic strength, and the solution pHs were adjusted to 5.0 ± 0.05 by adding 0.1 M $HNO_3$ solution for preventing the formation of precipitation. 50-mg CHR300 or CHR600 were added into 10 mL of the desired concentration of $Pb^{2+}$ and $Cd^{2+}$ solutions in 15-mL vials. All the vials were shaken at 150 rpm at 25 °C for over 48 h to reach sorption equilibrium [7], and centrifugated at 3800 rpm for 15 min and filtered through a 0.22 μm filter. The $Pb^{2+}$ and $Cd^{2+}$ concentration in the filtrates were analyzed by a flame atomic absorption spectrometer (FAAS, M6, Termo Elemental, USA) at 283.3 nm and 228.8 nm, respectively. All samples were run along with the blanks without the biochars or toxic metal ions by triplication. The removal rate of the toxic metals can be calculated using an equation of $(C_0\text{-}C_e)/C_0\times100\%$, where $C_0$ and $C_e$ is the initial and equilibrium aqueous concentration of $Pb^{2+}$ or $Cd^{2+}$, respectively.

The biochars after $Pb^{2+}$ or $Cd^{2+}$ removal ($C_0$ = 100 mg/L) were collected and washed by excess deionized water, and then dried at 80 °C for 48 h. The surface structure of $Pb^{2+}$ or $Cd^{2+}$-loaded biochars was observed by scanning electron microscopy (SEM, S4800, Hitachi, Japan), and the equipped electron dispersive X-ray analysis (EDX, 7593-H, Horiba, Japan) was employed to determine the sorbed $Pb^{2+}$ or $Cd^{2+}$ onto the biochars and other elemental composition.

### *2.3 Data analysis*

Microsoft office excel (2013) was used for data analysis. Significant difference was tested by Statistical Product and Service Solutions (SPSS 20.0, LSD, $P < 0.05$).

## 3. Results and discussion

### *3.1 $Pb^{2+}$ and $Cd^{2+}$ removal by biochars*

The removal rate of $Pb^{2+}$ and $Cd^{2+}$ by CHR300 and CHR600 is presented in Fig. 1. For $C_0$ of $Pb^{2+}$ and $Cd^{2+}$ is 10 mg/L, the removal rate by the two biochars was greater than 90.0%, suggesting that CHR300 and CHR600 could effectively remove the heavy metals from water. The removal rate of $Pb^{2+}$ by CHR600 was 97.5% at $C_0$ of 100 mg/L, significantly greater than that by CHR300 (88.6%), implying that the higher sorption affinity of $Pb^{2+}$ onto the biochar produced at high temperature. It should be noted that the adsorption of $Pb^{2+}$ onto the two biochars was both stronger than that of $Cd^{2+}$ at $C_0$ of 100 mg/L of the metal ions, which can well agree with the first stability constant of their associated metal hydroxide (log $K_1$ = 7.82 for $Pb(OH)^+$ and 4.17 for $Cd(OH)^+$) and metal acetate (log $K_1$ = 2.52 for $Pb(Ac)^+$ and 1.5 for $Cd(Ac)^+$). The same affinity order of $Pb^{2+}$ and $Cd^{2+}$ was detected for the oxidized-CNTs [8] and graphene oxide (GO) [9]. The surface oxygen-containing functional groups in these carbonaceous material plays a significant role in the adsorption process of the metal ions by an electrostatic attraction or complexation. In addition, the difference of sorption affinity of CHR300 and CHR600 for $Pb^{2+}$ and $Cd^{2+}$ may be related with the solubility product ($K$sp) of formed precipitation in the sorption of the toxic metals, such as some associated metal chloride, carbonate and phosphate, which was identified as a dominant mechanism for removing the metal ions from solution [10].

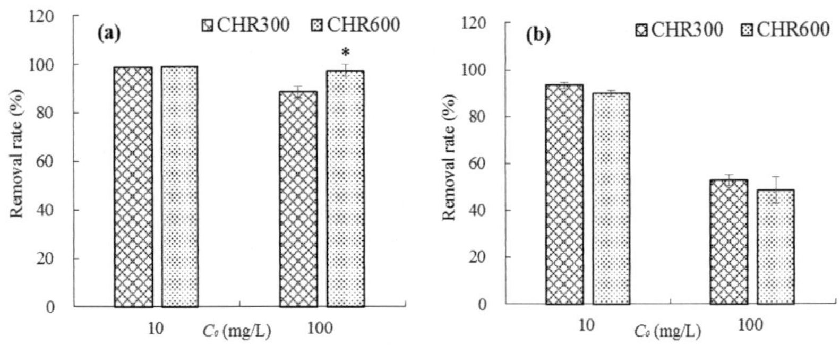


Fig. 1. The removal rate of (a) $Pb^{2+}$ and (b) $Cd^{2+}$ by the biochars in aqueous solution ($C_0$ = 10 or 100 mg/L, pH = 5.0, T = 25 °C, t = 48 h). CHR300 and CHR600 are the biochars from Chinese herb residues at 300 °C and 600 °C, respectively. For a given initial concentration of the toxic metal ions, significant difference about the removal rate between CHR300 and CHR600 was marked with "*" ($P$ < 0.05, n = 3).

*3.2 SEM-EDX of biochars after $Pb^{2+}$ and $Cd^{2+}$ sorption*

The SEM images of CHR300 and CHR600 after the adsorption of $Pb^{2+}$ and $Cd^{2+}$ ($C_0$ = 100 mg/L, pH = 5.0) are shown in Fig. 2a-d. As reported similarly in numerous previous studies [2, 4], CHR600, as a high-temperature biochar, has abundant pore structures, but little pore was formed during the charring process of CHRs into biochar at 300 °C.

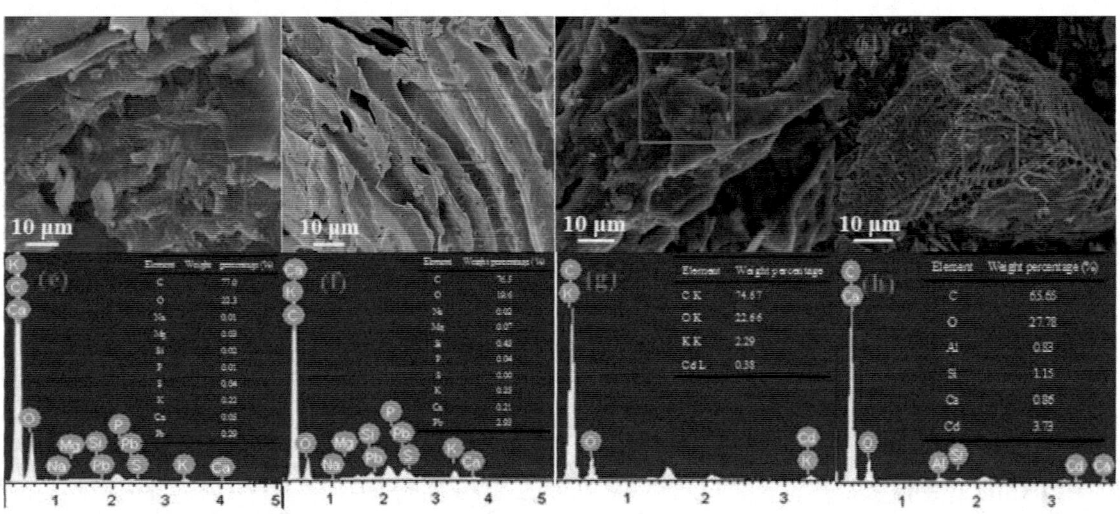


Fig. 2. SEM images for the biochars after the toxic metal ions adsorption: (a) $Pb^{2+}$ on CHR300; (b) $Pb^{2+}$ on CHR600; (c) $Cd^{2+}$ on CHR300; and (d) $Cd^{2+}$ on CHR600. (e-h) EDX spectra was collected from the labeled regions in the panels (a-d), respectively.

IOP Conf. Series: Earth and Environmental Science **61** (2017) 012147 doi:10.1088/1755-1315/61/1/012147

The EDX data of the $Pb^{2+}$- and $Cd^{2+}$-loaded biochars confirmed that both $Pb^{2+}$ and $Cd^{2+}$ could be removed by CHR300 and CHR600 from aqueous environment. But, the quantitative information about the removal of $Pb^{2+}$ and $Cd^{2+}$ by the biochars can not be accurately provided due to the outofflatness and unevenness of the surface of biochars, which have great influence on the EDX analysis. Moreover, the tested biochar particle for EDX analysis was randomly selected. The EDX analysis found that CHR300 and CHR600 contained the element of phosphorus (P) and sulfur (S), suggesting that the precipitation reaction between $PO_4^{3-}$ and $SO_4^{2-}$ released from the P- and S-minerals and the toxic metals may be an important mechanism of $Pb^{2+}$ and $Cd^{2+}$ removal by the biochars.

## 4. Conclusions

Both $Pb^{2+}$ and $Cd^{2+}$ can be effectively removed by CHR300 and CHR600 from water. For $Pb^{2+}$, CHR600 showed greater adsorption capacity than CHR300, but the $Cd^{2+}$ removal of CHR300 and CHR600 did not have significant difference. The higher removal of $Pb^{2+}$ by CHR300 and CHR600 than $Cd^{2+}$ suggeated that the stronger sorption favorite for $Pb^{2+}$ than $Cd^{2+}$ onto the CHRs-derived biochars. The SEM-EDX data confirmed the sorption of two testing toxic metals onto CHR300 and CHR600, and the P- and S-contained minerals might release some anions (i.e., $PO_4^{3-}$ and $SO_4^{2-}$), and thus result in the metals removal by the biochars through forming Pb/Cd-P and Pb/Cd-S precipitates.

## Acknowledgements

This work was financially supported by Open Foundation of Shandong Provincial Key Laboratory of Marine Ecology and Environment and Disaster Prevention and Mitigation (201511, 201512), Natural Science Foundation of China (41406085, 41503093), and Postdoctoral Applied Research Program in Qindao (861605040095).

## References

[1] Sun K, Ro K, Guo M, Novak J, Mashayekhi H and Xing B 2011 *Bioresour. Technol.* **102** 5757-5763
[2] Beesley L, Moreno-Jimenez E, Gomez-Eyles J L, Harris E, Robinson B and Sizmur T 2011 *Environ. Pollut.* **159** 3269-3282
[3] Mohan D, Sarswat A, Ok Y S and Pittman Jr. C U 2014 *Bioresour. Technol.* **160** 191-202
[4] Zheng H, Wang Z, Zhao J, Herbert S and Xing B 2013 *Environ. Pollut.* **181** 60-67
[5] Guo F, Dong Y, Dong L and Jing Y 2013 *Waste. Manag.* **33** 825-832
[6] Zheng H, Wang Z, Deng X, Zhao J, Luo Y, Novak J, Herbert S and Xing B 2013 *Bioresour. Technol.* **130** 463-471
[7] Wang Z, Liu G, Xing M, Li F and Zheng H 2014 *Environ. Sci.* **35** 4735-4744 (In Chinese)
[8] Gao Z, Bandosz T J, Zhao Z, Han M and Qiu J 2009 *J. Hazard. Mater.* **167** 357-365
[9] Sitko R, Turek E, Zawisza B, Malicka E, Talik R, Heimann J, Gagor A, Feist B and Wrzalik R 2013 *Dalton T.* **42** 5682-5689
[10] Lu H, Zhang W, Yang Y, Huang X, Wang S and Qiu R 2012 *Water Res.* **46**, 854-862

IOP Conf. Series: Earth and Environmental Science **61** (2017) 012148 doi:10.1088/1755-1315/61/1/012148

# Effects of polystyrene microplastics on the fitness of earthworms in an agricultural soil

**Dongdong Cao, Xiao Wang, Xianxiang Luo, Guocheng Liu and Hao Zheng***

Key Laboratory of Marine Environment and Ecology, Ministry of Education and College of Environmental Science and Engineering, Ocean University of China, Qingdao 266100, China

E-mail address: zhenghao2013@ouc.edu.cn

**Abstract.** Microplastics (MPs) pollution is widespread in the environment, while the effects of MPs on the soil organisms are poorly understood. In this study, we investigated the fitness of earthworms (*E. Foetida*) exposed to MPs (Polystyrene, 58 μm) in soils at the concentrations of 0, 0.25, 0.5, 1 and 2% (w/w). The results showed that MPs had little effects on the fitness of earthworms under low exposure concentrations ($\leq$ 0.5 % (w/w)), while MPs exposure with high concentrations (i.e., 1% and 2%) significantly inhibited the growth and increased the mortality of earthworms. The results indicated that the MPs pollution in soils have an adverse effect on the fitness of soil organisms, and implied the ecological risk of MPs in terrestrial ecosystems.

## 1. Introduction

The global production of plastics is around 280 million tons annually and large amount of plastic wastes accumulate in the marine and terrestrial ecosystem [1]. Large plastic items in waters and soils gradually become smaller pieces or particles through various environmental weathering processes, such as mechanical breakdown, microbiological decomposition and photodegradation [2]. When the plastic particles are less than 5 mm in diameter, they are defined as "microplastics (MPs)" by the National Oceanic and Atmospheric Administration (NOAA) [3].

MPs have adverse effects on organisms, mainly due to their great accumulation in the gut or stomach of the organisms [4], which further affect the feeding behavior and development of the organisms [5,6]. However, intense studies have been focused on the effects of MPs on the fitness of aquatic organisms in the marine ecosystem, and little is known about the effects of MPs on soil organisms. It has been reported that the concentration of MPs ranged from 1000 to 4000 particles per kilogram soil in European agricultural soils [7]. Therefore, it is essential to investigate the effects of MPs on soil organisms at a range of concentrations for ecological risk assessments. The objective of this study was to investigate the potential effects of MPs on the survival and growth of earthworms at different MPs exposure concentrations in soils.

## 2. Materials and methods

### *2.1. Earthworms, MPs and soil*

Earthworms (*E. Foetida)* were obtained from an earthworm farm, located in Jinan, Shandong province, China. The obtained earthworms were acclimated in a culture system (SPX-358, Jiangnan, Ningbo) with a constant temperature of 24 ± 1 °C for 7 days. Then, the healthy earthworms with similar weight (0.32 ± 0.02g) were selected and fasted for another 3 days before exposure. The polystyrene microspheres (PS-MPs, 58μm) was purchased from Sigma-Aldrich (USA). The soil was collected from an orchard in Jinling Town, Shandong Province, China. The soil was air dried at room temperature (20°C) and thoroughly sieved through 2mm for the exposure experiment. The sea sand was collected from Shilaoren beaches (Qingdao, China). Following the removal of visible gravel, the sea sand was washed with hydrochloric acid several times until its pH reached around 7. Afterwards, PS-MPs was thoroughly incorporated into the mixture of the air-dried sea sand and soil (40% sea sand, and 60% soil) at the rate of 0.25, 0.5, 1 and 2% (w/w), respectively.

### *2.2. Exposure experiment*

For PS-MPs exposure, 10 healthy pre-cultured earthworms were randomly selected, and then were moved into every 250-ml glass beaker which already contained PS-MPs (0, 0.25, 0.5, 1 or 2% (w/w)) in the mixed soil (500g) in the culture system under dark. During the exposure, 5g cow dung was uniformly placed on the surface of mixed soils as food in each beaker every 3 days. The moisture of soil mixture was kept at 20% during the experiment. After exposure for 30 days, the earthworms were rinsed for four times with ultrapure water to remove the adsorbed particles (e.g., soil and PS-MPs), dried with filter paper and then weighted by analytical balance to get the growth parameters of earthworms. The mortality of earthworms was assessed by counting the death adults in the beakers. Each treatment was replicated five times.

### *2.3. Statistical analysis*

All results were expressed as the mean values ($n$ = 5). Error bars presented in the results represent the standard deviation. Significant differences between the treatments were analyzed using one-way analysis of variance (ANOVA) with Duncan's multiple range test ($P = 0.05$) using Statistical Product and Service Solutions Software 20.0 (SPSS 20.0).

## 3. Results and discussion

The morphology the PS-MPs particles, which were observed using a scanning electronic microscope (SEM) (S-4800, Hitachi, Japan) is shown in figure 1. Imagery of the PS-MPs used in this study showed that they were flaky in nature and the major size of them was between 50 μm and 80 μm (figure 1).

The effects of PS-MPs on the weight and mortality of earthworms are shown in figure 2. PS-MPs exposure at lower rates (i.e., 0%, 0.25% and 0.5%) had no obvious effects on the weight of earthworms, while PS-MPs exposure at higher rates (i.e., 1% and 2%) caused a significant inhibition effect on the growth of earthworms ($P < 0.05$, figure 2a), which resulted in 27.6% and 29.8% decreases in the weight of earthworms in the 1% and 2% treatment, respectively, compared to the 0% treatment (figure 2a). Similarly, the PS-MPs exposure elevated the mortality of earthworms with the increasing exposure concentration (except for the 1% treatment) (figure 2b). Especially in the 2% treatment, around 40 % of earthworms were dead due to the PS-MPs exposure, implying that PS-MPs could have a lethal effect on earthworms. These results could be attributed to the serious damage in the self-defense system of earthworms caused by the PS-MPs exposure at the high concentrations (≥ 0.5%) [8].

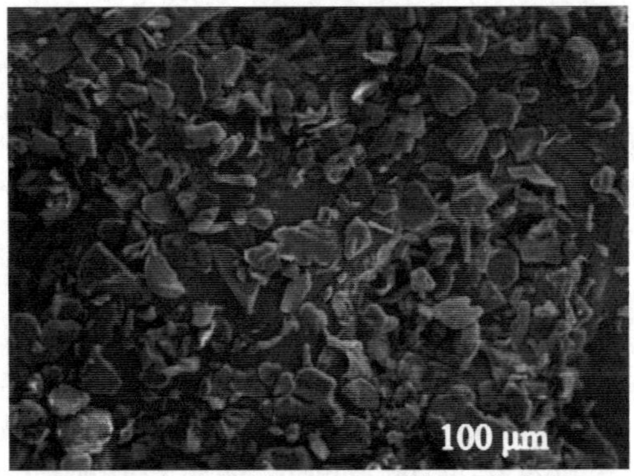


**Figure 1.** SEM images of the PS-MPs.

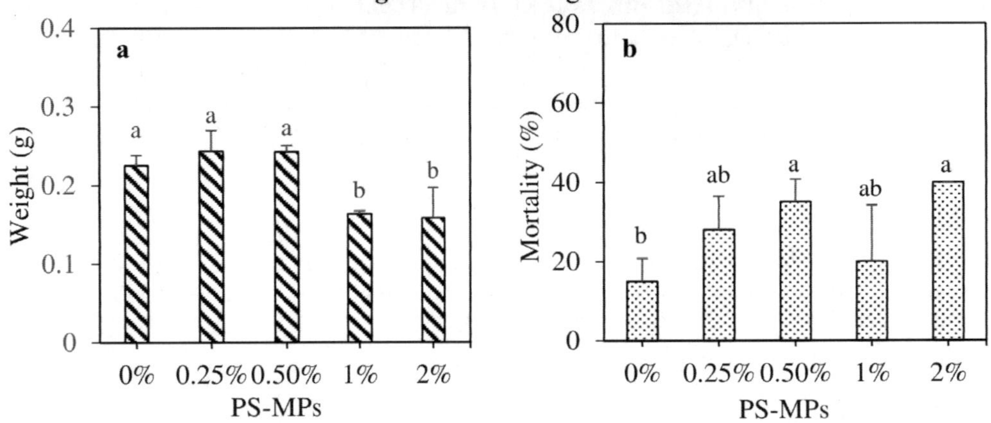


**Figure 2.** The weight (**a**) and mortality (**b**) of earthworms (*E. Foetida*) with PS-MPs concentration of 0%, 0.25%, 0.50%, 1% and 2% (w/w) under a 30-day exposure. The bars represent standard deviation of the means ($n$ = 5). Different letters among different treatments indicate significant differences, which were analyzed by Duncan's test ($P = 0.05$) using SPSS 20.0.

## 4. Conclusions

The results suggested that PS-MPs could inhibit the growth of earthworms and had an obvious lethal effect on earthworms at the high exposure concentrations (≥ 0.5%), which were possibly explained by the damage in the self-defense system of earthworms. This study will support the evidences that the MPs pollution in soils could pose a great adverse effect on the fitness of earthworms.

## Acknowledgments

This study was supported by National Natural Science Foundation of China (41325013, 41406085, 41503093) and Open Foundation of Shandong Provincial Key Laboratory of Marine Ecology and Environment and Disaster Prevention and Mitigation (201511, 201512).

## References

[1] Rillig M C 2012 *Environ. Sci. Technol.* **46** 6453-54
[2] Watts A J R, Lewis C, Goodhead R M, Beckett S J, Moger J, Tyler C R and Galloway T S 2014 *Environ. Sci. Technol.* **48** 8823-30
[3] Wright S L, Thompson R C and Galloway T S 2013 *Environ. Pollut.* **178** 483-492
[4] Bergami E, Bocci E, Vannuccini M L, Monopoli M, Salvati A, Dawson K A and Corsi I 2016 *Ecotoxicol. Environ. Saf.* **123** 18-25
[5] Watts A J, Urbina M A, Goodhead R, Moger J, Lewis C and Galloway T S 2016 *Environ. Sci. Technol.* **50** 5364-69.
[6] Von Moos N, Burkhardt-Holm P and Köhler A 2012 *Environ. Sci. Technol.* **46** 11327-35

[7] Barnes D K A, Galgani F, Thompson R C and Barlaz M 2009 *Philos. Trans. R. Soc., B* **364** 1985−98

[8] Huerta Lwanga E, Gertsen H, Gooren H, Peters P, Salánki T, van der Ploeg M, Besseling E, Koelmans A and Geissen V 2016 *Environ. Sci. Technol.* **50** 2685-91

IOP Conf. Series: Earth and Environmental Science **61** (2017) 012149 doi:10.1088/1755-1315/61/1/012149

# Effect of adding biochar with wood vinegar on the growth of cucumber

**Xin Pan, Yipeng Zhang, Xiao Wang, and Guocheng Liu***

Key Laboratory of Marine Environment and Ecology, Ministry of Education and College of Environmental Science and Engineering, Ocean University of China, Qingdao 266100, China
E-mail address: iceliugc@126.com

**Abstract**. The chemical fertilizers are abused to improve crop yields, which cause lots of soil problems (e.g., soil compaction and native nutrient loss). We thus investigated the potential of the combination of biochar and wood vinegar as a new type of fertilizer to replace the traditional fertilizer. The results in this study showed that the combined addition of biochar with wood vinegar had the greatest promotion effect on the plant growth of cucumber. Compared to the control treatment, the biochar addition with wood vinegar significantly increased the plant height, root length, root volume and root tips by 29.7%, 117%, 121% and 76.1%, respectively. These positive effects could be attributed to the benefits of the addition of biochar with wood vinegar treatment on improving soil fertility, increasing nutrient supply, and further stimulating plant growth. Overall, the combination of biochar and wood vinegar could be a promising fertilizer to promote plants growth and enhance crop yields.

## 1. Introduction

In the past decades, the demand of global food increased dramatically due to the rapid increased population. With the increasing environment pollution and climate issues, the development of China 's agricultural is facing a huge crisis. To solve this problem, chemical fertilizers are abused to improve crop yield, resulting in a series of soil environmental problems, such as soil compaction and the loss of soil nutrients [1]. Biochar acting as a promising soil amendment, which is produced from the pyrolysis of waste biomass under a limited or no oxygen condition [2], can improve soil physico-chemical properties, promote plant growth and ameliorate soil pollution [3]. Jeffery et al. reported that biochar could promote plant growth in soils by increasing soil organic carbon (SOC) content and water holding capacity [4]. However, the addition of biochar not always causes consistent yield increase. Vaccari et al. reported that the biochar addition increased the availability of $NH_4^+$, P and K, but not improved the yield of tomato [5]. Additionally, wood vinegar, a liquid by-product that obtained from the condensed vapors generated during the biomass pyrolysis, had been reported to enhance the crop yields and the nutrient uptake by plants followed by its application into agricultural soils [6]. And Meanwhile, there is limited study on the combine effects of biochar and wood vinegar on plant growth and the corresponding mechanisms are still unclear. Thereby, the major objective of this study is to investigate the effects of the biochar addition with or without wood vinegar into the agricultural soils on the growth of cucumber.

## 2. Materials and methods

### *2.1 The preparation of samples*

The surface soil (0-20cm) used in this study was collected from a farmland, located Qingdao, Shandong province, China. The field-moist soil samples were thoroughly mixed, gently passed through a 2-mm sieve, and air dried for 14 days. The poplar (*Populus trichocarpa*) was pyrolyzed at 500 °C for 5 h to produce the woody biochar and the wood vinegar (the liquid by-product collected from the cooling systems after pyrolysis process). All the biochar samples were ground to pass through a 0.5-mm sieve using a pulverizer (GM200, Retsch, Germany) for further analyses and pot experiment. The wood vinegar was standed for 3 months under darkness and then the upper component was collected for further analyses and pot experiment.

### *2.2 Characterization of biochar*

FTIR analysis was carried out to characterize the organic functional groups in the biochar sample. The biochar powders were mixed with spectrum pure KBr at ratio of 1:100 (w/w) fully, and the spectra of mixture were recorded in a 4000-400 $cm^{-1}$ region by a fourier transform infrared spectrometer (FTIR, GX, PerkinElmer, USA) . The surface morphology of the biocahr was observed using a scanning electronic microscope (SEM,S4800, Hitachi, Japan).

### *2.3 Experimental design*

Cucumber (*Cucumis sativus L*) were cultivated in a glasshouse, where the temperature was kept at 24 °C during the incubation. Afterwards, 75-g air-dried soil was packed into a nursery box for a 30 days of pot experiment. The prepared biochar was incorporated into the soils at a rate of 1.5% (w/w), hereafter referred to as the BC treatment. The wood vinegar used as a leaf fertilizer was fertilized on the cucumber at the rate of 0.02% (v/v), hereafter referred to as the WV treatment. The combined application of biochar with wood vinegar into the soils and the plant leaves at the rate of 1.5% (w/w) and 0.02% (v/v), respectively, which was referred to as the BCW treatment. The deionized water treatment was used as the control check (CK) and all the tests were performed in four replicates. After the 30 days of incubation, the root parameters including root length, root volume, root surface area and root tip number were measured by a root system scanner (Expression 1680, Epson, USA) and analyze with a WinRHIZO Pro 2005b (Canada).

### *2.4 Statistical analysis*

All the data in this study were expressed as mean values. Error bars presented in the results represent the standard deviation. Significant differences among the treatments were analyzed using one-way analysis of variance (ANOVA) with Duncan's multiple range test ($P = 0.05$) using Statistical Product and Service Solutions Software 20.0 (SPSS 20.0).

## 3. Result and discussion

### *3.1 Characterization of biochar*

Abundant adsorption peaks were found in biochar (Figure 1a), including -OH at 3400 $cm^{-1}$, aliphatic -$CH_2$- at 2950 $cm^{-1}$, 2882 $cm^{-1}$ and 1421 $cm^{-1}$, C-O-C stretching vibrations at 1050 $cm^{-1}$, ester C=O at 1746$cm^{-1}$, aromatic C=O at 1640 $cm^{-1}$ (C=O of quinones or conjugated ketones) and out-of-plane deformations of C-H at 796 $cm^{-1}$. The results indicated that the poplar-derived biochar contained abundant functional groups. Moreover, large amounts of pores can be observed in the biochar by the SEM micrograph (Figure 1b). Therefore, the biochar acting as a promising soil amendment, will potentially improve the physico-chemical properties of the soils, such as enhancing the porosity and cation exchange content [7].

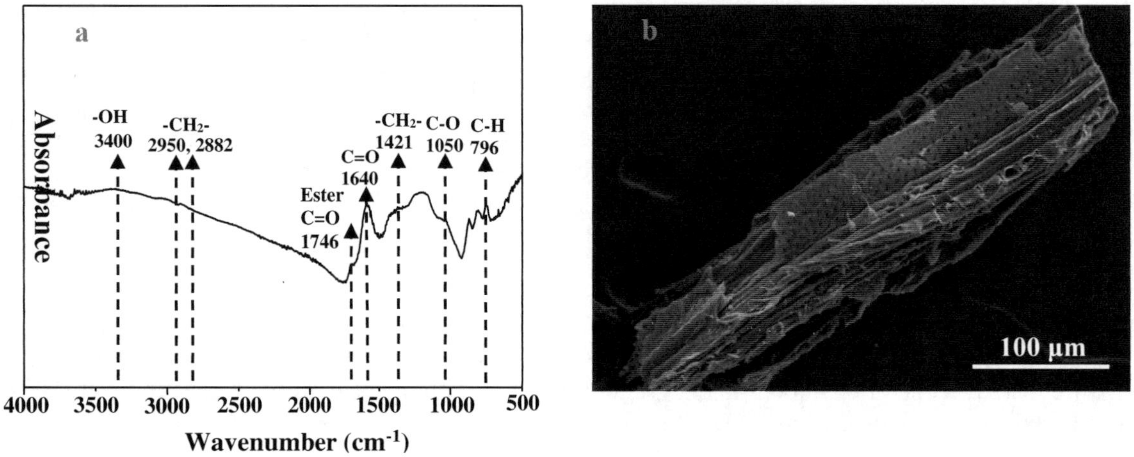


**Figure 1.** FTIR spectra (**a**) and SEM image (**b**) of the biochar from poplar at 500 °C.

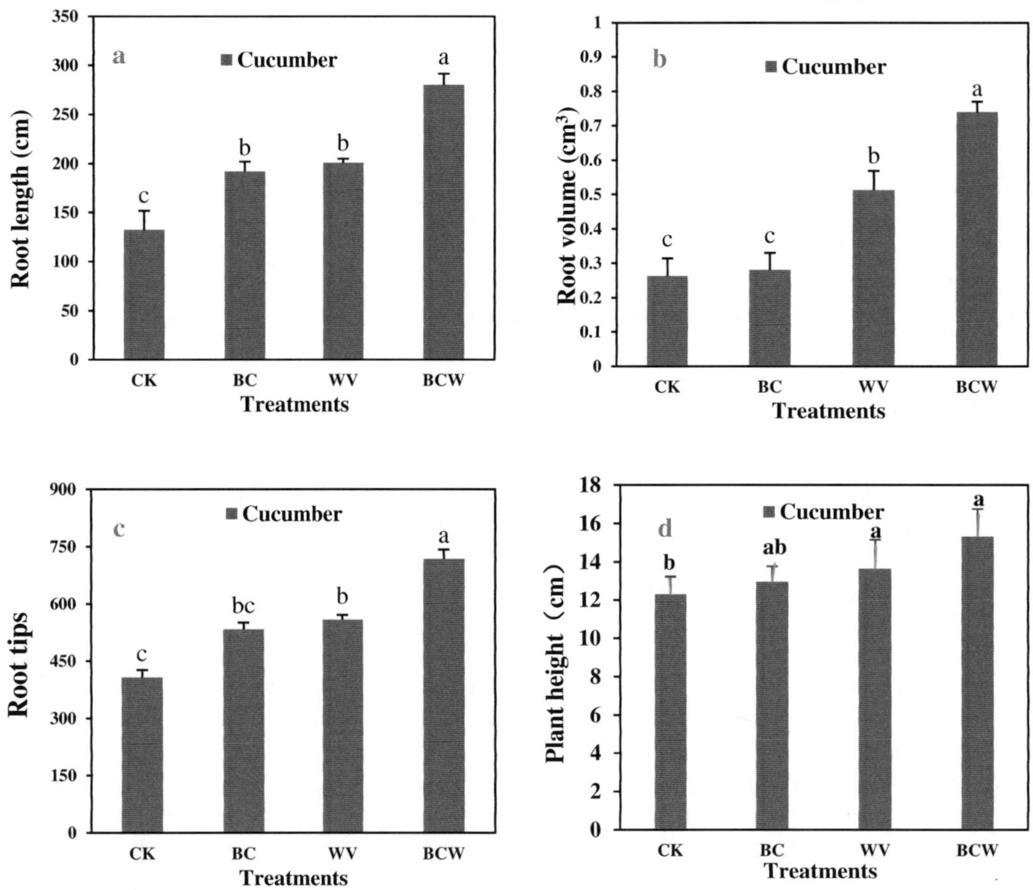


**Figure 2.** Effects of adding biochar and wood vinegar on the plant growth of cucumber.

CK, BC, WV and BCW indicate the plants was treated with deionized water, biochar, wood vinegar and the combination of biochar and wood vinegar, respectively. The different letters among different treatments indicate the significant differences, which were analyzed by Duncan's test ($P$ = 0.05) using SPSS 20.0. Error bars represent standard errors of the mean ($n$ = 4).

*3.2 Effects of adding biochar and wood vinegar on the cucumber growth*

As shown in Figure 2, the wood vinegar alone addition significantly increased the plant height, root length, root volume and root tips of cucumber by 20.2%, 45.2%, 7.8% and 30.9%, respectively, compared to that of the CK treatment. Consistently, the biochar alone addition also obviously increased the growth parameters of cucumber, compared to the CK treatment. Moreover, the combined addition of biochar with wood vinegar showed the greatest promotion effects on the growth of cucumber, which significantly increased the plant height, root length, root volume and root tips of cucumber by 29.7%, 117%, 121% and 76.1% than that in the CK treatment, respectively. These results showed that the combined addition of biochar with wood vinegar had the greatest promotion effects on the plant growth in contrast with the other treatments, which was mainly attributed to the following reason, i.e., improving the soil physico-chemical properties, such as loosening soil structure, improving soil acidity, elevating soil cation exchange capacity (CEC), and consequently benefiting the N and P translocation from soil to plant [7]. In addition, the wood vinegar could stimulate the growth of plants through increasing the chlorophyll content and enhancing photosynthetic rate and root vigor [8]. Thus, the combined addition of biochar with wood vinegar could significantly promote the plant growth over the biochar and wood vinegar adding alone.

## 4. Conclusion

The results from the pot experiments revealed that the combination of biochar with wood vinegar could maximize the promote effects of biochar and wood vinegar on the growth of plants. The mechanisms for promoting the plant growth may be attributed to the multiple benefits on the improvement of soil fertility, nutrient supply and translocation to the plant after the biochar and wood vinegar amendment. These effects offer an overall unique beneficial condition for the plantgrowth, and the specific mechanisms still deserve further research.

## Acknowledgements

This work was financially supported by Open Foundation of Shandong Provincial Key Laboratory of Marine Ecology and Environment and Disaster Prevention and Mitigation (201512, 201511) and Natural Science Foundation of China (41325013,41406085, 41503093).

## References

[1] Koch A, McBratney and Lai R 2012 *Nature* **492** 186
[2] Lehmann J 2007 *Nature* **447** 143-144
[3] Zheng H, Wang Z Y, Deng X, Zhao J,Luo Y , Novak J, Herbert S and Xing B S 2013 *Bioresour. Technol.* **130** 463-471
[4] Jeffery S, Verheijen F G, van der Velde M and Bastos A C 2011 *Agr. Ecosyst. Environ.* **144** 175-187
[5] Vaccari F P, Maienza A, Miglietta F, Baronti S, Di Lonnardo S, Giagnoni L and Valboa G 2015 *Agr. Ecosyst. Environ.* **207** 163-170
[6] Wei Q, Ma X H and Dong J E 2010 *J. Anal. Appl. Pyrolysis* **87** 21-28
[7] Zheng H, Wang Z Y, Deng X, Herbert S and Xing B S 2013 *Geoderma* **260** 32-39
[8] Bass A M. Bird M I, Kat G and Muirhwad B 2016 *Sci. Total Environ.* **550** 459-470

IOP Conf. Series: Earth and Environmental Science **61** (2017) 012150 doi:10.1088/1755-1315/61/1/012150

# Synthesis and application of water-soluble phenol-formaldehyde resin crosslinking agent

**Zhiqiang Wang[1], Shanshan Gao[1], Jing You[1], Jiliang.Yu[1], Tao Jiang[1], Dehai He[2], Hui Gong[2], Tiantian Zhang[1], JunjieWei[1], & Suzhen Guo[1]**

[1]Petroleum Production Engineering Research Institute of PetroChina Huabei Oilfield Company, Renqiu, Hebei, China
[2]The Third Exploit Factory of PetroChina Huabei Oilfield Company, Hejian, Hebei, China
cyy_wangzq@petrochina.com.cn

**Abstract**. Polyacrylamide and water-soluble phenol-formaldehyde resin can produce crosslinking reaction to generate the movable gel, which can be used in deep profile control to enhance oil recovery. In this paper, the preparation process of water-soluble phenol-formaldehyde resin crosslinking agent (polyhydroxymethyl phenolic) was presented, including the raw material mixture ratio and synthesis process. The water-soluble phenol-formaldehyde resin was synthesized by the technology of two-step base catalysis with NaOH and $Ba(OH)_2 \cdot 8H_2O$ composite catalyst. The movable gel prepared with this crosslinking agent and polyacrylamide has good stability and long shelf life through the performance evaluation experiment. The above movable gel was applied to deep profile control and obtained better flooding efficiency.

## 1. Introduction

The deep profile control technique is an effective measure for enhancing oil recovery and improving development effect. To some extent, the movable gel is the key of the deep profile control. Therefore, the movable gel has been applied in many oilfields around the country. Polyhydroxymethyl phenol solution could crosslink with the amide group of polyacrylamide to generate gels with net structure which is widely used in deep profile control. However, the application effect of the movable gel above is not good because of some reasons, such as the short storage time, the homopolymerization, etc.

Water-soluble phenol-formaldehyde resin crosslinking agent is the A-stage product of thermosetting phenolic resin. However, its properties are affected by many factors, such as the molar ratio of phenol to formaldehyde, the type of catalyst, the pH value, as well as the reaction temperature [1-3]. The molar ratio of phenol to formaldehyde is generally 1:2.5～3 in the synthesis, besides, the catalyst usually is NaOH. In this paper, the water-soluble phenol-formaldehyde resin crosslinking agent was synthetised by a new technology, in which NaOH and $Ba(OH)_2 \cdot 8H_2O$ was the composite catalyst[1]. Lu36 fault block of Huabei sandstone reservoir has obtained better flooding efficiency because of the successful application of the the crosslinking agent.

## 2. Experiment methods

### *2.1. Experimental materials*

Phenol and formaldehyde (36% aqueous solution), industrial products, Shandong Luhua Chemical Co.,Ltd.. NaOH and $Ba(OH)_2 \cdot 8H_2O$, chemical pure reagents. Polyacrylamide (the molecular weight is

IOP Conf. Series: Earth and Environmental Science **61** (2017) 012150 doi:10.1088/1755-1315/61/1/012150

25 million and the degree of hydrolysis is 25%～30%), Beijing Hengju Chemical Co.,Ltd.. Promoting cross agent[4], made in the laboratory by combining ammonium chloride with thiourea, etc.

### 2.2. *Experimental equipments*
Constant temperature water bath. Multi-speed mixer. Constant temperature oven, Shanghai Experimental Instrument Factory. Mars rheometer, Haake Company of Germany.

### 2.3. *Experimental methods*

*2.3.1. Synthesis of phenol-formaldehyde resin.* There are two synthetic methods commonly used in water-soluble phenolic resin synthesis. One is the one-step base catalysis and the other is two-step base catalysis. The technique of two-step base catalysis synthesis was adopted in this paper.

The steel stand was used to fix the 500mL three-necked flask, which was equipped with a thermometer, a mixer and a reflux condenser. An appropriate amount of phenol was added to the flask, then heated to 50°C to melt it into a liquid. NaOH was weighted at 4% of the amount of phenol and prepared as an aqueous solution at a concentration of 40%. Then the aqueous solution was divided into two portions according to the ratio of 7:3. The heavier portion was poured into the flask and the other one reserved. $Ba(OH)_2 \cdot 8H_2O$ was weighted at 10% of the total amount of NaOH and prepared as an aqueous solution at a concentration of 2%. Start the mixer and maintain a constant temperature for 20 minutes. At the same time the formaldehyde solution was weighted according to the molar ratio of phenol to formaldehyde 1:3. After that, divide the formaldehyde solution into two parts according to the ratio of 8:2. The portion of 80% was added slowly to the flask in 30 minutes and heated to 70°C holding for 30minutes. Then the solution of $Ba(OH)_2$, the remaining NaOH and formaldehyde were added into the flask. Heat the flask to 84～86°C slowly and maintain the constant temperature for 1.5～2h, then cool down. At last the tert butyl alcohol which accounted for 2% of the total solution volume was added. The final brown-red solution completely soluble in water was the water-soluble phenol-formaldehyde resin crosslinking agent and the concentration was 40%.

*2.3.2. Preparation and performance detection of polyacrylamide/water-soluble phenol-formaldehyde resin gels.* The polyacrylamide was firstly dissolved in a certain concentration of aqueous solution. Secondly, the right amount of water-soluble phenol-formaldehyde resin crosslinking agent and promoting cross agent were added. Then the solution was mixed evenly and filled into 100mL high temperature bottles. Place the bottles in the oven, the temperature of which was 76°C (the reservoir temperature of Lu36 fault block). The fluidity and elasticity of the gel were observed at regular intervals. The samples were taken out after 5 days and the viscosity was measured by using the Mars rheometer at a shear rate of $7.34s^{-1}$.

## 3. Experimental results and discussion

### 3.1. *Synthesis theory of water-soluble phenol-formaldehyde resin*
Phenol-formaldehyde resin is the condensation polymer synthesized from phenolic and aldehyde substances under the condition of acid or alkaline catalysis. The product is called thermosetting phenolic resin in the alkaline catalyst condition, and is called thermal shrinkage phenolic resin when the catalyst is alkaline. The properties of resin are affected significantly by the functionality, molar ratio, catalyst type and reaction temperature of raw materials. It takes two steps to synthesize water-soluble phenol-formaldehyde resin from phenol and formaldehyde.

*3.1.1. Addition reaction.* Phenol and formaldehyde firstly produce addition reaction to generate a mixture of polyhydroxymethyl phenol, which is the A-stage product of phenol-formaldehyde resin. The polyhydroxymethyl phenol not only has good water solubility and high reaction activity with the

IOP Conf. Series: Earth and Environmental Science **61** (2017) 012150 doi:10.1088/1755-1315/61/1/012150

amide group of polyacrylamide, but also can be used as the crosslinking agent to prepare the water-based gel.

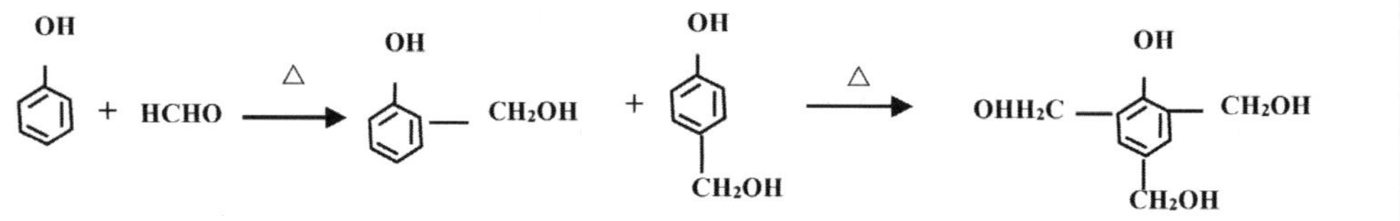


**Figure 1.** Schematic of the addition reaction

*3.1.2. Condensation reaction.* Hydroxymethyl phenol can be further polycondensation resulting in resin viscosity increases and generating viscous or solid phenolic resin, which is water-insoluble.

The reaction rate of addition is faster than that of the polycondensation under the alkaline catalyst condition. In general, the polycondensation rarely occurs, and the addition reaction rate is about 5 times of the polycondensation[1] when pH is high and the temperature is below 60°C. In addition, the reaction of formaldehyde with hydroxymethyl phenol is easier than that with phenol. The phenomenon will continue until 50% of formaldehyde is reacted. Therefore, the maximum reaction temperature must be strictly controlled below 90°C in order to obtain the A-stage phenol-formaldehyde resin solution (polyhydroxymethyl phenol), which has excellent water-solubility and crosslinking property. NaOH is the commonly used alkaline catalyst and $Ba(OH)_2 \cdot 8H_2O$ is added in this study. The composite catalyst can increase the content of dihydroxymethyl phenol and trihydroxymethyl phenol[5]. The technique of two-step base catalysis synthesis to synthesize water-soluble phenol-formaldehyde resin crosslinking agent is studied in this paper. The molar ratio of phenol to formaldehyde is chosen to be 1:3 in order to facilitate the hydroxylmethylation reaction occurring both in ortho-position and polarization of benzene ring.

The synthesized crosslinking agent has some advantages such as good solubility in water, no precipitation and flocculation, etc. Therefore this product can be further used for gelling experiments[6].

*3.2. Polyacrylamide/water-soluble phenol-formaldehyde resin gelling experiments*

*3.2.1. Influence of polyacrylamide concentration to the gel.* The concentration of crosslinking agent was 1500mg/L, the promoting cross agent was 750mg/L, and the concentration of polyacrylamide was varied. The experimental temperature was 76°C and the results were shown in Figure 2.

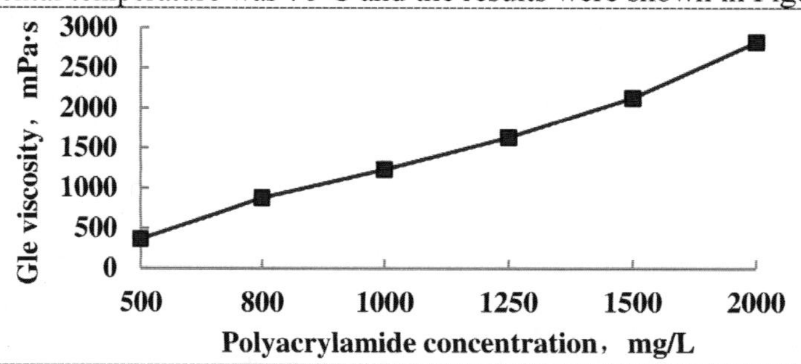


**Figure 2.** Influence of polyacrylamide concentration to the gel

The experimental results indicated that the gel viscosity increased with the increase of polyacrylamide concentration. Based on the experience of deep profile control, the gel stability is good when the gel viscosity is above 1000mPa·s. Therefore, the polyacrylamide concentration was preferred 1000mg/L～2000mg/L.

*3.2.2. Influence of crosslinking agent concentration to the gel.* The concentration of polyacrylamide was 1250mg/L, the promoting cross agent was 750mg/L, and the concentration of crosslinking agent was varied. The experimental temperature was 76°C and the results were shown in Figure 3.

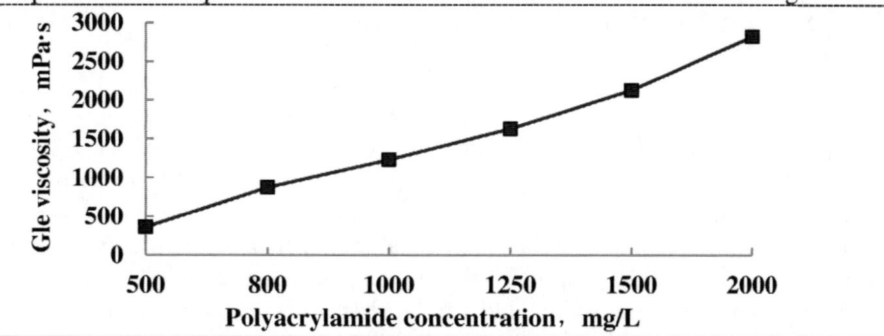


**Figure 3.** Influence of crosslinking agent concentration to the gel

The gel viscosity increased with the increase of crosslinking agent concentration. However, the increasing speed slowed down when the concentration reached 1500mg/L. Therefore, the crosslinking agent concentration was preferred 1500mg/L～2000mg/L.

*3.2.3. Influence of promoting cross agent concentration to the gel.* The concentration of polyacrylamide was 1250mg/L, the crosslinking agent was 1500mg/L, and the concentration of promoting cross agent was varied. The experimental temperature was 76°C and the results were shown in Figure 4.

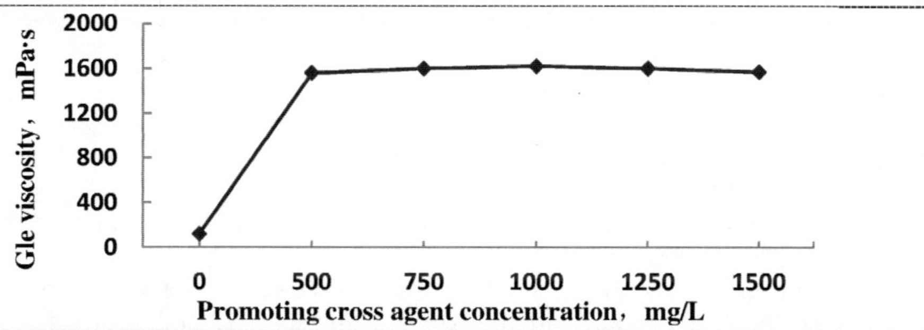


**Figure 4.** Influence of promoting cross agent concentration to the gel

The crosslinking reaction would not occur without promoting cross agent. The viscosity value was stable after the promoting agent concentration reached 500mg/L. Therefore, the promoting cross agent concentration was preferred 500mg/L～1000mg/L.

## 4. Field application

Water-soluble phenol-formaldehyde resin crosslinking agent was industrially produced on the foundation of this synthetic process. It was used in deep profile control[6] in Lu36 fault block of Huabei Oilfield and the corresponding wells obtained desirable stimulation result. During the field application, the crosslinking agent was stored for 3 months and still maintained good water-solubility, cross-linking property with polyacrylamide.

There are 4 injection wells and 5 production wells in NgII of North Lu36 fault block. The fault block is a high porosity and high permeability reservoir, and the reservoir connectivity is good. The water is injected by horizontal wells and the oil is produced by vertical wells. Based on the above experiment results, the polyacrylamide concentration applied in the field was 1200mg/L～1500mg/L, the crosslinking agent concentration was 1500mg/L, and the promoting cross agent was 750mg/L. The movable gel was injected in 4 injection wells simultaneously. The diurnal injection and formulation concentration were adjusted to keep the pressure of 4 wells as same as possible. The cumulative gel injection was 51200m$^3$ from June 17, 2014 to July 23, 2015. The production of the 5 oil wells increased significantly by October 2014, and the daily production increased from 5.2t to 17.1t, the composite water cut declined from 94.6% to 85.8%. The cumulative increase of oil production was up to 6483t and the term of validity was more than two years.

## 5. Conclusions

i. Water-soluble phenol-formaldehyde resin was synthesized by two-step base catalysis technology using composite catalyst. It has excellent water-solubility and crosslinking property. It can crosslink with polyacrylamide to generate the gel with good stability even if the polyacrylamide concentration is lower.

ii. The synthesized crosslinking agent has good stability which is conducive to massive injection. Polyacrylamide/water-soluble phenol-formaldehyde resin gels obtained better stimulation results in Lu36 fault block and can be introduced to similar reservoirs.

iii. It is feasible to use the technique of deep profile control in the reservoir, in which the water is injected by horizontal wells and the oil is produced by vertical wells. The study provides an effective way to increase the oil production in this type of reservoirs.

## Author information

Zhiqiang Wang was born in 1963 and is a senior engineer mainly working on the reaserch of EOR now. Phone: 0317-2756426, E-mail: cyy_wangzq@petrochina.com.cn.

## Acknowledgements

The study was supported by PetroChina Science and Technology Major Project "Study and application of key technologies for production stable and increase to 8 million tons in Huabei oilfield" under Grant No.2014E-3507.

## References

[1] Huang, F.R. & Wan, L.Q. (2011). Phenol-formaldehyde Resin and Application. Beijing: Chemical Industry Press.

[2] Zhang, H.T & Huang, J.X. (2011). Preparation and Application of Waterborne Resin. Beijing: Chemical Industry Press.

[3] Li, G. & Wang L.J, et al. (2002). Study on the synthesis and the performance of water soluble phenol-formaldehyde resin. Journal of Hebei University of Tchnology, 31(4), pp.37-41.

[4] Yu, J.L. & Wang, Z.Q. et al. (2016). Experiment and application of movable gel prepared with reinjected waste water in block Cha 39 of Huabei oilfield. Journal of Yangtze University (Nat Sci Edit), 13(7), pp.9-12.

[5] Chen, J.M. & Ao, W.L. et al. (2004). Review on synthesis of resole. Thermosetting Resin,19(6), pp.31-34.

[6] Gong, Z.B. & Luo, Q. et al. (2014). Porosity and permeability adaptability of organic phenolic polymer gel for deep profile control system. Journal of Oil and Gas Technology (J.JPI), 36(5), PP.136-140.

IOP Conf. Series: Earth and Environmental Science **61** (2017) 012151 doi:10.1088/1755-1315/61/1/012151

# Influence of lithium slag from lepidolite on the durability of concrete

**Luo Qi[1,2], Huang Shaowen[2] , Zhou Yuxuan[2], Li Jinyang[2], Peng Weiliang[2], Wen Yufeng[2]**

1.School of Materials Science and engineering, Tongji University, Shanghai, China;
2.School of Materials Science and engineering, Nanchang University, Jiangxi, China
chinalodger1@163.com

**Abstract**. This paper mainly studies the effect of lithium slag from lepidolite on the property of concrete including dry shrinkage, anti-carbonation, wear resistance and chloride ion resistance. Concrete interface structure has been observed with SEM. The results show that adding lithium slag to concrete can improve concrete property including dry shrinkage, wear resistance and chloride ion resistance. However, the wear resistance tends to decrease when the amount of lithium slag reach 20%. Lithium slag also has negative effect on anti-carbonation property. With the increasing amount of lithium slag, anti-carbonation property of concrete decrease gradually.

## 1. Introduction

Using lithium slag as concrete admixture can solve the pollution problem caused by excessive lithium slag in industrial production. Domestic experts have done a lot of researches on adding lithium slag to concrete. Zhang Lanfang, Chen Jianxiong and some other researchers[1~4] studied the effect of admixtures including lithium slag, fly ash, mineral powder on concrete property including working performance, strength, anti-carbonation, frost resistance. Zhao Ruopeng[5] produced high fluidity concrete compound lithium slag and silicon powder which strength can reach 100MPa. Wen Yong and Liu Guojun[6] studied the effect of lithium slag on fresh properties and surface area of concrete. WangGuoqiang[7] also studied the effect of lithium slag on strength and crack resistance. However, there is little researches on the application of lithium slag from Yichun lepidolite (lithium slag for short) on concrete, also no systematic researches on endurance of concrete adding lithium slag. Systematic researches about the effect of lithium slag on concrete property including dry shrinkage, anti-carbonation, wear resistance, chloride ion resistance and surface structure can be the theory reference in actual application.

## 2.Experienment

### *2.1. Raw material*

Cement: P.O42.5R cement produced by East Asia JiangXi Cement Co. Ltd. The apparent density of the cement is 3.1g/cm$^3$. Lithium slag: Lithium slag extract from lepidolite produced by Jiangxi Ganfeng Lithium Co. Ltd. The specific surface of lithium slag is equal to 550m$^2$/kg and the middle diameter of it equal to 12μm. The chemical constituents of lithium slag are given in Table 1. Fine aggregate: Sand in gan river with fineness modulus of 2.4 and apparent density of 2.6g/cm$^3$. Coarse aggregate: Continuous grading limestone artificial gravels with the maximum grain size of 20mm. Water reducing agent: Polycarboxylate superplasticizer (powder) saled in the market with water reducing ratio of 26%.

**Table 1. Chemical composition of lithium slag %**

| Raw material | $SiO_2$ | $Al_2O_3$ | CaO | $Fe_2O_3$ | MgO | $SO_3$ | $Na_2O$ | $K_2O$ | $TiO_2$ | Loss | Sum |
|---|---|---|---|---|---|---|---|---|---|---|---|
| Lithium slag | 47.62 | 21.56 | 2.02 | 0.48 | 0.12 | 0.03 | 10.68 | 3.05 | 3.46 | 0.14 | 90.63 |

*2.2. Testing method*

The testing of drying shrinkage, anti-carbonization and chloride penetration resistance of concrete refer to GB/T 50082-2009 Standard for test methods of long-term performance and durability of ordinary concrete. The wear resistance of concrete is tested in accordance with the JTG E30-2005 highway engineering and cement concrete testing rule. Meso-structure: Using QUANTA 200FEG Field Emission Environment Scanning Electron Microscope (FEI, America) analysis interface meso-structure of cement concrete. The mix proportion of cement concrete shows in Tab.2, water-binder ratio is 0.47 and the mixing value of the water reducer is 0.4% invariably.

**Table 2. The mix proportion of lithium slag cement concrete kg/m³**

| Sample | Water | Cement | Lithium slag | Fine aggregate | Coarse aggregate |
|---|---|---|---|---|---|
| L0 | 234 | 500 | 0 | 640 | 1139 |
| L1 | 234 | 450 | 50 | 640 | 1139 |
| L2 | 234 | 400 | 100 | 640 | 1139 |
| L3 | 234 | 350 | 150 | 640 | 1139 |

## 3.Results and discussion

*3.1. The effects of lithium slag on concrete drying shrinkage*

Drying shrinkage is referred to the irreversible shrinkage caused by the loss of adsorbed water in gel pores and capillary pores in unsaturated air after the concrete maintenance stopped. The drying shrinkage after curing may cause cracks in concrete without loading. To research the effects of lithium slag on concrete drying shrinkage, we designed two variables, the mount of lithium slag and concrete ages. The test results are shown in Fig.1. As depicted in Fig.1, the dry contraction percentage was steady and the increasing speed of it was slow after 60d. With the increment of the mixing mount of lithium slag, the dry contraction percentage was decreased. Comparing to the specimen without lithium slag (L0), the dry contraction of the specimen with 30% lithium slag decreased 23% on 180d.It may because Lithium slag need plenty of water causing that ,Compared with L0, there is less adsorbed water in gel pores and capillary pores in lithium slag concrete, making the dry contraction percentage of concrete decreases.

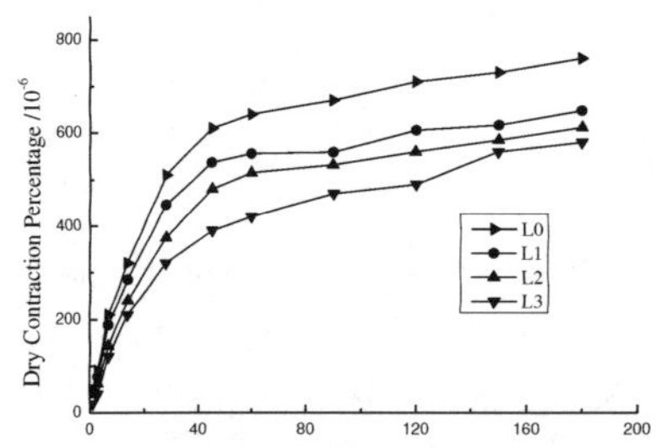


**Fig 1. The effects of lithium slag on concrete drying shrinkage**

### *3.2. The effects of lithium slag on concrete carbonation resistance*

Carbonation resistance is related to the concrete durability closely. It is used to evaluating the service life of reinforced concrete structures [8]. To study the effects of lithium slag on concrete carbonation resistance, we tested carbonation depth in different ratios of lithium slag and different concrete ages. The test results are shown in Fig.2. As depicted in Fig.2, with more lithium slag and high of concrete ages, the carbonation depth grows. The carbonation average depth of concrete without lithium slag L0 is 3mm on 28d. However, the carbonation average depth of concrete with 30% lithium slag L3 is 11mm. It may because that lithium slag can replace the cement and have secondary hydration with the production of cement hydration. So the alkalinity of concrete decreased. The neutralization resistance decreases in the process of $CO_2$ infiltration which causes that the carbonation resistance of concrete decreases gradually with the increasing of lithium slag.

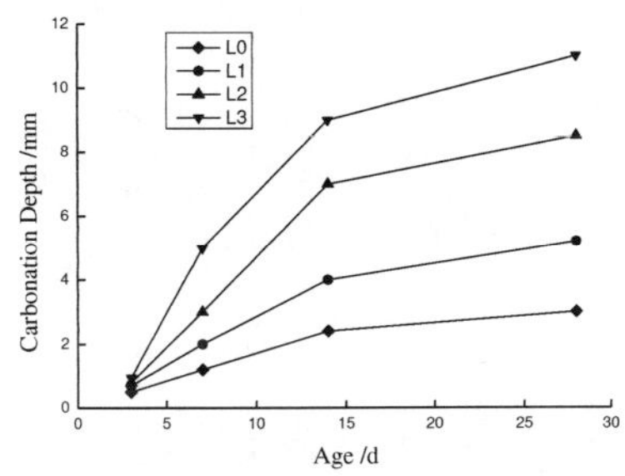


**Fig 2. The effects of lithium slag to concrete carbonation resistance**

### *3.3. The effects of lithium slag on concrete chloride-penetration resistance.*

When chloride ion infiltrate and form the galvanic cell in concrete, passive film on the surface of reinforcement will be destroyed easier and the reinforcement will rust faster. Fig.3 shows that there is a positive correlation between the amount of adding lithium slag and chloride-penetration resistance. Electric flux of concrete L3 adding 30% lithium slag decrease about 43% compared to concrete L0 without adding lithium slag after 6h at the condition of electrify. It may because that lithium slag and cement hydration product can have secondary hydration in a short time, translating the loose $Ca(OH)_2$ crystal which grow in concrete interface in specific direction into compact gelling layer. Thus give a rise to higher resistance of ion infiltrating, improving the concrete chloride-penetration resistance significantly.

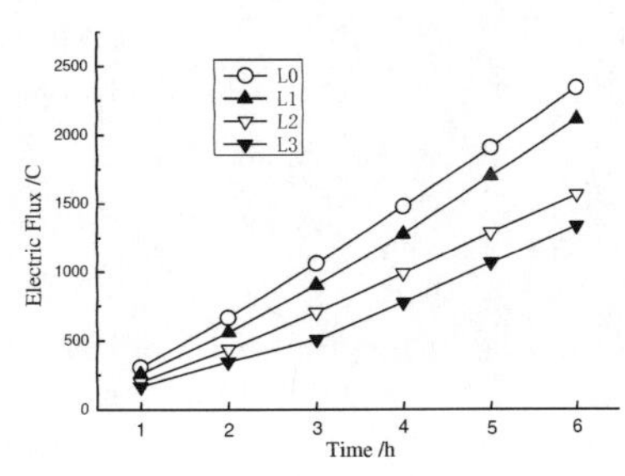


**Fig 3. The effects of lithium slag on concrete chloride-penetration resistance**

### *3.4. The effects of lithium slag on the wear resistance of concrete*

The test results of wear resistance of concrete can be seen in Fig.4. It shows that with the increasing of lithium slag mixing amount, the wear loss decreased first and then increased. Although the mixing amount of lithium slag reaches 30%, the wear loss of L3 lithium slag concrete is still less than L0 concrete without lithium slag. It may due to that adding lithium slag can improve the water retention and cohesiveness which increase the compactness of concrete. On the other hand, lithium slag can have secondary hydration with the production of cement hydration which also increases the compactness of concrete, improving the wear resistance. However, when the mixing amount of lithium slag is more than 20%, the wear resistance of lithium slag concrete will show decreasing trend. It is because with the increasing of lithium slag, the relevant hydration products will reduce. As lithium slag itself cannot hydrate, residual lithium slag can only fill in the concrete, making the compactness of concrete shows decreasing trend. Thus, it can decrease the improvement effect of lithium slag to compactness of concrete.

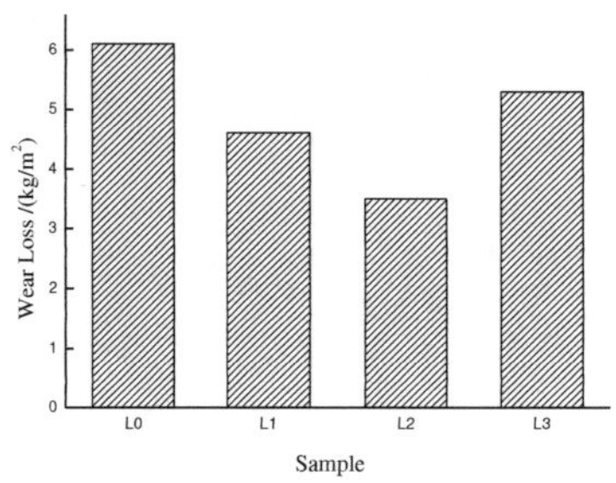


**Fig 4. The effects of lithium slag on the wear resistance of concrete**

*3.5. Effects of Lithium Slag on the interfacial microstructure of Cement Concrete*

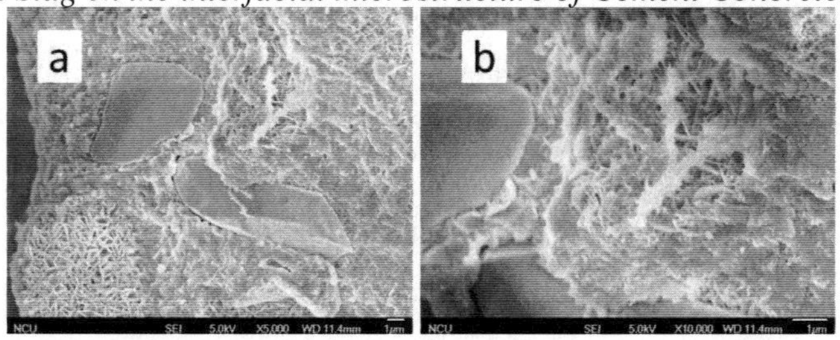


**Figure 5.SEM image of interfaces of L0 on 28d**

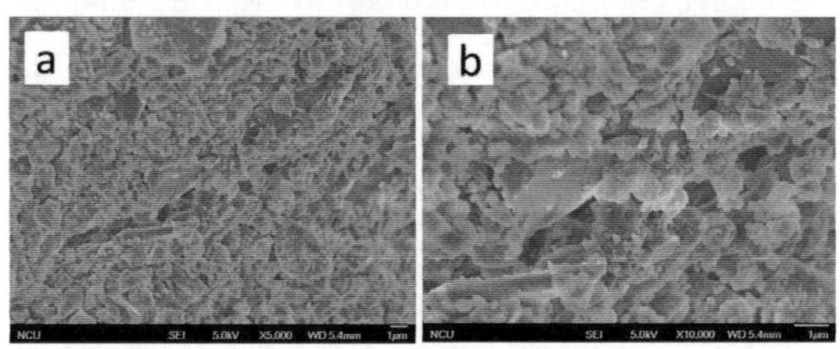


**Figure 6. SEM image of interfaces of L1 on 28d**

Fig.5 and Fig.6 are SEM images of interfaces of cement concrete without lithium slag and with 10% lithium slag respectively on 28d. As shown in Fig.5, without lithium slag, the interfaces of cement concrete appeared a lot flocculent C-S-H gels accompanied much acicular ettringite. As depicted in Fig.6, the interface structure of cement concrete with 10% lithium slag is much tighter. The hydration product mainly is flat C-S-H gels. The amount of ettringite was significantly decreased compared with the cement concrete without lithium slag. There was few ettringite in sample in Fig.6. It shows that the cement mixed with lithium slag consume more $Ca(OH)_2$ and reduce the production of ettringite due to the pozzolanic effect[9]. Meanwhile, as a result of the adding of lithium slag, the amounts of C-S-H

IOP Conf. Series: Earth and Environmental Science **61** (2017) 012151 doi:10.1088/1755-1315/61/1/012151

gels increased and the density of cement concrete was improved which was important reasons of strengthening cement concrete.

## 4.Conclusion

Dry contractibility, abrasion resistance and chloride-penetration resistance of concrete can be improved by adding lithium slag but the abrasion resistance presents the downward trend when adding over 20% lithium slag into the concrete.

Adding lithium slag has negative effect on carbonation resistance of concrete and the carbonation resistance of concrete decreases gradually with the increasing amount of the lithium slag.

Adding lithium slag can enhance the density of the concrete and inhibit the formation of ettringite, meanwhile, it also can improve the concrete chloride-penetration resistance so that dry shrinkage performance, abrasion resistance and chloride-penetration resistance of concrete enhanced obviously.

## References

[1] Zhang Lanfang, Chen Jianxiong, Yue yu, et al. Experimental study on the high strength concrete added with lithium slag[J]. New Building Materials, 2005, (3): 29-31.(in Chinese)

[2] Zhang Lanfang, Chen Jianxiong, Li Shiwei. Examination Study of Alkali-Activated Slag-Lithium Slag Concrete[J]. Journal of Building Materials, 2006, 9(4):488-492. (in Chinese)

[3] Zhang Lanfang, Yue Yu. The Research on Alkali-activated Mineral Slag-Flyash-Lithium Slag Concrete[J]. Fly Ash Comprehensive Utilization, 2008, (2): 21-23.(in Chinese)

[4] Zhang Lanfang, Xie Guihai, Chen Jianxiong, et al.Experimental Study on the Frost Resistance Properties of New Lithium salt Slag Concrete[J]. Low Temperature Architecture Technology, 2005, (4): 8-10.(in Chinese)

[5] Zhao Ruopeng, Guo Zili, Zhang Jing, et al. Study on High Strength and High Fluidity Concerte with Lithium Slag and Silica Fume[J]. Industrial Construction, 2004, 34(12): 61-63.(in Chinese)

[6] Wen Yong, Song Yafeng, Abula Kanjie, et al. Influence on Fresh Concrete with Ground Granulated Lithium Slag[J]. Concrete, 2011, (10): 58-60.(in Chinese)

[7] Wang Guowei, Nuerkaili·Yiziteliopu, Shi Ke-bin, et al. Influence and Fractal Evaluation of the Fineness Level of Lithium Slag on Early-Age Crack-Resistant Property of Concrete[J]. Fly Ash Comprehensive Utilization, 2010, (5): 23-28.(in Chinese)

[8] Luo Qi, Huang Shaowen, Hu Xiuxia, et al. Durability of Crumb Rubber-Modified Cement Concrete[J]. Journal of Nanchang University(Engineering Edition), 2007, 29(4):340-343.(in Chinese)

[9] Wang Ling. Ralation between the $C_3A$ content and the sulphate resistance property of concrete[J]. Cenment Technology,1998 , (4): 28.(in Chinese)

IOP Conf. Series: Earth and Environmental Science **61** (2017) 012152 doi:10.1088/1755-1315/61/1/012152

# Research and application on the horizontal tectonic stress influence on the stability of deep roadway

**CAO Jian-jun[1,2,3] , YAO Zhuang-zhuang[2,3]**

[1]Anhui University of Science and Technology, Huainan 232001, China
[2]China Coal Technology Engineering Group Chongqing Research Institute, Chongqing 400037, China
[3]National Key Laboratory of Gas Disaster Detecting, Preventing and Emergency Controlling, Chongqing 400037, China
[a]caojianjun207@126.com

**Abstract.** According to the test result of ground stress of -817.0m Horizontal shaft station in Huainan mine area, the stress field distribution characteristic of mainly with horizontal tectonic stress was analyzed, the deep roadway surrounding rock deformation and failure regularity under the situation of the maximum horizontal stress and roadway axis in the different angle were studied by using the $FLAC^{3D}$ simulation software, and then specific way of supporting and its parameters were designed combined with the shaft station layout conditions. The result shows that: the roadway stability is relatively good when its tunnel axis the maximum horizontal stress direction within 30° angle, surrounding rock deformation and the plastic zone at the top and bottom are significantly increasing along with the angle increased from 30° to 90°, the major form of deep soft rock deformation and failure is the floor heave. According to the simulation results, the conditions under the layout condition of the shaft station roadway , support methods and support parameters have been designed and determined .The supporting method can be referenced by the similar mines.

## 1. Introduction

The ground stress is the natural stress in the earth's crust, and its magnitude and direction have great influence on the stability of surrounding rock[1-3].In order to determine the mechanical properties of engineering rock mass, to analyze the stability of surrounding rock and to realize the scientific design of underground excavation, it is necessary to study the distribution law of surrounding rock stress[4]. A large number of in-situ stress measurement and research results show that the horizontal tectonic stress in deep strata is generally greater than the vertical stress, and the horizontal stress has obvious directivity. And it is one of the important factors that affect the stability of the roof and floor[5-10]. For mine engineering design, the determination of the general layout of the mine, the selection of mining methods, and the optimum section shape, section size, excavation steps, support form, support structure parameter, support time and so on, the ground stress condition is the prerequisite[11-12]. Therefore, it is important to study the influence of the axial and horizontal tectonic stress on the stability of surrounding rock of deep soft rock roadway when the original rock stress field is certain.

## 2. General situation of engineering geology

### *2.1. Geological conditions of surrounding rock in deep tunnel*

The coalfield in Huainan mining area is a totally hidden area covered by the loose layer of the

IOP Conf. Series: Earth and Environmental Science **61** (2017) 012152 doi:10.1088/1755-1315/61/1/012152

Cenozoic, which is a nearly east-west tectonic form. The north and south sides of the basin are hedged by imbricate-like nappe structure, and the basin is characterized by a gentle northwest-west syncline structure. The tectonic movement occurred mainly in the Indosinian and Yanshan period. The regional tectonics is formed by the push of the north-south direction. Syncline within the stratum dip gentle is generally 9°~19°, which is a series of gentle folds. In general, the mine strata formed in the Carboniferous and Permian sandstone, shale, mudstone lithology, experienced compressional and shear tectonic movement many times the role of the rock with mud cementation and katamorphic characteristics, and the occurrence of residual tectonic stress is higher.

According to the rock mechanical properties test results of long-term statistics, index of physical and mechanical properties of rock in the mine: Aluminum mudstone, mudstone and sandy mudstone and argillaceous sandstone in the natural water content under the condition of uniaxial compressive strength is 1.4~58.5MPa, is 3.3~7.9MPa, tensile strength 0.43~10.4MPa, shear strength was 2.1~10.9MPa; cements of sandstone is dominated by clay, so the strength is low.

*2.2 Gound stress distribution characteristics*[13]

In order to find out the law of stress distribution of the mine, we carried out in-situ stress test in the mine shaftbottom central air shaft -817.0m level, the test results are shown in table 1.

Table.1 -817.0m shaft station principal stress size and direction

| | Stress /MPa | Dip /° | Azimuth /° |
|---|---|---|---|
| First Main Stress $\sigma_1$ | 24.89 | 66.64 | 150.14 |
| Second Main Stress $\sigma_2$ | 19.64 | 20.19 | 1.77 |
| Third Main Stress $\sigma_3$ | 16.88 | 11.24 | 267.57 |

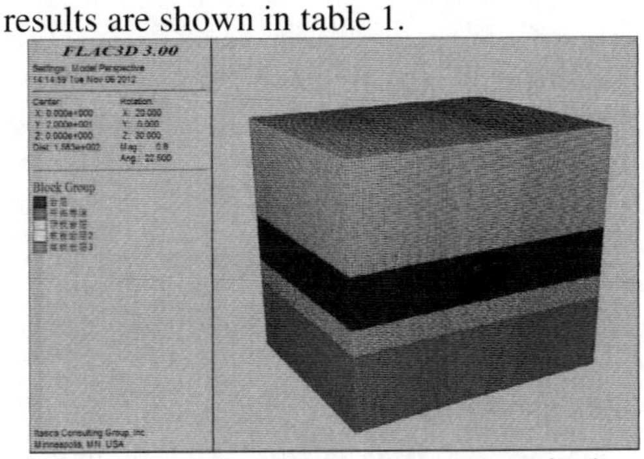

Figure1. Three dimensional numerical simulation model of FLAC$^{3D}$

The stress test results also show that the maximum horizontal principal stress is 19.95MPa, 26.01MPa, 23.40MPa, and the vertical stress is 24.01MPa, 23.11MPa, 22.03MPa, and the stress level is higher. The maximum horizontal principal stress direction is 350.9 degrees, 310 degrees, 330 degrees and 319 degrees, which is close to NWN. Through a series of stress monitoring, the pore region of lateral pressure coefficient is slightly greater than 1. We can deduce that the test area is mainly based on the horizontal tectonic stress.

## 3. Numerical analysis of horizontal stress on roadway stability

*3.1 The establishment of numerical calculation model*

In order to obtain the influence degree distribution of horizontal tectonic stress and the deformation of surrounding rock plastic zone and the extent of damage, based on the mine shaftbottom central air shaft -817.0m level as the engineering background, the simulation numerical model is established by using FLAC$^{3D}$ numerical software. Using the Mohr-Coulomb model as failure criterion, the model of maximum horizontal stress and the axial angle of roadway at 0, 30, 45, 60 and 90 degrees were established. Three dimensional numerical model is established as shown in figure 1.

*3.2 Numerical simulation results*

The distribution of surrounding rock deformation and plastic zone in the surrounding rock of the maximum horizontal stress and the axial direction is shown in Figure 2.

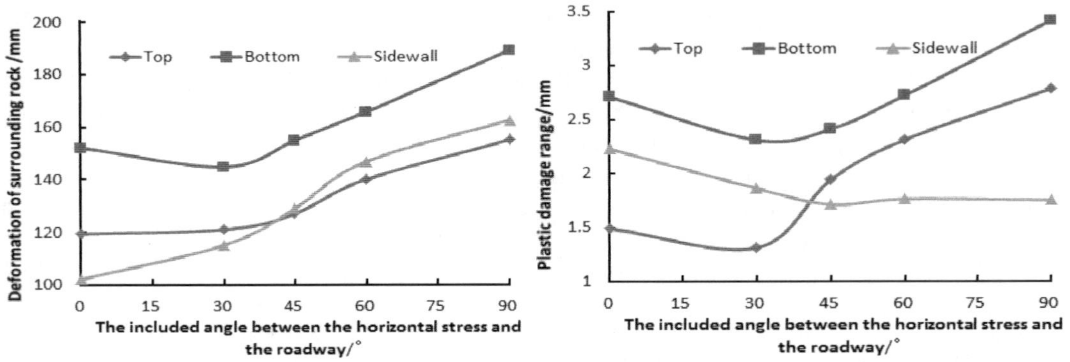


(a) Deformation of surrounding rock (b) Plastic damage range

Figure2. Displacement and plastic zone of surrounding rock with different maximum horizontal stress and axial direction of roadway

As shown in Figure 2 :

(1) When the angle between the maximum horizontal stress and the axial direction of the roadway is 0°, the two sides displacement is 102.14mm, and the damage range of plastic zone is 2.23m; the subsidence of roof is 119.32mm, and the damage range of plastic zone is 1.49m; the floor heave volume is 152.25 mm, and the damage range of plastic zone is 2.71 m. It is not difficult to see the deformation of roadway surrounding rock and plastic area is relatively small.

(2) When the angle between the maximum horizontal stress and the axial direction of the roadway is 30°, and the two sides displacement is 115.17mm, and the damage range of plastic zone is 1.86m; the subsidence of roof is 121mm the damage range of plastic zone is 1.31m; and the floor heave volume is 144.91mm, and the damage area of plastic zone is 2.31m.

(3) When the angle between the maximum horizontal stress and the axial direction of the roadway is 45°, and the two sides displacement is 129mm, and the damage range of plastic zone is 1.71m; the subsidence of roof is 127mm, and the damage range of plastic zone is 1.94m; the floor heave volume is 155 mm, and the damage range of plastic zone is 2.41 m.

(4) When the angle between the maximum horizontal stress and the axial direction of the roadway is 60°, and the two sides displacement is 146.87mm, and the damage range of plastic zone is 1.76m; the subsidence of roof is 140mm, and the damage range of plastic zone is 2.31m; the floor heave volume is 165.82 mm, and the damage range of plastic zone is 2.72 m. Compared with the maximum horizontal stress is 0 ° to the axial direction of the roadway, and the displacement of the roof and the bottom plate and the displacement of two sides are obviously increased

(5) When the angle between the maximum horizontal stress and the axial direction of the roadway is 90°, and the two sides displacement is 162.63mm, and the damage range of plastic zone is 1.75m; the subsidence of roof is 155.17mm, and the damage range of plastic zone is 2.78m; the floor heave volume is 189.15 mm, and the damage range of plastic zone is 3.41 m. The damage range of plastic zone has reached the maximum value.

As shown in the numerical simulation results:

(1) When the angle between the maximum horizontal stress and the axial direction of the roadway is less than 30°, and the deformation amount of the wall rock at the bottom is slightly decreased and the deformation amount of the top and the two sides is slightly increased with the increase of the included angle. The plastic zone of the top, bottom and two sides is slightly reduced with the increase of the included angle.

(2) When the maximum horizontal stress and the axial angle of the roadway are in the range of 30°~90°, the deformation of the top and bottom plate and the surrounding rock increases significantly with the increase of the included angle. Compared to the top and bottom of the plastic zone of the surrounding rock increased significantly, the deformation of two sides decreased slightly. When the angle between the maximum horizontal stress and the axial direction of the roadway is 90°, the deformation and plastic zone of roadway roof, bottom plate and wall rock have reached the maximum

value. Among them, the floor heave is more serious in the total deformation of the roadway, which shows that the floor heave is the main form of the deformation and failure of the deep high stress soft rock roadway.

## 4. Engineering practice

According to the ground stress test and rock physical mechanical properties test, the angle between -817 Horizontal track cross-cut direction and the direction of the average maximum principal stress is 48.5°.The ratio of horizontal stress to vertical stress is 0.97:1. The results of $FLAC^{3D}$ numerical simulation show that the horizontal stress in the included angle has a great influence on the surrounding rock stability. Therefore, in order to maintain the long-term stability of roadway surrounding rock, it is necessary to take into account the influence of horizontal stress on the stability of roadway surrounding rock when selecting roadway support mode and determining support parameters.

The results of numerical simulation show that the deformation of the roof and floor of the roadway is especially affected by the horizontal tectonic stress. Therefore, special strengthening method should be adopted to support the roof and floor of the roadway. Full-face ultra-high-strength pre-stressed anchor bolt with high-strength bolt-net support technology can effectively control the strong deformation of the soft rock roadway and maintain the stability of the roadway.

### *4.1 Parameters of roadway support*

Bolt layout: full-profile layout.

Support system: roadway surrounding rock support with a force of not less than 60kN ultra-high-strength prestressed bolt support system.

Treatment of special parts: The maximum plastic zone of the surrounding rock is the corner of the roadway where a row of bolts are arranged. Bolts are arranged in the straight wall and semicircular arch, the interval is 800mm × 800mm. The supporting section is shown in Figure 3.

Bolt :Diameter is 22mm, length is 2.5m, yield strength is more than 700MPa, elongation is more than 10%. The construction prestress of the bolt should reach 7 ~ 8t; using high-strength spherical type pallet with thickness of 8mm and area of 150mm × 150mm and standard high-strength torque nut with cover.

Steel mesh: reinforced with diameter of 8mm welding, the mesh grid size specifications: 110mm×110mm, mesh: 1650mm×850mm.

Injection method: a full cross-section injection, jet thickness: 120mm.

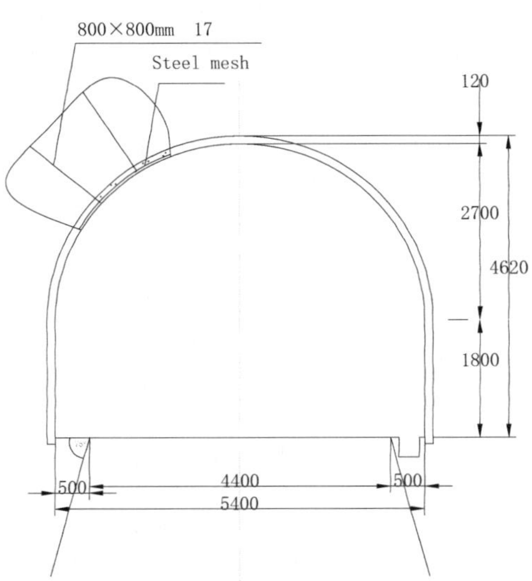


Figure 3. section of roadway support

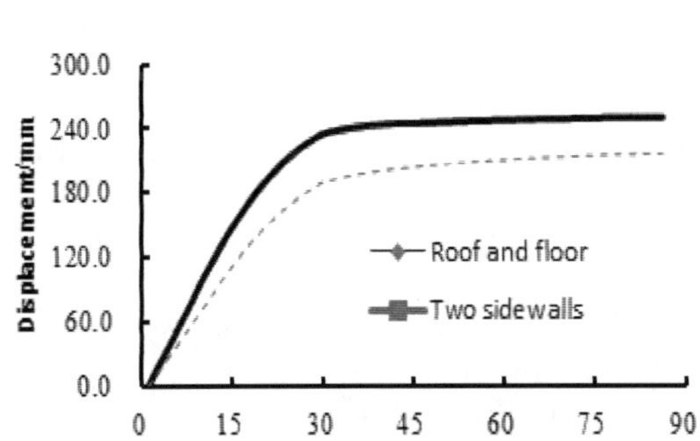


Figure. 4 curve of surface displacement of Roadway

*4.2 Analysis of supporting effect*

After supporting the roadway, the deformation of surrounding rock of roadway is monitored by 90d, and the observation results are shown in Figure 4.

As shown in Figure 4, after the roadway surrounding rock is supported, in the roadway surface displacement observation period, the surrounding rock deformation is not large, the maximum displacement of two to help the near 215.8mm, the top floor of the maximum displacement of 250mm, consistent with numerical simulation results After 30 days, the deformation speed of the surrounding rock began to slow down, and after 35 days, the deformation velocity was 0.6mm/d and finally stabilized. It can be explained that the support scheme effectively controls the strong deformation of high stress soft rock roadway and maintains the stability of roadway. It is also verified that the numerical simulation study on the maximum horizontal stress and the axial angle of the roadway can play a good guiding role.

## 5 Conclusion

Through Flac$^{3D}$ numerical simulation study, the deformation and failure law of surrounding rock of soft rock roadway under different angles are obtained. And successfully applied to the deep roadway support engineering practice, maintaining the stability of the roadway surrounding rock, the effect is good. The main conclusions of this paper are as follows:

(1) The distribution of stress field in deep mine, which is mainly caused by horizontal tectonic stress, is an important influence on the stability of deep roadway, especially in soft rock tunnel.

(2) The numerical results reveal that the maximum horizontal stress direction affects the stability of roadway surrounding rock, and when the angle between the roadway axial and maximum horizontal stress direction was 30°, the stability is relatively good. When the angle was in the range of 30°~90°, with the increase of the angle, the deformation of surrounding rock and plastic zone at the top and bottom are significantly increased. The floor heave is the main form of deformation and failure of deep high stress soft rock roadway.

(3) The engineering practice shows that the high strength bolt net support increasing the whole section of the corner bolt technology is adopted, which can effectively control the strong deformation of the high stress soft rock roadway and maintain the stability of the roadway

## Fund Project

This work was financially supported by the National Natural Science Foundation of China (51604298) and China Coal Technology Engineering Group Chongqing Research Institute of independent research and development project (2015YBXM39).

## Reference

[1] KANG Hong Pu. Stress distribution characteristics and strata control technology for roadways in deep coal mines [J]. Coal Science and Technology, 2013, 41(9): 12-17.
[2] LIU Chao Ru. Characteristics of in-situ stress distribution in deep coal mine and its influence on the stress field of surrounding rock [D]. Beijing: China Coal Research Institute, 2012.
[3] XIAO Tong Qiang. Study on stability and control of surrounding rock of thick coal seam roadway under deep tectonic stress [D]. Xuzhou: China University of Mining and Technology, 2011.
[4] CAI Meng Feng, HENG Mang Chao, LIU Dong Yan. Rock Mechanics and Engineering [M]. Beijing: Science Press, 2002
[5] Kong De Sen, JIANG Jin Quan. Stability analysis of deep roadway in tectonic stress Field [J]. Mine Pressure and Roof Management, 2007, 17(4): 56-58.
[6] Zhuang Yan Xin,CAI Mei Feng. Theoretical study on relationship between gound stress and roadway arrangement [J]. Geotechnical Engineering Technology, 2005, 19(2): 93-97.

[7] Gou Pan Feng, WEI Si Hong, ZHANG Sheng. Simulation study on the stability of the roadway with different horizontal stress [J]. Journal of Mining and Safety Engineering, 2010, 2(27): 143-148.
[8] Lu Yan, ZOU Xi Zheng, Liu Chang You. Roadway layout in tectonic stress field [J]. Journal of Mining and Safety Engineering, 2008, 25(2): 144-149.
[9] Lu Yan. Stability principle and control of roadway surrounding rock under the influence of tectonic stress field [D]. Xuzhou:China University of Mining and Technology, 2008.
[10] Gao Fu Qiang. Numerical analysis of the influence of the maximum horizontal principal stress on the stability of surrounding rock [J]. Mining Research and Development,2008, 1(28):62-64.
[11] Xu Yong qi. Coal Mining [M]. Xuzhou:China University of Mining and Technology press, 2009,7.
[12] WANG Qi Zhou, XIE Wen Bing, Jin Sheng Guo. Supporting technology of high stress soft rock roadway in fault structure zone [J]. Coal Mine Safety, 2012, 43(3): 51-53.
[13] China University of Mining and Technology. Study on the coupled deformation mechanism and control technology of deep roadway complex surrounding rock: analysis of in-situ stress field in Huainan mining area [R]. Huainan: Zhu Ji Xi coal mine. 2012:7-8.

IOP Conf. Series: Earth and Environmental Science **61** (2017) 012153 doi:10.1088/1755-1315/61/1/012153

# Strategy research of harbin city green transport and sustainable development from low carbon ecological perspective

**Wang Qiang, Qi Xiao-jie**

Heilongjiang Institute of Technology, School of Automobile and Traffic Engineering, Harbin
630702666@qq.com

**Abstract**. With quick development of urbanization and mechanization, there exist some problems in the cities, such as traffic jam, traffic safety, and traffic pollution and so on. It is extremely urgent for the city to develop green transport, in order to relieve these problems and push forward low carbon ecological construction in Harbin. Strategy research of Harbin city green transport and sustainable development is done from the eight aspects of building public transport system of integration, bicycle, walking, and slow-moving system and so on based on analyzing demands of low carbon ecology on city green transport development, and Harbin traffic development state.

## 1. Low carbon ecology demands on urban green transport development

Green transport indicates a kind of traffic concept which traffic tools with less pollution and suitable to urban environment are used to complete socially economical activities. It is put forward under the objective condition which urban and urban traffic face a series traffic problems and developing bottleneck. Green transport comes from harmonious transport system based on sustainable development transport idea. Developing green transport aims at decreasing urban air pollution and noise pollution from transport, and green transport should embody uniform combination of three aspects of smooth and order, safety and comfort, and low energy consumption and less pollution. Connotation can be shown in Table 1, its overall goal is to pursue sustainable economy, sustainable social development, sustainable environment and sustainable resource.

**Tab.1 Connotation of green transport**

| Content | Connotation | Specific requirement | Core realization pathway |
|---|---|---|---|
| Sustainable economy | High efficiency | Travel efficiency optimum of all social strata | Transport development mode of "public transport plus moving slowly" |
| Sustainably social development | Safety | Minimum traffic accident rate | |
| | Fair | Satisfying travel demand of all social strata | |
| Sustainable environment | Environment protection | Less traffic pollution | |
| Sustainable resource | Low consumption | Water conservation, field conservation, energy conservation, wealth conservation | |

## 2. Transport development state in Harbin

Traveling mode of citizen in Harbin is relatively unitary, at present, public traveling ways are only public buses on ground and line 1 of subway. Citizen's travel is influenced due to urban crowd, in the meanwhile, ecological environment and resource of city are also destroyed by fast mechanism. Under this condition, green transport is suitable to real needs of traffic development in Harbin . Harbin will develops its green transport because of its qualification and condition, however, traffic planning, management, supervision, infrastructure and so on should be further improved. The proposal of general planning 2030 of Harbin points out that estimation of carbon emission amount in Harbin will be calculated including carbon emission of the main aspects of industry, building industry, and traffic, and total carbon of vegetation, in which carbon emission of traffic comes from direct emission caused by energy consumption of automobile. By 2030, total amount of carbon emission in Harbin will be about 218950 thousand tons, but state standard of carbon emission is 28980 thousand tons at that time, which is 7.55 times of carbon emission standard. Moreover, traffic carbon emission will be estimated to be13860 thousand tons, which takes up 6% gross carbon emission. Therefore, low carbon, energy saving, and carbon emission reduction need to be realized in every walk of life, in order to realize overall goal of carbon emission. Among them is traffic domain, highway transportation must use automobiles burning clean energy, and citizens must choose travel mode of low carbon traffic. Development mainstream of future traffic in Harbin is low carbon, pollution-free, and high efficiency, and green "slow moving traffic" will become new trend of traffic system development in Harbin. It is predicted that contribution rate of public traffic in Harbin will have been 50% by 2030, and walking contribution rate 20%, in which trail traffic and BRT system will have accounted for 40% of contribution rate.

## 3 Development strategy of green transport and sustainable in Harbin

Lands for city are relatively stressful in Harbin, because of a large population and less land resources. So basic population density condition for developing public traffic is possessed by Harbin, then it is a modern road to develop public traffic with high efficiency and less pollution which is suitable to Harbin. Public traffic development in Harbin should be done according to actual situation, feasible strategies should be made combining near-term and long-term plans, for this reason, convenient, shortcut, multi-level and integrative public traffic system would be constructed. Division of labor of kinds of traffic tool modes is clear, link is close, and transfer is convenient, therefore, efficient network system is formed. The writer put forward eight strategies, in order to realize goals of low carbon, ecology, and green transport.

### *3.1 Building dynamic network with effective connection which comprises subway, light railway and conventional railway*

Subway and light railway will be backbone system of public traffic in Harbin downtown in the future, and be main strength to relieve traffic pressure of passenger transport within downtown, but only one or two lines can't play a role, only to form transport network will can bear large amount of public passenger transport. Light railway facilities should be planned and constructed as soon as possible in Harbin, subway, light railway and conventional railway are effectively connected for convenient transfer, thus integrative and dynamic network can be formed. Overall planning of subway construction in Harbin is "nine lines and one loop line", and twelve lines, one loop line, and two straight line are preinstalled. At present, subway line 1 has been operated in Harbin, and the whole construction cycle is about twenty years. Unified coordination and synchronous development are considered in process of construction of rail building, light railway, conventional railway and conventional public transport system. Moreover, connection and transfer among them must be thought out, so as to form dynamic network of effective connection.

### *3.2 Restoring and planning to use railroad and trolley bus*

Railroad and trolley bus as the conventional public bus can realize true "green traffic tools " with zero emission and pollution-free. Under the situation of emphasizing ecology, environment protection and sustainable development, railway and trolley buses should be paid more attention to and developed greatly. New style railway bus has been recovered in Shanghai, Tianjin, Changchun and so on, whose features of efficiency, economy, and environment protection have been generally accepted. They can not only bear public transport task, but also become a beautiful scenery in city. Railway and trolley

IOP Conf. Series: Earth and Environmental Science **61** (2017) 012153 doi:10.1088/1755-1315/61/1/012153

buses have long operating history in Harbin, recovery or planning of new railway and trolley bus operation will become key element for leading "ice city -Harbin" into green transport era.

### *3.3 Improving public bus attraction degree to citizen travelling*

Harbin became first batch demonstration city of "public transport metropolis" in China in 2012. Developing green public transport needs to use advanced and environment protecting new energy automobile, and public transport lines should be reasonably and scientifically planned, in order to form ground public transport network system with perfect function and obvious level. Application of intelligent transport technique should be paid more attention, and intelligent public transport system with humanization should be built. Public transport stations should be rationally distributed, boarding environment around public transport station should be improved, and enclosed waiting stations are proposed to construct because of much colder winter in Harbin. Drivers' driving technique and quality should be increased, and boarding environment inside public buses should be made better. Public transport ticket system with cheap price and great variety should be constructed, transfer cost of passengers' would be reduced, thus various and human-based service would be offered. BRT system and public transportation lane are set up, and precedence system of public transport signal is established, so as to increase public transport speed, reduce delay time, at last, service level of public transport can be constantly improved. Integration public transport of city and countryside and town and village should be built and completed, especially the last one kilometer public transport of "from door to door", therefore, public transport can increase attraction to traveling choice of citizen.

### *3.4 Carefully designed, convenient, and shortcut transfer pivot*

According to long-term development, land usage and fiscal capacity of Harbin, public transport is generally planned. Connection and transfer of kinds of traffic modes will directly influence quality of traffic network, so it is significant that transfer pivot should be carefully designed, in order to make passengers transfer at will. Priority system of public transport should be established, and public transportation lanes should be set up, so that idea priority, facility priority, efficiency priority, management priority, and safety priority of public transport will be gotten to. A set of complete strategy plan of transportation should be constructed, and effective transfer would be realized among rail transport and other traffic modes. According to features of different passenger flow volume and different passenger demand, all public transport tools would be combined by different level. Railway transport and fast public as backbone of public transport system are laid along passenger transport corridor of compact passenger flow. Conventional public transport lines should be designed by different level, such as conventional public transport backbone line, conventional public transport subsidiary line, conventional public transport branch line, and all these lines combine public transport pivots to realize that different lines can connect smoothly, and satisfy passengers' demands.

### *3.5 Paying more attention to protecting bikes and pedestrians and building comfortable environment for pedestrians*

Bicycle and walking traffic are the short-distance, effective, pollution-free traffic modes which are good at health and environment and should be greatly encouraged. However, there is a very cold and long winter in Harbin, ice roads are the common situation, so it is unwilling for citizens to choose bicycle to travel. Except this reason, safety by bicycle in Harbin can't be guaranteed as well. At present, there aren't bicycle lanes on many road, existing bicycle lanes are mixed with automobile lanes without physical partitions, such as guardrails or other facilities, moreover, continuity of running spaces for bicycles can't be well guaranteed, so some automobiles usually and illegally occupy bicycle lanes. For this reason, it is vital to attach importance to reasonable planning and construction of special lanes for bicycles and pedestrians and establishing bicycle channel and partitions, in order to develop green transport and realize sustainable development in Harbin zone. In the meantime, snow and ice on roads must be eliminated promptly and traffic lights for bicycles are set up at crossroads to ensure safety for traveling by bicycles. Public bicycle systems are built at main communities, commercial centers, and public transport bus stations, short-distance traveling superior of bicycle should be made full use of, and close connection between bicycles and public transport buses should be realized.

At the present time, planning of pedestrian street is not valued in Harbin and few planning is to aim at pedestrian street because of no more concerns for pedestrians, so that there aren't integration and continuity in current pedestrian street system, and there are some problems, for example: pedestrian spaces are occupied, and so on. In view of this, constructing comfortable walking environment should be attached importance to, and special pedestrian roads and overpasses at crossroads are built. Vegetation and afforestation should be reasonably arranged for pedestrian's sun-proof in summer, sceneries of pedestrian street are made taking advantage of ice and snow in winter, thus, more citizens are naturally absorbed to travel by foot because they would relax their mind and enjoy their life in such environment. In the view of this, water system and green pergola in Harbin, important scene sites, and green areas in parks of different level will be well made use of, furthermore, athletic facilities and culture of city should be connected to construct the integrate slow-moving system of "bicycle plus walking " which linking the whole city.

### *3.6 Establish mass participating system*

Planning, construction, and management of urban transport are closely interrelated to people's daily life, and implementation of green transport is unable to do without mass' participation. Traffic tool choice of green transport is a determinative question of comprehensive transport and life quality, it needs consensus of people in community to survey new human value again, so that green transport tools are chosen as one of their life styles. Mass' participating in green transport is different from their participating in other activities, or environment protection, and it is more profound and extensive. Green transport is the effective way to solve traveling quality and life quality which has close connection with mass themselves. It is very significant that people can set up temperate traffic traveling thought by improving their inner quality and idea, so as to change their behavior style through the way of complying with green transport.

## References

[1] Wang Wenyuan, Liu Yingshun, Ye Mao, Yin Fengjun. Strategy Research on Green Transport Development of Low Carbon Ecology New City[J]. Modern City Study, 2015(4):97-103.
[2] He Peichun, Chen Yueqing. Strategy Research on Green Transport Development in Guilin City[J]. Henan Urban Construction Institute Journal, 2013,22(4):24-27.
[3] Yu Yi. Strategy of Solving Urban Traffic Jam in Harbin[J]. Commercial Economy, 2014(8):18-19,119.
[4] Liu Zihe, Gong Xizhen. Green Transport Traveling Mode Analysis between Satellite Town and Centre City[J]. Highway and Motor Vehicle Transport, 2014(5):46-51.
[5] Zhang Xiaochun, Lu Rongjie, Tian Feng. Exploration and Practice of Green Transport Development in Shenzhen City[J]. Green Transport Planning, 2014(5):27-32.
[6] Ding Jinxue, Liang Yuelin. Reviews and Prospects of Urban Green Transport Development[J]. Transport Development, 2013(9):17-21.

IOP Conf. Series: Earth and Environmental Science **61** (2017) 012154 doi:10.1088/1755-1315/61/1/012154

# A Remote Registration Based on MIDAS

**JIN Xin**

Department of Computer Wuhan Polytechnic (Wuhan, China)
38390681@qq.com

**Abstract**. We often need for software registration to protect the interests of the software developers. This article narrated one kind of software long-distance registration technology. The registration method is: place the registration information in a database table, after the procedure starts in check table registration information, if it has registered then the procedure may the normal operation; Otherwise, the customer must input the sequence number and registers through the network on the long-distance server. If it registers successfully, then records the registration information in the database table. This remote registration method can protect the rights of software developers.

## 1. The realization principle The first section in your paper

MIDAS（ Multi-tier-Distributed Application Services Suite） is a multi-tier distributed data application technology, this technology is based on the DataSnap database. Programmer do not need to write a special program code for database access, the programmer can more easily and conveniently develop database application system based on C/S. The user of Delphi development platform can easily implement software registration program. The principle can be seen in Figure 1.

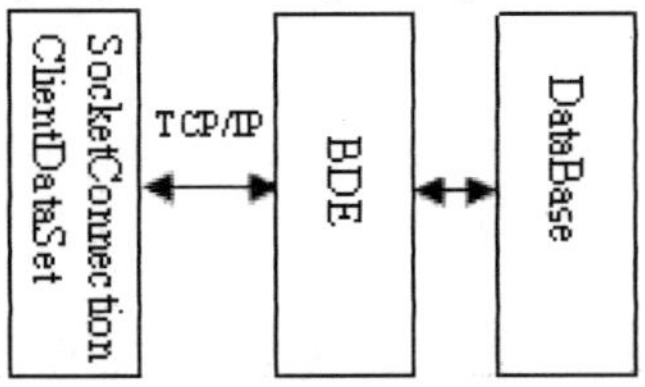


Figure 1. The principle of DataSnap database application system

It can be seen from figure 1, the application software using “BDE SocketConnection” to connections, which need to connect through TCP/IP protocol. “BDE Socket Connection” links to Database by SQL. It is easy to implement in Delphi. We are to realize the network software registration using the multilayer distributed database principle.

## 2. Steps to realize remote registration

After the customer application installed on the local host, will automatically put the table file “register.db” corresponding to the installation folder. Table register.db contains a field named "registered", the data_field type is Logical. In the remote login server, we install a table named server.db, server.db contains three fields: the "serialno", the "reged_num" and the total_num.

The realization of software DataSnap based on remote registration are as follows:

(1) After the client program start, We open the local host table file “c:\windows\ register.db”(for secrecy, in reality you can use file irrelevant, the extension does not need to be “db”, the file is hidden, placed in the irrelevant folder). Read the field reged_num in the file “server.db”. If the "registered"

value is true, it indicating that the user is a valid user, then it display module interface, hide the registration interface, you can use the client application.

(2) If the "registered" is the value of the false, it means the user has not registered. Program hiding function module interface, it displays the registration interface, the user must enter the serial number. The client software to the input serial number and the remote server in server.db each record serialno comparison. If the file "server.db" does not exist in the serial number of the input, this means that the client software is illegal. The Client software will be forcibly end run.

(3) If server.db exists in the input serial number. Is the serial number corresponding to the serialno and total_num. If the reged_num is equal to total_num, registered failure, end the client application. If the serialno we input is in the file "server.db". Put the serial number corresponding to the reged_num plus one, and the client in the file register.db "registered" set to true, prompt software registered successful. It display function module interface, hide the registration screen. At this point, based on DataSnap software registration process is complete.

## 3. Specific implementation based on datasnap software registration

### *3.1 Establish a remote login information table file "server.db"*

We establish the remote server registration information table file "server.db". The table has three fields: serialno、 reged_num and total_num. We save the file to the "d:\ remote login server \database" folder. The table file as shown in Figure 2.

Table : server.db

| server | Serino | Server | Reged_num | Total_num |
|---|---|---|---|---|
| 1 | 123456 | 1.00 | 0.00 | 1.00 |
| 2 | abcdef | 2.00 | 0.00 | 10.00 |
| 3 | hello | 3.00 | 0.00 | 20.00 |

Figure 2. The table file of server.db

The meaning of each field is as the follows:

Serialno: The services provided to customers, the strings for the use of register. If there is a serialno in the register file, the user is allowed to register. If the serialno of input is not exists in the register file, it means the user is illegal, will be prohibited to register the software.

reged_num: The serialno corresponding to the number of registered.

total_num: This software allows the serialno of the registration number.

In order to make registration more powerful software, different serialno allows registration number may vary. When the reged_num is less than the total_num, it allows the serialno to register. If the reged_num is equal to total_num, this serialno is not allowed to register, customers must buy new serialno.

A serialno is allowed to register multiple; this is in order to meet the needs of users to install software on multiple machines.

### *3.2 Establish the remote registered server*

The remote registered server is connected to the table file "server.db"; provide the software registration service for customers. The establishment of the registration server comprises the following steps:

(1) We using Delphi development platform, Establishment of an application. Add four controls on the form: Label、Table、DataSource、DBGrid and DBNavigator. The main properties of these controls such as table 1. Save the form and project, named "zhucefwqunit1.pas" and "zcfwq.dpr". Unit file and project files are saved to had built in folder "d:\ remote login server \database".

IOP Conf. Series: Earth and Environmental Science **61** (2017) 012154 doi:10.1088/1755-1315/61/1/012154

**Table 1.** The properties of the controls

| control | property | property value |
|---|---|---|
| Table1 | DataBaseName | d:\ remote login server \database |
| | TableName | server.db |
| | Active | True |
| DataSource1 | DataSet | Table1 |
| DBGrid1 | DataSource | DataSource1 |
| DBNavigator1 | DataSource | DataSource1 |

(2) We establishment of remote data module. In accordance with the order of execution of the menu command: [File]→[New]→[Others], open the New Item dialog, select "Multitier" page, select "Remote Data Module", save the unit named "socketzc.pas", save the class named "socketzhuce". The saving location is "d:\ remote login server \database".

(3) We add a "table" control and a "DataSetProvider" control for the remote data module. The main properties of these controls such as table 2.

**Table 2.** Remote data module control's properties

| control | property | property value |
|---|---|---|
| Table1 | DataBaseName | d:\ remote login server \database |
| | TableName | server.db |
| | Active | true |
| DataSetProvider1 | DataSet | Table1 |

Explain: The DatabaseName property Tabel1 must be set to an absolute path.

(4) After we run Socket proxy server(in the folder "Borland\Delphi7\Bin"). The file name is "scktsrvr.exe", this can be seen when the server has been running in the lower right corner of the icon windows.

(5) After we run the registration server, the registration server is registered in the proxy server.

So far, the registration server has been running normally, as shown in Figure 3 and 4.

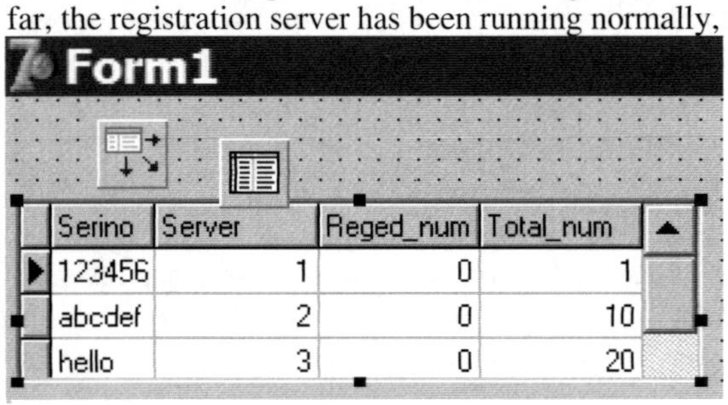


Figure 3. the remote register server

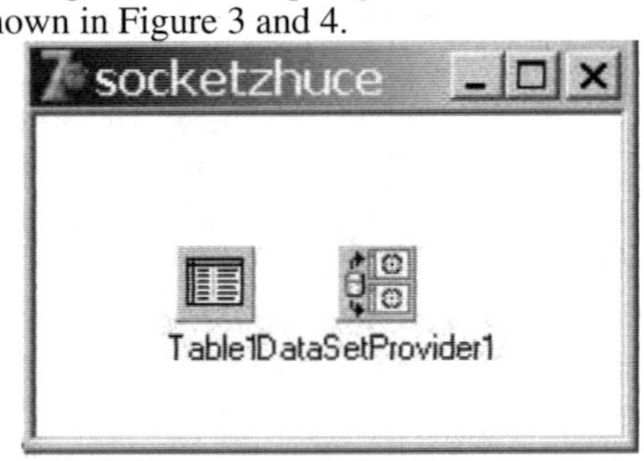


Figure 4. the remote data module

### *3.3 The establishment of the client application*

We first create a file "register.db" on the client application, this table contains only one field "registered", the data type is "logical", the establishment of the client application, this application contains two forms named "form1" and "form2", the user can use the form1 to software registration,form2 is a software function area. Add controls "SocketConnection", "ClientDataset", "Table" in form Form1. The table of "register.db" is shown in Figure 5, and the form1 is shown in Figure 6.

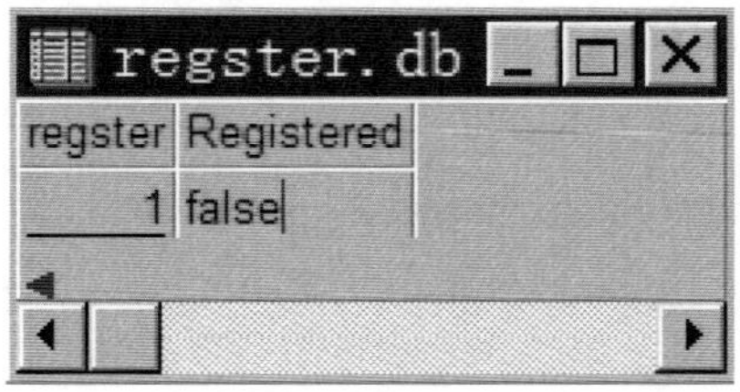


Figure 5. The table of "register.db"

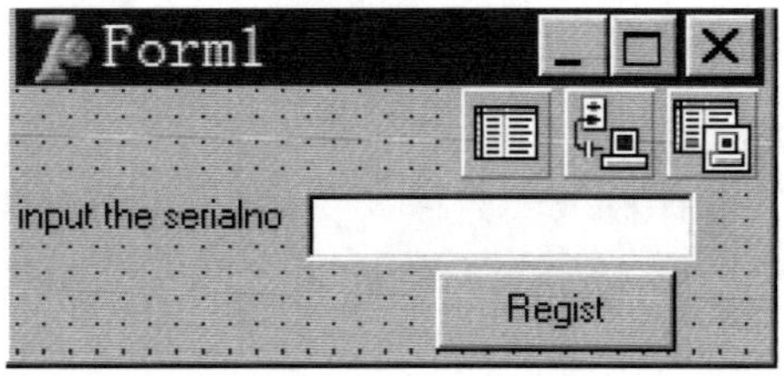


Figure 6. The registration form form1

Some important property of the controls is in Table 3.

**Table 3.** Some important property of the controls

| control | property | property value |
|---|---|---|
| SocketConnection1 | address | 127.0.0.1 |
| | Port | 211 |
| | ServerName | zcfwq.socketzhuce |
| | Connected | false |
| ClientDataSet1 | RemoteServer | SocketConnection1 |
| | ProviderName | DataSetProvider1 |
| | Active | false |

Explain: Control SocketConnection is responsible for connecting to the remote server; Control ClientDataset1 is responsible for connecting to the remote table file in server.db by SocketConnection1. The SocketConnection connected property and the ClientDataSet1's Active property is set to false. In the implementation of the registration operation and set them to true. The properties of Table do not need to set, its properties set to complete in the program through the code.

We save the program to the "d:\local". In the "procedure TForm1.FormActivate(Sender: TObject) " event process to determine whether the client application has registered, If no registration, registration is required before it can be normal use.

The key algorithm is as follows:

```
procedure TForm1.FormActivate(Sender: TObject);
 var
   path: array[0..MAX_PATH] of Char;
begin
  GetWindowsDirectory(path, MAX_PATH);
  table1.DatabaseName:=path;
  table1.TableName:='server.db';
  table1.Active:=true;
  table1.First;
  if table1.FieldByName(' You have successfully registered.').AsBoolean=false
  then begin showmessage(' You have not registered'); form1.Show;   form2.hide;  end
  else begin form2.show;  form1.Hide;  end;
end;
```

In the button1 write event process key code is as follows:

```
procedure TForm1.Button1Click(Sender: TObject);
begin
    socketconnection1.Connected:=true;
    clientdataset1.Active:=true;
    clientdataset1.First;
     if clientdataset1.Locate(' serialno ',edit1.Text,[loPartialKey]) then
      if clientdataset1.FieldByName('Reged_num').AsInteger
```

```
    <clientdataset1.FieldByName('Total_num').AsInteger then
    begin
     clientdataset1.Edit;
    clientdataset1.FieldByName(' Total_num ').AsInteger:=clientdataset1.FieldByName
     (' Total_num ').AsInteger+1;
      clientdataset1.Post;
      clientdataset1.ApplyUpdates(2);
      table1.Edit;
      table1.FieldByName(' registered ').AsBoolean:=true;
      table1.Post;
      showmessage(' Congratulations on your successful registration ');
     form1.Hide;
     form2.Show;
   end else
   begin
    showmessage(' The serial number of the registration number is full, Please buy a serial number ');
    application.Terminate;
   end
  else begin
     showmessage(' You are an illegal user ');
     application.Terminate;
    end;
end;
```

So far, the remote registry software has been implemented.

## 4. Conclusion

Based on the remote registry DataSnap software can solve a sequence number is arranged in a multi machine problems. Author throws away a brick in order to get a gem, hope that readers can infer other things from one fact. For example: to add a field in the table server.db, this field represents a registration allows the number of runs, The field "registered" of local table "register.db" changed into "remaining number". In this way, the registration system function will be stronger.

**Acknowledgment**

This work is supported by 2014 year Hubei Province Department of education science research project, the project name is “Design and implementation of online examination platform”).

## References

[1] HOU taiping,TONG aihong.Delphi Database programming[M].Beijing:Tsinghua University press.
[2] Product R & D Center of FEISI Science and technology. Delphi 6 distributed development[M].Beijing: Publishing House of electronics industry.
[3] JING Hai.The Delphi network programming[M].Beijing: China Railway Publishing House.
[4] Ray Liscbner.Delphi technical manuals[M].Bejing:China electric power press.
[5] LI Huan,Gaoling,LIU ling,XIN Bin. Improved algorithm for generating random CAPTCHA[J]. Journal of Computer Applications. Vol.30 No.6 June 2010:1501-1504.

IOP Conf. Series: Earth and Environmental Science **61** (2017) 012155 doi:10.1088/1755-1315/61/1/012155

# The application and value of virtual animation in Jinzhai revolutionary site

**Gang Liu, Xiaocheng Zhou**

College of Animation,Anhui Xinhua University, Hefei, 230088,china
729334333@qq.com,zhou_xiao_cheng@163.com

**Abstract**. A virtual interactive technology in Jinzhai revolutionary site virtual interactive animation, with "Internet plus Jinzhai revolutionary site virtual animation" spread display, to achieve a variety of ways to carry forward the patriotic education, Jinzhai red culture, but also promote benign circulation of red tourism and economic development of Jinzhai revolutionary site protection investment group.

## 1. Jinzhai revolutionary site introduction

Anhui city of Lu'an Province, Dabie Mountain is a national AAAA level Geological Park, located in the central region of China Anhui Province, Hubei Province, Henan Province, at the junction of three provinces, East Nanjing, west of Wuhan, is located in the middle position of Nanjing and Wuhan two city, with its unique geographical environment and geographical barrier, thus highlights of Dabie Mountain has an important military value of the Republic of China, the capital city of Nanjing and the central city of Wuhan, and Jinzhai County in the middle area of the Dabie Mountains Revolutionary Base, has a large number of revolutionary sites according to the. In 1924, Jiang Guangci, a famous revolutionary pioneer in our country, returned to his hometown from the Soviet Union to carry out the development of Party members on the basis of the local situation, and the Communist Party of China (CPC) was established in the area of swallow River and Nanxi. The 1929 outbreak of the famous "Lixia uprising and 6 suddenly uprise", then set up Chinese Red Army thirty-second division thirty-third division, as Chinese Hubei, Henan, Anhui open communist revolutionary base established immortal meritorious service [1]. In 1947, Liu Baicheng, led by the army of thousands of miles into the Dabie Mountains, the liberation of the whole territory, after this place name changed to Jinzhai county.

Jinzhai county is located in Anhui city of Lu'an Province, and adjacent to Henan, Hubei two provinces, in Heaven Village, Mazongling, Jingangtai, swallow River Canyon, Xiang Shan Temple, Lenin primary school site, Lixia uprising site such as nature reserves and cultural relics protection units. Is the largest county in Anhui province the largest area of red culture tourism resources, since ancient times is the points of battle, military transportation is extremely important. Jinzhai County revolutionary site reflects the magic of Mao Zedong thought to guide the rural revolutionary Chinese surrounding the city to victory and took the lead in the lead in establishing the party organization, party organization, led by agricultural workers and soldiers transported an armed uprising, to the countryside to encircle the city of victory, the establishment of democratic regime. Jinzhai County revolutionary site is treasure of Chinese revolutionary historical heritage, is the best place to carry out the education of patriotism and revolutionary tradition, every year attracts numerous audiences around the country to accept the revolutionary patriotic education, Burma revolutionary martyrs [2].

IOP Conf. Series: Earth and Environmental Science **61** (2017) 012155 doi:10.1088/1755-1315/61/1/012155

Our party and state leaders have always attached importance to the protection of cultural heritage, and has promulgated a series of laws and regulations for the protection of cultural relics, a large number of important cultural heritage value of the revolution has get better protection to sites and ancient buildings. In the late twentieth Century, China has accelerated the pace of the world cultural heritage declaration, many revolutionary base sites, cultural relics, ancient buildings and cities are listed or will be included in the world cultural heritage [3].

## 2. The application of virtual animation in Jinzhai revolutionary site roaming display

Virtual animation is the form of animation in the virtual reality, which is based on the concept of virtual reality and animation. Is the computer as the carrier, installation and operation of 3D animation software in the computer, based on the establishment of a virtual three-dimensional space, then according to the size of the object shape structure, the proportion of the construction of virtual scene model, combined with the predetermined trajectory angle setting model, according to the virtual camera movement and other animation parameters assigned on the model material and set the lights to simulate the visual, auditory, tactile and other sensory, as general view personally on the scene in the virtual 3D space of the people and things [1]. Virtual animation is widely used in many fields such as education, medicine, military, games, entertainment and so on. The development of the existing animation production technology in Jinzhai revolutionary site virtual roaming animation, through the original appearance of the site of reduction and revolutionary history allusions are well represented in the form of animation 3D animation modeling technology, and with the help of the audience for animated films, to meet the audience's visual needs.

### *2.1. Application of virtual animation*

The application of virtual animation technology in Jinzhai revolutionary site roaming display, can recover Jinzhai revolutionary site, combining with the audience favorite anime, is intuitive and interactive.

Intuitive, real scene simulation model using 1:1 proportional to build 3D modeling technology, combined with the animation of the map can be the performance of the real scene, application plus music, sound effects and effects of the 360 degree performance of Jinzhai revolutionary site three-dimensional sense of space, very real drawn on the inside and outside scenery, increasing its ornamental.

Interactivity is one of the most important aspects of modern virtual animation. The audience can be changed from passive viewing to active appreciation. Is no longer in the body to watch, but into the scene, there is an immersive tour.

### *2.2. Jinzhai mining revolutionary connotation*

Jinzhai revolutionary site an important position in the military area, the occurrence of some major revolutionary battle, Jinzhai revolutionary site layout and relevance and touching stories capable of evoking praises and tears full comb, so that these elements and modern animation art combined with visual effects and animation art and good 3D animation technology, virtual exhibition add glory to the Jinzhai revolutionary site virtual roaming animation effect, greatly show the cultural connotation of Jinzhai revolutionary site.

## 3. The application of virtual animation in Jinzhai revolutionary site roaming display

Virtual animation in the application value of Jinzhai revolutionary site With the acceleration of the process of information, people through computers, mobile phones, tablet PCs and other mobile communication devices to connect to the Internet, you can carry out activities in all aspects of resources. Jinzhai is a revolutionary site patriotic education, to commemorate the martyrs and the local tourism resources, if can "Internet plus revolutionary use of 3D animation technology and interactive technology of virtual roaming display", using the Internet can more show, let more people familiar with Jinzhai revolutionary site, has important significance.

### *3.1. Innovation of patriotism education mode*

The traditional patriotism education is the form of education groups to the scene to watch, hear, feel, etc.. If the group of minors by the school or organization to patriotism education base, it is necessary to use financial resources, material resources, human resources and to prevent the occurrence of various security incidents. Jinzhai revolutionary site culture, using Internet plus virtual animation display mode in the era of rapid development of information technology, carry forward the spirit of martyrs. Change the distance between the region is effectively reduced due to and from the patriotic education base cost of manpower, financial and material resources; changed due to a large number of people into the base, causing the destruction of the revolutionary site activities, through the existing media can achieve real-time education needs.

### *3.2. Jinzhai red cultural tourism*

Anhui provincial government issued the implementation of the Dabie old revolutionary base areas the revitalization and development planning and implementation plan (hereinafter referred to as the "plan"). Planning the development goal of our province Dabie mountain old revolutionary base areas during 13th Five-Year that promote Wanxi traditional villages in the overall protection and utilization, put the Dabie Mountain in Jinzhai County of Anhui developed to become well-known red tourist attractions and important cultural, ecological and leisure tourism destination. Jinzhai group is a revolutionary site in the Dabie Mountains, as the development and promotion of local red culture tourism resources, promoting local culture connotation construction and tourism brand effect, thus driving the rapid development of Dabie Mountain area economy, improve the local fiscal revenue and the improvement of living standards.

The realization of Jinzhai revolutionary site roaming virtual animation project, hand pull the Jinzhai tourism economic development and expand the influence of Jinzhai, increasing the local finance income; on the other hand to break the geographical restrictions, to achieve inter city, provincial and even cross-border spread, let more people familiar with Jinzhai Red Culture Revolution site.

### *3.3. Inherited the spirit of revolutionary martyrs*

Anhui Jinzhai revolutionary site has written numerous heroes, such as Liu Baicheng, Deng Xiaoping and other old revolutionary according to the CPC Central Committee's strategic deployment requirements, led by Shanxi and Hebei, Shandong and Henan and other army from Dabie Mountain, opened China people's war of liberation from strategic defense to strategic offensive prelude. In the Dabie Mountains as the center of the Jianghuai region, has wiped out 190 thousand people, according to the Central Plains to the re-establishment of the people's war of liberation, reversing the situation, in order to win the victory, achievements of the immortal exploits [1]; the Hubei Henan Anhui district will be set up in the Dabie Mountains, and to carry out the training of cadres, made one after another the great victory of the revolutionary struggle leading Henan area. These historical events through the virtual animation rendering technology, show in the world's eyes, the revolutionary deeds further inherit and carry forward the construction of spiritual civilization, is the indispensable resources for Chinese people, let the world know the spirit of martyrs and suggest that people learn from history, make people not easily won the peace.

## 4. Summary

Jinzhai County revolutionary site group because of various reasons by different degree of damage, some valuable cultural resources or even disappear very painful, thinking about the contemporary cultural workers, cultural workers, scholars and other groups, how to break the industry integration and development space. It is of great significance to apply the new technology to the protection and innovation, inheritance and development of traditional culture. Application of 3D animation technology, virtual reality technology and Jinzhai revolutionary site group combined with the development of virtual animation Jinzhai revolutionary site group roaming works, the characters, story, site restoration, environmental revolution landscape, every detail is not missing, are all displayed in front of. The use of the Internet virtual animation Jinzhai revolutionary site roaming works uploaded

IOP Conf. Series: Earth and Environmental Science **61** (2017) 012155 doi:10.1088/1755-1315/61/1/012155

to the network platform for more people to visit and study. In order to change the traditional school youth patriotism education form, inheritance martyrs will; can not only enhance the cultural tourism resources in Jinzhai propaganda, carry forward the revolutionary spirit; and then promote the development of local economy, thus protecting the revolution in Jinzhai to re-enter the site; the industry to achieve cross fusion for innovation the development of.

**References**

[1] Wu Yahui. Creation of 3D animation of product [J]. ship engineering, 2015 (2): 206-211.
[2] ho Hui. [J]. plays a important role of popular literature on "Jinzhai revolutionary site" in Henan area, 2015 (22): 171-171.
[3] Zhang Jiachen, Zhang Rong. Study on the protection and utilization of the existing buildings and historic buildings [C]. planning and Innovation: Proceedings of the 2014 annual meeting of Chinese urban planning, 2014 (22): 67-72.
[4] Excellent young talents in colleges and universities in anhui province support plan key projects (gxyqZD2016387)
[5] Colleges and universities in anhui province (SK2015A673) as a key project of humanities and social science research; Colleges and universities in anhui province the humanities and social sciences research project (SK2013B324) generally
[6] Anhui xinhua college humanities and social science research project (2015 rw003)
[7] Research on the application of virtual 3D animation technology in the design of human machine interface. (Item number: 2016td004)

IOP Conf. Series: Earth and Environmental Science **61** (2017) 012156 doi:10.1088/1755-1315/61/1/012156

# Analysis on policies text of air pollution control in Beijing

**Yujuan ZHANG*[1], Wen WANG[1], Wei ZHANG[1]**

*[1]Beijing Institute of Science and Technology Information, Beijing City, 100032
* zhangyujuan427@126.com

**Abstract:** Air pollution is one of the most serious environmental problems, and it is also the inevitable result of the extensive economic development mode. The matter of air pollution in Beijing is becoming more and more serious since 2010, which has a great impact on the normal social production, living and human health. These hazards have been highly valued by the whole society. More than 30 years have been pasted since controlling the air pollution and the system of policies was relatively complete. These policies have improved the quality of atmospheric and prevented environment further deterioration. The policies performance is not obvious. It is urgent trouble to improve policy performance. This paper analyzes the 103 policies text of air pollution control in Beijing, and researches status, history and problems, and put forward suggestions on policy improvement and innovation at last.

## 1. Background

Beijing Municipal Environmental Protection Bureau issued "Beijing Annual Air Quality Status Report in 2016" in January 3, 2017. Although the main pollutants, such as serious and frequent haze, PM2.5 index were downward in recent years, all these have critical threat people's normal life and health. Kinds of environmental accidents sounded the alarm: a grave problem of environmental pollution has imperiled people's health and life; the economic has suffered serious losses; we are facing severe situation of environmental pollution, environmental problem requires urgent attention and solution.

It is an objective requirement of Beijing as the capital and the vanguard in the developing countries which aim at the international high-end form and constantly improve the level of internationalization of the city. But the environment matter has become barrier and hindered the pace of the construction of the world city. The environmental pollution not only brings huge economic losses, but also threatens public health threat. It is fortunately the government has great pay attention to. The disposal needs to learn the advanced experience of developed countries, and to explore suitable countermeasures for the situation of our country and city, accurately holds the regional policy differences and orderliness. These are of practical significance to perfect the system of policy, promote the economy of Beijing and social sustainable. At the same time policies may present a useful reference to relieve.

## 2. Research State of Air Pollution Control Policies

### *2.1. Domestic Research*

With the increasing of air pollution in our country, many researchers have studied the protection and management of air pollution. Zhang Yongan's [1] research was that taking relative policies released by all ministries from 1988 to 2014 as data source, discovering the shortage, putting forward some measures. They deemed that policy standards and technological standards need further improvement, and policy implementation need coordination among different departments so as to avoid conflicts.

The analysis of Wu Liufen [2] pointed out that the formulation and implementation of governance policy was a result after action of public participation and government. The policy has a strong acute characteristic, lacked professional, and was facing potential risks.

Wang [3] analyzed the fiscal policy of air pollution prevention, believed that financial investment of air pollution control was not enough, scale of fiscal expenditure was small, tax policy was not perfect, standard of sewage charges was low, fund management was not standardized, and so on. The research proposed it should increase the financial investment of air pollution control, expand the scale of energy conservation and environmental protection, and optimize the tax policy of air pollution.

Zhao [4] studied policy from the perspective of regional cooperative governance on basis of the analysis framework, which was defining policy, making and operating policy, coordinating policy. The research made a detailed analysis of trouble among three governments of Beijing City, Tianjin City and Hebei Province. At last it tried to seek the implementation of atmosphere pollution of Beijing City, Tianjin City and Hebei Province from the four aspects of the value level, the organizational structure, the realization mechanism and the balance of interests. Zhang [5] reviewed the state, the measures and the deeper reasons of Beijing City, Tianjin City and Hebei Province, made the result that it should focus on deepening the reform of resource tax, to ensure the rationality of investment in industrial pollution control, let environmental tax payments to become economic means of sustainable development of three regions.

### *2.2. Aboard Research*

Researches on air pollution control policy in foreign countries mainly reflected in the developed countries. Peterson [6] pointed out the high standard for sewage could raise the cost of corporate, prompt enterprises to amplify equipment to reduce pollution emissions. Anthony investigated the policy affect of "Road Pricing" and "Vehicle Quota Scheme" actualized on traffic and air pollution caused by automobiles in Singapore. The discuss result was that the reasons for bettering traffic and air pollution were consummate legal and financial measures, intensive land use planning, public transportation system construction, road infrastructure construction and efficient traffic management, not above only two policies [7].

Entering twentieth Century, the air pollution has weakened in developed countries, but the researchers did not cut down the policy protection and measures research. Elietal [8] believed that the mechanism of "Polluter Pays" would debase the efficiency of air pollution control due to the existence of asymmetric information and costs of government regulation. Vinishet Kathuria [9] found that the government strictly sewage charges would not affect the economic effects of enterprises in the long-term. Technical strength and good social situation maybe aggrandize enterprise reputation and enhance competitiveness. Kanada and Fujita [10] looked into "Win-win Policy" outcome for abating carbon dioxide emissions and air pollution between the government and the Private Companies. They take Kawasaki as the research object, deemed that "Win-win Policy" not only making the local industry to reduce air pollution, but also enhancing the efficiency of energy utilization, suggested that preventive methods and enforcement tools would be more propitious to effective implementation of "Win-win Policy" of long-term.

## 3. Contents and Methods

There are two points in this paper. The one is the review of air pollution control policies of Beijing since 1989 to 2016. The other is the analysis on overall policies including the history of policy development, policy objectives and policy system. All these are the content of this article attempts to study.

The research methods and ideas are as follows: The first is in view of survey documents which reorganized the policy texts in the field of environmental protection since 1989 to 2016 using statistical method. The second is that research adopting to quantitative analysis on issued time, policy subject, policy content on the framework of process, policy organization and achieved the research conclusions and policy recommendations.

## 4. Analysis Texts on Air Pollution Policies

### *4.1. Analysis Framework of Policy Texts*

Policy is code of conduct drafted by national or party in order to achieve a task in a certain stage. Its contents is described by policy documents, it is a whole process from policy making to the end of the policy. Policy analysis should consider the whole process of policy activities, including policy formulation, implementation and termination. It is important task to analyze the contents in the paper. Organization survey is based on the organization, such as the legislative and the administrative organs. The process of policy analysis the corresponding policy life cycle from policy formulation to amendment, repeal, and the whole analysis frame. Policy text is the carrier of policy content and idea, it is not easy to discover significant difference from texts. So it needs to take many factors into account in discussing policy text, and reflect the connotation through exterior text.

Based on the consideration of the intrinsic structure and characteristics of policy texts, the article selected the date, the theme and the main body as the main research elements, and carried through quantitative analysis.

### *4.2. Collect Data and Select*

There has issued a lot of policies in order to promote the environmental since 1980s in China. Ministry of Environmental Protection of the People's Republic of China and regions Bureaus often released environmental policy. Beijing Municipal Environmental Protection Bureau follows the national policy, and presents environmental protection policies in the Bureau website. Then researcher may gain these policies from the website.

From the figure 4-1, the text mainly involves the laws and the regulations of environmental protection of national and bureau. There are total of 103 policies as the research subject. The notice class policy is the most, accounting for 67%. Next is circular, accounting for more than 18%. The minimum is the rule, advice and regulations, only one.

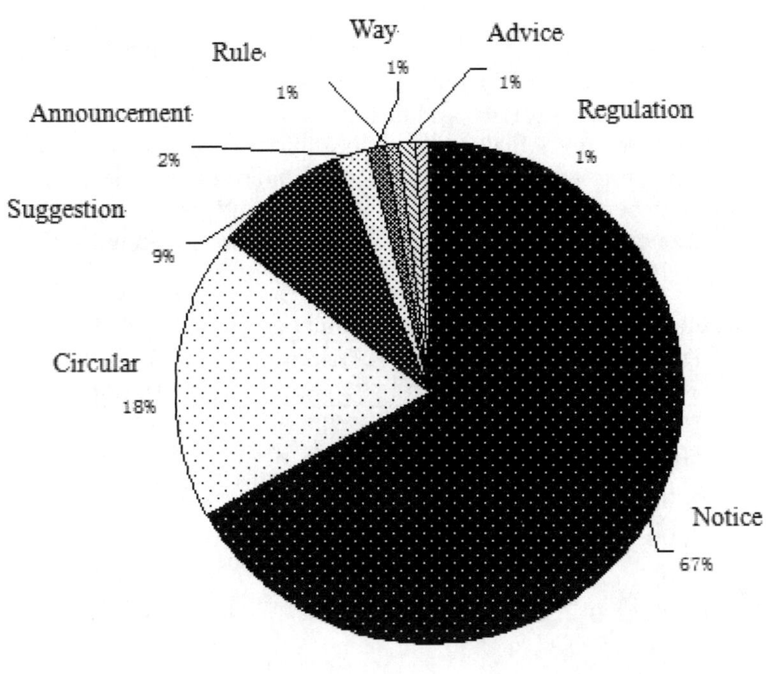


Figure 4-1 Kinds of policies ratio

IOP Conf. Series: Earth and Environmental Science **61** (2017) 012156 doi:10.1088/1755-1315/61/1/012156

**Table 4-1** List of policy of air pollution control

| List | Name | Year | Unit |
|---|---|---|---|
| 1 | Notice of the deadline for the treatment of excessive gasoline vehicles | 1989 | Beijing Government |
| 2 | Administrative punishment measures of the regulations the People's Republic of China on the prevention and control of air pollution | 1990 | Beijing Government |
| 3 | Notice on banning the sale of leaded gasoline to unleaded gasoline | 1997 | Beijing Government |
| 4 | Circular on emergency measures to control air pollution in Beijing | 1998 | Beijing Government |
| 5 | Announcement on the second phase of the city's air pollution control measures | 1999 | Beijing Government |
| … | …… | …… | …… |
| 100 | Notice of the Beijing Municipality on the administration of expert bases for the technical review of environmental impact assessment (for Trial Implementation) | 2016 | Beijing Municipal Environmental Protection Bureau |
| 101 | Circular on printing and distributing the construction plan of Beijing ecological environment monitoring network | 2016 | Beijing Government |
| 102 | Notice of Beijing emergency plan for heavy air pollution (revised in 2016) | 2016 | Beijing Government |
| 103 | Circular on traffic management measures for light duty gasoline vehicles in China I and national emission standards | 2016 | Beijing Municipal Environmental Protection Bureau, Beijing Municipal Commission of Transport, Beijing Traffic Management Bureau |

*4.3. Analysis Issued Time*

This paper has analyzed the promulgation, implementation, modification and abolition of policies, and it has discovered the policy continuity and forward-looking.

In this dissertation, 103 laws and regulations of air pollution control (including the revised version) were collected, and the issued time was investigated. From the table (4-2), it discovered that the number of policies have been greatly improved since 2011, especially after 2013. It deemed that there were two reasons prompting the number increase. The one was the rapid growth of Beijing City, Tianjin City and Hebei Province economy. The second was the PM2.5 index announced by Embassy of the United State in 2013.

**Table 4-2** Policies issued time and number

| Dispatch Time | Dispatch Number | Dispatch Time | Dispatch Number | Dispatch Time | Dispatch Number |
|---|---|---|---|---|---|
| 1989 | 1 | 2003 | 1 | 2011 | 8 |
| 1990 | 1 | 2004 | 2 | 2012 | 6 |
| 1997 | 1 | 2005 | 4 | 2013 | 11 |
| 1998 | 1 | 2006 | 3 | 2014 | 13 |
| 1999 | 5 | 2007 | 3 | 2015 | 16 |
| 2000 | 2 | 2008 | 5 | 2016 | 9 |
| 2001 | 5 | 2009 | 3 | | |
| 2002 | 1 | 2010 | 2 | | |

*4.4. Analysis on policy subject*

With the rapid development of capital economy, atmospheric environmental protection has become an indispensable part of people's life in the capital, and has attracted government more and more attention. The atmospheric environment protection policy plays an important role. More and more of the bureau have participated in policy formulation. The research classified and studied the 103 policies. There are sixteen departments involved in the formulation of policies. Most documents were published by Beijing Government and Beijing Municipal Environmental Protection Bureau. They respectively released 59 and 40. Beijing Municipal Finance Bureau drafted 7 file following above two from table 4-3.

**Table 4-3** Policies issued unit and number

| | Dispatch Unit | Joint Dispatch | Separate Dispatch |
|---|---|---|---|
| 1 | Beijing Municipal People's Congress | | 1 |
| 2 | Beijing Municipal Party Committee | 1 | 1 |
| 3 | Beijing Government | 1 | 58 |
| 4 | Beijing Municipal Environmental Protection Bureau | 14 | 26 |
| 5 | Beijing Municipal Finance Bureau | 7 | 0 |
| 6 | Beijing Municipal Commission of Development and Reform | 4 | 1 |
| 7 | Beijing Traffic Management Bureau | 4 | 0 |
| 8 | Beijing Municipal Commission of Transport | 3 | 0 |
| 9 | Beijing Municipal Administration of Quality and Technology Supervision | 3 | 0 |
| 10 | Beijing Statistical Information Bureau | 1 | 0 |
| 11 | Beijing Supervision Bureau | 1 | 0 |
| 12 | Beijing Municipal Human Resources and Social Security Bureau | 1 | 0 |
| 13 | Beijing Municipal Commission of Economy and Information Technology | 1 | 0 |
| 14 | Beijing Industrial Promotion Bureau | 1 | 0 |
| 15 | Beijing Municipal Commission of Commerce | 1 | 0 |
| 16 | Beijing Municipal Science and Technology Commission | 1 | 0 |

It found that about 88 policies have been determined by a department in all 103 policies. There are 7 documents announced by two branches, 8 documents draw up by three or more bureau or commission. It showed that the bureau cooperation was less in terms of environmental protection in Beijing. And it illustrates that the air pollution was becoming more and more important, the government should set up a special department to deal with and solve the matter of air pollution.

## 5. Whole Analysis of Policy of Air Pollution Control in Beijing

### *5.1. Analysis of Policy History*

The earliest policy of air pollution control was "The decision on prevention of silica dust factories harm" in 1956 in China. The earliest policy of air pollution control was "Regulations of the Beijing for Atmospheric Pollution Prevention and Control Law of the People's Republic" in 1988 in Beijing. Although it was later than the nation, the policies can be divided into four stages.

#### *5.1.1. Begin stage*

The initial stage of atmospheric pollution prevention and control was 1970s and 1980s, and it was also the beginning period of atmospheric environmental protection in Beijing. The main task of this phase

IOP Conf. Series: Earth and Environmental Science **61** (2017) 012156 doi:10.1088/1755-1315/61/1/012156

was to decrease excessive pollution emissions of enterprises and vehicles. "Regulations of the Beijing for Atmospheric Pollution Prevention and Control Law of the People's Republic" was published in the period. It planned set overall requirements of supervision and administration of pollution prevention and control and prevention smoke pollution. In this stage, manage aim was to transfer the boiler, to eliminate of smoke and dust, and to control source. The measure has played a key role. The whole supervise level and achievements were not high.

*5.1.2. Development stage*

The trouble of air pollution was more and more serious in 1990s which the economic development was on the right track, the consumption of energy was rising, especially sulfur dioxide pollution was exacerbating. The main way was No.11 Decree in 1990s drafted by Beijing Municipal People's Government-- "Administrative Punishment Measures of Regulations of the Beijing for Atmospheric Pollution Prevention and Control Law of the People's Republic". The work has achieved three transformations from decentralized to central management, from the end of administration to the whole process of governance, and from single concentration control to total control. The policy implementation performance improved.

*5.1.3. Apply stage*

The major contaminant of air was exhaust gas, acid rain, PM10 and photochemical pollution in 2000-2010s, and presented the characteristics of regional and compound. The government began to strengthen control contaminant. Beijing Government has also increased the intensity of this stage, and introduced about 31 policies of air pollution control. For instant, "Beijing Energy Structure Adjustment Plan" was released in 2001, "Interim Measures for Cleaner Production Audit" was published in 2006. In this period, the contradiction between economic development and environmental protection was highlighted. The national started to study the strategy of air pollution prevention and to control, to put forward the planning map and the overall goal of air pollution prevention and control, actualized sustainable development strategy. The manifestation of air pollution control policy has improved significantly compared with the previous period.

*5.1.4. Transfer stage*

There were four strategic transformations for air pollution prevention and control since 2010. The first was to regulate aim it has shifted to coordinate relationship between the total emission control and the environmental quality. The second was the target which has become coordinated control of multiple pollutants from focusing on coal pollutants. Third was the subject which has converted composite restriction of multiple pollutants from industrial point sources. Fourth was the supervise pattern which has be regional management from possession control. There were numerous policies what has taken in the period. "Beijing Environment Protection and Construction Plan in 12th Five-Year" and "Beijing Air Pollution Control Measures During 2012-2020" were issued in 2011, and "Beijing Clean Air Action Plan in 2013-2017" was released in 2013 and so on. The government began to comprehensively use the air pollution control policies and measures, and established air pollution control policy system at this stage. At the same time the nation strengthened air pollution control legislation, actively promoted economic management means, encouraged the development of environmental protection industry and pursuit of green GDP. However, the upgrading of the industrial structure, the transformation of the mode of economic development, and the reform of government functions has not yet been completed. The objective condition of the establishment of the system was not yet available. Then the administrative order still has played a leading role in air pollution control.

*5.2. Analysis of Policy Characteristic*

There shows the following characteristics of the development of air pollution control policy in Beijing:

*5.2.1. Restricted by economic development*

The reason air pollution control lagging behind the speed of economic development was related to the basic route of the primary stage of socialism which existing "One Center, Two Basic Points", and related to the extensive mode of economic development. During this period, people pursued material well-being living, the government stressed the economic development while ignoring the environmental problems. These maybe destructed the environment. But environment protection has lagged far behind in the environment damage rate.

*5.2.2. Policy means from single to complex change*

The means of air pollution control policy has experienced a transformation from a single administrative order to composite means, formed policy system which including the laws, the regulation, the administrative order, the market economy, the technology, and the public in China. Beijing's air pollution control policy followed the pace of the country. Beijing Government has take n politics solution, market instruments, and technical means, etc. in the framework of nation.

*5.2.3. Dominant Means of Administrative Command*

In a variety of air pollution control policy means, administrative orders have been playing a major role, but also it was the highest performance of government policy means. According to the different period of the atmospheric environmental pollution situation, Beijing Government has held a number of environmental protection conference, and formulated a series of strategic significance policies for the environmental protection. In addition, the government has continued to strengthen the authority and implementation of environmental policy. The government has adopted strong administrative orders to improve air quality, and let weather manifest a short "APEC blue" in 2014. The government has taken strong administrative method once again to create a "Two Sessions blue" in 2015

*5.3. Analysis on Policy Target*

According to the Ministry of public data, policy goal included the air quality standards, the pollutant emission standards, and the related testing and methods standards at present. But the goal has exited the following problems as an important public policy.

The aim system was imperfect. The policy target system based on the economic development has ignored the development of Tianjin City and Hebei Province. Not same as economic level of Beijing City, Tianjin City and Hebei Province demanded diverse methods to fit regional development. In addition, there are differences in resource security. Even if the air pollution situation is the same, it is impossible to achieve the same goal in the same time.

It was not timely of policy objectives update. Focusing on the control technology and norms of air pollution emissions were morn, and not matched the air pollution control strategy. Moreover, most of the standards of air pollution control has not been revised in a timely manner, and has been divorced from the actual situation.

*5.4. Analysis on Policy System*

The environmental policy system generally includes five levels: Law, administrative regulations, departmental rules, standards and planning. Laws, administrative regulations and plans are the basis for the formulation of departmental rules and regulations by the relevant functional departments which announce department standards for air pollution control. It was found that there existed the following questions in Beijing's air pollution control policy.

It lacked laws and regulations. The laws and regulations were shortage in the whole policy system, which lacked authority, not conducive to the implementation of the policy from the national rank. Little evidence for dependence from Beijing rank, it was easy to cause the phenomenon of unclear responsibilities and omission.

It was deficiency of policy authority. From the national degree, the policy grade was lower than aboard. Although "Atmospheric Pollution Prevention and Control Law of the People's Republic of China" was announced, its level was not as good as "Environmental Law". From the Beijing degree, policy show was vague. A great quantity of "should", "encouragement" and other vague words emerged in the policy text. The word "must" was less. These phenomenon leads to the lack of policy authority and it was difficult to effectively implement.

## 6. Suggestion of Air Pollution Control of Beijing

It is a systematic and comprehensive long-term project of air pollution control. There are formed a set of work system including administrative orders, market economy, and technology etc. Yet the stern situation testified that the policy effect has not achieved the desired results. Then it is necessary to further improve and better air pollution control policies to turn the state of air pollution. According to the defects of the current system, drawing lesions of the experience of foreign advanced governance policies, in line with the demand of coordinated development of Beijing City, Tianjin City, and Hebei Province, the paper put forward some suggestions in the next.

### *6.1. Perfect Target System of Air Pollution Control*

According to the defects of the current system, drawing lesions of the experience of foreign advanced governance policies, it may be further perfect the target system through the following aspects.

Setting detailed policy goal includes three aspects. The one is it should build a strategic plan, medium-term plan, and short-term plan. The second is it should utilize clear word in policy text, and put to use explicit quantitative criteria to define the aim of air pollution control. The third is it should pitch Beijing's goal in step with the situation of air pollution the condition of social and economic development.

The government should promote target standard. It should especially have strict requirement for air protection because of air is a special public goods which be bound up with human health. It should revise to conform to the standards to follow closely the international health organization.

The government should update policy goal. Air pollution situation is distinction in different, and harm is not same. It should reissue the air pollution control goal in time keep in step with the condition of air pollution.

### *6.2. Perfect Policy System of Air Pollution Control*

The premise of effective implementation of air pollution control policy is perfect air pollution control policy system. Based on the current air pollution control policy system, the article purposes the next ideas.

The government should perfect the local regulations of air pollution control. There are two sides. The one is to raise the policy grade of air pollution control, and to strengthen policy authority. The other one is to improve supporting policies. From the "Clean Air Act" of USA to introduce explain of supporting policies, it should learn to enhance the operation of laws and regulations in the light of "Atmospheric Pollution Prevention and Control Law of the People's Republic of China".

The government should perfect policy description. Management of air pollution is a comprehensive progress. There are some different subject, are easy to appear the phenomenon of cross function, unknown power and responsibility, which led to the omission of main body. So it should describe body's duty and right, especially a few of subject in the same policy. It may avoid to evade oneself obligation and to laze matters. Meanwhile it may prevent vague language happening in the policy text, and may definite policy implementation and evaluation criteria.

## Reference

[1] Zhang Yongan, Wu Long. On Control of Air Pollution in China Based on Econometric Analysis of Policies [J]. Productivity Research, 2015(1): 122-126.
[2] Wu Liufen, Hong Dayong. On China's Environmental Policy Making and Public Participation in Decision-making: A Case Study on Policy Making in Haze Governance [J]. Journal of Nanjing Tech University (Social Science Edition), 2015(2): 55-62.
[3] Wang Leying. The Analysis on Finance Policy of Air Pollution Control [J]. Manager, 2016(8): 31-31.
[4] Zhao Xinfeng, Yuan Zongwei. Study on the Jing-Jin-Ji Regional Intergovernmental Air

Pollution Control Policy Coordination [J]. Chinese Public Administration, 2014(11): 18-23.
[5] Zhang Yueqing. Select Study on Finance Policy on the Jing-Jin-Ji Regional Air Pollution [J]. China Market, 2016(38): 174-175.
[6] J.M Peterson.Estimating an effluent Charge: The Reserve Mining Case[J]. Land Economics, 1977, 53(3): 328-341.
[7] Anthony T. H. Chin. Containing Air Pollution and Traffic Congestion: Transport Policy and the Environment in Singapore [J]. Atmospheric Environment, 1996, 30(5): 787-801.
[8] Elietal Feinerman,Yakir Plessner. Recycled Effluent: Should the Polluter Pay[J]. American Journal ofAgricultural Economics, 2001, 83(4): 958-971
[9] Vinish Kathuria and Nisar A. Khan. Environmental Compliance versus Growth: Lessons from Malaysia's Regulations on Palm Oil Mills [J]. Economic & Political Weekly, 2002, 37(39): 3993-3999.
[10] Momoe Kanada, Tsuyoshi Fujita, Minoru Fujii, Satoshi Ohnishi. The Long-term Impacts of Air Pollution Control Policy: Historical Links between Municipal Actions and Industrial Energy Efficiency in Kawasaki City, Japan [J]. Journal of Cleaner Production, 2013, 58(11): 92-101.

IOP Conf. Series: Earth and Environmental Science **61** (2017) 012157 doi:10.1088/1755-1315/61/1/012157

# Construction and applied research of low-carbon building evaluation index system

**Zhongwen Liu[1], Xiaohong Dong[2], Pengzhao Gao[1]**

1 Shandong Women's University, Jinan, Shandong, 250300;
2 Qilu Normal University, Jinan, Shandong, 250200
Lzw3541@163.com

**Abstract.** Energy conservation in building is a key link on alleviating energy-deficient contradiction, improving the quality of human life environment, and realizing sustainable development in our country. In this paper, we construct low-carbon building evaluation index system and evaluation method from five aspects—low-carbon structure, low-carbon materials, low-carbon energy, low-carbon technology and low-carbon management. Finally, taking "Solar Valley" in Dezhou as an example, we make the evaluation to its situation of low-carbon building.

## 1. Introduction

At present, our country is in the period of rapid advance of urbanization. Obviously, urban construction promotes the rapid development of building industry. The increase of energy consumption of building and rise continuously of consumption ratio rise continuously bring great pressure to energy saving and emission reduction in our country. Hence, the low-carbon transformation in building industry is imperative.

China is a building giant. Building industry is not only large energy consumption, but also the pillar industry of national economy. Energy conservation in building is a key link on alleviating energy-deficient contradiction, improving the quality of human life environment, and realizing sustainable development in our country. The development of low-carbon building is an indispensable and important part in low-carbon urban construction. Building industry is one of the three major energy consumption industries. Low-carbon transformation is the urgent requirement and strategic choice to speed up the transformation of economic development patterns and adjustment of economic structure, having great significance for the development of low-carbon economy and energy conservation and emissions reduction.

## 2. Construction of low-carbon building evaluation index system

According to the evaluation index system of green building at home and abroad, the method of AHP fuzzy comprehensive evaluation was used to analyze the construction of low-carbon building evaluation index system comprehensively. Finally, a set of evaluation index system that is suitable to China's low-carbon buildings is put forward.

### *2.1. Low-carbon building Evaluation index system*

In this paper, the evaluation index system on low-carbon building could be divided three levels: object level, criterion level and index level. In addition, we establish 5 first class indexes—low-carbon structure, low-carbon materials, low-carbon energy, low-carbon technology, low-carbon management, and 20 second class indexes—energy measurement and monitoring, collective heating, utilization ratio of renewable energy and so on, constructing hierarchical structure model based on low-carbon building evaluation index system, as shown in Table 1. "Low-carbon building evaluation index system" belongs to object level, 5 first class indexes belong to criterion level and 20 second class indexes belong to index level.

**Table 1.** Low-carbon building evaluation index system

| Object level | Criterion level | Index level |
|---|---|---|
| Low- carbon building evaluation index system U | low-carbon structure index U1 | Lighting system U11 |
| | | Heating system U12 |
| | | Flooring system U13 |
| | | Wall system U14 |
| | low-carbon materials index U2 | Using local materials U21 |
| | | Building materials production U22 |
| | | Reuse of materials and waste disposal U23 |
| | low-carbon energy index U3 | Energy measurement and monitoring U31 |
| | | Collective heating U32 |
| | | Utilization ratio of renewable energy U33 |
| | | Utilization ratio of clean energy U34 |
| | low-carbon technology index U4 | Utilization of low-carbon and novel materials U41 |
| | | Introduction of novel construction technique U42 |
| | | Utilization of energy conservation and emissions reduction technique U43 |
| | | Investment of intelligent electronic equipment U44 |
| | | Specialization ratio of building construction team U45 |
| | low-carbon management index U5 | Greening rate U51 |
| | | Talent team construction of real estate U52 |
| | | Waste disposal U53 |
| | | Operations management U54 |

*2.1.1. Low-carbon structure index.*This index mainly refers to the integrated planning and structural system of building in the process of construction. It is mainly to make evaluation from 4 indexes—lighting system, heating system, floors and wall system. The low-carbon structure index evaluation criterion is shown in Table 2.

**Table 2.** Low-carbon structure index evaluation criterion

| Evaluation index | Excellent | Good | Fair | Poor |
|---|---|---|---|---|
| Lighting system | Fully coincident | Basically coincident | Partial coincident | Not coincident |
| Heating system | 4 items are qualified. | 3 items are qualified. | 2 items are qualified. | One or no item is qualified. |
| Flooring system | 4 items are qualified. | 3 items are qualified. | 2 items are qualified. | One or no item is qualified. |
| Wall system | 3 items are qualified. | 2 items are qualified. | 1 item is qualified. | No item is qualified. |

*2.1.2. Low-carbon materials index.*This index refers to whether the used building materials are low-carbon and recycled in the process of construction. It is mainly to make evaluation from 3 indexes—using local materials, building materials production, material recycling and waste management. The low-carbon materials index evaluation criterion is shown in Table 3.

**Table 3.** Low-carbon materials index evaluation criterion

| Evaluation index | Excellent | Good | Fair | Poor |
|---|---|---|---|---|
| Using local materials | 3 items are qualified. | 2 items are qualified. | 1 item is qualified. | No item is qualified. |
| Building materials production | 3 items are qualified. | 2 items are qualified. | 1 item is qualified. | No item is qualified. |
| Reuse of materials and waste disposal | 4 items are qualified. | 3 items are qualified. | 2 items are qualified. | One or no item is qualified. |

*2.1.3. Low-carbon energy index.* This index refers to the effective and sufficient utilization of energy in the process of building construction and operations management. It is mainly to make evaluation from 4 indexes—energy measurement and monitoring, collective heating, utilization ratio of renewable energy and utilization ratio of clean energy. The low-carbon energy index evaluation criterion is shown in Table 4.

**Table 4.** Low-carbon energy index evaluation criterion

| Evaluation index | Excellent | Good | Fair | Poor |
|---|---|---|---|---|
| Energy measurement and monitoring | 3 items are qualified. | 2 items are qualified. | 1 item is qualified. | No item is qualified. |
| Collective heating | 3 items are qualified. | 2 items are qualified. | 1 item is qualified. | No item is qualified. |

*2.1.4. Low-carbon technology index.* This index refers to a series of adopted energy conservation and emissions reduction techniques, the enhancement of energy utilization efficiency and reduction of waste emission in the process of building construction. It is mainly to make evaluation from 5 indexes—utilization of low-carbon and novel materials, introduction of novel construction technique, utilization of energy conservation and emissions reduction technique, investment of intelligent electronic equipment, specialization ratio of building construction team. The low-carbon technology index evaluation criterion is shown in Table 5.

**Table 5.** Low-carbon technology index evaluation criterion

| Evaluation index | Excellent | Good | Fair | Poor |
|---|---|---|---|---|
| Utilization of low-carbon and novel materials | 3 items are qualified. | 2 items are qualified. | 1 item is qualified. | No item is qualified. |
| Introduction of novel construction technique | 3 items are qualified. | 2 items are qualified. | 1 item is qualified. | No item is qualified. |
| Utilization of energy conservation and emissions reduction technique | 4 items are qualified | 3 items are qualified. | 2 items are qualified. | One or no item is qualified. |

*2.1.5. Low-carbon management index.* This index refers to the implemented low-carbon management in the process of construction completion, putting into normal use and operations maintenance. It is mainly to make evaluation from 4 indexes—greening rate, talent team construction of real estate, waste disposal and operations management. The low-carbon management index evaluation criterion is shown in Table 6.

**Table 6.** Low-carbon management index evaluation criterion

| Evaluation index | Excellent | Good | Fair | Poor |
|---|---|---|---|---|
| Talent team construction of real estate | 4 items are qualified | 3 items are qualified. | 2 items are qualified. | Two or no item is qualified. |
| Waste disposal | 3 items are | 2 items are | 1 item is | No item is |

| | | | | |
|---|---|---|---|---|
| | qualified. | qualified. | qualified. | qualified. |
| Operations management | 4 items are qualified | 3 items are qualified. | 2 items are qualified. | One or no item is qualified. |

*2.2. Evaluation method of low-carbon building*

*2.2.1. Establish evaluation factors set.* Generally, evaluation factors gather is expressed as $U=(u_1, u_2,\cdots, u_n)$. The evaluation element can be quantitative or qualitative. Evaluation factors in this paper refer to the 20 indexes established above.

*2.2.2. Establish level comment set.* Comment set refers to the set which is consisted of various evaluations to each sub-factor set made by the estimator according to the actual demand. Generally, it is expressed as $V=(v_1, v_2,\cdots, v_m)$, $v_i$ is evaluation level criterion, and $m$ is number of level comment. In this paper, according to the scores and votes made by experts, the level comment set of low-carbon building evaluation index system is identified as $V=$(excellent, good, fair, poor).

*2.2.3. Fuzzy comprehensive evaluation.* The goal of fuzzy comprehensive evaluation is to obtain more reasonable evaluation results, and it is essential to consider all factors comprehensively, not just analyze one factor. According to single factor evaluation known matrix $R$ and weight set $W$, comprehensive evaluation should be made through fuzzy transformation. Fuzzy comprehensive evaluation is expressed as $B$, fuzzy comprehensive evaluation index is expressed as $b_j$, namely evaluation index which could be explained as membership grade of evaluation object to evaluation element $v_i$ when consider all factors.

$$B=W\circ R=(w_1, w_2,\cdots, w_n)\circ\begin{bmatrix} r_{11} & r_{12} & \cdots & r_{1m} \\ r_{21} & r_{22} & \cdots & r_{2m} \\ \vdots & \vdots & \vdots & \vdots \\ r_{n1} & r_{n2} & \cdots & r_{nm} \end{bmatrix}=(b_1, b_2,\cdots, b_n)$$

## 3. Applied research of low-carbon building evaluation index system

"Solar Valley" in Dezhou is a joint name, covering solar production manufacturing, technology research, talent cultivation and relative supporting industries; it is a generalization of image of solar industrial cluster, planning to construct a solar "Sillcon Valley" that incorporate production, learning and research as a whole. "Solar Valley" in Dezhou is a green "Jerusalem" that reveal human wisdom. Represented by "Solar Valley" in Dezhou, Chinese Sun Ciy—Dezhou introduce an idea of low-carbon ecology starting from tourism project planning, implement energy saving and environmental protection building materials in the process of construction, using new energy products in lighting, landscaping, building in scenic spot whenever possible; guide accommodation and catering industry to implement low-carbon and green management model, insisting on clean production, advocating green consumption, protecting ecological environment and use the resources rationally; osmosis the idea of low-carbon into eating, accommodation, transportation, traveling, shopping, entertainment and other links of every tourist, launched a activity—"Access Chinese Sun City, Experience Low-carbon Trip", attracting tourists to experience low-carbon tourism. Above all, we should build brand of "Chinese Sun City—low-carbon Dezhou", and make efforts to become a pioneer of low-carbon green and practitioner of low-carbon tourism.

Through field researching and communicating with staff of "Solar Valley" in Dezhou, we manipulate collected data detailedly, and analyze the raw data of quantitative index, as is shown in Table 7.

**Table 7**.Raw data of quantitative index evaluation criterion

| Quantitative index | Unit | Data | Evaluation rating |
|---|---|---|---|
| Utilization ratio of renewable energy | % | 25 | Excellent |
| Utilization ratio of clean energy | % | 42 | Excellent |
| Investment ratio of intelligent electronic equipment | % | 28 | Good |
| Specialization ratio of building construction team | % | 95 | Excellent |
| Greening rate | % | 40 | Excellent |

Through experts' questionnaire, telephone interview and other forms, consulting the leaders and staff of "Solar Valley" in Dezhou, university professors and relative experts of Environmental Protection Agency in Dezhou, 15 qualitative indexes of low-carbon were got.

After calculating the scores of each evaluation index in low-carbon building evaluation system by evaluation panel of experts based on the method of membership grade of quantitative index and qualitative index, fuzzy matrix of low-carbon building evaluation index of "Solar Valley" in Dezhou could be obtained, as is shown in Table 8.

**Table 8**.Fuzzy matrix of low-carbon building evaluation index of "Solar Valley" in Dezhou

| Evaluation index | Excellent | Good | Fair | Poor |
|---|---|---|---|---|
| Lighting system | 0.533 | 0.267 | 0.200 | 0 |
| Heating system | 0.400 | 0.467 | 0.133 | 0 |
| Flooring system | 0 | 0.600 | 0.400 | 0 |
| Wall system | 0.400 | 0.400 | 0.200 | 0 |
| Using local materials | 0.200 | 0.400 | 0.400 | 0 |
| Building materials production | 0.800 | 0.200 | 0 | 0 |
| Reuse of materials and waste disposal | 0.133 | 0.600 | 0.267 | 0 |
| Energy measurement and monitoring | 0.544 | 0.456 | 0 | 0 |
| Collective heating | 0 | 0.942 | 0.058 | 0 |
| Utilization ratio of renewable energy | 1 | 0 | 0 | 0 |
| Utilization ratio of clean energy | 1 | 0 | 0 | 0 |
| Utilization of low-carbon and novel materials | 0.667 | 0.200 | 0.133 | 0 |
| Introduction of novel construction technique | 0.333 | 0.667 | 0 | 0 |
| Utilization of energy conservation and emissions reduction technique | 0.733 | 0.267 | 0 | 0 |
| Investment of intelligent electronic equipment | 0 | 1 | 0 | 0 |
| Specialization ratio of building construction team | 1 | 0 | 0 | 0 |
| Greening rate | 1 | 0 | 0 | 0 |
| Talent team construction of real estate | 0.200 | 0.667 | 0.133 | 0 |
| Waste disposal | 0.267 | 0.533 | 0.133 | 0.067 |
| Operations management | 0.133 | 0.600 | 0.267 | 0 |

Bring fuzzy evaluation matrix and weight to the model, the result can be obtained:

$$B = W \circ R = (w_1, \ w_2, \cdots, \ w_n) \circ \begin{bmatrix} r_{11} & r_{12} & \cdots & r_{1m} \\ r_{21} & r_{22} & \cdots & r_{2m} \\ \vdots & \vdots & \vdots & \vdots \\ r_{n1} & r_{n2} & \cdots & r_{nm} \end{bmatrix}$$

$$= (0.632, 0.217, 0.103, 0.048)$$

0.632+0.217+0.103+0.048=1, so the result of evaluation has been normalized.

According to the principle of maximum membership grade, evaluation result of building of "Solar Valley" in Dezhou that the relative performance being "excellent" is 0.632. 63.2 % of the total votes think the result of low-carbon building evaluation is "excellent", 21.7 % of the total votes think that is "good", 10.3 % of the total votes think that is "fair", and only 1.8 % of the total votes think that is "poor". Order scores set $F=(f_1, f_2, f_3, f_4)^T=(90,80,70,60)^T$, namely 80-90 scores is excellent, 70-80 scores is good, 60-70 scores is fair, below 60 scores is poor. Therefore, the calculated result of comprehensive evaluation value is:

$$Z=B\times F=\left(0.632,0.217,0.103,0.048\right)\times(90,80,70,60)^T=84.33$$

84.33 is between 80 and 90, so low-carbon building evaluation result of "Solar Valley" in Dezhou is "excellent".

The evaluation result corresponds to the actual situation, and the comprehensive evaluation of low-carbon building of "Solar Valley" in Dezhou is 84.33. It is mainly ascribed the efforts that "Solar Valley" in Dezhou made in utilization of energy conservation and emissions reduction technology, clean and renewable energy, adoption of modern management means, vigorously development new solar products and technique, and "Solar Valley" in Dezhou has acquired good benefits.

## 4. Conclusion

Investigating and evaluating carbon emission in the construction project, analyzing various influencing factors in the process of implementing low-carbon building, constructing low-carbon building evaluation index system and evaluation method are conductive to understanding the current situation of their own carbon emission for construction corporation, and accordingly improve the links and activities that have great impact on carbon emission in the process of construction effectively, as well as provide decision and gist in understanding the situation of low-carbon building development and taking specific carbon reduction measures for relative government departments.

Low-carbon building is in the early stage of development in our country, and the difficulties and challenges that we are faced with is huge. Therefore, it is necessary to give full play to the role of three main body—government, enterprise and society, strengthening the consciousness of developing low-carbon economy in the whole society, forming a national low-carbon green life style and consumption patterns, and jointly promoting the healthy development of low-carbon building.

## Acknowledgments

This article is a staged achievement of Social Science Planning research Project in Shandong province—"Research of Global Tourism Development under the Low-carbon Perspective" (No: 16CLYJ04) in 2016.

## References

[1] Di Gong, Xiao Yuan, Ming Lei. A simple analysis of green building system and low-carbon economy[J]. Environmental Science and Management, 2012, (10): 157.
[2] Changchun Liu, Fenghui Qin. Research of low-carbon buildings development[J]. Industries Forum, 2012, (9): 57.
[3] Lian Wu. Research of the scheme of urban buildings based on the concept of low-carbon and green[J]. Journal of Hunan University of Technology, 2012, (6):6.
[4] Baoxing Qiu. From green buildings to low-carbon eco-city[J]. Urban Studies, 2009, (7): 1-10.
[5] Manhong Shen, Longxiang Wang. Research summary and outlook of low-carbon buildings[J]. Construction Economy of China. 2012, (12): 67-70.

IOP Conf. Series: Earth and Environmental Science **61** (2017) 012157 doi:10.1088/1755-1315/61/1/012157

[6] Deli Yao, Tong Chen. Construction of low-carbon building evaluation index under the concept of eco-city[J]. China Population Resources and Environment, 2012, (S1): 268-271.

IOP Conf. Series: Earth and Environmental Science **61** (2017) 012158 doi:10.1088/1755-1315/61/1/012158

# Development strategy research of low-carbon tourist city

**Xiaohong Dong**

Qilu Normal University, Jinan, Shandong, 250200
Xiaohonghh@163.com

**Abstract.** Construction of low-carbon tourist city has become a strategic choice for the development of city construction in our country, becoming the direction and goal of future city development in China. In this paper, the development strategy of low-carbon tourist city is put forward from the aspects of building low-carbon tourism culture, strengthening and perfecting the relevant rules and regulations, establishing and perfecting the decision-making management mechanism of low-carbon tourist city construction, establishing accurate, timely, efficient and comprehensive ecological environment monitoring and supervision network, building economical resource utilization system, strengthening science and technology supporting of low-carbon city construction, establishing low-carbon tourism scenic spot, low-carbon community or low-carbon demonstration area, etc.

## 1. Introduction

With the rapid development of economy, urban tourism become a new growth point of city economy. At the same time of achievements of material and spiritual civilization provided by cities, the achievements advanced civilization in cities is suffering destruction constantly, such as continuous deterioration of urban environment, the urban ecological environmental unbance caused by tourist overload, the smaller and smaller living space of human being, sharply decline of biological diversity. As a result, the natural and cultural ecological system was damaged. In our country, a decision has been issued, that is "In our country, it is hoped to cut carbon dioxide emissions per unit of gross domestic product (GDP) by 40 %~45 % by 2020 from the 2005 level. Also, we should incorporate it into the long-term plan of national economic and social development as an obligatory target, and formulate relevant domestic statistics, monitoring and assessment measures." Therefore, it has profound significance to advocate low-carbon tourism and develop low-carbon tourist city. Under the double pressure of global climate change and the rapid development of urbanization in China, low-carbon tourism city construction has become a strategic choice for the development of city construction in our country, becoming the direction and goal of future city development in China.

## 2. Development strategies of low-carbon tourist city

### *2.1. Raise awareness and build low-carbon tourism culture*

The building of low-carbon city culture needs cooperation between various aspects strength, adhering to concept priotity is an important measure. On the one hand, we should advertise "low-carbon concept" to people using various propaganda means, enhancing cognition of tourists and tourism operators, building low-carbon tourism culture jointly. For example, we could creat the atmosphere of low-carbon tourism culture by means of making feature films, photos and information of indicator board in scenic spot. On the other hand, tourism product system should be reconstruction. Firstly, it needs to dig deeply and extract containing culture in tourism resources, promote and leisure the original tourism product brand, doing depth and fine in tourism product. Secondly, heavyweight tourist attractions that could attract foreign or remote tourists are digged based on a detailed investigation of tourism resources. Thirdly, leisure vacation products are packed and digged, and tourism experience of tourists should be diversified. We can build novel low-carbon tourism culture driven by all kinds of tourism projects, such as self-driving, camping ground, provincial low-carbon tourism demonstration zone, etc.

### *2.2. Improve and perfect the relevant rules and regulations, provide fundamental guarantee for the development of low-carbon city*

We need to earnestly implement a series of laws and regulations, including "Circulation Economy Promotion Law", "Environmental Protection Law", "Cleaner Production Promotion Law", "Energy Saving Law", "Environmental Impact Assessment Law", etc. Also, we need to implement green management vigorously, promote clean production and improve the added value of products and technical content continually, providing green products with long service life, less consumption of resources and environmental friendly.

It needs to issue some laws and regulations, such as "Low-carbon City Construction Regulation", "Circular Economy Development Guidelines of Low-carbon City", "Cleaner Production Verification Rules of Low-carbon City", "Building Energy Saving Regulation of Low-carbon City" and "Environment Protection Regulation of Low-carbon City Construction Project". Moreover, it is necessary to establish industry technical standards and technical specifications of low-carbon city, establishing and improving technical regulations system.

Policies and regulations of financial, tax and environmental protection that could promote the development of low-carbon city circular economy are formulated and improved. Mandatory standard of energy consumption for products is established, the standard of waste recycle, technical regulations and detection methods are improved. Moreover, low-carbon compensation mechanism is established and perfected, environmental symbols and green procurement system are established, the sustainable consumption is guided, and the market for green products is standardized.

### *2.3. Establish an perfect the decision-making management mechanism of low-carbon tourist city construction*

Organizational leadership is strengthen, target management responsibility system is perfected. Leading group of low-carbon tourist city construction is established, overall the action of departments and regional, and the major events in the low-carbon city construction are coordinated.

Decision-making mechanism of first making enbironmental impact assessment and then making decision is established. It needs to make environmental impact assessment and preliminary review for all kinds of onstruction projects in accordance with the law. As to the project not passing pre-qualification, the relevant departments shall not handle related formalities. Also, we need establish expert consultation committee of low-carbon city construction, and implement evaluation system of expert consultation on significant plans, policies and projects.

Assessing, monitoring and incentive mechanism of low-carbon city construction is formulate. Annual implementation scheme of low-carbon city construction is decomposed and implemented, the work of environmental protection and low-carbon city construction is listed into the important content of work performance appraisement of the leading cadres at all levels and accountability system is established. The work of low-carbon city construction should be incorporated into the supervision scope of major events. In addition, auditing system of environment protection should be built, and be incorporated into the scope of responsibility

IOP Conf. Series: Earth and Environmental Science **61** (2017) 012158 doi:10.1088/1755-1315/61/1/012158

auditing during the leader's tenure of office.

*2.4. Establish accurate, timely, efficient and comprehensive ecological environment monitoring and supervision network*

The network construction of environmental dynamic monitoring is strengthen. Environmental dyamic monitoring system is established, remote sensing, geographic information system and satellite positioning system are applied, the accuracy and timeliness of the monitoring are enhanced. Local area network of environmental protection is established to realize automatic monitored control system in city networking, and information about the ecological environmental quality and government information of environmental protection are released timely through environmental protection website.

Environmental warning system and fast feedback system are established and perfected. Construction of warning system, such as natural disasters, geological disasters, biological safety, public health is strengthen, forecast information timely is released, reducing the loss of all kinds of disasters. Emergency monitoring and disposal system dealing with unexpected pollution accidents is constructed, database of risk sources and emergency monitoring plan of key pollution sources are established, the ability of emergency monitoring is strengthen and the speed of making fast feedback is enhanced, reducing the pollution losses to the minimum.

The construction of environmental management institution and team are strengthen. environmental management department are perfected and strengthen according to the principle of unity, simplicity and effectiveness, and the funds of environmental monitoring, supervision, education and information personnel should be pulled into financial budget.

*2.5. Build economical resource utilization system*

On the basis of playing the market' role to the fullest extent, we need to optimize the utilization of land resource, water resource and energy, establish economical resource utilization system and stable supply system, enhance resource bearing capacity and eliminate the resource bottleneck of sustainable development effectively.

Resource management system is perfect. Natural resource management system with combining sufficiently with each other, efficient, administrating according to law and authoritative is established. The paid use of natural resources, competitive exploitation and asset-like management are promoted, and the reform of property right system of natural resources is promoted actively.

Resources are exploited and protected reasonably. It needs to make meeting the development of economic and social, and requirements of water quantity and water quality for ecological environment as the guidance and the unified management system of water resources is established and perfected. Land use control system is implemented strictly, the quantum of construction land is controled strictly, land arrangement and reclamation are developed actively, enhancing the intensive utilization level of land. Mining structure and layout are adjust, the number of mines is reduced, and scale exploition and intensive management is promoted.

*2.6. Strengthen science and technology supporting of low-carbon city construction*

Technical supporting system of low-carbon city is established. Based on developing high and new technology, developing ecological link technology of economy system as the key, our goal is to form a green technology supporting system of low-carbon city, including environmental engineering, resource utilization of waste, clean production technology as well as ecological restoration.

Technical innovation is strengthen, the development of low-carbon technologies is promoted. Energy saving, cost-reducing and pollutants discharge reduction are promoted using technologyof energy conservation and emissions reduction. New "carbon neutral" tourism activities are developed through promotion of low-carbon technology, advocating tourists compensating their own carbon emissions in the process of tourism. We could organize "carbon neutral" tourism activities by means of enhancing energy efficiency, research renewable energy and afforestation.

Industrial structure is optimized, cleaner production is implemented. On one hand, we should try our best to use toxic or harmful raw materials and intermediate products as few as possible or don't use that. On the other hand, we should recycle raw materials and intermediate products, improve management and enhance efficiency. At the same time, we should keep clean in the process of the whole production by mean of updating equipment, technology and process timely, scientific management of promoting production, maintenance and training.

*2.7. Strengthen of public participation ability construction*

Public participation system is established and perfected, conditions for the public participation are created, and public participation should be pulled into systematic orbit. Relevant trainings of low-carbon city construction at all levels of government, enterprise and community are developed. We should organize and guide the masses to carry out all kinds of mass constructive activities, such as rubbish classification, saving water and decreasing pollution, recycling and reusing resources, green communities, green courtyard, forming mass foundation of low-carbon city construction. Also, system of environmental integrity is established, the government information of environmental protection is made public, citizens' rights to know, to participate and to supervise of the low-carbon city construction are expanded.

*2.8. Increase investment of low-carbon city construction*

Governments at all levels should arrange sufficient funds used in ecological environment protection and construction in finacial budget in accordance with the requirements of establishing public finance. In addition, it is needed to establish and perfect funding system of environmental protection. Municipal public finance should arrange appropriate pollution control fund and special funds of supervision capability construction of environmental protection every year.

Pollution charges in full according to law are levied, and the charges should be used in low-carbon construction projects, on a priority basis. According to the new pollution charge regulation, it is essential to levy pollution charges in full comprehensively and in accordance with the law. Pollution charges are mainly used in helping enterprises to implement cleaner production, environmental management system certification, establishing environmental protection model enterprises and organizing low-carbon project construction.

Incentive finance and taxation policies are introduced. The government should build frequent supporting policies of direct financial subsidies, stimulating enthusiasm for developing low-carbon economy for enterprises by way of funding or discount loans. For example, in order to enhancing the enterprises' technology investment and product development of clean energy and strengthening the comprehensive utilization of resources, we could adopt a method that combines deduction from income tax, investment exempts ortax rebate, accelerated depreciation and tax deferral to provide tax support.

*2.9. Establish low-carbon tourism scenic spot, low-carbon community or low-carbon demonstration area*

Internal resources in scenic spots are integrated, and industry chain of low-carbon scenic spots is built. The basic elements of tourism scenic area mainly include six aspects—eating, accommodation, transportation, traveling, shopping, entertainment. Therefore, it is necessary to integrate the six key elements of scenic spot using the idea of integrated management scientifically and reasonably, forming the internal industry chain development model of low-carbon scenic spot.

Low-carbon community aims at not only minimizing carbon emissions produced by all the activities in the community, but also hoping to achieve the goal of zero carbon emission through some measures, for example, ecological afforestation. Community is a entity unit of a city. Establishing low-carbon city could start from new residential district, such as building low-carbon communit or low-carbon demonstration area. Moreover, we can learn the advanced experience of low-carbon community construction from abroad, pay attention to construct energy saving building, use environmental protection material, optimize community structure, advocate green transportation and advanced garbage disposal idea according to the local climate, terrain and other natural environment conditions.

*2.10. Give priority to the development of public transportation*

With the accelerating process of urbanization, rising of people's living standards, increasing number of private vehicles, city disease, such as traffic congestion, environmental pollution, arises at different levels in the process of urban development. As to low-carbon tourist city, it is necessary to optimize urban bus lines, increase the number of bus stations, increase the use of new energy public transport vehicles and strengthen investment of public transport facilities on the basis of the existing transport facilities. In addition, the corresponding travel policy should be set up, and the government should take the lead to take bus to commute to work, limiting the use of private cars. Strongle advocate walking, riding a bicycle or electric vehicles to commute to work. Increase rental stations of bicyclesand electric vehicles, decrease the times of transfering between buses in people's travel and increase the convenient degree of bus trip.

## 3. Conclusion

Constructing low-carbon tourist city needs extensive participation of the whole society. Government should set up development plan of low-carbon city and strengthen supervision; enterprises shoud strengthen technological innovation and production of low-carbon products; also, the public should cultivate low-carbon consumption patterns and consumption habits. Only vigorously implement low-carbon transport, low-carbon building, low-carbon production, low-carbon tourism, and create synergy in the whole society, can the development of low-carbon tourist city realize.

**Acknowledgments**

This article is a staged achievement of Social Science Planning research Project in Shandong province—"Research of Global Tourism Development under the Low-carbon Perspective" (No: 16CLYJ04) in 2016.

## Reference

[1] Qiong Lei, Research of Low-carbon Tourism Scenic Spot Management[J]. Modern Buaines Trade Industry, 2010, (16): 46-47.
[2] Huilin Cao, Discussion of Low-carbon Measures of Low-carbon tourism and tourism enterprises[J]. Value Engineering, 2010, (4): 30.
[3] Yongbo Zhou, Run Sha, Discussion of Intentional Strategy of Low-carbon Tourism in China[J]. Jiangsu Business and Trade, 2010, (5): 115-116.
[4] Wenliang Hou, Research of Low-carbon Tourism and Carbon Emission Reduction Countermeasure[D]. Henan City, Henan University, 2010.
[5] Xiao Liu, Low-carbon tourism—New Model of the Future Development of Suburban Tourism in Beijing[J]. Social Sciences of Beijing, 2010, (1): 42-46.
[6] Yun Pu, Constructing Proposal of Low-carbon Tourism in Sichuan[J]. Co-operative Economy & Science, 2010, (10): 10-12.

IOP Conf. Series: Earth and Environmental Science **61** (2017) 012159 doi:10.1088/1755-1315/61/1/012159

# Innovation in the management innovation of planting methods of vegetation in the southwestern edge of Mu Us Sandy land--from a sustainable development view

**Xing Li[1a], ZHU Yan-feng[2b], Jianmin He[1c], BingJie Hou[1d]**

[1]College of Economics and Management,LongDong University ,Gansu 745000, China
[2]Forestry Research Institute in Qingyang, Qingyang city, Gansu 745000, China
[a]lixing@ldxy.edu.cn, [b]gsqylkszyf@126.com, [c]ee_he96@126.com, [d]monkeyber@126.com

**Abstract**. All articles *must* contain an abstract. The abstract text should be formatted using 10 point Times or Times New Roman and indented 25 mm from the left margin. Leave 10 mm space after the abstract before you begin the main text of your article, starting on the same page as the abstract. The abstract should give readers concise information about the content of the article and indicate the main results obtained and conclusions drawn. The abstract is not part of the text and should be complete in itself; no table numbers, figure numbers, references or displayed mathematical expressions should be included. It should be suitable for direct inclusion in abstracting services and should not normally exceed 200 words in a single paragraph. Since contemporary information-retrieval systems rely heavily on the content of titles and abstracts to identify relevant articles in literature searches, great care should be taken in constructing both.

## 1. Introduction

Management innovation is an important means for an organization to pursue superiority and maintain their competitive edge. In China, the selection of trees for the prevention and control of sand is mainly based on the biological environment of the dunes, special properties of tree types and empirical judgment, all qualitative analysis based approaches. As a result, the outcome of planting the trees selected this way is not ideal and this tree type selection method needs improvement in terms of scientific and advanced properties. Therefore, like what was learned from past sand prevention studies which emphasized rational efficiency in the selection of trees to plant for prevent sands (Rogers, 1976), trees planted for restoring vegetation should be selected according to the land conditions and the restoration targets and with certain necessary sifting technologies. Only in this way can we make sure that the selection be scientific and the restoration of vegetation be a success, achieving sustainable development.

The northern part of Huan County is located in southwestern edge of the Mu Us Sandy Land and is in the transitional place between the Mengxi Plateau and the Huangtu Plateau of China, being a typical juncture of agricultural and husbandry areas. It's also the area most severely affected by wind and sand in Longdong, Gansu. The environmental conditions in the area is very unique and ecological problems are prominent and complex. The unique ecological environment of this area has determined the growth and distribution of the vegetation and also determined that the plants used for restoring the vegetation there should be special. Conducting a research and experiment on the method for selecting tree types for vegetation restoration in this are is of great significance for northwest China and places of similar ecology.

## 2. Research Background

In this study, we have selected 18 plants in the tree, shrub and grass categories. 6 are trees, which include, Pinus sylvestris L. var. mongolica Litv, Platycladus orientalis (L.) Franco, Populus alba var. pyramidalis Bge, Populus diversifolia Schrenk, Rhus typhina Nutt and Elaeagnus angustifolia Linn. 10 are shrubs, which include Sabina vulgaris Ant, Salix psammophila C. Wang et Ch. Y. Yang, Hedysarum fruticosum Pall. var. Mongolicum, Hedysarum scoparium Fisch. et Me, Amorpha fruticosa Linn, Caragana Korshinskii Kom, Caryopteris clandonensis, Sambucus williamsii Hance, Artemisia desterorum Spreng Syst. Veg, and Hippophae rhamnoides L. 2 are grasses, which include Astragalus adsurgens Pall and Medicago falcata L. 4 plants are locally grown in Qingyang City, Gansu Province of China, which are Platycladus orientalis (L.) Franco, Populus alba var. pyramidalis Bge, Caragana Korshinskii Kom and Hippophae rhamnoides L. The other 14 plants are all brought in from elsewhere.

We have selected fixed and semi-fixed dunes that cover a relatively large area as the research target. To be specific, we established a vegetation restoration experiment demonstration zone in the Hegou Shoal of Tianshuijie Village, Tianshui Township, Huan County and used the 25-mu broad-leaved mixed forest and the 25-mu shrub forest as the basic research target and respectively set fixed survey and monitoring sample lands.

In response to different restoration types, we set survey sample lands respectively. We set 10m×10m quadrats and survey 30 trees or shrubs in the sample lands at random. In the first year, we surveyed the survival rate. In the third year, we surveyed the preserving rate and measured the tree heights, length of branches that grow in the measuring year and the crown width with meter sticks and measured the diameter at breast height and the ground diameter of trees with vernier caliper.

By considering the adaptability to the environment, growth conditions and resilience of the selected trees, we classified the growth conditions of the trees into excellent, good, average, bad and worse grades for assessment. The specific classification and assessment standards are displayed in Table 1.

| Performance status | excellent | good | commonly | difference | Worst |
|---|---|---|---|---|---|
| adaptability | Fully adapt to local environmental conditions | It is suitable for the growing environment | Able to adapt to local environmental conditions | Not adapt to local environmental conditions and site conditions | Completely unable to adapt to local environmental conditions |
| Growth status | Vigorous growth, can blossom with luxuriant foliage | Growth is relatively strong, branches, leaves and other basic growth is good | Branches and leaves are sparse, can maintain the crown and tree | The branches and leaves are sparse, and the crown type is incomplete | Stems dysplasia, serious pests and diseases, crown scattered |

**Table1** Growth status evaluation of tree species classification table

## 3. Analysis

To narrow down the scope of trees that can be selected for vegetation restoration and reduce the difficulties of ensuing analysis, in accordance with the survival rate, preserving rate and growth of the plants, we did a preliminary sift of the 18 tree types experimented and had a qualitative assessment of their adaptability, survival rate, preserving rate and growth. (As shown in Table 2)

**Table2** Survival rate, preserving rate and growth status of different species

| varieties of trees | Number of seedling | Survival rate (%) | Preservation rate (%) | Average tree height (cm) | | | Ground diameter (cm) | | | crown width (cm) | | | shoot growth | | | Growth status |
|---|---|---|---|---|---|---|---|---|---|---|---|---|---|---|---|---|
| | | | | 2012 | 2013 | 2014 | 2012 | 2013 | 2014 | 2012 | 2013 | 2014 | 2012 | 2013 | 2014 | |
| Amorpha fruticosa Linn | 30 | 89 | 83 | 62.9 | 79.7 | 103.1 | 1.29 | 1.3 | 1.31 | 69.83 | 90.78 | 102.2 | 38.19 | 44.07 | 45.6 | good |
| Pinus sylvestris L. var. mongolica Litv | 30 | 95 | 89 | 98.3 | 109.25 | 116.7 | 2.83 | 3.26 | 3.42 | 49.65 | 57.98 | 62 | 6.94 | 7.82 | 8.1 | excellent |
| Populus alba var. pyramidalis Bge | 30 | 94 | 79 | 131.5 | 160.6 | 216.5 | 2.35 | 2.6 | 3.22 | 22.1 | 24 | 78.85 | 36.16 | 37.54 | 71.8 | good |
| Platycladus orientalis (L.) Franco | 30 | 91 | 86 | 97.9 | 111.1 | 123.6 | 1.81 | 2.06 | 2.28 | 41.9 | 46 | 49.25 | 11.05 | 13.25 | 14.35 | good |
| Elaeagnus angustifolia Linn | 30 | 85 | 85 | | 94.6 | 100.7 | | 1.06 | 1.08 | | 23.45 | 26.67 | | 8.95 | 12.47 | good |
| Populus diversifolia Schrenk | 30 | 3 | 0 | | | | | | | | | | | | | good |
| Sabina vulgaris Ant | 30 | 96 | 94 | 19.9 | 25.77 | 37.8 | 0.66 | 1.15 | 1.2 | 27.65 | 33.37 | 38 | 9.27 | 11.61 | 13.1 | excellent |
| Caragana Korshinskii Kom | 30 | 91 | 94 | 17.3 | 27.2 | 53.5 | 0.32 | 0.5 | 0.95 | 13.05 | 16.3 | 29.4 | 8.7 | 9.65 | 12.68 | excellent |
| Caryopteris clandonensis | 30 | 94 | 84 | 22.7 | 46.7 | 68.8 | | | | 26.1 | 32.55 | 55.45 | 16.78 | 17.38 | 21.8 | good |
| Hippophae rhamnoides L | 30 | 45 | 86 | 61.1 | 36.7 | 84.54 | 0.88 | 0.88 | 1.45 | 16.4 | 15.9 | 33.6 | 12 | 11.7 | 15.43 | commonly |
| Hedysarum fruticosum Pall. var. mongolicum (Turcz.) Turcz | 28 | 90 | 86 | 20.1 | 45.4 | 84.3 | | | | 18.3 | 39.8 | 64.4 | | 23.4 | 22.7 | excellent |
| Hedysarum scoparium Fisch. et Mey. | 26 | 85 | 87 | 17.8 | 41.3 | 75.2 | | | | 17.8 | 38.5 | 62.2 | | 21.6 | 20.4 | excellent |
| Sambucus williamsii Hance | 28 | 73 | 39 | | | 25.9 | | | | | | 35.9 | | | 20 | difference |
| Artemisia desterorum Spreng Syst. Veg | 25 | 84 | 88 | 51.6 | 95.4 | 145 | | | | 88.3 | 142.6 | 156.3 | 75.4 | 85.3 | 105.93 | excellent |
| Salix psammophila C | 27 | 27 | | | | 55.4 | | | | | | 32.1 | | | 20.3 | difference |
| Rhus typhina Nutt | 30 | 10 | 0 | | | | | | | | | | | | | difference |
| Astragalus adsurgens Pall | 20 | 66 | 90 | | | 94.4 | | | | | | 60.2 | | | 94 | good |
| Medicago falcata L | 30 | 95 | 93 | 27 | | 45.7 | | | | 25.15 | | 37.15 | 15375 | | 16.98 | good |

The result suggests that Populus alba var. pyramidalis Bge, Pinus sylvestris L. var. mongolica Litv, Artemisia desterorum Spreng Syst. Veg, Hedysarum fruticosum Pall. var. mongolicum (Turcz.) Turcz, and Hedysarum scoparium Fisch. et Mey have larger growth increment and better adaptability to the environment, suitable to be the pioneering tree seedlings for ecological restoration in the dunes in the north of Huan County. Medicago falcata L is used relatively more widely in the production and ecological restoration of the local region and can be used as the major candidates in the grass category.

## 4. Conclusion

This research adopts the southwestern edge of the Mu Us Sandy land of Huan County of Gansu Province as the subject, upholds scientific planting and sustainable development as the core and have discovered through management innovation that tree type selection is the key chain of ecological restoration and that the appropriateness of the tree types selected directly affects whether the introduction of trees can succeed and the overall restoration effects. The specific conclusion is as follows:

(1) According to the survival rate, the preserving rate and the growth, we had a comprehensive assessment of the 18 types of trees commonly used in ecological restoration and the result suggests that 7 shrubs (including Hippophae rhamnoides L, Caryopteris clandonensis, Artemisia desterorum Spreng Syst. Veg,Caragana Korshinskii Kom, Amorpha fruticosa Linn, Hedysarum fruticosum Pall. var. mongolicum (Turcz.) Turcz, and Hedysarum scoparium Fisch. et Mey.), 4 trees (including Pinus sylvestris L. var. mongolica Litv., Populus alba var. pyramidalis Bge, Platycladus orientalis (L.) Franco and Elaeagnus angustifolia Linn), and 1 grass (Medicago falcata L.), altogether 12 plants grow well in this region and are plants that can be selected for the local restoration work for sandy and deteriorated land.

(2) By ranking the competitive matrix according to appropriateness, we confirmed that shrubs are the types of plants that should be given first priority to when selecting plants for regional ecological restoration. And among all the plants in the experiment, the ranking of them in terms of appropriateness is: Caragana Korshinskii Kom＞Hedysarum fruticosum Pall. var. mongolicum (Turcz.) Turcz＞Hedysarum scoparium Fisch. et Mey.＞Artemisia desterorum Spreng Syst. Veg.＞sabina vulgaris＞Amorpha fruticosa Linn.＞Populus alba var. pyramidalis Bge＞Caryopteris clandonensis＞Pinus sylvestris L. var. mongolica Litv.＞Platycladus orientalis (L.) Franco. Therefore， when selecting plants for ecological restoration in the sandy areas of the southwestern edge of the Mu Us Sandy land, we should make the decision based on the ranking above. By doing so, we can significantly improve the status quo of sand prevention planting that upholds sustainable development as the core.

**Acknowledgements:**
Central finance forestry science and technology extension demonstration project [2015] ZYTG8
Qingyang science and technology support program (NK2011-41)
Longdong University doctoral research start-up funding project (No.: XYBE1605).

## References

[1] Wang Y. Preliminary study on landcape of green space affiliated to urban road system: a case study of Daxing district of Beijing.Journal of northwest forestry university, 2013,28(4):218-222.
[2] BO Y S, HAN E X, Han G, et al..A Study on Tree Species Selection for the Fire Protection Forest Belt in Shaanxi Province. Journal of northwest forestry university,1997,12(04):24-30.
[3] HUANG W D. Fuzzy probability model for selecting tree species and application. Journal of nanjing institute of forestry,1985,4:51-62.
[4] ZHANG Y S. Research into the Tree Choices for Desertification Prevention and Control in the Sources of Sand Storms in Northern Shanxi Province. Shanxi science and technology,2008,6:88-90.
[5] Xing Li, Qiu Wenhong, Zhang Peiyao, and. Discussion on service innovation of cultural and Creative Park: cultural value perspective [J]. macroeconomic management, 2017, (S1): 34-36.
[6] WANG H T, ZHANG G G, Liu X. Principles of silvicultural tree species selection for loess hilly areas.World Forestry Research, 2001,14(5): 74- 78.
[7] QIAN C J, ZHANG H H, LI T et al.. Composition and growth adaptability of introduce landscape tree species Guiyang. Guizhou Agricultural Sciences,2013, 41(12):184-188.

[8] Li X., Qu T., He J.M., Hou B.J.,Seek Excellence on the Hospital Quality Management Competition,Basic & Clinical Pharmacology & Toxicology,2016 ,December,Volume 119, Issue S4,HHME16-H24

IOP Conf. Series: Earth and Environmental Science **61** (2017) 012160 doi:10.1088/1755-1315/61/1/012160

# CFD-based optimization of truck fairing structure

**Jing Chen [1,a], Ning Deng[2], Yaguo Li [3,b,*], Chenghuan Liang[4], Beiqing Lin [4]**

[1]LiuZhou Vocational & Technical College, Liuzhou, Guangxi 545005, China;
[2]Liuzhou City Vocational College, Liuzhou, Guangxi 545036, China;
[3]Fine Optical Engineering Research Centre, Chengdu, Sichuan 610064, China ;
[4]Faw Jiefang Liuzhou Special Automotive Co., Ltd, Liuzhou, Guangxi 546100, China
[a]1936307477@qq.com, [b] yargolee@163.com.

**Abstract.** Computational Fluid Dynamics (CFD) technology was used to analyse a big truck with newly-designed head. Three models were designed to analyze: (1) Project 1: a trunk without fairing, (2) Project 2: a trunk with an old fairing, (3) Project 3: a trunk with a new fairing (the old fairing of Project 2 is particularly moved 500mm forehead). During the movement of vehicle, the more similar streamlined shape of the vehicle is, the more reduction of the wind resistance can achieve. The wind resistance could be thereby reduced in case of a truck having a fairing because of a more similar with streamlined shape for the truck structure. The challenge here is that the exactly optimal location of the fairing on the vehicle is not clear and generally was determined only by experience. However, here by the comparison of each model with CFD analysis, engineers can give the better design for the goal of reducing the resistance fuel consumption.

## 1. Introduction

The structure of vehicles directly affects fuel consumption. The body shape of the vehicles have to conform to streamlines shape as far as possible when considering the gas dynamic aspects e.g. wind resistance and safety coefficient demand. Drag coefficient, which refers to the wind resistance, exhibits real close relationship with the stability and safety of vehicles [1]. Generally, the wind resistance mainly comes from the front of the vehicle when it is on running. In this case, drag coefficient plays a key role on vehicle performance, whereby as reported that 60% of the fuel consumption is assumed to overcome the wind resistance when a car is running at the speed of 80 km/h [1-6].

In this paper, the professional Computational Fluid Dynamics (CFD) software named SolidWorks Flow Simulation was used to analyze a big truck with newly-designed long-headstocks. Three models were designed here: 1) Project 1, a truck without fairing; 2) Project 2, a truck with an old fairing, 3) Project 3: a truck with a new fairing (the old fairing is particularly moved 500mm forehead compared with Project 2). The smaller wind resistance is achieved during the movement of vehicle if its whole structure is more similar with the streamlined shape. In this context, the wind resistance could be reduced in case of a truck having a fairing because of in such way the truck structure being more similar with streamlined shape. It is difficult to fix the fairing the fairing on the vehicle at exactly optimal location and generally was determined only by experience. However, here by the comparison of three different designs with CFD analysis, the optimal location of the fairing could be fixed for the goal of reducing the resistance fuel consumption.

IOP Conf. Series: Earth and Environmental Science **61** (2017) 012160 doi:10.1088/1755-1315/61/1/012160

## 2. Premise & Simplification

*2.1 Premise of simulation*

This numerical simulation is based on the CFD turbulence model in an open state. The analysis object is the air fluent around the trunk inside the computational domain. To simplify the project, a standard κ-ε double equation model is chosen since the fluid is only air. Addition assumptions are as follows: (1) physical parameters between solid and fluent are set as constants; (2) fluent is set as turbulence; (3) carriage of the trunk is truncated to reduce the computational domain size and grid number; (4) ignore the influence of the fluent temperature [2-3, 7-12].

*2.2 Simplified model*

The exact model of trunk is very complicated. To accelerate the convergence rate and reduce processing workload, the model of the trunk should be simplified. Simplifications are as follows: 1. to remove small fillets and make sharp corners smooth; 2. to integrate the head as a whole; 3. To cut the length of the carriage. Simplified models are shown in Fig.1.

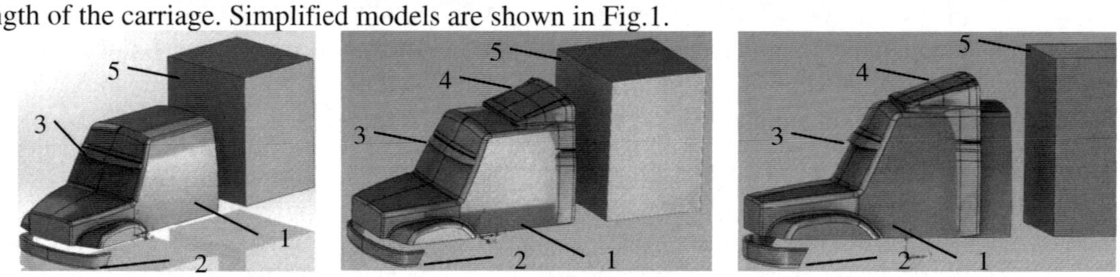


(a) Project 1 (b)Project 2 (c)Project 3

1-Cab; 2-Front-bumper; 3-Sun shield; 4-Fairing; 5-Carriage;

Notes: 1-4 is called head as a whole, 5 is just named carriage.

Fig.1 Simplified Model of Truck Body

## 3. Pre-treatment

*3.1. Definition of Drag coefficient & Boundary conditions setting*

Definition of Drag coefficient is: $C_t = 2F_t / \rho U^2 S$ , which is the resistance acting on the entity along the direction of the wind, S is the frontal projected area of the entity, ρ is air density, U is the relative velocity of truck to air.

In the SolidWorks Flow Simulation, the air is considered to be incompressible gas. Pressure-based coupled solver and implicit iterative method are chosen. The κ-ε model and first-order discrete form are chosen. Pressure and velocity coupling algorithms are selected to be the SIMPLE algorithm.

Note: The κ-εmodel has become the most popular of the turbulence models and is often used in many calculations of flow of practical interest.

Boundary conditions settings are as follows:

Inlet: velocity is 30m/s (the trunk is at around the speed of 100km/h)[13-14];

Outlet: pressure-outlet[8,15-16].

### 3.2 Mesh & Solve

Since the models are complicated, the option of 'Narrow channel' is used to refine the mesh. With the same setting, total grids of all the projects are close to 5 million. Details can be found at Table 1. A high configuration fluid analysis workstation is used for calculation (8 core processor +16G memory, solid state hard disk and professional graphics card)[17].

Table 1 Calculations

| Project name | Project 1 (without Fairing) | Project 2 (Fairing I) | Project 3 (Fairing I 500mm forward ) |
|---|---|---|---|
| Total grids | 4764505 | 4993732 | 4966173 |

IOP Conf. Series: Earth and Environmental Science **61** (2017) 012160 doi:10.1088/1755-1315/61/1/012160

| | | | |
|---|---|---|---|
| Fluent grids | 1926675 | 2147930 | 2135526 |
| Solid grids | 1422271 | 1317488 | 1311244 |
| Partial grids | 1415559 | 1528314 | 1519403 |
| CPU time/s | 130443 | 41847 | 44844 |
| Iterations | 1494 | 454 | 485 |

Table 2 Pressure Distribution

| Project names | Project 1 | Project 2 | Project 3 | Project 1-Project2 | Project 1-Project 3 |
|---|---|---|---|---|---|
| Glass window/Pa | 101843.50 | 101785.09 | 101768.05 | 17.04 | 75.45 |
| Area/m2 | 2.2021 | 2.2021 | 2.2021 | - | - |
| Hood/Pa | 101399.10 | 101562.57 | 101557.04 | 5.53 | -157.94 |
| Area/m2 | 2.0270 | 2.0270 | 2.0270 | - | - |
| Carriage front face /Pa | 101485.08 | 101404.37 | 101389.76 | 14.61 | 95.32 |
| Area/m2 | 6.4400 | 6.4400 | 6.4400 | - | - |
| Total Pressure/Pa | 304727.7 | 304752 | 304714.9 | 12.83 | 37.18 |

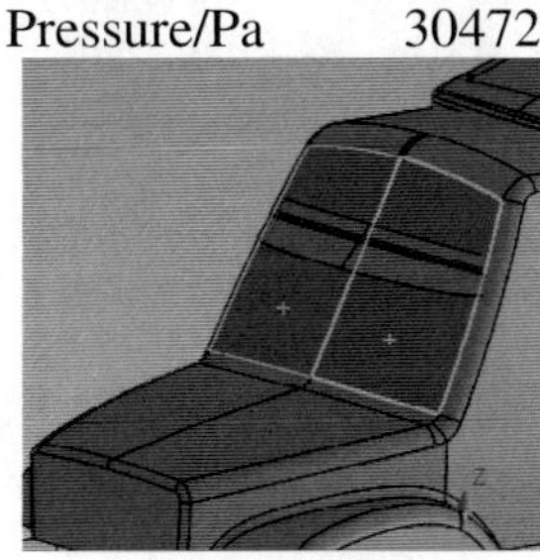
Glass window

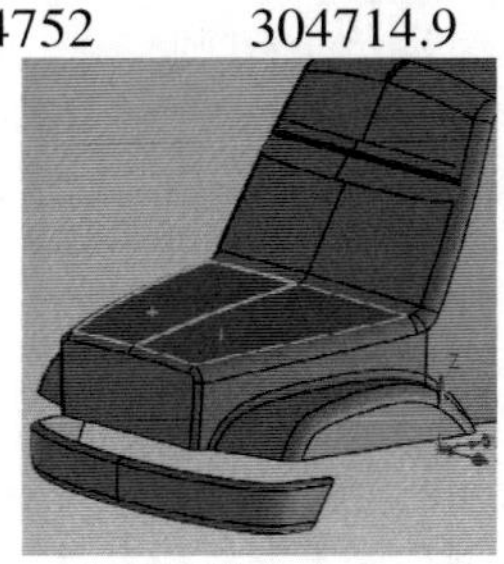
Hood

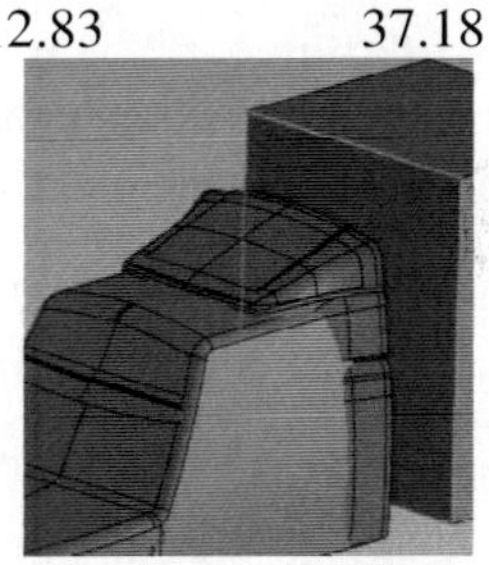
Carriage front face

## 4. Results

### *4.1 Surface Pressure Contours*

Fig.2 refers to the pressure contours of the three projects. (a1) & (a2), (b1) & (b2), (c1) & (c2)are almost the same, only with different color grades. (a1)- (c1) with 10 color grades, while (a2)-(c2) with 100 color grades. Note: high color grade means smooth color change but could hardly be provided insights into the design, while low color grade means poor color change but could easily be provided insights into the design. Compared with the 3 projects, judging from the entire trunk color (especially the circular area), Project 3 has the least red range (high pressure) indicating that this fairing is good for reducing the pressure on carriage. The pressure distribution (including glass window, hood and carriage front face) is listed in Table 2.

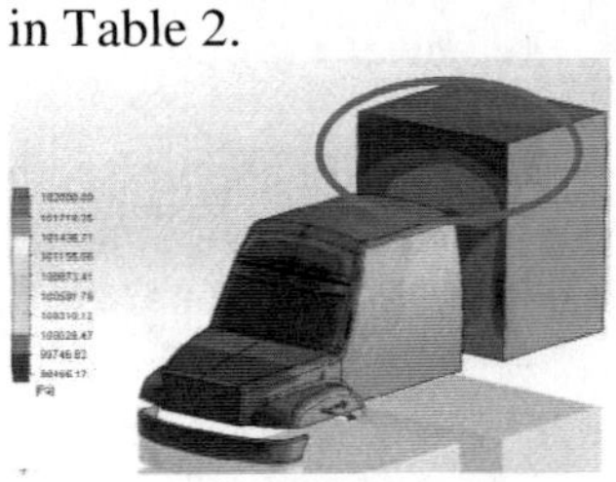
(a1) Project 1-grades 10

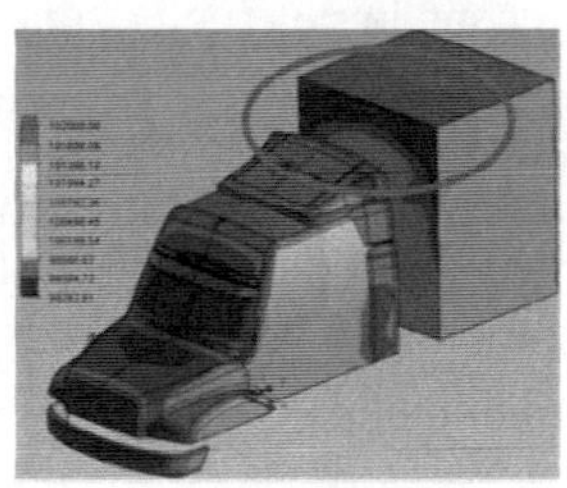
(b1)Project 2-grades 10

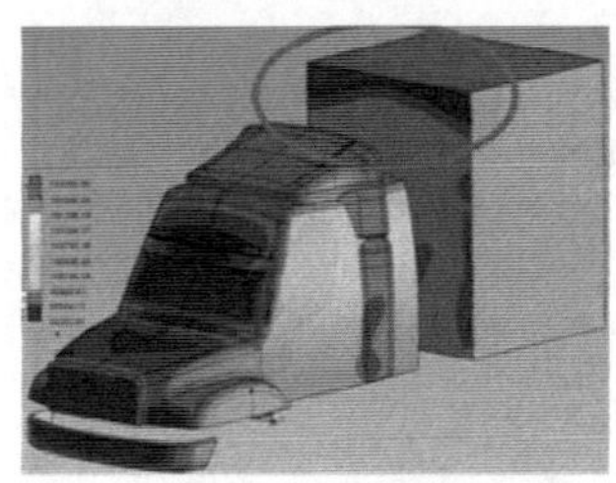
(c1)Project 3-grades 10

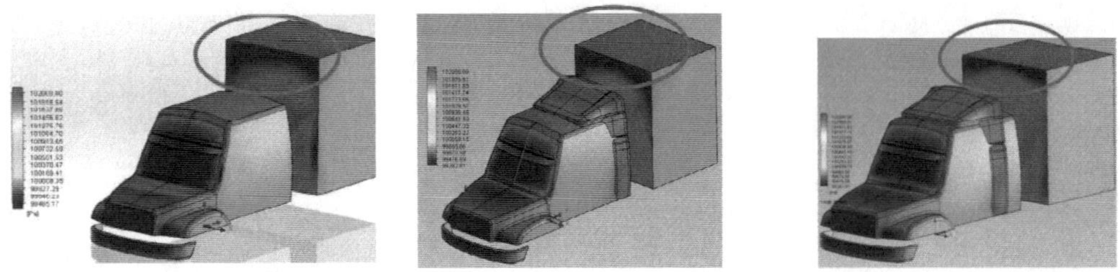

(a2)Project 1-grades 100 (b2)Project 2-grades 100 (c2)Project 3 -grades 100

Fig.2 Pressure contours of Truck Body

As shown in Table 2, the pressure of the glass window, hood and carriage face from big to small are: Project 1> Project 2> Project 3. Commonly speaking, the glass window is the weakest part. With the help of fairing, the pressure of the glass window is decreased by 75.45 Pa comparing Project 3 to Project 1, and by 17.04 Pa comparing Project 2 to Project 1. The pressure of hood is decreased by 5.53Pa compared Project 2 to Project 1, but increased by 157.94 Pa compared Project 3 to Project 1, which illustrated that the fairing separated the wind pressure from glass window to other parts, so that to release the pressure of the vulnerable glass window. The pressure of carriage front face is also reduced by fairing.

*4.2 Velocity contours (extract the middle section)*

Fig.3 refers to the velocity contours of the three projects (extract the middle section). Comparing with the 3 projects, we can observe that the streamline area between head and carriage of Project 1& 2 are separated, but the streamline area of Project 3 is merged.

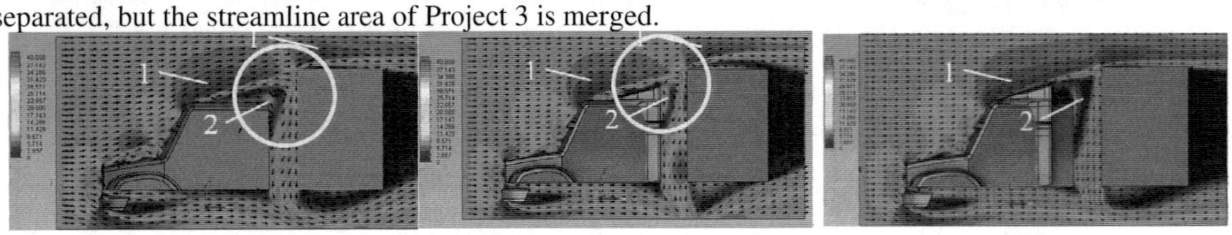

(a) Project 1-grades 10 (b)Project 2 -grades 10 (c)Project 3-grades 10

1-High velocity zone; 2-Low velocity zone;

Notes: the circular area of (a) (b) is separated area the circular area of c) is merged.

Fig.3 Velocity contours of Truck Body middle section

Judging from color of contours, blue means low velocity zone, while red means high velocity zone. The separated area is name for the area that the velocity changed rapidly.

As shown in Fig. 3, in Project 1, without the fairing, the separated area could be clearly observed above the trunk; in Project 2, since the fairing is not completely occupied the separated area, so the design do not conform to the streamline; in Project 3, the fairing almost completely covers the separated area, so that the design conforms to the streamline.

*4.3 Trunk flow trajectories*

Fig.4 shows the velocity 3D flow trajectories around Truck Body. In fact, 3D animation is also can be exported to display the motion state. However, the 3D trajectories temporarily have no direct information with value.

(a) Project 1 (without Fairing)-grades 10 (b)Project 2 (Fairing I)-grades 10 (c)Project 3 (Fairing I 500mm forward )-grades 10

Fig.4 Flow trajectories around Truck Body

### *4.4 Drag Coefficient*

Trunk in motion gets two kind of resistances： 1. Wind resistance coefficient (also named drag coefficient); 2. Wind resistance force. The definition of drag coefficient is shown in 2.1 [16-19].

After iterative calculation, Fig.5, Fig.6 and Table 3 are obtained. Fig.5 is the wind resistance coefficients of all the 3 projects. Fig.6 is the wind resistance forces of all the 3 projects. Combined with Fig.5 and Fig.6, we can see that the iterative curve of Project 1 fluctuated severely. The reason may be that the stable calculation of wind resistance force & coefficient are hard to obtain without fairing. On contrast, with the fairing, the calculations are relatively stable, according to curves of Project 2 &3. The wind resistance forces of Project 1 (without Fairing), Project 2 (Fairing I) and Project 3 (Fairing I 500mm forward) are 2358N, 2111N and 1761N, respectively. Accordingly, the drag coefficients of 3 projects are 0.784, 0.689, and 0.571, respectively. The drag coefficient and wind resistance force of Project 3 are the smallest, so the structure of Project 3 is the best one among the 3 projects [6, 18-20].

Table 3 Wind resistance force & coefficient

| Project names | Wind resistance force(N) | Drag coefficient |
|---|---|---|
| Project 1 | 2358.569 | 0.7836654 |
| Project 2 | 2111.374 | 0.6894526 |
| Project 3 | 1761.157 | 0.57082067 |

Note: in Table 3, the Wind resistance force & coefficient of Project 1 is the average between irritation 1000 and 1500. The wind resistance force & coefficient of Project 2 is the average between irritation 200 and 454. Because of good convergence, the curve of Project 3 has been descending before convergence, so the average is gotten from the last 5 irritations.

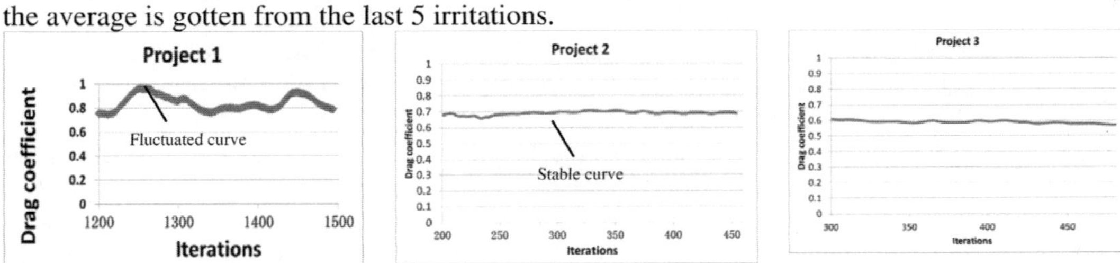


(a) Project 1 (without Fairing) (b)Project 2 (Fairing I) (c)Project 3 (Fairing I 500mm forward )

Fig.5 Contrast of Drag coefficient

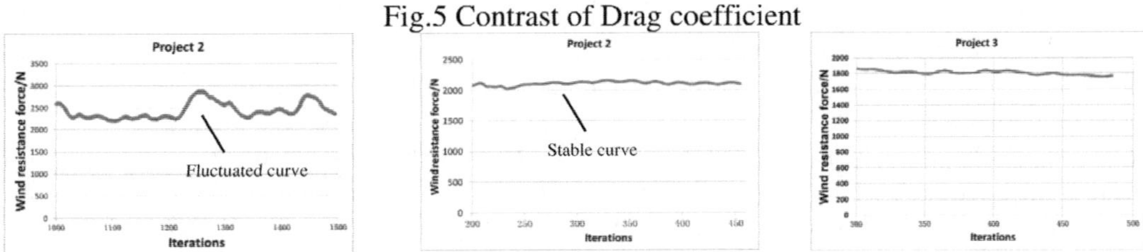


(a) Project 1 (without Fairing) (b)Project 2 (Fairing I) (c)Project 3 (Fairing I 500mm forward )

Fig.6 Contrast of wind resistance force

## 5. Conclusions

(1)The wind resistance forces of Project 1 (without Fairing), Project 2 (Fairing I), and Project 3 (Fairing I 500mm forward) are 2358N, 2111N, and 1761N, respectively. The drag coefficients of 3 projects are 0.784, 0.689, and 0.571, respectively.

(2)According to (1), the drag coefficient and wind resistance force of Project 3 are the smallest, so the structure of Project 3 is the best one among the 3 projects.

(3)The more similar streamlined shape of the vehicle can obtain the smaller wind resistance during the movement of vehicle. Therefore, the wind resistance could be reduced when a truck has a fairing because of the truck structure being more similar with streamlined shape. The challenge here is that the exactly optimal location of the fairing on the vehicle is not clear and generally was determined only by experience, which is not good. In this case, the optimal one could be fixed by the comparison of different design models with CFD analysis for the goal of reducing the resistance fuel consumption.

**Acknowledgement**

1.Enhance Project of Middle-aged and Young College Teachers of Guangxi, China (KY2016Y640);
2.The software of SolidWorks provided by New Well Technology Co. Ltd.

## References

[1] Li Cuiping, (2014) 'Optimization design of FSAE racer body based on CFD', Journal of Machine Design, Vol. 31, No.8, pp74-77.
[2] Wang Jun, (2013) 'Application of CFD analysis method in Vehicle design'. Automobile Technology, No.4, pp 17-20.
[3] Jian zhong Xiao-Ming-Cai-*-William-James-Bloss. (2016) 'Coupling dynamics and chemistry in the air pollution modelling of street canyons: A review'. Environmental Pollution, Vol.214.
[4] M. lateba R.N.-Meroneyb-M.-Yataghenec-H.-Fellouahd-F.-Saleha-M.C.-Boufadela. (2016) 'On the use of numerical modelling for near-field pollutant dispersion in urban environments ? A review'. Environmental Pollution, Vol. 208, pp 271-283.
[5] Jorge d. cambaa Manuel-Conterob-Pedro-Companyc. (2016) 'Parametric CAD modeling: An analysis of strategies for design reusability'. Computer-Aided Design, Vol.74, pp18-31.
[6] Hamid nawaz Yan-Sheng-Yuan. (2013) 'Thermal Comfort Analysis of a Ship Air-Conditioning System Using Solidworks Flow Simulation'. Advanced Materials Research , Vol.773, pp 883-888.
[7] Lee T,Filipi Z.( 2011) 'Synthesis of Real-world Driving Cycles Using Stochastic Process and Statistical Methodology'. International Journal of Vehicle Design, Vol.57, No.1, pp 17-36.
[8] Paramadayalan T,Pant A. (2015) 'Transient Cfd Model of Full Twc Converter'. International Journal of Vehicle Design, Vol.68, No.4.
[9] Saleh Saad-N.1,Barghi Shahzad2(-sbarghi2@eng.uwo.ca).(2016) 'Reduction of fine particle emission from a prilling tower using CFD simulation'. Chemical Engineering Research and Design, 109: 171-179.
[10] Paramadayalan Thiyagarajan,Pant Atul. (2016) CFD model of multi-converter HC adsorber-TWC system'. International Journal of Vehicle Design, Vol.70, No.4, pp 358-376.
[11] Chen Lei1,Chen Youhua,Huang Kai1. (2016) 'Investigation of effective thermal conductivity for pebble beds by one-way coupled CFD-DEM method for CFETR WCCB'. Fusion Engineering and Design, Vol.106, pp 1-8.
[12] Danca Calogero,Mancuso Antonio,Mariotti Gabriele-Virzi. (2005) 'Optimisation of a vehicle shape by CFD code'. International Journal of Vehicle Design, Vol.38, No.1, pp 26-41.
[13] Chen Jing,(2015) 'Analysis of CFD automobile Three-way catalytic converter'. Environmental Engineering, (12, pp 90-94.
[14] Chen Jing. (2016) 'Feeding ball valve optimized by CFD technology. China Adhesives, No.2, pp 16-20.

[15] Kiani M,Shiozaki H,Motoyama K. (2015) 'Simulation–based Design Optimisation to Develop a Lightweight Body–in–white Structure Focusing on Dynamic and Static Stiffness'. International Journal of Vehicle Design, Vol.67 No.3.
[16] Jing Chen. 'An analysis method of vehicle shape optimization based on CFD technology' China, 201510239905.9 Patent.
[17] Chen Jing. (2013) 'Dispersion plate selected by Flow Simulaiton plugin from SolidWorks software'. China Adhesives, No.12, pp 13-16.
[18] Matthew chua Chee-Kong-Chui-Constance-Teo. (2015) Computer aided design and experiment of a novel patient-specific carbon nanocomposite voice prosthesis'. Computer-Aided Design, Vol.59, pp 109-118.
[19] Min-feng sung Chin-Fu-Chen-Chao-Jung-Chen. (2015) 'Using the CFD Technique to Analyze Tire Tread Hydroplaning Effects'. Asian Journal of Engineering and Technology, Vol.3, No.3, pp 151-157.
[20] Yan luximon Roger-M.-Ball-&-Eric-H.C.-Chow. (2015) A design and evaluation tool using 3D head templates'. Computer-Aided Design and Applications, Vol. 13, No.2, pp 153-161.

IOP Conf. Series: Earth and Environmental Science **61** (2017) 012161 doi:10.1088/1755-1315/61/1/012161

# A technigue exploitation about anti-slide tire polyploid on ice-snow road in winter

**Qi Xiaojie, Wang Qiang*, Yang Zhao, Wang Yunlong, Wang Guotian, Lv Degang**

Heilongjiang Institute of Technology, School of Automobile and Traffic Engineering, Harbin
630702666@qq.com

**Abstract.** Present studies focus on improving anti-slide property of tyes on ice-snow road by changing material modification of tyre tread and designing groove. However, the basic reason causing starting slide, long braking distance, turning slide slip and so on of tyres used in winter is that tyre tread materials are unitary and homogenous rubber composite which can't coordinate driving demands of tyres in winter under muti-work condition, and can't exert their best property when starting, braking and sliding slip. In order to improve comprehensive anti-slide property of tyres, this paper discusses about changing structure, shape and distribution proportion among haploid materials of tyre tread rubber. Polyploid bubber tyre tread technique based on artificial neural network which is in favor of starting, braking and anti-slide slip is optimized and combined. Friction feature and anti-slide mechanism on ice-snow road of polyploid rubber tyre tread are studied using testing technique of low-temperature cabin and computer simulation. A set high anti-slide theories and realizing method systems of polyploid rubber composite formed from basic theory, models and technique method are developped which will be applied into solving anti-slide problem of winter tyres, provide theory instruction for studies on high anti-slide winter tyres, and promote development of application and usage safety of winter tyres.

## 1. Intruduction

High anti-slide is the important property of winter tires for safety. Friction and anti-slide mechanism of tire rubber on ice-snow road is the hot point studied on currently. Present studies focus on improving starting performance, braking performance, and anti-slide slip performance of tires on ice-snow roads by changing material modification of tire tread and designing tread groove. Whatever techniques mentioned above are used, winter tires usually have problems, such as sliding and idling due to less traction, long braking distance, turning slide slip, deviating and so on, which will cause serious traffic accidents. Certainly, except driving operation, basic reasons for these problems are that tire treads are made of unitary and homogenous rubber composite which can't coordinate imbalance among driving demands of winter tyres under the condition of multi-work state(such as : differences between non ice-snow road and ice-snow road, differences of demands on tyres when the properties of braking with anti-slide and starting improve, differences of demands on tyres when the properties of acceleration and anti-slide slip improve), traction and starting property, braking property and anti-slide slip property well. In order to improve comprehensively driving properties of tiers, i.e. their starting property, braking property and anti-slide slip property can all be exerted on ice-snow roads, it is difficult to achieve only relying on changing formulation of unitary and homogenous tire tread and structure of tread groove. Because unitary and homogenous rubber materials can't exert their best properties in different motion states of starting, braking, and sliding slip, then some motion states are possible out of control. So some attachments, such as anti-slide stud or anti-slide chain, which can't solve problems, are fixed on tires.

## 2. Technique exploitation about anti-slide polyploid

A lot of tread rubber standard specimens with different physical and mechanical properties are gotten from tread gross mix rubber of summer, winter, and all-weather tires with grooves and unitary, homogenous, and classical formulation through changing formulation system of gross mix rubber and vulcanization technique (temperature, depth, pressure and time and so on of vulcanization). Parameters of stretch strength, elasticity modulus, maximum elongation, stress at definite elongation, permanent deformation, tearing resistance and so on are achieved under the condition of low temperature cabin taking advantage of omnipotent stretch test machine. Moreover, physical and mechanical property parameters are obtained, such as hardness gotten by Sauer durometer, wearing resistance by Akron abrasion machine, elasticity by shock elasticity testing machine, fatigue durability by rubber fatigue and chap fracturing testing machine, brittleness temperature by rubber low temperature brittleness testing machine, heat and aging resistance by aging testing box, and so on. Friction property test platform is built under the condition of low temperature and many calculation numerical simulations are made using ABAQUS finite element software, in order to do research on friction properties of tread rubber standard specimens at different motion states (starting, braking and slide slip) and on different ice-snow roads. Mutual influence relation between friction property and physical and mechanical property is set up by experiment and computer simulation analysis. Which are main influence parameters on friction coefficient, traction (starting state), braking force (braking state), skid force (slide slip state) and so on among physical and mechanical property parameters will be clarified, considering conditions of load, turning radius, temperature and slippage speed. Then relativity between friction property and tread material property will be built.

In the research, butadiene styrene rubber, three-poly butadiene rubber, natural rubber, and cis-isoprene rubber are used as rubber base, then base rubbers of two kinds or several kinds from that mentioned above are chosen and optimized by changing corresponding amount, and then optimization of vulcanization system is made through changing species and usage amount of sulfur and accelerators, and some parameters, such as vulcanization temperature, vulcanization depth, vulcanization pressure, vulcanization time and so on, last optimization of reinforcement and packing system by means of changing species and usage amount of reinforcement, carbon soot, silica and so on. Optimized tread haploid will be gotten by combining application probability and statistics principle, ABAQUS finite element numerical value modeling, with thermostatic experiment in low temperature cabin, and integration of design, analysis, experiment, preparation and optimization is considered as principle.Three-stage mixing technique is used, and optimum mixing temperature, optimum time, rubber removal temperature, and rotor speed are controlled during mixing process. Flow chart of preparation technique of tread rubber haploid is shown as in the figure1. The key technique is to control rubber removal temperature of mixing rubber at the first stage and at the second stage which is higher than that of common tread rubber and rotor turning speed of banbury mixer is lower than that of the common one.

IOP Conf. Series: Earth and Environmental Science **61** (2017) 012161 doi:10.1088/1755-1315/61/1/012161

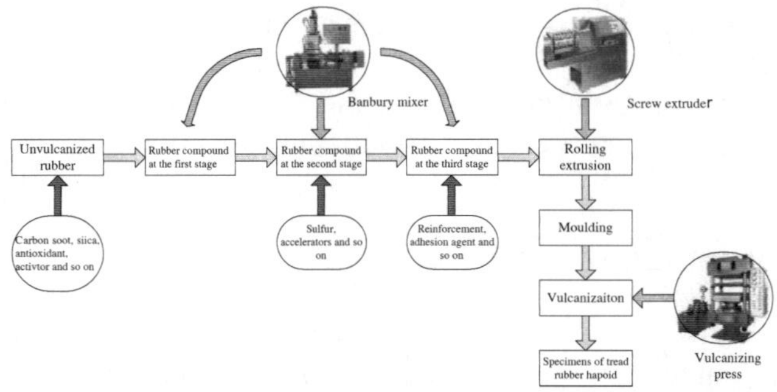

Fig.1 Preparation technique flow of tread rubber haploid

Distribution, combination, and optimization of tread rubber haploid is made aiming at being favor of three kinds of motion states of starting, braking and anti-slide, four kinds of classic working condition of snow roads, ice-snow roads, compacted ice roads and ice-water roads, and four kinds of condition of loading, environment temperature（-5～-40$^oC$），driving speeds（10～60km/h）and turning radius. So tread rubber polyploid with better comprehensive properties are achieved which are suitable to snow roads, ice-snow roads, compacted ice roads, ice-water roads, and staring, braking and anti-slide of vehicles under the condition of four kinds of working condition mentioned above.The optimum combination plan will be found by means of changing structure, shape and distribution of tread haploid which benefits to starting, braking and anti-slide. (distribution and combination are shown in figure 2). Multi-goal distribution and optimization problem is solved by using algorithm of multi-layer error opposite direction spread(BP algorithm) and genetic algorithm based on shared econiche technique(pareto GA) in artificial neural network and Matlab software, then multi-layer forward direction neural network constituted by input layer, the first , the second implicit layer, and output layer is constructed. It is demonstrated in figure3.

Preparation of polyploid is made taking advantage of cold feed extruding technique i.e. extruding technique using three compounding machine heads, tread die base planks of different distribution and combination, and by means of controlling different temperature(temperature of machine head, die base plank, rubber removal and heating area) and extruding speed. Structure characterization: Combination interface microstructure of tread rubber polyploid is characterized and analyzed through rubber processing analyser, scanning electron microscope (SEM), transmission electron microscope (TEM), sight difference heat checkout machine(DSC), thermogravimetic analyser(TGA), dynamic mechanical analyser(DMA), gas chromatograph analyser, and fourier transform infrared analyser. Property test: Tread rubber polyploid is tested about its properties of stretching, compressing bending, bonding strength, tearing, peeling and so on using universal tensile testing machine and cutting and stripping machine in low temperature cabin.

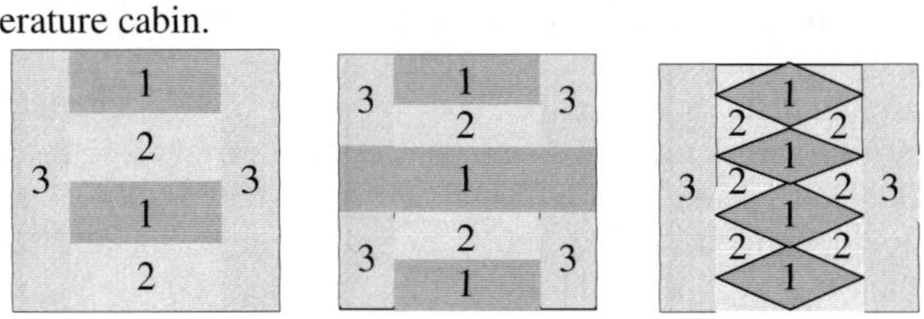

1-Starting haploid 2-Braking haploid 3-Sliding slip haploid
Fig.2 Distribution and combination plan of different proportions and shapes

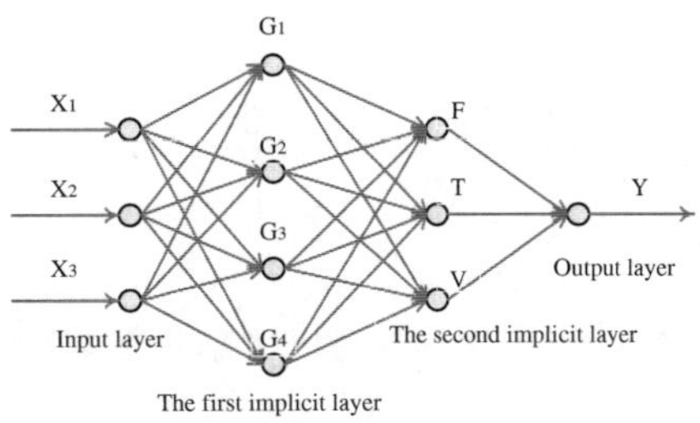


X1—Haploid benefit to starting X2—Haploid benefits to braking X3—Haploid benefits to anti-slide G1—Snow road G2—Ice-snow road G3—Compacted road G4—Ice-water road; F—Load T—Environment temperature V—Driving speed R—Turning radius Y—Polyploid after being combined and optimized

Fig.3 Combining and optimizing neural network structure

## 3. Conclusion

It is significant for vehicles on ice and snow roads in winter in northern area to reinforce to do research on safely driving of winter tire, reducing damage or having no damage to road surface, reducing consumption of tire driving energy, especially for safely driving and keeping roads clear, and decreasing traffic accidents on intermittent, thin ice, hard ice roads in winter in cities of north of China.This paper exploits anti-slide tire polyploid technique on winter ice-snow roads from the point of changing unitary and homogenous material of tires, and many study technique plans of friction property on different ice and snow roads indoor or outside, last anti-slide mechanism of tire tread rubber polyploid composite is made known through applying the technique to tires, in order to lay foundation for design and application of winter tires.

### References

[1] Xiang Wensen. Study on Cause and Occurrence Mechanism of Traffic Accident on City Ice-Snow Roads[J].Harbin Institute of Technology, 2010,07.
[2] He Haoming, Ye Liqing, Zhang Jianjun and so on. Development and Property Test of Winter Tyre[J].Tire Industry,2009, 29 (1),30-36.
[3] Deng Shitao, Tian Zhaoju, Houbo and so on. Research and Development of Car Radial tire Used in Winter[J]. Tire Industry,2013,33
[4] Yang Honghui, Zhao Jia. Winter Tire Market Saves up Strength to Start out[J].Chinese Rubber, 2013(7),11-15.
[5] Zhou Lichong, Wang Hongwei.Tire Groove Design Simulating Octopus Osculum Type and Finite Element[J]. Xian Science and Engineering University Journal,2013,29(2), 228-232.

IOP Conf. Series: Earth and Environmental Science **61** (2017) 012162 doi:10.1088/1755-1315/61/1/012162

# An image overall complexity evaluation method based on LSD line detection

**Jianan Li[1], Jin Duan[2*] and Xu Yang[1*] & Bo Xiao[2]**

[1]School of Electronic and Information Engineering, Changchun University of Science and Technology, Changchun,China

[2]Space Photoelectric Technology Research Institute, Changchun University of Science and Technology, Changchun, China

[*]Corresponding Author: Jin Duan: duanjin@vip.sina.com, Xu Yang: lbaomi@cust.edu.cn

**ABSTRACT.** In the artificial world, whether it is the city's traffic roads or engineering buildings contain a lot of linear features. Therefore, the research on the image complexity of linear information has become an important research direction in digital image processing field. This paper, by detecting the straight line information in the image and using the straight line as the parameter index, establishing the quantitative and accurate mathematics relationship. In this paper, we use LSD line detection algorithm which has good straight-line detection effect to detect the straight line, and divide the detected line by the expert consultation strategy. Then we use the neural network to carry on the weight training and get the weight coefficient of the index. The image complexity is calculated by the complexity calculation model. The experimental results show that the proposed method is effective. The number of straight lines in the image, the degree of dispersion, uniformity and so on will affect the complexity of the image.

## 1. Introduction

At present, image complexity has been widely used in many fields, including: military field for battlefield environmental assessment, military combat situation prediction; medical aspects for disease diagnosis, pathological analysis. In many academic articles, scholars use geometric information to analyze the image, the general will refer to edge information and geometric elements and other terms, usually with the information to describe the characteristics of these things. Geometric elements usually include linear, round, triangular and other shape features, in real life, these elements constitute the material world, increasing the diversity of objects. In this paper, the complexity of the image information from the linear aspects of the relevant research.

## 2. Image complexity characteristic parameter

There are many algorithms for straight line detection. In recent years, many scholars have been working on the algorithm of straight line detection. This paper will realize and contrast the classic Hough transform line detection and LSD (a Line Segment Detector) line detection algorithm. It is found that the Hough transform algorithm is not very accurate in locating the end points of the straight line, and it is easy to produce false line segments. The detection result is not very ideal. Fortunately, LSD linear extraction algorithm not only avoids the drawbacks of Hough transform algorithm, but also has the advantages of low computational complexity, good real-time performance and low storage space requirement. Therefore, considering this problem, this paper adopts LSD line detection algorithm to carry out experiments.

### *2.1 Based On LSD Linear Detection Algorithm*

(1) The scale factor of 80% (S=0.8), is the smallest image reduction that reasonably solves the staircase problem while producing almost the same result as a full scale analysis on images without artifacts. (A 80% scaling means here that the x and y axis are each reduced to 80%; the number of pixels is thus reduced to 64%. )

$$\sigma = \Sigma / s, \Sigma = 0.6 \tag{1}$$

(2)Gradient Computation:

$$g_x(x, y) = \frac{I(x+1, y) + I(x+1, y+1) - I(x, y) - I(x, y+1)}{2} \tag{2}$$

Note that the computed value corresponds to the image gradient at coordinates (x + 0:5; y + 0:5) and not (x; y) . This half-pixel offset is then added to the output rectangles coordinates to produce coherent results. Pixel $(x, y)$ gradient detection is defined as follows:

$$\theta_I = \arctan(\frac{g_x(x, y)}{-g_x(x, y)}) \tag{3}$$

the gradient magnitude as:

$$\theta_I = \arctan(\frac{g_x(x, y)}{-g_x(x, y)}) \tag{4}$$

$$G(x, y) = \sqrt{(g(x)^2 + g(y)^2)} \tag{5}$$

(3) Rectangular estimate that in the initial conditions, the support area as a single pixel, and sets the line direction of the support area to the pixel gradient direction, tags get straight line support area of the pixel, and then the linear area is rectangular simulation.

(4)NFA Computation:

$$NFA(r) = (MN)^{5/2} * \sum_{j=k}^{n} (\frac{n}{k}) p^j (1 - p^j)^{n-j} \tag{6}$$

The rectangles with $NFA(r) < \varepsilon$ ,are validated as detections.

Hough change algorithm and LSD line detection algorithm comparison test results are as follows:

Original image Hough LSD

### *2.2 Straight Line Characteristic Parameter*

In order to establish the mathematical relationship between the detected linear feature and the image complexity. In this paper, the straight lines extracted in each image are normalized without loss of generality. The length of straight line segment is straight to ignore. This article will each image is

IOP Conf. Series: Earth and Environmental Science **61** (2017) 012162 doi:10.1088/1755-1315/61/1/012162

extracted by a straight line within the length in do not break under the condition of generalized linear normalization processing, line length is:

$$l = len(l_i) = \sqrt{(x_{mi} - x_{li})^2 + (y_{mi} - y_{li})^2} \tag{7}$$

Where $x_{mi}$, $y_{mi}$, $x_{li}$, $y_{li}$ is the horizontal and vertical coordinates of the two end points of the ith straight line segment.

Line length of the normalized treatment, the formula is:

$$L = \frac{l}{l_{\max}} \tag{8}$$

$l_{\max}$ is the longest line detected in the image.

Then after the normalization processing range is divided into such as straight line [0, 0.2), [0.2, 0.4), [0.4, 0.6), [0.6, 0.8, [0.8, 1] five interval. Then the image complexity evaluation index X1, X2, X3, X4, X5 undertake unity to dimension, using interval value method to dimension :

$$W_i = \left( \frac{x_i(1) - \min(X_i)}{\max(X_i) - \min(X_i)}, \frac{x_i(2) - \min(X_i)}{\max(X_i) - \min(X_i)}, \cdots, \frac{x_i(n) - \min(X_i)}{\max(X_i) - \min(X_i)} \right) = \left( x_i^0(1), x_i^0(2), \cdots, x_i^0(n) \right) \tag{9}$$

That was five parametric image complexity of the model $W_{[0-0.2)}$, $W_{[0.2-0.4)}$, $W_{[0.4-0.6)}$, $W_{[0.6-0.8)}$, $W_{[0.8-1]}$.

## 3. Image complexity evaluation model

This paper adopted in the process of the construction of the model can reflect the change of the index weight, and can improve the scientific, effectiveness and evaluation of the effectiveness of the BP neural network training method [5-6].

Evaluation index for each neural network unit number: 5, the BP neural network input layer node number is five, the number of output layer neurons identified as 1, number of hidden layer neurons for four.

Network accuracy is achieved when all the sample training and, by right of the connection between the input layer to hidden layer matrix V, calculate the input layer node to the sum of the absolute value of the right of all the connection between the hidden layer nodes, and normalization, m the weight of indicators. The specific formula is as follows:

$$w_j = \frac{\sum_{l=1}^{k} |v_{jl}|}{\sum_{i=1}^{m} \sum_{l=1}^{k} |v_{il}|} \quad j = 1, 2, \cdots, m \tag{10}$$

$V_{ij}$ is the input layer to hidden layer between the connection weight matrix.

## 4. The experimental results and analysis

This paper established an image database of 800 images within 800 image contains four different levels of complexity (easy, complicated, very complicated) image, the experiment selected method is based on human visual image from the image database, chose 16 is representative of the image as a test sample, the figure 1 (figure 1 (a) ~ (l)) method based on human visual from the simple to the complex situation of sorting.

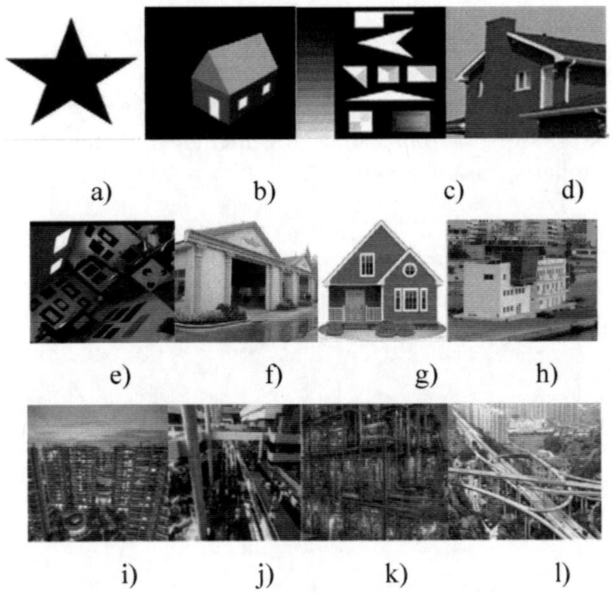


Figure 1 The method based on human visual image (figure b (a) ~ (l)) from simple to complex

800 images extracted from image library, using LSD linear detection algorithm for straight line detection, each image is calculated by the image complexity evaluation index X1, X2, X3, X4, X5 data stored in the database. All the line number within the range as shown in figure 2:

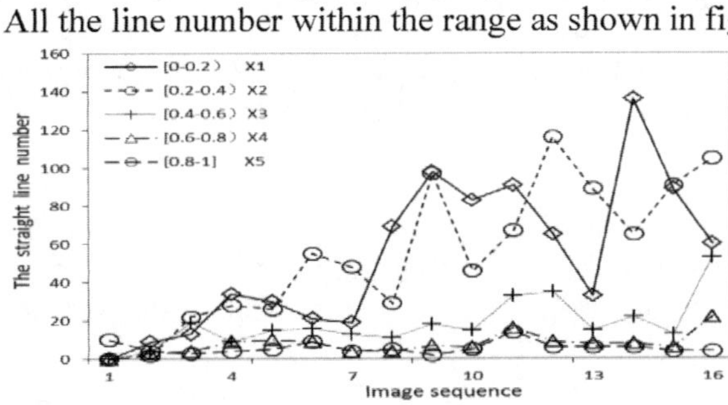


Figure 2 Image of each interval in the straight line number case diagram

From randomly selected 500 groups of data as training data network, 300 groups of data as test data, test network classification ability, parameter is obtained by the method of BP neural network training weight coefficient.

Table 1 Evaluation Index Weight Image Complexity

| Index | Weight |
|---|---|
| X1 | 0.2670 |
| X2 | 0.2700 |
| X3 | 0.2381 |
| X4 | 0.1417 |
| X5 | 0.0832 |

Image complexity calculation model for:

$$C = A_1W_{[0-0.2)} + A_2W_{[0.2-0.4)} + A_3W_{[0.4-0.6)} + A_4W_{[0.6-0.8)} + A_5W_{[0.8-1]} \tag{11}$$

In the formula, $A_i$ is the weight coefficient of every index , i=1,2,…,5.

In order to verify the validity of the results, the method calculated results and the method of reference literature [4] comparing calculated results, comparing the situation as shown in figure 3:

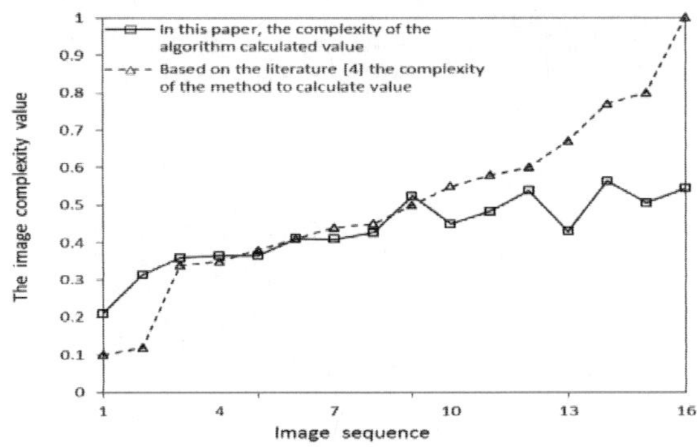


Figure 3 Complexity value contrast figure of each image

In this paper, compared with the images of the human visual complexity method is more objective and avoid the human subjective awareness of errors, from the perspective of other linear features susceptible to light, the influence of surface texture characteristics of condition, so the method from the perspective of the image complexity has opened up a new train of thought.

## 5. Conclusion

Geometric features on the analysis of the image of the role is of more and more important, geometric feature unique characteristics is based on the image, if it can be very good to extract the image geometric features, it will be the image research and has important significance. , of course, this article also has some shortcomings, such as linear interval dividing standards remains to be improved, the author will be added to modify in a follow-up study. The experimental results show the image to detect the straight line, and it can efficiently describe the image structure characteristics, and can express the complexity of the image.

## References

[1] R. Peters II, and R. Strickland, Image complexity metrics for automatic Target recognizers. 1990 ATR System and Technology Conference, pp.1-17, October 1990.
[2] Mario Ignacio Chacon M, Luis Enrique Aguilar D and Abdi Delgado S. A Fuzzy Approach on Image Complexity Measure. Computación y Sistemas, ISSN 1405-5546, Vol. 10, No. 3, 2007, accepted.
[3] Rafael Grompone von Gioi, Jérémie Jakubowicz, Jean-Michel Morel, and Gregory Randall, LSD: a Line Segment Detector, Image Processing On Line, 2 (2012), pp. 35–55.
[4] Matthieur Perreira Da Silva, Vincent COURBOULAY, Pascal Estraillier. Image Complexity Measure Based on Visual At-tention[C].18th IEEE International Conference on Image Processing,11-14 Sept. 2011: 3281-3284.

[5] Sun Huijun, et. Application of artificial neural network to determine the weight of evaluation index [J]. Journal of shandong university of science and technology (natural science edition), 2001, 3( 9) :84-86.

[6] Kai-li Zhou, Yao-hong Kang. Neural network model and MATLAB simulation program design [M]. Beijing: Tsinghua university press, 2005:73-74.

[7] Shao Feng, jiang ping, jiang fortitude, etc. Based on the analysis of the significance of three-dimensional image visual comfort prediction [J]. Optical precision engineering, 2014, 22 (6) : 1631-1638

IOP Conf. Series: Earth and Environmental Science **61** (2017) 012163 doi:10.1088/1755-1315/61/1/012163

# Detecting collapsed buildings from image data for estimation of disaster debris

**N. Yoshinari, M. Yamamoto, R. Tonegawa, Y. Oyanagi and A. Ohuchi**

Hokkaido University, Sapporo, Japan
Sumitomo Electric Industries, Ltd, Osaka, Japan
The Public Interest Incorporated Foundation for International Peace, Tokyo, Japan
Sendai Environmental Development, Inc., Sendai, Japan

**Abstract.** In the event of a massive earthquake, it is difficult to assess the situation in affected areas because access to these areas is usually blocked off. Tsunamis, in particular, create a large amount of debris and block many access paths. It is necessary to estimate the amount of disaster debris as early as possible in order to begin restoration in affected areas. This data helps to determine the required amount of temporary storage and arrange proper transportation. We propose a system that detects collapsed buildings after a disaster by using three aerial photographs. We verify the accuracy of the proposed method and discuss how detecting collapsed buildings aids in estimating total disaster debris.

## 1 Introduction

The Great East Japan Earthquake, which occurred on March 11, 2011 in Japan, caused enormous damage. The earthquake occurred with a Tsunami, destroying many houses and access paths. Additionally, the Tsunami left large amounts of debris in the affected areas. Media outlets all over the world reported this crisis. According to a report on the day after the earthquake, the area of collapsed buildings caused by the Tsunami was approximately 150 square kilometers. The collapsed buildings included infrastructure facilities such as stations. As a result, access to the affected areas was blocked by debris, flooding, and land subsidence. This forced the use of helicopters for damage surveying and rescue work.

One important aspect of restoration is assessing a situation to enable the removal and disposal of disaster debris. For disaster debris disposal, it is necessary to estimate the amount of disaster debris as early as possible. Furthermore, determining the required amount of temporary storage and arranging transportation is also essential. Both of these tasks also require an estimate of disaster debris. This is why estimating the amount of disaster debris is an essential task for restoration of the affected areas. However, little information is typically available from the affected areas immediately following a disaster. We focus on the use of aerial photographs, which can be taken immediately after a disaster, to accomplish debris estimation.

We propose a system that detects collapsed buildings after a disaster by using aerial photographs. This data is used to estimate the total amount of disaster debris. To detect collapsed buildings using aerial photographs, we create a classifier based on characteristics such as the number of straight lines in the images. We then apply the proposed method to a real aerial photograph taken after a disaster. The aerial photographs used were taken in the affected areas at the time of the Great East Japan Earthquake.

IOP Conf. Series: Earth and Environmental Science **61** (2017) 012163 doi:10.1088/1755-1315/61/1/012163

## 2 Related works

Disasters generate large amounts of debris. Some researchers have examined debris management plans and methods from real disasters such as Hurricane Katrina or the Great East Japan Earthquake. Some frameworks for methods to restore the affected areas have also been proposed. Luther et al. provided an overview of the types and amount of debris generated by Hurricane Katrina, as well as the unique complicating factors involved. Brown et al. examined various aspects of disaster debris management, such as debris management goals, prioritization and timing, environmental impact, economics, social factors, organizational and coordination structures, legislative issues, and financial aspects or funding mechanisms. Nojima et al. described a framework for estimating damage using various dimensions. The first dimension uses a combination of a fragility function and seismic information. The second dimension is data from affected areas, gathered using remote sensing methods. The third dimension is information obtained through audits of the disaster area. This final dimension is typically the most reliable. However, capturing this information takes large amount of time and manpower. Therefore, it is difficult to use this dimension immediately after a disaster. We focus on the second dimension because it is easier to obtain than the third dimension, and more specific than the first dimension.

Some researchers have focused on the estimation of the amount of disaster debris. They also used the previously described secondary information. Hirayama et al. proposed a system that estimates the amount of disaster debris in Tsunami-inundated areas. Mitomi et al. analyzed debris characteristics in aerial photographs to detect heaps of debris.

Some other studies that detect collapsed buildings have been conducted. Matsuoka et al. built a system that detects building damage in areas affected by an earthquake using satellite SAR intensity images. Urabe et al. analyzed road blockages after earthquakes in mountainous areas. They detected damaged areas by comparing two images with a digital elevation model. Sakurada et al. proposed a method for detecting temporal changes in three-dimensional structures from multi-view images captured at different times.

## 3 Detecting collapsed buildings

We target the detection of collapsed buildings in this paper. If collapsed buildings can be detected, the amount of disaster debris can be estimated. We focus on aerial photographs and map data to detect collapsed buildings. This is because they can be easily obtained at the time of a disaster. They are both taken in the areas affected by the disaster.

Image data, including aerial photographs, has standard image characteristics. We can use these to detect collapsed buildings. We focus on regions of the image likely to contain a building. A building that collapses after a disaster should create debris. On the other hand, a building that has not collapsed should appear as it did before the disaster. Our proposed processing method follows the process below:

1) Acquire image data.
2) Focus on region of image likely to contain a building.
3) Extract features from focus region.
4) Classify building as collapsed or not collapsed.

We describe the specific features and how extract them below.

### *3.1 Focus on Image Region*

A building that is captured in an aerial photograph and a building found on a map are in the same position in both images. We now discuss a method that uses this fact to focus on a particular image region.

Initially, the system focuses on a building on the map. Next, the system trims the aerial photograph to encompass the same area defined by the map coordinates. As a general rule, one focus region contains one building. If it is determined that the target building is damaged, the affected area is added to the size of the target building, which is obtained from the map data. Figure 1 shows an example of the process.

IOP Conf. Series: Earth and Environmental Science **61** (2017) 012163 doi:10.1088/1755-1315/61/1/012163

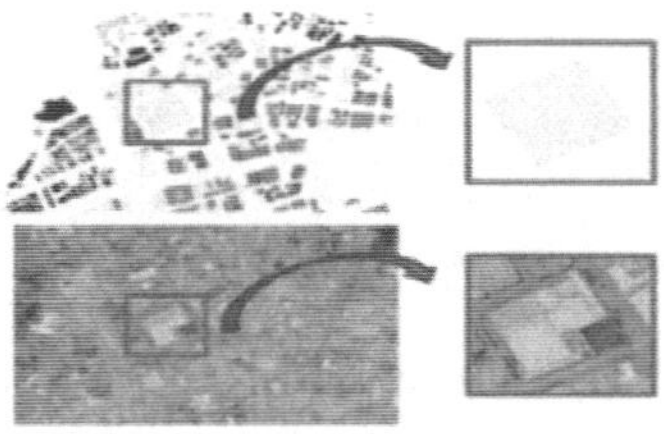


**Figure 1.** Process for determining a focus region.

*3.2 Feature Extraction*
It is necessary to analyze the aerial photographs and quantify which features indicate the presence of damage. To classify collapsed buildings using aerial photographs, we use extracted image features.

We focus on the roofs of houses in this paper. Typically, if a house is drawn on a map, the roof of the house is captured in an aerial photograph. However, if the target house has collapsed, the roof will appear differently, or be missing entirely in the aerial photograph. Therefore, the condition of roofs is a feature that can classify damage. The actual process is shown below.

*3.2.1 Convert color image into grayscale.* An aerial photograph is a color image. The first step is to convert $RGB$ color information into luminance.

Equation (1) is the conversion equation. The equation is applied to all pixels. The result is a color aerial photograph converted into a grayscale.

$$Y = 0.299r + 0.587g + 0.114b \tag{1}$$

Where $Y$ represents the 0-255 luminance value of each converted pixel. $r, g$, and $b$ are the $RGB$ values of each pixel before conversion. Figure 2 displays an example conversion result.

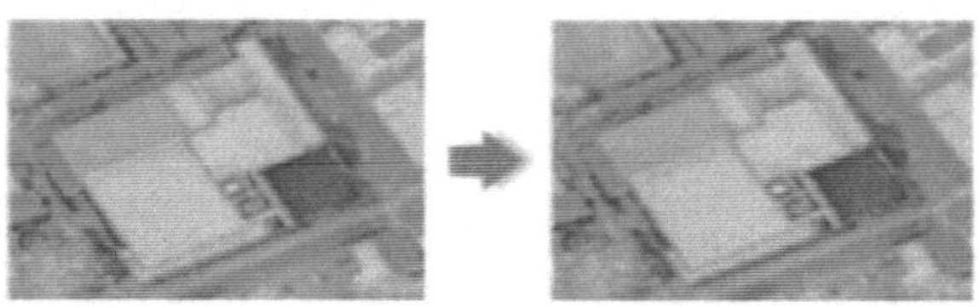

**Figure 2.** Conversion of color image into grayscale image.

*3.2.2 Creating a differential image: Sobel Filter.* By applying a differential filter, we can extract the portion of a grayscale image where brightness changes rapidly. This allows the detection of roof edges. For this research, we chose the widely used Sobel filter.

The Sobel filter is defined by equations (2) and (3).

$$g(x,y) = (-1)f(x-1,y-1) + (-2)f(x,y-1) + (-1)f(x+1,y-1) + f(x-1,y+1) + 2f(x,y+1) + f(x+1,y+1) \tag{2}$$

$$g(x,y) = (-1)f(x-1,y-1) + f(x+1,y-1) + (-2)f(x-1,y) + 2f(x+1,y) + (-1)f(x-1,y+1) + f(x+1,y+1) \tag{3}$$

Where $g(x,y)$ is a pixel value in the resulting image, $f(x,y)$ is a pixel value in the grayscale image, and $x, y$ is the pixel coordinate. Figure 3 shows an example of a differential image.

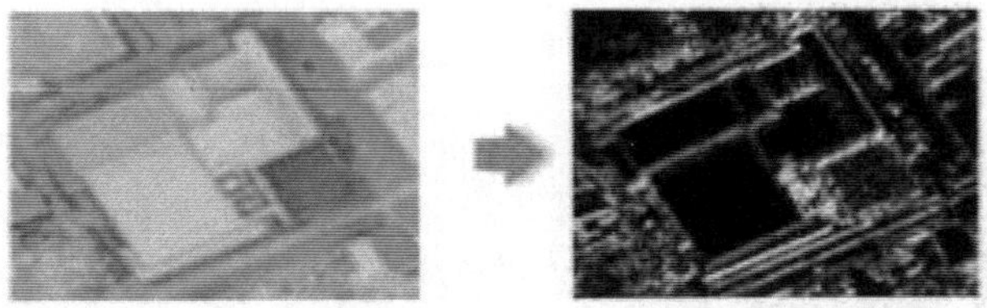

**Figure 3.** Conversion from grayscale into differential image.

A differential image is a grayscale image, meaning it is multi-valued. To carry out detection of line segments, it must be binarized using an automatic threshold selection method on the obtained image. Specifically, we search for a threshold value t that maximizes the following equation (4) for a luminance histogram.

$$t = \max\{\omega_1 \omega_2 (m_1 - m_2)^2\} \tag{4}$$

Where $t$ is a threshold used to divide two classes, $\omega_i$ is the average of a class, and $m_i$ is the divergence of a class.

*3.2.3 Detecting Straight Lines.* If a house has not collapsed, a differential image will contain defined roof edges. Typically, roof edges are straight lines, so it is necessary to detect straight lines in a differential image.

The Hough transformation is used to detect straight lines. We apply the widely used $\theta\rho$-Hough transformation to a differential image. If a house has not collapsed, the straight lines of a roof can be found by applying the Hough transformation. On the other hand, if a house has collapsed, straight lines will not be found. Furthermore, sand and debris will be present in their place.

The Hough transform is a method of converting XY-coordinates into polar coordinates. It is applied to all original pixel values using equation (5).

$$\rho = x \cos\theta + y \sin\theta \tag{5}$$

Where $x, y$ is a coordinate in an image. By applying equation (5), coordinates become a relationship between $\rho$ and $\theta$. As shown in Figure 4(a), similar processing is applied to all other coordinates. If the points are arranged in a straight line, the curves will intersect at one point as shown in Figure 4(b). To find straight lines, we look for many overlapping points in the polar plane. Then the point is converted back into a straight line in the XY plane. In this research, we determined a threshold parameter using equation (6). Coefficients for this equation were determined through preliminary experiments.

$$t = 0.0085S + 16.75 \tag{6}$$

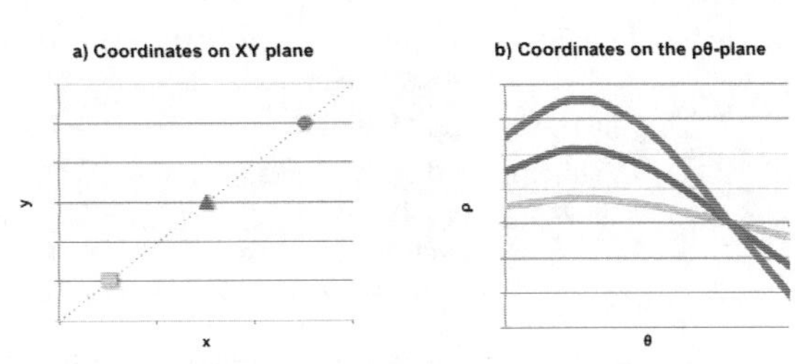


**Figure 4.** The principle of the $\rho\theta$-Hough transformation.

Figure 5 shows an example of line detection. Red lines are the results of the $\rho\theta$-Hough transformation

IOP Conf. Series: Earth and Environmental Science **61** (2017) 012163 doi:10.1088/1755-1315/61/1/012163

superimposed over the original image.

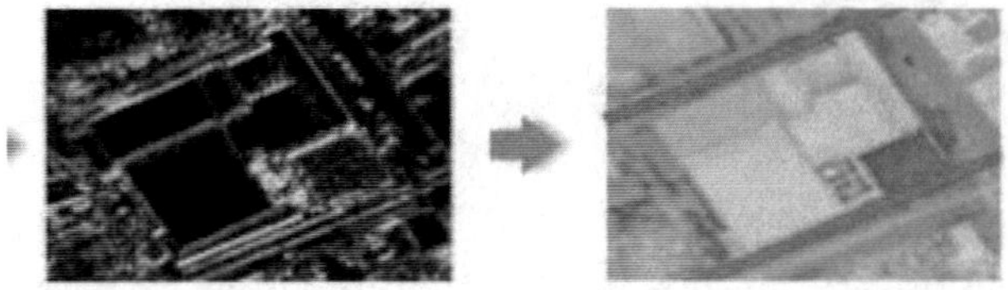

**Figure 5.** The results of the $\rho\theta$-Hough transformation.

## 4 Estimating disaster debris

We used equation (7) for estimating disaster debris.

$$W_D = \sum_{i=l} C_i N_i \tag{7}$$

Where $W_D$ is the amount of debris. $C_i$ is a basic unit for disasters that is defined by disaster types and building structures (e.g. Was collapse caused by fire or earthquake? Is a structure rebar or wooden?). $N_i$ represents the number of collapsed buildings. It can be used to indicate the physical area of collapsed buildings.

If a building is classified as "collapsed," the building area is added to the total damage area. At this time, building areas are converted into physical areas. We defined the basic unit of disaster using a wooden building collapse. The unit is debris created per square meter, 0.62[ton/$m^2$].

## 5 Experiment and evaluation

In this section, we describe the experimental results and evaluate the difference between real damage and our prediction. We performed the following experiment to verify the proposed method. Our proposed method detects collapsed buildings from image features. It then calculates the collapse area and the amount of disaster debris left by the building. We compare the prediction results with the actual damage for each building to evaluate the proposed method.

### *5.1 Aerial Photograph and Map Data*

A set of images is prepared for the experiment. The images are an aerial photograph taken after a disaster and a map of the affected areas. Information on the location and area of each building can be obtained from the map. Test images are required to satisfy the following conditions: collapsed building percentage in aerial photographs is nearly fifty percent, a map corresponding to the buildings in the aerial photographs exists, and real damage information is known.

We prepared images that satisfied all conditions. In this experiment, we used images of Minamisanriku Town, Rikuzentakata City, and Kamaishi City, in Japan. The images are all from areas affected by the Great East Japan earthquake.

Images scale is either 1:2500 or 1:4500. Pixel density is 24 dpi. The photographs come from the Geospatial Information Authority of Japan. They are shown in figure 6 below.

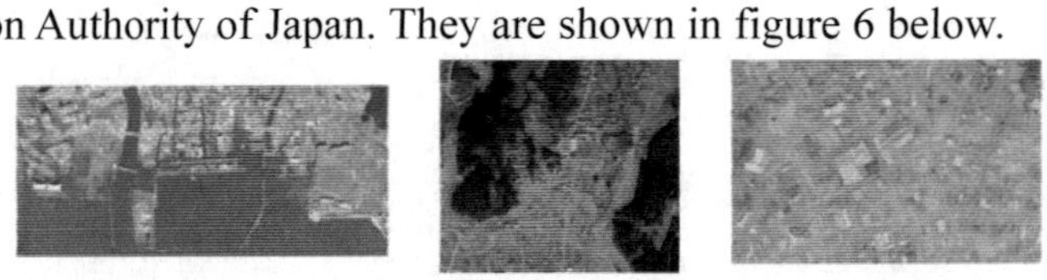

(a) Minamisanriku Town (b) Rikuzentakata City (c) Kamaishi City

**Figure 6.** Example aerial photographs of affected areas.

We use a Geographic Information System (GIS) for map data. The map data is publicly available from the Great East Japan Earthquake Reconstruction Support Research Archive. The available map information is the position of buildings, area of each building, and a classification of the damage. We use only the position and area of buildings.

### *5.2 Damage Data from the Actual Disaster*

To evaluate the validity of the experimental results, we gathered the actual damage data from the Great East Japan Earthquake Reconstruction Support Research Archive for comparison. This data was created through a huge investment of time and manpower. Figure 7 displays an example of this data. The image has been blurred to protect the privacy of the victims. By using the color data, we can verify the location, size, and damage level of every building.

**Figure 7.** Actual damage data from the Great East Japan Earthquake Reconstruction Support Research Archive.

Red indicates a collapsed building dragged away by the tsunami, pink indicates a collapsed house in place. We defined all collapsed buildings using these two colors. The colors used for buildings that have not completely collapsed are yellow, blue, and green. Yellow indicates half-collapsed, and blue indicates partially-collapsed. According to the Japanese Ministry of Internal Affairs and Communications, half-collapsed and partially-collapsed are defined as buildings that can be restored by repairs. Finally, green indicates buildings that were not damaged.

### *5.3 Evaluation of Experimental Results*

We use three metrics to evaluate experimental results: percentage of correct answers, precision, and recall. Figure 8 shows the legend, and the correct answers are $A_r$ and $A_b$.

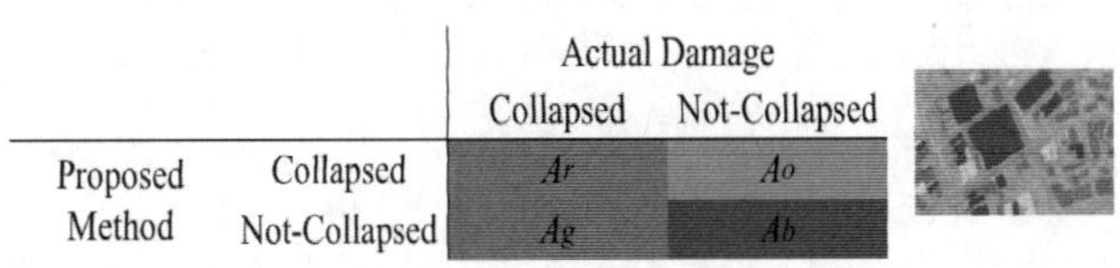


**Figure 8.** Legend for experiment results.

$A_r$ is the number of correct "collapsed" predictions. $A_b$ is the number of correct "non-collapsed" predictions. In this paper, we define the percentage of correct answers as the sum of $A_b$ and $A_r$ over total predictions.

The next metric used to measure the effectiveness of the proposed method is precision. Precision P is defined by equation (8).

$$P = A_r/(A_r + A_o) \quad (8)$$

This is the fraction of "collapsed" predictions over the actual number of collapses. Precision is used in conjunction with recall, the percentage of relevant "collapsed" predictions returned by the proposed method. Recall R is defined by equation (9).

$$R = A_r/(A_r + A_g) \tag{9}$$

*5.4 Experimental Results*

Figure 9 presents the results for estimation of collapsed buildings. We find that there is a large number of collapsed buildings and that our proposed method is able to detect most of them. Tables 1, 2, and 3 show the actual percentages of Figure 9.

(a) (b) (c)

**Figure 9.** The experimental results for estimation of collapsed buildings.

**Table 1.** Percentages for Minamisanriku Town (Figure 9-(a))

| | | Actual Damage | | Total |
|---|---|---|---|---|
| | | Collapsed | Not-Collapsed | |
| Proposed Method | Collapsed | 0.822 | 0.012 | 0.834 |
| | Not-Collapsed | 0.166 | 0.000 | 0.166 |
| Total | | 0.988 | 0.012 | 1.000 |

**Table 2.** Percentages for Rikuzentakata City (Figure 9-(b))

| | | Actual Damage | | Total |
|---|---|---|---|---|
| | | Collapsed | Not-Collapsed | |
| Proposed Method | Collapsed | 0.826 | 0.083 | 0.909 |
| | Not-Collapsed | 0.066 | 0.025 | 0.091 |
| Total | | 0.892 | 0.108 | 1.000 |

**Table 3.** Percentages for Kamaishi City (Figure 9-(c))

| | | Actual Damage | | Total |
|---|---|---|---|---|
| | | Collapsed | Not-Collapsed | |
| Proposed Method | Collapsed | 0.288 | 0.191 | 0.479 |
| | Not-Collapsed | 0.142 | 0.379 | 0.521 |
| Total | | 0.892 | 0.570 | 1.000 |

In tables 1 and 2, the results had high overall precision. In table 1, the percentage of correct answers is 0.823, precision is 0.986, and recall is 0.832. In table 2, the percentage of correct answers is 0.850, precision is 0.908, and recall is 0.926. However, the results in table 3 are not as good as in tables 1 and 2. The percentage of correct answers is 0.667, precision is 0.602, and recall is 0.670.

In this evaluation method, there are two different types of incorrect answers. One is that the proposed method classifies a building as collapsed, but the actual building has not collapsed. This error occurs when the threshold for detecting straight lines in the Hough transformation is not met. It is necessary to integrate a different technique for line-detection to solve this problem.

The other incorrect answer in when the proposed method classifies a building as non-collapsed, but the actual building has collapsed. The reason for this result is due to a difference in patterns of collapse between figure 9-(c) and the other two images. Figure 10 presents specific examples of incorrect answers in each image.

IOP Conf. Series: Earth and Environmental Science **61** (2017) 012163 doi:10.1088/1755-1315/61/1/012163

Figure 10-(a) is specific example of collapsed buildings from figure 9-(a) and 9-(b). The houses have collapsed without a trace. In this case, the error occurred because the road was falsely detected as straight lines of a roof. However, the roof of the collapsed building is clearly visible in figure 10-(b), which shows specific examples of collapsed buildings from figure 9-(c). This error occurred for 20 buildings out of 25 erroneous classifications. This error cannot be resolved when using our method with aerial photographs. However, these errors occur in less than 10% of the 361 total buildings in the image.

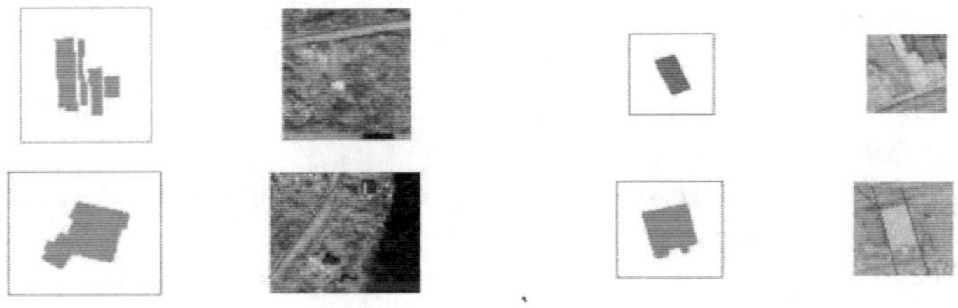


**Figure 10.** Specific examples of incorrect answers.

## 6 Discussion

In this section, we present a discussion from the perspective of evaluating disaster debris in affected areas.

Table 4 presents approximate estimations of total waste volume from the previous images using the proposed method.

**Table 4.** Results of estimating disaster debris in affected areas.

| | Minamisanriku Town | Rikuzentakata City | Kamaishi City |
|---|---|---|---|
| A) Actual damage[$\times 10^4$ ton] | 68.2 | 65.9 | 15.7 |
| B) Proposed method[$\times 10^4$ ton] | 57.6 | 67.2 | 17.5 |
| B/A | 0.843 | 1.012 | 1.113 |

In a massive disaster, it is difficult to estimate the amount of debris immediately. For example, we can examine the Great East Japan Earthquake. Some researchers attempted to estimate the total amount of debris. Hirayama researched a method of estimating debris extensively. However, the estimated amount of debris in that study was approximately 1.5 times the actual amount of debris.

If we can more exhaustively estimate debris, the system will become more practical and achieve better accuracy. Parallelization and high processing speeds will be required to achieve this task.

## 7 Conclusion

We proposed a system that detects collapsed buildings using aerial photographs taken after a disaster in order to estimate the amount of disaster debris. To detect collapsed buildings using aerial photograph, we extract straight lines (assuming they represent the roofs of buildings) from the photographs. Subsequently, we applied the proposed method to real aerial photographs that were taken in affected areas immediately after the Great East Japan Earthquake. Our proposed method achieved higher accuracy for estimating amounts of disaster debris when compared to existing methods. However, the performance of the extraction of roof features must improve for the system to become a practical application. In future research, we will consider image features other than straight lines. Parallelization and improved processing speed will be required to achieve this task.

## References

[1] Brown, C. & Milke, M.W. (2009). Planning for Disaster Debris Management. Christchurch, New

Zealand. *WasteMINZ 21st Annual Conference*. 14-16.
[2] Cabinet Office, Government of Japan. http://www.bousai.go.jp/taisaku/pdf/shishin002.pdf
[3] Geospatial Information Authority of Japan. http://www.gsi.go.jp
[4] Hirayama, N. & Kawata, Y & Okumura, Y. (2012). Quantitative Estimation of Disaster Debris for Operational Management after the Great East Japan Earthquake. Japan Society of Material Cycles and Waste Management Vol.23, No.1. 3-9.
[5] Hirayama, N. & Shimaoka, T. & Fujiwara, T. & Okayama, T. & Kawata, Y. (2010). Establishment of disaster debris management based on quantitative estimation using natural hazard maps. Waste Management and the Environment V, 40. 167–178.
[6] Luther, L. (2006). Disaster Debris Removal after Hurricane Katrina. Status and Associated Issues. Congressional Research Service, The Library of Congress.
[7] Marko, T. & Jurij, F. (2003). Colour Spaces - Perceptual, Historical and Applicational Background. EUROCON 2003 Ljubljana. 304-308.
[8] Matsuoka, M. & Yamazaki, F. (2002). Detection of Building Damage Areas due to Earthquakes using Satellite SAR Intensity Images. Architectural Institute of Japan, No.551. 53-60.
[9] Mitomi, H. & Matsuoka, M. & Yamazaki, F. (2002). ATTEMPT FOR AUTOMATED DETECTION OF BUILDING DAMAGE AREAS DUE TO RECENT EARTHQUAKES USING AERIAL TELEVISION IMAGES. Japan Society of Civil Engineering, No.703/I-59. 267-278.
[10] Nojima, N. & Sugito, M. & Kanazawa, N. (2001). Decision Support Model for Post-Earthquake Emergency Management Based on Sequential Data Processing of Damage Information. Japan Society of Civil Engineering, No682/I-56. 129-142.
[11] Otsu, N. (1979). Threshold selection method from gray-level histograms. IEEE TRANSACTIONS ON SYSTREMS, MAN, AND CYBERNETICS, SMC-9. 62–66.
[12] Sakurada, K. & Okatani, T. & Deguchi, K. (2013). Detecting Changes in 3D Structure of a Scene from Multi-view Images Captured by a Vehicle-mounted Camera. Computer Vision and Pattern Recognition (CVPR), 2013 IEEE Conference. 137-144.
[13] The Great East Japan Earthquake Reconstruction Support Research Archive. http://fukkou.csis.u-tokyo.ac.jp
[14] Urabe K. & Saji, H. (2009). Detection of Road Blockage Areas after Earthquake Disaster in Mountainous Districts Using Aerial Images. Japan Association for Earthquake Engineering. Vol 9/No.4. 26-38.

IOP Conf. Series: Earth and Environmental Science **61** (2017) 012164 doi:10.1088/1755-1315/61/1/012164

# Zircon U-Pb age and geochemistry of the late ordovician monzonitic granite in the Xiaoxilin from the lesser Khingan

**Feifei Zhao *, Fengyue Sun, Liang Li, Jian Wang**

College of Earth Sciences, Jilin University, Changchun 130061, China
*Email: terry_xp@163.com

**Abstract:** In this paper, we present new U-Pb zircon ages and major and trace elements for the Late Ordovician monzonitic granite rocks in Xiaoxilin area in the Eastern lesser khingan of Northeast China to elucidate the westward subduction of the Oceanic plate in Late Ordovician. U-Pb zircon dating results demonstrate that the monzonitic granites formed at 451±5Ma, belonging to the Late Ordovician.These monzonitic granites are characterized have $SiO_2$ of 66.45%~72.93%, $Na_2O$ of 2.41%~3.93%, $K_2O$ of 3.09%~5.58%, CaO of 2.02%~3.76%, ALK of 6.56%~7.99% and A/CNK=0.97~1.02, belonging to the high-K calc-alkaline series. The monzonitic granites is characterized by the enrichment of LILEs (e.g., Rb and K) and incompatible elements(e.g., Th and U), and depletion of HFSEs(e.g., Nb, Ta, Sr, P and Ti), as well as very lesser Eu anomalies (Eu/Eu*=0.49~1.06). The rock falls into the zone of Pre-Plate collision with the characteristics of active continental margin before the collision. Combine the mineral assemblage and geochemical characteristics, it is believed that the monzonitic granite in Xiaoxilin area was formed in the environment of oceanic crust subduction, which indicates that there was an westward subduction oceanic plate between Jiamusi Massif and Songnen-Zhangguangcailing Massif. The melt/fluid precipitated from the subducting oceanic plate confessed the overlying mantle wedge and caused the partial melting of it, which resulted in the arc magmatism. The Jiamusi Massif and Songnen-Zhangguangcailing Massif had not yet joined in the Late Ordovician.

## 1 Introduction

The study of the granites in the Xiaoxilin area mostly focus on the Mesozoic granite. In this paper we present new U-Pb zircon ages, and major and trace elements for poorly studied Early Paleozoic granitic rocks in Xiaoxilin area in the Lesser Khingan of Northeast China.

## 2 Geological background

The Xiaoxilin area is located in 40 km southeast of Yichun City, Heilongjiang Province, the geographic coordinates range between 129°07′56″~129°11′07″E and 47°21′59″~47°26′18″N. The tectonic location belongs to the Wuxing-Guansongzhen uplift belt in the central part of the Yichun-Yanshou fold orogenic belt which lactated in the outer edge with overlapping parts of Mongolian-Okhotsk collision orogenic belt and the continental Pacific subduction zone. In the east, it's bounded the Jiamusi Massif by Jiameng-Yilan-Mudanjiang fault and connected with the Songliao Basin in the western part (Fig. 1).

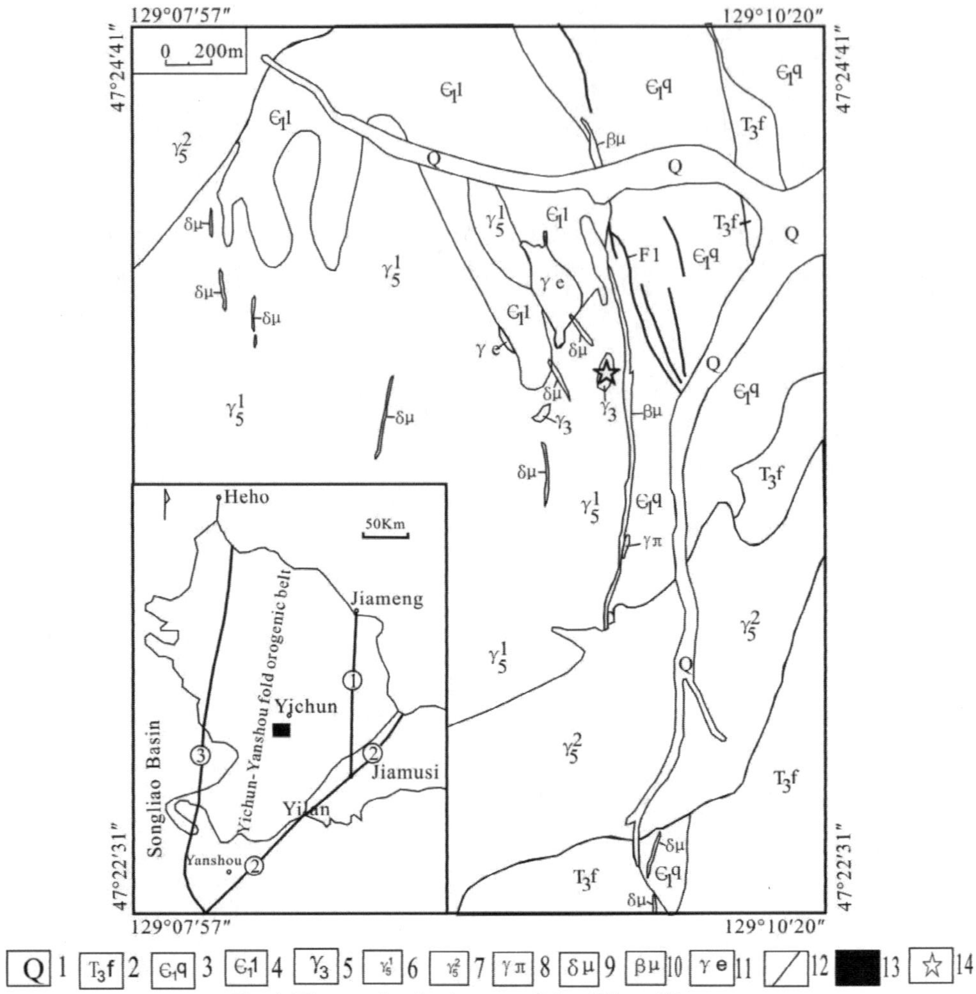


Fig.1 The geological map of the Xiaoxilin area

① Jiameng-Yilan-Mudanjiang fault; ② Jiamusi-Yisuhe Fault; ③ Xuhe-Shangzhi Fault.

1. Quaternary; 2. Fengshan Formation; 3. Qianshan Group; 4. Laodaomiaogou Formation; 5. Caledonian granite; 6. Indosinian granite; 7. Yanshanian granite 8. Granite porphyry; 9. granite diorite veins; 10. Diabase porphyry; 11. Granite fine crystal; 12. fault; 13. mine location; 14. sampling position.

The main strata in the area are the Laodaomiaogou Formation, the Qianshan Formation, the Fengshantun Formation of the Upper Triassic, and the quaternary system. The Laodaomiaogou Formation is mainly composed of metamorphic sandstone, siltstone, and gravel sandstone locally, the Qianshan Formation consists of dolomite, dolomitic marble folder carbonaceous slate. Fengshantun Formation mainly composed of acid volcanic rocks and their clastic rocks, local alteration and Pb - Zn mineralization. Separation fracture develops in the area, which is mainly in the longitudinal (near north-south) and the direction roughly coincides with the direction of strata or slightly oblique. F1 fracture is the largest vertical fracture in the area, exposed to the transition zone from Xilin syncline to Qianshan anticline, with the direction of strike near SN, trend E. The magmatism in the area is strong, mainly in the three periods of the Caledonian, the Indosinian and the Yanshan, on lithology it has adamellite, granite porphyry, granodiorite, granite aplite and allgovite and so on.

### 3 Petrography and sample descriptions

The samples of the monzonitic granite used during this study were collected from 47°23′57″N,

129°09′35″E. The sample contain Alkali feldspar(~40%), quartz (~30%), plagioclase (~20%), Sericite (~5%) and Opaque mineral(~5%)(Fig.2)..

Fig. 2 Microscopic photograph of the monzonitic granite (plane polars)
Qtz: quartz; Pl: Plagioclase; Al: Alkali feldspar

## 4 Analytical methods

### *4.1 U-Pb Zircon dating*

The samples used for zircon U-Pb dating were crushed using conventional crushing and were separated using heavy liquids and a magnetic separator at the Langfang Regional Geological Survey, Hebei Province, China. Laser ablation inductively coupled plasma mass spectrometry (LA–ICP–MS) U–Pb analyses were conducted using the Agilent 7500a ICP-MS instrument equipped with ComPex102 ArF (193 nm) at the State Key Laboratory of Continental Dynamics, Northwestern University, Xi'an, China, using a 30 μm spot diameter. Details of the analytical technique are described in Yuan et al. (2004)[1], common Pb concentrations were evaluated following Andersen (2002)[2], and ISOPLOT version 3.0 (Ludwig, 2003)[3] was used for age calculations and for constructing concordia diagrams.

### *4.2 Whole-rock geochemistry determination*

For geochemical analysis, whole-rock samples, after the removal of altered surfaces, were crushed in an agate mill to ~200 mesh. X-ray fluorescence (XRF; PW1401/10) using fused-glass disks and ICP-MS (Agilent 7500a with a shield torch) were used to measure the major and trace elements compositions, respectively, at the Testing Center of Jilin University, after acid digestion of samples in Teflon bombs.

## 5 Results

### *5.1 U-Pb Zircon ages*

The analytical results of U-Pb zircon dating in the Late Ordovician monzonitic granites are listed in Table 1. Zircons CL images from monzonitic granite (samples XXL-N2) reveal that the zircons display typical magmatic oscillatory zoning and rhythmically zoned textures (Fig.3) euhedral-subhedral in shape and the Th/U ratios of 0.22–0.69. The 12 analytical sites show that the ages range from 443 to 461 Ma, and the majority of ages are centralized at around 450 Ma. The weighted mean age is 451±5 Ma (1σ, MSWD= 0.43, n=12; Fig. 4).

IOP Conf. Series: Earth and Environmental Science **61** (2017) 012164 doi:10.1088/1755-1315/61/1/012164

Table.1 LA-ICPMS zircon U-Pb data of the monzonitic granite

| Sample | Th/U | $^{207}Pb/^{206}Pb$ | 1σ | $^{207}Pb/^{235}U$ | 1σ | $^{206}Pb/^{238}U$ | 1σ | $^{206}Pb/^{238}U$ | 1σ |
|---|---|---|---|---|---|---|---|---|---|
| XXL-N2-01 | 0.47 | 0.05824 | 0.00654 | 0.57043 | 0.06134 | 0.07121 | 0.0024 | 443 | 14 |
| XXL-N2-02 | 0.35 | 0.05544 | 0.0024 | 0.546 | 0.01951 | 0.07158 | 0.00136 | 446 | 8 |
| XXL-N2-03 | 0.22 | 0.05593 | 0.00177 | 0.54915 | 0.01152 | 0.07134 | 0.00123 | 444 | 7 |
| XXL-N2-04 | 0.69 | 0.056 | 0.00553 | 0.55127 | 0.05176 | 0.07152 | 0.00215 | 445 | 13 |
| XXL-N2-05 | 0.52 | 0.05542 | 0.00535 | 0.54517 | 0.05002 | 0.07145 | 0.0021 | 445 | 13 |
| XXL-N2-06 | 0.42 | 0.05693 | 0.00298 | 0.57109 | 0.02612 | 0.07285 | 0.00149 | 453 | 9 |
| XXL-N2-07 | 0.45 | 0.05659 | 0.00317 | 0.56444 | 0.0281 | 0.07243 | 0.00152 | 451 | 9 |
| XXL-N2-08 | 0.40 | 0.05597 | 0.00218 | 0.56451 | 0.01713 | 0.07323 | 0.00133 | 456 | 8 |
| XXL-N2-09 | 0.69 | 0.05926 | 0.00227 | 0.59659 | 0.01755 | 0.07304 | 0.00131 | 454 | 8 |
| XXL-N2-10 | 0.26 | 0.05632 | 0.00185 | 0.57556 | 0.01271 | 0.07412 | 0.00126 | 461 | 8 |
| XXL-N2-11 | 0.36 | 0.05634 | 0.0026 | 0.56178 | 0.0217 | 0.07231 | 0.00137 | 450 | 8 |
| XXL-N2-12 | 0.42 | 0.05589 | 0.00379 | 0.54911 | 0.03411 | 0.07124 | 0.00163 | 444 | 10 |

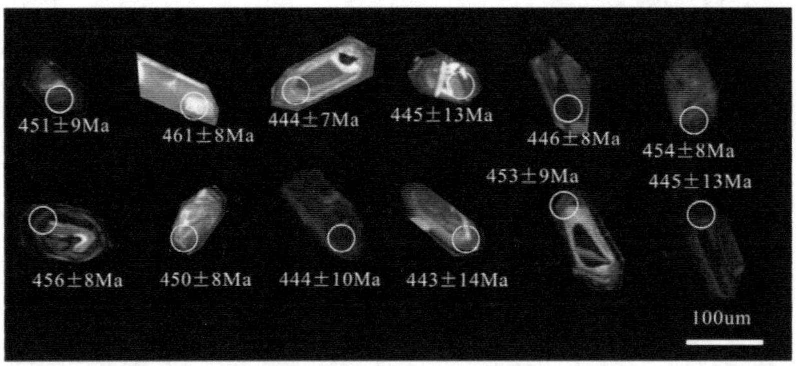


Fig.3 CL images of representative zircons from the monzonitic granite

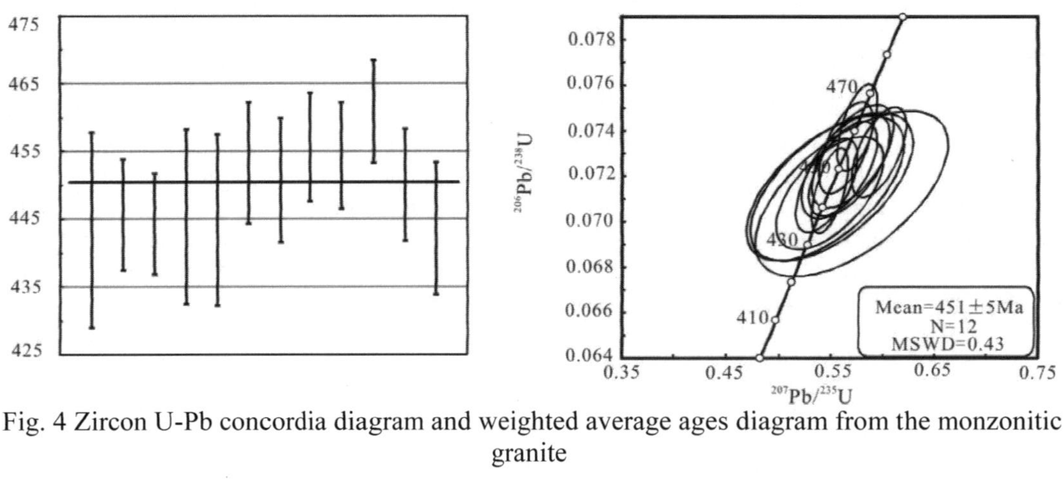


Fig. 4 Zircon U-Pb concordia diagram and weighted average ages diagram from the monzonitic granite

IOP Conf. Series: Earth and Environmental Science **61** (2017) 012164 doi:10.1088/1755-1315/61/1/012164

*5.2 Geochemistry characteristics*

The analytical results of major and trace elements in the Late Ordovician monzonitic granites are listed in Table 2. The monzonitic granite in Xiaoxilin area have $SiO_2$ of 66.45%~72.93%, $Na_2O$ of 2.41%~3.93%, $K_2O$ of 3.09%~5.58%,CaO of 2.02%～3.76%, ALK of 6.56%～7.99% and A/CNK=0.97~1.02. They are classified as granite in a total subalkalic versus $SiO_2$ (Fig.5a) diagram and are classified as high-K and calc-alkaline in a $K_2O$ versus $SiO_2$ diagram(Fig.5b).

Table 2. Major and trace element compositions of the monzonitic granite

| Sample | XXL-B1 | XXL-B2 | XXL-B3 | XXL-B4 |
|---|---|---|---|---|
| $SiO_2$ | 72.93 | 68.02 | 66.62 | 66.45 |
| $TiO_2$ | 0.27 | 0.42 | 0.46 | 0.47 |
| $Al_2O_3$ | 13.49 | 15.76 | 15.9 | 15.88 |
| $Fe_2O_3$ | 1.16 | 1.55 | 1.79 | 1.7 |
| MnO | 0.08 | 0.17 | 0.15 | 0.16 |
| MgO | 0.44 | 1.33 | 1.21 | 1.29 |
| CaO | 2.02 | 2.99 | 3.59 | 3.76 |
| $Na_2O$ | 3.93 | 2.41 | 3.3 | 3.47 |
| $K_2O$ | 3.43 | 5.58 | 3.76 | 3.09 |
| $P_2O_5$ | 0.1 | 0.18 | 0.23 | 0.24 |
| LOI | 0.64 | 0.46 | 1.08 | 1.14 |
| Total | 98.49 | 98.87 | 98.09 | 97.65 |
| $K_2O/Na_2O$ | 0.87 | 2.32 | 1.14 | 0.89 |
| $K_2O+Na_2O$ | 7.36 | 7.99 | 7.06 | 6.56 |
| A/CNK | 0.97 | 1.02 | 0.99 | 1 |
| La | 15.98 | 41.41 | 23.57 | 11.02 |
| Ce | 24.04 | 71.14 | 55.17 | 21.42 |
| Pr | 2.68 | 9.11 | 6.45 | 2.7 |
| Nd | 9.09 | 32.92 | 24.18 | 9.52 |
| Sm | 1.61 | 6.4 | 5.15 | 1.9 |
| Eu | 0.52 | 1.01 | 1.02 | 0.35 |
| Gd | 1.31 | 6.14 | 5.03 | 1.8 |
| Tb | 0.16 | 0.96 | 0.84 | 0.29 |
| Dy | 0.89 | 6.1 | 5.35 | 1.82 |
| Ho | 0.18 | 1.36 | 1.19 | 0.41 |
| Er | 0.52 | 4.4 | 3.79 | 1.29 |
| Tm | 0.08 | 0.73 | 0.62 | 0.21 |
| Yb | 0.59 | 5.08 | 4.43 | 1.5 |
| Lu | 0.11 | 0.78 | 0.69 | 0.23 |
| Cr | 8.94 | 10.91 | 4.92 | 4.68 |
| Ni | 5.08 | 6.2 | 3.96 | 2.22 |
| V | 21.86 | 31.56 | 45.39 | 10.85 |
| Rb | 112.7 | 222.7 | 54.56 | 44.19 |
| Sr | 294.4 | 201.4 | 202.8 | 81.52 |
| Ba | 296.9 | 935.5 | 605.2 | 192.1 |
| Pb | 32.76 | 237.3 | 68.92 | 18.19 |
| Ga | 16.82 | 19.77 | 19.09 | 5.48 |
| Y | 4.45 | 37.99 | 28.2 | 11.39 |

IOP Conf. Series: Earth and Environmental Science **61** (2017) 012164 doi:10.1088/1755-1315/61/1/012164

| | | | | |
|---|---|---|---|---|
| Zr | 123.3 | 262.3 | 173.6 | 56.57 |
| Nb | 5.19 | 18.26 | 13.83 | 3.81 |
| Ta | 0.49 | 1.57 | 1.21 | 0.36 |
| Hf | 4 | 7.53 | 5.05 | 1.62 |
| Th | 11.84 | 14.99 | 11.12 | 5.08 |
| ΣREE | 57.74 | 187.54 | 137.48 | 54.45 |
| ΣLREE | 53.92 | 161.99 | 115.54 | 46.91 |
| ΣHREE | 3.84 | 25.55 | 21.94 | 7.55 |
| LREE/HREE | 14.03 | 6.34 | 5.27 | 6.22 |
| （La/Yb）$_N$ | 18.3 | 5.49 | 3.58 | 4.97 |
| ΔEu | 1.09 | 0.49 | 0.61 | 0.59 |

A/CNK=$Al_2O_3$/(CaO+$Na_2O$+$K_2O$); δEu=2×(Eu/0.0735)/(Gd/0.259+Sm/0.195); LREE=La+Ce+Pr+Nd +Sm+ Eu; HREE=Gd+Tb+Dy+Ho+Er+Tm+Yb+Lu; $(La/Yb)_N$=(La/0.310)/(Yb/0.209)。

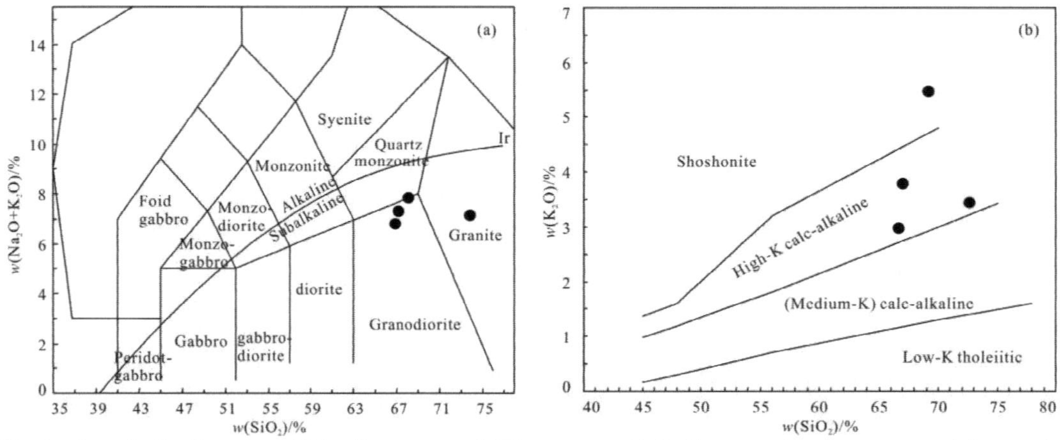


Fig.5 Total alkali versus $SiO_2$ diagram and $SiO_2$ versus $K_2O$ diagram for the monzonitic granite (a: base map modified after reference[4]; b:base map base map modified after reference[5])

The monzonitic granites in the Xiaoxilin area are weak enrichment in LREE (Fig.) and the $(La/Yb)_N$ values of 3.58–18.30, with the weak Eu anomalies (Eu/Eu*=0.49～1.09). The monzonitic granite is characterized by the enrichment of LILEs (e.g., Rb and K) and incompatible elements(e.g., Th and U), and depletion of HFSEs(e.g., Nb, Ta, Sr, P and Ti).(Fig.6),

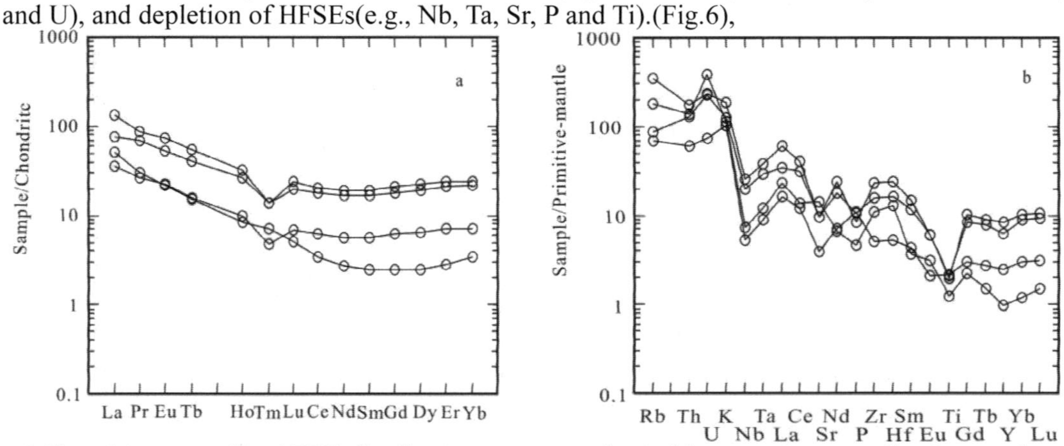


Fig.6 Chondrite-normalized REE distribution patternsand primitive mantle-normalized trace element spider diagramsof the monzonitic granite (a: base map modified after reference[6]; b:base map base map modified after reference[7])

## 6 Discussion

The monzonitic granites are high-k calc alkaline granites, which are mainly formed in the tectonic setting of the continental arc and the collision of the plate after the collision of the two. In the Rb-(Y+Nb) tectonic discriminant diagrams (Fig.7a), all the samples fall into the volcanic arc granite area. The ratio of Zr/Y (4.97~27.70), with an average value of 11.43, three in the four ratios are from 4.97 to 6.90, the scope of the main continental margin between the andesite (Zr/Y ratio ranged from 4 to 12, Condie 1989), shows the characteristics of active continental margin arc. The average La/Nb ratio (1.70~3.08) is 2.49, which is that of active continental margin La/Nb ratio (>2). In the R1-R2 diagram of granite (Fig.7b), the samples mainly fall into the zone of granite from the subduction zone before the collision, which is equivalent to the active plate margin. The above characteristics indicate that the late Ordovician granite is mainly formed in the tectonic environment of active continental margin.

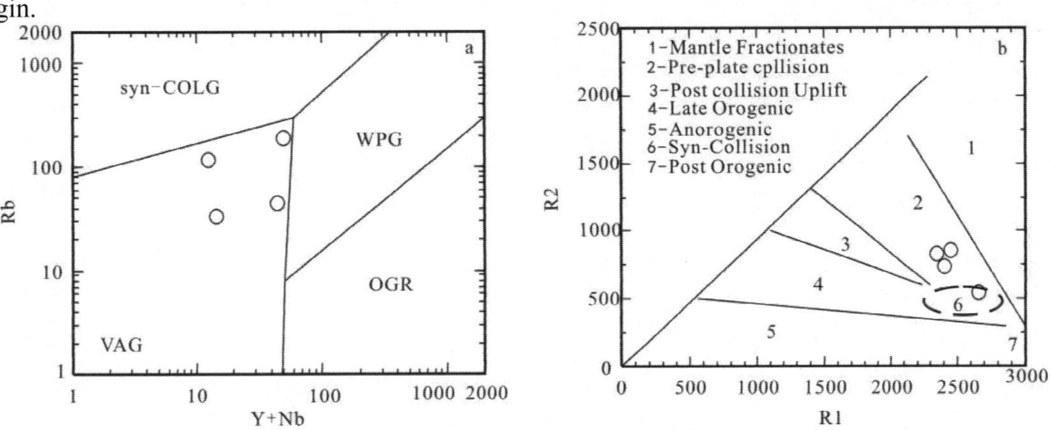


Fig. 7 Tectonic setting discrimination diagrams of the monzonitic granite
(ORG–ocean ridge granites; VAG–volcanic arc granites; WPG–within plate granites;
syn-COLG–syn-collision granites)

The Northeast orogenic belt is composed of many micro plates (XingKai Massif、Jiamusi Massif、Songnen-Zhangguangcai Massif、Erguna Massif and Xingan Massif).There are two points about the time of the Jiamusi Massif associated with Songnen-Zhangguangcai Massif. Some studies suggest that the Jiamusi Massif was associated with Songnen-Zhangguangcai Massif in Ordovician and the others suggest the time is Silurian.The monzonitic granite in study area formed at 451±5 Ma, it belongs to Late Ordovician. The above characteristics indicate that the late Ordovician granite is mainly formed in the tectonic environment of active continental margin. We conclude, therefore, that the the Jiamusi Massif was not yet associated with Songnen-Zhangguangcai Massif in in Late Ordovician.

Based on the research result and the previous studies, we summarize the evolution of the Northeast China Orogenic belt. In the Late Proterozoic, the Northeast China had been affected by the Rodinic Super-continent breakup. From the Late Mesoproterozoic to the Late Proterozoic, the Paleo Asian ocean basin formd in the Northeast China. From the Late Proterozoic to the Early Cambrian, there are many small oceans in the micro plates (XingKai Massif、Jiamusi Massif、Songnen-Zhangguangcai Massif、Erguna Massif and Xingan Massif). In the Ordovician, The Jiamusi Massif and Songnen-Zhangguangcailing Massif had not yet joined. In the Silurian, the Jiamusi Massif was associated with the Songnen-Zhangguangcailing Massif.

## 7 Conclusions

Based on the U-Pb zircon ages and geochemical data presented in this paper, we draw the following conclusions:

1. U-Pb zircon dating results demonstrate that the monzonitic granites formed in Late Ordovician (451±5 Ma).

2. These monzonitic granites are characterized have $SiO_2$ of 66.45%~72.93%, $Na_2O$ of 2.41%~3.93%, $k_2O$ of 3.09%~5.58%, CaO of 2.02%～3.76%, ALK of 6.56%～7.99% and A/CNK=0.97~1.02, belonging to the high-K calc-alkaline series

3. The late Ordovician monzonitic granites is mainly formed in the tectonic environment of active continental margin.

## References

[1] Yuan, H. L., Gao, S., Liu, X. M., Li, H. M., Gunther, D. and Wu, F. Y. (2004) Accurate U-Pb age and trace element determinations of zircon by laser ablation-inductively coupled plasma-mass spectrometry . *Geostandards and Geoanalytical Research*, 28(3), 353-370.

[2] Andersen T. 2002. Correction of common lead in U-Pb analyses that do not report $^{204}$Pb. *Chemical Geology*, 192(1-2): 59-79.

[3] Ludwig K R. 2003. User's manual for isoplot 3.00：A geochronological toolkit for Microsoft Excel. *Berkeley Geochronology Center Special Publication*, 4:70

[4] Irvine, T. H. and Baragar, W. R. A. (1971) A guide to the chemical classification of the common volcanic rocks. *Canadian Journal of Earth Sciences,* 8, 523-548.

[5] Taylor, S. R. and McLennan, S. M. (1995) The geochemical evolution of the continental crust. *Reviews of Geophysics,* 33(2), 241-265.

[6] Boynton, W. V. (1984) Geochemistry of the rare earth elements: meteorite studies. In: Henderson, P. (Ed.), Rare earth Element Geochemistry. *Elsevier, Amsterdam,* pp, 63-114.

[7] Mckenzie, D. P. (1989) Some remarks on the movement of small melt fractions in the mantle. Earth and Planetary Science Letters,95, 53-72.

**Institute of Physics**
Dirac House, Temple Back
Bristol BS1 6BE UK

ISSN: 1755-1307
ISBN 978-1-5108-3957-1

9 781510 839571

CURRAN ASSOCIATES INC.
proceedings.com